The HUTCHINSON
SCIENCE
DESK REFERENCE

The HUTCHINSON
SCIENCE
DESK REFERENCE

Helicon

Copyright © Helicon Publishing Ltd 1999

Helicon Publishing Ltd
42 Hythe Bridge Street
Oxford OX1 2EP
e-mail: admin@helicon.co.uk
Web site: http://www.helicon.co.uk

Printed and bound in Great Britain by
The Bath Press Ltd, Bath, Somerset

ISBN 1-85986-274-8

British Cataloguing in Publication Data
A catalogue record of this book is available from the British Library

Contents

Contributors

Tony Jones
Scott Kirsner
Julian Rowe

Chris Stringer
Gordon Woods
Edward Young

Editorial director
Hilary McGlynn

Managing editor
Elena Softley

Project editor
Catherine Thompson

Technical editor
Tracey Auden

Text editors
Jayne Harrison
Chris Holdsworth
Sara Jenkins-Jones
John Leonard
Edith Summerhayes
Karen Young

Production director
Tony Ballsdon

Art and design manager
Terence Caven

Typesetting and page make-up
TechType

INTRODUCTION

Science is one of the fastest changing areas of human knowledge and one of the main aims of *The Hutchinson Science Desk Reference* is to enable the reader to make sense of the vast amount of fascinating information and data that comprise modern science. The editors have produced a readable, accurate, and accessible reference book that will appeal to the specialist and general reader alike. In every chapter the reader will find a clear overview of the subject; easily accessible facts displayed as tables and diagrams; key vocabulary in the individual subject glossaries; historical context in the form of a chronology; outlines of the achievements of notable scientists; and numerous weblinks and bibliographies to aid further research.

Definition of science

Science is represented within this book at its very broadest. Pure science does not exist in a vacuum, but requires the back up of technology, computing science, and mathematics; in addition, no science desk reference would really be complete without some information on medicine and the natural world.

How to use The Hutchinson Science Desk Reference

The book is arranged thematically with each chapter covering one scientific discipline or an area within a discipline. Every chapter follows the same structure (except the final chapter 'Discoveries, Inventions, and Prizes', which consists of tables):

- **subject overview** with tables, illustrations, weblinks, topic features, interesting facts, and memory joggers to complement the main text;
- **chronology** to provide a brief history of the subject, listing milestones such as groundbreaking experiments and important publications;
- **biographies** of important figures, outlining their contributions to the field;
- **glossary** of key vocabulary used within the chapter;
- **further reading** for those who would appreciate some pointers for additional research.

cross references

Cross-references (indicated by ♉) are self-contained within each chapter and refer to glossary or biography entries at the end of that chapter only.

subject overlap

Some amount of subject overlap is inevitable when one attempts to carve up the interrelated world of science into chapters – indeed Austrian-born British scientist and mathematician Hermann Bondi summed this up very neatly with his observation that science is 'above all a co-operative enterprise'. For reasons of space, glossary and biography entries that have bearing in a number of areas normally appear in one chapter only. The extensive index ensures readers can locate all relevant information, even if they are unsure in which chapter to look.

MEASUREMENT

Measurement is the determination of the dimensions, extent, or quantity of something. A **unit** is a standard quantity in relation to which other quantities are measured. There have been many systems of units. Some ancient units, such as the day, the foot, and the pound, are still in use. SI units, the latest version of the metric system, are widely used in science.

A **scale** is an indexing system that shows the order of magnitude of a physical quantity through a one-to-one correspondence between a **number** and the physical quantity. Physical quantities like temperature and sound are measured using scales.

systems of measurement

imperial system

The imperial system is the traditional system of units developed in the UK, based largely on the foot, pound, and second (f.p.s.) system. In 1991 it was announced that the acre, pint, troy ounce, mile, yard, foot, and inch would remain in use indefinitely for beer, cider, and milk measures, and in road-traffic signs and land registration. Other units, including the fathom and therm, would be phased out by 1994. The imperial system is still in common usage in the USA, despite the fact that SI units have been adopted by the US scientific community.

> **imperial system**
>
> The foot, yard and inch owe their length to Henry I of England. He decreed that one yard was the length of his extended arm, measured from nose to fingertip.

c.g.s. system

The c.g.s. system of units is based on the centimetre, gram, and second, as units of length, mass, and time, respectively.

m.k.s. system

In the m.k.s. system the base units of the metre, kilogram, and second replace the centimetre, gram, and second of the c.g.s. system. As in the c.g.s. system, there is no standard unit for volume.

Units in the Imperial System

Length

1 foot	= 12 inches
1 yard	= 3 feet
1 rod	= $5\frac{1}{2}$ yards (=$16\frac{1}{2}$ feet)
1 chain	= 4 rods (= 22 yards)
1 furlong	= 10 chains (= 220 yards)
1 mile	= 5,280 feet
1 mile	= 1,760 yards
1 mile	= 8 furlongs

Nautical

1 fathom	= 6 feet
1 cable length	= 120 fathoms
1 nautical mile	= 6,076.11549 feet

Area

1 square foot	= 144 square inches
1 square yard	= 9 square feet
1 square rod	= 301/4 square yards
1 acre	= 4 roods
1 acre	= 4,840 square yards
1 square mile	= 640 acres

Volume

1 cubic foot	= 1,728 cubic inches
1 cubic yard	= 27 cubic feet
1 bulk barrel	= 5.8 cubic feet

Shipping

1 register ton	= 100 cubic feet

Capacity

1 fluid ounce	= 8 fluid drahms
1 gill	= 5 fluid ounces
1 pint	= 4 gills
1 quart	= 2 pints
1 gallon	= 4 quarts
1 peck	= 2 gallons
1 bushel	= 4 pecks
1 quarter	= 8 bushels
1 bulk barrel	= 36 gallons

Weight (avoirdupois)

1 ounce	= 437 1/2 grains
1 ounce	= 16 drams
1 pound	= 16 ounces
1 stone	= 14 pounds
1 quarter	= 28 pounds
1 hundredweight	= 4 quarters
1 ton	= 20 hundredweight

metric system

The metric system of weights and measures was developed in France in the 18th century and recognized by other countries in the 19th century. Based largely on the m.k.s. system, the metric system uses the metre as the base unit of length, the kilogram as the base unit of mass, the second as the base unit of time, and the litre as the base unit of volume.

The metric system has been replaced for scientific work by the SI units to avoid inconsistencies in definition of the thermal calorie and electrical quantities.

The metric system was made legal for most purposes in the UK and USA in the 19th century. Shops in the UK switched in October 1995 to the metric system for transactions involving liquid measures – other than for beer, cider, and milk – and linear measures.

Units in the Metric System

Length

1 centimetre	= 10 millimetres	
1 decimetre	= 10 centimetres	= 100 millimetres
1 metre	= 10 decimetres	= 1,000 millimetres
1 decametre	= 10 metres	
1 hectometre	= 10 decametres	= 100 metres
1 kilometre	= 10 hectometres	= 1,000 metres

Area

1 square centimetre	= 100 square millimetres	
1 square metre	= 10,000 square centimetres	= 1,000,000 square millimetres
1 are	= 100 square metres	

Mass (avoirdupois)

1 hectare	= 100 ares	= 10,000 square metres
1 square kilometre	= 100 hectares	= 1,000,000 square metres
1 centigram	= 10 milligrams	
1 decigram	= 10 centigrams	= 100 milligrams
1 gram	= 10 decigrams	= 1,000 milligrams
1 decagram	= 10 grams	
1 hectogram	= 10 decagrams	= 100 grams
1 kilogram	= 10 hectograms	= 1,000 grams
1 metric ton	= 1,000 kilograms	

Volume

1 cubic centimetre	= 1,000 cubic millimetres	
1 cubic decimetre	= 1,000 cubic centimetres	= 1,000,000 cubic millimetres
1 cubic metre	= 1,000 cubic decimetres	= 1,000,000,000 cubic millimetres

Capacity

1 centilitre	= 10 millilitres	
1 decilitre	= 10 centilitres	= 100 millilitres
1 litre	= 10 decilitres	= 1,000 millilitres
1 decalitre	= 10 litres	
1 hectolitre	= 10 decalitres	= 100 litres
1 kilolitre	= 10 hectolitres	= 1,000 litres

Imperial and Metric Conversion Factors

To convert from imperial to metric	Multiply by	Multiply by	To convert from metric to imperial
Length			
inches	25.4	0.0393701	millimetres
feet	0.3048	3.28084	metres
yards	0.9144	1.09361	metres
furlongs	0.201168	4.97097	kilometres
miles	1.609344	0.621371	kilometres
Area			
square inches	6.4516	0.1550	square centimetres
square feet	0.092903	10.7639	square metres
square yards	0.836127	1.19599	square metres
square miles	2.589988	0.386102	square kilometres
acres	4046.856422	0.000247	square metres
acres	0.404866	2.469955	hectares
Volume/capacity			
cubic inches	16.387064	0.061024	cubic centimetres
cubic feet	0.028317	35.3147	cubic metres
cubic yards	0.764555	1.30795	cubic metres
cubic miles	4.1682	0.239912	cubic kilometres
fluid ounces (US)	29.5735	0.033814	millilitres
fluid ounces (imperial)	28.413063	0.035195	millilitres
pints (US)	0.473176	2.113377	litres
pints (imperial)	0.568261	1.759754	litres
quarts (US)	0.946353	1.056688	litres
quarts (imperial)	1.136523	0.879877	litres
gallons (US)	3.785412	0.364172	litres
gallons (imperial)	4.54609	0.219969	litres
Mass/weight			
ounces	28.349523	0.035274	grams
pounds	0.453592	2.20462	kilograms
stone (14 lb)	6.350293	0.157473	kilograms
tons (US)	907.18474	0.001102	kilograms
tons (imperial)	1016.046909	0.000984	kilograms
tons (US)	0.907185	1.10231	metric tonnes
tons (imperial)	1.016047	0.984207	metric tonnes
Speed			
miles per hour	1.609344	0.621371	kilometres per hour
feet per second	0.3048	3.28084	metres per second
Force			
pound force	4.44822	0.224809	newton
kilogram force	9.80665	0.101972	newton
Pressure			
pound-force per square inch	6.89476	0.145038	kilopascals
tons-force per square inch (imperial)	15.4443	0.064779	megapascals
atmospheres	10.1325	0.098692	newtons per square centimetre
atmospheres	14.695942	0.068948	pounds-force per square inch
Energy			
calorie	4.1868	0.238846	joule
watt hour	3,600	0.000278	joule
Power			
horsepower	0.7457	1.34102	kilowatts
Fuel consumption			
miles per gallon (US)	0.4251	2.3521	kilometres per litre
miles per gallon (imperial)	0.3540	2.824859	kilometres per litre
gallons per mile (US)	2.3521	0.4251	litres per kilometre
gallons per mile (imperial)	2.824859	0.3540	litres per kilometre

SI units

SI units (French *Système International d'Unités*) comprise a standard system of scientific units used by scientists worldwide. In 1960 an international conference on weights and measures recommended the universal adoption of a revised international (or SI) system with seven prescribed 'base units': the metre (m) for length, kilogram (kg) for mass, second (s) for time, ampere (A) for electric current, kelvin (K) for thermodynamic temperature, candela (cd) for luminous intensity, and mole (mol) for quantity of matter.

SI Units

Quantity	SI unit	Symbol
absorbed radiation dose	gray	Gy
amount of substance	mole[1]	mol
electric capacitance	farad	F
electric charge	coulomb	C
electric conductance	siemens	S
electric current	ampere[1]	A
energy or work	joule	J
force	newton	N
frequency	hertz	Hz
illuminance	lux	lx
inductance	henry	H
length	metre[1]	m
luminous flux	lumen	lm
luminous intensity	candela[1]	cd
magnetic flux	weber	Wb
magnetic flux density	tesla	T
mass	kilogram[1]	kg
plane angle	radian	rad
potential difference	volt	V
power	watt	W
pressure	pascal	Pa
radiation dose equivalent	sievert	Sv
radiation exposure	roentgen	R
radioactivity	becquerel	Bq
resistance	ohm	Ω
solid angle	steradian	sr
sound intensity	decibel	dB
temperature	°Celsius	°C
temperature, thermodynamic	kelvin[1]	K
time	second[1]	s

[1] SI base unit.

Two *supplementary units* are included in the SI system – the radian (rad) and steradian (sr) – used to measure plane and solid angles. In addition, there are recognized derived units that can be expressed as simple products or divisions of powers of the basic units, with no other integers appearing in the expression; for example, the watt.

Some non-SI units, well established and internationally recognized, remain in use in conjunction with the SI system: minute, hour, and day in measuring time; multiples or submultiples of base or derived units which have long-established names, such as tonne for mass, the litre for volume; and specialist measures such as the metric carat for gemstones.

scientific notation

Scientific notation is a method of writing numbers often used by scientists, particularly for very large or very small numbers. The numbers are written with one digit before the decimal point and multiplied by a power of 10. The number of digits given after the decimal point depends on the accuracy required. For example, the speed of light is 2.9979×10^8 m/1.8628×10^5 mi per second.

types of measurement

length and distance

metre The metre (symbol m) is the SI unit of length, equivalent to 1.093 yards. It is defined by scientists as the length of the path travelled by light in a vacuum during a time interval of 1/299,792,458 of a second.

metre

The metre was originally (in 1791) defined as one ten-millionth of the distance from the North Pole to the equator, on a line through Paris.

foot The foot (symbol ft) is the imperial unit of length, equivalent to 0.3048 m, used in Britain since Anglo-Saxon times. It originally represented the length of a human foot. One foot contains 12 inches and is one-third of a yard.

yard A yard (symbol yd) is equivalent to 3 feet (0.9144 m). It is a commonly-used unit of length in the USA, where it is also sometimes used to denote a cubic yard (0.7646 cubic meters), as of topsoil.

area Area is the size of a surface. It is measured in square units, usually square centimetres (cm^2), square metres (m^2), or square kilometres (km^2). Surface area is the area of the outer surface of a solid.

hectare A hectare (symbol ha) is the metric unit of area equal to 100 ares or 10,000 square metres (2.47 acres). Trafalgar Square, London's only metric square, was laid out as one hectare.

SI Prefixes

Multiple	Prefix	Symbol	Example
1,000,000,000,000,000,000 (10^{18})	exa-	E	Eg (exagram)
1,000,000,000,000,000 (10^{15})	peta-	P	PJ (petajoule)
1,000,000,000,000 (10^{12})	tera-	T	TV (teravolt)
1,000,000,000 (10^{9})	giga-	G	GW (gigawatt)
1,000,000 (10^{6})	mega-	M	MHz (megahertz)
1,000 (10^{3})	kilo-	k	kg (kilogram)
100 (10^{2})	hecto-	h	hm (hectometre)
10 (10^{1})	deca-	da	daN (decanewton)
1/10 (10^{-1})	deci-	d	dC (decicoulomb)
1/100 (10^{-2})	centi-	c	cm (centimetre)
1/1,000 (10^{-3})	milli-	m	mm (millimetre)
1/1,000,000 (10^{-6})	micro-	(	(F (microfarad)
1/1,000,000,000 (10^{-9})	nano-	n	nm (nanometre)
1/1,000,000,000,000 (10^{-12})	pico-	p	ps (picosecond)
1/1,000,000,000,000,000 (10^{-15})	femto-	f	frad (femtoradian)
1/1,000,000,000,000,000,000 (10^{-18})	atto-	a	aT (attotesla)

acre The traditional English land measure, the acre, is equal to 4,840 square yards (4,047 sq m/0.405 ha). Originally meaning a field, it was the size that a yoke of oxen could plough in a day.

As early as Edward I's reign (1272–1307), the acre was standardized by statute for official use, although local variations in Ireland, Scotland, and some English counties continued. It may be subdivided into 160 square rods (one square rod equalling 25.29 sq m/30.25 sq yd).

volume and capacity

The space occupied by a three-dimensional solid object is called volume. Capacity, the alternative term for volume, is generally used to refer to the amount of liquid or gas that may be held in a container. Units of capacity include the litre and millilitre (metric) and the pint and gallon (imperial).

cubic measure Cubic measure is the measure of volume, indicated either by the word 'cubic' followed by a linear measure, as in 'cubic foot', or the word 'cubed' after a linear measure, as in 'metre cubed' (m^3). A **cubic decimetre** (symbol dm^3) corresponds to the volume of a cube whose edges are all 1 dm (10 cm) long; it is equivalent to a capacity of one litre.

litre The litre (symbol l) is the metric unit of volume and capacity, equal to one cubic decimetre (1.76 imperial pints/2.11 US pints). It was formerly defined as the volume occupied by one kilogram of pure water at 4°C at standard pressure, but this is slightly larger than one cubic decimetre.

gallon The imperial liquid or dry measure of capacity, the gallon, is equal to 4.546 litres and is subdivided into four quarts or eight pints. The US gallon is equivalent to 3.785 litres.

pint The pint is the imperial dry or liquid measure of capacity equal to 20 fluid ounces, half a quart, one-eighth of a gallon, or 0.568 litre. In the USA a liquid pint is equal to 0.473 litre, while a dry pint is equal to 0.550 litre.

weight and mass

Weight is the force exerted on an object by gravity. The weight of an object depends on its mass – the amount of material in it – and the strength of the Earth's gravitational pull, which decreases with height. Consequently, an object weighs less at the top of a mountain than at sea level. On the Moon, an object has only one-sixth of its weight on Earth, because the pull of the Moon's gravity is one-sixth that of the Earth.

If the mass of a body is m kilograms and the gravitational field strength is g newtons per kilogram, its weight W in newtons is given by $W = mg$.

Mass is the quantity of matter in a body as measured by its inertia. In the SI system, the base unit of mass is the kilogram. At a given place, equal masses experience equal gravitational forces, which are known as the weights of the bodies. Masses may, therefore, be compared by comparing the weights of bodies at the same place. The standard unit of mass to which all other masses are compared is a platinum-iridium cylinder of 1 kg, which is kept at the International Bureau of Weights and Measures in Sèvres, France.

kilogram SI unit (symbol kg) of mass equal to 1,000 grams (2.24 lb). It is defined as a mass equal to that of the international prototype held at the International Bureau of Weights and Measures in Sèvres, France.

pound The imperial unit of mass is the pound (abbreviation lb). The commonly used avoirdupois pound, also called the **imperial standard pound** (7,000 grains/0.45 kg), differs from the **pound troy** (5,760 grains/0.37 kg), which is used for weighing precious metals. It derives from the Roman *libra*, which weighed 0.327 kg.

ton and tonne A ton is an imperial unit of mass. The **long ton**, used in the UK, is 1,016 kg/2,240 lb; the **short ton**, used in the USA, is 907 kg/2,000 lb. The **metric ton** or **tonne** is 1,000 kg/ 2,204.6 lb.

In shipping, the ton is a unit of volume equal to 2.83 cubic metres/100 cubic feet. **Gross tonnage** is the total internal volume of a ship in tons; **net register tonnage** is the volume used for carrying cargo or passengers. **Displacement tonnage** is the weight of the vessel, in terms of the number of imperial tons of seawater displaced when the ship is loaded to its load line; it is used to describe warships.

The **metric tonne** of 1,000 kg/2,204.6 lb is equivalent to 0.9842 of an imperial ton.

mole A mole (symbol mol) is the SI unit of the amount of a substance. It is defined as the amount of a substance that contains as many elementary entities (atoms, molecules, and so on) as there are atoms in 12 g of the isotope carbon-12.

One mole of an element that exists as single atoms weighs as many grams as its atomic number (so one mole of carbon weighs 12 g), and it contains 6.022045 x 1023 atoms, which is Avogadro's number.

carat The carat is the unit for measuring the mass of precious stones; it is equal to 0.2 g/0.00705 oz, and is part of the ◊troy system of weights used for precious metals and gems. It is also the unit of purity in gold (US 'karat'). Pure gold is 24-carat; 22-carat (the purest used in jewellery) is 22 parts gold and two parts alloy (to give greater strength). Originally, one carat was the weight of a carob seed (Arabic *quirrat* 'seed').

time
The continuous passage of existence is recorded by division into hours, minutes, and seconds. Formerly the measurement of time was based on the Earth's rotation on its axis, but this was found to be irregular. Therefore the second, the standard SI unit of time, was redefined in 1956 in terms of the Earth's annual orbit of the Sun, and in 1967 in terms of a radiation pattern of the element caesium.

Time
http://physics.nist.gov/GenInt/Time/time.html
Designed by the National Institute of Standards and Technology (NIST) Physics Laboratory to provide an historic understanding of the evolution of time measurement. There are sections on Ancient Calendars, Early Clocks, World Time Scales, the Atomic Age, and the Revolution in Timekeeping.

---

year A year is a large unit of time measurement based on the orbital period of the Earth around the Sun. The **tropical year**, from one spring equinox to the next, lasts 365.2422 days. It governs the occurrence of the seasons, and is the period on which the calendar year is based. The **sidereal year** is the time taken for the Earth to complete one orbit relative to the fixed stars, and lasts 365.2564 days (about 20 minutes longer than a tropical year). The difference is due to the effect of precession, which slowly moves the position of the equinoxes. The **calendar year** consists of 365 days, with an extra day added at the end of February each leap year. **Leap years** occur in every year that is divisible by four, except that a century year is not a leap year unless it is divisible by 400. Hence 1900 was not a leap year, but 2000 will be.

second The second (symbol sec or s) is the basic SI unit of time, one-sixtieth of a minute. It is defined as the duration of 9,192,631,770 cycles of regulation (periods of the radiation corresponding to the transition between two hyperfine levels of the ground state) of the caesium-133 isotope.

minute A minute is a unit of time consisting of 60 seconds; 60 minutes make an hour.

hour
An hour is a period of time comprising 60 minutes; 24 hours make one calendar day.

day A day is the time taken for the Earth to rotate once on its axis. The **solar day** is the time that the Earth takes to rotate once relative to the Sun. It is divided into 24 hours, and is the basis of our calendar day. The **sidereal day** is the time that the Earth takes to rotate once relative to the stars. It is 3 minutes 56 seconds shorter than the solar day, because the Sun's position against the background of stars as seen from Earth changes as the Earth orbits it.

month

A month is a unit of time based on the motion of the Moon around the Earth. The time from one new or full Moon to the next (the **synodic** or **lunar month**) is 29.53 days. The time for the Moon to complete one orbit around the Earth relative to the stars (the **sidereal month**) is 27.32 days. The **solar month** equals 30.44 days, and is exactly one-twelfth of the solar or tropical year, the time taken for the Earth to orbit the Sun. The **calendar month** is a human invention, devised to fit the calendar year.

temperature

Celsius scale Previously called centigrade, the Celcius scale of temperature ranges from freezing to boiling of water and is divided into 100 degrees, freezing point being 0 degrees and boiling point 100 degrees.

The degree centigrade (°C) was officially renamed Celsius in 1948 to avoid confusion with the angular measure known as the centigrade (one-hundredth of a grade). The Celsius scale is named after the Swedish astronomer Anders Celsius who devised it in 1742 but in reverse (freezing point was 100°; boiling point 0°).

Celsius
http://users.bart.nl/~sante/engtemp.html
Brief but comprehensive description of how Swedish astronomer Anders Celsius devised his temperature scale.

Fahrenheit scale The Fahrenheit temperature scale was invented in 1714 by the Polish-born Dutch physicist Gabriel Fahrenheit and was commonly used in English-speaking countries until the 1970s, after which the Celsius scale was generally adopted, in line with the rest of the world. In the Fahrenheit scale, intervals are measured in degrees (°F); $°F = (°C \times {}^9/_5) + 32$.

Fahrenheit took as the zero point the lowest temperature he could achieve anywhere in the laboratory, and, as the other fixed point, body temperature, which he set at 96°F. On this scale, water freezes at 32°F and boils at 212°F.

Kelvin scale Introduced in the 19th century by the Irish physicist William Thomson Kelvin, the Kelvin scale of absolute temperature is used by scientists. It begins at absolute zero (-273.15°C) and increases by the same degree intervals as the Celsius scale; that is, 0°C is the same as 273.15 K and 100°C is 373.15 K.

sound

Loudness is the subjective judgement of the level or power of sound reaching the ear. The human ear cannot give an absolute value to the loudness of a single sound, but can only make comparisons between two different sounds. The precise measure of the power of a sound wave at a particular point is called its intensity.

Accurate comparisons of sound levels may be made using sound-level meters, which are calibrated in units called decibels.

decibel scale

The decibel scale is used for audibility measurements, as one decibel unit (symbol dB), representing an increase of about 25%, is about the smallest change the human ear can detect. A whisper has an intensity of 20 dB; 140 dB (a jet aircraft taking off nearby) is the threshold of pain.

Used originally to compare sound intensities, the decibel was used subsequently to measure electrical or electronic power outputs and is now also used to compare voltages. An increase of 10 dB is equivalent to a 10-fold increase in intensity or power, and a 20-fold increase in voltage. The difference in decibels between two levels of intensity (or power) L_1 and L_2 is $10 \log_{10}(L_1/L_2)$; a difference of 1 dB thus corresponds to a change of about 25%. For two voltages V_1 and V_2, the difference in decibels is $20 \log_{10}(V_1/V_2)$; 1 dB corresponding in this case to a change of about 12%.

Decibel measurements of noise are often 'A-weighted' to take into account the fact that some sound wavelengths are perceived as being particularly loud.

electricity

ampere The SI unit of electrical current is the ampere (symbol A). Electrical current is measured in a similar

Decibel Scale	
Decibels	**Typical sound**
0	threshold of hearing
10	rustle of leaves in gentle breeze
10	quiet whisper
20	average whisper
20–50	quiet conversation
40–45	hotel; theatre (between performances)
50–65	loud conversation
65–70	traffic on busy street
65–90	train
75–80	factory (light/medium work)
90	heavy traffic
90–100	thunder
110–140	jet aircraft at take-off
130	threshold of pain
140–190	space rocket at take-off

Table of Equivalent Temperatures

Celsius and Fahrenheit temperatures can be interconverted as follows: C = (F − 32) × 100/180; F = (C × 180/100) + 32.

°C	°F	°C	°F	°C	°F	°C	°F
100	212.0	70	158.0	40	104.0	10	50.0
99	210.2	69	156.2	39	102.2	9	48.2
98	208.4	68	154.4	38	100.4	8	46.4
97	206.6	67	152.6	37	98.6	7	44.6
96	204.8	66	150.8	36	96.8	6	42.8
95	203.0	65	149.0	35	95.0	5	41.0
94	201.2	64	147.2	34	93.2	4	39.2
93	199.4	63	145.4	33	91.4	3	37.4
92	197.6	62	143.6	32	89.6	2	35.6
91	195.8	61	141.8	31	87.8	1	33.8
90	194.0	60	140.0	30	86.0	0	32.0
89	192.2	59	138.2	29	84.2	−1	30.2
88	190.4	58	136.4	28	82.4	−2	28.4
87	188.6	57	134.6	27	80.6	−3	26.6
86	186.8	56	132.8	26	78.8	−4	24.8
85	185.0	55	131.0	25	77.0	−5	23.0
84	183.2	54	129.2	24	75.2	−6	21.2
83	181.4	53	127.4	23	73.4	−7	19.4
82	179.6	52	125.6	22	71.6	−8	17.6
81	177.8	51	123.8	21	69.8	−9	15.8
80	176.0	50	122.0	20	68.0	−10	14.0
79	174.2	49	120.2	19	66.2	−11	12.2
78	172.4	48	118.4	18	64.4	−12	10.4
77	170.6	47	116.6	17	62.6	−13	8.6
76	168.8	46	114.8	16	60.8	−14	6.8
75	167.0	45	113.0	15	59.0	−15	5.0
74	165.2	44	111.2	14	57.2	−16	3.2
73	163.4	43	109.4	13	55.4	−17	1.4
72	161.6	42	107.6	12	53.6	−18	-0.4
71	159.8	41	105.8	11	51.8	−19	-2.2

way to water current, in terms of an amount per unit of time; one ampere represents a flow of about 6.28 x 10^{18} electrons per second, or a rate of flow of charge of one coulomb per second.

The ampere is defined as the current that produces a specific magnetic force between two long, straight, parallel conductors placed 1 m/3.3 ft apart in a vacuum. It is named after the French physicist and mathematician André Ampère (1775–1836).

volt The SI unit of electromotive force or electric potential is the volt (symbol V). A small battery has a potential of 1.5 volts, whilst a high-tension transmission line may carry up to 765,000 volts. The domestic electricity supply in the UK is 230 volts (lowered from 240 volts in 1995); it is 110 volts in the USA.

The **absolute volt** is defined as the potential difference necessary to produce a current of one ampere through

an electric circuit with a resistance of one ohm (SI unit of electrical resistance). It can also be defined as the potential difference that requires one joule of work to move a positive charge of one coulomb from the lower to the higher potential. It is named after the Italian physicist Alessandro Volta (1745–1827).

watt The watt (symbol W) is the SI unit of power (the rate of expenditure or consumption of energy), defined as one joule (SI unit of work and energy, see below) per second. A light bulb, for example, may use 40, 60, 100, or 150 watts of power; an electric heater will use several kilowatts (thousands of watts). The watt is named after the Scottish engineer James Watt (1736–1819).

The **absolute watt** is defined as the power used when one joule of work is done in one second. In electrical terms, the flow of one ampere of current through a conductor whose ends are at a potential difference of one volt uses one watt of power (watts = volts x amperes).

energy

joule The joule (symbol J) is the SI unit of work and energy, replacing the calorie (one joule equals 4.2 calories). It is defined as the work done (energy transferred) by a force of one newton (SI unit of force) acting over one metre. It can also be expressed as the work done in one second by a current of one ampere at a potential difference of one volt. One watt is equal to one joule per second.

calorie The c.g.s. unit of heat, the calorie, is defined as the amount of heat required to raise the temperature of one gram of water by 1°C. It is now replaced by the joule (one calorie is approximately 4.2 joules). In dietetics, the Calorie or kilocalorie is equal to 1,000 calories.

The kilocalorie measures the energy value of food in terms of its heat output: 28 g/1 oz of protein yields 120 kilocalories; the same amount of carbohydrate yields 110 kilocalories; fat, 270 kilocalories; and alcohol, 200 kilocalories.

US Weights and Measures

A metric act was passed in the USA in 1975, but metric measurement has not been readily adopted, except by the scientific community.

Customary US weights and measures

Length

1 foot	= 12 inches
1 yard	= 3 feet
1 rod (pole or perch)	= $5\frac{1}{2}$ yards
	= $16\frac{1}{2}$ feet
1 chain	= 4 rods
	= 22 yards
1 furlong	= 10 chains
	= 220 yards
	= 40 rods
	= 660 feet
1 mile	= 8 furlongs
	= 1,760 yards
	= 5,280 feet
1 league	= 3 miles
	= 5,280 yards
	= 15,840 feet

Nautical

1 fathom	= 6 feet
1 cable length	= 120 fathoms
	= 720 feet
1 international nautical mile	= 6,076.11549 feet

Area

1 square foot	= 144 square inches
1 square yard	= 9 square feet
	= 1,296 square inches
1 square rod	= $30\frac{1}{4}$ square yards
	= $272\frac{1}{4}$ square feet
1 square mile acres	= 640
1 section (of land)	= 1 mile square[1]

Area

1 township	= 6 miles square[1]
	= 36 sections (of land)
	= 36 square miles[1]

Cubic measure

1 cubic foot	= 1,728 cubic inches
1 cubic yard	= 27 cubic feet
	= 46,656 cubic inches

Volume (dry measure)

1 quart	= 2 pints
	= 67.2006 cubic inches
1 peck	= 8 quarts
	= 16 pints
	= 537.605 cubic inches
1 bushel	= 4 pecks
	= 32 quarts
	= 2,150.42 cubic inches

Volume (liquid measure) and apothecaries' fluid measures

1 fluid dram	= 60 minims
	= 0.2256 cubic inches
1 fluid ounce	= 8 fluid drams
	= 1.8047 cubic inches
1 gill	= 32 fluid drams
	= 4 fluid ounces
	= 7.2188 cubic inches
1 pint	= 128 fluid drams
	= 16 fluid ounces
	= 4 gills
	= 28.875 cubic inches
1 quart	= 256 fluid drams
	= 32 fluid ounces
	= 8 gills

[1] Square miles are not the same as miles square.

Volume (liquid measure) and apothecaries' fluid measures

	= 2 pints
	= 57.75 cubic inches
1 gallon	= 1,024 fluid drams
	= 128 fluid ounces
	= 32 gills
	= 8 pints
	= 4 quarts
	= 231 cubic inches

Avoirdupois weight

1 dram	= $27\frac{11}{32}$ grains
1 ounce	= $437\frac{1}{2}$ grains
	= 16 drams
1 pound	= 7,000 grains
	= 256 drams
	= 16 ounces

Avoirdupois weight

1 hundredweight[2]	= 100 pounds
1 gross (or long) hundredweight[2]	= 112 pounds
1 ton[2]	= 2,000 pounds
	= 20 hundredweights
1 gross (or long ton)[2]	= 2,240 pounds
	= 20 gross (or long ton) hundredweights
1 scruple	= 20 grains
1 dram	= 60 grains
	= 3 scruples

Avoirdupois weight

1 apothecaries' ounce	= 480 grains
	= 24 scruples
	= 8 drams
1 apothecaries' pound	= 5,760 grains
	= 288 scruples
	= 96 apothecaries' drams

Apothecaries' weights

	= 12 apothecaries' ounces

Troy weights

1 pennyweight	= 24 grains
1 troy ounce	= 480 grains
	= 240 pennyweights
1 troy pound	= 5,760 grains
	= 240 pennyweights
	= 12 troy ounces

Gunter's or surveyor's chain measures

1 link	= 7.92 inches
1 chain	= 100 links
	= 4 rods
	= 66 feet
1 statute mile	= 88 chains
	= 320 rods
	= 5,280 feet

Units of circular measure

Second	
Minute	= 60 seconds
Degree	= 60 minutes
Right angle	= 90 degrees
Straight angle	= 180 degrees
Circle	= 360 degrees

[2] The terms 'hundredweight' and 'ton' are normally used to mean the 100 pound hundredweight and the 2,000 pound ton. These units are sometimes designated the terms 'net' or 'short' to differentiate them from the corresponding 'gross' or 'long' hundredweight and ton.

Miscellaneous Units

Unit	Definition
acoustic ohm	c.g.s. unit of acoustic impedance (the ratio of sound pressure on a surface to sound flux through the surface)
acre-foot	unit sometimes used to measure large volumes of water such as reservoirs; 1 acre-foot = 1,233.5 cu m/43,560 cu ft
astronomical unit	unit (symbol AU) equal to the mean distance of the Earth from the Sun: 149,597,870 km/92,955,808 mi
atmosphere	unit of pressure (abbreviation atm); 1 standard atmosphere = 101,325 Pa
barn	unit of area, especially the cross-sectional area of an atomic nucleus; 1 barn = 10^{-28} sq m
barrel	unit of liquid capacity; the volume of a barrel depends on the liquid being measured and the country and state laws. In the USA, 1 barrel of oil = 42 gal (159 l/34.97 imperial gal), but for federal taxing of fermented liquor (such as beer), 1 barrel = 31 gal (117.35 l/25.81 imperial gal). Many states fix a 36-gallon barrel for cistern measurement and federal law uses a 40-gallon barrel to measure 'proof spirits'. 1 barrel of beer in the UK = 163.66 l (43.23 US gal/36 imperial gal)
base box	imperial unit of area used in metal plating; 1 base box = 20.232 sq m/31,360 sq in
baud	unit of electrical signalling speed equal to 1 pulse per second
brewster	unit (symbol B) for measuring reaction of optical materials to stress
British thermal unit	imperial unit of heat (symbol Btu); 1 Btu = approximately 1,055 J

Unit	Definition
bushel	measure of dry and (in the UK) liquid volume. 1 bushel (struck measure) = 8 dry US gallons (64 dry US pt/35.239 l/2,150.42 cu in). 1 heaped US bushel = 1,278 bushels, struck measure (81.78 dry pt/45.027 l/2,747.715 cu in), often referred to a $1\frac{1}{4}$ bushels, struck measure. In the UK, 1 bushel = 8 imperial gallons (64 imperial pt); 1 UK bushel = 1.03 US bushels
cable	unit of length used on ships, taken as $\frac{1}{10}$ of a nautical mile (185.2 m/607.6 ft)
carcel	obsolete unit of luminous intensity
cental	name for the short hundredweight; 1 cental = 45.36 kg/100 lb
chaldron	obsolete unit measuring capacity; 1 chaldron = 1.309 cu m/46.237 cu ft
clausius	in engineering, a unit of entropy; defined as the ratio of energy to temperature above absolute zero
cleanliness unit	unit for measuring air pollution; equal to the number of particles greater than 0.5 (m in diameter per cu ft of air
clo	unit of thermal insulation of clothing; standard clothes have insulation of about 1 clo, the warmest have about 4 clo per 2.5 cm/1 in of thickness
clusec	unit for measuring the power of a vacuum pump
condensation number	in physics, the ratio of the number of molecules condensing on a surface to the number of molecules touching that surface
cord	unit for measuring the volume of wood cut for fuel; 1 cord = 3.62 cu m/128 cu ft, or a stack 2.4 m/8 ft long, 1.2 m/4 ft wide and 1.2 m/4 ft high
crith	unit of mass for weighing gases; 1 crith = the mass of 1 litre of hydrogen gas at standard temperature and pressure
cubit	earliest known unit of length; 1 cubit = approximately 45.7 cm/18 in, the length of the human forearm from the tip of the middle finger to the elbow
curie	former unit of radioactivity (symbol Ci); 1 curie = 3.7×10^{10} becquerels
dalton	international atomic mass unit, equivalent to $\frac{1}{12}$ of the mass of a neutral carbon-12 atom
darcy	c.g.s. unit (symbol D) of permeability, used mainly in geology to describe the permeability of rock
darwin	unit of measurement of evolutionary rate of change
decontamination factor	unit measuring the effectiveness of radiological decontamination; the ratio of original contamination to the radiation remaining
demal	unit measuring concentration; 1 demal = 1 gram-equivalent of solute in 1 cu dm of solvent
denier	unit used to measure the fineness of yarns; 9,000 m of 15 denier nylon weighs 15 g/0.5 oz
dioptre	optical unit measuring the power of a lens; the reciprocal of the focal length in metres
dram	unit of apothecaries' measure; 1 dram = 60 grains/3.888 g
dyne	c.g.s. unit of force; 10^5 dynes = 1 N
einstein unit	unit for measuring photoenergy in atomic physics
eotvos unit	unit (symbol E) for measuring small changes in the intensity of the Earth's gravity with horizontal distance
erg	c.g.s. unit of work; equal to the work done by a force of 1 dyne moving through 1 cm
erlang	unit for measuring telephone traffic intensity; for example, 90 minutes of carried traffic measured over 60 minutes = 1.5 erlangs ('carried traffic' refers to the total duration of completed calls made within a specified period)
fathom	unit of depth measurement in mining and seafaring; 1 fathom = 1.83 m/6 ft
finsen unit	unit (symbol FU) for measuring intensity of ultraviolet light
fluid ounce	measure of capacity; equivalent in the USA to $\frac{1}{16}$ of a pint ($\frac{1}{20}$ of a pint in the UK and Canada)
foot-candle	unit of illuminance, replaced by the lux; 1 foot-candle = 10.76391 lux
foot-pound	imperial unit of energy (symbol ft-lb); 1 ft-lb = 1.356 joule
frigorie	unit (symbol fg) used in refrigeration engineering to measure heat energy, equal to a rate of heat extraction of 1 kilocalorie per hour
furlong	unit of measurement, originating in Anglo-Saxon England, equivalent to 201.168 m/220 yd
galileo	unit (symbol Gal) of acceleration; 1 galileo = 10^{-2} m s^{-2}
gauss	c.g.s. unit (symbol) of magnetic flux density, replaced by the tesla; 1 gauss = 1×10^{-4} tesla
gill	imperial unit of volume for liquid measure; equal to $\frac{1}{4}$ of a pint (in the USA, 4 fl oz/0.118 l; in the UK, 5 fl oz/0.142 l)
grain	smallest unit of mass in the three English systems of measurement (avoirdupois, troy, apothecaries' weights) used in the UK and USA; 1 grain = 0.0648 g
hand	unit used in measuring the height of a horse from front hoof to shoulder (withers); 1 hand = 10.2 cm/4 in

Unit	Definition
hardness number	unit measuring hardness of materials. There are many different hardness scales: Brinell, Rockwell, and Vickers scales measure the degree of indentation or impression of materials; Mohs' scale measures resistance to scratching against a standard set of minerals
hartree	atomic unit of energy, equivalent to atomic unit of charge divided by atomic unit of length; 1 hartree = 4.850×10^{-18} J
haze factor	unit of visibility in mist or fog; the ratio of brightness of mist compared with that of the object
Hehner number	unit measuring concentration of fatty acids in oils; a Hehner number of 1 = 1 kg of fatty acid in 100 kg of oil or fat
hide	unit of measurement used in the 12th century to measure land; 1 hide = 60–120 acres/25–50 ha
horsepower	imperial unit (abbreviation hp) of power; 1 horsepower = 746 W
hundredweight	imperial unit (abbreviation cwt) of mass; 1 cwt = 45.36 kg/100 lb in the USA and 50.80 kg/112 lb in the UK
inferno	unit used in astrophysics for describing the temperature inside a star; 1 inferno = 1 billion K (degrees Kelvin)
iodine number	unit measuring the percentage of iodine absorbed in a substance, expressed as grams of iodine absorbed by 100 grams of material
jansky	unit used in radio astronomy to measure radio emissions or flux densities from space; 1 jansky = 10^{-26} Wm^{-2} Hz^{-1}. Flux density is the energy in a beam of radiation which passes through an area normal to the beam in a single unit of time. A jansky is a measurement of the energy received from a cosmic radio source per unit area of detector in a single time unit
kayser	unit used in spectroscopy to measure wave number (number of waves in a unit length); a wavelength of 1.0 cm has a wave number of 1 kayser
knot	unit used in navigation to measure a ship's speed; 1 knot = 1 nautical mile per hour, or about 1.15 miles per hour
league	obsolete imperial unit of length; 1 league = 3 nautical mi/5.56 km or 3 statute mi/4.83 km
light year	unit used in astronomy to measure distance; the distance travelled by light in one year, approximately 9.46×10^{12} km/5.88×10^{12} mi
mache	obsolete unit of radioactive concentration; 1 mache = 3.7×10^{-7} curies of radioactive material per cu m of a medium
maxwell	c.g.s. unit (symbol Mx) of magnetic flux, the strength of a magnetic field in an area multiplied by the area; 1 maxwell = 10^{-8} weber
megaton	measurement of the explosive power of a nuclear weapon; 1 megaton = 1 million tons of trinitrotoluene (TNT)
millimetre of mercury	unit of pressure (symbol mmHg) used in medicine for measuring blood pressure
morgan	arbitrary unit used in genetics; 1 morgan is the distance along the chromosome in a gene that gives a recombination frequency of 1%
nautical mile	unit of distance used in navigation, equal to the average length of 1 minute of arc on a great circle of the Earth; 1 international nautical mile =1.852 km/6,076 ft
neper	unit used in telecommunications; gives the attenuation of amplitudes of currents or powers as the natural logarithm of the ratio of the voltage between two points or the current between two points
oersted	c.g.s. unit (symbol Oe) of magnetic field strength, now replaced by amperes per metre (1 Oe = 79.58 amp per m)
parsec	unit (symbol pc) used in astronomy for distances to stars and galaxies; 1 pc = 3.262 light years, 2.063×10^5 astronomical units, or 3.086×10^{13} km
peck	obsolete unit of dry measure, equal to 8 imperial quarts or 1 quarter bushel (8.1 l in the USA or 9.1 l in the UK)
pennyweight	imperial unit of mass; 1 pennyweight = 24 grains = 1.555×10^{-3} kg
perch	obsolete imperial unit of length; 1 perch = $5^1/_2$ yards = 5.029 m, also called the rod or pole
point	metric unit of mass used in relation to gemstones; 1 point = 0.01 metric carat = 2×10^{-3} g
poise	c.g.s. unit of dynamic viscosity; 1 poise = 1 dyne-second per sq cm
poundal	imperial unit (abbreviation pdl) of force; 1 poundal = 0.1383 newton
quart	imperial liquid or dry measure; in the USA, 1 liquid quart = 0.946 l, while 1 dry quart = 1.101 l; in the UK, 1 quart =2 pt/1.137 l
rad	unit of absorbed radiation dose, replaced in the SI system by the gray; 1 rad = 0.01 joule of radiation absorbed by 1 kg of matter

Unit	Definition
relative biological effectiveness	relative damage caused to living tissue by different types of radiation
rood	imperial unit of area; 1 rood = $^1/_4$ acre = 1,011.7 sq m
roentgen	unit (symbol R) of radiation exposure, used for X- and gamma rays
rydberg	atomic unit of energy; 1 rydberg = 2.425×10^{-18} J
sabin	unit of sound absorption, used in acoustical engineering; 1 sabin = absorption of 1 sq ft (0.093 sq m) of a perfectly absorbing surface
scruple	imperial unit of apothecaries' measure; 1 scruple = 20 grains = 1.3×10^{-3} kg
shackle	unit of length used at sea for measuring cable or chain; 1 shackle = 15 fathoms (90 ft/27 m)
slug	obsolete imperial unit of mass; 1 slug = 14.59 kg/32.17 lb
snellen	unit expressing the visual power of the eye
sone	unit of subjective loudness
standard volume	in physics, the volume occupied by 1 kilogram molecule (molecular mass in kilograms) of any gas at standard temperature and pressure; approximately 22.414 cu m
stokes	c.g.s. unit (symbol St) of kinematic viscosity; 1 stokes = 10^{-4} m^2 s^{-1}
stone	imperial unit (abbreviation st) of mass; 1 stone = 6.35 kg/14 lb
strontium unit	measures concentration of strontium-90 in an organic medium relative to the concentration of calcium
tex	metric unit of line density; 1 tex is the line density of a thread with a mass of 1 gram and a length of 1 kilometre
tog	measure of thermal insulation of a fabric, garment, or quilt; the tog value is equivalent to 10 times the temperature difference (in °C) between the two faces of the article, when the flow of heat across it is equal to 1 W per sq m

Measurement Chronology

c. 3500 BC The gnomon – the first clock – is invented, probably in Egypt. It consists of a vertical stick or pillar inserted in the ground, the length of its shadow giving an idea of the time.

c. 3000 BC The cubit, the length of the arm from the elbow to the extended finger tips, is devised in Egypt as the standard unit of linear measure. A royal cubit of black granite serves as the standard for all other cubit sticks.

c. 2400 BC Sumerian scribes develop a calendar consisting of twelve 30-day months (360 days).

1800 BC The Babylonian Empire standardizes the year by adopting the lunar calendar of the Sumerian sacred city of Nippur. Previously, each city inserted intercalated months according to its own needs.

c. 1000 BC The Hindu calendar is developed in India. It is based on a solar year of 360 days divided into 12 lunar months of 27 or 28 days with a leap month intercalated every 60 months to bring it into line with the true solar year.

c. 975 BC The Gezer Calendar is devised. Based on a lunar cycle of 12 months and 354 days, it is tied into the solar year and forms the basis of the Hebrew calendar.

738 BC Romulus, traditionally the founder of Rome, devises a lunar calendar with 10 months, 6 of 30 days and 4 of 31 days. The year begins in March and ends in December and is followed by an uncounted winter gap.

c. 600 BC The Roman king Tarquinius Priscus introduces the Roman Republican calendar. It consists of 12 months with a total of 355 days. An intercalated month is added between February 23 and 24 every two years in order to keep step with the seasons. Intercalations, however, are made irregularly and it becomes hopelessly confused. The calendar forms the basis of the Gregorian calendar.

587 BC The Babylonians introduce their calendar to Jerusalem after their conquest of the city. It provides the Jews with a finite calendar with a New Year's day. The Babylonian names continue to be used.

c. 300 BC The Greek scholar Dicaearchus places the first orientation line on a map of the world; it runs through Gibraltar and Rhodes. The idea eventually leads to the system of parallels and meridians, and methods of projecting them.

237 BC The Egyptian king Ptolemy III improves the Egyptian calendar by introducing an extra day every four years to the basic 356 day calendar.

c. 140 BC The Greek Stoic philosopher Crates of Mallus, in Anatolia, makes the first globe.

105 BC	The Chinese historian Sima Qian reforms the Chinese calendar.
46 BC	The Roman consul and dictator Julius Caesar instructs Alexandrian astronomer Sosigenes to bring the Roman Republican calendar into line with the solar year. He creates the Julian calendar in which the year is 365 days long and begins 1 January. An extra day is inserted between 23 and 24 February every four years. The year 46 BC is 445 days long to bring it into line with the solar year.
c. 12 BC	A geographical commentary written by the Roman co-ruler (with Augustus) Marcus Vipsanius Agrippa, based on surveys of Rome's military roads, is used to produce a map of the world.
9 BC	The first calliper rule for measuring the thickness of objects is made by an anonymous Chinese inventor who inscribes the date of manufacture on it.
c. 800	Harun ar-Rashid presents an elaborate astronomical water-clock, built by the Arab engineer al-Jazari, to the Frankish emperor Charlemagne. The mechanism is driven by falling water, and the clock sounds the time by dropping bronze balls.
1120	The French-born Prior Walcher of Malvern Abbey, England, introduces the measurement of latitude and longitude in degrees, minutes, and seconds.
1326	The English astronomer Richard of Wallingford writes *Canones de instrumento*, which describes the astronomical clock he built at St Albans, England, in 1320.
1340	The avoirdupois weight system is introduced in England.
15 Dec 1582	The Gregorian calendar is adopted in the United Netherlands; the date advances by ten days.
1670	Gabriel Menton, the vicar of the Church of Saint-Paul, Lyon, France, invents the metre as a scientific measure of distance, based on a minute of arc on the Earth's surface.
1671	Respected French scientist Jean Picard is among several scientists who take up French clergyman Gabriel Menton's proposal of the metre as a standard measure of distance.
1700	The new Gregorian calendar is introduced in Germany and other Protestant European states, replacing the older, less accurate, Julian calendar.
10 Dec 1799	The metric system of measurement is officially adopted in France.
1822	The German scientist Friedrich Mohs introduces a scale for specifying the hardness of minerals.
1827	King George IV of Great Britain and Ireland standardizes the acre at 43,560 sq ft (1/640 of a sq mile).
1864	The metric system of measurement is officially adopted in England, though it does not supplant the imperial system until the 1970s.
1866	The US government legalizes the metric system of measurement.
20 May 1875	The International Bureau of Weights and Measures is established in France by a treaty signed in Paris. Located at Sèvres, its purpose is to unify systems of measurement, and to establish standards by providing a prototype metre and kilogram as the basis for all scientific and other measures.
1 Jan 1876	The International System of Weights and Measures comes into effect in France.
1927	The International Temperature Scale is adopted by countries subscribing to the International Bureau of Weights and Measures.
1916	The Royal Society sponsors a Board of Scientific Societies to promote cooperation in pure and applied science and to promote the application of science for the service of Britain.
1947	The British government sets up the Advisory Committee on Scientific Policy.
1949	The Chinese Academy of Sciences is founded in Shanghai.
Oct 1960	The 11th general Conference on Weights and Measures replaces the metric system with the International System (SI) of weights and measures. It redefines the seven basic units of measurement, from which all others are derived, in atomic terms. The metre, for instance is redefined as 1,650,763.73 wavelengths of the orange-red line in the krypton-86 spectrum.
1967	The 13th General Committee on Weights and Measures redefines the second in terms of the resonant frequency of the caesium atom.
1971	In its report to Congress, *A Metric America: A Decision Whose Time has Come*, the US National Bureau of Standards recommends that the USA change to the International Metric System over a period of ten years.
11 Dec 1975	The US Congress passes legislation calling for the voluntary conversion to the metric system in ten years.
20 Oct 1983	The General Conference on Weights and Measures at Sèvres, France, redefines the metre as the distance that light travels through a vacuum in 1/299,792,458 seconds.

Biographies

Ångström, Anders Jonas (1814–1874) Swedish astrophysicist who worked in spectroscopy and solar physics. His *Recherches sur le spectre solaire* (1868) presented an atlas of the solar spectrum with measurements of 1,000 spectral lines expressed in units of one-ten-millionth of a millimetre, the unit which later became the **angstrom**.

Arnold, John (1736–1799) English horologist. He first made his name as the maker of a very small half-quarter repeater for King George III, but later became known for his chronometers.

Atwater, Wilbur Olin (1844–1907) US agricultural chemist and educator. As professor of chemistry at Wesleyan University from 1873, he directed the first agricultural experiment station. He was also involved in fertilizer testing and the investigation of nutritional and calorimetric standards.

Banneker, Benjamin (1731–1806) American astronomer, surveyor, and mathematician who published almanacs 1792–97. In 1753, having studied only a pocket watch, Banneker constructed a striking clock, the first of its kind in America. He took part in the survey that prepared the establishment of the US capital, Washington, DC.

Bürgi, Jost (1552–1632) Swiss-born clockmaker and mathematician. One of the first clockmakers to use second hands, Bürgi also introduced a mechanism for providing the escapement with a constant driving force. In mathematics, he developed a comprehensive system of logarithms (about 30 years before the Scottish mathematician John Napier), but his work has been largely ignored.

Celsius, Anders (1701–1744) Swedish astronomer, physicist, and mathematician who introduced the **Celsius temperature scale**. In 1742 he presented a proposal to the Swedish Academy of Sciences that all scientific measurements of temperature should be made on a fixed scale based on two invariable (generally speaking) and naturally occurring points. His scale defined 0° as the temperature at which water boils, and 100° as that at which water freezes. This scale, in an inverted form devised eight years later by his pupil Martin Strömer, has since been used in almost all scientific work.

Edgeworth, Richard Lovell (1744–1817) Anglo-Irish inventor. He produced an early form of visual telegraphy, the velocipede, the perambulator land-measuring wheel, and various forms of carriage, including a phaeton and a sail-propelled version.

Fahrenheit, Gabriel Daniel (1686–1736) Polish-born Dutch physicist who invented the first accurate thermometer in 1724 and devised the **Fahrenheit temperature scale**. Using his thermometer, Fahrenheit was able to determine the boiling points of liquids and found that they vary with atmospheric pressure. His first thermometers contained a column of alcohol which expanded and contracted directly, as originally devised in 1701 by the Danish astronomer Ole Römer (1644–1710). Fahrenheit substituted mercury for alcohol because its rate of expansion, although less than that of alcohol, is more constant and could be used over a much wider temperature range. To reflect the greater sensitivity of his thermometer, Fahrenheit expanded Römer's scale so that blood heat was 90° and an ice-salt mixture was 0°; on this scale freezing point was 30°. Fahrenheit later adjusted the scale to ignore body temperature as a fixed point so that the boiling point of water came to 212° and freezing point was 32°. This is the Fahrenheit scale that is still in use today.

Harrison, John (1693–1776) English horologist and instrumentmaker. He made the first chronometers that were accurate enough to allow the precise determination of longitude at sea, and so permit reliable and safe navigation over long distances.

Ohm, Georg Simon (1789–1854) German physicist who studied electricity and discovered the fundamental law that bears his name. The SI unit of electrical resistance, the **ohm**, is named after him, and the unit of conductance (the inverse of resistance) was formerly called the **mho**, which is `ohm´ spelled backwards.

Vernier, Pierre (c. 1580–1637) French engineer and instrumentmaker who invented a means of making very precise measurements with what is now called the **vernier scale**. He realized the need for a more accurate way of reading angles on the surveying instruments he used in mapmaking whilst working as a military engineer. In 1631 he published *The Construction, Uses and Properties of a New Mathematical Quadrant*, in which he explained his method.

Glossary

absolute weight the weight of a body considered apart from all modifying influences such as the atmosphere. To determine its absolute weight, the body must, therefore, be weighed in a vacuum or allowance must be made for buoyancy.

accelerometer apparatus, either mechanical or electro-mechanical, for measuring acceleration or deceleration – that is, the rate of increase or decrease in the velocity of a moving object.

acoustic ohm c.g.s. unit of acoustic impedance (the ratio of the sound pressure on a surface to the sound flux through the surface). It is analogous to the ohm as the unit of electrical impedance.

acre-foot unit sometimes used to measure large volumes of water, such as the capacity of a reservoir (equal to its area in acres multiplied by its average depth in feet). One acre-foot equals 1,233.5 cu m/43,560 cu ft or the amount of water covering one acre to a depth of one foot.

altitude measurement of height, usually given in metres above sea level.

angstrom unit (symbol Å) of length equal to 10^{-10} metres or one-ten-millionth of a millimetre, used for atomic measurements and the wavelengths of electromagnetic radiation. It is named after the Swedish astrophysicist Anders Ångström.

ANSI abbreviation for **American National Standards Institute**, a US national standards body.

apothecaries' weights obsolete units of mass, formerly used in pharmacy: 20 grains equal one scruple; three scruples equal one dram; eight drams equal an apothecary's ounce (oz apoth.), and 12 such ounces equal an apothecary's pound (lb apoth.). There are 7,000 grains in one pound avoirdupois (0.454 kg).

are metric unit of area, equal to 100 square metres (119.6 sq yd); 100 ares make one hectare.

astronomical unit (symbol AU) unit equal to the mean distance of the Earth from the Sun: 149,597,870 km/92,955,800 mi. It is used to describe planetary distances. Light travels this distance in approximately 8.3 minutes.

atmosphere, or **standard atmosphere**, in physics, a unit (symbol atm) of pressure equal to 760 torr, 1013.25 millibars, or 1.01325×10^5 newtons per square metre. The actual pressure exerted by the atmosphere fluctuates around this value, which is assumed to be standard at sea level and 0°C/32°F, and is used when dealing with very high pressures.

atomic time time as given by atomic clocks, which are regulated by natural resonance frequencies of particular atoms, and display a continuous count of seconds.

Avogadro's number, or **Avogadro's constant**, the number of carbon atoms in 12 g of the carbon-12 isotope (6.022045×10^{23}). The relative atomic mass of any element, expressed in grams, contains this number of atoms. It is named after the Italian physicist Amedeo Avogadro (1776–1856).

avoirdupois system of units of mass based on the pound (0.45 kg), which consists of 16 ounces (each of 16 drams) or 7,000 grains (each equal to 65 mg).

bar unit of pressure equal to 10^5 pascals or 10^6 dynes/cm^2, approximately 750 mmHg or 0.987 atm. Its diminutive, the **millibar** (one-thousandth of a bar), is commonly used by meteorologists.

barrel unit of liquid capacity, the value of which depends on the liquid being measured. It is used for petroleum, a barrel of which contains 159 litres/35 imperial gallons; a barrel of alcohol contains 189 litres/41.5 imperial gallons.

baud in engineering, a unit of electrical signalling speed equal to one pulse per second, measuring the rate at which signals are sent between electronic devices such as telegraphs and computers; 300 baud is about 300 words a minute.

becquerel SI unit (symbol Bq) of radioactivity, equal to one radioactive disintegration (change in the nucleus of an atom when a particle or ray is given off) per second.

bel unit of sound measurement equal to ten decibels. It is named after the Scottish scientist Alexander Graham Bell (1847–1922).

bells nautical term applied to half-hours of watch. A day is divided into seven watches, five of four hours each and two, called dogwatches, of two hours. Each half-hour of each watch is indicated by the striking of a bell, eight bells signalling the end of the watch.

bolometer sensitive thermometer that measures the energy of radiation by registering the change in electrical resistance of a fine wire when it is exposed to heat or light. It was devised in 1880 by the US astronomer Samuel Langley (1834–1906) for measuring radiation from stars.

brewster unit (symbol B) for measuring the reaction of optical materials to stress, defined in terms of the slowing down of light passing through the material when it is stretched or compressed.

British Standards Institution (BSI) UK national standards body. Although government funded, the institution is independent. The BSI interprets international technical standards for the UK, and also sets its own.

British thermal unit imperial unit (symbol Btu) of heat, now replaced in the SI system by the ◊joule (one British thermal unit is approximately 1,055 joules). Burning one cubic foot of natural gas releases about 1,000 Btu of heat.

BSI abbreviation for ◊**British Standards Institution**.

BST abbreviation for **British Summer Time**.

bushel dry or liquid measure equal to eight gallons or four pecks (2,219.36 cu in/36.37 litres) in the UK; some US states have different standards according to the goods measured.

°C symbol for degrees Celsius, sometimes called centigrade.

cable unit of length, used on ships, originally the length of a ship's anchor cable or 120 fathoms (219 m/720 ft), but now taken as one-tenth of a nautical mile (185.3 m/608 ft).

calendar division of the year into months, weeks, and days and the method of ordering the years. From year one, an assumed date of the birth of Jesus, dates are calculated backwards (BC 'before Christ' or BCE 'before common era') and forwards (AD, Latin *anno Domini* 'in the year of the Lord', or CE 'common era'). The **lunar month** (period between one new Moon and the next) naturally averages 29.5 days, but the Western calendar uses for convenience a **calendar month** with a complete number of days, 30 or 31 (February has 28). For adjustments, since there are slightly fewer than six extra hours a year left over, they are added to February as a 29th day every fourth year (**leap year**), century years being excepted unless they are divisible by 400. For example, 1896 was a leap year; 1900 was not.

candela SI unit (symbol cd) of luminous intensity, which replaced the old units of candle and standard candle. It measures the brightness of a light itself rather than the amount of light falling on an object, which is called **illuminance** and measured in ◊lux.

cc symbol for **cubic centimetre**.

centigrade former name for the Celsius temperature scale.

CET abbreviation for **Central European Time**.

chronometer instrument for measuring time precisely, originally used at sea. It is designed to remain accurate through all conditions of temperature and pressure. The first accurate marine chronometer, capable of an accuracy of half a minute a year, was made in 1761 by the English horologist and instrumentmaker John Harrison.

clausius in engineering, a unit of entropy (the loss of energy as heat in any physical process). It is defined as the ratio of energy to temperature above absolute zero.

cleanliness unit unit for measuring air pollution: the number of particles greater than 0.5 micrometres in diameter per cubic foot of air. A more usual measure is the weight of contaminants per cubic metre of air.

cm symbol for **centimetre**.

comfort index estimate of how tolerable conditions are for humans in hot climates. It is calculated as the temperature in degrees Fahrenheit plus a quarter of the relative humidity, expressed as a percentage. If the sum is less than 95, conditions are tolerable for those unacclimatized to the tropics.

condensation number in physics, the ratio of the number of molecules condensing on a surface to the total number of molecules touching that surface.

cord unit for measuring the volume of wood cut for fuel. One cord equals 128 cubic feet (3.456 cubic metres), or a stack 8 feet (2.4 m) long, 4 feet (1.2 m) wide, and 4 feet high.

coulomb SI unit (symbol C) of electrical charge. One coulomb is the quantity of electricity conveyed by a current of one ampere in one second.

cu abbreviation for **cubic** (measure).

cubic decimetre metric measure (symbol dm^3) of volume corresponding to the volume of a cube whose edges are all 1 dm (10 cm) long; it is equivalent to a capacity of one litre.

cubit earliest known unit of length, which originated between 2800 and 2300 BC. It is approximately 50.5 cm/20.6 in long, which is about the length of the human forearm measured from the tip of the middle finger to the elbow.

curie former unit (symbol Ci) of radioactivity, equal to 3.7 x 10^{10} ◊becquerels. One gram of radium has a radioactivity of about one curie. It was named after the French physicist Pierre Curie (1859–1906).

cwt symbol for ◊**hundredweight**, a unit of weight equal to 112 pounds (50.802 kg); 100 lb (45.36 kg) in the USA.

decontamination factor in radiological protection, a measure of the effectiveness of a decontamination process. It is the ratio of the original contamination to the remaining radiation after decontamination: 1,000 and above is excellent; 10 and below is poor.

degree in mathematics, a unit (symbol °) of measurement of an angle or arc. A circle or complete rotation is divided into 360°. A degree may be subdivided into 60 minutes (symbol '), and each minute may be subdivided in turn into 60 seconds (symbol "). **Temperature** is also measured in degrees, which are divided on a decimal scale.

denier unit used in measuring the fineness of yarns, equal to the mass in grams of 9,000 metres of yarn. Thus 9,000 metres of 15 denier nylon, used in nylon stockings, weighs 15 g/0.5 oz, and in this case the thickness of thread would be 0.00425 mm/0.0017 in. The term is derived from the French silk industry; the *denier* was an old French silver coin.

DIN (abbreviation for Deutsches Institut für Normung) German national standards body, which has set internationally accepted standards for (among other things) paper sizes and electrical connectors.

dioptre optical unit in which the power of a lens is expressed as the reciprocal of its focal length in metres. The usual convention is that convergent lenses are positive and divergent lenses negative. Short-sighted people need lenses of power about –0.7 dioptre; a typical value for long sight is about +1.5 dioptre.

dyne c.g.s. unit (symbol dyn) of force. 10^5 dynes make one ◊newton. The dyne is defined as the force that will accelerate a mass of one gram by one centimetre per second per second.

electron volt unit (symbol eV) for measuring the energy of a charged particle (ion or electron) in terms of the energy of motion an electron would gain from a potential difference of one volt. Because it is so small, more usual units are mega- (million) and giga- (billion) electron volts (MeV and GeV).

eotvos unit unit (symbol E) for measuring small changes in the intensity of the Earth's gravity with horizontal distance.

erg c.g.s. unit of work, replaced in the SI system by the ◊joule. One erg of work is done by a force of one (dyne moving through one centimetre.

°F symbol for degrees Fahrenheit.

farad SI unit (symbol F) of electrical capacitance (how much electric charge a capacitor can store for a given voltage). One farad is a capacitance of one ◊coulomb per volt. For practical purposes the microfarad (one millionth of a farad, symbol (F) is more commonly used.

faraday unit of electrical charge equal to the charge on one mole of electrons. Its value is 9.648 x 10^4 ◊coulombs.

fathom unit of depth measurement (1.83 m/6 ft) used prior to metrication; it approximates to the distance between an adult man's hands when the arms are outstretched.

femtosecond SI unit of time. It is 10^{-15} seconds (one millionth of a billionth).

fermi unit of length equal to 10^{-15}m, used in atomic and nuclear physics. The unit is named after the Italian-born US physicist Enrico Fermi (1901–1954).

finsen unit unit (symbol FU) for measuring the intensity of ultraviolet (UV) light; for instance, UV light of 2 FUs causes sunburn in 15 minutes.

fire-danger rating unit index used by the UK Forestry Commission to indicate the probability of a forest fire. 0 means a fire is improbable, 100 shows a serious fire hazard.

foot-candle unit of illuminance, now replaced by the ◊lux. One foot-candle is the illumination received at a distance of one foot from an international candle. It is equal to 10.764 lux.

foot-pound imperial unit of energy (ft-lb), defined as the work done when a force of one pound moves through a distance of one foot. It has been superseded for scientific work by the ◊joule: one foot-pound equals 1.356 joule.

f.p.s. system system of units based on the foot, pound, and second as units of length, mass, and time, respectively. It has now been replaced for scientific work by the SI system.

ft symbol for ◊**foot**, a measure of distance.

furlong unit of measurement, originating in Anglo-Saxon England, equivalent to 220 yd (201.168 m).

g symbol for **gram**.

gal symbol for **gallon, galileo**.

galileo unit (symbol gal) of acceleration, used in geological surveying. One galileo is 10^{-2} metres per second per second. The Earth's gravitational field often differs by several milligals (thousandths of gals) in different places, because of the varying densities of the rocks beneath the surface.

gauge any scientific measuring instrument – for example, a wire gauge or a pressure gauge. The term is also applied to the width of a railway or tramway track.

gauss c.g.s. unit (symbol Gs) of magnetic induction or magnetic flux density, replaced by the SI unit the ◊tesla, but still commonly used. It is equal to one line of magnetic flux per square centimetre. The Earth's magnetic field is about 0.5 Gs, and changes to it over time are measured in gammas (one gamma equals 10^{-5} gauss).

giga- prefix signifying multiplication by 10^9 (1,000,000,000 or 1 billion), as in **gigahertz**, a unit of frequency equivalent to 1 billion hertz.

gill imperial unit of volume for liquid measure, equal to one-quarter of a pint or 5 fluid ounces (0.142 litre). It is used in selling alcoholic drinks.

GMT abbreviation for ◊**Greenwich Mean Time**.

grain smallest unit of mass in the three English systems (avoirdupois, troy, and apothecaries' weights) used in the UK and USA, equal to 0.0648 g. It was reputedly the weight of a grain of wheat. One pound avoirdupois equals 7,000 grains; one pound troy or apothecaries' weight equals 5,760 grains.

gray SI unit (symbol Gy) of absorbed radiation dose. It replaces the ◊rad (1 Gy equals 100 rad), and is defined as the dose absorbed when one kilogram of matter absorbs one joule of ionizing radiation. Different types of radiation cause different amounts of damage for the same absorbed dose; the SI unit of **dose equivalent** is the ◊sievert.

Greenwich Mean Time (GMT) local time on the zero line of longitude (the **Greenwich meridian**), which passes through the Old Royal Observatory at Greenwich, London. It was replaced in 1986 by coordinated universal time (UTC), but continued to be used to measure longitudes and the world's standard time zones.

g-scale scale for measuring force by comparing it with the force due to gravity (*g*), often called *g*-force.

ha symbol for **hectare**.

hand unit used in measuring the height of a horse from front hoof to shoulder (withers). One hand equals 10.2 cm/4 in.

haze factor unit of visibility in mist or fog. It is the ratio of the brightness of the mist compared with that of the object.

henry SI unit (symbol H) of inductance (the reaction of an electric current against the magnetic field that surrounds it). One henry is the inductance of a circuit that produces an opposing voltage of one volt when the current changes at one ampere per second.

hertz SI unit (symbol Hz) of frequency (the number of repetitions of a regular occurrence in one second). Radio waves are often measured in megahertz (MHz), millions of hertz, and the clock rate of a computer is usually measured in megahertz. The unit is named after the German physicist Heinrich Hertz (1857–1894).

hide, or **hyde**, Anglo-Saxon unit of measurement used to measure the extent of arable land; it varied from about 296 ha/120 acres in the east of England to as little as 99 ha/40 acres in Wessex. One hide was regarded as sufficient to support a peasant and his household; it was the area that could be ploughed in a season by one plough and one team of oxen.

horsepower (hp) imperial unit (abbreviation hp) of power, now replaced by the watt.

hundredweight imperial unit (abbreviation cwt) of mass, equal to 112 lb (50.8 kg). It is sometimes called the **long hundredweight**, to distinguish it from the **short hundredweight**, or **cental**, equal to 100 lb (45.4 kg).

IEEE abbreviation for **Institute of Electrical and Electronic Engineers**, the US institute which sets technical standards for electrical equipment and computer data exchange.

inch (in) imperial unit of linear measure, a twelfth of a foot, equal to 2.54 centimetres.

inferno in astrophysics, a unit for describing the temperature inside a star. One inferno is 1 billion K, or approximately 1 billion °C.

international biological standards drugs (such as penicillin and insulin) of which the activity for a specific mass (called the international unit, or IU), prepared and stored under specific conditions, serves as a standard for measuring doses. For penicillin, one IU is the activity of 0.0006 mg of the sodium salt of penicillin, so a dose of a million units would be 0.6 g.

International Organization for Standardization (ISO) international organization founded in 1947 to standardize technical terms, specifications, units, and so on. Its headquarters are in Geneva, Switzerland.

jansky unit of radiation received from outer space, used in radio astronomy. It is equal to 10^{-26} watts per square metre per hertz, and is named after the US radio engineer Karl Jansky (1905–1950).

joule SI unit (symbol J) of work and energy, replacing the calorie (one joule equals 4.2 calories)

k symbol for **kilo-**, as in kg (kilogram) and km (kilometre).

K symbol for **kelvin**, a scale of temperature.

kayser unit of wave number (number of waves in a unit length), used in spectroscopy. It is expressed as waves per centimetre, and is the reciprocal of the wavelength. A wavelength of 0.1 cm has a wave number of 10 kaysers.

kcal symbol for **kilocalorie**.

kilo- prefix denoting multiplication by 1,000, as in kilohertz, a unit of frequency equal to 1,000 hertz.

kilowatt unit (symbol kW) of power equal to 1,000 watts or about 1.34 horsepower.

kilowatt-hour commercial unit of electrical energy (symbol kWh), defined as the work done by a power of 1,000 watts in one hour and equal to 3.6 megajoules. It is used to calculate the cost of electrical energy taken from the domestic supply.

knot in navigation, unit by which a ship's speed is measured, equivalent to one ◊nautical mile per hour (one knot equals about 1.15 miles per hour). It is also sometimes used in aviation.

kph, or **km/h,** symbol for **kilometres per hour.**

kW symbol for ◊kilowatt.

l symbol for **litre,** a measure of liquid volume.

lambert unit of luminance (the light shining from a surface), equal to one ◊lumen per square centimetre. In scientific work the ◊candela per square metre is preferred.

light watt unit of radiant power (brightness of light). One light watt is the power required to produce a perceived brightness equal to that of light at a wavelength of 550 nanometres and 680 lumens.

lumen SI unit (symbol lm) of luminous flux (the amount of light passing through an area per second).

lux SI unit (symbol lx) of illuminance or illumination (the light falling on an object). It is equivalent to one ◊lumen per square metre or to the illuminance of a surface one metre distant from a point source of one ◊candela.

maxwell c.g.s. unit (symbol Mx) of magnetic flux (the strength of a magnetic field in an area multiplied by the area). It is now replaced by the SI unit, the ◊weber (one maxwell equals 10^{-8} weber).

mega- prefix denoting multiplication by a million. For example, a megawatt (MW) is equivalent to a million watts.

megaton one million (10^6) tons. Used with reference to the explosive power of a nuclear weapon, it is equivalent to the explosive force of one million tons of trinitrotoluene (TNT).

mg symbol for **milligram.**

mho SI unit of electrical conductance, now called the ◊siemens; equivalent to a reciprocal ohm.

mi symbol for **mile.**

micro- prefix (symbol () denoting a one-millionth part (10^{-6}). For example, a micrometre, (m, is one-millionth of a metre.

micrometre one-millionth of a metre (symbol (m).

micron obsolete name for the micrometre, one-millionth of a metre.

milli- prefix (symbol m) denoting a one-thousandth part (10^{-3}). For example, a millimetre, mm, is one-thousandth of a metre.

millibar unit of pressure, equal to one-thousandth of a ◊bar.

millilitre one-thousandth of a litre (ml), equivalent to one cubic centimetre (cc).

millimetre of mercury unit of pressure (symbol mmHg), used in medicine for measuring blood pressure defined as the pressure exerted by a column of mercury one millimetre high, under the action of gravity.

ml symbol for ◊millilitre.

mmHg symbol for ◊millimetre of mercury.

molar volume volume occupied by one mole (the molecular mass in grams) of any gas at standard temperature and pressure, equal to 2.24136×10^{-2} m³.

nano- prefix used in SI units of measurement, equivalent to a one-billionth part (10^{-9}). For example, a nanosecond is one-billionth of a second.

nautical mile unit of distance used in navigation, an internationally agreed-on standard (since 1959) equalling the average length of one minute of arc on a great circle of the Earth, or 1,852 m/6,076.12 ft. The term formerly applied to various units of distance used in navigation.

newton SI unit (symbol N) of force. One newton is the force needed to accelerate an object with mass of one kilogram by one metre per second per second. The weight of a medium size (100 g/3 oz) apple is one newton.

oersted c.g.s. unit (symbol Oe) of magnetic field strength, now replaced by the SI unit ampere per metre. The Earth's magnetic field is about 0.5 oersted; the field near the poles of a small bar magnet is several hundred oersteds; and a powerful electromagnet can have a field strength of 30,000 oersteds.

ohm SI unit (symbol () of electrical resistance (the property of a conductor that restricts the flow of electrons through it).

ounce (oz) unit of mass, one-sixteenth of a pound xavoirdupois, equal to 437.5 grains (28.35 g); also one-twelfth of a pound troy, equal to 480 grains.

paper sizes standard European sizes for paper, designated by a letter (A, B, or C) and a number (0–6). The letter indicates the size of the basic sheet at manufacture; the number is how many times it has been folded. A4 is obtained by folding an A3 sheet, which is half an A2 sheet, in half, and so on.

parsec in astronomy, a unit (symbol pc) used for distances to stars and galaxies. One parsec is equal to 3.2616 light years, 2.063×10^5 ◊astronomical units, and 3.086×10^{13} km.

pascal SI unit (symbol Pa) of pressure, equal to one ◊newton per square metre. It replaces ◊bars and millibars (10^5 Pa equals one bar). It is named after the French mathematician Blaise Pascal (1623–1662).

pedometer small portable instrument for counting the number of steps taken, and measuring the approximate distance covered by a person walking. Each step taken by the walker sets in motion a swinging weight within the instrument, causing the mechanism to rotate, and the number of rotations are registered on the instrument face.

phon unit of loudness, equal to the value in decibels of an equally loud tone with frequency 1,000 Hz. The higher the frequency, the louder a noise sounds for the same decibel value; thus an 80-decibel tone with a frequency of 20 Hz sounds as loud as 20 decibels at 1,000 Hz, and the phon value of both tones is 20. An aircraft engine has a loudness of around 140 phons.

Planck's constant in physics, a fundamental constant (symbol h) that relates the energy (E) of one quantum of electromagnetic radiation (the smallest possible 'packet' of energy; seequantum theory) to the frequency (f) of its radiation by $E = hf$. Its value is 6.6261×10^{-34} joule seconds.

planimeter simple integrating instrument for measuring the area of a regular or irregular plane surface. It consists of two hinged arms: one is kept fixed and the other is traced around the boundary of the area. This actuates a small graduated wheel; the area is calculated from the wheel's change in position.

poise c.g.s. unit (symbol P) of dynamic viscosity (the property of liquids that determines how readily they flow). It is equal to one dyne-second per square centimetre. For most liquids the centipoise (one-hundredth of a poise) is used. Water at 20°C/68°F has a viscosity of 1.002 centipoise.

poundal imperial unit (abbreviation pdl) of force, now replaced in the SI system by the ◊newton. One poundal equals 0.1383 newtons.

pt symbol for **pint**.

quart imperial liquid or dry measure, equal to two pints or 1.136 litres. In the USA, a liquid quart is equal to 0.946 litre, while a dry quart is equal to 1.101 litres.

rad unit of absorbed radiation dose, now replaced in the SI system by the ◊gray (one rad equals 0.01 gray), but still commonly used. It is defined as the dose when one kilogram of matter absorbs 0.01 joule of radiation energy (formerly, as the dose when one gram absorbs 100 ergs).

radian SI unit (symbol rad) of plane angles, an alternative unit to the ◊degree. It is the angle at the centre of a circle when the centre is joined to the two ends of an arc (part of the circumference) equal in length to the radius of the circle.

radiation units units of measurement for radioactivity and radiation doses. In SI units, the activity of a radioactive source is measured in becquerels (symbol Bq), where one becquerel is equal to one nuclear disintegration per second (an older unit is the curie). The exposure is measured in coulombs per kilogram ($C\ kg^{-1}$); the amount of ionizing radiation (X-rays or gamma rays) which produces one coulomb of charge in one kilogram of dry air (replacing the roentgen). The absorbed dose of ionizing radiation is measured in grays (symbol Gy) where one gray is equal to one joule of energy being imparted to one kilogram of matter (the rad is the previously used unit). The dose equivalent, which is a measure of the effects of radiation on living organisms, is the absorbed dose multiplied by a suitable factor which depends upon the type of radiation. It is measured in sieverts (symbol Sv), where one sievert is a dose equivalent of one joule per kilogram (an older unit is the rem).

rationalized units units for which the defining equations conform to the geometry of the system. Equations involving circular symmetry contain the factor 2π; those involving spherical symmetry 4π. SI units are rationalized, c.g.s. units are not.

relative biological effectiveness (RBE) the relative damage caused to living tissue by different types of radiation. Some radiations do more damage than others; alpha particles, for example, cause 20 times as much destruction as electrons (beta particles).

rem, acronym of **roentgen equivalent man**, unit of radiation dose equivalent.

rhe unit of fluidity equal to the reciprocal of the ◊poise.

rhm abbreviation for **roentgen–hour–metre**, the unit of effective strength of a radioactive source that produces gamma rays. It is used for substances for which it is difficult to establish radioactive disintegration rates.

roentgen, or **röntgen**, unit (symbol R) of radiation exposure, used for X-rays and gamma rays. It is defined in terms of the number of ions produced in one cubic centimetre of air by the radiation. Exposure to 1,000 roentgens gives rise to an absorbed dose of about 870 rads (8.7 grays), which is a dose equivalent of 870 rems (8.7 sieverts).

Rydberg constant in physics, a constant that relates atomic spectra to the spectrum of hydrogen. Its value is 1.0977×10^7 per metre.

sabin unit of sound absorption, used in acoustical engineering. One sabin is the absorption of one square foot (0.093 square metre) of a perfectly absorbing surface (such as an open window).

Saffir–Simpson damage-potential scale scale of potential damage from wind and sea when a hurricane is in progress: 1 is minimal damage, 5 is catastrophic.

sec, or **s**, abbreviation for **second**, a unit of time.

shackle obsolete unit of length, used at sea for measuring cable or chain. One shackle is 15 fathoms (90 ft/27 m).

SI abbreviation for **Système International d'Unités** (French 'International System of Metric Units').

siemens SI unit (symbol S) of electrical conductance, the reciprocal of the resistance of an electrical circuit. One siemens equals one ampere per volt. It was formerly called the **mho** or **reciprocal ohm**.

sievert SI unit (symbol Sv) of radiation dose equivalent. It replaces the rem (1 Sv equals 100 rem). Some types of radiation do more damage than others for the same absorbed dose – for example, an absorbed dose of alpha radiation causes 20 times as much biological damage as the same dose of beta radiation. The equivalent dose in sieverts is equal to the absorbed dose of radiation in grays multiplied by the relative biological effectiveness. Humans can absorb up to 0.25 Sv without immediate ill effects; 1 Sv may produce radiation sickness; and more than 8 Sv causes death.

speed of light speed at which light and other electromagnetic waves travel through empty space. Its value is 299,792,458 m/186,281 mi per second. The speed of light is the highest speed possible, according to the theory of relativity, and its value is independent of the motion of its source and of the observer. It is impossible to accelerate any material body to this speed because it would require an infinite amount of energy.

standard atmosphere alternative term for ◊atmosphere, a unit of pressure.

standard gravity acceleration due to gravity, generally taken as 9.81274 m/32.38204 ft per second per second. See also ◊g-scale.

standard illuminant any of three standard light intensities, A, B, and C, used for illumination when phenomena involving colour are measured. A is the light from a filament at 2,848K (2,575°C/4,667°F), B is noon sunlight, and C is normal daylight. B and C are defined with respect to A.

Standardization is necessary because colours appear different when viewed in different lights.

standard volume in physics, the volume occupied by one kilogram molecule (the molecular mass in kilograms) of any gas at standard temperature and pressure. Its value is approximately 22.414 cubic metres.

steradian SI unit (symbol sr) of measure of solid (three-dimensional) angles, the three-dimensional equivalent of the ⚬radian. One steradian is the angle at the centre of a sphere when an area on the surface of the sphere equal to the square of the sphere's radius is joined to the centre.

stokes c.g.s. unit (symbol St) of kinematic viscosity (a liquid's resistance to flow).

Système International d'Unités official French name for SI units.

t symbol for **tonne, ton**.

tesla SI unit (symbol T) of magnetic flux density. One tesla represents a flux density of one ⚬weber per square metre, or 10^4 ⚬gauss. It is named after the Yugoslavian-born US physicist and electrical engineer Nikola Tesla (1856–1943).

therm unit of energy defined as 10^5 British thermal units; equivalent to 1.055×10^8 J. It is no longer in scientific use.

torr unit of pressure equal to 1/760 of an ⚬atmosphere, used mainly in high-vacuum technology.

troy system system of units used for precious metals and gems. The pound troy (0.37 kg) consists of 12 ounces (each of 120 carats) or 5,760 grains (each equal to 65 mg).

UTC abbreviation for **coordinated universal time**, the standard measurement of time.

vernier device for taking readings on a graduated scale to a fraction of a division. It consists of a short divided scale that carries an index or pointer and is slid along a main scale. It was invented by the French engineer and instrumentmaker Pierre Vernier.

weber SI unit (symbol Wb) of magnetic flux (the magnetic field strength multiplied by the area through which the field passes). It is named after the German chemist Wilhelm Weber (1804–1891). One weber equals 10^8 ⚬maxwells.

wind-chill factor, or **wind-chill index**, estimate of how much colder it feels when a wind is blowing. It is the sum of the temperature (in °F below zero) and the wind speed (in miles per hour). So for a wind of 15 mph at an air temperature of −5°F, the wind-chill factor is 20.

Zhubov scale scale for measuring ice coverage, developed in the USSR. The unit is the **ball**; one ball is 10% coverage, two balls 20%, and so on.

Further Reading

Adam, Barbara E *Time and Social Theory* (1990)

Cooper, Frank *Sundials* (1969)

Fraser, J T *Time the Familiar Stranger* (1987)

Gell, Alfred *The Anthropology of Time: Cultural Construction of Temporal Maps and Images* (1992)

Hall, A Rupert *The Revolution in Science 1500–1750* (1983)

Howse, Derek *Greenwich Time and the Discovery of Longitude* (1980)

Mann, Wilfrid Basil *Radioactivity and Its Measurement* (1966)

Mendelssohn, K *The Quest for Absolute Zero: Meaning of Low Temperature Physics* (1977)

O'Neil, W M *Time and the Calendars* (1975)

Ronan, Colin *The Cambridge Illustrated History of the World's Science* (1983)

Sethares, William A *Tuning, Timbre, Spectrum, Scale* (1998)

Smith, C, and Wise, M N *Energy and Empire* (1989)

Taylor, E G R *The Haven Finding Art. A History of Navigation from Odysseus to Captain Cook* (1971)

MATHEMATICS

Mathematics is the science of relationships between numbers, between spatial configurations, and abstract structures. The main divisions of **pure mathematics** include arithmetic, algebra, geometry, trigonometry, and calculus. Mechanics, statistics, numerical analysis, computing, the mathematical theories of astronomy, electricity, optics, thermodynamics, and atomic studies come under the heading of **applied mathematics.**

In the 20th century mathematics has become diversified. Each specialist subject is being studied in far greater depth and advanced work in some fields may be unintelligible to researchers in other fields. Mathematicians working in universities have had the economic freedom to pursue the subject for its own sake. Nevertheless, new branches of mathematics have been developed which are of great practical importance and which have basic ideas simple enough to be taught in schools. Probably the most important of these is the mathematical theory of statistics in which much pioneering work was done by the English mathematician Karl Pearson. Another new development is operations research, which is concerned with finding optimum courses of action in practical situations, particularly in economics and management. Higher mathematics has a powerful tool in the high-speed electronic computer, which can create and manipulate mathematical 'models' of various systems in science, technology, and commerce.

Traditionally the subject of mathematics is divided into arithmetic, which studies numbers; algebra, which studies structures; geometry, which studies space; analysis, which studies infinite processes (in particular, calculus); and probability theory and statistics, which study random processes. Modern additions to school syllabuses such as sets, group theory, matrices, and graph theory are sometimes referred to as 'new' or 'modern' mathematics.

numbers

number systems

Numbers are symbols used in counting or measuring. Our everyday number system is the decimal ('proceeding by tens') system. The ancient Egyptians, Greeks, Romans, and Babylonians all evolved number systems, although none had a zero, which was introduced from India by way of Arab mathematicians in about the 8th century AD and allowed a place-value system to be devised on which the decimal system is based. Other systems are mainly used in computing and include the binary number system, octal number system, and hexadecimal number system.

what is a base? A base is the number of different single-digit symbols used in a particular number system. In our usual (decimal) counting system of numbers (with symbols 0, 1, 2, 3, 4, 5, 6, 7, 8, 9) the base is 10. In the binary number system (see below), which has only the symbols 1 and 0, the base is two.

A base is also a number that, when raised to a particular power (that is, when multiplied by itself a particular number of times as in $10^2 = 10 \times 10 = 100$), has a logarithm (see below) equal to the power. For example, the logarithm of 100 to the base ten is 2.

In general, any number system subscribing to a place-value system with base value b may be represented by $...b^4, b^3, b^2, b^1, b^0, b^{-1}, b^{-2}, b^{-3},...$.

Number Systems

Binary (Base 2)	Octal (Base 8)	Decimal (Base 10)	Hexadecimal (Base 16)
0	0	0	0
1	1	1	1
10	2	2	2
11	3	3	3
100	4	4	4
101	5	5	5
110	6	6	6
111	7	7	7
1000	10	8	8
1001	11	9	9
1010	12	10	A
1011	13	11	B
1100	14	12	C
1101	15	13	D
1110	16	14	E
1111	17	15	F
10000	20	16	10
11111111	377	255	FF
11111010001	3721	2001	7D1

Hence in base ten the columns represent $...10^4, 10^3, 10^2, 10^1, 10^0, 10^{-1}, 10^{-2}, 10^{-3}...$, in base two $...2^4, 2^3, 2^2, 2^1, 2^0, 2^{-1}, 2^{-2}, 2^{-3}...$, and in base eight $...8^4, 8^3, 8^2, 8^1, 8^0, 8^{-1}, 8^{-2}, 8^{-3}...$. For bases beyond 10, the denary numbers 10, 11, 12, and so on must be replaced by a single digit. Thus in base 16, all numbers up to 15 must be represented by single-digit 'numbers', since 10 in hexadecimal would mean 16 in decimal. Hence decimal 10, 11, 12, 13, 14, 15 are represented in hexadecimal by letters A, B, C, D, E, F.

Other number systems use other bases. For example, numbers to base two (binary numbers), using only 0 and 1, are commonly used in digital computers to represent the two-state 'on' or 'off' pulses of electricity. The Babylonians, however, used a complex base-sixty system, residues of which are found today in the number of minutes in each hour and in angular measurement (6×60 degrees). The Mayas used a base-twenty system.

Babylonian and Egyptian Mathematics
http://www-history.mcs.st-and.ac.uk/
~history/HistTopics/Babylonian_and_Egyptian.html
Description of the mathematics of these two early civilizations. The site includes maps of the area at the time, images of tablets and scrolls dating back to that era, and also a list of publications for further reference.

Roman numerals

Roman numerals are an ancient European number system using symbols different from Arabic numerals (the ordinary numbers 1, 2, 3, 4, 5, and so on). The seven key symbols in Roman numerals, as represented today, are I (1), V (5), × (10), L (50), C (100), D (500), and M (1,000). There is no zero, and therefore no place-value as is fundamental to the Arabic system. The first ten Roman numerals are I, II, III, IV (or IIII), V, VI, VII, VIII, IX, and X. When a Roman symbol is preceded by a symbol of equal or greater value, the values of the symbols are added (XVI = 16).

When a symbol is preceded by a symbol of less value, the values are subtracted (XL = 40). A horizontal bar over a symbol indicates a multiple of 1,000 ($\bar{X} = 10,000$). Although addition and subtraction are fairly straightforward using Roman numerals, the absence of a zero makes other arithmetic calculations (such as multiplication) clumsy and difficult.

Roman Numerals

Roman	Arabic
I	1
II	2
III	3
IV	4
V	5
VI	6
VII	7
VIII	8
IX	9
X	10
XI	11
XIX	19
XX	20
XXX	30
XL	40
L	50
LX	60
XC	90
C	100
CC	200
CD	400
D	500
CM	900
M	1,000

decimal number system

The decimal numeral system, or **denary number system**, evolved from the Hindu-Arabic number system and is the most commonly used number system today. It employs ten numerals (0, 1, 2, 3, 4, 5, 6, 7, 8, 9) and is said to operate in 'base ten'. In a base-ten number, each position has a value ten times that of the position to its immediate right; for example, in the number 23 the numeral 3 represents three units (ones), and the numeral 2 represents two tens. Decimal numbers do not necessarily contain a decimal point; 563, 5.63, and −563 are all decimal numbers.

Decimal numbers may be thought of as written under column headings based on the number ten. For example, the number 2,567 stands for 2 thousands, 5 hundreds, 6 tens, and 7 ones. Large decimal numbers may also be expressed in floating-point notation.

History of Numbers
http://www.islam.org/Mosque/ihame/Ref6.htm
Account of how numbers developed from their ancient Indian origins to the Modern Arabic numerals that are generally used today. There is a brief description of numbering systems used by the ancient Egyptians, Greeks, and Romans, and some of the problems these systems presented.

---

binary number system Binary numbers were first developed by the German mathematician Gottfried Leibniz in the late 17th century. The binary system is a system of numbers to base two, using combinations of the digits 1 and 0.

The value of any position in a binary number increases by powers of 2 (doubles) with each move from right to left (1, 2, 4, 8, 16, and so on). For example, 1011 in the binary number system represents $(1 \times 8) + (0 \times 4) + (1 \times 2) + (1 \times 1)$, which adds up to 11 in the decimal system.

The value of any position in a normal decimal, or base-ten, number increases by powers of 10 with each move from right to left (1, 10, 100, 1,000, 10,000, and so on). For example, the decimal number 2,567 stands for:

$$(2 \times 1,000) + (5 \times 100) + (6 \times 10) + (7 \times 1)$$

Codes based on binary numbers are used to represent instructions and data in all modern digital computers, the values of the binary digits (contracted to 'bits') being stored or transmitted as, for example, open/closed switches, magnetized/unmagnetized discs and tapes, and high/low voltages in circuits. Because the main operations of subtraction, multiplication, and division can be reduced mathematically to addition, digital computers carry out calculations by adding, usually in binary numbers in which the numerals 0 and 1 can be represented by off and on pulses of electric current.

types of numbers

Numbers can be categorized broadly as real and complex. Concepts such as negative numbers, rational numbers, and irrational numbers can be rigorously and precisely defined in terms of the natural numbers. There remains then the problem of defining the natural numbers. A modern approach defines the natural numbers in terms of sets (see below). Zero is defined to be the empty set: 0 = ø (i.e. the set with no elements). Then 1 is defined to be the union of 0 and the set that consists of 0 (which is a set with 1 element, zero). Now we can define 2 as the union of 1 and the set containing 1 (which is a set containing 2 elements, zero and one), and so on.

An alternative procedure for constructing a number system is to define the real numbers in terms of their algebraic and analytical properties.

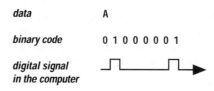

binary number system The capital letter A represented in binary form.

---

real numbers Real numbers include all rational numbers (integers, or whole numbers, and fractions) and irrational numbers (those not expressible as fractions). **Rational numbers** are whole numbers, or integers, and fractions. The whole numbers are represented by the natural numbers, 0, 1, 2, 3, 4, 5, 6, 7, 8, and 9, which give a counting system that, in the decimal system, continues 10, 11, 12, 13, and so on. Fractions of these numbers are represented as, for example, $\frac{1}{4}$, $\frac{1}{2}$, $\frac{3}{4}$, or as decimal fractions (0.25, 0.5, 0.75).

Irrational numbers cannot be represented in this way and require symbols, such as $\sqrt{2}$, π, and e. They can be expressed numerically only as the (inexact) approximations 1.414, 3.142, and 2.718 (to three places of decimals) respectively. The symbols π and e are also examples of **transcendental numbers**, because they (unlike $\sqrt{2}$) cannot be derived by solving a polynomial equation (an equation with one variable quantity) with rational coefficients (multiplying factors).

complex numbers Complex numbers include the real numbers described above *and* **imaginary numbers,**

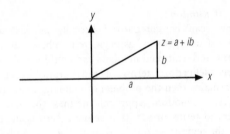

complex number A complex number can be represented graphically as a line whose end-point coordinates equal the real and imaginary parts of the complex number. This type of diagram is called an Argand diagram after the French mathematician Jean Argand (1768–1822) who devised it.

which are real-number multiples of the square root of −1. A complex number is a number written in the form $a + ib$, where a and b are real numbers and i is the square root of −1 (that is, $i^2 = -1$); i used to be known as the 'imaginary' part of the complex number. Some equations in algebra, such as those of the form

$$x^2 + 5 = 0$$

cannot be solved without recourse to complex numbers, because the real numbers do not include square roots of negative numbers.

The sum of two or more complex numbers is obtained by adding separately their real and imaginary parts, for example:

$$(a + bi) + (c + di) = (a + c) + (b + d)i$$

Complex numbers can be represented graphically on an Argand diagram, which uses rectangular Cartesian coordinates in which the x-axis represents the real part of the number and the y-axis the imaginary part. Thus the number $z = a + bi$ is plotted as the point (a, b). Complex numbers have applications in various areas of science, such as the theory of alternating currents in electricity.

arithmetic

Arithmetic is the branch of mathematics concerned with the study of numbers and their properties. The fundamental operations of arithmetic are addition, subtraction, multiplication, and division. Raising to powers (for example, squaring or cubing a number), the extraction of roots (for example, square roots), percentages, fractions, and ratios are developed from these operations.

Forms of simple arithmetic existed in prehistoric times. In China, Egypt, Babylon, and early civilizations

generally, arithmetic was used for commercial purposes, records of taxation, and astronomy. During the Dark Ages in Europe, knowledge of arithmetic was preserved in India and later among the Arabs. European mathematics revived with the development of trade and overseas exploration. Hindu-Arabic numerals replaced Roman numerals, allowing calculations to be made on paper, instead of by the abacus.

There have been many inventions and developments to make the manipulation of the arithmetic processes easier, such as the invention of logarithms by the Scottish mathematician John Napier in 1614 and of the slide rule in the period 1620–30. Since then, many forms of ready reckoners, mechanical and electronic calculators, and computers have been invented.

Modular or **modulo arithmetic**, sometimes known as residue arithmetic or clock arithmetic, can take only a specific number of digits, whatever the value. For example, in modulo 4 (mod 4) the only values any number can take are 0, 1, 2, or 3. In this system, 7 is written as 3 mod 4, and 35 is also 3 mod 4. Notice 3 is the residue, or remainder, when 7 or 35 is divided by 4. This form of arithmetic is often illustrated on a circle. It deals with events recurring in regular cycles, and is used in describing the functioning of petrol engines, electrical generators, and so on. For example, in the mod 12, the answer to a question as to what time it will be in five hours if it is now ten o'clock can be expressed 10 + 5 = 3.

properties of numbers
All the properties of numbers may be deduced from the **associative law**, which states that the sum of a set of numbers is the same whatever the order of addition, and that the product of a set of numbers is the same whatever the order of multiplication.

commutative law The **commutative law** is a special case of the associative law producing commutativity where there are only two numbers in the set. For example:

$$a + b = b + a$$
$$ab = ba$$

distributive law The **distributive law** for multiplication over addition states that, given a set of numbers a, b, c, ... and a multiplier m:

$$m(a + b + c + ...) = ma + mb + mc + ...$$

For example:

$$9 \times 132 = (9 \times 100) + (9 \times 30) + (9 \times 2)$$

The distributive law does not apply for addition over multiplication; for example:

$$7 + (3 \times 5) \; ((7 + 3) \times (7 + 5)$$

identities Zero is described as the identity for addition because adding zero to any number has no effect on that number.

$$n + 0 = 0 + n = n$$

One is the identity for multiplication because multiplying any number by one leaves that number unchanged.

$$n \times 1 = 1 \times n = n$$

negatives Every number has a negative $-n$ such that:

$$n + (-n) = 0$$

inverse Every number (except 0) has an inverse $\frac{1}{n}$ such that:

$$n \times \frac{1}{n} \;—\; = 1$$

prime numbers A prime number can be divided only by 1 and itself, that is, having no other factors. There is an infinite number of primes, the first ten of which are 2, 3, 5, 7, 11, 13, 17, 19, 23, and 29 (by definition, the number 1 is excluded from the set of prime numbers). The number 2 is the only even prime because all other even numbers have 2 as a factor.

Over the centuries mathematicians have sought general methods (algorithms) for calculating primes, from Eratosthenes' sieve to programs on powerful computers.

The largest prime, $2^{859433}-1$ (258,716 digits long) was discovered in 1993. It is the thirty-third Mersenne prime. All Mersenne primes are in the form 2^q-1, where q is also a prime.

Squares, Cubes, and Roots

No.	Square	Cube	Square root	Cube root
1	1	1	1.000	1.000
2	4	8	1.414	1.260
3	9	27	1.732	1.442
4	16	64	2.000	1.587
5	25	125	2.236	1.710
6	36	216	2.449	1.817
7	49	343	2.646	1.913
8	64	512	2.828	2.000
9	81	729	3.000	2.080
10	100	1000	3.162	2.154
11	121	1331	3.317	2.224
12	144	1728	3.464	2.289
13	169	2197	3.606	2.351
14	196	2744	3.742	2.410
15	225	3375	3.873	2.466
16	256	4096	4.000	2.520
17	289	4913	4.123	2.571
18	324	5832	4.243	2.621
19	361	6859	4.359	2.668
20	400	8000	4.472	2.714
25	625	15625	5.000	2.924
30	900	27000	5.477	3.107
40	1600	64000	6.325	3.420
50	2500	125000	7.071	3.684

Prime Numbers
http://www.utm.edu/research/primes/
largest.html#contents
Everything you ever wanted to know about prime numbers, including the top ten recorded primes and Euclid's proof that the largest prime can never be reached.

Prime Numbers

Prime numbers between 1 and 1,000

2	3	5	7	11	13	17	19	23	29
31	37	41	43	47	53	59	61	67	71
73	79	83	89	97	101	103	107	109	113
127	131	137	139	149	151	157	163	167	173
179	181	191	193	197	199	211	223	227	229
233	239	241	251	257	263	269	271	277	281
283	293	307	311	313	317	331	337	347	349
353	359	367	373	379	383	389	397	401	409
419	421	431	433	439	443	449	457	461	463
467	479	487	491	499	503	509	521	523	541
547	557	563	569	571	577	587	593	599	601
607	613	617	619	631	641	643	647	653	659
661	673	677	683	691	701	709	719	727	733
739	743	751	757	761	769	773	787	797	809
811	821	823	827	829	839	853	857	859	863
877	881	883	887	907	911	919	929	937	941
947	953	967	971	977	983	991	997		

arithmetic operations The four basic operations of arithmetic are addition, subtraction, multiplication, and division. Powers, roots, fractions, and percentages are developed from these operations. **Addition** is the operation of combining two numbers to form a sum; thus, 7 + 4 = 11. **Subtraction** involves taking one number or quantity away from another, or finding the difference between two quantities. Subtraction is neither commutative:

$$a - b \neq b - a$$

nor associative:

$$a - (b - c) \neq (a - b) - c$$

For example:

$$8 - 5 \neq 5 - 8$$
$$7 - (4 - 3) \neq (7 - 4) - 3$$

Multiplication Table

	2	3	4	5	6	7	8	9	10	11	12	13	14	15	16	17	18	19	20	21	22	23	24	25
2	4	6	8	10	12	14	16	18	20	22	24	26	28	30	32	34	36	38	40	42	44	46	48	50
3	6	9	12	15	18	21	24	27	30	33	36	39	42	45	48	51	54	57	60	63	66	69	72	75
4	8	12	16	20	24	28	32	36	40	44	48	52	56	60	64	68	72	76	80	84	88	92	96	100
5	10	15	20	25	30	35	40	45	50	55	60	65	70	75	80	85	90	95	100	105	110	115	120	125
6	12	18	24	30	36	42	48	54	60	66	72	78	84	90	96	102	108	114	120	126	132	138	144	150
7	14	21	28	35	42	49	56	63	70	77	84	91	98	105	112	119	126	133	140	147	154	161	168	175
8	16	24	32	40	48	56	64	72	80	88	96	104	112	120	128	136	144	152	160	168	176	184	192	200
9	18	27	36	45	54	63	72	81	90	99	108	117	126	135	144	153	162	171	180	189	198	207	216	225
10	20	30	40	50	60	70	80	90	100	110	120	130	140	150	160	170	180	190	200	210	220	230	240	250
11	22	33	44	55	66	77	88	99	110	121	132	143	154	165	176	187	198	209	220	231	241	253	264	275
12	24	36	48	60	72	84	96	108	120	132	144	156	168	180	192	204	216	228	240	252	264	276	288	300
13	26	39	52	65	78	91	104	117	130	143	156	169	182	195	208	221	234	247	260	273	286	299	312	325
14	28	42	56	70	84	98	112	126	140	154	168	182	196	210	224	238	252	266	280	294	308	322	336	350
15	30	45	60	75	90	105	120	135	150	165	180	195	210	225	240	255	270	285	300	315	330	345	360	375
16	32	48	64	80	96	112	128	144	160	176	192	208	224	240	256	272	288	304	320	336	352	368	384	400
17	34	51	68	85	102	119	136	153	170	187	204	221	238	255	272	289	306	323	340	357	374	391	408	425
18	36	54	72	90	108	126	144	162	180	198	216	234	252	270	288	306	324	342	360	378	396	414	432	450
19	38	57	76	95	114	133	152	171	190	209	228	247	266	285	304	323	342	361	380	399	418	437	456	475
20	40	60	80	100	120	140	160	180	200	220	240	260	280	300	320	340	360	380	400	420	440	460	480	500
21	42	63	84	105	126	147	168	189	210	231	252	273	294	315	336	357	378	399	420	441	462	483	504	525
22	44	66	88	110	132	154	176	198	220	242	264	286	308	330	352	374	396	418	440	462	484	506	528	550
23	46	69	92	115	138	161	184	207	230	253	276	299	322	345	368	391	414	437	460	483	506	529	552	575
24	48	72	96	120	144	168	192	216	240	264	288	312	336	360	384	408	432	456	480	504	528	552	576	600
25	50	75	100	125	150	175	200	225	250	275	300	325	350	375	400	425	450	475	500	525	550	575	600	625

Multiplication is usually written in the form $a \times b$ or ab, and involves repeated addition in the sense that a is added to itself b times. Multiplication obeys commutative, associative, and distributive laws (the latter over addition) and every number (except 0) has a multiplicative inverse. The number 1 is the identity for multiplication. The inverse of multiplication is **division**.

powers A power is represented by an exponent or index, denoted by a superior small numeral. A number or symbol raised to the power of 2 – that is, multiplied by itself – is said to be squared (for example, 3^2, x^2), and when raised to the power of 3, it is said to be cubed (for example, 2^3, y^3). Any number to the power zero always equals 1.

Powers can be negative. Negative powers produce fractions, with the numerator as one, as a number is divided by itself, rather than being multiplied by itself, so for example $2^{-1} = \frac{1}{2}$ and $3^{-3} = \frac{1}{27}$.

square roots A number that when squared (multiplied by itself) equals a given number is called a square root. For example, the square root of 25 (written $\sqrt{25}$) is ± 5, because $5 \times 5 = 25$, and $(-5) \times (-5) = 25$. As an exponent, a square root is represented by $\frac{1}{2}$, for example, $16^{\frac{1}{2}} = 4$.

Negative numbers (less than 0) do not have square roots that are ♭real numbers. Their roots are represented by complex numbers (described above), in which the square root of –1 is given the symbol i (that is, $\pm i^2 = -1$). Thus the square root of –4 is $\sqrt{[(-1) \times 4]} = \sqrt{-1} \times \sqrt{4} = 2i$.

fractions

A fraction (from Latin *fractus* 'broken') is a number that indicates one or more equal parts of a whole. Usually, the number of equal parts into which the unit is divided (denominator) is written below a horizontal line, and the number of parts comprising the fraction (numerator) is written above; thus $\frac{1}{2}$ or $\frac{3}{4}$. Such fractions are called **vulgar** or **simple fractions**. The denominator can never be zero.

A **proper fraction** is one in which the numerator is less than the denominator. An **improper fraction** has a numerator that is larger than the denominator, for example $\frac{3}{2}$. It can therefore be expressed as a mixed number, for example, $1\frac{1}{2}$. A combination such as $\frac{5}{0}$ is not regarded as a fraction (an object cannot be divided into zero equal parts), and mathematically any number divided by 0 is equal to infinity.

A **decimal fraction** has as its denominator a power of 10, and these are omitted by use of the decimal point and notation, for example 0.04, which is $\frac{4}{100}$.

Fractions as Decimals

Fraction	Decimal	Fraction	Decimal	Fraction	Decimal	Fraction	Decimal
$\frac{1}{2}$	0.5000	$\frac{7}{8}$	0.8750	$\frac{10}{11}$	0.9091	$\frac{19}{20}$	0.9500
$\frac{1}{3}$	0.3333	$\frac{1}{9}$	0.1111	$\frac{1}{12}$	0.0833	$\frac{1}{32}$	0.0312
$\frac{2}{3}$	0.6667	$\frac{2}{9}$	0.2222	$\frac{5}{12}$	0.4167	$\frac{3}{32}$	0.0938
$\frac{1}{4}$	0.2500	$\frac{4}{9}$	0.4444	$\frac{7}{12}$	0.5833	$\frac{5}{32}$	0.1562
$\frac{3}{4}$	0.7500	$\frac{5}{9}$	0.5556	$\frac{11}{12}$	0.9167	$\frac{7}{32}$	0.2188
$\frac{1}{5}$	0.2000	$\frac{7}{9}$	0.7778	$\frac{1}{16}$	0.0625	$\frac{9}{32}$	0.2812
$\frac{2}{5}$	0.4000	$\frac{8}{9}$	0.8889	$\frac{3}{16}$	0.1875	$\frac{11}{32}$	0.3438
$\frac{3}{5}$	0.6000	$\frac{1}{10}$	0.1000	$\frac{5}{16}$	0.3125	$\frac{13}{32}$	0.4062
$\frac{4}{5}$	0.8000	$\frac{3}{10}$	0.3000	$\frac{7}{16}$	0.4375	$\frac{15}{32}$	0.4688
$\frac{1}{6}$	0.1667	$\frac{7}{10}$	0.7000	$\frac{9}{16}$	0.5625	$\frac{17}{32}$	0.5312
$\frac{5}{6}$	0.8333	$\frac{9}{10}$	0.9000	$\frac{11}{16}$	0.6875	$\frac{19}{32}$	0.5938
$\frac{1}{7}$	0.1429	$\frac{1}{11}$	0.0909	$\frac{13}{16}$	0.8125	$\frac{21}{32}$	0.6562
$\frac{2}{7}$	0.2857	$\frac{2}{11}$	0.1818	$\frac{15}{16}$	0.9375	$\frac{23}{32}$	0.7188
$\frac{3}{7}$	0.4286	$\frac{3}{11}$	0.2727	$\frac{1}{20}$	0.0500	$\frac{25}{32}$	0.7812
$\frac{4}{7}$	0.5714	$\frac{4}{11}$	0.3636	$\frac{3}{20}$	0.1500	$\frac{27}{32}$	0.8438
$\frac{5}{7}$	0.7143	$\frac{5}{11}$	0.4545	$\frac{7}{20}$	0.3500	$\frac{29}{32}$	0.9062
$\frac{6}{7}$	0.8571	$\frac{6}{11}$	0.5455	$\frac{9}{20}$	0.4500	$\frac{31}{32}$	0.9688
$\frac{1}{8}$	0.1250	$\frac{7}{11}$	0.6364	$\frac{11}{20}$	0.5500		
$\frac{3}{8}$	0.3750	$\frac{8}{11}$	0.7273	$\frac{13}{20}$	0.6500		
$\frac{5}{8}$	0.6250	$\frac{9}{11}$	0.8182	$\frac{17}{20}$	0.8500		

The digits to the right of the decimal point indicate the numerators of vulgar fractions whose denominators are 10, 100, 1,000, and so on. Most fractions can be expressed exactly as decimal fractions ($\frac{1}{3}$ = 0.333...). Fractions are also known as **rational numbers**; that is, numbers formed by a ratio. **Integers** may be expressed as fractions with a denominator of 1, so 6 is $\frac{6}{1}$, for example.

addition and subtraction To add or subtract with fractions, a **common denominator** (a number divisible by both the bottom numbers) needs to be identified. For example, for $\frac{3}{4}$ + $\frac{5}{6}$ the smallest common denominator is 12. Both the numerators and denominators of the fractions to be added (or subtracted) are then multiplied by the number of times the denominator goes into the common denominator, so $\frac{3}{4}$ is multiplied by 3 and $\frac{5}{6}$ by 2. The numerators can then be simply added or subtracted.

$$\frac{9}{12} + \frac{10}{12} = \frac{19}{12} = 1\frac{7}{12}$$

If whole numbers appear in the calculation they can be added/subtracted separately first.

multiplication and division All whole numbers in a division or multiplication calculation must first be converted into improper fractions. For multiplication, the numerators are then multiplied together and the denominators are then multiplied to provide the solution. For example:

$$7\frac{2}{3} \times 4\frac{1}{2} = \frac{23}{3} \times \frac{9}{2} = \frac{207}{6} = 34\frac{1}{2}$$

In division, the procedure is similar, but the second fraction must be inverted before multiplication occurs. For example:

$$5\frac{5}{12} \div 1\frac{1}{8} = \frac{65}{12} \div \frac{9}{8} = \frac{65}{12} \times \frac{8}{9} = \frac{520}{108} = 4\frac{22}{27}$$

percentages

A percentage is way of representing a number as a fraction of 100. Thus 45 percent (45%) equals $\frac{45}{100}$, and 45% of 20 is $\frac{45}{100} \times 20 = 9$.

In general, if a quantity x changes to y, the percentage change is $\frac{100(x - y)}{x}$. Thus, if the number of people in a room changes from 40 to 50, the percentage increase is $(100 \times 10)/40 = 25\%$. To express

Percentages as Fractions or Decimals

%	Decimal	Fraction	%	Decimal	Fraction	%	Decimal	Fraction	%	Decimal	Fraction
1	0.01	$\frac{1}{100}$	16	0.16	$\frac{4}{25}$	32	0.32	$\frac{8}{25}$	48	0.48	$\frac{12}{25}$
2	0.02	$\frac{1}{50}$	$16\frac{2}{3}$	0.167	$\frac{1}{6}$	33	0.33	$\frac{33}{100}$	49	0.49	$\frac{49}{100}$
3	0.03	$\frac{3}{100}$	17	0.17	$\frac{17}{100}$	$33\frac{1}{3}$	0.333	$\frac{1}{3}$	50	0.50	$\frac{1}{2}$
4	0.04	$\frac{1}{25}$	18	0.18	$\frac{9}{50}$	34	0.34	$\frac{17}{50}$	55	0.55	$\frac{11}{20}$
5	0.05	$\frac{1}{20}$	19	0.19	$\frac{19}{100}$	35	0.35	$\frac{7}{20}$	60	0.60	$\frac{3}{5}$
6	0.06	$\frac{3}{50}$	20	0.20	$\frac{1}{5}$	36	0.36	$\frac{9}{25}$	65	0.65	$\frac{13}{20}$
7	0.07	$\frac{7}{100}$	21	0.21	$\frac{21}{100}$	37	0.37	$\frac{37}{100}$	$66\frac{2}{3}$	0.667	$\frac{2}{3}$
8	0.08	$\frac{2}{25}$	22	0.22	$\frac{11}{50}$	38	0.38	$\frac{19}{50}$	70	0.70	$\frac{7}{10}$
$8\frac{1}{3}$	0.083	$\frac{1}{12}$	23	0.23	$\frac{23}{100}$	39	0.39	$\frac{39}{100}$	75	0.75	$\frac{3}{4}$
9	0.09	$\frac{9}{100}$	24	0.24	$\frac{6}{25}$	40	0.40	$\frac{2}{5}$	80	0.80	$\frac{4}{5}$
10	0.10	$\frac{1}{10}$	25	0.25	—	41	0.41	$\frac{41}{100}$	85	0.85	$\frac{17}{20}$
11	0.11	$\frac{11}{100}$	26	0.26	$\frac{13}{50}$	42	0.42	$\frac{21}{50}$	90	0.90	$\frac{9}{10}$
12	0.12	$\frac{3}{25}$	27	0.27	$\frac{27}{100}$	43	0.43	$\frac{43}{100}$	95	0.95	$\frac{19}{20}$
$12\frac{1}{2}$	0.125	$\frac{1}{8}$	28	0.28	$\frac{7}{25}$	44	0.44	$\frac{11}{25}$	100	1.00	1
13	0.13	$\frac{13}{100}$	29	0.29	$\frac{29}{100}$	45	0.45	$\frac{9}{20}$			
14	0.14	$\frac{7}{50}$	30	0.30	$\frac{3}{10}$	46	0.46	$\frac{23}{50}$			
15	0.15	$\frac{3}{20}$	31	0.31	$\frac{31}{100}$	47	0.47	$\frac{47}{100}$			

a fraction as a percentage, its denominator must first be converted to 100 – for example, $\frac{1}{8}$ = 12.5/100 = 12.5%. The use of percentages often makes it easier to compare fractions that do not have a common denominator.

To convert a fraction to a percentage on a calculator, divide numerator by denominator. The percentage will correspond to the first figures of the decimal, for example $^7/_{12}$ = 0.5833333 = 58.3% correct to three decimal places.

The percentage sign is thought to have been derived as an economy measure when recording in the old counting houses; writing in the numeric symbol for $\frac{25}{100}$ of a cargo would take two lines of parchment, and hence the '100' denominator was put alongside the 25 and rearranged to '%'.

logarithms
The exponent or index of a number to a specified base – usually 10 – is called a logarithm, or **log**. For example, the logarithm to the base 10 of 1,000 is 3 because 10^3 = 1,000; the logarithm of 2 is 0.3010 because 2 = $10^{0.3010}$. The whole-number part of a logarithm is called the **characteristic**; the fractional part is called the **mantissa**.

Before the advent of cheap electronic calculators, multiplication and division could be simplified by being replaced with the addition and subtraction of logarithms.

For any two numbers $\times$ and y (where $x = b^a$ and $y = b^c$) $\times$ x $y = b^a \times b^c = b^{a+c}$; hence we would add the logarithms of x and y, and look up this answer in antilogarithm tables.

Tables of logarithms and antilogarithms are available that show conversions of numbers into logarithms, and vice versa. For example, to multiply 6,560 by 980, one looks up their logarithms (3.8169 and 2.9912), adds them together (6.8081), then looks up the antilogarithm of this to get the answer (6,428,800). **Natural** or **Napierian logarithms** are to the base e, an irrational number equal to approximately 2.7183.

The principle of logarithms is also the basis of the slide rule. With the general availability of the electronic pocket calculator, the need for logarithms has been reduced. The first log tables (to base e) were published by the Scottish mathematician John Napier in 1614. Base-ten logs were introduced by the Englishman Henry Briggs (1561–1631) and Dutch mathematician Adriaen Vlacq (1600–1667).

sets

A set, or **class,** is any collection of defined things (elements), provided the elements are distinct and that there is a rule to decide whether an element is a member of a set. It is usually denoted by a capital letter and indicated by curly brackets {}.

For example, *L* may represent the set that consists of all the letters of the alphabet. The symbol [member of] stands for 'is a member of'; thus *p* [member of] *L* means that *p* belongs to the set consisting of all letters, and 4 [not member of] *L* means that 4 does not belong to the set consisting of all letters.

There are various types of sets. A **finite set** has a limited number of members, such as the letters of the alphabet; an **infinite set** has an unlimited number of members, such as all whole numbers; an **empty** or **null** set has no members, such as the number of people

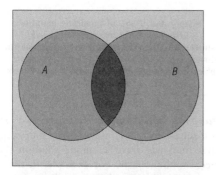

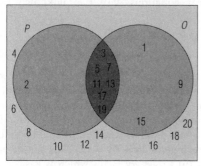

ξ = set of whole numbers from 1 to 20
O = set of odd numbers
P = set of prime numbers

Venn diagram The sets are drawn as circles – the area of overlap between the circles shows elements that are common to each set, and thus represent a third set. Here (top) is a Venn diagram of two intersecting sets and (bottom) a Venn diagram showing the set of whole numbers from 1 to 20 and the subsets P and O of prime and odd numbers, respectively. The intersection of P and O contains all the prime numbers that are also odd.

Set Theory
http://www-history.mcs.st-and.ac.uk/
~history/HistTopics/Beginnings_of_set_theory.html
Details of the beginnings of set theory.. It includes biographical details of the mathematicians involved and gives a brief description of the topic .

who have swum across the Atlantic Ocean, written as {} or ø; a **single-element set** has only one member, such as days of the week beginning with M, written as {Monday}. **Equal sets** have the same members; for example, if *W* = {days of the week} and *S* = {Sunday, Monday, Tuesday, Wednesday, Thursday, Friday, Saturday}, it can be said that *W* = *S*. Sets with the same number of members are **equivalent sets.** Sets with some members in common are **intersecting sets;** for example, if *R* = {red playing cards} and *F* = {face cards}, then *R* and *F* share the members that are red face cards. Sets with no members in common are **disjoint sets.** Sets contained within others are **subsets;** for example, *V* = {vowels} is a subset of *L* = {letters of the alphabet}. Sets and their interrelationships are often illustrated by a Venn diagram.

algebra

Algebra is the branch of mathematics in which the general properties of numbers are studied by using symbols, usually letters, to represent variables and unknown quantities. For example, the algebraic statement

$$(x + y)^2 = x^2 + 2xy + y^2$$

is true for all values of × and *y*. If × = 7 and *y* = 3, for instance, it becomes:

$$(7 + 3)^2 = 7^2 + 2(7 \times 3) + 3^2 = 100$$

An algebraic expression that has one or more variables (denoted by letters) is a polynomial equation. Algebra is used in many areas of mathematics – for example, matrix algebra and Boolean algebra (the latter is used in working out the logic for computers).

In ordinary algebra the same operations are carried on as in arithmetic, but, as the symbols are capable of a more generalized and extended meaning than the figures used in arithmetic, it facilitates calculation where the numerical values are not known, or are inconveniently large or small, or where it is desirable to keep them in an analysed form.

Within an algebraic equation the separate calculations involved must be completed in a set order. Any elements in brackets should always be calculated first, followed by multiplication, division, addition, and subtraction.

algebraic terms

variable A variable is a changing quantity (one that can take various values), as opposed to a constant. For example, in the algebraic expression $y = 4x^3 + 2$, the variables are x and y, whereas 4 and 2 are constants. Variables are generally represented as letters.

A variable may be dependent or independent. Thus if y is a function of x, written $y = f(x)$, such that $y = 4x^3 + 2$, the domain of the function includes all values of the **independent variable** $\times$ while the range (or codomain) of the function is defined by the values of the **dependent variable** y.

constant A constant is a fixed quantity or one that does not change its value in relation to variables (changing quantities). For example, in the algebraic expression $y^2 = 5x - 3$, the numbers 3 and 5 are constants. In physics, certain quantities are regarded as universal constants, such as the speed of light in a vacuum.

coefficient A coefficient is the number part in front of an algebraic term, signifying multiplication. For example, in the expression $4x^2 + 2xy - x$, the coefficient of x^2 is 4 (because $4x^2$ means $4 \times x^2$), that of xy is 2, and that of $\times$ is -1 (because $-1 \times x = -x$).

In general algebraic expressions, coefficients are represented by letters that may stand for numbers; for example, in the equation $ax^2 + bx + c = 0$, a, b, and c are coefficients, which can take any number.

polynomial Algebraic expressions that have one or more variables (denoted by letters) are called polynomial equations. A polynomial of degree one, that is, whose highest power of x is 1, as in $2x + 1$, is called a linear polynomial;

$$3x^2 + 2x + 1$$

is quadratic;

$$4x^3 + 3x^2 + 2x + 1$$

is cubic.

quadratic equation A quadratic equation is a polynomial equation of second degree (that is, an equation containing as its highest power the square of a variable, such as x^2). The general formula of such equations is:

$$ax^2 + bx + c = 0$$

in which a, b, and c are real numbers, and only the coefficient a cannot equal 0.

Some quadratic equations can be solved by factorization, or the values of x can be found by using the formula for the general solution

$$x = \frac{[-b \pm \sqrt{(b^2 - 4ac)}]}{2a}$$

Depending on the value of the discriminant $b^2 - 4ac$, a quadratic equation has two real, two equal, or two complex roots (solutions). When

$$b^2 - 4ac > 0$$

there are two distinct real roots. When

$$b^2 - 4ac = 0$$

there are two equal real roots. When

$$b^2 - 4ac < 0$$

there are two distinct complex roots.

History and usage of quadratic equations
http://www-history.mcs.st-and.ac.uk/
~history/HistTopics/Quadratic_etc_equations.html
The discovery and development of quadratic equations and also biographical background information on those mathematicians responsible.

---

simultaneous equations If there are two or more algebraic equations that contain two or more unknown quantities that may have a unique solution they can be solved simultaneously. For example, in the case of two linear equations with two unknown variables, such as:

(i) $x + 3y = 6$ and
(ii) $3y - 2x = 4$

the solution will be those unique values of x and y that are valid for both equations. Linear simultaneous equations can be solved by using algebraic manipulation to eliminate one of the variables. For example, both sides of equation (i) could be multiplied by 2, which gives $2x + 6y = 12$. This can be added to equation (ii) to get $9y = 16$, which is easily solved: $y = \frac{16}{9}$. The variable x can now be found by inserting the known y value into either original equation and solving for x.

history of algebra

'Algebra' was originally the name given to the study of equations. In the 9th century, the Arab mathematician Muhammad ibn-Musa al-Khwarizmi used the term *al-jabr* for the process of adding equal quantities to both sides of an equation. When his treatise was later translated into Latin, *al-jabr* became 'algebra' and the word was adopted as the name for the whole subject.

The basics of algebra were familiar in ancient Babylonia (*c.* 18th century BC). Numerous tablets giving sets of problems and their answers, evidently classroom exercises, survive from that period. The subject was also considered by mathematicians in ancient Egypt, China, and India. A comprehensive treatise on the subject, entitled *Arithmetica*, was written in the 3rd century AD by Diophantus of Alexandria. In the 9th century, al-Khwarizmi drew on Diophantus' work and on Hindu sources to produce his influential work *Hisab al-jabr wa'l-muqabalah/Calculation by Restoration and Reduction*.

the development of symbolism From ancient times until the Middle Ages, equation-solving depended on expressing everything in words or in geometric terms. It was not until the 16th century that the modern symbolism began to be developed (notably by François Viète) in response to the growing complexity of mathematical statements which were impossibly cumbersome when expressed in words. Further research in algebra was aided not only because the symbolism was a convenient 'shorthand' but also because it revealed the similarities between different problems and pointed the way to the discovery of generally applicable methods and principles.

quaternions and the idempotent law In the mid-19th century, algebra was raised to a completely new level of abstraction. In 1843, the Irish mathematician William Rowan Hamilton (1805–1865) discovered a three-dimensional extension of the number system, which he called 'quaternions', in which the commutative law of multiplication is not generally true; that is, $ab \neq ba$ for most quaternions a and b. In 1854 George Boole applied the symbolism of algebra to logic and found it fitted perfectly except that he had to introduce a 'special law' that $a^2 = a$ for all a (called the idempotent law).

algebraic structures Discoveries like this led to the realization that there are many possible 'algebraic structures', which can be described as one or more operations acting on specified objects and satisfying certain laws. (Thus the number system has the operations of addition and multiplication acting on numbers and obeying the commutative, associative, and distributive laws.)

In modern terminology, an algebraic structure consists of a set, A, and one or more binary operations

---
The History of Algebra
http://www-history.mcs.st-and.ac.uk/
~history/HistTopics/Fund_theorem_of_algebra.html
The history of the theorem of algebra. The site also provides biographical details on many of the main characters involved in its development.

(that is, functions mapping $A \times A$ into A) which satisfy prescribed 'axioms'. A typical example is a structure which had been studied from the 18th century onwards and is known as a group. This structure had turned up in the study of the solvability of polynomial equations, but it also appears in numerous other problems (for example, in geometry), and even has applications in modern physics.

modern algebra The objective of modern algebra is to study each possible structure in turn, in order to establish general rules for each structure which can be applied in any situation in which the structure occurs. Numerous structures have been studied, and since 1930 a greater level of generality has been achieved by the study of 'universal algebra', which concentrates on properties that are common to all types of algebraic structure.

geometry

Geometry is concerned with the properties of space, usually in terms of plane (two-dimensional) and solid (three-dimensional) figures. The subject is usually divided into **pure geometry**, which embraces roughly the plane and solid geometry dealt with in Greek mathematician Euclid's *Stoicheia/Elements*, and **analytical** or **coordinate geometry**, in which problems are solved using algebraic methods. A third, quite distinct, type includes the non-Euclidean geometries.

Introduction to Euclidean Geometry
http://ecomod.tamu.edu/~dcljr/euclid.html#intro
Thorough introduction to the principles of Euclidean geometry. There is an emphasis on the *Elements*, but other works associated with, or attributed to, Euclid are discussed.
---

pure geometry

Pure geometry is chiefly concerned with properties of figures that can be measured, such as lengths, areas, and angles, and is therefore of great practical use.

angle An angle is an amount of turn or rotation; it may be defined by a pair of rays (half-lines) that share a common endpoint but do not lie on the same line. Angles are measured in ◊degrees (°) or ◊radians (rads) – a complete turn or circle being 360° or 2π rads.

Angles are classified generally by their degree measures: **acute angles** are less than 90°; **right angles** are exactly 90° (a quarter turn); **obtuse angles** are greater than 90° but less than 180° (a straight line); **reflex angles** are greater than 180° but less than 360°.

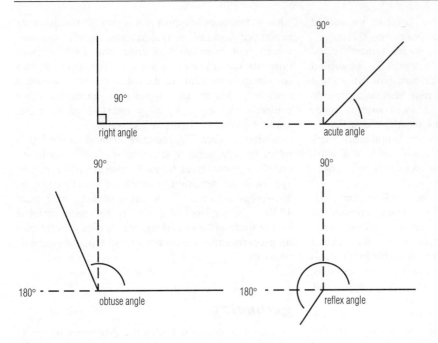

angle The four types of angle, as classified by their degree measures. No angle is classified as having a measure of 180°, as by definition such an 'angle' is actually a straight line.

Angles that add up to 180° are called **supplementary angles**.

angles in triangles A triangle has three interior angles which together add up to 180°. In an equilateral triangle these angles are equal (60°). The exterior angles of a triangle (those produced if one side is extended beyond the triangle) are equal to the sum of the opposite internal angles. Unknown angles in a right-angled triangle can be worked out using trigonometry (see below).

angles in polygons Regular polygons have three types of angle: interior, exterior, and the angle at the centre (produced when a triangle is drawn inside the polygon, with the centre as its apex and one side of the polygon as its base). The angle at the centre is equal to the exterior angle and it is found by dividing 360° by the number of sides in the polygon. For example, the angle at the centre of an octagon is 45° (360 ÷ 8).

An important idea in Euclidean geometry is the idea of **congruence**. Two figures are said to be congruent if they have the same shape and size (and area). If one figure is imagined as a rigid object that can be picked up, moved, and placed on top of the other so that they exactly coincide, then the two figures are congruent. Some simple rules about congruence may be stated: two line segments are congruent if they are of equal length; two triangles are congruent if their corresponding sides are equal in length or if two sides and an angle in one is equal to those in the other; two circles are congruent if they have the same radius; two polygons are congruent if they can be divided into

congruent triangles assembled in the same order.

The idea of picking up a rigid object to test congruence can be expressed more precisely in terms of elementary 'movements' of figures: a translation (or glide) in which all points move the same distance in the same direction (that is, along parallel lines); a rotation through a defined angle about a fixed point; a reflection (equivalent to turning the figure over).

Two figures are congruent to each other if one can be transformed into the other by a sequence of these elementary movements. In Euclidean geometry a fourth

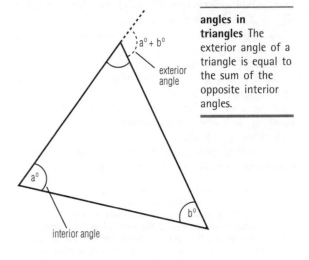

angles in triangles The exterior angle of a triangle is equal to the sum of the opposite interior angles.

triangle Types of triangle.

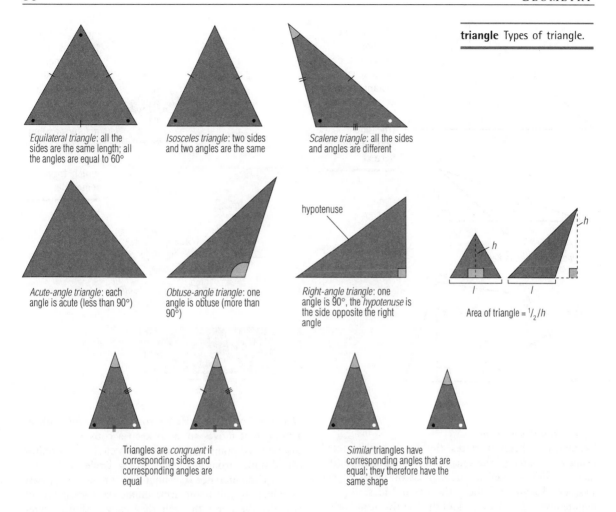

Equilateral triangle: all the sides are the same length; all the angles are equal to 60°

Isosceles triangle: two sides and two angles are the same

Scalene triangle: all the sides and angles are different

Acute-angle triangle: each angle is acute (less than 90°)

Obtuse-angle triangle: one angle is obtuse (more than 90°)

hypotenuse

Right-angle triangle: one angle is 90°, the *hypotenuse* is the side opposite the right angle

Area of triangle = $^1\!/_2\, lh$

Triangles are *congruent* if corresponding sides and corresponding angles are equal

Similar triangles have corresponding angles that are equal; they therefore have the same shape

kind of movement is also studied; this is the enlargement in which a figure grows or shrinks in all directions by a uniform scale factor. If one figure can be transformed into another by a combination of translation, rotation, reflection, and enlargement then the two are said to be **similar**. All circles are similar. All squares are similar. Triangles are similar if corresponding angles are equal.

triangle A triangle is a three-sided plane figure, the sum of whose interior angles is 180°. Triangles can be classified by the relative lengths of their sides. A **scalene triangle** has three sides of unequal length; an **isosceles triangle** has at least two equal sides; an **equilateral triangle** has three equal sides (and three equal angles of 60°).

A **right-angled triangle** has one angle of 90°. If the length of one side of a triangle is l and the perpendicular distance from that side to the opposite corner is h (the height or altitude of the triangle), its area $A = \frac{1}{2} lh$.

hypotenuse The longest side of a right-angled triangle, opposite the right angle, is the hypotenuse. It is of particular application in Pythagoras' theorem (the square of the hypotenuse equals the sum of the squares of the other two sides), and in trigonometry where the ratios sine and cosine (see under *trigonometry* below) are defined as the ratios opposite/hypotenuse and adjacent/hypotenuse respectively.

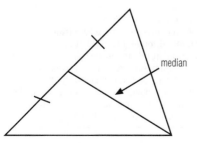

median

median The median is the name given to a line from the vertex (corner) of a triangle to the mid-point of the opposite side.

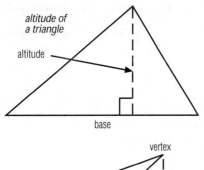

altitude of a triangle

altitude

base

vertex

altitude

base

altitude The altitude of a figure is the perpendicular distance from a vertex (corner) to the base (the side opposite the vertex).

two altitudes of a quadrilateral

vertex

vertex

altitudes

base

diameter
radius
centre
circumference
tangent

minor arc
minor segment
chord
major segment
major arc

major sector
minor sector

16b
16a
1 2 3 4 5 6 7 8 9 10 11 12 13 14 15

r
16a
1 3 5 7 9 11 13 15
2 4 6 8 10 12 14
16b
πr

circle Technical terms used in the geometry of the circle; the area of a circle can be seen to equal πr^2 by dividing the circle into segments which form a rectangle.

Pythagoras' theorem

Pythagorus' theorem states that in a right-angled triangle, the area of the square on the hypotenuse (the longest side) is equal to the sum of the areas of the squares drawn on the other two sides. If the hypotenuse is h units long and the lengths of the other sides are a and b, then $h^2 = a^2 + b^2$.

The theorem provides a way of calculating the length of any side of a right-angled triangle if the lengths of the other two sides are known. It is also used to determine certain trigonometrical relationships such as:

$$\sin^2 \theta + \cos^2 \theta = 1$$

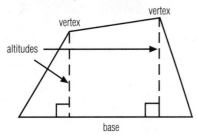

for right-angled triangles

a
c
a
c
c
b
b

Pythagoras' theorem
$a^2 = b^2 + c^2$

Pythagoras' theorem
Pythagoras' theorem for right-angled triangles is likely to have been known long before the time of Pythagoras. It was probably used by the ancient Egyptians to lay out the pyramids.

circle A circle is a perfectly round shape, the path of a point that moves so as to keep a constant distance from a fixed point (the centre). Each circle has a **radius** (the distance from any point on the circle to the centre), a **circumference** (the boundary of the circle, part of which is called an arc), **diameters** (straight lines crossing the circle through the centre), **chords** (lines joining two points on the circumference), **tangents** (lines that touch the circumference at one point only), **sectors** (regions inside the circle between two radii), and **segments** (regions between a chord and the circumference).

The ratio of the distance all around the circle (the circumference) to the diameter is an irrational number called π (**pi**), roughly equal to 3.1416. A circle of radius r and diameter d has a circumference $C = \pi d$, or $C = 2\pi r$, and an area $A = \pi r^2$. The area of a

Pi Through The Ages
http://www–history.mcs.st–and.ac.uk/~history/
HistTopics/Pi_through_the_ages.html
The history of the calculation of the number pi. This site gives the figure that mathematicians from the time of Ptolemy to the present day have used for pi, and describes exactly how they calculated their figure. Also included are many historical asides, such as how calculation of pi led to a racial attack on an eminent professor in pre-war Germany.

Pi

In 1853, the English mathematician William Shanks published the value of pi to 707 decimal places. The calculation had taken him 15 years and was surpassed only in 1945, when computations made on an early desk calculator showed that the last 180 decimal places he had calculated were incorrect.

circle can be shown by dividing it into very thin sectors and reassembling them to make an approximate rectangle. The proof of $A = \pi r^2$ can be done only by using integral calculus.

pi Pi, symbol π, is the ratio of the circumference of a circle to its diameter. The value of pi is 3.1415926, correct to seven decimal places. Common approximations to pi are $\frac{22}{7}$ and 3.14, although the value 3 can be used as a rough estimation.

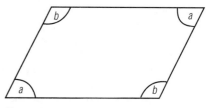

(i) opposite sides and angles are equal

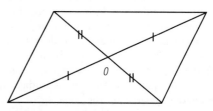

(ii) diagonals bisect each other at *0*

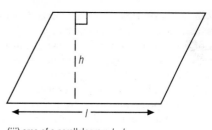

(iii) area of a parallelogram *l* x *h*

parallelogram Some properties of a parallelogram.

cylinder

A cylinder is a tubular solid figure with a circular base. In everyday use, the term applies to a **right cylinder**, the curved surface of which is at right angles to the base.

The volume V of a cylinder is given by the formula $V = \pi r^2 h$, where r is the radius of the base and h is the height of the cylinder. Its total surface area A has the formula $A = 2\pi r(h + r)$, where $2\pi rh$ is the curved surface area, and $2\pi r^2$ is the area of both circular ends.

area is the product of the length of one side and the perpendicular distance between this and the opposite side. In the special case when all four sides are equal in length, the parallelogram is known as a **rhombus**, and when the internal angles are right angles, it is a **rectangle** or **square**.

Pavilion of Polyhedreality
http://www.li.net/~george/pavilion.html
A collection of images and instructions on how to make various polyhedrals, some from quite surprising materials.

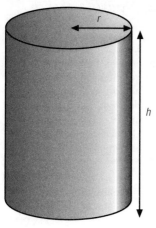

volume = $\pi r^2 h$
area or curved
surface = $2\pi rh$

total surface area
= $2\pi r(r + h)$

cylinder The volume and area of a cylinder are given by simple formulae relating the dimensions of the cylinder.

parallelogram Parallelograms are quadrilaterals (four-sided plane figures) with opposite pairs of sides equal in length and parallel, and opposite angles equal. The diagonals of a parallelogram bisect each other. Its

coordinate geometry

Coordinate geometry is a system of geometry in which points, lines, shapes, and surfaces are represented by algebraic expressions. In plane (two-dimensional) coordinate geometry, the plane is usually defined by two axes at right angles to each other, the horizontal x-axis and the vertical y-axis, meeting at O, the origin. A point on the plane can be represented by a pair of Cartesian coordinates, which define its position in terms of its distance along the x-axis and along the y-

axis from O. These distances are respectively the x and y coordinates of the point.

Lines are represented as equations; for example, $y = 2x + 1$ gives a straight line, and $y = 3x^2 + 2x$ gives a parabola (a curve). The graphs of varying equations can be drawn by plotting the coordinates of points that satisfy their equations, and joining up the points. One of the advantages of coordinate geometry is that geometrical solutions can be obtained without drawing but by manipulating algebraic expressions. For example, the coordinates of the point of intersection of two straight lines can be determined by finding the unique values of × and y that satisfy both of the equations for the lines, that is, by solving them as a pair of simultaneous equations. The curves studied in simple coordinate geometry are the conic sections (circle, ellipse, parabola, and hyperbola), each of which has a characteristic equation.

Cartesian coordinates Cartesian coordinates are components used in coordinate geometry to define the position of a point by its perpendicular distance from a set of two or more axes, or reference lines. For a two-dimensional area defined by two axes at right angles (a horizontal x-axis and a vertical y-axis), the coordinates of a point are given by its perpendicular distances from the y-axis and x-axis, written in the form (x,y). For example, a point P that lies three units from the y-axis and four units from the x-axis has Cartesian coordinates (3,4).

The Cartesian coordinate system can be extended to any finite number of dimensions (axes), and is used thus in theoretical mathematics. So coordinates can be negative numbers, or a positive and a negative, for example (–4, –7), where the point would be to the left of and below zero on the axes. In three-dimensional coordinate geometry, points are located with reference to a third, z-axis, mutually at right angles to the x and y axes.

Cartesian coordinates are named after the French mathematician René Descartes. The system is useful in creating technical drawings of machines or buildings, and in computer-aided design (CAD).

point A point is a basic element of geometry. Its position in the Cartesian system may be determined by its. Mathematicians have had great difficulty in defining the point, as it has no size, and is only the place where two lines meet. According to the Greek mathematician Euclid, (i) a point is that which has no part; (ii) the straight line is the shortest distance between two points.

abscissa The abscissa is the x-coordinate of a point – that is, the horizontal distance of that point from

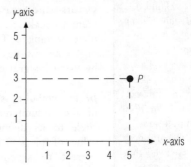

Cartesian coordinates

the Cartesian coordinates of *P* are (5,3)

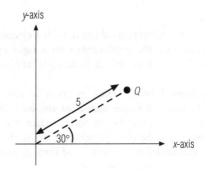

Polar coordinates

the Polar coordinates of *Q* are (5,30°)

coordinate Coordinates are numbers that define the position of points in a plane or in space. In the Cartesian coordinate system, a point in a plane is charted based upon its location along intersecting horizontal and vertical axes. In the polar coordinate system, a point in a plane is defined by its distance from a fixed point and direction from a fixed line.

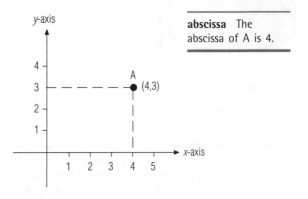

abscissa The abscissa of A is 4.

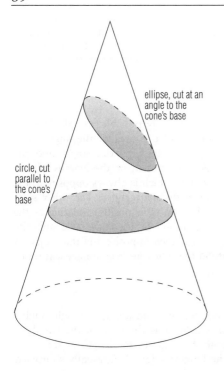

ellipse, cut at an angle to the cone's base

circle, cut parallel to the cone's base

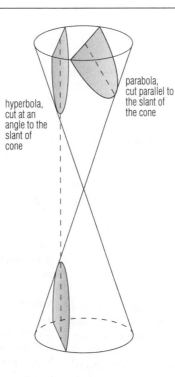

hyperbola, cut at an angle to the slant of cone

parabola, cut parallel to the slant of the cone

conic section The four types of curve that may be obtained by cutting a single or double right-circular cone with a plane (two-dimensional surface).

the vertical or y-axis. For example, a point with the coordinates (4, 3) has an abscissa of 4. The y-coordinate of a point is known as the **ordinate**.

Hyperbola
http://www-groups.dcs.st-andrews.ac.uk/~history/
Curves/Hyperbola.html
Introduction to hyperbola and links to associated curves. If your browser can handle Java code, you can experiment interactively with this curve and its associates. There are also links to mathematicians who have studied the hyperbola, to the hyperbola's particular attributes, and to related Web sites.

---

conic section A curve obtained when a conical surface is intersected by a plane is called a conic section. If the intersecting plane cuts both extensions of the cone, it yields a hyperbola; if it is parallel to the side of the cone, it produces a parabola. Other intersecting planes produce circles or ellipses. The Greek mathematician Apollonius wrote eight books with the title *Conic Sections*, which superseded previous work on the subject by Aristarchus and Euclid.

Parabola
http://www-groups.dcs.st-andrews.ac.uk/
~history/Curves/Parabola.html
Introduction to parabolas and links to associated curves. If your browser can handle Java code, you can experiment interactively with this curve and its associates.

---

topology

Topology is the branch of geometry that deals with those properties of a figure that remain unchanged even when the figure is transformed (bent, stretched) – for example, when a square painted on a rubber sheet is deformed by distorting the sheet.

Topology has scientific applications, as in the study of turbulence in flowing fluids. The topological theory,

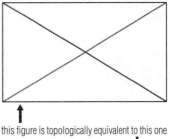

topology Despite distortion some properties, such as the intersection of the lines, remain unchanged.

this figure is topologically equivalent to this one

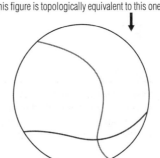

| dodecahedron | icosahedron | tetrahedron | cube | octahedron |

polyhedron The five regular polyhedra or Platonic solids.

proposed in 1880, that only four colours are required in order to produce a map in which no two adjoining countries have the same colour, inspired extensive research, and was proved in 1972 by the US mathematicians Kenneth Appel and Wolfgang Haken.

The map of the London Underground system is an example of the topological representation of a network; connectivity (the way the lines join together) is preserved, but shape and size are not.

trigonometry

Trigonometry solves problems relating to plane and spherical triangles. Its principles are based on the fixed proportions of sides for a particular angle in a right-angled triangle, the simplest of which are known as the sine, cosine, and tangent (so-called trigonometrical ratios). Trigonometry is of practical importance in navigation, surveying, and simple harmonic motion in physics.

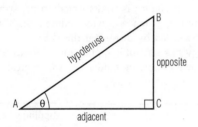

for any right-angled triangle with angle θ as shown the trigonometrical ratios are

$$\sin(e)\ \theta = \frac{BC}{AB} = \frac{opposite}{hypotenuse}$$

$$\cos\theta = \frac{AC}{AB} = \frac{adjacent}{hypotenuse}$$

$$\tan\theta = \frac{BC}{AC} = \frac{opposite}{adjacent}$$

trigonometry At its simplest level, trigonometry deals with the relationships between the sides and angles of triangles. Unknown angles or lengths are calculated by using trigonometrical ratios such as sine, cosine, and tangent.

Using trigonometry, it is possible to calculate the lengths of the sides and the sizes of the angles of a right-angled triangle as long as one angle and the length of one side are known, or the lengths of two sides. The longest side, which is always opposite to the right angle, is called the **hypotenuse**. The other sides are named depending on their position relating to the angle that is to be found or used: the side opposite this angle is always termed **opposite** and that adjacent is the **adjacent**. So the following trigonometrical ratios are used:

sine

The sine is the function of an angle in a right-angled triangle which is defined as the ratio of the length of the side opposite the angle to the length of the hypotenuse (the longest side). It is usually shortened to **sin**.

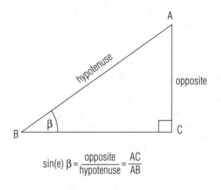

$$\sin(e)\ \beta = \frac{opposite}{hypotenuse} = \frac{AC}{AB}$$

sine The sine is a function of an angle in a right-angled triangle found by dividing the length of the side opposite the angle by the length of the hypotenuse (the longest side). Sine (usually abbreviated sin) is one of the fundamental trigonometric ratios.

Various properties in physics vary sinusoidally; that is, they can be represented diagrammatically by a sine wave (a graph obtained by plotting values of angles against the values of their sines). Examples include simple harmonic motion, such as the way alternating current (AC) electricity varies with time.

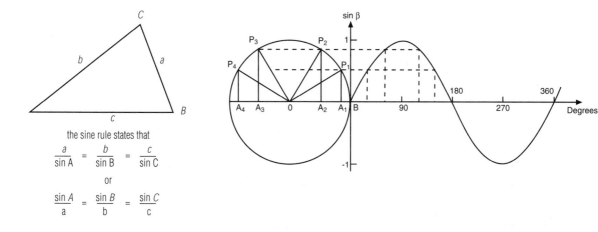

the sine rule states that

$$\frac{a}{\sin A} = \frac{b}{\sin B} = \frac{c}{\sin C}$$

or

$$\frac{\sin A}{a} = \frac{\sin B}{b} = \frac{\sin C}{c}$$

sine (left) The sine of an angle; (right) constructing a sine wave. The sine of an angle is a function used in the mathematical study of the triangle. If the sine of angle β is known, then the hypotenuse can be found given the length of the opposite side, or the opposite side can be found from the hypotenuse. Within a circle of unit radius (left), the height P_1A_1 equals the sine of angle P_1OA_1. This fact and the equalities below the circle allow a sine curve to be drawn, as on the right.

cosine

The cosine is the function of an angle in a right-angled triangle found by dividing the length of the side adjacent to the angle by the length of the hypotenuse (the longest side). It is usually shortened to **cos**.

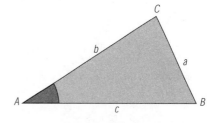

the cosine rule states that
$$a^2 = b^2 + c^2 - 2bc \cos A$$

cosine rule The cosine rule is a rule of trigonometry that relates the sides and angles of triangles. It can be used to find a missing length or angle in a triangle.

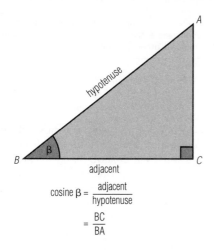

$$\text{cosine } \beta = \frac{\text{adjacent}}{\text{hypotenuse}}$$

$$= \frac{BC}{BA}$$

cosine The cosine of angle β is equal to the ratio of the length of the adjacent side to the length of the hypotenuse (the longest side, opposite to the right angle).

The two non-right angles of a right-angled triangle add up to 90° and are, therefore, described as **complementary angles** (or co-angles). If the two non-right angles are α and β, it may be seen that sin α = cos β and sin β = cos α. Therefore, the sine of each angle equals the cosine of its co-angle. For example, if the co-angles of a triangle are 30° and 60°, sin 30° = cos 60° = 0.5 sin 60° = cos 30° = 0.8660.

tangent

The tangent is a function of an acute angle in a right-angled triangle, defined as the ratio of the length of the side opposite the angle to the length of the side

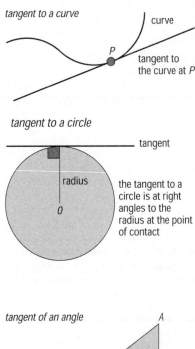

tangent to a curve

curve

P

tangent to
the curve at *P*

tangent to a circle

tangent

radius

0

the tangent to a
circle is at right
angles to the
radius at the point
of contact

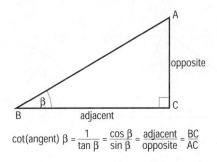

A

opposite

β

C

B   adjacent

$$\cot(\text{angent})\ \beta = \frac{1}{\tan\beta} = \frac{\cos\beta}{\sin\beta} = \frac{\text{adjacent}}{\text{opposite}} = \frac{BC}{AC}$$

cotangent The cotangent of angle β is equal to the ratio of the length of the adjacent side to the length of the opposite side.

the same as those employed for plane triangles.

Trigonometry arose out of the study of astronomy, and was originated by the Greek astronomer Hipparchus. It was also known to early Hindu and Arab mathematicians. Ptolemy, the Alexandrian astrologer, greatly extended the subject and German astronomer Regiomontanus made it a science independent of astronomy much later on, when he began compiling trigonometrical tables in 1467.

Trigonomic Functions
http://www-history.mcs.st-and.ac.uk/~history/
HistTopics/Trigonometric_functions.html
History of the development of the trigonometric functions in mathematics. The site also explains the basic principles of the use of these functions and details the role they have played in the advancement of other branches of mathematics.

calculus

Calculus (Latin 'pebble') is probably the most widely used part of mathematics. Many real-life problems are analysed by expressing one quantity as a function of another – position of a moving object as a function of time, temperature of an object as a function of distance from a heat source, force on an object as a function of distance from the source of the force, and so on – and calculus is concerned with such functions. Calculus uses the concept of a derivative to analyse the way in which the values of a function vary.

There are several branches of calculus. Differential and integral calculus, both dealing with small quantities which during manipulation are made smaller and smaller, compose the **infinitesimal calculus**. **Differential equations** relate to the derivatives of a set

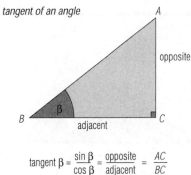

tangent of an angle   A

opposite

β

B   C
adjacent

$$\text{tangent}\ \beta = \frac{\sin\beta}{\cos\beta} = \frac{\text{opposite}}{\text{adjacent}} = \frac{AC}{BC}$$

tangent The tangent of an angle is a mathematical function used in the study of right-angled triangles. If the tangent of an angle β is known, then the length of the opposite side can be found given the length of the adjacent side, or vice versa.

adjacent to it; a way of expressing the gradient of a line. It is usually written **tan**.

three-dimensional and spherical trigonometry

The methods of elementary trigonometry can be used to solve problems in three dimensions by considering triangles in different planes that have a side in common. This may also involve the use of a **dropping perpendicular**, that is the envisaging of an imaginary line from a point above the base that will fall vertically to the base. Spherical triangles can be solved using the trigonometric functions, though the formulae are not of variables and may include the variables. Many give the mathematical models for physical phenomena such as simple harmonic motion. Differential equations are solved generally by integration, depending on their degree. If no analytical processes are available, integration can be performed numerically. Other branches of calculus include calculus of variations and calculus of errors.

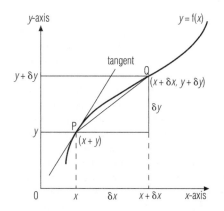

Calculus
http://www-history.mcs.st-and.ac.uk/~history/
HistTopics/The_rise_of_calculus.html
History of the development of calculus, from the time of the Greeks to the works of Cauchy in the 19th century. Also included are several mathematical demonstrations on the use of calculus.

---

function

A function f is a non-empty set of ordered pairs (x, $f(x)$) of which no two can have the same first element. Hence, if $f(x) = x^2$ two ordered pairs are (-2,4) and (2,4). The set of all first elements in a function's ordered pairs is called the **domain**; the set of all second elements is the **range**. In the algebraic expression

$$y = 4x^3 + 2$$

the dependent variable y is a function of the independent variable x, generally written as $f(x)$.

Functions are used in all branches of mathematics, physics, and science generally; for example, the formula $t = 2\pi\sqrt{(l/g)}$ shows that for a simple pendulum the time of swing t is a function of its length l and of no other variable quantity (π and g, the acceleration due to gravity, are constants).

limit

In an infinite sequence, the final value towards which the sequence is tending is termed the **limit**. For example, the limit of the sequence $\frac{1}{2}$, $\frac{3}{4}$, $\frac{7}{8}$, $\frac{15}{16}$ ... is 1, although no member of the sequence will ever exactly equal 1 no matter how many terms are added together.

derivative or differential coefficient

A derivative is the limit of the gradient of a chord linking two points on a curve as the distance between the points tends to zero; for a function of a single variable, $y = f(x)$, it is denoted by $f'(x)$, $Df(x)$, or dy/dx, and is equal to the gradient of the curve.

differentiation

The procedure for determining the derivative or gradient of the tangent to a curve $f(x)$ at any point x is called differentiation. The derivative may be regarded as the limit of the expression $[f(x + \delta x) -$

differentiation A mathematical procedure for determining the gradient, or slope, of the tangent to any curve $f(x)$ at any point x.

$f(x)]/\delta x$ as δx tends to zero. Graphically, this is equivalent to the gradient (slope) of the curve represented by $y = f(x)$ at any point x.

integration Integration is the method in calculus of determining the solutions of definite or indefinite integrals. An example of a **definite integral** can be thought of as finding the area under a curve (as represented by an algebraic expression or function) between particular values of the function's variable. In practice, integral calculus provides scientists with a powerful tool for doing calculations that involve a continually varying quantity (such as determining the position at any given instant of a space rocket that is accelerating away from Earth). Its basic principles were discovered in the late 1660s independently by the German philosopher Leibniz and the British scientist Isaac Newton.

statistics

Statistics is concerned with the collection and interpretation of data. For example, to determine the mean age of the children in a school, a statistically acceptable answer might be obtained by calculating an average based on the ages of a representative sample, consisting, for example, of a random tenth of the pupils from each class. Probability is the branch of statistics dealing with predictions of events.

One of the most important uses of statistical theory is in testing whether experimental data support hypotheses or not. For example, an agricultural researcher arranges for different groups of cows to be fed different diets and records the milk yields. The milk-yield data are analysed and the means and standard deviations of yields for different groups vary. The researcher can use statistical tests to assess whether the variation is of an amount that should be expected because of the natural variation in cows or whether it is larger than normal and therefore likely to be influenced by the difference in diet.

Correlation measures the degree to which two quantities are associated, in the sense that a variation in one quantity is accompanied by a predictable variation in the other. For example, if the pressure on a quantity of gas is increased then its volume decreases. If observations of pressure and volume are taken then statistical correlation analysis can be used to determine whether the volume of a gas can be completely predicted from a knowledge of the pressure on it.

mean, median, and mode

The mean, median, and mode are different ways of finding a 'typical' or 'central' value of a set of data. The mean is obtained by adding up all the observed values and dividing by the number of values; it is the number which is commonly used as an average value. The median is the middle value, that is, the value which is exceeded by half the items in the sample. The mode is the value which occurs with greatest frequency, the most common value. The mean is the most useful measure for the purposes of statistical theory. The idea of the median may be extended and a distribution can be divided into four quartiles. The first quartile is the value which is exceeded by three-quarters of the items; the second quartile is the same as the median; the third quartile is the value that is exceeded by one-quarter of the items.

mean The measure of the average of a number of terms or quantities is termed the mean. The simple **arithmetic mean** is the average value of the quantities, that is, the sum of the quantities divided by their number. The **weighted mean** takes into account the frequency of the terms that are summed; it is calculated by multiplying each term by the number of times it occurs, summing the results and dividing this total by the total number of occurrences. The **geometric mean** of n quantities is the nth root of their product. In statistics, it is a measure of central tendency of a set of data.

median The median is the middle number of an ordered group of numbers. If there is no middle number (because there is an even number of terms), the median is the mean (average) of the two middle numbers. For example, the median of the group 2, 3, 7,

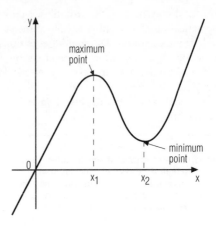

correlation Scattergraphs showing different kinds of correlation. In this way, a causal relationship between two variables may be proved or disproved, provided there are no hidden factors.

11, 12 is 7; that of 3, 4, 7, 9, 11, 13 is 8 (the average of 7 and 9).

In geometry, the term refers to a line from the vertex of a triangle to the midpoint of the opposite side.

standard deviation and other measures of dispersion

The mean is a very incomplete summary of a group of observations; it is useful to know also how closely the individual members of a group approach the mean, and this is indicated by various measures of dispersion. The **range** is the difference between the maximum and minimum values of the group; it is not very satisfactory as a measure of dispersion. The **mean deviation** is the arithmetic mean of the differences between the mean and the individual values, the differences all being taken as positive. However, the **mean deviation** also does not convey much useful information about a group of observations. The most useful measure of dispersion is the **variance**, which is the arithmetic mean of the squares of the deviations from the mean. The positive square root of the variance is called the **standard deviation**, a measure of the spread of data. It is usual to standardize the measurements by working in units of the standard deviation measured from the mean of the distributions, enabling statistical theories to be generalized. A standardized distribution has a mean of zero and a standard deviation of unity. Another useful measure of dispersion is the semi-interquartile range, which is one-half of the distance between the first and third quartiles, and can be considered as the average distance of the quartiles from the median. In many typical distributions the semi-interquartile range is about two-thirds of the standard deviation and the mean deviation is about four-fifths of the standard deviation.

standard deviation Standard deviation is a measure (symbol σ or s) of the spread of data. The deviation (difference) of each of the data items from the mean is found, and their values squared. The mean value of these squares is then calculated. The standard deviation is the square root of this mean.

If n is the number of items of data, $\bar{x}$ is the value of each item, and x is the mean value, the standard deviation σ may be given by the formula:

$$\sigma = \sqrt{\frac{\Sigma(x_i - x)^2}{n}}$$

where (indicates that the differences between the value of each item of data and the mean should be summed. To simplify the calculations, the formula may be rearranged to:

$$\sigma = \sqrt{\frac{\Sigma x^2}{n} - x^2}$$

As a result, it becomes necessary only to calculate Σx and Σx^2.

For example, if the ages of a set of children were 4, 4.5, 5, 5.5, 6, 7, 9, and 11, Σx would be 52, x would be $\frac{52}{n} = \frac{52}{8} = 6.5$, and Σx^2 would be:

$$378.5 \; (= 4^2 + 4.5^2 + 5^2 + 5.5^2$$
$$+ \; 6^2 + 7^2 + 9^2 + 11^2)$$

Therefore, the standard deviation σ would be:

$$\sqrt{\frac{378.5}{8} - (6.5)^2} = \sqrt{5.0625} = 2.25$$

probability

The likelihood, or chance, that an event will occur is its probability, which is often expressed as odds, or in mathematics, numerically as a fraction or decimal.

In general, the probability that n particular events will happen out of a total of m possible events is $\frac{n}{m}$. A certainty has a probability of 1; an impossibility has a probability of 0. Empirical probability is defined as the number of successful events divided by the total possible number of events.

In tossing a coin, the chance that it will land 'heads' is the same as the chance that it will land 'tails', that is, 1 to 1 or even; mathematically, this probability is expressed as $\frac{1}{2}$ or 0.5. The odds against any chosen number coming up on the roll of a fair die are 5 to 1; the probability is $\frac{1}{6}$ or 0.1666... .

If two dice are rolled there are $6 \times 6 = 36$ different possible combinations. The probability of a double (two numbers the same) is $\frac{6}{36}$ or $\frac{1}{6}$ since there are six doubles in the 36 events: (1,1), (2,2), (3,3), (4,4), (5,5), and (6,6).

Independent events are those which do not affect each other, for example rolling two dice are independent events, as the rolling of the first die does not effect the outcome of the rolling of the second die. If events are described as **mutually exclusive** it means that if one happens, then it prevents the other from happening. So tossing a coin is a mutually exclusive event as it can result in a head or a tail but not both. The sum of the probabilities of mutually exclusive events is always equal to one. For example, if one has a bag containing three marbles, each of a different colour, the probability of selecting each colour would be $\frac{1}{3}$.

To find out the probability of two or more mutually exclusive events occurring, their individual probabilities are added together. So, in the above example, the probability of selecting either a blue marble or a red marble is:

$$\frac{1}{3} + \frac{1}{3} + \frac{1}{3} = 1$$

The probability of two independent events both occurring is smaller than the probability of one such event occurring. For example, the probability of throwing a three when rolling a die is $\frac{1}{6}$, but the probability of throwing two threes when rolling two dice is $\frac{1}{36}$.

Probability theory was developed by the French mathematicians Blaise Pascal and Pierre de Fermat in the 17th century, initially in response to a request to calculate the odds of being dealt various hands at cards. Today probability plays a major part in the mathematics of atomic theory and finds application in insurance and statistical studies.

chaos theory

Chaos theory or **complexity theory**, attempts to describe irregular, unpredictable systems – that is, systems whose behaviour is difficult to predict because there are so many variable or unknown factors. Weather is an example of a chaotic system.

Chaos theory, which attempts to predict the *probable* behaviour of such systems, based on a rapid calculation of the impact of as wide a range of elements as possible, emerged in the 1970s with the development of sophisticated computers. First developed for use in meteorology, it has also been used in such fields as quantum physics and economics.

Playing Cards and Dice Chances

Poker

Hand	Number possible	Odds against
royal flush	4	649,739 to 1
straight flush	36	72,192 to 1
four of a kind	624	4,164 to 1
full house	3,744	693 to 1
flush	5,108	508 to 1
straight	10,200	254 to 1
three of a kind	54,912	46 to 1
two pairs	123,552	20 to 1
one pair	1,098,240	1.37 to 1
high card	1,302,540	1 to 1
total	2,598,960	

Bridge

Suit distribution in a hand	Odds against
4–4–3–2	4 to 1
5–4–2–2	8 to 1
6–4–2–1	20 to 1
7–4–1–1	254 to 1
8–4–1–0	2,211 to 1
13–0–0–0	158,753,389,899 to 1

Dice
(Chances with two dice and a single throw)

Total count	Odds against
2	35 to 1
3	17 to 1
4	11 to 1
5	8 to 1
6	31 to 5
7	5 to 1
8	31 to 5
9	8 to 1
10	11 to 1
11	17 to 1
12	35 to 1

Dice
(Chances of consecutive winning throws)

Number of consecutive wins	By 7, 11, or point
1	1 in 49
2	6 in 25
3	3 in 25
4	1 in 17
5	1 in 34
6	1 in 70
7	1 in 141
8	1 in 287
9	1 in 582

Fractal Patterns in Nature
http://polymer.bu.edu/museum/
Examination of fractal patterns in nature, including those of
bacterial growth, erosion, metal deposition, and termite trails.
Are they dominated by chance or are other factors involved?

Mathematics Chronology

c. 30000 BC Palaeolithic peoples record tallies on bone in central Europe and France; one wolf bone has 55 cuts arranged in groups of five – the earliest counting system.

c. 5000 BC A decimal number system is in use in Egypt.

c. 3400 BC The first symbols for numbers, simple straight lines, are used in Egypt.

c. 3000 BC The abacus, which uses rods and beads for making calculations, is developed in the Middle East and adopted throughout the Mediterranean. A form of the abacus is also used in China at this time.

The Sumerians of Babylon develop a sexagesimal (based on 60) numbering system. Used for recording financial transactions, the order of the numbers determines their relative, or unit value (place-value), although no zero value is used. It continues to be used for mathematics and astronomy until the 17th century AD, and is still used for measuring angles and time.

c. 1900 BC The Golenishev papyrus is written. It documents Egyptian knowledge of geometry.

c. 1750 BC The Babylonians under Hammurabi use the sexagesimal system to solve linear and quadratic algebraic equations, compile tables of square and cube roots. They are also aware of the Pythagorean property of the right-angled triangle.

876 BC The Hindus in India invent a symbol for zero – one of the greatest inventions in mathematics.

530 BC Pythagoras of Samos starts researching and teaching theories of mathematics, geometry, music, and reincarnation. A mystic as well as a mathematician, he argues that the key to the universe lies in numbers. His work leads to a number of important results, including Pythagoras' theorem of right-angled triangles and the discovery of irrational numbers (those that cannot be represented by fractions).

c. 440 BC Greek mathematician Hippocrates of Chios writes *Elements*, the first compilation of the elements of geometry.

c. 425 BC Greek mathematician Theodorus of Cyrene demonstrates that certain square roots cannot be written as fractions.

c. 360 BC Greek mathematician and astronomer Eudoxus of Cnidus develops the theory of proportion (dealing with irrational numbers), and the method of exhaustion (for calculating the area bounded by a curve) in mathematics.

c. 300 BC Alexandrian mathematician Euclid sets out the laws of geometry in his *Stoicheion/Elements*; it remains a standard text for 2,000 years.

287 BC–212 BC The prolific Greek mathematician Archimedes of Syracuse produces a number of works on two- and three-dimensional geometry, including circles, spheres, and spirals.

c. 250 BC Greek mathematician and inventor Archimedes provides the formulae for finding the volume of a sphere and a cylinder, arrives at an approximation of the value of pi, and creates a place-value system of notation for Greek mathematics.

c. 230 BC Alexandrian mathematician Apollonius of Perga, writes *Conics*, a systematic treatise on the principles of conics in which he introduces the terms, parabola, ellipse, and hyperbola.

c. 230 BC Greek scholar Eratosthenes of Cyrene develops a method of finding all prime numbers. Known as the sieve of Eratosthenes it involves striking out the number 1 and every nth number following the number n. Only prime numbers then remain.

c. 190 BC Chinese mathematicians use powers of 10 to express magnitudes.

127 BC Greek scientist Hipparchus of Bithynia makes an early formulation of trigonometry.

c. 100 BC Chinese mathematicians begin using negative numbers.

62 Greek mathematician and engineer Hero of Alexandria writes *Metrica/Measurements*, containing many formulae for working out areas and volumes.

100 Greek mathematician and inventor Hero of Alexandria devises a method of representing numbers and performing simple calculating tasks using a train of gears – a primitive computer.

100–150 The classical Chinese mathematics text *Jiuzhang Suanshu/Nine Chapters on the Mathematical Art* is assembled.

250 Greek mathematician Diophantus of Alexandria writes *Arithmetica*, a study of problems in which only whole numbers are allowed as solutions.

370 Greek mathematician Hypatia writes commentaries on Diophantus and Apollonius. She is the first recorded female mathematician.

516 The Indian astronomer and mathematician Aryabhata I produces his *Aryabhatiya*, a treatise on quadratic equations, the value of π, and other scientific problems, in which he adds tilted epicycles to the orbits of the planets to explain their movement.

595 Decimal notation is used for numbers in India. This is the system on which our current system is based.

598–665 Indian mathematician and astronomer Brahmagupta introduces negative numbers into mathematics.

780–850 Arab mathematician Muhammad ibn Musa al-

Khwârizma introduces the Indian system of numbers to the West and give us the word 'algebra', from *al-jabr*.

c. 970 The Muslim astronomer Abu al-Wafa' discovers, and plots tables for, several new trigonometrical functions.

1040 Ahmad al-Nasawi writes on fractions, square and cubic roots, and other mathematical phenomena using Hindu (or Arabic) numerals.

1175 Arabic numerals are introduced into Europe with Gerard of Cremona's translation of the Egyptian astronomer Ptolemy's astronomical work the *Almagest*.

1202 Italian mathematician Fibonacci writes *Liber abaci/The Book of the Abacus*, which introduces the famous sequence of numbers now called the Fibonacci sequence.

1335 The English abbot of St Albans Richard of Wallingford writes *Quadripartitum de sinibus demonstratis*, the first original Latin treatise on trigonometry.

1364 French astronomer and bishop Nicole d'Oresme writes *Latitudes of Forms*, an early work on co-ordinate systems.

1533 Flemish cartographer and mathematician Gemma Frisius publishes a method for accurate surveying using trigonometry.

1545 The Italian mathematician Girolamo Cardano publishes a formula that will solve any cubic equation, discovered by his student Niccolo Tartaglia.

1591 French mathematician François Viète uses letters of the alphabet to represent unknown quantities. Before this, equations had been written out in long descriptive sentences.

1614 Scottish mathematician John Napier invents logarithms, a method for doing difficult calculations quickly.

1617 Dutch mathematician and physicist Willibrord Snell establishes the technique of trigonometrical triangulation to improve the accuracy of cartographic measurements.

1617 Scottish mathematician John Napier, inventor of the log table, devises a system of numbered sticks, called Napier's bones, to aid complex calculations.

1622 English mathematician William Oughtred invents an early form of circular slide rule, adapting the principle behind Scottish mathematician John Napier's 'bones'.

1637 Pierre de Fermat claims to have proved a certain theorem, but leaves no details of his proof. Known as Fermat's last theorem, it is finally proved by US mathematician Andrew Wiles in 1994.

1639 French mathematician Girard Desargues begins the study of projective geometry, which considers what happens to shapes when they are projected onto a screen.

1642 French mathematician Blaise Pascal, aged only 19, builds an adding machine to help his father, the Intendant of Rouen, with tax calculations.

1642–1727 English scientist Isaac Newton lays the foundations of modern mathematics and physics throughout his lifetime.

1653 French mathematician Blaise Pascal publishes his 'triangle' of numbers. This has many applications in arithmetic, algebra, and combinatorics (the study of counting combinations).

1660 Newton begins work on the calculus, a fundamental tool in physics for studying rates of change.

1673 German mathematician Gottfried von Leibniz presents a calculating machine to the Royal Society, London, England. It is the most advanced yet, capable of multiplication, division, and extracting roots.

1676 Newton proves the binomial theorem, a basic tool in solving algebraic equations.

1679 German mathematician Gottfried von Leibniz introduces binary arithmetic, in which only two symbols are used to represent all numbers. It will eventually pave the way for computers.

1684 Leibniz invents the differential calculus, a fundamental tool in studying rates of change.

1690 Swiss mathematician Jacques Bernoulli of the uses the word 'integral' for the first time to refer to the area under a curve.

1694 Swiss mathematician Jean Bernoulli discovers L'Hôpital's rule for determining the correct value of certain ratios.

1707 French mathematician Abraham de Moivre uses trigonometric functions to understand complex numbers for the first time.

1713 Jacques Bernoulli's book *Ars conjectandi/The Art of Conjecture* is the first to deal with probability.

1717 Jean Bernoulli declares that the principle of virtual displacement is applicable to all cases of equilibrium.

1718 Jacques Bernoulli's work on the calculus of variations (the study of functions that are close to their maximum and minimum values) is published posthumously.

1722 De Moivre proposes an equation that is fundamental to the development of complex numbers.

1724 Italian mathematician Jacapo Riccati propounds his equation, an important type of differential equation.

1730 De Moivre propounds theorems of trigonometry concerning imaginary quantities.

1733 De Moivre describes the normal distribution curve.

1742	German mathematician Christian Goldbach conjectures that every even number greater than two can be written as the sum of two prime numbers. Goldbach's conjecture has not yet been proved.
1746	French mathematician Jean d'Alembert develops the theory of complex numbers.
1748	Swiss mathematician Leonhard Euler publishes *Analysis Infinitorum/Analysis of Infinities*, an introduction to pure analytical mathematics. He introduces a formula linking the value of pi to the square root of –1.
1763	The Reverend Thomas Bayes, the English mathematician and theologian, publishes 'An Essay Towards Solving a Problem in the Doctrine of Chances'. This includes Baye's Theorem, which is an important theorem in statistics.
1767	German mathematician Johann Heinrich Lambert proves that the value of pi cannot be written exactly as a fraction.
1772	Italian-born French mathematician Joseph-Louis Lagrange proves that every whole number can be written as the sum of four square numbers.
1789	The French mathematician Baron Augustine-Louis Cauchy completes his lifelong mathematical research, having made great advances in analysis, probability, and group theory.
1797	Lagrange introduces the modern notation for derivatives.
1798	Norwegian mathematician Caspar Wessel introduces the vector representation of complex numbers.
1799	The German mathematician Karl Friedrich Gauss proves the fundamental theorem of algebra: that every algebraic equation has as many solutions as the exponent of the highest term.
1810	French mathematician Jean-Baptiste-Joseph Fourier publishes his method of representing functions by a series of trigonometric functions.
1812	French mathematician Pierre-Simon Laplace describes the mathematical tools he has invented for predicting the probabilities of occurrence of natural events and is the first complete theoretical account of probability.
1815	English physician and philologist Peter Roget invents the 'log-log' slide rule.
c. 1820	Gauss introduces the normal distribution curve ('Gaussian distribution') – a basic statistical tool.
1820	French mathematician Charles-Thomas de Colmar develops the first mass-produced calculator – the 'arithmometer'.
1822	French mathematician Jean-Victor Poncelet systematically develops the principles of projective geometry.
1822	Fourier introduces a technique now known as Fourier analysis, which has widespread applications in mathematics, physics, and engineering.

1823	English mathematician Charles Babbage begins construction of the 'difference' engine, a machine for calculating logarithms and trigonometric functions.
1824	German mathematician and astronomer Friedrich Wilhelm Bessel discovers a class of functions, now called Bessel functions, that arise in many areas of physics.
1824	Swiss mathematician Jakob Steiner develops inversive geometry.
1827	Gauss introduces the subject of differential geometry that describes features of surfaces by analysing curves that lie on it – the intrinsic-surface theory.
1828	English mathematician George Green introduces a theorem that enables volume integrals to be calculated in terms of surface integrals.
1828	Norwegian mathematician Niels Abel begins the study of elliptic functions.
1829	French mathematician Evariste Galois invents group theory, which helps use ideas of symmetry to solve equations.
1829	Russian mathematician Nikolay Ivanovich Lobachevsky develops hyperbolic geometry, in which a plane is regarded as part of a hyperbolic surface shaped like a saddle. It is the beginning of non-Euclidean geometry.
1832	Swiss mathematician Jakob Steiner founds synthetic geometry with the publication of *Systematic Development of the Dependency of Geometrical Forms on One Another*.
1836	Poncelet introduces the use of mathematics to machine design.
1837	French mathematician Siméon-Denis Poisson establishes the rules of probability and describes the Poisson distribution.
1843	English mathematician Arthur Cayley is the first person to investigate spaces with more than three dimensions.
1843	Irish mathematician William Rowan Hamilton invents quaternions, which make possible the application of arithmetic to three-dimensional objects.
1844	French mathematician Joseph Liouville finds the first transcendental numbers – numbers that cannot be expressed as the roots of an algebraic equation with rational coefficients.
1845	Cayley publishes *Theory of Linear Transformations*, which lays the foundation of the school of pure mathematics.
1845	Augustine Cauchy proves the fundamental theorem of group theory, subsequently known as Cauchy's theorem.
1847	English mathematician Augustus de Morgan proposes two laws of logic that are now known as de Morgan's laws.
1847	The English mathematician George Boole publishes *The Mathematical Analysis of Logic*, in

which he shows that the rules of logic can be treated mathematically. Boole's work lays the foundation of computer logic.

1854 Boole outlines his system of symbolic logic now known as Boolean algebra.

1854 Cayley makes important advances in group theory.

1854 German mathematician Georg Friedrich Bernhard Riemann formulates his concept of non-Euclidean geometry in *On the Hypotheses forming the Foundation of Geometry*.

1857–1936 English mathematician Karl Pearson introduces a number of fundamental concepts to the study of statistics.

1858 Cayley invents matrices (rectangular arrays of numbers) and studies their properties.

1859 French artillery officer Amédée Mannheim invents the first modern slide rule that has a cursor or indicator.

1859 Riemann makes a conjecture about a function called the zeta function. Riemann's hypothesis is still unproved, but is an important key to understanding prime numbers.

1865 German mathematician and physicist Julius Plücker invents line geometry.

1871 German mathematician Karl Theodor Wilhelm Weierstrass discovers a curve that, while continuous, has no definable gradient at any point.

1872 German mathematician Richard Dedekind demonstrates how irrational numbers (those that cannot be written as a fraction) may be defined formally.

1874 German mathematician Georg Cantor is the first person rigorously to describe the notion of infinity.

1880 French mathematician Jules-Henri Poincaré publishes important results on automorphic functions, a subject of great importance in modern algebra.

1881 English mathematician John Venn introduces the idea of using pictures of circles to represents sets, subsequently known as Venn diagrams.

1881 US scientist Josiah Willard Gibbs develops the theory of vectors in three dimensions.

1882 Ferdinand Lindemann proves π is a transcendental number.

1882 German mathematician Carl Louis Ferdinand von Lindemann proves that it is impossible to construct a square with the same area as a given circle using a ruler and compass.

1892 German mathematician Georg Cantor demonstrates that there are different kinds of infinity.

1895 Poincaré publishes the first paper on topology, often referred to as 'rubber sheet geometry'.

1896 The prime number theorem is proved independently by mathematicians Jacques-Salomon Hadamard of France and Charles-Jean de la Vallée-Poussin of Belgium. This theorem gives an estimate of the number of primes there are up to a given number.

1899 German mathematician David Hilbert publishes *Grundlagen der Geometrie/Foundations of Geometry*, which provides a rigorous basis for geometry.

1902 US physicist J Willard Gibbs publishes *Elementary Principles of Statistical Mechanics*, in which he develops the mathematics of statistical mechanics.

1906 Russian mathematician Andrey Andreyevich Markov studies random processes that are subsequently known as Markov chains.

1908 German mathematician Ernst Zermelo publishes *Untersuchungen über die Grundlagen der Mengenlehre/Investigations on the Foundations of Set Theory*, which forms the basis of modern set theory.

1910 English philosophers Bertrand Russell and

Alfred North Whitehead publish the first volume of their three-volume *Principia Mathematica/Principles of Mathematics*, in which they attempt to derive the whole of mathematics from a logical foundation. The last volume appears in 1913.

1922 Polish mathematician Stefan Banach begins his work on a development of vector spaces, an important tool in general analysis.

1931 Austrian mathematician Kurt Gödel publishes 'Gödel's proof'. His proof questions the possibility of establishing dependable axioms in mathematics, showing that any formula strong enough to include the laws of arithmetic is either incomplete or inconsistent.

1931 Gödel proves that in any mathematical system such as arithmetic, there are statements that cannot be proved true or false.

1935 US mathematician Alonzo Church invents lambda calculus, a mathematical method for representing mechanical computations.

1936 British mathematician Alan Turing supplies the theoretical basis for digital computers by describing a machine, now known as the Turing machine, capable of universal rather than special-purpose problem solving.

1937–39 US mathematician and physicist John V Atanasoff invents an electromechanical digital computer for solving systems of linear equations. It uses punched cards and is the first electronic calculator using electronic vacuum tubes.

1948 US mathematician Claude Elwood Shannon invents information theory, a mathematical treatment of information that has important applications in computer science and communications.

1950 German-born US logician Rudolf Carnap publishes *Logical Foundations of Probability*.

1950 Russian mathematician Andrey Nikolaevich Kolmogorov presents the first formal treatment of probability in *Foundations of the Theory of Probability*.

1961 US meteorologist Edward Lorenz discovers a mathematical system with chaotic behaviour, leading to a new branch of mathematics known as chaos theory.

1972 French mathematician René Frédéric Thom formulates catastrophe theory, an attempt to describe biological processes mathematically.

1975 US mathematician Mitchell Fingelbaum discovers a new fundamental constant (approximately 4.6692016), which plays an important part in chaos theory.

1976 US mathematicians Kenneth Appel and Wolfgang Haken use a computer to prove the four-colour problem – that the minimum number of colours needed to colour a map such that no two adjacent sections have the same colour is four. The proof takes 1,000 hours of computer time and hundreds of pages.

1980 Mathematicians worldwide complete the classification of all finite and simple groups, a task that has taken over 100 mathematicians more than 35 years to complete. The results take up more than 14,000 pages in mathematical journals.

1980 Polish-born French mathematician Benoit Mandelbrot discovers fractals. The Mandelbrot set is a spectacular shape with a fractal boundary (a boundary of infinite length enclosing a finite area).

1994 US mathematician Andrew Wiles proves Fermat's last theorem, a problem that had remained unsolved since 1637.

1998 Cambridge University professors Richard Borchers and Tim Gowers win Fields Medals in Mathematics for work in vertex algebra and probabilistic number theory, at the International Congress of Mathematicians in Berlin, Germany.

Biographies

Alembert, Jean le Rond d' (1717–1783) French Fmathematician, encyclopedist, and theoretical physicist. He framed several theorems and principles – notably **d'Alembert's principle** – in dynamics and celestial mechanics, and devised the theory of partial differential equations. The principle that now bears his name was first published in his *Traité de dynamique* (1743), and was an extension of the third of Isaac Newton's laws of motion. From the early 1750s, together with other mathematicians such as Joseph ◊Lagrange and Pierre ◊Laplace, he applied calculus to celestial mechanics. In particular, he worked out in 1754 the theory needed to set Newton's discovery of the precession of the equinoxes on a sound mathematical basis, and explained the phenomenon of the oscillation of the Earth's axis.

Apollonius of Perga (*c.* 262–*c.* 190 BC) Greek mathematician, called 'the Great Geometer'. In his eight-volume work *Konica/The Conics*, he showed that a plane intersecting a cone will generate an ellipse, a parabola, or a hyperbola, depending on the angle of intersection. The first four books consisted of an introduction and a statement of the state of mathematics provided by his predecessors. In the last four volumes, Apollonius put forth his own important work on conic sections, the foundation of much of the geometry still used today in astronomy, ballistic science, and rocketry.

Archimedes (c. 287–212 BC) Greek mathematician who made major discoveries in geometry, hydrostatics, and mechanics, and established the sciences of statics and hydrostatics. He formulated a law of fluid displacement (Archimedes' Principle), and is credited with the invention of the Archimedes screw, a cylindrical device for raising water. His method of finding mathematical proof to substantiate experiment and observation became the method of modern science in the High Renaissance.

Archimedes wrote many mathematical treatises. His approximation for the value for π was more accurate than any previous estimate – the value lying between $^{223}/_{71}$ and $^{220}/_{70}$. The average of these two numbers is less than 0.0003 different from the modern approximation for π. He also examined the expression of very large numbers, using a special notation to estimate the number of grains of sand in the universe. Although the result, 10^{63}, was far from accurate, he demonstrated that large numbers could be considered and handled effectively. Archimedes also evolved methods to solve cubic equations and to determine square roots by approximation. His formulae for the determination of the surface areas and volumes of curved surfaces and solids anticipated the development of integral calculus, which did not come for another 2,000 years.

Barrow, Isaac (1630–1677) British mathematician, theologian, and classicist. His *Lectiones geometricae* (1670) contains the essence of the theory of calculus, which was later expanded by Isaac Newton and Gottfried Leibniz.

Bernoulli, Daniel (1700–1782) Swiss mathematical physicist. He made important contributions to trigonometry and differential equations (differentiation). Among his achievements in mathematics, he demonstrated how the differential calculus could be used in problems of probability. He did pioneering work in trigonometrical series and the computation of trigonometrical functions. He also showed the shape of the curve known as the lemniscate. In hydrodynamics he proposed Bernoulli's principle, which states that the pressure of a moving fluid decreases the faster it flows – an early formulation of the idea of conservation of energy.

Bernoulli, Jakob (1654–1705) Swiss mathematician who with his brother Johann pioneered the German mathematician Gottfried Leibniz's calculus. Jakob used calculus to study the forms of many curves arising in practical situations, and studied mathematical probability (*Ars conjectandi* 1713); **Bernoulli numbers** (a series of complex fractions) are named after him.

Bernoulli, Johann (1667–1748) Swiss mathematician who with his brother Jakob pioneered the German mathematician Gottfried Leibniz's calculus. He was the father of Daniel ◊Bernoulli. Johann also contributed to many areas of applied mathematics, including the problem of a particle moving in a gravitational field. He found the equation of the catenary in 1690 and developed exponential calculus in 1691.

Bolyai, János (1802–1860) Hungarian mathematician, one of the founders of non-Euclidean geometry. He was the first to see Euclidean geometry as only one case, and that others were possible. By about 1820 he had become convinced that a proof of Euclid's postulate about parallel lines was impossible; he began instead to construct a geometry which did not depend upon Euclid's axiom. He developed his formula relating the angle of parallelism of two lines with a term characterizing the line. In his new theory Euclidean space was simply a limiting case of the new space, and Bolyai introduced his formula to express what later became known as the space constant.

Boole, George (1815–1864) English mathematician. His work *The Mathematical Analysis of Logic* (1847) established the basis of modern mathematical logic, and his **Boolean algebra** can be used in designing computers. His system is essentially two-valued. By subdividing objects into separate classes, each with a given property, his algebra makes it possible to treat different classes according to the presence or absence of the same property. Hence it involves just two numbers, 0 and 1 – the binary system used in the computer.

Calderón, Alberto P (1920–1998) Argentine mathematician who specialized in Fourier analysis and partial differential equations. Together with mathematician Antoni Zygmund, he devised the Calderón-Zygmund theory of singular integrals.

Cantor, Georg Ferdinand Ludwig Philipp (1845–1918) German mathematician. He defined real numbers and produced a treatment of irrational numbers using a series of transfinite numbers. Investigating sets of the points of convergence of the Fourier series (which enables functions to be represented by trigonometric series), Cantor derived the theory of sets that is the basis of modern mathematical analysis. His work contains many definitions and theorems in topology. For the theory of sets, he had to arrive at a definition of infinity, and also therefore consider the transfinite; for this he used the ancient term 'continuum'. He showed that within the infinite there are countable sets and there are sets having the power of a continuum, and proved that for every set there is another set of a higher power.

Cauchy, Augustin Louis, Baron de (1789–1857) French mathematician who employed rigorous methods of analysis. His prolific output included work on complex functions, determinants, and probability, and on the convergence of infinite series. In calculus, he refined the concepts of the limit and the definite integral. Cauchy has the credit for 16 fundamental concepts and theorems in mathematics and mathematical physics, more than any other mathematician. His work provided a basis for the calculus. He provided the first comprehensive theory of complex numbers, which contributed to the development of mathematical physics and, in particular, aeronautics.

Condorcet, Marie Jean Antoine Nicolas de Caritat, Marquis de Condorcet (1743–1794) French philosopher, mathematician, and politician. As a mathematician he made important contributions to the theory of probability.

Descartes, René (1596–1650) French philosopher and mathematician. He believed that commonly accepted knowledge was doubtful because of the subjective nature of the senses, and attempted to rebuild human knowledge using as his foundation the dictum *cogito ergo sum* ('I think, therefore I am'). He also believed that the entire material universe could be explained in terms of mathematical physics, and founded coordinate geometry as a way of defining and manipulating geometrical shapes by means of algebraic expressions. **Cartesian coordinates**, the means by which points are represented in this system, are named after him. His great work in mathematics was *La Géométrie/ Geometry* (1637). Although not the first to apply algebra to geometry, he was the first to apply geometry to algebra. He was also the first to classify curves systematically, separating 'geometric curves' (which can be precisely expressed as an equation) from 'mechanical curves' (which cannot).

Diophantus (lived AD 250) Greek mathematician in Alexandria whose *Arithmetica* is one of the first known works on problem solving by algebra, in which both words and symbols are used. His main mathematical study was in the solution of what are now known as 'indeterminate' or 'Diophantine' equations – equations that do not contain enough facts to give a specific answer but enough to reduce the answer to a definite type. These equations have led to the formulation of the theory of numbers, regarded as the purest branch of present-day mathematics. In the solution of equations Diophantus was the first to abbreviate the expression of his calculations by means of a symbol representing the unknown quantity.

Eratosthenes (c. 276–c. 194 BC) Greek geographer and mathematician whose map of the ancient world was the first to contain lines of latitude and longitude, and who calculated the Earth's circumference with an error of about 10%. His mathematical achievements include a method for duplicating the cube, and for finding prime numbers (**Eratosthenes' sieve**).

Euclid (c. 330–c. 260 BC) Greek mathematician who wrote the *Stoicheia/ Elements* in 13 books, nine of which deal with plane and solid geometry and four with number theory. His great achievement lay in the systematic arrangement of previous mathematical discoveries and a methodology based on axioms (statements assumed to be true), definitions, and theorems. He used two main styles of presentation: the synthetic (in which one proceeds from the known to the unknown via

logical steps) and the analytical (in which one posits the unknown and works towards it from the known, again via logical steps). Both methods were based on axioms and from which mathematical propositions, or theorems, were deduced.

Euler, Leonhard (1707–1783) Swiss mathematician. He developed the theory of differential equations and the calculus of variations, developed spherical trigonometry, and demonstrated the significance of the coefficients of trigonometric expansions; **Euler's number** (e, as it is now called) has various useful theoretical properties and is used in the summation of particular series.

Fermat, Pierre de (1601–1665) French mathematician who, with Blaise ◊Pascal, founded the theory of probability and the modern theory of numbers. Fermat also made contributions to analytical geometry. In 1657 he published a series of problems as challenges to other mathematicians, in the form of theorems to be proved. **Fermat's last theorem** states that equations of the form $x^n + y^n = z^n$ where x, y, z, and n are all integers have no solutions if $n > 2$. In 1993, English mathematician Andrew Wiles announced a proof; this turned out to be premature, but he put forward a revised proof in 1994.

Fibonacci, Leonardo, also known as Leonardo of Pisa (c. 1170–c. 1250) Italian mathematician. He published *Liber abaci/The Book of the Calculator* (1202), which was instrumental in the introduction of Arabic notation into Europe. From 1960, interest increased in **Fibonacci numbers**, in their simplest form a sequence in which each number is the sum of its two predecessors (1, 1, 2, 3, 5, 8, 13, ...). Fibonacci also published *Practica geometriae* (1220), in which he used algebraic methods to solve many arithmetical and geometrical problems.

Fourier, Jean Baptiste Joseph (1768–1830) French applied mathematician whose formulation of heat flow in 1807 contains the proposal that, with certain constraints, any mathematical function can be represented by trigonometrical series. This principle forms the basis of **Fourier analysis**, used today in many different fields of physics. His idea is embodied in his *Théorie analytique de la chaleur/The Analytical Theory of Heat* (1822). Light, sound, and other wavelike forms of energy can be studied using Fourier's method, a developed version of which is now called harmonic analysis. Fourier laid the groundwork for the later development of dimensional analysis and linear programming. He also investigated probability theory and the theory of errors.

Galois, Evariste (1811–1832) French mathematician who originated the theory of groups and greatly extended the understanding of the conditions in which an algebraic equation is solvable. What has come to be known as the **Galois theorem** demonstrated the insolubility of higher-than-fourth-degree equations by radicals. **Galois theory** involved groups formed from the arrangements of the roots of equations and their subgroups, which he fitted into each other rather like Chinese boxes.

Gauss, Carl Friedrich (1777–1855) German mathematician who worked on the theory of numbers, non-Euclidean geometry, and the mathematical development of electric and magnetic theory. A method of neutralizing a magnetic field, used to protect ships from magnetic mines, is called 'degaussing'. In statistics, the normal distribution curve, which he studied, is sometimes known as the Gaussian distribution. *Disquisitiones* *arithmeticae* (1801) summed up Gauss's work in number theory and formulated concepts and questions that are still relevant today.

Gödel, Kurt (1906–1978) Austrian-born US mathematician and philosopher. In his paper, 'On formally undecidable propositions of *Principia Mathematica* and related systems' (1931), he proved that a mathematical system always contains statements that can be neither proved nor disproved within the system; in other words, as a science, mathematics can never be totally consistent and totally complete. He worked on relativity, constructing a mathematical model of the universe that made travel back through time theoretically possible.

Gunter, Edmund (1581–1626) English mathematician. He is reputed to have invented a number of surveying instruments as well as the trigonometrical terms 'cosine' and 'cotangent'.
 His measuring instruments include Gunter's Chain, the 22-yard-long, 100-link chain used by surveyors; Gunter's Line, the forerunner of the slide-rule; Gunter's Scale, a two-foot rule with scales of chords, tangents and logarithmic lines for solving navigational problems, and the portable Gunter's Quadrant.

Hilbert, David (1862–1943) German mathematician, philosopher, and physicist whose work was fundamental to 20th-century mathematics. He founded the formalist school with *Grundlagen der Geometrie/Foundations of Geometry* (1899), which was based on his idea of postulates. He attempted to put mathematics on a logical foundation through defining it in terms of a number of basic principles, which Kurt ◊Gödel later showed to be impossible. In 1900 he proposed a set of 23 problems for future mathematicians to solve, and gave 20 axioms to provide a logical basis for Euclidean geometry.
 Studying algebraic invariants, Hilbert had by 1892 not only solved all the known central problems of this branch of mathematics, he had introduced sweeping developments and new areas for research, particularly in algebraic topology.

Khwarizmi, al-, Muhammad ibn-Musa (*c.* 780– *c.* 850) Persian mathematician. He wrote a book on algebra, from part of whose title (*al-jabr*) comes the word 'algebra', and a book in which he introduced to the West the Hindu–Arabic decimal number system. He compiled astronomical tables and was responsible for introducing the concept of zero into Arab mathematics. The word 'algorithm' is a corruption of his name.

Kovalevskaia, Sofya Vasilevna (1850–1891) Russian mathematician who worked on partial differential equations and Abelian integrals. In 1886 she won the Prix Bordin of the French Academy of Sciences for a paper on the rotation of a rigid body about a point, a problem the 18th-century mathematicians Leonhard ◊Euler and Joseph ◊Lagrange had both failed to solve.

Lagrange, Joseph Louis (1736–1813) Italian-born French mathematician. His *Mécanique analytique* (1788) applied mathematical analysis, using principles established by Isaac Newton, to such problems as the movements of planets when affected by each other's gravitational force. He presided over the commission that introduced the metric system in 1793. Lagrange proved some of Pierre de GFermat's theorems, which had remained unproven for a century.

Laplace, Pierre Simon, Marquis de Laplace (1749–1827) French astronomer and mathematician. Among his mathematical achievements was the development of probability theory. In 'Théorie des attractions des sphéroïdes et de la figure des planètes' (1785) he introduced the potential function and the Laplace coefficients, both of them useful as a means of applying analysis to problems in physics.

Leibniz, Gottfried Wilhelm (1646–1716) German mathematician, philosopher, and diplomat. Independently of, but concurrently with, the English scientist Isaac Newton, he developed the branch of mathematics known as calculus and was one of the founders of symbolic logic. Free from all concepts of space and number, his logic was the prototype of future abstract mathematics. Leibniz is due the credit for first using the **infinitesimals** (very small quantities that were precursors of the modern idea of limits) as differences. He devised a notation for integration and differentiation that was so much more convenient than Newton's **fluxions** that it remains in standard use today. He also designed a calculating machine, completed about 1672, which was able to multiply, divide, and extract roots.

Lobachevsky, Nikolai Ivanovich (1792–1856) Russian mathematician who founded non-Euclidean geometry, concurrently with, but independently of, Karl ◊Gauss in Germany and János ◊Bolyai in Hungary. Lobachevsky published the first account of the subject in 1829, but his work went unrecognized until Georg ◊Riemann's system was published.
 In Euclid's system, two parallel lines will remain equidistant from each other, whereas in Lobachevskian geometry, the two lines will approach zero in one direction and infinity in the other. In Euclidean geometry the sum of the angles of a triangle is always equal to the sum of two right angles; in Lobachevskian geometry, the sum of the angles is always less than the sum of two right angles. In Lobachevskian space, also, two geometric figures cannot have the same shape but different sizes.

Mandelbrot, Benoit B (1924–) Polish-born French mathematician who coined the term **fractal** to describe geometrical figures in which an identical motif repeats itself on an ever-diminishing scale. The concept is associated with chaos theory. Mandelbrot's research has provided mathematical theories for erratic chance phenomena and self-similarity methods in probability. He has also carried out research on sporadic processes, thermodynamics, natural languages, astronomy, geomorphology, computer art and graphics, and the fractal geometry of nature.

Markov, Andrei Andreyevich (1856–1922) Russian mathematician, formulator of the **Markov chain**, an example of a stochastic (random) process. A Markov chain may be described as a chance process that possesses a special property, so that its future may be predicted from the present state of affairs just as accurately as if the whole of its past history were known. Markov chains are now used in the social sciences, atomic physics, quantum theory, and genetics.

Möbius, August Ferdinand (1790–1868) German mathematician and theoretical astronomer, discoverer of the **Möbius strip** and considered one of the founders of topology. In 1818 he formulated his **barycentric calculus**, a mathematical system in which numerical coefficients were assigned to points. The position of any point in the system could be expressed by varying the numerical coefficients of any four or more noncoplanar points. He also discovered the Möbius net, later of value in the development of projective geometry; the Möbius tetrahedra, two tetrahedra that mutually circumscribe and inscribe each other, which he described in 1828; and the Möbius function in number theory, published in 1832.

Napier, John, 8th Laird of Merchiston (1550–1617) Scottish mathematician who invented logarithms in 1614 and 'Napier's bones', an early mechanical calculating device for multiplication and division. It was Napier who first used and then popularized the decimal point to separate the whole number part from the fractional part of a number.

Oughtred, William (1575–1660) English mathematician, credited as the inventor of the slide rule in 1622. His major work *Clavis mathematicae/The Key to Mathematics* (1631) was a survey of the entire body of mathematical knowledge of his day. It introduced the 'x' symbol for multiplication, as well as the abbreviations 'sin' for sine and 'cos' for cosine.

Pascal, Blaise (1623–1662) French philosopher and mathematician. He contributed to the development of hydraulics, calculus, and the mathematical theory of probability. His work in mathematics widened general understanding of conic sections, introduced an algebraic notational system that rivalled that of ◊Descartes, and made use of the arithmetical triangle (called Pascal's triangle) in the study of probabilities. Between 1642 and 1645, Pascal constructed a machine to carry out the processes of addition and subtraction, and then organized the manufacture and sale of these first calculating machines.

Together with Pierre de ◊Fermat, Pascal studied two specific problems of probability: the first concerned the probability that a player will obtain a certain face of a die in a given number of throws; and the second was to determine the portion of the stakes returnable to each player of several if a game is interrupted. Pascal used the arithmetical triangle to derive combinational analysis. **Pascal's triangle** is a triangular array of numbers in which each number is the sum of the pair of numbers above it. In general the nth (n = 0, 1, 2, ...) row of the triangle gives the binomial coefficients ^nC_r, with r = 0, 1, ..., n. In 1657–59 Pascal also perfected his 'theory of indivisibles' – the forerunner of integral calculus –, which enabled him to study problems involving infinitesimals, such as the calculations of areas and volumes.

Pearson, Karl (1857–1936) English statistician who followed the English scientist Francis Galton (1822–1911) in introducing statistics and probability into genetics. He introduced the term 'standard deviation' into statistics. In 1900 he introduced the χ^2 (chi-squared) test to determine whether a set of observed data deviates significantly from what would have been predicted by a 'null hypothesis' (that is, totally at random). He demonstrated that it could be applied to examine whether two hereditary characteristics (such as height and hair colour) were inherited independently. Pearson's discoveries included the Pearson coefficient of correlation (1892), the theory of multiple and partial correlation (1896), the coefficient of variation (1898), work on errors of judgement (1902), and the theory of random walk (1905).

Poincaré, Jules Henri (1854–1912) French mathematician who developed the theory of differential equations and was a pioneer in relativity theory. He suggested that Isaac Newton's laws for the behaviour of the universe could be the exception rather than the rule. However, the calculation was so complex and time-consuming that he never managed to realize its full implication. He published the first paper devoted entirely to topology (the branch of geometry that deals with the unchanged properties of figures), and developed several new mathematical techniques, including the theories of asymptotic expansions and integral invariants, and the new subject of topological dynamics.

Poisson, Siméon Denis (1781–1840) French applied mathematician and physicist. In probability theory he formulated the **Poisson distribution**. **Poisson's ratio** in elasticity is the ratio of the lateral contraction of a body to its longitudinal extension. The ratio is constant for a given material. Much of Poisson's work involved applying mathematical principles in theoretical terms to contemporary and prior experiments in physics, particularly with reference to electricity, magnetism, heat, and sound. Poisson was also responsible for a formulation of the 'law of large numbers'.

Pythagoras
http://www.utm.edu/research/iep/p/pythagor.htm
Profile of the legendary mathematician, scientist, and philosopher Pythagoras. An attempt is made to disentangle the known facts of his life from the mass of legends about him.

Pythagoras (c. 580–500 BC) Greek mathematician and philosopher who formulated **Pythagoras' theorem** (the square of the hypotenuse equals the sum of the squares of the other two sides). Much of his work concerned numbers, to which he assigned mystical properties. For example, he classified numbers into triangular ones (1, 3, 6, 10, ...), which can be represented as a triangular array, and square ones (1, 4, 9, 16, ...), which form squares. He also observed that any two adjacent triangular numbers add to a square number (for example, 1 + 3 = 4; 3 + 6 = 9; 6 + 10 = 16).

Using geometrical principles, his followers, the Pythagoreans, were able to prove that the sum of the angles of any regular-sided triangle is equal to that of two right angles (using the theory of parallels), and to solve any algebraic quadratic equations having real roots. They formulated the theory of proportion (ratio), which enhanced their knowledge of fractions, and used it in their study of harmonics upon their stringed instruments.

Quetelet, Lambert Adolphe Jacques (1796–1874) Belgian statistician. He developed tests for the validity of statistical information, and gathered and analysed statistical data of many kinds. From his work on sociological data came the concept of the 'average person'.

Riemann, Georg Friedrich Bernhard (1826–1866) German mathematician whose system of non-Euclidean geometry, thought at the time to be a mere mathematical curiosity, was used by the physicist Albert Einstein to develop his general theory of relativity. Riemann made a breakthrough in conceptual understanding within several other areas of mathematics: the theory of functions, vector analysis, projective and differential geometry, and topology.

He developed the **Riemann surfaces** to study complex function behaviour. These multiconnected, many-sheeted surfaces can be dissected by cross-cuts into a singly connected surface. By means of these surfaces he introduced topological considerations into the theory of functions of a complex variable, and into general analysis. He showed, for example, that all curves of the same class have the same Riemann surface. He also published a paper on hypergeometric series, invented 'spherical' geometry as an extension of hyperbolic geometry, and in 1855–56 lectured on his theory of Abelian functions, one of his fundamental developments in mathematics.

Russell, Bertrand Arthur William, 3rd Earl Russell (1872–1970) British philosopher and mathematician who contributed to the development of modern mathematical logic. His works include *Principia Mathematica* (1910–13), with Alfred ◊Whitehead, in which he attempted to show that mathematics could be reduced to a branch of logic.

Shannon, Claude Elwood (1916–) US mathematician who founded the science of information theory. He argued that entropy is equivalent to a shortage of information content (a degree of uncertainty), and obtained a quantitative measure of the amount of information in a given message. He reduced the notion of information to a series of yes/no choices, which could be presented by a binary code. Each choice, or piece of information, he called a 'bit'. In this way, complex information could be organized according to strict mathematical principles.

Viète, François (1540–1603) French mathematician who developed algebra and its notation. He was the first mathematician to use letters of the alphabet to denote both known and unknown quantities, and is credited with introducing the term 'coefficient' into algebra. His mathematical achievements were the result of his interest in cosmology; for example, a table giving the values of six trigonometrical lines based on a method originally used by Egyptian astronomer Ptolemy. Viète was the first person to use the cosine law for plane triangles and he also published the law of tangents.

Whitehead, Alfred North (1861–1947) English philosopher and mathematician. In his 'theory of organism', he attempted a synthesis of metaphysics and science. His works include *Principia Mathematica* (1910–13), with Bertrand ◊Russell. Whitehead's research in mathematics involved a highly original attempt – incorporating the principles of logic – to create an extension of ordinary algebra to universal algebra (*A Treatise of Universal Algebra* (1898)), and a meticulous re-examination of the relativity theory of Albert Einstein.

Glossary

abacus ancient calculating device made up of a frame of parallel wires on which beads are strung. The method of calculating with a handful of stones on a 'flat surface' (Latin *abacus*) was familiar to the Greeks and Romans, and used by earlier peoples, possibly even in ancient Babylon; it survives in the more sophisticated bead-frame form of the Russian *schoty* and the Japanese *soroban*. The abacus has been superseded by the electronic calculator.

absolute value, or **modulus,** the value, or magnitude, of a number irrespective of its sign. The absolute value of a number n is written $|n|$ (or sometimes as mod n), and is defined as the positive square root of n^2. For example, the numbers –5 and 5 have the same absolute value:

$$|5| = |-5| = 5.$$

acute angle an angle between 0° and 90°; that is, an amount of turn that is less than a quarter of a circle.

algorithm procedure or series of steps that can be used to solve a problem.

arc a section of a curved line or circle. A circle has three types of arc: a **semicircle,** which is exactly half of the circle; **minor arcs,** which are less than the semicircle; and **major arcs,** which are greater than the semicircle.

arc minute, arc second units for measuring small angles, used in geometry, surveying, map-making, and astronomy. An arc minute (symbol ') is one-sixtieth of a degree, and an arc second (symbol ") is one-sixtieth of an arc minute. Small distances in the sky, as between two close stars or the apparent width of a planet's disc, are expressed in minutes and seconds of arc.

arithmetic mean the average of a set of n numbers, obtained by adding the numbers and dividing by n. For example, the arithmetic mean of the set of 5 numbers 1, 3, 6, 8, and 12 is

$$\frac{(1 + 3 + 6 + 8 + 12)}{5} = \frac{30}{5} = 6.$$

arithmetic progression, or **arithmetic sequence,** sequence of numbers or terms that have a common difference between any one term and the next in the sequence. For example, 2, 7, 12, 17, 22, 27, ... is an arithmetic sequence with a common difference of 5.

asymptote in coordinate geometry, a straight line that a curve approaches progressively more closely but never reaches. The x and y axes are asymptotes to the graph of xy = constant (a rectangular hyperbola).

average in statistics, a term used inexactly to indicate the typical member of a set of data. It usually refers to the ◊arithmetic mean. The term is also used to refer to the middle member of the set when it is sorted in ascending or descending order (the median), and the most commonly occurring item of data (the mode), as in 'the average family'.

axiom statement that is assumed to be true and upon which theorems are proved by using logical deduction; for example, two straight lines cannot enclose a space. The Greek mathematician Euclid used a series of axioms that he considered could not be demonstrated in terms of simpler concepts to prove his geometrical theorems.

axis (plural **axes**) in geometry, one of the reference lines by which a point on a graph may be located. The horizontal axis is usually referred to as the *x*-axis, and the vertical axis as the *y*-axis. The term is also used to refer to the imaginary line about which an object may be said to be symmetrical (**axis of symmetry**) – for example, the diagonal of a square – or the line about which an object may revolve (**axis of rotation**).

binomial expression consisting of two terms, such as $a + b$ or $a – b$.

cardinal number one of the series of numbers 0, 1, 2, 3, 4, … . Cardinal numbers relate to quantity, whereas ordinal numbers (first, second, third, fourth, … .) relate to order.

catastrophe theory mathematical theory developed by the French mathematician René Thom in 1972, in which he showed that the growth of an organism proceeds by a series of gradual changes that are triggered by, and in turn trigger, large-scale changes or 'catastrophic' jumps. It also has applications in engineering – for example, the gradual strain on the structure of a bridge that can eventually result in a sudden collapse – and has been extended to economic and psychological events.

chance likelihood, or probability, of an event taking place, expressed as a fraction or percentage. For example, the chance that a tossed coin will land heads up is 50%.

circumference curved line that encloses a curved plane figure, for example a circle or an ellipse. Its length varies according to the nature of the curve, and may be ascertained by the appropriate formula. The circumference of a circle is πd or $2\pi r$, where d is the diameter of the circle, r is its radius, and π is the constant pi, approximately equal to 3.1416.

concave polygon term used to describe any polygon that has an interior angle greater than 180°.

cone solid or surface consisting of the set of all straight lines passing through a fixed point (the vertex) and the points of a circle or ellipse whose plane does not contain the vertex.

congruent having the same shape and size, as applied to two-dimensional or solid figures. With plane congruent figures, one figure will fit on top of the other exactly, though this may first require rotation and/or rotation of one of the figures.

convex term used to describe any polygon possessing no interior angle greater than 180°.

coordinate number that defines the position of a point relative to a point or axis (reference line). **Cartesian coordinates** define a point by its perpendicular distances from two or more axes drawn through a fixed point mutually at right angles to each other. **Polar coordinates** define a point in a plane by its distance from a fixed point and direction from a fixed line.

correlation the degree of relationship between two sets of information. If one set of data increases at the same time as the other, the relationship is said to be positive or direct. If one set of data increases as the other decreases, the relationship is negative or inverse. Correlation can be shown by plotting a best-fit line on a scatter diagram.

cube regular solid figure whose faces are all squares. It has 6 equal-area faces and 12 equal-length edges.

cube to multiply a number by itself and then by itself again. For example, 5 cubed = 5^3 = $5 \times 5 \times 5$ = 125. The term also refers to a number formed by cubing; for example, 1, 8, 27, 64 are the first four cubes.

degree unit (symbol °) of measurement of an angle or arc. A circle or complete rotation is divided into 360°. A degree may be subdivided into 60 minutes (symbol ´), and each minute may be subdivided in turn into 60 seconds (symbol ´´).

digit any of the numbers from 0 to 9 in the decimal system. Different bases have different ranges of digits. For example, the hexadecimal system has digits 0 to 9 and A to F, whereas the binary system has two digits (or bits), 0 and 1.

dimension any directly measurable physical quantity such as mass (M), length (L), and time (T), and the derived units obtainable by multiplication or division from such quantities. For example, acceleration (the rate of change of velocity) has dimensions (LT^{-2}), and is expressed in such units as km s^{-2}. A quantity that is a ratio, such as relative density or humidity, is dimensionless.

equilateral of a geometrical figure, having all sides of equal length. For example, a square and rhombus are both equilateral four-sided figures. An equilateral triangle, to which the term is most often applied, has all three sides equal and all angles equal (at 60°).

Fibonacci numbers in their simplest form, a sequence in which each number is the sum of its two predecessors (1, 1, 2, 3, 5, 8, 13, …). They have unusual characteristics with possible applications in botany, psychology, and astronomy.

fractal irregular shape or surface produced by a procedure of repeated subdivision. Generated on a computer screen, fractals are used in creating models of geographical or biological processes (for example, the creation of a coastline by erosion or accretion, or the growth of plants).

golden section visually satisfying ratio, first constructed by the Greek mathematician Euclid and used in art and architecture. It is found by dividing a line AB at a point O such that the rectangle produced by the whole line and one of the segments is equal to the square drawn on the other segment. The ratio of the two segments is about 8:13 or 1:1.618, and a rectangle whose sides are in this ratio is called a **golden rectangle**. The ratio of consecutive ◊Fibonacci numbers tends to the golden ratio.

infinity mathematical quantity that is larger than any fixed assignable quantity; symbol ∞. By convention, the result of dividing any number by zero is regarded as infinity.

Möbius strip structure made by giving a half twist to a flat strip of paper and joining the ends together. It has certain remarkable properties, arising from the fact that it has only one edge and one side. If cut down the centre of the strip, instead of two new strips of paper, only one long strip is produced. It was discovered by the German mathematician August Möbius.

ordinal number one of the series first, second, third, fourth, … . Ordinal numbers relate to order, whereas cardinal numbers (1, 2, 3, 4, …) relate to quantity, or count.

parallel lines and parallel planes straight lines or planes that always remain a constant distance from one another no matter how far they are extended. This is a principle of Euclidean geometry. Some non-Euclidean geometries, such as elliptical and hyperbolic geometry, however, reject Euclid's parallel axiom.

perpendicular at a right angle; also, a line at right angles to another or to a plane. For a pair of skew lines (lines in three dimensions that do not meet), there is just one common perpendicular, which is at right angles to both lines; the nearest points on the two lines are the feet of this perpendicular.

polygon plane (two-dimensional) figure with three or more straight-line sides. Common polygons have names which define the number of sides (for example, triangle, quadrilateral, pentagon).

polyhedron solid figure with four or more plane faces. There are only five types of polyhedron (with all faces the same size and shape); they are the tetrahedron (four equilateral triangular faces), cube (six square faces), octahedron (eight equilateral triangles), dodecahedron (12 regular pentagons), and icosahedron (20 equilateral triangles).

progression sequence of numbers each occurring in a specific relationship to its predecessor. An **arithmetic progression** has numbers that increase or decrease by a common sum or difference (for example, 2, 4, 6, 8); a **geometric progression** has numbers each bearing a fixed ratio to its predecessor (for example, 3, 6, 12, 24); and a **harmonic progression** has numbers whose reciprocals are in arithmetical progression, for example $1, \frac{1}{2}, \frac{1}{3}, \frac{1}{4}$.

radian SI unit (symbol rad) of plane angles, an alternative unit to the ◊degree. It is the angle at the centre of a circle when the centre is joined to the two ends of an arc (part of the circumference) equal in length to the radius of the circle. There are 2π (approximately 6.284) radians in a full circle (360°).

ratio measure of the relative size of two quantities or of two measurements (in similar units), expressed as a proportion.

For example, the ratio of vowels to consonants in the alphabet is 5:21; the ratio of 500 m to 2 km is 500:2,000, or 1:4. Ratios are normally expressed as whole numbers, so 2:3.5 would become 4:7 (the ratio remains the same provided both numbers are multiplied or divided by the same number).

real number any of the rational numbers (which include the integers) or irrational numbers. Real numbers exclude imaginary numbers, found in complex numbers of the general form $a + bi$ where i = $\sqrt{-1}$, although these do include a real component a.

rhombus equilateral parallelogram. Its diagonals bisect each other at right angles, and its area is half the product of the lengths of the two diagonals. A rhombus whose internal angles are 90° is called a square.

root of an equation, a value that satisfies the equality. For example, $x = 0$ and $x = 5$ are roots of the equation $x^2 - 5x = 0$.

symmetry exact likeness in shape about a given line (axis), point, or plane. A figure has symmetry if one half can be rotated and/or reflected onto the other. (Symmetry preserves length, angle, but not necessarily orientation.)

tangram puzzle made by cutting up a square into seven pieces.

tetrahedron (plural **tetrahedra**) solid figure (polyhedron) with four triangular faces; that is, a pyramid on a triangular base. A regular tetrahedron has equilateral triangles as its faces.

theorem mathematical proposition that can be deduced by logic from a set of axioms (basic facts that are taken to be true without proof). Advanced mathematics consists almost entirely of theorems and proofs, but even at a simple level theorems are important.

trapezium (US **trapezoid**) in geometry, a four-sided plane figure (quadrilateral) with two of its sides parallel. If the parallel sides have lengths a and b and the perpendicular distance between them is h (the height of the trapezium), its area $A = \frac{1}{2}h(a + b)$.

Further Reading

Abbott, Percival, and Wardle, Michael Ernest *Trigonometry* (1991)

Anderson, Marlow *First Course in Abstract Algebra: Rings, Groups and Fields* (1995)

Ashurst, F Gareth *Founders of Modern Mathematics* (1982)

Balz, A G *Descartes and the Modern Mind* (1952)

Barnsley, Michael F *Fractals Everywhere* (1993)

Bell, R J T *Coordinate Geometry* (1959)

Belsom, Chris *Statistics* (1997)

Berger, Marcel; Cole, Michael; and Levy, Silvio *Geometry* (1989)

Boyer, Carl B *A History of Mathematics* (1968)

Brown, Stuart *Leibniz* (1984)

Butler, E M *Heinrich Heine: A Biography* (1956)

Casti, John L *Reality Rules* (1992)

Chaitin, Gregory J *Algorithmic Information Theory* (1987)

Chase, Warren, and Bown, Fred *General Statistics* (1997)

Cohen, Jack, and Stewart, Ian *The Collapse of Chaos* (1993)

Coxeter, H S M *Introduction to Geometry* (1969)

Crossley, J N; Ash, C J; Brickhill, C J; Stillwell, J C; and Williams, N H *What Is Mathematical Logic?* (1990)

Davidson, Hugh *Blaise Pascal* (1983)

Davis, Philip J, and Hersh, Reuben *The Mathematical Experience* (1981), *Descartes' Dream* (1986)

Davis, Philip J, and Chinn, William G *3.1416 and All That* (1985)

Devlin, Keith *Mathematics: The New Golden Age* (1988), *Logic and Information* (1991), *Mathematics, the Science of Patterns* (1994)

Dijksterhuis, E J *Archimedes* (1956)

Dudley, Underwood *A Budget of Trisections* (1987)

Ekeland, Ivar *Mathematics and the Unexpected* (1988)

Field, Michael, and Golubitsky, Martin *Symmetry in Chaos* (1992)

Fraleigh, John B *A First Course in Abstract Algebra* (1989)

Garfunkel, Solomon (ed) *For All Practical Purposes* (1994)

Gauk, Roger *Descartes: An Intellectual Biography* (1995)

Goldstein, Larry Joel *Algebra and Trigonometry and Their Applications* (1996)

Gorman, Peter *Pythagoras: A Life* (1978)

Graham, Ronald L; Knuth, Donald E; and Patashnik, Oren *Concrete Mathematics* (1994)

Grattan-Guinness, I *Joseph Fourier* (1972)

Greenberg, Martin Jay *Euclidean and Non-Euclidean Geometries* (1993)

Grene, Marjorie *Descartes* (1985)

Guillen, Michael *Bridges to Infinity* (1984)

Hansen, Vagn Lundsgaard *Geometry in Nature* (1993)

Harper, P G, and Weaire, D L *Introduction to Physical Mathematics* (1985)

Jacobs, Harold R *Mathematics, a Human Endeavour* (1994)

Jacobs, Konrad *Invitation to Mathematics* (1992)

Jesseph, Douglas M *Berkeley's Philosophy of Mathematics* (1993)

Kaufmann, Jerome E *Trigonometry* (1994)

Kitchens, Larry *Exploring Statistics: A Modern Introduction to Data Analysis and Inference* (1998)

Kitcher, Philip *The Nature of Mathematical Knowledge* (1984)

Kline, Morris *Mathematics and the Search for Knowledge* (1985)

Koosis, Donald J *Statistics: A Self-Teaching Guide* (1997)

Kostrikin, A I *Introduction to Algebra* (1982)

Lavine, Shaughan *Understanding the Infinite* (1994)

Legendre, Adrien Marie; Brewster, David; and Carlyle, Thomas *Elements of Geometry and Trigonometry: With Notes* (1824)

Lewin, Roger *Complexity* (1992)

Livingston, Charles *Knot Theory* (1993)

McLeish, John *Number* (1991)

Maor, Eli *The Story of a Number* (1994), *Trigonometric Delights* (1998)

Mendelson, Elliott *Introduction to Mathematical Logic* (1964)

Menhard, J *Pascal: His Life and Works* (1952)

Newmark, Joseph *Statistics and Probability in Modern Life* (1997)

O'Meara, Dominic *Pythagoras Revived: Mathematics and Philosophy in Late Antiquity* (1989)

Osserman, Robert *Poetry of the Universe* (1995)

Ore, Oystein, and Wilson, Robin James *Graphs and Their Uses* (1990)

Paulos, John Allen *Innumeracy* (1988), *Beyond Numeracy* (1991)

Peterson, Ivars *The Mathematical Tourist* (1988), *Islands of Truth* (1990)

Rosen, Kenneth H *Elementary Number Theory* (1988)

Salem, Lionel; Testard, Frédéric; and Salem, Coralie *The Most Beautiful Mathematical Formulas* (1992)

Saw, Ruth *Leibniz* (1956)

Schroeder, Manfred *Fractals, Chaos, Power Laws* (1991)

Siegel, Andrew F, and Morgan, Charles J *Statistics and Data Analysis: An Introduction* (1996)

Sigmund, Karl *Games of Life* (1993)

Smith, Geoff *Introductory Mathematics: Algebra and Analysis* (1998)

Stanley, Thomas *Pythagoras* (1989)

Stewart, Ian *Does God Play Dice?* (1989), *Galois Theory* (1989), *The Problems of Mathematics* (1992), *Nature's Numbers* (1995), *Concepts of Modern Mathematics* (1995)

Sved, Marta *Journey into Geometries* (1991)

Tignol, Jean-Pierre *Galois' Theory of Algebraic Equations* (1988)

Weeks, Jeffrey R *The Shape of Space* (1985)

Wells, David *The Penguin Dictionary of Curious and Interesting Numbers* (1986), *You Are a Mathematician: a Wise and Witty Introduction to the Joy of Numbers* (1998)

CHEMISTRY

Chemistry is the area of science concerned with the study of the structure and composition of the different kinds of matter, the changes which matter may undergo, and the phenomena which occur in the course of these changes.

Ancient civilizations were familiar with certain chemical processes – for example, extracting metals from their ores, and making alloys. The alchemists endeavoured to turn base (nonprecious) metals into gold, and chemistry evolved towards the end of the 17th century from the techniques and insights developed during alchemical experiments.

Robert Boyle defined elements as the simplest substances into which matter could be resolved. The alchemical doctrine of the four elements (earth, air, fire, and water) gradually lost its hold, and the theory that all combustible bodies contain a substance called phlogiston (a weightless 'fire element' generated during combustion) was discredited in the 18th century by the experimental work of Joseph Black, Antoine Lavoisier, and Joseph Priestley (who discovered the presence of oxygen in air).

Henry Cavendish discovered the composition of water, and John Dalton put forward the atomic theory, which ascribed a precise relative weight to the 'simple atom' characteristic of each element. Much research then took place leading to the development of biochemistry, chemotherapy, and plastics.

Chemistry is commonly divided into three main branches. **Organic chemistry** is the branch of chemistry that deals with carbon compounds. **Inorganic chemistry** deals with the description, properties, reactions, and preparation of all the elements and their compounds, with the exception of carbon compounds. **Physical chemistry** is concerned with the quantitative explanation of chemical phenomena and reactions, and the measurement of data required for such explanations. This branch studies in particular the movement of molecules and the effects of temperature and pressure, often with regard to gases and liquids.

molecules, atoms, and elements

All matter can exist in three states: gas, liquid, or solid. It is composed of minute particles termed **molecules**, which are constantly moving, and may be further divided into **atoms**.

Molecules that contain atoms of one kind only are known as **elements**; those that contain atoms of different kinds are called **compounds**. Chemical compounds are produced by a chemical action that alters the arrangement of the atoms in the reacting molecules. Heat, light, vibration, catalytic action, radiation, or pressure, as well as moisture (for ionization), may

Molecule of the Month
http://www.bris.ac.uk/Depts/Chemistry/
MOTM/motm.htm
Pages on interesting – and sometimes hypothetical – molecules, contributed by university chemistry departments throughout the world.

be necessary to produce a chemical change. Examination and possible breakdown of compounds to determine their components is **analysis**, and the building up of compounds from their components is **synthesis**. When substances are brought together without changing their molecular structures they are said to be **mixtures**.

the atom

The atom is the smallest unit of matter that can take part in a chemical reaction, and which cannot be broken down chemically into anything simpler. The atoms of the various elements differ in atomic number, relative atomic mass, and chemical behaviour.

Atoms are much too small to be seen by even the most powerful optical microscope (the largest, caesium, has a diameter of 0.0000005 mm/0.00000002 in), and they are in constant motion. However, modern electron microscopes, such as the scanning tunnelling microscope (STM) and the atomic force microscope (AFM), can produce images of individual atoms and molecules.

atomic structure The core of the atom is the **nucleus**, a dense body only one ten-thousandth the diameter of the atom itself. The simplest nucleus, that of hydrogen, comprises a single stable positively charged particle, the **proton**. Nuclei of other elements contain more protons and additional particles, called **neutrons**, of about the same mass as the proton but with no electrical charge. Each element has its own characteristic nucleus with a unique number of protons, the atomic number. The number of neutrons may vary. Where atoms of a single element have different numbers of neutrons, they are called **isotopes**. Although some isotopes tend to be unstable and exhibit radioactivity, they all have identical chemical properties.

The nucleus is surrounded by a number of moving **electrons**, each of which has a negative charge equal to the positive charge on a proton, but which weighs only 1/1,839 times as much. In a neutral atom, the nucleus is surrounded by the same number of electrons as it contains protons. According to quantum theory, the position of an electron is uncertain; it may be found at any point. However, it is more likely to be found in some places than others. The region of space in which an electron is most likely to be found is called an **orbital** or **shell**. The chemical properties of an element are determined by the ease with which its atoms can gain or lose electrons from its outer orbitals. These shells can be regarded as a series of concentric spheres, each of which can contain a certain maximum number of electrons; the **inert gases** have an arrangement in which every shell contains this number. The energy levels are usually numbered beginning with the shell nearest to the nucleus. The outermost shell is known as the **valency shell** as it contains the valence (bonding) electrons.

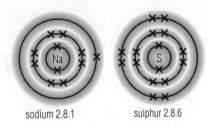

sodium 2.8.1   sulphur 2.8.6

atom, electronic structure The arrangement of electrons in a sodium atom and a sulphur atom. The number of electrons in a neutral atom gives that atom its atomic number: sodium has an atomic number of 11 and sulphur has an atomic number of 16.

The lowest energy level, or innermost shell, can contain no more than two electrons. Outer shells are considered to be stable when they contain eight electrons but additional electrons can sometimes be accommodated provided that the outermost shell has a stable configuration. Electrons in unfilled shells are available to take part in chemical bonding, giving rise to the concept of valency. In ions, the electron shells contain more or fewer electrons than are required for a neutral atom, generating negative or positive charges.

Look Inside the Atom
http://www.aip.org/history/electron/jjhome.htm
Part of the American Institute of Physics site, this page examines J J Thomson's 1897 experiments that led to the discovery of a fundamental building block of matter, the electron. The site includes images and quotes and a section on the legacy of his discovery.

atomic mass unit This is the unit of mass that is used to measure the relative mass of atoms and molecules. Sometimes called the **dalton unit**, it is equal to onetwelfth of the mass of a carbon-12 atom, which is equivalent to the mass of a proton or 1.66×10^{-27} kg. The relative atomic mass of an atom has no units; thus oxygen-16 has an atomic mass of 16 daltons but a relative atomic mass of 16.

atomic number This is the number (symbol Z) of protons in the nucleus of an atom (sometimes called the **proton number**). It is equal to the positive charge on the nucleus.

In a neutral atom, it is also equal to the number of electrons surrounding the nucleus. The chemical elements are arranged in the periodic table of the elements according to their atomic number.

Discovery of the Elements

(- = not applicable.)

Date	Element	Discoverer (symbol)	Date	Element	Discoverer (symbol)
Prehistoric knowledge	antimony (Sb) arsenic (As) bismuth (Bi) carbon (C) copper (Cu) gold (Au) iron (Fe) lead (Pb) mercury (Hg) silver (Ag) sulphur (S) tin (Sn) zinc (Zn)	–		iridium (Ir) osmium (Os) palladium (Pd) rhodium (Rh)	Smithson Tennant Smithson Tennant William Wollaston William Wollaston
			1807	potassium (K) sodium (Na)	Humphry Davy Humphry Davy
			1808	barium (Ba) boron (B)	Humphry Davy Humphry Davy, and independently by Joseph Gay-Lussac and Louis-Jacques Thénard
				calcium (Ca) strontium (Sr)	Humphry Davy Humphry Davy
1557	platinum (Pt)	Julius Scaliger	1811	iodine (I)	Bernard Courtois
1674	phosphorus (P)	Hennig Brand	1817	cadmium (Cd)	Friedrich Strohmeyer
1730	cobalt (Co)	Georg Brandt		lithium (Li)	Johan Arfwedson
1751	nickel (Ni)	Axel Cronstedt		selenium (Se)	Jöns Berzelius
1755	magnesium (Mg)	Joseph Black (oxide isolated	1823	silicon (Si)	Jöns Berzelius
		by Humphry Davy in 1808; pure form isolated by Antoine-Alexandre-Brutus Bussy in 1828)	1824	aluminium (Al)	Hans Oersted (also attributed to Friedrich Wöhler in 1827)
			1826	bromine (Br)	Antoine-Jérôme Balard
1766	hydrogen (H)	Henry Cavendish	1827	ruthenium (Ru)	G W Osann (isolated by Karl Klaus in 1844)
1771	fluorine (F)	Karl Scheele (isolated by Henri Moissan in 1886)	1828	thorium (Th)	Jöns Berzelius
1772	nitrogen (N)	Daniel Rutherford	1839	lanthanum (La)	Carl Mosander
1774	chlorine (Cl)	Karl Scheele	1843	erbium (Er)	Carl Mosander
	manganese (Mn)	Johann Gottlieb Gahn		terbium (Tb)	Carl Mosander
	oxygen (O)	Joseph Priestley and Karl Scheele, independently of each other	1860	caesium (Cs)	Robert Bunsen and Gustav Kirchhoff
1781	molybdenum (Mo)	named by Karl Scheele (isolated by Peter Jacob Hjelm in 1782)	1861	rubidium (Rb)	Robert Bunsen and Gustav Kirchhoff
1782	tellurium (Te)	Franz Müller		thallium (Tl)	William Crookes (isolated by William Crookes and Claude August Lamy, independently of each other in 1862)
1783	tungsten (W)	isolated by Juan José Elhuyar and Fausto Elhuyar			
1789	uranium (U)	Martin Klaproth (isolated by Eugène Péligot in 1841)	1863	indium (In)	Ferdinand Reich and Hieronymus Richter
	zirconium (Zr)	Martin Klaproth	1868	helium (He)	Pierre Janssen
1790	titanium (Ti)	William Gregor	1875	gallium (Ga)	Paul Lecoq de Boisbaudran
1794	yttrium (Y)	Johan Gadolin	1876	scandium (Sc)	Lars Nilson
1797	chromium (Cr)	Louis-Nicolas Vauquelin	1878	ytterbium (Yb)	Jean Charles de Marignac
1798	beryllium (Be)	Louis-Nicolas Vauquelin (isolated by Friedrich Wöhler and Antoine-Alexandre-Brutus Bussy in 1828)	1879	holmium (Ho)	Per Cleve
				samarium (Sm)	Paul Lecoq de Boisbaudran
				thulium (Tm)	Per Cleve
1801	vanadium (V)	Andrés del Rio (disputed), or Nils Sefström in 1830	1885	neodymium (Nd) praseodymium (Pr)	Carl von Welsbach Carl von Welsbach
	niobium (Nb)	Charles Hatchett	1886	dysprosium (Dy)	Paul Lecoq de Boisbaudran
1802	tantalum (Ta)	Anders Ekeberg		gadolinium (Gd)	Paul Lecoq de Boisbaudran
1804	cerium (Ce)	Jöns Berzelius and Wilhelm Hisinger, and independently by Martin Klaproth		germanium (Ge)	Clemens Winkler
			1894	argon (Ar)	John Rayleigh and William Ramsay

Date	Element	Discoverer (symbol)	Date	Element	Discoverer (symbol)
1898	krypton (Kr)	William Ramsay and Morris Travers	1952	einsteinium (Es)	Albert Ghiorso and co-workers
	neon (Ne)	William Ramsay and Morris Travers	1955	fermium (Fm)	Albert Ghiorso and co-workers
	polonium (Po)	Marie and Pierre Curie		mendelevium (Md)	Albert Ghiorso, Bernard G Harvey, Gregory Choppin, Stanley Thompson, and Glenn Seaborg
	radium (Ra)	Marie Curie			
	xenon (Xe)	William Ramsay and Morris Travers			
1899	actinium (Ac)	André Debierne	1958	nobelium (No)	Albert Ghiorso, Torbjørn Sikkeland, J R Walton, and Glenn Seaborg
1900	radon (Rn)	Friedrich Dorn			
1901	europium (Eu)	Eugène Demarçay			
1907	lutetium (Lu)	Georges Urbain and Carl von Welsbach, independently of each other	1961	lawrencium (Lr)	Albert Ghiorso, Torbjørn Sikkeland, Almon Larsh, and Robert Latimer
			1964	dubnium (Db)	claimed by Soviet scientist Georgii Flerov and co-workers (disputed by US workers)
1913	protactinium (Pa)	Kasimir Fajans and O Göhring			
	hafnium (Hf)	Dirk Coster and Georg von Hevesy	1967	unnilpentium (Unp)	claimed by Georgii Flerov and co-workers (disputed by US workers)
1925	rhenium (Re)	Walter Noddack, Ida Tacke, and Otto Berg			
1937	technetium (Tc)	Carlo Perrier and Emilio Segrè	1969	rutherfordium (Rf)	claimed by US scientist Albert Ghiorso and co-workers (disputed by Soviet workers)
1939	francium (Fr)	Marguérite Perey			
1940	astatine (At)	Dale R Corson, K R MacKenzie, and Emilio Segrè	1970	dubnium (Db)	claimed by Albert Ghiorso and co-workers (disputed by Soviet workers)
	neptunium (Np)	Edwin McMillan and Philip Abelson	1974	seaborgium (Sg)	claimed by Georgii Flerov and co-workers, and independently by Albert Ghiorso and co-workers
	plutonium (Pu)	Glenn Seaborg, Edwin McMillan, Joseph Kennedy, and Arthur Wahl			
1944	americium (Am)	Glenn Seaborg, Ralph James, Leon Morgan, and Albert Ghiorso	1976	bohrium (Bh)	Georgii Flerov and Yuri Oganessian (confirmed by German scientist Peter Armbruster and co-workers)
	curium (Cm)	Glenn Seaborg, Ralph James, and Albert Ghiorso	1982	meitnerium (Mt)	Peter Armbruster and co-workers
1945	promethium (Pm)	J A Marinsky, Lawrence Glendenin, and Charles Coryell	1984	hassium (Hs)	Peter Armbruster and co-workers
1949	berkelium (Bk)	Glenn Seaborg, Stanley Thompson, and Albert Ghiorso	1994	ununnilium (Uun)	team at GSI heavy-ion cyclotron, Darmstadt, Germany
1950	californium (Cf)	Glenn Seaborg, Stanley Thompson, Kenneth Street Jr, and Albert Ghiorso		unununium (Uuu)	team at GSI heavy-ion cyclotron, Darmstadt, Germany

chemical bonds

The principal types of bonding are ionic, covalent, metallic, and intermolecular (such as hydrogen bonding).

The type of bond formed depends on the elements concerned and their electronic structure. In an **ionic** or **electrovalent** bond, common in inorganic compounds, the combining atoms gain or lose electrons to become ions; for example, sodium (Na) loses an electron to form a sodium ion (Na^+) while chlorine (Cl) gains an electron to form a chloride ion (Cl^-) in the ionic bond of sodium chloride (NaCl).

In a **covalent bond** the atomic orbitals of two atoms overlap to form a molecular orbital containing two electrons, which are thus effectively shared between the two atoms. Covalent bonds are common in organic compounds, such as the four carbon–hydrogen bonds in methane (CH_4). In a dative covalent or coordinate bond, one of the combining atoms supplies both of the valence electrons in the bond.

A **metallic bond** joins metals in a crystal lattice; the atoms occupy lattice positions as positive ions, and

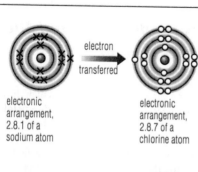

electronic
arrangement,
2.8.1 of a
sodium atom

electron
transferred

electronic
arrangement,
2.8.7 of a
chlorine atom

becomes a
sodium ion, Na$^+$,
with an electron
arrangement 2.8

becomes a
chloride ion, Cl$^-$,
with an electron
arrangement 2.8.8

ionic bond The formation of an ionic bond between a sodium atom and a chlorine atom to form a molecule of sodium chloride. The sodium atom transfers an electron from its outer electron shell (becoming the positive ion Na$^+$) to the chlorine atom (which becomes the negative chloride ion Cl$^-$). The opposite charges mean that the ions are strongly attracted to each other. The formation of the bond means that each atom becomes more stable, having a full quota of electrons in its outer shell.

valence electrons are shared between all the ions in an 'electron gas'.

In a **hydrogen bond** a hydrogen atom joined to an electronegative atom, such as nitrogen or oxygen, becomes partially positively charged, and is weakly attracted to another electronegative atom on a neighbouring molecule.

The strongest known noncovalent bond is the **superbond** that is formed between avidin, a protein found in egg white, and the growth factor biotin. It is almost impossible to separate the two molecules once bonded. The superbond is used in a number of biomedical research applications.

formulae and equations

Symbols are used to denote the elements. The symbol is usually the first letter or letters of the English or Latin name of the element – for example, C for carbon; Ca for calcium; Fe for iron (*ferrum*). These symbols represent one atom of the element; molecules containing more than one atom of an element are denoted by a subscript figure – for example, water is H$_2$O. In some substances a group of atoms acts as a single entity, and these are enclosed in parentheses in

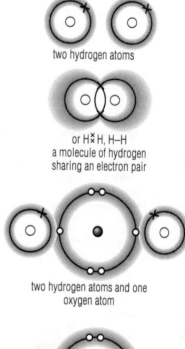

two hydrogen atoms

or H×H, H–H
a molecule of hydrogen
sharing an electron pair

two hydrogen atoms and one
oxygen atom

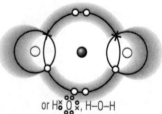

or H$^\times_\circ$O$^\times_\circ$, H–O–H
a molecule of water
showing the two covalent bonds

covalent bond The formation of a covalent bond between two hydrogen atoms to form a hydrogen molecule (H$_2$), and between two hydrogen atoms and an oxygen atom to form a molecule of water (H$_2$O). The sharing means that each atom has a more stable arrangement of electrons (its outer electron shells are full).

the symbol – for example (NH$_4$)$_2$SO$_4$ denotes ammonium sulphate. The symbolic representation of a molecule is known as a **formula**. A figure placed before a formula represents the number of molecules of a substance taking part in, or being produced by, a chemical reaction – for example, 2H$_2$O indicates two molecules of water.

Chemical reactions are expressed by means of **equations**. A chemical equation gives two basic pieces of information: (1) the reactants (on the left-hand side) and products (right-hand side); and (2) the reacting proportions (stoichiometry) – that is, how many units of each reactant and product are involved. The equation must balance; that is, the total number of atoms of a particular element on the left-hand side must be the same as the number of atoms of that element on the right-hand side.

$$Na_2CO_3 + 2HCl \rightarrow 2NaCl + CO_2 + H_2O$$
$$\textit{reactants} \rightarrow \textit{products}$$

This equation states that one molecule of sodium carbonate combines with two molecules of hydrochloric acid to form two molecules of sodium chloride, one of carbon dioxide, and one of water. Double arrows indicate that the reaction is reversible – in the formation of ammonia from hydrogen and nitrogen, the direction depends on the temperature and pressure of the reactants.

$$3H_2 + N_2 \leftrightarrow 2NH_3$$

metals, nonmetals, and the periodic system

Elements are divided into **metals**, which have lustre and conduct heat and electricity, and **nonmetals**, which usually lack these properties. The **periodic system**, developed by John Newlands in 1863 and established by Dmitri Mendeleyev in 1869, classified elements according to their relative atomic masses. Those elements that resemble each other in general properties were found to bear a relation to one another by weight, and these were placed in groups or families. Certain anomalies in this system were later removed by classifying the elements according to their atomic numbers. The latter is equivalent to the positive charge on the nucleus of the atom.

periodic table of the elements
The periodic table of the elements arranges the elements into horizontal rows (called periods) and vertical columns (called groups) according to their atomic numbers. The elements in a group or column all have similar properties – for example, all the elements in the far right-hand column are inert gases. .

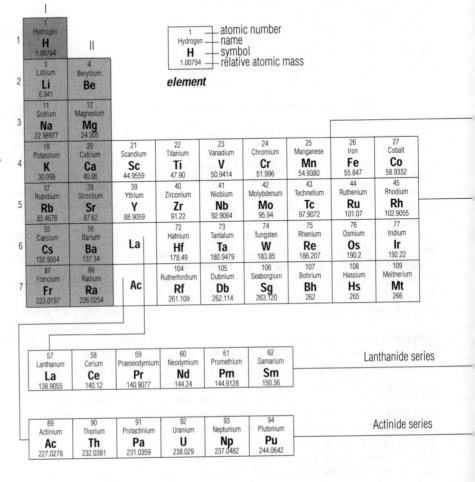

periodic table: first three periods

To remember the elements in their correct order in the first three periods of the table (hydrogen being omitted and potassium included):

Here **li**es **Be**njamin **B**old **c**ry **n**ot **o**ld **f**riend **ne**edlessly **Na**ture **m**agnifies **al**l **si**mple **p**eople **s**ometimes, **cl**ots **a**nd **k**ings.

helium, lithium, beryllium, boron, carbon, nitrogen, oxygen, fluorine, neon, sodium, magnesium, aluminium, silicon, phosphorus, sulphur, chlorine, argon, potassium

organic chemistry

Organic compounds, i.e. those based on linked carbon atoms, form the chemical basis of life and are more abundant than inorganic compounds. In a typical organic compound each carbon atom forms bonds covalently with each of its neighbouring carbon atoms in a chain or ring, and additionally with other atoms, commonly hydrogen, oxygen, nitrogen, or sulphur.

Many organic compounds (such as proteins and carbohydrates) are made only by living organisms, and it was once believed that organic compounds could not be made by any other means. This was disproved when Friedrich Wöhler synthesized urea in 1828, but the name 'organic' (that is, 'living') chemistry has remained in use. Many organic compounds are derived from petroleum, which represents the chemical remains of millions of microscopic marine organisms.

In inorganic chemistry, a specific formula usually represents one substance only, but in organic chemistry, it is exceptional for a molecular formula to represent only one substance, owing to the different ways that carbon chains can be branched and interlinked. Different chemical compounds having the same molecular composition and mass as one another, but with different physical or chemical properties owing to the

						0
						2 Helium **He** 4002.60

	III	IV	V	VI	VII	
	5 Boron **B** 10.81	6 Carbon **C** 12.011	7 Nitrogen **N** 14.0067	8 Oxygen **O** 15.9994	9 Fluorine **F** 18.99840	10 Neon **Ne** 20.179
	13 Aluminium **Al** 26.98154	14 Silicon **Si** 28.066	15 Phosphorus **P** 30.9738	16 Sulphur **S** 32.06	17 Chlorine **Cl** 35.453	18 Argon **Ar** 39.948

28 Nickel **Ni** 58.70	29 Copper **Cu** 63.546	30 Zinc **Zn** 65.38	31 Gallium **Ga** 69.72	32 Germanium **Ge** 72.59	33 Arsenic **As** 74.9216	34 Selenium **Se** 78.96	35 Bromine **Br** 79.904	36 Krypton **Kr** 83.80
46 Palladium **Pd** 106.4	47 Silver **Ag** 107.868	48 Cadmium **Cd** 112.40	49 Indium **In** 114.82	50 Tin **Sn** 118.69	51 Antimony **Sb** 121.75	52 Tellurium **Te** 127.75	53 Iodine **I** 126.9045	54 Xenon **Xe** 131.30
78 Platinum **Pt** 195.09	79 Gold **Au** 196.9665	80 Mercury **Hg** 200.59	81 Thallium **Tl** 204.37	82 Lead **Pb** 207.37	83 Bismuth **Bi** 207.2	84 Polonium **Po** 210	85 Astatine **At** 211	86 Radon **Rn** 222.0176
110 Ununnilium **Uun** 269	111 Unununium **Uuu** 272							

63 Europium **Eu** 151.96	64 Gadolinium **Gd** 157.25	65 Terbium **Tb** 158.9254	66 Dysprosium **Dy** 162.50	67 Holmium **Ho** 164.9304	68 Erbium **Er** 167.26	69 Thulium **Tm** 168.9342	70 Ytterbium **Yb** 173.04	71 Lutetium **Lu** 174.97
95 Americium **Am** 243.0614	96 Curium **Cm** 247.0703	97 Berkelium **Bk** 247	98 Californium **Cf** 251.0786	99 Einsteinium **Es** 252.0828	100 Fermium **Fm** 257.0951	101 Mendelevium **Md** 258.0986	102 Nobelium **No** 259.1009	103 Lawrencium **Lr** 260.1054

The Chemical Elements

An element is a substance that cannot be split chemically into simpler substances. The atoms of a particular element all have the same number of protons in their nuclei (their atomic number).
(- = not applicable.)

Name	Symbol	Atomic number	Atomic mass (amu)[1]	Relative density[2]	Melting or fusing point (°C)
Actinium	Ac	89	227[3]	-	-
Aluminium	Al	13	26.9815	2.58	658
Americium	Am	95	243[3]	-	-
Antimony	Sb	51	121.75	6.62	629
Argon	Ar	18	39.948	gas	-188
Arsenic	As	33	74.9216	5.73	volatile, 450
Astatine	At	85	210[3]	-	-
Barium	Ba	56	137.34	3.75	850
Berkelium	Bk	97	249[3]	-	-
Beryllium	Be	4	9.0122	1.93	1,281
Bismuth	Bi	83	208.9806	9.80	268
Bohrium	Bh	107	262[3]	-	-
Boron	B	5	10.81	2.5	2,300
Bromine	Br	35	79.904	3.19	-7.3
Cadmium	Cd	48	112.40	8.64	320
Caesium	Cs	55	132.9055	1.88	26
Calcium	Ca	20	40.08	1.58	851
Californium	Cf	98	251[3]	-	-
Carbon	C	6	12.011	3.52	infusible
Cerium	Ce	58	140.12	6.68	623
Chlorine	Cl	17	35.453	gas	-102
Chromium	Cr	24	51.996	6.5	1,510
Cobalt	Co	27	58.9332	8.6	1,490
Copper	Cu	29	63.546	8.9	1,083
Curium	Cm	96	247[3]	-	-
Dubnium	Db	105	262[3]	-	-
Dysprosium	Dy	66	162.50	-	-
Einsteinium	Es	99	254[3]	-	-
Erbium	Er	68	167.26	4.8	-
Europium	Eu	63	151.96	-	-
Fermium	Fm	100	253[3]	-	-
Fluorine	F	9	18.9984	gas	-223
Francium	Fr	87	223[3]	-	-
Gadolinium	Gd	64	157.25	-	-
Gallium	Ga	31	69.72	5.95	30
Germanium	Ge	32	72.59	5.47	958
Gold	Au	79	196.9665	19.3	1,062
Hafnium	Hf	72	178.49	12.1	2,500
Hassium	Hs	108	265[3]	-	-
Helium	He	2	4.0026	gas	-272
Holmium	Ho	67	164.9303	-	-
Hydrogen	H	1	1.0080	gas	-258
Indium	In	49	114.82	7.4	155
Iodine	I	53	126.9045	4.95	114
Iridium	Ir	77	192.22	22.4	2,375
Iron	Fe	26	55.847	7.86	1,525
Krypton	Kr	36	83.80	gas	-169
Lanthanum	La	57	138.9055	6.1	810
Lawrencium	Lr	103	260[3]	-	-
Lead	Pb	82	207.2	11.37	327
Lithium	Li	3	6.941	0.585	186
Lutetium	Lu	71	174.97	-	-
Magnesium	Mg	12	24.305	1.74	651
Manganese	Mn	25	54.9380	7.39	1,220
Meitnerium	Mt	109	266[3]	-	-

Name	Symbol	Atomic number	Atomic mass (amu)[1]	Relative density[2]	Melting or fusing point (°C)
Mendelevium	Md	101	256[3]	–	–
Mercury	Hg	80	200.59	13.596	–38.9
Molybdenum	Mo	42	95.94	10.2	2,500
Neodymium	Nd	60	144.24	6.96	840
Neon	Ne	10	20.179	gas	–248.6
Neptunium	Np	93	237[3]	–	–
Nickel	Ni	28	58.71	8.9	1,452
Niobium	Nb	41	92.9064	8.4	1,950
Nitrogen	N	7	14.0067	gas	–211
Nobelium	No	102	254[3]	–	–
Osmium	Os	76	190.2	22.48	2,700
Oxygen	O	8	15.9994	gas	–227
Palladium	Pd	46	106.4	11.4	1,549
Phosphorus	P	15	30.9738	1.8–2.3	44
Platinum	Pt	78	195.09	21.5	1,755
Plutonium	Pu	94	242[3]	–	–
Polonium	Po	84	210[3]	–	–
Potassium	K	19	39.102	0.87	63
Praseodymium	Pr	59	140.9077	6.48	940
Promethium	Pm	61	145[3]	–	–
Protactinium	Pa	91	231.0359	–	–
Radium	Ra	88	226.0254	6.0	700
Radon	Rn	86	222[3]	gas	–150
Rhenium	Re	75	186.2	21	3,000
Rhodium	Rh	45	102.9055	12.1	1,950
Rubidium	Rb	37	85.4678	1.52	39
Ruthenium	Ru	44	101.07	12.26	2,400
Rutherfordium	Rf	104	262[3]	–	–
Samarium	Sm	62	150.4	7.7	1,350
Scandium	Sc	21	44.9559	–	–
Seaborgium	Sg	106	263[3]	–	–
Selenium	Se	34	78.96	4.5	170–220
Silicon	Si	14	28.086	2.0–2.4	1,370
Silver	Ag	47	107.868	10.5	960
Sodium	Na	11	22.9898	0.978	97
Strontium	Sr	38	87.62	2.54	800
Sulphur	S	16	32.06	2.07	115–119
Tantalum	Ta	73	180.9479	16.6	2,900
Technetium	Tc	43	99[3]	–	–
Tellurium	Te	52	127.60	6.0	446
Terbium	Tb	65	158.9254	–	–
Thallium	Tl	81	204.37	11.85	302
Thorium	Th	90	232.0381	11.00	1,750
Thulium	Tm	69	168.9342	–	–
Tin	Sn	50	118.69	7.3	232
Titanium	Ti	22	47.90	4.54	1,850
Tungsten	W	74	183.85	19.1	2,900–3,000
Ununnilium	Uun[4]	110	269[3]	–	–
Unununium	Uuu[4]	111	272[3]	–	–
Uranium	U	92	238.029	18.7	–
Vanadium	V	23	50.9414	5.5	1,710
Xenon	Xe	54	131.30	gas	–140
Ytterbium	Yb	70	173.04	–	–
Yttrium	Y	39	88.9059	3.8	–
Zinc	Zn	30	65.37	7.12	418
Zirconium	Zr	40	91.22	4.15	2,130

[1] Atomic mass units.

[2] Also known as specific gravity.

[3] The number given is that for the most stable isotope of the element.

[4] Elements as yet unnamed; temporary identification assigned until a name is approved by the International Union of Pure and Applied Chemistry.

different structural arrangement of the constituent atoms are known as **isomers**. For example, the organic compounds butane ($CH_3(CH_2)_2CH_3$) and methyl propane ($CH_3CH(CH_3)CH_3$) are isomers, each possessing four carbon atoms and ten hydrogen atoms but differing in the way that these are arranged with respect to each other. **Structural isomers** such as these obviously have different constructions, but **geometrical** and **optical isomers** must be drawn or modelled in order to appreciate the difference in their three-dimensional arrangement. Geometrical isomers have a plane of symmetry and arise because of the restricted rotation of atoms around a bond; optical isomers are mirror images of each other. For instance, 1,1-dichloroethene ($CH_2=CCl_2$) and 1,2-dichloroethene (CHCl=CHCl) are structural isomers, but there are two possible geometric isomers of the latter (depending on whether the chlorine atoms are on the same side or on opposite sides of the plane of the carbon–carbon double bond).

Hydrocarbons form one of the most prolific of the many organic types; fuel oils are largely made up of hydrocarbons. Typical groups containing only carbon, hydrogen, and oxygen are alcohols, aldehydes, ketones, ethers, esters, and carbohydrates. Among groups containing nitrogen are amides, amines, nitrocompounds, amino acids, proteins, purines, alkaloids, and many others, both natural and artificial. Other organic types contain sulphur, phosphorus, or halogen elements.

The basis of organic chemistry is the ability of carbon to form long chains of atoms, branching chains, rings, and other complex structures. Compounds containing only carbon and hydrogen are known as **hydrocarbons**. Carbon bonds to adjacent atoms covalently.

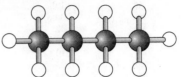

butane $CH_3(CH_2)_2CH_3$

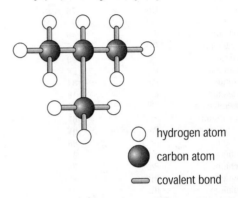

methyl propane $CH_3CH(CH_3)CH_3$

○ hydrogen atom

● carbon atom

⬭ covalent bond

isomer The chemicals butane and methyl propane are isomers. Each has the molecular formula $CH_3CH(CH_3)CH_3$, but with different spatial arrangements of atoms in their molecules.

In a **covalent bond** two atoms share one or more pairs of electrons (usually each atom contributes an electron). The bond is often represented by a single line drawn between the two atoms. Double bonds, seen, for example, in the alkenes, are formed when two atoms share two pairs of electrons (the atoms usually

aromatic compound
Compounds whose molecules contain the benzene ring, or variations of it, are called aromatic. The term was originally used to distinguish sweet-smelling compounds from others.

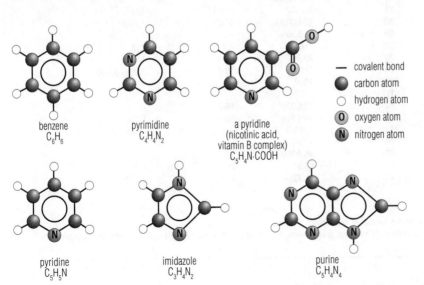

benzene
C_6H_6

pyrimidine
$C_4H_4N_2$

a pyridine
(nicotinic acid,
vitamin B complex)
$C_5H_4N \cdot COOH$

pyridine
C_5H_5N

imidazole
$C_3H_4N_2$

purine
$C_5H_4N_4$

— covalent bond

● carbon atom

○ hydrogen atom

Ⓞ oxygen atom

Ⓝ nitrogen atom

contribute a pair each); triple bonds, seen in the alkynes, are formed when atoms share three pairs of electrons. Such bonds are represented by a double or triple line, respectively, between the atoms concerned. Covalent compounds have the following general properties: they have low melting and boiling points; never conduct electricity; and are usually insoluble in water and soluble in organic solvents.

homologous series Organic chemistry is largely the chemistry of a great variety of **homologous series** – those in which the molecular formulae, when arranged in ascending order, form an arithmetical progression. The physical properties undergo a gradual change from one member to the next.

The linking carbon atoms that form the backbone of an organic molecule may be built up from beginning to end without branching, or may throw off branches at one or more points. Sometimes the ropes of carbon atoms curl round and form rings (**cyclic compounds**), usually of five, six, or seven atoms. Open-chain and cyclic compounds may be classified as aliphatic or aromatic depending on the nature of the bonds between their atoms. Compounds containing oxygen, sulphur, or nitrogen within a carbon ring are called **heterocyclic compounds.**

Organic chemical compounds in which the carbon atoms are joined in straight chains, as in hexane (C_6H_{14}), or in branched chains, as in 2-methylpentane ($CH_3CH(CH_3)CH_2CH_2 CH_3$) are known as **aliphatic compounds.** Aliphatic compounds have bonding electrons localized within the vicinity of the bonded atoms. Cyclic compounds that do not have delocalized electrons are also aliphatic, as in the alicyclic compound cyclohexane (C_6H_{12}) or the heterocyclic piperidine ($C_5H_{11}N$).

Among the principal aliphatic compounds are the alkanes (paraffins), which include methane, ethane, pentane, petrol, kerosene, lubricating oil, and paraffin wax; the alkenes (olefins), including ethene (C_2H_4); the ethynes (for example, acetylene C_2H_2); the alcohols, ethers, ethanoic and other acids, esters, certain classes of amines, the carbohydrates, such as starch, sugar, and cellulose; and the fats.

In **aromatic compounds** some of the bonding electrons are delocalized (shared among several atoms within the molecule and not localized in the vicinity of the atoms involved in bonding). The commonest aromatic compounds have ring structures, the atoms comprising the ring being either all carbon or containing one or more different atoms (usually nitrogen, sulphur, or oxygen). Typical examples are benzene (C_6H_6) and pyridine (C_6H_5N).

The most fundamental of all natural processes are oxidation, reduction, hydrolysis, condensation, polymerization, and molecular rearrangement. In nature, such changes are often brought about through the agency of promoters known as **enzymes**, which act as catalytic agents in promoting specific reactions. The most fundamental of all natural processes is **synthesis**, or building up. In living plant and animal organisms the energy stored in carbohydrate molecules, derived originally from sunlight, is released by slow oxidation and utilized by the organisms. The complex carbohydrates thereby revert to carbon dioxide and water, from where they were built up with absorption of energy. Thus, a carbon food cycle exists in nature. In a corresponding nitrogen food cycle, complex proteins are synthesized in nature from carbon dioxide, water, soil nitrates, and ammonium salts, and these proteins ultimately revert to the elementary raw materials from which they came, with the discharge of their energy of chemical combination.

polymer A compound made up of a large long-chain or branching matrix composed of many repeated

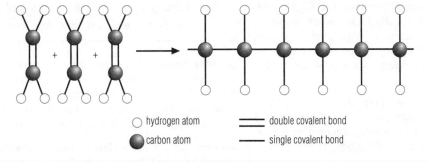

○ hydrogen atom   ══ double covalent bond
● carbon atom   —— single covalent bond

polymerization In polymerization small molecules (monomers) join together to make large molecules (polymers). In the polymerization of ethene to polyethene, electrons are transferred from the carbon–carbon double bond of the ethene molecule, allowing the molecules to join together as a long chain of carbon–carbon single bonds.

simple units (**monomers**) linked together by polymerization. There are many polymers, both natural (cellulose, chitin, lignin) and synthetic (polyethylene and nylon, types of plastic). Synthetic polymers belong to two groups: thermosoftening and thermosetting.

The size of the polymer matrix is determined by the amount of monomer used; it therefore does not form a molecule of constant molecular size or mass.

Polymers and Liquid Crystals
http://plc.cwru.edu/
Online tutorial about two modern physical wonders. The site is divided into a 'virtual textbook' and a 'virtual laboratory', with corresponding explanations and experiments.

---

some important organic compounds

alcohols These are organic compounds characterized by the presence of one or more OH (hydroxyl) groups in the molecule, and which form esters with acids. The main uses of alcohols are as solvents for gums, resins, lacquers, and varnishes; in the making of dyes; for essential oils in perfumery; and for medical substances in pharmacy. The alcohol produced naturally in the fermentation process and consumed as part of alcoholic beverages is called ethanol.

Alcohols may be liquids or solids, according to the size and complexity of the molecule. The five simplest alcohols form a series in which the number of carbon and hydrogen atoms increases progressively, each one having an extra CH_2 (methylene) group in the molecule: methanol or wood spirit (methyl alcohol, CH_3OH); ethanol (ethyl alcohol, C_2H_5OH); propanol (propyl alcohol, C_3H_7OH); butanol (butyl alcohol, C_4H_9OH); and pentanol (amyl alcohol, $C_5H_{11}OH$). The lower alcohols are liquids that mix with water; the higher alcohols, such as pentanol, are oily liquids immiscible with water; and the highest are waxy solids – for example, hexadecanol (cetyl alcohol, $C_{16}H_{33}OH$) and melissyl alcohol ($C_{30}H_{61}OH$), which occur in sperm-whale oil and beeswax, respectively. Alcohols containing the CH_2OH group are primary; those containing CHOH are secondary; while those containing COH are tertiary.

aldehydes Aldehydes are prepared by oxidation of primary alcohols, so that the OH (hydroxyl) group loses its hydrogen to give an oxygen joined by a double bond to a carbon atom (the aldehyde group, with the formula CHO).

The name is made up from **al**cohol **dehyd**rogenation – that is, alcohol from which hydrogen has been removed. Aldehydes are usually liquids and include methanal (formaldehyde), ethanal (acetaldehyde), and benzaldehyde.

alkanes Alkanes are molecules with the general formula C_nH_{2n+2}, and used to be known as **paraffins**. As they contain only single covalent bonds, alkanes are said to be saturated. Lighter alkanes, such as methane, ethane, propane, and butane, are colourless gases; heavier ones are liquids or solids. In nature they are found in natural gas and petroleum. Their principal reactions are combustion and bromination.

> **methane**
> The flatulence of a single sheep could power a small lorry for 40 km/25 mi a day. The digestive process produces methane gas, which can be burnt as fuel. According to one New Zealand scientist, the methane from 72 million sheep could supply the entire fuel needs of his country.

alkenes Alkenes are members of the group of hydrocarbons having the general formula C_nH_{2n}, formerly known as **olefins**. Alkenes are unsaturated compounds, characterized by one or more double bonds between adjacent carbon atoms. Lighter alkenes, such as ethene and propene, are gases, obtained from the cracking of oil fractions. Alkenes react by **addition**, and many useful compounds, such as polyethene and bromoethane, are made from them.

alkynes Alkynes are hydrocarbons with the general formula C_nH_{2n-2}, are known as alkynes (formerly **acetylenes**). They are unsaturated compounds, characterized by one or more triple bonds between adjacent carbon atoms. Lighter alkynes, such as ethyne, are gases; heavier ones are liquids or solids.

> **organic chemistry**
>
> To remember the prefixes for naming carbon chains:
> **Met Ethel properly but my pants had holes**
> (Meth, eth, prop, but, pent, hex, hept)

---

amides Amides are organic chemicals derived from a carboxylic acid (fatty acid) by the replacement of the hydroxyl group (OH) with an amino group (NH_2). One of the simplest amides is acetamide (CH_3CONH_2), which has a strong mousy odour.

amines Amines are a class of organic chemical compounds in which one or more of the hydrogen atoms of ammonia (NH_3) have been replaced by other groups of atoms. **Methyl amines** have unpleasant ammonia odours and occur in decomposing fish. They are all

Name	Molecular formula	Structural formula
methane	CH_4	

uses: domestic fuel (natural gas)

| ethane | C_2H_6 | |

uses: industrial fuel and chemical feedstock

| propane | C_3H_8 | |

uses: bottled gas (camping gas)

| butane | C_4H_{10} | |

uses: bottled gas (lighter fuel, camping gas)

alkanes The lighter alkanes methane, ethane, propane, and butane, showing the aliphatic chains, where a hydrogen atom bonds to a carbon atom at all available sites.

Formula	Name	Atomic bonding
CH_3	methyl	
CH_2CH_3	ethyl	
CC	double bond	
CHO	aldehyde	
CH_2OH	alcohol	
CO	ketone	
COOH	acid	
CH_2NH_2	amine	
C_6H_6	benzene ring	

gases at ordinary temperature. **Aromatic amine compounds** include aniline, which is used in dyeing.

amino acids Amino acids are the basic building blocks of life, being a vital constituent of proteins. They have a carboxyl group (COOH) and an amino group (NH_2) joined to the same carbon atom, and have the general formula $RCHNH_2$ (COOH). Their biological versitility derives from the fact that they have both acidic and basic properties. **Peptides** are molecules comprising two or more amino acid molecules (not necessarily different) joined by **peptide bonds**, whereby the acid group of one acid is linked to the amino group of the other (–CO.NH). The number of amino acid molecules in the peptide is indicated by referring to it as a di-, tri-, or polypeptide (two, three, or many amino acids).

organic chemistry Common organic molecule groupings. Organic chemistry is the study of carbon compounds, which make up over 90% of all chemical compounds. This diversity arises because carbon atoms can combine in many different ways with other atoms, forming a wide variety of loops and chains.

carboxylic acids Carboxylic acids (**fatty acids**) contain the carboxyl group COOH. They are weak acids which occur widely throughout nature. The simplest and best-known carboxylic acid is acetic acid, or vinegar, which has the formula CH_3COOH.

esters Esters are organic compounds formed by the reaction between an alcohol and an organic acid, with the elimination of water. Unlike salts, esters are covalent compounds.

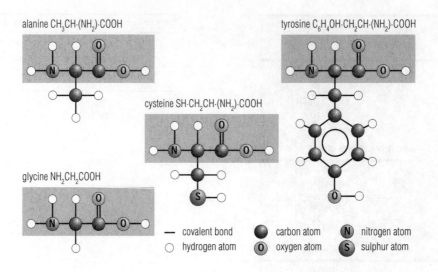

alanine $CH_3CH \cdot (NH_2) \cdot COOH$

tyrosine $C_6H_4OH \cdot CH_2CH \cdot (NH_2) \cdot COOH$

cysteine $SH \cdot CH_2CH \cdot (NH_2) \cdot COOH$

glycine NH_2CH_2COOH

— covalent bond   ● carbon atom   Ⓝ nitrogen atom
○ hydrogen atom   Ⓞ oxygen atom   Ⓢ sulphur atom

amino acid Amino acids are natural organic compounds that make up proteins and can thus be considered the basic molecules of life. There are 20 different common amino acids. They consist mainly of carbon, oxygen, hydrogen, and nitrogen. Each amino acid has a common core structure (consisting of two carbon atoms, two oxygen atoms, a nitrogen atom, and four hydrogen atoms) to which is attached a variable group, known as the R group. In glycine, the R group is a single hydrogen atom; in alanine, the R group consists of a carbon and three hydrogen atoms.

ethers Ethers are organic compounds with an oxygen atom linking the carbon atoms of two hydrocarbon radical groups (general formula R-O-R[prime]); also the common name for ethoxyethane $C_2H_5OC_2H_5$ (also called diethyl ether). This is used as an anaesthetic and as an external cleansing agent before surgical operations. It is also used as a solvent, and in the extraction of oils, fats, waxes, resins, and alkaloids.

Ethoxyethane is a colourless, volatile, inflammable liquid, slightly soluble in water, and miscible with ethanol. It is prepared by treatment of ethanol with excess concentrated sulphuric acid at 140°C/284°F.

hydrogen

carbon

oxygen

ester Molecular model of the ester ethyl ethanoate (ethyl acetate) $CH_3CH_2COOCH_3$.

ketones Ketones contain the carbonyl group (C=O) bonded to two atoms of carbon (instead of one carbon and one hydrogen as in aldehydes). They are liquids or low-melting-point solids, slightly soluble in water, often having a sweet smell. An example is propanone (acetone, CH_3COCH_3), which is used as a solvent.

petrol
Petrol is a mixture of hydrocarbons derived from petroleum, mainly used as a fuel for internal-combustion engines. It is colourless and highly volatile. **Leaded petrol** contains antiknock (a mixture of tetraethyl lead and dibromoethane), which improves the combustion of petrol and the performance of a car engine. The lead from the exhaust fumes enters the atmosphere, mostly as simple lead compounds. There is strong evidence that it can act as a nerve poison on young children and cause mental impairment. This has prompted a gradual switch to the use of **unleaded petrol** in the UK.

petroleum or crude oil natural mineral oil, a thick greenish-brown flammable liquid found underground in permeable rocks. Petroleum consists of hydrocarbons mixed with oxygen, sulphur, nitrogen, and other elements in varying proportions. It is thought to be derived from ancient organic material that has been converted by, first, bacterial action, then heat, and pressure (but its origin may be chemical also).

From crude petroleum, various products are made

by distillation and other processes; for example, fuel oil, petrol, kerosene, diesel, and lubricating oil. Petroleum products and chemicals are used in large quantities in the manufacture of detergents, artificial fibres, plastics, insecticides, fertilizers, pharmaceuticals, toiletries, and synthetic rubber.

inorganic chemistry

Inorganic chemistry is the branch of chemistry that deals with the chemical properties of the elements and their compounds, excluding the more complex covalent compounds of carbon, which are considered in organic chemistry. The origins of inorganic chemistry lay in observing the characteristics and experimenting with the uses of the substances (compounds and elements) that could be extracted from mineral ores. These could be classified according to their chemical properties: elements could be classified as metals or nonmetals; compounds as acids or bases, oxidizing or reducing agents, ionic compounds (such as salts), or covalent compounds (such as gases).

Web Elements
http://www.shef.ac.uk/~chem/web-elements/
web-elements-home.html
Periodic table on the Web, with 12 different categories of information available for each element – from its physical and chemical characteristics to its electronic configuration.

Elements are classified as metals, nonmetals, or metalloids (weakly metallic elements) depending on a combination of their physical and chemical properties; about 75% are metallic. Some elements occur abundantly (oxygen, aluminium); others occur moderately or rarely (chromium, neon); some, in particular the radioactive ones, are found in minute (neptunium, plutonium) or very minute (technetium) amounts.

Symbols (devised by Swedish chemist Jöns Berzelius) are used to denote the elements; the symbol is usually the first letter or letters of the English or Latin name (for example, C for carbon, Ca for calcium, Fe for iron, from the Latin *ferrum*). The symbol represents one atom of the element.

According to current theories, hydrogen and helium were produced in the Big Bang at the beginning of the universe. Of the other elements, those up to atomic number 26 (iron) are made by nuclear fusion within the stars. The more massive elements, such as lead and uranium, are produced when an old star explodes; as its centre collapses, the gravitational energy squashes nuclei together to make new elements.

The arrangement of elements into groups possessing similar properties led to Mendeleyev's periodic table of the elements, which prompted chemists to predict the properties of undiscovered elements that might occupy gaps in the table. This, in turn, led to the discovery of new elements, including a number of highly radioactive elements that do not occur naturally.

The periodic table today is the most recognizable logo of chemistry. Despite the range of formats they are all related to Mendeleyev's original formulation.

Elementistory
http://smallfry.dmu.ac.uk/chem/periodic/
elementi.html
Periodic table of elements showing historical rather than scientific information. The contents under the chemical links in the table are mainly brief in nature, commonly just giving names and dates of discovery.

---

metals

Metallic elements comprise about 75% of the 112 elements in the periodic table of the elements. Physical properties include a sonorous tone when struck, good conduction of heat and electricity, opacity but good reflection of light, malleability, which enables them to be cold-worked and rolled into sheets, ductility, which permits them to be drawn into thin wires, and the possible emission of electrons when heated (thermionic effect) or when the surface is struck by light (photoelectric effect).

The majority of metals are found in nature in a combined form only, as compounds or mineral ores; about 16 of them also occur in the elemental form, as native metals. Their chemical properties are largely determined by the extent to which their atoms can lose one or more electrons and form positive ions (cations).

In a metal the valency (bonding) electrons are able to move within the crystal and these electrons are said to be delocalized. Their movement creates short-lived, positively charged ions. The electrostatic attraction between the delocalized electrons and the ceaselessly forming ions constitutes the **metallic bond**.

The following metals are widely used in commerce: **precious metals** – gold, silver, and platinum, used principally in jewellery; **heavy metals** – iron, copper, zinc, tin, and lead, the common metals of engineering; **rarer heavy metals** – nickel, cadmium, chromium, tungsten, molybdenum, manganese, cobalt, vanadium, antimony, and bismuth, used principally for alloying with the heavy metals; **light metals** – aluminium and magnesium; **alkali metals** – sodium, potassium, and lithium; and **alkaline-earth metals** – calcium, barium, and strontium, used principally for chemical purposes.

Other metals have come to the fore because of special nuclear requirements – for example, technetium,

The Periodic Table

BY GORDON WOODS

Think how difficult it would be to assemble a jigsaw of which about 35 of the pieces were missing and roughly 20 of the 65 you had were too damaged to fit properly. Could you work out the shapes of some of the missing pieces? This is exactly what Mendeleyev did in 1869 when he produced the first periodic table.

Several 19th-century scientists had sought earlier to identify patterns in the properties of chemical elements linked to the weights of the atoms. Dobereiner had noticed sets of three similar elements (triads) for which the average weight of the lightest and heaviest was close to the weight of the middle one (try chlorine 35.5, bromine 80, and iodine 127). British chemist Newlands wrote the elements in order of increasing weight, noting that an element resembled the eight following. His 'Law of Octaves' soon broke down because of missing elements from a fundamentally correct law. He was ridiculed (why not list the elements alphabetically!), flung out of the Royal Society, Britain's top scientific institution, only to be reinstated when it was realized that he had nearly beaten Mendeleyev.

who was Mendeleyev?

Dmitri Ivanovitch Mendeleyev was born in Tobolsk (Siberia) in 1834 and brought to St Petersburg, capital of Tzarist Russia, for his secondary education by his ambitious mother who had realized her youngest son's potential. After research throughout Europe he was appointed chemistry professor at St Petersburg in 1865. His first marriage foundered because of all the time he spent researching; later he married a young student. After discontent among the university students Mendeleyev took their petition to the education minister who sacked him from his professorial chair. Later both Oxford and Cambridge awarded him honorary doctorates. Aged 53 he made a solo balloon ascent to view a solar eclipse. A Periodic Table was carried aloft in his funeral procession in February 1907.

Mendeleyev is said to have been playing patience when he suddenly visualized the arrangement of the elements in the patterns of the cards. It is certainly true that 20 years of previous education had equipped him to formulate the 'Periodic Law' which he developed for the remaining 40 years of his life. It stated that the elements display periodic (i.e. regularly repeating) properties when listed in order of increasing atomic weight.

Accurate atomic weights were needed for all 19th-century element patterns. Mendeleyev correctly recalculated some values better to fit his table. However iodine and tellurium provided a problem since iodine was chemically similar to bromine yet the heavier element tellurium fitted below bromine according to the weight order. Mendeleyev's solution was to have his research assistants redetermine tellurium's atomic weight to be the smaller. Thus he correctly positioned the two elements but for the wrong reason. Today we know that the elements are listed in order of increasing atomic number which only differs from the atomic weight order in three instances.

A stroke of genius was to leave spaces for elements yet to be discovered and to boldly predict properties for five such elements. Good scientists explain known information, great ones correctly predict unknown facts. Fortunately three of these miss-

ing elements were found within 20 years and their properties matched predictions to a remarkable degree.

modern periodic tables

Note the plural. There are many different formats, some are three dimensional but all show the elements in order of increasing atomic number. Most have vertical columns called groups and horizontal rows called periods. The underlying reason for the arrangement is the electron arrangement of the atoms of the elements. All elements in the same group have the same number of electrons in the outermost shell which governs the chemical nature of an element, hence their chemical similarity. Elements in the same period have the same number of electron shells, an extra electron being added for each increase in the atomic number.

metals and nonmetals

Crossing from left to right elements become more nonmetallic and descending they become more metallic. Thus moving diagonally down to the right, elements are comparable in their metallic/nonmetallic nature. This can be shown by a staircase, or better, as a diagonal line through the boxes of those elements which cannot be clearly classified either as a metal or a nonmetal. For example the use of silicon in computers stems from it being a semiconductor. Metals are conductors, nonmetals are insulators, so semiconductors are both. Science is not black and white but shades of grey.

One can copy Mendeleyev's work by making predictions about 'unknown' elements. Knowing that both sodium and potassium react with water producing hydrogen and an alkali, it is reasonable to predict the same behaviour for rubidium and caesium directly below them. Since potassium is more reactive than sodium, it is likely that caesium will be the most reactive of the four. It is! When added to a bowl of water it shatters the container! Incidentally caesium is one of three elements which have a different spelling in North America and the UK. Which are the others?

are there more elements to discover?

By 1950 all the elements for which Mendeleyev had left gaps had been discovered. Some are extremely rare and were identified from the radioactive decay of other elements, sometimes in the debris of atomic bomb tests (1944–50). However, elements with atomic numbers above 92 have been created by bombarding atoms of uranium with neutrons, carbon nuclei, and other subatomic particles. These synthetic elements are all identified from their radioactivity. As the atomic number increases it becomes harder to make the elements, so less of the element exists... and it is decaying all the time. It is possible to identify 10–18 g but there is less than 1 g of the elements with atomic number greater than 100.

The discoverer of an element has the right to name it. Some chose their country (gallium, francium), or a property (chromium has compounds in many colours), or the place of discovery (such as strontium, named after the Scottish village of Strontian). Only three places in the world have the facilities to make the synthetic elements. Both the USSR and the USA claimed initial discovery of elements 104 and 105 but each country gave them different names (for example 104 to the Russians was Kurschatovium (Ku) but to the Americans was Rutherfordium (Rf)). This scientific argument was initially solved by giving them artificial temporary names, unnilquadium (a hundredandfourium) and unnilpentium. A committee sat for years to decide officially who were the discoverers.

symbols and names
Element symbols are internationally agreed but the name differs with the language. This difference is only slight with elements discovered since 1850. Less reactive metals, isolated for hundreds of years may have very different names. Petrol is

unleaded in the UK, bleifrei in Germany, sans plomb in France, senza piomba in Italy, sin plomo in Spain. Note how the last three names relate to the Latin plumbum from which the symbol Pb is derived, as are plumber and plumbline.

unusual formats
Hundreds of versions exist of the periodic table. For example, some show the physical state (solid, liquid, or gas) of the element (in the UK only, the metal mercury and the nonmetal bromine are liquids) and three-dimensional varieties divide up the periodic table into its four blocks labelled s, p, d, and f according to the shape of the outer electron cloud round the nucleus.

produced in nuclear reactors, is corrosion-inhibiting; zirconium may replace aluminium and magnesium alloy in canning uranium in reactors.

reactions of metals with acids Metals replace the hydrogen in an acid to form a salt. For example, with magnesium and sulphuric acid the products are magnesium sulphate and hydrogen:

$$Mg + H_2SO_4 \rightarrow MgSO_4 + H_2$$

with oxygen

Most metals form oxides. For example magnesium will react with oxygen to form magnesium oxide:

$$Mg + O_2 \rightarrow 2MgO$$

with water

Some metals displace hydrogen in water to form hydroxides or oxides. For example when sodium is added to water sodium hydroxide is formed and hydrogen is given off:

$$2Na + 2H_2O \rightarrow 2NaOH + H_2$$

All metals except mercury are solid at ordinary temperatures, and all of them will crystallize under suitable conditions. The chief chemical properties of metals also include their strong affinity for certain non-metallic elements, for example sulphur and chlorine, with which they form sulphides and chlorides. Metals will, when fused, enter into the forming of alloys.

Metals have been put to many uses since prehistoric times, both structural and decorative, and the Copper Age, Bronze Age, and Iron Age are named after the metal that formed the technological base for that stage of human evolution. The science and technology of producing metals is called metallurgy.

alkali metals The alkili metals form a linked group (Group One) in the periodic table of the elements. They comprise six metallic elements with similar chemical

Flame Test

The flame test is a laboratory method used to ascertain the presence of positive ions in a sample. The sample is moistened with sulphuric acid and held in a Bunsen flame; the colour of the flame that results indicates which ions are present.

element	colour of flame
sodium	orange-yellow
potassium	lilac
calcium	red or yellow-red
strontium, lithium	crimson
barium, manganese	
(manganese chloride)	pale green
copper, thallium, boron (boric acid)	bright green
lead, arsenic, antimony	livid blue
copper (copper (II) chloride)	bright blue

properties: lithium, sodium, potassium, rubidium, caesium, and francium. They are univalent (have a valency of one) and of very low density (lithium, sodium, and potassium float on water); in general they are reactive, soft, low-melting-point metals. Because of their reactivity they are only found as compounds in nature.

sodium Sodium is a soft, waxlike, silver-white, metal (symbol Na from Latin *natrium*). It has a very low density, being light enough to float on water. It is the sixth most abundant element (the fourth most abundant metal) in the Earth's crust. Sodium is highly reactive, oxidizing rapidly when exposed to air and reacting violently with water. Sodium is important in the nervous systems of animals.

Its most familiar compound is sodium chloride (common salt), which occurs naturally in the oceans and in salt deposits left by dried-up ancient seas.

Other sodium compounds used industrially include sodium hydroxide (caustic soda, NaOH), sodium carbonate (washing soda, Na_2CO_3) and hydrogencarbonate (sodium bicarbonate, $NaHCO_3$), sodium nitrate (saltpetre, $NaNO_3$, used as a fertilizer), and sodium thiosulphate (hypo, $Na_2S_2O_3$, used as a photographic fixer). Thousands of tons of these are manufactured annually. Sodium metal is used to a limited extent in spectroscopy, in discharge lamps, and alloyed with potassium as a heat-transfer medium in nuclear reactors. It was isolated from caustic soda in 1807 by English chemist Humphry Davy.

potassium Potassium (symbol K from Latin *kalium*) is physically similar to sodium and it undergoes similar chemical reactions. It also has a very low density and is the second lightest metal (after lithium). It oxidizes rapidly when exposed to air and reacts violently with water. Of great abundance in the Earth's crust, it is widely distributed with other elements and is found in salt and mineral deposits in the form of potassium aluminium silicates. Potassium is the main base ion of the fluid in the body's cells. Along with sodium, it is important to the electrical potential of the nervous system and, therefore, for the efficient functioning of nerve and muscle. Shortage, which may occur with excessive fluid loss (prolonged diarrhoea, vomiting), may lead to muscular paralysis; potassium overload may result in cardiac arrest. It is also required by plants for growth. The element was discovered and named in 1807 by Humphry Davy, who isolated it from potash in the first instance of a metal being isolated by electric current.

Davy Discovers Sodium and Potassium
http://dbhs.wvusd.k12.ca.us/Chem-History/
Davy-Na&K-1808.html
Davy's paper to the Royal Society in 1808 titled 'On some new phenomena of chemical changes produced by electricity, particularly the decomposition of fixed alkalies, and the exhibition of the new substances which constitute their bases: and on the general nature of alkaline bodies'. Quite apart from it being the longest title ever, it seems astounding that only 190 years ago humankind did not know that common salt was sodium chloride, this paper marking the leap in knowledge which created a century of chemical discoveries.

---

alkaline-earth metals The six metals belonging to Group Two in the periodic table of the elements are termed alkaline-earth metals. Beryllium, magnesium, calcium, strontium, barium, and radium have similar bonding properties. They are strongly basic (see base), bivalent (have a valency of two), and occur in nature only in compounds. They and their compounds are used to make alloys, oxidizers, and drying agents. Examples of alkaline-earth metals are given below.

calcium Calcium is (Latin *calcis*, 'lime') a soft, silvery-white metallic element, symbol Ca. It is the fifth most abundant element (the third most abundant metal) in the Earth's crust, and is found mainly as its carbonate $CaCO_3$, which occurs in a fairly pure condition as chalk and limestone (see calcite). Calcium is an essential component of bones, teeth, shells, milk, and leaves, and it forms 1.5% of the human body by mass. Calcium ions in animal cells are involved in regulating muscle contraction, blood clotting, hormone secretion, digestion, and glycogen metabolism in the liver. It is acquired mainly from milk and cheese, and its uptake is facilitated by vitamin D. Calcium deficiency leads to chronic muscle spasms (tetany); an excess of calcium may lead to the formation of stones in the kidney or gall bladder.

The element was discovered and named by Humphry Davy in 1808. Its compounds include slaked lime (calcium hydroxide, $Ca(OH)_2$); plaster of Paris (calcium sulphate, $CaSO_4.2H_2O$); calcium phosphate ($Ca_3(PO_4)_2$), the main constituent of animal bones; calcium hypochlorite ($CaOCl_2$), a bleaching agent; calcium nitrate ($Ca(NO)_2.4H_2O$), a nitrogenous fertilizer; calcium carbide (CaC_2), which reacts with water to give ethyne (acetylene); calcium cyanamide ($CaCN_2$), the basis of many pharmaceuticals, fertilizers, and plastics, including melamine; calcium cyanide ($Ca(CN)_2$), used in the extraction of gold and silver and in electroplating; and others used in baking powders and fillers for paints.

magnesium Magnesium is a lightweight, very ductile and malleable, silver-white, metallic element (symbol Mg). It is the lightest of the commonly used metals. Magnesium silicate, carbonate, and chloride are widely distributed in nature. The metal is used in alloys and flash photography. It is a necessary trace element in the human diet, and green plants cannot grow without it since it is an essential constituent of the photosynthetic pigment chlorophyll ($C_{55}H_{72}MgN_4O_5$).

It was named after the ancient Greek city of Magnesia, near where it was first found. It was first recognized as an element by Scottish chemist Joseph Black in 1755 and discovered in its oxide by English chemist Humphry Davy in 1808. Pure magnesium was isolated in 1828 by French chemist Antoine-Alexandre-Brutus Bussy.

strontium Strontium is a soft, ductile, pale yellow, metallic element (symbol Sr). It is widely distributed in small quantities only as a sulphate or carbonate. Strontium salts burn with a crimson flame and are used in fireworks and signal flares. The radioactive isotopes Sr-89 and Sr-90 (half-life 25 years) are some of the most dangerous products of the nuclear industry; they are fission products in nuclear explosions and in the reactors of nuclear power plants. Strontium is chemically similar to calcium and deposits in bones and other tissues, where the radioactivity is damaging. The element was named in 1808 by Humphry Davy, who isolated it by electrolysis, after Strontian, a mining location in Scotland where it was first found.

other important metals A few of the more commonly used metals are listed below.

iron Iron is a hard, malleable and ductile, silver-grey, metal (symbol Fe from Latin *ferrum*). It is the fourth most abundant element (the second most abundant metal, after aluminium) in the Earth's crust. Iron occurs in concentrated deposits as the ores hematite (Fe_2O_3), spathic ore ($FeCO_3$), and magnetite (Fe_3O_4). It sometimes occurs as a free metal, occasionally as fragments of iron or iron-nickel meteorites.

Iron is the most common and most useful of all metals; it is strongly magnetic and is the basis for **steel**, an alloy with carbon and other elements. In electrical equipment iron is used in all permanent magnets and electromagnets, and forms the cores of transformers and magnetic amplifiers. In the human body, iron is an essential component of haemoglobin, the molecule in red blood cells that transports oxygen to all parts of the body. A deficiency in the diet causes a form of anaemia.

Iron is a member of the group known as the **transition metals**. Transition metals bond in a more complex way than other metals, having the ability to bond covalently, and have more than one valency: in its compounds iron may exist as Fe(II) (e.g. $FeCO_3$) or Fe(III) (e.g. Fe_2O_3). Other transition metals include nickel, copper, and gold.

transition metals

To remember the first row of transition metals in the periodic table:

Scandinavian **TV** **c**orrupts **m**any **F**rench **co**almen's **ni**eces and **cou**zins

(**Sc**, **Ti**, **V**, **Cr**, **Mn**, **Fe**, **Co**, **Ni**, **Cu**, **Zn**)

aluminium Aluminium is a lightweight, silver-white, ductile and malleable metallic element (symbol Al). It is the third most abundant element (and the most abundant metal) in the Earth's crust, of which it makes up about 8.1% by mass. It is nonmagnetic, an excellent conductor of electricity, and oxidizes easily, the layer of oxide on its surface making it highly resistant to tarnish. In the USA the original name suggested by the scientist Humphry Davy, 'aluminum', is retained. Pure aluminium is a reactive element with stable compounds, so a great deal of energy is needed in order to separate aluminium from its ores, and the pure metal was not readily obtainable until the middle of the 19th century. Commercially, it is prepared by the electrolysis of alumina (aluminium oxide), which is obtained from the ore bauxite. In its pure state aluminium is a weak metal, but when combined with elements such as copper, silicon, or magnesium it forms alloys of great strength.

Aluminium is widely used in the shipbuilding and aircraft industries because of its light weight (relative density 2.70). It is also used in making cooking utensils, cans for beer and soft drinks, and foil. It is much used in steel-cored overhead cables and for canning uranium slugs for nuclear reactors. Aluminium is an essential constituent in some magnetic materials; and, as a good conductor of electricity, is used as foil in electrical capacitors. A plastic form of aluminium, developed in 1976, which moulds to any shape and extends to several times its original length, has uses in electronics, cars, and building construction.

commercial production The method now used for its commercial production is the electrolysis of alumina. An iron pot, lined with carbon, is charged with cryolite and heated to about 800°C/1,470°F by the electric current. For the electrolysis, a bundle of carbon rods is used as the anode, while the pot itself forms the cathode. The oxygen liberated combines with the carbon of the anode to form carbon dioxide, while the aluminium falls to the bottom of the vessel.

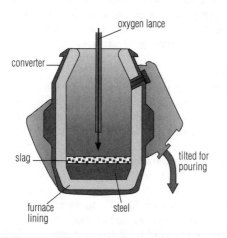

oxygen lance

converter

slag

furnace
lining

tilted for
pouring

steel

basic–oxygen process The basic-oxygen process is the primary method used to produce steel. Oxygen is blown at high pressure through molten pig iron and scrap steel in a converter lined with basic refractory materials. The impurities, principally carbon, quickly burn out, producing steel.

More alumina is added and the process continued, the molten metal being drawn off from time to time.

copper Copper is an orange-pink, very malleable and ductile transition metal (symbol Cu from Latin *cuprum*). It is used for its durability, pliability, high thermal and electrical conductivity, and resistance to corrosion. It was the first metal used systematically for tools by humans; when mined and worked into utensils it formed the technological basis for the Copper Age in prehistory. When alloyed with tin it forms bronze, which is stronger than pure copper and may hold a sharp edge; the systematic production and use of this alloy was the basis for the prehistoric Bronze Age. Brass, another hard copper alloy, includes zinc.

zinc Zinc is a hard, brittle, bluish-white, metallic element (symbol Zn). The principal ore is sphalerite or zinc blende (zinc sulphide, ZnS). Zinc is hardly affected by air or moisture at ordinary temperatures; its chief uses are in alloys such as brass and in coating metals (for example, galvanized iron). Its compounds include zinc oxide, used in ointments (as an astringent) and cosmetics, paints, glass, and printing ink. Zinc is a transition metal.

Zinc is an essential trace element in most animals; adult humans have 2–3 g/0.07–0.1 oz zinc in their bodies. There are more than 300 known enzymes that contain zinc. Zinc has been used as a component of brass since the Bronze Age, but it was not recognized as a separate metal until 1746, when it was described by German chemist Andreas Sigismund Marggraf (1709–1782). The name derives from the shape of the crystals on smelting. The zinc industry in Europe generates about 80,000 tons of zinc waste each year.

nickel Nickel is a hard, malleable and ductile, silver-white transition metal (symbol Ni). It occurs in igneous rocks and as a free metal (native metal), occasionally occurring in fragments of iron-nickel meteorites. It is a component of the Earth's core, which is held to consist principally of iron with some nickel. It has a high melting point, low electrical and thermal conductivity, and can be magnetized. It does not tarnish and therefore is much used for alloys, electroplating, and for coinage.

It was discovered in 1751 by Swedish mineralogist Axel Cronstedt (1722–1765) and the name given as an abbreviated form of *kopparnickel* (Swedish for 'false copper'), since the ore in which it is found resembles copper but yields none.

silver Silver is a white, lustrous, extremely malleable and ductile transition metal (symbol Ag from Latin *argentum*). It occurs in nature in ores and as a free metal; the chief ores are sulphides, from which the metal is extracted by smelting with lead. It is one of the best metallic conductors of both heat and electricity; its most useful compounds are the chloride and bromide, which darken on exposure to light and are the basis of photographic emulsions.

Silver is used ornamentally, for jewellery and tableware, for coinage, in electroplating, electrical contacts, and dentistry, and as a solder. It has been mined since prehistory; its name is an ancient non-Indo-European one, *silubr*, borrowed by the Germanic branch as *silber*.

gold Gold is a heavy, yellow precious metal (symbol Au). It is unaffected by temperature changes and is highly resistant to acids. For manufacture, gold is alloyed with another strengthening metal (such as copper or silver), its purity being measured in carats on a scale of 24. It is a transition metal.

In 1990 the three leading gold-producing countries were South Africa, 605.4 tonnes; USA, 295 tonnes; and Russia, 260 tonnes. In 1989 gold deposits were found in Greenland with an estimated yield of 12 tonnes per year.

Gold occurs naturally in veins, but following erosion it can be transported and redeposited. It has long been valued for its durability, malleability, and ductility, and its uses include dentistry and jewellery. As it will not corrode, it is also used in the manufacture of electric contacts for computers and other electrical devices.

A Japanese company produced a malleable form of gold in 1995, made of fine gold powder mixed with water and a secret binder. Designers can work with the putty in the same way as clay, but once the putty is fired (at 1,000°C), the water and binder evaporate, leaving the fused gold particles.

alloys

Some metals can be blended with some other metallic or nonmetallic substances to give them special qualities, such as resistance to corrosion, greater hardness, or tensile strength. Useful alloys include bronze, brass, cupronickel, duralumin, German silver, gunmetal, pewter, solder, steel, and stainless steel.

Among the oldest alloys is bronze (mainly an alloy of copper and tin), the widespread use of which ushered in the Bronze Age. Complex alloys are now common; for example, in dentistry, where a cheaper alternative to gold is made of chromium, cobalt, molybdenum, and titanium. Among the most recent alloys are superplastics: alloys that can stretch to double their length at specific temperatures, permitting, for example, their injection into moulds as easily as plastic.

Alloys are usually made by melting the metals

together. (Certain elements which will not melt together, for example copper and graphite may be combined using techniques of powder metallurgy.) Before adding the alloying element to the principal metal in the molten state it is necessary to ensure that it is free from oxygen, which would otherwise react with the alloying element, reducing the amount which would be dissolved and so causing an error in the composition. For this purpose a deoxidizer is added; this is often another metal.

Compositions made only for the purpose of melting with other metals to form alloys are called **master alloys** or **foundry alloys**. They are used to overcome the problems of alloying metals of widely differing melting points, or to facilitate closer control over the final composition, or as deoxidizers. **Shape memory alloys** are imprinted with a shape so that even after distortion, a threshold temperature will bring about a return to the original shape. Nitinol, an alloy of titanium and nickel, and brass are examples.

brass Brass is a metal alloy of copper and zinc, with not more than 5% or 6% of other metals. The zinc content ranges from 20% to 45%, and the colour of brass varies accordingly from coppery to whitish yellow. Brasses are characterized by the ease with which they may be shaped and machined; they are strong and ductile, resist many forms of corrosion, and are used for electrical fittings, ammunition cases, screws, household fittings, and ornaments.

nonmetals

There are about 20 elements that have certain physical and chemical properties opposite to those of metals. Nonmetals accept electrons and are sometimes called **electronegative** elements. Nonmetals include the halogens, carbon, oxygen, sulphur, and phosphorus. Their typical reactions are given below.

reactions of nonmetals with acids and alkalis
Nonmetals do not react with dilute acids but may react with alkalis.

$$2NaOH + Cl_2 \rightarrow NaCl + NaOCl + H_2O$$

with air or oxygen They form acidic or neutral oxides.

$$S + O_2 \rightarrow SO_2$$

with chlorine They react with chlorine gas to form covalent chlorides.

$$2P(s) + 3Cl_2 \rightarrow 2PCl_3$$

with reducing agents Nonmetals act as oxidizing agents.

$$2FeCl_2 + Cl_2 \rightarrow 2FeCl_2$$

Chemistry of Carbon
http://cwis.nyu.edu/pages/mathmol/modules/
carbon/carbon1.html
Introduction to carbon, the element at the heart of life as we know it. This site is illustrated throughout and explains the main basic forms of carbon and the importance of the way scientists choose to represent these various structures.

---

carbon Carbon (Latin *carbo, carbonaris* 'coal') is a nonmetallic element, atomic number 6 (symbol C). It occurs on its own as the **allotropes** diamond, graphite, and as fullerenes. (Allotropy is the property whereby an element can exist in two or more forms (allotropes), each possessing different physical properties but the same state of matter - gas, liquid, or solid.) Its compounds are found in carbonaceous rocks such as chalk and limestone, as carbon dioxide in the atmosphere, as hydrocarbons in petroleum, coal, and natural gas, and as a constituent of all organic substances.

In its amorphous form, it is familiar as coal, charcoal, and soot. The atoms of carbon can link with one another in rings or chains, giving rise to innumerable complex compounds. Of the inorganic carbon compounds, the chief ones are **carbon dioxide**, a colourless gas formed when carbon is burned in an adequate supply of air; and **carbon monoxide** (CO), formed when carbon is oxidized in a limited supply of air. **Carbon disulphide** (CS_2) is a dense liquid with a sweetish odour. Another group of compounds is the **carbon halides**, including carbon tetrachloride (tetrachloromethane, CCl_4).

Nobel Prize in Chemistry 1996
http://www.nobel.se/announcement-96/ chemistry96.html
Description of the discovery of fullerene carbon molecules that led to the award of the 1996 Nobel prize. The description on this page covers all the basic aspects of the discovery, including the historical background and brief biographical information on the chemists involved. The structure of the molecule is also shown in a diagram, and some of the chemistry used in their manufacture is also described here.

When added to steel, carbon forms a wide range of alloys with useful properties. In pure form, it is used as a moderator in nuclear reactors; as colloidal graphite it is a good lubricant and, when deposited on a surface in a vacuum, reduces photoelectric and secondary emission of electrons. Carbon is used as a fuel in the form of coal or coke. The **radioactive isotope** carbon-14 (half-life 5,730 years) is used as a tracer in biological research and in radiocarbon dating. Analysis of interstellar dust has led to the discovery of discrete carbon molecules, each containing 60 carbon atoms. The C_{60} molecules have been named buckminster-fullerenes because of their structural similarity to the geodesic domes designed by US architect and engineer Buckminster Fuller.

halogens The halogens are a group of five non-metallic elements with similar chemical bonding properties: fluorine, chlorine, bromine, iodine, and astatine. They form a linked group in the periodic table of the elements, descending from fluorine, the most reactive, to astatine, the least reactive. They combine directly with most metals to form salts, such as common salt (NaCl). Each halogen has seven electrons in its valence shell, which accounts for the chemical similarities displayed by the group.

chlorine

Chlorine (Greek *chloros* 'green') is a greenish-yellow, gaseous, nonmetallic element with a pungent odour (symbol Cl). It is a member of the halogen group and is widely distributed, in combination with the alkali metals, as chlorates or chlorides.

Chlorine is obtained commercially by the electrolysis of concentrated brine and is an important bleaching agent and germicide, used for sterilizing both drinking water and swimming pools. As an oxidizing agent it finds many applications in organic chemistry.

chlorine
Fritz Haber developed chlorine gas for use by the Germans in World War I. Unable to live with this, his wife committed suicide in 1915.

The pure gas (Cl_2) is a poison and was used in gas warfare in World War I, where its release seared the membranes of the nose, throat, and lungs, producing pneumonia. Chlorine is a component of chlorofluorocarbons (CFCs) and is partially responsible for the depletion of the ozone layer; it is released from the CFC molecule by the action of ultraviolet radiation in the upper atmosphere, making it available to react with and destroy the ozone. The concentration of chlorine in the atmosphere in 1997 reached just over three parts per billion. It is expected to reach its peak in 1999 and then start falling rapidly due to international action to curb ozone-destroying chemicals.

Some typical reactions are given below.

with metals
When dry chlorine is passed over a heated metal, the chloride is formed.

$$Zn + Cl_2 \rightarrow ZnCl_2$$

$$2Fe + 3Cl_2 \rightarrow 2FeCl_3$$

with nonmetals
The same reaction occurs with certain nonmetals, when the dry gas is passed over the heated element.

$$2P + 5Cl_2 \rightarrow 2PCl_5$$

with compounds
With water, chlorine forms a bleaching solution.

$$H_2O + Cl_2 \rightarrow HCl + 2HOCl_2OCl \rightarrow 2Cl^- + O_2$$

Iron (II) salts are oxidized to iron (III) salts.

$$2FeCl_2 + Cl_2 \rightarrow 2FeCl_3$$

Organic compounds undergo halogenation.

$$C_2H_6 + Cl_2 \rightarrow C_2H_5Cl + HCl \quad C_2H_4 + Cl_2 \rightarrow C_2H_4Cl_2$$

Alkalis form chlorides, chlorates, and water.

$$2NaOH + Cl_2 \rightarrow NaCl + NaOCl + H_2O$$

Other halogens are displaced in a **redox** reaction.

$$2KBr + Cl_2 \rightarrow 2KCl + Br_2$$

On a Combination of Oxymuriatic Gas and Oxygene Gas
http://dbhs.wvusd.k12.ca.us/Chem-History/avy-Chlorine-1811.html
Transcript of Humphry Davy's submission to the Royal Society naming chlorine and labelling it as an element.

---

oxygen Oxygen is a colourless, odourless, tasteless, nonmetallic, gaseous element (symbol O). It is the most abundant element in the Earth's crust (almost 50% by mass), forms about 21% by volume of the atmosphere, and is present in combined form in water and many other substances. Oxygen is a by-product of photosynthesis and the basis for respiration in plants and animals.

Oxygen is very reactive and combines with all other elements except the inert gases and fluorine. It is present in carbon dioxide, silicon dioxide (quartz), iron ore, and calcium carbonate (limestone). As a gas it exists as a molecule composed of two atoms (O_2) or as **ozone** (O_3) a highly reactive pale blue gas with a penetrating odour. Ozone is an **allotrope** (see allotropy) of oxygen, made up of three atoms of oxygen. It is formed when the molecule of the stable form of oxygen (O_2) is split by ultraviolet radiation or electrical discharge. It forms the ozone layer in the upper atmosphere, which protects life on Earth from ultraviolet rays, a cause of skin cancer.

Oxygen is obtained for industrial use by the fractional distillation of liquid air, by the electrolysis of water, or by heating manganese (IV) oxide with potassium chlorate. It is essential for combustion, and is used with ethyne (acetylene) in high-temperature oxy-acetylene welding and cutting torches.

The element was first identified by English chemist Joseph Priestley in 1774 and independently in the same year by Swedish chemist Karl Scheel. It was named by French chemist Antoine Lavoisier in 1777.

inert gases The elements helium, neon, argon, krypton, xenon, and radon are known as the inert (or noble) gases, so named because they were originally thought not to enter into any chemical reactions. This is now known to be incorrect: in 1962, xenon was made to combine with fluorine, and since then, compounds of argon, krypton, and radon with fluorine and/or oxygen have been described.

The extreme unreactivity of the inert gases is due to the stability of their electronic structure. All the electron shells (energy levels) of inert gas atoms are full and, except for helium, they all have eight electrons in their outermost (valency) shell. The apparent stability of this electronic arrangement led to the formulation of the octet rule to explain the different types of chemical bond found in simple compounds.

Inert Gases: Electronic Structure

name	symbol	atomic number	electronic arrangement
helium	He	2	2.
neon	Ne	10	2.8.
argon	Ar	18	2.8.8.
krypton	Kr	36	2.8.18.8.
xenon	Xe	54	2.8.18.18.8.
radon	Rn	86	2.8.18.32.18.8.

inert gases

To remember the inert gases:

He neatly arranges Kremlin executive ranks
(helium, neon, argon, krypton, xenon, radon)

argon

Argon is grouped with the inert gases, since it was long believed not to react with other substances, but observations now indicate that it can be made to combine with boron fluoride to form compounds. It constitutes almost 1% of the Earth's atmosphere, and was discovered in 1894 by British chemists John Rayleigh and William Ramsay after all oxygen and nitrogen had been removed chemically from a sample of air. It is colourless and odourless, and used in electric discharge tubes and argon lasers.

ions

An **ion** is an atom, or group of atoms, that is either positively charged (**cation**) or negatively charged (**anion**), as a result of the loss or gain of electrons during chemical reactions or exposure to certain forms of radiation. In solution or in the molten state, ionic compounds such as salts, acids, alkalis, and metal oxides conduct electricity. These compounds are known as **electrolytes**.

cation

To remember that **cat**ions are atoms that have
lost an electron:

Cat lost an eye

Ions are produced during electrolysis, for example the salt zinc chloride ($ZnCl_2$) dissociates into the positively charged Zn^{2+} and negatively charged Cl^- when electrolysed.

tests for negative ions The presence of negative ions can be determined by performing a number of different tests.

bromide (Br^-): addition of dilute nitric acid to bromide solution immediately yields a whitish precipitate of silver bromide, which is partially soluble in concentrated ammonia solution, for example:

$$KBr + AgNO_3 \rightarrow AgBr + KNO_3$$

carbonate ($CO_3{}^{2-}$): a solid carbonate treated with dilute hydrochloric acid gives off carbon dioxide gas, which turns limewater milky:

$$CaCO_3 + 2HCl \rightarrow CaCl_2 + H_2O + CO_2$$

chloride (Cl^-): treatment of a chloride with concentrated sulphuric acid produces colourless hydrogen chloride gas, which forms thick white fumes of ammonium chloride on mixing with gaseous ammonia:

$$NH_3 + HCl \rightarrow NH_4Cl(s)$$

hydrogencarbonate (HCO^{3-}): heating a solution of a hydrogencarbonate produces carbon dioxide, which turns limewater milky:

$$Ca(HCO_3)_2 \rightarrow CaCO_3 + H_2O + CO_2$$

Hydrogencarbonates react with dilute hydrochloric acid giving off carbon dioxide, in a similar way to carbonates.

iodide (I^-): on addition of silver nitrate solution to an acidified solution of an iodide, a yellow precipitate of silver iodide is formed immediately, which is insoluble in ammonia solution:

$$KI + AgNO_3 \rightarrow AgI + KNO_3$$

nitrate (NO_3^-): there are two tests for the nitrate ion in solution. Sodium hydroxide solution and aluminium powder (or Devarda's alloy, which contains aluminium) are added to a solution of the nitrate. The mixture is warmed and the ammonia gas produced turns red litmus paper blue:

$$3NO_3^- + 5OH^- + 2H_2O \rightarrow 3NH_3 + 8AlO_2^-$$

The brown ring test: an equal volume of iron(II) sulphate solution (acidified with dilute sulphuric acid) is added to the nitrate solution in a test tube. Concentrated sulphuric acid is carefully poured down the side of the test tube, so that it forms a separate layer at the bottom of the tube. A brown ring is formed at the junction of the two layers. This is $FeSO_4.NO$, which is produced by the reaction of nitrate ions to nitrogen monoxide by the iron(II) ions:

$$NO_3^- + 4H^+ + 3Fe^{2+} \rightarrow NO(g) + 3Fe^{3+} + 2H_2O$$

Care should be taken with this test, as nitrites and bromides can give similar results.

nitrite (NO^{2-}): addition of dilute sulphuric acid to a nitrite produces brown nitrogen dioxide gas, which turns blue litmus paper red without bleaching it. The solution turns pale blue. No heating is required.

sulphate (SO_4^{2-}): addition of dilute hydrochloric acid and barium sulphate solution to a solution of a sulphate results in the immediate precipitation of barium sulphate:

$$Na_2SO_4 + BaCl_2 \rightarrow BaSO_4 + 2NaCl$$

sulphide (S^{2-}): addition of dilute hydrochloric acid to a sulphide results in the production of colourless hydrogen sulphide gas, which smells of rotten eggs and turns lead nitrate (soaked into filter paper) black.

$$Na_2S + 2HCl \rightarrow 2Na_2Cl + H_2S$$

sulphite (SO_3^{2-}): addition of dilute hydrochloric acid to a sulphite, with heating, produces colourless sulphur dioxide gas. This turns potassium dichromate from orange to green, but does not change the colour of lead nitrate solution.

$$K_2SO_3 + 2HCl \rightarrow 2KCl + SO_2 + H_2O.$$

Tests for common positive ions are given above in the Flame Test table.

free radicals A free radical is an atom or molecule that has an unpaired electron and is therefore highly reactive. Most free radicals are very short-lived. They are by-products of normal cell chemistry and rapidly oxidize other molecules they encounter. Free radicals are thought to do considerable damage. They are neutralized by protective enzymes.

Free radicals are often produced by high temperatures and are found in flames and explosions.

The action of ultraviolet radiation from the Sun splits chlorofluorocarbon (CFC) molecules in the upper atmosphere into free radicals, which then break down the ozone layer.

A very simple free radical is the methyl radical CH_2 produced by the splitting of the covalent carbon-to-carbon bond in ethane.

$$CH_3CH_3 \rightarrow 2CH_3$$

compounds
The elements react with one another to form an enormous variety of different compounds. Some common classes of compound are illustrated below.

acids Any compound that releases hydrogen ions (H^+ or protons) in the presence of an ionizing solvent (usually water) is called an **acid**. Acids react with bases to form salts, and they act as solvents. Strong acids are corrosive; dilute acids have a sour or sharp taste, although in some organic acids this may be partially masked by other flavour characteristics. The strength of an acid is measured by its hydrogen-ion concentration, indicated by the pH value (see below). All acids have a pH below 7.0.

Acids can be classified as monobasic, dibasic, tribasic, and so on, according to their basicity (the number of hydrogen atoms available to react with a base) and degree of ionization (how many of the available hydrogen atoms dissociate in water). Dilute sulphuric acid is classified as a strong (highly ionized), dibasic acid.

Inorganic acids include boric, carbonic, hydrochloric, hydrofluoric, nitric, phosphoric, and sulphuric. Organic acids include acetic (vinegar), benzoic, citric, formic, lactic, oxalic, and salicylic, as well as complex substances such as nucleic acids and amino acids.

Sulphuric, nitric, and hydrochloric acid are sometimes referred to as the mineral acids. Most naturally occurring acids are found as organic compounds, such as the fatty acids R-COOH and sulphonic acids R-SOOH, where R is an organic group.

All acids produce hydrogen ions when dissolved in water; for example hydrochloric acid is produced when hydrogen chloride gas dissolves in water.

---

acid

To remember how to mix acid and water safely:

Add acid to water, just as you oughter!

The reactions of acids are the reactions of the H^+ ion. These are as follows.

with indicators They give a specific colour reaction with indicators; for example, litmus turns red.

with alkalis They react with alkalis to form a salt and water (neutralization). For example, hydrochloric acid added to sodium hydroxide gives the salt sodium chloride plus water.

$$HCl(aq) + NaOH(aq) \rightarrow NaCl(aq) + H_2O(l)$$

Acids react with many bases, such as oxides and hydroxides, but the product is not always soluble in water so the reaction soon ceases, as when sulphuric acid reacts with calcium oxide, hydroxide, or carbonate.

$$H_2SO_4(aq) + CaO(Aq) \rightarrow CaSO_4(s) + H_2O(l)$$

with carbonates With carbonates and hydrogencarbonates, acids form a salt and displace carbon dioxide. For example, as with nitric acid added to sodium hydrogencarbonate:

$$NaHCO_3(aq) + HNO_3(aq) \rightarrow NaNO_3(aq) + H_2O(l) + CO_2(g)$$

with metals Acids react with metals to give off hydrogen and form a salt. For example, with magnesium and sulphuric acid the products are magnesium sulphate and hydrogen.

$$H_2SO_4(aq) + Mg(s) \rightarrow MgSO_4(aq) + H_2(g)$$

pH scale from 0 to 14 for measuring acidity or alkalinity. A pH of 7.0 indicates neutrality, below 7 is acid, while above 7 is alkaline. Strong acids, such as those used in car batteries, have a pH of about 2; strong alkalis such as sodium hydroxide are pH 13.

Acidic fruits such as citrus fruits are about pH 4. Fertile soils have a pH of about 6.5 to 7.0, while weak alkalis such as soap are 9 to 10.

The pH of a solution can be measured by using a broad-range indicator, either in solution or as a paper strip. The colour produced by the indicator is compared with a colour code related to the pH value. An alternative method is to use a pH meter fitted with a glass electrode.

Sören Sörenson and pH
http://dbhs.wvusd.k12.ca.us/Chem-History/
Sorenson-article.html
Excerpt from a paper on enzymatic processes in which Sörenson defined pH as the relative concentration of hydrogen ions in a solution.

---

bases and alkalis A base is the chemical opposite of an acid: it accepts protons. Bases can contain negative ions such as the hydroxide ion (OH^-), which is the strongest base, or be molecules such as ammonia (NH_3). Ammonia is a weak base, as only some of its molecules accept protons. Bases that dissolve in water are called alkalis.

Inorganic bases are usually oxides or hydroxides of metals, which react with dilute acids to form a salt and water. Many carbonates also react with dilute acids, additionally giving off carbon dioxide.

Alkalis neutralize acids and are soapy to the touch. The strength of an alkali is measured by its hydrogen-ion concentration, indicated by the pH value. They may be divided into strong and weak alkalis: a strong alkali (for example, potassium hydroxide, KOH) ionizes completely when dissolved in water, whereas a weak alkali (for example, ammonium hydroxide, NH_4OH) exists in a partially ionized state in solution. All alkalis have a pH above 7.0.

The hydroxides of metals are alkalis. Those of sodium and potassium are chemically powerful; both were historically derived from the ashes of plants.

The four main alkalis are sodium hydroxide (caustic soda, NaOH); potassium hydroxide (caustic potash, KOH); calcium hydroxide (slaked lime or limewater, $Ca(OH)_2$); and aqueous ammonia (NH_3 (aq)). Their solutions all contain the hydroxide ion OH^-, which gives them a characteristic set of properties.

with acids
Alkalis react with acids to form a salt and water (neutralization). For example potassium hydroxide and nitric acid gives potassium nitrate and water:

$$KOH(aq) + HNO_3(aq) \rightarrow HNO_3(aq) + H_2O(l)$$

with indicators
They give a specific colour reaction with indicators; for example, litmus turns blue.

with ammonium salts

Alkalis displace ammonia gas from ammonium salts:

$$NaOH(aq) + NH_4Cl(aq) \rightarrow NaCl(aq)$$
$$+ NH_3 + H_2O(l)$$

with soluble salts

Alkalis precipitate the insoluble hydroxides of most metals from soluble salts. For example iron chloride:

$$FeCl_2 + 2NaOH \rightarrow Fe(OH)_2 + 2NaCl$$

$$Fe^{2+}(aq) + 2OH^-(aq) \rightarrow Fe(OH)_2(s)$$

salts A salt is any compound formed from an acid and a base through the replacement of all or part of the hydrogen in the acid by a metal or electropositive radical. **Common salt** is sodium chloride.

A salt may be produced by a chemical reaction between an acid and a base, or by the displacement of hydrogen from an acid by a metal (see displacement activity). As a solid, the ions normally adopt a regular arrangement to form crystals. Some salts only form stable crystals as hydrates (when combined with water). Most inorganic salts readily dissolve in water to give an electrolyte (a solution that conducts electricity).

As all salts are electrically neutral, the **formula** of a salt can be worked out by making sure that the total numbers of positive and negative charges arising from the ions are equal.

Common Ions That Form Salts

positive ions	negative ions
silver Ag^+	bromide Br^-
aluminium Al^{3+}	chloride Cl^-
barium Ba^{2+}	carbonate CO_3^{2-}
calcium Ca^{2+}	fluoride F^-
copper Cu^{2+}	hydrogencarbonate HCO_3^-
iron(II) Fe^{2+}	hydrogensulphate HSO_4^-
iron(III) Fe^{3+}	iodide I^-
hydrogen H^+	nitrate NO_3^-
potassium K^+	oxide O^{2-}
lithium Li^+	hydroxide OH^-
magnesium Mg^{2+}	sulphide S^{2-}
sodium Na^+	sulphite SO_3^{2-}
ammonium NH_4^+	sulphate SO_4^{2-}
lead Pb^{2+}	
zinc $Zn2+$	

preparation

Various methods can be used to prepare salts in the laboratory; the choice is dictated by the starting materials available and by whether the required salt is soluble or insoluble.

Methods include:

(i) **acid + metal** for salts of magnesium, iron, and zinc;

(ii) **acid + base** for salts of magnesium, iron, zinc, and calcium;

(iii) **acid + carbonate** for salts of all metals;

(iv) **acid + alkali** for salts of sodium, potassium, and ammonium;

(v) **direct combination** for sulphides and chlorides;

(vi) **double decomposition** for insoluble salts.

In methods (i)–(iii) an excess of the solid reactant is added to the acid to ensure that no acid remains. The excess solid is filtered from the salt solution and the filtrate is boiled to a much smaller volume; it is then allowed to cool and crystallize. The salt crystals are filtered and dried on filter paper.

In method (iv) an indicator is used to determine the volume of acid needed to neutralize the alkali (or vice versa). The colour can then be removed by charcoal treatment, or alternatively the experiment can be repeated without the indicator. The solution is boiled to a smaller volume, cooled to crystallize the salt, and the crystals filtered and dried as in (i)–(iii) above.

In method (v) the salt is made in one step and does not require drying.

In method (vi) the two solutions are mixed and stirred. The precipitated salt is filtered, washed well with water to remove the soluble impurities, and allowed to dry in air or an oven at 60–80°C/ 140–176°F.

salts

To create a salt:

If a soluble salt you wish to provide,
You first on the acid settle;
Then neutralize with the proper oxide,
hydroxide, carbonate, or metal
But if the salt will not dissolve,
A simpler means you'll try:
Precipitate it, you resolve,
Then filter, wash, and dry

oxides Oxygen is a highly reactive element and combines with a variety of other elements to form oxides, frequently by burning the element or a compound of it in air.

Oxides of metals are normally bases and will react with an acid to produce a salt in which the metal forms the cation. Some of them will also react with a strong alkali to produce a salt in which the metal is part of a complex anion. Most oxides of nonmetals are acidic (dissolve in water to form an acid). Some oxides display no pronounced acidic or basic properties.

---

oxidation and reduction: principles

To remember the principles of oxidation and reduction:

Remember that Leo the lion goes 'ger'

(**L**ose **e**lectrons - **o**xidation, **g**ain **e**lectrons - **r**eduction)

---

oxidizing and reducing agents Oxidation is the loss of electrons, gain of oxygen, or loss of hydrogen by an atom, ion, or molecule during a chemical reaction. An **oxidizing agent** brings about oxidation by accepting electrons or hydrogen from a compound, or donating oxygen to that compound, and is simultaneously **reduced** in the reaction. Oxidation may also be brought about electrically at the anode (positive electrode) of an electrolytic cell.

Reduction is the opposite of oxidation, that is the gain of electrons or hydrogen, or the loss of oxygen. Reduction of a compound may be brought about by reaction with a **reducing agent**, which is simultaneously oxidized, or electrically at the cathode (negative electrode) of an electric cell. Examples include the reduction of iron(III) oxide to iron by carbon monoxide:

$$Fe_2O_3 + 3CO \rightarrow 2Fe + 3CO_2$$

the hydrogenation of ethene to ethane:

$$CH_2=CH_2 + H_2 \rightarrow CH_3-CH_3$$

and the reduction of a sodium ion to sodium.

$$Na^+ + e^- \rightarrow Na$$

A **redox reaction** is said to occur where one reactant is reduced and the other reactant oxidized. The reaction can only occur if both reactants are present and each changes simultaneously. For example, hydrogen reduces copper (II) oxide to copper while it is itself oxidized to water:

$$CuO + H_2 \rightarrow Cu + H_2O$$

---

oxidation and reduction: electrons

To remember the difference between oxidation and reduction with relation to electrons:

OILRIG

(**O**xidation **i**s **l**oss; **R**eduction **i**s **g**ain)

---

physical chemistry

Most chemical reactions exhibit some physical phenomena (change of state, temperature, pressure, or volume, or the use or production of electricity or light). Physical chemistry is the branch of chemistry concerned with examining the relationships between the chemical compositions of substances and the physical properties that they display. The measurement and study of such phenomena has led to many chemical theories and laws.

physical states of matter
gas A gas is a form of matter, such as air, in which the molecules move randomly in otherwise empty space, filling any size or shape of container into which the gas is put.

liquid A liquid is a state of matter between a solid and a gas. A liquid forms a level surface and assumes the shape of its container. Its atoms do not occupy fixed positions as in a crystalline solid, nor do they have freedom of movement as in a gas. Unlike a gas, a liquid is difficult to compress since pressure applied at one point is equally transmitted throughout (Pascal's principle). Hydraulics makes use of this property.

solid A solid is a a state of matter that holds its own shape. According to kinetic theory, the atoms or molecules in a solid are not free to move but merely vibrate about fixed positions, such as those in crystal lattices.

solution A solution is two or more substances mixed to form a single, homogenous phase. One of the substances is the **solvent** and the others (**solutes**) are said to be dissolved in it.

The constituents of a solution may be solid, liquid, or gaseous. The solvent is normally the substance that is present in greatest quantity; however, if one of the constituents is a liquid this is considered to be the solvent even if it is not the major substance. Although the commonest solvent is water, in popular use the term refers to low-boiling-point organic liquids, which are harmful if used in a confined space. They can give rise to respiratory problems, liver damage, and neurological complaints.

Typical organic solvents are petroleum distillates (in glues), xylol (in paints), alcohols (for synthetic and natural resins such as shellac), esters (in lacquers, including nail varnish), ketones (in cellulose lacquers and resins), and chlorinated hydrocarbons (as paint stripper and dry-cleaning fluids). The fumes of some solvents, when inhaled (glue-sniffing), affect mood and

Densities of Some Common Substances

Substance	Density in kg m^{-3}	Substance	Density in kg m^{-3}
Solids		**Liquids**	
balsa wood	200	milk	1,030
oak	700	sea water	1,030
butter	900	glycerine	1,260
ice	920	Dead Sea brine	1,800
ebony	120	**Gases**	
sand (dry)	1,600	(at standard temperature and pressure	
concrete	2,400	of 0°C and 1 atm)	
aluminium	2,700	air	1.30
steel	7,800	hydrogen	0.09
copper	8,900	helium	0.18
lead	11,300	methane	0.72
uranium	19,000	nitrogen	1.25
Liquids		oxygen	1.43
water	1,000	carbon dioxide	1.98
petrol, paraffin	800	propane	2.02
olive oil	900	butane (iso)	2.60

perception. In addition to damaging the brain and lungs, repeated inhalation of solvent from a plastic bag can cause death by asphyxia.

suspension A suspension is a mixture consisting of small solid particles dispersed in a liquid or gas, which will settle on standing. An example is milk of magnesia, which is a suspension of magnesium hydroxide in water.

colloid A colloid is a substance composed of extremely small particles of one material (the dispersed phase) evenly and stably distributed in another material (the continuous phase). The size of the dispersed particles (1–1,000 nanometres across) is less than that of particles in suspension but greater than that of molecules in true solution. Colloids involving gases include **aerosols** (dispersions of liquid or solid particles in a gas, as in fog or smoke) and **foams** (dispersions of gases in liquids).

Those involving liquids include **emulsions** (in which both the dispersed and the continuous phases are liquids) and **sols** (solid particles dispersed in a liquid). Sols in which both phases contribute to a molecular three-dimensional network have a jellylike form and are known as **gels**; gelatin, starch 'solution', and silica gel are common examples.

gel A gel is a solid produced by the formation of a three-dimensional cage structure, commonly of linked large-molecular-mass polymers, in which a liquid is trapped. It is a form of ◊colloid. A gel may be a jellylike mass (pectin, gelatin) or have a more rigid structure (silica gel).

kinetic theory of matter

According to the molecular or kinetic theory of matter, matter is made up of molecules that are in a state of constant motion, the extent of which depends on their temperature. Molecules also exert forces on one another. The nature and strength of these forces depends on the temperature and state of the matter (solid, liquid, or gas).

The existence of molecules was first inferred from the Italian physicist Amedeo Avogadro's hypothesis in 1811. He observed that when gases combine, they do so in simple proportions. For example, exactly one volume of oxygen and two volumes of hydrogen combine to produce water. He hypothesized that equal volumes of gases at the same temperature and pressure contain equal numbers of molecules. Avogadro's hypothesis only became generally accepted in 1860, when proposed by the Italian chemist Stanislao Cannizzaro.

Avogadro's Hypothesis of 1811
http://dbhs.wvusd.k12.ca.us/
Chem–History/Avogadro.html
Transcript of a translation of the essay containing Avogadro's hypothesis.

The movement of some molecules can be observed in a microscope. As early as 1827, Robert Brown observed that very fine pollen grains suspended in water move about in a continuously agitated manner. This continuous, random motion of particles in a fluid medium (gas or liquid) as they are subjected to impact from the molecules of the medium is known as **Brownian movement**.

The spontaneous and random movement of molecules or particles in a fluid can also be observed as **diffusion** occurs from a region in which they are at a high concentration to a region of lower concentration, until a uniform concentration is achieved throughout. No mechanical mixing or stirring is involved. For example, if a drop of ink is added to water, its molecules will diffuse until the colour becomes evenly distributed.

In biological systems, diffusion plays an essential role in the transport, over short distances, of molecules such as nutrients, respiratory gases, and neurotransmitters. It provides the means by which small molecules pass into and out of individual cells and microorganisms, such as amoebae, that possess no

before   after

sugar and water molecules become evenly mixed

gas exchange in amoeba

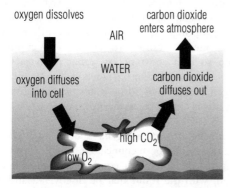

oxygen dissolves

carbon dioxide enters atmosphere

AIR

WATER

oxygen diffuses into cell

carbon dioxide diffuses out

high CO₂

low O₂

diffusion Diffusion is the movement of molecules from a region of high concentration into a region of lower concentration.

circulatory system. Plant and animal organs whose function depends on diffusion – such as the lung – have a large surface area. Diffusion of water across a semipermeable membrane is termed **osmosis**.

One application of diffusion is the separation of isotopes, particularly those of uranium. When uranium hexafluoride diffuses through a porous plate, the ratio of the 235 and 238 isotopes is changed slightly. With sufficient number of passages, the separation is nearly complete. There are large plants in the USA and UK for obtaining enriched fuel for fast nuclear reactors and the fissile uranium-235, originally required for the first atom bombs. Another application is the diffusion pump, used extensively in vacuum work, in which the gas to be evacuated diffuses into a chamber from which it is carried away by the vapour of a suitable medium, usually oil or mercury.

Laws of diffusion were formulated by Thomas Graham in 1829 (for gases) and Adolph Fick 1829–1901 (for solutions).

kinetic theory of gases

The effects of pressure, temperature, and volume on a gas were investigated during the 17th and 18th centuries. **Boyle's law** states that for a fixed mass of gas the volume of the gas is inversely proportional to the pressure at constant temperature. **Charles's law** states that for a fixed mass of gas the volume of the gas is proportional to the absolute temperature at constant pressure. The pressure law states that the pressure of a fixed mass of gas at constant volume is directly proportional to its absolute temperature.

These statements together give the gas laws which can be expressed as: A plot of the volume of a gas against its temperature gives a straight line, showing that the two are proportional. The line intercepts the x axis at –273°C/–459°F. This suggests that, if the gas did not liquefy first, it would occupy zero volume at a temperature of –273°C. This temperature is referred to as absolute zero, or zero Kelvin (0K) on the Kelvin scale, and is the lowest temperature theoretically possible.

This behaviour applies only to ideal gases, which are assumed to occupy negligible volume and contain negligible forces between particles. A real gas often behaves rather differently, and the van der Waals' law contains a correction to the gas laws to account for the nonideal behaviour of real gases.

change of state

As matter is heated its temperature may rise or it may cause a change of state. As the internal energy of matter increases the energy possessed by each particle increases too. This can be visualized as the kinetic energy of the molecules increasing, causing them to move more quickly. This movement includes both vibration within the molecule (assuming the substance is made of more than one atom) and rotation.

A solid is made of particles that are held together by forces. As a solid is heated, the particles vibrate more vigorously, taking up more space, and causing the material to expand. As the temperature of the solid increases, it reaches its melting point and turns into a liquid. The particles in a liquid can move around more freely but there are still forces between them. As further energy is added, the particles move faster until they are able to overcome the forces between them. When the boiling point is reached the liquid boils and becomes a gas. Gas particles move around independently of one another except when they collide.

Different objects require different amounts of heat energy to change their temperatures by the same amount. The heat capacity of an object is the quantity of heat required to raise its temperature by one degree. The specific heat capacity of a substance is the heat capacity per unit of mass, measured in joules per kilogram per kelvin. As a substance is changing state while being heated, its temperature remains constant, provided that thermal energy is being added. For example, water boils at a constant temperature as it turns to steam. The energy required to cause the change of state is called latent heat. This energy is used to break down the forces holding the particles together so that the change in state can occur. Specific latent heat is the thermal energy required to change the state of a certain mass of that substance without any temperature change. Evaporation causes cooling as a liquid vaporizes.

Heat is transferred by the movement of particles (that possess kinetic energy) by conduction, convection, and radiation. Conduction involves the movement of heat through a solid material by the movement of free electrons. Convection involves the transfer of energy by the movement of fluid particles. Convection currents are caused by the expansion of a liquid or gas as its temperature rises. The expanded material, being less dense, rises above colder and therefore denser material.

attraction and repulsion

Atoms are held together by the electrical forces of attraction between each negative electron and the positive protons within the nucleus. The latter repel one another with enormous forces; a nucleus holds together only because an even stronger force, called the **strong nuclear force**, attracts the protons and neutrons to one another. The strong force acts over a very short range – the protons and neutrons must be in virtual contact with one another. If, therefore, a fragment of a complex nucleus, containing some protons, becomes only slightly loosened from the main group of neutrons and protons, the natural repulsion between the protons will cause this fragment to fly apart from the rest of the nucleus at high speed. It is by such fragmentation of atomic nuclei (nuclear fission) that nuclear energy is released.

two Greek theories   Among the ancient Greeks there were two theories as to the nature of matter, or substance. Some, such as Anaxagoras and Aristotle, held that matter was infinite and continuous, and that therefore any substance could theoretically be divided and subdivided to an infinite extent. Others, such as Democritus and Epicurus, taught that matter was *grained*, that is, consisted of minute particles which could not be divided. Both theories were based on naturally slender experimental evidence.

the conservation of matter   Towards the end of the 18th century, the development of experimental chemistry demanded greater quantitative exactness, and experimental evidence, primarily from studies in combustion, led to the principle of the conservation of matter. The value of this principle has been enormous, particularly in the direction of detecting new elements.

Dalton's theory   John Dalton, in the 19th century, believed that gases consisted of particles or 'corpuscles'. He appears to have reasoned that, as all the particles of the same substances are alike, any chemical action between two substances means a corresponding change in the individual particles of the substances concerned. Particles of a compound must therefore be divisible into atomic particles of the atoms combined. Dalton enunciated the law of constant proportions, which states that when two elements unite to form a compound they do so in a constant ratio that is characteristic of that compound. For instance, when oxygen and hydrogen combine to form water, the weights combining always take the same ratio.

determining atomic weights   Shortly after Dalton's atomic theory had been enunciated, Joseph Gay-Lussac investigated the volumetric conditions of gases in combination, with the result that he discovered and published the law that when gases combine together they do so in volumes which bear a simple ratio to one another and to that of their product (if gaseous). In 1811 Amadeo Avogadro published his hypothesis on the molecular constitution of gases, which asserts that under the same conditions of temperature and pressure equal volumes of all gases contain the same number of molecules whether those molecules consist of single atoms or many atoms in combination. Both hypotheses were well supported by experimental evidence, and were used to determine the relative atomic masses of the elements. Much of the progress in chemistry has been based on quantitative analysis using atomic weights.

Rutherford and Moseley Around 1900 it became apparent that atoms themselves have structure and are not indivisible. From his experiments with alpha particles, Ernest Rutherford and others (1911–13) showed that practically the whole mass of any atom is concentrated in an extremely small central nucleus bearing a positive electrical charge. With Henry Moseley in 1913 he showed that the nucleus contains a number of positive charges dependent on the element, and called the atomic number of the element. Around the nucleus move an equal number of electrons at a relatively great distance. The lightest nucleus, the hydrogen nucleus, contains a single positive charge, and is called a proton.

Bohr In 1913 Niels Bohr proposed that the electrons move in orbits around the nucleus like planets round the Sun, and suggested how atoms might emit or absorb light. These ideas were developed and applied with great success by Bohr and others using quantum theory, to the full elucidation of atomic structure, and the explanation of the properties of matter in bulk, and of the substructure of the nucleus itself.

Chadwick In 1932 James Chadwick discovered that the bombardment of beryllium by alpha particles produced neutral particles which he called neutrons. From the atomic weights of atoms, and the known weights of the proton and the electron it became clear that (1) protons and neutrons have essentially equal masses, and (2) that atomic nuclei contain approximately equal numbers of protons and neutrons, the protons carrying the nuclear charge.

subatomic particles High-energy physics research has discovered the existence of subatomic particles other than the proton, neutron, and electron. More than three hundred kinds of particle are now known, and these are classified into several classes according to their mass, electric charge, spin, magnetic moment, and interaction. The elementary particles, which include the electron, are indivisible and may be regarded as the fundamental units of matter; the **hadrons**, such as the proton and neutron, are composite particles made up of either two or three elementary particles called quarks.

electronic structure of the atom

Electrons are arranged around the nucleus of an atom in distinct energy levels, also called **orbitals** or **shells**. These shells can be regarded as a series of concentric spheres, each of which can contain a certain maximum number of electrons; the noble gases have an arrangement in which every shell contains this number. The energy levels are usually numbered beginning with the shell nearest to the nucleus. The outermost shell is known as the **valency shell** as it contains the valence electrons.

The lowest energy level, or innermost shell, can contain no more than two electrons. Outer shells are considered to be stable when they contain eight electrons but additional electrons can sometimes be accommodated provided that the outermost shell has a stable configuration. Electrons in unfilled shells are available to take part in chemical bonding, giving rise to the concept of **valency**. In ions, the electron shells contain more or fewer electrons than are required for a neutral atom, generating negative or positive charges.

The atomic number of an element indicates the number of electrons in a neutral atom. From this it is possible to deduce its electronic structure. For example,

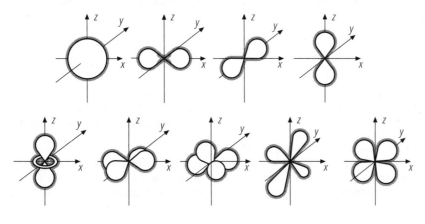

orbital The shapes of atomic orbitals. An atomic orbital is a picture of the 'electron cloud' that surrounds the nucleus of an atom. There are four basic shapes for atomic orbitals: spherical, dumbbell, clover-leaf, and complex (shown at bottom left).

Valency Shell

group number	I	II	III	IV	V	VI	VII
element	Na	Mg	Al	Si	P	S	Cl
atomic number	11	12	13	14	15	16	17
electron arrangement	2.8.1	2.8.2	2.8.3	2.8.4	2.8.5	2.8.6	2.8.7
valencies	1	2	3	4(2)	5(3)	6(2)	7(1)

sodium has atomic number 11 ($Z = 11$) and its electronic arrangement (configuration) is two electrons in the first energy level, eight electrons in the second energy level and one electron in the third energy level – generally written as 2.8.1. Similarly for sulphur ($Z = 16$), the electron arrangement will be 2.8.6. The electronic structure dictates whether two elements will combine by ionic or covalent bonding or not at all.

chemical reactions Physical chemistry is interested in the mechanics of how chemical change takes place. Chemical equations show the reactants and products of a chemical reaction by using chemical symbols and formulae.

State symbols and the energy symbol (ΔH) can be used to show whether reactants and products are solids, liquids, or gases, and whether energy has been released or absorbed during the reaction. Elements, compounds, and ions may react with each other in many different ways.

addition reactions: two or more compounds react together to form one compound. For example hydrogen chloride reacts with ethene to give chloroethane:

$$CH_2=CH_2 + HCl \rightarrow CH_3CH_2Cl$$

chain reactions produce very fast, exothermic reactions, as in the formation of flames and explosions.

displacement reactions: a less reactive element is replaced in a compound by a more reactive one. For example, the addition of powdered zinc to a solution of copper (II) sulphate displaces copper metal, which can be detected by its characteristic colour.

endothermic reactions: there is a physical or chemical change where energy is absorbed by the reactants from the surroundings. The energy absorbed is represented by the symbol $+\Delta H$. Photosynthesis is an example.

exothermic reactions: chemical reactions during which heat is given out. (ΔH is negative)

heterogeneous reactions: there is an interface between the different components or reactants. Examples of heterogeneous reactions are those between a gas and a solid or between two immiscible liquids.

homogeneous reactions: there is no interface between the components. The term applies to all reactions where only gases are involved or where all the components are in solution.

photochemical reactions: light is produced or light initiates the reaction. Light can initiate reactions by exciting atoms or molecules and making them more reactive: the light energy becomes converted to chemical energy.

redox reactions: one reactant is reduced and the other reactant oxidized. The reaction can only occur if both reactants are present and each changes simultaneously. For example, hydrogen reduces copper (II) oxide to copper while it is itself oxidized to water:

$$CuO + H_2 \rightarrow Cu + H_2O$$

reversible reactions: proceed in both directions at the same time, as the product decomposes back into reactants as it is being produced. Such reactions do not run to completion, provided that no substance leaves the system. The manufacture of ammonia from hydrogen and nitrogen is an example:

$$N_2 + 3H_2 \rightarrow 2NH_3$$

The term is also applied to those reactions that can be made to go in the opposite direction by changing the conditions, but these run to completion because some of the substances escape from the reaction. An example is the decomposition of calcium hydrogen-carbonate on heating:

$$Ca(HCO_3)_2 \rightarrow CaCO_3 + CO_2 + H_2O$$

substitution reactions: one atom or functional group in an organic molecule is replaced by another.

catalysts A catalyst is a substance that alters the speed of, or makes possible, a chemical or biochemical reaction but remains unchanged at the end of the reaction. Enzymes are natural biochemical catalysts. In practice most catalysts are used to speed up reactions.

electrolysis

If an electric current is passed through a solution or molten salt (the electrolyte), ions will migrate to the electrodes: positive ions (cations) to the negative electrode (cathode) and negative ions (anions) to the positive electrode (anode).

During electrolysis, the ions react with the electrode, either receiving or giving up electrons (see oxidation and reduction). The resultant atoms may be liberated as a gas, or deposited as a solid on the electrode, in amounts that are proportional to the amount of current passed, as discovered by English chemist Michael Faraday. For instance, when acidified water is electrolysed, hydrogen ions (H^+) at the cathode receive electrons to form hydrogen gas; hydroxide ions (OH^-) at the anode give up electrons to form oxygen gas and water.

One application of electrolysis is **electroplating**, in

The Transuranic Elements

A transuranic element is a chemical element with an atomic number of 93 or more – that is, with a greater number of protons in the nucleus than uranium. All transuranic elements are radioactive.
(– = not applicable.)

Atomic number	Name	Symbol	Year discovered	Source of first preparation identified	Isotope	Half-life of first isotope identified
Actinide Series						
93	neptunium	Np	1940	irradiation of uranium-238 with neutrons	Np-239	2.35 days
94	plutonium	Pu	1941	bombardment of uranium-238 with deuterons	Pu-238	86.4 years
95	americium	Am	1944	irradiation of plutonium-239 with neutrons	Am-241	458 years
96	curium	Cm	1944	bombardment of plutonium-239 with helium nuclei	Cm-242	162.5 days
97	berkelium	Bk	1949	bombardment of americium-241 with helium nuclei	Bk-243	4.5 h
98	californium	Cf	1950	bombardment of curium-242 with helium nuclei	Cf-245	44 min
99	einsteinium	Es	1952	irradiation of uranium-238 with neutrons in first thermonuclear explosion	Es-253	20 days
100	fermium	Fm	1953	irradiation of uranium-238 with neutrons in first thermonuclear explosion	Fm-235	20 h
101	mendelevium	Md	1955	bombardment of einsteinium-253 with helium nuclei	Md-256	76 min
102	nobelium	No	1958	bombardment of curium-246 with carbon nuclei	No-255	2.3 sec
103	lawrencium	Lr	1961	bombardment of californium-252 with boron nuclei	Lr-257	4.3 sec
Transactinide Elements						
104	rutherfordium	Rf	1969	bombardment of californium-249 with carbon-12 nuclei	Db-257	3.4 sec
105	dubnium	Db	1970	bombardment of californium-249 with nitrogen-15 nuclei	Unp-260	1.6 sec
106	seaborgium	Sg	1974	bombardment of californium-249 with oxygen-18 nuclei	Rf-263	0.9 sec
107	bohrium	Bh	1977	bombardment of bismuth-209 with nuclei of chromium-54	Uns	102 millisec
108	hassium	Hs	1984	bombardment of lead-208 with nuclei of iron-58	Uno-265	1.8 millisec
109	meitnerium	Mt	1982	bombardment of bismuth-209 with nuclei of iron-58	Une	3.4 millisec
110	ununnilium[1]	Uun	1994	bombardment of lead nuclei with nickel nuclei	–	–
111	unununium[1]	Uuu	1994	bombardment of bismuth-209 with nickel nuclei	–	–

[1] Temporary names as proposed by the International Union for Pure and Applied Chemistry.

which a solution of a salt, such as silver nitrate ($AgNO_3$), is used and the object to be plated acts as the negative electrode, thus attracting silver ions (Ag^+). Electrolysis is used in many industrial processes, such as coating metals for vehicles and ships, and refining bauxite into aluminium; it also forms the basis of a number of electrochemical analytical techniques, such as polarography.

radioactivity

The nuclei of many large atoms may disintegrate spontaneously, emitting energy as alpha particles (helium nuclei), beta particles (electrons), or gamma radiation. Many of the elements with high atomic numbers exist as mixtures of **isotopes**, atoms that have the same number of protons in the nucleus but different numbers of neutrons. The average time required for the radioactivity of a sample to drop to half of its original value is known as the **half-life** and is a measure of the stability of that isotope.

Elements with atomic numbers 43, 61, and from 84 up, are radioactive. Those elements with atomic numbers above 96 do not occur in nature and are synthesized only, produced in particle accelerators. Elements 110 and 111 were discovered in 1994 and 1995. Element 110 was detected for a millisecond at the GSI heavy-ion cyclotron in Darmstadt, Germany, while lead atoms were bombarded with nickel atoms. It has an atomic mass of 269. Element 111 was later detected at GSI, when bismuth-209 was bombarded with nickel. It has an atomic mass of 272. Element 112 was discovered there in 1996. It has an atomic mass of 277. After firing 5 billion billion zinc ions at a speed of 30,000 kps/18,640 mps at lead, the German scientists created a single atom of 112 that survived for a third of a millisecond.

uranium Uranium is a hard, lustrous, silver-white, malleable and ductile, radioactive, metallic element of the actinide series, symbol U, with atomic number 92 and relative atomic mass 238.029. It is the most abundant radioactive element in the Earth's crust, its decay giving rise to essentially all radioactive elements in nature; its final decay product is the stable element lead. Uranium combines readily with most elements to form compounds that are extremely poisonous. The chief ore is pitchblende, in which the element was discovered by German chemist Martin Klaproth in 1789; he named it after the planet Uranus, which had been discovered in 1781.

Small amounts of certain compounds containing uranium have been used in the ceramics industry to make orange-yellow glazes and as mordants in dyeing; however, this practice was discontinued when the dangerous effects of radiation became known.

Uranium is one of three **fissile** elements (i.e. it will undergo nuclear fission; the others are thorium and plutonium). It was long considered to be the element with the highest atomic number to occur in nature. The isotopes U-238 and U-235 have been used to help determine the age of the Earth.

Uranium-238, which comprises about 99% of all naturally occurring uranium, has a half-life of 4.51 x 10^9 years. Because of its abundance, it is the isotope from which fissile plutonium is produced in breeder nuclear reactors. The fissile isotope U-235 has a half-life of 7.13 x 10^8 years and comprises about 0.7% of naturally occurring uranium; it is used directly as a fuel for nuclear reactors and in the manufacture of nuclear weapons.

Many countries mine uranium; large deposits are found in Canada, the USA, Australia, and South Africa.

What is Uranium?
http://www.uic.com.au/uran.htm
Comprehensive and informative page on uranium, its properties and uses, mainly in nuclear reactors and weapons, provided by the Uranium Information Council.

---

analytical chemistry

Analytical chemistry is the branch of chemistry that deals with the determination of the chemical composition of substances. **Qualitative analysis** determines the identities of the substances in a given sample; **quantitative analysis** determines how much of a particular substance is present.

Simple qualitative techniques exploit the specific, easily observable properties of elements or compounds – for example, the flame test makes use of the different flame colours produced by metal cations when their compounds are held in a hot flame. More sophisticated methods, such as those of spectroscopy, are required where substances are present in very low concentrations or where several substances have similar properties.

Most quantitative analyses involve initial stages in which the substance to be measured is extracted from the test sample, and purified. The final analytical stages (or 'finishes') may involve measurement of the substance's mass (gravimetry) or volume (volumetry, titrimetry), or a number of techniques initially developed for qualitative analysis, such as fluorescence and absorption spectroscopy, chromatography, electrophoresis, and polarography. Many modern methods enable quantification by means of a detecting device that is integrated into the extraction procedure (as in gas–liquid chromatography).

spectroscopy

Spectroscopy is the study of spectra associated with atoms or molecules in solid, liquid, or gaseous phase. Spectroscopy can be used to identify unknown compounds and is an invaluable tool in science, medicine, and industry (for example, in checking the purity of drugs).

Emission spectroscopy is the study of the characteristic series of sharp lines in the spectrum produced when an element is heated. Thus an unknown mixture can be analysed for its component elements. Related is **absorption spectroscopy**, dealing with atoms and molecules as they absorb energy in a characteristic way. Again, dark lines can be used for analysis. More detailed structural information can be obtained using **infrared spectroscopy** (concerned with molecular vibrations) or **nuclear magnetic resonance (NMR) spectroscopy** (concerned with interactions between adjacent atomic nuclei). **Supersonic jet laser beam spectroscopy** enables the isolation and study of clusters in the gas phase. A laser vaporizes a small sample, which is cooled in helium, and ejected into an evacuated chamber. The jet of clusters expands supersonically, cooling the clusters to near absolute zero, and stabilizing them for study in a mass spectrometer.

chromatography

Chromatography (from the Greek *chromos* 'colour') is a technique for separating or analysing a mixture of gases, liquids, or dissolved substances. This is brought about by means of two immiscible substances, one of which (**the mobile phase**) transports the sample mixture through the other (**the stationary phase**). The mobile phase may be a gas or a liquid; the stationary phase may be a liquid or a solid, and may be in a column, on paper, or in a thin layer on a glass or plastic support. The components of the mixture are absorbed or impeded by the stationary phase to different extents and therefore become separated. The technique is used for both qualitative and quantitive analyses in biology and chemistry.

In **paper chromatography**, the mixture separates because the components have differing solubilities in the solvent flowing through the paper and in the chemically bound water of the paper.

In **thin-layer chromatography**, a wafer-thin layer of adsorbent medium on a glass plate replaces the filter paper. The mixture separates because of the differing solubilities of the components in the solvent flowing up the solid layer, and their differing tendencies to stick to the solid (adsorption). The same principles apply in **column chromatography**.

In **gas–liquid chromatography**, a gaseous mixture is passed into a long, coiled tube (enclosed in an oven) filled with an inert powder coated in a liquid. A carrier gas flows through the tube. As the mixture proceeds along the tube it separates as the components dissolve in the liquid to differing extents or stay as a gas. A detector locates the different components as they emerge from the tube. The technique is very powerful, allowing tiny quantities of substances (fractions of parts per million) to be separated and analysed.

Preparative chromatography is carried out on a large scale for the purification and collection of one or more of a mixture's constituents; for example, in the recovery of protein from abattoir wastes.

Analytical chromatography is carried out on far smaller quantities, often as little as one microgram (one-millionth of a gram), in order to identify and quantify the component parts of a mixture. It is used to determine the identities and amounts of amino acids in a protein, and the alcohol content of blood and urine samples. The technique was first used in the separation of coloured mixtures into their component pigments.

crystallography

Crystallography is the scientific study of crystals. In 1912 it was found that the shape and size of the repeating atomic patterns (unit cells) in a crystal could be determined by passing X-rays through a sample. This method, known as **X-ray diffraction**, opened up an entirely new way of 'seeing' atoms. It has been found that many substances have a unit cell that exhibits all the symmetry of the whole crystal; in table salt (sodium chloride, $NaCl$), for instance, the unit cell is an exact cube.

applications of chemistry

There can be few aspects of the modern world that have remained unaffected by the discoveries of chemists and the industrial and commercial application of these discoveries. Medical practice has been revolutionized in the past century by the development of modern pharmaceutical drugs, allowing effective treatment of an ever wider set of ailments. The petrochemical industry has come up with countless new materials with properties suited to specialized applications, as well as fertilizers and pesticides that have radically changed the way that food is produced, preserved, and presented. The ability to produce highly complex electrical circuitry on silicon chips, and the development of miniature electrical power sources have led to the computer revolution that has had a huge impact on many aspects of design, production, finance, and entertainment.

A few applications of chemistry are outlined below.

the battery

An **electrical battery** or **electrical cell** is a device in which chemical energy is converted into electrical energy (the popular name 'battery' actually refers to a collection of cells in one unit). The reactive chemicals of a **primary cell** cannot be replenished, whereas **secondary cells** – such as storage batteries – are rechargeable: their chemical reactions can be reversed and the original condition restored by applying an electric current. Primary-cell batteries are an extremely uneconomical form of energy, since they produce only 2% of the power used in their manufacture. It is dangerous to attempt to recharge a primary cell.

Each cell contains two conducting electrodes immersed in an electrolyte, in a container. A spontaneous chemical reaction within the cell generates a negative charge (excess of electrons) on one electrode, and a positive charge (deficiency of electrons) on the other. The accumulation of these equal but opposite charges prevents the reaction from continuing unless an outer connection (external circuit) is made between the electrodes allowing the charges to dissipate.

When this occurs, electrons escape from the cell's negative terminal and are replaced at the positive, causing a current to flow. After prolonged use, the cell will become flat (ceases to supply current). The first cell was made by Italian physicist Alessandro Volta in 1800. Types of primary cells include the Daniell, Lalande, Leclanché, and so-called 'dry' cells; secondary cells include the Planté, Faure, and Edison. Newer types include the Mallory (mercury depolarizer), which has a very stable discharge curve and can be made in very small units (for example, for hearing aids), and the Venner accumulator, which can be made substantially solid for some purposes. Rechargeable nickel–cadmium dry cells are available for household use.

The reactions that take place in a simple cell depend on the fact that some metals are more

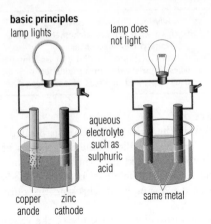

basic principles

lamp lights

lamp does not light

aqueous electrolyte such as sulphuric acid

copper anode   zinc cathode

same metal

a simple cell

electron flow

zinc rod

salt bridge (KCl)

copper rod

porous plugs

zinc salt solution

copper salt solution

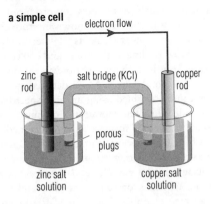

cell, electrical When electrical energy is produced from chemical energy using two metals acting as electrodes in an aqueous solution, it is sometimes known as a galvanic cell or voltaic cell. Here the two metals copper (+) and zinc (–) are immersed in dilute sulphuric acid, which acts as an electrolyte. If a light bulb is connected between the two, an electric current will flow with bubbles of gas being deposited on the electrodes in a process known as polarization.

reactive than others. If two different metals are joined by an electrolyte and a wire, the more reactive metal loses electrons to form ions. The ions pass into solution in the electrolyte, while the electrons flow down the wire to the less reactive metal. At the less reactive metal the electrons are taken up by the positive ions in the electrolyte, which completes the circuit. If the two metals are zinc and copper and the electrolyte is dilute sulphuric acid, the following cell reactions occur. The zinc atoms dissolve as they lose electrons (oxidation):

$$Zn - 2e^- \rightarrow Zn^{2+}$$

The two electrons travel down the wire and are taken up by the hydrogen ions in the electrolyte (reduction):

$$2H^+ + 2e^- \rightarrow H_2$$

The overall cell reaction is obtained by combining these two reactions; the zinc rod slowly dissolves and bubbles of hydrogen appear at the copper rod:

$$Zn + 2H^+ \rightarrow Zn^{2+} + H_2$$

If each rod is immersed in an electrolyte containing ions of that metal, and the two electrolytes are joined by a salt bridge, metallic copper deposits on the copper rod as the zinc rod dissolves in a redox reaction, just as if zinc had been added to a copper salt solution:

$$Zn - 2e^- \rightarrow Zn^{2+}$$

$$Cu^{2+} + 2e^- (Cu \ Zn(s) +$$
$$Cu^{2+}(aq) \rightarrow Zn^{2+} (aq) + Cu(s)$$

adhesives

Natural adhesives (glues) include gelatin in its crude industrial form (made from bones, hide fragments, and fish offal) and vegetable gums. Synthetic adhesives include thermoplastic and thermosetting resins, which are often stronger than the substances they join; mixtures of epoxy resin and hardener that set by chemical reaction; and elastomeric (stretching) adhesives for flexible joints. Superglues are fast-setting adhesives used in very small quantities.

natural water-based adhesives Typical natural substances used for water-based adhesives are starch, casein, tree exudates, skin, and bones.

starch, usually provided from maize, potato, tapioca, or sago, is extracted from the vegetable matter by disintegration and extraction with cold water. Dextrin is prepared from starch by roasting it in the presence of acid or by acid hydrolysis. Its adhesive properties vary widely with the source of the starch and the degree of hydrolysis.

casein is prepared from milk by precipitating its protein with acid or rennet. After further purification the casein is dissolved in a solution of alkali or urea to form a strong adhesive.

Among **tree exudates**, natural rubber latex is obtained from incisions in the bark of *Hevea brasiliensis*. Gum arabic (acacia) and gum tragacanth are solid exudates from *Acacia leguminosae* and *Astragalus leguminosae* respectively. These natural substances (except latex) are brittle when dry and support mould growth when wet. Adhesives are formulated from them by the additions of water, plasticizers, fungicides, tack, and wetting agents. Their adhesive action is due to the formation of physicochemical bonds and penetration to give mechanized keying. Setting and hardening of

this type of adhesive depends on loss of water; therefore they are mainly used for joining porous substrates such as paper, board, and wood. As a group these natural adhesives are less waterproof and strong, but cheaper than the synthetic resin-based adhesives.

natural nonwater-based adhesives Bitumen, derived from asphalt; **shellac**, produced by parasitic tree insects; and **resin**, an exudate from pine trees, are examples of naturally occurring resins used both as hot melt adhesives and as spirit-based cements.

The marine bacterium *Shewanell colwellii* secretes a natural glue, PAVE (polysaccharide adhesive viscous exopolymer), in large quantities. Because PAVE can be used in wet conditions and is resistant to sea water, it has been developed commercially since 1994 as a sealant for ships' hulls.

synthetic adhesives Many modern adhesives are based on the increasing number of synthetic resins available. Thanks to the chemists' closer control over these polymers, a wider variety of type and nomenclature of adhesives has evolved. Definitions, common uses, and the names of some resins used in these adhesives are listed below.

Solution adhesives are resins dissolved in a volatile organic solvent, used for bonding porous materials. They include natural and synthetic rubbers, nitrocellulose, polyvinylacetate, and polymethylmethacrylate.

Emulsion adhesives are resins dispersed in an aqueous base, used for bonding porous materials. Natural and synthetic rubbers, polyvinylacetate, and polymethylmethacrylate are used.

Contact adhesives are emulsions or solutions formulated to bond impervious materials. Both faces are covered with adhesive and allowed to dry before being brought into contact. They are used for bonding plastics, sticking rubber shoe soles, and on self-seal envelopes.

Pressure-sensitive adhesives are used on tapes or sheet material, sometimes with a nonstick backing paper. Applied pressure forms a bond. Some adhesives are formulated for ease and cleanliness of removal (permanently tacky adhesive), and some to give a permanent bond. The materials used include modified natural and synthetic rubbers and polyisobutylene.

Thermoplastic adhesives are solventless adhesives, softened by the application of heat before bonding. They may be remelted after bonding and are used for high-speed packaging, labelling, and unsewn bookbinding. Polyamides and polyvinylacetate and its copolymers are used.

Thermosetting adhesives are solventless, and are cured by heat to form a bond that, once cured, can not be resoftened by heat. They are used for exterior plywood and for bonding brake linings to shoes. They are usually made from epoxy resins.

Two-part or chemical-cure adhesives consist of resin and hardener, which are mixed together shortly before use and set by chemical action without the necessity for the application of heat. They are used for bonding aluminium alloys in the aircraft industry and varied domestic applications. Resins used include epoxypolyamide and resorcinol.

Structural adhesives are adhesives of high strength, toughness, and creep resistance, used for bonding load-bearing members. They are mainly confined to chemical-cure and thermosetting adhesives.

plastics

Plastics are stable synthetic materials that are fluid at some stage in their manufacture, when they can be shaped, and that later set to rigid or semirigid solids. Plastics today are chiefly derived from petroleum. Most are polymers, made up of long chains of identical molecules.

environmental influence Since plastics have afforded an economical replacement for ivory in the manufacture of piano keys and billiard balls, the industrial chemist may well have been responsible for the survival of the elephant.

Most plastics cannot be broken down by microorganisms, so cannot easily be disposed of. Incineration leads to the release of toxic fumes, unless carried out at very high temperatures.

Processed by extrusion, injection-moulding, vacuum-forming, and compression, plastics emerge in consistencies ranging from hard and inflexible to soft and rubbery. They replace an increasing number of natural substances, being lightweight, easy to clean, durable, and capable of being rendered very strong – for example, by the addition of carbon fibres – for building aircraft and other engineering projects.

thermoplastics Thermoplastics soften when warmed, then reharden as they cool. Examples of thermoplastics include polystyrene, a clear plastic used in kitchen utensils or (when expanded into a 'foam' by gas injection) in insulation and ceiling tiles; polyethylene (polythene), used for containers and wrapping; and polyvinyl chloride (PVC), used for drainpipes, floor tiles, audio discs, shoes, and handbags.

thermosets

Thermosets remain rigid once set, and do not soften when warmed. They include Bakelite, used in electrical insulation and telephone receivers; epoxy resins, used in paints and varnishes, to laminate wood, and

as adhesives; polyesters, used in synthetic textile fibres and, with fibreglass reinforcement, in car bodies and boat hulls; and polyurethane, prepared in liquid form as a paint or varnish, and in foam form for upholstery and in lining materials (where it may be a fire hazard). One group of plastics, the silicones, are chemically inert, have good electrical properties, and repel water. Silicones find use in silicone rubber, paints, electrical insulation materials, laminates, waterproofing for walls, stain-resistant textiles, and cosmetics.

polyamides Polyamides are widely used for the production of film, sheet, and injection-moulded articles. Nylon, the first polyamide, was synthesized in 1934 by Wallace Carothers at the du Pont laboratories in the USA and was intended to have many of the properties possessed of natural silk. Although it does have other applications, nylon is known principally for its applications in the textile field. Nylon yarn, once it has been stretched during the filament-forming process, has a combination of properties unique among textile fibres. One of the most notable is remarkable tensile strength, combined with lightness in weight and a high degree of resilience.

shape-memory polymers Shape-memory polymers are plastics that can be crumpled or flattened and will resume their original shape when heated. They include *trans*-polyisoprene and polynorbornene. The initial shape is determined by heating the polymer to over 35°C/95°F and pouring it into a metal mould. The shape can be altered with boiling water and the substance solidifies again when its temperature falls below 35°C/95°F.

biodegradable plastics Biodegradable plastics are increasingly in demand: Biopol was developed in 1990. Soil microorganisms are used to build the plastic in their cells from carbon dioxide and water (it constitutes 80% of their cell tissue). The unused parts of the microorganism are dissolved away by heating in water. The discarded plastic can be placed in landfill sites where it breaks back down into carbon dioxide and water. It costs three to five times as much as ordinary plastics to produce. Another plastic digested by soil microorganisms is polyhydroxybutyrate (PHB), which is made from sugar.

celluloid Celluloid is a transparent or translucent, highly flammable, plastic material (a thermoplastic) made from cellulose nitrate and camphor. It was once used for toilet articles, novelties, and photographic film, but has now been replaced by the nonflammable substance cellulose acetate.

intelligent gels Intelligent gels are polymer gels that respond 'intelligently' to their environments. Most gels shrink or swell in fairly strict proportion to the quality of solvent added to them, but some undergo a sudden change in dimension in response to relatively small fluctuations. This rapid response could make gels suitable for use as 'muscle' for robots, or as valves in engineering. They are also likely to have medical applications; for example, in long-term drug administration. The gel could sense conditions inside the body and vary drug delivery rate to maintain suitable levels in the bloodstream.

anodizing

The natural resistance to corrosion of a metal, such as aluminium, may be increased by building up a protective oxide layer on the surface. Anodizing increases the thickness of this film and thus the corrosion protection.

It is so called because the metal becomes the anode in an electrolytic bath containing a solution of, for example, sulphuric or chromic acid as the electrolyte. During electrolysis oxygen is produced at the anode, where it combines with the metal to form an oxide film.

dyes

A dye is a substance that imparts colour to a fabric and is resistant to washing. There are three main types of dye: **direct dyes** combine with the material of the fabric, yielding a coloured compound; **indirect dyes** require the presence of another substance (a mordant), with which the fabric must first be treated; **vat dyes** are colourless soluble substances that on exposure to air yield an insoluble coloured compound.

Naturally occurring dyes include indigo, madder (alizarin), logwood, and cochineal, but industrial dyes (introduced in the 19th century) are usually synthetic: acid green was developed in 1835 and bright purple in 1856. Synthetic dyes now allow an almost infinite range of colours to be applied to a wide variety of materials.

Industrial dyes include azo dyestuffs, acridine, anthracene, and aniline.

Chemistry Chronology

c. 1500 BC The liquid metal mercury is known to the Egyptians who place it in a tomb about this date. It is also known to the Chinese and Hindus.

c. 1000 BC The following elements are known by this date: carbon, copper, gold, iron, lead, mercury, silver, sulphur, tin, zinc.

c. 230 BC Copper-lined pottery jars, with asphalt plugs, containing metal rods – the first electric battery – are used in Baghdad to coat objects with thin layers of gold or silver – the first example of electroplating.

c. 100 BC The Romans produce mercury by heating the sulphide mineral cinnabar and condensing the vapours.

c. 100 Mary the Jewess, an alchemist, succeeds in her laboratory inventions with metals and lays the foundation for later work in chemistry. She creates the world's first distillation device, a double boiler, a way to capture vapours of metals, and a metal alloy called Mary's Black.

297 The tomb of a Chinese military commander of this date contains metal belt ornaments made of aluminium, not isolated by Western scientists until 1827.

742 The most famous alchemist of the period, Jabir ibn-Hayyan of Kufa (Geber in Iraq) practises as a physician at Kufa, Persia. He becomes court physician to the caliph Harun ar-Rashid. He is said to have been the first person to manufacture mineral acids (nitric acid, etc.).

1044 The Chinese text *Wu Ching Tsung Yao* is written, including a recipe for black powder which uses a mix of saltpetre (potassium nitrate), charcoal, and sulphur to produce the earliest form of gunpowder.

1126 Gunpowder is first recorded in military use during the siege of Kaifeng, China, capital of the Chinese Sung dynasty. It is used by troops of the rival Jin dynasty.

1249 The English monk and scholar Roger Bacon records the use of explosives, and documents a recipe for gunpowder (possibly his own invention). He writes this dangerous information in coded form.

c. 1313 The German Grey Friar Berthold der Schwarze is traditionally credited with the independent invention of gunpowder. He is also said to have cast the first bronze cannon.

1597 The German scientist and alchemist Andreas Libavius publishes his *Alchymia/Alchemy*, an outline of chemistry as used in medicine at the time, which also describes the metal zinc.

1648 German chemist Johann Glauber creates nitric acid from the reaction of potassium nitrate and sulphuric acid.

1649 Arsenic is first isolated and identified as an element by German pharmacist Johann Schroeder.

1661 Anglo-Irish chemist and physicist Robert Boyle publishes *The Sceptical Chymist*, in which he proposes a corpuscular or atomic theory of matter, introducing the modern concept of chemical elements, and distinguishing alkali and acid properties.

1665 Using vacuum pumps, Boyle proves that air is necessary for candles to burn and for animals to live.

1669 Phosphorus is isolated from urine by German alchemist Hennig Brand. He names it 'light-bearer' because it glows in the dark.

1670 Boyle discovers hydrogen, produced when certain metals react with acid, although he does not identify it as an element.

March 1676 French physicist Edmé Mariotte discovers the relationship between volume and pressure in a fixed mass of gas, independently of Robert Boyle.

1680 Boyle obtains the element phosphorus by evaporating urine and distilling the residue with sand.

1680 Boyle invents the match, striking a sulphur-tipped splinter of wood against phosphorus-coated paper.

1700 French chemist and physician Nicolas Lemery shows that a mixture of hydrogen and air detonates on the application of a spark or flame, although he does not identify hydrogen as a separate element.

1718 French chemist Etienne Geoffroy presents his *tables des rapports* ('tables of affinities') to the French Academy of Sciences – the first systematic record of the chemical reactivity of elements and compounds.

1724 Dutch chemist Herman Boerhaave publishes a pioneer study of organic chemistry.

1730 Zinc smelting is first practised in England, by William Champion of Bristol.

1750 British chemist William Brownrigg first identifies the metal platinum as a separate and distinct element.

1751 Swedish mineralogist Axel Cronstedt isolates nickel from its ore niccolite. He names the pure material *Kupfernickel*, or 'devil's copper'.

1755 Scottish chemist Joseph Black discovers carbon dioxide, which he calls 'fixed air'.

1766 English natural philosopher Henry Cavendish discovers hydrogen and delivers papers to the Royal Society, London, England, on the chemistry of gases.

1772 Scottish chemist Daniel Rutherford discovers nitrogen.

1772 Swedish chemist Karl Wilhelm Scheele advances the concept of oxygen two years before the English chemist Joseph Priestley.

1 Aug 1774 English clergyman, chemist, and natural philosopher Joseph Priestley discovers oxygen, which he calls 'dephlogisticated air'.

1774 French chemist Antoine-Laurent Lavoisier demonstrates the conservation of mass in chemical reactions.

1774 Scheele discovers chlorine and baryta (barium oxide).

1777 Lavoisier shows that air is made up of a mixture of gases, and that one of them (oxygen) is the substance necessary for combustion and rusting to take place. He also assigns the name 'oxygen' to Joseph Priestley's dephlogisticated air.

1777 Scheele discovers that silver nitrate, when exposed to light, results in a blackening effect, an important discovery for the development of photography.

1787 French physicist Jacques Charles discovers Charles's law, stating that the volume of a given mass of gas at constant pressure is directly proportional to its absolute temperature (temperature in kelvin).

1789 German chemist Martin Heinrich Klaproth discovers uranium and zirconium.

9 April 1800 English chemist Humphry Davy details the effects of nitrous oxide, later used as the first anaesthetic.

1800 Italian physicist Alessandro Volta invents the voltaic pile made of discs of silver and zinc – the first battery.

1801 English chemist and physicist John Dalton formulates the law of partial pressure in gases – Dalton's Law – that states that each component of a gas mixture produces the same pressure as if it occupied the container alone.

1802 French chemist and physicist Joseph-Louis Gay-Lussac demonstrates that all gases expand by the same fraction of their volume when subjected to the same temperature increase; it permits the establishment of a new temperature scale.

1803 Dalton formulates his atomic theory of matter: that all elements are made of minute indestructible particles, called atoms, that are all identical.

1803 Dalton devises a system of chemical symbols and arranges the relative weights of atoms in a table.

1804 Dalton proposes the law of multiple proportions that states that when two elements combine to form more than one compound the weights of one element combine with a fixed weight of the other in a ratio of small whole numbers. The law provides strong support for his atomic theory.

1805 Gay-Lussac determines the relative proportions of hydrogen and oxygen in water by measuring the proportions of the gases that combine.

31 Dec 1808 French chemist Joseph-Louis Gay-Lussac, in *The Combination of Gases*, announces that gases combine chemically in simple proportions of volumes, and that the contraction in volume observed when they combine is a simple relation to the original volume of the gases – Gay-Lussac's Law.

1808 Davy isolates the alkaline-earth metals magnesium, calcium, strontium, and barium.

1811 Italian physicist Amedeo Avogadro hypothesizes that equal volumes of gases at the same temperature and pressure contain equal numbers of molecules.

1811 Swedish chemist Jöns Jakob Berzelius introduces the modern system of chemical symbols.

1819 French physicists Pierre-Louis Dulong and Alexis-Thérèse Petit formulate the Dulong–Petit Law that states that the specific heat of an element, times its atomic weight, is a constant. It proves useful in establishing atomic weights.

1825 Danish scientist Hans Christian Oersted isolates aluminium in powdered form.

1827 Robert Brown observed that very fine pollen grains suspended in water move about in a continuously agitated manner. This became known as Brownian movement.

1828 German chemist Friedrich Wöhler synthesizes urea from ammonium cyanate. It is the first synthesis of an organic substance from an inorganic compound and signals the beginning of organic chemistry.

1829 Scottish chemist Thomas Graham formulates the law named after him on the diffusion rates of gases. He also devises a dialysis method of separating colloids from crystalloids and thereby establishes the science of colloidal chemistry.

1830 French chemist Jean-Baptiste-André Dumas discovers a method of burning organic compounds to determine their nitrogen content.

1831 Peregrine Phillips develops the contact process for producing sulphuric acid.

1836 English chemist John Frederic Daniell invents the Daniell cell, a battery that generates a steady current during continuous operation – an improvement over the voltaic cell which loses power over time.

1837 German chemist Karl Friedrich Mohr enunciates the theory of conservation of energy.

1845 German chemist Hermann Kolbe synthesizes acetic acid from carbon disulphide – the first organic compound to be synthesized from inorganic materials.

1858	English chemists W H Perkin and B F Duppa synthesize glycine, the first amino acid to be manufactured.
1858	German chemist Friedrich Kekulé shows that carbon atoms can link together to form long chains – the basis of organic molecules.
1858	Italian chemist Stanislao Cannizzaro differentiates the atomic and molecular weight of an element. It becomes generally accepted in 1860.
1861	Belgian chemist Ernest Solvay patents a method for the economic production of sodium carbonate (washing soda) from sodium chloride, ammonia, and carbon dioxide. Used to make paper, glass, and bleach, and to treat water and refine petroleum, it is a key development in the Industrial Revolution. The first production plant is established in 1863.
1863	English chemist John Newlands develops the first periodic system. It is later adapted by Russian chemist Dmitry Ivanovich Mendeleyev.
1865	Kekulé suggests that the benzene molecule has a six-carbon ring structure. His theory refines current knowledge of organic chemistry.
1867	Swedish chemist Alfred Nobel patents dynamite. It consists of 75% nitroglycerin and 25% of an absorbent material known as ghur which makes the explosive safe and easy to handle.
1869	Based on the fact that the elements exhibit recurring patterns of properties when placed in order of increasing atomic weight, Russian chemist Dmitry Ivanovich Mendeleyev develops the periodic classification of the elements. He leaves gaps for elements yet to be discovered.
1869	US scientist John Wesley Hyatt, in an effort to find a substitute for the ivory in billiard balls, invents (independently of Alexander Parkes) celluloid. The first artificial plastic, it can be produced cheaply in a variety of colours, is resistant to water, oil, and weak acids, and quickly finds use in making such things as combs, toys, and false teeth.
1871	Austrian physicist Ludwig Boltzmann describes the general statistical distribution of energies among the molecules in a gas.
1876	US physicist Josiah Willard Gibbs publishes 'On the Equilibrium of Heterogeneous Substances', which lays the theoretical foundation of physical chemistry.
1881	Dutch physicist Johannes van der Waals develops a version of the gas law, now known as the van der Waals equation, which takes into account the size and attraction of atoms and molecules.
1883	English physicist and chemist Joseph Wilson Swan patents a method of creating nitrocellulose (cellulose nitrate) fibre by squeezing it though small holes. It becomes a basic process in the artificial textile industry.
May 1884	Swedish chemist Svante August Arrhenius

suggests that electrolytes (solutions or molten compounds that conduct electricity) disassociate into ions (atoms or groups of atoms that carry an electrical charge).

1885	French horticulturist Pierre-Marie-Alexis Millardet develops Bordeaux Mixture, a blend of copper sulphate and hydrated lime. The first successful fungicide, it rapidly achieves worldwide usage.
1886	Arrhenius introduces the idea that acids are substances that disassociate in water to yield hydrogen ions ($H+$) and that bases are substances that disassociate to yield hydroxide ions ($OH-$), thus explaining the properties of acids and bases through their ability to yield ions.
1886	US chemist Charles Martin Hall and French chemist Paul-Louis-Toussaint Héroult, working independently, each develop a method for the production of aluminium by the electrolysis of aluminium oxide. The process reduces the price of the metal dramatically and brings it into widespread use.
1898	French chemists Pierre and Marie Curie discover the radioactive elements radium and polonium. Radium is discovered in pitchblende and is the first element to be discovered radiochemically.
1901	Dutch chemist Jacobus van't Hoff receives the first Nobel Prize for Chemistry for his discovery of the laws of chemical dynamics and osmotic pressure.
1901	German engineer Carl von Linde separates liquid oxygen from liquid air. It leads to the widespread use of oxygen in industry.
1902	German chemists Emil Fischer and Franz Hofmeister discover that proteins are polypeptides consisting of amino acids.
1903	Scottish chemist William Ramsay shows that helium is produced during the radioactive decay of radium – an important discovery for the understanding of nuclear reactions.
1906	English physicist Frederick Soddy discovers that ionium and radiothorium, are chemically indistinguishable variants of thorium but have different radioactive properties. He later calls them isotopes.
1907	German chemist Emil Fischer, describes the synthesis of amino acid chains in proteins.
1908	Belgian-born US chemist Leo H Baekeland invents the plastic Bakelite: its insulating and malleable properties, combined with the fact that it does not bend when heated, ensures it has many uses.
1909	Danish biochemist Søren Sørensen devises the pH scale for measuring acidity and alkalinity.
1911–13	Ernest Rutherford and Henry Moseley showed that practically the whole mass of any atom is concentrated in an extremely small central nucleus bearing a positive electrical charge. Around the nucleus move an equal number of electrons at a relatively great distance. Mosely draws up the periodic table, based on atomic numbers, that is in use today.

1913   Danish scientist Niels Bohr proposes that the electrons move in orbits around the nucleuslike planets round the Sun, and suggests how atoms might emit or absorb light.

1916   US chemist Gilbert Lewis states a new valency theory, in which electrons are shared between atoms.

1923   Danish chemist Johannes Brønsted and British chemist Thomas Martin Lowry simultaneously and independently introduce the idea that an acid tends to lose a proton and a base tends to gain a proton.

1923   Dutch chemist Peter Debye and German chemist Erich Hückel demonstrate that the disassociation of positive and negative ions of salts in solution is complete and not partial.

1925   Austrian physicist Wolfgang Pauli discovers the exclusion principle.

1926   Debye proposes a method of obtaining temperatures a fraction of a degree above absolute zero by removing their magnetic field. Canadian-born US scientist William Giauque independently proposes the same idea the following year.

1928   Polyvinyl chloride (PVC) is developed, simultaneously, by the US companies Carbide and Carbon Corporation and Du Pont and the German firm I G Farben.

1932   British physicist James Chadwick bombards beryllium with alpha-particles and produces neutral particles, which he called neutrons.

1934   French physicists Frédéric and Irène Joliot-Curie bombard boron, aluminium, and magnesium with alpha particles and obtain radioactive isotopes of nitrogen, phosphorus, and aluminium – elements that are not normally radioactive. They are the first radioactive elements to be prepared artificially.

1935   Chemists working for the British company Imperial Chemical Industries (ICI) polymerize ethylene to make polyethylene, the first true plastic.

c. 1936   Catalytic cracking, a chemical process in which long-chain hydrocarbon molecules are broken down into smaller ones, is introduced to produce gasoline from low-grade crude oil by the US Sun Oil Company and Socony-Vacuum Company.

1938   German physicists Lise Meitner, Otto Hahn, and Fritz Strassmann conclude that bombarding uranium atoms with neutrons splits the atom and releases huge amounts of energy by the conversion of some of the mass of the uranium atom into energy.

1938   The Soviet physicist Pyotr Kapitza discovers that liquid helium exhibits superfluidity, the ability to flow over its containment vessel without friction, when cooled below 2.18K/–270.97°C.

1940   US physicist J R Dunning leads a research team that uses a gaseous diffusion technique to isolate uranium-235 from uranium-238. Because uranium-235 readily undergoes fission into two atoms, and in doing so releases large amounts of energy, it is used for fuelling nuclear reactors.

1940   US physicists Edwin McMillan and Philip Abelson synthesize the first transuranic element, neptunium, by bombarding uranium with neutrons at the cyclotron at Berkeley, California.

1944   British chemists Archer J P Martin and Richard L M Synge separate amino acids by using a solvent in a column of silica gel. The beginnings of partition chromatography, the technique leads to further advances in chemical, medical, and biological research.

1946   US physicists Edward Mills Purcell and Felix Bloch independently discover nuclear magnetic resonance, which is used to study the structure of pure metals and composites.

1947   US physicist Willard Libby develops carbon-14 dating.

1962   English chemist Neil Bartlett prepares xenon hexafluoroplatinate, the first compound of an inert gas.

1973   US biochemists Stanley Cohen and Herbert Boyer develop the technique of recombinant DNA (deoxyribonucleic acid). Strands of DNA are cut by restriction enzymes from one species and then inserted into the DNA of another; this marks the beginning of genetic engineering.

1974   Mexican chemist Mario Molina and US chemist F Sherwood Rowland warn that the chlorofluorocarbons (CFCs) used in fridges and as aerosol propellants may be damaging the atmosphere's ozone layer that filters out much of the Sun's ultraviolet radiation.

28 Aug 1976   Indian-born US biochemist Har Gobind Khorana and his colleagues announce the construction of the first artificial gene to function naturally when inserted into a bacterial cell. This is a major breakthrough in genetic engineering.

1977   English biochemist Frederick Sanger describes the full sequence of 5,386 bases in the DNA (deoxyribonucleic acid) of virus phiX174 in Cambridge, England; the first sequencing of an entire genome.

1981   French researchers Claude Michel and Bernard Reveau synthesize some metallic oxides that have excellent conducting properties; the materials prove invaluable in achieving superconductivity at relatively high temperatures.

1985   Harold Kroto and David Walton at the University of Sussex, England, discover a new unusually stable elemental form of solid carbon made up of closed cages of 60 carbon atoms shaped like soccer balls; they call them buckminsterfullerines or 'buckyballs'.

1988	Dutch firm CCA Biochem develops the biodegradable polymer polyactide; capable of being broken down by human metabolism, it is ideal for use in suturing threads, bone platelets, and artificial skin.	Oct 1994	Ununnilium is discovered by researchers at the heavy-ion cyclotron based at Darmstadt, Germany. It lasts for a millisecond.
1990	The British company Imperial Chemical Industries (ICI) develops the first practical biodegradable plastic, Biopal.	Dec 1994	Unununium is discovered by researchers at the heavy-ion cyclotron based in Darmstadt, Germany.
1993	The first pictures of individual atoms, obtained by the use of a scanning tunnelling microscope, are published.	Feb 1996	Element no. 112 is discovered at the GSI heavy-ion research centre, Darmstadt, Germany. A single atom is created, which lasts for a third of a millisecond.

Biographies

Accum, Friedrich Christian (1769–1838) German chemist who introduced illumination by gas in 1815.

Achard, Franz Karl (1753–1821) German chemist who was largely responsible for developing the industrial process by which table sugar (sucrose) is extracted from sugar beet.

Adams, Roger (1889–1971) US organic chemist, known for his painstaking analytical work to determine the composition of naturally occurring substances such as complex vegetable oils and plant alkaloids.

Alder, Kurt (1902–1958) German organic chemist who with Otto Diels developed the diene synthesis in 1928, a fundamental process that has become known as the **Diels–Alder reaction**. It is used in organic chemistry to synthesize cyclic (ring) compounds.

Altman, Sidney (1939–) Canadian-born US biochemist who shared the Nobel Prize for Chemistry in 1989 with Thomas Cech for his research on the catalytic activities of RNA.

Andrews, Thomas (1813–1885) Irish physical chemist, best known for postulating the idea of critical temperature and pressure from his experimental work on the liquefaction of gases.

Anfinsen, Christian Boehmer (1916–) US biochemist who shared the Nobel Prize for Physiology or Medicine in 1972 with Stanford Moore and William Stein for his work on the shape and primary structure of ribonuclease (the enzyme that hydrolyses RNA).

Dissociation of Substances Dissolved in Water by Svante Arrhenius
http://dbhs.wvusd.k12.ca.us/Chem-History/
Arrhenius-dissociation.html
Extract from the above paper Arrhenius discusses the dissociation of certain substances in water, an observation which led to deductions on electrolysis and his Nobel prize in 1903.

Arrhenius, Svante August (1859–1927) Swedish scientist, the founder of physical chemistry. For his study of electrolysis, he received the Nobel Prize for Chemistry in 1903. In 1905 he predicted global warming as a result of carbon dioxide emission from burning fossil fuels.

Baekeland, Leo Hendrik (1863–1944) Belgian-born US chemist. He invented Bakelite, the first commercial plastic, made from formaldehyde (methanal) and phenol. He also made a photographic paper, Velox, which could be developed in artificial light.

Baeyer, Johann Friedrich Wilhelm Adolf von (1835–1917) German organic chemist who synthesized the dye indigo in 1880.

Baker, Henry (1698–1774) English scientist who wrote two popular instructional books on the use of the microscope in natural history, and made observations on the crystallization of salts in 1744.

Bartlett, Neil (1932–) British-born US chemist. In 1962 he prepared the first compound of one of the inert gases, which were previously thought to be incapable of reacting with anything.

Barton, Derek Harold Richard (1918–) English organic chemist who investigated the stereochemistry of natural compounds.

Berg, Paul (1926–) US molecular biologist. In 1972, using gene-splicing techniques developed by others, Berg spliced and combined into a single hybrid the DNA from an animal tumour virus (SV40) and the DNA from a bacterial virus.

Bergius, Friedrich Karl Rudolph (1884–1949) German research chemist who invented processes for converting coal into oil and wood into sugar. He shared a Nobel prize in 1931 with Carl Bosch for his part in inventing and developing high-pressure industrial methods.

Bergstrom, Sune Karl (1916–) Swedish biochemist who shared the Nobel Prize for Physiology or Medicine in 1982 with John Vane and Bengt Samuelsson for the purification of prostaglandins.

Berthelot, Pierre Eugène Marcellin (1827–1907) French chemist and politician who carried out research into dyes and explosives, proving that hydrocarbons and other organic compounds can be synthesized from inorganic materials.

Berthollet, Claude Louis, Count (1748–1822) French chemist who carried out research into dyes and bleaches (introducing the use of chlorine as a bleach) and determined the composition of ammonia.

Berzelius, Jöns Jakob (1779–1848) Swedish chemist. He accurately determined more than 2,000 relative atomic and molecular masses. In 1813–14 he devised the system of chemical symbols and formulae now in use.

Bloch, Konrad (1912–) German-born US chemist whose research concerned cholesterol.

Bosch, Carl (1874–1940) German metallurgist and chemist. He developed the Haber process from a small-scale technique for the production of ammonia into an industrial high-pressure process.

Boyle, Robert (1627–1691) Irish chemist and physicist who published the seminal *The Sceptical Chymist* in 1661. He formulated (**Boyle's law** in 1662. He was a pioneer in the use of experiment and scientific method.

Boyle questioned the alchemical basis of the chemical theory of his day and taught that the proper object of chemistry was to determine the compositions of substances. The term 'analysis' was coined by Boyle and many of the reactions still used in qualitative work were known to him. He introduced certain plant extracts, notably litmus, for the indication of acids and bases. He was also the first chemist to collect a sample of gas.

Bredig, Georg (1868–1944) German physical chemist who devised a method of preparing colloidal solutions in 1898

Brønsted, Johannes Nicolaus (1879–1947) Danish physical chemist whose work in solution chemistry, particularly electrolytes, resulted in a new theory of acids and bases, the theory of proton donors and proton acceptors, published in 1923.

Brown, Herbert Charles (1912–) US inorganic chemist who is noted for his research on boron compounds.

Buchner, Eduard (1860–1917) German chemist who researched the process of fermentation.

Bunsen, Robert Wilhelm (1811–1899) German chemist credited with the invention of the **Bunsen burner**. His name is also given to the carbon–zinc electric cell, which he invented in 1841 for use in arc lamps.

Butenandt, Adolf Friedrich Johann (1903–1995) German biochemist who isolated the first sex hormones (oestrone, androsterone, and progesterone), and determined their structure. He shared the 1939 Nobel Prize for Chemistry with Leopold Ruzicka (1887–1976).

Calvin, Melvin (1911–1997) US chemist who, using radioactive carbon-14 as a tracer, determined the biochemical processes of photosynthesis, in which green plants use chlorophyll to convert carbon dioxide and water into sugar and oxygen.

Cannizzaro, Stanislao (1826–1910) Italian chemist who revived interest in the work of Avogadro that had, in 1811, revealed the difference between atoms and molecules, and so established atomic and molecular weights as the basis of chemical calculations.

Carothers, Wallace Hume (1896–1937) US chemist who carried out research into polymerization and, with Paul Flory, invented nylon.

Carr, Emma Perry (1880–1972) US chemist who in the USA pioneered techniques to synthesize and analyse the structure of complex organic molecules using absorption spectroscopy.

Cavendish, Henry (1731–1810) English physicist and chemist. He discovered hydrogen (which he called 'inflammable air') in 1766, and determined the compositions of water and of nitric acid.

Cavendish demonstrated in 1784 that water is produced when hydrogen burns in air, thus proving that water is a compound and not an element. He also worked on the production of heat and determined the freezing points for many materials, including mercury.

Cech, Thomas (1947–) US biochemist who discovered the catalytic activity of RNA.

Chardonnet, (Louis-Marie) Hilaire Bernigaud, comte de (1839–1924) French chemist who developed artificial silk in 1883, the first artificial fibre.

Chargaff, Erwin (1905–) Czech-born US biochemist, best known for his work on the base composition of deoxyribonucleic acid (DNA).

Chevreul, Michel-Eugène (1786–1889) French chemist who studied the composition of fats and identified a number of fatty acids, including 'margaric acid', which became the basis of margarine.

Claude, Georges (1870–1960) French industrial chemist, responsible for inventing neon signs.

Cleve, Per Teodor (1840–1905) Swedish chemist and geologist who discovered the elements holmium and thulium in 1879.

Corey, Elias James (1928–) US organic chemist who received the Nobel Prize for Chemistry in 1990 for the development of **retrosynthetic analysis**, a method of synthesizing complex substances.

Cornforth, John Warcup (1917–) Australian chemist. Using radioisotopes as markers, he found out how cholesterol is manufactured in the living cell and how enzymes synthesize chemicals that are mirror images of each other (optical isomers).

Coulson, Charles Alfred (1910–1974) English theoretical chemist. He developed a molecular orbital theory, which is an extension of atomic quantum theory.

Cram, Donald James (1919–) US chemist who shared the 1987 Nobel Prize for Chemistry with Jean-Marie Lehn and Charles J Pedersen for their work on molecules with highly selective structure-specific interactions.

Crookes, William (1832–1919) English scientist whose many chemical and physical discoveries include the metallic element thallium in 1861, the radiometer in 1875, and the Crookes high-vacuum tube used in X-ray techniques. Knighted in 1897.

Curie, Marie (1867–1934), born Manya Sklodowska, Polish scientist who, with husband Pierre Curie, discovered in 1898 two new radioactive elements in pitchblende ores: polonium and radium. They isolated the pure elements in 1902. Both scientists refused to take out a patent on their discovery and were jointly awarded the Nobel Prize for Physics in 1903, with Henri Becquerel. Marie Curie was also awarded the Nobel Prize for Chemistry in 1911.

Curie, Pierre (1859–1906) French scientist. He shared the Nobel Prize for Physics in 1903 with his wife Marie Curie and Henri Becquerel. From 1896 the Curies had worked together on radioactivity, discovering two radioactive elements.

Dalton, John (1766–1844) English chemist who proposed the theory of atoms, which he considered to be the smallest parts of matter. He produced the first list of relative atomic masses in 'Absorption of Gases' in 1805 and put forward the law of partial pressures of gases (**Dalton's law**).

Davy, Humphry (1778–1829) English chemist. He discovered, by electrolysis, the metallic elements sodium and potassium in 1807, and calcium, boron, magnesium, strontium, and barium in 1808. In addition, he established that chlorine is an element and proposed that hydrogen is present in all acids. He invented the safety lamp for use in mines where methane was present, enabling miners to work in previously unsafe conditions.

Deisenhofer, Johann (1943–) German chemist who was the first to apply the technique of X-ray crystallography (the use of X-rays to discern atomic structure) to biological molecules.

Dewar, James (1842–1923) Scottish chemist and physicist who invented the vacuum flask.

Diels, Otto Paul Hermann (1876–1954) German chemist. In 1950 he and his former assistant, Kurt Alder, were jointly awarded the Nobel Prize for Chemistry for their research into the synthesis of organic chemical compounds.

Domagk, Gerhard Johannes Paul (1895–1964) German pathologist, discoverer of antibacterial sulphonamide drugs.

Dulong, Pierre Louis (1785–1838) French chemist and physicist. In 1819 he discovered, together with physicist Alexis Petit, the law that now bears their names.

Dumas, Jean Baptiste André (1800–1884) French chemist. He made contributions to organic analysis and synthesis, and to the determination of atomic weights (relative atomic masses) through the measurement of vapour densities.

Eigen, Manfred (1927–) German chemist who worked on extremely rapid chemical reactions (those taking less than 1 millisecond).

Emeléus, Harry Julius (1903–) English chemist. He made wide-ranging investigations in inorganic chemistry, studying particularly nonmetallic elements and their compounds.

Ernst, Richard Robert (1933–) Swiss physical chemist who improved the technique of nuclear magnetic resonance (NMR) spectroscopy.

Eyde, Samuel (1866–1940) Norwegian industrial chemist. He helped to develop a commercial process for the manufacture of nitric acid that made use of comparatively cheap hydroelectricity.

Fajans, Kasimir (1887–1975) Polish-born US chemist. He did pioneering work on radioactivity and isotopes, he also formulated rules that help to explain valence and chemical bonding.

Faraday, Michael (1791–1867) English chemist and physicist. Faraday isolated benzene from gas oils and produced the basic laws of electrolysis in 1834. Faraday's laws of electrolysis established the link between electricity and chemical affinity, one of the most fundamental concepts in science.

Electrical Decomposition by Michael Faraday
http://dbhs.wvusd.k12.ca.us/Chem-History/
Faraday-electrochem.html
Transcript of Faraday's paper in Philosophical *Transactions of the Royal Society, 1834,* in which Faraday describes for the first time the phenomenon of electrolysis.

Fischer, Edmond (1920–) US biochemist who shared the 1992 Nobel Prize for Physiology or Medicine with Edwin Krebs for isolating and describing the action of the enzymes responsible for reversible protein phosphorylation.

Fischer, Emil Hermann (1852–1919) German chemist who produced synthetic sugars and, from these, various enzymes.

Fischer, Ernst Otto (1918–) German inorganic chemist. He showed that transition metals can bond chemically to carbon.

Fischer, Hans (1881–1945) German chemist awarded a Nobel prize in 1930 for his work on haemoglobin, the oxygen-carrying red colouring matter in blood.

Fischer, Hermann Otto Laurenz (1888–1960) German organic chemist. He carried out research into the synthetic and structural chemistry of carbohydrates, glycerides, and inositols.

Flory, Paul John (1910–1985) US polymer chemist. He was awarded the 1974 Nobel Prize for Chemistry for his investigations of synthetic and natural macromolecules. With Wallace Carothers, he developed nylon, the first synthetic polyamide.

Freundlich, Herbert Max Finlay (1880–1941) German physical chemist. He worked on the nature of colloids, particularly sols and gels and introduced the term 'thixotropy' to describe the behaviour of gels.

Friedel, Charles (1832–1899) French organic chemist and mineralogist. Together with US chemist James Mason Crafts (1839–1917) he discovered the **Friedel–Crafts reaction** which uses aluminium chloride as a catalyst to facilitate the addition of an alkyl halide (halogenoalkane) to an aromatic compound, orbenne.

Fröhlich, Herbert (1905–1991) German-born British physicist who helped lay the foundations for modern theoretical physics in the UK.

Fukui, Kenichi (1918–1998) Japanese industrial chemist who shared the Nobel Prize for Chemistry in 1981 with Roald Hoffman for his work on 'frontier orbital theory', predicting the change in molecular orbitals (the arrangement of electrons around the nucleus during chemical reactions).

Funk, Casimir (1884–1967) Polish-born US biochemist who pioneered research into vitamins.

Geber, Latinized form of Jabir ibn Hayyan (c. 721–c. 776) Arabian alchemist. His influence lasted for more than six hundred years.

Giauque, William Francis (1895–1982) Canadian-born US physical chemist who specialized in chemical thermodynamics, in particular the behaviour of matter at extremely low temperatures.

Gibbs, Josiah Willard (1839–1903) US theoretical physicist and chemist who developed a mathematical approach to thermodynamics and established vector methods in physics. He devised the phase rule and formulated the Gibbs adsorption isotherm.

Gibbs, Josiah
http://www-history.mcs.st-and.ac.uk/~history/
Mathematicians/Gibbs.html
Photograph and biography of the 19th-century US mathematician. This site, run by St Andrews University, also provides information on Gibbs's constant along with literature references for further study.

Gilbert, Walter (1932–) US molecular biologist who studied genetic control, seeking the mechanisms that switch genes on and off.

Gilman, Henry (1893–1986) US organic chemist. He made a comprehensive study of methods of high-yield synthesis, quantitative and qualitative analysis, and uses of organometallic compounds, particularly Grignard reagents.

Goldschmidt, Victor Moritz (1888–1947) Swiss-born Norwegian chemist. He did fundamental work in geochemistry, particularly on the distribution of elements in the Earth's crust.

Graham, Thomas (1805–1869) Scottish chemist who laid the foundations of physical chemistry by his work on the diffusion of gases and liquids.

Grignard, (François Auguste) Victor (1871–1935) French chemist. In 1900 he discovered a series of organic compounds, the **Grignard reagents**, that found applications as some of the most versatile reagents in organic synthesis.

Haber, Fritz (1868–1934) German chemist whose conversion of atmospheric nitrogen to ammonia opened the way for the synthetic fertilizer industry.

Hall, Charles Martin (1863–1914) US chemist who developed a process for the commercial production of aluminium in 1886.

Hammick, Dalziel Llewellyn (1887–1966) English chemist whose major contributions were in the fields of theoretical and synthetic organic chemistry.

Harden, Arthur (1865–1940) English biochemist who investigated the mechanism of sugar fermentation and the role of enzymes in this process.

Hassel, Odd (1897–1981) Norwegian physical chemist who established the technique of conformational analysis – the determination of the properties of a molecule by rotating it around a single bond – and received the Nobel Prize for Chemistry in 1969.

Hauptman, Herbert A (1917–) US mathematician who shared the 1985 Nobel Prize for Chemistry with Jerome Karle for discovering a general method of determining crystal structures by X-ray diffraction.

Haworth, (Walter) Norman (1883–1950) English organic chemist who was the first to synthesize a vitamin (ascorbic acid, vitamin C) in 1933, for which he shared a Nobel prize in 1937.

Helmont, Jean Baptiste van (1579–1644) Flemish physician who was the first to realize that there are gases other than air, and claimed to have coined the word 'gas' (from Greek *cháos*).

Henry, William (1774–1836) English chemist and physician. In 1803 he formulated **Henry's law**, which states that when a gas is dissolved in a liquid at a given temperature, the mass that dissolves is in direct proportion to the pressure of the gas.

Herschbach, Dudley R (1932–) US chemist who shared the 1986 Nobel Prize for Chemistry with Yuan T Lee and John C Polanyi for their researches into the dynamics of the processes which occur when atoms and molecules react.

Hess, Germain Henri (1802–1850) Swiss-born Russian chemist, a pioneer in the field of thermochemistry. The law of constant heat summation is named after him.

Hevesy, Georg Karl von (1885–1966) Hungarian-born Swedish chemist, discoverer of the element hafnium. He was the first to use a radioactive isotope to follow the steps of a biological process.

Heyrovsky, Jaroslav (1890–1967) Czech chemist who was awarded the 1959 Nobel prize for his invention and development of polarography, an electrochemical technique of chemical analysis.

Hinshelwood, Cyril Norman (1897–1967) English chemist who shared the 1956 Nobel prize for his work on chemical chain reactions.

Hodgkin, Dorothy Mary Crowfoot (1910–1994) English biochemist who analysed the structure of penicillin, insulin, and vitamin B12.

Hoffman, Roald (1937–) Polish chemist who worked on **molecular orbital theory** with Robert Woodward and developed the Woodward–Hoffman rules for the conservation of orbital symmetry.

Hofmann, August Wilhelm von (1818–1892) German chemist who studied the extraction and exploitation of coal-tar derivatives, mainly for dyes.

Huber, Robert (1937–) German chemist who shared the Nobel Prize for Chemistry in 1988 with Hartmut Michel and

Johann Deisenhofer for his use of high resolution X-ray crystallography.

Hückel, Erich Armand Arthur Joseph (1896–1980) German physical chemist who, with Peter Debye, developed in 1923 the modern theory that accounts for the electrochemical behaviour of strong electrolytes in solution.

Hyatt, John Wesley (1837–1920) US inventor who in 1869 invented celluloid, the first artificial plastic, intended as a substitute for ivory.

Ingold, Christopher Kelk (1893–1970) English organic chemist who specialized in the concepts, classification, and terminology of theoretical organic chemistry.

Ipatieff, Vladimir Nikolayevich (1867–1952) Russian-born US organic chemist who developed catalysis in organic chemistry, particularly in reactions involving hydrocarbons.

Karle, Jerome (1918–) US chemist who, with colleague Herbert Hauptman, tested the range of available diffraction techniques, such as X-ray diffraction and the 'heavy atom' technique.

Karrer, Paul (1889–1971) Russian-born Swiss organic chemist who synthesized various vitamins.

Kekulé von Stradonitz, Friedrich August (1829–1896) German chemist whose 1858 theory of molecular structure revolutionized organic chemistry.

Kendrew, John Cowdery (1917–1997) English biochemist who determined the structure of the muscle protein myoglobin.

Kenyon, Joseph (1885–1961) English organic chemist who studied optical activity, particularly of secondary alcohols.

Kipping, Frederic Stanley (1863–1949) English chemist who pioneered the study of the organic compounds of silicon; he invented the term 'silicone', which is now applied to the entire class of oxygen-containing polymers.

Klaproth, Martin Heinrich (1743–1817) German chemist who first identified the elements uranium and zirconium, in 1789.

Klug, Aaron (1926–) South African molecular biologist who improved the quality of electron micrographs by using laser lighting.

Kolbe, (Adolf Wilhelm) Hermann (1818–1884) German chemist, generally regarded as the founder of modern organic chemistry with his synthesis of acetic acid (ethanoic acid) – an organic compound – from inorganic starting materials.

Kornberg, Arthur (1918–) US biochemist. In 1956 he discovered the enzyme DNA-polymerase, which enabled molecules of the genetic material DNA to be synthesized for the first time.

Langmuir, Irving (1881–1957) US scientist who invented the mercury vapour pump for producing a high vacuum, and the atomic hydrogen welding process; he was also a pioneer of the thermionic valve.

Lapworth, Arthur (1872–1941) British chemist, one of the founders of modern physical-organic chemistry. He formulated the electronic theory of organic reactions (independently of Robert Robinson).

Lavoisier, Antoine Laurent (1743–1794) French chemist. He proved that combustion needs only a part of the air, which he called oxygen, thereby destroying the theory of phlogiston (an imaginary 'fire element' released during combustion). With astronomer and mathematician Pierre de Laplace, he showed in 1783 that water is a compound of oxygen and hydrogen.

Lavoisier, Antoine Laurent
http://www.knight.org/advent/cathen/09052a.htm
Account of the life and achievements of the French chemist, philosopher, and economist.

Leblanc, Nicolas (1742–1806) French chemist who in the 1780s developed a process for making soda ash (sodium carbonate, Na_2CO_3) from common salt (sodium chloride, $NaCl$).

Le Châtelier, Henri Louis (1850–1936) French physical chemist who formulated the principle now named after him, which states that if any constraint is applied to a system in chemical equilibrium, the system tends to adjust itself to counteract or oppose the constraint.

Lee, Yuan Tseh (1936–) Taiwanese chemist who contributed much to the field of chemical reaction dynamics.

Lehn, Jean-Marie (1939–) French chemist who demonstrated for the first time how metal ions could be made to exist in a nonplanar structure, tightly bound into the cavity of a crown ether molecule.

Leloir, Luis Frederico (1906–1987) Argentinian chemist who studied glucose metabolism.

G N Lewis and the Covalent Bond
http://dbhs.wvusd.k12.ca.us/Chem-History/
Lewis-1916/Lewis-1916.html
Transcript of one of the most important papers in the history of chemistry. In the paper Lewis forwards his ideas on the shared electron bond, later to become known as the covalent bond.

Lewis, Gilbert Newton (1875–1946) US theoretical chemist who defined a base as a substance that supplies a pair of electrons for a chemical bond, and an acid as a substance that accepts such a pair. He also set out the electronic theory of valency and in thermodynamics listed the free energies of 143 substances.

Libby, Willard Frank
http://kroeber.anthro.mankato.msus.edu/
bio/Libby.htm
Profile of the Nobel prizewinning US chemist. It traces his academic career and official appointments and the process which led to his discovery of the technique of radiocarbon dating.

Libby, Willard Frank (1908–1980) US chemist whose development in 1947 of radiocarbon dating as a means of determining the age of organic or fossilized material won him a Nobel prize in 1960.

Liebig, Justus, Baron von (1803–1873) German organic chemist who extended chemical research into other scientific fields, such as agricultural chemistry and biochemistry.

Lipmann, Fritz Albert (1899–1986) German-born US biochemist. He investigated the means by which the cell acquires energy and highlighted the crucial role played by the energy-rich phosphate molecule adenosine triphosphate (ATP).

Lipscomb, William Nunn (1919–) US chemist who studied the relationships between the geometric and electronic structures of molecules and their chemical and physical behaviour.

Longuet-Higgins, Hugh Christopher (1923–) English theoretical chemist whose main contributions have involved the application of precise mathematical analyses, particularly statistical mechanics, to chemical problems.

Lonsdale, Kathleen (1903–1971), born Yardley, Irish X-ray crystallographer who was among the first to determine the structures of organic molecules.

Marcus, Rudolph Arthur (1923–) Canadian chemist who advanced the theory of electron-transfer reactions (involving soluble molecules and/or ions) which drive many biological processes.

Martin, Archer John Porter (1910–) British biochemist who received the 1952 Nobel Prize for Chemistry for work with Richard Synge on paper chromatography in 1944.

Mendeleyev, Dmitri Ivanovich (1834–1907) Russian chemist who framed the periodic law in chemistry in 1869, which states that the chemical properties of the elements depend on their relative atomic masses. This law is the basis of the periodic table of the elements, in which the elements are arranged by atomic number and organized by their related groups.

Merrifield, R Bruce (1921–) US chemist who was awarded the 1984 Nobel Prize for Chemistry for his development of a method for synthesizing large organic molecules using a solid support or matrix.

Meyer, (Julius) Lothar (1830–1895) German chemist who, independently of his Russian contemporary Dmitri Mendeleyev, produced a periodic law describing the properties of the chemical elements.

Meyer, Viktor (1848–1897) German organic chemist who invented an apparatus for determining vapour densities (and hence molecular weights), now named after him.

Michel, Hartmut (1948–) German biochemist who worked on determining the molecular structure of photosynthetic reaction centres.

Mitscherlich, Eilhard (1794–1863) German chemist who discovered isomorphism (the phenomenon in which substances of analogous chemical composition crystallize in the same crystal form). He also synthesized many organic compounds for the first time.

Molina, Mario (1943–) Mexican chemist who shared the 1995 Nobel Prize for Chemistry with Paul Crutzen and F Sherwood Rowland for their work in atmospheric chemistry,

particularly concerning the formation and decomposition of ozone.

Müller, Paul Herman (1899–1965) Swiss chemist who discovered the first synthetic contact insecticide, DDT, in 1939.

Natta, Giulio (1903–1979) Italian chemist who worked on the production of polymers. He shared a Nobel prize in 1963 with German chemist Karl Ziegler.

Nernst, (Walther) Hermann (1864–1941) German physical chemist who won a Nobel prize in 1920 for work on heat changes in chemical reactions.

Alfred Nobel – His Life and Work
http://www.nobel.se/alfred/biography.html
Presentation of the life and work of Alfred Nobel. The site includes references to Nobel's life in Paris, as well as his frequent travels, his industrial occupations, his scientific discoveries and especially his work on explosives, which led to the patenting of dynamite, his numerous chemical inventions which included materials such as synthetic leather and artificial silk, his interest in literature and in social and peace-related issues, and of course the Nobel prizes which came as a natural extension of his lifelong interests.

Nobel, Alfred Bernhard (1833–1896) Swedish chemist and engineer. He invented dynamite in 1867, gelignite in 1875, and ballistite, a smokeless gunpowder, in 1887. Having amassed a large fortune from the manufacture of explosives and the exploitation of the Baku oilfields in Azerbaijan, near the Caspian Sea, he left this in trust for the endowment of five Nobel prizes.

Olah, George Andrew (1927–) Hungarian-born US chemist who was awarded the 1992 Nobel Prize for Chemistry for his isolation of carbocations, electrically charged fragments of hydrocarbon molecules.

Ostwald, (Friedrich) Wilhelm (1853–1932) Latvian-born German chemist who devised the Ostwald process (the oxidation of ammonia over a platinum catalyst to give nitric acid).

Pauling, Linus Carl (1901–1994) US theoretical chemist and biologist. His ideas on chemical bonding are fundamental to modern theories of molecular structure.

Dr Linus Pauling Profile
http://www.achievement.org/autodoc/page/pau0pro-1
Description of the life and works of the multiple Nobel prizewinner also holds a lengthy interview with Dr Pauling from 1990.

Pedersen, Charles (1904–1990) US organic chemist who shared the Nobel Prize for Chemistry in 1987 with Jean Lehn and Donald Cram for his discovery of 'crown ether', a cyclic polyether.

Pelletier, Pierre-Joseph (1788–1842) French chemist whose extractions of a range of biologically active compounds from plants founded the chemistry of the alkaloids. The most important of his discoveries was quinine, used against malaria.

Perey, Marguérite (Catherine) (1909–1975) French nuclear chemist who discovered the radioactive element francium in 1939.

Perutz, Max Ferdinand (1914–) Austrian-born British biochemist who shared the 1962 Nobel Prize for Chemistry with his coworker John Kendrew for work on the structure of the haemoglobin molecule.

Polanyi, John Charles (1929–) German physical chemist whose research on infrared light given off during chemical reactions (infrared chemical luminescence) laid the foundations for the development of chemical lasers.

Porter, George (1920–) English chemist. From 1947 he and Ronald Norrish developed a technique by which flashes of high energy are used to bring about extremely fast chemical reactions.

Pregl, Fritz (1869–1930) Austrian chemist who, during his research on bile acids, devised new techniques for microanalysis (the analysis of very small quantities).

Prelog, Vladimir (1906–1998) Bosnian-born Swiss organic chemist who studied alkaloids and antibiotics.

Priestley, Joseph (1733–1804) English chemist and Unitarian minister. He identified oxygen in 1774 and several other gases. Dissolving carbon dioxide under pressure in water, he began a European craze for soda water.

Prigogine, Ilya, Viscount Prigogine (1917–) Russian-born Belgian chemist who, as a highly original theoretician, has made major contributions to the field of thermodynamics.

Proust, Joseph Louis (1754–1826) French chemist. He was the first to state the principle of constant composition of compounds – that compounds consist of the same proportions of elements wherever found.

Prout, William (1785–1850) British physician and chemist. In 1815 Prout published his hypothesis that the relative atomic mass of every atom is an exact and integral multiple of the mass of the hydrogen atom.

Ramsay, William (1852–1916) Scottish chemist who, with Lord Rayleigh, discovered argon in 1894.

Raoult, François Marie (1830–1901) French chemist. In 1882, while working at the University of Grenoble, Raoult formulated one of the basic laws of chemistry. **Raoult's law** enables the relative molecular mass of a substance to be determined by noting how much of it is required to depress the freezing point of a solvent by a certain amount.

Regnault, Henri Victor (1810–1878) German-born French physical chemist who showed that Boyle's law applies only to ideal gases.

Richards, Theodore William (1868–1928) US chemist who determined as accurately as possible the relative atomic masses of a large number of elements.

Robertson, Robert (1869–1949) Scottish chemist who worked on explosives for military use, such as TNT.

Robinson, Robert (1886–1975) English chemist, Nobel prizewinner in 1947 for his research in organic chemistry on the structure of many natural products, including flower pigments and alkaloids. He formulated the electronic theory now used in organic chemistry.

Rodbell, Martin (1925–) US molecular biochemist who shared the 1994 Nobel Prize for Physiology or Medicine with Alfred Gilman for their discovery of a family of proteins that translate messages from outside a cell into action inside cells.

Rowland, F Sherwood (1927–) US chemist who shared the 1995 Nobel Prize for Chemistry with Mario Molina and Paul Crutzen for their work in atmospheric chemistry.

Ruzicka, Leopold Stephen (1887–1976) Swiss chemist who began research on natural compounds such as musk and civet secretions. Ruzicka shared the 1939 Nobel Prize for Chemistry with Adolf Butenandt.

Sabatier, Paul (1854–1941) French chemist. He found in 1897 that if a mixture of ethylene and hydrogen was passed over a column of heated nickel, the ethylene changed into ethane.

Saint-Claire Deville, Henri Etienne (1818–1881) French inorganic chemist who worked on high-temperature reactions and was the first to extract metallic aluminium in any quantity.

Sanger, Frederick (1918–) English biochemist. He was the first person to win a Nobel Prize for Chemistry twice: the first in 1958 for determining the structure of insulin, and the second in 1980 for work on the chemical structure of genes.

Scheele, Karl Wilhelm (1742–1786) Swedish chemist and pharmacist who isolated many elements and compounds for the first time, including oxygen in about 1772, and chlorine in 1774.

Seaborg, Glenn Theodore (1912–) US nuclear chemist. For his discovery of plutonium and research on the transuranic elements, he shared a Nobel prize in 1951 with his coworker Edwin McMillan.

Semenov, Nikolai Nikolaevich (1896–1986) Russian physical chemist who studied chemical chain reactions, particularly branched-chain reactions.

Smalley, Richard E (1943–) US chemist who, with colleagues Robert Curl and Harold Kroto, discovered buckminsterfullerene (carbon 60) in 1985.

Soddy, Frederick (1877–1956) English physical chemist who pioneered research into atomic disintegration and coined the term isotope. He was awarded a Nobel prize in 1921 for investigating the origin and nature of isotopes.

Solvay, Ernest (1838–1922) Belgian industrial chemist who in the 1860s invented the ammonia-soda process, also known as the **Solvay process**, for making the alkali sodium carbonate.

Sørensen, Søren Peter Lauritz (1868–1939) Danish chemist who in 1909 introduced the concept of using the pH scale as a measure of the acidity of a solution.

Stahl, Georg Ernst (1660–1734) German chemist who developed the theory that objects burn because they contain a combustible substance, phlogiston.

Stanley, Wendell Meredith (1904–1971) US biochemist who crystallized the tobacco mosaic virus (TMV) in 1935.

Stas, Jean Servais (1813–1891) Belgian analytical chemist who made the first accurate determinations of atomic weights (relative atomic masses).

Staudinger, Hermann (1881–1965) German organic chemist, founder of macromolecular chemistry, who carried out pioneering research into the structure of albumen and cellulose.

Stein, William Howard (1911–1980) US biochemist who determined the amino acid sequence of the enzyme ribonuclease.

Sumner, James Batcheller (1887–1955) US biochemist. In 1926 he succeeded in crystallizing the enzyme urease and demonstrating its protein nature. For this work Sumner shared the 1946 Nobel Prize for Chemistry with John Northrop and Wendell Stanley.

Svedberg, Theodor (1884–1971) Swedish chemist. In 1924 he constructed the first ultracentrifuge.

Synge, Richard Laurence Millington (1914–1994) British biochemist who improved paper chromatography (a means of separating mixtures) to the point where individual amino acids could be identified.

Taube, Henry (1915–) US chemist who established the basis of inorganic chemistry through his study of the loss or gain of electrons by atoms during chemical reactions.

Tiselius, Arne Wilhelm Kaurin (1902–1971) Swedish chemist who developed a powerful method of chemical analysis known as electrophoresis.

Todd, Alexander Robertus (1907–), Baron Todd, Scottish organic chemist who won a Nobel prize in 1957 for his work on the role of nucleic acids in genetics.

Travers, Morris William (1872–1961) English chemist who, with Scottish chemist William Ramsay, between 1894 and 1908 first identified what were called the inert or noble gases: krypton, xenon, and radon.

Tswett, Mikhail Semyonovich (1872–1919) Italian-born Russian scientist who made an extensive study of plant pigments and developed the technique of chromatography to separate them.

Urey, Harold Clayton (1893–1981) US chemist. In 1932 he isolated heavy water and discovered deuterium, for which he was awarded the 1934 Nobel Prize for Chemistry.

van't Hoff, Jacobus Henricus (1852–1911) Dutch physical chemist. He explained the 'asymmetric' carbon atom occurring in optically active compounds and developed the concept of chemical affinity as the maximum work obtainable from a reaction.

Vauquelin, Louis Nicolas (1763–1829) French chemist who worked mainly in the inorganic field, analysing minerals. He discovered the elements chromium (1797) and beryllium.

Virtanen, Artturi Ilmari (1895–1973) Finnish chemist who from 1920 made discoveries in agricultural chemistry.

Volhard, Jacob (1834–1910) German chemist who devised various significant methods of organic synthesis.

Wald, George (1906–1997) US biochemist who explored the chemistry of vision.

Wallach, Otto (1847–1931) German analytic chemist who isolated a new class of compounds, called terpenes, from essential oils (oils extracted from plants and used in medicine, aromatherapy, and perfume).

Werner, Alfred (1866–1919) French-born Swiss chemist. He was awarded a Nobel prize in 1913 for his work on valency theory, which gave rise to the concept of coordinate bonds and coordination compounds.

Wieland, Heinrich Otto (1877–1957) German organic chemist who determined the structures of steroids and related compounds.

Wilkinson, Geoffrey (1921–1996) English inorganic chemist who shared a Nobel prize in 1973 for his pioneering work on the organometallic compounds of the transition metals.

Willstätter, Richard (1872–1942) German organic chemist who investigated plant pigments – such as chlorophyll – and alkaloids, determining the structure of cocaine, tropine, and atropine.

Windaus, Adolf Otto Reinhold (1876–1959) German chemist who was awarded the Nobel Prize for Chemistry in 1928 for his research on the structure of cholesterol, its relationship to vitamin D, and his discovery that steroids are precursors of vitamins.

Wittig, Georg (1897–1987) German chemist whose method of synthesizing olefins (alkenes) from carbonyl compounds is a reaction often termed the **Wittig synthesis**. For this achievement he shared the 1979 Nobel Prize for Chemistry.

Wöhler, Friedrich (1800–1882) German chemist who in 1828 became the first person to synthesize an organic compound (urea) from an inorganic compound (ammonium cyanate).

Wollaston, William Hyde (1766–1828) English chemist and physicist who discovered in 1804 how to make malleable platinum.

Woodward, Robert Burns (1917–1979) US chemist who worked on synthesizing a large number of complex molecules.

Wurtz, Charles Adolphe (1817–1884) French organic chemist who discovered a method of producing paraffin hydrocarbons (alkanes) using alkyl halides (haloalkanes) and sodium in ether. The method was named the **Wurtz reaction**.

Ziegler, Karl (1898–1973) German organic chemist. In 1963 he shared the Nobel Prize for Chemistry with Giulio Natta of Italy for his work on the chemistry and technology of large polymers.

Zsigmondy, Richard Adolf (1865–1929) Austrian-born German chemist who devised and built an ultramicroscope in 1903.

Glossary

absolute zero lowest temperature theoretically possible according to kinetic theory, zero kelvin (0 K), equivalent to $-273.15°C/-459.67°F$, at which molecules are in their lowest energy state. Near absolute zero, the physical properties of some materials change substantially; for example, some metals lose their electrical resistance and become superconducting.

acid in chemistry, compound that releases hydrogen ions (H+ or protons) in the presence of an ionizing solvent (usually water). Acids react with bases to form salts, and they act as solvents. Strong acids are corrosive; dilute acids have a sour or sharp taste, although in some organic acids this may be partially masked by other flavour characteristics. The strength of an acid is measured by its hydrogen-ion concentration, indicated by the pH value. All acids have a pH below 7.0.

actinide any of a series of 15 radioactive metallic chemical elements with atomic numbers 89 (actinium) to 103 (lawrencium). Elements 89 to 95 occur in nature; the rest of the series are synthesized elements only. Actinides are grouped together because of their chemical similarities (for example, they are all bivalent), the properties differing only slightly with atomic number. The series is set out in a band in the periodic table of the elements, as are the lanthanides.

affinity in chemistry, the force of attraction (see bond) between atoms that helps to keep them in combination in a molecule. The term is also applied to attraction between molecules, such as those of biochemical significance (for example, between enzymes and substrate molecules). This is the basis for affinity chromatography, by which biologically important compounds are separated.

aliphatic compound any organic chemical compound in which the carbon atoms are joined in straight chains, as in hexane (C_6H_{14}), or in branched chains, as in 2-methylpentane ($CH_3CH(CH_3)CH_2CH_2CH_3$).

alkali a base that is soluble in water. Alkalis neutralize acids and are soapy to the touch. The strength of an alkali is measured by its hydrogen-ion concentration, indicated by the pH value. They may be divided into strong and weak alkalis: a strong alkali (for example, potassium hydroxide, KOH) ionizes completely when disssolved in water, whereas a weak alkali (for example, ammonium hydroxide, NH_4OH) exists in a partially ionized state in solution. All alkalis have a pH above 7.0.

The hydroxides of metals are alkalis. Those of sodium and potassium are chemically powerful; both were historically derived from the ashes of plants.

alkali metal any of a group of six metallic elements with similar chemical properties: lithium, sodium, potassium, rubidium, caesium, and francium.

alkaline-earth metal any of a group of six metallic elements with similar bonding properties: beryllium, magnesium, calcium, strontium, barium, and radium.

allotropy property whereby an element can exist in two or more forms (allotropes), each possessing different physical properties but the same state of matter (gas, liquid, or solid). The allotropes of carbon are diamond and graphite. Sulphur has several allotropes (flowers of sulphur, plastic, rhombic, and monoclinic). These solids have different crystal structures, as do the white and grey forms of tin and the black, red, and white forms of phosphorus.

alpha particle positively charged, high-energy particle emitted from the nucleus of a radioactive atom. It is identical with the nucleus of a helium atom – that is, it consists of two protons and two neutrons. The process of emission, **alpha decay**, transforms one element into another, decreasing the atomic (or proton) number by two and the atomic mass (or nucleon number) by four.

aromatic compound organic chemical compound in which some of the bonding electrons are delocalized (shared among several atoms within the molecule and not localized in the vicinity of the atoms involved in bonding).

assay the determination of the quantity of a given substance present in a sample. Usually it refers to determining the purity of precious metals.

atom smallest unit of matter that can take part in a chemical reaction, and which cannot be broken down chemically into anything simpler. An atom is made up of protons and neutrons in a central nucleus surrounded by electrons. The atoms of the various elements differ in atomic number, relative atomic mass, and chemical behaviour.

atomic mass unit, or **dalton unit**, (symbol amu or u) unit of mass that is used to measure the relative mass of atoms and molecules. It is equal to one-twelfth of the mass of a carbon-12 atom, which is equivalent to the mass of a proton, or $1.66 \times 10^{(27}$ kg. The relative atomic mass of an atom has no units; thus oxygen-16 has an atomic mass of 16 daltons but a relative atomic mass of 16.

atomic number, or **proton number**, the number (symbol Z) of protons in the nucleus of an atom. It is equal to the positive charge on the nucleus.

In a neutral atom, it is also equal to the number of electrons surrounding the nucleus. The chemical elements are arranged in the periodic table of the elements according to their atomic number. See also nuclear notation.

Avogadro's hypothesis the law stating that equal volumes of all gases, when at the same temperature and pressure, have the same numbers of molecules. It was first propounded by Amedeo Avogadro.

base a substance that accepts protons. Bases react with acids to give water and a salt. They may be a soluble oxide, hydroxide (such as sodium hydroxide, NaOH), or compound (such as ammonia, NH_3) that dissolves in water to give the hydroxide ion ($OH^($), or an insoluble oxide or hydroxide (such as copper(II) oxide, CuO) that reacts with an acid. A base that is soluble in water is called an alkali.

battery any energy-storage device allowing release of electricity on demand. It is made up of one or more electrical cells.

beta particle electron ejected with great velocity from a radioactive atom that is undergoing spontaneous disintegration. Beta particles do not exist in the nucleus but are created on disintegration, beta decay, when a neutron converts to a proton to emit an electron.

bond in chemistry, the result of the forces of attraction that hold together atoms of an element or elements to form a molecule. The principal types of bonding are ionic, covalent, metallic, and intermolecular (such as hydrogen bonding).

Boyle's law law stating that the volume of a given mass of gas at a constant temperature is inversely proportional to its pressure. For example, if the pressure of a gas doubles, its volume will be reduced by a half, and vice versa. The law was discovered in 1662 by Irish physicist and chemist (Robert Boyle.

catalyst substance that alters the speed of, or makes possible, a chemical or biochemical reaction but remains unchanged at the end of the reaction. Enzymes are natural biochemical catalysts. In practice most catalysts are used to speed up reactions.

cathode in chemistry, the negative electrode of an electrolytic cell, towards which positive particles (cations), usually in solution, are attracted.

cation ion carrying a positive charge. During electrolysis, cations in the electrolyte move towards the cathode (negative electrode).

cell, electrical, or **voltaic cell** or **galvanic cell,** device in which chemical energy is converted into electrical energy; the popular name is 'battery', but this actually refers to a collection of cells in one unit. The reactive chemicals of a **primary cell** cannot be replenished, whereas **secondary cells** – such as storage batteries – are rechargeable: their chemical reactions can be reversed and the original condition restored by applying an electric current. It is dangerous to attempt to recharge a primary cell.

chain reaction in chemistry, a succession of reactions, usually involving free radicals, where the products of one stage are the reactants of the next. A chain reaction is characterized by the continual generation of reactive substances.

Charles's law law stating that the volume of a given mass of gas at constant pressure is directly proportional to its absolute temperature (temperature in kelvin). It was discovered by French physicist Jacques Charles in 1787, and independently by French chemist Joseph Gay-Lussac in 1802.

chromatography technique for separating or analysing a mixture of gases, liquids, or dissolved substances. This is brought about by means of two immiscible substances, one of which (**the mobile phase**) transports the sample mixture through the other (**the stationary phase**). The mobile phase may be a gas or a liquid; the stationary phase may be a liquid or a solid, and may be in a column, on paper, or in a thin layer on a glass or plastic support. The components of the mixture are absorbed or impeded by the stationary phase to different extents and therefore become separated. The technique is used for both qualitative and quantitive analyses in biology and chemistry.

combustion burning, defined in chemical terms as the rapid combination of a substance with oxygen, accompanied by the evolution of heat and usually light. A slow-burning candle flame and the explosion of a mixture of petrol vapour and air are extreme examples of combustion. Combustion is an exothermic reaction as heat energy is given out.

compound chemical substance made up of two or more elements bonded together, so that they cannot be separated by physical means. Compounds are held together by ionic or covalent bonds.

colloid substance composed of extremely small particles of one material (the dispersed phase) evenly and stably distributed in another material (the continuous phase). The size of the dispersed particles (1–1,000 nanometres across) is less than that of particles in suspension but greater than that of molecules in true solution.

covalent bond chemical bond produced when two atoms share one or more pairs of electrons (usually each atom contributes an electron). The bond is often represented by a single line drawn between the two atoms. Covalently bonded substances include hydrogen (H_2), water (H_2O), and most organic substances.

crystal substance with an orderly three-dimensional arrangement of its atoms or molecules, thereby creating an external surface of clearly defined smooth faces having characteristic angles between them. Examples are table salt and quartz.

crystallography the scientific study of crystals. In 1912 it was found that the shape and size of the repeating atomic patterns (unit cells) in a crystal could be determined by passing X-rays through a sample. This method, known as X-ray diffraction, opened up an entirely new way of 'seeing' atoms. It has been found that many substances have a unit cell that exhibits all the symmetry of the whole crystal; in table salt (sodium chloride, NaCl), for instance, the unit cell is an exact cube.

dative bond covalent bond in which one atom supplies both bonding electrons.

density measure of the compactness of a substance; it is equal to its mass per unit volume and is measured in kg per cubic metre/lb per cubic foot. Density is a scalar quantity. The average density D of a mass m occupying a volume V is given by the formula:

$$D = m/V$$

Relative density is the ratio of the density of a substance to that of water at 4°C/32.2°F.

diffusion spontaneous and random movement of molecules or particles in a fluid (gas or liquid) from a region in which they are at a high concentration to a region of lower concentration, until a uniform concentration is achieved throughout. The difference in concentration between two such regions is called the **concentration gradient**. No mechanical mixing or stirring is involved. For instance, if a drop of ink is added to water, its molecules will diffuse until their colour becomes evenly distributed throughout. Diffusion occurs more rapidly across a higher concentration gradient and at higher temperature.

dilution process of reducing the concentration of a solution by the addition of a solvent.

electrode any terminal by which an electric current passes in or out of a conducting substance; for example, the anode or cathode in a battery or the carbons in an arc lamp. The terminals that emit and collect the flow of electrons in thermionic valves (electron tubes) are also called electrodes: for example, cathodes, plates, and grids.

electron stable, negatively charged elementary particle; it is a constituent of all atoms, and a member of the class of particles known as leptons. The electrons in each atom surround the nucleus in groupings called shells; in a neutral atom the number of electrons is equal to the number of protons in the nucleus. This electron structure is responsible for the chemical properties of the atom (see atomic structure).

element substance that cannot be split chemically into simpler substances. The atoms of a particular element all have the same number of protons in their nuclei (their atomic number).

fatty acid, or **carboxylic acid**, organic compound consisting of a hydrocarbon chain, up to 24 carbon atoms long, with a carboxyl group (–COOH) at one end. The covalent bonds between the carbon atoms may be single or double; where a double bond occurs the carbon atoms concerned carry one instead of two hydrogen atoms. Chains with only single bonds have all the hydrogen they can carry, so they are said to be **saturated** with hydrogen. Chains with one or more double bonds are said to be **unsaturated**. Fatty acids are produced in the small intestine when fat is digested.

formula representation of a molecule, radical, or ion, in which the component chemical elements are represented by their symbols. An **empirical formula** indicates the simplest ratio of the elements in a compound, without indicating how many of them there are or how they are combined. A **molecular formula** gives the number of each type of element present in one molecule. A **structural formula** shows the relative positions of the atoms and the bonds between them. For example, for ethanoic acid, the empirical formula is CH_2O, the molecular formula is $C_2H_4O_2$, and the structural formula is CH_3COOH.

free radical in chemistry, an atom or molecule that has an unpaired electron and is therefore highly reactive. Most free radicals are very short-lived. They are by-products of normal cell chemistry and rapidly oxidize other molecules they encounter. Free radicals are thought to do considerable damage to living organisms. They are neutralized by protective enzymes.

functional group the part of a molecule that actively takes part in the normal reactions of that molecule.

gamma radiation high-energy electromagnetic radiation emitted by some nucleides in the course of radioactive decay.

gas a form of matter, such as air, in which the molecules move randomly in otherwise empty space, filling any size or shape of container into which the gas is put.

gel solid produced by the formation of a three-dimensional cage structure, commonly of linked large-molecular-mass polymers, in which a liquid is trapped. It is a form of colloid. A gel may be a jellylike mass (pectin, gelatin) or have a more rigid structure (silica gel).

half-life during radioactive decay, the time in which the strength of a radioactive source decays to half its original value. In theory, the decay process is never complete and there is always some residual radioactivity. For this reason, the half-life of a radioactive isotope is measured, rather than the total decay time. It may vary from millionths of a second to billions of years.

halogen any of a group of five nonmetallic elements with similar chemical bonding properties: fluorine, chlorine, bromine, iodine, and astatine. They form a linked group in the periodic table of the elements, descending from fluorine, the most reactive, to astatine, the least reactive. They combine directly with most metals to form salts, such as common salt (NaCl). Each halogen has seven electrons in its valence shell, which accounts for the chemical similarities displayed by the group.

hydrocarbon any of a class of chemical compounds containing only hydrogen and carbon (for example, the alkanes and alkenes). Hydrocarbons are obtained industrially principally from petroleum and coal tar.

inert gas, or **noble gas**, any of a group of six elements (helium, neon, argon, krypton, xenon, and radon), so named because they were originally thought not to enter into any chemical reactions. This is now known to be incorrect: in 1962, xenon was made to combine with fluorine, and since then, compounds of argon, krypton, and radon with fluorine and/or oxygen have been described.

ion atom, or group of atoms, that is either positively charged (cation) or negatively charged (anion), as a result of the loss or gain of electrons during chemical reactions or exposure to certain forms of radiation. In solution or in the molten state, ionic compounds such as salts, acids, alkalis, and metal oxides conduct electricity. These compounds are known as electrolytes.

ionic bond, or **electrovalent bond**, bond produced when atoms of one element donate electrons to atoms of another element, forming positively and negatively charged ions, respectively. The attraction between the oppositely charged ions constitutes the bond. Sodium chloride (Na^+Cl^-) is a typical ionic compound.

isomer chemical compound having the same molecular composition and mass as another, but with different physical or chemical properties owing to the different structural arrangement of its constituent atoms. For example, the organic compounds butane ($CH_3(CH_2)_2CH_3$) and methyl propane ($CH_3CH(CH_3)CH_3$) are isomers, each possessing four carbon atoms and ten hydrogen atoms but differing in the way that these are arranged with respect to each other.

isotope one of two or more atoms that have the same atomic number (same number of protons), but which contain a different number of neutrons, thus differing in their atomic mass (see relative atomic mass). They may be stable or radioactive, naturally occurring or synthesized. For example, hydrogen has the isotopes ^2H (deuterium) and ^3H (tritium). The term was coined by English chemist Frederick Soddy, pioneer researcher in atomic disintegration.

lanthanide any of a series of 15 metallic elements (also known as rare earths) with atomic numbers 57 (lanthanum) to 71 (lutetium).

One of its members, promethium, is radioactive. All occur in nature. Lanthanides are grouped because of their chemical similarities (most are trivalent, but some can be divalent or tetravalent), their properties differing only slightly with atomic number.

liquid state of matter between a solid and a gas. A liquid forms a level surface and assumes the shape of its container. Its atoms do not occupy fixed positions as in a crystalline solid, nor do they have freedom of movement as in a gas. Unlike a gas, a liquid is difficult to compress since pressure applied at one point is equally transmitted throughout (Pascal's principle). Hydraulics makes use of this property.

lone pair a pair of electrons in the outermost shell of an atom that are not used in bonding. In certain circumstances, they will allow the atom to bond with atoms, ions, or molecules (such as boron trifluoride, BF_3) that are deficient in electrons, forming coordinate covalent (dative) bonds in which they provide both of the bonding electrons.

metallic bond the force of attraction operating in a metal that holds the atoms together. In the metal the valency electrons are able to move within the crystal and these electrons are said to be delocalized. Their movement creates short-lived, positively charged ions. The electrostatic attraction between the delocalized electrons and the ceaselessly forming ions constitutes the metallic bond.

molecule molecules are the smallest particles of an element or compound that can exist independently. Hydrogen atoms, at room temperature, do not exist independently. They are bonded in pairs to form hydrogen molecules. A molecule of a compound consists of two or more different atoms bonded together. Molecules vary in size and complexity from the hydrogen molecule (H_2) to the large macromolecules of proteins. They may be held together by ionic bonds, in which the atoms gain or lose electrons to form ions, or by covalent bonds, where electrons from each atom are shared in a new molecular orbital.

monomer chemical compound composed of simple molecules from which polymers can be made. Under certain conditions the simple molecules (of the monomer) join together (polymerize) to form a very long chain molecule (macromolecule) called a polymer. For example, the polymerization of ethene (ethylene) monomers produces the polymer polyethene (polyethylene).

neutralization in chemistry, a process occurring when the excess acid (or excess base) in a substance is reacted with added base (or added acid) so that the resulting substance is neither acidic nor basic.

nonmetal one of a set of elements (around 20 in total) with certain physical and chemical properties opposite to those of metals. Nonmetals accept electrons and are sometimes called electronegative elements.

nucleon in particle physics, either a proton or a neutron, when present in the atomic nucleus.

nucleus in physics, the positively charged central part of an atom, which constitutes almost all its mass. Except for hydrogen nuclei, which have only protons, nuclei are composed of both protons and neutrons. Surrounding the nuclei are electrons, of equal and opposite charge to that of the protons, thus giving the atom a neutral charge.

orbital the area of space around an atom or molecule where an electron is likely to be found.

oxidation in chemistry, the loss of electrons, gain of oxygen, or loss of hydrogen by an atom, ion, or molecule during a chemical reaction.

oxide compound of oxygen and another element, frequently produced by burning the element or a compound of it in air or oxygen.

peptide molecule comprising two or more amino acid molecules (not necessarily different) joined by **peptide bonds**, whereby the acid group of one acid is linked to the amino group of the other (–CO.NH). The number of amino acid molecules in the peptide is indicated by referring to it as a di-, tri-, or polypeptide (two, three, or many amino acids).

periodic table of the elements in chemistry, a table in which the elements are arranged in order of their atomic number. The table summarizes the major properties of the elements and enables predictions to be made about their behaviour.

pH scale from 0 to 14 for measuring acidity or alkalinity. A pH of 7.0 indicates neutrality, below 7 is acid, while above 7 is alkaline. Strong acids, such as those used in car batteries, have a pH of about 2; strong alkalis such as sodium hydroxide are pH 13.

plastic any of the stable synthetic materials that are fluid at some stage in their manufacture, when they can be shaped, and that later set to rigid or semirigid solids. Plastics today are chiefly derived from petroleum. Most are polymers, made up of long chains of identical molecules.

polymer compound made up of a large long-chain or branching matrix composed of many repeated simple units (**monomers**) linked together by polymerization. There are many polymers, both natural (cellulose, chitin, lignin) and synthetic (polyethylene and nylon, types of plastic). Synthetic polymers belong to two groups: thermosoftening and thermosetting.

polysaccharide long-chain carbohydrate made up of hundreds or thousands of linked simple sugars (monosaccharides) such as glucose and closely related molecules.

precipitation in chemistry, the formation of an insoluble solid in a liquid as a result of a reaction within the liquid between two or more soluble substances. If the solid settles, it forms a **precipitate**; if the particles of solid are very small, they will remain in suspension, forming a **colloidal precipitate** (see colloid).

radical in chemistry, a group of atoms forming part of a molecule, which acts as a unit and takes part in chemical reactions without disintegration, yet often cannot exist alone for any length of time; for example, the methyl radical $-CH_3$, or the carboxyl radical –COOH.

radioactive decay process of disintegration undergone by the nuclei of radioactive elements, such as radium and various isotopes of uranium and the transuranic elements. This changes the element's atomic number, thus transmuting one element into another, and is accompanied by the emission of radiation. Alpha and beta decay are the most common forms.

radioactivity spontaneous alteration of the nuclei of radioactive atoms, accompanied by the emission of radiation. It is the property exhibited by the radioactive isotopes of stable elements and all isotopes of radioactive elements, and can be either natural or induced. See radioactive decay.

reaction in chemistry, the coming together of two or more atoms, ions, or molecules with the result that a chemical change takes place; that is, a change that occurs when two or more substances interact with each other, resulting in the production of different substances with different chemical compositions. The nature of the reaction is portrayed by a chemical equation.

reduction in chemistry, the gain of electrons, loss of oxygen, or gain of hydrogen by an atom, ion, or molecule during a chemical reaction.

relative atomic mass the mass of an atom relative to one-twelfth the mass of an atom of carbon-12. It depends primarily on the number of protons and neutrons in the atom, the electrons having negligible mass. If more than one isotope of the element is present, the relative atomic mass is calculated by taking an average that takes account of the relative proportions of each isotope, resulting in values that are not whole numbers. The term **atomic weight**, although commonly used, is strictly speaking incorrect.

salt in chemistry, any compound formed from an acid and a base through the replacement of all or part of the hydrogen in the acid by a metal or electropositive radical. **Common salt** is sodium chloride.

semiconductor material with electrical conductivity intermediate between metals and insulators and used in a wide range of electronic devices. Certain crystalline materials, most notably silicon and germanium, have a small number of free electrons that have escaped from the bonds between the atoms. The atoms from which they have escaped possess vacancies, called holes, which are similarly able to move from atom to atom and can be regarded as positive charges. Current can be carried by both electrons (negative carriers) and holes (positive carriers). Such materials are known as **intrinsic semiconductors**.

soap mixture of the sodium salts of various fatty acids: palmitic, stearic, and oleic acid. It is made by the action of sodium hydroxide (caustic soda) or potassium hydroxide (caustic potash) on fats of animal or vegetable origin. Soap makes grease and dirt disperse in water in a similar manner to a detergent.

solid a state of matter that holds its own shape (as opposed to a liquid, which takes up the shape of its container, or a gas, which totally fills its container). According to kinetic theory, the atoms or molecules in a solid are not free to move

but merely vibrate about fixed positions, such as those in crystal lattices.

solution two or more substances mixed to form a single, homogenous phase. One of the substances is the **solvent** and the others (**solutes**) are said to be dissolved in it.

solvent substance, usually a liquid, that will dissolve another substance. Although the commonest solvent is water, in popular use the term refers to low-boiling-point organic liquids, which are harmful if used in a confined space. They can give rise to respiratory problems, liver damage, and neurological complaints.

spectroscopy study of spectra associated with atoms or molecules in solid, liquid, or gaseous phase. Spectroscopy can be used to identify unknown compounds and is an invaluable tool in science, medicine, and industry (for example, in checking the purity of drugs).

standard temperature and pressure, STP, in chemistry, a standard set of conditions for experimental measurements, to enable comparisons to be made between sets of results. Standard temperature is 0°C/32°F (273K) and standard pressure 1 atmosphere (101,325 Pa).

sublimation in chemistry, the conversion of a solid to vapour without passing through the liquid phase.

substitution reaction in chemistry, the replacement of one atom or functional group in an organic molecule by another.

suspension mixture consisting of small solid particles dispersed in a liquid or gas, which will settle on standing. An example is milk of magnesia, which is a suspension of magnesium hydroxide in water.

trace element chemical element necessary in minute quantities for the health of a plant or animal. For example, magnesium, which occurs in chlorophyll, is essential to photosynthesis, and iodine is needed by the thyroid gland of mammals for making hormones that control growth and body chemistry.

transition metal metal with an unfilled inner electron shell. Such elements are good conductors of heat and electricity and in their compounds can have variable valency. An example is copper, which in its compounds can have a charge of +1 or +2. The compounds of transition elements are usually coloured.

valency in chemistry, the measure of an element's ability to combine with other elements, expressed as the number of atoms of hydrogen (or any other standard univalent element) capable of uniting with (or replacing) its atoms. The number of electrons in the outermost shell of the atom dictates the combining ability of an element.

volatile in chemistry, term describing a substance that readily passes from the liquid to the vapour phase. Volatile substances have a high vapour pressure.

Further Reading

Adloff, Jean-P (ed) *Handbook of Hot Atom Chemistry* (1992)

Alexander, W and Street, A *Metals in the Service of Man* (1972)

Ansell, M R *Rodd's Chemistry of Carbon Compounds* (1990)

Asimov, Isaac *Asimov on Chemistry* (1975)

Atkins, Peter William *Atoms, Electrons and Change* (1991)

Atkins, Peter William *General Chemistry* (1992, Second edition)

Atkins, Peter William *Molecules* (1987)

Atkins, Peter William *Physical Chemistry* (1994, Fifth edition)

Bowser, J *Inorganic Chemistry* (1993)

Brock, W H *The Fontana History of Chemistry* (1992)

Burgess, John *Ions in Solution: Basic Principles of Chemical Interactions* (1988)

Cox, P A *Introduction to Quantum Theory and Atomic Structure* (1996)

Crawford, Elisabeth *Arrhenius: From Ionic Theory to the Greenhouse Effect* (1997)

Donovan, A *Antoine Lavoisier: Science, Administration and Revolution* (1994)

Faraday, Michael *The Chemical History of the Candle* (1861)

Frausto da Silva, J and Williams, R *The Biological Chemistry of the Elements* (1993)

Gillam, John *The Crucible: The Story of Joseph Priestley* (1954)

Guerlac, Henry *Antoine-Laurent Lavoisier: Chemist and Revolutionary* (1975)

Hand, Clifford W and Blewitt, Harry Lyon *Acid-Base Chemistry* (1986)

Hibbert, D *Introduction to Electrochemistry* (1993)

Hill, G and Holman, J *Chemistry in Context* (1989)

Holum, T (ed) *Fundamentals of General Organic and Biological Chemistry* (1994)

Hornby, M and Peach, J *Foundations of Organic Chemistry* (1993)

Hutton, Kenneth *Chemistry: The Conquest of Materials* (1957)

Jaffe, B *Crucibles: The Story of Chemistry From Ancient Alchemy to Nuclear Fission* (1976)

Kieft, Lester and Willeford, Bennett R, Jr (eds) *Joseph Priestley: Scientist, Theologian, and Metaphysician* (1979)

Kroto, H W and Walton, D R *The Fullerenes: New Horizons for the Chemistry, Physics and Astrophysics of Carbon* (1993)

Marsh, Jerry *Advanced Organic Chemistry: Reactions, Mechanisms, and Structure* (1992, Fourth edition)

Morago, Guillermo *Cluster Chemistry: Introduction to the Chemistry of Transition Metal and Main Group Element Molecular Clusters* (1993)

Murrel, J and Jenkins, A (eds) *Properties of Liquids and Solutions* (1994)

Nicolaou, K C and Sorensen, E J *Classics in Total Synthesis* (1996)

Olah, George A and Molnar, Arpad *Hydrocarbon Chemistry* (1995)

Owen, S M and Brooker, A T *A Guide to Modern Inorganic Chemistry* (1991)

Partington, J R *A Short History of Chemistry* (1937)

Pauling, Linus Carl *The Nature of the Chemical Bond and the Structure of Molecules and Crystals: An Introduction to Modern Structural Chemistry* (1939)

Phillips, J C *Bonding and Structure in Solids* (1992); *Encyclopedia of Physical Science and Technology,* (Second edition)

Richards, W G *The Problems of Chemistry* (1986)

Roberts, R M *Serendipity: Accidental Discoveries in Science* (1989)

Serafini, Anthony *Linus Pauling: A Man and His Science* (1989)

Skoog, D A, West, D M, and Holler, F J *Analytical Chemistry: An Introduction* (1994)

Suppan, Paul *Chemistry and Light* (1994)

Thackrey, A *John Dalton: Critical Assessments of His Life and Science* (1973)

Treneer, Anne *The Mercurial Chemist: A Life of Sir Humphry Davy* (1963)

Vollhardt, K P C and Schore, N E *Organic Chemistry* (1994)

von Baeyer, H C *Taming the Atom: The Emergence of the Visible Microworld* (1992)

Warren, W S *The Physical Basis of Chemistry* (1994)

Wills, Christine and Wills, Martin *Organic Synthesis* (1995)

Winter, Mark J *Chemical Bonding* (1994)

PHYSICS

Physics is the branch of science concerned with the laws that govern the structure of the universe, and the properties of matter and energy and their interactions. For convenience, physics is often divided into branches such as atomic physics, nuclear physics, particle physics, solid-state physics, molecular physics, electricity and magnetism, optics, acoustics, heat, thermodynamics, quantum theory, and relativity. Before the 20th century, physics was known as **natural philosophy.**

the atom

An atom is the smallest unit of matter that can take part in a chemical reaction, and which cannot be broken down chemically into anything simpler. An atom is made up of protons and neutrons in a central nucleus surrounded by electrons. The atoms of the various elements differ in atomic number, relative atomic mass, and chemical behaviour.

Thomson on the Number of Corpuscles in an Atom
http://dbhs.wvusd.k12.ca.us/Chem-History/
Thomson-1906/Thomson-1906.html
J J Thomson's paper on the number of particles, or as he called them corpuscles, in an atom. The Web site is a reproduction of Thomson's publication in *Philosophical Magazine* vol 11, in June 1906.

---

nucleus The core of the atom is the nucleus, a dense body only one ten-thousandth the diameter of the atom itself. The simplest nucleus, that of hydrogen, comprises a single stable positively charged particle, the **proton.** Nuclei of other elements contain more protons, and additional particles called **neutrons**, of about the same mass as the proton but with no electrical charge. Each element has its own characteristic nucleus with a unique number of protons – the atomic number. The number of neutrons may vary. Atoms of a single element with different numbers of neutrons are called isotopes. Although some isotopes tend to be unstable and exhibit radioactivity, they all have identical chemical properties.

electron The nucleus is surrounded by a number of moving electrons, each of which has a negative charge equal to the positive charge on a proton, but which weighs only $\frac{1}{1839}$ times as much. In a neutral atom, the nucleus is surrounded by the same number of electrons as it contains protons. According to quantum theory, the position of an electron is uncertain; it may be found at any point. It is more likely, however, to be found in some places than others. The region of space in which an electron is most likely to be found is called an orbital. The chemical properties of an element are determined by the ease with which its atoms can gain or lose electrons from its outer orbitals.

orbitals

orbitals

To remember the order of atomic orbitals:

Spin pairs don't form - go higher.

The Pauli selection rule states that two electrons of like spin may not be present in any orbital if they have the same set of quantum numbers. Since the electronic orbitals in atoms are listed by the letters s, p, d, f, g, h, in order of ascending energy, this mnemonic acts as a useful reminder.

---

proton A proton is a positively charged subatomic particle, a constituent of the nucleus of all atoms. It belongs to the baryon subclass of the ɸhadrons. A proton is extremely long-lived, with a lifespan of at least 10^{32} years. It carries a unit positive charge equal to the negative charge of an electron. Its mass is almost 1,836 times that of an electron, or 1.67×10^{-27} kg. Protons are composed of two up quarks and one down quark held together by gluons (see section of quarks below). The number of protons in the atom of an element is equal to the atomic number of that element.

neutron The neutron is a composite particle, being made up of three quarks, and therefore belongs to the baryon group of the ◊hadrons. Neutrons have about the same mass as protons but no electric charge, and occur in the nuclei of all atoms except hydrogen. They contribute to the mass of atoms but do not affect their chemistry.

isotope

An isotope is one of two or more atoms that have the same atomic number (same number of protons), but which contain a different number of neutrons, thus differing in their atomic mass. They may be stable or radioactive (see radioisotope below), naturally occurring or synthesized. Hydrogen, for example, has the isotopes ^{2}H (deuterium) and ^{3}H (tritium). Elements at the lower end of the periodic table have atoms with roughly the same number of protons as neutrons. These elements are called **stable isotopes**. The stable isotopes of oxygen include ^{16}O, ^{17}O, and ^{18}O.

Fajans on the Concept of Isotopes
http://dbhs.wvusd.k12.ca.us/Chem-History/
Fajans-Isotope.html
Extract from Kasimir Fajans paper of 1913 titled *Radioactive Transformations and the Periodic System of The Elements*. The paper describes Fajans's discovery of isotopes, and goes on to offer examples of radioactive transformations leading to their production.

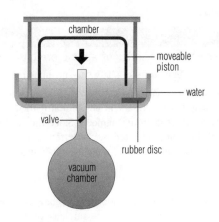

cloud chamber The cloud chamber devised by C T R Wilson was the first instrument to detect the tracks of atomic particles. It consisted originally of a cylindrical glass chamber fitted with a hollow piston, which was connected, via a valve, to a large evacuated flask. The piston falls rapidly when the valve is opened, and water vapour condenses along the tracks of any particles in the chamber.

radioisotope

A radioisotope or **radioactive isotope** is a naturally occurring or synthetic radioactive form of an element. Elements with high atomic mass numbers have many more neutrons than protons and are therefore less stable. It is these isotopes that are more prone to radioactive decay. One example is ^{238}U, uranium-238. Most natural isotopes of relative atomic mass below 208 are not radioactive. Those from 210 and up are all radioactive. Most radioisotopes are made by bombarding a stable element with neutrons in the core of a nuclear reactor (see fission below). The radiations given off by radioisotopes are easy to detect (hence their use as tracers), and in some instances can penetrate substantial thicknesses of materials, and can have profound effects (such as genetic mutation) on living matter.

splitting the atom

accelerators

An accelerator is a device to bring charged particles (such as protons and electrons) up to high speeds and energies, at which they can be of use in industry, medicine, and pure physics. At low energies, accelerated particles can be used to produce the image on a television screen and generate X-rays (by means of a ◊cathode-ray tube), destroy tumour cells, or kill bacteria. When high-energy particles collide with other particles, the fragments formed reveal the nature of the fundamental forces (described below).

Early accelerators directed the particle beam onto a stationary target; large modern accelerators usually collide beams of particles that are travelling in opposite directions. This arrangement doubles the effective energy of the collision. The world's most powerful accelerator is the 2-km/1.25-mi diameter machine at Fermilab near Batavia, Illinois, USA. This machine, the Tevatron, accelerates protons and antiprotons and then collides them at energies up to a thousand billion electron volts (or 1 TeV, hence the name of the machine). The largest accelerator is the Large Electron Positron Collider at CERN near Geneva, which has a circumference of 27 km/16.8 mi around which electrons and positrons are accelerated before being allowed to collide.

ALEPH Experiment
http://alephwww.cern.ch/Public.html
Home page of one of the four high energy particle physics experiments which uses the LEP at CERN. This Web site displays images of the components of ALEPH, including a cutaway schematic diagram of the huge detector assembly.

linear accelerator The linear accelerator, or **linac**, consists of a line of metal tubes, called drift tubes, through which the particles travel. The particles are accelerated by electric fields in the gaps between the drift tubes. The world's longest linac is also a colliding beam machine: the Stanford Linear Collider, in California, in which electrons and positrons are accelerated along a straight track, 3.2 km/2 mi long, and then steered to a head-on collision with other particles, such as protons and neutrons

cyclotron In a cyclotron an electric field is used to bend the path of a particle into a circle so that it passes repeatedly through the same electric field. A cyclotron consists of an electromagnet with two hollow, metal semicircular structures, called dees, supported between the poles of an electromagnet. Particles such as protons are introduced at the centre of the machine and travel outwards in a spiral path, being accelerated by an oscillating electric field each time they pass through the gap between the dees. Cyclotrons can accelerate particles up to energies of 25 MeV (25 million electron volts). To produce higher energies, new techniques are needed.

synchrotron In the synchrotron, particles travel in a circular path of constant radius, guided by electromagnets. The strengths of the electromagnets are varied to keep the particles on an accurate path. Electric fields at points around the path accelerate the particles.

Accelerator Physics Page
http://www-laacg.atdiv.lanl.gov/accphys.html
Virtual library dedicated to accelerator physics, with pages on design and components, as well as direct links to laboratories throughout the world.

---

fission Fission is the splitting of a heavy atomic nucleus into two or more major fragments. It is accompanied by the emission of two or three neutrons and the release of large amounts of nuclear energy. Fission occurs spontaneously in nuclei of uranium-235, the main fuel used in nuclear reactors. However, the process can also be induced by bombarding nuclei with neutrons because a nucleus that has absorbed a neutron becomes unstable and soon splits. The neutrons released spontaneously by the fission of uranium nuclei may therefore be used in turn to induce further fissions, setting up a ▷chain reaction that must be controlled if it is not to result in a nuclear explosion. The minimum amount of fissile material that can undergo a continuous chain reaction is referred to as the critical mass.

fusion

In nuclear fusion, the nuclei of light elements, such as hydrogen, combine to form the bigger nucleus of a heavier element, such as helium. The resultant loss in their combined mass is converted into energy. Stars and thermonuclear weapons are powered by nuclear fusion. Very high temperatures and pressures are thought to be required in order for fusion to take place. Under these conditions the atomic nuclei can approach each other at high speeds and overcome the mutual repulsion of their positive charges. At very close range another force, the strong nuclear force, comes into play, fusing the particles together to form a larger nucleus.

Fusion
http://www.pppl.gov/~rfheeter/
All-text site, but packed with information about fusion research and its applications. It is quite well organized and it includes a glossary of commonly used terms to aid the uninitiated.

As fusion is accompanied by the release of large amounts of energy, the process might one day be harnessed to form the basis of commercial energy production. So far no successful fusion reactor – one able to produce the required conditions and contain the reaction – has been built. An important step was taken in 1991, however, when, in an experiment that lasted 2 seconds, a 1.7 megawatt pulse of power was produced by the Joint European Torus (JET) at Culham, Oxfordshire, UK. This was the first time that a substantial amount of fusion power had been produced in a controlled experiment, as opposed to a bomb. In 1997 JET produced a record 21 megajoule of fusion power, and tested the first large-scale plant of the type needed to supply and process tritium in a future fusion power station.

subatomic particles

Subatomic particles are particles that are smaller than an atom. Such particles may be indivisible elementary particles, such as the electron and quark, or they may be composites, such as the proton, neutron, and alpha particle.

elementary particle An elementary particle is a subatomic particle that is not made up of smaller particles, and so can be considered one of the fundamental units of matter. There are three groups of elementary particles: quarks, leptons, and gauge bosons.

antiparticle An antiparticle is a particle corresponding in mass and properties to a given elementary

Principal Subatomic Particles

	Group	Particle	Symbol	Charge	Mass (MeV)	Spin	Lifetime (sec)
elementary particle	quark	up	u	$\frac{2}{3}$	336	$\frac{1}{2}$	?
		down	d	$-\frac{1}{3}$	336	$\frac{1}{2}$	?
		(top)	t	$(\frac{2}{3})$	(<600,000)	$(\frac{1}{2})$	?
		bottom	b	$-\frac{1}{3}$	4,700	$\frac{1}{2}$	?
		strange	s	$-\frac{1}{3}$	540	$\frac{1}{2}$	?
		charm	c	$\frac{2}{3}$	1,500	$\frac{1}{2}$	?
	lepton	electron	e^-	−1	0.511	$\frac{1}{2}$	stable
		electron neutrino	ν_e	0	(0)	$\frac{1}{2}$	stable
		muon	μ^-	−1	105.66	$\frac{1}{2}$	2.2×10^{-6}
		muon neutrino	ν_μ	0	(0)	$\frac{1}{2}$	stable
		tau	τ^-	−1	1,784	$\frac{1}{2}$	3.4×10^{-13}
		tau neutrino	ν_τ	0	(0)	$\frac{1}{2}$	?
	gauge boson	photon	γ	0	0	1	stable
		graviton	g	0	(0)	2	stable
		gluon	g	0	0	1	?
		weakon	$W^\pm$	±1	81,000	1	?
			Z	0	94,000	1	?
hadron	meson	pion	Π^+	1	139.57	0	2.6×10^{-8}
			Π^0	0	134.96	0	8.3×10^{-17}
		kaon	K^+	1	493.67	0	1.2×10^{-8}
			K^{0S}	0	497.67	0	8.9×10^{-11}
			K^{0L}	0	497.67	0	5.18×10^{-8}
		psi	ψ	0	3,100	1	6.3×10^{-2}
		upsilon	Υ	0	9,460	1	$\sim 1 \times 10^{-20}$
	baryon	nucleon					
		proton	p	1	938.28	$\frac{1}{2}$	stable
		neutron	n	0	939.57	$\frac{1}{2}$	920
		hyperon:					
		lambda	Λ	0	1,115.6	$\frac{1}{2}$	2.63×10^{-10}
		sigma	Σ^+	1	1,189.4	$\frac{1}{2}$	8.0×10^{-11}
			Σ^-	−1	1,197.3	$\frac{1}{2}$	1.5×10^{-10}
			Σ^0	0	1,192.5	$\frac{1}{2}$	5.8×10^{-20}
		xi	X^-	−1	1,321.3	$\frac{1}{2}$	1.64×10^{-10}
			X^0	0	1,314.9	$\frac{1}{2}$	2.9×10^{-10}
		omega	Ω	−1	1,672.4	$\frac{3}{2}$	8.2×10^{-11}

? indicates that the paricle's lifetime has yet to be determined
( ) indicates that the property has been deduced but not confirmed
MeV = million electron volts

particle but with the opposite electrical charge, magnetic properties, or coupling to other fundamental forces. For example, an electron carries a negative charge whereas its antiparticle, the positron, carries a positive one. When a particle and its antiparticle collide, they destroy each other, in the process called 'annihilation', their total energy being converted to lighter particles and/or photons. A substance consisting entirely of antiparticles is known as antimatter.

quarks

There are six types or 'flavours' of quarks (up, down, charm, strange, top, and bottom), each of which has three varieties or 'colours': red, green, and blue (visual colour is not meant, although the analogy is useful in many ways). To each quark there is an antiparticle, called an antiquark. Quarks combine in groups of three to produce heavy particles called **baryons**, and in groups of two to produce particles with masses between those of electrons and protons, called **mesons**. Baryons and mesons together are known as **hadrons**, and they and their composite particles are influenced by the strong nuclear force. Quarks have electric charges that are fractions of the electronic charge ($+\frac{2}{3}$ or $-\frac{1}{3}$ of the electronic charge).

Leaping Leptoquarks
http://www.sciam.com/explorations/032497lepto/
032497horgan.html

Part of a larger site maintained by *Scientific American*, this page follows the trail of two German physicists in search of the elusive leptoquark. Find out about events that led them to believe they were witnessing a new phenomenon: a particle that combined aspects of the two elementary particles that make up atoms, leptons and quarks.

---

baryon A baryon is a heavy subatomic particle made up of three quarks. The baryons form a subclass of the hadrons and comprise the nucleons (protons and neutrons) and hyperons. Baryons have half-integral spins.

meson A meson is a group of unstable subatomic particles made up of two indivisible elementary particles. It has a mass intermediate between that of the electron and that of the proton, is found in cosmic radiation, and is emitted by nuclei under bombardment by very high-energy particles. There are believed to be 15 ordinary types. The last of these to be found was identified by physicists at Fermilab, USA in 1998. Mesons have whole-number or zero spins.

leptons

Leptons are light particles. There are six types: the electron, muon, and tau; and their neutrinos, the electron neutrino, muon neutrino, and tau neutrino. Each also has a corresponding antiparticle. These particles are influenced by the weak nuclear force but not the strong force, and do not interact strongly with other particles or nuclei.

electron The electron is a stable, negatively charged elementary particle and is a constituent of all atoms. The electrons in each atom surround the nucleus in groupings called shells; in a neutral atom the number of electrons is equal to the number of protons in the nucleus. This electron structure is responsible for the chemical properties of the atom.

positron The positron is the antiparticle of the electron and has the same mass as an electron but an equal and opposite charge. The positron was discovered in 1932 by the US physicist Carl Anderson at Caltech, USA, its existence having been predicted by the British physicist Paul Dirac in 1928. This was the first example of antimatter.

muon The muon is an elementary particle (found by the US physicist Carl Anderson in cosmic radiation in 1937) similar to the electron except for its mass which is 207 times greater than that of the electron. It has a half-life of 2 millionths of a second, decaying into electrons and neutrinos. The muon produces the muon neutrino when it decays. The muon was originally thought to be a meson and is thus sometimes called a mu meson, although current opinion is that it is a lepton.

tau The tau is an elementary particle with the same electric charge as the electron but a mass nearly double that of a proton. It has a lifetime of around 3 x 10^{-13} seconds. The tau produces the tau neutrino when it decays.

neutrino Neutrinos are any of three uncharged elementary particles (and their antiparticles) of the lepton class, having a mass too close to zero to be measured. The most familiar type, the antiparticle of the electron neutrino, is emitted in the beta decay of a nucleus. The other two are the muon and tau neutrinos. The existence of the tau neutrino has never been conclusively confirmed but it is believed to cause a less stable particle, tau, to be emitted from the nucleus of an atom when the atom is struck by a tau neutrino. The tau decays almost instantaneously. Researchers at Fermilab believe they found traces of the tau released during an experiment in 1997.

gauge bosons

Gauge bosons carry the four fundamental forces between other particles. There are four types: gluon, photon, weakon, and graviton. The gluon carries the strong nuclear force, the photon the electromagnetic force, the weakons the weak nuclear force, and the graviton the force of gravity (see fundamental forces below). Gravitons have yet to be discovered.

gluon A gluon is a gauge boson that carries the strong nuclear force, responsible for binding quarks together

European Laboratory for Particle Physics
http://www.cern.ch/
Information about CERN, the world class physics laboratory in
Geneva. As well as presenting committees, groups, and
associations hosted by the Laboratory, this official site offers
important scientific material and visual evidence on several
activities and projects currently undertaken by the various
groups.

to form the strongly interacting subatomic particles known as þhadrons. There are eight kinds of gluon. Gluons cannot exist in isolation; they are believed to exist in balls ('glueballs') that behave as single particles. Glueballs may have been detected at CERN in 1995 but further research is required to confirm their existence.

photon The photon is the elementary particle or 'package' (quantum) of energy in which light and other forms of electromagnetic radiation are emitted. The photon has both particle and wave properties; it has no charge, is considered massless but possesses momentum and energy. It is the carrier of the electromagnetic force. According to quantum theory the energy of a photon is given by the formula $E = hf$, where h is Planck's constant and f is the frequency of the radiation emitted.

Beam Me Up: Photons and Teleportation
http://www.sciam.com/explorations/122297teleport/
Part of a larger site maintained by *Scientific American*, this
page reports on the amazing research conducted by physicists at
the University of Innsbruck who have turned science fiction into
reality by teleporting the properties of one photon particle to
another.

weakon The weakon, or **intermediate vector boson**, carries the weak nuclear force, one of the fundamental forces of nature. There are three types of weakon, the positive and negative W particle and the neutral Z particle.

radioactivity

Radioactivity is the spontaneous alteration of the nuclei of radioactive atoms, accompanied by the emission of radiation. It is the property exhibited by the radioactive isotopes of stable elements and all isotopes of radioactive elements, and can be either natural or induced. There are three types of radioactive radiation: alpha particles, beta particles, and gamma rays.

When alpha, beta, and gamma radiation pass through matter they tend to knock electrons out of atoms, ionizing them. They are therefore called ionizing radiation. Alpha particles are the most ionizing, being heavy, slow moving, and carrying two positive charges. Gamma rays are weakly ionizing as they carry no charge. Beta particles fall between alpha and gamma radiation in ionizing potential. Alpha, beta, and gamma radiation are dangerous to body tissues because of their ionizing properties, especially if a radioactive substance is ingested or inhaled.

radioactive decay Radioactive decay occurs when the unstable nuclei of radioactive elements, such as radium and various isotopes of uranium and the transuranic elements, disintegrate to become more stable. This changes the element's atomic number, thus transmuting one element into another. The energy given out by disintegrating atoms is called atomic radiation and consists either of alpha, beta, or gamma rays. Alpha and beta decay are the most common forms. Certain lighter, artificially created, isotopes also undergo radioactive decay. Radioactive decay takes place at a constant rate expressed as a specific half-life, which is the time taken for half of any mass of that particular isotope to decay completely.

Radioactive decay can take place either as a one-step decay, or through a series of steps that transmute one element into another. This is called a decay series or chain, and sometimes produces an element more radioactive than its predecessor. For example, uranium 238 decays by alpha emission to thorium 234; thorium 234 is a beta emitter and decays to give protactinium 234. This emits a beta particle to form uranium 234, which in turn undergoes alpha decay to form thorium 230. A further alpha decay yields the isotope radium 226.

half-life During radioactive decay, the time in which the strength of a radioactive source decays to half its original value is known as its half-life. In theory, the decay process is never complete and there is always some residual radioactivity. For this reason, the half-life

Half-Life	
isotope	**half-life**
(least stable)	
lithium-5	4.4×10^{-22} sec
polonium-213	4.2×10^{-6} sec
lead-211	36 min
lead-209	3.3 hours
uranium-238	4.551×10^{9} years
thorium-232	1.39×10^{10} years
tellurium-128	1.5×10^{24} years
(most stable)	

of a radioactive isotope is measured, rather than the total decay time. It may vary from millionths of a second to billions of years. Radioactive substances decay exponentially; thus the time taken for the first 50% of the isotope to decay will be the same as the time taken by the next 25%, and by the 12.5% after that, and so on. For example, carbon-14 takes about 5,730 years for half the material to decay; another 5,730 for half of the remaining half to decay; then 5,730 years for half of that remaining half to decay, and so on. Plutonium-239, one of the most toxic of all radioactive substances, has a half-life of about 24,000 years. The final product in all modes of decay is a stable element

---

Rutherford's Discovery of Half-Life
http://dbhs.wvusd.k12.ca.us/
Chem-History/Rutherford-half-life.html
Transcript of Ernest Rutherford's paper describing his discovery of the half life of radioactive materials.

---

alpha decay In alpha decay an alpha particle (two protons and two neutrons) is emitted from a nucleus and the atomic number decreases by two to form a new nucleus. For example, an atom of uranium isotope of mass 238, on emitting an alpha particle, becomes an atom of thorium, mass 234.

beta decay Beta decay is the disintegration of the nucleus of an atom to produce a beta particle, or high-speed electron, and an electron-antineutrino. During beta decay a neutron in the nucleus changes into a proton, thereby increasing the atomic number by one while the mass number stays the same. For example, the decay of the carbon 314 isotope results in the formation of an atom of nitrogen (mass 14, atomic number 7) and the emission of an electron. The mass lost in the change is converted into kinetic (movement) energy of the beta particle. Beta decay is caused by the weak nuclear force, one of the fundamental forces of nature operating inside the nucleus.

Rutherford on the Discovery of Alpha and Beta Radiation
http://dbhs.wvusd.k12.ca.us/
Chem-History/Rutherford-Alpha&Beta.html
Transcript of Rutherford's paper describes the nature of the two types of radiation he discovered to be emitted from uranium as it decays.

---

alpha particle Alpha particles are positively charged, high-energy particles emitted from the nucleus of a radioactive atom. They consist of two neutrons and two protons and are thus identical to the nucleus of a helium atom and are one of the products of the spontaneous disintegration of radioactive elements such as radium and thorium. The process of emission, **alpha decay**, transforms one element into another, decreasing the atomic (or proton) number by two and the atomic mass (or nucleon number) by four. Because of their large mass, alpha particles have a short range of only a few centimetres in air, and can be stopped by a sheet of paper. They have a strongly ionizing effect on the molecules that they strike, and are therefore capable of damaging living cells. Alpha particles travelling in a vacuum are deflected slightly by magnetic and electric fields.

beta particle The beta particle is an electron emitted at high velocity from a radioactive atom that is undergoing spontaneous disintegration. Beta particles do not exist in the nucleus but are created on disintegration – beta decay – when a neutron converts to a proton to emit an electron. Beta particles are more penetrating than alpha particles, but less so than gamma radiation; they can travel several metres in air, but are stopped by 2–3 mm of aluminium. They are less strongly ionizing than alpha particles and, like cathode rays, are easily deflected by magnetic and electric fields.

gamma radiation Gamma rays comprise very high-frequency electromagnetic radiation, similar in nature to X-rays but of shorter wavelength (wavelengths of less than 10^{-10}) emitted by the nuclei of radioactive substances during decay or by the interactions of high-energy electrons with matter. Gamma rays are stopped only by direct collision with an atom and are therefore very penetrating; they can, however, be stopped by about 4 cm/1.5 in of lead or by a very thick concrete shield. Gamma emission usually occurs as part of alpha or beta emission. They are less ionizing in their effect than alpha and beta particles, but are dangerous nevertheless because they can penetrate deeply into body tissues such as bone marrow. They are not deflected by either magnetic or electric fields. Gamma radiation is used to kill bacteria and other microorganisms, sterilize medical devices, and change the molecular structure of plastics to modify their properties (for example, to improve their resistance to heat and abrasion). Cosmic gamma rays have been identified as coming from pulsars, radio galaxies, and quasars, although they cannot penetrate the Earth's atmosphere.

electricity and magnetism

electricity
Electricity is all phenomena caused by electric charge, whether static or in motion.

electric charge Electric charge is the property of some bodies that causes them to exert forces on each other and is caused by an excess or deficit of electrons in the charged substance and is therefore either positive or negative. Objects with a like charge always repel one another while objects with an unlike charge attract each other. In atoms, electrons possess a negative charge, and protons an equal positive charge. Atoms have no charge but can sometimes gain electrons to become negative ions or lose them to become positive ions. A coulomb (C), named after the French scientist Charles Augustin de Coulomb (1736–1806), is the unit of charge, and is defined as the charge passing a point in a wire each second when the current is exactly 1 amp.

static electricity Static electricity is an electric charge that is stationary, usually acquired by a body by means of electrostatic induction or friction. Rubbing different materials can produce static electricity or cause them to be electrically charged so that it they have an excess or deficit of electrons, as seen in the sparks produced on combing one's hair or removing a nylon shirt. This charge on the object exerts an electric field in the space around itself that can attract or repel other objects. In some processes static electricity is useful, as in paint spraying where the parts to be sprayed are charged with electricity of opposite polarity to that on the paint droplets, and in xerography.

electric current An electric current is the movement of electrically charged particles through a conducting material. For charge to flow in a circuit there must be a ◊potential difference (pd) applied across the circuit. Conventionally, current is regarded as a movement of positive electricity from points at high potential to points at a lower potential. Potential difference is often supplied in the form of a battery that has a positive terminal and a negative terminal. Under the influence of the potential difference, the electrons are repelled from the negative terminal side of the circuit and attracted to the positive terminal of the battery. A steady flow of electrons around the circuit is produced. Current flowing through a circuit can be measured using an ammeter and is measured in ◊amperes (or

standard potential divider

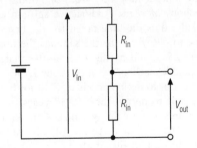

potentiometer used as a potential divider

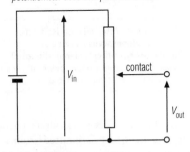

potential divider A potential divider is a resistor or a chain of resistors connected in series in an electrical circuit. It is used to obtain a known fraction of the total voltage across the whole resistor or chain. When a variable resistor, or potentiometer, is used as a potential divider, the output voltage can be varied continuously by sliding a contact along the resistor. Devices like this are used in electronic equipment to to vary volume, tone, and brightness control.

amps). **Direct current** (DC) flows continuously in one direction; **alternating current** (AC) flows alternately in each direction. The flow of current is measured in amperes (symbol A).

In a circuit the battery provides energy to make charge flow through the circuit. The amount of energy supplied to each unit of charge is called the electromotive force (emf). The unit of emf is the volt (V). A battery has an emf of 1 volt when it supplies 1 joule

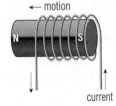

← motion

current

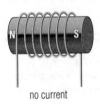

no current

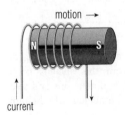

motion →

current

induced current Movement of a magnet in a coil of wire induces a current.

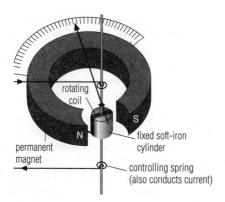

moving-coil meter A simple moving-coil meter. Direct electric current (DC) flowing through the wire coil combined with the presence of a magnetic field causes the coil to rotate; this in turn moves a pointer across a calibrated scale so that the degree of rotation can be related to the magnitude of the current.

electronics

To remember whether current leads voltage or lags it in reactive circuits:

Think of 'Eli the Ice man'. In inductive ('L') circuits, voltage ('E') leads current ('I'), hence 'E L I'. In capacitive ('C') circuits, it is the other way, so 'I C E'

of energy to each coulomb of charge flowing through it. The energy carried by flowing charges can be used to do work, for example to light a bulb, to cause current to flow through a resistor, to emit radiation, or to produce heat. When the energy carried by a current is made to do work in this way, a potential difference can be measured across the circuit component concerned by a voltmeter or a cathode-ray oscilloscope. The potential difference is also measured in volts. Power, measured in watts, is the product of current and voltage. Although potential difference and current measure different things, they are related to

circuit

To remember the order of items in an AC circuit:

The **voltage (V)** across a **capacitor (C)** lags the **current (I)** by 90 degrees. The voltage across the **inductor (L)** leads the current by 90 degrees. This is given by the word CIVIL, when split into CIV and VIL.

one another. This relationship was discovered by the German physicist Georg Ohm, and is expressed by Ohm's law: the current through a wire is proportional to the potential difference across its ends. The potential difference divided by the current is a constant for a given piece of wire. This constant for a given material is called the resistance. When current flows in a component possessing resistance, electrical energy is converted into heat energy.

Ohm's law

To remember one expression of this:

Vampires **a**re **r**are.

(volts = amps x resistance)

conductors Electrical conduction is the flow of charged particles through a material giving rise to electric current. Electrical conductors are substances, such as metals, that allow the passage of electricity through them readily. In metals and other conducting materials, the charge is carried by negatively charged free electrons that are not bound tightly to the atoms and are thus able to move through the material.

Conduction in many liquids involves a flow not merely of electrons, but of atoms or groups of atoms as well. When a salt such as sodium chloride is dissolved in water, the chlorine atoms each gain an electron and become negatively charged, while the sodium atoms each lose one and become positively charged. These charged atoms, or ions, can move through the liquid and transport electricity. Current flows by the movement of charged ions through a solution or molten salt (the electrolyte), resulting in the migration of ions to the electrodes: positive ions (cations) to the negative electrode (cathode) and negative ions (anions) to the positive electrode (anode). This process is called **electrolysis** and represents bi-directional flow of charge as opposite charges move to oppositely charged electrodes. In metals, charges are only carried by free electrons and therefore move in only one direction. Gases are, under normal circumstances, almost completely nonconducting. They may be ionized by irradiation with X-rays or by radioactive radiations. They are more readily maintained in a conducting state at high temperatures, as in the electric arc, or at low pressures, as in electric discharge lamps.

A magnetic field is created around all conductors that carry a current. When a current-bearing conductor is made into a coil it forms an electromagnet with a magnetic field that is similar to that of a bar magnet, but which disappears as soon as the current is switched off. The strength of the magnetic field is

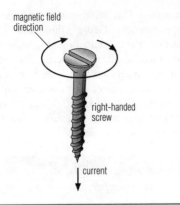

magnetic field
direction

right-handed
screw

current

Maxwell's screw rule Maxwell's screw rule, named after the physicist James Maxwell, predicts the direction of the magnetic field produced around a wire carrying electric current. If a right-handed screw is turned so that it moves forward in the same direction as the current, its direction of rotation will give the direction of the magnetic field.

directly proportional to the current in the conductor – a property that allows a small electromagnet to be used to produce a pattern of magnetism on recording tape or disc that accurately represents the sound or data to be stored. The direction of the field created around a conducting wire may be predicted by using Maxwell's screw rule.

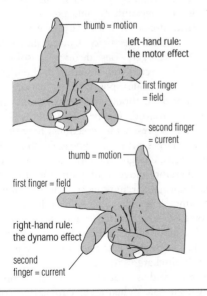

thumb = motion

left-hand rule:
the motor effect

first finger
= field

second finger
= current

thumb = motion

first finger = field

right-hand rule:
the dynamo effect

second
finger = current

Fleming's rules Fleming's rules give the direction of the magnetic field, motion, and current in electrical machines. The left hand is used for motors, and the right hand for generators and dynamos.

Electrical Decomposition by Michael Faraday
http://dbhs.wvusd.k12.ca.us/Chem-History/
Faraday-electrochem.html
Transcript of Faraday's paper in *Philosophical Transactions of the Royal Society*, 1834 in which Faraday describes for the first time the phenomena of electrolysis.

A conductor carrying current in a magnetic field experiences a force, and is impelled to move in a direction perpendicular to both the direction of the current and the direction of the magnetic field. The direction of motion may be predicted by Fleming's left-hand rule. The magnitude of the force experienced depends on the length of the conductor and on the strengths of the current and the magnetic field, and is greatest when the conductor is at right angles to the field. A conductor wound into a coil that can rotate between the poles of a magnet forms the basis of an electric motor.

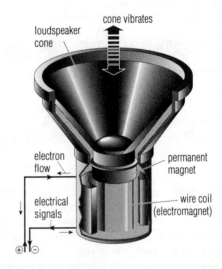

cone vibrates

loudspeaker
cone

electron
flow

electrical
signals

permanent
magnet

wire coil
(electromagnet)

loudspeaker A moving-coil loudspeaker. Electrical signals flowing through the wire coil turn it into an electromagnet, which moves as the signals vary. The attached cone vibrates, producing sound waves.

insulators Insulators, such as rubber, are extremely poor conductors in which the electrons are more tightly bound to the atoms and conduction is less.

semiconductors Semiconductors are substances with electrical conductivity intermediate between metals and insulators and used in a wide range of electronic devices. Certain crystalline materials, most notably silicon and germanium, have a small number of free electrons that have escaped from the bonds between the

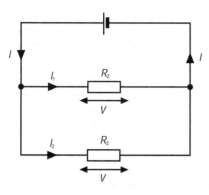

parallel circuit In a parallel circuit, the components are connected side by side, so that the current is split between two or more parallel paths or conductors.

atoms. The atoms from which they have escaped possess vacancies, called holes, which are similarly able to move from atom to atom and can be regarded as positive charges. Current can be carried by both electrons (negative carriers) and holes (positive carriers). Such materials are known as **intrinsic semiconductors**.

Conductivity of a semiconductor can be enhanced by doping the material with small numbers of impurity atoms which either release free electrons (making an **n-type semiconductor** with more electrons than holes) or capture them (a **p-type semiconductor** with more holes than electrons). When p-type and n-type materials are brought together to form a p–n junction, an electrical barrier is formed which conducts current more readily in one direction than the other. This is the basis of the semiconductor diode, used for rectification, and numerous other devices including ◊transistors, rectifiers, and ◊integrated circuits (silicon chips). The conductivity of semi-conductors can also be improved by the addition of heat or light. Increase

of temperature frees more electrons, so the conductivity of nonmetals increases with rising temperature.

superconductivity Superconductivity is the increase in electrical conductivity at low temperatures. The resistance of some metals and metallic compounds decreases uniformly with decreasing temperature until at a critical temperature (the superconducting point), within a few degrees of absolute zero (0K/ –273.15°C/–459.67°F), the resistance suddenly falls to zero. In the superconducting state, an electric current will continue indefinitely once started, provided that the material remains below the superconducting point

In 1986 IBM researchers achieved superconductivity with some ceramics at –243°C/ –405°F, opening up the possibility of 'high-temperature' superconductivity; Paul Chu at the University of Houston, Texas, achieved superconductivity at –179°C/ –290°F, a temperature that can be sustained using liquid nitrogen. In 1993 Swiss researchers produced an alloy of mercury, barium, and copper which becomes superconducting at 133 K (–140°C/–220°F). A high-temperature semiconductor material, called bismuth ceramic, which is superconducting at 100 K, became commercially available in 1997.

Some metals, such as platinum and copper, do not become superconductive; as the temperature decreases, their resistance decreases to a certain point but then rises again. Superconductivity can be nullified by the application of a large magnetic field.

Introduction to High Temperature Superconductivity
http://phycmt2.sogang.ac.kr/~smshin/physics/
curriculum/superconductor/superconductor.html
Texas Centre for Superconductivity at the University of Houston, USA, offers a beginners guide to the phenomenon of high temperature superconductivity. The site, run by one of the world centres for superconductor research, includes some of the classic images of superconductors at work, and describes in reasonably understandable terms the history and theory of the materials and their bizarre properties.

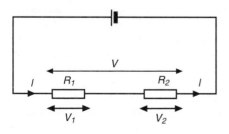

series circuit In a series circuit, the components of the circuit are connected end to end, so that the current passes through each component one after the other, without division or branching into parallel circuits.

electromagnetism Magnetic fields are produced either by current-carrying conductors or by permanent magnets. In current-carrying wires, the magnetic field lines are concentric circles around the wire. Their direction depends on the direction of the current and their strength on the size of the current. If a conducting wire is moved within a magnetic field, the magnetic field acts on the free electrons within the conductor, displacing them and causing a current to flow. The force acting on the electrons and causing them to move is greatest when the wire is perpendicular to the magnetic field lines. The direction of the current is given by the left-hand rule. The generation of a current by the relative movement of a conductor in a magnetic field is called electromagnetic induction. This is the basis of how a dynamo works.

magnetism

magnet A magnet is any object that forms a magnetic field (displays magnetism), either permanently or temporarily through induction, causing it to attract materials such as iron, cobalt, nickel, and alloys of these. It always has two magnetic poles, called north and south. The world's most powerful magnet was built in 1997 at the Lawrence Berkeley National Laboratory, California. It produces a field 250,000 times stronger than the Earth's magnetic field (13.5 teslas). The coil magnet is made of an alloy of niobium and tin.

magnetism and magnetic fields Magnetism refers to phenomena associated with magnetic fields. A magnetic field is the region around a permanent magnet, or around a conductor carrying an electric current, in which a force acts on a moving charge or on a magnet placed in the field. Magnetic fields are produced by moving charged particles: in electromagnets, electrons flow through a coil of wire connected to a battery; in permanent magnets, spinning electrons within the atoms generate the field. The field can be represented by lines of force, which by convention link north and south poles and are parallel to the directions of a small compass needle placed on them. A magnetic field's magnitude and direction are given by the magnetic flux density, expressed in teslas.

All substances are magnetic to a greater or lesser degree, and their magnetic properties, however feeble, may be observed when they are placed in an intense magnetic field. Materials that can be strongly magnetized, such as iron, cobalt, and nickel, are said to be **ferromagnetic**; this is due to the formation of areas called domains in which atoms, weakly magnetic because of their spinning electrons, align to form areas of strong magnetism. Magnetic materials lose their magnetism if heated to the ◊Curie temperature. Furthermore, if the magnetizing force is increased, a stage is reached when the magnet becomes saturated, that is, its pole strength reaches a maximum value. Most substances are **paramagnetic**, being only weakly pulled towards a strong magnet. This is because their atoms have a low level of magnetism and do not form ◊domains. **Diamagnetic** materials, notably bismuth, are weakly repelled by a magnet since electrons within their atoms act as electromagnets and oppose the applied magnetic force. **Antiferromagnetic** materials have a very low susceptibility that increases with temperature.

molecular magnets When a magnet is broken in two, we do not obtain two halves, one with a north pole, the other with a south pole. Two new poles appear at the point of fracture. However often this process is repeated the same result is obtained: every magnet has two poles.

The German physicist Wilhelm Weber suggested that every magnet was really composed of magnetic particles or magnetized domains that are now believed to be of molecular dimensions. The Scottish physicist J Alfred Ewing (1855–1935) developed his theory and suggested that, since the act of magnetization did not change the chemical character nor the weight of the specimen, but simply endowed it with magnetic properties, magnetizable substances consisted of molecular magnets. According to this theory, an ordinary piece of iron is made up of molecular magnets arranged in haphazard fashion, so that they neutralize each other's effects on external bodies. This disorder disappears when the iron is placed in a magnetic field and the molecular magnets are set with their axes parallel to the field: free poles appear at the ends of the magnet, while the central portions exhibit only feeble magnetic powers because equal and opposite poles neutralize each other's effects. This theory accounts for the appearance of new poles wherever the magnet is broken, and the state of saturation is reached when all the molecular magnets have been arranged in order. Subsequent loss of magnetism is explained by the partial return to disordered array.

magneton theory Early in the 20th century, Pierre Weiss suggested the existence of the magneton or elementary magnet, an analogue of the electron, the elementary charge of electricity. An electric current flowing around a circular coil has a magnetic field similar to that of a magnet whose axis coincides with that of the coil: the electrical theory of matter attempts to ascribe the magnetic properties of bodies to the orbital motions of the electrons in the atom. The ◊quantum theory of the atom developed by Niels Bohr supported the magneton theory, and subsequently direct experimental evidence of the existence of the magnetic moment associated with electron orbits was obtained by Otto Stern and Walther Gerlach in 1921.

matter

Matter is anything that has mass. All matter is made up of atoms, which in turn are made up of elementary particles; it ordinarily exists in one of three physical states: solid, liquid, or gas. The state it exists in depends on its temperature and the pressure on it. ◊Kinetic theory describes how the state of a material depends on the movement and arrangement of its atoms or molecules. In a solid, the atoms or molecules vibrate in a fixed position. In a liquid, they do not occupy fixed positions as in a solid, and yet neither do they have the freedom of random movement that occurs within a gas, so the atoms or molecules within a liquid will always follow the shape of their container. The transition between states takes place at definite temperatures, called melting point and boiling point. In chemical reactions matter is conserved, so no matter is lost or gained and the sum of the mass of the reactants will always equal the sum of the end products.

antimatter

Antimatter is a form of matter in which most of the attributes (such as electrical charge, magnetic moment, and spin) of elementary particles are reversed. Such particles (antiparticles) can be created in particle accelerators, such as those at CERN in Geneva, Switzerland, and at Fermilab in the USA. In 1996 physicists at CERN created the first atoms of antimatter: nine atoms of antihydrogen survived for 40 nanoseconds.

mass

Mass is the quantity of matter in a body as measured by its inertia (tendency to remain in a state of rest or uniform motion until an external force is applied).

mass and volume

To remember that if an object floats, it displaces water equal to its mass, but if it sinks, it displaces water equal to its volume.

Think of a pebble, made of neutronium. It is small, but it weighs a lot. If it were to displace water equal to its mass, then when you threw this little pebble into a swimming pool, all the water would have to jump out of the swimming pool. So it must only displace water equal to its volume.

Mass determines the acceleration produced in a body by a given force acting on it, the acceleration being inversely proportional to the mass of the body. The mass also determines the force exerted on a body by gravity on Earth, although this attraction varies slightly from place to place. In the SI system, the base unit of mass is the kilogram. At a given place, equal masses experience equal gravitational forces, which are known as the weights of the bodies. Masses may, therefore, be compared by comparing the weights of bodies at the same place.

density Density is the measure of the compactness of a substance; it is equal to its mass per unit volume and is measured in kg per cubic metre/lb per cubic foot. Density is a scalar quantity. The average density D of a mass m occupying a volume V is given by the formula: $D = m/V$. Relative density is the ratio of the density of a substance to that of water at 4°C/32.2°F.

solid

A solid is a state of matter that holds its own shape (as opposed to a liquid, which takes up the shape of

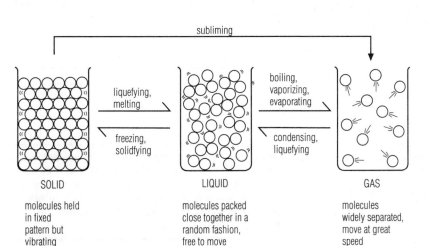

SOLID	molecules held in fixed pattern but vibrating
LIQUID	molecules packed close together in a random fashion, free to move
GAS	molecules widely separated, move at great speed

change of state The state (solid, liquid, or gas) of any substance is not fixed but varies with changes in temperature and pressure.

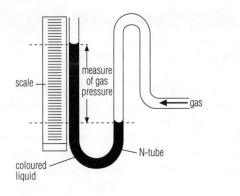

scale

measure
of gas
pressure

gas

N-tube

coloured
liquid

manometer The manometer indicates gas pressure by the rise of liquid in the tube.

its container, or a gas, which totally fills its container). According to ◊kinetic theory, the atoms or molecules in a solid are not free to move but merely vibrate about fixed positions, such as those in crystal lattices.

liquid

A liquid is a state of matter between a solid and a gas. A liquid forms a level surface and assumes the shape of its container. Its atoms do not occupy fixed positions as in a crystalline solid, nor do they have freedom of movement as in a gas. Unlike a gas, a liquid is difficult to compress since pressure applied at one point is equally transmitted throughout (Pascal's principle). Hydraulics makes use of this property.

gas

Gas is a form of matter, such as air, in which the molecules move randomly in otherwise empty space, filling any size or shape of container into which the gas is put. A sugar-lump sized cube of air at room temperature contains 30 trillion molecules moving at an average speed of 500 metres per second (1,800 kph/1,200 mph). **Plasma** is an ionized gas produced at extremely high temperatures, as in the Sun and other stars, which contains positive and negative charges in equal numbers. It is a good electrical conductor. In thermonuclear reactions the plasma produced is confined through the use of magnetic fields.

mechanics

Mechanics is the branch of physics dealing with the motions of bodies and the forces causing these motions, and also with the forces acting on bodies in equilibrium. It is usually divided into **dynamics**, or **kinetics**, and **statics**. Dynamics is the mathematical and physical study of the behaviour of bodies under the action of forces that produce changes of motion in them. Statics is concerned with the behaviour of bodies at rest and forces or moving with constant velocity where the forces acting on the body cancel each other out; that is, the forces are in equilibrium.

machines

As well as dealing with the direct action of forces on bodies, mechanics studies the nature and action of forces when they act on bodies by the agency of machinery. This gives the origin of the word 'mechanics': in its early stages it was the science of making machines. A machine in mechanics means any contrivance in which a force applied at one point is made to raise weight or overcome a resisting force acting at another point. All machines can be resolved into three primary machines: the lever, the inclined plane, and the wheel and axle.

mechanical advantage

To remember the definition of mechanical advantage:

Men **a**lways **l**ike **e**ating.
(MA - load over effort [l/e])

inertia Inertia is the tendency of an object to remain in a state of rest or uniform motion until an external force is applied, as described by Isaac Newton's first law of motion.

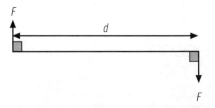

F

d

F

couple Two equal but opposite forces (F) will produce a turning effect on a rigid body, provided that they do not act through the same straight line. The turning effect, or moment, is equal to the magnitude of one of the turning forces multiplied by the perpendicular distance (d) between those two forces.

Newton's laws of motion

Isaac Newton's three laws of motion form the basis of Newtonian mechanics. (1) Unless acted upon by an external force, a body at rest stays at rest, and a moving body continues moving at the same speed in the same straight line. (2) An external force applied to a body gives it an acceleration proportional to the force (and in the direction of the force) and inversely proportional to the mass of the body. (3) When a body A exerts a force on a body B, B exerts an equal and opposite force on A; that is, to every action there is an equal and opposite reaction.

force

Force is any influence that tends to change the state of rest or the uniform motion in a straight line of a body. The action of an unbalanced or resultant force results in the acceleration of a body in the direction of action of the force, or it may, if the body is unable to move freely, result in its deformation (see ⟡Hooke's

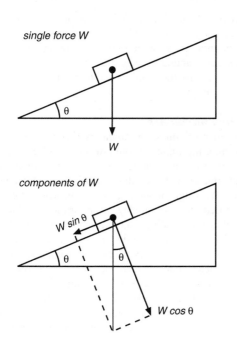

single force W

W

components of W

W sin θ

W cos θ

resolution of forces In mechanics, the resolution of forces is the division of a single force into two parts that act at right angles to each other. In the diagram, the weight W of an object on a slope, tilted at an angle θ, can be resolved into two parts or components: one acting at a right angle to the slope, equal to Wcosθ, and one acting parallel to and down the slope, equal to Wsinθ.

law). Force is a vector quantity, possessing both magnitude and direction; its SI unit is the newton.

speed and distance

In order to understand movement and what causes it, we need to be able to describe it. Speed is a measure of how fast something is moving. Speed is measured by dividing the distance travelled by the time taken to travel that distance. Hence speed is distance moved in unit time. Speed is a scalar quantity in which the direction of travel is not important, only the rate of travel.

velocity

Velocity is the speed of an object in a given direction. Velocity is therefore a vector quantity, in which both magnitude and direction of movement must be taken into account. The velocity at any instant of a particle travelling in a curved path is in the direction of the tangent to the path at the instant considered. The velocity v of an object travelling in a fixed direction may be calculated by dividing the distance s it has travelled by the time t taken to do so, and may be expressed as: $v = s/t$.

acceleration

Acceleration is the rate of change of velocity of a moving body with time. This is also a vector quantity. Acceleration is usually measured in metres per second per second (m s^{-2}) or feet per second per second (ft s^{-2}). Because velocity is a vector quantity (possessing both magnitude and direction), a body travelling at constant speed may be said to be accelerating if its direction of motion changes. According to Isaac Newton's law of gravitation, all objects fall to Earth with the same acceleration, regardless of mass. According to Newton's second law of motion, a body will accelerate only if it is acted upon by an unbalanced, or resultant, force. Acceleration due to gravity is the acceleration of a body falling freely under the influence of the Earth's gravitational field; it varies slightly at different latitudes and altitudes. The value adopted internationally for gravitational acceleration is 9.806 m s^{-2}/32.174 ft s^{-2}. The average acceleration a of an object travelling in a straight line over a period of time t may be calculated using the formula:

> **gravity**
> The maximum speed with which a falling raindrop can hit you is about 29 kmph/18 mph. In a vacuum, the further an object falls, the more speed it gains, but in the real world, air resistance eventually balances out the accelerating effect of gravity.

Physical Constants

Physical constants, or fundamental constants, are standardized values whose parameters do not change.

Constant	Symbol	Value in SI units
acceleration of free fall	g	9.80665 m s^{-2}
Avogadro's constant	N_A	6.0221367×10^{23} mol^{-1}
Boltzmann's constant	k	1.380658×10^{-23} J K^{-1}
elementary charge	e	$1.60217733 \times 10^{-19}$ C
electronic rest mass	m_e	$9.1093897 \times 10^{-31}$ kg
Faraday's constant	F	9.6485309×10^4 C mol^{-1}
gas constant	R	8.314510 J K^{-1} mol^{-1}
gravitational constant	G	6.672×10^{-11} N m^2 kg^{-2}
Loschmidt's number	N_L	2.686763×10^{25} m^{-3}
neutron rest mass	m_n	$1.6749286 \times 10^{-27}$ kg
Planck's constant	h	$6.6260755 \times 10^{-34}$ J s
proton rest mass	m_p	$1.6726231 \times 10^{-27}$ kg
speed of light in a vacuum	c	2.99792458×10^8 m s^{-1}
standard atmosphere	atm	1.01325×10^5 Pa
Stefan–Boltzmann constant	σ	5.67051×10^{-8} W m^{-2} K^{-4}

$a = v - \frac{u}{t}$, where u is its initial velocity and v its final velocity.

A negative answer shows that the object is slowing down (decelerating).

momentum Momentum is a function both of the mass of a body and of its velocity and is the product of the mass of a body and its velocity. If the mass of a body is m kilograms and its velocity is v m s^{-1} then its momentum is given by: momentum = mv. Its unit is the kilogram metre-per-second (kg ms^{-1}) or the newton second. The momentum of a body does not change unless a resultant or unbalanced force acts on that body. The law of conservation of momentum is one of the fundamental concepts of classical physics. It states that the total momentum of all bodies in a closed system is constant and unaffected by processes occurring within the system.

forces and motion Galileo discovered that a body moving on a perfectly smooth horizontal surface would neither speed up nor slow down. All moving bodies continue moving with the same velocity unless a force is applied to cause an acceleration. The reason we appear to have to push something to keep it moving with constant velocity is because of frictional forces acting on all moving objects on Earth. Friction occurs when two solid surfaces rub on each other; for example, a car tyre in contact with the ground. Friction opposes the relative motion of the two objects in contact and acts to slow the velocity of the moving object. A force is required to push the moving object and to cancel out the frictional force. If the forces combine to give a net force of zero, the object will not accelerate but will continue moving at constant velocity. A resultant force is a single force acting on a particle or body whose effect is equivalent to the combined effects of two or more separate forces.

quantum mechanics
Quantum mechanics, or **quantum theory**, superseded Newtonian mechanics in the interpretation of physical phenomena on the atomic scale and is the theory that energy (the capacity for doing work) does not have a continuous range of values, but is, instead, absorbed or radiated discontinuously, in multiples of definite, indivisible units called quanta. Just as earlier theory showed how light, generally seen as a wave motion, could also in some ways be seen as composed of discrete particles (photons), quantum theory shows how atomic particles such as electrons may also be seen as having wavelike properties. Quantum theory is the basis of particle physics, modern theoretical chemistry, and the solid-state physics that describes the behaviour of the silicon chips used in computers.

Subatomic Logic
http://www.sciam.com/explorations/
091696explorations.html
Part of a larger site maintained by *Scientific American*, this page provides information on the recent progress of scientists who are attempting to harness quantum physics to run a lightning fast, super-charged 'quantum computer'. This article explains how a quantum computer would work and why it would be so much faster than silicon-based computer systems.

Quantum Age Begins
http://www-history.mcs.st-and.ac.uk/~history/
HistTopics/The_Quantum_age_begins.html
St Andrews University-run Web site chronicling the discovery of
quantum theory.

Explanation of Temperature Related Theories
http://www.unidata.ucar.edu/staff/blynds/t
mp.html#KT
Detailed explanatory site on the laws and theories of
temperature. It explains what temperature actually is, what a
thermometer is, and the development of both, complete with
illustrations and links to pioneers in the field. There is a
temperature conversion facility and explanations of associated
topics such as kinetic theory and thermal radiation.

thermodynamics

Thermodynamics deals with the transformation of heat into and from other forms of energy. It is the basis of the study of the efficient working of engines, such as the steam and internal combustion engines. The three laws of thermodynamics are: (1) energy can be neither created nor destroyed, heat and mechanical work being mutually convertible; (2) it is impossible for an unaided self-acting machine to convey heat from one body to another at a higher temperature; and (3) it is impossible by any procedure, no matter how idealized, to reduce any system to the absolute zero of temperature (0K/–273°C/–459°F) in a finite number of operations. Put into mathematical form, these laws have widespread applications in physics and chemistry.

Schrödinger's Cation
http://www.sciam.com/explorations/
061796explorations.html
Part of a larger site maintained by *Scientific American*,
this page features an explanation of the quantum
mechanics paradox known as 'Schrödinger's Cat', an
experiment devised by Erwin Schrödinger to illustrate
the difference between the quantum and macroscopic
worlds.

heat

Heat is a form of energy possessed by a substance by virtue of the vibrating movement (kinetic energy) of its molecules or atoms. Heat energy is transferred by conduction, convection, and radiation. It always flows from a region of higher temperature (heat intensity) to one of lower temperature. Its effect on a substance may be simply to raise its temperature, or to cause it to expand, melt (if a solid), vaporize (if a liquid), or increase its pressure (if a confined gas).

Quantities of heat are usually measured in units of energy, such as joules (J) or calories (cal). The **specific heat** of a substance is the ratio of the quantity of heat required to raise the temperature of a given mass of the substance through a given range of temperature to the heat required to raise the temperature of an equal mass of water through the same range. It is measured by a calorimeter.

conduction

Conduction is flow of heat energy through a material without the movement of any part of the material itself (compare electrical conduction, described above) – for example, when the whole length of a metal rod is heated when one end is held in a fire. Heat energy is present in all materials in the form of the kinetic energy of their vibrating molecules, and may be conducted from one molecule to the next in the form of this mechanical vibration. In the case of metals, which are particularly good conductors of heat, the free electrons within the material carry heat around very quickly.

convection

Convection is the transmission of heat through a fluid (liquid or gas) in currents – for example, when the air in a room is warmed by a fire or radiator.

radiation

Radiation is heat transfer by infrared rays. All objects radiate heat; hotter objects emit more energy than cooler objects. Infrared radiation can pass through a

heat transfer

To remember the principles of heat transfer:

Conduction – imagine a line of passengers on a bus being asked to move down by the **conductor**, each passenger causing the next to bustle along (analogy for the movement/vibration of atoms that is passed along, causing heat to be transferred)

Convection – consider **vector**, a disease-carrying insect, e.g. a mosquito, which travels in swarms (very much like the movement of convection currents)

Radiation – heat radiation is a form of radiation – (think of nuclear fallout or the Sun's radiation) and thus travels in waves undetected until they fall upon another body vacuum, travels at the same speed as light, can be reflected and refracted, and does not affect the medium through which it passes. For example, heat reaches the Earth from the Sun by radiation.

energy

Energy is the capacity for doing work. Energy can exist in many different forms. For example, potential energy (PE) is energy deriving from position; thus a stretched spring has elastic PE, and an object raised to a height above the Earth's surface, or the water in an elevated reservoir, has gravitational PE. Moving bodies possess kinetic energy (KE). All atoms and molecules possess some amount of kinetic energy because they are all in some state of motion (see ◊kinetic theory). Adding heat energy to a substance increases the mean kinetic energy and hence the mean speed of its constituent molecules – a change that is reflected as a rise in the temperature of that substance.

Brownian Motion
http://dbhs.wvusd.k12.ca.us/Chem-History/
Brown-1829.html
Transcript of 'Remarks on Active Molecules' by Robert Brown from *Additional Remarks on Active Molecules* (1829). The text describes Robert Brown's observations of the random motion of particles.

Energy can be converted from one form to another, but the total quantity in a system stays the same (in accordance with the conservation of energy principle). Energy cannot be created or destroyed. For example, as an apple falls it loses gravitational PE but gains KE. Although energy is never lost, after a number of conversions it tends to finish up as the kinetic energy of random motion of molecules (of the air, for example) at relatively low temperatures. This is 'degraded' energy that is difficult to convert back to other forms. All forms of energy tend to be transformed into heat and can not then readily be converted into other, useful forms of energy.

A body with no energy can do no work. For example, a flat battery in a torch will not light the torch. If the battery is fully charged, it should contain enough chemical energy to do the work involved in illuminating the torch bulb. When one body A does work on another body B, A transfers energy to B. The energy transferred is equal to the work done by A on B. Energy is therefore measured in joules. The rate of doing work or consuming energy is called power and is measured in watts (joules per second).

It is now recognized that mass can be converted into energy under certain conditions, according to Einstein's theory of relativity. This conversion of mass into energy is the basis of atomic power. Einstein's special theory of relativity (1905; see below) correlates any gain, E, in energy with a gain, m, in mass, by the equation $E = mc^2$, in which c is the speed of light. The conversion of mass into energy in accordance with this equation applies universally, although it is only for nuclear reactions that the percentage change in mass is large enough to detect.

radiation

Radiation is the emission of radiant energy as particles or waves – for example, heat, light, alpha particles, and beta particles (see under electromagnetic waves below and radioactivity above). Of the radiation given off by the Sun, only a tiny fraction of it, called insolation, reaches the Earth's surface; much of it is absorbed and scattered as it passes through the atmosphere. The radiation given off by the Earth itself is called **ground radiation**.

Radiation Reassessed
http://whyfiles.news.wisc.edu/020radiation/index.html
Part of the Why Files project, published by the National Institute for Science Education (NISE) and funded by the National Science Foundation, this page provides insight into the controversy concerning the health effects of ionizing radiation.

---

background radiation Background radiation is radiation that is always present in the environment. By far the greater proportion (87%) of it is emitted from natural sources. Alpha and beta particles and gamma radiation are radiated by the traces of radioactive minerals that occur naturally in the environment and even in the human body, and by radioactive gases such as radon and thoron, which are found in soil and may seep upwards into buildings. Radiation from space (cosmic radiation) also contributes to the background level.

Radioactivity in Nature
http://www.sph.umich.edu/group/eih/UMSCHPS/
natural.htm
Detailed explanation of the different types of radiation found naturally on Earth and in its atmosphere, as well as those produced by humans. It includes tables of the breakdown of nuclides commonly found in soil, the oceans, the air, and even the human body.

---

solar radiation Solar radiation is radiation given off by the Sun and consists mainly of visible light, ultraviolet radiation, and infrared radiation although the

electromagnetic spectrum

To remember the different categories of radiation, in order of increasing wavelength:

Cary **G**rant **e**xpects **u**nanimous **v**otes **i**n **m**ovie **r**eviews **t**onight

(**C**osmic, **g**amma, **X**-rays, **u**ltraviolet, **v**isible, **i**nfrared, **m**icrowave, **r**adio, **t**elevision)

whole spectrum of electromagnetic waves is present, from radio waves to X-rays. High-energy charged particles, such as electrons, are also emitted, especially from solar flares. When these reach the Earth, they cause magnetic storms (disruptions of the Earth's magnetic field), which interfere with radio communications.

ultraviolet radiation Ultraviolet radiation is electromagnetic radiation near the short wavelength–high frequency end of the electromagnetic spectrum with wavelengths from about 400 to 4 nm (where the X-ray range begins). Physiologically, ultraviolet radiation is extremely powerful, producing sunburn and causing the formation of vitamin D in the skin. Ultraviolet rays are also strongly germicidal and may be produced artificially by mercury vapour and arc lamps for therapeutic use.

infrared radiation Infrared radiation is invisible electromagnetic radiation of wavelength between between 10^{-4} m and 7×10^{-7} m – that is, between the limit of the red end of the visible spectrum and the shortest microwaves. All bodies above the absolute zero of temperature ($0K/-273.15°C/-459.67°F$) absorb and radiate infrared radiation. Infrared absorption spectra are used in chemical analysis, particularly for organic compounds.

X-rays X-rays are a band of electromagnetic radiation in the wavelength range 10^{-11} to 10^{-9} m (between gamma rays and ultraviolet radiation. Applications of

X-rays make use of their short wavelength (as in X-ray diffraction) or their penetrating power (as in medical X-rays of internal body tissues). X-rays are dangerous and can cause cancer. The X-rays used in radiotherapy have very short wavelengths that penetrate tissues deeply and destroy them.

waves

Waves are oscillations that are propagated from a source. **Mechanical waves** require a medium through which to travel. **Electromagnetic waves** do not; they can travel through a vacuum. Waves carry energy but they do not transfer matter.

Introduction to Waves
http://id.mind.net/~zona/wintro.html
Interactive site that begins with the basics – explaining and allowing you to manipulate wavelength, amplitude, and phase shift of a simple wave. Further into the site there are more complex examinations of such things as Huygen's principle, interference, and wave propagation. You will need to have a Java-enabled browser to get the most out of this site.

---

amplitude The amplitude is the maximum displacement of an oscillation from the equilibrium position (the height of a crest or the depth of a trough). With a sound wave, for example, amplitude corresponds to the intensity (loudness) of the sound. If a mechanical system is made to vibrate by applying oscillations to it, the system vibrates. As the frequency of the oscillations is varied, the amplitude of the vibrations reaches a maximum at the natural frequency of the system. If a force with a frequency equal to the natural frequency is applied, the vibrations can become violent, a phenomenon known as resonance.

longitudinal wave In a longitudinal wave, such as a sound wave, the disturbance of the medium is parallel to the wave's direction of travel. A longitudinal

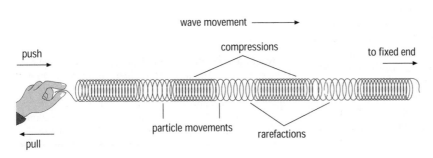

longitudinal wave The motion of a longitudinal wave. Sound, for example, travels through air in longitudinal waves: the waves vibrate back and forth in the direction of travel. In the compressions the particles are pushed together, and in the rarefactions they are pulled apart.

wave consists of a series of compressions and rarefactions (states of maximum and minimum density and pressure, respectively). Such waves are always mechanical in nature and thus require a medium through which to travel.

transverse wave

In a transverse wave, such as an electromagnetic wave, the displacement of the medium is perpendicular to the direction in which the wave travels. The directions of the electric and magnetic fields in electromagnetic waves are perpendicular to the wave motion. The medium (for example the Earth, for seismic waves) is not permanently displaced by the passage of a wave.

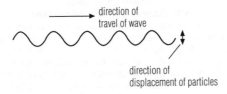

direction of
travel of wave

direction of
displacement of particles

transverse wave The diagram illustrates the motion of a transverse wave. Light waves are examples of transverse waves: they undulate at right angles to the direction of travel and are characterized by alternating crests and troughs. Simple water waves, such as the ripples produced when a stone is dropped into a pond, are also examples of transverse waves.

polarization

polarization Transverse waves can exhibit polarization. If the oscillations of the wave take place in lots of different directions (all at right angles to the directions of the wave) the wave is unpolarized. If the oscillations occur in one plane only, the wave is polarized. Light, which consists of transverse waves, can be polarized.

wavelength

wavelength Wavelength is the distance between successive crests of a wave. This is measured as the distance between successive crests (or successive troughs) of the wave. It is given the Greek symbol λ. The frequency of a wave is the number of vibrations per second. The reciprocal of this is the wave period. This is the time taken for one complete cycle of the wave oscillation. The speed of the wave is measured by multiplying wave frequency by the wavelength. The wavelength of a light wave determines its colour; red light has a wavelength of about 700 nanometres, for example. The complete range of wavelengths of

electromagnetic waves is called the electromagnetic spectrum.

frequency

frequency Frequency refers to the number of periodic oscillations, vibrations, or waves occurring per unit of time. The SI unit of frequency is the hertz (Hz), one hertz being equivalent to one cycle per second. Frequency is related to wavelength and velocity by the relationship $f = v/\lambda$, where f is frequency, v is velocity and λ is wavelength.

refraction

refraction When a wave moves from one medium to another (for example a light wave moving from air to glass) it moves with a different speed in the second medium. This change in speed causes it to change direction. This property is called refraction. The amount of refraction depends on the densities of the media, the angle at which the wave strikes the surface of the second medium, and the amount of bending and change of velocity corresponding to the wave's frequency (dispersion). Refraction differs from reflection (see below), which involves no change in velocity. The refractive index of a material indicates by how much a wave is bent. It is found by dividing the velocity of

refraction Refraction is the bending of a light beam when it passes from one transparent medium to another. This is why a spoon appears bent when standing in a glass of water and pools of water appear shallower than they really are. The quantity sin i/sin r has a constant value, for each material, called the refractive index.

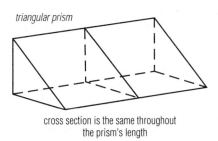

triangular prism

cross section is the same throughout
the prism's length

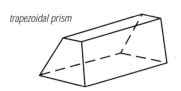

trapezoidal prism

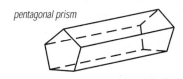

pentagonal prism

prism The volume of a prism is determined by multiplying the area of the cross section by the length of the prism.

the wave in the first medium by the velocity of the wave in the second medium.

dispersion Dispersion is a particular property of refraction in which the angle and velocity of waves passing through a dispersive medium depends upon their frequency. In the case of visible light the frequency corresponds to colour. The splitting of white light into a spectrum (see under electromagnetic waves below) when it passes through a prism occurs because each component frequency of light moves through at a slightly different angle and speed. A rainbow is formed when sunlight is dispersed by raindrops.

spectrum

To remember the order of colours:

Richard **o**f **Y**ork **g**ained **b**attles **i**n **v**ain.

(**r**ed, **o**range, **y**ellow, **g**reen, **b**lue, **i**ndigo, **v**iolet)

reflection Whenever a wave hits a barrier, reflection, occurs. The wave is sent back, or reflected, into the

medium at a different angle. The **law of reflection** states that the angle of incidence (the angle between the ray and a perpendicular line drawn to the surface) is equal to the angle of reflection (the angle between the reflected ray and a perpendicular to the surface).

When light passes from a denser medium to a less dense medium, such as from water to air, both refraction (see above) and reflection can occur. If the angle of incidence is small, the reflection will be relatively weak compared to the refraction. But as the angle of incidence increases the relative degree of reflection will increase. At some **critical angle of incidence** the angle of refraction is 90°. Since refraction cannot occur above 90°, the light is totally reflected at angles above this critical angle of incidence. This condition is known as **total internal reflection**. Total internal reflection is used in fibre optics to transmit data over long distances, without the need of amplification

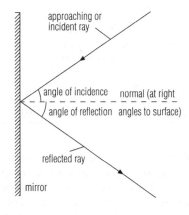

approaching or incident ray

angle of incidence   normal (at right
angle of reflection   angles to surface)

reflected ray

mirror

reflection The law of reflection: the angle of incidence of a light beam equals the angle of reflection of the beam.

diffraction Diffraction is the spreading out of waves when they pass through a small gap or around a small object, resulting in some change in their direction. The degree of diffraction depends on the relationship between the wavelength and the size of the object or gap through which the wave travels. In order for this effect to be observed the size of the object or gap must be comparable to, or smaller than, the wavelength of the waves. Diffraction occurs with all forms of progressive waves – electromagnetic, sound, and water waves – and explains such phenomena as why long-wave radio waves can bend round hills better than short-wave radio waves. Large objects cast shadows because the difference between their size and the wavelength is so large that light waves are not diffracted round the object.

The wavelength of light ranges from 4×10^{-7} to 7×10^{-7}, a few orders of magnitude smaller than radio waves. The slight spreading of a light beam through a narrow slit causes the different wavelengths of light to interfere with each other to produce a pattern of light and dark bands. A **diffraction grating** is a plate of glass or metal ruled with close, equidistant parallel lines used for separating a wave train such as a beam of incident light into its component frequencies (white light results in a spectrum). The wavelength of sound is between 0.5 m/1.6 ft and 2.0 m/6.6 ft. When sound waves travel through doorways or between buildings they are diffracted significantly, so that the sound is heard round corners. The regular spacing of atoms in crystals are used to diffract X-rays, and in this way the structure of many substances has been elucidated, including that of proteins.

Moseley Articles
http://dbhs.wvusd.k12.ca.us/Chem-History/
Moseley-article.html

Transcript of Henry Moseley's article *The High Frequency Spectra Of The Elements*. In the article, Moseley lays the foundations for X-ray spectroscopy, and forms a relationship between atomic number and the frequency of the emitted spectra.

---

interference When two or more waves meet at a point, they interact and combine to produce a resultant wave of larger or smaller amplitude (depending on whether the combining waves are in or out of phase with each other). This is known as interference.

Professor Bubbles' Official Bubble Home Page
http://bubbles.org/

Lively information about how to blow the best bubbles, and answers to frequently asked questions about bubbles – such as 'why are bubbles always round?' and 'why do bubbles have colour?'

Interference of white light (multiwavelength) results in spectral coloured fringes; for example, the iridescent colours of oil films seen on water or soap bubbles. Interference of sound waves of similar frequency produces the phenomenon of beats, often used by musicians when tuning an instrument. With monochromatic light (of a single wavelength), interference produces patterns of light and dark bands. This is the basis of holography, for example. Interferometry can also be applied to radio waves, and is a powerful tool in modern astronomy

sound wave Sound is the physiological sensation received by the ear, originating in a vibration that communicates itself as a pressure variation in the air and travels in every direction, spreading out as an expanding sphere. All sound waves in air travel with a speed dependent on the temperature; under ordinary conditions, this is about 330 m/1,070 ft per second. The pitch of the sound depends on the number of vibrations imposed on the air per second (frequency), but the speed is unaffected. The loudness of a sound is dependent primarily on the amplitude of the vibration of the air. Sound travels as a longitudinal wave, that is, its compressions and rarefactions are in the direction of propagation. Reflection of a sound wave is heard as an echo. Diffraction explains why sound can be heard round doorways. Sound travels faster in denser materials, such as solids and liquids.

Sound waves, unlike light, travel faster in denser materials, such as solids and liquids, than they travel in air. When sound waves enter a solid, their velocity and wavelength *increase* and they are bent away from the normal to the surface of the solid.

electromagnetic waves
Electromagnetic waves are oscillating electric and magnetic fields travelling together through space. All electromagnetic waves travel at the same speed, the speed of light – nearly 300,000 km/186,000 mi per second. The (limitless) range of possible wavelengths and frequencies of electromagnetic waves, which can be thought of as making up the **electromagnetic spectrum**, includes radio waves, infrared radiation, visible light, ultraviolet radiation, X-rays, and gamma rays. The different wavelengths and frequencies lend specific properties to electromagnetic waves.

---

speed of light

To remember the speed of light:

The speed of light is 299,792,458 m/s, which can be remembered from the number of letters in each word of the following phrase:

We guarantee certainty, clearly referring to this light mnemonic

---

radio wave A radio wave is an electromagnetic wave possessing a long wavelength (ranging from about 10^{-3} to 10^4 m) and a low frequency (from about 10^5 to 10^{11} Hz). Included in the radio-wave part of the spectrum are microwaves, used for both communications and for cooking; ultra high- and very high-frequency waves, used for television and FM (frequency modulation) radio communications; and short, medium, and

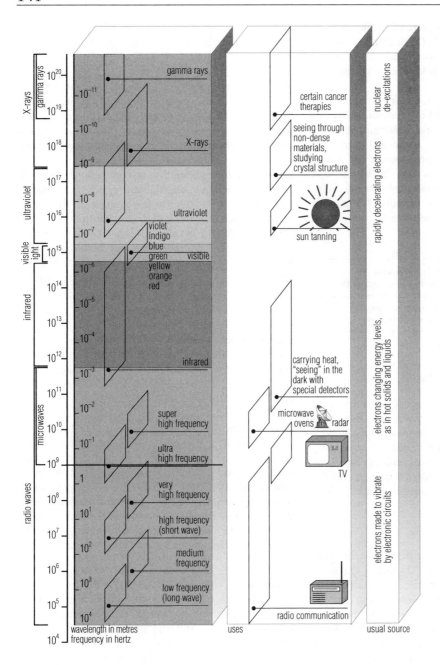

electromagnetic waves Radio waves have the lowest frequency. Infrared radiation, visible light, ultraviolet radiation, X-rays, and gamma rays have progressively higher frequencies.

long waves, used for AM (amplitude modulation) radio communications. Radio waves that are used for communications have all been modulated to carry information. Certain astronomical objects emit radio waves, which may be detected and studied using radio telescopes.

light wave A light wave is an electromagnetic wave in the visible range, having a wavelength from about 400 nanometres in the extreme violet to about 770 nanometres in the extreme red. Light is considered to exhibit particle and wave properties, and the fundamental particle, or quantum, of light is called the photon. The speed of light (and of all electromagnetic radiation) in a vacuum is approximately 300,000 km/186,000 mi per second, and is a universal constant denoted by *c*.

For all practical purposes light rays travel in straight

lines, although Einstein demonstrated that they may be 'bent' by a gravitational field. On striking a surface they are reflected or refracted with some absorption of energy, and the study of this is known as geometrical optics.

Sources of light have a characteristic spectrum or range of wavelengths. Hot solid objects emit light with a broad range of wavelengths, the maximum intensity being at a wavelength which depends on the temperature. The hotter the object, the shorter the wavelengths emitted, as described by Wien's displacement law. Hot gases, such as the vapour of sodium street lights, emit light at discrete wavelengths. The pattern of wavelengths emitted is unique to each gas and can be used to identify the gas.

Introduction to Mass Spectrometry
http://www.scimedia.com/chem-ed/ms/ms-intro.htm
Good introduction to the mass spectrometer and how it works. Different mass analyser designs are described, together with sections on ionization and ion detectors.

---

fundamental forces

Four fundamental interactions are believed to be at work in the physical universe. There are two long-range forces – **gravity** and the **electromagnetic force** – and two very short-range forces that operate only inside the atomic nucleus: the **weak nuclear force** and the **strong nuclear force**. The relative strengths of the four forces are: strong, 1; electromagnetic, 10^{-2}; weak, 10^{-6}; gravitational, 10^{-40}.

gravity
Gravity is the force of attraction that arises between objects by virtue of their masses. On Earth, gravity is the force of attraction between any object in the Earth's gravitational field and the Earth itself. It is the force which keeps the planets in orbit around the Sun. The gravitational force is the weakest of the four forces, but it acts over great distances. The particle that is postulated as the carrier of the gravitational force is the graviton.

---

astronaut

Astronauts in space cannot belch. It is gravity that causes bubbles to rise to the top of a liquid, so space shuttle crews were forced to request less gas in their fizzy drinks to avoid discomfort.

---

electromagnetic force The electromagnetic force stops solids from falling apart, and acts between all particles with electric charge. The elementary particle that is the carrier for the electromagnetic force is the photon.

weak nuclear force The weak nuclear force is responsible for the reactions that fuel the Sun and for the emission of beta particles from certain nuclei. The **weakon**, or **intermediate vector boson**, is the elementary particle that carries the weak nuclear force.

strong nuclear force The strong nuclear force was first described by the Japanese physicist Hideki Yukawa in 1935. It is the strongest of all the forces, acts only over very small distances (within the nucleus of the atom), and is responsible for binding together quarks to form hadrons, and for binding together protons and neutrons in the atomic nucleus. The particle that is the carrier of the strong nuclear force is the gluon, of which there are eight kinds, each with zero mass and zero charge

unified field theory
Unified field theory is a sought-for theory that would explain the four fundamental forces (strong nuclear, weak nuclear, electromagnetic, and gravity) in terms of a single unified force. By 1971 a theory developed by the US physicists Steven Weinberg and Sheldon Glashow, the Pakistani physicist Abdus Salam, and others, had demonstrated the link between the weak nuclear and electromagnetic forces. Called the **electroweak force**, experimental support came from observation at CERN in the 1980s. The next stage is to develop a theory (called the **grand unified theory**) that combines the strong nuclear force with the electroweak force. The final stage will be to incorporate gravity into the scheme.

superstring theory
Superstring theory is a mathematical theory developed in the 1980s to explain the properties of elementary particles and the forces between them (in particular, gravity and the nuclear forces) in a way that combines relativity and quantum theory. In string theory, the fundamental objects in the universe are not pointlike particles but extremely small stringlike objects. These objects exist in a universe of ten dimensions, although, for reasons not yet understood, only three space dimensions and one dimension of time are discernible. There are many unresolved difficulties with superstring theory, but some physicists think it may be the ultimate 'theory of everything' that explains all aspects of the universe within one framework.

relativity

Relativity is the theory of the relative rather than absolute character of motion and mass, and the interdependence of matter, time, and space, as developed by German-born US physicist Albert Einstein in two phases.

In his **special theory of relativity**, developed in 1905, starting with the premises that (1) the laws of nature are the same for all observers in unaccelerated motion, and (2) that the speed of light is independent of the motion of its source, Einstein arrived at some rather unexpected consequences. Intuitively familiar concepts, like mass, length, and time, had to be modified. For example, an object moving rapidly past an observer will appear to be both shorter and heavier than when it is at rest (that is, at rest relative to the observer), and a clock moving rapidly past the observer will appear to be running slower than when it is at rest. These changes are quite negligible at speeds less than about 1,500 km s^{-1}, and only become appreciable at speeds approaching the speed of light.

Einstein's **general theory of relativity**, developed in 1915, treats gravitation not as a force but as the curvature of space-time around a body. A planet's orbit around the Sun (as observed in three-dimensional space) arises from its natural trajectory in modified space-time; there is no need to invoke, as Isaac Newton did, a force of gravity coming from the Sun and acting on the planet. Einstein's general theory accounts for a peculiarity in the behaviour of the motion of the perihelion of the orbit of the planet Mercury that cannot be explained in Newton's theory.

Relativity also predicts that light rays should bend when they pass by a massive object, and that light should shift towards the red in the spectra of the Sun or star in a gravitational field; both have been observed. Another prediction of relativity is **gravitational waves**, which should be produced when massive bodies are violently disturbed. These waves are so weak that they have not yet been detected with certainty, although observations of a pulsar (which emits energy at regular intervals) in orbit around another star have shown that the stars are spiralling together at the rate that would be expected if they were losing energy in the form of gravitational waves.

Einstein showed that, for consistency with the above premises (1) and (2), the principles of dynamics as established by Newton needed modification; the most celebrated new result was the equation $E = mc^2$, which expresses an equivalence between mass (m) and energy (E), c being the speed of light in a vacuum. In 'relativistic mechanics', conservation of mass is replaced by the new concept of conservation of 'mass-energy'.

General relativity is central to modern astrophysics and cosmology; it predicts, for example, the possibility of black holes. General relativity theory was inspired by the simple idea that it is impossible in a small region to distinguish between acceleration and gravitation effects (as in a lift one feels heavier when the lift accelerates upwards), but the mathematical development of the idea is formidable. Such is not the case for the special theory, which a nonexpert can follow up to $E = mc^2$ and beyond.

relativity
The manuscript for Einstein's special theory of relativity sold for $5 million in 1996.

Usenet Relativity FAQ
http://math.ucr.edu/home/baez/physics/relativity.html
Concise answers to some of the most common questions about relativity. The speed of light and its relation to mass, dark matter, black holes, time travel, and the Big Bang are some of the things covered by this illuminating series of articles based both on Usenet discussions and good reference sources. The site also directs the visitors to appropriate discussion groups where they can pose more questions.

---

space-time

Space-time is the combination of space and time used in the theory of relativity. Einstein showed that time was in many respects like an extra dimension (or direction) to space. Space and time can thus be considered as entwined into a single entity, rather than two separate things.

Space-time is considered to have four dimensions: three of space and one of time. In relativity theory, events are described as occurring at points in space-time. The general theory of relativity describes how space-time is distorted by the presence of material bodies, an effect that we observe as gravity.

Einstein's Legacy
http://www.ncsa.uiuc.edu/Cyberia/NumRel/
EinsteinLegacy.html
Illustrated introduction to the man and his greatest legacy – relativity and the concept of space-time. There is a film and audio clip version of the page courtesy of a US scientist and details about how current research is linked to Einstein's revolutionary ideas.

Physics Chronology

c. 435 BC   Greek philosopher Leucippus is the first to propose the atomic theory. It is developed later by his pupil Democritus.

c. 420 BC   Greek philosopher Democritus of Abdera develops Leucippus' atomic theory and states that space is a vacuum and that all things consist of eternal, invisible and indivisible atomon (atoms). He also posits necessary laws by which they interact.

c. 250 BC   Greek mathematician and inventor Archimedes discovers the principle that bears his name – that submerged bodies are acted upon by an upward or buoyant force equal to the weight of the fluid displaced.

1039   The Muslim scientist Abu'Ali al-Hasan (Alhazen) writes treatise on optics explaining the function of lenses, curved mirrors, refraction, and other phenomena.

1590   Italian scientist Galileo publishes *De Motu/On Motion*, in which he discusses his ideas and discoveries about the motion of objects.

1600   The English physician William Gilbert writes *De magnete/On Magnetism*, a pioneering study of electricity and magnetism, which distinguishes between electrostatic and magnetic effects.

1604   Italian scientist Galileo Galilei discovers his law of falling bodies, proving that gravity acts with the same strength on all objects, independent of their mass. Traditionally, he is believed to have demonstrated this by dropping balls of the same size but different masses from the top of the Leaning Tower at Pisa.

1642   French mathematician Blaise Pascal puts forward the principles of hydraulics, the use of liquids to transmit force.

1647   Pascal demonstrates the pressure exerted by the atmosphere, using it to raise water and wine 12 m/40 ft up tubes fastened to a ship's mast.

1660   English physicist Robert Hooke discovers the law now named after him – that the extension of an elastic material such as a spring is in proportion to the force exerted on it.

1660   English mathematician Isaac Newton begins work on the calculus, a fundamental tool in physics for studying rates of change.

1662   Anglo-Irish chemist and physicist Robert Boyle describes the law that will bear his name, stating that, for a fixed mass of gas in a container, the volume occupied by the gas is inversely proportional to the pressure it exerts.

1663   German physicist Otto von Guericke makes a machine for generating static electricity by friction, consisting of a ball of sulphur isolated from earth, and turned by an axle and winch.

1664   Hooke suggests that planetary orbits may be maintained by the constant attractive force of gravity between two bodies.

1675   Newton proposes a corpuscular theory of light.

March 1676   French physicist Edmé Mariotte discovers the relationship between volume and pressure in a fixed mass of gas, independently of Boyle.

1678   Dutch physicist and astronomer Christiaan Huygens records his discovery of the polarization of light, responsible for phenomena such as double refraction.

1679   Hooke proposes an inverse-square law of gravity, pre-empting Newton's law of gravitation.

1680   Newton calculates that an inverse-square law of gravitational attraction between the Sun and planets would explain the elliptical orbits discovered by Kepler. He also puts forward a theory that the air resistance encountered by a body increases in proportion to the square of its speed.

1687   Newton publishes *Philosophiae naturalis principia mathematica/The Mathematical Principles of Natural Philosophy*, his most important work. It presents his theories of motion, gravity, and mechanics, which form the basis of much of modern physics.

1690   Dutch physicist Christiaan Huygens propounds a theory of light as a longitudinal wave with vibration in the direction of its travel.

1745   German scientist Ewald Georg von Kleist invents the Leyden jar, a simple capacitor that accumulates and preserves electricity. The following year Dutch scientist Pieter von Musschenbroek makes the same discovery independently.

1750   Scandinavian physicist Martin Stromer modifies the temperature scale devised by his mentor, the Swedish astronomer Anders Celsius. He inverts it, setting freezing point as 0°C and boiling point as 100°C, creating the Celsius scale still used today.

1767   English physicist Joseph Priestley publishes his *History and Present State of Electricity*, which suggests that electrical forces follow an inverse-square law, as does gravity.

1785   French scientist Charles Augustin Coulomb makes the first precise measurements of the electric forces of attraction and repulsion between charged bodies.

1787   French physicist Jacques-Alexandre Charles demonstrates that different gases expand by the same amount for the same temperature rise. It later becomes known as Charles's law.

1792   Italian physicist Alessandro Volta demonstrates the electrochemical series.

1798   US-born British physicist and inventor Benjamin Thompson, Count Rumford, demonstrates experimentally the theory that heat is the increased motion of particles.

1800 English physicist Thomas Young proposes a wave theory of light.

1800 Volta invents the voltaic pile made of discs of silver and zinc – the first battery.

1801 English chemist and physicist John Dalton formulates the law of partial pressure in gases – Dalton's Law – which states that each component of a gas mixture produces the same pressure as if it occupied the container alone.

1801 Young discovers the interference of light when he observes that light passing through two closely spaced pinholes produces alternating bands of light and dark in the area of overlap. He thereby establishes the wave theory of light.

1802 French chemist and physicist Joseph-Louis Gay-Lussac demonstrates that all gases expand by the same fraction of their volume when subjected to the same temperature increase; it permits the establishment of a new temperature scale.

1807 Young enunciates 'Young's modulus', a measurement of the elasticity of a material defined as the stress divided by the strain.

1811 Italian physicist Amedeo Avogadro proposes Avogadro's law which states that equal volumes of different gases under the same temperature and pressure conditions will contain the same number of molecules.

25 July 1814 German physicist Joseph von Fraunhofer plots more than 500 absorption lines (Fraunhofer lines) and discovers that the relative positions of the lines is constant for each element. His work forms the basis of modern spectroscopy.

1815 French physicist Augustin-Jean Fresnel shows that light has transverse waves – he thus explains the diffraction of light.

June 1819 Danish physicist Hans Christian Oersted discovers electromagnetism when he observes that a magnetized compass needle is deflected by an electric current.

1820 French physicist André Ampère formulates Ampère's law, which states the relationship between a magnetic field and the electric current that produces it.

1821 English physicist Michael Faraday builds an apparatus that transforms electrical energy into mechanical energy – the principle of the electric motor.

1821 German physicist Thomas Seebeck discovers thermoelectricity – the conversion of heat into electricity – when he generates a current by heating one end of a metal strip comprising two metals joined together.

1824 French scientist Sadi Carnot publishes a pioneering study of thermodynamics in which he explains that a steam engine's power results from the decrease in temperature from the boiler to the condenser. He also describes the 'Carnot cycle' whereby heat is converted into mechanical motion and mechanical motion converted into heat – the basis of the second law of thermodynamics.

1825 Ampère publishes *Electrodynamics*, in which he formulates the mathematical laws governing electric currents and magnetic fields. It lays the foundation for electromagnetic theory.

1827 German physicist Georg Ohm formulates Ohm's Law, which states that the current flowing through an electric circuit is directly proportional to the voltage, and indirectly proportional to the resistance.

1828 Scottish botanist Robert Brown observes the continuous motion of tiny particles in a liquid solution, now known as Brownian motion.

1829 French mathematician Gustave-Gaspard Coriolis is the first to use the term 'kinetic energy'.

1830 US scientist Joseph Henry discovers electromagnetic induction – the production of an electric current by change in magnetic intensity – but does not publish his discovery.

29 Aug 1831 Faraday discovers electromagnetic induction – the production of an electric current by change in magnetic intensity (and also the principle of the electric generator).

July 1832 Henry discovers the phenomenon of self-induction – the production of electric current when a conductor is disconnected from a battery.

1833 Faraday announces the basic laws of electrolysis: that the amount of a substance deposited on an electrode is proportional to the amount of electric current passed through the cell, and that the amounts of different elements are proportional to their atomic weights.

1834 French physicist Benoît-Pierre Clapeyron develops the second law of thermodynamics: entropy always increases in a closed system.

1836 French physicist Edmund Becquerel discovers the photovoltaic effect when he observes the creation of a voltage between two electrodes, one of which is exposed to light.

1839 Faraday discovers that each element has a specific electrical inductive capacity.

1840 English physicist James Joule states his law that the amount of heat produced per second in any conductor by an electric current is proportional to the product of the square of the current and the resistance of the conductor.

1842 Austrian physicist Christian Doppler describes how the frequency of sound and light waves changes with the motion of their source relative to the observer – the 'Doppler effect'.

1843 Joule determines the value for the mechanical equivalent of heat (now known as the joule), that is the amount of work required to produce a unit of heat.

1848 Scottish physicist William Thomson (Lord Kelvin) devises the absolute temperature scale. He defines absolute zero as –273°C/–459.67°F, where the molecular energy of molecules is zero. He also defines the quantities currently used to describe magnetic forces: magnitude of magnetic flux, beta, and H the magnetizing force.

1850 French physicist Jean Foucault establishes that light travels slower in water than in air. He also measures the velocity of light to within 1% of its true speed.

1851 Kelvin states that energy in a closed system tends to become unusable waste heat – the second law of thermodynamics.

1859 German chemists Robert Wilhelm von Bunsen and Gustav Kirchhoff discover that each element emits a characteristic wavelength of light. It initiates spectrum analysis, a valuable tool for both chemist and astronomer.

1864 Scottish physicist James Clerk Maxwell introduces mathematical equations that describe the electromagnetic field, and predict the existence of radio waves.

1868 Swedish physicist Anders Ångström expresses the wavelengths of Fraunhofer lines in units of 10^{-10} m, a unit now known as the angstrom.

1874 Austrian physicist Ludwig Boltzmann develops the basic principles of statistical mechanics when he demonstrates how the laws of mechanics and the theory of probability, when applied to the motions of atoms, can explain the second law of thermodynamics.

1874 Irish physicist George Johnstone Stoney names the electron and estimates the value of its charge.

1879 British-born US electrical engineer Elihu Thomson shows how induction coils can be used to increase current and step down voltage – the basic principle of the transformer developed a few years later.

1880 French physicists Pierre and Paul-Jacques Curie discover that electricity is produced when pressure is placed on certain crystals including quartz – the 'piezoelectric' effect.

1882 German-born US physicist Albert Michelson determines the speed of light to be 299,853 kps/186,329 mps.

1882 Scottish physicist Balfour Stewart postulates the existence of an electrically conducting layer of the outer atmosphere (now known as the ionosphere) to account for the daily variation in the Earth's magnetic field.

1883 Irish physicist George Francis FitzGerald suggests that electromagnetic waves (radio waves) can be created by oscillating an electric current. A later demonstration of such waves by the German physicist Heinrich Hertz leads to the development of wireless telegraphy.

1883 US inventor Thomas Alva Edison observes the flow of current between a hot electrode and a cold electrode in one of his vacuum bulbs. Known as the 'Edison effect', it results from the thermionic emission of electrons from the hot electrode, and is the principle behind the working of the electron tube, which is to form the basis of the electronics industry.

1885 US physicist Henry Augustus Rowland invents the concave diffraction grating, in which 20,000 lines to the inch are engraved on spherical concave mirrored surfaces. The grating revolutionizes spectrometry by dispersing light and permitting spectral lines to be focused.

1886 US astronomer and physicist Samuel Pierpont Langley begins the first systematic aerodynamic research. He measures lift and drag on models of wings and other objects, which he attaches to a counterweighted beam, mounted on a pivot, that may be rotated at a speed of up to 112 kph/70 mph.

1887 Hertz discovers the photoelectric effect, in which a material gives off charged particles when it absorbs radiant energy, when he observes that ultraviolet light affects the voltage at which sparking between two metal plates takes place. Later work on this phenomenon leads to the conclusion that light is composed of particles called photons.

1887 US physicist Albert Michelson and US chemist Edward Williams Morley fail in an attempt to measure the velocity of the Earth through the 'ether' by measuring the speed of light in two directions. Their failure discredits the idea of the ether and leads to the conclusion that the speed of light is a universal constant, a fundamental premise of Einstein's theory of relativity.

1891 Serbian-born US inventor Nikola Tesla invents the Tesla coil, which produces a high-frequency high-voltage current.

1893 British physicist Oliver Heaviside theorizes that as the velocity of an electric charge increases so does its mass. It presages Einstein's special theory of relativity.

1893 German physicist Wilhelm Wien states that the maximum wavelength emitted by a hot body is inversely proportional to the absolute temperature of the body.

8 Nov 1895 German physicist Wilhelm Conrad Röntgen discovers X-rays. Named because of their unknown origin, they revolutionize medicine and usher in the age of modern physics.

1896 British physicist Ernest Rutherford discovers that magnetic fields can be used to detect electromagnetic or radio waves.

1896 Scottish physicist Charles Thomson Rees Wilson develops the first cloud chamber.

1897 English physicist John Joseph Thomson demonstrates the existence of the electron, the first known subatomic particle. It revolutionizes knowledge of atomic structure by indicating that the atom can be subdivided.

1898 German physicist Gerhard Carl Schmidt and French physicist Marie Curie demonstrate, independently, that thorium is radioactive; it stimulates interest in radioactivity.

1898	German physicist Wilhelm Wien discovers the proton.
1899	Rutherford discovers alpha and beta rays, produced by the radioactivity of uranium.
1899	Thomson measures the charge of the electron.
1900	Canadian-born US scientist Reginald Aubrey Fessenden discovers the principle of amplitude modulation (AM) of radio waves.
1900	French physicist Antoine–Henri Becquerel demonstrates that the beta particle is the same thing as the electron.
1900	French physiologist Paul Ulrich Villard discovers gamma rays.
1900	German physicist Max Planck suggests that black bodies (perfect absorbers) radiate energy in packets or quanta, rather than continuously. He thus begins the science of quantum physics, which revolutionizes the understanding of atomic and subatomic processes.
1902	British physicist Oliver Heaviside and US electrical engineer Arthur Kennelly independently predict the existence of a conducting layer in the atmosphere that reflects radio waves.
1903	Rutherford discovers that a beam of alpha particles is deflected by electric and magnetic fields. From the direction of deflection he is able to prove that they have a positive charge and from their velocity he determines the ratio of their charge to their mass. He also names the high-frequency electromagnetic radiation escaping from the nuclei of atoms as gamma rays.
1903	Scottish chemist William Ramsay shows that helium is produced during the radioactive decay of radium – an important discovery for the understanding of nuclear reactions.
1904	English physicist Charles Glover Barkla demonstrates that each element can be made to emit X-rays of a characteristic frequency.
1904	Japanese physicist Hantaro Nagaoka proposes a model of the atom in which the electrons are located in an outer ring and orbit the positive charge which is located in a central nucleus. The model is ignored because it is thought the electrons would fall into the nucleus.
1905	Kelvin, proposes a model of the atom in which positive and negatively charged spheres alternate.
1906	English physicist Frederick Soddy discovers that ionium and radiothorium are chemically indistinguishable variants of thorium but have different radioactive properties. He later calls them isotopes.
1906	German physicist Walther Herman Nernst formulates the third law of thermodynamics, which states that matter tends towards random motion and that energy tends to dissipate at a temperature above absolute zero.
1908	German physicist Hans Geiger and Rutherford develop the Geiger counter, which counts individual alpha particles emitted by radioactive substances.

1908	US physicist Percy Williams Bridgman invents equipment that can create atmospheric pressures of 100,000 atmospheres (later 400,000) creating a new field of investigation.
1909	German physicist Albert Einstein introduces his idea that light exhibits both wave and particle characteristics.
1910	Thomson discovers the proton.
1910	French chemists Marie Curie and A Diebierne isolate radium.
1910	German physicist Wolfgang Gaede develops the molecular vacuum pump, which can generate a vacuum of 0.00001 mm of mercury.
1911	Austrian physicist Victor Francis Hess discovers cosmic radiation using crewed balloons.
1911	Dutch physicist Heike Kamerlingh-Onnes discovers superconductivity, the characteristic of a substance to display zero electrical resistance when cooled to just above absolute zero.
1911	German physicist Albert Einstein calculates the deflection of light caused by the Sun's gravitational field.
1911	Rutherford proposes the concept of the nuclear atom, in which the mass of the atom is concentrated in a nucleus occupying one ten-thousandth of the space of the atom and which has a positive charge balanced by surrounding electrons.
1911	Rutherford and Soddy devise a scheme for the 'transmutation' of the elements, producing a simpler atom from a complex one.
1911	US physicist Robert Millikan measures the electric charge on a single electron in his oil-drop experiment, in which the upward force of the electric charge on an oil droplet precisely counters the known downward gravitational force acting on it.
1912	German physicist Max von Laue demonstrates that crystals are composed of regular, repeated arrays of atoms by studying the patterns in which they diffract X-rays. It is the beginning of X-ray crystallography.
1913	Danish physicist Niels Bohr proposes that electrons orbit the atomic nucleus in fixed orbits thus upholding Rutherford's model proposed in 1911.
1913	Thomson develops a mass spectrometer, called a parabola spectrograph. A beam of charged ions is deflected by a magnetic field to produce parabolic curves on a photographic plate.
1913	Thomson discovers neon-22, an isotope of neon. It is the first isotope of a nonradioactive element to be discovered.
1913	English physicists William and Lawrence Bragg develop X-ray crystallography by establishing that the orderly arrangement of atoms in crystals display interference and diffraction patterns. They also demonstrate the wave nature of X-rays.
1913	Einstein formulates the law of photochemical equivalence, which states that for every

quantum of radiation absorbed by a substance one molecule reacts.

1914 German physicists James Franck and Gustav Hertz provide the first experimental evidence for the existence of discrete energy states in atoms and thus verify Bohr's atomic model.

1916 Einstein publishes *The Foundation of the General Theory of Relativity*, in which he postulates that space is a curved field modified locally by the existence of mass and that this can be demonstrated by observing the deflection of starlight around the Sun during a total eclipse. This replaces previous Newtonian ideas which invoke a force of gravity. Einstein also derives the basic equations for the exchange of energy between matter and radiation.

29 May 1919 English astrophysicist Arthur Eddington and others observe the total eclipse of the Sun on Principe Island (West Africa), and discover that the Sun's gravity bends the light from the stars beyond the edge of the eclipsed Sun, thus confirming Einstein's theory of relativity.

1919 English physicist Francis Aston builds the first mass-spectrograph, which allows him to separate ions or isotopes of the same element.

1919 Rutherford splits the atom by bombarding a nitrogen nucleus with alpha particles, discovering that it ejects hydrogen nuclei (protons). It is the first artificial disintegration of an element and inaugurates the development of nuclear energy.

1920 Rutherford recognizes the hydrogen nucleus as the fundamental particle and names it the 'proton'.

1922 US physicist Arthur Holly Compton discovers that X-rays scattered by an atom have a shift in frequency. He explains the phenomenon, known as the Compton effect, by treating the X-rays as a stream of particles, thus confirming the wave–particle idea of light.

1924 English physicist Edward Appleton discovers that radio emissions are reflected by an ionized layer of the atmosphere.

1924 French physicist Louis de Broglie argues that particles can also behave as waves, laying the foundations for wave mechanics. He demonstrates that a beam of electrons has a wave motion with a short wavelength. The discovery permits the development of the electron microscope.

1926 Austrian physicist Erwin Schrödinger develops wave mechanics.

1927 German physicist Werner Heisenberg propounds the 'uncertainty principle' in quantum physics, which states that it is impossible to simultaneously determine the position and momentum of an atom. It explains why Newtonian mechanics is inapplicable at the atomic level.

1928 English physicist Paul Dirac describes the electron by four wave equations. The equations imply that the electron must spin on its axis and that negative states of matter must exist.

1928 Germany physicist Rolf Wideröe develops the resonance linear accelerator, which he uses to accelerate potassium and sodium to an energy of 710 keV to split the lithium atom.

1928 Russian physicist George Gamow shows that the atom can be split using low-energy ions. It stimulates the development of particle accelerators.

1929 Irish physicist Ernest Walton and English physicist Douglas Cockcroft develop the first particle accelerator.

1931 US physicists Ernest Lawrence and M Stanley Livingston build a cyclotron (particle accelerator).

1932 British physicist James Chadwick discovers the neutron, an important discovery in the development of nuclear reactors.

1932 British physicists John D Cockcroft and Ernest Walker develop a high-voltage particle accelerator, which they use to split lithium atoms.

1932 US scientist Carl David Anderson, while analysing cosmic rays, discovers positive electrons ('positrons'), the first form of antimatter to be discovered.

1933 German physicists Walter Meissner and R Ochensfeld discover that superconducting materials expel their magnetic fields when cooled to superconducting temperatures – the Meissner effect.

1934 French physicists Frédéric and Irène Joliot-Curie bombard boron, aluminium, and magnesium with alpha particles and obtain radioactive isotopes of nitrogen, phosphorus, and aluminium – elements that are not normally radioactive. They are the first radioactive elements to be prepared artificially.

1934 Italian physicist Enrico Fermi suggests that neutrons and protons are the same fundamental particles in two different quantum states. He bombards uranium with neutrons and discovers the phenomenon of atomic fission, the basic principle of atomic bombs and nuclear power.

1935 Japanese physicist Hideki Yukawa proposes the existence of a new particle, the meson, to explain nuclear forces.

1936 Anderson discovers the muon, an electron-like particle over 200 times more massive than an electron.

1936 US physicists George Gamow and Edward Teller develop the theory of beta decay – the nuclear process of electron emission.

1938 The Soviet physicist Pyotr Kapitza discovers that liquid helium exhibits superfluidity, the ability to flow over its containment vessel

	without friction, when cooled below 2.18 K/ –270.97°C.
1939	French physicists Frédéric Joliot and Irène Curie-Joliot demonstrate the possibility of a chain reaction when they split uranium nuclei.
1940	US physicist J R Dunning leads a research team that uses a gaseous diffusion technique to isolate uranium-235 from uranium-238. Because uranium-235 readily undergoes fission into two atoms, and in doing so releases large amounts of energy, it is used for fuelling nuclear reactors.
1940	US physicists Edwin McMillan and Philip Abelson synthesize the first transuranic element, neptunium, by bombarding uranium with neutrons at the cyclotron at Berkeley, California.
1945	US physicist Edwin M McMillan and Soviet physicist V I Veksler (1943) independently describe the principle of phase stability. By removing an apparent limitation on the energy of particle accelerators for protons, it makes possible the construction of magnetic-resonance accelerators, or synchrotrons. Synchrocyclotrons are soon built at the University of California and in England.
1946	UK physicist Edward Appleton discovers the 'Appleton layer' in the ionosphere, which reflects radio waves; it makes long-range radio communication possible and also aids the development of radar.
1947	US physicist Willard Libby develops carbon-14 dating.
1948	Hungarian-British physicist Dennis Gabor invents holography, the production of three-dimensional images.
1948	US physicists Richard Feynman and Julian S Schwinger, and Japanese physicist Shin'ichiro Tomonaga, independently develop quantum electrodynamics, the theory that accounts for the interactions between radiation, electrons, and positrons.
1951	US physicist Edward Purcell discovers line radiation (radiation emitted at only one specific wavelength) at 21 cm/8 in emitted by hydrogen in space. It allows the distribution of hydrogen clouds in galaxies and the speed of the Milky Way's rotation to be determined.
1952	US nuclear physicist Donald Glaser develops the bubble chamber to observe the behaviour of subatomic particles. It uses a superheated liquid instead of a vapour to track particles.
1953	US physicist Murray Gell-Mann introduces the concept of 'strangeness', a property of subatomic particles, to explain their behaviour.
1956	US physicists B Cook, G R Lambertson, O Piconi, and W A Wentzel discover the antineutron by passing an antiproton beam through matter.
1956	US physicists Clyde Cowan and Fred Reines detect the existence of the neutrino, a particle with no electric charge and no mass, at the Los Alamos Laboratory.

1957	Japanese physicist Leo Esaki discovers tunnelling, the ability of electrons to penetrate solids by acting as radiating waves.
1957	US physicists John Bardeen, Leon Cooper, and John Schrieffer formulate the theory of superconductivity, the characteristic of a solid material to lose its resistance to electric current when cooled below a certain extremely low temperature.
1961	Gell-Mann and Israeli physicist Yuval Ne'eman independently propose a classification scheme for subatomic particles that comes to be known as the Eightfold Way.
1961	US physicist Robert Hofstadter discovers that protons and neutrons have an internal structure.
1962	Welsh physicist Brian Josephson discovers the Josephson effect, the high-frequency oscillation of a current between two superconductors across an insulating layer.
1964	Gell-Mann and George Zweig independently suggest the existence of the quark, a subatomic particle and the building block of hadrons, a subatomic particle that experiences the strong nuclear force.
1967	US nuclear physicists Sheldon Lee Glashow and Steven Weinberg and Pakistani nuclear physicist Abdus Salam separately develop the electroweak unification theory, which explains 'electromagnetic' interactions and the 'weak' nuclear force.
1970	US physicist Sheldon Glashow and associates postulate the existence of a fourth quark, which they name 'charm'.
1971	English theoretical physicist Stephen Hawking suggests that after the Big Bang, mini black holes no bigger than a proton but containing more than a billion tonnes of mass were formed and that they were governed by both the laws of relativity and of quantum mechanics.
1973	Researchers at the European Centre for Particle Research (CERN) find some confirmation for the electroweak force – one of the four fundamental forces – when they discover neutral currents in neutrino reactions.
1974	Hawking suggests that black holes emit subatomic particles until their energy is diminished to the point where they explode.
4 July 1978	Scientists at the Princeton Large Torus test reactor achieve a temperature of 60 million degrees Fahrenheit, and maintain it for one-twentieth of a second. It is hailed as a breakthrough for nuclear fusion.
1979	Physicists in Hamburg at DESY (Deutsches Elektron Synchroton) observe gluons – particles that carry the strong nuclear force which holds quarks together.
1980	The Tevatron at Fermilab located at the Fermi National Accelerator Laboratory in Batavia, Illinois, USA, is completed; the most powerful proton synchrotron in the world, it is designed

	to operate at 1,000 GeV or 1 TeV (teraelectron volt).
June 1983	The W and Z subatomic particles are detected in experiments at the European Centre for Nuclear Research (CERN), Switzerland, by Italian physicist Carlo Rubbia and Dutch physicist Somin van der Meer; the existence of these particles had been predicted as carriers of the weak nuclear force.
1984	A team of international physicists at CERN in Geneva, Switzerland, discovers the sixth (top) quark; its discovery completes the theoretical scheme of subatomic building blocks.
1986	German physicist Johannes Bednorz and Swiss physicist Karl Alex Müller announce the discovery of a superconducting ceramic material in which superconductivity occurs at a much higher temperature (30 K) than hitherto known, increasing the potential for use of superconductivity for more energy-efficient motors and computers. They receive the Nobel Prize for Physics – in record time – for their discovery.
1986	Scientists use 10 laser beams, which deliver a total energy of 100 trillion watts during one-billionth of a second, to convert a small part of the hydrogen nuclei contained in a glass sphere to helium at the Lawrence Livermore National Laboratory in California; it is the first fusion reaction induced by a laser.
12 Feb 1987	Chinese physicist Paul Ching-Wu Chu and associates at the University of Huston, Texas, make a material that is superconducting at the temperature of liquid nitrogen – 77 K or –196°C/–321°F.
March 1989	US physicist Stanley Pons and English physicist Martin Fleischmann announce that they have achieved nuclear fusion at room temperature (cold fusion); other scientists fail to replicate their experiment.
14 July 1989	The LEP (Large Electron Positron Collider) is

	inaugurated at the CERN research centre in Switzerland; the new accelerator has a circumference of 27 km/16.8 mi and is the largest scientific apparatus in the world.
9 Nov 1991	The Joint European Torus (JET) at Culham, near Oxford, England, produces a 1.7 megawatt pulse of power in an experiment that lasts 2 seconds. It is the first time that a substantial amount of fusion power has been produced in a controlled experiment, as opposed to an atomic bomb.
1992	The Hadron Electron Ring Accelerator (HERA) particle accelerator is built under the streets of Hamburg, Germany. Occupying a tunnel 6.3 km/3.9 mi in length, it is the world's most powerful particle accelerator, accelerating protons to energies of 820 GeV (billion electron volts) and electrons to 30 GeV.
June 1995	US physicists announce the discovery of a new form of matter, called a Bose–Einstein condensate (because its existence had been predicted by Einstein and Indian physicist Satyendra Bose), created by cooling rubidium atoms to just above absolute zero.
1995	The Omega laser is developed at the University of Rochester, New York State. It generates 60 trillion watts of ultraviolet light in pulses that last for 0.65 billionths of a second, and is used in researching the civil applications of nuclear fusion.
1995	US scientists at Fermilab, near Chicago, Illinois, announce the discovery of the top quark, an elementary particle almost as heavy as a gold atom.
4 Jan 1996	A team of European physicists at the CERN research centre in Switzerland create the first atoms of antimatter: nine atoms of antihydrogen survive for 40 nanoseconds.
11 July 1998	Researchers at the Fermi National Accelerator laboratory, announce the discovery of the tau neutrino.

Biographies

Ampère, André Marie (1775–1836) French physicist and mathematician who made many discoveries in electromagnetism and electrodynamics. He followed up the work of Hans Oersted on the interaction between magnets and electric currents, developing a rule for determining the direction of the magnetic field associated with an electric current. The unit of electric current, the **ampere**, is named after him.

Ampère's law is an equation that relates the magnetic force produced by two parallel current-carrying conductors to the product of their currents and the distance between the conductors. Today Ampère's law is usually stated in the form of calculus: the line integral of the magnetic field around an arbitrarily chosen path is proportional to the net electric current enclosed by the path.

Ångström, Anders Jonas (1814–1874) Swedish astrophysicist who worked in spectroscopy and solar physics. In 1861 he identified the presence of hydrogen in the Sun. His outstanding *Recherches sur le spectre solaire* (1868) presented an atlas of the solar spectrum with measurements of 1,000 spectral lines expressed in units of one-ten-millionth of a millimetre, the unit which later became the angstrom. He also investigated the conduction of heat and devised a method of determining thermal conductivity in 1863. His 'Optical investigations' (1853) contains his principle of spectrum analysis, demonstrating that a hot gas emits light at the same frequency as it absorbs it when it is cooled. In 1867 he investigated the spectrum of the aurora borealis, the first person to do so.

Avogadro, Amedeo, Conte di Quaregna (1776–1856) Italian physicist, one of the founders of physical chemistry, who proposed **Avogadro's hypothesis** on gases in 1811. His work enabled scientists to calculate **Avogadro's number**, or **constant**, and still has relevance for atomic studies. Avogadro made it clear that the gas particles need not be individual atoms but might consist of molecules., the term he introduced to describe combinations of atoms. No previous scientist had made this fundamental distinction between the atoms of a substance and its molecules.

Bardeen, John (1908–1991) US physicist. He became the first double winner of a Nobel prize in the same subject (with Leon Cooper (1930–) and Robert Schrieffer (1931–)) in 1972 for his theory of superconductivity, which states that superconductivity arises when electrons travelling through a metal interact with the vibrating atoms of the metal.

Becquerel, (Antoine) Henri (1852–1908) French physicist. The discovery of X-rays in 1896 prompted Becquerel to investigate fluorescent crystals for the emission of X-rays, and in so doing he accidentally discovered radioactivity in uranium salts. He subsequently investigated the radioactivity of radium, and in 1900 showed that it consists of a stream of electrons. In the same year, he also obtained evidence that radioactivity causes the transformation of one element into another.

Biography of A H Becquerel
http://www.nobel.se/laureates/
physics-1901-1-bio.html
Presentation of the life and discoveries of Becquerel, who was awarded the Nobel Prize in Physics jointly with Pierre and Marie Curie.

Bohr, Niels Henrik David (1885–1962) Danish physicist whose theoretical work in 1913 established the structure of the atom and the validity of quantum theory by showing that the nuclei of atoms are surrounded by shells of electrons, each assigned particular sets of quantum numbers according to their orbits. Bohr's atomic theory was validated in 1922 by the discovery of an element he had predicted, hafnium. He explained the structure and behaviour of the nucleus, as well as the process of nuclear fission. He also proposed the doctrine of **complementarity**, the theory that a fundamental particle is neither a wave nor a particle, because these are complementary modes of description. In 1939 Bohr proposed his liquid-droplet model for the nucleus, in which nuclear particles are pulled together by short-range forces, similar to the way in which molecules in a drop of liquid are attracted to one another. The extra energy produced by the absorption of a neutron causes the nuclear particles to separate into two groups of approximately the same size, thus breaking the nucleus into two smaller nuclei – as happens in nuclear fission. The model was vindicated when Bohr correctly predicted the differing behaviour of nuclei of uranium-235 and uranium-238 from the fact that the number of neutrons in each nucleus is odd and even respectively.

Born, Max (1882–1970) German-born British physicist. In 1924 Born coined the term 'quantum mechanics' and in 1925 he devised a system called matrix mechanics that accounted mathematically for the position and momentum of the electron in the atom. He received a Nobel prize in 1954 for fundamental work on quantum theory, especially his 1926 discovery that the wave function of an electron is linked to the probability that the electron is to be found at any point. He also devised a technique, called the Born approximation method, for computing the behaviour of subatomic particles, which is of great use in high-energy physics. In 1953 Born was also able to determine the energies involved in lattice formation, from which the properties of crystals may be derived, and thus laid one of the foundations of solid-state physics.

Born, Max
http://www-history.mcs.st-and.ac.uk/~history/
Mathematicians/Born.html
Biography of the German-born British physicist. The Web site details the work of Born, and his relationships with his contemporaries and colleagues.

Broglie, Louis Victor Pierre Raymond de, 7th duc de Broglie (1892–1987) French theoretical physicist. He established that all subatomic particles can be described either by particle equations or by wave equations, thus laying the foundations of wave mechanics. De Broglie's discovery of wave–particle duality enabled physicists to view Einstein's conviction that matter and energy are interconvertible as being fundamental to the structure of matter. The study of matter waves led not only to a much deeper understanding of the nature of the atom but also to explanations of chemical bonds and the practical application of electron waves in electron microscopes.

De Broglie, Louis
http://www-history.mcs.st-and.ac.uk/~history/
Mathematicians/Broglie.html
Biographical details and a photograph of Louis de Broglie, the famous French physicist and mathematician. There are also links to de Broglie's most famous work on quantum mechanics and to many of his contemporaries.

Carnot, (Nicolas Leonard) Sadi (1796–1832) French scientist and military engineer who founded the science of thermodynamics. His pioneering work was *Reflexions sur la puissance motrice du feu/On the Motive Power of Fire*, which considered the changes that would take place in an idealized, frictionless steam engine. Carnot's theorem showed that the amount of work that an engine can produce depends only on the temperature difference that occurs in the engine. In formulating his theorem, Carnot considered the case of an ideal heat engine following a reversible sequence known as the **Carnot cycle**. This cycle consists of the isothermal expansion and adiabatic expansion of a quantity of gas, producing work and consuming heat, followed by isothermal compression and adiabatic compression, consuming work and producing heat to restore the gas to its original state of pressure, volume, and temperature. Carnot's law states that no engine is more efficient than a reversible engine working between the same temperatures.

Charles, Jacques Alexandre César (1746–1823) French physicist who studied gases and made the first ascent in a hydrogen-filled balloon, in 1783. His work on the expansion of gases led to the formation of Charles's law.

Curie, Marie (born Manya Sklodowska) (1867–1934) Polish scientist who, with her husband Pierre Curie, discovered in 1898, two new radioactive elements in pitchblende ores: polonium and radium. They isolated the pure elements in 1902 and were jointly awarded the Nobel Prize for Physics in 1903. In 1910 with André Debierne (1874–1949), who had discovered actinium in pitchblende in 1899, Marie Curie isolated pure radium metal; she was awarded the Nobel Prize for Chemistry in 1911.

Marie and Pierre Curie
http://www.nobel.se/essays/curie/index.html
Biographies of the Curies. This full account of their lives not only provides a wealth of personal details but also places their work alongside others working to increase understanding of radiation. Marie Curie's life after the death of Pierre and the hostility she suffered from the French press and scientific establishment is movingly described.

Dirac, Paul Adrien Maurice (1902–1984) British physicist who worked out a version of quantum mechanics consistent with special relativity. In 1928 he formulated the relativistic theory of the electron. The model was able to describe many quantitative aspects of the electron, including such properties as the half-quantum spin and magnetic moment. The existence of antiparticles, such as the positron (positive electron), was one of its predictions.

Doppler, Christian Johann (1803–1853) Austrian physicist who in 1842 described the **Doppler effect** (change of observed frequency (or wavelength) of waves due to relative motion between the wave source and observer) and derived the observed frequency mathematically in **Doppler's principle.**

Einstein, Albert (1879–1955) German-born US physicist whose theories of relativity revolutionized our understanding of matter, space, and time. Einstein suggested that packets of light energy are capable of behaving as particles called 'light quanta' (later called photons). Einstein used this hypothesis to explain the photoelectric effect, proposing that light particles striking the surface of certain metals cause electrons to be emitted and deduced the **photoelectric law**, for which he was awarded the Nobel Prize for Physics in 1921. Einstein went on to show in 1907 that mass is related to energy by the famous equation $E=mc^2$, which indicates the enormous amount of energy that is stored as mass, some of which is released in radioactivity and nuclear reactions, for example in the Sun. He also investigated Brownian motion, explaining the phenomenon as being due to the effect of large numbers of molecules (in this case, water molecules) bombarding the particles. Einstein's explanation of Brownian motion and its subsequent experimental confirmation was one of the most important pieces of evidence for the hypothesis that matter is composed of atoms. His **special theory of relativity** started with the premises that the laws of nature are the same for all observers in unaccelerated motion, and that the speed of light is independent of the motion of its source. In the **general theory of relativity**, the properties of space-time were to be conceived as modified locally by the presence of a body with mass; and light rays should bend when they pass by a massive object. His last conception of the basic laws governing the universe was outlined in his unified field theory, made public in 1953.

Life and Theories of Albert Einstein
http://www.pbs.org/wgbh/nova/einstein/index.html
Heavily illustrated site on the life and theories of Einstein. There is an illustrated biographical chart, including a summary of his major achievements and their importance to science. The theory of relativity gets an understandably more in-depth coverage, along with photos and illustrations. The pages on his theories on light and time include illustrated explanations and an interactive test. There is also a 'time-traveller' game demonstrating these theories.

Faraday, Michael (1791–1867) English chemist and physicist. In 1821 he began experimenting with electromagnetism and discovered the induction of electric currents and made the first dynamo, the first electric motor, and the first transformer. Faraday isolated benzene from gas oils and produced the basic laws of electrolysis in 1834. Faraday's laws of electrolysis established the link between electricity and chemical affinity, one of the most fundamental concepts in science. It was Faraday who coined the terms anode, cathode, cation, anion, electrode, and electrolyte. He demonstrated in 1837 that electrostatic force consists of a field of curved lines of force, and that different substances have specific inductive capacities – that is, they take up different amounts of electric charge when subjected to an electric field. He also pointed out that the energy of a magnet is in the field around it and not in the magnet itself, extending this basic conception of field theory to electrical and gravitational systems.

Fermi, Enrico (1901–1954) Italian-born US physicist who proved the existence of new radioactive elements produced by bombardment with neutrons, and discovered nuclear reactions produced by low-energy neutrons. This research was the basis for studies leading to the atomic bomb and nuclear energy. Fermi built the first nuclear reactor in 1942 at Chicago University and later took part in the Manhattan Project to construct an atom bomb.

Feynman, Richard Phillips
http://www-groups.dcs.st-and.ac.uk/~history/
Mathematicians/Feynman.html
Part of an archive containing the biographies of the world's greatest mathematicians, this site is devoted to the life and contributions of physicist Richard Feynman.

Feynman, Richard P(hillips) (1918–1988) US physicist whose work laid the foundations of quantum electrodynamics, developing a simple and elegant system of **Feynman diagrams**

to represent interactions between particles and how they moved from one space-time point to another. He had rules for calculating the probability associated with each diagram. He also contributed to the theory of superfluidity and to many aspects of particle physics, including quark theory and the nature of the weak nuclear force.

Foucault, Jean Bernard Léon (1819–1868) French physicist who used a pendulum to demonstrate the rotation of the Earth on its axis, and invented the gyroscope in 1852. In 1862 he made the first accurate determination of the velocity of light.

Franklin, Benjamin (1706–1790) US scientist, statesman, writer, printer, and publisher. He proved that lightning is a form of electricity, distinguished between positive and negative charges, and invented the lightning conductor. Franklin also made a fundamental discovery when he realized that the gain and loss of electricity must be balanced – the concept of conservation of charge. Franklin's interest in atmospheric electricity led him to recognize the aurora borealis as an electrical phenomenon, postulating good conditions in the rarefied upper atmosphere for electrical discharges, and speculating on the existence of what we now call the ionosphere.

Autobiography of Benjamin Franklin, The
http://www.inform.umd.edu/EdRes/ReadingRoom/
HistoryPhilosophy/BenFranklin/
Contains the text of Franklin's autobiography, complete with an introduction and notes.

Fresnel, Augustin Jean (1788–1827) French physicist who refined the theory of polarized light. Fresnel realized in 1821 that light waves do not vibrate like sound waves longitudinally, in the direction of their motion, but transversely, at right angles to the direction of the propagated wave.

Galileo, properly Galileo Galilei (1564–1642) Italian mathematician, astronomer, and physicist. Galileo discovered that freely falling bodies, heavy or light, have the same, constant acceleration and that this acceleration is due to gravity. He also determined that a body moving on a perfectly smooth horizontal surface would neither speed up nor slow down. He invented a thermometer, a hydrostatic balance, and a compass, and discovered that the path of a projectile is made up of two components: one component consists of uniform motion in a horizontal direction, and the other component is vertical motion under acceleration or deceleration due to gravity. Galileo used this explanation to refute objections to the Sun-centred theory of Polish astronomer Nicolaus Copernicus. Galileo's work founded the modern scientific method of deducing laws to explain the results of observation and experiment.

Geiger, Hans (Wilhelm) (1882–1945) German physicist who produced the Geiger counter. He spent the period 1906–12 in Manchester, England, working with Ernest ◊Rutherford on radioactivity. In 1908 they designed an instrument to detect and count alpha particles, positively charged ionizing particles produced by radioactive decay.

Hawking, Stephen (William) (1942–) English physicist whose work in general relativity – particularly gravitational field theory – led to a search for a quantum theory of gravity to explain black holes and the Big Bang, singularities that classical relativity theory does not adequately explain. His book *A Brief History of Time* (1988) gives a popular account of cosmology and became an international bestseller. His latest book is *The Nature of Space and Time*, written with Roger Penrose.

Hawking, Stephen
http://www.damtp.cam.ac.uk/DAMTP/user/
hawking/home.html
Stephen Hawking's own home page, with a brief biography, disability advice, and a selection of his lectures, including 'the beginning of time' and a series debating the nature of space and time.

Heaviside, Oliver (1850–1925) English physicist. In 1902 he predicted the existence of an ionized layer of air in the upper atmosphere, which was known as the Kennelly–Heaviside layer but is now called the E-layer of the ionosphere. Deflection from it makes possible the transmission of radio signals around the world, which would otherwise be lost in outer space. Heaviside's theoretical work had implications for radio transmission. His studies of electricity published in *Electrical Papers* (1892) had considerable impact on long-distance telephony, and he added the concepts of inductance, capacitance, and impedance to electrical science.

Heaviside, Oliver
http://www-history.mcs.st-
and.ac.uk/~history/Mathematicians/Heaviside.html
Extensive biography of the English physicist. The site contains a description of his contribution to physics, and in particular his simplification of Maxwell's 20 equations in 20 variables, replacing them by two equations in two variables. Today we call these 'Maxwell's equations' forgetting that they are in fact 'Heaviside's equations'.

Hertz, Heinrich Rudolf (1857–1894) German physicist who studied electromagnetic waves, showing their behaviour resembles that of light and heat waves. He confirmed James Clerk Maxwell's theory of electromagnetic waves. In 1888 he realized that electric waves could be produced and would travel through air, and he confirmed this experimentally. He went on to determine the velocity of these waves (later called radio waves) and, on showing that it was the same as that of light, devised experiments to show that the waves could be reflected, refracted, and diffracted. The unit of frequency, the **hertz**, is named after him.

Heisenberg, Werner (Karl) (1901–1976) German physicist who developed quantum theory and formulated the uncertainty principle, which states that there is a theoretical limit to the precision with which a particle's position and momentum can be measured. In other words, it is impossible to specify precisely both the position and the simultaneous momentum (mass multiplied by velocity) of a particle. There is always a degree of uncertainty in either, and as one is determined with greater precision, the other can only be found less exactly.

Heisenberg, Werner Karl
http://www-groups.dcs.st-and.ac.uk/~history/
Mathematicians/Heisenberg.html
Part of an archive containing the details of the world's greatest mathematicians, this site is devoted to the life and contributions of physicist Werner Heisenberg.

Huygens (or Huyghens), Christiaan (1629–1695) Dutch mathematical physicist and astronomer. He proposed the wave theory of light, developed the pendulum clock in 1657, discovered polarization, and observed Saturn's rings. He made important advances in pure mathematics, applied mathematics, and mechanics, which he virtually founded.

Huygens, Christiaan
http://www-history.mcs.st-and.ac.uk/~history/
Mathematicians/Huygens.html
Extensive biography of the Dutch astronomer, physicist, and mathematician. The site contains a description of his contributions to astronomy, physics, and mathematics. Also included are the title page of his book *Horologium Oscillatorium* (1673) and the first page of his book *De Ratiociniis in Ludo Aleae* (1657).

Joliot-Curie, Frédéric (Jean) Joliot (1900–1958) and Irène (born Curie) (1897–1956) French physicists. They made the discovery of artificial radioactivity and the transmutation of elements. In 1934, while bombarding light elements with alpha particles, they noticed that although proton production stopped when the alpha particle bombardment stopped, another form of radiation continued. The alpha particles had produced an isotope of phosphorus not found in nature. This isotope was radioactive and was decaying through beta-decay.

Josephson, Brian David (1940–) Welsh physicist, a leading authority on superconductivity. In 1973 he shared a Nobel prize for his theoretical predictions of the properties of a supercurrent through a tunnel barrier (the **Josephson effect**), which led to the development of the **Josephson junction.**

Joule, James Prescott (1818–1889) English physicist. His work on the relations between electrical, mechanical, and chemical effects led to the discovery of the first law of thermodynamics. He determined the mechanical equivalent of heat (Joule's equivalent) in 1843, and the SI unit of energy, the joule, is named after him. He also discovered Joule's law, which defines the relation between heat and electricity; and with Irish physicist Lord Kelvin in 1852 the Joule–Kelvin (or Joule–Thomson) effect.

Kelvin, William Thomson, 1st Baron Kelvin (1824–1907) Irish physicist who introduced the **Kelvin scale**, the absolute scale of temperature. His work on the conservation of energy in 1851 led to the second law of thermodynamics. In 1847 he concluded that electrical and magnetic fields are distributed in a manner analogous to the transfer of energy through an

elastic solid. From 1849 to 1859, Kelvin also developed the work of English scientist Michael ◊Faraday into a full theory of magnetism, arriving at an expression for the total energy of a system of magnets.

Maxwell, James Clerk (1831–1879) Scottish physicist. His main achievement was in the understanding of electromagnetic waves: **Maxwell's equations** bring electricity, magnetism, and light together into one set of relations. He studied gases, optics, and the sensation of colour, and his theoretical work in magnetism prepared the way for wireless telegraphy and telephony. In developing the kinetic theory of gases, Maxwell gave the final proof that heat resides in the motion of molecules.

Maxwell, James Clerk
http://www-history.mcs.st-and.ac.uk/~history/
Mathematicians/Maxwell.html
Extensive biography of the Scottish physicist and mathematician. The site contains a description of his contribution to physics, in particular his work on electricity and magnetism.

Michelson, Albert Abraham (1852–1931) German-born US physicist. With his colleague Edward Morley (1838–1923), he performed in 1887 the **Michelson–Morley experiment** to detect the motion of the Earth through the postulated ether (a medium believed to be necessary for the propagation of light). The negative result of the experiment demonstrated that the velocity of light is constant whatever the motion of the observer. The failure of the experiment indicated the nonexistence of the ether, and led Einstein to his theory of relativity. Michelson invented the **Michelson interferometer** to detect any difference in the velocity of light in two directions at right angles.

Newton, Isaac (1642–1727) English physicist and mathematician who laid the foundations of physics as a modern discipline. During 1665–66, he discovered the binomial theorem, differential and integral calculus, and that white light is composed of many colours. He developed the three standard laws of motion (Newton's laws of motion) and the universal law of gravitation, set out in Philosophiae naturalis principia mathematica (1687, usually referred to as the Principia). Newton's greatest achievement was to demonstrate that scientific principles are of universal application. He clearly defined the nature of mass, weight, force, inertia, and acceleration. In 1679 he calculated the Moon's motion on the basis of his theory of gravity and also found that his theory explained the laws of planetary motion that had been derived by the German astronomer Johannes Kepler on the basis of observations of the planets.

Oersted, Hans Christian (1777–1851) Danish physicist who founded the science of electromagnetism. In 1820 he discovered the magnetic field associated with an electric current.

Ohm, Georg Simon (1789–1854) German physicist who studied electricity and discovered the fundamental law that bears his name. The SI unit of electrical resistance, the **ohm**, is named after him, and the unit of conductance (the inverse of resistance) was formerly called the **mho**, which is 'ohm' spelled backwards.

Planck, Max Karl Ernst (1858–1947) German physicist who framed the quantum theory in 1900. His research into the manner in which heated bodies radiate energy led him to report that energy is emitted only in indivisible amounts, called 'quanta', the magnitudes of which are proportional to the frequency of the radiation. His discovery ran counter to classical physics that radiation consisted of waves and is held to have marked the commencement of the modern science.

Planck, Max Karl Ernst Ludwig
http://www-history.mcs.st-and.ac.uk/~history/
Mathematicians/Planck.html
Biography of the eminent German physicist Max Planck. Planck is thought of by many to have been more influential than any other in the foundation of quantum physics, and partly as a consequence has a fundamental constant named after him.

Röntgen (or Roentgen), Wilhelm Konrad (1845–1923) German physicist. He discovered X-rays in 1895. While investigating the passage of electricity through gases, he noticed the fluorescence of a barium platinocyanide screen. This radiation passed through some substances opaque to light, and affected photographic plates. Developments from this discovery revolutionized medical diagnosis.

Röntgen, Wilhelm Conrad
http://www.nobel.se/laureates/
physics-1901-1-bio.html
Biography of Wilhelm Conrad Röntgen, the German physicist who first realized the huge potential of the electromagnetic field of X-rays. The presentation includes sections on Röntgen's early years and education, his academic career and scientific experiments, and the miraculous coincidences that led him to his great discovery of X-rays.

Rutherford, Ernest, 1st Baron Rutherford of Nelson (1871–1937) New Zealand–born British physicist. His main research was in the field of radioactivity, and he discovered alpha, beta, and gamma rays. He was the first to recognize the nuclear nature of the atom in 1911. In 1919 he produced the first artificial transformation, changing one element to another by bombarding nitrogen with alpha particles and getting hydrogen and oxygen. After further research he announced that the nucleus of any atom must be composed of hydrogen nuclei; at Rutherford's suggestion, the name 'proton' was given to the hydrogen nucleus in 1920. He speculated that uncharged particles (neutrons) must also exist in the nucleus. In 1934, using heavy water, Rutherford and his coworkers bombarded deuterium with deuterons and produced tritium. This may be considered the first nuclear fusion reaction.

Schrödinger, Erwin
http://www-groups.dcs.st-
and.ac.uk/~history/Mathematicians/Schrodinger.html
Part of an archive containing the biographies of the world's greatest mathematicians, this site is devoted to the life and contributions of physicist Erwin Schrödinger.

Schrödinger, Erwin (1887–1961) Austrian physicist. He advanced the study of wave mechanics to describe the behaviour of electrons in atoms. In 1926 he produced a solid mathematical explanation of the quantum theory and the structure of the atom.

Tesla, Nikola (1856–1943) Serbian-born US physicist and electrical engineer who invented fluorescent lighting, the Tesla induction motor (1882–87), and the Tesla coil, and developed the alternating current (AC) electrical supply system. The **Tesla coil** is an air core transformer with the primary and secondary windings tuned in resonance to produce high-frequency, high-voltage electricity.

Tesla, Nikola
http://www.neuronet.pitt.edu/~bogdan/tesla/
Short biography of the electrical inventor plus quotations by and about Tesla, anecdotes, a photo gallery, and links to other sites of interest.

Thomson, J(oseph) J(ohn) (1856–1940) English physicist. He discovered the electron in 1897. His work inaugurated the electrical theory of the atom, and his elucidation of positive rays and their application to an analysis of neon led to the discovery of isotopes. Using magnetic and electric fields to deflect positive rays, Thomson found in 1912 that ions of neon gas are deflected by different amounts, indicating that they consist of a mixture of ions with different charge-to-mass ratios. English chemist Frederick Soddy had earlier proposed the existence of isotopes and Thomson proved this idea correct when he identified, also in 1912, the isotope neon-22.

Volta, Alessandro Giuseppe Antonio Anastasio, Count (1745–1827) Italian physicist who invented the first electric cell (the voltaic pile, in 1800), the electrophorus (an early electrostatic generator, in 1775), and an electroscope. Volta also produced a list of metals in order of their electricity production based on the strength of the sensation they made on his tongue, thereby deriving the electromotive series. In about 1795, Volta recognized that the vapour pressure of a liquid is independent of the pressure of the atmosphere and depends only on temperature. The **volt** is named after him.

Weber, Wilhelm Eduard (1804–1891) German physicist who studied magnetism and electricity. Working with the German mathematician Karl Gauss (1777–1855), he made sensitive magnetometers to measure magnetic fields, and instruments to measure direct and alternating currents. He also built an electric telegraph. The SI unit of magnet flux, the **weber**, is named after him.

Young, Thomas (1773–1829) British physicist who revived the wave theory of light and identified the phenomenon of interference in 1801. He also established many important concepts in mechanics. Young assumed that light waves are propagated in a similar way to sound waves, and proposed that different colours consist of different frequencies. He obtained experimental proof for the principle of interference by passing light through extremely narrow openings and observing the interference patterns produced. In mechanics, Young was the first to use the terms 'energy' for the product of the mass of a body with the square of its velocity and 'labour expended' for the product of the force exerted on a body 'with the distance through which it moved'. He also stated that these two products are proportional to each other. He introduced an absolute measurement in elasticity, now known as **Young's modulus**.

Biographies of Physicists
http://hermes.astro.washington.edu/scied/
physics/physbio.html
Valuable compilation of biographies of the most famous physicists of all time, including Aristotle, Da Vinci, Kepler, Galilei, Newton, Franklin, Curie, Feynman, and Oppenheimer. Be careful to bookmark it properly because you can easily get lost in the many pages of this site.

Glossary

absorption phenomenon by which a substance retains radiation of particular wavelengths; for example, a piece of blue glass absorbs all visible light except the wavelengths in the blue part of the spectrum; it also refers to the partial loss of energy resulting from light and other electromagnetic waves passing through a medium. In nuclear physics, absorption is the capture by elements, such as boron, of neutrons produced by fission in a reactor.

acoustics experimental and theoretical science of sound and its transmission; in particular, that branch of the science that has to do with the phenomena of sound in a particular space such as a room or theatre.

adiabatic term used to describe a process that occurs without loss or gain of heat, especially the expansion or contraction of a gas in which a change takes place in the pressure or volume, although no heat is allowed to enter or leave.

adsorption taking up of a gas or liquid at the surface of another substance, most commonly a solid (for example, activated charcoal adsorbs gases). It involves molecular attraction at the surface, and should be distinguished from absorption (in which a uniform solution results from a gas or liquid being incorporated into the bulk structure of a liquid or solid).

aerodynamics branch of fluid physics that studies the forces exerted by air or other gases in motion. Examples include the airflow around bodies moving at speed through the atmosphere (such as land vehicles, bullets, rockets, and aircraft), the behaviour of gas in engines and furnaces, air conditioning of buildings, the deposition of snow, the operation of air-cushion vehicles (hovercraft), wind loads on buildings and bridges, bird and insect flight, musical wind instruments, and meteorology. For maximum efficiency, the aim is usually to design the shape of an object to produce a streamlined flow, with a minimum of turbulence in the moving air.

alternating current (AC) electric current that flows for an interval of time in one direction and then in the opposite direction, that is, a current that flows in alternately reversed directions through or around a circuit. Electric energy is usually generated as alternating current in a power station, and alternating currents may be used for both power and lighting. The advantage of alternating current over direct current (DC), as from a battery, is that its voltage can be raised or lowered economically by a transformer: high voltage for generation and transmission, and low voltage for safe utilization.

ammeter instrument that measures electric current (flow of charge per unit time), usually in amperes, through a conductor. It should not to be confused with a voltmeter, which measures potential difference between two points in a circuit.

ampere SI unit (symbol A) of electrical current. Electrical current is measured in a similar way to water current, in terms of an amount per unit time; one ampere represents a flow of about $6.28 \_ 10^{18}$ electrons per second, or a rate of flow of charge of one coulomb per second.

anion ion carrying a negative charge. During electrolysis, anions in the electrolyte move towards the anode (positive electrode).

Archimedes' principle principle that the weight of the liquid displaced by a floating body is equal to the weight of the body. The principle is often stated in the form: 'an object totally or partially submerged in a fluid displaces a volume of fluid that weighs the same as the apparent loss in weight of the object (which, in turn, equals the upwards force, or upthrust, experienced by that object).' It was discovered by the Greek mathematician Archimedes.

atmosphere, or **standard atmosphere**, unit (symbol atm) of pressure equal to 760 torr, 1013.25 millibars, or $1.01325 \_ 10^5$ newtons per square metre. The actual pressure exerted by the atmosphere fluctuates around this value, which is assumed to be standard at sea level and 0°C/32°F, and is used when dealing with very high pressures.

Avogadro's number, or **Avogadro's constant**, the number of carbon atoms in 12 g of the carbon-12 isotope (6.022045×10^{23}). The relative atomic mass of any element, expressed in grams, contains this number of atoms.

ballistics study of the motion and impact of projectiles such as bullets, bombs, and missiles. For projectiles from a gun, relevant exterior factors include temperature, barometric pressure, and wind strength; and for nuclear missiles these extend to such factors as the speed at which the Earth turns.

becquerel SI unit (symbol Bq) of radioactivity, equal to one radioactive disintegration (change in the nucleus of an atom when a particle or ray is given off) per second.

binding energy the amount of energy needed to break the nucleus of an atom into the neutrons and protons of which it is made.

boiling point for any given liquid, the temperature at which the application of heat raises the temperature of the liquid no further, but converts it into vapour.

Boltzmann constant the constant (symbol k) that relates the kinetic energy (energy of motion) of a gas atom or molecule to temperature. Its value is 1.38066×10^{-23} joules per kelvin. It is equal to the gas constant R, divided by ◊Avogadro's number.

boson elementary particle whose spin can only take values that are whole numbers or zero. Bosons may be classified as gauge bosons (carriers of the four fundamental forces) or mesons. All elementary particles are either bosons or fermions. Unlike fermions, more than one boson in a system (such as an atom) can possess the same energy state.

buckminsterfullerene form of carbon, made up of molecules (buckyballs) consisting of 60 carbon atoms arranged in 12 pentagons and 20 hexagons to form a perfect sphere. It was named after the US architect and engineer Richard Buckminster Fuller (1895–1983) because of its structural similarity to the geodesic dome that he designed.

capacitor, or **condenser**, device for storing electric charge, used in electronic circuits; it consists of two or more metal plates separated by an insulating layer called a dielectric.

cathode the part of an electronic device in which electrons are generated. In a thermionic valve, electrons are produced by the heating effect of an applied current; in a photocell, they are produced by the interaction of light and a semiconducting material. The cathode is kept at a negative potential relative to the device's other electrodes (anodes) in order to ensure that the liberated electrons stream away from the cathode and towards the anodes.

cathode ray stream of fast-moving electrons that travel from a cathode (negative electrode) towards an anode (positive electrode) in a vacuum tube. They carry a negative charge and can be deflected by electric and magnetic fields. Cathode rays focused into fine beams of fast electrons are used in cathode-ray tubes, the electrons' kinetic energy being converted into light energy as they collide with the tube's fluorescent screen.

centrifugal force useful concept in physics, based on an apparent (but not real) force. It may be regarded as a force that acts radially outwards from a spinning or orbiting object, thus balancing the centripetal force (which is real). For an object of mass m moving with a velocity v in a circle of radius r, the centrifugal force F equals mv^2/r (outwards).

centripetal force force that acts radially inwards on an object moving in a curved path. For example, with a weight whirled in a circle at the end of a length of string, the centripetal force is the tension in the string. For an object of mass m moving with a velocity v in a circle of radius r, the centripetal force F equals mv^2/r (inwards). The reaction to this force is the ◊centrifugal force.

chain reaction in nuclear physics, a fission reaction that is maintained because neutrons released by the splitting of some atomic nuclei themselves go on to split others, releasing even more neutrons. Such a reaction can be controlled (as in a nuclear reactor) by using moderators to absorb excess neutrons. Uncontrolled, a chain reaction produces a nuclear explosion (as in an atom bomb).

circuit arrangement of electrical components through which a current can flow. There are two basic circuits, series and parallel. In a **series circuit**, the components are connected end to end so that the current flows through all components one after the other. In a **parallel circuit**, components are connected side by side so that part of the current passes through each component. A **circuit diagram** shows in graphical form how components are connected together, using standard symbols for the components.

cloud chamber apparatus for tracking ionized particles. It consists of a vessel fitted with a piston and filled with air or other gas, saturated with water vapour. When the volume of the vessel is suddenly expanded by moving the piston outwards, the vapour cools and a cloud of tiny droplets forms on any nuclei, dust, or ions present. As fast-moving ionizing particles collide with the air or gas molecules, they show as visible tracks.

cold fusion in nuclear physics, the fusion of atomic nuclei at room temperature. If cold fusion were possible it would provide a limitless, cheap, and pollution-free source of energy, and it has therefore been the subject of research around the world. In 1989, Martin Fleischmann (1927–) and Stanley Pons (1943–) of the University of Utah, USA, claimed that they had achieved cold fusion in the laboratory, but their results could not be substantiated and in 1998 their patent was allowed to lapse.

colour quality or wavelength of light emitted or reflected from an object. Visible white light consists of electromagnetic radiation of various wavelengths, and if a beam is refracted through a prism, it can be spread out into a spectrum, in which the various colours correspond to different wavelengths. From long to short wavelengths (from about 700 to 400 nanometres) the colours are red, orange, yellow, green, blue, indigo, and violet.

cosmic radiation streams of high-energy particles from outer space, consisting of protons, alpha particles, and light nuclei, which collide with atomic nuclei in the Earth's atmosphere, and produce secondary nuclear particles (chiefly mesons, such as pions and muons) that shower the Earth.

coulomb SI unit (symbol C) of electrical charge. One coulomb is the quantity of electricity conveyed by a current of one ◊ampere in one second.

critical mass in nuclear physics, the minimum mass of fissile material that can undergo a continuous ◊chain reaction. Below this mass, too many neutrons escape from the surface for a chain reaction to carry on; above the critical mass, the reaction may accelerate into a nuclear explosion.

cryogenics science of very low temperatures (approaching ◊absolute zero), including the production of very low temperatures and the exploitation of special properties associated with them, such as the disappearance of electrical resistance (superconductivity).

Curie temperature temperature above which a magnetic material cannot be strongly magnetized. Above the Curie temperature, the energy of the atoms is too great for them to join together to form the small areas of magnetized material, or ◊domains, which combine to produce the strength of the overall magnetization.

dimension any directly measurable physical quantity such as mass (M), length (L), and time (T), and the derived units obtainable by multiplication or division from such quantities. For example, acceleration (the rate of change of velocity) has dimensions (LT^{-2}), and is expressed in such units as km s^{-2}. A quantity that is a ratio, such as relative density or humidity, is dimensionless.

diode combination of a cold anode and a heated cathode, or the semiconductor equivalent, which incorporates a p–n junction. Either device allows the passage of direct current in one direction only, and so is commonly used in a rectifier to convert alternating current (AC) to direct current (DC).

direct current (DC) electric current that flows in one direction, and does not reverse its flow as ◊alternating current does. The electricity produced by a battery is direct current.

domain small area in a magnetic material that behaves like a tiny magnet. The magnetism of the material is due to the movement of the electrons in the atoms of the domain. In an unmagnetized sample of material, the domains point in random directions, or form closed loops, so that there is no overall magnetization of the sample. In a magnetized sample, the domains are aligned so that their magnetic effects combine to produce a strong overall magnetism.

Doppler effect change in the observed frequency (or wavelength) of waves due to relative motion between the wave source and the observer. The Doppler effect is responsible for the perceived change in pitch of a siren as it approaches and then recedes, and for the red shift of light from distant galaxies. It is named after the Austrian physicist Christian Doppler.

echo repetition of a sound wave, or of a radar or sonar signal, by reflection from a surface. By accurately measuring the time taken for an echo to return to the transmitter, and by knowing the speed of a radar signal (the speed of light) or a sonar signal (the speed of sound in water), it is possible to calculate the range of the object causing the echo (echolocation).

electric field region in which a particle possessing electric charge experiences a force owing to the presence of another electric charge. The strength of an electric field, E, is measured in volts per metre (V m^{-1}). It is a type of ◊electromagnetic field.

electrode any terminal by which an electric current passes in or out of a conducting substance; for example, the anode or ◊cathode in a battery or the carbons in an arc lamp. The terminals that emit and collect the flow of electrons in thermionic valves (electron tubes) are also called electrodes: for example, cathodes, plates, and grids.

electromagnetic field region in which a particle with an electric charge experiences a force. If it does so only when moving, it is in a pure **magnetic field**; if it does so when stationary, it is in an **electric field**. Both can be present simultaneously.

electromotive force (emf) loosely, the voltage produced by an electric battery or generator in an electrical circuit or, more precisely, the energy supplied by a source of electric power in driving a unit charge around the circuit. The unit is the volt.

element substance that cannot be split chemically into simpler substances. The atoms of a particular element all have the same number of protons in their nuclei (their atomic number). Elements are classified in the periodic table of the elements. Of the known elements, 92 are known to occur in nature (those with atomic numbers 1–92). Those elements with atomic numbers above 96 do not occur in nature and are synthesized only, produced in particle accelerators. Of the elements, 81 are stable; all the others, which include atomic numbers 43, 61, and from 84 up, are radioactive.

entropy in thermodynamics, a parameter representing the state of disorder of a system at the atomic, ionic, or molecular level; the greater the disorder, the higher the entropy. Thus the fast-moving disordered molecules of water vapour have higher entropy than those of more ordered liquid water, which in turn have more entropy than the molecules in solid crystalline ice. In a closed system undergoing change, entropy is a measure of the amount of energy unavailable for useful work. At absolute zero (–273.15°C/–459.67°F/0 K), when all molecular motion ceases and order is assumed to be complete, entropy is zero.

evaporation process in which a liquid turns to a vapour without its temperature reaching boiling point. A liquid left to stand in a saucer eventually evaporates because, at any time, a proportion of its molecules will be fast enough (have enough kinetic energy) to escape through the attractive intermolecular forces at the liquid surface into the atmosphere. The temperature of the liquid tends to fall because the evaporating molecules remove energy from the liquid. The rate of evaporation rises with increased temperature because as the mean kinetic energy of the liquid's molecules rises, so will the number possessing enough energy to escape.

expansion increase in size of a constant mass of substance caused by, for example, increasing its temperature (thermal expansion) or its internal pressure. The **expansivity**, or coefficient of thermal expansion, of a material is its expansion (per unit volume, area, or length) per degree rise in temperature.

farad SI unit (symbol F) of electrical capacitance (how much electric charge a ◊capacitor can store for a given voltage). One farad is a capacitance of one ◊coulomb per volt. For practical purposes the microfarad (one millionth of a farad, symbol mF) is more commonly used.

fermion subatomic particle whose spin can only take values that are half-integers, such as $^{1}/_{2}$ or $^{3}/_{2}$. Fermions may be classified as leptons, such as the electron, and baryons, such as the proton and neutron. All elementary particles are either fermions or ◊bosons.

fibre optics branch of physics dealing with the transmission of light and images through glass or plastic fibres known as optical fibres.

field region of space in which an object exerts a force on another separate object because of certain properties they both possess. For example, there is a force of attraction between any two objects that have mass when one is in the gravitational field of the other. Other fields of force include ◊electric fields (caused by electric charges) and magnetic fields (caused by circulating electric currents), either of which can involve attractive or repulsive forces.

force any influence that tends to change the state of rest or the uniform motion in a straight line of a body. The action of an unbalanced or resultant force results in the acceleration of a body in the direction of action of the force, or it may, if the body is unable to move freely, result in its deformation (see ◊Hooke's law). Force is a vector quantity, possessing both magnitude and direction; its SI unit is the newton.

freezing change from liquid to solid state, as when water becomes ice. For a given substance, freezing occurs at a definite temperature, known as the **freezing point**, that is invariable under similar conditions of pressure, and the temperature remains at this point until all the liquid is frozen. The amount of heat per unit mass that has to be removed to freeze a substance is a constant for any given substance, and is known as the latent heat of fusion.

friction the force that opposes the relative motion of two bodies in contact. The **coefficient of friction** is the ratio of the force required to achieve this relative motion to the force pressing the two bodies together.

fundamental constant physical quantity that is constant in all circumstances throughout the whole universe. Examples are the electric charge of an electron, the speed of light, Planck's constant, and the gravitational constant.

Geiger counter any of a number of devices used for detecting nuclear radiation and/or measuring its intensity by counting the number of ionizing particles produced. It detects the momentary current that passes between ◊electrodes in a suitable gas when a nuclear particle or a radiation pulse causes the ionization of that gas. The electrodes are connected to electronic devices that enable the number of particles passing to be measured. The increased frequency of measured particles indicates the intensity of radiation. The device is named after the German physicist Hans Geiger.

hadron subatomic particle that experiences the strong nuclear force. Each is made up of two or three indivisible particles called quarks. The hadrons are grouped into the baryons (protons, neutrons, and hyperons) and the mesons (particles with masses between those of electrons and protons).

heat capacity quantity of heat required to raise the temperature of an object by one degree. The **specific heat capacity** of a substance is the heat capacity per unit of mass, measured in joules per kilogram per kelvin ($J kg^{-1} K^{-1}$).

Hooke's law law stating that the deformation of a body is proportional to the magnitude of the deforming force, provided that the body's elastic limit is not exceeded. If the elastic limit is not reached, the body will return to its original size once the force is removed. The law was discovered by the English scientist Robert Hooke (1635–1703) in 1676.

inductance phenomenon where a changing current in a circuit builds up a magnetic field which induces an ◊electromotive force either in the same circuit and opposing the current (self-inductance) or in another circuit (mutual inductance). The SI unit of inductance is the henry (symbol H). A component designed to introduce inductance into a circuit is called an inductor (sometimes inductance) and is usually in the form of a coil of wire. The energy stored in the magnetic field of the coil is proportional to its inductance and the current flowing through it.

inductor device included in an electrical circuit because of its inductance.

integrated circuit (IC), or **silicon chip**, miniaturized electronic circuit produced on a single crystal, or chip, of a semiconducting material – usually silicon.

joule SI unit (symbol J) of work and energy, replacing the calorie (one joule equals 4.2 calories).

Kelvin scale temperature scale used by scientists. It begins at –absolute zero (–273.15°C) and increases by the same degree intervals as the Celsius scale; that is, 0°C is the same as 273.15 K and 100°C is 373.15 K.

kinetic energy the energy of a body resulting from motion. It is contrasted with ◊potential energy.

kinetics branch of dynamics dealing with the action of forces producing or changing the motion of a body; **kinematics** deals with motion without reference to force or mass.

kinetic theory theory describing the physical properties of matter in terms of the behaviour – principally movement – of its component atoms or molecules. The temperature of a substance is dependent on the velocity of movement of its constituent particles, increased temperature being accompanied by increased movement. A gas consists of rapidly moving atoms or molecules and, according to kinetic theory, it is their continual impact on the walls of the containing vessel that accounts for the pressure of the gas. The slowing of molecular motion as temperature falls, according to kinetic theory, accounts for the physical properties of liquids and solids, culminating in the concept of no molecular motion at ◊absolute zero (0K/–273°C). By making various assumptions about the nature of gas molecules, it is possible to derive from the kinetic theory the various gas laws (such as Avogadro's hypothesis, Boyle's law, and Charles's law).

latent heat heat absorbed or released by a substance as it changes state (for example, from solid to liquid) at constant temperature and pressure.

luminescence emission of light from a body when its atoms are excited by means other than raising its temperature. Short-lived luminescence is called **fluorescence**; longer-lived luminescence is called **phosphorescence**. When exposed to an external source of energy, the outer electrons in atoms of a luminescent substance absorb energy and 'jump' to a higher energy level. When these electrons 'jump' back to their former level they emit their excess energy as light.

melting point temperature at which a substance melts, or changes from solid to liquid form. A pure substance under standard conditions of pressure (usually one atmosphere) has a definite melting point. If heat is supplied to a solid at its melting point, the temperature does not change until the melting process is complete. The melting point of ice is 0°C or 32°F.

newton SI unit (symbol N) of ◊force. One newton is the force needed to accelerate an object with mass of one kilogram by one metre per second per second. The weight of a medium size (100 g/3 oz) apple is one newton.

nucleon in particle physics, either a proton or a neutron, when present in the atomic nucleus. **Nucleon number** is an alternative name for the mass number of an atom.

nuclear physics study of the properties of the nucleus of the atom, including the structure of nuclei; nuclear forces; the interactions between particles and nuclei; and the study of radioactive decay.

ohm SI unit (symbol W) of electrical ◊resistance (the property of a conductor that restricts the flow of electrons through it).

optics study of light and vision – for example, shadows and mirror images, lenses, microscopes, telescopes, and cameras.

particle physics study of the particles that make up all atoms, and of their interactions. More than 300 subatomic particles have now been identified by physicists, categorized into several classes according to their mass, electric charge, spin, magnetic moment, and interaction.

Peltier effect a change in temperature at the junction of two different metals produced when an electric current flows through them. The extent of the change depends on what the conducting metals are, and the nature of change (rise or fall in temperature) depends on the direction of current flow. It is the reverse of the ◊Seebeck effect. It is named after the French physicist Jean Charles Peltier (1785–1845) who discovered it in 1834.

perpetual motion idea that a machine can be designed and constructed in such a way that, once started, it will continue in motion indefinitely without requiring any further input of energy (motive power). Such a device would contradict at least one of the two laws of thermodynamics that state that (1) energy can neither be created nor destroyed (the law of conservation of energy) and (2) heat cannot by itself flow from a cooler to a hotter object. As a result, all practical (real) machines require a continuous supply of energy, and no heat engine is able to convert all the heat into useful work.

piezoelectric effect property of some crystals (for example, quartz) to develop an electromotive force or voltage across opposite faces when subjected to tension or compression, and, conversely, to expand or contract in size when subjected to an electromotive force. Piezoelectric crystal oscillators are used as frequency standards (for example, replacing balance wheels in watches), and for producing ultrasound.

Planck's constant a fundamental constant (symbol h) that relates the energy (E) of one quantum of electromagnetic radiation (the smallest possible 'packet' of energy) to the frequency (f) of its radiation by E = hf. Its value is $6.6260755 \times 10^{-34}$.

potential difference (pd) difference in the electrical potential (see ◊potential, electric) of two points, being equal to the electrical energy converted by a unit electric charge moving from one point to the other. The SI unit of potential difference is the volt (V). The potential difference between two points in a circuit is commonly referred to as voltage. In equation terms, potential difference V may be defined by: V = W/Q, where W is the electrical energy converted in joules and Q is the charge in coulombs.

potential, electric the relative electrical state of an object. The potential at a point is equal to the energy required to bring a unit electric charge from infinity to the point. The SI unit of potential is the volt (V). Positive electric charges will flow

'downhill' from a region of high potential to a region of low potential. The difference in potential – ◊potential difference (pd) – is expressed in volts so, for example, a 12 V battery has a pd of 12 volts between its negative and positive terminals.

potential energy (PE) energy possessed by an object by virtue of its relative position or state (for example, as in a compressed spring or a muscle). It is contrasted with kinetic energy, the form of energy possessed by moving bodies. An object that has been raised up is described as having gravitational potential energy.

power rate of doing work or consuming energy. It is measured in watts (joules per second) or other units of work per unit time. If the work done or energy consumed is W joules and the time taken is t seconds, then the power P is given by the formula P = W/t.

pressure in a fluid, the force that would act normally (at right angles) per unit surface area of a body immersed in the fluid. The SI unit of pressure is the pascal (Pa), equal to a pressure of one newton per square metre. In the atmosphere, the pressure declines with height from about 100 kPa at sea level to zero where the atmosphere fades into space. Pressure is commonly measured with a barometer, manometer, or Bourdon gauge. Other common units of pressure are the bar and the torr.

prism triangular block of transparent material (plastic, glass, silica) commonly used to 'bend' a ray of light or split a beam into its spectral colours. Prisms are used as mirrors to define the optical path in binoculars, camera viewfinders, and periscopes. The dispersive property of prisms is used in the spectroscope.

quantum theory theory that energy does not have a continuous range of values, but is, instead, absorbed or radiated discontinuously, in multiples of definite, indivisible units called quanta. Just as earlier theory showed how light, generally seen as a wave motion, could also in some ways be seen as composed of discrete particles (photons), quantum theory shows how atomic particles such as electrons may also be seen as having wavelike properties. Quantum theory is the basis of particle physics, modern theoretical chemistry, and the solid-state physics that describes the behaviour of the silicon chips used in computers.

resistance that property of a conductor that restricts the flow of electricity through it, associated with the conversion of electrical energy to heat; also the magnitude of this property. Resistance depends on many factors, such as the nature of the material, its temperature, dimensions, and thermal properties; degree of impurity; the nature and state of illumination of the surface; and the frequency and magnitude of the current. The SI unit of resistance is the ohm. Resistance = voltage/current.

resonance rapid amplification of a vibration when the vibrating object is subject to a force varying at its natural frequency. In a trombone, for example, the length of the air column in the instrument is adjusted until it resonates with the note being sounded. Resonance effects are also produced by many electrical circuits. Tuning a radio, for example, is done by adjusting the natural frequency of the receiver circuit until it coincides with the frequency of the radio waves falling on the aerial.

Seebeck effect generation of a voltage in a circuit containing two different metals, or semiconductors, by keeping the junctions between them at different temperatures. Discovered by the German physicist Thomas Seebeck (1770–1831), it is also called the thermoelectric effect, and is the basis of the thermocouple. It is the opposite of the ◊Peltier effect (in which current flow causes a temperature difference between the junctions of different metals).

short circuit unintended direct connection between two points in an electrical circuit. Resistance is proportional to the length of wire through which current flows. By bypassing the rest of the circuit, the short circuit has low resistance and a large current flows through it. This may cause the circuit to overheat dangerously.

siemens SI unit (symbol S) of electrical conductance, the reciprocal of the ◊resistance of an electrical circuit. One siemens equals one ampere per volt. It was formerly called the mho or reciprocal ohm.

sievert SI unit (symbol Sv) of radiation dose equivalent. It replaces the rem (1 Sv =100 rem). Some types of radiation do more damage than others for the same absorbed dose – for example, an absorbed dose of alpha radiation causes 20 times as much biological damage as the same dose of beta radiation. The equivalent dose in sieverts is equal to the absorbed dose of radiation in grays multiplied by the relative biological effectiveness. Humans can absorb up to 0.25 Sv without immediate ill effects; 1 Sv may produce radiation sickness; and more than 8 Sv causes death.

solenoid coil of wire, usually cylindrical, in which a magnetic field is created by passing an electric current through it This field can be used to move an iron rod placed on its axis.

specific gravity alternative term for relative density.

specific heat capacity quantity of heat required to raise unit mass (1 kg) of a substance by one kelvin (1 K). The unit of specific heat capacity in the SI system is the ◊joule per kilogram kelvin (J kg^{-1} K^{-1}).

spectroscopy study of spectra (see ◊spectrum) associated with atoms or molecules in solid, liquid, or gaseous phase. Spectroscopy can be used to identify unknown compounds and is an invaluable tool in science, medicine, and industry (for example, in checking the purity of drugs).

spectrum (plural **spectra**) arrangement of frequencies or wavelengths when electromagnetic radiations are separated into their constituent parts. Visible light is part of the electromagnetic spectrum and most sources emit waves over a range of wavelengths that can be broken up or 'dispersed'; white light can be separated into red, orange, yellow, green, blue, indigo, and violet. There are many types of spectra, both emission and absorption, for radiation and particles, used in spectroscopy. An incandescent body gives rise to a **continuous spectrum** where the dispersed radiation is distributed uninterruptedly over a range of wavelengths. An element gives a **line spectrum** – one or more bright discrete lines at characteristic wavelengths. Molecular gases give **band spectra** in which there are groups of close-packed lines. In an **absorption spectrum** dark lines or spaces replace the characteristic bright lines of the absorbing medium. The **mass spectrum** of an element is obtained from a mass spectrometer and shows the relative proportions of its constituent isotopes.

speed of light speed at which light and other electromagnetic waves travel through empty space. Its value is 299,792,458 m/186,281 mi per second. The speed of light is the highest speed possible, according to the theory of relativity, and its value is independent of the motion of its source and of the observer. It is impossible to accelerate any material body to this speed because it would require an infinite amount of energy.

speed of sound speed at which sound travels through a medium, such as air or water. In air at a temperature of 0°C/32°F, the speed of sound is 331 m/1,087 ft per second. At higher temperatures, the speed of sound is greater; at 18°C/64°F it is 342 m/1,123 ft per second. It is greater in liquids and solids; for example, in water it is around 1,440 m/4,724 ft per second, depending on the temperature.

spin intrinsic angular momentum of a subatomic particle, nucleus, atom, or molecule, which continues to exist even when the particle comes to rest. A particle in a specific energy state has a particular spin, just as it has a particular electric charge and mass. According to ◊quantum theory, this is restricted to discrete and indivisible values, specified by a spin quantum number. Because of its spin, a charged particle acts as a small magnet and is affected by magnetic fields.

supercooling cooling of a liquid below its freezing point without freezing taking place; or the cooling of a saturated solution without crystallization taking place, to form a supersaturated solution. In both cases supercooling is possible because of the lack of solid particles around which crystals can form. Crystallization rapidly follows the introduction of a small crystal (seed) or agitation of the supercooled solution.

surface tension property that causes the surface of a liquid to behave as if it were covered with a weak elastic skin; this is why a needle can float on water. It is caused by the exposed surface's tendency to contract to the smallest possible area because of cohesive forces between ◊molecules at the surface. Allied phenomena include the formation of droplets, the concave profile of a meniscus, and the capillary action by which water soaks into a sponge.

temperature degree or intensity of heat of an object and the condition that determines whether it will transfer heat to another object or receive heat from it, according to the laws of thermodynamics. The temperature of an object is a measure of the average kinetic energy possessed by the atoms or molecules of which it is composed. The SI unit of temperature is the kelvin (symbol K) used with the Kelvin scale. Other measures of temperature in common use are the Celsius scale and the Fahrenheit scale.

thermal conductivity ability of a substance to conduct heat. Good thermal conductors, like good electrical conductors, are generally materials with many free electrons (such as metals). Thermal conductivity is expressed in units of joules per second per metre per kelvin (J s^{-1} m^{-1} K^{-1}). For a block of material of cross-sectional area a and length l, with temperatures T_1 and T_2 at its end faces, the thermal conductivity l equals $Hl/at(T_2 - T_1)$, where H is the amount of heat transferred in time t.

transducer device that converts one form of energy into another. For example, a thermistor is a transducer that converts heat into an electrical voltage, and an electric motor is a transducer that converts an electrical voltage into mechanical energy. Transducers are important components in many types of sensor, converting the physical quantity to be measured into a proportional voltage signal.

transformer device in which, by electromagnetic induction, an alternating current (AC) of one voltage is transformed to another voltage, without change of frequency. Transformers are widely used in electrical apparatus of all kinds, and in particular in power transmission where high voltages and low currents are utilized.

transistor solid-state electronic component, made of semiconductor material, with three or more electrodes, that can regulate a current passing through it. A transistor can act as an amplifier, oscillator, photocell, or switch, and usually operates on a very small amount of power.

ultrasonics branch of physics dealing with the theory and application of ultrasound: sound waves occurring at frequencies too high to be heard by the human ear (that is, above about 20 kHz).

uncertainty principle, or **indeterminacy principle**, in quantum mechanics, the principle that it is impossible to know with unlimited accuracy the position and momentum of a particle. The principle arises because in order to locate a particle exactly, an observer must bounce light (in the form of a photon) off the particle, which must alter its position in an unpredictable way. It was established by the German physicist Werner Heisenberg, and gave a theoretical limit to the precision with which a particle's momentum and position can be measured simultaneously: the more accurately the one is determined, the more uncertainty there is in the other.

vacuum in general, a region completely empty of matter; in physics, any enclosure in which the gas pressure is considerably less than atmospheric pressure (101,325 pascals).

vapour density density of a gas, expressed as the mass of a given volume of the gas divided by the mass of an equal volume of a reference gas (such as hydrogen or air) at the same temperature and pressure. It is equal approximately to half the relative molecular weight (mass) of the gas.

vapour pressure pressure of a vapour given off by (evaporated from) a liquid or solid, caused by vibrating atoms or molecules continuously escaping from its surface. In an enclosed space, a maximum value is reached when the number of particles leaving the surface is in equilibrium with those returning to it; this is known as the **saturated vapour pressure** or **equilibrium vapour pressure**.

viscosity resistance of a fluid to flow, caused by its internal friction, which makes it resist flowing past a solid surface or other layers of the fluid. It applies to the motion of an object moving through a fluid as well as the motion of a fluid passing by an object.

volt SI unit (symbol V) of electromotive force or electric potential. A small battery has a potential of 1.5 volts, whilst a high-tension transmission line may carry up to 765,000 volts. The domestic electricity supply in the UK is 230 volts (lowered from 240 volts in 1995); it is 110 volts in the USA.

voltage commonly used term for ◊potential difference (pd) or ◊electromotive force (emf).

watt SI unit (symbol W) of power (the rate of expenditure or consumption of energy) defined as one joule per second. A light bulb, for example, may use 40, 60, 100, or 150 watts of power; an electric heater will use several kilowatts (thousands of watts). The watt is named after the Scottish engineer James Watt (1736–1819).

weber SI unit (symbol Wb) of magnetic flux (the magnetic field strength multiplied by the area through which the field passes). It is named after German chemist Wilhelm Weber. One weber equals 10^8 maxwells.

weight force exerted on an object by gravity. The weight of an object depends on its mass – the amount of material in it – and the strength of the Earth's gravitational pull, which decreases with height. Consequently, an object weighs less at the top of a mountain than at sea level. On the surface of the Moon, an object has only one-sixth of its weight on Earth, because the Moon's surface gravity is one-sixth that of the Earth. If the mass of a body is m kilograms and the gravitational field strength is g newtons per kilogram, its weight W in newtons is given by $W = mg$.

W particle elementary particle, one of the weakons responsible for transmitting the weak nuclear force.

Z particle elementary particle, one of the weakons responsible for carrying the weak nuclear force.

Further Reading

Baeyer, Hans Taming the Atom: The Emergence of the Visible Microworld (1994)

Bailey, George The Making of Andrei Sakharov (1989)

Balfour, Mark The Sign of the Serpent: the Key to Creative Physics (1990)

Bernstein, Jeremy Einstein (1973)

Berry, A J Henry Cavendish: His Life and Scientific Work (1960)

Bowe, Frank Comeback (1981)

Brown, Andrew The Neutron and the Bomb: A Biography of Sir James Chadwick (1997)

Bunge, Mario, and Shea, William (eds) Rutherford and Physics at the Turn of the Century (1979)

Burt, Philip Barnes Quantum Mechanics and Nonlinear Waves (1981)

Calder, Nigel Einstein's Universe (1979)

Carrigan, Richard, and Trower, Peter (eds) Particles and Forces: At the Heart of the Matter (1990) Cassidy, D C Uncertainty (1992)

Chadwick, J (ed) The Collected Papers of Lord Rutherford of Nelson (1962–65)

Christianson, G E In the Presence of the Creator: Isaac Newton and His Times (1984)

Close, Frank The Cosmic Onion: Quarks and the Nature of the Universe (1983)

Close, Frank; Marten, Michael; and Sutton, Christine The Particle Explosion (1987)

Coleman, James A Relativity for the Layman (1969)

Cox, P A Introduction to Quantum Theory and Atomic Structure (1996)

Croll, J G A Force Systems and Equilibrium (1974)

Davies, D A Waves, Atoms, and Solids (1978)

Davies, P C W Superforce: the Search for a Grand Unified Theory of Nature (1995)

Davies, Paul The Forces of Nature (1986)

Draganic, Ivan G Radiation and Radioactivity on Earth and Beyond (second edition, 1993)

Drake, Ellen Tan Restless Genius: Robert Hooke and his Earthly Thoughts (1996)

Drake, J Electrochemistry and Clean Energy (1994)

Einstein, Albert, and Infeld, L The Evolution of Physics (1938)

Everitt, C W F James Clerk Maxwell: Physicist and Natural Philosopher (1975)

Faraday, Michael Experimental Researches in Electricity (1839–55)

Feather, N Lord Rutherford (1973)

Fermi, Laura Atoms in the Family: My Life with Enrico Fermi (1954)

Feynman, Richard Surely You're Joking, Mr Feynman! (memoirs) (1985)

Fogden, Edward Energy (1990)

Fölsing, Alexander Albert Einstein (1997)

Fritsch, Harald Quarks: The Stuff of Matter (1984)

Giroud, F Marie Curie: A Life (trs 1986)

Gjertson, Derek The Newton Handbook (1987)

Gleick, James Genius (1992), The Life and Science of Richard Feynman (1992)

Goddard, Peter (ed) Paul Dirac: the Man and His Work (1998)

Goldstein, M, and Goldstein, I F The Refrigerator and the Universe (1993)

Goodchild, P J Robert Oppenheimer: Shatterer of Worlds (1985)

Gooding, David, and James, Frank (eds) Faraday Rediscovered (1986)

Goodstein, David L States of Matter (1975)

Gribbin, John Schrödinger's Kittens (1995), Q is for Quantum: the A–Z of Particle Physics (1998)

Harbison, James P, and Nahory, Robert E Lasers: Harnessing the Atom's Light (1998)

Hesse, Mary B Forces and Fields: the Concept of Action at a Distance in the History of Physics (1970)

Hoffmann, Banesh Albert Einstein: Creator and Rebel (1973)

Jammer, Max The Conceptual Development of Quantum Mechanics (1966)

Kane, Gordon The Particle Garden (1994)

Lederman, Leon, and Schramm, David From Quarks to the Cosmos: Tools of Discovery (1989)

MacDonald, D K C Faraday, Maxwell and Kelvin (1965)

Maxwell, James Clerk Matter and Motion (1996)

Mendelssohn, K The Quest for Absolute Zero: Meaning of Low Temperature Physics (1977)

Michelmore, Peter Einstein: Profile of the Man (1963)

Milgrom, Lionel R The Colours of Life (1998)

Moore, Ruth E Niels Bohr: The Man and the Scientist (1967)

Pais, Abraham 'Subtle is the Lord': The Life and Science of Albert Einstein (1982), Niels Bohr's Times (1991)

Peierls, R E The Laws of Nature (1955)

Planck, Max Scientific Autobiography and Other Papers (trs 1949)

Quinn, Susan Marie Curie: A Life (1995)

Rabi, I, and others Oppenheimer (1969)

Rayleigh, R J The Life of Sir J J Thomson (1942)

Reid, Robert Marie Curie (1974)

Rhodes, R The Making of the Atomic Bomb (1986)

Rogers, E M Physics for the Inquiring Mind (1960)

Romer, Alfred Restless Atom (1983)

Rozental, S (ed) Niels Bohr: His Life and Work (1967)

Schwartz, Joseph, and McGuinness, Michael Einstein for Beginners (1993)

Segrè, Emilio Enrico Fermi, Physicist (1970), From Falling Bodies to Radio Waves (1984)

Sharlin, H and T Lord Kelvin: The Dynamic Victorian (1978)

Smith, A K, and Weiner, C (eds) Robert Oppenheimer: Letters and Recollections (1981)

Smith, C, and Wise, M N Energy and Empire (1989)

Stayer, Marcia Newton's Dream (1988)

Stehle, P Order, Chaos, Order (1994)

Sykes, Christopher (ed) No Ordinary Genius (1994)

Tabor, David Gases, Liquids, and Solids: and Other States of Matter (third edition, 1991)

Thompson, S P The Life of Lord Kelvin (1977)

Thomson, George J J Thomson and the Cavendish Laboratory of His Day (1965)

Thorning, William Harris Motion and Forces (1974)

Tolstoy, Ivan James Clerk Maxwell (1982)

Walton, Alan John Three Phases of Matter (second edition, 1985)

Weinberg, Steven The Discovery of Subatomic Particles (1983), Dreams of a Final Theory (1993)

Weissbluth, Mitchel Atoms and Molecules (1978)

Westfall, R The Life of Isaac Newton (1993)

White, Michael Stephen Hawking: a Life in Science (1992), Isaac Newton: The Last Sorcerer (1998)

Williams, D J Force, Matter, and Energy (1974)

Williams, Leslie Michael Faraday: A Biography (1965)

Wilson, David Rutherford: Simple Genius (1983)

Zohar, Danah The Quantum Self (1990)

Zohar, Danah, and Marshall, Ian The Quantum Society (1994)

ASTRONOMY

stronomy is the science of the celestial bodies: the Sun, the Moon, and the planets; the stars and galaxies; and all other objects in the universe. It is concerned with their positions, motions, distances, and physical conditions and with their origins and evolution. It is divided into fields such as astrophysics, celestial mechanics, and cosmology. Astronomy is perhaps the oldest recorded science; there are observational records from ancient Babylonia, China, Egypt, and Mexico. The first true astronomers were the Greeks, who deduced the Earth to be a sphere and attempted to measure its size. A summary of Greek astronomy is provided in Ptolemy of Alexandria's *Almagest*. The Arabs developed the astrolabe and produced good star catalogues. In 1543 the Polish astronomer Copernicus demonstrated that the Sun, not the Earth, is the centre of our planetary system. The Italian scientist Galileo was the first to use a telescope for astronomical study, in 1609–10. In the 17th and 18th centuries astronomy was mostly concerned with positional measurements. The British astronomer William Herschel's suggestions on the shape of our galaxy were verified in 1923 by the US astronomer Edwin Hubble's telescope at the Mount Wilson Observatory, California. The most remarkable recent extension of the powers of astronomy to explore the universe is in the use of rockets, satellites, space stations, and space probes, while the launching of the Hubble Space Telescope into permanent orbit in 1990 has enabled the detection of celestial phenomena seven times more distant than any Earth-based telescope.

the universe

The universe embraces all of space and its contents, the study of which is called **cosmology**. The universe is thought to be between 10 billion and 20 billion years old, and is mostly empty space, dotted with stars collected into vast aggregations called galaxies for as far as telescopes can see. The most distant detected galaxies and quasars (quasi-stellar objects) lie 10 billion light years or more from Earth, and are moving farther apart as the universe expands. Several theories attempt to explain how the universe came into being and evolved: for example, the **Big Bang theory** of an expanding universe originating in a single explosive event, and the contradictory **steady-state theory**.

Big Bang

This is the hypothetical 'explosive' event that marked the origin of the universe as we know it. At the time of the Big Bang, the entire universe was squeezed into a hot, superdense state. The Big Bang explosion threw this compact material outwards, producing the expanding universe. The cause of the Big Bang is unknown; observations of the current rate of expansion of the universe suggest that it took place about 10–20 billion years ago. The Big Bang theory began modern cosmology.

According to a modified version of the Big Bang, called the **inflationary theory**, the universe underwent a rapid period of expansion shortly after the Big Bang, which accounts for its current large size and uniform nature. The inflationary theory is supported by recent observations of the cosmic background radiation (see below).

Scientists have calculated that one 10^{-36} second (equivalent to one million-million-million-million-million-millionth of a second) after the Big Bang, the universe was the size of a pea, and the temperature was 10 billion million million million°C/18 billion million million million°F. One second after the Big Bang, the temperature was about 10 billion°C/18 billion°F.

steady-state theory

This rival theory to that of the Big Bang claims that the universe has no origin but is expanding because new matter is being created continuously throughout the universe. The theory was proposed in 1948 by Austrian-born British cosmologist Hermann Bondi, Austrian-born US astronomer and physicist Thomas Gold (1920–), and English astronomer Fred Hoyle, but was dealt a severe blow in 1965 by the discovery of cosmic background radiation and is now largely rejected.

Unsolved Mysteries
http://www.pbs.org/wnet/hawking/mysteries/html/myst.html
Part of a larger site on cosmology provided by PBS Online, this page features seven articles that address the most difficult questions regarding the mysteries of our universe. Some of the articles include 'Where does matter come from?', 'Is time travel possible?', 'Where is the missing matter?', 'An inhabited universe?', and 'Is there a theory for everything?' You can follow the link to the main site to explore other aspects of the cosmos.

---

cosmic background radiation This is electromagnetic radiation left over from the original formation of the universe in the Big Bang. It corresponds to an overall background temperature of 3 K (–270°C/–454°F), or 3°C above absolute zero. Cosmic background radiation was first detected in 1965 by US physicists Arno Penzias (1933–) and Robert Wilson (1936–). In 1992 the Cosmic Background Explorer satellite, COBE, detected slight 'ripples' in the strength of the background radiation that are believed to mark the first stage in the formation of galaxies.

Brief History of Cosmology
http://www-history.mcs.st-and.ac.uk/~history/HistTopics/Cosmology.html
Based at St Andrews University, Scotland, a site chronicling the history of cosmology from the time of the Babylonians to the Hubble Space Telescope.

---

the celestial sphere and constellations

celestial sphere

The celestial sphere is an imaginary sphere surrounding

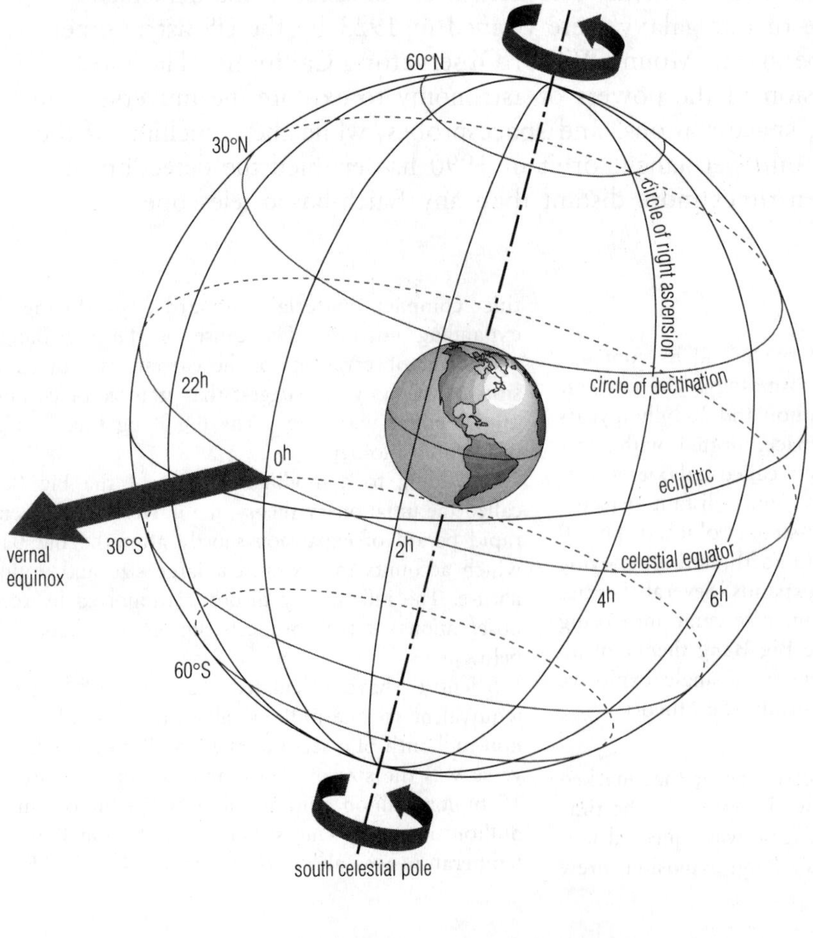

celestial sphere The main features of the celestial sphere. Declination, the equivalent of latitude on Earth, runs from 0° at the celestial equator to 90° at the celestial poles. Right ascension, the equivalent of longitude, is measured in hours eastwards from the vernal equinox, one hour corresponding to 15° of longitude.

the Earth, on which the celestial bodies seem to lie. The positions of bodies such as stars, planets, and galaxies are specified by their coordinates on the celestial sphere. The equivalents of latitude and longitude on the celestial sphere are called **declination** (measured in degrees) and **right ascension** (measured in hours from 0 to 24). The **celestial poles** lie directly above the Earth's poles, and the **celestial equator** lies over the Earth's Equator. The celestial sphere appears to rotate once around the Earth each day, actually a result of the rotation of the Earth on its axis.

constellations

A constellation is one of the 88 areas into which the sky is divided for the purposes of identifying and naming celestial objects. The first constellations were simple, arbitrary patterns of stars in which early civilizations visualized gods, sacred beasts, and mythical heroes.

The constellations in use today are derived from a list of 48 known to the ancient Greeks, who inherited some from the Babylonians. The current list of 88 constellations was adopted by the International Astronomical Union, astronomy's governing body, in 1930. Within the current system the definitive boundaries are arcs of constant right ascension or declination for the equinox of 1875. Each constellation includes not only the historic star grouping but also all the variable stars that have become associated with it. The genitive, or possessive, form of the name is used when an object is being identified by its constellation letter or number. For example, the star Polaris is Alpha Ursae Minoris, which is usually contracted to alpha UMi.

Constellations and Their Stars
http://www.vol.it/mirror/constellations/
Notes on the constellations (listed alphabetically, by month, and by popularity), plus lists of the 25 brightest stars and the 32 nearest stars, and photographs of the Milky Way.

---

zodiac

The zodiac is the zone of sky containing the 12 zodiacal constellations (after which the astrological signs of the zodiac are named) through which the paths of the Sun, Moon, and planets appear to move. The 12 astronomical constellations are uneven in size and do not between them cover the whole zodiac, or even the line of the ecliptic (the path that the Sun appears to follow each year as the Earth orbits the Sun), much of which lies in the constellation of Ophiuchus.

The sequence of the astrological signs (which are

each 30° wide and do not cover the same areas of sky as the astronomical constellations) is eastwards, following the motions of the Sun and Moon through the constellations, and is regarded as beginning at the point that marks the position of the Sun at the time of the March equinox. This point is sometimes called 'The First Point of Aries'.

The **equinoxes** occur as the Sun enters the signs of Aries and Libra, at which times the Sun passes directly above the Equator. The **solstices** occur as the Sun enters the signs of Cancer and Capricorn and passes directly over the Tropic of Cancer and Tropic of Capricorn, circles of latitude 23° 27' north and south of the Equator.

Zodiac constellations

To remember the constellations of the zodiac:

A tense grey cat lay very low, sneaking slowly, contemplating a pounce

(Aries, Taurus, Gemini, Cancer, Leo, Virgo, Libra, Scorpio, Sagittarius, Capricorn, Aquarius, Pisces)

---

stars

A star is a luminous globe of gas, mainly hydrogen and helium, which produces its own heat and light by nuclear reactions. Although stars shine for a very long time – many billions of years – they are not eternal, and have been found to change in appearance at different stages in their lives.

The smallest mass possible for a star is about 8% that of the Sun (80 times that of Jupiter, the largest planet in the Solar System), otherwise nuclear reactions cannot occur. Objects with less than this critical mass shine only dimly, and are termed brown dwarfs.

Stars are born when nebulae (giant clouds of dust and gas) contract under the influence of gravity. These clouds consist mainly of hydrogen and helium, with traces of other elements and dust grains. A huge volume of interstellar matter gradually separates from the cloud, and the temperature and pressure in its core rises as the star grows smaller and denser. As the star is forming, it is surrounded by evaporating gaseous globules (EGGs), the oldest of which was photographed in the Eta Carina Nebula in 1996 by the Hubble Space Telescope.

At first the temperature of the star scarcely rises, as dust grains radiate away much of the heat, but as it grows denser less of the heat generated can escape,

Constellations

A constellation is one of the 88 areas into which the sky is divided for the purposes of identifying and naming celestial objects. The first constellations were simple, arbitrary patterns of stars in which early civilizations visualized gods, sacred beasts, and mythical heroes.
(- = not applicable.)

Constellation	Abbreviation	Popular name	Constellation	Abbreviation	Popular name
Andromeda	And	-	Lacerta	Lac	Lizard
Antlia	Ant	Airpump	Leo	Leo	Lion
Apus	Aps	Bird of Paradise	Leo Minor	LMi	Little Lion
Aquarius	Aqr	Water-bearer	Lepus	Lep	Hare
Aquila	Aqi	Eagle	Libra	Lib	Balance
Ara	Ara	Altar	Lupus	Lup	Wolf
Aries	Ari	Ram	Lynx	Lyn	-
Auriga	Aur	Charioteer	Lyra	Lyr	Lyre
Boötes	Boo	Herdsman	Mensa	Men	Table
Caelum	Cae	Chisel	Microscopium	Mic	Microscope
Camelopardalis	Cam	Giraffe	Monoceros	Mon	Unicorn
Cancer	Cnc	Crab	Musca	Mus	Southern Fly
Canes Venatici	CVn	Hunting Dogs	Norma	Nor	Rule
Canis Major	CMa	Great Dog	Octans	Oct	Octant
Canis Minor	CMi	Little Dog	Ophiuchus	Oph	Serpent-bearer
Capricornus	Cap	Sea-goat	Orion	Ori	-
Carina	Car	Keel	Pavo	Pav	Peacock
Cassiopeia	Cas	-	Pegasus	Peg	Flying Horse
Centaurus	Cen	Centaur	Perseus	Per	-
Cepheus	Cep	-	Phoenix	Phe	Phoenix
Cetus	Cet	Whale	Pictor	Pic	Painter
Chamaeleon	Cha	Chameleon	Pisces	Psc	Fishes
Circinus	Cir	Compasses	Piscis Austrinus	PsA	Southern Fish
Columba	Col	Dove	Puppis	Pup	Poop
Coma Berenices	Com	Berenice's Hair	Pyxis	Pyx	Compass
Corona Australis	CrA	Southern Crown	Reticulum	Ret	Net
Corona Borealis	CrB	Northern Crown	Sagitta	Sge	Arrow
Corvus	Crv	Crow	Sagittarius	Sgr	Archer
Crater	Crt	Cup	Scorpius	Sco	Scorpion
Crux	Cru	Southern Cross	Sculptor	Scl	-
Cygnus	Cyn	Swan	Scutum	Sct	Shield
Delphinus	Del	Dolphin	Serpens	Ser	Serpent
Dorado	Dor	Goldfish	Sextans	Sex	Sextant
Draco	Dra	Dragon	Taurus	Tau	Bull
Equuleus	Equ	Foal	Telescopium	Tel	Telescope
Eridanus	Eri	River	Triangulum	Tri	Triangle
Fornax	For	Furnace	Triangulum Australe	TrA	Southern Triangle
Gemini	Gem	Twins	Tucana	Tuc	Toucan
Grus	Gru	Crane	Ursa Major	UMa	Great Bear
Hercules	Her	-	Ursa Minor	UMi	Little Bear
Horologium	Hor	Clock	Vela	Vel	Sails
Hydra	Hya	Watersnake	Virgo	Vir	Virgin
Hydrus	Hyi	Little Snake	Volans	Vol	Flying Fish
Indus	Ind	Indian	Vulpecula	Vul	Fox

and it gradually warms up. At about 10 million°C/18 million°F the temperature is hot enough for a nuclear reaction to begin, and hydrogen nuclei fuse to form helium nuclei; vast amounts of energy are released, contraction stops, and the star begins to shine. Stars at this stage are called **main-sequence stars**.

types of stars

binary star Binary stars are pairs of stars moving in orbit around their common centre of mass. Observations show that most stars are binary, or even multiple – for example, the nearest star system to the Sun, Alpha Centauri.

One of the stars in the pair called Epsilon Aurigae may be the largest star known. Its diameter is 2,800 times that of the Sun. If it were in the position of the Sun, it would engulf Mercury, Venus, Earth, Mars, Jupiter, and Saturn. A spectroscopic binary is a binary in which two stars are so close together that they cannot be seen separately, but their separate light spectra can be distinguished by a spectroscope. Another type is the eclipsing binary (see variable star below).

brown dwarf A brown dwarf is an object less massive than a star, but heavier than a planet. Brown dwarfs do not have enough mass to ignite nuclear reactions at their centres, but shine by heat released during their contraction from a gas cloud. Some astronomers believe that vast numbers of brown dwarfs exist throughout the galaxy. Because of the difficulty of detection, none were spotted until 1995, when US astronomers discovered a brown dwarf,

GI229B, in the constellation Lepus. It is about 20–40 times as massive as Jupiter but emits only 1% of the radiation of the smallest known star.

neutron star A neutron star is a very small, 'superdense' star composed mostly of neutrons. Neutron stars are thought to form when massive stars explode as supernovae, during which the protons and electrons of the star's atoms merge, owing to intense gravitational collapse, to make neutrons. A neutron star may have the mass of up to three Suns, compressed into a globe only 20 km/12 mi in diameter.

If its mass is any greater, its gravity will be so strong that it will shrink even

> **neutron star**
> Neutron stars are so condensed that a fragment the size of a sugar cube would weigh as much as all the people on Earth put together.

further to become a black hole. Being so small, neutron stars can spin very quickly. The rapidly flashing radio stars called pulsars are believed to be neutron stars. The flashing is caused by a rotating beam of radio energy similar in behaviour to a lighthouse beam of light.

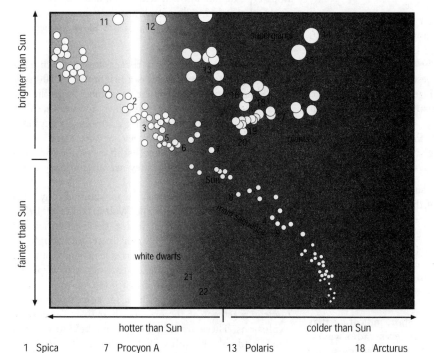

Hertzsprung–Russell diagram
The Hertzsprung–Russell diagram relates the brightness (or luminosity) of a star to its temperature. Most stars fall within a narrow diagonal band called the main sequence. A star moves off the main sequence when it grows old. The Hertzsprung–Russell diagram is one of the most important diagrams in astrophysics.

1 Spica	7 Procyon A	13 Polaris	18 Arcturus
2 Regulus	8 Tau Ceti	14 Betelgeuse	19 Pollux
3 Vega	9 61 Cygni A	15 Antares	20 Capella
4 61 Cygni B	10 Proxima Centauri	16 Mira	21 Sirius B
5 Sirius A	11 Rigel	17 Aldebaran	22 Procyon B
6 Altair	12 Deneb		

nova A nova is a faint star that suddenly increases in brightness by a factor of 10,000 or more, remains bright for a few days, and then fades away and is not seen again for very many years, if at all. Novae are believed to occur in close binary star systems, where gas from one star flows to a companion white dwarf. The gas ignites and is thrown off in an explosion at speeds of 1,500 kps/930 mps or more. Unlike a supernova, the star is not completely disrupted by the outburst. After a few weeks or months it subsides to its previous state, though it may erupt many more times.

Two or three such stars are detected in our galaxy each year, but on average, a nova is visible to the naked eye only about once every ten years. Novae very similar to those appearing in our own galaxy have also been observed in other galaxies.

pulsar Pulsars are celestial sources that emit pulses of energy at regular intervals, ranging from a few seconds to a few thousandths of a second. They are thought to be rapidly rotating neutron stars, which flash at radio and other wavelengths as they spin. They were discovered in 1967 by British astronomers Jocelyn Bell Burnell and Antony Hewish at the Mullard Radio Astronomy Observatory, Cambridge, England. Over 500 radio pulsars are now known in our galaxy, although a million or so may exist.

Pulsars slow down as they get older, and eventually the flashes fade. Of the 500 known radio pulsars, 20 are millisecond pulsars (flashing 1,000 times a second). Such pulsars are thought to be more than a billion years old. Two pulsars, one (estimated to be 1,000 years old) in the Crab nebula and one (estimated to be 11,000 years old) in the constellation Vela, give out flashes of visible light.

The closest pulsar to the Earth is named PSR J0108−1431 and lies 280 light years away in the constellation of Cetus; it was discovered in 1994.

quasar Quasars (the name is derived from **quasi**-stellar object, or QSO) are some of the most distant extragalactic objects known, and were discovered in 1963. Quasars appear starlike, but each emits more energy than 100 giant galaxies. They are thought to be at the centre of galaxies, their brilliance emanating from the stars and gas falling towards an immense black hole at their nucleus. Most quasars are found in elliptical galaxies.

Quasar light shows a large red shift, indicating that they are very distant. Some quasars emit radio waves, which is how they were first identified, but most are radio-quiet. The furthest are over 10 billion light years away.

red dwarf Any star that is cool, faint, and small (about one-tenth the mass and diameter of the Sun) is a red dwarf. Red dwarfs burn slowly, and have estimated lifetimes of 100 billion years. They may be the most abundant type of star, but are difficult to see because they are so faint. Two of the closest stars to the Sun, Proxima Centauri and Barnard's Star, are red dwarfs.

red giant A red giant is any large bright star with a cool surface. It is thought to represent a late stage in the evolution of a star like the Sun, as it runs out of hydrogen fuel at its centre and begins to burn heavier elements, such as helium, carbon, and silicon. Because of more complex nuclear reactions that then occur in the red giant's interior, it eventually becomes gravitationally unstable and begins to collapse and heat up. The result is either explosion of the star as a supernova, leaving behind a neutron star, or loss of mass by more gradual means to produce a white dwarf.

Red giants have diameters between 10 and 100 times that of the Sun. They are very bright because they are so large, although their surface temperature is lower than that of the Sun, about 2,000–3,000 K (1,700–2,700°C/3,000–5,000°F).

supergiant Supergiants are the largest and most luminous type of star known, with a diameter up to 1,000 times that of the Sun and absolute magnitudes of between −5 and −9. Supergiants are likely to become supernovae.

supernova A supernova is the explosive death of a star, which temporarily attains a brightness of 100 million Suns or more, so that it can shine as brilliantly as a small galaxy for a few days or weeks. Very approximately, it is thought that a supernova explodes in a large galaxy about once every 100 years. Many supernovae – astronomers estimate some 50% – remain undetected because they are obscured by interstellar dust.

The name 'supernova' was coined by US astronomers Fritz Zwicky and Walter Baade in 1934. Zwicky was also responsible for their classification into types I and II. **Type I** supernovae are thought to occur in binary star systems, in which gas from one star falls on to a white dwarf, causing it to explode. **Type II** supernovae occur in stars ten or more times as massive as the Sun, which suffer runaway internal nuclear reactions at the ends of their lives, leading to explosions. These are thought to leave behind neutron stars and black holes. Gas ejected by such an explosion causes an expanding radio source, such as the Crab nebula. Supernovae are thought to be the main source of elements heavier than hydrogen and helium.

Nearest Stars

Star	Distance (light years)
Proxima Centauri	4.2
Alpha Centauri A	4.3
Alpha Centauri B	4.3
Barnard's Star	6.0
Wolf 359	7.7
Lalande 21185	8.2
UV Ceti A	8.4
UV Ceti B	8.4
Sirius A	8.6
Sirius B	8.6
Ross 154	9.4
Ross 249	10.4

variable star A variable star's brightness changes, either regularly or irregularly, over a period ranging from a few hours to months or years. The Cepheid variables regularly expand and contract in size every few days or weeks.

Stars that change in size and brightness at less precise intervals include **long-period variables**, such as the

The 20 Brightest Stars

A star's brightness is referred to as its 'magnitude'. 'Apparent magnitude' is brightness as seen from Earth. 'Absolute magnitude' is measured at a standard distance of 32.6 light-years or 10 parsecs from the star.

	Scientific name	Common name	Distance (light years)
1	Alpha Canis Majoris	Sirius	9
2	Alpha Carinae	Canopus	1,170
3	Alpha Centauri	Rigil Kent	4
4	Alpha Boötis	Arcturus	36
5	Alpha Lyrae	Vega	26
6	Alpha Aurigae	Capella	42
7	Beta Orionis	Rigel	910
8	Alpha Canis Minoris	Procyon	11
9	Alpha Eridani	Achernar	85
10	Alpha Orionis	Betelgeuse	310
11	Beta Centauri	Hadar	460
12	Alpha Aquilae	Altair	17
13	Alpha Tauri	Aldebaran	25
14	Alpha Crucis	Acrux	360
15	Alpha Scorpii	Antares	330
16	Alpha Virginis	Spica	260
17	Beta Geminorum	Pollux	36
18	Alpha Piscis Austrini	Fomalhaut	22
19	Alpha Cygni	Deneb	1,830
20	Beta Crucis	Mimosa	420

red giant Mira in the constellation Cetus (period about 330 days), and **irregular variables**, such as some red supergiants. **Eruptive variables** emit sudden outbursts of light. Some suffer flares on their surfaces, while others, such as a nova, result from transfer of gas between a close pair of stars. In an **eclipsing binary**, the variation is due not to any change in the star itself, but to the periodic eclipse of a star by a close companion. The different types of variability are closely related to different stages of stellar evolution.

Brightest star

The brightest known star is the Pistol Star, discovered near the centre of the Milky Way in 1997 by the Hubble Space Telescope. It emits as much energy in seconds as the Sun does in one year, making it 10 million times brighter and 100 times larger.

white dwarf A white dwarf is a small, hot star, the last stage in the life of a star such as the Sun. White dwarfs make up 10% of the stars in our galaxy; most have a mass 60% of that of the Sun, but, with only 1% of the Sun's diameter, are similar in size to the Earth. Most have surface temperatures of 8,000°C/14,400°F or more, hotter than the Sun. Yet, being so small, their overall luminosities may be less than 1% of that of the Sun. The Milky Way contains an estimated 50 billion white dwarfs.

White dwarfs consist of degenerate matter in which gravity has packed the protons and electrons together as tightly as is physically possible, so that a spoonful of it weighs several tonnes. White dwarfs are thought to be the shrunken remains of stars that have exhausted their internal energy supplies. They slowly cool and fade over billions of years.

Stars and Constellations
http://www.astro.wisc.edu/~dolan/constellations/constellations.html
Hugely informative site on stars and constellations. It includes star charts of all major stars and constellations, details of the origins of the various names, photographs of our galaxy, and details on what stars can be seen at any given time.

some notable stars

Alpha Centauri, or **Rigil Kent**, is the brightest star in the constellation Centaurus and the third brightest star in the sky. It is actually a triple star (see binary star above); the two brighter stars orbit each other every 80 years, and the third, Proxima Centauri, is the closest star to the Sun, 4.2 light years away, 0.1 light years closer than the other two.

Arcturus, or **Alpha Boötis,** is the brightest star in the constellation Boötes and the fourth brightest star in the sky. It is a red giant about 28 times larger than the Sun and 70 times more luminous, 36 light years away from Earth. Arcturus was the first star for which a proper motion was detected. In 1718 the English astronomer Edmond Halley noticed that, relatively to the surrounding stars, it had moved by about one degree from the position recorded by Ptolemy in the *Almagest*. Its name is derived from the Greek and means 'the Guardian of the Bear'.

Canopus, or **Alpha Carinae,** is the second brightest star in the night sky (after Sirius), lying in the southern constellation Carina. It is a first-magnitude yellow–white supergiant about 120 light years from Earth, and thousands of times more luminous than the Sun. It may have been named in honour of the chief pilot of the Greek fleet that sailed against Troy, or the name may be derived from two Coptic words signifying 'Golden Earth'.

Capella, or **Alpha Aurigae,** is the brightest star in the constellation Auriga and the sixth brightest star in the night sky. It is a visual and spectroscopic binary that consists of a pair of yellow-giant stars 45 light years from Earth, orbiting each other every 104 days. It is a first-magnitude star, whose Latin name means the 'the Little Nanny Goat': its kids are the three adjacent stars Epsilon, Eta, and Zeta Aurigae.

Deneb, or **Alpha Cygni,** is the brightest star in the constellation Cygnus, and the 19th brightest star in the night sky. It is one of the greatest supergiant stars known, with a true luminosity about 60,000 times that of the Sun. Deneb is about 1,800 light years from Earth. Its name is derived from the Arabic word for 'tail'.

Mira, or **Omicron Ceti,** is the brightest long-period pulsating variable star, located in the constellation Cetus. Mira was the first star discovered to vary periodically in brightness. It has a periodic variation between third or fourth magnitude and ninth magnitude over an average period of 331 days. It can sometimes reach second magnitude and once almost attained first magnitude in 1779. At times it is easily visible to the naked eye, being the brightest star in that part of the sky, while at others it cannot be seen without a telescope. It was named 'Stella Mira', 'the wonderful star', by Hevelius, who observed it 1659–82.

Polaris, or the **Pole Star** or **North Star,** is the bright star closest to the north celestial pole, and the brightest star in the constellation Ursa Minor. Its position is indicated by the 'pointers' in Ursa Major. Polaris is a yellow supergiant about 500 light years away. It is a Cepheid variable, whose magnitude varies between 2.1 and 2.2 over 3.97 days. It is also known as **Alpha Ursae Minoris.**

Polaris currently lies within 1° of the north celestial pole; precession (Earth's axial wobble) will bring it closest to the celestial pole (less than 0.5° away) in about AD 2100. Then its distance will start to increase, reaching 1° in 2205 and 47° in 28000.

Procyon, or **Alpha Canis Minoris,** is the brightest star in the constellation Canis Minor and the eighth brightest star in the sky. Procyon is a first-magnitude white star 11.4 light years from Earth, with a mass of 1.7 Suns. It has a white dwarf companion that orbits it every 40 years. The name, derived from Greek, means 'before the dog', and reflects the fact that in midnorthern latitudes Procyon rises shortly before Sirius, the Dog Star. Procyon and Sirius are sometimes called 'the Dog Stars'.

Rigel, or **Beta Orionis,** is the brightest star in the constellation Orion. It is a blue–white supergiant, with an estimated diameter 50 times that of the Sun. It is 900 light years from Earth, and is intrinsically the brightest of the first-magnitude stars, its luminosity being about 100,000 times that of the Sun. It is the seventh brightest star in the sky. Its name is derived from the Arabic for 'foot'.

Sirius, or the **Dog Star** or **Alpha Canis Majoris,** is the the brightest star in the night sky, 8.6 light years from Earth in the constellation Canis Major. Sirius is a white star with a mass 2.3 times that of the Sun, a diameter 1.8 times that of the Sun, and a luminosity of 23 Suns. It is orbited every 50 years by a white dwarf, Sirius B, also known as the Pup. The name 'Sirius' is derived from the Greek word for 'sparkling'. In ancient Egypt, where its hieroglyph was a dog, its reappearance in the early morning sky heralded the annual rising of the River Nile.

Vega, or **Alpha Lyrae,** is the brightest star in the constellation Lyra and the fifth brightest star in the night sky. It is a blue–white star, 25 light years from Earth, with a luminosity 50 times that of the Sun. In 1983 the Infrared Astronomy Satellite (IRAS) discovered a ring of dust around Vega, possibly a disc from which a planetary system is forming. As a result of precession (Earth's axial wobble), Vega will become the north polar star about the year 14000.

galaxies

Each galaxy is a congregation of millions or billions of stars, held together by gravity. **Spiral galaxies,** such as the Milky Way, are flattened in shape, with a central bulge of old stars surrounded by a disc of younger stars, arranged in spiral arms like a Catherine wheel. **Barred spirals** are spiral galaxies that have a straight bar of stars across their centre, from the ends of which the spiral arms emerge. The arms of spiral galaxies contain gas and dust from which new stars are still forming. **Elliptical galaxies** contain old stars and very little gas. They include the most massive galaxies known, containing a trillion stars. At least some elliptical galaxies are thought to be formed by mergers between spiral galaxies. There are also irregular galaxies. Most galaxies occur in clusters, containing anything from a few to thousands of members.

Our own galaxy is the **Milky Way Galaxy**, with the Sun lying in one of its spiral arms. Only two galaxies, the **Magellanic Clouds**, are easily visible to the naked eye. The next brightest, the **Andromeda galaxy**, is just visible; it is 2.2 million light years away, in the constellation Andromeda. About 20 galaxies are known to be within 2.5 million light years away, and several thousand within 50 million light years. More than 100 million can be photographed with modern telescopes.

Milky Way

The Milky Way appears as a faint band of light crossing the night sky, consisting of stars in the plane of our galaxy. The name 'Milky Way' is also used for the galaxy itself. It is a spiral galaxy, 100,000 light years in diameter and 2,000 light years thick, containing at least 100 billion stars. The Sun is in one of its spiral arms, about 25,000 light years from the centre, not far from its central plane.

The densest parts of the Milky Way, towards the galaxy's centre, lie in the constellation Sagittarius. In places, the Milky Way is interrupted by lanes of dark dust that obscure light from the stars beyond, such as the Coalsack nebula in Crux (the Southern Cross). It is because of these that the Milky Way is irregular in width and appears to be divided into two between Centaurus and Cygnus.

The Milky Way passes through the constellations of Cassiopeia, Perseus, Auriga, Orion, Canis Major, Puppis, Vela, Carina, Crux, Centaurus, Norma, Scorpius, Sagittarius, Scutum, Aquila, and Cygnus.

Magellanic Clouds

These are the two galaxies nearest to our own galaxy. They are irregularly shaped, and appear as detached parts of the Milky Way, in the southern constellations Dorado, Tucana, and Mensa. The **Large Magellanic Cloud** spreads over the constellations of Dorado and Mensa. The **Small Magellanic Cloud** is in Tucana. The Large Magellanic Cloud is 169,000 light years from Earth, and about a third the diameter of our galaxy; the Small Magellanic Cloud, 180,000 light years away, is about a fifth the diameter of our galaxy. They are named after the Portuguese navigator Ferdinand

Magellan, who first described them. Being the nearest galaxies to ours, the Clouds are especially useful for studying stellar populations and objects such as supergiant stars.

radio galaxies

All galaxies, including our own, emit some radio waves, but those emitted by radio galaxies are up to a million times more powerful. In many cases the strongest radio emission comes not from the visible galaxy but from two clouds, invisible through an optical telescope, that can extend for millions of light years either side of the galaxy. This double structure is also shown by some quasars, suggesting a close relationship between the two types of object. In both cases, the source of energy is thought to be a massive black hole at the centre. Some radio galaxies are thought to result from two galaxies in collision or recently merged.

other astronomical objects

black holes

A black hole is an object in space whose gravity is so great that nothing can escape from it, not even light. Thought to form when massive stars shrink at the end of their lives, a black hole sucks in more matter, including other stars, from the space around it. Matter that falls into a black hole is squeezed to infinite density at the centre of the hole. Black holes can be detected because gas falling towards them becomes so hot that it emits X-rays.

Black holes containing the mass of millions of stars are thought to lie at the centres of quasars. Satellites have detected X-rays from a number of objects that may be black holes, but only four likely black holes in our galaxy had been identified by 1994.

Microscopic black holes may have been formed in the chaotic conditions of the Big Bang. The English physicist Stephen Hawking has shown that such tiny black holes could 'evaporate' and explode in a flash of energy.

nebulae

A nebula is a cloud of gas and dust in space. Nebulae are the birthplaces of stars, but some nebulae (**planetary nebulae** and **supernovae remnants**) are produced by gas thrown off from dying stars. Nebulae are classified according to on whether they emit, reflect, or absorb light.

An **emission nebula**, such as the Orion nebula, glows brightly because its gas is energized by the stars that have formed within it. In a **reflection nebula**, such as the one surrounding the stars of the Pleiades cluster, starlight reflects off the grains of dust within it. A **dark nebula** is a dense cloud, composed of molecular hydrogen, which partially or completely absorbs the light behind it. Examples include the Coalsack nebula in Crux and the Horsehead nebula in Orion.

Web Nebulae
http://www.vol.it/MIRROR/EN/ftp.seds.org/
html/billa/twn/
Images of nebulae, plus a short account of the different types of nebulae, and a glossary of related terms.

Crab nebula

The Crab nebula, the most famous supernova remnant (named after its crablike shape), is a cloud of gas 6,000 light years from Earth, in the constellation Taurus. It is the remains of a star that, according to Chinese records, exploded on 4 July 1054 as a brilliant point of light (now known to be a supernova). At its centre is a pulsar that flashes 30 times a second.

It is a powerful radio and X-ray source. Optically it appears as a diffuse elliptical area on which is superimposed an intricate network of bright filaments. Observations show that it is increasing in size; its present dimensions are of the order of 10 light years. The light is highly polarized, suggesting the presence of strong magnetic fields. This suggestion is strengthened by the fact that the diffuse portion is emitting radiation throughout the whole electromagnetic spectrum from radio to gamma waves, the energy coming from a pulsar near the centre.

the Solar System

The Solar System comprises the Sun (a star) and all the bodies orbiting it: the nine planets (Mercury, Venus, Earth, Mars, Jupiter, Saturn, Uranus, Neptune, and Pluto), their moons, the asteroids, and the comets. The Sun contains 99.86% of the mass of the Solar System.

Solar System
http://www.hawastsoc.org/solar/eng/
Educational tour of the Solar System. It contains information and statistics about the Sun, Earth, planets, moons, asteroids, comets, and meteorites found within the Solar System, supported by images.

The Solar System gives every indication of being a strongly unified system having a common origin and development. It is isolated in space; all the planets go round the Sun in orbits that are nearly circular and coplanar, and in the same direction as the Sun itself rotates; moreover this same pattern is continued in the regular system of satellites that accompany Jupiter, Saturn, and Uranus. It is thought to have formed by condensation from a cloud of gas and dust about 4.6 billion years ago.

the Sun

The Sun is the star at the centre of the Solar System, and the source of daylight on Earth. Its diameter is 1,392,000 km/865,000 mi; its temperature at the surface is about 5,800 K (5,500°C/9,900°F), and at

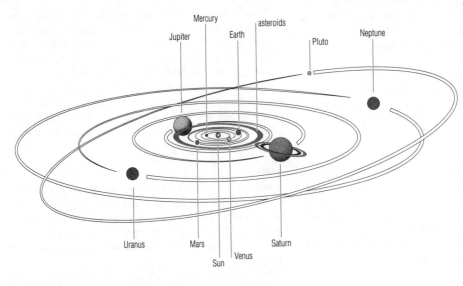

Solar System Most of the objects in the Solar System lie close to the plane of the ecliptic. The planets are tiny compared to the Sun (not to scale). If the Sun were the size of a basketball, the planet closest to the Sun, Mercury, would be the size of a mustard seed 15 m/48 ft from the Sun. The most distant planet, Pluto, would be a pinhead 1.6 km/1 mi away from the Sun. The Earth, which is the third planet out from the Sun, would be the size of a pea 32 m/100 ft from the Sun.

Sun The structure of the Sun. Nuclear reactions at the core releases vast amounts of energy in the form of light and heat that radiate out to the photosphere and corona. Surges of glowing gas rise as prominences from the surface of the Sun and cooler areas, known as sunspots appear as dark patches on the giant stars surface.

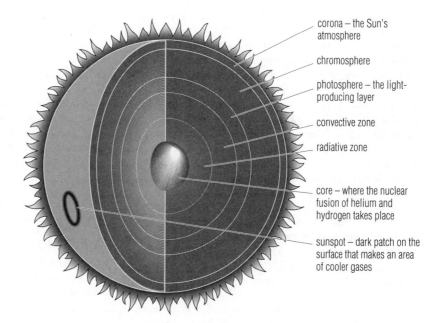

corona – the Sun's atmosphere

chromosphere

photosphere – the light-producing layer

convective zone

radiative zone

core – where the nuclear fusion of helium and hydrogen takes place

sunspot – dark patch on the surface that makes an area of cooler gases

the centre 15,000,000 K (about 15,000,000°C/27,000,000°F). It is composed of about 70% hydrogen and 30% helium, with other elements making up less than 1%. The Sun's energy is generated by nuclear-fusion reactions that turn hydrogen into helium at its centre. The gas core is far denser than mercury or lead on Earth. The Sun is about 4.7 billion years old, with a predicted lifetime of 10 billion years.

At the end of its life, it will expand to become a red giant the size of Mars's orbit, then shrink to become a white dwarf. The Sun spins on its axis every 25 days near its equator, but more slowly towards its poles. Its rotation can be followed by watching the passage of dark sunspots across its disc. Sometimes bright eruptions called flares occur near sunspots. Above the Sun's **photosphere**, or visible surface, lies a layer of thinner gas called the **chromosphere**, visible only by means of special instruments or at eclipses. Tongues of gas called prominences extend from the chromosphere into the **corona**, a halo of hot, tenuous gas surrounding the Sun. Gas boiling from the corona streams outwards through the Solar System, forming

the **solar wind**. Activity on the Sun, including sunspots, flares, and prominences, waxes and wanes during the **solar cycle**, which peaks every 11 years or so, and seems to be connected with the solar magnetic field.

the Moon

The Moon is Earth's closest neighbour and its natural satellite. It is 3,476 km/2,160 mi in diameter, with a mass 0.012 (approximately one-eightieth) that of Earth. Its surface gravity is only 0.16 (one-sixth) that of Earth. Its average distance from Earth is 384,400 km/238,855 mi, and it orbits in a west-to-east direction every 27.32 days (the **sidereal month**). It spins on its axis with one side permanently turned towards Earth. The Moon has no atmosphere and was thought to have no water until ice was discovered on its surface in 1998.

The Moon is illuminated by sunlight, and goes through a cycle of **phases** of shadow, waxing from **new** (dark) via **first quarter** (half Moon) to **full**, and waning back again to new every 29.53 days (the **synodic month**, also known as a **lunation**). On its sunlit side, temperatures reach 110°C/230°F, but during the two-week lunar night the surface temperature drops to –170°C/–274°F.

The origin of the Moon is still open to debate. Scientists suggest the following theories: that it split from the Earth; that it was a separate body captured by Earth's gravity; that it formed in orbit around Earth; or that it was formed from debris thrown off when a body the size of Mars struck Earth. Future exploration of the Moon may detect water permafrost, which could be located at the permanently shadowed lunar poles.

The Moon's composition is rocky, with a surface heavily scarred by meteorite impacts that have formed craters up to 240 km/150 mi across. Seismic observations indicate that the Moon's surface extends downwards for tens of kilometres; below this crust is a solid mantle about 1,100 km/688 mi thick, and below that a silicate core, part of which may be molten. Rocks brought back by astronauts show the Moon is 4.6 billion years old, the same age as Earth. It is made up of the same chemical elements as Earth, but in different proportions, and differs from Earth in that most of the Moon's surface features were formed within the first billion years of its history when it was hit repeatedly by meteorites.

The youngest craters are surrounded by bright rays of ejected rock. The largest scars have been filled by dark lava to produce the lowland plains called seas, or **maria** (plural of **mare**). These dark patches form the so-called 'man-in-the-Moon' pattern. One of the Moon's easiest features to observe is the mare **Plato**, which is about 100 km/62 mi in diameter and 2,700 m/8,860 ft deep, and at times is visible with the naked eye alone.

The far side of the Moon was first photographed from the Soviet space probe *Lunik 3* in October 1959. Much of our information about the Moon has been derived from this and other photographs and measurements taken by US and Soviet Moon probes, from geological samples brought back by US Apollo astronauts and by Soviet Luna probes, and by experiments set up by the US astronauts in 1969–72. The US probe *Lunar Prospector*, launched in January 1998, is intended to examine the composition of the lunar crust, record gamma rays, and map the lunar magnetic field.

eclipse An eclipse is the passage of an astronomical body through the shadow of another. The term is usually employed for solar and lunar eclipses, which may be either partial or total, but also, for example, for eclipses by Jupiter of its satellites. An eclipse of a star by a body in the Solar System is called an **occultation**.

Moon
http://www.hawastsoc.org/solar/eng/moon.htm
Detailed description of the Moon. It includes statistics and information about the surface, eclipses, and phases of the Moon, along with details of the Apollo landing missions.

A **solar eclipse** occurs when the Moon passes in front of the Sun as seen from Earth, and can happen only at new Moon. During a total eclipse the Sun's corona can be seen. A total solar eclipse can last up to 7.5 minutes. When the Moon is at its farthest from Earth it does not completely cover the face of the Sun, leaving a ring of sunlight visible. This is an **annular**

Phases of the Moon 2000

Phases of the Moon shown to the nearest hour with timings given in Greenwich Mean Time (GMT).

New moon			First quarter			Full moon			Last quarter		
Month	Day	Time	Month	Day	Time	Month	Day	Time	Month	Day	Time
January	6	18:14	January	14	13:34	January	21	04:40	January	28	07:57
Frebruary	5	13:03	February	12	23:21	February	19	16:27	February	27	03:53
March	6	05:17	March	13	06:59	March	20	04:44	March	28	00:21
April	4	18:12	April	11	13:30	April	18	17:41	April	26	19:30
May	4	04:12	May	10	20:00	May	18	07:34	May	26	11:55
June	2	12:14	June	9	03:29	June	16	22:27	June	25	01:00
July	1	19:20	July	8	12:53	July	16	13:55	July	24	11:02
July	31	02:25	August	7	01:02	August	15	05:13	August	22	18:51
August	29	10:19	September	5	16:27	September	13	19:37	September	21	01:28
September	27	19:53	October	5	10:59	October	13	08:53	October	20	07:59
October	27	07:58	November	4	07:27	November	11	21:15	November	18	15:24
November	25	23:11	December	4	03:55	December	11	09:03	December	18	00: 41
December	25	17:22									

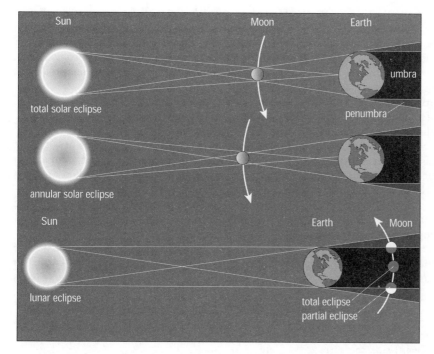

Eclipse The two types of eclipse: lunar and solar. A lunar eclipse occurs when the Moon passes through the shadow of the Earth. A solar eclipse occurs when the Moon passes between the Sun and the Earth, blocking out the Sun's light. During a total solar eclipse, when the Moon completely covers the Sun, the Moon's shadow sweeps across the Earth's surface from west to east at a speed of 3,200kph/ 2,000mph.

eclipse (from the Latin word *annulus* 'ring'). Between two and five solar eclipses occur each year.

A **lunar eclipse** occurs when the Moon passes into the shadow of the Earth, becoming dim until emerging from the shadow. Lunar eclipses may be partial or total, and they can happen only at full Moon. Total lunar eclipses last for up to 100 minutes; the maximum number each year is three.

Solar and Lunar Eclipses

Table does not include partial eclipses of the Moon.

Month	Day	Type of eclipse	Duration of maximum eclipse	Region for observation
2000				
January	21	lunar total	4 hr 44 min	the Americas, Europe, Africa, western Asia
February	5	solar partial	12 hr 50 min	Antarctica
July	1	solar partial	19 hr 34 min	southeastern Pacific Ocean
July	16	lunar total	13 hr 56 min	southeastern Asia, Australasia
July	31	solar partial	2 hr 14 min	Arctic regions
December	25	solar partial	17 hr 36 min	USA, eastern Canada, Central America, Caribbean
2001				
January	9	lunar total	20 hr 21 min	Africa, Europe, Asia
June	21	solar total	12 hr 4 min	central and southern Africa
December	14	solar annular	20 hr 52 min	Pacific Ocean
2002				
June	10	solar annular	23 hr 44 min	Pacific Ocean
December	4	solar total	7 hr 31 min	Southern Africa, Indian Ocean, Australia
2003				
May	16	lunar total	3 hr 40 min	Americas, E. W. Africa
May	31	solar annular	4 hr 8 min	Iceland, Greenland
November	9	lunar total	1 hr 18 min	Americas, Africa, Europe
November	23	solar total	22 hr 49 min	Antarctica

planets

A planet is a large celestial body in orbit around a star, composed of rock, metal, or gas. The nine planets in the Solar System are: Mercury, Venus, Earth, Mars, Jupiter, Saturn, Uranus, Neptune, and Pluto. The inner four, called the terrestrial planets, are small and rocky, and include the planet Earth. The outer planets, with the exception of Pluto, are called the major planets, and consist of large balls of rock, liquid, and gas; the largest is Jupiter, which contains a mass equivalent to 70% of all the other planets combined. Planets do not produce light, but reflect the light of their parent star.

As seen from the Earth, all the historic planets – Mercury, Venus, Mars, Jupiter, and Saturn – are conspicuous objects moving in looped paths against the stellar background. The size of these loops, which are caused by the Earth's own motion round the Sun, are inversely proportional to the planet's distance from the Earth.

Mercury

Mercury is the closest planet to the Sun. Its mass is 0.056 that of Earth. On its sunward side the surface temperature reaches over 400°C/752°F, but on the 'night' side it falls to −170°C/−274°F.
mean distance from the Sun: 58 million km/36 million mi
equatorial diameter: 4,880 km/3,030 mi
rotation period: 59 Earth days
year: 88 Earth days
atmosphere: Mercury has an atmosphere with minute traces of argon and helium.
surface: composed of silicate rock often in the form of lava flows. In 1974 the US space probe *Mariner 10* showed that Mercury's surface is cratered by meteorite impacts.
satellites: none

Mercury
http://www.hawastsoc.org/solar/eng/mercury.htm
Detailed description of the planet Mercury. It includes statistics and information about the planet, along with a chronology of its exploration supported by a good selection of images.

Its largest known feature is the Caloris Basin, 1,400 km/870 mi wide. There are also cliffs hundreds of kilometres long and up to 4 km/ 2.5 mi high, thought to have been formed by the cooling of the planet billions of years ago. Inside is an iron core three-quarters of the planet's diameter, which produces a magnetic field 1% the strength of Earth's.

Venus

Venus is the second planet from the Sun. It can approach Earth to within 38 million km/24 million mi, closer than any other planet. Its mass is 0.82 that of Earth. Venus rotates on its axis more slowly than any other planet, from east to west, the opposite direction to the other planets (except Uranus and possibly Pluto).
mean distance from the Sun: 108.2 million km/ 67.2 million mi
equatorial diameter: 12,100 km/7,500 mi
rotation period: 243 Earth days
year: 225 Earth days
atmosphere: Venus is shrouded by clouds of sulphuric acid droplets that sweep across the planet from east to west every four days. The atmosphere is almost entirely carbon dioxide, which traps the Sun's heat by the greenhouse effect and raises the planet's surface temperature to 480°C/900°F, with an atmospheric pressure 90 times that at the surface of the Earth.
surface: consists mainly of silicate rock and may have an interior structure similar to that of Earth: an iron–nickel core, a mantle composed of mafic rocks (rocks made of one or more dark-coloured, ferromagnesian minerals), and a thin siliceous outer crust. The surface is dotted with deep impact craters. Some of Venus's volcanoes may still be active.
satellites: no moons

The largest highland area is Aphrodite Terra near the equator, half the size of Africa. The highest mountains are on the northern highland region of Ishtar Terra, where the massif of Maxwell Montes rises to 10,600 m/35,000 ft above the average surface level. The highland areas on Venus were formed by volcanoes.

Venus has an ion-packed tail, 45 million km/28 million mi in length, that stretches away from the Sun and is caused by the bombardment of the ions in Venus's upper atmosphere by the solar wind.

The first human-made object to hit another planet was the Soviet probe *Venera 3*, which crashed on Venus on 1 March 1966. Later Venera probes parachuted down through the atmosphere and landed successfully on its surface, analysing surface material and sending back information and pictures. In December 1978 a US Pioneer Venus probe went into orbit around the planet and mapped most of its surface by radar, which penetrates clouds. In 1992 the US space probe *Magellan* mapped 99% of the planet's surface to a resolution of 100 m/ 330 ft.

Earth

Earth is the third planet from the Sun. It is almost spherical, flattened slightly at the poles, and is composed of three concentric layers: the core, the mantle, and the crust. About 70% of the surface (including the north and south polar icecaps) is covered with water. The Earth is surrounded by a life-supporting atmosphere and is the only planet on which life is known to exist.

Discovery of the Major Planets

BY TONY JONES

the first five

Other than the Earth, there are eight major planets in the Solar System. Five of them – Mercury, Venus, Mars, Jupiter, and Saturn – have been known since antiquity. Looking like bright stars, they reveal their true nature by moving slowly from night to night against the background of the constellations. Indeed, the word 'planet' comes from the ancient Greek for 'wanderer'.

Venus is unmistakable. Shining brilliant white, it is by far the brightest object in the sky after the Sun and Moon and is often visible in daylight. Every few months it is an arresting sight in the evening or morning sky. Jupiter, Mars and Saturn are also prominent, and Mercury may escape attention only because it is close to the Sun and never seen in darkness. These five, together with the Sun and Moon, were familiar sights in the heavens throughout recorded history and have been woven into religious and astrological myth since ancient times.

Uranus

The world was taken by surprise then, when a humble music teacher discovered a new member of the Sun's family of planets in 1781. William Herschel was a committed amateur astronomer and a skilled telescope maker. He had built a series of superb telescopes at his home in Bath, England. On 13 March, while conducting a survey of the night sky, he came across a curious star that appeared as a disc rather than a point of light. A few nights later it had changed position and within months the astronomical world had confirmed that Herschel's object was a new planet far beyond the orbit of Saturn, then the outermost known member of the Solar System.

Herschel wanted to call it George's Star, in honour of King George III, but the name Uranus, after the Greek sky god and father of Cronos (Saturn), was eventually accepted. The king was dazzled none the less, and Herschel, the first person ever to discover a new planet, became a favourite at the royal court and soon the most influential astronomer of his day.

Uranus, though faint, is actually visible to the naked eye under good conditions. It was later found that the planet had been recorded on at least 20 occasions, as far back as 1690, and had been mistaken for a star each time. The earlier observers had not had the benefit of Herschel's powerful telescopes, which were able to discern the greenish disc of the planet.

Neptune

Some years later astronomers were having problems with the orbit of Mercury, the planet closest to the Sun. Its movements could not be completely accounted for by Newton's laws of gravity. A French astronomer, Urbain Leverrier, proposed in 1845 that the discrepancy could be explained by the gravitational attraction of an undiscovered planet, which he called Vulcan, orbiting within the orbit of Mercury only 30 million km/18.5 million mi from the Sun. All attempts to find Vulcan failed, and the Mercury problem remained unsolved until 1915, when Albert Einstein (1879–1955) showed that the discrepancies in the orbit were a consequence of the general theory of relativity.

Leaving astronomers to search for Vulcan, Leverrier turned his attention to Uranus, which was also deviating from its predicted path. Again, he attributed the perturbation to another planet, this time beyond the orbit of Uranus, and predicted where in the sky it would be found. Johann Galle at the Berlin Observatory pointed a telescope at Leverrier's position on 23 September 1846 and almost at once discovered the new planet only one degree from the predicted location. It became known as Neptune, after the Roman god of the sea.

It turned out that a young Cornish mathematician, John Couch Adams had predicted the position of Neptune a year earlier but had been unable to persuade the Astronomer Royal, George Airy to take it seriously. When Airy finally asked James Challis (1803–1882) at Cambridge Observatory to search for the planet, Challis saw it on two occasions more than a month before Galle but failed to recognize it.

Pluto

Buoyed by this outstanding triumph of Newtonian mechanics, astronomers soon suspected that Neptune in turn was being affected by the pull of a still more distant planet. One of the scientists who attempted to calculate its position was US businessman and astronomer Percival Lowell. He built his own observatory in 1895 at Flagstaff in Arizona but failed to track down the mystery planet. It was not until 18 February 1930, 14 years after Lowell's death, that Clyde Tombaugh, an assistant at the observatory, finally stumbled across the planet. He had ignored the predictions of the mathematicians and had systematically worked his way through the zodiac, the band of sky in which the planets move, comparing pairs of photographs taken on different nights and looking for a 'star' that had moved.

The planet, which was named Pluto after the god of the underworld, was much fainter than expected and was later found on several photographs taken during earlier searches.

Though astronomers continue to discover asteroids, comets and other small bodies in the outer Solar System, modern search techniques are so thorough that there is little chance of more major planets lying unseen beyond the orbit of Pluto.

mean distance from the Sun: 149,500,000 km/ 92,860,000 mi

equatorial diameter: 12,756 km/7,923 mi

circumference: 40,070 km/24,900 mi

rotation period: 23 hr 56 min 4.1 sec

year: (complete orbit, or sidereal period) 365 days 5 hr 48 min 46 sec. Earth's average speed around the Sun is 30 kps/18.5 mps; the plane of its orbit is inclined to its equatorial plane at an angle of 23.5°, the reason for the changing seasons

atmosphere: nitrogen 78.09%; oxygen 20.95%; argon 0.93%; carbon dioxide 0.03%; and less than 0.0001% neon, helium, krypton, hydrogen, xenon, ozone, and radon

---

Earth and Moon Viewer
http://www.fourmilab.ch/earthview/vplanet.html
View a map of the Earth showing the day and night regions at this moment, or view the Earth from the Sun, the Moon, or any number of other locations. Alternatively, take a look at the Moon from the Earth or the Sun, or from above various formations on the lunar surface.

---

surface: land surface 150,000,000 sq km/ 57,500,000 sq mi (greatest height above sea level 8,872 m/29,118 ft Mount Everest); water surface 361,000,000 sq km/139,400,000 sq mi (greatest depth 11,034 m/36,201 ft Mariana Trench in the Pacific). The interior is thought to be an inner core about 2,600 km/1,600 mi in diameter, of solid iron and nickel; an outer core about 2,250 km/1,400 mi thick, of molten iron and nickel; and a mantle of mostly solid rock about 2,900 km/1,800 mi thick, separated from the Earth's crust by the Mohorovicic discontinuity. The crust and the topmost layer of the mantle form about 12 major moving plates, some of which carry the continents. The plates are in constant, slow motion, called tectonic drift. US geophysicists announced in

1996 that they had detected a difference in the spinning time of the Earth's core and the rest of the planet; the core is spinning slightly faster.

satellite: the Moon

age: 4.6 billion years. The Earth was formed with the rest of the Solar System by consolidation of interstellar dust. Life began 3.5–4 billion years ago.

Mars

The fourth planet from the Sun. It is much smaller than Venus or Earth, with a mass 0.11 that of Earth. Mars is slightly pear-shaped, with a low, level northern hemisphere, which is comparatively uncratered and geologically 'young', and a heavily cratered 'ancient' southern hemisphere.

mean distance from the Sun: 227.9 million km/141.6 million mi

equatorial diameter: 6,780 km/4,210 mi

rotation period: 24 hr 37 min

year: 687 Earth days

atmosphere: 95% carbon dioxide, 3% nitrogen, 1.5% argon, and 0.15% oxygen. Red atmospheric dust from the surface whipped up by winds of up to 450 kph/280 mph accounts for the light pink sky. The surface pressure is less than 1% of the Earth's atmospheric pressure at sea level.

surface: The landscape is a dusty, red, eroded lava plain. Mars has white polar caps (water ice and frozen carbon dioxide) that advance and retreat with the seasons.

satellites: two small satellites: Phobos and Deimos.

There are four enormous volcanoes near the equator, of which the largest is Olympus Mons 24 km/15 mi high, with a base 600 km/375 mi across, and a crater 65 km/40 mi wide. To the east of the four volcanoes lies a high plateau cut by a system of valleys, Valles Marineris, some 4,000 km/2,500 mi long, up to 200 km/120 mi wide and 6 km/4 mi deep; these features

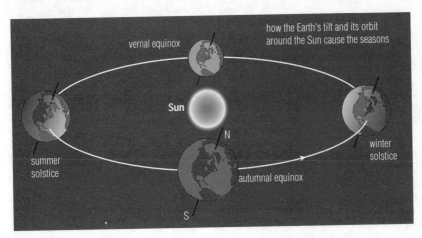

how the Earth's tilt and its orbit around the Sun cause the seasons

vernal equinox

Sun

N

summer solstice

winter solstice

autumnal equinox

S

season The cause of the seasons. As the Earth orbits the Sun, its axis of rotation always points in the same direction. This means that, during the northern hemisphere summer solstice (21 June), the Sun is overhead in the northern hemisphere. At the northern hemisphere winter solstice (22 December), the Sun is overhead in the southern hemisphere.

are apparently caused by faulting and wind erosion. Recorded temperatures vary from −100°C/−148°F to 0°C/32°F.

Mars
http://www.hawastsoc.org/solar/eng/mars.htm
Detailed description of the planet Mars, commonly referred to as 'the red planet'. It includes statistics and information about the surface, volcanoes, satellites, and clouds of the planet, supported by a good selection of images.

Mars may approach Earth to within 54.7 million km/34 million mi. The first human-made object to orbit another planet was *Mariner 9*. *Viking 1* and *Viking 2*, which landed, also provided much information. Studies in 1985 showed that enough water might exist to sustain prolonged missions by space crews.

In January 1997 NASA launched the *Mars Pathfinder*, which made a successful landing on Mars in July 1997 on a flood plain called Ares Vallis. After initial technical problems, its 0.3-m/1-ft rover, 'Sojourner', began to explore the Martian landscape and to transmit data back to Earth. Photographs from the *Mars Pathfinder* indicated that the planet is rusting. NASA announced this in July 1997 and said that a supercorrosive force was eroding rocks on the surface due to iron oxide in the soil.

In May 1997 American scientists announced that Mars is becoming increasingly colder and cloudier. Images from the Hubble Space Telescope showed that dust storms had covered areas of the planet that had been dark features in the early century, including one section as large as California.

The *Global Surveyor*, which entered Martian orbit in September 1997, revealed that Mars's magnetic field is a mere 800th that of the Earth.

Welcome to the Mars Missions, Year 2000 and Beyond!
http://marsweb.jpl.nasa.gov/
Well presented NASA site with comprehensive information on current and future missions to Mars. There are fascinating and well written accounts of *Pathfinder* and *Global Surveyor*, and large numbers of images of the red planet.

Jupiter

Jupiter is the fifth planet from the Sun, and the largest in the Solar System, with a mass equal to 70% of all the other planets combined, 318 times that of Earth. It is largely composed of hydrogen and helium, liquefied by pressure in its interior, and probably with a rocky core larger than Earth. Its main feature is the **Great Red Spot**, a cloud of rising gases, 14,000 km/ 8,500 mi wide and 30,000 km/20,000 mi long, revolving anticlockwise.

mean distance from the Sun: 778 million km/ 484 million mi
equatorial diameter: 142,800 km/88,700 mi
rotation period: 9 hr 51 min
year: (complete orbit) 11.86 Earth years
atmosphere: consists of clouds of white ammonia crystals, drawn out into belts by the planet's high speed of rotation (the fastest of any planet). Darker orange and brown clouds at lower levels may contain sulphur, as well as simple organic compounds. Further down still, temperatures are warm, a result of heat left over from Jupiter's formation, and it is this heat that drives the turbulent weather patterns of the planet. In 1995 the *Galileo* probe revealed Jupiter's atmosphere to consist of 0.2% water, less than previously estimated.
surface: although largely composed of hydrogen and helium, Jupiter probably has a rocky core larger than Earth.
satellites: Jupiter has 16 moons. The four largest moons, Io, Europa (which is the size of our Moon), Ganymede, and Callisto, are the **Galilean satellites**, discovered in 1610 by Galileo (Ganymede, which is about the size of Mercury, is the largest moon in the Solar System). Three small moons were discovered in 1979 by the Voyager space probes, as was a faint ring of dust around Jupiter's equator 55,000 km/34,000 mi above the cloud tops.

The Great Red Spot was first observed in 1664. Its top is higher than the surrounding clouds; its colour is thought to be due to red phosphorus. Jupiter's strong magnetic field gives rise to a large surrounding magnetic 'shell', or magnetosphere, from which bursts of radio waves are detected. The Southern Equatorial Belt in which the Great Red Spot occurs is subject to unexplained fluctuation. In 1989 it sustained a dramatic and sudden fading.

Jupiter
http://www.hawastsoc.org/solar/eng/jupiter.htm
Full details of the planet and its moons, including a chronology of exploration, various views of the planet and its moons, and links to other planets.

Saturn

Saturn is the second largest planet in the Solar System, sixth from the Sun, and encircled by bright and easily visible equatorial rings. Viewed through a telescope, it is ochre in colour. Its polar diameter is 12,000 km/ 7,450 mi smaller than its equatorial diameter, a result of its fast rotation and low density, the lowest of any planet. Its mass is 95 times that of Earth, and its magnetic field 1,000 times stronger.

mean distance from the Sun: 1.427 billion km/ 0.886 billion mi

equatorial diameter: 120,000 km/75,000 mi

rotational period: 10 hr 14 min at equator, 10 hr 40 min at higher latitudes

year: 29.46 Earth years

atmosphere: visible surface consists of swirling clouds, probably made of frozen ammonia at a temperature of −170°C/−274°F, although the markings in the clouds are not as prominent as Jupiter's. The space probes *Voyager 1* and *2* found winds reaching 1,800 kph/1,100 mph

surface: Saturn is believed to have a small core of rock and iron, encased in ice and topped by a deep layer of liquid hydrogen

satellites: 18 known moons, more than for any other planet. The largest moon, Titan, has a dense atmosphere. Other satellites include Epimetheus, Janus, Pandor, and Prometheus. The rings visible from Earth begin about 14,000 km/9,000 mi from the planet's cloudtops and extend out to about 76,000 km/ 47,000 mi. Made of small chunks of ice and rock (averaging 1 m/3 ft across), they are 275,000 km/170,000 mi rim to rim, but only 100 m/300 ft thick. The Voyager probes showed that the rings actually consist of thousands of closely spaced ringlets, looking like the grooves in a gramophone record.

From Earth, Saturn's rings appear to be divided into three main sections. Ring A, the outermost, is separated from ring B, the brightest, by the Cassini division, named after its discoverer Italian astronomer Giovanni Cassini, which is 3,000 km/2,000 mi wide; the inner, transparent ring C is also called the Crepe Ring. Each ringlet of the rings is made of a swarm of icy particles like snowballs, a few centimetres to a few metres in diameter. Outside the A ring is the narrow and faint F ring, which the Voyagers showed to be twisted or braided. The rings of Saturn could be the remains of a shattered moon, or they may always have existed in their present form.

Saturn
http://www.hawastsoc.org/solar/eng/saturn.htm
How many rings does Saturn have? How many satellites? Find out this and more at this site, which also features a video of a storm in the planet's atmosphere.

Uranus
Uranus is the seventh planet from the Sun, discovered by the British astronomer William Herschel in 1781. It is twice as far out as the sixth planet, Saturn. Uranus

---

Uranian satellites
To remember the Uranian satellites visible from Earth:

MAUTO

(**M**iranda, **A**riel, **U**mbriel, **T**itania, **O**beron)

---

has a mass 14.5 times that of Earth. The spin axis of Uranus is tilted at 98°, so that one pole points towards the Sun, giving extreme seasons.

mean distance from the Sun: 2.9 billion km/1.8 billion mi

equatorial diameter: 50,800 km/31,600 mi

rotation period: 17.2 hr

year: 84 Earth years

atmosphere: deep atmosphere composed mainly of hydrogen and helium

surface: composed primarily of hydrogen and helium but may also contain heavier elements, which might account for Uranus's mean density being higher than Saturn's

satellites: 17 moons (two discovered in 1997); 11 thin rings around the planet's equator were discovered in 1977.

Uranus has a peculiar magnetic field, whose axis is tilted at 60° to its axis of spin, and is displaced about a third of the way from the planet's centre to its surface. Uranus spins from east to west, the opposite of the other planets, with the exception

Uranus
All Uranus' moons are named after characters from Shakespeare.

of Venus and possibly Pluto. The rotation rate of the atmosphere varies with latitude, from about 16 hours in mid-southern latitudes to longer than 17 hours at the equator.

Uranus's equatorial ring system comprises 11 rings. The ring furthest from the planet centre, Epsilon, is 100 km/62 mi at its widest point. In 1995 US astronomers determined that the ring particles contained long-chain hydrocarbons. Looking at the brightest region of Epsilon, they were also able to calculate the precession (axial wobble) of Uranus as 264 days, the fastest known precession in the Solar System.

The space probe *Voyager 2* detected 11 rings, composed of rock and dust, around the planet's

---

Uranus
http://www.hawastsoc.org/solar/eng/uranus.htm
Did you know that Uranus is tipped on its side? Find out more about Uranus, its rings, and its moons at this site. Also included are a table of statistics about the planet, photographs, and animations of it rotating.

equator, and found 10 small moons in addition to the 5 visible from Earth. Titania, the largest moon, has a diameter of 1,580 km/980 mi. The rings are charcoal black, and may be debris of former 'moonlets' that have broken up.

Neptune

Neptune is the eighth planet in average distance from the Sun. It is a giant gas planet, with a mass 17.2 times that of Earth. It has the highest winds in the Solar System.

mean distance from the Sun: 4.4 billion km/2.794 billion mi
equatorial diameter: 48,600 km/30,200 mi
rotation period: 16 hr 7 min
year: 164.8 Earth years
atmosphere: methane in its atmosphere absorbs red light and gives the planet a blue colouring. Consists primarily of hydrogen (85%) with helium (13%) and methane (1–2%).
surface: hydrogen, helium and methane. Its interior is believed to have a central rocky core covered by a layer of ice.
satellites: of Neptune's eight moons, two (Triton and Nereid) are visible from Earth. Six were discovered by the *Voyager 2* probe in 1989, of which Proteus (diameter 415 km/260 mi) is larger than Nereid (300 km/200 mi).
rings: there are four faint rings: Galle, Le Verrier, Arago, and Adams (in order from Neptune). Galle is the widest at 1,700 km/ 1,060 mi. Leverrier and Arago are divided by a wide diffuse particle band called the plateau.

Neptune was located in 1846 by German astronomers Johan Galle and Heinrich d'Arrest (1822–1875) after calculations by English astronomer John Couch Adams and French mathematician Urbain Leverrier had predicted its existence from disturbances in the movement of Uranus. *Voyager 2*, which passed Neptune in August 1989, revealed various cloud features, notably an Earth-sized oval storm cloud, the Great Dark Spot, similar to the Great Red Spot on Jupiter, but images taken by the Hubble Space Telescope in 1994 show that the Great Dark Spot has since disappeared. A smaller dark spot, DS2, has also gone.

Neptune
http://www.hawastsoc.org/solar/eng/neptune.htm
Detailed description of the planet Neptune. It includes a chronology of the exploration of the planet, along with statistics and information on its rings, moons, and satellites.

Pluto

Pluto is the smallest and, usually, the outermost planet of the Solar System. The existence of Pluto was predicted by calculation by the US astronomer Percival Lowell and the planet was located by US astronomer Clyde Tombaugh in 1930. Its highly elliptical orbit occasionally takes it within the orbit of Neptune, as in 1979–99. Pluto has a mass about 0.002 of that of Earth.
mean distance from the Sun: 5.8 billion km/3.6 billion mi
equatorial diameter: 2,300 km/1,438 mi
rotation period: 6.39 Earth days
year: 248.5 Earth years
atmosphere: thin atmosphere with small amounts of methane gas
surface: low density, composed of rock and ice, primarily frozen methane; there is an ice cap at Pluto's north pole
satellites: one moon, Charon

Charon, Pluto's moon, was discovered in 1978 by James Walter Christy (1938–   ). It is about 1,200 km/750 mi in diameter, half the size of Pluto, making it the largest moon in relation to its parent planet in the Solar System. It orbits about 20,000 km/12,500 mi from the planet's centre every 6.39 days – the same time that Pluto takes to spin on its axis. Charon is composed mainly of ice. Some astronomers have suggested that Pluto was a former moon of Neptune that escaped, but it is more likely that it was an independent body that was captured. The Hubble Space Telescope photographed Pluto's surface in 1996.

Pluto and Charon
http://www.hawastsoc.org/solar/eng/pluto.htm
Site devoted to our most distant planet and its satellite. It contains a table of statistics, photographs, and an animation of their rotation. You can also find out about NASA's planned mission to Pluto and Charon in 2010.

---

asteroids

An asteroid, or **minor planet**, is any one of many thousands of small bodies, composed of rock and iron, that orbit the Sun. Most lie in a belt between the orbits of Mars and Jupiter, and are thought to be fragments left over from the formation of the Solar System. About 100,000 may exist, but their total mass is only a few hundredths the mass of the Moon.

They include Ceres (the largest asteroid, 940 km/584 mi in diameter), Vesta (which has a light-coloured surface, and is the brightest as seen from Earth), Eros, and Icarus. Some asteroids are in orbits that bring them close to Earth, and some, such as the Apollo asteroids, even cross Earth's orbit; at least some of these may be remnants of former comets. One group, the Trojans, moves along the same orbit as Jupiter, 60° ahead and behind the planet. One unusual asteroid, Chiron, orbits beyond Saturn.

The Largest Asteroids

An asteroid is a small body, composed of rock and iron, that orbits the Sun. Most lie in a belt between the orbits of Mars and Jupiter, and are thought to be fragments left over from the formation of the Solar System.

Name	Diameter	Average distance from Sun (Earth = 1)		Orbital period (yrs)
		km	mi	
Ceres	940	584	2.77	4.6
Pallas	588	365	2.77	4.6
Vesta	576	358	2.36	3.6
Hygeia	430	267	3.13	5.5
Interamnia	338	210	3.06	5.4
Davida	324	201	3.18	5.7

The first asteroid was discovered by the Italian astronomer Giuseppe Piazzi at the Palermo Observatory, Sicily, on 1 January 1801. The first asteroid moon was observed by the space probe *Galileo* in 1993, orbiting asteroid Ida.

comets

A comet is a small, icy body orbiting the Sun, usually on a highly elliptical path. It consists of a central

Asteroid Introduction
http://www.hawastsoc.org/solar/eng/asteroid.htm
What is the difference between an asteroid and a meteorite? Find out at this site, which contains a table of statistics about asteroids, a chronology of asteroid exploration, and images of selected asteroids.

nucleus a few kilometres across, and has been likened to a dirty snowball because it is mostly made up of ice mixed with dust. As a comet approaches the Sun its nucleus heats up, releasing gas and dust, which form a tenuous coma, up to 100,000 km/ 60,000 mi wide, around the nucleus. Gas and dust stream away from the coma to form one or more tails, which may extend for millions of kilometres. US astronomers concluded in 1996 that there are two distinct types of comet: one rich in methanol and one low in methanol. Evidence for this comes in part from observations of the spectrum of Comet Hyakutake.

Comets are believed to have been formed at the birth of the Solar System. Billions of them may reside in a halo (the **Oort cloud**) beyond the planet Pluto. The gravitational effect of passing stars pushes some towards the Sun, when they eventually become visible from Earth. Most comets swing around the Sun and return to distant space, never to be seen again for thou-

Major Comets

(– = not applicable.)

Name	First recorded sighting	Orbital period (yrs)	Interesting facts
Halley's comet	240 BC	76	parent of Eta Aquarid and Orionid meteor showers
Comet Tempel-Tuttle	AD 1366	33	parent of Leonid meteors
Biela's comet	1772	6.6	broke in half in 1846; not seen since 1852
Encke's comet	1786	3.3	parent of Taurid meteors
Comet Swift-Tuttle	1862	130	parent of Perseid meteors; reappeared 1992
Comet Ikeya-Seki	1965	880	so-called 'Sun-grazing' comet, passed 500,000 km/300,000 mi above surface of the Sun on 21 October 1965
Comet Kohoutek	1973	–	observed from space by Skylab astronauts
Comet West	1975	500,000	nucleus broke into four parts
Comet Bowell	1980	–	ejected from Solar System after close encounter with Jupiter
Comet IRAS-Araki-Alcock	1983	–	passed only 4.5 million km/2.8 million mi from the Earth on 11 May 1983
Comet Austin	1989	–	passed 32 million km/20 million mi from the Earth in 1990
Comet Shoemaker-Levy 9	1993	–	made up of 21 fragments; crashed into Jupiter in July 1994
Comet Hale-Bopp	1995	1,000	spitting out of gas and debris produced a coma, a surrounding hazy cloud of gas and dust, of greater volume than the Sun; the bright coma is due to an outgassing of carbon monoxide; clearly visible with the naked eye in March 1997
Comet Hyakutake	1996	–	passed 15 million km/9,300,000 mi from the Earth

Heaviest Meteorites

(N/A = not available.)

Name and location	Weight (tonnes)	Year found	Dimensions	Composition
Hoba West, Grootfontein, Namibia	60	1920	9 x 9 x 3.5 ft	nickel-rich iron
Ahnighito, Greenland	30	–	–	–
Bacuberito, Mexico	27	1863	12 ft long	iron
Mbosi, Tanzania	26	1930	13.5 x 4 x 4 ft	iron
Agpalik, Greenland	20	–	–	–
Armanty, Mongolia	20	1935 (known in 1917)	–	iron
Chupaderos, Mexico	14	known for centuries; first mentioned in 1852	2 masses	iron
Willamette (OR), USA	14	1902	–	iron
Campo del Cielo, Argentina	13	–	–	–
Mundrabilla, Western Australia	12	–	–	–
Morito, Mexico	11	known in 1600	–	iron

sands or millions of years, although some, called **periodic comets,** have their orbits altered by the gravitational pull of the planets so that they reappear every 200 years or less. Periodic comets are thought to come from the Kuiper belt, a zone just beyond the planet Neptune. Of the 800 or so comets whose orbits have been calculated, about 160 are periodic. The one with the shortest known period is **Encke's comet,** which orbits the Sun every 3.3 years. The brightest periodic comet, **Hale-Bopp,** flew past the Earth in March 1997. A dozen or more comets are discovered every year, some by amateur astronomers.

Asteroid and Comet-Impact Hazards
http://impact.arc.nasa.gov/index.html
Overview and the latest news on asteroid- and comet-impact hazards from NASA's Ames Space Science Division, with the last Spaceguard Survey Report and a list of future Near Earth Objects (NEOs).

---

meteors

A meteor is seen as a flash of light in the sky and is popularly known as a **shooting** or **falling star.** It is caused by a particle of dust, a **meteoroid,** entering the atmosphere at speeds up to 70 kps/45 mps and burning up by friction at a height of around 100 km/60 mi. On any clear night, several **sporadic meteors** can be seen each hour. Several times each year the Earth encounters swarms of dust shed by

---

meteorite

An explosion in Northern Ireland in December 1997 was blamed on terrorists, but was later discovered to be caused by a meteorite. It left a 1.2-m/4-ft crater.

---

comets, which give rise to a **meteor shower.** This appears to radiate from one particular point in the sky, after which the shower is named; the Perseid meteor shower in August appears in the constellation Perseus. A brilliant meteor is termed a **fireball.** Most meteoroids are smaller than grains of sand. The Earth sweeps up an estimated 16,000 tonnes of meteoric material every year.

Meteors, Meteorites, and Impacts
http://www.seds.org/billa/tnp/meteorites.html
Informative collection of facts about meteorites: how they are formed, classification, and what happens when one hits Earth. It includes images of a selection of meteorites.

---

telescopes

A telescope is traditionally an optical instrument that magnifies images of faint and distant objects; it can also be any of a variety of devices for collecting and focusing light and other forms of electromagnetic radiation – for example, **radio telescopes** (see below). Telescopes are major research tools in astronomy. An optical telescope with a large aperture, or opening, can distinguish finer detail and fainter objects than one with a small aperture. The **refracting telescope** uses lenses, and the **reflecting telescope** uses mirrors. A third type, the **catadioptric telescope,** is a combination of lenses and mirrors.

In a refracting telescope, or refractor, light is collected by a lens called the **object glass** or **objective,** which focuses light down a tube, forming an image magnified by an **eyepiece.** The largest refracting telescope in the world, at Yerkes Observatory, Wisconsin, USA, has an aperture of 102 cm/40 in.

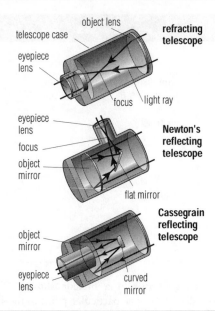

telescope Three kinds of telescope. The refracting telescope uses a large objective lens to gather light and form an image which the smaller eyepiece lens magnifies. A reflecting telescope uses a mirror to gather light. The Cassegrain telescope uses a corrective lens to achieve a wide field of view. It is one of the most widely used tools of astronomy.

In a reflecting telescope, or reflector, light is collected and focused by a concave mirror. Large mirrors are cheaper to make and easier to mount than large lenses, so all the largest telescopes are reflectors. The largest reflector with a single mirror, 6 m/236 in, is at Zelenchukskaya, Russia. Telescopes with larger apertures composed of numerous smaller segments have been built, such as the Keck Telescope on Mauna Kea. A **multiple-mirror telescope** was installed on Mount Hopkins, Arizona, USA, in 1979. It consists of six mirrors of 1.8 m/72 in aperture, which perform like a single 4.5-m/176-in mirror.

Schmidt telescopes are used for taking wide-field photographs of the sky. They have a main mirror plus a thin lens at the front of the tube to increase the field of view.

The **liquid-mirror telescope** is a reflecting telescope constructed with a rotating mercury mirror. In 1995 NASA completed a 3-m/9.8-ft liquid mirror telescope at its Orbital Debris Observatory in New Mexico, USA.

Large telescopes can now be placed in orbit above the distorting effects of the Earth's atmosphere. Telescopes in space have been used to study infrared, ultraviolet, and X-ray radiation that does not penetrate the atmosphere but carries a great deal of information about the births, lives, and deaths of stars and galaxies. The 2.4-m/94-in **Hubble Space Telescope**, launched in 1990, can see the sky more clearly than can any telescope on Earth.

In 1996 an X-ray telescope was under development by UK, US, and Australian astronomers, based on the structure of a lobster's eye, which has thousands of square tubes reflecting light onto the retina. The £4 million **Lobster Eye telescope** will contain millions of tubes 10–20 micrometres across and is intended for use on a satellite.

FAQ on Telescope Buying and Usage
http://www-personal.umich.edu/ ~dnash/saafaq/faq.html
Extensive 'plain English' guide to buying a telescope. For anybody interested in astronomy and contemplating buying a telescope or setting up an observatory, this is an indispensable source of non-commercial advice.

Hubble Space Telescope

The Hubble Space Telescope (HST) is a space-based astronomical observing facility, orbiting the Earth at an altitude of 610 km/380 mi. It consists of a 2.4-m/94-in telescope and four complimentary scientific instruments, is roughly cylindrical, 13 m/43 ft long, and 4 m/ 13 ft in diameter, with two large solar panels. The HST produces a wealth of scientific data, and allows astronomers to observe the birth of stars, find planets around neighbouring stars, follow the expanding remnants of exploding stars, and search for black holes in the centre of galaxies. The HST is a cooperative programme between the European Space Agency (ESA) and the US agency NASA, and is the first spacecraft specifically designed to be serviced in orbit as a permanent space-based observatory. It was launched in 1990.

Before the HST could reach its full potential, a flaw in the shape of its main mirror, discovered two months after the launch, had to be corrected. In 1993, as part of a planned servicing and instrument upgrade mission, NASA astronauts aboard the space shuttle *Endeavor* installed a set of corrective lenses to compensate for the error in the mirror figure.

Hubble discoveries In December 1995 the Hubble Space Telescope was trained on an 'empty' area of sky near the Plough, now termed the Hubble Deep Field. Around 1,500 galaxies, mostly new discoveries, were photographed.

In May 1997, three months after astronauts had installed further new equipment, US scientists reported that Hubble had made an extraordinary finding. Within 20 minutes of searching, it discovered evidence

Major Ground-Based Telescopes and Observatories

Observatory/ Telescope	Location	Description	Year opened	Run by
Algonquin Radio Observatory	Ontario, Canada	radio telescope, 46 m/150 ft in diameter	1966	National Research Council (NRC) of Canada
Arecibo Observatory	Puerto Rico	home of the largest radar-radio telescope in the world; a 305-m/1,000-ft diameter spherical reflector with a surface made up of nearly 40,000 perforated aluminium panels. Each panel can be adjusted to maintain a precise spherical shape that varies less than 3 mm/0.12 in over the whole 20 acre surface.	inaugurated in 1963; upgraded in 1974, and again in the mid-1990s	National Astronomy and Ionosphere Center, which is operated by Cornell University and the National Science Foundation
Australia Telescope National Facility	New South Wales, Australia	giant radio telescope consisting of six 22-m/72-ft antennae at Culgoora, a similar antenna at Siding Spring Mountain, and the 64-m/210-ft Parkes radio telescope; together they simulate a dish 300 km/186 mi across	1993	Commonwealth Scientific and Industrial Research Organization (CSIRO)
Cerro Tololo Inter-American Observatory	Cerro Tololo mountain in the Chilean Andes	main instrument is a 4-m/158-in reflector, a twin of that at Kitt Peak, Arizona, USA	1974	Association of Universities for Research into Astronomy (AURA)
David Dunlap Observatory	Richmond Hill, Ontario, Canada	1.88-m/74-in reflector, the largest optical telescope in Canada	1935	University of Toronto
Dominion Astrophysical Observatory	near Victoria, British Columbia, Canada	1.85-m/73-in reflector	1918	NRC of Canada
Dominion Radio Astrophysical Observatory	Penticton, British Columbia, Canada	26-m/84-ft radio dish and an aperture synthesis radio telescope	1996	NRC of Canada through its Herzborg Institute of Astrophysics
Effelsberg Radio Telescope	near Bonn, Germany	the world's largest fully steerable radio telescope; 100-m/328-ft radio dish	1971	Max Planck Institute for Radio Astronomy
European Southern Observatory	La Silla, Chile	telescopes include: 3.6-m/142-in reflector 3.58-m/141-in New Technology Telescope	1976 1990	operated jointly by Belgium, Denmark, France, Germany, Italy, the Netherlands, Sweden, and Switzerland with headquarters near Munich, Germany
Gemini 8-Meter Telescopes Project		Very Large Telescope (VLT), consisting of four 8-m/315-in reflectors mounted independently but capable of working in combination	1999	international partnership of USA, UK, Canada, Chile, Argentina, and Brazil
	Mauna Kea (HI), USA; Cerro	two 8-m/26-ft Aperture Optical/Infrared Telescopes	1999; 2001	

Observatory/ Telescope	Location	Description	Year opened	Run by
Green Bank Telescope	Green Bank, Pocahontas County (WV), USA	largest fully steerable radio telescope in the world, with a 110 m _ 110 m/300 ft x 340 ft surface	under construction; expected inauguration in 1999	National Radio Astronomy Observatory (NRAO)
Jodrell Bank	Cheshire, UK	Lovell Telescope, a 76-m/250-ft radio dish	1957, modified 1970	Nuffield Radio Astronomy Laboratories of the University of Manchester
		elliptical radio dish 38 m _ 25 m/125 ft _ 82 ft capable of working at shorter wavelengths	1964	
Keck I	Mauna Kea (HI), USA	world's largest optical telescope, with a primary mirror 10- m/33-ft in diameter, with 36 hexagonal sections, each controlled by a computer to generate single images of the objects observed	first images 1990	jointly owned by the California Institute of Technology and the University of California/Lick Observatory
Keck II	Mauna Kea (HI), USA	identical to Keck I: primary mirror 10 m/33 ft in diameter, with 36 hexagonal sections, each controlled by a computer	1996	jointly owned by the California Institute of Technology and the University of California/Lick Observatory
Kitt Peak National Observatory	Quinlan Mountains near Tucson (AZ), USA	numerous telescopes including the 4-m/ 158-in Mayall reflector	1973	AURA in agreement with the National Science Foundation of the USA
		McMath-Pierce Solar Telescope, the world's largest of its type	1960	National Solar Observatory (NSO)
		3.5-m/138-in reflecting telescope	1994	WIYN consortium comprising the University of Wisconsin, Indiana University, Yale University, and National Optical Astronomy Observatories (NOAO)
La Palma Observatory or the Observatorio del Roque de los Muchachos	La Palma, Canary Islands	Isaac Newton Group of telescopes, including the 4.2- m/165-in William Herschel Telescope	1987	Royal Greenwich Observatories (RGO)
Las Campanas Observatory	Chile	2.5-m/100-in Irénée du Pont telescope	1977	Carnegie Institution of Washington
Lick Observatory or the University of California/Lick Observatory	Mount Hamilton (CA), USA	several instruments including: 3.04 m/120 in Shane reflector	1959	University of California and Mount Hamilton
		91-cm/36-in refractor, the second-largest refractor in the world	1888	
Lowell Observatory	Flagstaff (AZ), USA	8 telescopes, including the 61- cm/24-in Alvan Clark refractor	1894	Lowell Observatory staff

Observatory/ Telescope	Location	Description	Year opened	Run by
	Anderson Mesa	several telescopes, including the 1.83-m/ 72-in Perkins reflector		Ohio State and Ohio Wesleyan universities
McDonald Observatory	Davis Mountains (TX), USA	2.72-m/107-in reflector	1969	University of Texas
		2.08-m/82-in reflector	1939	
		9.2-m/30-ft Hobby-Eberly telescope (HET) for spectral analysis	1997	Penn State University, Stanford University, and the Ludwig-Maximilians Universität
Magellan	Las Campanas, Chile	6.5-m/21.3-ft honeycomb-back optical mirror	1997	Carnegie Institution of Washington, University of Arizona, and Harvard College Observatory
Mauna Kea	Mauna Kea (HI), USA	telescopes include: the 2.24-m/88-in University of Hawaii reflector	1970	University of Hawaii
		3.8-m/150-in United Kingdom Infrared Telescope (UKIRT) (also used for optical observations)	1978	Royal Observatory, Edinburgh
		3-m/120-in NASA Infrared Telescope Facility (IRTF)	1979	NASA
		3.6-m/142-in Canada-France–Hawaii Telescope (CFHT), designed for optical and infrared work	1979	NRC of Canada, University of Hawaii, and Centre National de la Recherche Scientifique of France
		15-m/50-ft diameter UK/Netherlands James Clerk Maxwell Telescope (JCMT), the world's largest telescope specifically designed to observe millimetre wave radiation from nebulae, stars, and galaxies	1987	Joint Astronomy Centre in Hilo (HI) for the NRC of Canada, Particle Physics and Astronomy Research Council (PPARC) of the UK, and Netherlands Organization for Scientific Research (NOSR)
		Keck I and Keck II telescopes	1990 and 1996	California Institute of Technology and the University of California/Lick Observatory
Mount Palomar	80 km/50 mi northeast of San Diego (CA), USA	5-m/200-in diameter reflector called the Hale; 1.2-m/48-in Schmidt telescope; it was the world's premier observatory during the 1950s	1948	California Institute of Technology and the University of California/Lick Observatory
Mount Wilson Observatory	San Gabriel Mountains, near Los Angeles (CA), USA	several telescopes including: 2.5-m/100-in Hooker telescope, with which Edwin Hubble discovered the expansion of the universe	1917	Mount Wilson Institute
		solar telescopes in towers 18.3 m/60 ft and 45.7 m/150 ft tall	1912	operated presently by University of California (UCLA)

Observatory/ Telescope	Location	Description	Year opened	Run by
Mullard Radio Astronomy Observatory	Cambridge, UK	Ryle Telescope, eight dishes 12.8 m/42 ft wide in a line 5 km/3 mi long	1972	University of Cambridge, UK
Multiple Mirror Telescope	Mt Hopkins (AZ), USA	6.5-m/21.3-ft honeycomb-back optical telescope conversion	1979	University of Arizona and the Smithsonian Institute
		6 mirrors of 1.8-m/72-in aperture, which perform as a single 4.5 m/176 in mirror	1996	
New Technology Telescope	La Silla, Chile	optical telescope that forms part of the European Southern Observatory; it has a thin, lightweight mirror, 3.38 m/141 in across with active optics, which is kept in shape by computer-adjustable supports	1991	operated jointly by Belgium, Denmark, France, Germany, Italy, the Netherlands, Sweden, and Switzerland with headquarters near Munich
Royal Greenwich Observatory (RGO)	Cambridge, UK	operates Isaac Newton Group of telescopes, including:	founded 1675 (moved to Cambridge in 1990)	RGO
	La Palma, Canary Islands	4.2-m/165-in William Herschel Telescope	1987	Anglo-Australian Observatory
Siding Spring Mountain	400 km/250 mi northwest of Sydney, Australia	1.2-m/48-in UK Schmidt Telescope	1973	
		3.9-m/154-in Anglo-Australian Telescope	1975	
South African Astronomical Observatory	Sutherland, South Africa	main telescope is a 1.88-m/74-in reflector	founded 1973	Council for Scientific and Industrial Research of South Africa
Subaru ('Pleiades')	Mauna Kea (HI), USA	8-m/26.4-ft optical-infrared telescope	1999	National Astronomical Observatory (NAO) of the University of Tokyo
United Kingdom Infrared Telescope (UKIRT)	Mauna Kea (HI), USA	3.8-m/150-in reflecting telescope for observing at infrared wavelengths	1979	Royal Observatory, Edinburgh
US Naval Observatory	Washington DC, USA	several telescopes including: 66-cm/26-in refracting telescope	1873	US Naval Observatory
	Flagstaff (AZ), USA	1.55-m/61-in reflector for measuring positions of celestial objects	1964	
Very Large Array (VLA)	Plains of San Augustine (NM), USA	largest and most complex single-site radio telescope in the world, comprising 27 dish antennae, each 25 m/82 ft in diameter, forming a Y-shaped array	1981	NRAO
Very Large Telescope (VLT)	Cerro Paranal, Chile	4 × 8 m/26 ft optical array	1999	European Southern Observatory consisting of eight European countries

Observatory/ Telescope	Location	Description	Year opened	Run by
Very Long Baseline Array (VLBA)	St Croix (VI); Hancock (NH); North Liberty (IA); Fort davis (TX); Los Alamos (NM); Pie Town (NM); Kitt Peak (AZ); Owens Valley (CA); Brewster (WA); Mauna Kea (HI), USA	system of 10 radio telescopes, each a 25-m/82-ft diameter dish antenna, controlled remotely from the Array Operations Center in Socorro, New Mexico, that work together as the world's largest dedicated, full-time astronomical instrument		NRAO
Yerkes Observatory	Wisconsin, USA	houses the world's largest refracting optical telescope, with a lens of diameter 102 cm/40 in	observatory founded 1897	University of Chicago Department of Astronomy and Astrophysics
Zelenchukskaya	Caucasus Mountains of Russia	site of the world's largest single mirror optical telescope, with a mirror of 6 m/236 in diameter Radio Astronomy Telescope of the Russian Academy of Sciences (RATAN) 600 radio telescope, consisting of radio reflectors in a circle 600 m/2,000 ft in diameter	1976	Special Astrophysical Observatory of the Russian Academy of Sciences (RAS) in St Petersburg

of a black hole 300 million times the mass of the Sun. It is located in the middle of galaxy M84 about 50 million light-years from Earth. Further findings in December 1997 concerned different shapes of dying stars. Previously, astronomers had thought that most stars die with a round shell of burning gas expanding into space. The photographs taken by the HST show shapes such as pinwheels and jet exhaust.

Space Telescope Electronic Information Service
http://www.stsci.edu/
Home page of the Hubble Space Telescope, which includes an archive of past observations, a description of the instruments aboard, and a section for educators, students, and the general public – with pictures, audio clips, and press releases.

---

radio telescopes

This type of telescope is used in **radio astronomy** to detect radio waves emitted naturally by objects in space. Radio emission comes from hot gases (**thermal radiation**); electrons spiralling in magnetic fields (**synchrotron radiation**); and specific wavelengths (**lines**) emitted by atoms and molecules in space, such as the 21-cm/8-in line emitted by hydrogen gas. Radio telescopes usually consist of a metal bowl that collects and focuses radio waves the way a concave mirror collects and focuses light waves. Radio telescopes are much larger than optical telescopes, because the wavelengths they are detecting are much longer than the wavelength of light. The largest single dish is 305 m/1,000 ft across, at Arecibo, Puerto Rico.

A large dish such as that at Jodrell Bank, England, can see the radio sky less clearly than a small optical telescope sees the visible sky. **Interferometry** is a technique in which the output from two dishes is combined to give better resolution of detail than with a single dish. **Very long baseline interferometry** (VBLI) uses radio telescopes spread across the world to resolve minute details of radio sources.

In **aperture synthesis**, several dishes are linked together to simulate the performance of a very large single dish. This technique was pioneered by Martin Ryle at Cambridge, England, site of a radio telescope consisting of eight dishes in a line 5 km/3 mi long. The Very Large Array in New Mexico, USA, consists of 27 dishes arranged in a Y-shape, which simulates the performance of a single dish 27 km/17 mi in diameter. Other radio telescopes are shaped like long troughs, and some consist of simple rod-shaped aerials.

Radio astronomy has greatly improved our understanding of the evolution of stars, the structure of galaxies, and the origin of the universe. Astronomers have mapped the spiral structure of the Milky Way from the radio waves given out by interstellar gas, and they have detected many individual radio sources within our galaxy and beyond. Radio astronomy is also used in the search for extraterrestrial intelligence, which has been ongoing since 1983.

the exploration of space

rockets

Rockets are used as the means of propelling satellites into geostationary orbit and spacecraft out of the Earth's atmosphere into and through space. To escape from Earth's gravity rockets must reach an **escape velocity** of 11.2 kps/6.9 mps. They are driven by the reaction of gases produced by a fast-burning fuel, such as liquid hydrogen and kerosene. Unlike jet engines, which are also reaction engines, modern rockets carry their own oxygen supply (usually in liquid form) to burn their fuel and do not require any surrounding atmosphere. Being the only form of propulsion available that can function in a vacuum, rockets are essential to exploration in outer space. Multistage rockets have to be used, consisting of a number of rockets joined together.

One of the largest rockets ever built, the *Saturn V* Moon rocket, was a three-stage design, standing 111 m/365 ft high, as tall as a 30-storey skyscraper. It weighed more than 2,700 tonnes on the launch pad, developed a takeoff thrust of some 3.4 million kg/7.5 million lb, and could place almost 140 tonnes into low Earth orbit.

satellites

Artificial satellites, used for scientific purposes, communications, weather forecasting, and military applications, have been rocket-launched into orbit around the Earth since the late 1950s. The first, *Sputnik 1*, was launched by the USSR in 1957. Most satellites are in **geostationary orbit**, following a circular path 35,900 km/22,300 mi above the Earth's Equator and taking 24 hours, moving from west to east, to complete an orbit, thus appearing to hang stationary over one place on the Earth's surface. The brightest artificial satellites can be seen by the naked eye.

At any time, there are several thousand artificial satellites orbiting the Earth, including active satellites, satellites that have ended their working lives, and discarded sections of rockets. Artificial satellites eventually re-enter the Earth's atmosphere. Usually they burn up by friction, but sometimes debris falls to the Earth's

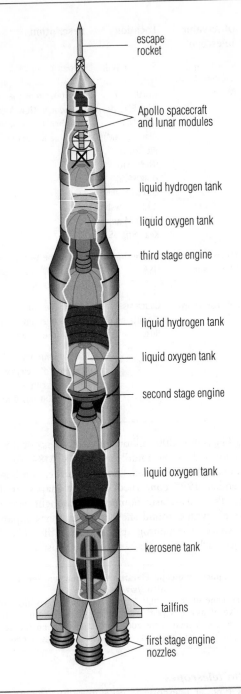

escape rocket

Apollo spacecraft and lunar modules

liquid hydrogen tank

liquid oxygen tank

third stage engine

liquid hydrogen tank

liquid oxygen tank

second stage engine

liquid oxygen tank

kerosene tank

tailfins

first stage engine nozzles

rocket The three-stage *Saturn V* rocket used in the Apollo moonshots of the 1960s and 1970s. It developed a power equivalent to 50 Boeing 747 jumbo jets.

surface, as with *Skylab* and *Salyut 7*. In 1997 there were 300 active artificial satellites in orbit around Earth, the majority used in communications.

J-Track Satellite Tracker
http://liftoff.msfc.nasa.gov/RealTime/
JTrack/Spacecraft.html
Real-time tracking system that displays on a world map the
current position and orbit information for the space shuttle,
Mir, Hubble Space Telescope, and the UARS and COBE
satellites.

---

space programmes

Mercury project US project to put a human in space in the one-seat Mercury spacecraft 1961–63. The first two Mercury flights, on Redstone rockets, were short flights to the edge of space and back. The orbital flights, beginning with the third in the series (made by John Glenn), were launched by Atlas rockets.

Project Mercury
http://www.ksc.nasa.gov/history/mercury/ mercury.html
Official NASA archive of the programme that led to the first
crewed US space flight. There are comprehensive details
(technical and of general interest) on all the crewed and
uncrewed flights included in the project.

Gemini project US space programme 1965–66 in which astronauts practised rendezvous and docking of spacecraft, and working outside their spacecraft, in preparation for the Apollo Moon landings. Gemini spacecraft carried two astronauts and were launched by Titan rockets.

Project Gemini
http://www.ksc.nasa.gov/history/gemini/gemini.html
Official NASA archive of the project that paved the way for the
first Moon landing. There are comprehensive details (technical
and of general interest) on all the missions included in the
project.

Apollo project US space project to land a person on the Moon, achieved on 20 July 1969, when Neil Armstrong was the first to set foot there. He was accompanied on the Moon's surface by Buzz Aldrin; Michael Collins remained in the orbiting command module.

The programme was announced in 1961 by President Kennedy. The world's most powerful rocket, *Saturn V*, was built to launch the Apollo spacecraft, which carried three astronauts. When the spacecraft was in orbit around the Moon, two astronauts would descend to the surface in a lunar module to take samples of rock and set up experiments that would send data back to Earth. After three other preparatory

flights, *Apollo 11* made the first lunar landing. Five more crewed landings followed.

Apollo 1: during a preliminary check on the ground,the three crew were killed by a fire on 27 January 1967

Apollo 4: launched on 9 November 1967 into an orbit around the Earth; the first time the *Saturn V* rocket was used

Apollo 7: the first Apollo mission carrying a crew, *Apollo 7* was a test flight sent into orbit around the Earth on 11 October 1968

Apollo 8: launched on 21 December 1968; the first rocket to take a crew round the Moon

Apollo 9: launched on 3 March 1969; the lunar module was tested in orbit around the Earth

Apollo 10: launched on 18 May 1969; the lunar module was successfully tested 14.5 km/9 mi above the surface of the Moon

Apollo 11: launched on 16 July 1969; Armstrong and Aldrin landed the lunar module (named 'Eagle') in an area called the Sea of Tranquillity on the Moon's surface on 20 July 1969. Armstrong had to land manually because the automatic navigation system was heading for a field of boulders. On landing, Armstrong announced 'Tranquillity base here. The Eagle has landed.' The module remained on the Moon for 22 hours during which time they collected rocks, set up experiments, and mounted a US flag. Apart from a slight wobble when rejoining the command module, the return flight went without a hitch. After splashdown the astronauts were quarantined as a precaution against unknown illnesses from the Moon

Apollo 12: launched on 14 November 1969; in spite of twice being struck by lightning, another successful Moon landing was achieved.

Apollo 13: intended to be the third Moon landing, *Apollo 13* was launched on 11 April 1970 with the crew of John Swigert, Fred Haise, and James Lovell. On the third day of the mission Swigert reported to Houston 'We seem to have a problem.' An electrical fault had caused an explosion in one of the oxygen tanks, cutting off supplies of power and oxygen to the command module. The planned landing was abandoned and the rocket was sent round the Moon before heading back to Earth. The crew used the lunar module Aquarius as a 'lifeboat', though they had to endure near-freezing temperatures to save power, making sleep almost impossible. Attempting re-entry in the crippled ship almost led to disaster, but the crew splashed down safely on 17 April

Apollo 15: launched on 26 July 1971; the first surface vehicle, 'Lunar Rover', was used

Apollo 17: launched on 7 December 1972, this was the last of the Apollo Moon landings. Detailed geological studies were carried out and large amounts of rock and soil were brought back.

Notable Crewed Space Flights

Launch date	Spacecraft	Crew	Duration	Remarks
12 April 1961	*Vostok* 1	Yuri Gagarin	1 hr 58 min	first man in space; Gagarin landed separately from the spacecraft after ejecting at 1 hr 48 min, in a procedure followed by all *Vostok* pilots
5 May 1961	*Freedom* 7	Alan Shepard	15 min 28 sec	first American in space; suborbital flight ended in planned splashdown
6 August 1961	*Vostok* 2	Gherman Titov	1 day 1 hr 18 min	at 25, Titov was the youngest person in space; he was spacesick and the first to sleep in space
20 February 1962	*Friendship* 7	John Glenn	4 hr 55 min	first American to orbit the Earth
11 August 1962	*Vostok* 3	Andrian Nikolyev	3 days 22 hr 25 min	long duration flight
14 June 1963	*Vostok* 5	Valeri Bykovsky	4 days 23 hr 7 min	solo flight record-holder
16 June 1963	*Vostok* 6	Valentina Tereshkova	2 days 22 hr 50 min	first woman in space, flew close to *Vostok* 5
12 October 1964	*Voskhod* 1	Vladimir Komarov, Konstantin Feoktistov, Boris Yegerov	1 day 17 min	first multi-crewed spaceflight without spacesuits or ejection seats, flying in stripped down one-man *Vostok*
18 March 1965	*Voskhod* 2	Pavel Belyayev, Alexei Leonov	1 day 2 hr 2 min	Leonov made first walk in space
25 March 1965	*Gemini* 3	Gus Grissom, John Young	4 hr 52 min	first US two-man flight; Grissom first man to return to space
3 June 1965	*Gemini* 4	James McDivitt, Edward White	4 days 1 hr 56 min	White makes first American spacewalk
4 December 1965	*Gemini* 7	Frank Borman, James Lovell	13 days 18 hr 35 min	acted as rendezvous target for *Gemini* 6; broke endurance record
15 December 1965	*Gemini* 6	Wally Schirra, Tom Stafford	1 day 1 hr 51 min	first rendezvous in space
12 September 1966	*Gemini* 11	Charles Conrad, Richard Gordon	2 days 23 hr 17 min	docking on first orbit; spacewalk; re-boost to altitude of 1368 km/850 mi; automatic landing
11 November 1966	*Gemini* 12	James Lovell, Edwin Aldrin	3 days 22 hr 34 min	record spacewalk of over 2 hr by Aldrin
23 April 1967	*Soyuz* 1	Vladimir Komarov	1 day 2 hr 47 min	Komarov killed when parachute failed after emergency landing; intended to dock with *Soyuz* 2
11 October 1968	*Apollo* 7	Wally Schirra, Donn Eisele, Walt Cunningham	10 days 20 hr 9 min	Earth orbit maiden flight of Apollo Command/service modules
21 December 1968	*Apollo* 8	Frank Borman, James Lovell, William Anders	6 days 3 hr	first crewed craft to orbit the Moon
15 January 1969	*Soyuz* 5	Boris Volynov, Alexei Yeleseyev, Yevgeny Khrunov	3 days 54 min	Yeleseyev and Khrunov spacewalk to *Soyuz* 4 after docking with it
18 May 1969	*Apollo* 10	Tom Stafford, John Young, Eugene Cernan	8 days 3 min	lunar module tested in lunar orbit and flew to 14.5 km/9 mi from the Moon
16 July 1969	*Apollo* 11	Neil Armstrong, Michael Collins, Edwin Aldrin	8 days 3 hr 18 min	first crewed lunar landing; Armstrong and Aldrin walked on the Moon for over 2 hr
14 November 1969	*Apollo* 12	Charles Conrad, Richard Gordon, Alan Bean	10 days 4 hr 36 min 25 sec	pinpoint landing near uncrewed *Surveyor* craft on the Moon; two moonwalks
11 April 1970	*Apollo* 13	James Lovell, Jack Swigert, Fred Haise	5 days 22 hr 54 min	service module exploded 55 hr into mission; crew limped home using lunar module as lifeboat
6 June 1971	*Soyuz* 11	Georgi Dobrovolsky, Vladislav Volkov, Viktor Patsayev	23 days 18 hr 21 min	crew died as craft depressurized before re-entry; they were not wearing spacesuits
26 July 1971	*Apollo* 15	David Scott, Alfred Worden, James Irwin	12 days 7 hr 11 min	use of first lunar rover
16 April 1972	*Apollo* 16	John Young, Ken Mattingly, Charles Duke	11 days 1 hr 51 min	Mattingly in lunar orbit makes longest solo US flight; three moonwalks were taken
25 May 1973	*Skylab* 2	Charles Conrad, Joe Kerwin, Paul Weitz	28 days 49 min	spacewalk to repair severely disabled *Skylab* 1 space station
16 November 1973	*Skylab* 4	Gerry Carr, Edward Gibson, Bill Pogue	84 days 1 hr 15 min	longest US crewed spaceflight until 1996
24 May 1975	*Soyuz* 18	Pyotr Klimuk, Vitali Sevastyanov	62 days 23 hr 20 min	record stay on *Salyut* 4
15 July 1975	*Soyuz* 19	Alexei Leonov, Valeri Kubasov	5 days 22 hr 30 min	docked with *Apollo 18* in joint Apollo-Saturn Test Project (ASTP) mission
5 July 1975	*Apollo* 18	Tom Stafford, Vance Brand, Donald 'Deke' Slayton	9 days 1 hr 28	docked with *Soyuz* 19; first US-Soviet space link-up
10 December 1977	*Soyuz* 26	Yuri Romanenko, Georgi Grechko	96 days 10 hr	aboard *Salyut* 6; broke endurance record
9 April 1980	*Soyuz* 35	Leonid Popov, Valeri Ryumin	184 days 20 hr 11 min	record *Salyut* 6 mission, with Ryumin achieving 361 days' space experience
12 April 1981	*Columbia* STS 1	John Young, Bob Crippen	2 days 6 hr 20 min	maiden flight of space shuttle
13 May 1982	*Soyuz* T5	Anatoli Berezevoi, Valentin Lebedev	211 days 9 hr 4 min	first record-breaking visit to *Salyut* 7
27 June 1982	*Columbia* STS 4	Ken Mattingly, Hank Hartsfield	7 days 1 hr 9 min	military flight; final shuttle test flight
11 November 1982	*Columbia* STS 5	Vance Brand, Robert Overmyer, Joe Allen, William Lenoir	5 days 2 hr 14 min	first commercial mission of Shuttle; deployed two communications satellites; first four-person flight
18 June 1983	*Challenger* STS 7	Bob Crippen, Rick Hauck, John Fabian, Sally Ride, Norman Thagard	6 days 2 hr 24 min	satellite deployment mission; first with five-person crew and first US woman in space

Launch date	Spacecraft	Crew	Duration	Remarks
28 November 1983	*Columbia* STS 9	John Young, Brewster Shaw, Owen Garriott, Robert Parker, Byron Lichtenberg, Ulf Merbold	10 days 7 hr 47 min	flight of European *Spacelab* 1; Merbold from West Germany; first six-person flight
3 February 1984	*Challenger* STS 41B	Vance Brand, Robert Gibson, Bruce McCandless, Robert Stewart, Ronald McNair	7 days 23 hr 15 min	first independent spacewalk using Manned mission Manoeuvring Unit (MMU) by McCandless; first space mission to end at launch site (Kennedy/Canaveral);
8 February 1984	*Soyuz* T10	Leonid Kizim, Vladimir Solovyov, Oleg Atkov	236 days 22 hr 9 min	longest crewed space mission to date; Kizim and Solovyov made record six spacewalks
6 April 1984	*Challenger* STS 41C	Bob Crippen, Dick Scobee, George Nelson, Terry Hart, James van Hoften	6 days 23 hr 40 min	repaired Solar Max; with *Soyuz* T10 and T11 crews in space; 11 people in space at once
17 July 1984	*Soyuz* T12	Vladimir Dhzanibekov, Svetlana Savitskaya, Oleg Volk	11 days 19 hr 14 min	Savitskaya became first woman space-walker, outside *Salyut* 7
17 September 1985	*Soyuz* T14	Vladimir Vasyutin, Georgi Grechko, Alexander Volkov	64 days 21 hr 52 min	mission cut short after Vasyutin suffered depression and anxiety
28 January 1986	*Challenger* STS 51	Dick Scobee, Mike Smith, Judith Resnik, Ronald McNair, Ellison Onizuka, Christa McAuliffe, Gregory Jarvis	73 sec	broke apart at 14,325 m/47,000 ft; crew killed; first flight to take off but not to reach space; first American in-flight fatalities
13 March 1986	*Soyuz* T15	Leonid Kizim, Vladimir Solovyov	125 days 1 min	first mission to new space station *Mir* 1; also docked with *Salyut* 7; Kizim achieved over a year in space experience
21 December 1987	*Soyuz* TM4	Vladimir Titov, Musa Manarov, Anatoli Levchenko	365 days 22 hr 39 min	longest duration mission by Titov and Manarov
26 November 1988	*Soyuz* TM7	Alexander Volkov, Sergei Krikalev, Jean-Loup Chretien	151 days 11 hr 10 min	visit to *Mir*; Frenchman Chretien is first non-US, non-USSR astronaut to make two space flights, to make spacewalk, and the oldest spacewalker at 50; Chretien returned in TM6 after 25 days
4 May 1989	*Atlantis* STS 30	David Walker, Ron Grabe, Norman Thagard, Mary Cleave, Mark Lee	4 days 57 min	deployed *Magellan* for its journey to orbit the planet Venus; first deployment of a planetary spacecraft from a crewed spacecraft
18 October 1989	*Atlantis* STS 34	Donald Williams, Michael McCulley, Shannon Lucid, Franklin Chang-Diaz, Ellen Baker	4 days 23 hr 39 min	deployed Jupiter orbiter *Galileo*
24 April 1990	*Discovery* STS 31	Loren Shriver, Charles Bolden, Steven Hawley, Bruce McCandless, Kathryn Sullivan	5 days 1 hr 16 min 532 km/319 mi	deployed Hubble Space Telescope; reached record Shuttle altitude
18 May 1991	*Soyuz* TM12	Anatoli Artsebarski, Sergei Krikalev, Helen Sharman	144 days 15 hr 22 min	Sharman first non-Soviet, non-US woman and first Briton in space, returned in TM11 after seven days; Krikalev stayed aboard *Mir* and returned after 311 days; Artsebarski and Krikalev made a record six spacewalks in 33 days
22 January 1992	*Discovery* STS 42	Ronald Grabe, Stephen Oswald, Norman Thagard, David Hilmers, William Readdy, Roberta Bondar, Ulf Merbold	8 days 1 hr 14 min	International Microgravity Laboratory mission; Bondar from Canada and Merbold from Germany; Thagard achieves record 25-day Shuttle flight time on fourth mission
7 May 1992	*Endeavour* STS 49	Dan Brandenstein, Kevin Chilton, Rick Heib, Bruce Melnick, Pierre Thuot, Kathryn Thornton, Tom Akers	8 days 21 hr 17 min	retrieved *Intelsat* 6 and reboosted it into geostationary orbit; record breaking 8 hr 29 min extra-vehicular activity (EVA) by Thuot, Hieb, and Akers
31 July 1992	*Atlantis* STS 46	Loren Shriver, Andrew Allen, Claude Nicollier, Marsha Ivins, Jeff Hoffman, Franklin Chang-Diaz, Franco Malerba	7 days 23 hr 15 min	deployed *Eureca* and tethered satellites; Nicollier first Swiss astronaut and first non-US NASA mission specialist; Malerba first Italian in space; 12 people in space at once, with record five space nations being represented, with *Mir* mission
2 December 1993	*Endeavour* STS 61	Richard Covey, Ken Bowersox, Claude Nicollier, Story Musgrave, Jeff Hoffman, Tom Akers, Kathryn Thornton	10 days 19 hr 58 min	Hubble Space Telescope servicing and repair mission; Musgrave, first to fly five Shuttle missions, achieves record 35 days' flight time
8 January 1994	*Soyuz* TM18	Viktor Afanasyev, Yuri Usachev, Valeri Poliakov	182 days 27 min	new residency aboard *Mir* 1 space station; Poliakov remained on *Mir* and landed on 22 March 1995 with a flight time of 437 days
3 February 1994	*Discovery* STS 60	Charles Bolden, Kenneth Reightler, Franklin Chang-Diaz, Jan Davis, Ron Sega, Sergei Krikalev	8 days 7 hr 9 min	Krikalev first Russian to fly on US rocket
3 October 1994	*Soyuz* TM20	Alexander Viktorenko, Yelena Kondakova, Ulf Merbold	169 days 5 hr 21 min	new crew to *Mir*, including Kondakova, the first woman to make a long duration flight, and German European Space Agency (ESA) visitor Merbold, the first non-US, non-Soviet spaceperson to make three flights and first Western European to fly both US and Russian rockets; with Shuttle in space, 12 people in orbit at once; Merbold landed in TM19 after 31 days

Launch date	Spacecraft	Crew	Duration	Remarks
3 February 1995	*Discovery* STS 63	James Wetherbee, Eileen Collins, Michael Foale, Bernard Harris, Janice Ford, Vladimir Titov	8 days 6 hr 28 min	*Spacelab* science rendezvous mission with *Mir* space station and spacewalk – first by British-born astronaut, Foale; Collins first female Shuttle pilot; Titov from Russia
14 March 1995	*Soyuz* TM21	Vladimir Dezhurov, Gennady Strekalov, Norman Thagard	115 days 8 hr 44 min	mission to *Mir 1* with first US astronaut to ride a Russian rocket; record 13 people in space at same time on 14–18 March; crew landed in STS 71
27 June 1995	*Atlantis* STS 71	Robert Gibson, Charles Precourt, Ellen Baker, Bonnie Dundar, Gregory Harbaugh, Anatoli Solovyov, Nikolai Budarin	9 days 19 hr 23 min	100th US crewed flight including Thagard's *Soyuz* to TM21 launch; Shuttle/*Mir 1* mission 1; 5 days joined *Mir*; delivered Solovyov and Budarin and returned with the TM21 crew; first time ten people on board one spacecraft (223 tonnes) in orbit
12 November 1995	*Atlantis* STS 74	Ken Cameron, James Halsall, Jerry Ross, Bill McArthur, Chris Hadfield	8 days 4 hr 30 min	Shuttle/*Mir* mission 2; carried docking module to be left at Mir; Hadfield NASA mission specialist from Canada
22 February 1996	*Columbia* STS 75	Andrew Allen, Scott Horowitz, Maurizio Cheli, Claude Nicollier, Jeff Hoffman, Franklin Chang-Diaz, Umberto Guidoni	15 days 17 hr 40 min	tethered satellite system reflight, satellite lost when tether broke; 12 people, five nations (two from Italy) in space with TM22 and TM23 crews also orbiting
22 March 1996	*Atlantis* STS 76	Kevin Chilton, Richard Searfoss, Ronald Sega, Ric hr Clifford, Linda Godwin, Shannon Lucid	9 days 5 hr 15 min	Shuttle/*Mir* mission 3, delivered Shannon Lucid for extended stay on *Mir*; returned 26 September aboard STS 79; after stay of 188 days, a record for a woman; spacewalk
17 August 1996	*Soyuz* TM24	Valeri Korzun, Alexander Kaleri, Claudie Andre-Deshays	196 days 16 hr 26 min	new crew for *Mir 1* with Deshays, the first French woman in space, as commercial crew-person on 15-day flight, landing in TM23; Korzun and Kaleri first back-up crew to fly since *Soyuz 11* in 1971 after prime commander Gennady Manakov hospitalized (if one crew member unable to fly, back-up crew takes over) with heart attack
12 January 1997	*Atlantis* STS 81	Mike Baker, Brent Jett, John Grunsfield, Jeff Wisoff, Marsha Ivins, Jerry Linenger	10 days 4 hr 55 min	Shuttle/*Mir* mission 5, delivered Jerry Linenger and returned John Blaha from *Mir* after 128 days; Lineger made first US-Russian spacewalk with Tsiblyev, wearing Russian spacesuit
10 February 1997	*Soyuz* TM25	Vasili Tsiblyev, Alexander Lazutkin, Reinhold Ewald	184 days 22 hr 7 min	new crew for *Mir* with German Ewald flying shorter commercial mission; this crew experienced a fire on the space station and a collision with the *Progress* M34 supply ship
11 February 1997	*Discovery* STS 82	Ken Bowersox, Scott Horowitz, Steven Hawley, Mark Lee, Joe Tanner, Greg Harbaugh, Steve Smith	9 days 23 hr 37 min	second mission to service the Hubble Space Telescope; featured five spacewalks
5 August 1997	*Soyuz* TM26	Anatoli Solovyov, Pavel Vinogradev,	9 days 23 hr 37 min	200th launched crewed spaceflight in history; new crew to *Mir* to carry out major repair work

---

Apollo 11
http://www.nasa.gov/hqpao/apollo_11.html
This NASA page relives the excitement of the *Apollo 11* mission, with recollections from the participating astronauts, images, audio clips, access to key White House documents, and a bibliography.

Soyuz project The Soviet Soyuz series of spacecraft, capable of carrying up to three cosmonauts, was launched in 1967. Soyuz spacecraft consist of three parts: a rear section containing engines; the central crew compartment; and a forward compartment that gives additional room for working and living space. They are now used for ferrying crews up to space stations, though they were originally used for independent space flight.

Soyuz 1 crashed on its first flight in April 1967, killing the lone pilot, Vladimir Komarov. *Soyuz 11* had three deaths on re-entry in 1971. In 1975 the Apollo–Soyuz test project resulted in a successful docking in orbit.

space shuttles The first space shuttle, a reusable crewed spacecraft, was launched on 12 April 1981 by the USA. It was developed by NASA to reduce the cost of using space for commercial, scientific, and military purposes.

The space-shuttle orbiter, the part that goes into space, is 37.2 m/122 ft long and weighs 68 tonnes. Two to eight crew members occupy the orbiter's nose section, and missions last up to 30 days. In its cargo bay the orbiter can carry up to 29 tonnes of satellites, scientific equipment, Spacelab (see below), or military payloads. After leaving its payload in space, the orbiter can be flown back to Earth to land on a runway, and is then available for reuse.

Four orbiters were built: *Columbia*, *Challenger*, *Discovery*, and *Atlantis*. *Challenger* was destroyed in a midair explosion just over a minute after its tenth launch on 28 January 1986, killing all seven crew members, the result of a failure in one of the solid rocket boosters. Flights resumed with redesigned boosters in September 1988. A replacement orbiter, *Endeavour*, was built, which had its maiden flight in May 1992.

The USSR produced a shuttle of similar size and

Notable Uncrewed Space Flights

Date	Remarks	Date	Remarks
September 1959	first craft to reach the Moon, by USSR's *Luna 2*	January 1958	first US satellite, first science satellite, Explorer 1
October 1959	first pictures of lunar far side from *Luna 3*	December 1958	first experimental communications satellite, Score 1
December 1962	first fly-by of Venus by US *Mariner 2*	April 1959	first military spy satellite, Discoverer 2
July 1964	first close-up images of the Moon from *Ranger 7*	April 1960	first weather satellite, Tiros 1
July 1965	first successful fly-by of Mars with first images of the planet from *Mariner 4*		first navigation satellite, Transit 1B
January 1966	first 'soft' lunar landing and surface images from *Luna 9*	August 1960	first recovery of craft from orbit, Discoverer 13
April 1966	first lunar orbiter, *Luna 10*		first recovery of living creatures (two dogs) from orbit, Sputnik 5
October 1967	first successful exploration of Venus's atmosphere by *Venera 4*	July 1962	first commercial communications satellite, Telstar 1
September 1968	first spacecraft to fly round the Moon and return to Earth, *Zond 5*	July 1963	first geostationary orbiting communications satellite, Syncom 2
September 1970	first return of sample from Moon by unmanned craft, *Luna 16*	April 1966	first astronomical satellite, OAO 1
November 1970	first lunar rover, *Lunakhod 1*	December 1966	first French satellite launch, A1
December 1970	first successful Venus landing by *Venera 7*	October 1967	first automatic, unmanned docking in orbit, Cosmos 186 and 188
November 1971	first Mars orbiter, *Mariner 9*	February 1970	first Japanese satellite launch, Ohsumi
December 1973	first craft to explore Jupiter, *Pioneer 10*	April 1970	first Chinese satellite launch, Tungfanghung
March 1974	first fly-by of Mercury by *Mariner 10*		
October 1975	first craft to orbit Venus and return surface pictures, *Venera 9*	October 1971	first British satellite launch, Prospero
July 1976	first soft landing on Mars by *Viking 1*	July 1972	first Earth resources remote sensing satellite, Landsat 1
September 1979	first exploration of Saturn by *Pioneer 11*	February 1976	first maritime mobile communications satellite, Marisat 1
October 1983	first mapping of Venus by radar, *Venera 15*	July 1980	first Indian satellite launch, Rohini
January 1986	first exploration of Uranus, *Voyager 2*	June 1981	first European operational satellite launch by Ariane, Meteosat 2
March 1986	first encounter with coma of comet (Halley) by *Giotto*	April 1984	first satellite to be captured, repaired, and redeployed, SMM 1
August 1989	first exploration of Neptune by *Voyager 2*	May 1984	first fully commercial satellite launch by Arianespace, Spacenet1
October 1991	first fly-by of a comet, Gaspra, by *Galileo*	November 1985	first satellite capture and return to Earth, Palapa and Westar
December 1995	first craft to enter Jupiter's atmosphere, *Galileo* probe	February 1986	first privately operated, commercial, remote-sensing craft, Spot 1
	first Jupiter orbiter, *Galileo*	September 1988	first Israeli satellite launch, Ofeq 1
October 1957	first artificial Earth satellite, Soviet Union's Sputnik 1	December 1988	first privately operated, commercial TV satellite, Astra 1A
November 1957	first living being (dog, Laika) in orbit aboard Sputnik 2	April 1990	first optical telescope in orbit, Hubble Space Telescope

appearance to the US one. The first Soviet shuttle, *Buran*, was launched without a crew by the Energiya rocket on 15 November 1988. In Japan, development of a crewless shuttle began in 1986.

Spacelab Spacelab is a small space station built by the European Space Agency, carried in the cargo bay of the US space shuttle, in which it remains throughout each flight, returning to Earth with the shuttle.

Spacelab consists of a pressurized module in which astronauts can work, and a series of pallets, open to the vacuum of space, on which equipment is mounted.

Spacelab is used for astronomy, Earth observation, and experiments utilizing the conditions of weightlessness and vacuum in orbit. The pressurized module can be flown with or without pallets, or the pallets can be flown on their own, in which case the astronauts remain in the shuttle's crew compartment. All the sections of Spacelab can be reused many times. The first Spacelab mission, consisting of a pressurized module and pallets, lasted ten days November–December 1983.

space probes

Space probes are uncrewed, instrumented objects sent beyond Earth to collect data from other parts of the Solar System and from deep space. The first probe was the Soviet *Lunik 1*, which flew past the Moon in 1959. The first successful planetary probe was the US *Mariner 2*, which flew past Venus in 1962, using transfer orbit. The first space probe to leave the Solar System was *Pioneer 10* in 1983. Space probes include *Galileo, Giotto, Magellan, Mars Observer, Ulysses*, the Moon probes, and the Mariner, Pioneer, Viking, and Voyager series.

space stations

Space stations are large structures designed for human occupation in space for extended periods of time. They are used for carrying out astronomical observations and surveys of Earth, as well as for biological studies and the processing of materials in weightlessness. The first space station was *Salyut 1*, and the USA has launched *Skylab*.

NASA plans to build a larger space station, to be called *Alpha* in cooperation with other countries, including the European Space Agency, which is building a module called *Columbus*; Russia and Japan are also building modules.

Salyut *Salyut* was cylindrical in shape, 15 m/50 ft long, and weighed 19 tonnes. It housed two or three cosmonauts at a time, for missions lasting up to eight months. Seven *Salyut* space stations were launched by the USSR 1971–82.

Salyut 1 was launched on 19 April 1971. It was occupied for 23 days in June 1971 by a crew of three, who died during their return to Earth when their Soyuz ferry craft depressurized. In 1973 *Salyut 2* broke up in orbit before occupation. The first fully successful Salyut mission was a 14-day visit to *Salyut 3* in July 1974. In 1984–85 a team of three cosmonauts endured a record 237-day flight in *Salyut 7*. In 1986 the *Salyut* series was superseded by *Mir*, an improved design capable of being enlarged by additional modules sent up from Earth. *Salyut 7* crashed to Earth in February 1991, scattering debris in Argentina.

Mir Soviet space station, the core of which was launched on 20 February 1986. It is intended to be a permanently occupied space station. *Mir* weighs almost 21 tonnes, is approximately 13.5 m/44 ft long, and has a maximum diameter of 4.15 m/13.6 ft. It carries a number of improvements over the earlier *Salyut* series of space stations, including six docking ports; four of these can have scientific and technical modules attached to them. The first of these was the *Kvant* (quantum) astrophysics module, launched in 1987. This had two main sections: a main experimental module, and a service module that would be separated in orbit. The experimental module was 5.8 m/19 ft long and had a maximum diameter matching that of *Mir*. When attached to the *Mir* core, *Kvant* added a further 40 cu m/1,413 cu ft of working space to that already there. Among the equipment carried by *Kvant* were several X-ray telescopes and an ultraviolet telescope. In June 1995 the US space shuttle *Atlantis* docked with *Mir*, exchanging crew members. In 1997, *Mir* suffered a series of problems, culminating in its collision with an crewless cargo ship in June.

Skylab

This US space station, launched on 14 May 1973, was made from the adapted upper stage of a *Saturn V* rocket. At 75 tonnes/82.5 tons, it was the heaviest object ever put into space, and was 25.6 m/84 ft long. *Skylab* contained a workshop for carrying out experiments in weightlessness, an observatory for monitoring the Sun, and cameras for photographing the Earth's surface.

Project Skylab
http://www.ksc.nasa.gov/history/skylab/skylab.html
Official NASA archive of the project that launched the USA's first experimental space station. There are comprehensive details (technical and of general interest) on all the experiments included in the project.

Damaged during launch, it had to be repaired by the first crew of astronauts. Three crews, each of three astronauts, occupied *Skylab* for periods of up to 84 days, at that time a record duration for human spaceflight. *Skylab* finally fell to Earth on 11 July 1979, dropping debris on Western Australia.

NASA Home Page
http://www.nasa.gov/
Latest news from NASA, plus the most recent images from the Hubble Space Telescope, answers to questions about NASA resources and the space programme, and a gallery of video, audio clips, and still images.

directory of world space agencies

Argentina Comision Nacional de Investig-aciones Espaciales, Avenida Torrego 4010, Buenos Aires; phone: +54 1 776 2913

Australia Australian Space Office, 40 Allara Street, Canberra, ACT 2601; phone: +61 6 276 1000; fax: +61 6 276 1942)

Austria Austrian Space Agency, Garnisongasse 7, 1090 Wien; phone: +43 1 40 38 177; fax: +43 1 42 82 28

Belgium Science Policy Office, Rue de la Science 8, 1040 Brussels; phone: +32 2 238 3411; fax: +32 2 230 5912

Brazil Instituto de Atividades Espaciales, c/o Secretaria de Assuntos Estratégicos, Palácio do Planalto 4 andar, 70150-900 Brasilia-DF; phone: +55 61 211 1410; fax: +55 61 321 2466)

Canada Canadian Space Agency, 500 René-Lévesque Boulevard West, Montréal, Quebec H2Z 1Z7; phone: +1 514 496 4000; fax: +1 514 496 4039

China Ministry of Aero-Space Industry, 8 Fucheng Road, Haidian District, Beijing 100830; phone: +86 10 683 72 221; fax: +86 10 683 70 849

Denmark Danish Space Board, Bredgade 43, 1260 Copenhagen K; phone: +45 33 929 700; fax: +45 33 323 501

Europe European Space Agency, 8-10 rue Mario-Nikis, 75015 Paris; phone: +33 1 42 73 76 54; fax: +33 1 42 73 75 60

Finland Finnish Space Committee, c/o Technology Development Centre, PO Box 69, 00101 Helsinki; phone: +358 0 693 691; fax: +358 0 694 9196

France Centre National d'Etudes Spatiales, 2 place Maurice Quentin, 75039 Paris; phone: +33 1 45 08 75 00; fax: +33 1 45 08 76 76

Germany Deutsche Argentur für Raumfahrtangelegenheiten, Königswinterer Strasse 522-524, 5300 Bonn 3; phone: +49 228 447 0; fax: +49 228 447 700

India Indian Space Research Organization, Antariksh Bhavan, New BEL Road, Bangalore 560094; phone: +91 80 333 4474; fax: +91 80 333 2253

Indonesia National Institute for Aeronautics and Space, JL Pemuda Persil 1, Jakarta; phone: +62 21 489 4941; fax: +62 21 489 4815

Ireland, Republic of Forbairt Science and Technology Directorate, Glasnevin, Dublin 9; phone: +353 1 808 2000; fax: +353 1 808 2587

Israel Israeli Space Agency, Kiryat Hamenshala, 3rd Building PO Box 18195, Jerusalem 91181; phone: +972 2 584 7096; fax: + 972 2 582 5581

Italy Agenzia Spaziale Italiana, Via di Villa Patrizi 13, 00161 Rome; phone: +39 6 85 679; fax: +39 6 44 04 212

Japan Space Activities Commission, 2-2-1 Kasumigaseki, Chiyoda-ku, Tokyo 100; phone: +81 3 35 81 15 59. National Space Develop-ment Agency, World Trade Center Building, 2-4-1 Hamamatau-cho, Minato-ku, Tokyo 105; fax: +81 3 54 02 79 34

Korea, South Korea Aerospace Research Institute, c/o Korea Advanced Institute of Science and Technology Headquarters, 373-1 Kusong-Dong, Yusong-Gu, Taegu; phone: +82 53 869 2114

Netherlands The Netherlands Institute for Air and Space Development, Kluyverweg 1, PO Box 35, 2600 AA Delft; phone: +31 15 78 80 25; fax: +31 15 62 30 96

Norway Norsk Romesenter, Hoffsveien 65A, PO Box 85, Smestad, 0309 Oslo 3; phone: +47 2 523 800; fax: +47 2 522 397

Pakistan Pakistan Space and Upper Atmosphere Research Commission, Sector 28, Gulaz-e-Hujri, Off University Road, PO Box 8402, Karachi 75270; phone: +92 21 472 630; fax: +92 21 466 902

Russia Russian Space Agency, Shchepkin Street 42, 129857 Moscow; phone: +7 095 971 9176; fax: +7 095 975 6936

Spain Centro para el Desarollo Tecnologico Industrial, Paseo de la Castellana 141, 28046 Madrid; phone: +34 1 581 5500; fax: +34 1 581 5584

Sweden Swedish Board for Space Activities, Rymdstyrelsen, Box 4006, 171 04 Solna; phone: +46 8 627 6480; fax: +46 8 627 5014

Switzerland Federal Space Affairs Commission, Federal Department of Foreign Affairs, 3003 Berne; phone: +41 31 324 1065; fax: +41 31 324 1073

Taiwan National Science Council, No. 106 Ho-Ping East Road, Sector 2, Taipei 10636; phone: +886 2 737 7501; fax: +886 2 737 7668

UK British National Space Centre, Dean Bradley House, 52 Horseferry Road, London SW1P 2AG; phone: +44 171 276 2688; fax: +44 171 276 2377

USA National Aeronautics and Space Administration, Headquarters, Washington, DC 20546; phone: +1 202 358 0000; fax: +1 202 358 0037

Astronomy Chronology

c. 1300 BC The Egyptians have identified 43 constellations and are familiar with those planets visible to the naked eye: Mercury, Venus, Mars, Jupiter, and Saturn.

c. 366 BC Greek mathematician and astronomer Eudoxus of Cnidus builds an observatory and constructs a model of 27 nested spheres to give the first systematic explanation of the motion of the Sun, Moon, and planets around the Earth.

c. 350 BC Aristotle defends the doctrine that the Earth is a sphere, in *De caelo/Concerning the Heavens*, and estimates its circumference to be about 400,000 stadia (one stadium varied from 154 m/505 ft to 215 m/705 ft). It is the first scientific attempt to estimate the circumference of the Earth.

c. 280 BC Greek astronomer Aristarchus of Samos writes *On the Size and Distances of the Sun and the Moon*. He is the first to maintain that the Earth rotates and revolves around the Sun.

c. 200 BC The Greeks invent the astrolabe – the first scientific instrument. It is used for observing the positions and altitudes of stars.

129 BC Greek scientist Hipparchus of Bithynia completes the first known star catalogue. It gives the latitude and longitude and brightness of nearly 850 stars and is later used by Ptolemy.

150 Greek astronomer Ptolemy publishes the work known as the *Almagest*, a highly influential astronomical textbook that outlines a theory of a geocentric (Earth-centred) universe based on years of observations.

772 Muslim astronomer Al-Fazari translates the Indian astronomical compendium *Mahasiddhanta/Treatise on Astronomy*, and begins the establishment of a uniquely Arabic astronomy.

c. 970 The Muslim astronomer Abu al-Wafa' invents the wall quadrant for the accurate measurement of the declination of stars in the sky.

1150 The Hebrew scholar Solomon Jarchus produces the first almanac, a table for the calculation of celestial movements, and a calendar.

1424 Mongolian ruler and astronomer Ulugh Beg, Prince of Samarkand, builds a great observatory, including a 40 m/132 ft sextant, which enables extremely accurate measurements to be made, cataloguing over 1,000 stars.

1543 Copernicus publishes *De revolutionibus orbium coelestium/On the Revolutions of the Celestial Sphere*, detailing his theory that the Earth and other planets orbit the Sun on circular paths.

1609 Italian astronomer Galileo Galilei, having obtained a Dutch telescope, makes his own instruments, including one that magnifies objects 32 times. They are the first telescopes that can be used for astronomical observation.

1613 Convinced by his telescopic observations of the Solar System, Galileo promotes the heliocentric system devised by Polish astronomer Copernicus.

1647 German astronomer Johannes Hevelius first charts the lunar surface accurately in his *Selenographia/Moon Map*.

1668 English physicist and mathematician Isaac Newton constructs the first reflecting telescope, using a series of mirrors instead of lenses to bring light to a focus, and creating a clearer, brighter image.

Aug 1675 The Royal Greenwich Observatory is established by British king Charles II on the outskirts of London, England. English astronomer John Flamsteed is appointed Astronomical Observator (later Astronomer Royal).

1705 English astronomer Edmund Halley conjectures that a comet seen in 1682 was identical with comets observed in 1607, 1531, and earlier; he correctly predicts its return in 1758.

13 March 1781 German-born English astronomer William Herschel discovers the planet Uranus.

1798 French astronomer Pierre-Simon Laplace predicts the existence of black holes.

1 Jan 1801 The Italian astronomer Giuseppe Piazzi discovers the first asteroid, Ceres.

1819 German astronomer Johann Encke discovers the short-period comet (Encke's comet), which returns every 3.29 years.

11 Oct 1838 Using the method of parallax, German astronomer Friedrich Bessel calculates the star 61 Cygni to be 10.3 light years away from Earth. It is the first determination of the distance of a star other than the Sun.

1840 Harvard College professor William Cranch Bond erects the first astronomical observatory in the USA.

1840 US astronomer John William Draper takes the first photograph of the Moon.

1842 Bessel accurately explains that the wavy course of Sirius is due to the existence of a companion star – the first binary star to be discovered.

23 Sept 1846 German astronomer Johann Gotfried Galle discovers the planet Neptune on the basis of French astronomer Urbain Leverrier's calculations of its position.

1865 Maria Mitchell of Massachusetts becomes the first woman professor of astronomy when she receives an appointment to Vassar College in Poughkeepsie, New York.

1905–1907 Danish astronomer Ejnar Hertzsprung discovers that there is a relationship between the colour and absolute brightness of stars and classifies them according to this relationship. The relationship is used to determine the distances of stars and forms the basis of theories of stellar evolution.

1914	British astronomer John Franklin publishes the Franklin–Adams Charts, the first photographic star charts of the entire sky.
1917	Dutch astronomer Willem de Sitter shows that Einstein's theory of general relativity implies that the universe must be expanding.
1917	The 2.5-m/100-in Hooker reflecting telescope is installed at Mount Wilson Observatory, California. It is the world's largest reflecting telescope to date.
1919	The International Astronomical Union (IAU) is founded to promote international cooperation in astronomy.
1923	German mathematician Hermann Oberth is the first to provide the mathematics of how to achieve escape velocity.
1924	US astronomer Edwin Hubble demonstrates that certain Cepheid variable stars are several hundred thousand light years away and thus outside the Milky Way Galaxy. The nebulae they are found in are the first galaxies to be discovered that are proved to be independent of the Milky Way.
1927	Belgian astronomer Georges Lemaître proposes that the universe was created by an explosion of energy and matter from a 'primaeval atom' – the beginning of the Big Bang theory.
c. 1928	English mathematician and physicist James Hopwood Jeans proposes the steady-state hypothesis, which states that the universe is constantly expanding and maintaining a constant average density through the continuous creation of new matter.
1929	Edwin Hubble publishes Hubble's Law, which states that the ratio of the speed of a galaxy to its distance from Earth is a constant (now known as Hubble's constant).
18 Feb 1930	US astronomer Clyde Tombaugh, at the Lowell Observatory, Arizona, discovers the ninth planet, Pluto.
1931	US engineer Karl Jansky discovers that the interference in telephone communications is caused by radio emissions from the Milky Way. He thus begins the development of radio astronomy.
1937	US astronomer Grote Reber builds the first radio telescope. It has a parabolic reflector 9.4 m/31 ft in diameter and begins service in Wheaton, Illinois.
1942	Reber makes the first radio maps of the sky, locating individual radio sources.
1946	Cygnus A, the first radio galaxy, and the most powerful cosmic source of radio waves, is discovered.
1947	The first recorded sighting of an unidentified flying object (UFO) is made in the sky over Kansas, USA, by Kenneth Arnold.
3 June 1948	The 5-m/200-in Hale reflector telescope is installed at Mount Palomar Observatory, California; it remains the world's largest and most powerful telescope until 1974.
1948	Austrian-born British mathematician Hermann Bondi and Austrian astronomer Thomas Gold publish *The Steady State Theory of the Expanding Universe* in which they argue that the universe is constantly expanding, but maintaining a constant density through the continual creation of new stars and galaxies at a rate equal to the rate at which old ones become unobservable because of their increasing distance.
1950	Cape Canaveral, Florida, is established as a rocket assembly and launching facility.
1951	US astronomer Gerard Kuiper proposes the existence of a ring of small, icy bodies orbiting the Sun beyond Pluto, thought to be the source of comets. It is discovered in the 1990s and named the Kuiper belt.
1953–1956	The British Royal Observatory is moved from Greenwich, London, to Herstmonceux, Sussex.
1955	English radio astronomer Martin Ryle builds the first radio interferometer. Consisting of three antennae spaced 1.6 km/1 mi apart, it increases the resolution of radio telescopes, permitting the diameter of a radio source to be determined, or two closely spaced sources to be separated.
19 Aug 1957	US astronomers, using a 33 cm/12 in telescope on board the uncrewed balloon-telescope *STRATOSCOPE I*, take the first clear photographs of the Sun from 24,384 m/80,000 ft.
4 Oct 1957	The USSR launches the first artificial satellite, *Sputnik 1*, to study the cosmosphere. It weighs 84 kg/184 lb and circles the Earth in 95 minutes, inaugurating the space age.
3 Nov 1957 –13 April 1958	The Soviet spacecraft *Sputnik 2* is placed in orbit carrying a dog, Laika. It is the first vehicle to carry a living organism into orbit. Laika dies in space.
1957	The Jodrell Bank observatory, located near Manchester, England, begins operating. The first large radio telescope, it has a 76-m/250-ft diameter reflector, which can be rotated horizontally at 20° per minute and vertically at 24° per minute.
31 Jan 1958	The US Army launches the first US satellite, *Explorer 1*, into Earth orbit. It is used to study cosmic rays.
15 May 1958	The USSR places *Sputnik 3* in orbit. It contains the first multipurpose space laboratory and transmits data about cosmic rays, the composition of the Earth's atmosphere, and ion concentrations.
May 1958	Using data from the *Explorer* rockets, US physicist James Van Allen discovers a belt of radiation around the Earth. Now known as the Van Allen belts (additional belts were discovered later), they consist of charged particles from the Sun trapped by the Earth's magnetic field.
29 July 1958	The US National Aeronautics and Space Administration (NASA) is created for the research and development of vehicles and activities involved in space exploration.

Nov 1958 The USA launches *Atlas*, from Cape Canaveral, Florida. It is a one-half-stage rocket and has a range of 14,400 km/9,000 mi. It was originally designed as an intercontinental ballistic missile (ICBM).

18 Dec 1958 The USA launches *PROJECT SCORE* (Signal Communications by Orbiting Relay Equipment), the first US communications satellite. It functions for 13 days relaying messages stored on magnetic tape.

2 Jan 1959 The USSR launches *Lunik 1*. The first spacecraft to escape Earth's gravity, it passes within 6,400 km/4,000 mi of the Moon.

28 Feb 1959 The US Air Force launches *Discoverer 1* into a low polar orbit where it photographs the entire surface of the Earth every 24 hours.

Feb 1959 The US Navy launches *Vanguard 2*, the first weather satellite.

14 Sept 1959 The Soviet spacecraft *Luna 2* (launched on 12 September) becomes the first spacecraft to strike the Moon.

11 March 1960 The USA launches *Pioneer 5*, which relays the first measurements of deep space.

12 April 1961 Soviet cosmonaut Yury Gagarin, in *Vostok 1*, is the first person to enter space. His flight lasts 108 minutes.

5 May 1961 US astronaut Alan Shepard in the Mercury capsule *Freedom 7* makes a 14.8-minute single suborbital flight. He is the first US astronaut into space.

21 May 1961 US president John F Kennedy commits the country to 'landing a man on the Moon and returning him safely to Earth before this decade is out'.

26 April 1962 The USA and UK launch the Earth satellite *Ariel*. Designed to study the ionosphere, it is the first international cooperative launch.

14 June 1962 The European Space Research Organization is established in Paris, France.

10 July 1962 The US telecommunications company AT&T launches *Telstar*, the first communications satellite, and begins experimental trans-Atlantic transmissions.

26 Aug 1962 US space probe *Mariner 2* is launched. It makes a flyby of Venus (14 December) passing within 34,000 km/21,600 mi of the planet's surface and takes measurements of temperature and atmospheric density.

5 Dec 1962 The USA and the USSR sign an agreement on cooperation for the peaceful use of outer space.

1962 The USA launches the Orbiting Solar Observatory (OSO). The first of a series of solar observatories, it collects and transmits data on the Sun's electromagnetic radiation.

16 June 1963 Soviet cosmonaut Valentina Tereshkova, the first woman in space, is launched into a three-day orbital flight aboard *Vostok 6*, to study the problem of weightlessness.

1963 Dutch-born US astronomer Maarten Schmidt discovers the first quasar (3C 273), an extraordinarily distant object brighter than the largest known galaxy yet with a star-like image.

1963 The Arecibo radio telescope in Puerto Rico begins operation; its 300-m/1,000-ft reflector is built into a naturally occurring parabola and is the largest single-reflector telescope in the world.

Jan 1964 In the first US–Soviet joint space venture, the satellite *Echo 2* takes off from California's Vandenberg Air Force Base. It will transmit messages around the world.

28 Nov 1964 The USA launches *Mariner 4* to Mars. Passing within 1,865 km/6,118 mi of the planet's surface (14 July 1965), it relays the first close-up photographs of the planet's surface as well as information on the Martian atmosphere.

18 March 1965 Soviet cosmonaut Alexey Leonov leaves spacecraft *Voskhod 2* and floats in space for 20 minutes – the first space walk.

June 1965 The US National Aeronautics and Space Administration (NASA) launches *Gemini 4*, whose four-day mission sees the first US space walk by astronaut Edward White.

1965 The first Soviet *Molniya* communications satellite is launched.

1965 US astronomers Arno Penzias and Robert Wilson detect microwave background radiation in the universe and suggest that it is the residual radiation from the Big Bang.

3 Feb 1966 Soviet spacecraft *Luna 9* (launched 31 January) makes the first soft landing on the Moon and transmits photographs and soil data for three days.

1 March 1966 Soviet probe *Venera 3* (launched 16 November 1965) crash-lands on Venus, the first artificial object to land on another planet.

16 March 1966 US astronauts Neil Armstrong and David Scott, aboard *Gemini 8*, achieve the first link-up of a crewed spacecraft with another object, an *Agena* rocket.

11 Nov 1966 *Gemini 12*, the last of the *Gemini* two-person space missions, is launched. It makes the first fully automatically controlled re-entry.

27 Jan 1967 A treaty banning nuclear weapons from outer space is signed by 60 countries, including the USA and USSR. It will be effective from 10 October.

27 Jan 1967 Three US astronauts, Virgil ('Gus') I Grissom, Edward H White, II, and Roger B Chaffee, die in a fire during a countdown rehearsal on the *Apollo 1* spacecraft at Cape Kennedy, Florida. They are the first human casualties of the US space programme.

24 April 1967 Soviet cosmonaut Vladimir Komarov dies

during the descent of his *Soyuz 1* spacecraft when his parachute fails to open properly. He is the first fatality of the Soviet space programme.

7 July 1967 British astronomer Jocelyn Bell and English astronomer Anthony Hewish discover the first pulsar (announced in 1968). A new class of stars, they are later shown to be collapsed neutron stars emitting bursts of radio energy.

18 Oct 1967 The Soviet spacecraft *Venera 4* (launched 12 June) lands on Venus. The first soft landing on another planet, its instrument-laden capsule transmits information about Venus' atmosphere.

22 Oct 1967– The US spacecraft *Cosmos 186* and *Cosmos*
29 Oct 1967 *188* complete the first automatic docking.

7 Nov 1967 US spacecraft *Surveyor 6* photographs one area of the Moon then lifts off, repositions itself 2.4 m/8 ft away and resumes photographing. It is the first lift-off from an extraterrestrial body.

1967 The Outer Space Treaty that bans military activities in space is ratified by 63 members of the United Nations .

14 Sept 1968– The Soviet spacecraft *Zond 5* flies around the
21 Sept 1968 Moon and returns to Earth – the first spacecraft to do so.

11 Oct 1968– *Apollo 7*, the first US Apollo space mission
22 Oct 1968 with a crew, tests the Command module used on subsequent flights to the Moon, during 163 orbits of the Earth. The crew make the first live transmission from space on 13 October.

1968 US astronauts Frank Borman, James Lovell, and William Anders become the first astronauts to orbit the Moon during the *Apollo 8* space mission. They complete 10 orbits.

16 Jan 1969 Two cosmonauts aboard Soviet spacecraft *Soyuz 5* (launched 15 January) dock and transfer to *Soyuz 4* (launched 14 January). Locked together for four hours they form the first experimental space station.

20 July 1969 US astronaut Neil Armstrong, onboard *Apollo 11* (launched 16 July), is the first person to walk on the Moon, famously saying 'That's one small step for man, one giant leap for mankind.' He and astronaut Buzz Aldrin also install and operate the first Moon seismograph at Tranquillity base, spending a total of 21 hours 37 minutes on the Moon's surface.

13–17 April NASA narrowly diverts a disaster aboard the
1970 Moon-bound spaceship *Apollo 13*, after a canister of liquid oxygen explodes in the command module. The crew of James Lovell, John Swigert, and Fred Haise enter the lunar module, which they use as a 'lifeboat' to return safely to Earth.

1970 The Effelsberg radio telescope near Bonn, Germany, begins operating; its 100-m/328-ft moveable dish is the largest fully steerable dish in the world.

26 July–7 *Apollo 15* astronauts use a Lunar Roving

Aug 1971 Vehicle (LRV) to explore the Moon's surface.

12 Nov 1971 The US space probe *Mariner 9* (launched in May) becomes the first artificial object to orbit another planet (Mars); it transmits 7,329 photographs of the planet and its two moons, Deimos and Phobos.

1971 The binary X-ray system Cygnus X-1 is discovered; the centre is believed to contain a black hole.

23 July 1972 The USA launches *Landsat 1*, the first of a series of satellites for surveying the Earth's resources from space.

14 May 1973 The USA launches the first *Skylab* space
–8 Feb 1974 station. It contains a workshop for carrying out experiments in weightlessness. It is visited by three three-person crews and astronauts make observations of the Sun, manufacture superconductors, and conduct other scientific and medical experiments. The third mission lasts a record 84 days and gathers data about long space flights.

July 1975 The launch of the Soviet spaceship *Soyuz 19* signals the start of a joint US–Soviet space mission. US and Soviet astronauts meet in space on July 17 when *Soyuz 19* docks with its NASA counterpart, *Apollo 18*.

1 Aug 1975 The European Space Agency is founded in Paris, France, to undertake research and develop technologies for use in space.

22 Oct 1975– The Soviet spacecraft *Venera 9* and *10* land on
25 Oct 1975 Venus and transmit the first pictures from the surface of another planet.

1976 The 6-m/19.7-ft UTR-Z telescope is completed at Zelenchukskaya, USSR; it is the largest reflecting telescope in the world.

1976 The Bol'shoi Teleskop Azimutal'nyi on Mount Pastukhov, USSR, begins operating; its 6-m/19.7-ft reflector makes it the largest optical telescope in the world.

1976 The US spacecraft *Viking 1* and *Viking 2* (launched in 1975) soft-land on Mars (20 July, 3 September). They make meteorological readings of the Martian atmosphere and search for traces of bacterial life which prove inconclusive.

4 Dec 1979 The European Space Agency's first Ariane rocket is launched from the Guiana Space Centre in Kourou, French Guiana; it is designed to deploy satellites into orbit.

1979 The 3-m/9.8-ft NASA Infrared Telescope Facility (IRTF) telescope and the 3.8- m/12.5-ft UK Infrared Telescope both begin operation on Mauna Kea, Hawaii.

1979 The Multiple Mirror Telescope begins operation on Mount Hopkins, Arizona; it focuses the light from six 180-cm/70-in telescopes to form one image, giving the light-gathering power of a single 4.5-m/15.7-ft telescope; it becomes the prototype for larger optical telescopes.

1980	A thin layer of iridium-rich clay, about 65 million years old, is found around the world. US physicist Luis Alvarez suggests that it was caused by the impact of a large asteroid or comet which threw enough dust into the sky to obscure the Sun and cause the extinction of the dinosaurs.	
1980	Alan Guth proposes the theory of the inflationary universe – that the universe expanded very rapidly for a short time after the Big Bang.	
1980	The Multi-Element Radio-Linked Interferometer Network (MERLIN) radio telescope begins operation in the UK. It has five dishes measuring 25 m/82 ft, one measuring 32 m/105 ft, and one measuring 76 m/249 ft, making it the largest radio telescope in the world.	
1980	The Very Large Array (VLA) radio telescope at Socorro, New Mexico, enters service; its 27 25-m/81-ft dishes are steerable and moveable on railway tracks and are equivalent to one dish 27 km/17 mi in diameter; together they provide high-resolution radio images.	
12 April 1981 –14 April 1981	The US reusable space shuttle, using the orbiter *Columbia*, makes its first flight (second shuttle flight 12–14 November). It achieves the first US spacecraft landing on land (instead of water).	
1982	The US rocket *Conestoga 1* makes a suborbital flight; it is the first private space craft.	
13 June 1983	The US space probe *Pioneer 10* becomes the first artificial object to leave the Solar System.	
18 June 1983	Astronauts on board the space shuttle *Challenger* first use the Remote Manipulating Structure ('arm') to deploy and retrieve a satellite.	
1983	The Search for Extraterrestrial Intelligence (SETI) programme is established.	
19 Feb 1986	The Soviet space station *Mir 1* is launched; it is intended to be permanently occupied.	
1986	Scientists at Arizona State University conduct computer simulations that strongly	

suggest that a Mars-sized object struck the Earth a glancing blow about 4.6 billion years ago and was then captured by the Earth; by the end of the year the impact theory is the leading hypothesis about the Moon's origin.

1986	The California Submillimetre Observatory telescope begins operation on Mauna Kea, Hawaii; its 10.4-m/34-ft disc makes it the largest submillimetre telescope in the world.
24 Feb 1987	The first supernova (explosion of a star) visible to the naked eye since 1604 is seen.
1987	Radio waves are observed from 3C326 – believed to be a galaxy in the process of formation.
1987	The James Clerk Maxwell Telescope, operated by the Royal Observatory, based in Edinburgh, Scotland, begins operation on Mauna Kea, Hawaii; its 15-m/49-ft dish makes it the largest submillimetre telescope in the world.
25 Aug 1989	The US space probe *Voyager 2* reaches Neptune and transmits pictures; it discovers a great dark spot on the planet and six new moons.
1989	The US Cosmic Background Explorer (COBE) satellite is launched to study microwave background radiation, thought to be a vestige of the Big Bang.
1989	The US Delta Star 'Star Wars' satellite is launched; it successfully detects and tracks test missiles shortly after they are launched.
Feb 1990	The US space probe *Voyager 1*, now near the edge of the Solar System, turns and takes the first photograph of the entire Solar System from space.
24 April 1990	The space shuttle *Discovery* places the Hubble Space Telescope in Earth orbit; the main mirror proves to be defective.
10 Aug 1990	The US *Magellan* radar mapper arrives in orbit around Venus; it transmits the most detailed pictures of the planet's surface yet produced.
Nov 1990	The Keck 1 Telescope on Mauna Kea volcano, Hawaii, is erected; its 10-m/ 32.8-ft

reflector, composed of 36 segments, makes it the largest optical telescope in the world.

2 Dec 1990–12 Dec 1990 The Soviet spacecraft *Soyuz TM-11* is launched, marking the first paying-passenger space flight.

1990 The British Royal Observatory is transferred from Herstmonceaux, Sussex, to Cambridge.

18–26 May 1991 English chemist Helen Sharman becomes the first Briton to go into space, as a participant in a Soviet space mission launched in *Soyuz TM-12*. She spends six days with Soviet cosmonauts aboard the *Mir* space station.

29 Oct 1991 The US space probe *Galileo* takes the closest ever picture of an asteroid – Gaspra – at a distance of 26,071 km/ 16,200 mi.

8 Feb 1992 The gravity of Jupiter is used to swing the US space probe *Ulysses* towards the Sun.

1992 The Cosmic Background Explorer (COBE) satellite detects ripples in the microwave background radiation, thought to originate from the formation of galaxies.

1992 The US space probe *Magellan* maps 99% of the surface of Venus to a resolution of 100 m/330 ft.

7 Dec 1993 The Hubble Space Telescope is repaired and reboosted into a nearly circular orbit by five US astronauts operating from the US space shuttle *Endeavour*.

6 Dec 1994 Pictures taken by the Hubble Space Telescope of galaxies in their infancy are published.

Dec 1994 The Apollo asteroid (an asteroid with an orbit that crosses that of Earth) 1994 XM1 passes within 100,000 km/60,000 mi of Earth, the closest observed approach of any asteroid.

7 Dec 1995 The US spacecraft *Galileo*'s probe enters Jupiter's atmosphere while its parent continues to orbit the planet. The probe radios information back about the chemical composition of the atmosphere to the orbiter for 57 minutes before being destroyed by atmospheric pressure.

1995 The largest liquid-mirror telescope (a reflecting telescope constructed with a rotating mercury mirror) is completed for NASA's Orbital Debris Observatory in New Mexico. It is 3 m/9.8 ft across.

1995 US astronomers discover the first brown dwarf, an object larger than a planet but not massive enough to ignite into a star, in the constellation Lepus. It is about 20–40 times as massive as Jupiter. Four other brown dwarfs are discovered in 1996.

1995 US astronomers discover water in the Sun – in the form of superheated steam – in two sunspots where the temperature is only 3,000°C/5,400°F.

24 March 1996 Comet Hyakutake makes its closest approach, passing within 15.4 million km/9.5 million mi of Earth. It is the brightest comet for decades, with a tail extending over 12 degrees of the sky.

4 Dec 1996 NASA launches the *Mars Pathfinder*. Its main goal is to demonstrate the feasibility of low-cost landings on, and exploration of, Mars. The spacecraft carries a roving machine to explore the surface.

6 Dec 1996 Cosmonauts aboard the space station *Mir* successfully harvest a small wheat crop, the first plants to be successfully cultivated from seed in space.

23 March 1997 The comet Hale-Bopp comes to within 196 million km/122 million mi of Earth, becoming the closest comet since 2000 BC. NASA launches rockets to study the comet. Its icy nucleus is estimated to be 40 km/25 mi wide, making it at least ten times larger than that of Comet Hyakutake and twice the size of Comet Halley.

4 July 1997 The US spacecraft *Mars Pathfinder* lands on Mars. Two days later the probe's rover *Sojourner*, a six-wheeled vehicle controlled by an Earth-based operator, begins to explore the area around the spacecraft.

12 Sept 1997 The US spacecraft *Mars Global Surveyor* goes into orbit around Mars to conduct a detailed photographic survey of the planet, commencing in March 1998.

Jan 1998 The US spaceprobe *Lunar Prospector* is launched to examine the composition of the lunar crust, record gamma rays, and map the lunar magnetic field.

1998 The European Southern Observatory's Very Large Telescope on Mount Paranal, Chile, transmits its first images.

Biographies

Adams, John Couch (1819–1892) English astronomer. He mathematically deduced the existence of the planet Neptune in 1845 from the effects of its gravitational pull on the motion of Uranus, although it was not found until 1846 by J G Galle. Adams also studied the Moon's motion, the Leonid meteors, and terrestrial magnetism. He was Lowndean professor of astronomy and geometry at Cambridge 1859–92 and director of Cambridge observatory 1861–92.

Airy, George Biddell (1801–1892) English astronomer. He became the seventh Astronomer Royal in 1835. He installed a transit telescope at the Royal Observatory at Greenwich, England, and accurately measured Greenwich Mean Time by the stars as they crossed the meridian. Greenwich Mean Time was adopted as legal time in Britain in 1880. Airy's *Mathematical Tracts on Physical Astronomy* (1826) became a standard work.

Anaximander (c. 610–c. 546 BC) Greek astronomer and philosopher. He claimed that the Earth was a cylinder three times wider than it is deep, motionless at the centre of the universe, and that the celestial bodies were fire seen through holes in the hollow rims of wheels encircling the Earth. He is thought to have been the first to determine solstices and equinoxes, by means of a sundial, and he is credited with drawing the first geographical map of the whole known world.

Aristarchus of Samos (c. 320–c. 250 BC) Greek astronomer. The first to argue that the Earth moves around the Sun, he was ridiculed for his beliefs. He was also the first astronomer to estimate (quite inaccurately) the sizes of the Sun and Moon and their distances from the Earth. His model of the universe described the Sun and the fixed stars as stationary in the cosmos, and the planets – including the Earth – as travelling in circular orbits around the Sun. His only surviving work is *Magnitudes and Distances of the Sun and Moon*.

Baade, (Wilhelm Heinrich) Walter (1893–1960) German-born US astronomer who made observations that doubled the distance, scale, and age of the universe. He discovered that stars are in two distinct populations according to their age, known as Population I (the younger, bluish in colour) and Population II (the older, reddish). Later, he found that Cepheid variable stars of Population I are brighter than had been supposed and that distances calculated from them were wrong.

Bell Burnell, (Susan) Jocelyn (1943–) British astronomer. In 1967, while a research student at Cambridge, she discovered the first pulsar (rapidly flashing star) with Antony Hewish and colleagues at Cambridge University, England.

Bessel, Friedrich Wilhelm (1784–1846) German astronomer and mathematician. He was the first person to find the approximate distance to a star by direct methods when he measured the parallax (annual displacement) of the star 61 Cygni in 1838. Bessel's work laid the foundations for a more accurate calculation of the scale of the universe and the sizes of stars, galaxies, and clusters of galaxies. He was a pioneer of very precise observation and reduction in astronomy, and published a catalogue of 3,222 star positions under the title *Fundamenta Astronomiae* (1818).

Bode, Johann Elert (1747–1826) German astronomer and mathematician. He contributed greatly to the popularization of astronomy. He published the first atlas of all stars visible to the naked eye, *Uranographia* (1801), and devised **Bode's law**, a numerical sequence that gives the approximate distances in astronomical units, of the planets from the Sun by adding 4 to each term of the series 0, 3, 6, 12, 24, ... and then adding 10.

Bondi, Hermann (1919–) Austrian-born British cosmologist. In 1948 he joined Fred Hoyle and Thomas Gold in developing the steady-state theory of cosmology, which suggested that matter is continuously created in the universe. Bondi also described the likely characteristics and physical properties of gravitational waves, and demonstrated that such waves are compatible with and are indeed a necessary consequence of the general theory of relativity.

Bradley, James (1693–1762) English astronomer. In 1728 he discovered the aberration of starlight. From the amount of aberration in star positions, he was able to calculate the speed of light. In 1748 he announced the discovery of nutation (variation in the Earth's axial tilt). He became the third Astronomer Royal in 1742.

Brahe, Tycho (1546–1601) Danish astronomer. His accurate observations of the planets enabled German astronomer and mathematician Johannes Kepler to prove that planets orbit the Sun in ellipses. Brahe's discovery and report of the 1572 supernova brought him recognition, and his observations of the comet of 1577 proved that it moved in an orbit among the planets, thus disproving Aristotle's view that comets were in the Earth's atmosphere.

Cannon, Annie Jump (1863–1941) US astronomer. She carried out revolutionary work on the classification of stars by examining their spectra. Her system, still used today, has spectra arranged according to temperature into categories labelled O, B, A, F, G, K, M, R, N, and S. O-type stars are the hottest, with surface temperatures of over 25,000K. She discovered 300 new variable stars and classified the spectra of over 300,000 stars.

Cassini, Giovanni Domenico (1625–1712) Italian-born French astronomer. During 1664–67 he determined the rotation periods of Mars, Jupiter, and Venus. He discovered four moons of Saturn and the gap in the rings of Saturn now called the **Cassini division**.

Chandrasekhar, Subrahmanyan (1910–1995) Indian-born US astrophysicist who made pioneering studies of the structure and evolution of stars. The **Chandrasekhar limit** is the maximum mass of a white dwarf before it turns into a neutron star. He also investigated the transfer of energy in stellar atmospheres by radiation and convection, and the polarization of light emitted from particular stars. He was awarded the Nobel Prize for Physics in 1983.

Copernicus, Nicolaus (1473–1543) Polish astronomer who believed that the Sun, not the Earth, is at the centre of the Solar System, thus defying the Christian church doctrine of the time. For 30 years, he worked on the hypothesis that the rotation and the orbital motion of the Earth are responsible for the apparent movement of the heavenly bodies. His great work *De revolutionibus orbium coelestium/On the Revolutions of the Heavenly Spheres* was the important first step to the more accurate picture of the Solar System built up by Tycho Brahe, Kepler, Galileo, and later astronomers.

Eddington, Arthur Stanley (1882–1944) British astrophysicist. He studied the motions, equilibrium, luminosity, and atomic structure of the stars. In 1919 his observation of stars during a solar eclipse confirmed Albert Einstein's prediction that light is bent when passing near the Sun, in accordance with the general theory of relativity. In *The Expanding Universe* (1933) he expressed the theory that in the spherical universe the outer galaxies or spiral nebulae are receding from one another.

Eudoxus of Cnidus (c. 400–c. 347 BC) Greek mathematician and astronomer. He devised the first system to account for the motions of celestial bodies, believing them to be carried around the Earth on sets of spheres. Work attributed to Eudoxus includes methods to calculate the area of a circle and to derive the volume of a pyramid or a cone. Eudoxus is said to have been the one who first fixed the length of the year as 365.25 days, and to have invented the sundial.

Evershed, John (1864–1956) English astronomer who made solar observations. In 1909 he discovered the radial movements of gases in sunspots (the **Evershed effect**). He also gave his name to a spectroheliograph, the Evershed spectroscope.

Flamsteed, John (1646–1719) English astronomer. He began systematic observations of the positions of the stars, Moon, and planets at the Royal Observatory he founded at Greenwich, London, in 1676. His observations were published in *Historia Coelestis Britannica* (1725).

As the first Astronomer Royal of England, Flamsteed determined the latitude of Greenwich, the slant of the ecliptic, and the position of the equinox. He also worked out a method of observing the absolute right ascension (a coordinate of the position of a heavenly body) that removed all errors of parallax, refraction, and latitude. Having obtained the positions of 40 reference stars, he then computed positions for the rest of the 3,000 stars in his catalogue.

Galileo Galilei (1564–1642) Italian mathematician, astronomer, and physicist. He developed the astronomical telescope and was the first to see sunspots, the four main satellites of Jupiter, and the appearance of Venus going through phases, thus proving it was orbiting the Sun. Galileo discovered that freely falling bodies, heavy or light, have the same, constant acceleration and that this acceleration is due to gravity. He also determined that a body moving on a perfectly smooth horizontal surface would neither speed up nor slow down. He invented a thermometer, a hydrostatic balance, and a compass, and discovered that the path of a projectile is a parabola. Galileo's work founded the modern scientific method of deducing laws to explain the results of observation and experiment.

Galle, Johann Gottfried (1812–1910) German astronomer. He located the planet Neptune in 1846, close to the position predicted by French mathematician Urbain Leverrier and the

English astronomer J C Adams. He also suggested a method of measuring the scale of the Solar System by observing the parallax of asteroids, first applying his method to the asteroid Flora in 1873. The method was employed with great success after Galle's death.

Gamow, George (Georgi Antonovich) (1904–1968) Russian-born US cosmologist, nuclear physicist, and popularizer of science. His work in astrophysics included a study of the structure and evolution of stars and the creation of the elements. He explained how the collision of nuclei in the solar interior could produce the nuclear reactions that power the Sun. With the 'hot Big Bang' theory, he indicated the origin of the universe.

Hale, George Ellery (1868–1938) US astronomer. He made pioneer studies of the Sun and founded three major observatories. In 1889 he invented the spectroheliograph, a device for photographing the Sun at particular wavelengths. In 1917 he established on Mount Wilson, California, a 2.5-m/100-in reflector, the world's largest telescope until superseded in 1948 by the 5-m/200-in reflector on Mount Palomar, which Hale had planned just before he died.

He, more than any other, was responsible for the development of observational astrophysics in the USA. He also founded the Yerkes Observatory in Wisconsin 1897, with the largest refractor, 102 cm/40 in, ever built at that time.

Halley, Edmond (1656–1742) English astronomer. He identified Halley's Comet, compiled a star catalogue, detected the proper motion of stars, using historical records, and began a line of research that, after his death, resulted in a reasonably accurate calculation of the astronomical unit.

Halley calculated that the comet sightings reported in 1456, 1531, 1607, and 1682 all represented reappearances of the same comet. He reasoned that the comet would follow a parabolic path and announced in 1705 in his *Synopsis Astronomia Cometicae* that it would reappear in 1758. When it did, public acclaim for the astronomer was such that his name was irrevocably attached to it. He was Astronomer Royal from 1720.

Herschel, Caroline (Lucretia) (1750–1848) German-born English astronomer, sister of William Herschel, and from 1772 his assistant. She discovered eight comets and worked on her brother's catalogue of star clusters and nebulae.

Herschel, John Frederick William (1792–1871) English scientist, astronomer, and photographer who discovered thousands of close double stars, clusters, and nebulae. Together with James South, he systematically remeasured, from 1821 to 1823, the double stars discovered by his father and, in 1824, published a catalogue of double stars. Herschel went on to revise his father's survey of the northern heavens, and mapped the southern skies from the Cape of Good Hope Observatory in South Africa 1834–38.

Herschel, (Frederick) William (1738–1822) German-born English astronomer. He was a skilled telescopemaker, and pioneered the study of binary stars and nebulae. He discovered the planet Uranus in 1781 and infrared solar rays in 1801. He catalogued over 800 double stars, and found over 2,500 nebulae, catalogued by his sister Caroline

Herschel; this work was continued by his son John Herschel. By studying the distribution of stars, William established the basic form of our galaxy, the Milky Way.

Hewish, Antony (1924–) English radio astronomer who was awarded, with Martin Ryle, the Nobel Prize for Physics in 1974 for his work on pulsars, rapidly rotating neutron stars that emit pulses of energy.

The discovery by Jocelyn Bell Burnell of a regularly fluctuating signal, which turned out to be the first pulsar, began a period of intensive research. Hewish discovered another three straight away, and more than 170 pulsars have been found since 1967.

Hipparchus (c. 190–c. 120 BC) Greek astronomer and mathematician. He calculated the lengths of the solar year and the lunar month, discovered the precession of the equinoxes, made a catalogue of 850 fixed stars, and advanced Eratosthenes' method of determining the situation of places on the Earth's surface by lines of latitude and longitude.

In 134 BC he noticed a new star in the constellation Scorpio, which inspired him to put together his star catalogue – the first of its kind. He entered his observations of stellar positions using a system of celestial latitude and longitude, and taking the precaution wherever possible to state the alignments of other stars as a check on present position. He classified the stars by magnitude (brightness). His finished work, completed in 129 BC, was used by Edmond Halley some 1,800 years later.

Hoyle, Fred(erick) (1915–) English astronomer, cosmologist, and writer. His astronomical research has dealt mainly with the internal structure and evolution of the stars. In 1948 he developed with Hermann Bondi and Thomas Gold the steady-state theory of the universe. In 1957, with William Fowler, he showed that chemical elements heavier than hydrogen and helium may be built up by nuclear reactions inside stars.

Hubble, Edwin (Powell) (1889–1953) US astronomer. He discovered the existence of galaxies outside our own, and classified them according to their shape. His theory that the universe is expanding is now generally accepted. It has been said that Hubble opened up the observable region of the universe in the same way that Galileo opened up the Solar System and the Herschels the Milky Way.

His data on the speed at which galaxies were receding (based on their red shifts) were used to determine the portion of the universe that we can ever come to know, the radius of which is called the **Hubble radius**. Beyond this limit, any matter will be travelling at the speed of light, so communication with it will never be possible. The ratio of the velocity of galactic recession to distance has been named the **Hubble constant**.

Hubble discovered Cepheid variable stars in the Andromeda galaxy in 1923, proving it to lie far beyond our own Galaxy. In 1925 he introduced the classification of galaxies as spirals, barred spirals, and ellipticals. In 1929 he announced **Hubble's law**, stating that the galaxies are moving apart at a rate that increases with their distance.

Huygens, Christiaan (1629–1695) Dutch mathematical physicist and astronomer. He proposed the wave theory of light, developed the pendulum clock in 1657, discovered polarization, and observed Saturn's rings. His work in astronomy was an impressive defence of the Copernican view of the Solar System.

Jansky, Karl Guthe (1905–1950) US radio engineer who in 1932 discovered that the Milky Way galaxy emanates radio waves; he did not follow up his discovery, but it marked the birth of radioastronomy.

Kepler, Johannes (1571–1630) German mathematician and astronomer. He formulated what are now called **Kepler's laws** of planetary motion: (1) the orbit of each planet is an ellipse with the Sun at one of the foci; (2) the radius vector of each planet sweeps out equal areas in equal times; (3) the squares of the periods of the planets are proportional to the cubes of their mean distances from the Sun. Kepler's laws are the basis of our understanding of the Solar System, and such scientists as Isaac Newton built on his ideas.

His *Rudolphine Tables* (1627) were based on Tycho Brahe's observations, whose assistant he became in 1600. These were the first modern astronomical tables, enabling astronomers to calculate the positions of the planets at any time in the past, present or future.

Korolev, Sergei Pavlovich (1906–1966) Russian designer of the first Soviet intercontinental missile, used to launch the first *Sputnik* satellite in 1957 and the *Vostok* spacecraft, also designed by Korolev, in which Yuri Gagarin made the world's first space flight in 1961.

Korolev and his research team built the first Soviet liquid-fuel rocket, launched in 1933. His innovations in rocket and space technology include ballistic missiles, rockets for geophysical research, launch vehicles, and crewed spacecraft. Korolev was also responsible for the *Voskhod* spaceship, from which the first space walks were made.

Kuiper, Gerard Peter (1905–1973) Dutch-born US astronomer who made extensive studies of the Solar System. His discoveries included the atmosphere of the planet Mars and that of Titan, the largest moon of the planet Saturn. His spectroscopic studies of Uranus and Neptune led to the discovery of features subsequently named **Kuiper bands**, which indicate the presence of methane. He was an adviser to many NASA exploratory missions, and pioneered the use of telescopes on high-flying aircraft.

Laplace, Pierre Simon, Marquis de Laplace (1749–1827) French astronomer and mathematician. In 1796 he theorized that the Solar System originated from a cloud of gas (the nebular hypothesis). He studied the motion of the Moon and planets, and published a five-volume survey of celestial mechanics, *Traité de méchanique céleste* (1799–1825). This work contained the law of universal attraction – the law of gravity as applied to the Earth – and explanations of such phenomena as the ebb and flow of tides and the precession of the equinoxes.

Leavitt, Henrietta Swan (1868–1921) US astronomer who in 1912 discovered the **period–luminosity law**, which links the brightness of a Cepheid variable star to its period of variation. This law allows astronomers to use Cepheid variables as 'standard candles' for measuring distances in space. She discovered a total of 2,400 new variable stars and four novae.

Lemaître, Georges Edouard (1894–1966) Belgian cosmologist. He proposed the Big Bang theory of the origin of the universe in 1933. The US astronomer Edwin Hubble had shown that the universe was expanding, but it was Lemaître who suggested that the expansion had been started by an initial explosion, the Big Bang, a theory that is now generally accepted.

Leverrier, Urbain Jean Joseph (1811–1877) French astronomer. He predicted the existence and position of the planet Neptune from its influence on the orbit of the planet Uranus. It was discovered in 1846.

The possibility that another planet might exist beyond Uranus, influencing its orbit, had already been suggested. Leverrier calculated the orbit and apparent diameter of the hypothetical planet, and wrote to a number of observatories, asking them to test his prediction of its position. Johann Galle at the Berlin Observatory found it immediately, within 1° of Leverrier's coordinates.

Lovell, (Alfred Charles) Bernard (1913–) \English radio astronomer, director 1951–81 of Jodrell Bank Experimental Station (now Nuffield Radio Astronomy Laboratories). He showed that radar could be a useful tool in astronomy, and lobbied for the setting-up of a radio-astronomy station. Jodrell Bank was built near Manchester 1951–57. Although its high cost was criticized, its public success after tracking the Soviet satellite *Sputnik I* in 1957 assured its future.

Lowell, Percival (1855–1916) US astronomer who predicted the existence of a planet beyond Neptune, starting the search that led to the discovery of Pluto in 1930. In 1894 he founded the Lowell Observatory in Flagstaff, Arizona, where he reported seeing 'canals' (now known to be optical effects and natural formations) on the surface of Mars. He led an expedition to the Chilean Andes in 1907, which produced the first high-quality photographs of the planet.

Maskelyne, Nevil (1732–1811) English astronomer. He made observations to investigate the reliability of the lunar distance method for determining longitude at sea. In 1774 he estimated the mass of the Earth by noting the deflection of a plumb line near Mount Schiehallion in Perthshire, Scotland. He was Astronomer Royal 1765–1811.

Messier, Charles (1730–1817) French astronomer. He discovered 15 comets and in 1784 published a list of 103 star clusters and nebulae. Objects on this list are given M (for Messier) numbers, which astronomers still use today, such as M1 (the Crab nebula) and M31 (the Andromeda galaxy).

Messier's search was continually hampered by rather obscure forms which he came to recognize as nebulae. During the period 1760–84, therefore, he compiled a list of these nebulae and star clusters, so that he and other astronomers would not confuse them with possible new comets.

Oort, Jan Hendrik (1900–1992) Dutch astronomer. In 1927 he calculated the mass and size of our galaxy, the Milky Way, and the Sun's distance from its centre, from the observed movements of stars around the Galaxy's centre. In 1950 Oort proposed that comets exist in a vast swarm, now called the **Oort cloud**, at the edge of the Solar System. He established radio observatories at Dwingeloo and Westerbork, which put the Netherlands in the forefront of radio astronomy.

Ptolemy (Claudius Ptolemaeus) (c. AD 100–c. AD 170) Greek astronomer and geographer. His *Almagest* developed the theory that Earth is the centre of the universe, with the Sun, Moon, and stars revolving around it. The *Almagest* contains all his works on astronomical themes, the only authoritative works until the time of Copernicus. Ptolemy began with the premise that the Earth was a perfect sphere. All planetry orbits were circular, but those of Mercury and Venus, and possibly Mars (Ptolemy was not sure), were epicyclic (the planets orbited a point that itself was orbiting the Earth). The sphere of the stars formed a dome with points of light attached or pricked through. In 1543 the Polish astronomer Copernicus proposed an alternative to the **Ptolemaic system**.

Ryle, Martin (1918–1984) English radio astronomer. At the Mullard Radio Astronomy Observatory, Cambridge, he developed the technique of sky-mapping using 'aperture synthesis', combining smaller dish aerials to give the characteristics of one large one. His work on the distribution of radio sources in the universe brought confirmation of the Big Bang theory. He was Astronomer Royal 1972–82, and, with his co-worker Antony Hewish, won the Nobel Prize for Physics in 1974.

Schiaparelli, Giovanni Virginio (1835–1910) Italian astronomer. He drew attention to linear markings on Mars, which gave rise to the popular belief that they were canals. The markings were soon shown by French astronomer Eugène Antoniadi to be optical effects and not real lines. Schiaparelli also gave observational evidence for the theory that all meteor showers are fragments of disintegrating comets.

Shapley, Harlow (1885–1972) US astronomer. He established that our galaxy was much larger than previously thought. His work on the distribution of globular clusters showed that the Sun was not at the centre of the galaxy as assumed, but two-thirds of the way out to the rim; globular clusters were arranged in a halo around the Galaxy. His surveys recorded tens of thousands of galaxies in both hemispheres.

Tombaugh, Clyde William (1906–1997) US astronomer who discovered the planet Pluto in 1930. As an assistant at the Lowell Observatory in Flagstaff, Arizona, from 1929, he photographed the sky in search of the predicted remote planet, discovering it on 18 February 1930. His failure to find any other planets placed strict limits on the possible existence of planets beyond Pluto.

Van Allen, James Alfred (1914–) US physicist whose instruments aboard the first US satellite *Explorer 1* in 1958 led to the discovery of the Van Allen belts, two zones of intense radiation around the Earth. He pioneered high-altitude research with rockets after World War II and was responsible for the instrumentation of the first US satellites.

Whipple, Fred Lawrence (1906–) US astronomer whose hypothesis in 1949 that the nucleus of a comet is like a dirty snowball was confirmed in 1986 by space-probe studies of Halley's comet. He discovered six new comets and worked on ascertaining cometary orbits and defining the relationship between comets and meteors.

Whipple proposed that the nucleus of a comet consisted of a frozen mass of water, ammonia, methane, and other hydrogen compounds together with silicates, dust, and other materials. As the comet's orbit brought it nearer to the Sun, solar radiation would cause the frozen material to evaporate, thus producing a large amount of silicate dust which would form the comet's tail.

Zwicky, Fritz (1898–1974) Swiss astronomer. He predicted the existence of neutron stars in 1934. He discovered 18 supernovae and determined that cosmic rays originate in them. Beginning in 1936, he compiled a catalogue of galaxies and galaxy clusters in which he listed 10,000 clusters.

Glossary

aberration of starlight apparent displacement of a star from its true position, due to the combined effects of the speed of light and the speed of the Earth in orbit around the Sun (about 30 kps/18.5 mps).

Aquarius zodiacal constellation a little south of the celestial equator near Pegasus. Aquarius is represented as a man pouring water from a jar. The Sun passes through Aquarius from late February to early March.

arc minute, arc second units for measuring small angles, used in astronomy. An arc minute (symbol ') is one-sixtieth of a degree, and an arc second (symbol ") is one-sixtieth of an arc minute. Small distances in the sky, as between two close stars or the apparent width of a planet's disc, are expressed in minutes and seconds of arc.

Aries zodiacal constellation in the northern hemisphere between Pisces and Taurus, near Auriga, represented as the legendary ram whose golden fleece was sought by Jason and the Argonauts. Its most distinctive feature is a curve of three stars of decreasing brightness. The brightest of these is Hamal or Alpha Arietis, 65 light years from Earth.

astrolabe ancient navigational instrument, forerunner of the sextant. Astrolabes usually consisted of a flat disc with a sighting rod that could be pivoted to point at the Sun or bright stars. From the altitude of the Sun or star above the horizon, the local time could be estimated.

astronomical unit unit (symbol AU) equal to the mean distance of the Earth from the Sun: 149,597,870 km/92,955,800 mi. It is used to describe planetary distances. Light travels this distance in approximately 8.3 minutes.

astrophotography use of photography in astronomical research. The first successful photograph of a celestial object was the daguerreotype plate of the Moon taken by John W Draper (1811–1882) of the USA in March 1840. The first photograph of a star, Vega, was taken by US astronomer William C Bond (1789–1859) in 1850. Modern-day astrophotography uses techniques such as charge-coupled devices (CCDs).

astrophysics study of the physical nature of stars, galaxies, and the universe. It began with the development of spectroscopy in the 19th century, which allowed astronomers to analyse the composition of stars from their light. Astrophysicists view the universe as a vast natural laboratory in which they can study matter under conditions of temperature, pressure, and density that are unattainable on Earth.

Auriga constellation of the northern hemisphere, represented as a charioteer. Its brightest star is the first-magnitude Capella, about 45 light years from Earth; Epsilon Aurigae is an eclipsing binary star with a period of 27 years, the longest of its kind (last eclipse 1983).

aurora coloured light in the night sky near the Earth's magnetic poles, called aurora borealis ('northern lights') in the northern hemisphere and aurora australis in the southern hemisphere. Although aurorae are usually restricted to the polar skies, fluctuations in the solar wind occasionally cause them to be visible at lower latitudes. Aurorae are caused at heights of over 100 km/60 mi by a fast stream of charged particles from solar flares and low-density 'holes' in the Sun's corona.

Bear, Great and Little common names (and translations of the Latin) for the constellations Ursa Major and Ursa Minor respectively.

Betelgeuse, or Alpha Orionis, red supergiant star in the constellation of Orion. It is the tenth brightest star in the night sky, although its brightness varies. It is 1,100 million km/700 million mi across, about 800 times larger than the Sun, roughly the same size as the orbit of Mars. It is over 10,000 times as luminous as the Sun, and lies 650 light years from Earth. Light takes 60 minutes to travel across the giant star.

blue shift a manifestation of the Doppler effect in which an object appears bluer when it is moving towards the observer or the observer is moving towards it (blue light is of a higher frequency than other colours in the spectrum). The blue shift is the opposite of the red shift.

Boötes constellation of the northern hemisphere represented by a herdsman driving a bear (Ursa Major) around the pole. Its brightest star is Arcturus (or Alpha Boötis), which is about 37 light years from Earth. The herdsman is assisted by the neighbouring Canes Venatici, 'the Hunting Dogs'.

Cancer faintest of the zodiacal constellations (its brightest stars are fourth magnitude). It lies in the northern hemisphere between Leo and Gemini, and is represented as a crab. The Sun passes through the constellation during late July and early August.

Cape Canaveral promontory on the Atlantic coast of Florida, USA, 367 km/228 mi north of Miami, used as a rocket launch site by NASA.

Capricornus zodiacal constellation in the southern hemisphere next to Sagittarius. It is represented as a fish-tailed goat, and its brightest stars are third magnitude. The Sun passes through it late January to mid-February.

Cassiopeia prominent constellation of the northern hemisphere, named after the mother of Andromeda. It has a distinctive W-shape, and contains one of the most powerful radio sources in the sky, Cassiopeia A. This is the remains of a supernova (star explosion) that occurred c. AD 1702, too far away to be seen from Earth.

celestial mechanics the branch of astronomy that deals with the calculation of the orbits of celestial bodies, their gravitational attractions (such as those that produce the Earth's tides), and also the orbits of artificial satellites and space probes. It is based on the laws of motion and gravity laid down by Isaac Newton.

Cepheid variable yellow supergiant star that varies regularly in brightness every few days or weeks as a result of pulsations. The time that a Cepheid variable takes to pulsate isdirectly related to its average brightness; the longer the period, the brighter the star.

charge-coupled device (CCD) device for forming images electronically, using a layer of silicon that releases electrons when struck by incoming light. The electrons are stored in pixels and read off into a computer at the end of the exposure. CCDs have now almost entirely replaced photographic film for applications such as astrophotography where extreme sensitivity to light is paramount.

chromosphere layer of mostly hydrogen gas about 10,000 km/6,000 mi deep above the visible surface of the Sun (the photosphere). It appears pinkish red during eclipses of the Sun.

Comet Hale-Bopp (C/1995 01) large and exceptionally active comet, which in March 1997 made its closest approach to Earth since 2000 BC, coming within 190 million km/118 million mi. It has a diameter of approximately 40 km/25 mi and an extensive gas coma (when close to the Sun, Hale-Bopp released 10 tonnes of gas every second). Unusually, Hale-Bopp has three tails: one consisting of dust particles, one of charged particles, and a third of sodium particles. Comet Hale-Bopp was discovered independently in July 1995 by two amateur US astronomers, Alan Hale and Thomas Bopp.

communications satellite relay station in space for sending telephone, television, telex, and other messages around the world. Messages are sent to and from the satellites via ground stations. Most communications satellites are in geostationary orbit, appearing to hang fixed over one point on the Earth's surface.

conjunction the alignment of two celestial bodies as seen from Earth. A superior planet (or other object) is in conjunction when it lies behind the Sun. An inferior planet (or other object) comes to **inferior conjunction** when it passes between the Earth and the Sun; it is at **superior conjunction** when it passes behind the Sun. **Planetary conjunction** takes place when a planet is closely aligned with another celestial object, such as the Moon, a star, or another planet.

corona faint halo of hot (about 2,000,000°C/ 3,600,000°F) and tenuous gas around the Sun, which boils from the surface. It is visible at solar eclipses or through a **coronagraph**, an instrument that blocks light from the Sun's brilliant disc. Gas flows away from the corona to form the solar wind.

cosmic radiation streams of high-energy particles from outer space, consisting of protons, alpha particles, and light nuclei, which collide with atomic nuclei in the Earth's atmosphere, and produce secondary nuclear particles (chiefly mesons, such as pions and muons) that shower the Earth.

ecliptic path, against the background of stars, that the Sun appears to follow each year as it is orbited by the Earth. It can be thought of as the plane of the Earth's orbit projected on to the celestial sphere (imaginary sphere around the Earth). The ecliptic is tilted at about 23.5° with respect to the celestial equator, a result of the tilt of the Earth's axis relative to the plane of its orbit around the Sun.

equinox the points in spring and autumn at which the Sun's path, the ecliptic, crosses the celestial equator, so that the day and night are of approximately equal length. The **vernal equinox** occurs about 21 March and the **autumnal equinox**, 23 September.

Eridanus the sixth largest constellation, which meanders from the celestial equator deep into the southern hemisphere of the sky. Eridanus is represented as a river. Its brightest star is Achernar, a corruption of the Arabic for 'the end of the river'.

ESA abbreviation for **European Space Agency**.

escape velocity minimum velocity with which an object must be projected for it to escape from the gravitational pull of a planetary body. In the case of the Earth, the escape velocity is 11.2 kps/6.9 mps; the Moon, 2.4 kps/1.5 mps; Mars, 5 kps/3.1 mps; and Jupiter, 59.6 kps/37 mps.

European Space Agency (ESA) organization of European countries (Austria, Belgium, Denmark, Finland, France, Germany, Ireland, Italy, the Netherlands, Norway, Spain, Sweden, Switzerland, and the UK) that engages in space research and technology. It was founded in 1975, with headquarters in Paris.

falling star popular name for a meteor.

Gemini prominent zodiacal constellation in the northern hemisphere represented as the twins Castor and Pollux. Its brightest star is Pollux; Castor is a system of six stars. The Sun passes through Gemini from late June to late July. Each December, the Geminid meteors radiate from Gemini.

Great Bear popular name for the constellation Ursa Major.

Hale-Bopp, Comet see Comet Hale-Bopp.

Halley's comet comet that orbits the Sun about every 76 years, named after Edmond Halley, who calculated its orbit. It is the brightest and most conspicuous of the periodic comets. Recorded sightings go back over 2,000 years. It travels around the Sun in the opposite direction to the planets. Its orbit is inclined at almost 20° to the main plane of the Solar System and ranges between the orbits of Venus and Neptune. It will next reappear in 2061.

Jodrell Bank site in Cheshire, England, of the Nuffield Radio Astronomy Laboratories of the University of Manchester. Its largest instrument is the 76-m/250-ft radio dish (the Lovell Telescope), completed in 1957 and modified in 1970. An elliptical radio dish, measuring 38 x 25 m/125 x 82 ft and capable of working at shorter wave lengths , was introduced in 1964.

Kennedy Space Center NASA launch site on Merritt Island, near Cape Canaveral, Florida, used for *Apollo* and space-shuttle launches. The first flight to land on the Moon (1969) and *Skylab*, the first orbiting laboratory (1973), were launched here.

Leo zodiacal constellation in the northern hemisphere, represented as a lion. The Sun passes through Leo from mid-

August to mid-September. Its brightest star is first-magnitude Regulus at the base of a pattern of stars called the Sickle.

Libra faint zodiacal constellation on the celestial equator adjoining Scorpius, and represented as the scales of justice. The Sun passes through Libra during November. The constellation was once considered to be a part of Scorpius, seen as the scorpion's claws.

light year the distance travelled by a beam of light in a vacuum in one year, approximately 9.46 trillion (million million) km/5.88 trillion miles.

Local Group a cluster of about 30 galaxies that includes our own, the Milky Way. Like other groups of galaxies, the Local Group is held together by the gravitational attraction among its members, and does not expand with the expanding universe. Its two largest galaxies are the Milky Way and the Andromeda galaxy; most of the others are small and faint.

magnetosphere volume of space, surrounding a planet, controlled by the planet's magnetic field, and acting as a magnetic 'shell'. The Earth's magnetosphere extends 64,000 km/40,000 mi towards the Sun, but many times this distance on the side away from the Sun.

magnitude measure of the brightness of a star or other celestial object. The larger the number denoting the magnitude, the fainter the object. Zero or first magnitude indicates some of the brightest stars. Still brighter are those of negative magnitude, such as Sirius, whose magnitude is −1.46. **Apparent magnitude** is the brightness of an object as seen from Earth; **absolute magnitude** is the brightness at a standard distance of 10 parsecs (32.6 light years).

meteorite piece of rock or metal from space that reaches the surface of the Earth, Moon, or other body. Most meteorites are thought to be fragments from asteroids, although some may be pieces from the heads of comets. Most are stony, although some are made of iron and a few have a mixed rock-iron composition.

moon any natural satellite that orbits a planet. Mercury and Venus are the only planets in the Solar System that do not have moons.

NASA (acronym for National Aeronautics and Space Administration) US government agency for space flight and aeronautical research, founded in 1958 by the National Aeronautics and Space Act. Its headquarters are in Washington, DC, and its main installation is at the Kennedy Space Center in Florida. NASA's early planetary and lunar programmes included Pioneer spacecraft from 1958, which gathered data for the later crewed missions, the most famous of which took the first people to the Moon in *Apollo 11* on 16–24 July 1969.

nutation a slight 'nodding' of the Earth in space, caused by the varying gravitational pulls of the Sun and Moon. Nutation changes the angle of the Earth's axial tilt (average 23.5°) by about 9 seconds of arc to either side of its mean position, a complete cycle taking just over 18.5 years.

observatory site or facility for observing astronomical or meteorological phenomena. The earliest recorded observatory was in Alexandria, north Africa, built by Ptolemy Soter in about 300 BC. The modern observatory dates from the invention of the telescope. Observatories may be ground-based, carried on aircraft, or sent into orbit as satellites, in space stations, and on the space shuttle.

opposition the moment at which a body in the Solar System lies opposite the Sun in the sky as seen from the Earth and crosses the meridian at about midnight.

orbit path of one body in space around another, such as the orbit of Earth around the Sun, or the Moon around Earth. When the two bodies are similar in mass, as in a binary star, both bodies move around their common centre of mass. The movement of objects in orbit follows Johann Kepler's laws, which apply to artificial satellites as well as to natural bodies.

Orion very prominent constellation in the equatorial region of the sky, identified with the hunter of Greek mythology.

parallax the change in the apparent position of an object against its background when viewed from two different positions. In astronomy, nearby stars show a shift owing to parallax when viewed from different positions on the Earth's orbit around the Sun. A star's parallax is used to deduce its distance from the Earth. Nearer bodies such as the Moon, Sun, and planets also show a parallax caused by the motion of the Earth. **Diurnal parallax** is caused by the Earth's rotation.

parsec a unit (symbol pc) used for distances to stars and galaxies. One parsec is equal to 3.2616 light years, 2.063 x 10^5 astronomical units, and 3.086 x 10^{13} km.

photosphere visible surface of the Sun, which emits light and heat. About 300 km/200 mi deep, it consists of incandescent gas at a temperature of 5,800K (5,530°C/9,980°F). Rising cells of hot gas produce a mottling of the photosphere known as **granulation**, each granule being about 1,000 km/620 mi in diameter. The photosphere is often marked by large, dark patches called sunspots.

Pisces inconspicuous zodiac constellation, mainly in the northern hemisphere between Aries and Aquarius, near Pegasus. It is represented as two fish tied together by their tails. The Circlet, a delicate ring of stars, marks the head of the western fish in Pisces. The constellation contains the **vernal equinox**, the point at which the Sun's path around the sky (the **ecliptic**) crosses the celestial equator. The Sun reaches this point around 21 March each year as it passes through Pisces from mid-March to late April.

planetary nebula shell of gas thrown off by a star at the end of its life. Planetary nebulae have nothing to do with planets. They were named by William Herschel, who thought their rounded shape resembled the disc of a planet. After a star such as the Sun has expanded to become a red giant, its outer layers are ejected into space to form a planetary nebula, leaving the core as a white dwarf at the centre.

Pleiades an open star cluster about 400 light years away in the constellation Taurus, represented as the Seven Sisters of Greek mythology. Its brightest stars (highly luminous, blue-white giants only a few million years old) are visible to the naked eye, but there are many fainter ones.

precession slow wobble of the Earth on its axis, like that of a spinning top. The gravitational pulls of the Sun and Moon on the Earth's equatorial bulge cause the Earth's axis to trace out a circle on the sky every 25,800 years. The position of the celestial poles is constantly changing owing to precession, as are the positions of the equinoxes (the points at which the celestial equator intersects the Sun's path around the sky). The **precession of the equinoxes** means that there is a gradual westward drift in the ecliptic – the path that the Sun appears to follow – and in the coordinates of objects on the celestial sphere.

radar astronomy bouncing of radio waves off objects in the Solar System, with reception and analysis of the 'echoes'. Radar contact with the Moon was first made in 1945 and with Venus in 1961. The travel time for radio reflections allows the distances of objects to be determined accurately. Analysis of the reflected beam reveals the rotation period and allows the object's surface to be mapped. The rotation periods of Venus and Mercury were first determined by radar. Radar maps of Venus were obtained first by Earth-based radar and subsequently by orbiting space probes.

red shift the lengthening of the wavelengths of light from an object as a result of the object's motion away from us. It is an example of the Doppler effect. The red shift in light from galaxies is evidence that the universe is expanding.

Royal Greenwich Observatory the national astronomical observatory of the UK, founded in 1675 at Greenwich, SE London, England, to provide navigational information for sailors. After World War II it was moved to Herstmonceux Castle, Sussex; in 1990 it was transferred to Cambridge. It also operates telescopes on La Palma in the Canary Islands, including the 4.2-m/165-in William Herschel Telescope, commissioned in 1987.

Sagittarius bright zodiac constellation in the southern hemisphere, represented as a centaur aiming a bow and arrow at neighbouring Scorpius. The Sun passes through Sagittarius from mid-December to mid-January, including the winter solstice, when it is farthest south of the Equator. The constellation contains many nebulae and globular clusters, and open star clusters. Kaus Australis and Nunki are its brightest stars. The centre of our galaxy, the Milky Way, is marked by the radio source Sagittarius A.

satellite any small body that orbits a larger one, either natural or artificial. Natural satellites that orbit planets are called moons.

Scorpius bright zodiacal constellation in the southern hemisphere between Libra and Sagittarius, represented as a scorpion. The Sun passes briefly through Scorpius in the last week of November. The heart of the scorpion is marked by the bright red supergiant star Antares. Scorpius contains rich Milky Way star fields, plus the strongest X-ray source in the sky, Scorpius X-1. The whole area is rich in clusters and nebulae.

sextant navigational instrument for determining latitude by measuring the angle between some heavenly body and the horizon. It was invented in 1730 by John Hadley (1682–1744) and can be used only in clear weather.

shooting star popular name for a meteor.

singularity in astrophysics, the point in space–time at which the known laws of physics break down. Singularity is predicted to exist at the centre of a black hole, where infinite gravitational forces compress the infalling mass of a collapsing star to infinite density. It is also thought, according to the Big Bang model of the origin of the universe, to be the point from which the expansion of the universe began.

solar wind stream of atomic particles, mostly protons and electrons, from the Sun's corona, flowing outwards at speeds of between 300 kps/200 mps and 1,000 kps/600 mps.

solstice either of the days on which the Sun is farthest north or south of the celestial equator each year. The **summer solstice**, when the Sun is farthest north, occurs around 21 June; the **winter solstice** around 22 December.

space, or **outer space**, the void that exists beyond Earth's atmosphere. Above 120 km/75 mi, very little atmosphere remains, so objects can continue to move quickly without extra energy. The space between the planets is not entirely empty, but filled with the tenuous gas of the solar wind as well as dust specks.

sunspot dark patch on the surface of the Sun, actually an area of cooler gas, thought to be caused by strong magnetic fields that block the outward flow of heat to the Sun's surface. Sunspots consist of a dark central **umbra**, about 4,000K (3,700°C/6,700°F), and a lighter surrounding **penumbra**, about 5,500K (5,200°C/9,400°F). They last from several days to over a month, ranging in size from 2,000 km/1,250 mi to groups stretching for over 100,000 km/62,000 mi.

Taurus conspicuous zodiacal constellation in the northern hemisphere near Orion, represented as a bull. The Sun passes through Taurus from mid-May to late June.

tektite small, rounded glassy stone, found in certain regions of the Earth, such as Australasia. Tektites are probably the scattered drops of molten rock thrown out by the impact of a large meteorite.

Titan largest moon of the planet Saturn, with a diameter of 5,150 km/3,200 mi and a mean distance from Saturn of 1,222,000 km/759,000 mi. It was discovered in 1655 by Dutch mathematician and astronomer Christiaan Huygens, and is the second largest moon in the Solar System (Ganymede, of Jupiter, is larger).

UFO abbreviation for unidentified flying object.

unidentified flying object, or **UFO**, any light or object seen in the sky whose immediate identity is not apparent. Despite unsubstantiated claims, there is no evidence that UFOs are alien spacecraft. On investigation, the vast majority of sightings turn out to have been of natural or identifiable objects, notably bright stars and planets, meteors, aircraft, and satellites, or to have been perpetrated by pranksters. The term **flying saucer** was coined in 1947.

Ursa Major (Latin 'Great Bear') the third largest constellation in the sky, in the north polar region. Its seven brightest stars make up the familiar shape, or asterism, of the **Big Dipper** or **Plough**. The second star of the handle of the dipper, called Mizar, has a companion star, Alcor.   Two stars forming the far side of the bowl act as pointers to the north star, Polaris. Dubhe, one of them, is the constellation's brightest star.

Ursa Minor (Latin 'Little Bear') small constellation of the northern hemisphere. It is shaped like a dipper, with the bright north pole star Polaris at the end of the handle.

Van Allen radiation belts two zones of charged particles around the Earth's magnetosphere, discovered in 1958 by US physicist James Van Allen. The atomic particles come from the Earth's upper atmosphere and the solar wind, and are trapped by the Earth's magnetic field. The inner belt lies 1,000–5,000 km/620–3,100 mi above the Equator, and contains protons and electrons. The outer belt lies 15,000–25,000 km/9,300–15,500 mi above the Equator, but is lower around the magnetic poles. It contains mostly electrons from the solar wind.

Virgo zodiacal constellation of the northern hemisphere, the second largest in the sky. It is represented as a maiden holding an ear of wheat, marked by first-magnitude Spica, Virgo's brightest star. The Sun passes through Virgo from late September to the end of October.

zenith uppermost point of the celestial horizon, immediately above the observer; the nadir is below, diametrically opposite.

Further Reading

Chown, Marcus *Afterglow of Creation* (1993)

Clark, Stuart *Universe in Focus* (1997)

Gleiser, Marcelo *The Dancing Universe: From Creation Myths to the Big Bang* (1998)

Gribbin, John *In Search of the Big Bang* (1986), In the Beginning (1993), *In Search of the Edge of Time* (1992), *Companion to the Cosmos* (1996), *Time and Space* (with Mary Gribbin, 1994)

Gribbin, John, and Rees, Martin *The Stuff of the Universe* (1990)

Harrington, Philip *Eclipse* (1997)

Hogan, Craig *The Little Book of the Big Bang: a Cosmic Primer* (1998)

Illingworth, Valerie *Macmillan Dictionary of Astronomy* (1988)

King, Henry *History of the Telescope* (1955)

Lewis, H A *Times Atlas of the Moon* (1969)

Lovell, Bernard *The Story of Jodrell Bank* (1968)

Mather, John C *The Very First Light: The True Inside Story of the Scientific Journey Back to the Dawn of the Universe* (1996)

Mitton, Jacqueline *Concise Dictionary of Astronomy* (1991), *Penguin Dictionary of Astronomy* (1993)

Moore, Patrick *Atlas of the Universe* (1970), *The Amateur Astronomer* (1970), *Guide to the Planets* (1976), *Guide to Comets* (1977), *Exploring the Night Sky with Binoculars* (1986), *The Guinness Book of Astronomy* (1992)

Muirden, James *Astronomy with a Small Telescope* (1985)

Nahin, Paul *Time Machines* (1993)

Nicolson, Iain *Heavenly Bodies: Beginner's Guide to Astronomy* (1995)

North, John *Astronomy and Cosmology* (1994)

Norton, Arthur Philip *Norton's 2000* (1989)

Overbye, Dennis *Lonely Hearts of the Cosmos* (1991)

Phillips, Kenneth *Guide to the Sun* (1992)

Plant, Malcolm *Dictionary of Space* (1986)

Ronan, C A *The Natural History of the Universe* (1994)

Rycroft, Michael (ed) *The Cambridge Encyclopedia of Space* (1990)

Schweighauser, Charles A *Astronomy from A to Z: A Dictionary of Celestial Objects and Ideas* (1991)

Sheehan, William *The Planet Mars: A History of Observation and Discovery* (1996)

Smith, Robert *The Expanding Universe 1900–31* (1982)

Sullivan, Walter *We Are Not Alone* (1970)

Tayler, Roger *The Hidden Universe* (1991)

Whipple, Fred *Orbiting the Sun: Planets and Satellites of the Solar System* (1981)

Will, Clifford *Was Einstein Right?* (1986)

EARTH SCIENCE: EVOLUTION AND STRUCTURE OF PLANET EARTH

Earth is the third planet from the Sun. It was formed with the rest of the Solar System by consolidation of interstellar dust approximately 4.6 billion years ago. Earth is the only planet on which life is known to exist. Its life-supporting atmosphere is composed of nitrogen (78.09%), oxygen (20.95%), argon (0.93%), carbon dioxide (0.03%), and less than 0.0001% neon, helium, krypton, hydrogen, xenon, ozone, and radon. Life here began an estimated 3.5–4.0 billion years ago.

About 70% of Earth's surface, including the north and south polar ice caps, is covered with water, totalling 361,000,000 sq km/139,400,000 sq mi. The greatest depth of Earth's hydrosphere is the Mariana Trench in the Pacific Ocean, at a depth of 11,034 m/36,201 ft. Earth's land surface covers 150,000,000 sq km/57,500,000 sq mi. The greatest height above sea level is Mount Everest at 8,872 m/29,118 ft.

Earth science is the scientific study of the planet Earth as a whole. The mining and extraction of minerals, weather prediction, earthquakes, the pollution of the atmosphere, and the forces that shape the physical world all fall within its scope of study. The term 'earth science' reflects a broadening of the traditional field of geology to include any physical or life sciences applied to the study of the Earth. The broadening of the discipline reflects scientists' concern that an understanding of the global aspects of the Earth's structure and its past will hold the key to how humans affect its future, ensuring that its resources are used in a sustainable way. Similarly, we now know that our understanding of Earth is linked to our knowledge of the other planets in our Solar System and the evolution of the universe.

Earth science includes geology, mineralogy, and palaeontology, as well as geophysics, geochemistry, mineral physics, palaeomagnetics, meteorology, and oceanography. Meteorology, oceanography, and the study of Earth's hydrosphere and atmosphere will be discussed in the following chapter.

structure and composition of the Earth

Earth is almost spherical in shape, flattened slightly at the poles, and is composed roughly of three concentric layers: the crust, the mantle, and the core.

crust

The outermost part of the structure of Earth is the crust. There are two distinct types of crust, **oceanic crust** and **continental crust**. The oceanic crust is on average about 10 km/6.2 mi thick and consists mostly of basaltic types of rock. By contrast, the continental crust is largely made of granite and is more complex in its structure. Because of the movements of plate tectonics (see below), the oceanic crust is in no place older than about 200 million years. However, parts of the continental crust are over 3 billion years old.

Beneath a layer of surface sediment, the oceanic crust is made up of a layer of basalt, followed by a layer of gabbro. The composition of the oceanic crust overall shows a high proportion of silicon and **magnesium** oxides, hence named **sima** by geologists. The continental crust varies in thickness from about 40 km/25 mi to 70 km/45 mi, being deeper beneath mountain ranges. The surface layer consists of many

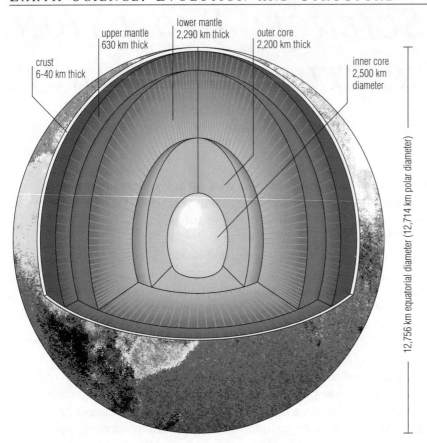

crust
6-40 km thick

upper mantle
630 km thick

lower mantle
2,290 km thick

outer core
2,200 km thick

inner core
2,500 km
diameter

12,756 km equatorial diameter (12,714 km polar diameter)

Earth Inside the Earth. The surface of the Earth is a thin crust about 6 km/4 mi thick under the sea and 40 km/25 mi thick under the continents. Under the crust lies the mantle about 2,900 km/1,800 mi thick and with a temperature of 1,500–3,000°C/2,700–5,400°F. The outer core is about 2,250 km/1,400 mi thick, of molten iron and nickel. The inner core is probably solid iron and nickel at about 5,000°C/9,000°F.

kinds of sedimentary and igneous rocks. Beneath lies a zone of metamorphic rocks built on a thick layer of granodiorite. Silicon and aluminium oxides dominate the composition and the name **sial** is given to continental crustal material.

Earth
To remember the most common elements of the planet's crust, in descending order:
Only **s**illy **a**sses **i**n **c**ollege **s**tudy **p**ast **m**idnight
(**O**xygen, **s**ilicon, **a**luminium, **i**ron, **c**alcium, **s**odium, **p**otassium, **m**agnesium)

lithosphere
The crust is sometimes confused with the lithosphere, Earth's brittle outer skin forming the jigsaw of plates that take part in the movements of plate tectonics. The crust is only the topmost layer of the lithosphere. The lithosphere is comprised of crust underlain by a rigid portion of the upper mantle. The lithosphere is about 100 km/63 mi thick and moves about on the more elastic and less rigid asthenosphere. The movements of the lithospheric plates are responsible for Earth's major

physical features, such as mountains, islands, and deep-sea trenches.

continent
The seven large land masses of Earth's crust, as distinct from the oceans, are called continents. They are Asia, Africa, North America, South America, Europe, Australia, and Antarctica. Because continents are part of lithospheric plates, they are constantly moving and evolving due to plate tectonics. A continent does not end at the coastline; its boundary is the edge of the shallow continental shelf, which may extend several hundred kilometres out to sea.

At the centre of each continental mass lies a shield or **craton**, a deformed mass of old metamorphic rocks

continents
To remember the seven continents:
Eat **an** **as**pirin **af**ter **a** **n**ight-time **s**nack
(**E**urope, **An**tarctica, **As**ia, **Af**rica, **A**ustralia, **N**orth America, **S**outh America)
The second letter in the first three 'A' words helps to remember the 'A' continents.

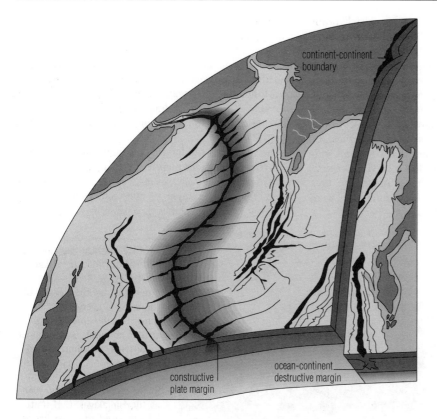

continent-continent boundary

constructive plate margin

ocean-continent destructive margin

crust The crust of the Earth is the top layer of the tectonic plates. These plates are bounded by different kinds of margins. In mid-ocean, there are constructive plate margins, where magma wells up from the Earth's interior, forming new crust. On continent–continent destructive margins, mountain ranges are flung up by the collision of two continents. At an ocean–continent destructive margin, the plate containing denser oceanic crust is forced beneath the plate with less dense continental crust, forming an area of volcanic instability.

dating from Precambrian times. The shield is thick, compact, and solid (the Canadian Shield is an example), having undergone all the mountain-building activity it is ever likely to, and is usually worn flat. Around the shield is a concentric pattern of mountains comprised of folded rock, with older ranges, such as the Rockies, closest to the shield, and younger ranges, such as the coastal ranges of North America, farther away. This general concentric pattern is modified when two continental masses have moved together and they become welded with a great mountain range along the join, the way Europe and northern Asia are joined along the Urals. If a continent is torn apart, the new continental edges have no mountains formed by wrinkling and folding of the crust; for instance, the western coast of Africa, which rifted apart from South America 200 million years ago.

isostasy

The theoretical balance in buoyancy of all parts of the Earth's crust, as though they were floating on a denser layer beneath, is called isostasy. There are two

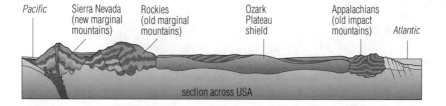

Pacific   Sierra Nevada (new marginal mountains)   Rockies (old marginal mountains)   Ozark Plateau shield   Appalachians (old impact mountains)   *Atlantic*

section across USA

continent The North American continent is growing in the west as a result of collision with the Pacific plate. On the east of the wide area of the Ozark Plateau shield lie the Appalachian Mountains, showing where the continent once collided with another continent. The eastern coastal rifting formed when the continents broke apart. On the western edge, new impact mountains have formed.

theories of the mechanism of isostasy, the **Airy hypothesis** and the **Pratt hypothesis**, both of which have validity. In the Airy hypothesis crustal blocks have the same density but different depths: like ice cubes floating in water, higher mountains have deeper roots. In the Pratt hypothesis, crustal blocks have different densities allowing the depth of crustal material to be the same.

There appears to be more geological evidence to support the Airy hypothesis of isostasy. During an ice age the weight of the ice sheet pushes that continent into the Earth's mantle; once the ice has melted, the continent rises again. This accounts for shoreline features being found some way inland in regions that were heavily glaciated during the Pleistocene period.

mantle

The mantle is the intermediate zone of the Earth between the crust and the core, accounting for 82% of Earth's volume. The boundary between the mantle and the crust above is the Mohorovicic discontinuity, or **Moho**, located at an average depth of 32 km/20 mi. The lower boundary with the core is the Gutenburg discontinuity at an average depth of 2,900 km/ 1,813 mi.

The mantle is subdivided into **upper mantle, transition zone,** and **lower mantle,** based upon the different velocities with which seismic waves travel through these regions. The upper mantle includes a zone characterized by low velocities of seismic waves, called the

isostasy Isostasy explains the vertical distribution of Earth's crust. George Bedell Airy proposed that the density of the crust is everywhere the same and the *thickness* of crustal material varies. Higher mountains are compensated by deeper roots. This explains the high elevations of most major mountain chains, such as the Himalayas. G H Pratt hypothesized that the *density* of the crust varies, allowing the base of the crust to be the same everywhere. Sections of crust with high mountains, therefore, would be less dense than sections of crust where there are lowlands. This applies to instances where density varies, such as the difference between continental and oceanic crust.

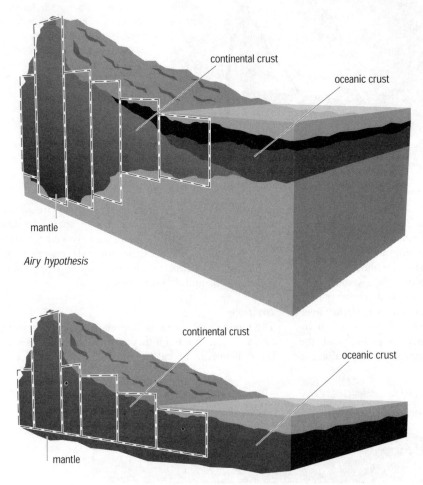

continental crust

oceanic crust

mantle

Airy hypothesis

continental crust

oceanic crust

mantle

Pratt hypothesis

Highest and Lowest Elevations by Continent

Continent	Highest elevation	Height		Deepest elevation	Depth below sea level	
		m	ft		m	ft
Africa	Kilimanjaro, Tanzania	5,895	19,340	Lake Assal, Djibouti	153	502
Antarctica	Vinson Massif	5,140	16,863	Hollick-Keyon plateau[1]	2,468	8,097
Asia	Everest, China–Nepal	8,848	29,029	Dead Sea, Israel/Jordan	400	1,312
Europe	Elbrus, Russia	5,642	18,510	Caspian Sea, Azerbaijan/Russia/		
				Kazakhstan/Turkmenistan/Iran	28	92
North America	McKinley (AK), USA	6,194	20,321	Death Valley (CA), USA	86	282
Oceania	Jaya, New Guinea	5,030	16,502	Lake Eyre, South Australia	16	52
South America	Cerro Aconcagua, Argentina	6,960	22,834	Valdés Peninsula, Argentina	40	131

[1] This is the deepest estimated depression, beneath the Marie Byrd Land ice cap. Vast areas of western Antarctica would be below sea level if stripped of their ice sheet; lower points beneath the ice may exist.

low velocity zone, at 72 km/45 mi to 250 km/155 mi depth. This zone corresponds to the aesthenosphere (see below) upon which Earth's tectonic plates of lithosphere glide. Seismic velocities in the upper mantle are overall less than those in the transition zone and those of the transition zone are in turn less than those of the lower mantle. Faster propagation of seismic waves in the lower mantle implies that the lower mantle is more dense than the upper mantle.

The mantle is composed primarily of magnesium, silicon, and oxygen in the form of silicate minerals. In the upper mantle, the silicon in silicate minerals, such as olivine, is surrounded by four oxygen atoms. Deeper in the transition zone, greater pressures promote denser packing of oxygen such that some silicon is surrounded by six oxygen atoms, resulting in magnesium silicates with garnet and pyroxene structures. Deeper

still in the lower mantle, all silicon is surrounded by six oxygen atoms so that the new mineral $MgSiO_3$-perovskite predominates.

asthenosphere

The asthenosphere is the rigid layer of the upper mantle that underlies the lithosphere, typically beginning at a depth of approximately 100 km/63 mi and extending to depths of approximately 260 km/160 mi.

Sometimes referred to as the 'weak sphere', the asthenosphere is characterized by being weaker and more elastic than the surrounding mantle. Its lack of shear strength results from the high temperature of the rocks approaching the melting point. Since seismic waves travel more slowly in the asthenosphere; it is also referred to as the 'low velocity zone'.

The asthenosphere's elastic behaviour and low

Deepest Geographical Depressions in the World

Depression	Location	Maximum depth below sea level	
		m	ft
Dead Sea	Israel/Jordan	400	1,312
Turfan Depression	Xinjiang, China	154	505
Lake Assal	Djibouti	153	502
Qattâra Depression	Egypt	133	436
Poloustrov Mangyshlak	Kazakhstan	131	430
Danakil Depression	Ethiopia	120	394
Death Valley	California, USA	86	282
Salton Sink	California, USA	71	233
Zapadnyy Chink Ustyurta	Kazakhstan	70	230
Priaspiyskaya Nizmennost	Russia/Kazakhstan	67	220
Ozera Sarykamysh	Uzbekistan/Kazakhstan	45	148
El Faiyûm	Egypt	44	144
Valdés Peninsula	Argentina	40	131

viscosity allow the overlying plates to move laterally and also allow overlying crust and mantle to move vertically in response to gravity to achieve **isostatic equilibrium.**

core

The innermost part of Earth is the core. It is divided into an outer core, which begins at a depth of 2,898 km/1,800 mi, and an inner core, which begins at a depth of 4,982 km/3,095 mi. Both parts are thought to consist of iron-nickel alloy. The outer core is liquid and the inner core is solid.

The fact that seismic shear waves (see seismic wave below) disappear at the mantle–outer core boundary indicates that the outer core is molten, since shear waves cannot travel through fluid. Scientists infer the iron-nickel rich composition of the core from Earth's density and its moment of inertia. The temperature of the core, as estimated from the melting point of iron at high pressure, is thought to be at least 4,000°C/7,232°F, but remains controversial. Earth's magnetic field is believed to be the result of the motions involving the inner and outer cores.

plate tectonics

Plate tectonics is the theory formulated in the 1960s to explain the phenomena of ♢continental drift and seafloor spreading, and the formation of the major physical features of the Earth's surface. The Earth's outermost layer, the lithosphere, is regarded as a jigsaw puzzle of rigid plates that move relative to each other, probably under the influence of convection currents in the mantle beneath. At the margins of the plates, where they collide or move apart, major landforms such as mountains, volcanoes, ocean trenches, and mid-ocean ridges are created. The rate of plate movement is at most 15 cm/6 in per year.

New plate material is generated between two plates along the mid-ocean ridges, where basaltic lava is poured out by submarine volcanoes. Basaltic lava spreads outwards away from the ridge crest at 1–6 cm/0.5–2.5 in per year. Plate material is consumed at a rate of 5–15 cm/2–6 in per year at the site of the deep ocean trenches, for example, along the Pacific coast of South America. These trenches are sites, called subduction zones, where two plates of lithosphere meet; the one bearing denser ocean-floor basalts plunges beneath the adjacent continental mass at an angle of 45°, giving rise to shallow earthquakes near the coast and progressively deeper earthquakes inland. In places the sinking plate may descend beneath an island arc of offshore islands, as in the Aleutian Islands and Japan, and in this case the shallow earthquakes will occur beneath the island arc. The destruction of

ocean crust in this way accounts for another well-known geological fact – that there are no old rocks found in the ocean basins. The oldest sediments found are 150 million years old, but the vast majority are less than 80 million years old. This suggests that plate tectonics has been operating for at least the last 200 million years. In other areas plates slide past each other along transform faults, giving rise to shallow earthquakes. Sites where three plates meet are known as triple junctions.

Plate Tectonics
http://www.seismo.unr.edu/ftp/pub/louie/class/ 100/ plate-tectonics.html
Well-illustrated site on this geological phenomenon. As well as the plentiful illustrations, this site also has a good clear manner of explaining the way the plates of the Earth's crust interact to produce seismic activity.

---

plate

A plate (or tectonic plate) is one of several sections of lithosphere approximately 100 km/60 mi thick and at least 200 km/120 mi across, which together comprise the outermost layer of the Earth like the pieces of the cracked surface of a hard-boiled egg.

The plates are made up of two types of crustal material: **oceanic crust** (sima) and **continental crust** (sial), both of which are underlain by a solid layer of mantle. Dense oceanic crust lies beneath Earth's oceans and consists largely of basalt. Continental crust, which underlies the continents and their continental shelves, is thicker, less dense, and consists of rocks rich in silica and aluminum.

Due to convection in the Earth's mantle these pieces of lithosphere are in motion, riding on a more plastic layer of the mantle, called the aesthenosphere. Mountains, volcanoes, earthquakes, and other geological features and phenomena all come about as a result of interaction between the plates.

There are three types of plate margins: constructive, destructive, and conservative.

constructive margins Where two plates are moving apart from each other, molten rock from the mantle wells up in the space between the plates and hardens to form new crust, usually in the form of a mid-ocean ridge (such as the Mid-Atlantic Ridge). This is called a constructive margin because the newly formed crust accumulates on either side of the ridge, causing the seafloor to spread; the floor of the Atlantic Ocean is growing by 5 cm/2 in each year because of the welling-up of new material at the Mid-Atlantic Ridge.

destructive margins Destructive margins occur where two plates are moving towards each other. When plates

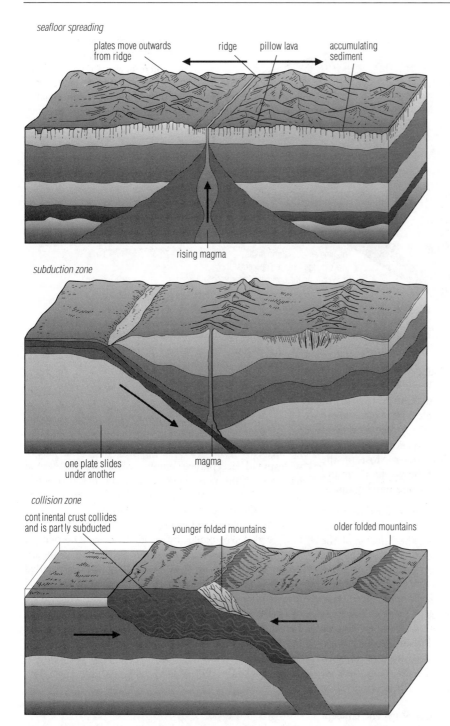

seafloor spreading

plates move outwards from ridge | ridge | pillow lava | accumulating sediment

rising magma

subduction zone

one plate slides under another | magma

collision zone

continental crust collides and is partly subducted | younger folded mountains | older folded mountains

plate tectonics The three main types of action in plate tectonics. (top) Seafloor spreading. The upwelling of magma forces apart the plates, producing new crust at the joint. Rapid extrusion of magma produces a domed ridge; more gentle spreading produces a central valley. (middle) The drawing downwards of an oceanic plate beneath a continent produces a range of volcanic mountains parallel to the plate edge. (bottom) Collision of continental plates produces immense collisional mountains, such as the Himalayas.

Younger mountains are found near the coast with older ranges inland. The plates of the Earth's lithosphere are always changing in size and shape of each plate as material is added at constructive margins and removed at destructive margins. The process is extremely slow, but it means that the tectonic history of the Earth cannot be traced back further than about 200 million years.

with denser oceanic crust meet plates with less dense continental crust, the denser of the two plates may be forced under the other into a region called the **subduction zone**. The descending plate melts to form a body of ⋄magma, which may then rise to the surface through cracks and faults to form volcanoes. If the two plates consist of more buoyant continental crust, subduction does not occur. Instead, the crust crumples gradually to form ranges of young mountains, such as the Himalayas in Asia, the Andes in South America, and the Rockies

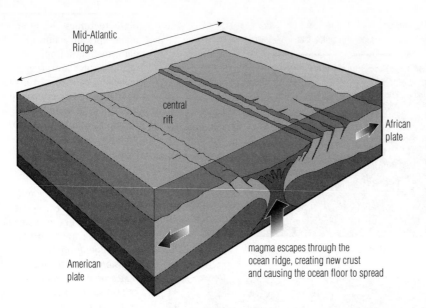

Mid-Atlantic Ridge

central rift

African plate

American plate

magma escapes through the ocean ridge, creating new crust and causing the ocean floor to spread

constructive margin A rift in the lithospheric plates where new material is being formed causes the plates to be pushed apart. This usually occurs as a result of volcanic action.

in North America. This process of mountain building is termed orogeny, or orogenesis.

conservative margins Plate boundaries in which two plates slide past each other are called conservative margins because crust is neither created nor destroyed. An example is the ◊San Andreas Fault, California, where the movement of the plates sometimes takes the form of sudden jerks, causing the earthquakes common in the San Francisco–Los Angeles area. Most of the earthquake and volcanic zones of the world are found in regions where two plates meet or are moving apart.

causes of plate movement The causes of plate movement are poorly understood. It has been known for some time that heat flow from the interior of the Earth is high over the ocean ridges, and so thermal convection in the mantle has been proposed as a driving mechanism for the plates. The rising limbs of the convective mantle cells may be the plumes of hot, molten material that rise beneath the ocean ridges to be extruded as basaltic lava. The descending limbs of the convective cells may be linked to subduction zones. Oceanic crust is continually produced by, and returned to, the mantle, but the continental crust rocks, because of their buoyancy, remain on the surface.

development of plate tectonics theory The concept of continental drift was first put forward in 1915 in a book entitled *The Origin of Continents and Oceans* by the German meteorologist Alfred Wegener, who recognized that continental plates rupture, drift apart, and eventually collide with one another. Wegener's theory explained why the shape of the east coast of

the Americas and that of the west coast of Africa seem to fit together like pieces of a jigsaw puzzle; evidence for the drift came from the presence of certain rock deposits which indicated that continents have changed position over time. In the early 1960s scientists discovered that most earthquakes occur along lines parallel to ocean trenches and ridges, and in 1965 the theory of plate tectonics was formulated by Canadian geophysicist John Tuzo Wilson; it has now gained widespread acceptance among earth scientists who have traced the movements of tectonic plates millions of years into the past. The widely accepted belief is that all the continents originally formed part of an enormous single land mass, known as Pangaea. This land was surrounded by a giant ocean known as Panthalassa. About 200 million years ago, Pangaea began to break up into two large masses called Gondwanaland and Laurasia, which in turn separated into the continents as they are today, and which have drifted to their present locations. In 1995 US and French geophysicists produced the first direct evidence that the Indo-Australian plate has split in two in the middle of the Indian Ocean, just south of the Equator. They believe the split began about 8 million years ago.

major physical features of the Earth's surface

fault A fracture in the Earth either side of which rocks have moved past one another is called a fault. Faults involve displacements, or offsets, ranging from the microscopic scale to hundreds of kilometres. Large offsets along a fault are the result of the accumulation of smaller movements (metres or less) over long

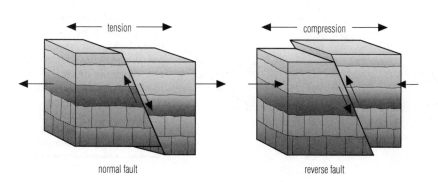

normal fault   reverse fault

fault Faults are caused by the movement of rock layers, producing such features as block mountains and rift valleys. A normal fault is caused by a tension or stretching force acting in the rock layers. A reverse fault is caused by compression forces. Faults can continue to move for thousands or millions of years.

periods of time. Large motions cause detectable earthquakes.

Faults are planar features. Fault orientation is described by the inclination of the fault plane with respect to the horizontal (see ⟠dip) and its direction in the horizontal plane (see ⟠strike). Faults at a high angle with respect to the horizontal (in which the fault plane is steep) are classified as either **normal faults**, where one block has apparently moved downhill along the inclined fault plane, or **reverse faults**, where one block appears to have moved uphill along the fault plane. Normal faults occur where rocks on either side have moved apart. Reverse faults occur where rocks on either side have been forced together. A reverse fault that forms a low angle with the horizontal plane is called a **thrust fault**.

A **lateral fault**, or **tear fault**, occurs where the relative movement along the fault plane is sideways. A particular kind of fault found only in ocean ridges is the **transform fault**. On a map, an ocean ridge has a stepped appearance. The ridge crest is broken into sections, each section offset from the next. Between each section of the ridge crest the newly generated plates are moving past one another at different rates. Transform faults form to accommodate these variations in spreading rates.

Faults produce lines of weakness on the Earth's surface (along their strike) that are often exploited by processes of weathering and erosion. Coastal caves and geos (narrow inlets) often form along faults and, on a larger scale, rivers may follow the line of a fault.

San Andreas Fault
http://sepwww.stanford.edu/oldsep/joe/fault_images/Bay
AreaSanAndreasFault.html
Detailed tour of the San Andreas Fault and the San Francisco Bay area, with information on the origination of the fault.

mid-ocean ridge A mid-ocean ridge is a mountain range on the seabed indicating the presence of a constructive plate margin (where tectonic plates are moving apart and magma rises to the surface; see plate tectonics above). Ocean ridges, such as the Mid-Atlantic Ridge, consist of many segments offset along transform faults, and can rise thousands of metres above the surrounding seabed.

Ocean ridges usually have a rift valley along their crests, indicating where the flanks are being pulled apart by the growth of the plates of the lithosphere beneath. The crests are generally free of sediment; increasing depths of sediment are found with increasing distance down the flanks.

ocean trench Ocean trenches are deep trenches in the seabed indicating the presence of a destructive margin (produced by the movements of plate tectonics). The subduction, or dragging downwards of one plate of the lithosphere beneath another, means that the ocean floor is pulled down. Ocean trenches are found around the edge of the Pacific Ocean and the northeastern Indian Ocean; minor ones occur in the Caribbean and near the Falkland Islands.

Ocean trenches represent the deepest parts of the ocean floor, the deepest being the Mariana Trench which has a depth of 11,034 m/36,201 ft. At depths of below 6 km/3.6 mi there is no light and very high pressure; ocean trenches are inhabited by crustaceans, coelenterates (for example, sea anemones), polychaetes (a type of worm), molluscs, and echinoderms.

mountain Mountains are natural upward projections of the Earth's surface that are higher and steeper than hills. Mountains are at least 330 m/1,000 ft above the surrounding topography. The process of mountain building (⟠orogeny) consists of volcanism, folding, faulting, and thrusting, resulting from the collision of two tectonic plates at a **convergent margin**. The existing rock is also subjected to high temperatures and

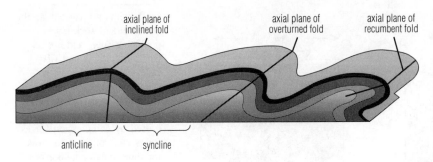

anticline syncline

axial plane of inclined fold

axial plane of overturned fold

axial plane of recumbent fold

fold The folding of rock strata occurs where compression causes them to buckle. Over time, folding can assume highly complicated forms, as can sometimes be seen in the rock layers of cliff faces or deep cuttings in the rock. Folding contributed to the formation of great mountain chains such as the Himalayas.

pressures causing metamorphism. Plutonic activity also can accompany mountain building.

island An island is an area of land surrounded entirely by water. Australia is classed as a continent rather than an island, because of its size.

Islands can be formed in many ways. **Continental islands** were once part of the mainland, but became isolated (by tectonic movement, erosion, or a rise in sea level, for example). **Volcanic islands**, such as Japan, were formed by the explosion of underwater volcanoes. **Coral islands** consist mainly of coral, built up over many years. An **atoll** is a circular coral reef surrounding a lagoon; atolls were formed when a coral reef grew up around a volcanic island that

subsequently sank or was submerged by a rise in sea level. **Barrier islands** are found by the shore in shallow water, and are formed by the deposition of sediment eroded from the shoreline.

Earth phenomena: earthquakes, volcanism, and magnetism

earthquake
An abrupt motion that propagates through the Earth and along its surfaces is called an earthquake. Earthquakes are caused by the sudden release in rocks

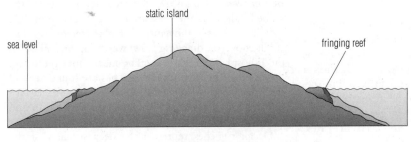

sea level

static island

fringing reef

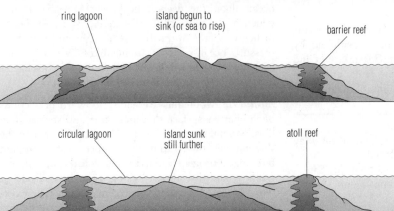

ring lagoon

island begun to sink (or sea to rise)

barrier reef

circular lagoon

island sunk still further

atoll reef

coral atoll The formation of a coral atoll by the gradual sinking of a volcanic island. The reefs fringing the island build up as the island sinks, eventually producing a ring of coral around the spot where the island sank.

Highest Mountains in the World, by Region

Region/mountain	Location	Height m	ft
Africa			
Kilimanjaro	Tanzania	5,895	19,337
Kenya (Batian)	Kenya	5,199	17,057
Ngaliema (formerly Mt Stanley and Margherita Peak)	Democratic Republic of Congo/Uganda	5,110	16,765
Duwoni (formerly Umberto Peak)	Uganda	4,896	16,063
Baker (Edward Peak)	Uganda	4,843	15,889
Alpine Europe			
Mont Blanc	France/Italy	4,807	15,771
Monte Rosa	Switzerland	4,634	15,203
Dom	Switzerland	4,545	14,911
Liskamm	Switzerland/Italy	4,527	14,852
Weisshorn	Switzerland	4,505	14,780
Antarctica			
Vinson Massif		5,140	16,863
Tyree		4,965	16,289
Shinn		4,800	15,748
Gardner		4,690	15,387
Epperley		4,511	14,800
Asia			
Everest	China/Nepal	8,848	29,029
K2	Kashmir/Jammu	8,611	28,251
Kangchenjunga	India/Nepal	8,598	28,208
Lhotse	China/Nepal	8,511	27,923
Yalung Kang	India/Nepal	8,502	27,893
Australia			
Kosciusko	Snowy Mountains, New South Wales	2,230	7,316
Carpathians			
Gerlachvka	Slovak Republic	2,655	8,711
Moldoveanu	Romania	2,544	8,346
Negoiu	Romania	2,535	8,317
Mindra	Romania	2,518	8,261
Peleaga	Romania	2,509	8,232
Caucasia			
Elbrus, West Peak	Russia	5,642	18,510
Dykh Tau	Russia/Georgia	5,203	17,070
Shkhara	Russia/Georgia	5,201	17,063
Kashtan Tau	Russia/Georgia	5,144	16,876
Dzanghi Tau	Russia	5,049	16,565
New Zealand			
Cook (called Aorongi in Maori)	west coast, South Island	3,754	12,316
North and Central America			
McKinley	Alaska, USA	6,194	20,321
Logan, Yukon	Canada	6,050	19,849
North and Central America			
Citlaltépetl (Orizaba)	Mexico	5,610	18,405
St Elias	Alaska, USA/Yukon, Canada	5,489	18,008
Popocatépetl	Mexico	5,452	17,887

Region/mountain	Location	Height	
		m	ft
Oceania[1]			
Jaya	West Irian, Papua New Guinea	5,030	16,502
Daam	West Irian, Papua New Guinea	4,922	16,148
Oost Carstensz (also known as Jayakusumu Timur)	West Irian, Papua New Guinea	4,840	15,879
Trikora	West Irian, Papua New Guinea	4,730	15,518
Enggea	West Irian, Papua New Guinea	4,717	15,476
Polynesia			
Mauna Kea	Hawaii, USA	4,205	13,796
Mauna Loa	Hawaii, USA	4,170	13,681
Pyrenees			
Pico de Aneto	Spain	3,404	11,168
Pico de Posets	Spain	3,371	11,060
Monte Perdido	Spain	3,348	10,984
Pico de la Maladeta	Spain	3,312	10,866
Pic de Vignemale	France/Spain	3,298	10,820
Scandinavia			
Glittertind	Norway	2,472	8,110
Galdhøpiggen	Norway	2,469	8,100
Skagastølstindane	Norway	2,405	7,890
Snøhetta	Norway	2,286	7,500
South America			
Cerro Aconcagua	Argentina	6,960	22,834
Ojos del Salado	Argentina/Chile	6,908	22,664
Bonete	Argentina	6,872	22,546
Nevado de Pissis	Argentina/Chile	6,779	22,241
Huascarán Sur	Peru	6,768	22,204

[1] Including all of Papua New Guinea.

Largest Islands in the World

Island	Location	Area	
		sq km	sq mi
Greenland	northern Atlantic	2,175,600	840,000
New Guinea	southwestern Pacific	800,000	309,000
Borneo	southwestern Pacific	744,100	287,300
Madagascar	Indian Ocean	587,041	226,657
Baffin	Canadian Arctic	507,258	195,928
Sumatra	Indian Ocean	424,760	164,000
Honshu	northwestern Pacific	230,966	89,176
Great Britain	northern Atlantic	218,078	84,200
Victoria	Canadian Arctic	217,206	83,896
Ellesmere	Canadian Arctic	196,160	75,767
Sulawesi	Indian Ocean	189,216	73,057
South Island, New Zealand	southwestern Pacific	149,883	57,870
Java	Indian Ocean	126,602	48,900
North Island, New Zealand	southwestern Pacific	114,669	44,274
Cuba	Caribbean Sea	110,860	42,804
Newfoundland	northwestern Atlantic	108,860	42,030
Luzon	western Pacific	104,688	40,420

Island	Location	Area	
		sq km	sq mi
Iceland	northern Atlantic	103,000	39,800
Mindanao	western Pacific	94,630	36,537
Ireland (Northern Ireland and the Republic of Ireland)	northern Atlantic	84,406	32,590
Hokkaido	northwestern Pacific	83,515	32,245
Sakhalin	northwestern Pacific	76,400	29,500
Hispaniola – Dominican Republic and Haiti	Caribbean Sea	76,192	29,418
Banks	Canadian Arctic	70,028	27,038
Tasmania	southwestern Pacific	67,800	26,200
Sri Lanka	Indian Ocean	65,610	25,332
Devon	Canadian Arctic	55,247	21,331

of strain accumulated over time as a result of tectonics. The study of earthquakes is called seismology. Most earthquakes occur along faults (fractures or breaks) and ◊Benioff zones. Plate tectonic movements generate the major proportion: as two plates move past each other they can become jammed. When sufficient strain has accumulated, the rock breaks, releasing a series of elastic waves (seismic waves) as the plates spring free. The force of earthquakes (magnitude) is measured on the Richter scale, and their effect (intensity) on the Mercalli scale. The point at which an earthquake originates is the **seismic focus** or **hypocentre**; the point on the Earth's surface directly above this is the **epicentre**.

Most earthquakes happen at sea and cause little damage. However, when severe earthquakes occur in

---

Earthquake Image Information System
http://www.eerc.berkeley.edu/
cgi-bin/eqiis_form?eq=4570&count=1
EqIIS – Earthquake Image Information System – is a fully searchable library of almost 8,000 images from more than 80 earthquakes.

---

highly populated areas they can cause great destruction and loss of life. The Alaskan earthquake of 27 March 1964 ranks as one of the greatest ever recorded. The ◊San Andreas fault in California, where the North American and Pacific plates move past each other, is a notorious site of many large earthquakes. The 1906 San Francisco earthquake is among the most famous

Richter Scale

The Richter scale is based on measurement of seismic waves, used to determine the magnitude of an earthquake at its epicentre. The magnitude of an earthquake differs from its intensity, measured by the Mercalli scale, which is subjective and varies from place to place for the same earthquake. The Richter scale was named after the US seismologist Charles Richter.

Magnitude	Relative amount of energy released	Examples	Year
1	1		
2	31		
3	960		
4	30,000	Carlisle, England (4.7)	1979
5	920,000	Wrexham, Wales (5.1)	1990
6	29,000,000	San Fernando, California, USA (6.5)	1971
		northern Armenia (6.8)	1988
7	890,000,000	Loma Prieta, California, USA (7.1)	1989
		Kobe, Japan (7.2)	1995
		Rasht, Iran (7.7)	1990
		San Francisco, California, USA (7.7–7.9)[1]	1906
8	28,000,000,000	Tangshan, China (8.0)	1976
		Gansu, China (8.6)	1920
		Lisbon, Portugal (8.7)	1755
9	850,000,000,000	Prince William Sound, Alaska, USA (9.2)	1964

[1] Richter's original estimate of a magnitude of 8.3 has been revised by two recent studies carried out by the California Institute of Technology and the US Geological Survey.

Earthquake Locator Site
http://www.geo.ed.ac.uk/quakes/quakes.html
Edinburgh University, Scotland, runs this worldwide earthquake locator site that allows visitors to search for the world's latest earthquakes. The locator works on a global map on which perspective can be zoomed in or out. There are normally around five or six earthquakes a day; you'll find it surprising how few make the news. The site also has some general information on earthquakes.

in history. The deadliest, most destructive earthquake in historical times is thought to have been in China in 1556.

A reliable form of earthquake prediction has yet to be developed, although the seismic gap theory has had some success in identifying likely locations. In 1987 a California earthquake was successfully predicted by measurement of underground pressure waves; prediction attempts have also involved the study of such phenomena as the change in gases issuing from the crust, the level of water in wells, slight deformation of the rock surface, a sequence of minor tremors, and the behaviour of animals.

The possibility of earthquake prevention is remote. However, rock slippage might be slowed at movement points, or promoted at stoppage points, by the extraction or injection of large quantities of water underground, since water serves as a lubricant. This would ease overall pressure.

seismic wave A seismic wave is an energy wave generated by an earthquake or an artificial explosion. There are two types of seismic wave: **body waves** that travel through the Earth's interior, and **surface waves** that travel through the surface layers of the crust and can be felt as the shaking of the ground, as in an earthquake.

body wave There are two types of body wave: **P-waves** and **S-waves**, so-named because they are the primary and secondary waves detected by a seismograph. P-waves are longitudinal waves (wave motion in the direction the wave is travelling), whose compressions and rarefactions resemble those of a sound wave. S-waves are transverse waves or shear waves, involving a back-and-forth shearing motion at right angles to the direction the wave is travelling.

Most Destructive Earthquakes in the World

Source: US Geological Survey National Earthquake Information Center
(N/A = not available.)

Date	Location	Estimated number of deaths	Magnitude (Richter scale)
23 January 1556	Shaanxi, China	830,000	N/A
11 October 1737	Calcutta, India	300,000	N/A
27 July 1976	Tangshan, China	255,000 [1]	8.0
9 August 1138	Aleppo, Syria	230,000	N/A
22 May 1927	near Xining, China	200,000	8.3
22 December 856	Damghan, Iran	200,000	N/A
16 December 1920	Gansu, China	200,000	8.6
23 March 893	Ardabil, Iran	150,000	N/A
1 September 1923	Kwanto, Japan	143,000	8.3
30 December 1730	Hokkaido, Japan	137,000	N/A
September 1290	Chihli, China	100,000	N/A
November 1667	Caucasia, Russia	80,000	N/A
18 November 1727	Tabriz, Iran	77,000	N/A
28 December 1908	Messina, Italy	70,000–100,000	7.5
1 November 1755	Lisbon, Portugal	70,000	8.7
25 December 1932	Gansu, China	70,000	7.6
31 May 1970	northern Peru	66,000	7.8
1268	Cilicia, Asia Minor	60,000	N/A
11 January 1693	Sicily, Italy	60,000	N/A
4 February 1783	Calabria, Italy	50,000	N/A
20 June 1990	Iran	50,000	7.7
30 May 1935	Quetta, India	30,000–60,000	7.5

[1] This is the official casualty figure; the estimated death toll is as high as 750,000.

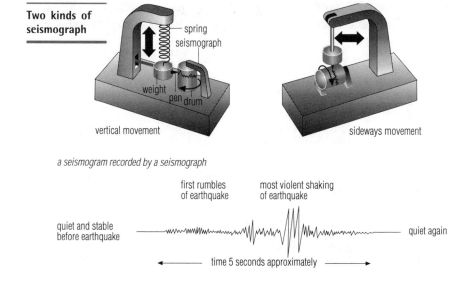

Two kinds of seismograph

spring
seismograph

weight
pen drum

vertical movement

sideways movement

a seismogram recorded by a seismograph

first rumbles
of earthquake

most violent shaking
of earthquake

quiet and stable
before earthquake

quiet again

time 5 seconds approximately

seismograph A seismogram, or recording made by a seismograph. Such recordings are used to study earthquakes and in prospecting.

Because liquids have no resistance to shear and cannot sustain a shear wave, S-waves cannot travel through liquid material. The Earth's outer core is believed to be liquid because S-waves disappear at the mantle-core boundary, while P-waves do not.

surface wave Surface waves travel in the surface and subsurface layers of the crust. **Rayleigh waves** travel along the free surface (the uppermost layer) of a solid material. The motion of particles is elliptical, like a water wave, creating the rolling motion often felt during an earthquake. **Love waves** are transverse waves trapped in a subsurface layer due to different densities in the rock layers above and below. They have a horizontal side-to-side shaking motion transverse (at right angles) to the direction the wave is travelling.

Earth's volcanism A volcano is a crack in the Earth's crust through which hot magma (molten rock) and gases well up. The magma is termed lava when it reaches the surface. A volcanic mountain, usually cone shaped with a crater on top, is formed around the opening, or vent, by the build-up of solidified lava and ashes (rock fragments). Most volcanoes arise on plate margins (see plate tectonics above), where the move-

ments of plates generate magma or allow it to rise from the mantle beneath. However, a number are found far from plate-margin activity, on 'hot spots' where the Earth's crust is thin.

volcano There are two main types of volcano, but three distinctive cone shapes. Composite volcanoes emit a stiff, rapidly solidifying lava which forms high, steep-sided cones. Volcanoes that regularly throw out ash build up flatter domes known as cinder cones. The lava from a shield volcano is not ejected violently, flowing over the crater rim forming a broad low profile.

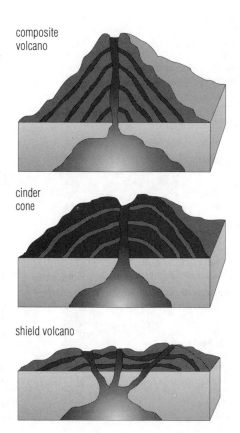

composite
volcano

cinder
cone

shield volcano

Major Volcanoes Active in the 20th Century, by Region

| Volcano | Height | | Location | Date of last |
	m	ft		eruption or activity
Africa				
Cameroon	4,096	13,353	isolated mountain, Cameroon	1986
Nyiragongo	3,470	11,385	Virungu, Democratic Republic of Congo	1994
Nyamuragira	3,056	10,028	Democratic Republic of Congo	1994
Ol Doinyo Lengai	2,886	9,469	Tanzania	1993
Lake Nyos	918	3,011	Cameroon	1986
Erta-Ale	503	1,650	Ethiopia	1995
Antarctica				
Erebus	4,023	13,200	Ross Island, McMurdo Sound	1995
Deception Island	576	1,890	South Shetland Island	1970
Asia				
Kerinci	3,800	12,467	Sumatra, Indonesia	1987
Rindjani	3,726	12,224	Lombok, Indonesia	1966
Semeru	3,676	12,060	Java, Indonesia	1995
Slamet	3,428	11,247	Java, Indonesia	1989
Raung	3,322	10,932	Java, Indonesia	1993
Agung	3,142	10,308	Bali, Indonesia	1964
On-Taka	3,063	10,049	Honshu, Japan	1991
Merapi	2,911	9,551	Java, Indonesia	1998
Marapi	2,891	9,485	Sumatra, Indonesia	1993
Asama	2,530	8,300	Honshu, Japan	1990
Nigata Yake-yama	2,475	8,111	Honshu, Japan	1989
Mayon	2,462	8,084	Luzon, Philippines	1993
Canlaon	2,459	8,070	Negros, Philippines	1993
Chokai	2,225	7,300	Honshu, Japan	1974
Galunggung	2,168	7,113	Java, Indonesia	1984
Azuma	2,042	6,700	Honshu, Japan	1977
Sangeang Api	1,935	6,351	Lesser Sunda Island, Indonesia	1988
Pinatubo	1,759	5,770	Luzon, Philippines	1995
Kelut	1,730	5,679	Java, Indonesia	1990
Unzen	1,360	4,462	Japan	1996
Krakatoa	818	2,685	Sumatra, Indonesia	1996
Taal	300	984	Philippines	1977
Atlantic Ocean				
Pico de Teide	3,716	12,192	Tenerife, Canary Islands, Spain	1909
Fogo	2,835	9,300	Cape Verde Islands	1995
Beerenberg	2,277	7,470	Jan Mayen Island, Norway	1985
Hekla	1,491	4,920	Iceland	1991
Krafla	654	2,145	Iceland	1984
Helgafell	215	706	Iceland	1973
Surtsey	174	570	Iceland	1967
Caribbean				
La Grande Soufrière	1,467	4,813	Basse-Terre, Guadeloupe	1977
Pelée	1,397	4,584	Martinique	1932
La Soufrière St Vincent	1,234	4,048	St Vincent and the Grenadines	1979
Soufrière Hills/Chances Peak	968	3,176	Montserrat	1997
Central America				
Acatenango	3,960	12,992	Sierra Madre, Guatemala	1972
Fuego	3,835	12,582	Sierra Madre, Guatemala	1991
Tacana	3,780	12,400	Sierra Madre, Guatemala	1988
Santa Maria	3,768	12,362	Sierra Madre, Guatemala	1993
Irazú	3,452	11,325	Cordillera Central, Costa Rica	1992
Turrialba	3,246	10,650	Cordillera Central, Costa Rica	1992
P992	2,721	8,930	Cordillera Central, Costa Rica	1994
Pacaya	2,543	8,346	Sierra Madre, Guatemala	1996
San Miguel	2,131	6,994	El Salvador	1986
Arenal	1,552	5,092	Costa Rica	1996

Volcano	Height		Location	Date of last
	m	ft		eruption or activity
Europe				
Kliuchevskoi	4,750	15,584	Kamchatka Peninsula, Russia	1997
Koryakskaya	3,456	11,339	Kamchatka Peninsula, Russia	1957
Sheveluch	3,283	10,771	Kamchatka Peninsula, Russia	1997
Etna	3,236	10,625	Sicily, Italy	1998
Bezymianny	2,882	9,455	Kamchatka Peninsula, Russia	1997
Alaid	2,335	7,662	Kurile Islands, Russia	1986
Tiatia	1,833	6,013	Kurile Islands, Russia	1981
Sarychev Peak	1,512	4,960	Kurile Islands, Russia	1989
Vesuvius	1,289	4,203	Italy	1944
Stromboli	931	3,055	Lipari Islands, Italy	1996
Santorini (Thera)	584	1,960	Cyclades, Greece	1950
Indian Ocean				
Karthala	2,440	8,000	Comoros	1991
Piton de la Fournaise (Le Volcan)	1,823	5,981	Réunion Island, France	1998
Mid-Pacific				
Mauna Loa	4,170	13,681	Hawaii, USA	1984
Kilauea	1,247	4,100	Hawaii, USA	1998
North America				
Popocatépetl	5,452	17,887	Altiplano de México, Mexico	1997
Colima	4,268	14,003	Altiplano de México, Mexico	1994
Spurr	3,374	11,070	Alaska Range (AK) USA	1953
Lassen Peak	3,186	10,453	California, USA	1921
Redoubt	3,108	10,197	Alaska Range (AK) USA	1991
Iliamna	3,052	10,016	Alaska Range (AK) USA	1978
Shishaldin	2,861	9,387	Aleutian Islands, Alaska, USA	1997
St Helens	2,549	8,364	Washington, USA	1995
Pavlof	2,517	8,261	Alaska Range (AK) USA	1997
Veniaminof	2,507	8,225	Alaska Range (AK) USA	1995
Novarupta (Katmai)	2,298	7,540	Alaska Range (AK) USA	1931
El Chichon	2,225	7,300	Altiplano de México, Mexico	1982
Makushin	2,036	6,680	Aleutian Islands, Alaska, USA	1987
Oceania				
Ruapehu	2,796	9,175	New Zealand	1997
Ulawun	2,296	7,532	Papua New Guinea	1993
Ngauruhoe	2,290	7,515	New Zealand	1977
Oceania				
Bagana	1,998	6,558	Papua New Guinea	1993
Manam	1,829	6,000	Papua New Guinea	1997
Lamington	1,780	5,844	Papua New Guinea	1956
Karkar	1,499	4,920	Papua New Guinea	1979
Lopevi	1,450	4,755	Vanuatu	1982
Ambrym	1,340	4,376	Vanuatu	1991
Tarawera	1,149	3,770	New Zealand	1973
Langila	1,093	3,586	Papua New Guinea	1996
Rabaul	688	2,257	Papua New Guinea	1997
Pagan	570	1,870	Mariana Islands	1993
White Island	328	1,075	New Zealand	1995
South America				
San Pedro	6,199	20,325	Andes, Chile	1960
Guallatiri	6,060	19,882	Andes, Chile	1993
Lascar	5,990	19,652	Andes, Chile	1995
San José	5,919	19,405	Andes, Chile	1931
Cotopaxi	5,897	19,347	Andes, Ecuador	1975
Tutupaca	5,844	19,160	Andes, Ecuador	1902
Ubinas	5,710	18,720	Andes, Peru	1969
Tupungatito	5,640	18,504	Andes, Chile	1986
Islunga	5,566	18,250	Andes, Chile	1960
Nevado del Ruiz	5,435	17,820	Andes, Colombia	1992
Tolima	5,249	17,210	Andes, Colombia	1943
Sangay	5,230	17,179	Andes, Ecuador	1996

Composite volcanoes, such as Stromboli and Vesuvius in Italy, are found at destructive plate margins (areas where plates are being pushed together), usually in association with island arcs and coastal mountain chains. The magma is mostly derived from plate material and is rich in silica. This makes a very stiff lava such as andesite, which solidifies rapidly to form a high, steep-sided volcanic mountain. The magma often clogs the volcanic vent, causing violent eruptions as the blockage is blasted free, as in the eruption of Mount St Helens, USA, in 1980. The crater may collapse to form a caldera.

Shield volcanoes, such as Mauna Loa in Hawaii, are found along the rift valleys and ocean ridges of constructive plate margins (areas where plates are moving apart), and also over hot spots. The magma is derived from the Earth's mantle and is quite free-flowing. The lava formed from this magma – usually basalt – flows for some distance over the surface before it sets and so forms broad low volcanoes. The lava of a shield volcano is not ejected violently but simply flows over the crater rim.

Vesuvius
http://volcano.und.nodak.edu/vwdocs/volc_images/
img_vesuvius.html
Site examining the complex geology of Vesuvius and its famous eruption of AD 79. There are images of the volcano and historical drawings. There is also a link to a local site campaigning for an improved civil defence plan as the volcano prepares once more to explode.

The type of volcanic activity is also governed by the age of the volcano. The first stages of an eruption are usually vigorous as the magma forces its way to the surface. As the pressure drops and the vents become established, the main phase of activity begins,

Volcanoes
http://www.geo.mtu.edu/volcanoes/
Provided by Michigan Technological University, this site includes a world map of volcanic activity with information on recent eruptions, the latest research in remote sensing of volcanoes, and many spectacular photographs.

---

composite volcanoes giving pyroclastic debris and shield volcanoes giving lava flows. When the pressure from below ceases, due to exhaustion of the magma chamber, activity wanes and is confined to the emission of gases and in time this also ceases. The volcano then enters a period of quiescence, after which activity may resume after a period of days, years, or even thousands of years. Only when the root zones of a volcano have been exposed by erosion can a volcano be said to be truly extinct.

Many volcanoes are submarine and occur along mid-ocean ridges. The chief terrestrial volcanic regions are around the Pacific rim (Cape Horn to Alaska); the central Andes of Chile (with the world's highest volcano, Guallatiri, 6,060 m/19,900 ft); North Island, New Zealand; Hawaii; Japan; and Antarctica. There are more than 1,300 potentially active volcanoes on Earth. Volcanism has helped shape other members of the Solar System, including the Moon, Mars, Venus, and Jupiter's moon Io.

There are several methods of monitoring volcanic activity. They include seismographic instruments on the ground, aircraft monitoring, and space monitoring using remote sensing satellites.

Earth's magnetism The change in polarity of Earth's magnetic field is called a **polar reversal**. Like all magnets, Earth's magnetic field has two opposing regions, or poles, one of attraction and one of repulsion,

Major Pre-20th Century Volcanic Eruptions

Volcano	Location	Year	Estimated number of deaths
Santorini (Thera)	Greece	c. 1470 BC	[1]
Vesuvius	Italy	AD 79	16,000[2]
Kelut	Java, Indonesia	1586	10,000
Etna	Sicily, Italy	1669	20,000
Vesuvius	Italy	1631	4,000
Papandayan	Java, Indonesia	1772	3,000
Laki	Iceland	1783	9,350
Unzen	Japan	1792	14,500
Tambora	Sumbawa, Indonesia	1815	10,000[3]
Krakatoa	Indonesia	1883	36,000

[1] Number of deaths unknown; the explosion was four times more powerful than Krakatoa.

[2] Estimates vary greatly, from 2,000 (lowest) to 20,000 (highest).

[3] A further 82,000 deaths were caused by starvation and disease brought on by the eruption.

Major Volcanic Eruptions in the 20th Century

Volcano	Location	Year	Estimated number of deaths
Santa María	Guatemala	1902	1,000
Pelée	Martinique	1902	30,000
Taal	Philippines	1911	1,400
Kelut	Java, Indonesia	1919	5,500
Vulcan	Papua New Guinea	1937	500
Lamington	Papua New Guinea	1951	3,000
St Helens	USA	1980	57
El Chichon	Mexico	1982	1,880
Nevado del Ruiz	Colombia	1985	23,000
Lake Nyos	Cameroon	1986	1,700
Pinatubo	Luzon, Philippines	1991	639
Unzen	Japan	1991	39
Mayon	Philippines	1993	70
Loki	Iceland	1996	0
Soufriere	Montserrat	1997	23
Merapi	Java, Indonesia	1998	38

positioned approximately near the geographical North and South Poles. During a period of normal polarity the region of attraction corresponds with the North Pole. Today, a compass needle, like other magnetic materials, aligns itself parallel to the magnetizing force and points to the North Pole. During a period of reversed polarity, the region of attraction would change to the South Pole and the needle of a compass would point south.

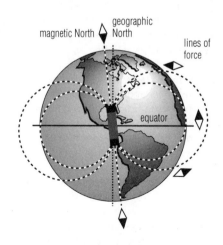

magnetic field The Earth's magnetic field is similar to that of a bar magnet with poles near, but not exactly at, the geographic poles. Compass needles align themselves with the magnetic field, which is horizontal near the equator and vertical at the magnetic poles.

Studies of the magnetism retained in rocks at the time of their formation (like little compasses frozen in time) have shown that the polarity of the magnetic field has reversed repeatedly throughout geological time.

Polar reversals are a random process. Although the average time between reversals over the last 10 million years has been 250,000 years, the rate of reversal has changed continuously over geological time. The most recent reversal was 700,000 years ago; scientists have no way of predicting when the next reversal will occur. The reversal process takes about a thousand years. Movements of Earth's molten core are thought to be responsible for the magnetic field and its polar reversals. Dating rocks using distinctive sequences of magnetic reversals is called palaeomagnetic stratigraphy.

secular variation The changes in the position of Earth's magnetic poles measured with respect to geographical positions, such as the North pole, throughout geological time is called secular variation.

rocks

Earth's crust is composed of rocks, aggregates of minerals or materials of organic origin that have consolidated into hard masses. There are three types of rocks: igneous, sedimentary, and metamorphic. The property of a rock will depend on its components and the conditions of its formation. **Igneous rocks** are formed by the cooling and solidification of magma, the molten rock material that originates in the lower part of the Earth's crust, or mantle, where it reaches temperatures

oldest known rocks

The oldest known rocks are found near the Great Slave and Great Bear lakes in Canada. The rocks, which include granite, are 4.03 billion years old.

as high as 1,000°C. The rock may form on or below the Earth's surface and is usually crystalline in texture. Larger crystals are more common in rocks such as granite which have cooled slowly within the Earth's crust; smaller crystals form in rocks such as basalt which have cooled more rapidly on the surface.

Sedimentary rocks are formed by the compression of particles deposited by water, wind, or ice. They may be created by the erosion of older rocks, the deposition of organic materials, or they may be formed from chemical precipitates. For example, sandstone is derived from sand particles, limestone from the remains of sea creatures, and gypsum is precipitated from evaporating salt water. Sedimentary rocks are typically deposited in distinct layers or strata and many contain fossils.

Metamorphic rocks are formed through the action of high pressure or heat on existing igneous or sedimentary rocks, causing changes to the composition, structure, and texture of the rocks. For example, marble is formed by the effects of heat and pressure on limestone, while granite may be metamorphosed into gneiss, a coarse-grained foliated rock.

rock identification

Rocks can often be identified by their location and appearance. For example, **sedimentary rocks** lie in stratified, or layered, formations and may contain fossils; many have markings such as old mud cracks or ripple marks caused by waves. Except for volcanic glass, all **igneous rocks** are solid and crystalline. Some appear dense, with microscopic crystals, and others have larger, easily seen crystals. They occur in volcanic areas, and in intrusive formations that geologists call batholiths, laccoliths, sills, dykes, and stocks. Many **metamorphic rocks** have characteristic bands, and are easily split into sheets or slabs. Rock formations and strata are often apparent in the cliffs that line a seashore, or where rivers have gouged out deep channels to form gorges and canyons. They are also revealed when roads are cut through hillsides, or by excavations for quarrying and mining. Rock and fossil collecting has been a popular hobby since the 19th century and such sites can provide a treasure trove of finds for the collector.

Where deposits of economically valuable minerals occur rocks are termed ores. Rocks break down as a result of weathering into very small particles that combine with organic materials from plants and animals

to form soil. In geology the term 'rock' can also include unconsolidated materials such as sand, mud, clay, and peat.

rock studies

The study of the Earth's crust and its composition fall under a number of interrelated sciences, each with its own specialists. Among these are **geologists**, who identify and survey rock formations and determine when and how they were formed, **petrologists**, who identify and classify the rocks themselves, and **mineralogists**, who study the mineral contents of the rocks. **Palaeontologists** study the fossil remains of plants and animals found in rocks.

applications of rock studies Data from rock studies and surveys enable scientists to trace the history of the Earth and learn about the kind of life that existed here millions of years ago. The data are also used in locating and mapping deposits of fossil fuels such as coal, oil, and natural gas, and valuable mineral-containing ores providing metals such as aluminium, iron, lead, and tin, and radioactive elements such as radium and uranium. These deposits may lie close to the Earth's surface or deep underground, often under oceans. In some regions, entire mountains are composed of deposits of iron or copper ores, while in other regions rocks may contain valuable nonmetallic minerals such as borax and graphite, or precious gems such as diamonds and emeralds.

rocks as construction materials In addition to the mining and extraction of fuels, metals, minerals, and gems, rocks provide useful building and construction materials. Rock is mined through quarrying, and cut into blocks or slabs as building stone, or crushed or broken for other uses in construction work. For instance, cement is made from limestone and, in addition to its use as a bonding material, it can be added to crushed stone, sand, and water to produce strong, durable concrete, which has many applications, such as the construction of roads, runways, and dams.

Among the most widely used building stones are granite, limestone, sandstone, marble, and slate. **Granite** provides one of the strongest building stones and is resistant to weather, but its hardness makes it difficult to cut and handle. **Limestone** is a hard and lasting stone that is easily cut and shaped and is widely used for public buildings. The colour and texture of the stone can vary with location; for instance, Portland stone from the Jurassic rocks of Dorset is white, even-textured, and durable, while Bath stone is an oolitic limestone that is honey-coloured and more porous. **Sandstone** varies in colour and texture; like limestone, it is relatively easy to quarry and work and is used for

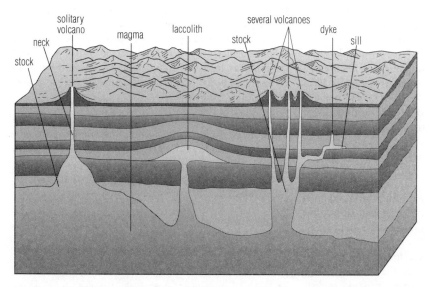

solitary volcano · neck · stock · magma · laccolith · stock · several volcanoes · dyke · sill

similar purposes. **Marble** is a classic stone, worked by both builders and sculptors. Pure marble is white, streaked with veins of black, grey, green, pink, red, and yellow. **Slate** is fine-grained rock that can be split easily into thin slabs and used as tiles for roofing and flooring. Its colour varies from black to green and red.

igneous rocks

Igneous rocks are formed from cooling magma or lava, solidifying from a molten state. Igneous rocks are largely composed of silica (SiO_2) and they are classified according to their crystal size, texture, method of formation, or chemical composition, for example by the proportions of light and dark minerals.

plutonic or intrusive igneous rocks

Igneous rocks that crystallize from magma below the Earth's surface are called plutonic or intrusive, depending on the depth of formation. They have large crystals produced by slow cooling; examples include gabbro and granite.

gabbro

A gabbro is a mafic (consisting primarily of dark-coloured crystals) igneous rock formed deep in the Earth's crust. It contains pyroxene and calcium-rich feldspar, and may contain small amounts of olivine and amphibole. Its coarse crystals of dull minerals give it a speckled appearance.

Gabbro is the plutonic version of basalt (that is, derived from magma that has solidified below the Earth's surface), and forms in large, slow-cooling intrusions.

granite

Granite is a coarse-grained intrusive igneous rock, typically consisting of the minerals quartz, feldspar, and biotite mica. It may be pink or grey, depending on the composition of the feldspar. Granites

are chiefly used as building materials.

Granites often form large intrusions in the core of mountain ranges, and they are usually surrounded by zones of metamorphic rock (rock that has been altered by heat or pressure). Granite areas have characteristic moorland scenery. In exposed areas the bedrock may be weathered along joints and cracks to produce a tor, consisting of rounded blocks that appear to have been stacked upon one another.

Britain's largest quarry, Glensanda, near Fort William, Scotland, produces 5 million tonnes of granite a year.

extrusive or volcanic igneous rocks

Igneous rocks extruded at the surface from lava are called **extrusive** or **volcanic**. Rapid cooling results in small crystals; basalt is an example.

basalt

Basalt is the commonest extrusive igneous rock in the Solar System. Much of the surfaces of the terrestrial planets Mercury, Venus, Earth, and Mars, as well as the Moon, are composed of basalt. Earth's ocean floor is virtually entirely made of basalt. Basalt is mafic, that is, it contains relatively little ⬦silica: about 50% by weight. It is usually dark grey but can also be green, brown, or black. Its essential constituent minerals are calcium-rich feldspar and calcium and magnesium-rich pyroxene.

The groundmass may be glassy or finely crystalline, sometimes with large crystals embedded. Basaltic lava tends to be runny and flows for great distances before solidifying. Successive eruptions of basalt have formed the great plateaus of Colorado and the Deccan plateau region of southwest India. In some places, such as Fingal's Cave in the Inner Hebrides of Scotland and the Giant's Causeway in Antrim, Northern Ireland, shrinkage during the solidification of the molten lava caused the formation of hexagonal columns.

The dark-coloured lowland maria regions of the Moon are underlain by basalt. Lunar mare basalts have higher concentrations of titanium and zirconium and lower concentrations of volatile elements like potassium and sodium relative to terrestrial basalts. Martian basalts are characterized by low ratios of iron to manganese relative to terrestrial basalts, as judged from some martian meteorites and spacecraft analyses of rocks and soils on the Martian surface.

sedimentary rocks

Sedimentary rocks are formed by the accumulation and cementation of deposits that have been laid down by water, wind, ice, or gravity. They cover more than two-thirds of the Earth's surface and comprise three major categories: **clastic, chemically precipitated,** and **organic (or biogenic)**. Clastic sediments are the largest group and are composed of fragments of pre-existing rocks; they include clays, sands, and gravels. Chemical precipitates include some limestones and evaporated deposits such as gypsum and halite (rock salt). Coal, oil shale, and limestone made of fossil material are examples of organic sedimentary rocks.

Most sedimentary rocks show distinct layering (stratification), caused by alterations in composition or by changes in rock type. These strata may become folded or fractured by the movement of the Earth's crust, a process known as deformation.

chalk Chalk is a soft, fine-grained, whitish sedimentary rock composed of calcium carbonate ($CaCO_3$), extensively quarried for use in cement, lime, and mortar, and in the manufacture of cosmetics and toothpaste. **Blackboard chalk** in fact consists of gypsum (calcium sulphate, $CaSO_4.2H_2O$).

Chalk was once thought to derive from the remains of microscopic animals or foraminifera. In 1953, however, it was seen under the electron microscope to be composed chiefly of coccolithophores, unicellular lime-secreting algae, and hence primarily of plant origin. It is formed from deposits of deep-sea sediments called oozes.

Chalk was laid down in the later Cretaceous period (see below) and covers a wide area in Europe. In England it stretches in a belt from Wiltshire and Dorset continuously across Buckinghamshire and Cambridgeshire to Lincolnshire and Yorkshire, and also forms the North and South Downs, and the cliffs of southern and southeast England.

limestone Limestone is a sedimentary rock composed chiefly of calcium carbonate ($CaCO_3$), either derived from the shells of marine organisms or precipitated from solution, mostly in the ocean. Various types of limestone are used as building stone.

Marble is metamorphosed limestone. Certain so-called marbles are not in fact marbles but fine-grained fossiliferous limestones that take an attractive polish. Caves commonly occur in limestone. Karst is a type of limestone landscape.

sandstone Sandstones are sedimentary rocks formed from the consolidation of sand, with sand-sized grains (0.0625–2 mm/0.0025–0.08 in) in a matrix or cement. Their principal component is quartz. Sandstones are commonly permeable and porous, and may form fresh-water aquifers. They are mainly used as building materials.

Sandstones are classified according to the matrix or cement material (whether derived from clay or silt; for example, as calcareous sandstone, ferruginous sandstone, siliceous sandstone).

shale A fine-grained and finely layered sedimentary rock composed of silt and clay is called shale. It is a weak rock, splitting easily along bedding planes to form thin, even slabs (by contrast, mudstone splits into irregular flakes). Oil shale contains kerogen, a solid bituminous material that yields petroleum when heated.

fossils A fossil (from Latin *fossilis*, 'dug up') is a cast, impression, or the actual remains of an animal or plant preserved in rock. Fossils were created during periods of rock formation, caused by the gradual accumulation of sediment over millions of years at the bottom of the sea bed or an inland lake. Fossils may include footprints, an internal cast, or external impression. A few fossils are preserved intact, as with mammoths fossilized in Siberian ice, or insects trapped in tree resin that is today amber. The study of fossils is called palaeontology. Palaeontologists are able to deduce much of the geological history of a region from fossil remains.

About 250,000 fossil species have been discovered – a figure that is believed to represent less than 1 in 20,000 of the species that ever lived. **Microfossils** are so small they can only be seen with a microscope. They include the fossils of pollen, bone fragments, bacteria, and the remains of microscopic marine animals and plants, such as foraminifera and diatoms.

Trace Fossils Left By Dinosaurs
http://www.emory.edu/GEOSCIENCE/HTML/
Dinotraces.htm
All about the trace fossils left by dinosaurs. The site is divided into several categories of dinosaur fossils: tracks, eggs and nests, tooth marks, gastroliths, and coprolites and also includes images and descriptions of each one.

---

metamorphic rocks

Metamorphic rocks are those formed when either igneous or sedimentary rocks are altered in structure and composition by pressure, heat, or chemically active fluids after original formation. (If heat is sufficient to melt the original rock, technically it becomes an igneous rock upon cooling.) The term was coined in 1833 by the Scottish geologist Charles Lyell (1797–1875).

The mineral assemblage present in a metamorphic rock depends on the composition of the starting material (which may be sedimentary or igneous) and the temperature and pressure conditions to which it is subjected. There are two main types of metamorphism. **Thermal metamorphism**, or **contact metamorphism**, is brought about by the baking of solid rocks in the vicinity of an igneous intrusion (molten rock, or magma, in a crack in the Earth's crust). It is responsible, for example, for the conversion of limestone to marble. **Regional metamorphism** results from the heat and intense pressures associated with the movements and collision of tectonic plates (see plate tectonics above). It brings about the conversion of shale to slate, for example.

gneiss A gneiss is a coarse-grained metamorphic rock, formed under conditions of high temperature and pressure, and often occurring in association with schists and granites. It has a foliated, or layered, structure consisting of thin bands of micas and/or amphiboles, dark in colour, alternating with bands of granular quartz and feldspar that are light in colour. Gneisses are formed during regional metamorphism; **paragneisses** are derived from metamorphism of sedimentary rocks and **orthogneisses** from metamorphism of granite or similar igneous rocks.

marble Marble is formed by metamorphosis of sedimentary limestone. It takes and retains a good polish, and is used in building and sculpture. In its pure form it is white and consists almost entirely of the mineral calcite ($CaCO_3$). Mineral impurities give it various colours and patterns. Carrara, Italy, is known for white marble.

Mohs' scale of hardness

To remember the order of hardness:

Tall **gy**roscopes **ca**n **fl**y **ap**art, **or**biting **qu**ickly **to** **co**mplete **di**sintegration.

(**ta**lc, **gy**psum or rock salt, **ca**lcite, **fl**uorite, **ap**atite, **or**thaclase, **qu**artz, **to**paz, **co**orundum, **di**amond)

---

minerals

A mineral is any naturally formed inorganic substance with a particular chemical composition and a regularly repeating internal structure. Either in their perfect crystalline form or otherwise, minerals are the constituents of rocks. In more general usage, a mineral is any substance economically valuable for mining (including coal and oil, despite their organic origins).

Mineral forming processes include: melting of pre-existing rock and subsequent crystallization of a mineral to form magmatic or volcanic rocks; weathering

---

Mohs' Scale

Number	Defining mineral	Other substances compared
1	talc	
2	gypsum	$2\frac{1}{2}$ fingernail
3	calcite	$3\frac{1}{2}$ copper coin
4	fluorite	
5	apatite	$5\frac{1}{2}$ steel blade
6	orthoclase	$5\frac{3}{4}$ glass
7	quartz	7 steel file
8	topaz	
9	corundum	
10	diamond	

Note that the scale is not regular; diamond, at number 10 the hardest natural substance, is 90 times harder in absolute terms than corundum, number 9

Metamorphic Rocks

Main primary material (before metamorphism)

Typical depth and temperature	Shale with several minerals	Sandstone with only quartz	Limestone with only calcite
50,000 ft/570°F	slate	quartzite	marble
65,000 ft/750°F	schist		
82,000 ft/930°F	gneiss		
98,500 ft/1,100°F	hornfels	quartzite	marble

Crystal System

Crystal system	Minimum symmetry	Possible shape	Mineral examples
cubic	4 threefold axes	cube, octahedron, dodecahedron	diamond, garnet, pyrite
tetragonal	1 fourfold axis	square-based prism	zircon
orthorhombic	3 twofold axes or mirror planes	matchbox shape	baryte
monoclinic	1 twofold axis or mirror plane	matchbox distorted in one plane	gypsum
triclinic	no axes or mirror planes	matchbox distorted in three planes	plagioclase feldspar
trigonal	1 threefold axis	triangular prism, rhombohedron	calcite, quartz
hexagonal	1 sixfold axis	hexagonal prism	beryl

of rocks exposed at the land surface, with subsequent transport and grading by surface waters, ice, or wind to form sediments; and recrystallization through increasing temperature and pressure with depth to form metamorphic rocks.

Minerals are usually classified as **magmatic, sedimentary,** or **metamorphic.** The magmatic minerals include the feldspars, quartz, pyroxenes, amphiboles, micas, and olivines that crystallize from silica-rich rock melts within the crust or from extruded lavas. The most commonly occurring sedimentary minerals are either pure concentrates or mixtures of sand, clay minerals, and carbonates (chiefly calcite, aragonite, and dolomite). Minerals typical of metamorphism include andalusite, cordierite, garnet, tremolite, lawsonite, pumpellyite, glaucophane, wollastonite, chlorite, micas, hornblende, staurolite, kyanite, and diopside.

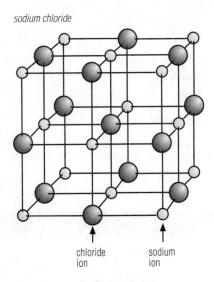

sodium chloride

chloride ion   sodium ion

crystal The sodium chloride, or common salt, crystal is a regular cubic array of charged atoms (ions) – positive sodium atoms and negative chlorine atoms. Repetition of this structure builds up into cubic salt crystals.

One of the ways to distinguish one mineral from another is by hardness. Mohs' scale of hardness is an established scale of hardness for minerals. The scale is useful in mineral identification because any mineral will scratch any other mineral lower on the scale than itself, and similarly it will be scratched by any other mineral higher on the scale.

crystals and crystallography

A mineral can often be identified by the shape of its crystals and the system of crystallization determined. A **crystal** is any substance with an orderly three-dimensional arrangement of its atoms or molecules, thereby creating an external surface of clearly defined smooth faces having characteristic angles between them. Examples are table salt and quartz.

Each geometrical form, many of which may be combined in one crystal, consists of two or more faces – for example, dome, prism, and pyramid. A single crystal can vary in size from a submicroscopic particle to a mass some 30 m/100 ft in length. Crystals fall into seven crystal systems or groups, classified on the basis of the relationship of three or four imaginary axes that intersect at the centre of any perfect, undistorted crystal.

Three common crystalline forms are: (1) the simple cubic structure of ionic crystals, such as those of sodium chloride (NaCl); (2) the face-centred cubic structure of metals such as aluminium, copper, gold, silver, and lead; and (3) the hexagonal close-packed structure of metals such as cadmium and zinc.

Crystallography
http://www.iumsc.indiana.edu/docs/crystmin.htm
Understand the shapes and symmetries of crystallography, with these interactive drawings of cubic, tetrahedral, octahedral, and dodecahedral solids..

The scientific study of crystals is called **crystallography.** In 1912 it was found that the shape and size of the repeating atomic patterns (unit cells) in a crystal could be determined by passing X-rays through a sample. This method, known as X-ray diffraction, opened up an entirely new way of 'seeing' atoms. It has been found that many substances have a unit cell that exhibits all the symmetry of the whole crystal; in table salt (sodium chloride, NaCl), for instance, the unit cell is an exact cube.

Many materials were not even suspected of being crystals until they were examined by X-ray crystallography. It has been shown that purified biomolecules, such as proteins and DNA, can form crystals, and such compounds may now be studied by this method. Other applications include the study of metals and their alloys, and of rocks and soils.

calcite Calcite is a colourless, white, or light-coloured common rock-forming mineral, calcium carbonate ($CaCO_3$). It is the main constituent of **limestone** and **marble** and forms many types of invertebrate shell. Calcite often forms stalactites and stalagmites in caves and is also found deposited in veins through many rocks because of the ease with which it is dissolved and transported by groundwater; **oolite** is a rock consisting of spheroidal calcite grains. It rates 3 on the Mohs' scale of hardness. Large crystals up to 1 m/3 ft have been found in Oklahoma and Missouri, USA. **Iceland spar** is a transparent form of calcite used in the optical industry; as limestone it is used in the building industry.

diamond A diamond is a generally colourless, transparent mineral, an allotrope of carbon. It is regarded as a precious gemstone, and is the hardest substance known (10 on the Mohs' scale). Industrial diamonds, which may be natural or synthetic, are used for cutting, grinding, and polishing.

Diamond crystallizes in the cubic system as octahedral crystals, some with curved faces and striations. The high refractive index of 2.42 and the high dispersion of light, or 'fire', account for the spectral displays seen in polished diamonds.

Diamonds may be found as alluvial diamonds on or close to the Earth's surface in riverbeds or dried watercourses; on the sea bottom (off southwest Africa); or, more commonly, in diamond-bearing volcanic pipes composed of 'blue ground', kimberlite or lamproite, where the original matrix has penetrated the Earth's crust from great depths. They are sorted from the residue of crushed ground by X-ray and other recovery methods.

feldspar Feldspar is actually a group of silicate minerals. Feldspars are the most abundant mineral type in the Earth's crust. They are the chief constituents of igneous rock and are present in most metamorphic and sedimentary rocks. All feldspars contain silicon, aluminium, and oxygen, linked together to form a framework. Spaces within this framework structure are occupied by sodium, potassium, calcium, or occasionally barium, in various proportions. Feldspars form white, grey, or pink crystals and rank 6 on the Mohs' scale of hardness.

The four extreme compositions of feldspar are represented by the minerals **orthoclase** ($KAlSi_3O_8$); **albite** ($NaAlSi_3O_8$); **anorthite** ($CaAl_2Si_2O_8$); and **celsian** ($BaAl_2Si_2O_8$). **Plagioclase feldspars** contain variable amounts of sodium (as in albite) and calcium (as in anorthite) with a negligible potassium content. **Alkali feldspars** (including orthoclase) have a high potassium content, less sodium, and little calcium.

The type known as moonstone has a pearl-like effect and is used in jewellery. Approximately 4,000 tonnes of feldspar are used in the ceramics industry annually.

quartz Quartz is the crystalline form of ◊silica (SiO_2), one of the most abundant minerals of the Earth's crust (12% by volume). Quartz occurs in many different kinds of rock, including sandstone and granite. It ranks 7 on the Mohs' scale of hardness and is resistant to chemical or mechanical breakdown. Quartzes vary according to the size and purity of their crystals. Crystals of pure quartz are coarse, colourless, transparent, show no cleavage, and fracture unevenly; this form is usually called rock crystal. Impure coloured varieties, often used as gemstones, include ◊agate, citrine quartz, and ◊amethyst. Quartz is also used as a general name for the cryptocrystalline and noncrystalline varieties of silica, such as chalcedony, chert, and opal.

Quartz is used in ornamental work and industry, where its reaction to electricity makes it valuable in electronic instruments. Quartz can also be made synthetically.

Crystals that would take millions of years to form naturally can now be 'grown' in pressure vessels to a standard that allows them to be used in optical and scientific instruments and in electronics, such as quartz wristwatches.

talc Talc ($Mg_3Si_4O_{10}(OH)_2$) is a hydrous magnesium silicate mineral. It occurs in tabular crystals, but the massive impure form, known as **steatite** or **soapstone**, is more common. It is formed by the alteration of magnesium compounds and is usually found in metamorphic rocks. Talc is very soft, ranked 1 on the Mohs' scale of hardness. It is used in powdered form in cosmetics, lubricants, and as an additive in paper manufacture.

French chalk and potstone are varieties of talc. Soapstone has a greasy feel to it, and is used for carvings such as Inuit sculptures.

Mineral Gallery
http://mineral.galleries.com/
This Mineral Gallery is a collection of descriptions and images of minerals, organized by mineral name, class (sulphides, oxides, carbonates, and so on), and grouping (such as gemstones, birth stones, and fluorescent minerals).

geological record: divisions of geological time and dating

The **geological time scale** embraces the history of the Earth from its physical origin to the present day. Geological time is traditionally divided into eons (Precambrian and Phanerozoic, in ascending chronological order), which in turn are subdivided into eras, periods, epochs, ages, and finally chrons.

The terms eon, era, period, epoch, age and chron are **geochronological units** representing intervals of geological time. Rocks representing an interval of geological time comprise a **chronostratigraphic unit**. Each of the hierarchical geochronological terms has a chronostratigraphic equivalent. Thus, rocks formed during an eon (a geochronological unit) are members of an eonothem (the chronostratigraphic unit equivalent of eon). Rocks of an era belong to an erathem. The chronostratigraphic equivalents of period, epoch, age, and chron are system, series, stage, and chronozone, repectively.

eon Eons are the largest units of geological time and include the Precambrian eon, spanning from 4.6 billion years to 570 million years, and the Phanerozoic eon, lasting from 570 million years to the present. Rocks representing an eon of geological time comprise and **eonothem**.

era Eras are subdivisions of eons. The currently recognized eras all fall within the Phanerozoic eon – or the vast span of time, starting about 570 million years ago, when fossils are found to become abundant. The eras in ascending order are the **Palaeozoic, Mesozoic,** and **Cenozoic.** We are living in the Recent epoch of

geological periods

To remember the geological periods:

Camels often sit down carefully. Perhaps their joints creak? Early oiling might prevent permanent rheumatism.

(Cambrian, Ordovician, Silurian, Devonian, Carboniferous, Permian, Triassic, Jurassic, Cretaceous, Eocene, Oligocene, Miocene, Pliocene, Pleistocene [Recent])

the Quaternary period of the Cenozoic era. Rocks representing an era of geological time comprise an **erathem**. Eras are further subdivided into **periods**.

period Periods are subdivisions of eras and are themselves subdivided into epochs. They are the basic time units of the geological timescale. However, the periods of the Mesozoic era – the Triassic, Jurassic, and Cretaceous periods – are not divided further into epochs.

epoch Epochs are subdivisions of geological periods in the geological time scale. Epochs are sometimes given their own names (such as the Palaeocene, Eocene, Oligocene, Miocene, and Pliocene epochs comprising the Tertiary period), or they are referred to as the late, early, or middle portions of a given period (such as the Late Cretaceous or the Middle Triassic epoch). Rocks representing an epoch of geological time comprise a **series**. The subdivisions of epochs are **ages**.

dating

Dating is the process of determining the age of geological structures, rocks, and fossils, and placing them in the context of geological time. The techniques are of two types: **relative dating** and **absolute dating**. Relative dating can be carried out by identifying fossils of creatures that lived only at certain times (marker fossils), and by looking at the physical relationships of rocks to other rocks of a known age.

Absolute dating is achieved by measuring how much of a rock's radioactive elements have changed since the rock was formed, and is called **radiometric dating**. Radiometric dating was discovered in the first part of the 20th century by the US radiochemist Bertram Boltwood, who used the ratio of the radioactive element uranium to the product of its decay, the element lead, to date rocks. Using his uranium–lead technique he determined that the Earth was at least 2.2 billion years old. Many radiometric dating methods have been developed since Boltwood's time, such as rubidium–strontium (Rb–Sr) and potassium–argon (K–Ar) methods, enabling earth scientists to establish the age of the Earth as 4.5 billion years.

radiocarbon dating Radiocarbon dating, or **carbon dating**, is a method of dating more recent organic materials (for example, bone or wood), which used in geology and archaeology. Plants take up carbon dioxide gas from the atmosphere and incorporate it into their tissues, and some of that carbon dioxide contains the radioactive isotope of carbon, ^{14}C or carbon-14. As this decays at a known rate (half of it decays every 5,730 years), the time elapsed since the plant died can be measured in a laboratory. Animals take carbon-14 into their bodies from eating plant tissues and their remains can be similarly dated. After 120,000 years so little carbon-14 is left that no measure is possible.

Radiocarbon dating was first developed in 1949 by the US chemist Willard Libby. The method yields reliable ages back to about 50,000 years, but its results require correction since Libby's assumption that the concentration of carbon-14 in the atmosphere was constant through time has subsequently been proved wrong. Discrepancies were noted between carbon-14 dates for Egyptian tomb artefacts and construction dates recorded in early local texts. Radiocarbon dates from tree rings showed that material before 1000 BC had been exposed to greater concentrations of carbon-14. Now radiocarbon dates are calibrated against calendar dates obtained from tree rings, or, for earlier periods, against uranium/thorium dates obtained from coral. The carbon-14 content is determined by counting beta particles with either a proportional gas or a liquid scintillation counter for a period of time. A new advance, accelerator mass spectrometry, requires only tiny samples and counts the atoms of carbon-14 directly, disregarding their decay.

palaeomagnetic stratigraphy The use of distinctive sequences of magnetic polarity reversals to date rocks is called palaeomagnetic stratigraphy. Magnetism retained in rocks at the time of their formation are matched with known dated sequences of polar reversals or with known patterns of secular variation.

geological timescale

Precambrian Eon representing the time from the formation of Earth (4.6 billion years ago) up to 570 million years ago. Its boundary with the succeeding Cambrian period marks the time when animals first developed hard outer parts (exoskeletons) and so left abundant fossil remains. The Precambrian eon comprises about 85% of geological time and is divided into two eras: the Archaean, in which no life existed, and the Proterozoic, in which there was life in some form.

Archaean, or Archaeozoic The widely used term for the earliest era of geological time is the Archaean; the first part of the Precambrian eon, spanning the interval from the formation of Earth to about 3.5 billion years ago.

Proterozoic The Proterozoic era of geological time is the second division of the Precambrian eon from 3.5 billion to 570 million years ago. It is defined as the time of simple life, since many rocks dating from this era show traces of biological activity, and some contain the fossils of bacteria and algae.

Phanerozoic Eon consisting of the most recent 570 million years. It comprises the Palaeozoic, Mesozoic, and Cenozoic eras. The vast majority of fossils come from this eon, owing to the evolution of hard shells and internal skeletons. The name means 'interval of well-displayed life'.

Palaeozoic The era of geological time 570–245 million years ago is called the Palaeozoic. It is comprised of the Cambrian, Ordovician, Silurian, Devonian, Carboniferous, and Permian periods. The Cambrian, Ordovician, and Silurian constitute the **Lower** or **Early Palaeozoic**; the Devonian, Carboniferous, and Permian make up the **Upper** or **Late Palaeozoic**. The era includes the evolution of hard-shelled multicellular life forms in the sea; the invasion of land by plants and animals; and the evolution of fish, amphibians, and early reptiles. The earliest identifiable fossils date from this era.

The climate at this time was mostly warm with short ice ages. The continents were very different from the present ones but, towards the end of the era, all were joined together as a single world continent called Pangaea.

Cambrian The Cambrian period of geological time lasted from 570–510 million years ago; it is the first period of the Palaeozoic era. All invertebrate animal life appeared, and marine algae were widespread. The **Cambrian Explosion** 530–520 million years ago saw the first appearance in the fossil record of all modern animal phyla; the earliest fossils with hard shells, such as trilobites, date from this period. The name comes from Cambria, the medieval Latin name for Wales, where Cambrian rocks are typically exposed and were first described.

Kirschvink Cambrian Explosion Research
http://www.sciam.com/explorations/082597cambrian/
powell.html
Part of a larger site maintained by Scientific American, this page reports on the research of Joseph Kirschvink of the California Institute of Technology that suggests the so-called 'Cambrian Explosion' resulted from a sudden shifting of the earth's crust.

Ordovician The Ordovician period of geological time is the second period of the Palaeozoic era, from 510–439 million years ago. Animal life was confined to the sea: reef-building algae and the first jawless fish

EON	ERA	PERIOD	EPOCH	TIME (my)	PLANT EVOLUTION
PHANEROZOIC	CENOZOIC *Age of mammals*	QUATERNARY *Age of man*	HOLOCENE	0.01	
			PLEISTOCENE	1.64	
		TERTIARY	PLIOCENE	5.20	flora and fauna similar to present day
			MIOCENE	23.5	prairie grasses flourish
			OLIGOCENE	35.5	flowering plants flourish
			EOCENE	56.5	flowering plants flourish
			PALAOCENE	65.0	flowering plants flourish; grasses appear
	MESOZOIC	CRETACEOUS		146	first flowering plants (angiosperms), cycads, and ginkgoes decline
		JURASSIC *Age of Cycads*		208	on land conifers, ginkgoes, cycads dominate; in the sea planktonic plants, such as diatoms and coccoliths, bloom
		TRIASSIC		245	conifers, ginkgoes, and cycads dominate
	PALAEOZOIC	PERMIAN *Age of Amphibians*		290	
		CARBONIFEROUS *Age of Coal* *Age of Amphibians*		363	first seeded land plants (gymnosperms) appear, such as conifers
		DEVONIAN *Age of Fishes*		409	cycads, ginkgoes seed, and ferns
		SILURIAN *Age of Fishes*		439	first seedless, vascular land plants (Psilopsida) and trees (Lycopsida), first ferns
		ORDOVICIAN *Age of Marine Invertebrates*		510	first non-vascular land plants, mosses and liverworts
		CAMBRIAN *Age of Marine Invertebrates*		570	green algae adapt to fresh water environment
PRECAMBRIAN	PROTEROZOIC			3500	multi-celled marine life – blue green algae (stromatolites) and later, green algae
	ARCHAEOZOIC			4600	single-celled marine life emerges, bacteria

ANIMAL EVOLUTION	TECTONIC HISTORY	ATMOSPHERE	ICE AGES
Homo sapiens *Homo erectus* *Australopithecus*	continents in present positions	glaciers retreated, climate became warmer ice ages begin	Pleistocene Ice Age
Ramapithecus, primates with jaws and teeth like man		cooler climates	
bears, raccoons, hyenas appear; monkeys and apes are the dominant primates; first homonids		slight warming trend	
first apes; first cats and dogs	formation of Himalayan – Alpine mountain system begins		
hoofed mammals abundant; first horses (*Hyracotherium*)	Europe separates from Greenland	warming, atmospheric CO_2 6 times modern values	
first primates (prosimians) appear, hand-grasp begins to develop	Australia begins to separate from Antarctica		
CRETACEOUS – TERTIARY (K – T) BOUNDARY – MAJOR EXTINCTION OF LAND FAUNA; INTENSE VOLCANISM dinosaurs, marine ammonoids and swimming reptiles become extinct			
dinosaurs dominate (*Iguanodon*, *Tyrannosaurus rex*, diplodocus, triceratops); shrew-like placental mammals (insectivores) develop	opening of South Atlantic Ocean begins	planktonic plant boom decreases atmospheric CO_2, climates cool	
dinosaurs flourish (stegosaurus, allosaurus); first birds (*Archaeopteryx*); modern amphibians develop (frogs, salamanders)	opening of North Atlantic Ocean begins	mild uniform climate	
original amphibians (labyrinthodonts) become extinct; modern reptiles (turtles, snakes, lizards) begin to develop from cotylosaurs; ammonites flourish in the sea	Pangaea begins to break apart into Laurasia and Gondwana Pangaea fully formed		
PERMO – TRIASSIC BOUNDARY – MAJOR EXTINCTION OF MARINE FAUNA 90% of all species become extinct, including placoderms, trilobytes and some corals			
first dinosaurs (Thecodonta); mammal-like reptiles (Therapsida) such as *Lystrosaurus* appear	northern Africa, Europe and Siberia converge: formation of Ural mountains and Hercynides		Permo-Carboniferous Ice Age
first reptiles (Cotylosauria) appear with the development of eggs with hard shells; amphibians diversify, labyrinthodonts dominate	late Carboniferous: formation of the supercontinent Pangaea begins		
placoderms reach peak; crossopterygians give rise to first amphibians (Labyrinthodontia); sharks; bony fishes; winged-insects			
first vertibratres which were jawed-fish (placoderms) appear; first insects (scorpion-like arachnids, appear			Ordovician Ice Age
jawless fish (Agnathas)		ozone layer lifts	
marine animals develop hard shells (trilobytes, corals, crustaceans)			
		oxygen-rich atmosphere	Veragian, Sturtian Gnejso, Huronian Ice Ages
		anoxic, carbon dioxide-rich atmosphere	

are characteristic. The period is named after the Ordovices, an ancient Welsh people, because the system of rocks formed in the Ordovician period was first studied in Wales.

Silurian The third period of the Palaeozoic era is the Silurian, lasting from 439–409 million years ago. Silurian sediments are mostly marine and consist of shales and limestone. Luxuriant reefs were built by coral-like organisms. The first land plants began to evolve during this period, and there were many ostracoderms (armoured jawless fishes). The first jawed fishes (called acanthodians) also appeared.

Devonian The Devonian is the fourth period of the Palaeozoic era representing 408–360 million years ago. Many desert sandstones from North America and Europe date from this time. The first land plants flourished in the Devonian period, corals were abundant in the seas, amphibians evolved from air-breathing fish, and insects developed on land. The name comes from the county of Devon in southwest England, where Devonian rocks were first studied.

Carboniferous The fifth period of the Palaeozoic era – 362.5 to 290 million years ago – is the Carboniferous. In the USA it is represented by two periods: the Mississippian (lower Carboniferous) and the Pennsylvanian (upper Carboniferous). Typical of the lower-Carboniferous rocks are shallow-water limestones, while upper-Carboniferous rocks have delta deposits with coal (hence the name). Amphibians were abundant, and reptiles evolved during this period.

Permian The sixth and last period of the Palaeozoic era, from 290–245 million years ago, is the Permian. Its end was marked by a significant change in marine life, including the extinction of many corals and trilobites. Deserts were widespread, terrestrial amphibians and mammal-like reptiles flourished, and cone-bearing plants (gymnosperms) came to prominence. In the oceans, 49% of families and 72% of genera vanished in the late Permian. On land, 78% of reptile families and 67% of amphibian families disappeared.

The transition from the Permian period to the Triassic period of the Mesozoic era is referred to as the **Permo-Triassic boundary**. This boundary represents the largest mass extinction in geological history; 90% of all species became extinct.

Mesozoic Following the Palaeozoic era is the Mesozoic era of geological time, spanning 245–65 million years ago. The Mesozoic era consists of the Triassic, Jurassic, and Cretaceous periods. At the beginning of the era, the continents were joined together as Pangaea; dinosaurs and other giant reptiles

dominated the sea and air; and ferns, horsetails, and cycads thrived in a warm climate worldwide. By the end of the Mesozoic era, the continents had begun to assume their present positions, flowering plants were dominant, and many of the large reptiles and marine fauna were becoming extinct.

Triassic The period of geological time 245–208 million years ago is the Triassic, the first period of the Mesozoic era. The continents were fused together to form the world continent Pangaea. Triassic sediments contain remains of early dinosaurs and other reptiles now extinct. By late Triassic times, the first mammals had evolved. The climate was generally dry; desert sandstones are typical Triassic rocks.

Jurassic The middle period of the Mesozoic era, 208–146 million years ago, is the Jurassic. Climates worldwide were equable, creating forests of conifers and ferns; dinosaurs were abundant, birds evolved, and limestones and iron ores were deposited. The name comes from the Jura Mountains in France and Switzerland, where the rocks formed during this period were first studied.

Cretaceous The Cretaceous (Latin *creta* 'chalk') is the period of geological time approximately 144.2–65 million years ago. It is the last period of the Mesozoic era, during which angiosperm (seed-bearing) plants evolved, and dinosaurs reached a peak before their extinction at the end of the period. The northern European chalk, which forms the white cliffs of Dover, was deposited during the latter half of the Cretaceous.

K-T boundary The geologists' shorthand for the boundary between the rocks of the Cretaceous and the Tertiary periods 65 million years ago is the K-T boundary. It coincides with the end of the extinction of the dinosaurs and in many places is marked by a layer of clay or rock enriched in the element iridium. Extinction of the dinosaurs at the K-T boundary and deposition of the iridium layer are thought to be the result of either the impact of a meteorite (or comet) that crashed into the Yucatán Peninsula (forming the Chicxulub crater) or the result of intense volcanism on the continent of India.

Cenozoic, or Caenozoic The Cenozoic is the era of geological time that began 65 million years ago and continues to the present day. It is divided into the Tertiary and Quaternary periods. The Cenozoic marks the emergence of mammals as a dominant group, including humans, and the formation of the mountain chains of the Himalayas and the Alps.

Tertiary The Tertiary period of geological time, spanning 65–1.64 million years ago, is divided into five

What Killed the Dinosaurs?

BY EDWARD YOUNG

Sixty-five million years ago dinosaurs disappeared. With them went 70% of all species of the time. This 'mass extinction', one of many such events throughout Earth's history, marks the end of the Cretaceous period and the beginning of the Tertiary period, an interval of geological time known as the K-T boundary (K for Kreide, German for Cretaceous). What killed the dinosaurs? Two rival hypotheses dominate. One is that a huge asteroid or comet collided with Earth during K-T time, the other is that voluminous volcanism in what is now western India was responsible. Distinguishing between these two hypotheses has proven difficult. In either case, the possible link between these events and the demise of the dinosaurs is forcing scientists to re-examine the role of catastrophe in Earth's history.

from catastrophism to uniformitarianism

Geology arose as a modern scientific discipline in the early 19th century with the fall of catastrophism, the notion that the history of our planet was shaped by successive catastrophic events, and the rise of uniformitarianism, which holds that everyday processes, operating however slowly, are sufficient to explain the principle features of Earth's history.

Naturalists of the late 18th century understood that fossil marine animals and the sedimentary rock in which they were found represented ancient sea floors. But the cast accumulations of fossil-bearing rock revealed in the sides of mountains posed a time problem. The Earth is just 6,000 years old, it was believed, and any casual observer of the oceans could see that sediment did not accumulate fast enough to yield kilometre thicknesses of marine sediment (later turned to rock) unless normal processes were somehow accelerated by cataclysmic events. The great flood described in the Old Testament of the Bible was one catastrophe commonly invoked. Catastrophic upheavals were also called upon to explain mountains.

Near the end of the 18th century a Scottish scientist named James Hutton reasoned that all of the Earth's geological features could be explained by the slow and unchanging forces operating all around us if the Earth was very much older than had been previously imagined. This new view, uniformitarianism, proved immensely successful in explaining geological observations, though not at first. Several decades later another Scot, Charles Lyell, refined and popularized Hutton's uniformitarianism. Lyell emphasized that humankind had existed for sufficient time as to bear witness to all kinds of processes that affect Earth history. Hutton and Lyell imagined Earth where the past looked much like the present, with no beginning nor an end. But it was their principle of uniformitarianism that allowed Charles Darwin to conceive of the evolution of species by natural selection and show that Earth's inhabitants, at least, had changed irrevocably over geological time.

Without uniformitarianism there could have been no modern geology and no Darwinian theory of evolution and geologists have been understandably reluctant to dismiss the premise that 'the present is the key to the past'.

the new catastrophism

Was Lyell right? Have human beings actually witnessed all of the important agents of Earth change? Since the early years of uniformitarianism we have learned that Earth is 4.5 billion years old and that our Solar System formed from the collapse of a fragment of molecular interstellar cloud approximately 4.6 billion years ago. Human history spans a mere one-tenth of one-thousandth of this interval and we would have been fortunate indeed if our existence had coincided with all of the forces that periodically effect our planet over millions and even billions of years. Therefore the most likely answer to the question: was Lyell right? must be no. Meteorite impact or volcanism on a massive scale may have killed off the dinosaurs. But both of these events, as it turns out, are normal in the course of the evolution of our planet. Should they be considered catastrophes in the context of Earth history?

In July 1994 comet Shoemaker–Levy 9 collided with Jupiter. The impact was a catastrophe (for Jupiter) the like of which we have not seen before. Despite the fact that this event was unique in terms of human experiences, such colossal collisions are business-as-usual in astronomical terms. Numerous impact craters scar the rocky surfaces of most bodies in our Solar System. These craters reveal that prior to 3.8 billion years ago, planets were routinely bombarded by metre to kilometre-sized objects. The collisions constituted the final stages of rocky accretion as the planets swept up the debris from which they were made. After 3.8 billion years accretion was essentially complete and impacts upon the planets became less frequent, but as Shoemaker–Levy 9 reminded us, they have not ceased entirely.

Impact structures dating back to the time of frequent bombardment have been destroyed on Earth by erosion and tectonic processes that constantly deform and make over the crust. Nonetheless, scientists have thus far managed to identify more than 100 younger impact craters on Earth. It is estimated that, on average, asteroids or comets measuring 5–10 km/3–6 mi in diameter impact our planet every 50 to 100 million years. Kilometre-sized bodies are thought to hit roughly every million years. Objects in the order of 50 m/165 ft in diameter strike about once every 1,000 years. For comparison, Shoemaker–Levy 9 was composed of several objects ranging from one to several kilometres in size, their collision with Jupiter released more energy than all of the nuclear weapons on Earth combined.

At present 150 asteroids are known to pass within one Earth orbit of the Sun. They range in size from a few metres to 8 km/5 mi. A working group under the auspices of the United States National Aeronautics and Space Administration (NASA) suggests that there may be 2,100 such bodies in all. The impact of a single asteroid of 2 or more kilometres/1.3 or more miles would have enough energy to profoundly affect Earth's biosphere, hydrosphere, and atmosphere. Fortunately, the likelihood of a collision of severe consequence in the next few hundred years is remote.

Not all catastrophes arrive from space. In 1783, just as catastrophism was about to give way to uniformitarianism, the Laki volcano erupted on Iceland. Unusually harsh winter conditions ensued in North America and Europe. Sulphuric acid rain destroyed Iceland's crops and livestock. Several events of comparable magnitude have confirmed that volcanic eruptions can bring about changes in climate. Have humans experienced the full extent of the influence of volcanism on Earth history? The geological record suggests probably not. the impact hypothesis

In 1980, physicist and Nobel Prize laureate Luis Alvarez and his colleagues Walter Alvarez, Frank Asaro, and Helen Michel suggested that the K-T mass extinction resulted from the impact of an asteroid or comet (the term bolide describes an impactor of any sort). A bolide greater than 10 km/6 mi in diameter and travelling 10 km/6 mi per second, they submitted, would liberate enough energy to trigger environmental disaster across the globe. The Earth would be plunged into months of darkness as the dust propelled into the air by the collision blocked the Sun's rays. Photosynthesis would cease and plant-eating animals would be deprived of food, touching off a breakdown in the food chain. Temperatures would fall and as the shroud of dust prevented the Sun from warming Earth's surface, a phenomenon referred to as 'impact winter' would occur. While land-dwelling animals were fighting starvation and the cold, marine creatures would have to battle acidification of the oceans; aerosols of debris and chemical reactions triggered by an atmospheric shock wave would result in nitric and sulphuric acid rain on a grand scale.

Alvarez and co-workers made their proposal on the basis of an unusual enrichment of the rare element iridium in a layer of clay marking the K-T boundary in Italy. Iridium, chemically similar to platinum and gold, is rare on Earth but much more abundant in primitive rocky meteorites. Since their original report, the 'iridium anomaly', as it has come to be known, has since been found in rocks and sediments deposited at the K-T boundary around the world. There is other evidence supportive of an impact. Ratios of the isotope of the element osmium, a platinum-group element like iridium, from some K-T samples are similar to primitive meteorites. Large amounts of soot in K-T boundary clay is thought to have come from large-scale burning of vegetation ignited by debris ejected from the atmosphere that fell back to Earth like a hail of red-hot meteors. Quartz grains shocked by pressures exceeding 100,000 atmospheres have been found in K-T deposits. Previously, shocked quartz had only been found at impact craters (Arizona's Meteor Crater for example) or sites where underground nuclear weapons were tested.

In 1991 a circular impact structure of K-T age was found buried beneath 1 km/0.6 mi of Tertiary carbonate rock in Mexico's Yucatan peninsula. The structure is referred to as the Chicxulub crater after the nearby town of Chicxulub Puerto. Experts agree that Chicxulub is the most spectacular crater on Earth and is the best candidate for the K-T impact site envisioned by Alvarez and co-workers. There are only two other impact structures of comparably large size and they are about 2 billion years old and poorly preserved. The precise size of Chicxulub has been debated. Estimates range from 180 to 300 km/112 to 186 mi in diameter. Most recent studies indicate that the hole upon impact was approximately 120 km/75 mi wide.

Despite its obvious attraction, the impact hypothesis is not without problems. Palaeontologists find evidence that the K-T extinction was not instantaneous and may have begun up to a million years before K-T time and continued for tens of thousands of years after. Shocked quartz and even the iridium anomaly are found to occur just above and below the K-T sediment. Disastrous environmental conditions that would have besieged the Earth after impact of a large bolide should have lasted about a decade and so none of these observations is easily explained by the crash of a single asteroid or comet. To redress this apparent shortcoming of the hypothesis, some proponents have argued that more than one bolide struck Earth at about K-T time.

the volcanism hypothesis

Another competing catastrophe theory has been put forward to explain not only the chemical and physical features of the K-T

boundary but are also the protracted nature of the mass extinction. Beginning in the 1970s it was observed that the largest episode of volcanism in the past 200 million years coincided with the K-T mass extinction. Remains of this volcanism are exposed today in the Deccan plateau region of western India. Here, vast quantities of basalt (a type of dark volcanic rock) known as the Deccan Traps are exposed. Combined, the ancient flows are more than 2 km/1.2 mi thick and comprise roughly 3% of the entire Indian continent. Deccan volcanism lasted several hundreds of thousands of years and was most active at the time of the K-T mass extinction. American geologist Dewey McLean and French geophysicist Vincent Courtillot, among others, argue that it was the voluminous Deccan volcanism that was responsible for the K-T mass extinction. Recall the climatic effects of the Laki eruption? Imagine the effects of continuous large-scale volcanism for hundreds of thousands of years.

The chemical composition of Earth's mantle is more like that of primitive meteorites than the crust. Deccan lavas came from the mantle. Thus instead of being derived from dust thrown up by an asteroid, the iridium anomaly and other chemical signatures of the K-T boundary could be the result of a deluge of volcanic dust. Similarly, shocked quartz, cited as critical evidence for bolide impact, can apparently be formed by eruption of some types of explosive volcanoes.

The environmental effects of large-scale volcanism and bolide impact may be similar. Volcanism, like impact, would loft large amounts of dust high into the atmosphere where it would be transported around the world. As with the impact hypothesis, the Earth might cool as the dust blocked the Sun's rays. Sulphur released by the volcanoes would cause rain to be acidic. Alternatively, McLean has argued that it was carbon dioxide, a greenhouse gas released during volcanism, that was the killer. Large amounts of carbon dioxide released in to the atmosphere would have changed the chemistry of the oceans and caused global warming, both of which would be harmful to life.

an unlucky coincidence

New theoretical studies suggest that without a vulnerability to extinction, catastrophe may have little influence on the diversity of living organisms. Mass extinctions occur with a crude 30 million year periodicity. The largest, in which 90% of species disappeared, was at the end of the Palaeozoic era approximately 250 million years ago. Smaller extinctions are more common. Theoreticians have shown recently that a pattern of frequent smaller extinctions and less frequent larger extinctions is the inevitable consequence of the dependence of living organisms on one another. The disappearance of an animal's prey, for example, may make the animal less fit for survival. The predator species might then vacate its ecological niche and in turn affect fitness of another species and so it goes until eventually an 'avalanche' of linked extinctions occurs. During an extinction 'avalanche' large numbers of species enter new habitats for which they are not well adapted. Mathematical simulations of such processes show that mass extinction requires the coincidence of both large numbers of vulnerable species due to an extinction avalanche, the result of normal evolutionary change, and an unusual amount of stress caused by an environmental catastrophe. Thus it seems the dinosaurs that lived during the Cretaceous period must have been ripe for expiration. Unfortunately for them, their vulnerability coincided with a particularly unlucky time in Earth's history when volcanism was rampant and a collision of the sort that occurs just once every 50 million years actually happened.

epochs: Palaeocene, Eocene, Oligocene, Miocene, and Pliocene. During the Tertiary period mammals took over all the ecological niches left vacant by the extinction of the dinosaurs, and became the prevalent land animals. The continents took on their present positions, and climatic and vegetation zones as we know them became established.

epochs

To remember the different epochs:

Please eliminate old men playing poker honestly

(**P**alaeocene, **E**ocene, **O**ligocene, **M**iocene, **P**liocene, **P**leistocene, **H**olocene)

Palaeocene The first epoch of the Tertiary period of geological time, 65–56.5 million years ago, is the Palaeocene (Greek 'old' + 'recent'). Many types of mammals spread rapidly after the disappearance of the great reptiles of the Mesozoic. Flying mammals replaced the flying reptiles, swimming mammals replaced the swimming reptiles, and all the ecological niches vacated by the reptiles were adopted by mammals.

At the end of the Palaeocene there was a mass extinction that caused more than half of all bottom-dwelling organisms in the oceans to disappear worldwide, over a period of around one thousand years. Surface dwelling organisms remained unaffected, as did those on land. The cause of this extinction remains unknown, though US palaeontologists have found evidence (released in 1998) that it may have been caused by the Earth releasing tonnes of methane into the oceans causing increased water temperatures.

Eocene The second epoch of the Tertiary period, 56.5–35.5 million years ago, is the Eocene. Originally considered the earliest division of the Tertiary, the name means 'early recent', referring to the early forms of mammals evolving at the time, following the extinction of the dinosaurs.

Oligocene The Oligocene is the third epoch of the Tertiary period, 35.5–23.5 million years ago. The name, from Greek, means 'a little recent', referring to the presence of the remains of some modern types of animals existing at that time.

Miocene The Miocene ('middle recent') is the fourth epoch of the Tertiary period, 23.5–5.2 million years ago. At this time grasslands spread over the interior of continents, and hoofed mammals rapidly evolved.

Pliocene The fifth and last epoch of the Tertiary period, 5.2–1.64 million years ago, is the Pliocene ('almost recent'). The earliest hominid, the humanlike ape *Australopithecines*, evolved in Africa.

Quaternary The Quaternary period of geological time began 1.64 million years ago and is still in process. It is divided into the Pleistocene and Holocene epochs.

Pleistocene The Pleistocene is the first epoch of the Quaternary period, beginning 1.64 million years ago and ending 10,000 years ago. The polar ice caps were extensive and glaciers were abundant during the ice age of this period, and humans evolved into modern *Homo sapiens sapiens* about 100,000 years ago.

Holocene The Holocene epoch began 10,000 years ago and is the second and current epoch of the Quaternary period. During this epoch the glaciers retreated, the climate became warmer, and humans developed significantly.

Extinction
http://www.museum.state.il.us/exhibits/larson/
LP_extinction.html
Exploration of possible causes of the Late Pleistocene extinction of most large mammals in North America.

Earth Science: Evolution and Structure Chronology

c. 1470 BC The island of Thera is destroyed in a volcanic eruption. It causes a tidal wave and subsequent famine in Egypt, and destroys the Minoan civilization on the island of Crete over 120 km/75 mi away. It may be the source of the Atlantis myth.

c. 1226 BC The first record of Mount Etna in Italy erupting is made. It has erupted 190 times since.

c. 780 BC Chinese scholars make the first record of an earthquake.

c. 560 BC The Greek philosopher Xenophanes correctly recognizes the nature of fossils when he suggests that fossil seashells are the result of a great flood that buried them in the mud.

c. 530 BC Pythagoras of Samos proposes the notion of a spherical Earth.

c. 350 BC Aristotle defends the doctrine that the Earth is a sphere, in *De caelo/Concerning the Heavens*, and estimates its circumference to be about 400,000 stadia (one stadium varied from 154 m/505 ft to 215 m/705 ft). It is the first scientific attempt to estimate the circumference of the Earth.

c. 295 BC Greek philosopher Theophrastus writes *De lapidibus/On Stones*, a classification of 70 different minerals. It is the oldest known work on rocks and minerals and is the best treatise on the subject for nearly 2,000 years.

c. 100 The geographer Marinus of Tyre develops a system of equal spacing for lines of latitude and longitude. Maps by Marinus are the first in the Roman Empire to show China.

114 A Chinese sculpture of this date shows an early form of compass – a polished magnetite spoon that spins to align with the Earth's magnetic field when placed on a smooth surface.

132 Chinese engineer Zhang Heng develops the first seismograph for determining the position of an earthquake's epicentre. It uses a series of balls suspended in the mouths of eight carved dragons. The balls that are dislodged indicate the direction of the earthquake centre.

1154 The Egyptian Muslim scholar al-Tifashi writes his pioneering work on mineralogy *Flowers of Knowledge of Precious Stones*.

1282 The Italian scholar Ristoro d'Arezzo writes his encyclopedia *Della composizione del mondo/On the Composition of the World*. It discusses the structure of the world and develops some sound geological theories despite its links with astrology.

1492 German navigator Martin Behaim, with painter Goerg Glockendon, constructs a terrestrial globe at Nuremberg, the earliest still in existence.

1492 The explorer Christopher Columbus, crossing the Atlantic, notes that the deviation of his compass from true north has changed from the east to the west, an early discovery of the variation of the Earth's magnetism.

1513 A new edition of Ptolemy's *Geography* shows the New World as two separate continents.

26 Jan 1531 An earthquake destroys the Portuguese capital of Lisbon, killing 50,000 people.

1536 Flemish cartographer Gerardus Mercator collaborates with geographer and mathematician Gemma Frisius to construct a terrestrial globe.

1544 German mineralogist Agricola (Georg Bauer) writes *De ortu et causis subterraneis/On Subterranean Origin and Causes*, a founding work in geology, identifying the erosive power of water, and the origin of mineral veins as depositions from solution.

1546 Flemish cartographer Gerardus Mercator states that the Earth must have a magnetic pole separate from its 'true' pole, in order to explain the deviation of a compass needle from true North.

1576 English scientist Robert Norman discovers the magnetic 'dip' or inclination in a compass needle that is caused by the Earth's magnetic field not running exactly parallel to the surface.

1580 An earthquake shakes London, England.

1622 British scientist Edward Gunter discovers that the magnetic needle does not retain the same declination in the same place all the time – the first evidence for variation in Earth's magnetic field.

1650 German geographer Bernhardus Varenius publishes his *Georgraphia generalis/General Geography*, establishing scientific principles for general and regional geography.

1691 German scientist Gottfried von Leibniz publishes *Protogaea/The Primordial Earth*, a study of geology containing many new insights, including a belief that the Earth was formed from a molten state.

1736 Swedish scientist Anders Celsius leads a French-sponsored expedition to Lapland. By measuring the length of a degree of meridian at high latitude, they prove the Earth is flattened at the poles.

1745 Russian mineralogist Mikhail Vasileyevich Lomonosov publishes a catalogue of over 3,000 minerals.

1746 French naturalist Jean-Etienne Guettard makes the first mineralogical map (of France). Amongst his discoveries, he identifies extinct volcanoes in the Auvergne region, and volcanic rocks in large deposits across France.

1749	French naturalist Georges-Louis de Buffon publishes the first book of of his 36-volume *Histoire naturelle, genérale et particulière/Natural History, General and Particular*, the first attempt to bring together the various fields of natural history.
1774	French naturalist Nicolas Desmarest's essays on extinct volcanoes demonstrate the volcanic origin of basalt, disproving the theory that all rocks are formed by sedimentation from primeval seas.
1776	English chemist James Keir suggests that some rocks may have formed as molten material that cooled and crystallized.
1778	In *Epoques de la nature/Epochs of Nature*, French scientist George-Louis Leclerc, comte de Buffon, reconstructs geological history as a series of stages - the first to recognize such stages. It contradicts the doctrine that the Earth is only 6,000 years old.
1779	Swiss geologist Horace Bénédict de Saussure coins the word 'geology'.
1783–84	Mount Skaptar in Iceland erupts, killing 9,000 people, about one-fifth of the population.
1788	Scottish geologist James Hutton's paper *Theory of the Earth* expounds his uniformitarian theory of continual change in the Earth's geological features and marks a turning point in geology.
1798	English natural philosopher Henry Cavendish determines the mean density of the Earth; it is 5.5 times as dense as water.
1801	French mineralogist René-Just Haüy publishes *Traité de minéralogie/Treatise on Mineralogy*, a theory of the crystal structure of minerals that establishes him as one of the founders of crystallography.
1809	René-Just Haüy publishes *Tableau comparatif/Comparative Tables*, one of the first classifications of minerals.
16 Dec 1811	The first, and largest, earthquake recorded in the USA destroys the city of New Madrid, Missouri. Two other earthquakes hit the town on 23 January and 7 February 1812.
1817–59	German geographer Carl Ritter publishes *Earth Science in its Relation to Nature and the History of Man*. The first of 19 volumes surveying world geography, it establishes him as co-founder, along with Alexander von Humboldt, of modern geography.
1822	Mary Anning discovers the first fossil to be recognized as that of a dinosaur - an *Iguanodon* - in Devon, England.
July 1830–April 1833	Scottish geologist Charles Lyell publishes his three-volume work *Principles of Geology* in which he argues that geological formations are the result of presently observable processes acting over millions of years. It creates a new time frame for other sciences such as biology and palaeontology.

1837	US geologist James Dana publishes *System of Mineralogy*, which is still the standard text on mineralogy.
1839	Scottish geologist Robert Murchison publishes *The Silurian System*, a geological treatise which establishes the geological sequence of the Early Palaeozoic rocks.
1841	German astronomer Friedrich Bessel deduces the elliptical distortion of the Earth - the amount it departs from a perfect sphere - to be $1/_{299}$.
c. 1850	French naturalist Antonio Snider-Pellegrini suggests that the similarities between European and North American plant fossils could be explained if the two continents were once in contact.
1854	English astronomer George Biddell Airy calculates the mass of the Earth by swinging a pendulum at the top and bottom of a deep coal mine and measuring the different gravitational effects on it.
1862	Scottish physicist William Thomson (Lord Kelvin), using the Earth's temperature, estimates the Earth to be between 20 and 400 million years old.
1862	Scottish-born German astronomer Johann von Lamont discovers the electrical current within the Earth's crust.
1877	Cotopaxi volcano in Ecuador erupts; over 1,000 people are killed by mudflows.
1877	The Como Bluff palaeontological site is discovered in Wyoming. It contains a large number and variety of dinosaur remains, including the first specimens of *Stegosaurus*, *Brontosaurus* (now known as *Apatosaurus*), and *Allosaurus*.
1880	British geologist John Milne invents the modern seismograph for measuring the strength of earthquakes.
1883–88	Austrian geologist Eduard Suess publishes *Das Antlitz der Erde/Face of the Earth*, in which he postulates the existence of an ancient supercontinent in the southern hemisphere called Gondwanaland, and discusses how various processes are responsible for the present features of the Earth's surface.
1888	Swedish geologist Alfred Elis Törnebohm presents the theory that mountain ranges are the result of overthrusting, in which the upper surface of a fault plane moves over the rocks of the lower surface.
1889	US geologist Clarence Edward Dutton discovers a method of determining the epicentre of earthquakes and accurately measuring the speed of seismic waves.
1892	US geologist Clarence Edward Dutton publishes the paper 'On Some of the Greater Problems of Physical Geology', in which he advances the idea of isostasy, whereby lighter material in the Earth's crust rises to form continents and mountains, while heavier material sinks, to form basins and oceans.

1902 New Zealand-born British physicist Ernest Rutherford and British physicist Frederick Soddy discover thorium X and publish *The Cause and Nature of Radioactivity*, which outlines the theory that radioactivity involves the disintegration of atoms of one element into atoms of another, laying the foundation for radiometric dating of natural materials.

1905 US chemist Bertram Boltwood suggests that lead is the final decay product of uranium. His work will eventually lead to the uranium–lead dating method.

1905 The Carnegie Institution of Washington establishes the Geophysical Laboratory in Washington, DC.

1906 Scientists discover two distinct forms of rock magnetism: some rocks magnetized with their north 'poles' parallel to Earth's present magnetic field and others magnetized with their poles reversed with respect to Earth's magnetic field.

1906 Irish geologist Richard Oldham proves that the Earth has a molten core, by studying seismic waves.

1907 French physicist Pierre-Ernest Weiss develops the domain theory of ferromagnetism, which suggests that in a ferromagnet, such as loadstone, there are regions, or domains, where the molecules are all magnetized in the same direction. His theory leads to a greater understanding of rock magnetism.

1907 US chemist Bertram Boltwood uses the ratio of lead and uranium in some rocks to determine their age. He estimates his samples to be 410 million to 2.2 billion years old.

1909 Croatian physicist Andrija Mohorovicic discovers the Mohorovicic discontinuity in the Earth's crust. Located about 30 km/18 mi below the surface, it forms the boundary between the crust and the mantle.

1909 US geologist and secretary of the Smithsonian Institution Charles D Walcott discovers fossils of soft parts of organisms in the Cambrian Burgess Shale of the Canadian Rockies. The discovery provides unprecedented evidence pertaining to the rapid evolution of life that started in the Cambrian period.

1910 US physicist Percy Bridgman invents a device, called the 'collar', that allows him to squeeze all kinds of materials to pressures comparable to the base of Earth's crust, giving rise to the new fields of high-pressure physics and mineral physics.

1910 Swedish archaeologist Gerhard de Geer publishes *A Geochronology of the Last 12,000 Years*, setting out his influential system for dating rock strata.

1912 German meteorologist Alfred Wegener suggests the idea of continental drift and proposes the existence of a supercontinent (Pangaea) in the distant past.

1912 German physicist Max von Laue demonstrates that crystals are composed of regular, repeated arrays of atoms by studying the patterns in which they diffract X-rays. It is the beginning of X-ray crystallography.

1913 English geologist Arthur Holmes uses radioactivity to date rocks, establishing that the Earth is 4.6 billion years old.

1914 German-born US geologist Beno Gutenberg discovers the discontinuity that marks the boundary between the Earth's lower mantle and outer core, about 2,800 km/1,750 mi below the surface.

1920 English physicist Frederick Soddy suggests that isotopes can be used to determine geological age.

1926 The Scott Polar Research Institute is opened in Cambridge, England, to conduct Antarctic research.

1928 US geochemist Norman L Bowen publishes *The Evolution of the Igneous Rocks* in which he suggests that Earth's crust is the product of melting of parts of the mantle, a process known as differentiation. His work firmly establishes the potential of the physical chemical approach to geology.

1929 By studying the magnetism of rocks, the Japanese geologist Motonori Matuyama shows that the Earth's magnetic field periodically reverses direction.

1929 Norwegian chemist Victor Moritz Goldschmidt produces the first table of ionic radii useful for predicting crystal structures.

1930 The Indian physicist Chandrasekhara Venkata Raman receives the Nobel Prize for Physics for his work on the scattering of light and for the Raman effect. Raman spectroscopy later becomes an important tool in mineral physics.

1931 US chemist Harold C Urey discovers deuterium, a heavy isotope of hydrogen, ushering in the modern field of stable isotope geochemistry.

1936 US archaeologist Andrew E Douglass develops dendrochronology; a dating system based on the measurement of growth rings in trees.

1937 Finnish chemist and geologist Victor Moritz Goldschmidt tabulates the absolute abundances of chemical elements in Earth from solar and meteorite chemical data.

1939 US chemist A O Nier with E A Gulbransen report natural variations in the isotopes of carbon. Geochemists will later use the stable isotopes of carbon to study fossil bones, teeth, and rocks.

1939 US chemist Linus Pauling consolidates his theory of the chemical bond in *The Nature of the Chemical Bond, and the Structure of Molecules and Crystals* in which he sets out rules for understanding mineral structures. 'Pauling's rules' revolutionize the understanding of the structure and chemistry of minerals.

1939 US geophysicist Walter Maurice Elsasser formulates the 'dynamo model' of the Earth, which proposes that eddy currents in the Earth's molten iron core cause its magnetism.

1946 Carbon-13, a stable isotope of carbon, is discovered. Its abundance will ultimately be used for deciphering the geochemistry of carbon.

1947 US physicist Willard Libby develops carbon-14 dating.

1948 US chemists L T Aldrich and A O Nier find argon from decay of potassium in four geologically old minerals, confirming predictions by German physicist C F von Weizsacker made in 1937. The basis for potassium-argon dating is established.

1949 US seismologist Hugo Benioff identifies planes of earthquake foci extending from deep ocean trenches to beneath adjacent continents at approximately 45° as faults. These faults, later called Benioff zones or Benioff–Wadati zones, will later be identified as the tops of plates of lithosphere.

1953 US chemists Stanley Miller and Harold Urey demonstrate experimentally how organic compounds might have begun on Earth, by exposing a warm, gaseous mixture of inorganic compounds to electrical discharges. The gases are meant to represent the pre-biotic atmosphere, the 'primordial soup' out of which life evolved.

1953 US geophysicist William Ewing announces that there is a crack, or rift, running along the middle of the Mid-Atlantic Ridge.

1954 Researchers at General Electric produce the first synthetic diamonds.

1954 The existence of pre-metazoan life becomes widely accepted when diverse fossil microscopic organisms are discovered in the Gunflint rocks along the north shore of Lake Superior, northern USA/southern Canada. The Gunflint biota prove that there had been significant evolutionary activity by at least about 2 billion years ago.

1956 US geologists Bruce Heezen and William Ewing discover a global network of oceanic ridges and rifts 60,000 km/37,000 mi long that divide the Earth's surface into 'plates'.

28 Feb 1959 The US Air Force launches *Discoverer 1* into a low polar orbit where it photographs the entire surface of the Earth every 24 hours.

1960 US geophysicist Harry Hess develops the theory of seafloor spreading (the term is coined by R S Dietz in 1961), in which molten material wells up along the mid-oceanic ridges forcing the seafloor to spread out from the ridges. The flow is thought to be the cause of continental drift.

1961 L O Nicolaysen invents the 'isochron' method of rubidium–strontium and uranium–lead dating of geological materials.

1961 US researchers establish Arctic Research Lab Ice Station II (Arliss II), a drifting sea ice station.

1962 US geologist Harry Hess publishes *History of Ocean Basin*, in which he formally proposes seafloor spreading, the idea that the ocean crust is like a giant conveyor belt produced by volcanism at the mid-ocean ridges, pushed or pulled away from the ridge axis, and eventually destroyed by plunging down into the deep-sea trenches.

1962 Part of the north summit of Mount Huascaran, Peru's highest mountain, breaks off during a thaw; 3,500 people are killed.

1963 British geophysicists Fred Vine and Drummond Matthews analyse the magnetism of rocks in the Atlantic Ocean floor, which assume a magnetization aligned with the Earth's magnetic field at the time of their creation. It provides concrete evidence of seafloor spreading.

1965 Canadian geologist John Tuzo Wilson publishes *A New Class of Faults and Their Bearing on Continental Drift*, in which he formulates the theory of plate tectonics to explain continental drift and seafloor spreading. He suggests that the mid-ocean ridges, deep-sea trenches, and the faults that connect them combine to divide the Earth's outer layer into rigid independent plates. The connecting faults are shown by Wilson to have a unique geometry, and are called 'transform faults'.

1965 The Large Aperture Seismic Array is established in Montana, USA. The signals from 525 seismometers, dispersed over an area of 30,000 sq km/11,600 sq mi, are combined to record seismic events with a high degree of sensitivity.

1966 US geologists Allan Cox, Richard Doell, and Brent Dalrymple publish a chronology of magnetic polarity reversals going back 3 million years. It is useful in dating rocks and fossils.

1967 Geophysicist Lynn Sykes uses first-motion seismic studies to establish that mid-ocean ridges form with offsets rather than being offset later, a major advance in understanding the formation of ocean basins.

1967 British geophysicist D McKenzie and US geophysicist Jason Morgan describe the motions of plates across Earth's surface. Morgan calls the plates 'tectosphere'. They are later referred to as plates of lithosphere.

1968 French geophysicist Xavier Le Pichon, working at the Lamont Observatory in New York, describes the motions of Earth's six largest plates using poles of rotation derived from the patterns of magnetic anomalies and fracture zones about mid-ocean ridges.

1968 The US survey ship *Glomar Challenger* starts drilling cores in the sea bed as part of the Deep Sea Drilling Project. Capable of drilling in water up to 6,000 m/20,000 ft deep, it can return core samples from 750 m/2,500 ft below the sea floor.

1969	The Joint Oceanographic Institutions Deep Earth Sampling (JOIDES) project begins. It makes boreholes in the ocean floor and confirms the theory of seafloor spreading and that the oceanic crust everywhere is less than 200 million years old.
1970	British geologist John F Dewey with J M Bird relate the positions of Earth's mountain belts to the motions of lithospheric plates.
1972	US palaeontologists Stephen Jay Gould and Nils Eldridge propose the punctuated equilibrium model – the idea that evolution progresses in fits and starts rather than at a uniform rate.
23 July 1972	The USA launches *Landsat 1*, the first of a series of satellites for surveying the Earth's resources from space.
1974	Earth scientist John Liu discovers that the lower mantle, comprising most of the Earth, is likely composed of silicate perovskite, a mineral with a structure wildly different from the minerals found in Earth's upper mantle and crust.
1975	David Mao and Peter Bell of the Geophysical Laboratory use a diamond-anvil cell to produce pressures exceeding a million atmospheres.
1975	The USA launches *Landsat 2*; located 180° away from *Landsat 1*, the two together provide a view of the same geographic area with the same Sun angle every nine days.
Oct 1975	The USA launches the first *Geostationary Operational Environmental Satellite* (GOES); it provides 24-hour coverage of US weather.
1976	The US *Lageos* (Laser Geodynamic Satellite) is launched; it uses laser beams to make precise measurements of the Earth's movements in an attempt to improve the prediction of earthquakes. Placed in an orbit 9,321 km/5,793 mi high, it is expected to remain in orbit for 8 million years.
1978	Core samples from the seabed are collected by the US research vessel *Glomar Challenger* from a record depth of 7,042 m/23,104 ft.
1980	A thin layer of iridium-rich clay, about 65 million years old, is found around the world. US physicist Luis Alvarez suggests that it was caused by the impact of a large asteroid or comet which threw enough dust into the sky to obscure the Sun and cause the extinction of the dinosaurs.
1980	M Ikeya and T Liki of Yamaguchi University, Japan, announce a new method of dating fossil remains: electron spin resonance

spectroscopy, which measures the amount of natural radiation received by such remains.

1980	The US *Magsat* satellite completes its mapping of the Earth's magnetic field.
1983	Studies from the *Lageos* satellite (launched in 1976) indicate that the Earth's gravitational field is changing.
1984	Australian geologists Bod Pidgeon and Simon Wilde discover zircon crystals in the Jack Hills north of Perth, Australia, that are estimated to be 4.2 billion years old – the oldest minerals ever discovered.
1990	Canadian scientists discover fossils of the oldest known multicellular animals, dating from 600 million years ago.
1991	A circular impact structure of Cretaceous–Tertiary (K–T) age is found buried beneath in Mexico's Yucatan peninsula. Called the Chicxulub crater, it is the best candidate for the K–T meteor impact site envisioned by Luis Alvarez and others.
1995	US and French geophysicists discover that the Indo-Australian plate split in two in the middle of the Indian Ocean about 8 million years ago.
1996	US geophysicists discover that the Earth's core spins slightly faster than the rest of the planet.
25 June 1997	A volcanic eruption in the Soufrière Hills on the British dependency of Montserrat in the Leeward Islands, West Indies, kills 23 people.
24 July 1997	Canadian researcher Richard Bottomley and colleagues date the 100-km/62-mi-wide Popigai impact crater in Siberia, thought to be the fifth largest impact crater on Earth, to 35.7 million years old. They suggest that the meteorite that created it may be responsible for the mass extinction that occurred at the end of the Eocene and the start of the Oligocene geological periods, which is dated to about the same time.
25 July 1997	US researcher Joseph L Kirschvink and colleagues, by examining the record of remnant magnetism in very ancient rocks, discover that the outer layers of the Earth shifted by 90 degrees relative to the core between about 535 and 520 million years ago. This major reorganization of the continents they suggest may have led to the Cambrian Explosion – the rapid appearance of abundant fossils in the geological record in the Cambrian period.

Biographies

Agassiz, (Jean) Louis Rodolphe (1807–1873) Swiss-born US palaeontologist and geologist who established his name through his work on the classification of fossil fishes. His book *Researches on Fossil Fish* (1833–44) described and classified over 1,700 species. Unlike Charles Darwin, he did not believe that individual species themselves changed, but that new species were created from time to time.

Alvarez, Luis Walter (1911–1988) US physicist. In 1980 he was responsible for the theory that dinosaurs disappeared because a meteorite crashed into Earth 70 million years ago, producing a dust cloud that blocked out the sun for several years, causing dinosaurs and plants to die.

Bakker, Robert T (1945–) US palaeontologist who in 1975 published the theory that dinosaurs were warm-blooded. He also views them as fast, agile, and colourful. He believes that the extinction of the dinosaurs was caused by disease carried across the new land bridges that appeared at the end of the Cretaceous period. He has popularized his work on television and in books such as *The Dinosaur Heresies* (1986).

Boltwood, Bertram (1870–1927) US radiochemist who pioneered the use of radioactive elements as tools for dating rocks. By studying abundances of radioactive elements in ores, he deduced that the radium present in an ore was the product of the breakdown of uranium in the ore and that uranium ultimately would decay to lead. In 1907 he demonstrated that by knowing the rate at which uranium decays (its half-life) he could calculate the age of a mineral by measuring the relative proportions of its uranium and lead. He dated rocks from several localities using his uranium–lead technique, obtaining ages between 410 million to 2.2 billion years old, and determined the age of the Earth to be at least 2.2 billion years old, significantly older than previously thought.

Bowen, Norman Levi (1887–1956) Canadian geologist whose work helped found modern petrology. His findings on the experimental melting and crystallization behaviour of silicates and similar mineral substances were published from 1912 onwards. He demonstrated the principles governing the formation of magma by partial melting, and the fractional crystallization of magma; his book *The Evolution of Igneous Rocks* (1928) deals particularly with magma.

Bridgman, Percy Williams (1882–1961) US physicist. His research into machinery producing high pressure led in 1955 to the creation of synthetic diamonds by General Electric. His technique for synthesizing diamonds was used to synthesize many more minerals, and a new school of geology developed, based on experimental work at high pressure and temperature. Because the pressures and temperatures that Bridgman achieved simulated those deep in the Earth, his discoveries gave an insight into the geophysical processes that take place within the Earth. His book *Physics of High Pressure* (1931) remains a basic work. He was awarded the Nobel Prize for Physics in 1946.

Brongniart, Alexandre (1770–1847) French naturalist and geologist who first used fossils to date strata of rock and was the first scientist to arrange the geological formations of the Tertiary period in chronological order. He introduced the idea of geological dating according to the distinctive fossils found in each stratum. From 1804–11 Brongniart and Georges ◊Cuvier studied the fossils deposited in the Paris basin and showed that the fossils had been laid down during alternate fresh and salt water conditions.

Buckland, William (1784–1856) English geologist and palaeontologist, a pioneer of British geology. He contributed to the descriptive and historical stratigraphy of the British Isles, inferring from the vertical succession of the strata a stage-by-stage temporal development of the Earth's crust. His interest in catastrophic transformations of the Earth's surface in the geologically recent past, as indicated by such features as fossil bones and erratic boulders, led him to become an early British exponent of the glacial theory of Louis ◊Agassiz.

Buffon, Georges-Louis Leclerc, comte de Buffon (1707–1788) French naturalist and author of the 18th century's most significant work of natural history, the 44-volume *Histoire naturelle génerale et particulière* (1749–67). In *The Epochs of Nature*, one of the volumes, he questioned biblical chronology for the first time, and raised the Earth's age from the traditional figure of 6,000 years to the seemingly colossal estimate of 75,000 years.

Bullard, Edward Crisp (1907–1980) English geophysicist who, with US geologist Maurice ◊Ewing, founded the discipline of marine geophysics. He pioneered the application of the seismic method to study the sea floor. He also studied continental drift before the theory became generally accepted. Bullard's earliest work was to devise a technique to measure minute gravitational variations in the East African Rift Valley. He then investigated the rate of efflux (outflow) of the Earth's interior heat through the land surface; later he devised apparatus for measuring the flow of heat through the deep sea floor. According to his 'dynamo' theory of geomagnetism, the Earth's magnetic field results from convective movements of molten material within the Earth's core.

Chamberlin, Thomas Chrowder (1843–1928) US geophysicist who asserted that the Earth was far older than then believed. He developed the planetesimal hypothesis for the origin of the Earth and other planetary bodies – that they had been formed gradually by accretion of particles.

Cuvier, Georges Léopold Chrétien Frédéric Dagobert, Baron Cuvier (1769–1832) French comparative anatomist, the founder of palaeontology. In 1799 he showed that some species have become extinct by reconstructing extinct giant animals that he believed were destroyed in a series of giant deluges. These ideas are expressed in *Recherches sur les ossiments fossiles de quadrupèdes/Researches on the Fossil Bones of Quadrupeds* (1812) and *Discours sur les révolutions de la surface du globe* (1825). He was the first to relate the structure of fossil animals to that of their living relatives.

De la Beche (born Beach), Henry Thomas (1796–1855) English geologist. He secured the founding of the Geological Survey in 1835, a government-sponsored geological study of Britain, region by region. He wrote books of descriptive stratigraphy, above all on the Jurassic and Cretaceous rocks of the Devon and Dorset area. He also conducted important fieldwork on the Pembrokeshire coast and in Jamaica. His main work is *The Geological Observer* (1851).

Desmarest, Nicolas (1725–1815) French champion of volcanist geology, who countered the widely held belief that all rocks were sedimentary. Studying the large basalt deposits of central France, he traced their origin to ancient volcanic activity in the Auvergne region. In 1768 he produced a detailed study of the geology and eruptive history of the volcanoes responsible. However, he did not believe that all rocks had igneous origin, and emphasized the critical role of water in the shaping of the Earth's history.

Du Toit, Alexander Logie (1878–1948) South African geologist. His work was to form one of the foundations for the synthesis of continental drift theory and plate tectonics that created the geological revolution of the 1960s. The theory of continental drift put forward by Alfred ◊Wegener inspired Du Toit's book *A Geological Comparison of South America and South Africa* (1927), in which he suggested that they had probably once been joined. In *Our Wandering Continents* (1937) he maintained that the southern continents had, in earlier times, formed the supercontinent of Gondwanaland, which was distinct from the northern supercontinent of Laurasia.

Edinger, Tilly (Johanna Gabrielle Ottilie) (1897–1967) German-born US palaeontologist. Her work in vertebrate palaeontology laid the foundations for the study of palaeoneurology. She demonstrated that the evolution of the brain could be studied directly from fossil cranial casts.

Elsasser, Walter Maurice (1904–1991) German-born US geophysicist. He pioneered analysis of the Earth's former magnetic fields, which are frozen in rocks. His research in the 1940s yielded the dynamo model of the Earth's magnetic field. The field is explained in terms of the activity of electric currents flowing in the Earth's fluid metallic outer core. The theory premises that these currents are magnified through mechanical motions, rather as currents are sustained in power-station generators.

Eskola, Pentti Eelis (1883–1964) Finnish geologist. He was one of the first to apply physicochemical postulates on a far-reaching basis to the study of metamorphism, thereby laying the foundations of most subsequent studies in metamorphic petrology. His approach enabled comparison of rocks of widely differing compositions in respect of the pressure and temperature under which they had originated.

Ewing, (William) Maurice (1906–1974) US geologist. His studies of the ocean floor provided crucial data for the plate tectonics revolution in geology in the 1960s. Ewing ascertained that the crust of the Earth under the ocean is much thinner (5–8 km/3–5 mi thick) than the continental shell (about 40 km/25 mi thick). He demonstrated that mid-ocean ridges, with deep central canyons, are common to all oceans, and his studies of ocean sediment showed that its depth increases with distance from the mid-ocean ridge, which gave clear support for the hypothesis of seafloor spreading.

Gardner, Julia (Anna) (1882–1960) US geologist and palaeontologist. Her work was important for petroleum geologists establishing standard stratigraphic sections for Tertiary rocks in the southern Caribbean.

Goldring, Winifred (1888–1971) US palaeontologist. Her research focused on Devonian fossils and during the late 1920s and the 1930s, as well as geologically mapping the Coxsackie and Berne quadrangles of New York, she developed and maintained the State Museum's public programme in palaeontology. Her works include *The Devonian Crinoids of the State of New York* (1923) and *Handbook of Paleontology for Beginners and Amateurs* (1929–31). She did much to popularize geology.

Guettard, Jean-Etienne (1715–1786) French pioneer of geological mapping, who studied the origin of various types of rock. Research in the field suggested that the rocks of the Auvergne district of central France were volcanic in nature, and Guettard boldly identified several peaks in the area as extinct volcanoes, though he later had doubts about this hypothesis. Of basalt, he originally took the view that it was not volcanic in origin, but changed his mind after visits to Italy in the 1770s.

Gutenberg, Beno (1889–1960) German-born US geophysicist who determined the depth of Earth's core and contributed to the understanding of Earth's deep interior. As a student in 1914, he used velocities of seismic waves to calculate the depth of Earth's core at 2,900 km/1,812 mi. This boundary between Earth's mantle and its core is called the **Gutenberg discontinuity**. In 1948 he suggested the existence of a low-velocity zone approximately 60–150 km/38–94 mi below the Earth's surface in which seismic waves travel more slowly. This zone is now known to be the asthenosphere, the more ductile layer of the mantle on which the Earth's lithospheric plates ride.

Hall, James (1761–1832) Scottish geologist. He was one of the founders of experimental geology and provided evidence in support of the theories of James ◊Hutton regarding the formation of the Earth's crust. By means of furnace experiments, he showed that Hutton had been correct to maintain that igneous rocks would generate crystalline structures if cooled very slowly. Hall also demonstrated that there was a degree of interconvertibility between basalt and granite rocks, and that, even though subjected to immense heat, limestone would not decompose if sustained under suitable pressure.

Haüy, René-Just (1743–1822) French mineralogist, the founder of modern crystallography. He regarded crystals as geometrically structured assemblages of units (integrant molecules), and developed a classification system on this basis. He proposed six primary forms: parallelepiped, rhombic dodecahedron, hexagonal dipyramid, right hexagonal prism, octahedron, and tetrahedron. His two major works are *Traité de minéralogie/Treatise of Mineralogy* (1801) and *Treatise of Crystallography* (1822).

Hess, Harry Hammond (1906–1969) US geologist who in 1962 proposed the notion of seafloor spreading. This played a key part in the acceptance of plate tectonics as an explanation of how the Earth's crust is formed and moves. Building on the recognition that certain parts of the ocean floor were anomalously young, and the discovery of the global distribution of mid-ocean ridges and central rift valleys, Hess suggested that convection within the Earth was continually creating new ocean floor, rising at mid-ocean ridges, and then flowing horizontally to form new oceanic crust. It would follow that the further from the mid-ocean ridge, the older would be the crust – an expectation confirmed by research in 1963.

Holmes, Arthur (1890–1965) English geologist who helped develop interest in the theory of continental drift. In 1928 he proposed that convection currents within the Earth's mantle, driven by radioactive heat, might furnish the mechanism for the continental drift theory broached a few years earlier by Alfred ◊Wegener. He also pioneered the use of radioactive decay methods for rock dating, giving the first reliable estimate of the age of the Earth.

Hutton, James (1726–1797) Scottish geologist, known as the 'founder of geology', who formulated the concept of uniformitarianism, suggesting that past events could be explained in terms of processes that work today. For example, the kind of river current that produces a certain settling pattern in a bed of sand today must have been operating many millions of years ago, if that same pattern is visible in ancient sandstones. In 1785 he developed a theory of the igneous origin of many rocks. His *Theory of the Earth* (1788) proposed that the Earth was incalculably old.

Knopf (born Bliss), Eleanora Frances (1883–1974) US geologist who studied metamorphic rocks. She introduced the technique of petrofabrics, which had been developed in Austria, to the USA.

Lyell, Charles (1797–1875) Scottish geologist. In his *Principles of Geology* (1830–33), he opposed Georges ◊Cuvier's theory that the features of the Earth were formed by a series of catastrophes, and expounded James ◊Hutton's concept of uniformitarianism, that past events were brought about by the same processes that occur today – a view that influenced Charles Darwin's theory of evolution. He suggested that the Earth was as much as 240 million years old (in contrast to the 6,000 years of prevalent contemporary theory), and provided the first detailed description of the Tertiary period, dividing it into the Eocene, Miocene, and older and younger Pliocene periods.

Le Pichon, Xavier (1937–) French geophysicist who worked out the motions of Earth's six major lithospheric plates. His work was instrumental in the development of plate tectonics. In 1968 he published *Sea-floor spreading and continental drift* in which he depicted Earth's lithosphere divided into six major plates. The boundaries between the plates were shown to have high seismic activity and occurred along mid-ocean ridges, island arcs, active orogenic (mountain-building) belts and transform faults.

Libby, Willard Frank (1908–1980) US chemist whose development in 1947 of radiocarbon dating as a means of determining the age of organic or fossilized material won him a Nobel prize in 1960.

Matsuyama, Motonori (1884–1958) Japanese geophysicist who determined that Earth's magnetic field reverses its polarity periodically throughout its geological history. The **Matsuyama Epoch**, a major polar reversal occurring approximately 0.5–2.5 million years ago, is named after him. He also pioneered the use of gravimetry in finding geological structures below the Earth's surface.

Mercator, Gerardus (Latinized form of Gerhard Kremer) (1512–1594) Flemish mapmaker who devised the first modern atlas, showing **Mercator's projection** in which the parallels and meridians on maps are drawn uniformly at 90°. It is often used for navigational charts, because compass courses can be drawn as straight lines, but the true area of countries is increasingly distorted the further north or south they are from the Equator.

Mohorovicic, Andrija (1857–1936) Croatian seismologist and meteorologist who discovered the **Mohorovicic discontinuity**, the boundary between the Earth's crust and the mantle. In 1909, after a strong earthquake occurred in the Kulpa Valley south of Zagreb, he discovered two distinct sets of P and S seismic waves – one set arriving earlier than the other. He deduced that one set of waves was slower than the other because it had travelled through denser material. He proposed that the Earth's surface consists of an outer layer of rocky material approximately 30 km/19 mi thick, which overlies a denser mantle.

Murchison, Roderick Impey (1792–1871) Scottish geologist responsible for naming the Silurian period, based on studies of slate rocks in south Wales. Expeditions to Russia 1840–45 led him to define another worldwide system, the Permian, named after the strata of the Perm region. He believed in a universal order of the deposition of strata, indicated by fossils rather than solely by lithological features. With Charles ◊Lyell, he also established the Devonian system in southwest England.

Oldham, Richard Dixon (1858–1936) Irish seismologist who discovered the Earth's core and first distinguished between primary and secondary seismic waves. While analysing seismic records in 1906, he noticed an area on the globe in which P-waves were not detected. Every time an earthquake occurred, this P-wave 'shadow zone' appeared on the opposite side of the globe. Oldham demonstrated that the Earth had a core that was causing the primary waves to refract (bend) away, leaving a seismic shadow. In 1919 he suggested that the core may be liquid.

Richter, Charles Francis (1900–1985) US seismologist, deviser of the **Richter scale** used to measure the strength of the waves from earthquakes.

Saussure, Horace Bénédict de (1740–1799) Swiss geologist who made the earliest detailed and first-hand study of the Alps. The results of his Alpine survey appeared in his classic work *Voyages des Alpes/Travels in the Alps* (1779–86).

Sedgwick, Adam (1785–1873) English geologist who contributed greatly to understanding the stratigraphy of the British Isles, using fossils as an index of relative time. In the 1830s he unravelled the stratigraphic sequence of fossil-bearing rocks in north Wales, naming the oldest of them the Cambrian period. In south Wales, Roderick ◊Murchison had concurrently developed the Silurian system. The question of where the boundary lay between the older Cambrian and the younger Silurian sparked a dispute that was not resolved until 1879, when Charles Lapworth (1842–1920) coined the term Ordovician for the middle ground. Together with Murchison, Sedgwick identified the Devonian system in southwest England.

Smith, William (1769–1839) English geologist. He produced the first geological maps of England and Wales, setting the pattern for geological cartography. Often called the founder of stratigraphical geology, he determined the succession of English strata across the whole country, from the Carboniferous up to the Cretaceous. He also established their fossil specimens.

Sorby, Henry Clifton (1826–1908) English geologist whose discovery in 1863 of the crystalline nature of steel led to the study of metallography. Thin-slicing of hard minerals enabled him to study the constituent minerals microscopically in transmitted light. He later employed the same techniques in the study of iron and steel under stress. He published *On the Microscopical Structure of Crystals* (1858).

Steno, Nicolaus (Latinized form of Niels Steensen) (1638–1686) Danish anatomist and naturalist, one of the founders of stratigraphy. To illustrate his ideas, he sketched what are probably the earliest geological sections. His examination of quartz crystals disclosed that, despite differences in the shapes, the angle formed by corresponding faces is invariable for a particular mineral. This constancy is known as **Steno's law**. In his *Sample of the Elements of Myology* (1667) Steno championed the organic origin of fossils.

Strabo (c. 63 BC–AD c. 24) Greek geographer and historian who travelled widely to collect first-hand material for his *Geography*.

Suess, Eduard (1831–1914) Austrian geologist who helped pave the way for the theories of continental drift. He suggested that there had once been a great supercontinent, made up of the present southern continents; this he named Gondwanaland, after a region in India. In his book *The Face of the Earth* (1885–1909) he offered an encyclopedic view of crustal movement, the structure and grouping of mountain chains, sunken continents, and the history of the oceans. He also made significant contributions to rewriting the structural geology of each continent.

Wegener, Alfred Lothar (1880–1930) German meteorologist and geophysicist whose theory of continental drift, expounded in *Origin of Continents and Oceans* (1915), was originally known as 'Wegener's hypothesis'. He supposed that a united supercontinent, Pangaea, had existed in the Mesozoic era. This had developed numerous fractures and had drifted apart some 200 million years ago. During the Cretaceous period, South America and Africa had largely been split, but not until the end of the Quaternary had North America and Europe finally separated; the same was true of the break between South America and Antarctica. Australia had been severed from Antarctica during the Eocene. His ideas can now be explained in terms of plate tectonics, the idea that the Earth's crust consists of a number of plates, all moving with respect to one another.

Werner, Abraham Gottlob (1749–1817) German geologist, one of the first to classify minerals systematically. He also developed the later discarded theory of neptunism – that the Earth was initially covered by water, with every mineral in suspension; as the water receded, layers of rocks 'crystallized'. His geology was particularly important for establishing a physically based stratigraphy, grounded on precise mineralogical knowledge. He linked the order of the strata to the history of the Earth, and related studies of mineralogy and strata.

Wilson, John Tuzo (1908–1993) Canadian geologist and geophysicist who established and brought about a general understanding of the concept of plate tectonics. He pioneered hands-on interactive museum exhibits, and could explain complex subjects like the movement of continents, the spreading of ocean floors, and the creation of island chains by using astonishingly simple models.

Glossary

abrasive substance used for cutting and polishing or for removing small amounts of the surface of hard materials. There are two types: natural and artificial abrasives, and their hardness is measured using the Mohs' scale. Natural abrasives include quartz, sandstone, pumice, diamond, emery, and corundum; artificial abrasives include rouge, whiting, and carborundum.

aclinic line the magnetic equator, an imaginary line near the Equator, where a compass needle balances horizontally, the attraction of the north and south magnetic poles being equal.

agate cryptocrystalline (with crystals too small to be seen with an optical microscope) ◊silica composed of cloudy and banded ◊chalcedony, sometimes mixed with ◊opal, that forms in rock cavities.

alabaster naturally occurring fine-grained white or light-coloured translucent form of gypsum, often streaked or mottled. A soft material, it is easily carved, but seldom used for outdoor sculpture.

amethyst variety of quartz, coloured violet by the presence of small quantities of impurities such as manganese or iron; used as a semiprecious stone. Amethysts are found chiefly in the Ural Mountains, India, the USA, Uruguay, and Brazil.

andalusite aluminium silicate (Al_2SiO_5), a white to pinkish mineral crystallizing as square- or rhombus-based prisms. It is common in metamorphic rocks formed from clay sediments under low pressure conditions. Andalusite, kyanite, and sillimanite are all polymorphs (see ◊polymorphism) of Al_2SiO_5.

anthracite hard, dense, shiny variety of ◊coal, containing over 90% carbon and a low percentage of ash and impurities, which causes it to burn without flame, smoke, or smell. Because of its purity, anthracite gives off relatively little sulphur dioxide when burnt.

antipodes places at opposite points on the globe.

aquamarine blue variety of the mineral beryl. A semiprecious gemstone, it is used in jewellery.

archipelago group of islands, or an area of sea containing a group of islands. The islands of an archipelago are usually volcanic in origin, and they sometimes represent the tops of peaks in areas around continental margins flooded by the sea.

asbestos any of several related minerals of fibrous structure that offer great heat resistance because of their nonflammability and poor conductivity. Commercial asbestos is generally either made from serpentine ('white' asbestos) or from sodium iron silicate ('blue' asbestos). The fibres are woven together or bound by an inert material. Over time the fibres can work loose and, because they are small enough to float freely in the air or be inhaled, asbestos usage is now strictly controlled; exposure to its dust can cause cancer.

asphalt mineral mixture containing semisolid brown or black ◊bitumen, used in the construction industry. Asphalt is mixed with rock chips to form paving material, and the purer varieties are used for insulating material and for waterproofing masonry. It can be produced artificially by the distillation of ◊petroleum.

atoll continuous or broken circle of ◊coral reef and low coral islands surrounding a lagoon.

batholith large, irregular, deep-seated mass of intrusive igneous rock, usually granite, with an exposed surface of more than 100 sq km/40 sq mi. The mass forms by the intrusion or upswelling of magma (molten rock) through the surrounding rock. Batholiths form the core of some large mountain ranges like the Sierra Nevada of western North America.

bauxite principal ore of aluminium, consisting of a mixture of hydrated aluminium oxides and hydroxides, generally contaminated with compounds of iron, which give it a red colour. It is formed by the chemical weathering of rocks in tropical climates. Chief producers of bauxite are Australia, Guinea, Jamaica, Russia, Kazakhstan, Suriname, and Brazil.

bed in geology, a single sedimentary rock unit with a distinct set of physical characteristics or contained fossils, readily distinguishable from those of beds above and below. Well-defined partings called **bedding planes** separate successive beds or strata.

Benioff zone seismically active zone inclined from a deep sea trench beneath a continent or continental margin. Earthquakes along Benioff zones define the top surfaces of plates of lithosphere that descend in to the mantle beneath another, overlying plate. The zone is named after Hugo Benioff (1899–1968), a US seismologist who first described this feature.

beryl mineral, beryllium aluminium silicate ($Be_3Al_2Si_6O_{18}$), which forms crystals chiefly in granite. It is the chief ore of the metallic element beryllium. Two of its gem forms are aquamarine (light-blue crystals) and emerald (dark-green crystals).

bitumen impure mixture of hydrocarbons, including such deposits as petroleum, asphalt, and natural gas, although sometimes the term is restricted to a soft kind of pitch resembling asphalt.

caldera very large basin-shaped crater. Calderas are found at the tops of volcanoes, where the original peak has collapsed into an empty chamber beneath. The basin, many times larger than the original volcano vent, may be flooded, producing a crater lake, or the flat floor may contain a number of small volcanic cones, produced by volcanic activity after the collapse.

cartography art and practice of drawing ◊maps.

cement any bonding agent used to unite particles in a single mass or to cause one surface to adhere to another. In geology, cement refers to a chemically precipitated material such as carbonate that occupies the interstices of clastic rocks.

chalcedony form of the mineral quartz in which the crystals are so fine-grained that they are impossible to distinguish with a microscope (cryptocrystalline). Agate, onyx, and carnelian are ◊gem varieties of chalcedony.

chalk soft, fine-grained, whitish sedimentary rock composed of calcium carbonate ($CaCO_3$), extensively quarried for use in cement, lime, and mortar, and in the manufacture of cosmetics and toothpaste. **Blackboard chalk** in fact consists of gypsum (calcium sulphate, $CaSO_4.2H_2O$).

cinnabar mercuric sulphide mineral (HgS), the only commercially useful ore of mercury. It is deposited in veins and impregnations near recent volcanic rocks and hot springs. The mineral itself is used as a red pigment, commonly known as **vermilion**. Cinnabar is found in the USA (California), Spain (Almadén), Peru, Italy, and Slovenia.

clay very fine-grained sedimentary deposit that has undergone a greater or lesser degree of consolidation. When moistened it is plastic, and it hardens on heating, which renders it impermeable. It may be white, grey, red, yellow, blue, or black, depending on its composition. Clay minerals consist largely of hydrous silicates of aluminium and magnesium together with iron, potassium, sodium, and organic substances. The crystals of clay minerals have a layered structure, capable of holding water, and are responsible for its plastic properties. According to international classification, in mechanical analysis of soil, clay has a grain size of less than 0.002 mm/0.00008 in.

coal black or blackish mineral substance formed from the compaction of ancient plant matter in tropical swamp conditions. It is used as a fuel and in the chemical industry. Coal is classified according to the proportion of carbon it contains. The main types are ◊anthracite (shiny, with about 90% carbon), **bituminous coal** (shiny and dull patches, about 75% carbon), and **lignite** (woody, grading into peat, about 50% carbon). Coal burning is one of the main causes of acid rain.

continental drift in geology, the theory that, about 250–200 million years ago the Earth consisted of a single large continent (Pangaea), which subsequently broke apart to form the continents known today. The theory was proposed in 1912 by German meteorologist Alfred Wegener, but such vast continental movements could not be satisfactorily explained until the study of plate tectonics in the 1960s.

coral marine invertebrate of the class Anthozoa in the phylum Cnidaria, which also includes sea anemones and jellyfish. It has a skeleton of lime (calcium carbonate) extracted from the surrounding water. Corals exist in warm seas, at moderate depths with sufficient light. Some coral is valued for decoration or jewellery, for example, Mediterranean red coral *Corallum rubrum*.

corundum native aluminium oxide (Al_2O_3), the hardest naturally occurring mineral known apart from diamond (corundum rates 9 on the Mohs' scale of hardness); lack of cleavage also increases its durability. Its crystals are barrel-shaped prisms of the trigonal system. Varieties of gem-quality corundum are **ruby** (red) and **sapphire** (any colour other than red, usually blue). Poorer-quality and synthetic corundum is used in industry, for example as an ◊abrasive.

crater bowl-shaped depression in the ground, usually round and with steep sides. Craters are formed by explosive events such as the eruption of a volcano or the impact of a meteorite.

deep-sea trench another term for ocean trench.

dip the angle at which a structural surface, such as a fault or a bedding plane, is inclined from the horizontal. Measured at right angles to the ◊strike of that surface, it is used together with strike to describe the orientation of geological features in the field. Rocks that are dipping have usually been affected by folding.

dolomite white mineral with a rhombohedral structure, calcium magnesium carbonate ($CaMg(CO_3)_2$). Dolomites are common in geological successions of all ages and are often formed when limestone is changed by the replacement of the mineral calcite with the mineral dolomite.

dyke sheet of igneous rock created by the intrusion of magma (molten rock) across layers of pre-existing rock. (By contrast, a sill is intruded *between* layers of rock.) It may form a ridge when exposed on the surface if it is more resistant than the rock into which it intruded.

emerald clear, green gemstone variety of the mineral beryl. It occurs naturally in Colombia, the Ural Mountains in Russia, Zimbabwe, and Australia. The green colour is caused by the presence of the element chromium in the beryl.

emery black to greyish form of impure ◊corundum that also contains the minerals magnetite and hematite. It is used as an ◊abrasive.

Equator, or **terrestrial equator**, the great circle whose plane is perpendicular to the Earth's axis (the line joining the poles). Its length is 40,092 km/24,901.8 mi, divided into 360 degrees of longitude. The Equator encircles the broadest part of the Earth, and represents 0° latitude. It divides the Earth into two halves, called the northern and the southern hemispheres.

extrusive rock, or **volcanic rock**, igneous rock formed on the surface of the Earth; for example, basalt. It is usually fine-grained (having cooled quickly), unlike the more coarse-grained ◊intrusive rocks. The magma (molten rock) that cools to form extrusive rock may reach the surface through a crack such as the constructive margin at the Mid-Atlantic Ridge, or through the vent of a volcano. Extrusive rock can be either **lava** (solidified from a flow) or a **pyroclastic deposit** (hot rocks or ash).

felsic rock plutonic rock composed chiefly of light-coloured minerals, such as quartz, feldspar, and mica. It is derived from **feldspar**, **lenad** (meaning feldspathoid), and **silica**. The term **felsic** also applies to light-coloured minerals as a group, especially quartz, feldspar, and feldspathoids.

flint compact, hard, brittle mineral (a variety of chert), brown, black, or grey in colour, found as nodules in limestone or shale deposits. It consists of cryptocrystalline (grains too small

to be visible even under a light microscope) ◊silica, principally in the crystalline form of quartz. Implements fashioned from flint were widely used in prehistory.

fold in geology, a bend in ◊beds or layers of rock. If the bend is arched in the middle it is called an **anticline**; if it sags downwards in the middle it is called a **syncline**. The line along which a bed of rock folds is called its axis. The axial plane is the plane joining the axes of successive beds.

garnet any of a group of silicate minerals with the formula $X_3Y_3(SiO_4)_3$, where X is calcium, magnesium, iron, or manganese, and Y is usually aluminium or sometimes iron or chromium. Garnets are used as semiprecious gems (usually pink to deep red) and as abrasives. They occur in metamorphic rocks such as gneiss and schist.

gem mineral valuable by virtue of its durability (hardness), rarity, and beauty, cut and polished for ornamental use, or engraved. Of 120 minerals known to have been used as gemstones, only about 25 are in common use in jewellery today; of these, the diamond, emerald, ruby, and sapphire are classified as precious, and all the others semiprecious; for example, the topaz, amethyst, opal, and aquamarine.

geodesy methods of surveying the Earth for making maps and correlating geological, gravitational, and magnetic measurements. Geodesic surveys, formerly carried out by means of various measuring techniques on the surface, are now commonly made by using radio signals and laser beams from orbiting satellites.

geography study of the Earth's surface; its topography, climate, and physical conditions, and how these factors affect people and society. It is usually divided into **physical geography**, dealing with landforms and climates, and **human geography**, dealing with the distribution and activities of peoples on Earth.

geology science of the Earth, its origin, composition, structure, and history. It is divided into several branches: **mineralogy** (the minerals of Earth), **petrology** (rocks), **stratigraphy** (the deposition of successive beds of sedimentary rocks), **palaeontology** (fossils), and **tectonics** (the deformation and movement of the Earth's crust).

Gondwanaland, or **Gondwana**, southern landmass formed 200 million years ago by the splitting of the single world continent ◊Pangaea. (The northern landmass was Laurasia.) It later fragmented into the continents of South America, Africa, Australia, and Antarctica, which then moved slowly to their present positions. The baobab tree found in both Africa and Australia is a relic of this ancient landmass.

graphite blackish-grey, laminar, crystalline form of carbon. It is used as a lubricant and as the active component of pencil lead.

gravel coarse ◊sediment consisting of pebbles or small fragments of rock, originating in the beds of lakes and streams or on beaches. Gravel is quarried for use in road building, railway ballast, and for an aggregate in concrete. It is obtained from quarries known as gravel pits, where it is often found mixed with sand or clay.

hematite principal ore of iron, consisting mainly of iron(III) oxide, Fe_2O_3. It occurs as **specular hematite** (dark, metallic lustre), **kidney ore** (reddish radiating fibres terminating in smooth, rounded surfaces), and a red earthy deposit.

hot spot isolated rising plume of molten mantle material that may rise to the surface of the Earth's crust creating features such as volcanoes, chains of ocean islands, sea mounts, and rifts in continents. Hot spots occur beneath the interiors of tectonic plates and so differ from areas of volcanic activity at plate margins. Examples of features made by hot spots are Iceland in the Atlantic Ocean, and in the Pacific Ocean the Hawaiian Islands and Emperor Seamount chain, and the Galápagos Islands.

International Date Line (IDL) imaginary line that approximately follows the 180° line of longitude. The date is put forward a day when crossing the line going west, and back a day when going east. The IDL was chosen at the International Meridian Conference in 1884.

intrusive rock igneous rock formed beneath the Earth's surface. Magma, or molten rock, cools slowly at these depths to form coarse-grained rocks, such as granite, with large crystals. A mass of intrusive rock is called an **intrusion**. Intrusion features include vertical cylindrical structures such as stocks, pipes, and necks; sheet structures such as dykes that cut across the strata and sills that push between them; laccoliths, which are blisters that push up the overlying rock; and batholiths, which represent chambers of solidified magma and contain vast volumes of rock.

island arc curved chain of islands produced by volcanic activity at a destructive margin (where one tectonic plate slides beneath another). Island arcs are common in the Pacific where they ring the ocean on both sides; the Aleutian Islands off Alaska are an example.

jade semiprecious stone consisting of either jadeite ($NaAlSi_2O_6$, a pyroxene), or nephrite ($Ca_2(Mg,Fe)_5Si_8O_{22}(OH,F)_2$, an amphibole), ranging from colourless through shades of green to black according to the iron content. Jade ranks 5.5–6.5 on the Mohs' scale of hardness.

kaolinite white or greyish ◊clay mineral, hydrated aluminium silicate ($Al_2Si_2O_5(OH)_4$), formed mainly by the decomposition of feldspar in granite. It is made up of platelike crystals, the atoms of which are bonded together in two-dimensional sheets, between which the bonds are weak, so that they are able to slip over one another, a process made more easy by a layer of water. China clay (kaolin) is derived from it. It is mined in France, the UK, Germany, China, and the USA.

kyanite aluminium silicate (Al_2SiO_5), a pale-blue mineral occurring as blade-shaped crystals. It is an indicator of high-pressure conditions in metamorphic rocks formed from clay sediments. Andalusite, kyanite, and sillimanite are all polymorphs (see ◊polymorphism).

lahar mudflow formed of a fluid mixture of water and volcanic ash. During a volcanic eruption, melting ice may combine with ash to form a powerful flow capable of causing great destruction. The lahars created by the eruption of Nevado del Ruiz in Colombia, South America, in 1985 buried 22,000 people in 8 m/26 ft of mud.

lapis lazuli rock containing the blue mineral lazurite in a matrix of white calcite with small amounts of other

minerals. It occurs in silica-poor igneous rocks and metamorphic limestones found in Afghanistan, Siberia, Iran, and Chile. Lapis lazuli was a valuable pigment of the Middle Ages, also used as a gemstone and in inlaying and ornamental work.

latitude and longitude imaginary lines used to locate position on the globe. Lines of latitude are drawn parallel to the Equator, with 0° at the Equator and 90° at the north and south poles. Lines of longitude are drawn at right angles to these, with 0° (the Prime Meridian) passing through Greenwich, England.

lava molten rock (usually 800–1,100°C/ 1,500– 2,000°F) that erupts from a volcano and cools to form extrusive igneous rock. It differs from magma in that it is molten rock on the surface; **magma** is molten rock below the surface. Lava that is viscous and sticky does not flow far; it forms a steep-sided conical composite volcano. Less viscous lava can flow for long distances and forms a broad flat shield volcano.

loam type of fertile soil, a mixture of sand, silt, clay, and organic material. It is porous, which allows for good air circulation and retention of moisture.

lode geological deposit rich in certain minerals, generally consisting of a large vein or set of veins containing ore minerals. A system of veins that can be mined directly forms a lode, for example the mother lode of the California gold rush.

longitude see ◊latitude and longitude.

mafic rock plutonic rock composed chiefly of dark-coloured minerals containing abundant magnesium and iron, such as olivine and pyroxene. It is derived from **magnesium** and **ferric** (iron). The term **mafic** also applies to dark-coloured minerals rich in iron and magnesium as a group.

magma molten rock material beneath the Earth's (or any of the terrestrial planets) surface from which igneous rocks are formed. ◊Lava is magma that has extruded onto the surface.

map diagrammatic representation of an area – for example, part of the Earth's surface or the distribution of the stars. Modern maps of the Earth are made using satellites in low orbit to take a series of overlapping stereoscopic photographs from which a three-dimensional image can be prepared. The earliest accurate large-scale maps appeared about 1580.

mass extinction an event that produces the extinction of many species at about the same time. One notable example is the boundary between the Cretaceous and Tertiary periods (known as the K-T boundary) that saw the extinction of the dinosaurs and other big reptiles, and many of the marine invertebrates as well. Mass extinctions have taken place frequently during Earth's history.

meridian half a great circle drawn on the Earth's surface passing through both poles and thus through all places with the same longitude. Terrestrial longitudes are usually measured from the Greenwich Meridian.

mica any of a group of silicate minerals that split easily into thin flakes along lines of weakness in their crystal structure (perfect basal cleavage). They are glossy, have a pearly lustre, and are found in many igneous and metamorphic rocks. Their good thermal and electrical insulation qualities make them valuable in industry.

mineralogy study of minerals. The classification of minerals is based chiefly on their chemical compostion and the kind of chemical bonding that holds these atoms together. The mineralogist also studies their crystallographic and physical characters, occurrence, and mode of formation.

nitre, or **saltpetre**, potassium nitrate (KNO_3), a mineral found on and just under the ground in desert regions; used in explosives. Nitre occurs in Bihar, India, Iran, and Cape Province, South Africa. The salt was formerly used for the manufacture of gunpowder, but the supply of nitre for explosives is today largely met by making the salt from nitratine (also called Chile saltpetre, $NaNO_3$). Saltpetre is a preservative and is widely used for curing meats.

nuée ardente rapidly flowing, glowing white-hot cloud of ash and gas emitted by a volcano during a violent eruption. The ash and other pyroclastics in the lower part of the cloud behave like an ash flow. In 1902 a nuée ardente produced by the eruption of Mount Pelée in Martinique swept down the volcano in a matter of seconds and killed 28,000 people in the nearby town of St Pierre.

obsidian black or dark-coloured glassy volcanic rock, chemically similar to granite, but formed by cooling rapidly on the Earth's surface at low pressure.

olivine greenish mineral, magnesium iron silicate (($Mg,Fe)_2SiO_4$). It is a rock-forming mineral, present in, for example, peridotite, gabbro, and basalt. Olivine is called **peridot** when pale green and transparent, and used in jewellery.

onyx semiprecious variety of chalcedonic ◊silica in which the crystals are too fine to be detected under a microscope, a state known as cryptocrystalline. It has straight parallel bands of different colours: milk-white, black, and red.

opal form of hydrous ◊silica, ($SiO_2.nH_2O$), often occurring as stalactites and found in many types of rock. The common opal is translucent, milk-white, yellow, red, blue, or green, and lustrous. Precious opal is opalescent, the characteristic play of colours being caused by close-packed silica spheres diffracting light rays within the stone.

ore body of rock, a vein within it, or a deposit of sediment, worth mining for the economically valuable mineral it contains. The term is usually applied to sources of metals. Occasionally metals are found uncombined (native metals), but more often they occur as compounds such as carbonates, sulphides, or oxides. The ores often contain unwanted impurities that must be removed when the metal is extracted.

orogeny, or **orogenesis**, the formation of mountains. It is brought about by the movements of the rigid plates making up the Earth's crust and upper-most mantle (described by plate tectonics). Where two plates collide at a destructive margin rocks become folded and lifted to form chains of mountains (such as the Himalayas).

palaeontology study of ancient life, encompassing the structure of ancient organisms and their environment, evolution, and ecology, as revealed by their fossils. The practical aspects of palaeontology are based on using the presence of different fossils to date particular rock strata and to identify rocks that were laid down under particular conditions; for instance, giving rise to the formation of oil.

Pangaea, or **Pangea** single land mass, made up of all the present continents, believed to have existed between 300 and 200 million years ago; the rest of the Earth was covered by the Panthalassa ocean. Pangaea split into two land masses – Laurasia in the north and ◊Gondwanaland in the south – which subsequently broke up into several continents. These then drifted slowly to their present positions.

peridotite rock consisting largely of the mineral olivine; pyroxene and other minerals may also be present. Peridotite is an ultramafic rock containing less than 45% silica by weight. It is believed to be one of the rock types making up the Earth's upper mantle, and is sometimes brought from the depths to the surface by major movements, or as inclusions in lavas.

perovskite yellow, brown, or greyish-black orthorhombic mineral ($CaTiO_3$), which sometimes contains cerium. Other minerals that have a similar structure are said to have the **perovskite structure**. The term also refers to $MgSiO_3$ with the perovskite structure, the principle mineral that makes up the Earth's lower mantle.

petroleum, or **crude oil**, natural mineral oil, a thick greenish-brown flammable liquid found underground in permeable rocks. Petroleum consists of hydrocarbons mixed with oxygen, sulphur, nitrogen, and other elements in varying proportions. It is thought to be derived from ancient organic material that has been converted by, first, bacterial action, then heat, and pressure (but its origin may be chemical also).

From crude petroleum, various products are made by distillation and other processes; for example, fuel oil, petrol, kerosene, diesel, and lubricating oil. Petroleum products and chemicals are used in large quantities in the manufacture of detergents, artificial fibres, plastics, insecticides, fertilizers, pharmaceuticals, toiletries, and synthetic rubber.

petrology branch of geology that deals with the study of rocks, their mineral compositions, and their origins.

pole either of the geographic north and south points of the axis about which the Earth rotates. The geographic poles differ from the magnetic poles, which are the points towards which a freely suspended magnetic needle will point.

polymorphism in mineralogy, the ability of a substance to adopt different internal structures and external forms, in response to different conditions of temperature and/or pressure. For example, diamond and graphite are both forms of the element carbon, but have very different properties and appearance.

pyroclastic in geology, pertaining to fragments of solidified volcanic magma, ranging in size from fine ash to large boulders, that are extruded during an explosive volcanic eruption; also the rocks that are formed by consolidation of such material. Pyroclastic rocks include tuff (ash deposit) and agglomerate (volcanic breccia).

pyroxene any one of a group of minerals, silicates of calcium, iron, and magnesium with a general formula X,YSi_2O_6, found in igneous and metamorphic rocks. The internal structure is based on single chains of silicon and oxygen. Diopside (X = Ca, Y = Mg) and augite (X = Ca, Y = Mg,Fe,Al) are common pyroxenes.

rhyolite igneous rock, the fine-grained volcanic (extrusive) equivalent of granite.

ruby red transparent gem variety of the mineral ◊corundum. Small amounts of chromium oxide (Cr_2O_3), substituting for aluminium oxide, give ruby its colour. Natural rubies are found mainly in Myanmar (Burma), but rubies can also be produced artificially and such synthetic stones are used in lasers.

salt, common, or **sodium chloride** (NaCl), white crystalline solid found dissolved in sea water and as rock salt (the mineral halite) in large deposits and salt domes. Common salt is used extensively in the food industry as a preservative and for flavouring, and in the chemical industry in the making of chlorine and sodium.

San Andreas fault geological fault stretching for 1,125 km/700 mi northwest–southeast through the state of California, USA. It marks a conservative plate margin, where two plates slide past each other.

sand loose grains of rock, sized 0.0625–2.00 mm/ 0.0025–0.08 in in diameter, consisting most commonly of quartz, but owing their varying colour to mixtures of other minerals. Sand is used in cement-making, as an abrasive, in glass-making, and for other purposes.

sapphire deep-blue, transparent gem variety of the mineral ◊corundum. Small amounts of iron and titanium give it its colour. A corundum gem of any colour except red (which is a ruby) can be called a sapphire; for example, yellow sapphire.

scarp and dip in geology, the two slopes formed when a sedimentary bed outcrops as a landscape feature. The scarp is the slope that cuts across the bedding plane; the dip is the opposite slope which follows the bedding plane. The scarp is usually steep, while the dip is a gentle slope.

schist metamorphic rock containing mica or another platy or elongate mineral, whose crystals are aligned to give a foliation (planar texture) known as schistosity. Schist may contain additional minerals such as ◊garnet.

seafloor spreading growth of the ocean crust outwards (sideways) from ocean ridges. The concept of seafloor spreading has been combined with that of continental drift and incorporated into plate tectonics.

sediment any loose material that has 'settled' – deposited from suspension in water, ice, or air, generally as the water current or wind speed decreases. Typical sediments are, in order of increasing coarseness, clay, mud, silt, sand, gravel, pebbles, cobbles, and boulders.

seismology study of earthquakes and how their shock waves travel through the Earth. By examining the global pattern of waves produced by an earthquake, seismologists can deduce the nature of the materials through which they have passed. This leads to an understanding of the Earth's internal structure.

silica silicon dioxide (SiO_2), the composition of the most common mineral group, of which the most familiar form is quartz. Other silica forms are ◊chalcedony, chert, opal, tridymite, and cristobalite. Common sand consists largely of silica in the form of quartz.

sill sheet of igneous rock created by the intrusion of magma (molten rock) between layers of pre-existing rock. An example of a sill in the UK is the Great Whin Sill, which forms the ridge along which Hadrian's Wall was built.

sillimanite aluminium silicate (Al_2SiO_5), a mineral that occurs either as white to brownish prismatic crystals or as minute white fibres. It is an indicator of high temperature conditions in metamorphic rocks formed from clay sediments. Andalusite, kyanite, and sillimanite are all polymorphs (see ◊polymorphism) of Al_2SiO_5.

strike the compass direction of a horizontal line on a planar structural surface, such as a fault plane, bedding plane, or the trend of a structural feature, such as the axis of a fold. Strike is 90° from ◊dip.

stratigraphy branch of geology that deals with the sequence of formation of sedimentary rock layers and the conditions under which they were formed. Its basis was developed by the English geologist William Smith. Stratigraphy in the interpretation of archaeological excavations provides a relative chronology for the levels and the artefacts within them. The basic principle of superimposition establishes that upper layers or deposits have accumulated later in time than the lower ones.

stromatolite mound produced in shallow water by mats of algae that trap mud particles. Another mat grows on the trapped mud layer and this traps another layer of mud and so on. The stromatolite grows to heights of a metre or so. They are uncommon today but their fossils are among the earliest evidence for living things – over 2,000 million years old.

subduction zone region where two plates of the Earth's rigid lithosphere collide, and one plate descends below the other into the weaker asthenosphere. Subduction occurs along ocean trenches, most of which encircle the Pacific Ocean; portions of the ocean plate slide beneath other plates carrying continents.

surveying accurate measuring of the Earth's crust, or of land features or buildings. It is used to establish boundaries, and to evaluate the topography for engineering work. The measurements used are both linear and angular, and geometry and trigonometry are applied in the calculations.

tectonics in geology, the study of the movements of rocks on the Earth's surface. On a small scale tectonics involves the formation of folds and faults, but on a large scale plate tectonics deals with the movement of the Earth's surface as a whole.

topaz mineral, aluminium fluorosilicate ($Al_2(F_2SiO_4)$). It is usually yellow, but pink if it has been heated, and is used as a gemstone when transparent. It ranks 8 on the Mohs' scale of hardness.

topography surface shape and composition of the landscape, comprising both natural and artificial features, and its study. Topographical features include the relief and contours of the land; the distribution of mountains, valleys, and human settlements; and the patterns of rivers, roads, and railways.

tremor minor earthquake.

turquoise mineral, hydrous basic copper aluminium phosphate $(CuAl_6(PO_4)_4(OH)_85H_2O)$. Blue-green, blue, or green, it is a gemstone. Turquoise is found in Australia, Egypt, Ethiopia, France, Germany, Iran, Turkestan, Mexico, and the southwestern USA. It was originally introduced into Europe through Turkey, from which its name is derived.

volcanic rock another name for ◊extrusive rock, igneous rock formed on the Earth's surface.

zircon zirconium silicate $(ZrSiO_4)$, a mineral that occurs in small quantities in a wide range of igneous, sedimentary, and metamorphic rocks. It is very durable and is resistant to erosion and weathering. It is usually coloured brown, but can be other colours, and when transparent may be used as a gemstone.

Further Reading

Ager, D V *The Nature of the Stratigraphical Record* (1992)

Allen, K, and Briggs, D (eds) *Evolution and the Fossil Record* (1989)

Bailey, E B *James Hutton* (1967)

Barnes, J W *Ores and Minerals: Introducing Economic Geology* (1988)

Benton, M J *Vertebrate Palaeontology* (1990)

Borchardt-Ott, W *Crystallography* (1993)

Briggs, David *Fundamentals of the Physical Environment* (1992)

Brown, G A, and others *Understanding the Earth* (1992)

Brown, G C, and Mussett, A E *The Inaccessible Earth* (1981)

Calder, Nigel *Restless Earth* (1972)

Carroll, R L *Vertebrate Palaeontology and Evolution* (1988)

Chisholm, Michael *Human Geography: Evolution or Revolution?* (1975)

Cocks, L R M *The Evolving Earth* (1981)

Condie, Kent C *Plate Tectonics and Crustal Evolution* (1997, 4th edition)

Cowen, Richard *History of Life* (1990)

Cox, Allan *Plate Tectonics: How It Works* (1986)

Cresser, M; Killham, K; and Edwards, T *Soil Chemistry and its Applications* (1993)

Decker, R W, and Decker, B B *Mountains of Fire: The Nature of Volcanoes* (1991)

Dietrich, R V, and Skinner, B J *Gems ,Granites and Gravels: Knowing and Using Rocks and Minerals* (1990)

Dineen, Jacqueline *Natural Disasters: Volcanoes* (1991)

Dixon, Dougal *The Practical Geologist* (1992)

Dunning, F W, and others *Britain Before Man* (1978), *The Story of the Earth* (1981)

Emery, Dominic, and Meyers, Keith *Sequence Stratigraphy* (1996)

Emery, D, and Robinson, A *Inorganic Geochemistry* (1994)

Erickson, Jon *Plate Tectonics: Unravelling the Mysteries of the Earth* (1992), *Rock Formations and Unusual Geologic Structures: Exploring the Earth's Surface* (1993)

Farndon, J *How the Earth Works* (1992)

Fortey, Richard *Fossils: The Key to the Past* (1991)

Fowler, C M R *The Solid Earth: An Introduction to Global Geophysics* (1990)

Fuller, Sue *Rocks and Minerals* (1995)

Goudie, Andrew, and Gardner, Rita *Discovering Landscape* (1985)

Gould, Stephen Jay *Wonderful Life* (1989), (ed) *The Book of Life* (1993)

Gubbins, David *Seismology and Plate Tectonics* **(1990)**

Hall, C *Gemstones* (1994)

Halstead, L B *Hunting the Past* (1982)

Harper, David, and Owen, Alan *Fossils of the Upper Ordovician* (1998)

Holdsworth, R V; Strachan, R A; and Dewey, J F (eds) *Continental Transpressional Tectonics and Transtensional Tectonics* (1998)

Holmes, Arthur *The Principles of Physical Geology* (1965)

Killops, S D, and Killops, V J *An Introduction to Organic Geochemistry* (1994)

Klein, C, and Hurlbutt, C S *Manual of Mineralogy* (1994)

MacKenzie, W S, and Adams, A E *A Colour Atlas of Rocks and Minerals in Thin Section* (1994)

Mayr, H *Fossils* (1992)

Milner, Angela *The Natural History Museum Book of Dinosaurs* (1995)

Myers, Norman *The Gaia Atlas of Planet Management* (1994)

Navrotsky, Alexandra *Physics and Chemistry of Earth Materials* (1994)

Nelson, C R (ed) *Chemistry of Coal Weathering* (1989)

Nield, E W *Drawing and Understanding Fossils: A Theoretical and Practical Guide for Beginners with Self-Assessment* (1987)

Payne, K R *Chemicals from Coal: New Processes* (1987)

Pellant, Chris *Rocks, Minerals and Fossils of the World* (1990), *Rocks and Minerals* (1992)

Playfair, John *Illustrations of the Huttonian Theory of the Earth* (1802; reprinted 1956)

Raup, David *The Nemesis Affair. A Story of the Death of Dinosaurs and Ways of Science* (1986)

Reading, Harold *Sedimentary Environments: Processes, Facies and Stratigraphy* (1996)

Roberts, J L *The Macmillan Field Guide to Geological Structures* (1989)

Roberts, Willard L, and others *Encyclopedia of Minerals* (1990)

Robinson, George *Minerals: An Illustrated Exploration of the Dynamic World of Minerals and Their Properties* (1995)

Roger, Jacques *Buffon: a Life in Natural History* (1998)

Ross, Simon *Hazard Geography* (1987), *The Challenge of the Natural Environment* (1989), *Exploring Geography* (1991)

Rudwick, Martin *The Meaning of Fossils* (1972), *Georges Cuvier, Fossil Bones, and Geological Catastrophes* (1998)

Russell, D A *An Odyssey in Time: The Dinosaurs of North America* (1989)

Sabins, Floyd *Remote Sensing* (1996)

Smith, James *Introduction to Geodesy: the History and Concepts of Modern Geodesy* (1998)

Stanley, Steven M *Earth and Life Through Time* (1986)

Stefoff, Rebecca *Maps and Mapmaking* (1995)

Tassy, Pascal *The Message of Fossils* (1991; translated 1993)

Thackray, John *The Age of the Earth* (1980)

Tooley, R V *Maps and Map-Makers* (1987)

Trueman, A E *Geology and Scenery in England and Wales* (1971)

van Rose, Susanna *Earthquakes* (1983), *Dorling Kindersley Eyewitness Guides: Volcano* (1992)

Walker, C, and Ward, D *Fossils, A Visual Guide to over 500 Fossils Genera from Around the World* (1992)

Waltham, A C *Foundations of Engineering Geology* (1994)

Wendt, Herbert *Before the Deluge* (1968; translated 1970)

Whittow, John *Disasters* (1980)

EARTH SCIENCE: THE ATMOSPHERE AND THE HYDROSHPERE

The earth sciences include the study of Earth's **atmosphere,** the layer of gases surrounding the Earth, and the **hydrosphere,** the water contained in oceans, rivers, lakes, and groundwater. Sub-disciplines such as atmospheric physics, meteorology, oceanography, and hydrology study the processes and phenomena of Earth's atmosphere and hydrosphere.

Earth science is also concerned with how the atmosphere and hydrosphere interact with Earth's lithosphere, affecting the ◊geomorphology of the planet's surface. Weathering, running water, waves, glacial ice, and wind – which result in the erosion, transportation, and deposition of rocks and soils – influence the form of Earth's surface. For example, the shape of a mountain range is largely the result of erosive processes that progressively remove material from the mountain range.

erosion

The wearing away of the Earth's surface is called erosion. It is caused by the breakdown and transportation of particles of rock or soil. By contrast, weathering does not involve transportation. **Water,** consisting of sea waves and currents, rivers, and rain; **ice,** in the form of glaciers; and **wind,** hurling sand fragments against exposed rocks and moving dunes along, are the most potent forces of erosion. There are four processes of erosion : ◊hydraulic action, ◊corrasion, ◊attrition, and ◊solution.

The form that erosion takes, and the degree to which it takes place, can vary with rock type. Unconsolidated sands and gravel are more easily eroded than solid granite, while rocks such as limestone are worn down by chemical processes rather than by physical forces.

weathering

Weathering is the process by which exposed rocks are broken down on the spot by the action of rain, frost, wind, and other elements of the weather. It differs from erosion in that no movement or transportation of the broken-down material takes place. Two types of weathering are recognized: physical (or mechanical) and chemical. They usually occur together.

Weathering

Physical weathering

temperature changes	weakening rocks by expansion and contraction
frost	wedging rocks apart by the expansion of water on freezing
unloading	the loosening of rock layers by release of pressure after the erosion and removal of those layers above

Chemical weathering

carbonation	breakdown of calcite by reaction with carbonic acid in rainwater
hydrolysis	breakdown of feldspar into china clay by reaction with carbonic acid in rainwater
oxidation	breakdown of iron-rich minerals due to rusting
hydration	expansion of certain minerals due to the uptake of water

transportation and deposition

After materials have been broken down and loosened by weathering, they are transported by mass movement, wind action, or running water, and deposited to new locations; glaciers transport materials embedded in them, winds lift dust particles and carry them over great distances, precipitation falling on sloping land shifts soils downhill, water currents carry materials along a riverbed or out to sea. Through deposition, the particles accumulate elsewhere; rivers and glaciers carve valleys and deposit eroded material in plains and deltas, desert winds wear away rock and form huge sand dunes, waves erode rocky shorelines and create sandy beaches. River ▷deltas such as those of the Nile and the Ganges, which are formed by the build-up of silt at the point where the river meets the sea, demonstrate the cumulative effects on landform of transportation and deposition.

Earth's atmosphere

An atmosphere is mixture of gases surrounding a planet. Planetary atmospheres are prevented from escaping by the pull of gravity. On Earth, atmospheric pressure decreases with altitude. In its lowest layer, Earth's atmosphere consists of nitrogen (78%) and oxygen (21%), both in molecular form (two atoms bonded together) and 1% argon. Small quantities of other gases are important to the chemistry and physics of the Earth's atmosphere, including water and carbon dioxide. The atmosphere plays a major part in the various cycles of nature (the water cycle, the ▷carbon cycle, and the nitrogen cycle). It is the principal industrial source of nitrogen, oxygen, and argon, which are obtained by fractional distillation of liquid air.

Earth's atmosphere is divided into four regions of atmosphere classified by temperature. The thermal structure of the Earth's atmosphere is the result of a complex interaction between the electromagnetic radiation from the Sun, radiation reflected from the Earth's surface, and molecules and atoms in the atmosphere.

layers of the atmosphere

troposphere The troposphere is the lowest level of the atmosphere (altitudes from 0 to 10 km/6 mi) and is heated to an average temperature of 15°C/59°F by the Earth, which in turn is warmed by infrared and visible radiation from the Sun. Warm air cools as it rises in the troposphere and this rising of warm air causes rain and most other weather phenomena. The top of the troposphere is approximately –60°C/–76°F. The temperature minimum between the troposphere and the stratosphere above marks the influence of the Earth's warming effects and is called the **tropopause**.

stratosphere

Temperature *increases* with altitude in the stratosphere, the layer (from 10 km/6 mi to 50 km/31 mi) above the troposphere. Tempera-tures rise in the stratosphere from –60°C/–76°F to near 0°C/32°F. The temperature maximum near the top of the stratosphere is called the **stratopause**. Here, temperatures rise as ultraviolet photons are absorbed by heavier molecules to form new gases. An important example is the production of ozone molecules (oxygen atom triplets, O_3) from oxygen molecules. Ozone is a better absorber of ultraviolet radiation than ordinary (two-atom) oxygen, and it is the ozone layer within the stratosphere that prevents lethal amounts of ultraviolet from reaching the Earth's surface.

mesosphere Temperature again decreases with altitude through the mesosphere (50 km/31 mi to 80 km/ 50 mi), from 0°C/32°F to below –100°C/–212°F.

thermosphere The thermosphere is the highest layer of the atmosphere (80 km/50 mi to about 700 km/435 mi). Temperature rises with altitude to extreme values of thousands of degrees. The meaning of these extreme temperatures can be misleading, however. High thermosphere temperatures represent little heat because they are defined by motions among so few atoms and molecules spaced widely apart from one another.

High in the thermosphere temperatures are high because of collisions between ultraviolet (UV) photons and atoms of the atmosphere. Temperature decreases at lower levels in the thermosphere because there are fewer UV photons available, having been absorbed by collisions higher up. The thermal minimum that results at the base of the thermosphere is called the **mesopause**.

ionosphere At altitudes above the ozone layer and above the base of the mesosphere (50 km/31 mi), ultraviolet photons collide with atoms, knocking out electrons to create a plasma of electrons and positively charged ions. The resulting **ionosphere** acts as a reflector of radio waves, enabling radio transmissions to 'hop' between widely separated points on the Earth's surface.

Van Allen radiation belts Far above the atmosphere lie the Van Allen radiation belts. These are regions in which high-energy charged particles travelling outwards from the Sun (the solar wind) have been captured by the Earth's magnetic field. The outer belt (about 1,600 km/1,000 mi) contains mainly protons, the inner belt (about 2,000 km/1,250 mi) contains mainly electrons. Sometimes electrons spiral down towards the Earth, noticeably at polar latitudes, where the magnetic field is strongest. When such particles collide with atoms and ions in the thermosphere, light is emitted. This is the origin of the glows visible in the sky as the **aurora borealis** (northern lights) and the **aurora australis** (southern lights).

During periods of intense solar activity, the atmosphere swells outwards; there is a 10–20% variation in atmosphere density. One result is to increase drag on satellites. This effect makes it impossible to predict exactly the time of re-entry of satellites.

chemistry of the atmosphere

The chemistry of atmospheres is related to the geology of the planets they envelop. Unlike Earth, Venus's dense atmosphere is predominantly carbon dioxide (CO_2). The carbon dioxide-rich atmosphere of Venus absorbs infrared radiation emanating from the planet's

atmosphere All but 1% of the Earth's atmosphere lies in a layer 30 km/19 mi above the ground. At a height of 5,500 m/18,000 ft, air pressure is half that at sea level. The temperature of the atmosphere varies greatly with height; this produces a series of layers, called the troposphere, stratosphere, mesosphere, and thermosphere.

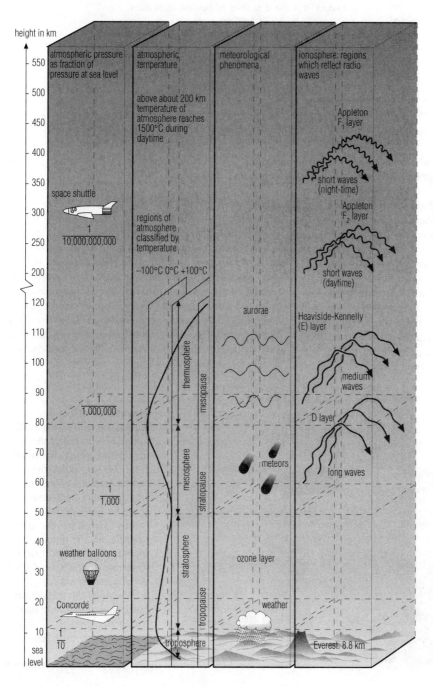

Atmosphere: Composition

Gas	Symbol	Volume (%)	Role
nitrogen	N_2	78.08	cycled through human activities and through the action of microorganisms on animal and plant waste
oxygen	O_2	20.94	cycled mainly through the respiration of animals and plants and through the action of photosynthesis
carbon dioxide	CO_2	0.03	cycled through respiration and photosynthesis in exchange reactions with oxygen. It is also a product of burning fossil fuels
argon	Ar	0.093	chemically inert and with only a few industrial uses
neon	Ne	0.0018	as argon
helium	He	0.0005	as argon
krypton	Kr	trace	as argon
xenon	Xe	trace	as argon
ozone	O_3	0.00006	a product of oxygen molecules split into single atoms by the Sun's radiation and unaltered oxygen molecules
hydrogen	H_2	0.00005	unimportant

surface, causing the very high surface temperatures capable of melting lead (see greenhouse effect below). If all of the carbon dioxide that has gone to form carbonate rock (limestone) on Earth were liberated into the troposphere, our atmosphere would be similar to that of Venus. It is the existence of liquid water that enables carbonate rock to form on Earth that has caused our atmosphere to differ substantially from the Venusian atmosphere.

Other atmospheric ingredients are found in particular localities: gaseous compounds of sulphur and nitrogen in towns, salt over the oceans, and everywhere dust composed of inorganic particles, decaying organic matter, tiny seeds and pollen from plants, and bacteria. Of particular importance are the anthropogenic chlorofluorocarbons (CFCs) that destroy stratospheric ozone.

greenhouse effect

The greenhouse effect is the phenomenon of the Earth's atmosphere by which solar radiation, trapped by the Earth and re-emitted from the surface as infrared radiation, is prevented from escaping by various gases in the air. Greenhouse gases trap heat because they readily absorb infrared radiation. The result is a rise in the Earth's temperature (global warming). The main greenhouse gases are carbon dioxide, methane, and chlorofluorocarbons (CFCs) as well as water vapour. Fossil-fuel consumption and forest fires are the principal causes of carbon

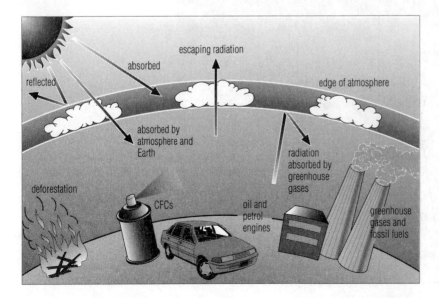

greenhouse effect The warming effect of the Earth's atmosphere is called the greenhouse effect. Radiation from the Sun enters the atmosphere but is prevented from escaping back into space by gases such as carbon dioxide (produced for example, by the burning of fossil fuels), nitrogen oxides (from car exhausts), and CFCs (from aerosols and refrigerators). As these gases build up in the atmosphere, the Earth's average temperature is expected to rise.

dioxide build-up; methane is a byproduct of agriculture (rice, cattle, sheep).

Dubbed the 'greenhouse effect' by the Swedish scientist Svante Arrhenius (1859–1927), it was first predicted in 1827 by the French mathematician Joseph Fourier (1768–1830).

The concentration of carbon dioxide in the atmosphere is estimated to have risen by 25% since the Industrial Revolution, and 10% since 1950; the rate of increase is now 0.5% a year. Chlorofluorocarbon levels are rising by 5% a year, and nitrous oxide levels by 0.4% a year, resulting in a global warming effect of 0.5% since 1900, and a rise of about 0.1°C/3°F a year in the temperature of the world's oceans during the 1980s. Arctic ice was 6–7 m/20–23 ft thick in 1976 and had reduced to 4–5 m/ 13–17 ft by 1987.

Dan's Wild Wild Weather Page
http://www.whnt19.com/kidwx/index.html
An introduction to the weather for kids. It has pages dealing with everything from climate to lightning, from satellite forecasting to precipitation; all explained in a lively style with plenty of pictures.

Earth's weather

The day-to-day variation of atmospheric and climatic conditions at any one place over a short period of time is termed **weather**. Such conditions include atmospheric pressure, temperature, humidity, wind, cloud cover, and precipitation such as rain, snow, and hail, together with extreme phenomena such as storms and blizzards. Weather differs from climate in that the latter is a composite of the average weather conditions of a locality or region over a long period of time (at least 30 years).

Meteorology is the study of short-term weather patterns and data within a circumscribed area, while **climatology** is the study of weather over longer timescales on a zonal or global basis. At meteorological stations readings are taken of the factors determining weather conditions. Satellites are used either to relay information transmitted from the Earth-based stations, or to send pictures of cloud development, indicating wind patterns and snow and ice cover.

World Meteorological Organization
http://www.wmo.ch/
Internet voice of the World Meteorological Organization, a UN division coordinating global scientific activity related to climate and weather. The site offers ample material on the long term objectives and immediate policies of the organization. It also disseminates important information on WMO's databases, training programmes, and satellite activities, as well as its projects related to the protection of the environment.

A traditional rhyme to predict the weather:

Red sky at night, shepherds' delight, red sky in the morning, shepherds' warning

(in other words, if the sky is red in the evening, the next day will be fine, but if it is red in the morning, expect bad weather)

---

atmospheric pressure

The weight of the air above any point presses downwards and the force this produces in all directions is called the air pressure; it is measured by barometers and barographs, the unit of measurement being the millibar (mb).

With increase in height there is less air above and therefore pressure decreases with height by about a factor of 10 for every increase of height of 16 km/10 mi. If the temperature of the air is known, the decrease in pressure can be calculated: near sea level it amounts to about 1 mb in every 10 m/33 ft. In order to compare pressures between many stations at a constant level, pressures are reduced to sea level, that is, barometer readings are adjusted to show what the pressure would be at sea level.

---

weather prediction: pressure

A weather prediction:
When smoke descends, good weather ends
(low pressure forces smoke down)

---

depression A depression, or **cyclone** or **low**, is a region of low atmospheric pressure. In mid latitudes a depression forms as warm, moist air from the tropics mixes with cold, dry polar air, producing warm and cold boundaries (fronts, see below) and unstable weather – low cloud and drizzle, showers, or fierce storms. The warm air, being less dense, rises above the cold air to produce the area of low pressure on the ground. Air spirals in towards the centre of the depression in an anticlockwise direction in the northern hemisphere, clockwise in the southern hemisphere, generating winds up to gale force. Depressions tend to travel eastwards and can remain active for several days.

A deep depression is one in which the pressure at the centre is very much lower than that round about; it produces very strong winds, as opposed to a shallow depression, in which the winds are comparatively light. A severe depression in the tropics is called a hurricane, tropical cyclone, or typhoon, and is a great

Summary of World Weather Records

Source: National Weather Service; National Oceanic and Atmospheric Administration
(Data as of April 1997. N/A = not available.)

Record	Location	Details
Highest amount of rainfall in the northern hemisphere in 24 hours	Paishih, Taiuan	.124 cm/49 in
Highest amount of rainfall in 24 hours (not induced by the presence of mountains)	Dharampuri, India	99 cm/39 in
Highest amount of rainfall in 24 hours	Cilaos, La Réunion Island	188 cm/74 in
Highest amount of rainfall over 5 days	Cilaos, La Réunion Island	386 cm/152 in
Highest amount of rainfall in 12 hours	Belouve, La Réunion Island	135 cm/53 in
Highest amount of rainfall in 20 minutes	Curtea-de-Arge, Romania	21 cm/8.1 in
Highest yearly number of days of rainfall	Bahia Felix, Chile	325 days
Longest period without rainfall	Arica, Chile	14 years
Highest yearly average period of thunderstorms	Kampala, Uganda	242 days
Highest sustained yearly average period of thunderstorms	Bogor, Indonesia	322 days per year from 1916 to 1919
Longest snowfall	Bessans, France	19 hours with 173 cm/68 in of snow
Highest yearly average rainfall in Africa	Debundscha, Cameroon	1,029 cm/405 in (with an average variability of 191 cm/75 in)
Lowest yearly average rainfall in Africa	Wadt Halfa, Sudan	3 mm/0.1 in
Lowest yearly average rainfall in Asia	Aden, South Yemen	5 cm/1.8 in
Highest yearly average rainfall in Europe	Crkvice, Yugoslavia	465 cm/183 in
Lowest yearly average rainfall in Europe	Astrakhan, Russia	16 cm/6.4 in
Highest amount of rainfall in Australia in 24 hours	Crohamhurst, Queensland	91 cm/36 in
Highest yearly average rainfall in Australia	Tully, Queensland	455 cm/179 in
Lowest yearly average rainfall in Australia	Mulka, South Australia	10 cm/4.1 in
Highest yearly average rainfall in South America	Quibdo, Colombia	899 cm/354 in
Lowest yearly average rainfall in South America	Arica, Chile	0.7 mm/0.03 in
Highest yearly average rainfall in North America	Henderson Lake, British Columbia, Canada	665 cm/262 in
Lowest yearly average rainfall in North America	Bataques, Mexico	3 cm/1.2 in
Highest temperature ever recorded in the world	El Aisisa, Libya	58°C/136°F
Lowest temperature ever recorded in the world	Vostok, Antarctica	–88°C/–127°F
Highest yearly average temperature in world	Dallol, Ethiopia	34°C/94°F
Highest yearly average temperature range	Eastern Sayan Region, Russia	through 63°C/146°F
Highest average temperature sustained over a long period	Marble Head, Australia	38°C/100°F for 162 consecutive days
Fastest temperature rise in short period	Edinburgh, Scotland	through –6°C/21°F
Highest temperature in Antarctica	N/A	near 16°C/60°F
Lowest temperature in Antarctica	South Pole	–77°C/–107°F
Lowest temperature in Africa	Ifrane, Morocco	–24°C/–11°F
Highest temperature in Asia	Tirat Tsvi, Israel	54°C/129°F
Highest temperature in Australia	Cloncurry, Queensland	53°C/128°F
Lowest temperature in Australia	Charlotte Press	–22°C/–8°F
Highest temperature in Europe	Seville, Spain	50°C/122°F
Lowest temperature in Europe	Ust 'Shchugor, Russia	–55°C/–67°F
Lowest temperature in Greenland	Northice	–66°C/–87°F
Lowest temperature in North America (excluding Greenland)	Snag, Yukon Territory, Canada	–63°C/–81°F
Lowest temperature in northern hemisphere	Verkhoyansk, Oimekon, Russia	–68°C/–90°F
Highest temperature in South America	Rivadavia, Argentina	49°C/120°F
Lowest temperature in South America	Sarmiento, Argentina	–33°C/–27°F
Highest temperature in western hemisphere	Death Valley, California, USA	57°C/134°F
Highest peak wind	Thule Air Base, Greenland	333 kph/207 mph
Highest average wind speed in 24 h	Port Martin, Antarctica	173 kph/108 mph
Highest peak wind gust	Mount Washington, New Hampshire, USA	372 kph/231 mph
Highest monthly average wind speed	Port Martin, Antarctica	104 kph/65 mph

weather system

A formula for finding the centre of a weather system (applies in the northern hemisphere only; reverse for the southern hemisphere):

When the wind is at your back, the low is on your left

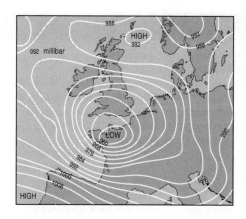

danger to shipping; a tornado is a very intense, rapidly swirling depression, with a diameter of only a few hundred metres or so.

isobar The isobars around a low-pressure area or depression. In the northern hemisphere, winds blow anticlockwise around lows, approximately parallel to the isobars, and clockwise around highs. In the southern hemisphere, the winds blow in the opposite directions.

wind Wind is the lateral movement of the Earth's atmosphere from high-pressure areas (anticyclones) to low-pressure areas (depressions). Its speed is measured using an anemometer or by studying its effects on, for example, trees by using the Beaufort scale (see below). Although modified by features such as land and water, there is a basic worldwide system of ◊**trade winds**, ◊**Westerlies**, and polar **easterlies**.

A belt of low pressure (the ◊**doldrums**) lies along the Equator. The trade winds blow towards this from the horse latitudes (areas of high pressure at about 30° N and 30° S of the Equator), blowing from the northeast in the northern hemisphere, and from the southeast in the southern. The Westerlies (also from the horse latitudes) blow north of the Equator from the southwest and south of the Equator from the northwest.

Cold winds blow outwards from high-pressure areas at the poles. More local effects result from land masses heating and cooling faster than the adjacent sea, producing onshore winds in the daytime and offshore winds at night.

The **monsoon** is a seasonal wind of southern Asia, blowing from the southwest in summer and bringing the rain on which crops depend. It blows from the northeast in winter.

Famous or notorious warm winds include the **chinook** of the eastern Rocky Mountains, North America; the **föhn** of Europe's Alpine valleys; the **sirocco** (Italy)/**khamsin** (Egypt)/**sharav** (Israel), spring winds that bring warm air from the Sahara and Arabian deserts across the Mediterranean; and the **Santa Ana**, a periodic warm wind from the inland deserts that strikes the California coast.

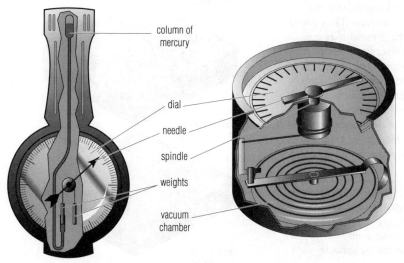

column of mercury
dial
needle
spindle
weights
vacuum chamber

mercury barometer

aneroid barometer

barometer The mercury barometer (left) and the aneroid barometer (right). In the mercury barometer, the weight of the column of mercury is balanced by the pressure of the atmosphere on the lower end. A change in height of the column indicates a change in atmospheric pressure. In the aneroid barometer, any change of atmospheric pressure causes the metal box which contains the vacuum to be squeezed or to expand slightly. The movements of the box sides are transferred to a pointer and scale via a chain of levers.

Beaufort Scale

Number and description	Features	Air speed kph	mph
0 calm	smoke rises vertically; water smooth	0–2	0–1
1 light air	smoke shows wind direction; water ruffled	2–5	1–3
2 light breeze	leaves rustle; wind felt on face	6–11	4–7
3 gentle breeze	loose paper blows around	12–19	8–12
4 moderate breeze	branches sway	20–29	13–18
5 fresh breeze	small trees sway, leaves blown off	30–39	19–24
6 strong breeze	whistling in telephone wires; sea spray from waves	40–50	25–31
7 near gale	large trees sway	51–61	32–38
8 gale	twigs break from trees	62–74	39–46
9 strong gale	branches break from trees	75–87	47–54
10 storm	trees uprooted; weak buildings collapse	88–101	55–63
11 violent storm	widespread damage	102–117	64–73
12 hurricane	widespread structural damage	above 118	above 74

The dry northerly **bise** (Switzerland) and the **mistral**, which strikes the Mediterranean area of France, are unpleasantly cold winds.

Wind is responsible for the erosion and deposition of sand and dust particles, forming **dunes**.

The **Beaufort scale** is a system of recording wind velocity (speed), devised by Francis Beaufort in 1806. It is a numerical scale ranging from 0 to 17, calm being indicated by 0 and a hurricane by 12; 13–17 indicate degrees of hurricane force.

In 1874 the scale received international recognition; it was modified in 1926. Measurements are made at 10 m/33 ft above ground level.

temperature Measuring the temperature of the air can be difficult, because a thermometer measures its own temperature, not necessarily that of its surroundings. In addition, temperature varies irregularly with height, particularly in the first few metres. Thus the temperature at 1 m/3.3 ft above the ground may easily be 5°C/41°F lower than the temperature closer to the ground, whereas on a following clear night the reverse usually occurs. On the other hand the temperature of air which rises 100 m/330 ft only cools about 1°C/33.8°F by reason of its change of height and consequent expansion because of decrease in pressure. Temperatures are therefore read at a standard height (usually 1.2 m/3.9 ft) above the ground, and

dune The shape of a dune indicates the prevailing wind pattern. Crescent-shaped dunes (barchans) form in sandy desert with winds from a constant direction. Seif dunes form on bare rocks, parallel to the wind direction. Irregular star dunes are formed by variable winds.

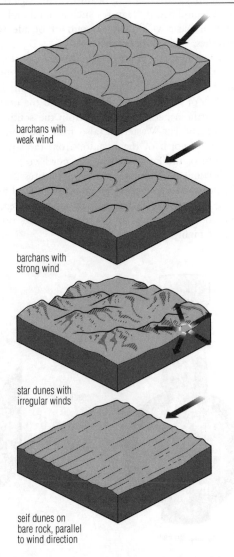

barchans with weak wind

barchans with strong wind

star dunes with irregular winds

seif dunes on bare rock, parallel to wind direction

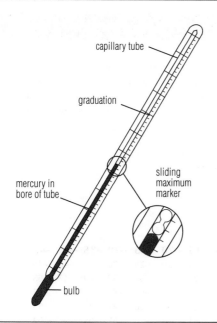

capillary tube

graduation

mercury in
bore of tube

sliding
maximum
marker

bulb

thermometer Maximum and minimum thermometers are universally used in weather-reporting stations. The maximum thermometer, shown here, includes a magnet that fits tightly inside a capillary tube and is moved up it by the rising mercury. When the temperature falls, the magnet remains in position, thus enabling the maximum temperature to be recorded.

not reduced to sea level. A thermometer is kept at the same temperature as the surrounding air by sheltering it in a Stevenson screen.

air mass An air mass is a large body of air with particular characteristics of temperature and humidity. An air mass forms when air rests over an area long enough to pick up the conditions of that area. When an air mass moves to another area it affects the weather of that area, but its own characteristics become modified in the process. For example, an air mass formed over the Sahara will be hot and dry, becoming cooler as it moves northwards. Air masses that meet form **fronts.**

There are four types of air masses. **Tropical continental** (Tc) air masses form over warm land and **Tropical maritime** (Tm) masses form over warm seas. Air masses that form over cold land are called **Polar continental** (Pc) and those forming over cold seas are called **Polar** or **Arctic maritime** (Pm or Am).

front The boundary between two air masses of different temperature or humidity is a front. A **cold front** marks the line of advance of a cold air mass from below, as it displaces a warm air mass; a **warm front** marks the advance of a warm air mass as it rises up over a cold one. Frontal systems define the weather of the mid-latitudes, where warm tropical air is constantly meeting cold air from the poles.

Warm air, being lighter, tends to rise above the cold; its moisture is carried upwards and usually falls as rain or snow, hence the changeable weather conditions at fronts. Fronts are rarely stable and move with the air mass. An **occluded front** is a composite form, where a cold front catches up with a warm front and merges with it.

humidity The humidity of the air is found by comparing readings taken by an ordinary thermometer with readings from a thermometer whose bulb is covered with moist muslin (a hygrometer). Temperature and humidity in the upper atmosphere are measured by attaching instruments to an aircraft, to a small bal-

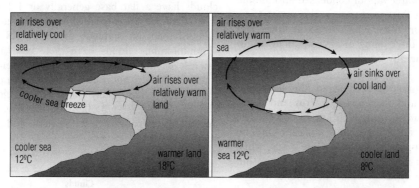

air rises over relatively cool sea

air rises over relatively warm land

cooler sea breeze

cooler sea 12°C

warmer land 18°C

air rises over relatively warm sea

air sinks over cool land

warmer sea 12°C

cooler land 8°C

sea breeze The direction of a sea breeze can vary depending on which is warmer, the land or the sea. Warm air rises and is replaced by cool air. When the land is warmer than the sea, the warm air rises drawing in the cooler air from the sea, creating an onshore breeze. The opposite happens when the sea is warmer than the land, drawing cooler air from the land creating an offshore or land breeze.

Wind Chill Index

Source: US National Weather Service

Wind speed can affect temperature, causing the outside air temperature to feel colder. To determine wind chill, find the outside air temperature on the top line, then read down the column to the measured wind speed in miles per hour. The point at which the two axes intersect provides the wind chill. For example, if the outside temperature is 0°F and the wind speed is 20 mph, the rate of heat loss is equivalent to –39°F. If the temperature is 0°F and there is no wind, the wind chill is between 0 and 4 mph.

A wind chill value of –20°F presents little danger.

A wind chill value between –21 and –74°F may cause flesh to freeze within a minute.

A wind chill value of –75°F and below may cause flesh to freeze within 30 seconds.

Wind chill chart

Wind speed (mph) / **Temperature (°F)**

	45	40	35	30	25	20	15	10	5	0	–5	–10	–15	–20	–25	–30	–35	–40	–45
0	45	40	35	30	25	20	15	10	5	0	–5	–10	–15	–20	–25	–30	–35	–40	–45
5	43	37	32	27	22	16	11	6	1	–5	–10	–16	–21	–26	–31	–37	–42	–47	–53
10	34	28	22	16	10	4	–3	–9	–15	–22	–28	–34	–40	–46	–52	–59	–65	–71	–77
15	29	22	16	9	2	–5	–12	–19	–25	–32	–39	–45	–52	–59	–66	–72	–79	–86	–93
20	25	18	11	4	–3	–11	–18	–25	–32	–39	–47	–54	–61	–68	–75	–82	–89	–96	–104
25	23	15	8	1	–7	–15	–22	–30	–37	–45	–52	–60	–67	–74	–82	–89	–97	–104	–112
30	21	13	6	–3	–11	–18	–26	–33	–41	–49	–56	–64	–72	–79	–87	–95	–102	–110	–117
35	19	11	4	–5	–13	–20	–28	–36	–44	–52	–59	–67	–75	–83	–91	–98	–106	–114	–122
40	18	10	3	–6	–14	–22	–30	–38	–46	–54	–61	–69	–77	–85	–93	–101	–109	–117	–124
45	17	9	2	–7	–15	–23	–31	–39	–47	–55	–63	–71	–79	–87	–95	–103	–111	–119	–127

loon, or even to a rocket. With aircraft, measurements have been made up to more than 15 km/9.3 mi above the Earth, with balloons to 30 km/18.5 mi, and with rockets to more than 150 km/93 mi. Initially, instruments carried by balloon had to be recovered before readings could be obtained; now, radiosonde balloons transmit observations to Earth.

clouds A cloud is water vapour condensed into minute water particles that float in masses in the atmosphere. Clouds, like fogs or mists, which occur at lower levels, are formed by the cooling of air containing water vapour, which generally condenses around tiny dust particles.

Clouds are classified according to the height at which they occur and their shape. **Cirrus** and **cirrostratus** clouds occur at around 10 km/33,000 ft. The former, sometimes called mares'-tails, consist of minute specks of ice and appear as feathery white wisps, while cirrostratus clouds stretch across the sky as a thin white sheet. Three types of cloud are found at 3–7 km/10,000–23,000 ft: cirrocumulus, altocumulus, and altostratus. **Cirrocumulus** clouds occur in small or large rounded tufts, sometimes arranged in the pattern called mackerel sky. **Altocumulus** clouds are similar, but larger, white clouds, also arranged in lines. **Altostratus** clouds are like heavy cirrostratus clouds

and may stretch across the sky as a grey sheet. **Stratocumulus** clouds are generally lower, occurring at 2–6 km/6,500–20,000 ft. They are dull grey clouds that give rise to a leaden sky that may not yield rain. Two types of clouds, **cumulus** and **cumulonimbus**, are placed in a special category because they are produced by daily ascending air currents, which take moisture into the cooler regions of the atmosphere. Cumulus clouds have a flat base generally at 1.4 km/4,500 ft where condensation begins, while the upper part is dome-shaped and extends to about 1.8 km/6,000ft. Cumulonimbus clouds have their base at much the same level, but extend much higher, often up to over 6 km/20,000 ft. Short heavy showers and sometimes thunder may accompany them. **Stratus** clouds, occurring below 1–2.5 km/3,000–8,000 ft, have the appearance of sheets parallel to the horizon and are like high fogs.

Clouds
http://ww2010.atmos.uiuc.edu/(Gh)/guides/mtr/cld/home.rxml

Illustrated guide to how clouds form and to the various different types. The site contains plenty of images and a glossary of key terms in addition to further explanations of the various types of precipitation.

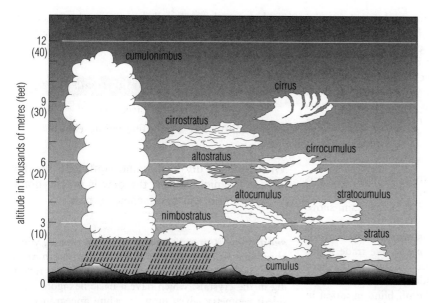

cloud Standard types of cloud. The height and nature of a cloud can be deduced from its name. Cirrus clouds are at high levels and have a wispy appearance. Stratus clouds form at low level and are layered. Middle-level clouds have names beginning with 'alto'. Cumulus clouds, ball or cottonwool clouds, occur over a range of height.

In addition to their essential role in the water cycle, clouds are important in the regulation of radiation in the Earth's atmosphere. They reflect short-wave radiation from the Sun, and absorb and re-emit long-wave radiation from the Earth's surface.

Clouds
http://www.hawastsoc.org/solar/eng/cloud1.htm
This site offers a unique look at clouds, containing photographs of various cloud types taken from space including thunderstorms over Brazil, jet stream cirrus clouds, and a description of how clouds form.

fog Fog is a cloud that collects at the surface of the Earth, composed of water vapour that has condensed on particles of dust in the atmosphere. Cloud and fog are both caused by the air temperature falling below dew point. The thickness of fog depends on the number of water particles it contains. Officially, fog refers to a condition when visibility is reduced to 1 km/0.6 mi or less, and mist or haze to that giving a visibility of 1–2 km or about 1 mi.

There are two types of fog. An **advection fog** is formed by the meeting of two currents of air, one cooler than the other, or by warm air flowing over a cold surface. Sea fogs commonly occur where warm and cold currents meet and the air above them mixes. A **radiation fog** forms on clear, calm nights when the land surface loses heat rapidly (by radiation); the air above is cooled to below its dew point and condensation takes place. A **mist** is produced by condensed water particles, and a haze by smoke or dust.

In drought areas, for example, Baja California, the Canary Islands, the Cape Verde Islands, the Namib Desert, Peru, and Chile, coastal fogs enable plant and

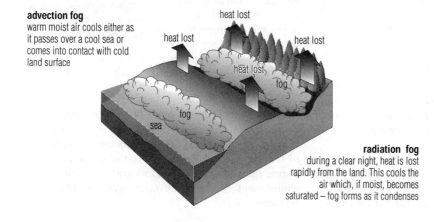

advection fog
warm moist air cools either as it passes over a cool sea or comes into contact with cold land surface

heat lost

heat lost

heat lost

heat lost

fog

fog

sea

radiation fog
during a clear night, heat is lost rapidly from the land. This cools the air which, if moist, becomes saturated – fog forms as it condenses

fog Advection fog occurs when two currents of air, one cooler than the other meet, or by warm air flowing over a cold surface. Radiation fog forms through rapid heat loss from the land, causing condensation to take place and a mist to appear.

animal life to survive without rain and are a potential source of water for human use (by means of water collectors exploiting the effect of condensation).

Industrial areas uncontrolled by pollution laws have a continual haze of smoke over them, and if the temperature falls suddenly, a dense yellow smog forms. At some airports since 1975 it has been possible for certain aircraft to land and take off blind in fog, using radar navigation.

precipitation

Water that falls to the Earth from the atmosphere is called precipitation. It is part of the hydrological or water cycle. Forms of precipitation include rain, snow, sleet, hail, dew, and frost.

The amount of precipitation in any one area depends on climate, weather, and phenomena like trade winds and ocean currents. The cyclical change in the Peru current off the coasts of Ecuador and Peru, El Niñor certain aircraft to land and take off blind in tation in South and Central America and throughout the Pacific region.

Precipitation can also be influenced by people. In urban areas dust, smoke, and other types of particulate pollution that comprise **condensation nuclei** cause water in the air to condense more readily. Fog is one example. Precipitation also can react chemically with air-borne pollutants producing acid rain.

rain Rain is a form of precipitation in which separate drops of water fall to the Earth's surface from clouds. The drops are formed by the accumulation of fine droplets that condense from water vapour in the air. The condensation is usually brought about by rising and subsequent cooling of air. Rain can form in three main ways – frontal (or cyclonic) rainfall, orographic (or relief) rainfall, and convectional rainfall.

Frontal rainfall takes place at the boundary, or front, between a mass of warm air from the tropics and a mass of cold air from the poles. The water vapour in the warm air is chilled and condenses to form clouds and rain.

Orographic rainfall occurs when an airstream is forced to rise over a mountain range. The air becomes cooled and precipitation takes place. In the UK, the Pennine hills, which extend southwards from Northumbria to Derbyshire in northern England, interrupt the path of the prevailing southwesterly winds, causing orographic rainfall. Their presence is partly responsible for the west of the UK being wetter than the east.

Convectional rainfall, associated with hot climates, is brought about by rising and abrupt cooling of air that has been warmed by the extreme heat of the ground surface. The water vapour carried by the air condenses and so rain falls heavily. Convectional rainfall is usually accompanied by a thunderstorm, and it can be intensified over urban areas due to higher temperatures.

weather prediction: rain

A weather prediction:

Rain before seven, fine before eleven.

or

When grass is dry in morning light, look for rain before the night. When the dew is on the grass, rain will never come to pass.

snow Snow is precipitation in the form of soft, white, crystalline flakes caused by the condensation in air of excess water vapour below freezing point. Light reflecting in the crystals, which have a basic hexagonal (six-sided) geometry, gives snow its white appearance.

Snow
http://www-nsidc.colorado.edu/NSIDC/EDUCATION/SNOW/
snow_FAQ.html
Comprehensive information about snow from the US National Snow and Ice Data Centre. Among the interesting subjects discussed are why snow is white, why snow flakes can be up to two inches across, what makes some snow fluffy, why sound travels farther across snowy ground, and why snow is a good insulator.

hail Precipitation in the form of pellets of ice (hailstones) is called hail. It is caused by the circulation of moisture in strong convection currents, usually within cumulonimbus clouds.

Water droplets freeze as they are carried upwards. As the circulation continues, layers of ice are deposited around the droplets until they become too heavy to be supported by the currents and they fall as a hailstorm. Hail is usually associated with thunderstorms.

> **hailstones**
> Hailstones can kill. In the Gopalganj region of Bangladesh in 1988, 92 people died after being hit by huge hailstones weighing up to 1 kg/2.2 lb.

dew Precipitation in the form of moisture that collects on the ground is called dew. It forms after the temperature of the ground has fallen below the dew point of the air in contact with it. As the temperature falls during the night, the air and its water vapour become chilled, and condensation takes place on the cooled surfaces.

extreme weather phenomena

thunderstorm Thunderstorms are severe storms of very heavy rain, thunder, and lightning. They are usually caused by the intense heating of the ground surface during summer. The warm air rises rapidly to form tall cumulonimbus clouds with a characteristic anvil-shaped top. Electrical charges accumulate in the clouds and are discharged to the ground as flashes of lightning. Air in the path of lightning becomes heated and expands rapidly, creating shock waves that are heard as a crash or rumble of thunder.

> **thunderstorms**
> There are an estimated 44,000 thunderstorms worldwide every day.

lightning Lightning is a high-voltage electrical discharge between two charged rainclouds or between a cloud and the earth, caused by the build-up of electrical charges. Air in the path of lightning ionizes (becomes conducting), and expands; the accompanying noise is heard as thunder. Currents of 20,000 amperes and temperatures of 30,000°C/54,000°F are common. Lightning causes nitrogen oxides to form in the atmosphere and approximately 25% of the atmospheric nitrogen oxides are formed in this way.

> **lightning**
> There are an estimated 8 million bolts of lightning worldwide every day.

According to a 1997 US survey on lightning strength and frequency, using information gathered from satellite images and data from the US Lightning Detection Network, there are 70–100 lightning flashes per second worldwide, with an average peak current of 36 kiloamps.

Lightning kills more than 200 people per year in the USA and causes extensive damage to property and lost revenue due to power cuts.

hurricane A hurricane, or **tropical cyclone**, is a severe depression (region of very low atmospheric pressure) in tropical regions, called a **typhoon** in the north Pacific. It is a revolving storm originating at latitudes between 5° and 20° N or S of the Equator, when the surface temperature of the ocean is above 27°C/80°F. A central calm area, called the eye, is surrounded by inwardly spiralling winds (anticlockwise in the northern hemisphere) of up to 320 kph/200 mph. A hurricane is accompanied by lightning and torrential rain, and can cause extensive damage. In meteorology, a hurricane is a wind of force 12 or more on the Beaufort scale.

During 1995 the Atlantic Ocean region suffered 19 tropical storms, 11 of them hurricanes. This was the third-worst season since 1871, causing 137 deaths. The most intense hurricane recorded in the Caribbean/Atlantic sector was Hurricane Gilbert in 1988, with sustained winds of 280 kph/175 mph and gusts of over 320 kph/200 mph.

Hurricanes
http://hurricane.terrapin.com/
Follow the current paths of Pacific and mid-Atlantic hurricanes and tropical storms at this site. Java animations of storms in previous years can also be viewed, and data sets for these storms may be downloaded. Current satellite weather maps can also be accessed, but only for the USA and surrounding region.

In October 1987 and January 1990, winds of near-hurricane strength were experienced in southern England. Although not technically hurricanes, they were the strongest winds there for three centuries.

The naming of hurricanes began in the 1940s with female names. Owing to public opinion that using female names was sexist, the practice was changed in 1978 to using both male and female names alternately.

tornado A tornado is an extremely violent revolving storm with swirling, funnel-shaped clouds, caused by

Hurricane: Worst of the 20th Century

Year	Date	Location	Deaths
1900	Aug–Sept	Galveston, Texas	6,000
1926	20 Oct	Cuba	600
1928	6–20 Sept	Southern Florida	1,836
1930	3 Sept	Dominican Republic	2,000
1938	21–22 Sept	Long Island, New York, New England	600
1942	15–16 Oct	Bengal, India	40,000
1963	4–8 Oct	(Flora) Caribbean	6,000
1974	19–20 Sept	(Fifi) Honduras	2,000
1979	30 Aug–7 Sept	(David) Caribbean, eastern USA	1,100

a rising column of warm air propelled by strong wind. A tornado can rise to a great height, but with a diameter of only a few hundred metres or less. Tornadoes move with wind speeds of 160–480 kph/100–300 mph, destroying everything in their path. They are common in the central USA and Australia.

Tornado Project Online
http://www.tornadoproject.com/
All about tornadoes – including myths and oddities, personal experiences, tornado chasing, tornado safety, and tornadoes past and present.

Earth's hydrosphere

Earth's hydrosphere encompasses the oceans, seas, rivers, streams, swamps, lakes, and groundwater. In some definitions atmospheric water vapour is included as part of the hydrosphere.

The **water cycle**, or **hydrological cycle**, is natural circulation of water between Earth's surface and its atmosphere. It is a complex system involving a number of physical and chemical processes (such as evaporation, precipitation, and infiltration) and stores (such as rivers, oceans, and soil).

Water is lost from the Earth's surface to the atmosphere by evaporation caused by the Sun's heat on the surface of lakes, rivers, and oceans, and through the transpiration of plants. This atmospheric water is carried by the air moving across the Earth, and **condenses** as the air cools to form clouds, which in turn deposit moisture on the land and sea as precipitation. The water that collects on land flows to the ocean overland – as streams, rivers, and glaciers – or through the soil (infiltration) and rock (groundwater). The boundary that marks the upper limit of groundwater is called the water table.

The oceans, which cover around 70% of the Earth's surface, are the source of most of the moisture in the atmosphere.

Earth's oceans

An ocean is great mass of salt water. Strictly speaking three oceans exist – the Atlantic, Indian, and Pacific – to which the Arctic is often added. They cover approximately 70% or 363,000,000 sq km/140,000,000 sq mi of the total surface area of the Earth. Water levels recorded in the world's oceans have shown an increase of 10–15 cm/4–6 in over the past 100 years.

distribution and depth The oceans cover about 70% of the Earth's surface. The proportion of land to water is 2:3 in the northern hemisphere and 1:4.7 in the southern hemisphere. The average depth of the oceans is about 4,000 m/13,125 ft, but the greatest depth sounded is 11,524 m/37,807 ft in the Mindanao Trench, east of the Philippines (for comparison, Mount Everest is 8,872 m/29,118 ft high). Some 76% of the ocean basins have a depth of 3–6 km/1.8–3.7 mi and only 1% is deeper. The deepest parts of the ocean occur mainly in Pacific trenches (see ◊Mariana Trench). The deepest sounding in the Atlantic Ocean is 9,560 m/31,365 ft in the Puerto Rico Trench. In some ocean basins the sea floor is relatively smooth, and stretches of the abyssal plain in the northwestern Atlantic have been found to be flat within 2 m/6.5 ft over distances of 100 km/62 mi. Comparing the average depth of

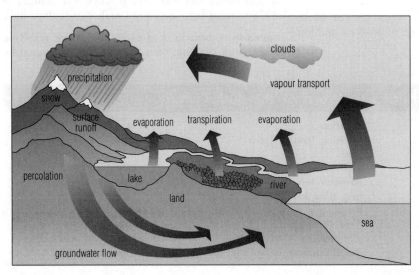

water cycle About one-third of the solar energy reaching the Earth is used in evaporating water. About 380,000 cubic km/95,000 cubic mi is evaporated each year. The entire contents of the oceans would take about 1 million years to pass through the water cycle.

about 4 km/2.5 mi with the horizontal dimensions, which are of the order of 5,000–15,000 km/3,100–9,320 mi, gives a ratio similar to that of the width and thickness of a single sheet of paper.

As the continents are approached, the abyssal plain rises through the continental slope to the continental shelf. The width of the continental shelf varies enormously, but its average width is 65 km/40 mi and its average depth about 130 m/425 ft. These are important areas for fishing and petroleum deposits and have a great influence on local tides. The mass of the oceans is calculated to be 1.43×10^{18} t, with a mean density of 1.045 gm/cc and a mean temperature of 3.9°C/39.02°F.

features The physical features of oceans include deep-sea trenches (off eastern and southeast Asia, and western South America), volcanic belts (in the western Pacific and eastern Indian Ocean), seamounts, and ocean ridges (in the mid-Atlantic, eastern Pacific, and Indian Ocean).

temperature On average, temperature distribution in the oceans has three distinct layers. From the surface to a depth of usually less than 500 m/ 1,640 ft, the water is quite uniformly warm. The temperature decreases comparatively rapidly in a layer 500–1,000 m/1,640–3,280 ft thick to about 5°C/41°F. This region is called the main thermocline and beneath it lie the deep ocean waters, where temperature decreases slowly with depth. Towards higher latitudes the thermocline becomes less deep and in subpolar regions the water column is uniformly cold.

The temperature beneath the main thermocline is fairly uniform throughout the oceans, but the temperature above the thermocline depends on latitude and the predominant currents. The mean annual surface temperature in the tropics is about 30°C/86°F; towards the poles this may drop to –1.7°C/28.9°F, the freezing point of sea water.

Except in the tropics the amount of heat the ocean receives at a given latitude varies with the seasons. In late spring and summer the surface temperature increases and heat is mixed downwards by turbulence, to form a mixed surface layer bounded underneath by a seasonal thermocline. This mixed layer is rarely thicker than 100 m/330 ft, and the seasonal thermocline has been shown to consist of many layers several metres deep at a uniform temperature, separated by thinner regions where temperature changes rapidly with depth. During the winter the surface temperature of the layer slowly decreases and the seasonal thermocline is eroded. The annual variation in surface temperature is at most about 10°C/50°F and often less.

The capacity of tropical waters to store heat has a great influence on the global circulation of the atmosphere, because this is driven by large-scale convection in the tropical atmosphere.

---

composition and salinity Salinity is usually expressed as the amount of dissolved salts contained in 1,000 parts of water, with an average value of about 35 parts per thousand (ppT). Average salinity of the oceans is about 3%. Areas of particularly heavy precipitation, such as the tropics, and those with slight evaporation or a great inflow of fresh water, have a low salinity. In the Baltic Sea, for instance, the salinity is always below 29 ppT.

Regions of the ◊trade winds and permanent anticyclonic conditions show high salinity, but the enclosed seas (such as the Mediterranean and the Red Sea) have the highest. The most striking contrasts of salinity are only a surface feature, and are greatly reduced in deeper waters. In the case of the Dead Sea, river water has been pouring down for thousands of years into a comparatively small lake where evaporation has been consistently high, and, as a result, the very high salinity of 200 ppT has been reached.

Minerals commercially extracted include bromine, magnesium, potassium, salt; those potentially recoverable include aluminium, calcium, copper, gold, manganese, silver.

pressure The pressure at any depth is due to the weight of the overlying water (and atmosphere). For every 10 m/33 ft of depth the pressure increases by about one standard atmosphere (atm), which is the pressure at sea level due to the weight of the atmosphere. Thus, at a depth of 4,000 m/13,125 ft the pressure is 400 atm, and is 1,000 atm or more at the bottom of the deepest trenches. Even at such enormous pressures marine life has been found. One of the effects of living at these great pressures is revealed when animals are brought up quickly in a trawl only to break into pieces on account of the sudden reduction in pressure. In the laboratory small unicellular creatures have been subjected to pressures as high as 600 atm without suffering any apparent harm.

density The density of a sample of sea water depends first on its temperature, then on its salinity, and lastly on its pressure. Increasing salinity and pressure cause an increase in density, and higher temperatures reduce density. The range of density values found in sea water is remarkably small, varying from about 1.025 at the surface to about 1.046 at 4,000 m/13,125 ft.

As density is so closely linked to temperature, the density distribution in the oceans tends to mirror the temperature distribution. Surface density is minimal in the tropics with a value of 1.022 and increases towards higher latitudes with values of about 1.026 at 60°N or S. Vertical sections show the same three-layer structure as temperature, with a less dense, mixed upper layer lying on the thermocline which surmounts the bottom layer. In subpolar regions the density may be quite uniform throughout the water column. Small though these density differences may be, they have a profound effect on vertical motions in the ocean and large-scale circulation.

pollution Oceans have always been used as a dumping area for human waste, but as the quantity of waste increases, and land areas for dumping it diminish, the problem is exacerbated. Today ocean pollutants include airborne emissions from land (33% by weight of total marine pollution); oil from both shipping and land-based sources; toxins from industrial, agricultural, and domestic uses; sewage; sediments from mining, forestry, and farming; plastic litter; and radioactive isotopes. Thermal pollution by cooling water from power plants or other industry is also a problem, killing coral and other temperature-sensitive sedentary species.

light The colour of the sea is a reflection in its surface of the colour of the sky above. The penetration of light into the ocean is of crucial importance to marine plants, which occur as plankton. Thus they are able to survive only in the top 100 m/330 ft, or less if the water is polluted. The amount of light to be found at any depth of the ocean depends on the altitude of the Sun, on the weather and surface conditions, and on the turbidity of the water. Sunlight can penetrate the sea's surface only when the Sun is vertically overhead and the sea is calm. Light that does not pass through the surface penetrates no deeper than about 150 m/500 ft. Not all the colours of the spectrum are absorbed equally: red rays are quickly absorbed, but blue and violet light penetrate much further. In clear waters, such as the Sargasso Sea, violet light may be present at 150 m/500 ft, though at a very low strength.

sound The speed at which sound travels underwater depends on density, which in turn depends on temperature, salinity, and pressure. In water at 0°C/32°F with salinity of 35 ppT and pressure of 1 atm the speed of sound is 1,445 m/4,740 ft s^{-1}. The speed of sound increases by about 4 m/13 ft s^{-1} for an increase in temperature of 1°C/33.8°F, by 1.5 m/4.9 ft s^{-1} for an increase of 1 ppT in salinity and by 1.7 m/5.6 ft s^{-1} for an increase in pressure of 10 atm. This dependence on density means that the distribution of sound–speed displays the three vertical regions shown by temperature and density.

Below the mixed upper layer the velocity of sound decreases with depth in the region of greatest temperature change (thermocline). From about 1,000–1,500 m/3,280–4,920 ft the pressure dependence is the most important factor and the velocity of propagation increases with depth.

oceanography
Oceanography is the study of the oceans. Its subdivisions deal with each ocean's extent and depth, the water's evolution and composition, its physics and chemistry, the bottom topography, currents and wind effects, tidal ranges, the biology, and the various aspects of human use.

Oceanography involves the study of water movements – currents, waves, and tides – and the chemical and physical properties of the seawater. It deals with the origin and topography of the ocean floor – ocean trenches and ridges formed by plate tectonics, and continental shelves from the submerged portions of the continents. Computer simulations are widely used in oceanography to plot the possible movements of the waters, and many studies are carried out by remote sensing.

current A current is a flow of a body of water or air, or of heat, moving in a definite direction. Ocean currents are fast-flowing currents of seawater generated by the wind or by variations in water density between two areas. They are partly responsible for transferring heat from the Equator to the poles and thereby evening out the global heat imbalance.

Ocean currents can be divided into two groups: wind-driven currents and thermohaline currents. Wind-driven currents are primarily horizontal and occur in the upper few hundred metres of the ocean. Trade winds blowing from the east in low latitudes and the westerly winds of mid-latitudes along with the ◊Coriolis effect produce a great clockwise gyre in the oceans of the northern hemisphere and an anticlockwise gyre in the southern hemisphere. The consequent return currents, or stream currents, such as the ◊Gulf Stream from the Equator towards the poles are relatively narrow and strong and occur on the western boundaries of the oceans. The Japan (or Kuroshio) Current is another example of a stream current.

Major Oceans and Seas in the World

Ocean/sea	Area[1]		Average depth	
	sq km	sq mi	m	ft
Pacific Ocean	166,242,000	64,186,000	3,939	12,925
Atlantic Ocean	86,557,000	33,420,000	3,575	11,730
Indian Ocean	73,429,000	28,351,000	3,840	12,598
Arctic Ocean	13,224,000	5,106,000	1,038	3,407
South China Sea	2,975,000	1,149,000	1,464	4,802
Caribbean Sea	2,754,000	1,063,000	2,575	8,448
Mediterranean Sea	2,510,000	969,000	1,501	4,926
Bering Sea	2,261,000	873,000	1,491	4,893
Sea of Okhotsk	1,580,000	610,000	973	3,192
Gulf of Mexico	1,544,000	596,000	1,614	5,297
Sea of Japan	1,013,000	391,000	1,667	5,468
Hudson Bay	730,000	282,000	93	305
East China Sea	665,000	257,000	189	620
Andaman Sea	565,000	218,000	1,118	3,667
Black Sea	461,000	178,000	1,190	3,906
Red Sea	453,000	175,000	538	1,764
North Sea	427,000	165,000	94	308
Baltic Sea	422,000	163,000	55	180
Yellow Sea	294,000	114,000	37	121
Persian Gulf	230,000	89,000	100	328
Gulf of California	153,000	59,000	724	2,375
English Channel	90,000	35,000	54	177
Irish Sea	89,000	34,000	60	197

[1] All figures are approximate, as boundaries of oceans and seas cannot be exactly determined.

There are also westerly counter-currents along the Equator, cold northerly currents from the Arctic, and a strong circumpolar current around Antarctica, which is driven by the 'roaring forties' (westerly winds). The currents in the northern Indian Ocean are more complicated and change direction with the ◊monsoon. Some parts of the centres of the main oceanic gyres have very little current, for example the Sargasso Sea, but other areas have recently been shown to contain large eddies that are similar to, but smaller than, depressions and anticyclones in the atmosphere. The source of these eddies is uncertain but one possibility is large meanders that break off from strong western boundary currents such as the Gulf Stream. These eddies are the subject of recent research, as are the deep ocean currents, which have recently been shown to be faster than was previously thought.

Thermohaline currents are caused by changes in the density of sea water due to changes in temperature and salinity. They are mainly vertical currents affecting the deep oceans. Deep ocean currents mostly run in the reverse direction to surface currents, in that dense water sinks off Newfoundland and Tierra del Fuego and then drifts towards the Equator in deep western boundary currents. In the Arctic the sinking occurs because of the cooling of Arctic water in winter, whereas in the Antarctic it is because of an increase in salinity, and hence density, due to surface sea water freezing. The return vertical flow of this thermohaline circulation occurs as a very slow rise in dense deep water towards the surface over most of the ocean. This rise is a result of winds blowing over the water which increases the depth of the wind-mixed layer and the thermocline bringing up denser water from below. These upwelling currents, such as the Gulf of Guinea Current and the Peru (Humboldt) Current, provide food for plankton, which in turn supports fish and sea birds. At approximate five-to-eight-year intervals, the Peru Current that runs from the Antarctic up the west coast of South America, turns warm, with heavy rain and rough seas, and has disastrous results for Peruvian wildlife and for the anchovy industry. The phenomenon is called El Niño (see below).

Ocean circulation and currents play a dominant role in the climate of oceanic land margins. The ocean is warmer than the land in winter and cooler in summer, so that the climate of coastal regions is equable in that annual temperature variations are smaller than in the centre of large land masses, where the climate is extreme. Warm and cold ocean currents can produce climatic contrasts in places at similar latitudes, for example western Europe and eastern Canada.

El Niño El Niño (Spanish 'the child') is a warm ocean surge of the Peru Current, so called because it tends to occur at Christmas, recurring about every five to eight years in the eastern Pacific off South America. It involves a change in the direction of ocean currents, which prevents the upwelling of cold, nutrient-rich waters along the coast of Ecuador and Peru, killing fishes and plants. It is an important factor in global weather.

El Niño is believed to be caused by the failure of trade winds and, consequently, of the ocean currents normally driven by these winds. Warm surface waters then flow in from the east. The phenomenon can disrupt the climate of the area disastrously, and has played a part in causing famine in Indonesia, drought and bush fires in the Galápagos Islands, rainstorms in California and South America, and the destruction of Peru's anchovy harvest and wildlife in 1982–1983. El Niño contributed to algal blooms in Australia's drought-stricken rivers and an unprecedented number of typhoons in Japan in 1991. It is also thought to have caused the 1997 drought in Australia and contributed to certain ecological disasters such as bush fires in Indonesia.

El Niño

http://www.pmel.noaa.gov/toga-tao/ el-nino/home.html

Wealth of scientific information about El Niño with animated views of the monthly sea changes brought about by it, El Niño-related climate predictions, and forecasts from meteorological centres around the world.

El Niño usually lasts for about 18 months, but the 1990 occurrence lasted until June 1995; US climatologists estimated this duration to be the longest in 2,000 years. The last prolonged El Niño of 1939–41 caused extensive drought and famine in Bengal. It is understood that there might be a link between El Niño and global warming.

In a small way, El Niño effects the entire planet. The wind patterns of the the 1998 El Niño have slowed the Earth's rotation, adding 0.4 milliseconds to each day, an effect measured on the Very Long Baseline Interferometer (VLBI).

By examining animal fossil remains along the west coast of South America, US researchers estimated in 1996 that El Niño began 5,000 years ago.

tide The rhythmic rise and fall of the sea level in the Earth's oceans and their inlets and estuaries are called

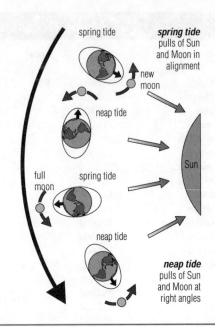

tide The gravitational pull of the Moon is the main cause of the tides. Water on the side of the Earth nearest the Moon feels the Moon's pull and accumulates directly under the Moon. When the Sun and the Moon are in line, at new and full Moon, the gravitational pull of Sun and Moon are in line and produce a high spring tide. When the Sun and Moon are at right angles, lower neap tides occur.

---

tides. Tides are due to the gravitational attraction of the Moon and, to a lesser extent, the Sun, affecting regions of the Earth unequally as it rotates. Water on the side of the Earth nearest the Moon feels the Moon's pull and accumulates directly below it, producing **high tide**.

High tide occurs at intervals of 12 hr 24 min 30 sec. The maximum high tides, or **spring tides**, occur at or near new and full Moon when the Moon and Sun are in line and exert the greatest combined gravitational pull. Lower high tides, or **neap tides**, occur when the Moon is in its first or third quarter and the Moon and Sun are at right angles to each other.

wave An ocean wave is a ridge or swell of water formed by wind or other causes. The power of a wave is determined by the strength of the wind and the distance of open water over which the wind blows (the fetch). Waves are the main agents of coastal erosion and deposition: sweeping away or building up beaches, creating spits and berms, and wearing down cliffs by their hydraulic action and by the corrosion of the sand and shingle that they carry. A tsunami (misleadingly

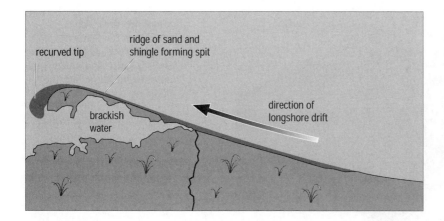

spit Longshore drift carries sand and shingle up coastlines. Deposited material gradually builds up over time at headlands forming a new stretch of land called a spit. A spit that extends across a bay is known as a bar.

called a 'tidal wave') is formed after a submarine earthquake.

As a wave approaches the shore it is forced to break. Friction with the sea bed causes the wavelenth to decrease, while the shallow depth causes the wave height to increase. The wave eventually becomes unstable and breaks. When it breaks on a beach, water and sediment are carried up the beach as **swash**; the water then drains back as **backwash**.

A **constructive wave** causes a net deposition of material on the shore because its swash is stronger than its backwash. Such waves tend be low and have crests that spill over gradually as they break. The backwash of a **destructive wave** is stronger than its swash, and therefore causes a net removal of material from the shore. Destructive waves are usually tall and have peaked crests that plunge downwards as they break, trapping air as they do so.

Atmospheric instability caused by the greenhouse effect and global warming appears to be increasing the severity of Atlantic storms and the heights of the ocean waves. Waves in the south Atlantic are shrinking – they are on average half a metre smaller than in the mid-1980s – and those in the northeast Atlantic have doubled in size over the last 40 years. As the height of waves affects the supply of marine food, this could affect fish stocks, and there are also implications for shipping and oil and gas rigs in the north Atlantic, which will need to be strengthened if they are to avoid damage.

Freak or 'episodic' waves form under particular weather conditions at certain times of the year, travelling long distances in the Atlantic, Indian, and Pacific oceans. They are considered responsible for the sudden disappearance, without distress calls, of many ships.

Freak waves become extremely dangerous when they reach the shallow waters of the continental shelves at 100 fathoms (180 m/600 ft), especially when they meet currents: for example, the Agulhas Current to the east of South Africa, and the Gulf Stream in the north Atlantic. A wave height of 34 m/112 ft has been recorded.

beach A beach is a strip of land bordering the sea, normally consisting of boulders and pebbles on exposed coasts or sand on sheltered coasts. It is usually defined by the high- and low-water marks. A berm, a ridge of sand and pebbles, may be found at the farthest point that the water reaches.

The unconsolidated material of the beach consists of a rocky debris eroded from exposed rocks and headlands by the processes of coastal erosion, or material carried in by rivers. The material is transported to the beach, and along the beach, by longshore drift. Incoming waves (swash) hit the beach at an angle and carry sand onto the beach. Outgoing waves (backwash) draw back at right angles to the beach carrying sand with them. This zigzag pattern results in a net movement of the material in one particular direction along the coast.

When the energy of the waves decreases due to interaction with currents or changes in the coastline, more sand is deposited than is transported, building up to create depositional features such as spits, bars, and tombolos.

Attempts often are made to artificially halt longshore drift and increase deposition on a beach by erecting barriers (groynes) at right angles to the beach. These barriers cause sand to build up on their upstream side but deplete the beach on the downstream side, causing beach erosion. The finer sand can be moved about by the wind, forming sand dunes.

coastal erosion The erosion of the land by the constant battering of the sea's waves is called coastal erosion. Coastal erosion occurs primarily by the processes of (hydraulic action, (corrasion, (attrition, and (solution. Frost shattering (or freeze-thaw), caused by the

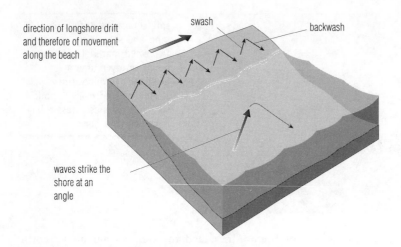

direction of longshore drift and therefore of movement along the beach

swash

backwash

waves strike the shore at an angle

longshore drift Waves sometimes hit the beach at an angle. The incoming waves (swash) carry sand and shingle up onto the shore and the outgoing wave takes some material away with it. Gradually material is carried down the shoreline in the same direction as the longshore current.

expansion of frozen sea water in cavities, and biological weathering, caused by the burrowing of rock-boring molluscs, also result in the breakdown of the coastal rock.

Where resistant rocks form headlands, the sea erodes the coast in successive stages. First it exploits weaknesses, such as faults and cracks, in cave openings and then gradually wears away the interior of the caves until their roofs are pierced through to form blowholes. In time, caves at either side of a headland may unite to form a natural arch. When the roof of the arch collapses, a (stack is formed. The Old Man of Hoy (137 m/449 ft high), in the Orkney Islands, is a fine example of a stack. Stacks may be worn down further to produce a stump and a wave-cut platform.

Beach erosion occurs when more sand is eroded and carried away from the beach than is deposited by longshore drift. Beach erosion can occur due to the construction of artificial barriers, such as groynes, or due to the natural periodicity of the **beach cycle**, whereby high tides and the high waves of winter storms tend to carry sand away from the beach and deposit it offshore in the form of bars. During the calmer summer season some of this sand is redeposited on the beach.

tsunami A tsunami (Japanese 'harbour wave') is an ocean wave generated by vertical movements of the sea floor resulting from earthquakes or volcanic activity. Unlike waves generated by surface winds, the entire depth of water is involved in the wave motion. In the open ocean the tsunami takes the form of several successive waves, rarely in excess of 1 m/3 ft in height but travelling at speeds of 650–800 kph/400–500 mph.

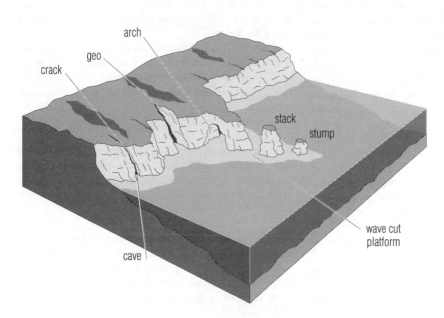

arch

geo

crack

stack

stump

cave

wave cut platform

coastal erosion Typical features of coastal erosion: from the initial cracks in less resistant rock through to arches, stacks, and stumps that can occur as erosion progresses.

Major Floods and Tsunamis of the 20th Century

Year	Event	Location	Number of deaths	Year	Event	Location	Number of deaths
1911	floods	Chang Jiang River, China	100,000		floods	Tanzania	283
1913	floods	Ohio/Indiana, USA	732	1991	floods	Afghanistan	1,367
1915	floods	Galveston, Texas, USA	275		floods	Bangladesh	150,450
1918	tsunami	Kuril Islands/Russia/Japan/	23		floods	Benin	30
		Hawaii			floods	Chad	39
1923	tsunami	Kamchatka/Hawaii	3		floods/storm	Chile	199
1931	floods	Huang Ho River, China	3,700,000		floods	China	6,728
1939	floods	northern China	200,000		floods	India	2,024
1944	tsunami	Japan	998		floods	Malawi	1,172
1946	tsunami	Japan	1,997		floods	Peru	40
	tsunami	Aleutian Islands/Hawaii/	159		floods/	Philippines	8,890
		California			typhoon		
1949	floods	China	57,000		floods	Romania	138
1952	tsunami	Kamchatka/Kuril Islands/Hawaii	many		floods	South Korea	54
1953	floods	northwestern Europe	1,794		floods	Sudan	2,000
1954	floods	China	40,000		floods	Turkey	30
	floods	Farhzad, Iran	2,000		floods	Texas, USA	33
1955	floods	India/Pakistan	1,700		floods	Vietnam	136
1959	floods	western Mexico	2,000	1992	floods	Afghanistan	450
	flood	Fréjus, France	412		floods	Argentina	104
1960	tsunami	Chile/Hawaii/Japan	5,000		floods	Chile	41
	floods	Bangladesh	10,000		floods	China	197
1962	floods	North Sea coast, Germany	343		floods	India	551
	floods	Barcelona, Spain	445		floods	Pakistan	1,446
1964	tsunami	Alaska/Aleutian Islands/California	122		floods	Vietnam	55
1966	floods	Brazil	200	1993	floods	Indonesia	18
	floods	Florence, Italy	113		floods	Midwestern USA	48
1967	floods	Brazil	>300	1994	floods	Moldova	47
1969	floods	southern California, USA	100		floods	southern China	1,400
	floods	South Korea	250		floods	India	>600
	floods	Tunisia	500		floods	Vietnam	>175
1970	floods	Romania	160		floods	northern Italy	>60
	floods	Himalayas, India	500	1995	floods	Benin	10
1971	floods	Rio de Janeiro, Brazil	130		floods	Bangladesh	>200
1972	floods	West Virginia, USA	118		floods	Somalia	20
	floods	South Dakota, USA	237		floods	northwestern Europe	40
1974	floods	Tubaro, Brazil	1,000		floods	Hunan Province, China	1,200
	floods	Bangladesh	2,500		floods	southwestern Morocco	136
1976	tsunami	Philippines	5,000		floods	Pakistan	>120
	floods	Colorado, USA	139		floods	South Africa	147
1978	floods	northern India	1,200		floods	Vietnam	85
1979	tsunami	Indonesia	539	1996	floods	southern and western India	>300
	floods	Morvi, India	15,000		floods	Tuscany, Italy	30
1981	floods	northern China	550		floods	North and South Korea	86
1982	floods	Peru	600		floods	Pyrenees, France/Spain	84
	floods	Guangdong, China	430		floods	Yemen	324
	floods	El Salvador/Guatemala	>1,300		floods	central and southern China	2,300
1983	tsunami	Japan/South Korea	107	1997	floods	west coast, USA	36
1984	floods	South Korea	>200		floods	Sikkim, India	>50
1987	floods	northern Bangladesh	>1,000		floods	Germany/Poland/Czech Republic	>100
1988	floods	Brazil	289		floods	Somalia	>1,700
	floods	Bangladesh	>1,300		floods	eastern Uganda	>30
1990	tsunami	Bangladesh	370		floods	Spain and Portugal	70
	floods	Mexico	85				

In the coastal shallows tsunamis slow down and build up, producing huge swells over 15 m/45 ft high in some cases and over 30 m/90 ft in rare instances. The waves sweep inland causing great loss of life and property. On 26 May 1983 an earthquake in the Pacific Ocean caused tsunamis up to 14 m/42 ft high, which killed 104 people along the west coast of Japan near Minehama, Honshu.

Before each wave there may be a sudden withdrawal of water from the beach. Used synonymously with tsunami, the popular term 'tidal wave' is misleading: tsunamis are not caused by the gravitational forces that affect tides.

Tsunami!
http://www.geophys.washington.edu/tsunami/
intro.html
Description of many aspects of tsunamis. Included are details on how a tsunami is generated and how it propagates, how they have affected humans, and how people in coastal areas are warned about them. The site also discusses if and how you may protect yourself from a tsunami and provides 'near real-time' tsunami information bulletins.

---

rivers

A river is a large body of water that flows down a slope along a channel restricted by adjacent banks and levées (naturally formed raised banks). A river originates at a point called its **source**, and enters a sea or lake at its **mouth**. Along its length it may be joined by smaller rivers called **tributaries**; a river and its tributaries are contained within a **drainage basin**. The point at which two rivers join is called the **confluence**.

Rivers are formed and moulded over time chiefly by the processes of erosion, and by the transport and deposition of sediment (see below). Rivers are able to work on the landscape because the energy stored in the water, or potential energy, is converted as it flows downhill into the kinetic energy used for erosion, transport, and deposition. The amount of potential energy available to a river is proportional to its initial height above sea level. A river follows the path of least resistance downhill, and deepens, widens, and lengthens its channel by erosion.

One way of classifying rivers is by their stage of development. A youthful stream is typified by a narrow V-

river The course of a river from its source of a spring or melting glacier, through to maturity where it flows into the sea.

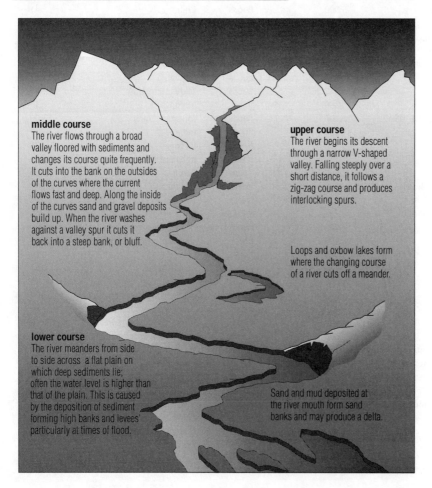

middle course
The river flows through a broad valley floored with sediments and changes its course quite frequently. It cuts into the bank on the outsides of the curves where the current flows fast and deep. Along the inside of the curves sand and gravel deposits build up. When the river washes against a valley spur it cuts it back into a steep bank, or bluff.

upper course
The river begins its descent through a narrow V-shaped valley. Falling steeply over a short distance, it follows a zig-zag course and produces interlocking spurs.

Loops and oxbow lakes form where the changing course of a river cuts off a meander.

lower course
The river meanders from side to side across a flat plain on which deep sediments lie; often the water level is higher than that of the plain. This is caused by the deposition of sediment forming high banks and levees particularly at times of flood.

Sand and mud deposited at the river mouth form sand banks and may produce a delta.

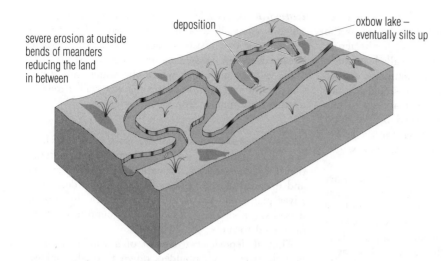

severe erosion at outside bends of meanders reducing the land in between

deposition

oxbow lake – eventually silts up

oxbow lake The formation of an oxbow lake. As a river meanders across a flood plain, the outer bends are gradually eroded and the water channel deepens; as the loops widen, the neck of the loop narrows and finally gives way, allowing the water to flow in a more direct route, isolating the old water channel and forming an oxbow lake.

shaped valley with numerous waterfalls, lakes, and rapids. Because of the steep gradient of the topography and the river's height above sea level, the rate of erosion is greater than the rate of deposition and down-

cutting occurs by **vertical corrasion**. These characteristics may also be said to typify a river's **upper course**.

In a mature river, the topography has been eroded down over time and the river's course has a shallow

Longest Rivers in the World

River	Location	km	mi	River	Location	km	mi
Nile	Africa	6,695	4,160	São Francisco	Brazil	3,199	1,988
Amazon	South America	6,570	4,083	Yukon	USA/Canada	3,185	1,979
Chang Jiang (Yangtze)	China	6,300	3,915	Rio Grande	USA/Mexico	3,058	1,900
				Indus	Tibet/Pakistan	2,897	1,800
Mississippi–Missouri–Red Rock	USA	6,020	3,741	Danube	eastern Europe	2,858	1,776
				Japura	Brazil	2,816	1,750
Huang He (Yellow River)	China	5,464	3,395	Salween	Myanmar/China	2,800	1,740
				Brahmaputra	Asia	2,736	1,700
Ob–Irtysh	China/Kazakhstan/ Russia	5,410	3,362	Euphrates	Iraq	2,736	1,700
				Tocantins	Brazil	2,699	1,677
Amur–Shilka	Asia	4,416	2,744	Zambezi	Africa	2,650	1,647
Lena	Russia	4,400	2,734	Orinoco	Venezuela	2,559	1,590
Congo–Zaire	Africa	4,374	2,718	Paraguay	Paraguay	2,549	1,584
Mackenzie–Peace–Finlay	Canada	4,241	2,635	Amu Darya	Tajikistan/Turkmenistan/ Uzbekistan	2,540	1,578
Mekong	Asia	4,180	2,597	Ural	Russia/Kazakhstan	2,535	1,575
Niger	Africa	4,100	2,548	Kolyma	Russia	2,513	1,562
Yenisei	Russia	4,100	2,548	Ganges	India/Bangladesh	2,510	1,560
Parana	Brazil	3,943	2,450	Arkansas	USA	2,344	1,459
Mississippi	USA	3,779	2,348	Colorado	USA	2,333	1,450
Murray–Darling	Australia	3,751	2,331	Dnieper	eastern Europe	2,285	1,420
Missouri	USA	3,726	2,315	Syr Darya	Asia	2,205	1,370
Volga	Russia	3,685	2,290	Irrawaddy	Myanmar	2,152	1,337
Madeira	Brazil	3,241	2,014	Orange	South Africa	2,092	1,300
Purus	Brazil	3,211	1,995				

gradient. This mature river is said to be graded. Erosion and deposition are delicately balanced as the river meanders (gently curves back and forth) across the extensive flood plain (sometimes called an inner (delta). **Horizontal corrasion** is the dominant erosive process. The flood plain is an area of periodic flooding along the course of river valleys made up of fine silty material called alluvium deposited by the flood water. Features of a the mature river (or the **lower course** of a river) include extensive meanders, ox-bow lakes, and braiding.

Many important flood plains, such as the inner Niger delta in Mali, occur in arid areas where their exceptional fertility has great importance for the local economy. However, using flood plains as the site of towns and villages involves a certain risk, and it is safer to use flood plains for other uses, such as agriculture and parks. Water engineers can predict when flooding is likely and take action to prevent it by studying hydrographs, graphs showing how the discharge of a river varies with time.

Flood!
http://www.pbs.org/wgbh/nova/flood/

As a companion to the US Public Broadcasting Service (PBS) television programme Nova, this page concerns many aspects of flooding. It takes an historical look at floods and examines the measures that engineers have taken to combat them. Three major rivers are discussed: the Yellow, Nile, and Mississippi. In addition to learning about the negative effects of floods, you can also find out about the benefits that floods bestow on farmland.

---

waterfall A cascade of water in a river or stream is called a waterfall. It occurs when a river flows over a bed of rock that resists erosion; weaker rocks downstream are worn away, creating a steep, vertical drop and a plunge pool into which the water falls. Over time, continuing erosion causes the waterfall to retreat upstream forming a deep valley, or gorge.

alluvial deposit An alluvial deposit is a layer of broken rocky matter, or sediment, formed from material that has been carried in suspension by a river or stream and dropped as the velocity of the current decreases. River plains and deltas are made entirely of alluvial deposits, but smaller pockets can be found in the beds of upland torrents.

Alluvial deposits can consist of a whole range of particle sizes, from boulders down through cobbles, pebbles, gravel, sand, silt, and clay. The raw materials are the rocks and soils of upland areas that are loosened by erosion and washed away by mountain streams. Much of the world's richest farmland lies on alluvial deposits. These deposits can also provide an economic source of minerals. River currents produce a sorting action, with particles of heavy material deposited first while lighter materials are washed downstream. Hence heavy minerals such as gold and tin, present in the original rocks in small amounts, can be concentrated and deposited on stream beds in commercial quantities. Such deposits are called 'placer ores'.

---

waterfall When water flows over hard rock and soft rock, the soft rocks erode creating waterfalls. As the erosion processes continue, the falls move backwards, in the opposite direction of the water

---

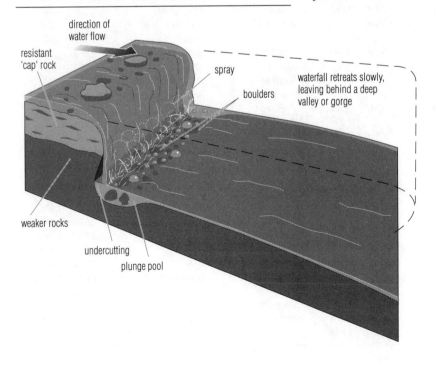

direction of water flow

resistant 'cap' rock

spray

boulders

waterfall retreats slowly, leaving behind a deep valley or gorge

weaker rocks

undercutting

plunge pool

Highest Waterfalls in the World

Waterfall	Location	Total drop m	ft	Waterfall	Location	Total drop m	ft
Angel Falls	Venezuela	979	3,212	Glass Falls	Brazil	404	1,325
Yosemite Falls	USA	739	2,425	Krimml	Austria	400	1,312
Mardalsfossen–South	Norway	655	2,149	Trummelbach Falls	Switzerland	400	1,312
Tugela Falls	South Africa	614	2,014	Takkakaw Falls	Canada	366	1,200
Cuquenan	Venezuela	610	2,000	Silver Strand Falls, Yosemite	USA	357	1,170
Sutherland	New Zealand	580	1,903	Wallaman Falls	Australia	346	1,137
Ribbon Fall, Yosemite	USA	491	1,612	Wollomombi	Australia	335	1,100
Great Karamang River Falls	Guyana	488	1,600	Cusiana River Falls	Colombia	300	984
Mardalsfossen–North	Norway	468	1,535	Giessbach	Switzerland	300	984
Della Falls	Canada	440	1,443	Skykkjedalsfossen	Norway	300	984
Gavarnie Falls	France	422	1,385	Staubbach	Switzerland	300	984
Skjeggedal	Norway	420	1,378				

sediment Sediment is any loose material that has 'settled' – deposited from suspension in water, ice, or air, generally as the water current or wind speed decreases. Typical sediments are, in order of increasing coarseness, clay, mud, silt, sand, gravel, pebbles, cobbles, and boulders.

Sediments differ from sedimentary rocks in which deposits are fused together in a solid mass of rock by a process called lithification. Pebbles are cemented into conglomerates; sands become sandstones; muds become mudstones or shales; peat is transformed into coal.

sand The term 'sand' refers to loose grains of rock, sized 0.0625–2.00 mm/0.0025–0.08 in in diameter, consisting most commonly of quartz, but owing their varying colour to mixtures of other minerals. Sand is used in cement-making, as an abrasive, in glass-making, and for other purposes. Sands are classified into marine, freshwater, glacial, and terrestrial. Some 'light' soils contain up to 50% sand. Sands may eventually consolidate into sandstone.

groundwater
Water collected underground in porous rock strata and soils is termed groundwater; it emerges at the surface as springs and streams. The groundwater's upper level is called the **water table**. Sandy or other kinds of beds that are filled with groundwater are called **aquifers**. Recent estimates are that usable ground water amounts to more than 90% of all the fresh water on Earth; however, keeping such supplies free of pollutants entering the recharge areas is a critical environmental concern.

Most groundwater near the surface moves slowly through the ground while the water table stays in the same place. The depth of the water table reflects the balance between the rate of **infiltration**, called recharge, and the rate of discharge at springs or rivers or pumped water wells. The force of gravity makes underground water run 'downhill' underground just as it does on the surface. The greater the slope and the permeability, the greater the speed. Velocities vary from 100 cm/40 in per day to 0.5 cm/0.2 in.

Hydrology
http://wwwdmorll.er.usgs.gov/~bjsmith/outreach/
hydrology.primer.html
Information from the US Geological Survey about all aspects of hydrology. The clickable chapters include facts about surface water and ground water, the work of hydrologists, and careers in hydrology. For answers to further questions click on 'ask a hydrologist', which provides links to other US national and regional sources.

---

water table The upper level of ground water is called the water table. Water that is above the water table will drain downwards; a spring forms where the water table cuts the surface of the ground. The water table rises and falls in response to rainfall and the rate at which water is extracted, for example, for irrigation and industry.

In many irrigated areas the water table is falling due to the extraction of water. Below northern China, for example, the water table is sinking at a rate of 1 m/3 ft a year. Regions with high water tables and dense industrialization have problems with pollution of the water table. In the USA, New Jersey, Florida, and Louisiana have water tables contaminated by both industrial wastes and saline seepage from the ocean.

aquifer A body of rock through which appreciable amounts of water can flow is called an aquifer. The rock of an aquifer must be porous and permeable (full

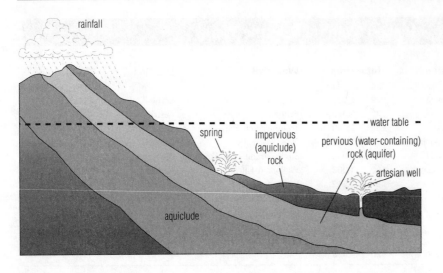

rainfall

water table

spring

impervious (aquiclude) rock

pervious (water-containing) rock (aquifer)

artesian well

aquiclude

artesian well In an artesian well, water rises from an underground water-containing rock layer under its own pressure. Rain falls at one end of the water-bearing layer, or aquifer, and percolates through the layer. The layer fills with water up to the level of the water table. Water will flow from a well under its own pressure if the well head is below the level of the water table.

of interconnected holes) so that it can conduct water. Aquifers are an important source of fresh water, for example, for drinking and irrigation, in many arid areas of the world, and are exploited by the use of artesian wells.

Edwards Aquifer
http://www.txdirect.net/users/eckhardt/
Guide to the Edwards Aquifer (a rock formation containing water) in Texas – one of the most prolific artesian aquifers in the world.

An aquifer may be underlain, overlain, or sandwiched between less permeable layers, called **aquicludes** or **aquitards**, which impede water movement. Sandstones and porous limestones make the best aquifers.

artesian well An artesian well is a well that is supplied with water rising naturally from an underground water-saturated rock layer (an aquifer). The water rises from the aquifer under its own pressure. Such a well may be drilled into an aquifer that is confined by impermeable rocks both above and below. If the water table (the top of the region of water saturation) in that aquifer is above the level of the well head, hydrostatic pressure will force the water to the surface.

Artesian wells are often overexploited because their water is fresh and easily available, and they eventually become unreliable. There is also some concern that pollutants such as pesticides or nitrates can seep into the aquifers.

Much use is made of artesian wells in eastern Australia, where aquifers filled by water in the Great Dividing Range run beneath the arid surface of the Simpson Desert. The artesian well is named after Artois, a French province, where the phenomenon was first observed.

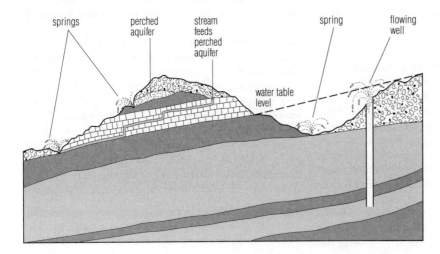

springs

perched aquifer

stream feeds perched aquifer

spring

flowing well

water table level

spring Springs occur where water-laden rock layers (aquifers) reach the surface. Water will flow from a well whose head is below the water table.

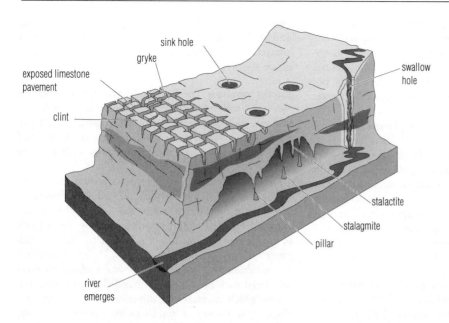

exposed limestone
pavement

sink hole

gryke

clint

swallow
hole

stalactite

stalagmite

pillar

river
emerges

cave A cave is a roofed-over cavity in the Earth's crust. It is usually produced by the action of underground water, or by waves on a seacoast. Inland caves commonly occur in areas underlain by limestone, such as Kentucky and many Balkan regions, where the rocks are soluble in water. A **pothole** is a vertical hole in rock caused by water descending a crack; it is thus open to the sky.

Most inland caves are found in karst regions, because limestone is soluble when exposed to ground water. As the water makes its way along the main joints, fissures, and bedding planes, they are constantly enlarged into potential cave passages, which ultimately join to form a complex network. ◊Stalactites and stalagmites and columns form due to water rich in calcium carbonate dripping from the roof of the cave. The collapse of the roof of a cave produces features such as **natural arches** and **steep-sided gorges**. Limestone caves are usually found just below the water table, wherever limestone outcrops on the surface. The biggest cave in the world is over 70 km/43 mi long, at Holloch, Switzerland.

During the ice age, humans began living in caves leaving many layers of debris that archaeologists have unearthed and dated in the Old World and the New. They also left cave art, paintings of extinct animals often with hunters on their trail. Cave animals often show loss of pigmentation or sight, and under isolation, specialized species may develop. The scientific study of caves is called **speleology**.

Celebrated caves include the Mammoth Cave in Kentucky, USA, 6.4 km/4 mi long and 38 m/125 ft high; the Caverns of Adelsberg (Postumia) near Trieste,

Italy, which extend for many miles; Carlsbad Cave, New Mexico, the largest in the USA; the Cheddar Caves, England; Fingal's Cave, Scotland, which has a range of basalt columns; and Peak Cavern, England.

lakes

A lake is a body of still water lying in depressed ground without direct communication with the sea. Lakes are common in formerly glaciated regions, along the courses of slow rivers, and in low land near the sea. The main classifications are by origin: **glacial lakes**, formed by glacial scouring; **barrier lakes**, formed by landslides and glacial moraines; **crater lakes**, found in volcanoes; and **tectonic lakes**, occurring in natural fissures.

Crater lakes form in the calderas of extinct volcanoes, for example Crater Lake, Oregon, USA. Subsidence of the roofs of limestone caves in a karst landscape exposes the subterranean stream network and provides a cavity in which a lake can develop. Tectonic lakes form during tectonic movement, as when a rift valley is formed. Lake Tanganyika was created in conjunction with the East African Great Rift Valley. Glaciers produce several distinct types of lake, such as the lochs of Scotland and the Great Lakes of North America.

Lakes are mainly freshwater, but salt and bitter lakes are found in areas of low annual rainfall and little surface runoff, so that the rate of evaporation exceeds the rate of inflow, allowing mineral salts to accumulate. The Dead Sea has a salinity of about 250 parts per 1,000 and the Great Salt Lake, Utah, about 220 parts per 1,000. Salinity can also be caused by volcanic gases or fluids, for example Lake Natron, Tanzania.

In the 20th century large artificial lakes have been created in connection with hydroelectric and other works. Some lakes have become polluted as a result of human activity. Sometimes eutrophication (a state of overnourishment) occurs, when agricultural fertilizers leaching into lakes cause an explosion of aquatic life, which then depletes the lake's oxygen supply until it is no longer able to support life.

glaciers

A glacier is a tongue of ice, originating in mountains in snowfields above the snowline, which moves slowly downhill and is constantly replenished from its source. Glaciers form where annual snowfall exceeds annual melting and drainage.The area at the top of the glacier is called the **zone of accumulation**. The lower area of the glacier is called the **ablation zone**. In the zone of accumulation, the snow compacts to ice under the weight of the layers above and moves downhill under the force of gravity. The ice moves plastically under pressure, changing its shape and crystalline structure permanently. Partial melting of ice at the sole of the glacier also produces a sliding component of glacial movement, as the ice travels over the bedrock. In the ablation zone, melting occurs and glacial till (boulder clay) is deposited.

When a glacier moves over an uneven surface, deep **crevasses** are formed in rigid upper layers of the ice mass; if it reaches the sea or a lake, it breaks up to form **icebergs**. A glacier that is formed by one or several valley glaciers at the base of a mountain is called a **piedmont** glacier. A body of ice that covers a large land surface or continent, for example Greenland or Antarctica, and flows outward in all directions is called an **ice sheet**.

glacial erosion The wearing-down and removal of rocks and soil by a glacier forms impressive landscape features, including glacial troughs (U-shaped valleys), arêtes (steep ridges), corries (enlarged hollows), and pyramidal peaks (high mountain peaks with concave faces).

Erosional landforms result from abrasion and plucking of the underlying bedrock. Abrasion is caused by the lodging of rock debris in the sole of the glacier, followed by friction and wearing away of the bedrock as the ice moves. The action is similar to that of sandpaper attached to a block of wood. The results include the polishing and scratching of rock surfaces to form powdered rock flour, and sub-parallel scratches or striations which indicate the direction of ice movement. Plucking is a form of glacial erosion restricted to the lifting and removal of blocks of bedrock already loosened by freeze-thaw activity in joint fracture.

The most extensive period of recent glacial erosion was the Pleistocene epoch in the Quaternary period when, over 2 to 3 million years, the polar ice caps repeatedly advanced and retreated. More ancient glacial episodes are also preserved in the geological record, the earliest being in the middle Precambrian and the most extensive in Permo-Carboniferous times.

Larger landforms caused by glacial erosion generally possess a streamlined form, as in ◊roche moutonnée and ◊corries. A common feature of glacial erosion is the glacial trough. Hanging valleys, with their tributary waterfalls, indicate extensive glacial erosion. The amount of lowering accomplished by glacial erosion has been estimated at 0.05–2.8 mm/0.002–0.11 in per year, a rate of lowering 10 to 20 times that associated with the action of rivers. Depositional landforms cover

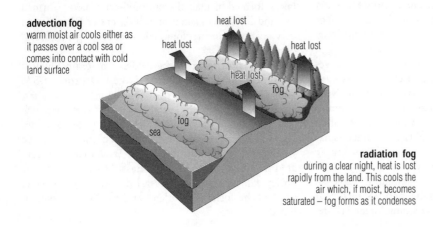

advection fog
warm moist air cools either as it passes over a cool sea or comes into contact with cold land surface

heat lost

heat lost

heat lost

heat lost

fog

fog

sea

radiation fog
during a clear night, heat is lost rapidly from the land. This cools the air which, if moist, becomes saturated – fog forms as it condenses

glacial deposition A glacier picks up large boulders and rock debris from the valley and deposits them at the snout of the glacier when the ice melts. Some deposited material is carried great distances by the ice to form erratics.

10% of the Earth's surface.

Periglacial processes result from frost and snow activity in areas on the margins of an ice-sheet. Among the most important periglacial processes are frost-weathering and ☿solifluction. Frost-weathering (the alternate freezing and thawing of ice in cracks in the rock) etches the outlines of rock outcrops, exploiting joints and areas of weakness, and results in aprons of scree.

Glaciers
http://www-nsidc.colorado.edu/ NSIDC/EDUCATION/GLACIERS/
Comprehensive information about glaciers from the US National Snow and Ice Data Center. There are explanations of why glaciers form, different kinds of glaciers, and what they may tell us about climate change. There are a number of interesting facts and a bibliography about the compacted tongues of ice which cover 10% of the land surface of our planet.

---

glacial deposition When glacial ice melts, it deposits the material that it has been carrying. The material deposited by a glacier is called **till**, or **boulder clay**. It comprises angular particles of all sizes from boulders to clay that are unsorted and lacking in stratification.

Unstratified till can be moulded by ice to form **drumlins**, egg-shaped hills. At the snout of the glacier, till piles up to form a ridge called a **terminal moraine**. Small depositional landforms may also result from glacial deposition, such as **kames** (small mounds) and **kettle holes** (small depressions, often filled with water).

Stratified till that has been deposited by meltwater is termed **fluvio-glacial**, because it is essentially deposited by running water. Meltwater flowing away from a glacier will carry some of the till many kilometres away. This sediment will become rounded (by the water) and, when deposited, will form a gently sloping area called an **outwash plain**. Several landforms owe their existence to meltwater (**fluvioglacial landforms**) and include the long ridges called eskers, which form parallel to the direction the ice flows. Meltwater may fill depressions eroded by the ice to form **ribbon lakes**.

Glacial deposits occur in many different locations beneath the ice (subglacial), inside it (englacial), on top of it (supraglacial), at the side of it (marginal), and in front of it (proglacial).

Nova
http://www.pbs.org/wgbh/nova/ice/
Companion to the US Public Broadcasting Service (PBS) television programme Nova, this page provides information about glaciation, the natural changes in climate over the past 60 million years, the greenhouse effect, global warming, and continental movement.

Ice Age: Major Ice Ages

Name	Date (years ago)
Pleistocene	1.64 million–10,000
Permo-Carboniferous	330–250 million
Ordovician	440–430 million
Verangian	615–570 million
Sturtian	820–770 million
Gnejso	940–880 million
Huronian	2,700–1,800 million

ice age The term 'ice age' refers to any period of glaciation occurring in Earth's history, but particularly that in the Pleistocene epoch, immediately preceding historic times. On the North American continent, glaciers reached as far south as the Great Lakes, and an ice sheet spread over northern Europe, leaving its remains as far south as Switzerland.

There were several glacial advances separated by interglacial stages during which the ice melted and temperatures were higher than today. Formerly there were thought to have been only three or four glacial advances, but recent research has shown about 20 major incidences. For example, ocean-bed cores record the absence or presence in their various layers of such cold-loving small marine animals as radiolaria, which indicate a fall in ocean temperature at regular intervals. Other ice ages have occurred throughout geological time: there were four in the Precambrian era, one in the Ordovician, and one at the end of the Carboniferous and beginning of the Permian.

The occurrence of an ice age is governed by a combination of factors (the **Milankovitch hypothesis**): (1) the Earth's change of attitude in relation to the Sun, that is, the way it tilts in a 41,000-year cycle and at the same time wobbles on its axis in a 22,000-year cycle, making the time of its closest approach to the Sun come at different seasons; and (2) the 92,000-year cycle of eccentricity in its orbit round the Sun, changing it from an elliptical to a near circular orbit, the severest period of an ice age coinciding with the approach to circularity. There is a possibility that the Pleistocene ice age is not yet over. It may reach another maximum in another 60,000 years.

The theory of ice ages was first proposed in the 19th century by, among others, the Swiss civil engineer Ignace Venetz in 1821 and the Swiss-born US geologist Louis Agassiz in 1837. (Before, most geologists had believed that the rocks and sediment they left behind were caused by the biblical flood.) The term 'ice age' was first used by the German botanist Karl Schimper in 1837.

Earth Science Chronology

c. 15000 BC	Rising temperatures in the northern hemisphere cause glacial ice sheets to melt and sea levels to rise by 36.6–39.6 m/120–130 ft, causing widespread flooding.
c. 7500 BC	The ice covering much of the Earth has by now melted to roughly modern levels. This destroys many land bridges, for instance across the Mediterranean. The separation of the British Isles from the European mainland occurs about this time.
c. 450 BC	The Greek historian Herodotus concludes, correctly, that the Nile delta is caused by the deposition of mud carried by the Nile. He declares the Caspian to be an inland sea and not part of the northern ocean as most scholars of the time believe.
c. 400 BC	Crude rain gauges 46 cm/18 in wide are used in India.
c. 350 BC	Aristotle writes of the weather and climate in *Meteorologica/Meteorology*.
c. 250 BC	Chinese scholars explicitly describe the hydrologic cycle. The idea has been known in China for a century.
c. 240 BC	The Greek scholar Eratosthenes of Cyrene makes a map of the Nile valley and correctly explains the reasons for the Nile's annual floods as being due to heavy rains in the upper reaches.
c. 100 BC	The Greek philosopher Poseidonius correlates tides with the lunar cycle.
c. 100 BC	The Greeks are the first to measure wind direction. They install a wind vane on the Acropolis.
440	The Gaulish town of Ys, in Armorica (modern Brittany), is overwhelmed by a great flood, and submerged beneath the sea – the foundation of many such legends of lost cities.
602	The Yellow River floods in China, with catastrophic results. In order to prevent a repetition of the disaster, the emperor orders a massive project to recut the riverbed channel.
700	Persian scientists make measurements of windspeed using the rotation of windmills.
764	A severe winter freezes the Bosporus linking Europe and Asia Minor and allows the channel to be crossed on foot.
1108	The first embankments are built on the Red River of Dai Viet in order to protect the rice fields from being flooded.
1130	The English monk Adelard of Bath writes *Quaestiones naturales/Inquiries into Nature*, comprising a series of 76 dialogues discussing scientific topics, including meteorology.
1307	The German scholar Dietrich of Freiburg writes his treatise *De iride et radialibus impressionibus/On the Rainbow and Radial Impressions*, on optical effects in meteorology, and particularly the rainbow.
1320	The Muslim scientist Al-Farisi writes *Tanqih al-Manazur/Correction of the Optics*, which includes his theory on the origin of the rainbow.
1340	William Merle of Oxford, England, writes *De futura aeris intemperie/On the Future Storm*, a treatise on various methods of predicting the weather.
c. 1450	German philosopher Cardinal Nicholas de Cusa devises the first hygrometer for measuring air humidity.
1450	Italian scientist Leone Alberti devises a vane anemometer and makes the first measurements of wind speeds.
1613	Italian scientist Giovan-Francesco Sagredo begins using a Galilean air thermometer to take daily temperature readings in Venice.
1654	Duke Ferdinand II of Tuscany establishes the world's first meteorological office, with Luigi Antinori in charge. Daily reports and temperature readings are provided by observers in Parma, Milan, Bologna, and Florence.
1660	German physicist Otto Von Guericke makes weather forecasts using a water barometer to measure changes in air pressure. He proposes a series of meteorological stations to improve forecasting. and discovers the sudden drop in air pressure preceding a violent storm – a discovery that will revolutionize weather forecasting.
1664	German philosopher Athanasius Kircher publishes *Mundus subterraneus/The Subterranean World*, suggesting that the tides are caused by water moving to and from a subterranean ocean.
1677	English scientist Richard Towneley devises a rain gauge to capture water falling on the roof of his house, and starts to maintain accurate weather records.
1686	English scientist Edmond Halley publishes a map of the world showing the directions of prevailing winds in different regions – the first meteorological chart.
1702	French scientist Guillaume Amontons, in his experiments on thermometers, notes for the first time that, for a constant mass of air, the change in pressure is proportional to the change in temperature.
1735	English lawyer George Hadley describes the circulation of the atmosphere in terms of

large-scale convection cells moving hot air away from the equator and towards the poles.

1740 Scottish mathematician Colin Maclaurin publishes a widely-praised gravitational theory to explain the tides.

June 1752 North American scientist and statesman Benjamin Franklin performs his most famous experiment, flying a kite during a thunderstorm and charging a Leyden jar to which it is connected. He thereby demonstrates the electrical nature of lightning.

1755 The Golden Horn around Constantinople in the Ottoman Empire freezes over in one of the coldest winters on record – the beginning of a 'mini Ice Age' that will last several decades.

1760 Russian scientist Mikhail Vasileivich Lomonosov explains the formation of icebergs.

1771 Swiss meteorologist Jean-André Deluc establishes rules for using the barometer to measure altitude and the height of mountains.

1783 Swiss geologist Horace Bénédict de Saussure invents the hair hygrometer to measure humidity. It is based on the principle that hair lengthens when wet.

1803 English meteorologist Luke Howard gives names to the cloud types (cirrus, cumulus, stratus, and nimbus), and recognizes that their shapes reflect their causes.

1804–07 German naturalist and explorer Alexander von Humboldt draws maps with isotherms (lines joining locations with the same mean temperature) and isobars (lines joining locations with the same mean barometric pressure).

1805 British navy commander Francis Beaufort devises the Beaufort wind force scale.

Sep 1821 US meteorologist William Redfield discovers that during a hurricane trees are toppled toward the northwest in Connecticut, and toward the southeast 80 km/50 mi further west, demonstrating that tropical storms are cyclones.

1828 Prussian meteorologist Heinrich Wilhelm Dove discovers that winds in tropical storms circulate counterclockwise in the northern hemisphere and clockwise in the southern hemisphere.

1835 French mathematician Gustave Coriolis describes the inertial forces acting on a rotating body that act at right angles to the direction of rotation. The Coriolis force causes wind and current systems to rotate to the right in the northern hemisphere and to the left in the southern.

1835 US meteorologist James Pollard Espy studies the energy sources of storms and explains frontal systems with attendant condensation and precipitation.

1840 Swiss-born US naturalist Jean-Louis Agassiz describes the motion and behaviour of glacier. He argues that Europe was covered by great sheets of ice in the geologically recent past.

1844 French scientist Lucien Vidie invents the aneroid barometer. It detects changes in air pressure through the deformation of an evacuated metal tube.

1845–58 Von Humboldt lays the basis of modern geography with the publication of *Kosmos/Cosmos*, in which he arranges geographic knowledge in a systematic fashion.

1848 The first wind and current charts for the North Atlantic are compiled by US naval officer Matthew Fontaine Maury.

1849 British meteorologist William Reid demonstrates that hurricanes in the northern hemisphere rotate and curve along paths opposite in direction to those in the southern hemisphere.

c. 1850 US physicist Joseph Henry arranges for telegraph offices to have meteorological equipment installed in exchange for weather information telegraphed to the Smithsonian Institution. By 1860 there are over 500 stations involved, and the information allows the production of the first daily weather maps.

1854 US naval officer Matthew Fontaine Maury maps the depth of the North Atlantic to 4,000 fathoms (7,300 m/23,950 ft).

1855 US navy officer Matthew Fontaine Maury publishes *Physical Geography of the Sea*, the first textbook on oceanography.

1856 French hydraulic engineer Henri Darcy develops a law to estimate the flow of groundwater.

1868 A tsunami over 21 m/66 ft high kills over 25,000 people in Hawaii and Chile.

1868 English meteorologist Alexander Buchan begins the use of weather maps for forecasting, with the publication of a map showing the movement of a cyclonic depression across North America and Europe. It marks the start of modern meteorology.

1872 Scottish physicist William Thomson (Lord Kelvin) develops a sounding-machine (the 'Kelvin') for determining depth at sea.

7 Dec 1872– 26 May 1876 The British ship *Challenger* undertakes the world's first major oceanographic survey. Under the command of the Scottish naturalist Wyville Thomson, it discovers hundreds of new marine animals, and finds that at 2,000 fathoms the temperature of the ocean is a constant 2.5°C/36.5°F.

1881 Nearly 300,000 people die when a typhoon hits Haiphong, Vietnam.

1882 Scottish physicist Balfour Stewart postulates the existence of an electrically conducting layer of the outer atmosphere (now known as

	the ionosphere) to account for the daily variation in the Earth's magnetic field.
1882	US meteorologist Elias Loomis publishes the first precipitation map, which shows mean annual precipitation throughout the world, using isohyets connecting places having equal rainfall.
1883	Austrian meteorologist Julius von Hann publishes *Handbook of Climatology*, which is a compilation of meterological knowledge and world climate.
1884	German meteorologist Vladimir Peter Köppen introduces a classification of the world's climate zones based on the number of months whose temperatures remain above or below certain means.
1895	US meteorologist Jeanette Picard launches the first balloon to be used for atmospheric research.
March 1895	Russian physicist Aleksandr Stepanovich Popov constructs a lightning detector to register atmospheric electrical disturbances, and suggests that it can be used to detect radio waves.
1897	Norwegian meteorologist Vilhelm Bjerknes develops mathematical theorems applicable to the motions of large-scale air masses, which are essential to weather forecasting.
1900	French meteorologist Léon-Philippe Teisserenc de Bort discovers that the Earth's atmosphere consists of two main layers: the troposphere, where the temperature continually changes and is responsible for the weather, and the stratosphere, where the temperature is invariant.
1900	German meteorologist Vladimir Peter Köppen develops a mathematical system for classifying climatic types, based on temperature and rainfall. It serves as the basis for subsequent classification systems.
1913	French physicist Charles Fabry discovers the ozone layer in the upper atmosphere.
1919	Norwegian meteorologists Vilhelm and Jakob Bjerknes introduce the term 'front' in meteorology (after the military front), which describes the transition zone between two masses of air differing in density and temperature.
1919	The US Navy develops the Hayes sonic depth finder. It consists of a device that generates sound waves and receives their echo from the ocean floor. A timing device indicates the depth of the water.
1920	Yugoslavian meteorologist and mathematician Milutin Milankovitch shows that the amount of energy, or heat, received by Earth from the Sun varies with long-term changes in Earth's orbit. Decades later, scientists will correlate fluctuations in global temperature to his 'Milankovitch cycles'.

1922	British meteorologist Lewis Fry Richardson publishes *Weather Prediction by Numerical Process*, in which he applies the first mathematical techniques to weather forecasting.
1925	The US Navy develops a pulse modulation technique to measure the distance above the Earth of the ionizing layer in the atmosphere.
1930	The Woods Hole Oceanographic Institution is established in Massachusetts, USA.
11 June 1930	The first bathysphere, a spherical steel craft for undersea exploration, built by US zoologist William Beebe and US engineer Otis Baron, descends to 435 m/1,428 ft.
18 Aug 1934	US explorers and biologists William Beebe and Otis Baron descend in a bathysphere to a record 923 m/3,028 ft in the Atlantic off Bermuda.
c. 1935	Aeroplanes begin to be used for weather reporting.
1935	The US Weather Bureau establishes hurricane forecasting centres.
1935	US seismologist Charles Richter introduces the Richter scale for measuring the magnitude of earthquakes at their epicentre.
1936	The radio meteorograph (radiosonde) is developed by the US Weather Service; it transmits information on temperature, humidity, and barometric pressure from uncrewed balloons. A network of stations is also inaugurated.
c. 1940	Information from uncrewed weather balloons indicates that columns of warm air rise more than 1.6 km/1mi above the Earth and winds form layers in the lower atmosphere, often blowing in different directions.
1945	Single-stage sounding rockets, reaching speeds of 4,800–8,000 kph/3,000–5,000 mph, and a maximum altitude of 160 km/100 mi, are launched carrying instrumentation to gather information about the upper atmosphere.
1946	The first cloud-seeding experiments are conducted in the USA in an attempt to produce rain.
1947	US meteorologist Irving Langmuir carries out the first hurricane-seeding experiment; 91 kg/200 lb of dry ice is distributed in a storm.
1952	Norwegian-US meteorologist Jacob Bjerknes discovers that centres of low pressure, or cyclones, develop at the fronts that separate different air masses. It leads to improved weather forecasting. He is also the first to use photographs, taken from high-altitude rockets, as a tool in weather analysis and forecasting.

1952	US meteorologists establish the first weather station in the Arctic, at Ice Island T-3.
1958	The Equatorial Undercurrent is discovered in the equatorial Pacific Ocean. It has a width of 320–480 km/ 200–300 mi, a height of 200–300 m/ 650–1,000 ft, and flows 50–150 m/165–500 ft below the surface.
May 1958	Using data from the *Explorer* rockets, US physicist James Van Allen discovers a belt of radiation around the Earth. Now known as the Van Allen belts (additional belts were discovered later), they consist of charged particles from the Sun trapped by the Earth's magnetic field.
c. 1960	Meteorologists begin to study storm systems using doppler radar, which can detect the speed and direction of moving storms because of the change in frequency of the reflected radar waves.
23 Jan 1960	Swiss engineer Jacques Piccard and US Navy lieutenant Don Walsh descend to the bottom of Challenger Deep (10,916 m/35,810 ft), off the Pacific island of Guam, in the bathyscaph *Trieste*, setting a new undersea record.
1 April 1960	The USA launches *TIROS 1* (Television and Infra-Red Observation Satellite). A weather satellite, it is equipped with television cameras, infrared detectors, and videotape recorders. It provides a worldwide weather observation system, along with subsequently launched *Tiros* satellites.
18–20 Aug 1961	The US Navy and the US Environmental Sciences Service Administration initiate Project Stormfury, an attempt to modify hurricanes through seeding, by heavily seeding Hurricane Debbie with silver iodide. Wind speeds drop markedly.
19 June 1963	The US satellite *TIROS 7* (Television and Infra-Red Observation Satellite) is launched. It is used by meteorologists to track, forecast, and analyse storms.
1965	French oceanographer Jacques Cousteau heads the Conshelf Saturation Dive Programme, which sends six divers 100 m/328 ft down in the Mediterranean for 22 days.
1965	NASA launches *GEOS 1* (Geodynamics Experimental Ocean Satellite). Its aim is to provide a three-dimensional map of the world accurate to within 10 m/ 30 ft.
1966	The National Science Foundation in the USA puts Scripps Institute of Oceanography in charge of the Joint Oceanographic Institutions Deep Earth Sampling (JOIDES) project and establishes the Deep Sea Drilling Project (DSDP).
1974	Mexican chemist Mario Molina and US chemist F Sherwood Rowland warn that the chlorofluorocarbons (CFCs) used in fridges and as aerosol propellants may be damaging the

	atmosphere's ozone layer that filters out much of the Sun's ultraviolet radiation.
1974	The Global Atmospheric Research Programme (GARP) is launched. An international project, its aim is to provide a greater understanding of the mechanisms of the world's weather by using satellites and by developing a mathematical model of the Earth's atmosphere.
May 1974	The US launches the world's first *Synchronous Meteorological Satellite* (SMS).
1975	Five new nations, USSR, West Germany, France, Japan, and the United Kingdom join the Deep Sea Drilling Project (DSDP) to form the International Phase of Ocean Drilling (IPOD).
1976–79	The International Magnetosphere Study conducts a three-year observation of the Earth's magnetosphere and its effects on the lower stratosphere including the disruptive effects of magnetic storms on communications.
1977	Scientists from the project FAMOUS (French-American Mid-Ocean Undersea Study) in their deep sea submersible vehicle ALVIN discover a host of strange life forms, such as large red and white tube worms, near undersea hot springs heated by ocean-ridge volcanism . The discovery proves the existence of life in extreme conditions.
27 June 1978	The US satellite *Seasat 1* is launched to measure the temperature of sea surfaces, wind and wave movements, ocean currents, and icebergs; it operates for 99 days before its power fails.
1983	US chemist Mark Thiemens and his colleagues demonstrate that the production of ozone in the upper atmosphere causes separation of the two different heavy isotopes of oxygen *independent* of their masses. The discovery of this non-mass dependent kinetic partitioning of oxygen isotopes provides a new means of tracing the mixing of gases between Earth's different layers of atmosphere.
1987	US researchers prove that thunderstorm systems can propel pollutants into the lower stratosphere when they observe high levels of carbon monoxide and nitric acid at high altitude during a thunderstorm.
1991	The European Space Agency's first remote-sensing satellite (*ERS-1*) is launched into polar orbit to monitor the Earth's temperature from space.
1991	The World Ocean Experiment (WOCE) programme is set up to monitor ocean temperatures, circulation, and other parameters.
1993	US scientist Albert Bradley develops the Autonomous Benthic Explorer (ABE), a robotic submersible that can descend to depths of 6.4 km/4 mi and remain at such depths for up to one year.

4 July 1994	Electrical flashes known as 'Sprites' – upper atmosphere optical phenomena associated with thunderstorms – are first examined by plane, by a team from the University of Alaska Statewide System, Fairbanks, Alaska, USA.		that up to a billion tonnes of dust a year are blown off the arid drought-prone lands surrounding the Sahara Desert in north Africa and carried as far as the UK and the Caribbean. The amount has more than doubled in the past 30 years.
1996	Scientists from the Scott Polar Institute, using data from the European Space Agency's ERS-1 satellite, discover a 14,000-sq-km/5,400-sq-mi, 125-m/410-ft-deep lake, 4 km/2.5 mi under the Antarctic ice sheet. Called Lake Vostok after the Russian ice-drilling station it lies beneath, the ice sheet, which acts as a blanket, and a pressure of 300–400 atmospheres allow the water to remain liquid.	4 Aug 1997	Using computer models, British meteorologist Alan O'Neill demonstrates a connection between the collapse of anchovy fishing in Peru, drought in Australia, and the late arrival of India's monsoons.
11 June 1997	French meteorologist Cyril Moulin shows	17 July 1998	A 10-m/30-ft tidal wave hits the north coast of Papua New Guinea, inundating several villages and killing an estimated 6,000 people. Of the survivors 70% are adults; a generation of children is wiped out.

Biographies

Agassiz, (Jean) Louis Rodolphe (1807–1873) Swiss-born US palaeontologist and geologist who developed the idea of the ice age. He proposed that glaciers, far from being static, were in a constant state of almost imperceptible motion. Finding rocks that had been shifted or abraded, presumably by glaciers, he inferred that in earlier times much of northern Europe had been covered with ice sheets. *Etudes sur les glaciers/Studies on Glaciers* (1840) developed the original concept of the ice age.

Beaufort, Francis (1774–1857) British admiral, hydrographer to the Royal Navy from 1829. He drew up the Beaufort scale for measuring wind speed. The Beaufort Sea in the Arctic Ocean is named after him.

Bjerknes, Vilhelm Firman Koren (1862–1951) Norwegian scientist whose theory of polar fronts formed the basis of all modern weather forecasting and meteorological studies. He also developed hydrodynamic models of the oceans and the atmosphere and showed how weather prediction could be carried out on a statistical basis, dependent on the use of mathematical models.

Coriolis, Gustave Gaspard de (1792–1843) French physicist. In 1835 he discovered the **Coriolis effect**, which governs the movements of winds in the atmosphere and currents in the ocean. Investigating the movements of moving parts in machines and other systems relative to the fixed parts, Coriolis explained how the rotation of the Earth causes objects moving freely over the surface to follow a curved path relative to the surface. Coriolis was also the first to derive formulas expressing and mechanical work.

Crutzen, Paul (1933–) Dutch meteorologist who shared the 1995 Nobel Prize for Chemistry with the Mexican chemist Mario Molina and the US chemist F Sherwood Rowland for their work in atmospheric chemistry, particularly concerning the formation and decomposition of ozone. They explained the chemical reactions which are destroying the ozone layer.

In 1970 Crutzen discovered that the nitrogen oxides NO and NO_2 speed up the breakdown of atmospheric ozone into molecular oxygen. These gases are produced in the atmosphere from nitrous oxide (N_2O) which is released by microorganisms in the soil. He showed that this process is the main natural method of ozone breakdown. He also discovered that ozone-depleting chemical reactions occur on the surface of cloud particles in the stratosphere.

Davis, William Morris (1850–1934) US physical geographer who analysed landforms. In the 1890s he developed the organizing concept of a regular cycle of erosion, a theory that dominated geomorphology and physical geography for half a century. He proposed a standard stage-by-stage life cycle for a river valley, marked by youth (steep-sided V-shaped valleys), maturity (flood-plain floors), and old age, as the river valley was imperceptibly worn down into the rolling landscape which he termed a 'peneplain'. On occasion these developments could be punctuated by upthrust, which would rejuvenate the river and initiate new cycles.

Fitzroy, Robert (1805–1865) British vice admiral and meteorologist. In 1828 he succeeded to the command of HMS *Beagle*, then engaged on a survey of the Patagonian coast of South America, and in 1831 was accompanied by naturalist Charles Darwin on a five-year survey. Fitzroy was governor of New Zealand 1843–45. In 1855 the Admiralty founded the Meteorological Office, which issued weather forecasts and charts, under his charge.

Forbes, Edward (1815–1854) British naturalist who studied molluscs and made significant contributions to oceanography. He discounted the contemporary conviction that marine life subsisted only close to the sea surface, spectacularly dredging a starfish from a depth of 400 m/1,300 ft in the Mediterranean. His *The Natural History of European Seas* (1859) was a pioneering oceanographical text. He proposed that Britain had once been joined to the continent by a land bridge.

Humboldt, (Friedrich Wilhelm Heinrich) Alexander, Baron von Humboldt (1769–1859) German geophysicist, botanist, geologist, and writer who was a founder of ecology. He aimed to erect a new science, a 'physics of the globe', analysing the deep physical interconnectedness of all terrestrial phenomena. He believed that geological phenomena were to be understood in terms of basic physical causes (for example, terrestrial magnetism or rotation). One of the first popularizers of science, he gave a series of lectures later published as *Kosmos/Cosmos* (1845–62), an account of the relations between physical environment and flora and fauna.

In meteorology, he introduced isobars and isotherms on weather maps, made a general study of global temperature and pressure, and instituted a worldwide programme for compiling magnetic and weather observations. His studies of American volcanoes demonstrated they corresponded to underlying geological faults; on that basis he deduced that volcanic action had been pivotal in geological history and that many rocks were igneous in origin. In 1804 he discovered that the Earth's magnetic field decreased from the poles to the Equator.

Lovelock, James Ephraim (1919–   ) British scientist who began the study of CFCs in the atmosphere in the 1960s (though he did not predict the damage they cause to the ozone layer) and who later elaborated the Gaia hypothesis. He invented the electron capture detector in the 1950s, a device for measuring minute traces of atmospheric gases.

Maury, Matthew Fontaine (1806–1873) US naval officer, founder of the US Naval Oceanographic Office. His system of recording oceanographic data is still used today.

Mercator, Gerardus, Latinized form of Gerhard Kremer (1512–1594) Flemish mapmaker who devised the first modern atlas, showing **Mercator's projection** in which the parallels and meridians on maps are drawn uniformly at 90°. It is often used for navigational charts, because compass courses can be drawn as straight lines, but the true area of countries is increasingly distorted the further north or south they are from the Equator.

Nuttall, Thomas (1786–1859) English-born US naturalist who explored the Arkansas, Red, and Columbia rivers. Between 1809 and 1811 he participated in an expedition up the Missouri River. In 1818–20 he explored the Arkansas and Red rivers, and 1834–35 the mouth of the Columbia River. In 1820 he studied the geology of the Mississippi Valley.

Powell, John Wesley (1834–1902) US geologist whose enormous and original studies produced lasting insights into erosion by rivers, volcanism, and mountain formation. His greatness as a geologist and geomorphologist stemmed from his capacity to grasp the interconnections of geological and climatic causes. He was appointed director of the US Geological Survey in 1881. He drew attention to the aridity of the American southwest, and campaigned for irrigation projects and dams, for the geological surveys necessary to implement adequate water strategies, and for changes in land policy and farming techniques.

Saussure, Horace Bénédict de (1740–1799) Swiss geologist who made the earliest detailed and first-hand study of the Alps. The results of his Alpine survey appeared in his classic work *Voyages des Alpes/Travels in the Alps* (1779–86).

Shaw, (William) Napier (1854–1945) English meteorologist who introduced the millibar as the meteorological unit of atmospheric pressure (in 1909, but not used internationally until 1929). He also invented the tephigram, a thermodynamic diagram widely used in meteorology, in about 1915. He pioneered the study of the upper atmosphere by using instruments carried by kites and high-altitude balloons. His work on pressure fronts formed the basis of a great deal of later work in the field. Shaw's *Manual of Meteorology* (1926–31) is still a standard reference work.

Glossary

abrasion the effect of ◊corrasion, a type of erosion in which rock fragments scrape and grind away a surface. The rock fragments may be carried by rivers, wind, ice, or the sea. Striations, or grooves, on rock surfaces are common abrasions, caused by the scratching of rock debris embedded in glacier ice.

abyssal plain broad expanse of sea floor lying 3–6 km/2–4 mi below sea level. Abyssal plains are found in all the major oceans, and they extend from bordering continental rises to mid-oceanic ridges.

abyssal zone dark ocean region 2,000–6,000 m/ 6,500–19,500 ft deep; temperature 4°C/39°F. Three-quarters of the area of the deep-ocean floor lies in the abyssal zone, which is too far from the surface for photosynthesis to take place. Some fish and crustaceans living there are blind or have their own light sources. The region above is the bathyal zone; the region below, the hadal zone.

alluvial fan roughly triangular sedimentary formation found at the base of slopes. An alluvial fan results when a sediment-laden stream or river rapidly deposits its load of gravel and silt as its speed is reduced on entering a plain.

anticyclone area of high atmospheric pressure caused by descending air, which becomes warm and dry. Winds radiate from a calm centre, taking a clockwise direction in the northern hemisphere and an anticlockwise direction in the southern hemisphere. Anticyclones are characterized by clear weather and the absence of rain and violent winds. In summer they bring hot, sunny days and in winter they bring fine, frosty spells, although fog and low cloud are not uncommon in the UK. **Blocking anticyclones**, which prevent the normal air circulation of an area, can cause summer droughts and severe winters.

Appleton layer, or F layer, band containing ionized gases in the Earth's upper atmosphere, at a height of 150–1,000 km/94–625 mi, above the E layer (formerly the Kennelly–Heaviside layer). It acts as a dependable reflector of radio signals as it is not affected by atmospheric conditions, although its ionic composition varies with the sunspot cycle.

arch any natural bridgelike land feature formed by erosion. A **sea arch** is formed from the wave erosion of a headland where the backs of two caves have met and broken through. The roof of the arch eventually collapses to leave part of the headland isolated in the sea as a ◊stack. A **natural bridge** is formed on land by wind or water erosion and spans a valley or ravine.

arête North American **combe-ridge** sharp narrow ridge separating two ◊glacial troughs (valleys), or ◊corries. The typical U-shaped cross sections of glacial troughs give arêtes very steep sides. Arêtes are common in glaciated mountain regions such as the Rockies, the Himalayas, and the Alps.

attrition process by which particles of rock being transported by river, wind, or sea are rounded and gradually reduced in size by being struck against one another. The rounding of particles is a good indication of how far they have been transported. This is particularly true for particles carried by rivers, which become more rounded as the distance downstream increases.

avalanche fall or flow of a mass of snow and ice down a steep slope under the force of gravity. Avalanches occur because of the unstable nature of snow masses in mountain areas.

barometer instrument that measures atmospheric pressure as an indication of weather. Most often used are the **mercury barometer** and the **aneroid barometer**.

barrier island long island of sand, lying offshore and parallel to the coast. Some barrier islands are over 100 km/60 mi in length. Most are derived from marine sands piled up by shallow ◊longshore currents that sweep sand parallel to the seashore. Others are derived from former ◊spits, connected to land and built up by drifted sand, that were later severed from the mainland.

barrier reef coral reef that lies offshore, separated from the mainland by a shallow lagoon.

bathyal zone upper part of the ocean, which lies on the continental shelf at a depth of between 200 m/650 ft and 2,000 m/6,500 ft.

bearing the direction of a fixed point, or the path of a moving object, from a point of observation on the Earth's surface, expressed as an angle from the north. Bearings are taken by compass and are measured in degrees (°), given as three-digit numbers increasing clockwise. For instance, north is 000°, northeast is 045°, south is 180°, and southwest is 225°.

biological weathering form of ◊weathering caused by the activities of living organisms – for example, the growth of roots or the burrowing of animals. Tree roots are probably the most significant agents of biological weathering as they are capable of prising apart rocks by growing into cracks and joints.

California Current cold ocean current in the eastern Pacific Ocean flowing southwards down the west coast of North America. It is part of the North Pacific gyre (a vast, circular movement of ocean water).

Canaries Current cold ocean current in the north Atlantic Ocean flowing southwest from Spain along the northwest coast of Africa. It meets the northern equatorial current at a latitude of 20° N.

canyon deep, narrow valley or gorge running through mountains. Canyons are formed by stream down-cutting, usually in arid areas, where the rate of down-cutting is greater than the rate of weathering, and where the stream or river receives water from outside the area.

chemical weathering form of ◊weathering brought about by chemical attack of rocks, usually in the presence of water. Chemical weathering involves the 'rotting', or breakdown, of the original minerals within a rock to produce new minerals (such as clay minerals, bauxite, and calcite). Some chemicals are dissolved and carried away from the weathering source.

chlorofluorocarbon (CFC) synthetic chemical that is odourless, nontoxic, nonflammable, and chemically inert. When CFCs are released into the atmosphere they drift up slowly into the stratosphere where, under the influence of ultraviolet radiation from the Sun, they break down into chlorine atoms which destroy the ◊ozone layer and allow harmful radiation from the Sun to reach the Earth's surface.

cirque French and North American name for a ◊corrie, a steep-sided hollow in a mountainside.

clay very fine-grained sedimentary deposit that has undergone a greater or lesser degree of consolidation. When moistened it is plastic, and it hardens on heating, which renders it impermeable. It may be white, grey, red, yellow, blue, or black, depending on its composition. Clay minerals consist largely of hydrous silicates of aluminium and magnesium together with iron, potassium, sodium, and organic substances. The crystals of clay minerals have a layered structure, capable of holding water, and are responsible for its plastic properties. According to international classification, in mechanical analysis of soil, clay has a grain size of less than 0.002 mm/0.00008 in.

condensation process by which water vapour turns into fine water droplets to form clouds. Condensation in the atmosphere occurs when the air becomes completely saturated and is unable to hold any more water vapour. As air rises it cools and contracts – the cooler it becomes the less water it can hold. Rain is frequently associated with warm weather fronts because the air rises and cools, allowing the water vapour to condense as rain. The temperature at which the air becomes saturated is known as the **dew point**. Water vapour will not condense in air if there are not enough condensation nuclei (particles of dust, smoke or salt) for the droplets to form on. It is then said to be supersaturated. Condensation is an important part of the water cycle.

continental rise the portion of the ocean floor rising gently from the abyssal plain toward the steeper continental slope. The continental rise is a depositional feature formed from sediments transported down the slope mainly by turbidity currents. Much of the continental rise consists of coalescing submarine alluvial fans bordering the continental slope.

continental shelf the submerged edge of a continent, a gently sloping plain that extends into the ocean. It typically has a gradient of less than 1°. When the angle of the sea bed increases to 1°–5° (usually several hundred kilometres away from land), it becomes known as the **continental slope**.

continental slope sloping, submarine portion of a continent. It extends downward from the edge of the continental shelf. In some places, such as south of the Aleutian Islands of Alaska, continental slopes extend directly to the ocean deeps or abyssal plain. In others, such as the east coast of North America, they grade into the gentler continental rises that in turn grade into the abyssal plains.

Coriolis effect the effect of the Earth's rotation on the atmosphere and on all objects on the Earth's surface. In the northern hemisphere it causes moving objects and currents to be deflected to the right; in the southern hemisphere it causes deflection to the left. The effect is named after its discoverer, Gaspard de Coriolis.

corrasion the grinding away of solid rock surfaces by particles carried by water, ice, and wind. It is generally held to be the most significant form of erosion. As the eroding particles are carried along they become eroded themselves due to the process of ◊attrition.

corrie (Welsh **cwm**, French and North American **cirque**) Scottish term for a steep-sided hollow in the mountainside of a glaciated area. The weight and movement of the ice has ground out the bottom and worn back the sides. A corrie is open at the front, and its sides and back are formed of ◊arêtes. There may be a lake in the bottom, called a **tarn**.

corrosion alternative term for ◊solution, the process by which water dissolves rocks such as limestone.

crevasse deep crack in the surface of a glacier; it can reach several metres in depth. Crevasses often occur where a glacier flows over the break of a slope, because the upper layers of ice are unable to stretch and cracks result. Crevasses may also form at the edges of glaciers owing to friction with the bedrock.

cyclone alternative name for a depression, an area of low atmospheric pressure. A severe cyclone that forms in the tropics is called a tropical cyclone or hurricane.

delta tract of land at a river's mouth, composed of silt deposited as the water slows on entering the sea. Familiar examples of large deltas are those of the Mississippi, Ganges and Brahmaputra, Rhône, Po, Danube, and Nile; the shape of the Nile delta is like the Greek letter *delta* [Delta], and thus gave rise to the name.

devil wind minor form of tornado, usually occurring in fine weather; formed from rising thermals of warm air (as is a ◊cyclone). A fire creates a similar updraught.

dew precipitation in the form of moisture that collects on the ground. It forms after the temperature of the ground has fallen below the dew point of the air in contact with it. As the temperature falls during the night, the air and its water vapour become chilled, and condensation takes place on the cooled surfaces.

doldrums area of low atmospheric pressure along the Equator, in the intertropical convergence zone where the northeast and southeast ◊trade winds converge. The doldrums are characterized by calm or very light winds, during which there may be sudden squalls and stormy weather. For this reason the areas are avoided as far as possible by sailing ships.

dune mound or ridge of wind-drifted sand common on coasts and in deserts. Loose sand is blown and bounced along by the wind, up the windward side of a dune. The sand particles then fall to rest on the lee side, while more are blown up from the windward side. In this way a dune moves gradually downwind.

drumlin long streamlined hill created in formerly glaciated areas. Rocky debris (till) is gathered up by the glacial icesheet and moulded to form an egg-shaped mound, 8–60 m/25–200 ft in height and 0.5–1 km/0.3–0.6 mi in length. Drumlins commonly occur in groups on the floors of ◊glacial troughs, producing a 'basket-of-eggs' landscape. They are important indicators of the direction of ice flow, as their blunt end points upstream and their gentler slopes trail off downstream.

Ekman spiral effect in oceanography, theoretical description of a consequence of the ◊Coriolis effect on ocean currents, whereby currents flow at an angle to the winds that drive them. It derives its name from the Swedish oceanographer Vagn Ekman (1874–1954).

esker narrow, steep-walled ridge, often sinuous and sometimes branching, formed beneath a glacier. It is made of sands and gravels, and represents the course of a subglacial river channel. Eskers vary in height from 3–30 m/10–100 ft and can be up to 160 km/100 mi or so in length.

exosphere uppermost layer of the atmosphere. It is an ill-defined zone above the thermosphere, beginning at about 700 km/435 mi and fading off into the vacuum of space. The gases are extremely thin, with hydrogen as the main constituent.

fjord, or **fiord**, narrow sea inlet enclosed by high cliffs. Fjords are found in Norway, New Zealand, and western parts of Scotland. They are formed when an overdeepened U-shaped glacial valley is drowned by a rise in sea-level. At the mouth of the fjord there is a characteristic lip causing a shallowing of the water. This is due to reduced glacial erosion and the deposition of ◊moraine at this point.

flooding inundation of land that is not normally covered with water. Flooding from rivers commonly takes place after heavy rainfall or in the spring after winter snows have melted. The river's discharge (volume of water carried in a given period) becomes too great, and water spills over the banks onto the surrounding flood plain. Small floods may happen once a year – these are called **annual floods** and are said to have a one-year return period. Much larger floods may occur on average only once every 50 years.

flood plain area of periodic flooding along the course of river valleys. When river discharge exceeds the capacity of the channel, water rises over the channel banks and floods the adjacent low-lying lands. As water spills out of the channel some alluvium (silty material) will be deposited on the banks to form ◊levees (raised river banks). This water will slowly seep into the flood plain, depositing a new layer of rich fertile alluvium as it does so. Many important floodplains, such as the inner Niger delta in Mali, occur in arid areas where their exceptional productivity has great importance for the local economy.

geomorphology branch of geology that deals with the nature and origin of surface landforms such as mountains, valleys, plains, and plateaus.

glacial trough, or **U-shaped valley**, steep-sided, flat-bottomed valley formed by a glacier. The erosive action of the glacier and of the debris carried by it results in the formation not only of the trough itself but also of a number of associated features, such as truncated spurs (projections of rock that have been sheared off by the ice) and ◊hanging valleys. Features characteristic of glacial deposition, such as ◊drumlins and ◊eskers, are commonly found on the floor of the trough, together with linear lakes called ribbon lakes.

glacier budget in a glacier, the balance between accumulation (the addition of snow and ice to the glacier) and ablation (the loss of snow and ice by melting and evaporation). If accumulation exceeds ablation the glacier will advance; if ablation exceeds accumulation it will probably retreat.

geyser natural spring that intermittently discharges an explosive column of steam and hot water into the air due to the build-up of steam in underground chambers. One of the most remarkable geysers is Old Faithful, in Yellowstone National Park, Wyoming, USA. Geysers also occur in New Zealand and Iceland.

gorge narrow steep-sided valley (or canyon) that may or may not have a river at the bottom. A gorge may be formed as a waterfall retreats upstream, eroding away the rock at the base of a river valley; or it may be caused by rejuvenation, when a river begins to cut downwards into its channel once again (for example, in response to a fall in sea level). Gorges are common in limestone country, where they may be formed by the collapse of the roofs of underground caverns.

gravel coarse sediment consisting of pebbles or small fragments of rock, originating in the beds of lakes and streams or on beaches. Gravel is quarried for use in road building, railway ballast, and for an aggregate in concrete. It is obtained from quarries known as gravel pits, where it is often found mixed with sand or clay.

Gulf Stream warm ocean current that flows north from the warm waters of the Gulf of Mexico along the east coast of America, from which it is separated by a channel of cold water originating in the southerly Labrador current. Off Newfoundland, part of the current is diverted east across the Atlantic, where it is known as the **North Atlantic Drift**, dividing to flow north and south, and warming what would otherwise be a colder climate in the British Isles and northwest Europe.

gyre circular surface rotation of ocean water in each major sea (a type of current). Gyres are large and permanent, and occupy the northern and southern halves of the three major oceans. Their movements are dictated by the prevailing winds and the ◊Coriolis effect. Gyres move clockwise in the northern hemisphere and anticlockwise in the southern hemisphere.

hadal zone the deepest level of the ocean, below the ◊abyssal zone, at depths greater than 6,000m/19,500 ft. The ocean trenches are in the hadal zone. There is no light in this zone and pressure is over 600 times greater than atmospheric pressure.

hanging valley valley that joins a larger ◊glacial trough at a higher level than the trough floor. During glaciation the ice in the smaller valley was unable to erode as deeply as the ice in the trough, and so the valley was left perched high on the side of the trough when the ice retreated. A river or stream flowing along the hanging valley often forms a waterfall as it enters the trough.

headward erosion backwards erosion of material at the source of a river or stream. Broken rock and soil at the source are carried away by the river, causing erosion to take place in the opposite direction to the river's flow. The resulting lowering of the land behind the source may, over time, cause the river to cut backwards into a neighbouring valley to 'capture' another river (see ◊river capture).

heat island large town or city that is warmer than the surrounding countryside. The difference in temperature is most pronounced during the winter, when the heat given off by the city's houses, offices, factories, and vehicles raises the temperature of the air by a few degrees.

hydraulic action erosive force exerted by water (as distinct from the forces exerted by rocky particles carried by water). It can wear away the banks of a river, particularly at the outer curve of a meander (bend in the river), where the current flows most strongly.

Hydraulic action occurs as a river tumbles over a waterfall to crash onto the rocks below. It will lead to the formation of a plunge pool below the waterfall. The hydraulic action of ocean waves and turbulent currents forces air into rock cracks, and therefore brings about erosion by cavitation.

hydrology study of the location and movement of inland water, both frozen and liquid, above and below ground. It is applied to major civil engineering projects such as irrigation schemes, dams, and hydroelectric power, and in planning water supply.

hygrometer instrument for measuring the humidity, or water vapour content, of a gas (usually air). A wet and dry bulb hygrometer consists of two vertical thermometers, with one of the bulbs covered in absorbent cloth dipped into water. As the water evaporates, the bulb cools, producing a temperature difference between the two thermometers. The amount of evaporation, and hence cooling of the wet bulb, depends on the relative humidity of the air.

Ice Age, Little period of particularly severe winters that gripped northern Europe between the 13th and 17th centuries. Contemporary writings and paintings show that Alpine glaciers were much more extensive than at present, and rivers such as the Thames, which do not ice over today, were so frozen that festivals could be held on them.

iceberg floating mass of ice, about 80% of which is submerged, rising sometimes to 100 m/300 ft above sea level. Glaciers that reach the coast become extended into a broad foot; as this enters the sea, masses break off and drift towards temperate latitudes, becoming a danger to shipping.

isobar line drawn on maps and weather charts linking all places with the same atmospheric pressure (usually measured in millibars). When used in weather forecasting, the distance between the isobars is an indication of the barometric gradient (the rate of change in pressure).

jet stream narrow band of very fast wind (velocities of over 150 kph/95 mph) found at altitudes of 10–16 km/6–10 mi in the upper troposphere or lower stratosphere. Jet streams usually occur about the latitudes of the ◊Westerlies (35°–60°).

karst landscape characterized by remarkable surface and underground forms, created as a result of the action of water on permeable limestone. Limestone is soluble in the weak acid of rainwater. Erosion takes place most swiftly along cracks and joints in the limestone and these open up into gullies called grikes. The rounded blocks left upstanding between them are called clints.

Kuroshio, or **Japan Current**, warm ocean current flowing from Japan to North America.

lagoon coastal body of shallow salt water, usually with limited access to the sea. The term is normally used to describe the shallow sea area cut off by a coral reef or barrier islands.

levée naturally formed raised bank along the side of a river channel. When a river overflows its banks, the rate of flow is less than that in the channel, and silt is deposited on the banks. With each successive flood the levée increases in size so that eventually the river may be above the surface of the surrounding flood plain. Notable levées are found on the lower reaches of the Mississippi in the USA and the Po in Italy.

longshore drift movement of material along a beach. When a wave breaks obliquely, pebbles are carried up the beach in the direction of the wave (swash). The wave draws back at right angles to the beach (backwash), carrying some pebbles with it. In this way, material moves in a zigzag fashion along a beach

magnetic storm in meteorology, a sudden disturbance affecting the Earth's magnetic field, causing anomalies in radio transmissions and magnetic compasses. It is probably caused by sunspot activity.

Mariana Trench lowest region on the Earth's surface; the deepest part of the sea floor. The trench is 2,400 km/1,500 mi long and is situated 300 km/200 mi east of the Mariana Islands, in the northwestern Pacific Ocean. Its deepest part is the gorge known as the Challenger Deep, which extends 11,034 m/36,210 ft below sea level.

marsh low-lying wetland. Freshwater marshes are common wherever groundwater, surface springs, streams, or run-off cause frequent flooding or more or less permanent shallow water. A marsh is alkaline whereas a bog is acid. Marshes develop on inorganic silt or clay soils. Large marshes with standing water throughout the year are commonly called swamps. Near the sea, salt marshes may form.

meander loop-shaped curve in a mature river flowing sinuously across flat country. As a river flows, any curve in its course is accentuated by the current. On the outside of the curve the velocity, and therefore the erosion, of the current is greatest. Here the river cuts into the outside bank, producing a **cutbank** or **river cliff** and the river's deepest point,

or **thalweg**. On the curve's inside the current is slow and deposits any transported material, building up a gentle slip-off slope. As each meander migrates in the direction of its cutbank, the river gradually changes its course across the flood plain.

meridian half a great circle drawn on the Earth's surface passing through both poles and thus through all places with the same longitude. Terrestrial longitudes are usually measured from the Greenwich Meridian.

midnight sun constant appearance of the Sun (within the Arctic and Antarctic circles) above the horizon during the summer.

monsoon wind pattern that brings seasonally heavy rain to South Asia; it blows towards the sea in winter and towards the land in summer. The monsoon may cause destructive flooding all over India and Southeast Asia from April to September, leaving thousands of people homeless each year.

moraine rocky debris or till carried along and deposited by a glacier. Material eroded from the side of a glaciated valley and carried along the glacier's edge is called a **lateral moraine**; that worn from the valley floor and carried along the base of the glacier is called a **ground moraine**. Rubble dropped at the snout of a melting glacier is called a **terminal moraine**.

oasis area of land made fertile by the presence of water near the surface in an otherwise arid region. The occurrence of oases affects the distribution of plants, animals, and people in the desert regions of the world.

ozone layer thin layer of the gas ozone in the upper atmosphere that shields the Earth from harmful ultraviolet rays.

Peru Current (formerly known as the **Humboldt Current**) cold ocean current flowing north from the Antarctic along the west coast of South America to southern Ecuador, then west. It reduces the coastal temperature, making the western slopes of the Andes arid because winds are already chilled and dry when they meet the coast.

physical weathering type of weathering involving such effects as: frost wedging, in which water trapped in a crack in a rock expands on freezing and splits the rock; sand blasting, in which exposed rock faces are worn away by sand particles blown by the wind; and soil creep, in which soil particles gradually move downhill under the influence of gravity.

playa temporary lake in a region of interior drainage. Such lakes are common features in arid desert basins fed by intermittent streams. The streams bring dissolved salts to the lakes, and when the lakes shrink during dry spells, the salts precipitate as evaporite deposits.

rainbow arch in the sky displaying the colours of the spectrum; it is formed by the refraction and reflection of the Sun's rays through rain or mist. The countless drops of water in the air each act as a tiny prism, splitting sunlight into the wavelengths of the spectrum.

ridge of high pressure elongated area of high atmospheric pressure extending from an ◊anticyclone. On a synoptic weather chart it is shown as a pattern of lengthened ◊isobars. The weather under a ridge of high pressure is the same as that under an anticyclone.

river capture diversion of the headwaters of one river into a neighbouring river. River capture occurs when a stream is carrying out rapid ◊headward erosion (backwards erosion at its source). Eventually the stream will cut into the course of a neighbouring river, causing the headwaters of that river to be diverted, or 'captured'.

roche moutonée outcrop of tough bedrock having one smooth side and one jagged side, found on the floor of a ◊glacial trough. It may be up to 40 m/132 ft high. A roche moutonée is a feature of glacial erosion – as a glacier moved over its surface, ice and debris eroded its upstream side by ◊corrasion, smoothing it and creating long scratches or striations. On the sheltered downstream side fragments of rock were plucked away by the ice, causing it to become steep and jagged.

season period of the year having a characteristic climate. The change in seasons is mainly due to the change in attitude of the Earth's axis in relation to the Sun, and hence the position of the Sun in the sky at a particular place. In temperate latitudes four seasons are recognized: spring, summer, autumn (fall), and winter. Tropical regions have two seasons – the wet and the dry. Monsoon areas around the Indian Ocean have three seasons: the cold, the hot, and the rainy.

solar pond natural or artificial 'pond', such as the Dead Sea, in which salt becomes more soluble in the Sun's heat. Water at the bottom becomes saltier and hotter, and is insulated by the less salty water layer at the top. Temperatures at the bottom reach about 100°C/212°F and can be used to generate electricity.

solifluction downhill movement of topsoil that has become saturated with water. Solifluction is common in periglacial environments (those bordering glacial areas) during the summer months, when the frozen topsoil melts to form an unstable soggy mass. This may then flow slowly downhill under gravity to form a **solifluction lobe** (a tonguelike feature).

solution process by which the minerals in a rock are dissolved in water. It is also referred to as **corrosion**. Solution is one of the processes of erosion as well as weathering (in which the dissolution of rock occurs without transport of the dissolved material). An example of this is when weakly acidic rainfall causes carbonation.

Solution commonly affects limestone and chalk, both forms of calcium carbonate. It can occur in coastal environments along with corrasion and hydraulic action, producing features like the White Cliffs of Dover, as well as fluvial (river) environments, like the one that formed Cheddar Gorge. In groundwater environments of predominantly limestone, solution produces **karst topography**, forming features such as sink holes, caves, and limestone pavement.

spit ridge of sand or shingle projecting from the land into a body of water. It is deposited by waves carrying material from one direction to another across the mouth of an inlet (◊longshore drift). Deposition in the brackish water behind a spit may result in the formation of a salt marsh.

spring natural flow of water from the ground, formed at the point of intersection of the water table and the ground's surface. The source of water is rain that has percolated through the overlying rocks. During its underground passage, the water may have dissolved mineral substances that may then be precipitated at the spring (hence, a mineral spring).

stack isolated pillar of rock that has become separated from a headland by coastal erosion. It is usually formed by the collapse of an ◊arch. Further erosion will reduce it to a stump, which is exposed only at low tide.

stalactite and stalagmite cave structures formed by the deposition of calcite dissolved in ground water. **Stalactites** grow downwards from the roofs or walls and can be icicle-shaped, straw-shaped, curtain-shaped, or formed as terraces. **Stalagmites** grow upwards from the cave floor and can be conical, fir-cone-shaped, or resemble a stack of saucers. Growing stalactites and stalagmites may meet to form a continuous column from floor to ceiling.

tidal wave common name for a tsunami.

till, or **boulder clay**, deposit of clay, mud, gravel, and boulders left by a glacier. It is unsorted, with all sizes of fragments mixed up together, and shows no stratification; that is, it does not form clear layers or beds.

trade wind prevailing wind that blows towards the Equator from the northeast and southeast. Trade winds are caused by hot air rising at the Equator and the consequent movement of air from north and south to take its place. The winds are deflected towards the west because of the Earth's west-to-east rotation. The unpredictable calms known as the ◊doldrums lie at their convergence.

tropics the area between the tropics of Cancer and Capricorn, defined by the parallels of latitude approximately 23°30[prime] north and south of the Equator. They are the limits of the area of Earth's surface in which the Sun can be directly overhead. The mean monthly temperature is over 20°C/68°F.

tufa, or **travertine**, soft, porous, limestone rock, white in colour, deposited from solution from carbonate-saturated ground water around hot springs and in caves.

turbidity current gravity-driven current in air, water, or another fluid resulting from accumulation of suspended material, such as silt, mud, or volcanic ash, and imparting a density greater than the surrounding fluid. Marine turbidity currents originate from tectonic movement, storm waves, tsunamis (tidal waves), or earthquakes and move rapidly downwards, like underwater avalanches, leaving distinctive deposits called **turbidites**. They are thought to be one of the mechanisms by which submarine canyons are formed.

typhoon violent revolving storm, a hurricane in the western Pacific Ocean.

wadi in arid regions of the Middle East, a steep-sided valley containing an intermittent stream that flows in the wet season.

weathering see ◊biological weathering, ◊chemical weathering, and ◊physical weathering.

Westerlies prevailing winds from the west that occur in both hemispheres between latitudes of about 35° and 60°. Unlike the ◊trade winds, they are very variable and produce stormy weather.

World Meteorological Organization agency, part of the United Nations since 1950, that promotes the international exchange of weather information through the establishment of a world-wide network of meteorological stations. It was founded as the International Meteorological Organization in 1873, and its headquarters are now in Geneva, Switzerland.

Further Reading

Abrahamson, D E (ed) *The Challenge of Global Warming* (1989)

Asimov, I *Atom Journey Across the Sub-Atomic Cosmos* (1992)

Barry, Roger G, and Chorley, Richard J *Atmosphere, Weather and Climate* (1992)

Bolin, B, and others (eds) *The Greenhouse Effect, Climate and Ecosystems* (1986)

Bridges, Edwin Michael *World Geomorphology* (1990)

Bryant, Edward *Climate: Process and Change* (1997)

Calder, Nigel *The Weather Machine* (1974)

Calow, Peter, and Petts, Geoffrey E *The Rivers Handbook: Hydrological and Ecological Principles* (1992–94; in two volumes)

Carter, Bill *Coastal Environments: An Introduction to the Physical, Ecological, and Cultural Systems of Coastlines* (1988)

Cline, W *The Economics of Global Warming* (1992)

Collard, Roy *The Physical Geography of Landscape* (1992)

Couper, A *The Times Atlas of the Oceans* (1989)

Dawson, Alastair *Ice Age Earth* (1992)

Elsom, D M *Atmospheric Pollution: A Global Problem* (1992)

Embleton, Clifford *Glaciers and Glacial Erosion* (1972)

Goudie, A *Climate* (1997)

Hambrey, M J *Glacial Environments* (1994)

Hardy, Ralph; Wright, Peter; Gribbin, John; and Kington,

John *The Weather Book* (1982)

Hare, F Kenneth *The Restless Atmosphere* (1953–1967; various reprints)

Hester, Nigel *The Living River* (1991)

Holford, Ingrid *The Guinness Book of Weather Facts and Feats* (1982, 2nd edition)

Knighton, David *Fluvial Forms and Processes* (1998, 2nd edition)

Komar, Paul D *CRC Handbook of Coastal Processes and Erosion* (1983)

Leopold, Luna Bergese *Water, Rivers, and Creeks* (1997)

MacInnis, Joseph *Saving the Oceans* (1996)

Manley, Gordon *Climate and the British Scene* (1975)

Moore, George W *Speleology: Caves and the Cave Environment* (1997)

Robinson, Peer John, and Henderson-Sellers, Ann *Climatology* (1998, 2nd edition)

Schneider, Stephen H *Encyclopedia of Climate and Weather* (1996)

Sharp, Robert Phillip *Living Ice: Understanding Glaciers and Glaciation* (1988)

Sparks, Bruce Wilfred *Geomorphology* (1986, 3rd edition)

Walker, H J, and Graban, W E *The Evolution of Geomorphology: A Nation-By-Nation Summary of Development* (1993)

World Survey of Climatology (1969– series)

ENVIRONMENT AND ECOLOGY

Ecology is the study of the relationship between organisms and the environments in which they live, including all living and nonliving components. The chief environmental factors governing the distribution of plants and animals are temperature, humidity, soil, light intensity, day length, food supply, and interaction with other organisms. The term was coined by the German biologist Ernst Haeckel in 1866. Ecology may be concerned with individual organisms (for example, behavioural ecology or feeding strategies), with populations (for example, population dynamics), or with entire communities (for example, competition between species for access to resources in an ecosystem, or predator–prey relationships). Applied ecology is concerned with the management and conservation of habitats and the consequences and control of pollution.

elements of an ecosystem

The narrow zone that supports life on our planet, the **biosphere,** is limited to the waters of Earth, a fraction of its crust, and the lower regions of the atmosphere. The biosphere is made up of all of Earth's ecosystems. It is affected by external forces such as the Sun's rays, which provide energy, the gravitational effects of the Sun and Moon, and cosmic radiations.

ecosystems

An ecosystem is an integrated unit consisting of a **community of living organisms** – bacteria, animals, and plants – and the physical **environment** – air, soil, water, and climate – that they inhabit. Individual organisms interact with each other and with their environment, or habitat, in a series of relationships that depends on the flow of energy and nutrients through the system. These relationships are usually complex and finely balanced, and in theory natural ecosystems are self-sustaining. However, major changes to an ecosystem, such as climate change, overpopulation, or the removal of a species, may threaten the system's sustainability and result in its eventual destruction. For instance, the removal of a major carnivore predator can result in the destruction of an ecosystem through overgrazing by herbivores.

system diversity Ecosystems can be identified at different scales or levels, ranging from macrosystems to microsystems. The global ecosystem (the ecosphere), for instance, consists of all the Earth's physical features – its land, oceans, and enveloping atmosphere (the geosphere) – together with all the biological organisms living on Earth; on a smaller scale, a freshwater-pond ecosystem includes the plants and animals living in the pond, the pond water and all the substances dissolved or suspended in that water, together with the rocks, mud, and decaying matter at the bottom of the pond. Thus, ecosystems can contain smaller systems and be contained within larger ones.

equilibrium and succession The term 'ecosystem' was first coined in 1935 by a British ecologist, Arthur Tansley (1871– 1955), to refer to a community of interdependent organisms with dynamic relationships between consumer levels, that can respond to change without altering the basic characteristics of the system. For example, cyclical changes in populations can sometimes result in large fluctuations in the numbers of a species, and are a fundamental part of most ecosystems, but because of the interdependence of all the components, any change in one part of its nature will result in a reaction in other parts of the community. In most cases, these reactions work to restore the equilibrium or balance of nature, but on occasions the overall change or disruption will be so great as to alter the system's balance irreversibly and result in the replacement of one type of ecosystem with another. Where this occurs as a natural process, as in the colonization of barren rock by living organisms, or the conversion of forest to grassland as a result of fires

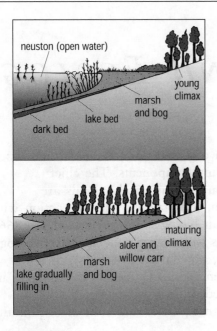

succession The succession of plant types along a lake. As the lake gradually fills in, a mature climax community of trees forms inland from the shore. Extending out from the shore, a series of plant communities can be discerned with small, rapidly growing species closest to the shore.

started by lightning, it is known as ecosystem development or ecological **succession**.

carrying capacity The maximum number of organisms that can be supported by a particular environment is called its carrying capacity. If the carrying capacity of an ecosystem is exceeded by overpopulation, there will be insufficient resources and one or more species will decline until an equilibrium, or balance of nature, is restored. Similarly, if the number of species in an environment is less than the carrying capacity, the population will tend to increase until it balances the available resources. Human interference frequently causes disruption to the carrying capacity of an area, for instance by the establishment of too many grazing animals on grassland, the over-culling of a species, or the introduction of a non-indigenous species into an area.

food chains

One of the main features of an ecosystem is its biodiversity. Members are usually classified either as **producers** (those that can synthesize the organic materials they need from inorganic compounds in the environment) or **consumers** (those that are unable to manufacture their own food directly from these

sources and depend upon producers to meet their needs). Thus plants, as producers, capture energy originating from the Sun through the process of photosynthesis and absorb nutrients from the soil and water; these stores of energy and nutrients then become available to the consumers – for example, they are passed via the herbivores that eat the plants, to the carnivores that feed on the herbivores. The sequence in which energy and nutrients pass through the system is known as a **food chain**, and the energy levels within a food chain are termed **trophic levels**. At each stage of assimilation, energy is lost by consumer functioning, and so there are always far fewer consumers at the end of the chain. This can be represented diagrammatically by a pyramid with the primary producers at the base; it is termed a pyramid of numbers. At each level of the chain, nutrients are returned to the soil through the decomposition of excrement and dead organisms, thus becoming available once again to plants, and completing a cycle crucial to the stability and survival of the ecosystem.

natural cycles

The biosphere is an interactive layer incorporating elements of the atmosphere, hydrosphere, and lithosphere, and involving natural cycles such as carbon, nitrogen, and water cycles. Human interference in the Earth's natural systems, which began with the transition of human society from nomadic hunter–gatherer tribes into settled agriculture-based communities, gathered pace in the 18th and 19th centuries with the coming of the agricultural revolution and the Industrial Revolution. The technological revolution of the 20th century, with its programmes of industrialization and urbanization and intensive farming practices, has become a major threat, damaging the planet's ecosystems at all levels.

carbon cycle The carbon cycle is the circulation of carbon through the natural world. The movements of carbon from one reservoir to another (carbon fluxes) controls the amount of carbon dioxide in the atmosphere. Carbon dioxide is released into the atmosphere by volcanism, metamorphism, decay of organic matter, burning of fossil fuels, and respiration by living organisms. It is drawn out of the atmosphere by chemical weathering of silicate rock, reactions between the atmosphere and the oceans (eventually forming carbonate rocks like limestone), burial of organic matter, and plant photosynthesis. The oceans absorb 25–40% of all carbon dioxide released into the atmosphere.

Today, the carbon cycle is in danger of being disrupted by the increased burning of fossil fuels and the burning of large tracts of tropical forests, both of

carbon cycle The carbon cycle is necessary for the continuation of life. Since there is only a limited amount of carbon in the Earth and its atmosphere, carbon must be continuously recycled if life is to continue. Other chemicals necessary for life – nitrogen, sulphur, and phosphorus, for example – also circulate in natural cycles.

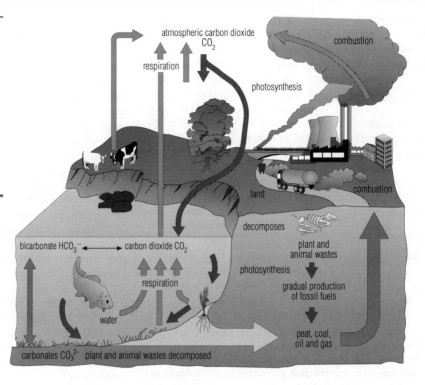

which can release large amounts of carbon into the atmosphere, contributing to the greenhouse effect (described in Chapter 7).

nitrogen cycle The nitrogen cycle is the process of nitrogen passing through the ecosystem. Nitrogen, in the form of inorganic compounds (such as nitrates) in the soil, is absorbed by plants and turned into organic compounds (such as proteins) in plant tissue. A proportion of this nitrogen is eaten by herbivores, with some of this in turn being passed on to the carnivores, which feed on the herbivores. The nitrogen is ultimately returned to the soil both as excrement and when organisms die and are converted back to inorganic form by decomposers.

Although about 78% of the atmosphere is nitrogen, this cannot be used directly by most organisms. However, certain bacteria and cyanobacteria (blue-green algae) are capable of nitrogen fixation. Some nitrogen-fixing bacteria live mutually with leguminous plants (peas and beans) or other plants (for example, alder), where they form characteristic nodules on the roots. The presence of such plants increases the nitrate content, and hence the fertility, of the soil.

water or hydrological cycle The water cycle, or hydrological cycle, is the natural circulation of water through the biosphere. It is a complex system involving a number of physical and chemical processes (such as evaporation, precipitation, and infiltration) and stores (such as rivers, oceans, and soil).

soil

Soil, an environmental feature of many ecosystems, is the loose covering of broken rocky material and decaying organic matter overlying the bedrock of the Earth's surface. It is comprised of minerals, organic matter (called humus) derived from decomposed plants and organisms, living organisms, air, and water. Soils differ according to climate, parent material, rainfall, relief of the bedrock, and the proportion of organic material. The study of soils is **pedology**.

A soil can be described in terms of its **soil profile**, that is, a vertical cross-section from ground-level to the bedrock on which the soils sits. The profile is

---

earthworms

Earthworms produce nitrous oxide as a by-product from digesting soil nitrates and nitrites. Soil with earthworms contains five times as much nitrous oxide as soil without.

---

soil

To remember the chief constituents of soil:

All hairy men will buy razors.

(air, humus, mineral salts, water, bacteria, rock particles)

soil

A handful of soil contains up to 5,000 species of bacteria.

divided into layers called horizons. The **A horizon**, or **topsoil**, is the uppermost layer, consisting primarily of humus and living organisms and some mineral material. Most soluble material has been leached from this layer or washed down to the B horizon. The **B horizon**, or **subsoil**, is the layer where most of the nutrients accumulate and is enriched in clay minerals. The **C horizon** is the layer of weathered parent material at the base of the soil.

Two common soils are the podzol and the chernozem soil. The **podzol** is common in coniferous forest regions where precipitation exceeds evaporation. The A horizon consists of a very thin litter of organic material producing a poor humus. The needles of coniferous trees take a long time to decompose. The relatively heavy precipitation causes leaching of minerals, as nutrients are washed downwards.

Chernozem soils are found in grassland regions, where evaporation exceeds precipitation. The A horizon is rich in humus due to decomposition of a thick litter of dead grass at the surface. Minerals and moisture migrate upward due to evaporation, leaving the B and A horizons enriched.

The organic content of soil is widely variable, ranging from zero in some desert soils to almost 100% in peats.

Soils influence the type of agriculture employed in a particular region – light well-drained soils favour arable farming, whereas heavy clay soils give rise to lush pasture land.

soil erosion   Soil erosion, the wearing away and redistribution of the Earth's soil layer, is caused by the action of water, wind, and ice, and also by improper methods of agriculture. If unchecked, soil erosion results in the formation of deserts (desertification). It has been estimated that 20% of the world's cultivated topsoil was lost between 1950 and 1990.

If the rate of erosion exceeds the rate of soil formation (from rock and decomposing organic matter), then the land will become infertile. The removal of forests (deforestation) or other vegetation often leads to serious soil erosion, because plant roots bind soil, and without them the soil is free to wash or blow away, as in the American dust bowl. The effect is worse on hillsides, and there has been devastating loss of soil where forests have been cleared from mountainsides, as in Madagascar.

Improved agricultural practices such as contour ploughing are needed to combat soil erosion. Windbreaks, such as hedges or strips planted with coarse grass, are valuable, and organic farming can reduce soil erosion by as much as 75%.

climate

Climate is another environmental factor in an ecosystem. A climate is the combination of weather conditions at a particular place over a period of time – usually a minimum of 30 years. A classification of climate encompasses the averages, extremes, and frequencies of all meteorological elements such as temperature, atmospheric pressure, precipitation, wind, humidity, and sunshine, together with the factors that influence them. The primary factors involved are: the Earth's rotation and latitudinal effects; ocean currents; large-scale movements of wind belts and air masses over the Earth's surface; temperature differences between land and sea surfaces; and topography. **Climatology**, the scientific study of climate, includes the construction of computer-generated models, and considers not only present-day climates, their effects and their classification, but also long-term climate changes, covering both past climates (paleoclimates) and future predictions. Climatologists are especially concerned with the influence of human activity on climate change.

climate classification   The word 'climate' comes from the Greek *klima*, meaning an inclination or slope (referring to the angle of the Sun's rays, and thus latitude) and the earliest known classification of climate was that of the ancient Greeks, who based their system on latitudes. In recent times, many different systems of classifying climate have been devised, most of which follow that formulated by the German climatologist Wladimir Köppen (1846–1940) in 1900. These systems use vegetation-based classifications such as desert, tundra, and rainforest. Classification by air mass is used in conjunction with this method. This idea was first introduced in 1928 by the Norwegian meteorologist Tor Bergeron, and links the climate of an area with the movement of the air masses it experiences.

In the 18th century, the British scientist George Hadley developed a model of the general circulation of atmosphere based on convection. He proposed a simple pattern of cells of warm air rising at the Equator and descending at the poles. In fact, due to the rotation of the Earth, there are three such cells (now called Hadley cells) in each hemisphere. The first two of these consist of air that rises at the Equator and sinks at latitudes north and south of the tropics; the second two exist at the mid-latitudes where the rising air from the sub-tropics flows towards the cold air masses of the third pair of cells circulating from the two polar regions. Thus, in this model, there are six main circulating cells of air above ground producing seven terrestrial zones – three rainy regions (at the Equator and the temperate latitudes) resulting from the moisture-laden rising air, interspersed and bounded by four dry or desert regions (at the poles and subtropics) resulting from the dry descending air.

prevailing winds Regions are also affected by different wind systems, which result from the rotation of the Earth and the uneven heating of surface air. As air is heated by radiation from the Sun, it expands and rises, and cooler air flows in to take its place. This movement of air produces belts of prevailing winds. Because of the rotation of the Earth, these are deflected to the right in the northern hemisphere and to the left in the southern hemisphere. This effect, which is greater in the higher latitudes, is known as the **Coriolis effect**. Because winds transport heat and moisture, they affect the temperature, humidity, precipitation, and cloudiness of an area. As a result, regions with different prevailing wind directions may have different climates.

temperature variations The amount of heat received by the Earth from the Sun varies with latitude and season. In equatorial regions, there is no large seasonal variation in the mean daily temperature of the air near the ground, while in the polar regions, temperatures in the long winters, when there is little incoming solar radiation, fall far below summer temperatures. The temperature of the sea, and of the air above it, varies little in the course of day or night, whereas the surface of the land is rapidly cooled by lack of solar radiation. Similarly, the annual change of temperature is relatively small over the sea but much greater over the land. Thus, continental areas are colder than maritime regions in winter, but warmer in summer. This results in winds blowing from the sea which, relative to the land, are warm in winter and cool in summer, while winds originating from the central parts of continents are hot in summer and cold in winter. On average, air temperature drops with increasing land height at a rate of 1°C/1.8°F per 90 m/300 ft – so that even in equatorial regions, the tops of mountains can be snow-covered throughout the year.

climate changes Changes in climate can occur naturally or as a result of human activity. Natural variations can be caused by fluctuations in the amount of solar radiation reaching the Earth – for example, sunspot activity is thought to produce changes in the Earth's climate. Variations in the Earth's orbit around the Sun, known as the **Milankovitch hypothesis**, is also thought to bring about climatic changes. Natural events on the surface of the Earth, such as volcanic

eruptions and the effects of **El Niño** (the warm ocean surge of the Peru Current, occurring in the eastern Pacific, off South America), can result in temporary climate changes on a worldwide scale, sometimes extending over several months or even years. Human influences on the climate range from localized effects such as cloud seeding to produce rain, to the global effects of acid rain from industrial emissions; pollution; and the destruction of the rainforests. The study of past climates, **paleoclimatogy**, involves the investigation of climate changes from the ice ages to the beginning of instrumental recording in the 19th century.

biomes

A biome is a broad natural assemblage of plants and animals shaped by common patterns of vegetation and climate. The varied distribution of land and sea areas produce the complexity of the general circulation of the atmosphere; this, in turn, directly affects the distribution of climate. Centred on the Equator is a belt of tropical rainforest, which may be either constantly wet or monsoonal (having wet and dry seasons in each year). On either side of this is a belt of savanna, with lighter seasonal rainfall and less dense vegetation, largely in the form of grasses. Then there is usually a transition through steppe (semi-arid) to desert (arid), with a further transition through steppe to what is termed Mediterranean climate with dry summers. Beyond this is the moist, temperate climate of middle latitudes, and then a zone of cold climate with moist winters. Where the desert extends into middle latitudes, however, the zones of Mediterranean and moist temperate climates are missing, and the transition is from desert to a cold climate with moist winters. In the extreme east of Asia a cold climate with dry winters extends from about 70° N to 35° N. The polar caps have tundra and glacial climates, with little or no precipitation.

rainforest Rainforests are areas of dense forest usually found on or near the Equator where the climate is hot and wet. Moist air brought by the converging tradewinds rises because of the heat, producing heavy rainfall. Over half the tropical rainforests are in Central and South America, primarily the lower Amazon and the coasts of Ecuador and Columbia. The rest are in Southeast Asia (Malaysia, Indonesia, and New Guinea) and in West Africa and the Congo.

Tropical rainforests once covered 14% of the Earth's land surface, but are now being destroyed at an increasing rate as their valuable timber is harvested and the land cleared for agriculture, causing problems

Countries Losing Greatest Areas of Forest

Source: Food and Agriculture Organization (FAO) of the United Nations

1990–95

Rank	Country	Area of lost forest (hectares)
1	Brazil	2,554,000
2	Indonesia	1,084,000
3	Congo, Democratic Republic of	740,000
4	Bolivia	581,000
5	Mexico	508,000
6	Venezuela	503,000
7	Malaysia	400,000
8	Myanmar	387,000
9	Sudan	353,000
10	Thailand	329,000
11	Paraguay	327,000
12	Tanzania	323,000
13	Zambia	264,000
14=	Colombia	262,000
	Philippines	262,000
16	Angola	237,000
17	Peru	217,000
18	Ecuador	189,000
19	Cambodia	164,000
20	Nicaragua	151,000
21	Laos	148,000
22	Vietnam	135,000
23	Papua New Guinea	133,000
24	Madagascar	130,000
25	Cameroon	129,000
26	Central African Republic	128,000
27	Nigeria	121,000
28	Afghanistan	118,000
29	Ghana	117,000
30	Mozambique	116,000
31	Guinea-Bissau	114,000
32	Honduras	102,000
33	Chad	94,000
34	Gabon	91,000
35	Argentina	89,000
36	Hong Kong, China	87,000
37	Guatemala	82,000
38	Guinea	75,000
39	Botswana	71,000
40	Panama	64,000

and harbour at least 40% of the Earth's species (plants and animals).

The vegetation in tropical rainforests typically includes an area of dense forest called **selva**. A **canopy** formed by high branches of tall trees provides shade for lower layers. Underneath there is an intermediate layer of shorter trees and tree roots, and lianas. The ground cover consists mainly of mosses and ferns. The lack of **seasonal rhythm** causes adjacent plants to flower and shed leaves simultaneously. Chemical weathering and leaching take place in the iron-rich soil due to the high temperatures and humidity.

Rainforests comprise some of the most complex and diverse ecosystems on the planet, deriving their energy from the sun and photosynthesis. The trees are the main **producers**. Herbivores such as insects, caterpillars, and monkeys feed on the plants and trees and in turn are eaten by the carnivores, such as ocelots and puma. Fungi and bacteria, the primary **decomposers**, break down the dead material from the plants, herbivores, and carnivores with the help of heat and humidity. This decomposed material provides the **nutrients** for the plants and trees.

The rainforest ecosystem helps to regulate global weather patterns – especially by taking up carbon dioxide (CO_2) from the atmosphere – and stabilizes the soil. Rainforests provide the bulk of the oxygen needed for plant and animal respiration. When deforestation occurs, the microclimate of the mature forest disappears; soil erosion and flooding become major problems since rainforests protect the shallow tropical soils. Once an area is cleared it is very difficult for shrubs and bushes to re-establish because the soils are poor in nutrients. This causes problems for plans to convert rainforests into agricultural land – after two or three years the crops fail and the land is left bare. Clearing of the rainforests may lead to a global warming of the atmosphere, and contribute to the greenhouse effect.

Tropical rainforests are characterized by a great diversity of species, usually of tall broad-leaved evergreen trees, with many climbing vines and ferns, some of which are a main source of raw materials for medicines. A tropical forest, if properly preserved, can yield medicinal plants, oils (from cedar, juniper, cinnamon, sandalwood), spices, gums, resins (used in inks, lacquers, linoleum), tanning and dyeing materials, forage for animals, beverages, poisons, green manure, rubber, and animal products (feathers, hides, honey). Other types of rainforest include montane, upper montane or cloud, mangrove, and subtropical.

Traditional ways of life in tropical rainforests are disappearing. The practice of **shifting cultivation**, in which small plots of forest are cultivated and

of deforestation. Although by 1991 over 50% of the world's rainforest had been removed, they still comprise about 50% of all growing wood on the planet,

desert slopes, Argentina
Eroded desert slopes near Tres Cruces in the Quebrada da Humahuaca in Jujuy province, Argentina.

abandoned after two or three harvests, is being replaced by slash-and-burn cultivation on such a large scale that the rainforests cannot regenerate. As a result, hunting and gathering as a way of life also is becoming less viable. In the last 30 years Central America has lost almost two-thirds of its rainforests to cattle ranching.

savanna
Savannas, or savannahs, are extensive open tropical grasslands with scattered trees and shrubs. They cover large areas of Africa, North and South America, and northern Australia. The soil is acidic and sandy and generally considered suitable only as pasture for low-density grazing. The name was originally given by Spaniards to the treeless plains of the tropical South American prairies. Most of North America's savannas have been built over.

A new strain of rice suitable for savanna conditions was developed in 1992. It not only grew successfully under test conditions in Colombia but also improved pasture quality so grazing numbers could be increased twentyfold.

steppe
Steppe is the temperate grasslands of Europe and Asia. Sometimes the term refers to other temperate grasslands and semi-arid desert edges.

desert
Deserts are arid areas with sparse vegetation (or, in rare cases, almost no vegetation). Soils are poor, and many deserts include areas of shifting sands. Deserts can be either hot or cold. Almost 33% of the Earth's land surface is desert, and this proportion is increasing.

The **tropical desert** belts of latitudes from 5° to 30° are caused by the descent of air that is heated over the warm land and therefore has lost its moisture. Other natural desert types are the **continental deserts**, such as the Gobi, which are too far from the sea to receive any moisture; **rain-shadow deserts**, such as California's Death Valley, which lie in the lee of mountain ranges, where the ascending air drops its rain only on the windward slopes; and **coastal deserts**, such as the Namib, where cold ocean currents cause local dry air masses to descend. Desert surfaces are usually rocky or gravelly, with only a small proportion being covered with sand. Deserts can be created by changes in climate, or by the human-aided process of desertification.

Characteristics common to all deserts include irregular rainfall of less than 250 mm/19.75 in per year, very high evaporation rates often 20 times the annual precipitation, and low relative humidity and cloud cover. Temperatures are more variable; tropical deserts have a big diurnal temperature range and very high daytime temperatures (58°C/136.4°F has been recorded at Azizia in Libya), whereas mid-latitude deserts have a wide annual range and much lower winter temperatures (in the Mongolian desert the mean temperature is below freezing point for half the year).

Desert soils are infertile, lacking in humus and generally grey or red in colour. The few plants capable of surviving such conditions are widely spaced, scrubby, and often thorny. Long-rooted plants (phreatophytes) such as the date palm and musquite commonly grow along dry stream channels. Salt-loving plants

Largest Deserts in the World

Desert	Location	Area[1]	
		sq km	sq mi
Sahara	northern Africa	9,065,000	3,500,000
Gobi	Mongolia/northeastern China	1,295,000	500,000
Patagonian	Argentina	673,000	260,000
Rub al-Khali	southern Arabian peninsula	647,500	250,000
Chihuahuan	Mexico/southwestern USA	362,600	140,000
Taklimakan	northern China	362,600	140,000
Great Sandy	northwestern Australia	338,500	130,000
Great Victoria	southwestern Australia	338,500	130,000
Kalahari	southwestern Africa	260,000	100,000
Kyzyl Kum	Uzbekistan	259,000	100,000
Thar	India/Pakistan	259,000	100,000
Sonoran	Mexico/southwestern USA	181,300	70,000
Simpson	Australia	103,600	40,000
Mojave	southwestern USA	51,800	20,000

[1]desert areas are very approximate because clear physical boundaries may not occur

(halophytes) such as saltbushes grow in areas of highly saline soils and near the edges of playas (dry saline lakes). Others, such as the xerophytes, are drought-resistant and survive by remaining leafless during the dry season or by reducing water losses with small waxy leaves. They frequently have shallow and widely branching root systems and store water during the wet season (for example, succulents and cacti with pulpy stems).

woodland

Woodlands are areas in which trees grow more or less thickly; they are generally smaller than forests.

Temperate climates, with four distinct seasons a year, tend to support a mixed woodland habitat, with some conifers but mostly broad-leaved and deciduous trees, shedding their leaves in autumn and regrowing them in spring. In the Mediterranean region and parts of the southern hemisphere, the trees are mostly evergreen.

Temperate woodlands grow in the zone between the cold coniferous forests and the tropical forests of the hotter climates near the Equator. They develop in areas where the closeness of the sea keeps the climate mild and moist.

Old woodland can rival tropical rainforest in the

evergreen forest Evergreen forest with Indian elephants in the foreground, in Periyar national park, Kerala, India. Its canopy is multistorey, with older trees rising above younger ones, indicating that it is an old established forest and a rich habitat containing many species of plants and animals.

number of species it supports, but most of the species are hidden in the soil. A study in Oregon, USA, in 1991 found that the soil in a single woodland location contained 8,000 arthropod species (such as insects, mites, centipedes, and millipedes), compared with only 143 species of reptile, bird, and mammal in the woodland above.

In England in 1900, about 2.5% of land was woodland, compared to about 3.4% in the 11th century. An estimated 33% of ancient woodland has been destroyed since 1945.

The trees determine the character of the wood. Sometimes a single species dominates, as in a pine or beech wood, but there is often a mixture of two or more co-dominants, as in a mixed oak and ash wood. Beneath the tree canopy there is frequently a layer of shrubs and beneath these the herbaceous plants. Woodland herbs grow in shady conditions and are adapted in various ways to make the best possible use of the available sunlight. The woodland floor provides moist conditions in which mosses and liverworts thrive and many fungi grow in the soil or on rotting wood. The trees themselves provide habitats for climbing plants, mosses, liverworts, lichens, fungi, and microscopic algae.

wetland

A permanently wet land area or habitat is called a wetland. Wetlands include areas of marsh, fen, bog, flood plain, and shallow coastal areas. Wetlands are extremely fertile. They provide warm, sheltered waters for fisheries, lush vegetation for grazing livestock, and an abundance of wildlife. Estuaries and seaweed beds are more than 16 times as productive as the open ocean.

The term is often more specifically applied to a naturally flooding area that is managed for agriculture or wildlife. A water meadow, where a river is expected to flood grazing land at least once a year thereby replenishing the soil, is a traditional example.

The largest area of tidal wetlands in Japan, a 3,550-hectare/8,800-acre area in Isahaya Bay, near Nagasaki, is threatened by a 7-km/4.3-mi-long barrier erected across the bay in 1997 as part of a flood control project. The wetlands, cut off from the tidal waters, are drying out rapidly and the ecosystem is likely to die if the barrier is not removed. The mudflats are home to 282 species of birds, crustaceans, and fish, including 21 endangered species.

taiga

Taiga is the Russian name for the forest zone south of the tundra, found across the northern hemisphere. Here, dense forests of conifers (spruces and hemlocks), birches, and poplars occupy glaciated regions punctuated with cold lakes, streams, bogs, and marshes.

Winters are prolonged and very cold, but the summer is warm enough to promote dense growth.

The varied fauna and flora are in delicate balance because the conditions of life are so precarious. This ecology is threatened by mining, forestry, and pipeline construction.

tundra

Tundra is a region of high latitude almost devoid of trees, resulting from the presence of permafrost. The vegetation consists mostly of grasses, sedges, heather, mosses, and lichens. Tundra stretches in a continuous belt across northern North America and Eurasia. Tundra is also used to describe similar conditions at high altitudes. The term was originally applied to the topography of part of northern Russia, but is now used for all such regions.

environmental issues

Environmental issues are matters relating to the detrimental effects of human activity on the biosphere, their causes, and the search for possible solutions. Since the Industrial Revolution, the demands made by both the industrialized and developing nations are increasingly affecting the balance of the Earth's natural resources. Over a period of time, some of these resources are renewable – trees can be replanted, soil nutrients can be replenished – but many resources, such as fossil fuels and minerals, are nonrenewable and in danger of eventual exhaustion. In addition, humans are creating many other problems which may endanger not only their own survival, but also that of other species. For instance, deforestation and air pollution are not only damaging and radically altering many natural environments, they are also affecting the Earth's climate by adding to the greenhouse effect and global warming, while water pollution is seriously affecting aquatic life, including fish populations, as well as human health.

public awareness Concern for the environment is not just a late-20th-century issue. In England, the first smoke-abatement law dates from 1273, while in 1306 the burning of coal was prohibited in London because of fears of air pollution. However, the inspiration for the creation of the modern environmental movements came about from the publication in 1962 of the US biologist Rachel Carson's book *Silent Spring*, in which she attacked the indiscriminate use of pesticides. This, combined with the increasing affluence of Western nations, which allowed people to look beyond their everyday needs, triggered an awareness of environmental issues on a global scale and resulted in the formation of the Green movement. In the mid-1960s, the detection of chlorofluorocarbons (CFCs) in the atmosphere by the British scientist James Lovelock led to a realization of the damaging effects of ozone depletion and added to public concern for the environment, as did his development of the Gaia hypothesis, which views the Earth as a single integrated and self-sustaining organism.

Gaia hypothesis The concept of the Earth as a single organism, or ecosystem, was formulated in the mid-1960s by the British scientist James Lovelock, while researching the possibility of life on Mars for NASA's space programme. The Gaia hypothesis, named after an ancient Greek earth goddess, views the planet as a self-regulating system in which all the individual elements coexist in a symbiotic relationship. In developing this hypothesis, Lovelock realized that the damage effected by humans on many of the Earth's ecosystems was posing a threat to the viability of the planet itself. The effects of this disruption are now becoming apparent in the changing landscapes and climates of almost every region or biome of the planet. They can be seen in the desertification of the Sahel, the shrinking of the Aral Sea in Central Asia, the destruction of tropical rainforests, and the creation of the holes in the ozone layer over the Arctic and Antarctic because of the pollution of the atmosphere with greenhouse gases. These gases include carbon and sulphur-dioxide emissions from the combustion of fossil fuels, and the CFCs widely used as propellants and refrigerants. The thinning of the protective ozone layer surrounding the planet, with its consequent threat of global warming, affects the basic functioning of energy flow within every ecosystem of the planet, from microorganisms to the ecosphere itself.

Friends of the Earth Home Page
http://www.foe.co.uk/
The Friends of the Earth Home Page appeals for raised awareness of environmental issues with masses of information and tips for action from Friends of the Earth. The site hosts lengthy accounts of several campaigns undertaken by FoE on climate, industry, transport, and sustainable development. It also maintains an archive of press releases from FoE on some of the most controversial environmental problems encountered in the course of last year around the world.

---

international measures In 1972 the United Nations Environment Program (UNEP) was formed to coordinate international measures for monitoring and protecting the environment, and in 1985 the Vienna Convention for the Protection of the Ozone Layer, which promised international cooperation in research, monitoring, and the exchange of information on the problem of ozone depletion, was signed by 22 nations. Discussions arising out of this convention led to the

Carbon Dioxide Emissions for OECD Countries

Source: Organization for Economic Cooperation and Development (OECD)
The data are from energy use only; waste in international marine bunkers is excluded.
(Data as of January 1997.)

Rank	Country	Carbon dioxide per capita (tonnes)
1	Luxembourg	27
2	USA	20
3=	Australia	16
	Canada	16
5=	Belgium	12
	Denmark	12
	Finland	12
	Czech Republic	12
9=	Netherlands	11
	Germany	11
11=	UK	10
	Ireland, Republic of	10
13=	Iceland	9
	Japan	9
	Poland	9
16=	New Zealand	8
	Norway	8
18=	Austria	7
	Italy	7
	South Korea	7
	Greece	7
22=	Sweden	6
	Spain	6
	Switzerland	6
	France	6
	Hungary	6
27	Portugal	5
28	Mexico	4
29	Turkey	2

signing in 1987 of the Montréal Protocol. In 1992, representatives of 178 nations met in Rio de Janeiro for the United Nations Conference on Environment and Development. Known as the 'Earth Summit', this was one of the most important conferences ever held on environmental issues. UN members signed agreements on the prevention of global warming and the preservation of forests and endangered species, along with many other environmental issues.

global warming
Average global temperature has increased approximately 1°F/0.5°C over the past century. Global

temperature has been highly variable in Earth history and many fluctuations in global temperature have occurred in historical times, but this most recent episode of warming coincides with the spread of industralization, prompting the hypothesis that it is the result of an accelerated greenhouse effect caused by atmospheric pollutants, especially carbon dioxide gas. Recent melting and collapse of the Larsen Ice Shelf, Antarctica, is a consequence of global warming. Melting of ice is expected to raise sea level in the coming decades.

In addition to a rise in average global temperature, global warming has caused seasonal variations to be more pronounced in recent decades. Examples are the most severe winter on record in the eastern USA in 1976–77 and the record heat waves in the Netherlands and Denmark the following year.

Global Warming
http://pooh.chem.wm.edu/chemWWW/courses/
chem105/projects/group1/page1.html
Interesting step-by-step explanation of the chemistry behind global warming. There is information on the causes of global warming, the environmental effects, and the social and economic consequences. The views of those who challenge the assertion that the world is warming up are also presented.

Natural, perhaps chaotic, climatic variations have not been ruled out as the cause of the current global rise in temperature, but scientists are still assessing the likely influence of anthropogenic (human-made) pollutants. Assessing the impact of humankind on global climate is complicated by the natural variability of global temperature on both geological and human time scales. The present episode of global warming has thus far still left England approximately 1°C *cooler* than during the peak of the so-called

Medieval Warm Period (1000 to 1400 AD). The latter was part of a purely natural climatic fluctuation on a global scale. With respect to historical times, the interval between the Medieval Warm Period and the rise in temperatures we see today was unusually cold throughout the world.

deforestation

Deforestation is the destruction of forest for timber, fuel, charcoal burning, and clearing for agriculture and extractive industries, such as mining, without planting new trees to replace those lost (reafforestation) or working on a cycle that allows the natural forest to regenerate. Deforestation causes fertile soil to be blown away or washed into rivers, leading to soil erosion, drought, flooding, and loss of wildlife. It may also increase the carbon dioxide content of the atmosphere and intensify the greenhouse effect, because there are fewer trees absorbing carbon dioxide from the air for photosynthesis.

Many people are concerned about the rate of deforestation as great damage is being done to the habitats of plants and animals. Deforestation ultimately leads to famine, and is thought to be partially responsible for the flooding of lowland areas – for example, in Bangladesh – because trees help to slow down water movement.

Soil degradation and erosion are becoming as serious as the loss of the rainforest. It is estimated that more than 10% of the world's soil lost a large amount of its natural fertility during the latter half of the 20th century. Some of the worst losses are in Europe, where 17% of the soil is damaged by human activity such as mechanized farming and fallout from acid rain. Mexico and Central America have 24% of soil highly degraded, mostly as a result of deforestation.

Deforestation of Tropical and Temperate Forests Worldwide

Source: Food and Agriculture Organization (FAO) of the United Nations
(In hectares.)

Region	Forest area	Total change 1990–95	Annual change 1990	Annual change 1995	Annual change (%)
Africa	538,978,000	520,237,000	−18,741,000	−3,748,000	−0.7
Asia	490,812,000	474,172,000	−16,640,000	−3,328,000	−0.7
Europe1	144,044,000	145,988,000	1,944,000	389,000	0.3
Former USSR1	813,381,000	816,167,000	2,786,000	557,000	0.1
North and Central America	537,898,000	536,529,000	−1,369,000	−274,000	−0.1
Oceania	91,149,000	90,695,000	−454,000	−91,000	−0.1
South America	894,466,000	870,594,000	−23,872,000	−4,774,000	−0.5

[1] No tropical forests exist in these regions, thus totals represent only temperate forests.

The first deforestation occurred more than 2,000 years ago in areas surrounding the Mediterranean as wood was increasingly used for fuel, building materials, and the construction of ships. Throughout the next two millennia most of the woodland in Europe was destroyed as demands increased. The current wave of deforestation in the tropics dates back only 30 years, but even so has reduced the amount of intact forest ecosystem from 34% of total land in the affected areas to 12%. Deforestation in the tropics is especially serious because such forests do not regenerate easily and because they are such a rich source of biodiversity. The Food and Agriculture Organization (FAO) of the United Nations estimated that the deforestation rate for the tropics between 1990 and 1995 was 130,000 sq km/50,200 sq mi a year.

desertification

The destruction of fertile topsoil, and consequent soil erosion, as a result of human activity is becoming a worldwide problem. About 25% of the planet's land surface is now thought to be at risk due to increased demand from expanding populations. This damage and destruction results not only from increased demand for food, but also as a result of changes in agricultural practices. Desertification of vast areas, such as in the Sahel in northern Africa, have resulted from the replacement of traditional farming methods in these marginal lands for the present-day cultivation of cash crops such as groundnuts and cotton. The consequence has been that the soil has lost its fertility and the land has become arid. Similarly, changes in agricultural practices produced the dust bowl in the USA in the 1930s and, more recently, the move from mixed farming to arable and the removal of hedges in order to enlarge fields for the use of modern agricultural machinery has resulted in the loss of topsoil in the large areas of the English Fenlands.

Sahel The Sahel (Arabic *sahil* 'coast') is the marginal area to the south of the Sahara, Africa, from Senegal to Somalia, which experiences desert-like conditions during periods of low rainfall. The desertification is partly due to climatic fluctuations but has also been caused by the pressures of a rapidly expanding population, which has led to overgrazing and the destruction of trees and scrub for fuelwood. In recent years many famines have taken place in the area.

The average rainfall in the Sahel ranges from 100 mm/4 in to 500 mm/20 in per year, but the rainfall over the past 30 years has been significantly below average. The resulting famine and disease are further aggravated by civil wars. The areas most affected are Ethiopia and the Sudan.

pollution

The term pollution refers to the harmful effect on the environment of by-products of human activity, principally industrial and agricultural processes – for example, noise; smoke; car emissions; chemical and radioactive effluents in air, seas, and rivers; pesticides; radiation; sewage; and household waste. Pollution contributes to the greenhouse effect.

Environmental pollution can have natural sources, for example volcanic activity, which can cause major air pollution or water pollution and destroy flora and fauna. In terms of environmental issues, however, environmental pollution relates to human actions. The extraction, transportation, utilization, and waste products of nonrenewable resources for fuel all give rise to pollutants of one form or another. The effects of these pollutants can have consequences not only for the local environment, but also at a global level.

Pollution control involves higher production costs for the industries concerned, but failure to implement adequate controls may result in irreversible environmental damage and an increase in the incidence of diseases such as cancer. Radioactive pollution results from inadequate nuclear safety.

Many people think of air, water, and soil pollution as distinctly separate forms of pollution. However, each part of the global ecosystem – air, water, and soil – depends upon the others, and upon the plants and animals living within the environment. Thus, pollution that might appear to affect only one part of the environment is also likely to affect other parts.

air pollution Air pollution is any contamination of the atmosphere caused by the discharge, accidental or deliberate, of a wide range of toxic airborne substances. Often the amount of the released substance is relatively high in a certain locality, so the harmful effects become more noticeable. The cost of preventing any discharge of pollutants into the air is prohibitive, so attempts are more usually made to reduce the amount of discharge gradually and

> ### hydrocarbons
> In summer, in a city the size of Melbourne, Australia, the mowing of lawns increases atmospheric hydrocarbons by 10%.

to disperse it as quickly as possible by using a very tall chimney, or by intermittent release.

The emission of vehicle exhausts or acidic gases from a power plant might appear to harm only the surrounding atmosphere. But once released into the air they are carried by the prevailing winds, often for several hundred kilometres, before being deposited as acid rain.

Sulphur Oxides, Nitrogen Oxides, and Carbon Dioxide Emissions for OECD Countries

Source: Organization for Economic Cooperation and Development (OECD)
(Data as of January 1997. N/A = not available.)

Country	Emissions per capita			Country	Emissions per capita		
oxides	Sulphur oxides (kg)	Nitrogen dioxide (kg)	Carbon (tonnes)	oxides	Sulphur oxides (kg)	Nitrogen dioxide (kg)	Carbon (tonnes)
Australia	N/A	N/A	16	South Korea	34	26	7
Austria	9	23	7	Luxembourg	26	N/A	27
Belgium	25	35	12	Mexico	N/A	N/A	4
Canada	91	68	16	Netherlands	9	35	11
Czech Republic	125	36	12	New Zealand	N/A	43	8
Denmark	30	53	12	Norway	8	51	8
Finland	22	54	12	Poland	68	29	9
France	17	26	6	Portugal	27	26	9
Germany	37	27	11	Spain	53	31	6
Greece	50	33	7	Sweden	11	45	6
Hungary	72	18	6	Switzerland	5	19	6
Iceland	29	81	9	Turkey	29	9	2
Ireland, Republic of	53	37	10	UK	47	38	10
Italy	25	37	7	USA	63	74	20
Japan	7	12	9				

smog Smog is a natural fog containing impurities, mainly nitrogen oxides and volatile organic compounds (VOCs) from domestic fires, industrial furnaces, certain power stations, and internal-combustion engines (petrol or diesel). It can cause substantial illness and loss of life, particularly among chronic bronchitics, and damage to wildlife.

acid rain Acid rain is acidic precipitation thought to be caused principally by the release into the atmosphere of sulphur dioxide (SO_2) and oxides of nitrogen, which dissolve in pure rainwater making it acidic. Sulphur dioxide is formed by the burning of fossil fuels, such as coal, that contain high quantities of sulphur; nitrogen oxides are contributed from various industrial activities and from car exhaust fumes.

Acidity is measured on the pH scale, where the value of 0 represents liquids and solids that are completely acidic and 14 represents those that are highly alkaline. Distilled water is neutral and has a pH of 7. Normal rain has a value of 5.6. It is slightly acidic due to the presence of carbonic acid formed by the mixture of carbon dioxide (CO_2) and rainwater. Acid rain has values of 5.6 or less on the pH scale.

Acid deposition occurs not only as wet precipitation (mist, snow, or rain), but also comes out of the atmosphere as dry particles (**dry deposition**) or is absorbed directly by lakes, plants, and masonry as gases. Acidic

gases can travel over 500 km/310 mi a day so acid rain can be considered an example of transboundary pollution.

Acid rain is linked with damage to and the death of forests and lake organisms in Scandinavia, Europe, and eastern North America. It also results in damage to buildings and statues. US and European power stations that burn fossil fuels release about 8 g/0.3 oz of sulphur dioxide and 3 g/0.1 oz of nitrogen oxides per kilowatt-hour. According to the UK Department of the Environment figures, emissions of sulphur dioxide from power stations would have to be decreased by 81% in order to arrest damage.

A 1993–95 Irish study found Manchester to be the European city worst affected by acid rain. Its rainfall was the most acidic, causing building stone to be destroyed faster there than anywhere else in Europe (other cities included Athens, Copenhagen, Amsterdam, Padua, and Donegal). The other British test site, Liphook, Hampshire, also fared badly.

The main effect of acid rain is to damage the chemical balance of soil, causing leaching of important minerals including magnesium and aluminium. Plants living in such soils, particularly conifers, suffer from mineral loss and become more prone to infection. The minerals from the soil pass into lakes and rivers, disturbing aquatic life, for instance, by damaging the gills of young fish and killing plant life. Lakes affected by acid rain are virtually clear due to the absence of green plankton. Lakes and rivers suffer more direct damage as well because they become acidified by rainfall draining directly from their catchment.

Owing to reductions in sulphur emissions, the amount of sulphur deposited per hectare of farm land has fallen from around 50 kg/110 lb in 1979 to 10 kg/22 lb in 1995. According to German research in 1995, although acid rain is harmful to crops, it also provides essential sulphur. There has been an increased incidence of sulphur-deficiency diseases in plants in the period 1990–95.

catalytic converter A catalytic converter is a device fitted to the exhaust system of a motor vehicle in order to reduce toxic emissions from the engine. It converts harmful exhaust products to relatively harmless ones by passing the exhaust gases over a mixture of catalysts coated on a metal or ceramic honeycomb (a structure that increases the surface area and therefore the amount of active catalyst with which the exhaust gases will come into contact). **Oxidation catalysts** (small amounts of precious palladium and platinum metals) convert hydrocarbons (unburnt fuel) and carbon monoxide into carbon dioxide and water, but do not affect nitrogen oxide emissions. **Three-way catalysts** (platinum and rhodium metals) convert nitrogen oxide gases into nitrogen and oxygen.

Over the lifetime of a vehicle, a catalytic converter can reduce hydrocarbon emissions by 87%, carbon monoxide emissions by 85%, and nitrogen oxide emissions by 62%, but will cause a slight increase in the amount of carbon dioxide emitted. Catalytic converters are standard in the USA, where a 90% reduction in pollution from cars was achieved without loss of engine performance or fuel economy. Only 10% of cars in Britain had catalytic converters in 1993.

Catalytic converters are destroyed by emissions from leaded petrol and work best at a temperature of 300°C. The benefits of catalytic converters are offset by any increase in the number of cars in use.

Unfortunately, catalytic converters emit nitrous oxide, which is itself a potent greenhouse gas. Vehicle emissions of nitrous oxide increased in the USA by 49% in the period 1990–96 reflecting an increase in the number of vehicles fitted with catalytic converters (vehicles with catalytic converters emit five times the amount of nitrous oxide of those without).

pesticides
Chemicals used in farming, gardening, or indoors to combat pests cause a number of pollution problems through spray drift onto surrounding areas, pollution of the groundwater, direct contamination of users or the public, and as residues on food. Pesticides are of

three main types: **insecticides** (to kill insects), **fungicides** (to kill fungal diseases), and **herbicides** (to kill plants, mainly those considered weeds). The safest pesticides include those made from plants, such as the insecticides pyrethrum and derris.

Pyrethrins are safe and insects do not develop resistance to them. Their impact on the environment is very small as the ingredients break down harmlessly.

More potent are synthetic products, such as chlorinated hydrocarbons. These products, including DDT and dieldrin, are highly toxic to wildlife and often to human beings, so their use is now restricted by law in some areas and is declining. Safer pesticides such as malathion are based on organic phosphorus compounds, but they still present hazards to health. The aid organization Oxfam estimates that pesticides cause about 10,000 deaths worldwide every year.

> **Pesticide poisoning**
>
> There are around 4,000 cases of acute pesticide poisoning a year in the UK.

Pesticides were used to deforest SE Asia during the Vietnam War, causing death and destruction to the area's ecology and lasting health and agricultural problems.

marine pollution Rubbish deposited in the sea by waste disposal from land or ships.

Many pesticides remain in the soil, since they are not biodegradable, and are then passed on to foods. In the UK, more than half of all potatoes sampled in 1995 contained residues of a storage pesticide; seven different pesticides were found in carrots, with concentrations up to 25 times the permitted level; and 40% of bread contained pesticide residues.

water pollution

This consists of any addition to fresh or sea water that disrupts biological processes or causes a health hazard. Common pollutants include nitrates, pesticides, and sewage, although a huge range of industrial contaminants, such as chemical by-products and residues created in the manufacture of various goods, also enter water – legally, accidentally, and through illegal dumping.

oil spills The leakage of oil from an ocean-going tanker, a pipeline, or other source can kill all shore life, clogging up the feathers of birds and suffocating other creatures. At sea, toxic chemicals leach from the oil into the water below, poisoning sea life. Mixed with dust, the oil forms globules that sink to the seabed, poisoning sea life there as well. Oil spills are broken up by the use of detergents but such chemicals can themselves damage wildlife. The annual spillage of oil is 8 million barrels a year. At any given time tankers are carrying 500 million barrels.

groundwater pollution

In 1980 the UN launched the 'Drinking Water Decade', aiming for cleaner water for all by 1990. However, in 1994 it was estimated that approximately half of all people in the developing world did not have safe drinking water. A 1995 World Bank report estimated that some 10 million deaths in developing countries were caused annually by contaminated water.

Major Oilspills Throughout the World

(Data as of January 1997.)

Date	Location	Description	Amount	
			tonnes	millions of gallons
March 1967	off Cornwall, England	grounding of Torrey Canyon	118,000	35.4
June 1968	off South Africa	Hull failure of World Glory	37,000	11.0
December 1972	Gulf of Oman	collision of Sea Star with another ship	103,500	31.0
May 1976	La Caruña, Spain	grounding of the Urquioia	60–70,000	18.0–21.0
December 1976	Nantucket, Massachusetts, USA	grounding of Argo Merchant	25,000	7.5
February 1977	mid-Pacific	Haiwaiian Patriot develops leak and catches fire	100,000	30.0
April 1977	North Sea	blow-out of well in Ekofisk oil field	270,000	81.0
March 1978	Portsall, Brittany, France	grounding of the Amoco Cadiz	226,000	68.0
June 1979	Gulf of Mexico	blow-out of well in Ixtoc 1 oil field	600,000	180.0
July 1979	off Tobago, Caribbean	collision of the Atlantic Empress and Aegean Captain	370,000	111.0
February 1983	Persian Gulf	blow-out of well in Nowruz oil field	600,000	180.0
August 1983	off Cape Town, South Africa	fire on board the Castillo de Beliver	250,000	75.0
September 1985	Delaware River, Delaware, USA	grounding of Grand Eagle	1,500	0.5
January 1988	Floreffe, Pennsylvania, USA	collapsing of Ashland oil storage tank	2,400–2,500	0.7–0.8
March 1989	Prince William Sound, off Alaskan Coast	grounding of Exxon Valdez	37,000	11.0
June 1989	Canary Islands	fire on board the Kharg 5	65,000	19.5
January 1991	Sea Island Terminal of Persian Gulf	deliberate release of oil by Iraqi troops at end of Persian Gulf War	799,120	240.0
January 1993	Shetland Islands, Scotland	grounding of Braer	130,000	39.0
August 1993	Tampa Bay, Florida, USA	collision of two barges and a Philippine freighter	984	0.3
January 1996	Pembrokeshire coastline of Wales, British Isles	grounding of Sea Empress	>100,000	19.0
January 1996	south shore of Rhode Island, USA	grounding of the tugboat Scandia and North Cape tanker it was towing	1,000	0.3

ozone depletion

Ozone (O_3), a highly reactive pale-blue gas, is formed when the molecule of the stable form of oxygen (O_2) is split by ultraviolet radiation or electrical discharge. It forms the ozone layer in the upper atmosphere, which protects life on Earth from ultraviolet rays, a cause of skin cancer.

Chemicals that destroy the ozone in the stratosphere are called **ozone depleters**. Most ozone depleters are chemically stable compounds containing chlorine or bromine, which remain unchanged for long enough to drift up to the upper atmosphere. The best known are chlorofluorocarbons (CFCs), but many other ozone depleters are known, including halons, used in some fire extinguishers; methyl chloroform and carbon tetrachloride, both solvents; some CFC substitutes; and the pesticide methyl bromide.

chlorofluorocarbons

Chlorofluorocarbons (CFCs) are a class of synthetic chemicals that are odourless, nontoxic, nonflammable, and chemically inert. The first CFC was synthesized in 1892, but no use was found for it until the 1920s. Since then their stability and apparently harmless properties have made CFCs popular as propellants in aerosol cans, as refrigerants in refrigerators and air conditioners, as degreasing agents, and in the manufacture of foam packaging. They are partly responsible for the destruction of the ozone layer. In June 1990 representatives of 93 nations, including the UK and the USA, agreed to phase out production of CFCs and various other ozone-depleting chemicals by the end of the 20th century.

The Chemistry of the Ozone Layer
http://pooh.chem.wm.edu/chemWWW/courses/chem105/projec
ts/group2/page1.html
Interesting step-by-step introduction to the ozone layer for those wishing to understand the chemistry of ozone depletion, the role of chlorofluorocarbons, the consequences of increased radiation for life on Earth, and actions to tackle the problem.

When CFCs are released into the atmosphere, they drift up slowly into the stratosphere, where, under the influence of ultraviolet radiation from the Sun, they react with ozone (O_3) to form free chlorine (Cl) atoms and molecular oxygen (O_2), thereby destroying the ozone layer that protects Earth's surface from the Sun's harmful ultraviolet rays. The chlorine liberated during ozone breakdown can react with still more ozone, making the CFCs particularly dangerous to the environment. CFCs can remain in the atmosphere for more than 100 years.

Replacements for CFCs are being developed, and research into safe methods for destroying existing CFCs is being carried out. In 1996 it was reported that chemists at Yale University, USA, had developed a process for breaking down freons and other gases containing CFCs into nonhazardous compounds.

resources

A resource is a commodity that can be used to satisfy human needs. Resources can be categorized into **human resources**, such as labour, supplies, and skills, and **natural resources**, such as climate, fossil fuels, and water. Natural resources may be further divided into nonrenewable resources and renewable resources.

A **nonrenewable resource** is a natural resource, such as coal or oil, that takes thousands or millions of years to form naturally and can therefore not be replaced once it is consumed. The main energy sources used by humans are nonrenewable. **Renewable** natural resources are replaced by natural processes in a reasonable amount of time. Soil, water, forests, plants, and animals are all renewable resources as long as they are properly conserved. Solar, wind, wave, and geothermal energies are based on renewable resources.

water supply

The water supply is the distribution of water for domestic, municipal, or industrial consumption. Water supply in sparsely populated regions usually comes from underground water rising to the surface in natural springs, supplemented by pumps and wells. Urban sources are deep artesian wells, rivers, and reservoirs, usually formed from enlarged lakes or dammed and flooded valleys, from which water is conveyed by pipes, conduits, and aqueducts to filter beds. As water seeps through layers of shingle, gravel, and sand, harmful organisms are removed and the water is then distributed by pumping or gravitation through mains and pipes.

water treatment Often other substances are added to the water, such as chlorine and fluoride; aluminium sulphate, a clarifying agent, is the most widely used chemical in water treatment. In towns, domestic and municipal (road washing, sewage) needs account for about 135 l/30 gal per head each day. In coastal desert areas, such as the Arabian peninsula, desalination plants remove salt from sea water. The Earth's waters, both fresh and saline, have been polluted by industrial and domestic chemicals, some of which are toxic and others radioactive.

drought A period of prolonged dry weather can disrupt water supply and lead to drought. The area of the world subject to serious droughts, such as the Sahara, is increasing because of destruction of forests,

lake A lake near Morondava in Madagascar. The water lilies are *Nymphaea stellata* and the trees in the distance, growing on dry land, are baobabs *Adansonia grandidieri*. Most lakes are the focal point of an ecosystem, drawing together a broad range of animal and plant life.

overgrazing, and poor agricultural practices. A World Bank report in 1995 warned that a global crisis was imminent: chronic water shortages were experienced by 40% of the world's population, notably in the Middle East, northern and sub-Saharan Africa, and central Asia. In 1997, 1.4 billion people (25% of the

The Largest Reservoirs by Volume in the World

Source: Institute of Civil Engineers, London

Reservoir	Location	Year completed	Volume cubic m (millions)	cubic yd (millions)
Owen Falls[1]	Uganda	1954	204,800	266,240
Bratsk	Russia	1964	169,000	219,700
High Aswan	Egypt	1970	162,000	210,600
Kariba	Zimbabwe/Zambia	1959	160,368	208,478
Akosombo	Ghana	1965	147,960	192,348
Daniel Johnson	Canada	1968	141,851	184,406
Guri	Venezuela	1986	135,000	175,500
Krasnoyarsk	Russia	1967	73,300	95,290
W A C Bennett	Canada	1967	70,309	91,402
Zeya	Russia	1978	68,400	88,920
Cabora Bassa	Mozambique	1974	63,000	81,900
La Grande 2 Barrage	Canada	1978	61,715	80,230
La Grande 3 Barrage	Canada	1981	60,020	78,026
Ust–Ilim	Russia	1977	59,300	77,090
Boguchany	Russia[2]		58,200	75,660
Kuibyshev	Russia	1955	58,000	75,400
Serra de Mesa	Brazil	–[2]	54,400	70,720
Caniapiscau Barrage K A 3	Canada	1980	53,790	69,927
Bukhtarma	Kazakhstan	1960	49,800	64,740
Atatürk	Turkey	1992	48,700	63,310

[1] This volume is not fully obtainable by the dam: the major part of it is the natural capacity of the lake. Owen Falls is not the largest artificial lake.
[2] Under construction.

Water Scarcity

Water-scarce countries, 1992, with projections for 2010[1] per capita renewable water supplies

region/country	1992	2010	change (%)
	(cubic metres/cubic feet per person)		
Africa			
Algeria	730/25,779	500/17,657	-32
Botswana	710/25,072	420/14,832	-41
Burundi	620/21,894	360/12,713	-42
Cape Verde	500/17,657	290/10,241	-42
Djibouti	750/26,485	430/15,185	-43
Egypt	30/1,059	20/706	-33
Kenya	560/19,775	330/11,653	-41
Libya	160/5,650	100/3,531	-38
Mauritania	190/6,709	110/3,884	-42
Rwanda	820/28,957	440/15,538	-46
Tunisia	450/15,891	330/11,653	-27
Middle East			
Israel	330/11,653	250/8,828	-24
Jordan	190/6,709	110/3,884	-42
Qatar	40/1,412	30/1,059	-25
Saudi Arabia	140/4,944	70/2,472	-50
Syria	550/19,422	300/10,594	-45
United Arab Emirates	120/4,238	60/2,119	-50
Yemen	240/8,475	130/4,591	-46
Other			
Barbados	170/6,003	170/6,003	0
Belgium	840/29,663	870/30,723	+4
Hungary	580/20,482	570/20,129	-2
Malta	80/2,825	80/2,825	0
Netherlands	660/23,307	600/21,188	-9
Singapore	210/7,416	190/6,709	-10
Additional countries by 2010			
Malawi	1,030/36,373	600/21,188	-42
Sudan	1,130/39,904	710/25,072	-37
Morocco	1,150/40,610	830/29,310	-28
South Africa	1,200/42,376	760/26,838	-37
Oman	1,250/44,142	670/23,660	-46
Somalia	1,390/49,086	830/29,310	-40
Lebanon	1,410/49,792	980/34,607	-30
Niger	1,690/59,680	930/32,841	-45

[1] Countries with per capita renewable water supplies of less than 1,000 cubic meters per year; does not include water flowing in from neighbouring countries .

population) had no access to safe drinking water.

In 1992, the town of Cgungungo in the Atacama Desert, South America, began using a system to convert water from fog as a public water supply. The system supplies 11,000 l/2,400 gal of water per day.

waste

Materials that are no longer needed and are discarded are termed waste. Examples are household waste, industrial waste (which often contains toxic chemicals), medical waste (which may contain organisms

Waste Generated in OECD Countries

Source: Organization for Economic Corporation and Development (OECD)
(Data as of January 1997. N = nil or negligible. N/A = not available.)

Country	Industrial waste per unit of GDP[1] (tonnes per $ millions)	Municipal waste (kg per capita)	Nuclear waste per unit of energy[2] (tonnes per Mtoe[3])
Australia	125	N/A	N
Austria	53	430	N
Belgium	76	470	2.0
Canada	N/A	630	7.4
Czech Republic	232	230	1.1
Denmark	24	520	N
Finland	201	410	2.2
France	100	560	5.1
Germany	60	360	1.4
Greece	44	310	N
Hungary	104	420	N/A
Iceland	2	660	N
Ireland, Republic of	N/A	N/A	N
Italy	22	470	N
Japan	61	410	1.9
South Korea	67	390	1.8
Luxembourg	164	530	N
Mexico	70	320	0.1
Netherlands	30	540	0.2
New Zealand	N/A	N/A	N
Norway	39	620	N
Poland	117	290	N/A
Portugal	N/A	350	N
Spain	28	370	1.6
Sweden	95	440	4.8
Switzerland	5	380	3.0
Turkey	86	390	N
UK	59	350	5.2
USA	142	730	1.2

[1] GDP = gross domestic product.

[2] Wastes from spent fuel in nuclear power plants, in tonnes of heavy metal per million tonnes of oil equivalent (primary energy supply).

[3] Mtoe = million tonnes.

that cause disease), and nuclear waste (which is radioactive). By recycling, some materials in waste can be reclaimed for further use. In 1990 the industrialized nations generated 2 billion tonnes of waste. In the USA, 40 tonnes of solid waste are generated annually per person, roughly twice as much as in Europe or Japan.

There has been a tendency to increase the amount of waste generated per person in industrialized countries, particularly through the growth in packaging and disposable products, creating a 'throwaway society'.

In Britain, the average person throws away about ten times their own body weight in household refuse each year. Collectively the country generates about 50 million tonnes of waste per year. In principle, over 50% of UK household waste could be recycled, although less then 5% is currently recovered.

recycling Some industrial and household waste (such as paper, glass, and some metals and plastics) can be reprocessed so that the materials can be reused. This saves expenditure on scarce raw materials, slows down the depletion of nonrenewable resources, and helps to reduce pollution. Aluminium is frequently recycled because of its value and special properties that allow it to be melted down and re-pressed without loss of

Municipal Waste Generation for OECD Countries: Top 20

Source: Organization for Economic Cooperation and Development (OECD)
(Data as of January 1997.)

Rank	Country	Municipal waste per capita (kg)	Rank	Country	Municipal waste per capita (kg)
1	USA	730	11	Sweden	440
2	Iceland	660	12	Austria	430
3	Canada	630	13	Hungary	420
4	Norway	620	14=	Finland	410
5	France	560		Japan	410
6	Netherlands	540	16=	Turkey	390
7	Luxembourg	530		South Korea	390
8	Denmark	520	18	Switzerland	380
9=	Italy	470	19	Spain	370
	Belgium	470	20	Germany	360

quality, unlike paper and glass, which deteriorate when recycled.

The USA recycles only around 25% of its waste (1998), compared to around 33% in Japan. However, all US states encourage or require local recycling programmes to be set up. It was estimated in 1992 that 4,000 cities collected waste from 71 million people for recycling. Most of these programmes were set up between 1989 and 1992. Around 33% of newspapers, 22% of office paper, 64% of aluminium cans, 3% of plastic containers, and 20% of all glass bottles and jars were recycled.

Most British recycling schemes are voluntary, and rely on people taking waste items to a central collection point. However, some local authorities now ask householders to separate waste before collection, making recycling possible on a much larger scale.

In 1998 Britain was recycling only 6% of its domestic waste, compared with 25% for the USA and up to 70% in parts of Canada. Britain recycles only 27% of its glass compared with 80% for the Netherlands; and 16% of steel packaging compared with 67% for Germany.

Recycle City
http://www.epa.gov/recyclecity/
Recycle City is a child-friendly site of the US Environmental Protection Agency designed to help people to live more ecologically. The site includes a host of fun games and activities to encourage children to think about waste disposal issues.

---

energy
Fossil fuels, such as coal, oil, and natural gas, formed from the fossilized remains of plants that lived hundreds of millions of years ago, are nonrenewable

---

Bioenergy Information Network
http://www.esd.ornl.gov/bfdp/
US government site about the possibilities and research into producing energy from rapidly replaced trees and grasses. There are sections on 'biopower basics', 'the biopower industry', and a library of photos and video stills.

---

resources and will eventually run out. Extraction of coal and oil causes considerable environmental pollution, and burning coal contributes to problems of acid rain and the greenhouse effect. **Alternative energy**, or **renewable energy**, is energy from sources that are renewable and ecologically safe, as opposed to sources that are nonrenewable with toxic by-products, such as fossil fuels and uranium (for nuclear power). The most important alternative energy source is flowing water, harnessed as hydroelectric power (described in Chapter 15, together with other alternative, renewable, energy technologies). Other sources include the oceans' tides and waves (harnessed by tidal power stations, floating booms, and other technologies; see Chapter 15), wind power (harnessed by windmills and wind turbines), the Sun (solar energy), and the heat trapped in the Earth's crust (geothermal energy).

Geothermal Technologies Programme
http://www.eren.doe.gov/geothermal/
Interesting geothermal home page of the US Department of Energy. There is a description of various geothermal energy sources and technologies being developed to exploit them. There is information about a number of energy authorities across the world tapping geothermal energy.

---

geothermal energy Geothermal energy is extracted for heating and electricity generation from natural steam, hot water, or hot dry rocks in the Earth's crust. It is an important source of energy in volcanically active areas such as Iceland and New Zealand. (See Chapter 15 for more details.)

solar energy Solar energy is derived from the Sun's radiation. The amount of energy falling on just 1 sq km/0.4 sq mi is about 4,000 megawatts, enough to heat and light a small town. In one second the Sun gives off 13 million times more energy than all the electricity used in the USA in one year. **Solar heaters** have industrial or domestic uses. They usually consist of a black (heat-absorbing) panel containing pipes through which air or water, heated by the Sun, is circulated, either by thermal convection or by a pump.

Solar energy may also be harnessed indirectly using **solar cells** (photovoltaic cells) made of panels of semiconductor material (usually silicon), which generate electricity when illuminated by sunlight. Although it is difficult to generate a high output from solar energy compared to sources such as nuclear or fossil fuels, it is a major nonpolluting and renewable energy source used as far north as Scandinavia as well as in the southwestern USA and in Mediterranean countries.

A solar furnace, such as that built in 1970 at Odeillo in the French Pyrenees, has thousands of mirrors to focus the Sun's rays; it produces uncontaminated intensive heat (up to 3,000°C/5,4000°F) for industrial and scientific or experimental purposes. The world's first solar power station connected to a national grid opened in 1991 at Adrano in Sicily. Scores of giant mirrors move to follow the Sun throughout the day, focusing the rays into a boiler. Steam from the boiler drives a conventional turbine. The plant generates up to 1 megawatt.

In March 1996 the first solar power plant capable of storing heat was switched on in California's Mojave Desert. Solar 2, part of a three-year government-sponsored project, consists of 2,000 motorized mirrors that will focus the Sun's rays on to a 91-m/300 ft metal tower containing molten nitrate salt. When the salt reaches 565°C/1049°F it boils water to drive a 10-megawatt steam turbine. The molten salt retains its heat for up to 12 hours.

Despite their low running costs, their high installation cost and low power output have meant that solar cells have found few applications outside space probes and artificial satellites. Solar heating is, however, widely used for domestic purposes in many parts of the world, and is an important renewable source of energy.

wind power Wind energy can be harnessed to produce power (see Chapter 15 for more details). In 1995 the UK generated nearly 350 megawatts from wind energy, a 10% increase on 1994.

---

wind turbine

The wind turbine is the modern counterpart of the windmill. The rotor blades are huge – up to 100 m/330 ft across – in order to extract as much energy as possible from the wind. Inside the turbine head, gears are used to increase the speed of the turning shaft so that the electricity generation is as efficient as possible.

---

conservation

Conservation is the action taken to protect and preserve the natural world, usually from pollution, over-exploitation, and other harmful features of human activity. The late 1980s saw a great increase in public concern for the environment, with membership of conservation groups, such as Friends of the Earth, Greenpeace, and the US Sierra Club, rising sharply. Globally the most important issues include the depletion of atmospheric ozone by the action of chlorofluorocarbons (CFCs), the build-up of carbon dioxide in the atmosphere (thought to contribute to an intensification of the greenhouse effect), and deforestation.

Action by governments has been prompted and supplemented by private agencies, such as the World Wide Fund for Nature (formerly the World Wildlife Fund). In attempts to save particular species or habitats, a distinction is often made between **preservation** – that is, maintaining the pristine state of nature exactly as it was or might have been – and **conservation**, the management of natural resources in such a way as to integrate the requirements of the local human population with those of the animals, plants, or the habitat being conserved.

biodiversity

Biodiverstiy is the measure of the variety of the Earth's animal, plant, and microbial species; of genetic differences within species; and of the ecosystems that support those species. Its maintenance is important for ecological stability and as a resource for research into, for example, new drugs and crops. In the 20th century the destruction of habitats is believed to have resulted in the most severe and rapid loss of biodiversity in the history of the planet.

Estimates of the number of species vary widely because many species-rich ecosystems, such as tropical forests, contain unexplored and unstudied habitats. Especially among small organisms, many are unknown; for instance, it is thought that only 1–10% of the world's bacterial species have been identified.

Protected Areas of the World

Source: Fish and Wildlife Service

The International Union for the Conservation of Nature and Natural Resources (IUCN) identifies a protected area as an area of land and/or sea dedicated to the protection and maintenance of biological diversity and natural and associated cultural resources that are managed through legal or other effective means. The main purposes of management of a region or area as a protected area are: for scientific research; wilderness protection; preservation of species and genetic diversity; maintenance of environmental services; protection of specific natural and cultural features; tourism and recreation; education; sustainable use of resources from natural ecosystems; and maintenance of cultural and traditional attributes.The table only includes those areas greater than 10 sq km/4 sq mi in extent, or completely protected islands of more than 1 sq km/0.4 sq mi, classified in any of the IUCN management categories. (N = nil or negligible.)

1996

Country	Area		Number of protected areas	Area protected		Land area protected
	sq km	sq mi		sq km	sq mi	(%)[1]
Afghanistan	652,090	251,707	6	2,184	843	0.33
Albania	28,748	11,097	11	340	131	1.18
Algeria	2,381,741	919,352	19	119,192	46,020	5.00
Angola	1,246,700	481,226	6	26,412	10,198	2.12
Antarctica	14,266,827	5,508,422	19	2,425	936	0.02
Antigua and Barbuda	441	170	2	61	24	13.86
Argentina	2,780,092	1,073,116	86	43,731	16,885	1.57
Armenia	29,800	11,500	4	2,139	826	7.18
Australia	7,682,300	2,966,136	892	935,455	361,180	12.18
Austria	83,500	32,374	170	20,058	7,744	23.92
Azerbaijan	86,600	33,400	12	1,909	737	2.20
Bahamas	13,864	5,352	10	1,244	480	8.97
Bangladesh	144,000	55,585	8	968	374	0.67
Belarus	207,600	80,100	10	2,425	936	1.17
Belgium	30,510	11,784	3	771	298	2.53
Belize	22,963	8,864	14	3,231	1,247	14.07
Benin	112,622	43,472	2	7,775	3,002	6.90
Bhutan	46,500	17,954	9	9,661	3,730	20.72
Bolivia	1,098,581	424,052	25	92,330	35,649	8.40
Bosnia-Herzegovina	51,129	19,745	5	251	97	0.49
Botswana	582,000	225,000	9	106,633	41,171	18.54
Brazil	8,511,965	3,285,618	273	321,898	124,285	3.78
Brunei	5,765	2,225	10	1,151	444	19.97
Bulgaria	110,912	42,812	46	3,699	1,428	3.34
Burkina Faso	274,122	105,811	12	26,619	10,278	9.71
Burundi	27,834	10,744	3	889	343	3.19
Cambodia	181,035	69,898	20	29,978	11,575	16.56
Cameroon	475,440	183,638	14	20,504	7,917	4.31
Canada	9,970,610	3,849,674	640	825,455	318,708	8.32
Central African Republic	622,436	240,260	13	61,060	23,575	9.77
Chad	1,284,000	495,624	9	114,940	44,378	8.95
Chile	756,950	292,257	66	137,251	52,992	18.26
China	9,596,960	3,599,975	463	580,666	224,195	6.05
Colombia	1,141,748	440,715	79	93,580	36,131	8.22
Congo, Democratic Republic of	2,344,900	905,366	8	99,166	38,288	4.23
Congo, Republic of the	342,000	132,012	10	11,774	4,546	3.44
Costa Rica	51,100	19,735	29	6,386	2,466	12.55

Country	Area		Number of protected areas	Area protected		Land area protected
	sq km	sq mi		sq km	sq mi	(%)[1]
Côte d'Ivoire	322,463	124,471	12	19,929	7,695	6.18
Croatia	56,538	21,824	29	3,853	1,488	6.82
Cuba	110,860	42,820	53	8,928	3,447	7.80
Cyprus	9,251	3,571	4	753	290	8.14
Czech Republic	78,864	30,461	34	10,668	4,119	13.53
Denmark	43,075	16,627	113	13,888	5,362	32.24
Djibouti	23,200	8,955	1	100	39	0.43
Dominica	751	290	1	69	27	9.15
Dominican Republic	48,442	18,700	17	10,483	4,047	21.64
Ecuador	461,475	178,176	15	111,139	42,910	24.08
Egypt	1,001,450	386,990	12	7,932	3,063	0.79
El Salvador	21,393	8,258	2	52	20	0.24
Estonia	45,000	17,000	39	4,402	1,700	9.76
Ethiopia	1,096,900	423,403	23	60,226	23,253	5.45
Fiji	18,333	7,078	5	189	73	1.03
Finland	338,145	130,608	82	27,286	10,535	8.10
France	543,965	209,970	110	56,015	21,627	10.30
Gabon	267,667	103,319	6	10,450	4,035	3.90
Gambia	10,402	4,018	5	229	88	2.15
Georgia	69,700	26,911	15	1,869	722	2.68
Germany	357,041	137,853	504	91,957	35,505	25.77
Ghana	238,305	91,986	9	11,036	4,261	4.63
Greece	131,957	50,935	24	2,231	861	1.69
Greenland	2,186,000	844,014	2	982,500	379,344	44.95
Guatemala	108,889	42,031	17	8,330	3,216	7.65
Guinea	245,857	94,901	3	1,635	631	0.67
Guyana	214,969	82,978	1	586	226	0.27
Haiti	27,750	10,712	3	97	37	0.35
Honduras	112,100	43,282	44	8,628	3,331	7.70
Hungary	93.032	35,910	53	5,740	2,216	6.17
Iceland	103,000	39,758	22	9,159	3,536	8.91
India	3,166,829	1,222,396	374	143,507	55,408	4.53
Indonesia	1,191,443	740,905	175	185,653	71,680	9.67
Iran	1,648,000	636,128	68	82,996	32,045	5.04
Ireland, Republic of	70,282	27,146	12	468	181	0.68
Israel	20,800	8,029	15	3,078	1,188	14.82
Italy	301,300	116,332	172	22,748	8,783	7.55
Jamaica	10,957	4,230	1	15	6	0.13
Japan	377,535	145,822	80	27,582	10,649	7.46
Jordan	89,206	34,434	10	2,903	1,121	3.02
Kazakhstan	2,717,300	1,049,150	9	8,915	3,442	0.33
Kenya	582,600	224,884	36	35,038	13,528	6.01
Kiribati	717	277	3	266	103	38.93
Korea, North	120,538	46,528	2	579	224	0.47
Korea, South	98,799	38,161	28	6,938	2,679	7.05
Kuwait	17,819	6,878	2	270	104	1.11
Kyrgyzstan	198,500	76,641	5	2,841	1,097	1.43
Laos	236,790	91,400	17	24,400	9,421	10.30
Latvia	63,700	24,595	45	7,747	2,991	12.16
Lebanon	10,452	4,034	1	35	14	0.34
Lesotho	30,355	11,717	1	68	26	0.22
Liberia	111,370	42,989	1	1,292	499	1.16
Libya	1,759,540	679,182	6	1,730	668	0.10

Country	Area		Number of protected areas	Area protected		Land area protected (%)[1]
	sq km	sq mi		sq km	sq mi	
Liechtenstein	160	62	1	60	23	37.50
Lithuania	65,200	25,174	76	6,347	2,451	9.73
Luxembourg	2,586	998	1	360	139	13.93
Macedonia, Former Yugoslav Republic of	25,700	9,920	16	2,165	836	8.42
Madagascar	587,041	226,598	37	11,153	4,306	1.88
Malawi	118,000	45,560	9	10,585	4,087	11.25
Malaysia	329,759	127,287	54	14,848	5,733	4.46
Mali	1,240,142	478,695	11	40,120	15,490	3.24
Mauritania	1,030,700	397,850	4	17,460	6,741	1.69
Mauritius	1,865	720	3	40	15	2.16
Mexico	1,958,201	756,198	65	97,287	37,563	4.93
Moldova	33,700	13,012	2	62	24	0.18
Mongolia	1,565,000	604,480	15	61,678	23,814	3.94
Morocco	458,730	177,070	10	3,621	1,398	0.79
Mozambique	799,380	308,561	1	20	8	N
Myanmar	676,577	261,228	2	1,733	669	0.26
Namibia	824,300	318,262	12	102,178	39,451	12.40
Nepal	147,181	56,850	12	11,085	4,280	7.84
Netherlands	41,863	16,169	79	3,885	1,500	9.44
New Zealand	268,680	103,777	206	61,478	23,737	23.19
Nicaragua	127,849	49,363	59	9,035	3,488	6.10
Niger	1,186,408	457,953	5	84,162	32,495	7.09
Nigeria	923,773	356,576	19	29,713	11,472	3.22
Northern Marianas	477	184	4	15	6	3.23
Norway	387,000	149,421	114	55,365	21,376	17.09
Oman	272,000	105,000	29	37,363	14,426	13.74
Pakistan	796,100	307,295	55	37,209	14,366	4.63
Palau	508	196	1	12	5	2.44
Panama	77,100	29,768	15	13,263	5,121	16.89
Papua New Guinea	462,840	178,656	5	820	317	0.18
Paraguay	406,752	157,006	19	14,830	5,726	3.65
Peru	1,285,200	496,216	22	41,762	16,124	3.25
Philippines	300,000	115,800	27	6,059	2,339	2.02
Poland	312,683	120,628	111	30,636	11,829	9.80
Portugal	92,000	35,521	25	5,826	2,249	6.31
Qatar	11,400	4,402	1	16	6	0.14
Romania	237,500	91,699	39	10,849	4,189	4.57
Russia	17,075,500	6,591,100	199	655,368	253,038	3.84
Rwanda	26,338	10,173	2	3,270	1,263	12.42
St Kitts and Nevis	269	104	1	26	10	10.00
St Lucia	617	238	1	15	6	2.41
St Vincent and the Grenadines	388	150	2	83	32	21.30
Samoa	2,840	1,097	3	101	39	3.55
Saudi Arabia	2,200,518	849,400	10	62,014	23,944	2.58
Senegal	196,200	75,753	10	21,807	8,420	11.09
Seychelles	453	175	3	379	146	93.79
Sierra Leone	71,740	27,710	2	820	317	1.13
Singapore	622	240	1	28	11	4.54
Slovak Republic	49,035	18,940	40	10,155	3,921	72.36
Slovenia	20,251	7,817	10	1,081	417	5.34
Somalia	637,700	246,220	1	1,800	695	0.29
South Africa	1,222,081	471,723	237	69,283	26,750	5.85

Country	Area		Number of protected	Area protected		Land area protected
	sq km	sq mi	areas	sq km	sq mi	(%)[1]
Spain	504,750	194,960	215	42,456	16,392	8.41
Sri Lanka	65,600	25,328	56	7,960	3,073	12.13
Sudan	2,505,815	967,489	16	93,825	36,226	3.74
Suriname	163,820	63,243	13	7,360	2,842	4.49
Swaziland	17,400	6,716	4	459	177	2.64
Sweden	450,000	173,745	214	29,890	11,541	6.78
Switzerland	41,300	15,946	109	7,307	2,821	17.70
Taiwan	36,179	13,965	14	4,266	1,647	11.54
Tajikistan	143,100	55,251	3	857	331	0.60
Tanzania	945,000	364,865	30	138,900	53,629	14.78
Thailand	513,115	198,108	111	70,203	27,105	13.66
Togo	56,800	21,930	11	6,469	2,498	11.39
Trinidad and Tobago	5,130	1,981	6	157	61	3.07
Tunisia	164,150	63,378	7	449	173	0.27
Turkey	779,500	300,965	44	8,194	3,164	1.05
Turkmenistan	488,100	188,406	8	11,116	4,292	2.28
Uganda	236,600	91,351	31	10,987	4,242	8.07
UK	244,100	94,247	191	51,280	19,799	20.94
Ukraine	603,700	233,089	20	5,224	2,017	0.87
Uruguay	176,200	68,031	8	3,209	1,239	0.17
USA	9,368,900	3,618,770	1,494	1,042,380	402,463	11.12
Uzbekistan	447,400	172,741	10	2,442	943	0.55
Venezuela	912,100	352,162	100	263,223	101,631	28.86
Vietnam	329,600	127,259	59	13,298	5,134	4.03
Yugoslavia	58,300	22,503	21	3,470	1,340	3.40
Zambia	752,600	290,579	21	63,636	24,570	8.46
Zimbabwe	390,300	150,695	25	30,678	11,845	7.86

[1] The figures in the percentage of land area protected column may be inflated due to the inclusion of protected marine areas in the calculation.

threats to biodiversity The most significant threat to biodiversity comes from the destruction of rainforests and other habitats in the southern hemisphere. It is estimated that 7% of the Earth's surface hosts 50–75% of the world's biological diversity. Costa Rica, for example, has an area less than 10% of the size of France but possesses three times as many vertebrate species.

In 1992 an international convention for the preservation of biodiversity was signed by over 100 world leaders. The convention called on industrialized countries to give financial and technological help to developing countries to allow them to protect and manage their natural resources, and profit from growing commercial demand for genes and chemicals from wild species. However, the convention was weakened by the USA's refusal to sign because of fears it would undermine the patents and licences of biotechnology companies.

A report issued in 1995, after almost a decade of research sponsored by the World Bank and international organization groups, designated 1,306 sites worldwide as marine protection areas (MPAs); of these, 155 were singled out for further protection, including the Bering Strait and Kachemak Bay, both in Alaska. The MPAs were selected on the basis of genetic diversity, biological productivity, and the extent to which they provided habitats for endangered species.

wildlife trade The international trade in live plants and animals, and in wildlife products such as skins, horns, shells, and feathers, has made some species virtually extinct, and threatens whole ecosystems (for

---

Welcome to Coral Forest
http://www.blacktop.com/coralforest/
Welcome to Coral Forest is a site dedicated to explaining the importance of coral reefs for the survival of the planet. It is an impassioned plea on behalf of the world's endangered coral reefs and includes a full description of their biodiversity, maps of where coral reefs are to be found (no less than 109 countries), and many photos.

example, coral reefs). Wildlife trade is to some extent regulated by CITES (Convention on International Trade in Endangered Species).

Species almost eradicated by trade in their products include many of the largest whales, crocodiles, marine turtles, and some wild cats. Until recently, some 2 million snakeskins were exported from India every year. Populations of black rhino and African elephant have collapsed because of hunting for their horns and tusks (ivory), and poaching remains a problem in cases where trade is prohibited.

whaling

Whales have been killed by humans since at least the Middle Ages. There were hundreds of thousands of whales at the beginning of the 20th century, but the invention of the harpoon gun in 1870 and improvements in ships and mechanization have led to the near-extinction of several species of whale. Commercial whaling was largely discontinued in 1986, although Norway and Japan have continued commercial whaling. Traditional whaling areas include the coasts of Greenland and Newfoundland, but the Antarctic, in the summer months, supplies the bulk of the catch.

Whales are killed for whale oil (made from the thick layer of fat under the skin called 'blubber'), which is used as a lubricant, or for making soap, candles, and margarine; for the large reserve of oil in the head of the sperm whale, used in the leather industry; and for **ambergris**, a waxlike substance from the intestines of the sperm whale, used in making perfumes. Whales have also been killed for use in petfood manufacture in the USA and Europe, and as a food in Japan. The flesh and ground bones are used as soil fertilizers.

The International Whaling Commission (IWC) was established in 1946 to enforce quotas on whale killing. It was largely unsuccessful until the 1970s, when world concern about the possible extinction of the whale mounted. By the end of the 1980s, 90% of blue, fin, humpback, and sperm whales had been wiped out, and their low reproduction rates mean that populations are slow to recover.

After 1986 only Iceland, Japan, Norway, and the USSR continued with limited whaling for 'scientific purposes', and proposals made by Japan, Norway, and the USSR for further scientific whaling were rejected by the IWC in 1990, when a breakaway whaling club was formed by Norway, Greenland, the Faeroes, and Iceland. In 1991 Japan, which has been repeatedly implicated in commercial whaling, held a 'final' whale feast before conforming to the regulations of the IWC. In 1992 the IWC established the Revised Management Procedure (RMP), to provide a new basis for regulating the exploitation of whales.

In 1994 it was revealed that Soviet whaling fleets had been systematically killing protected whales for over 40 years and exceeding their permitted quotas by more than 90%, leading to a revision of estimates of the remaining number of whales. The IWC's decision in May 1994 to create a vast Southern Ocean whale sanctuary was supported by 23 member states; Japan voted against it, and Norway abstained.

Environment and Ecology Chronology

c. 3100 BC	The Egyptian king Menes has a large (15 m/ 49 ft high) masonry dam built on the Nile south of Memphis (Cairo) to provide water for irrigation and for the city. It is the first large-scale dam.
c. 900 BC	Natural gas begins to be used in China.
c. 800 BC	Long underground aqueducts called 'qanats' are driven horizontally into hillsides in Persia to tap groundwater. They are still a major source of water in Iran.
c. 400 BC	The Chinese begin to use bitumen for cooking food and burning in lamps – the first use of oil as a source of energy.
c. 325 BC	Greek scholar Theophrastus writes about the relationship between organisms, and between organisms and their environment – the first ecological study.
c. 200 BC	Coal is first used in China as fuel.
100	The Romans begin the mining of coal in Britain.
536	Dust from volcanic eruptions in Southeast Asia is flung high in the atmosphere, blocking out sunlight and cooling the climate, causing a severe winter as far away as Europe, where the Mediterranean is covered by a 'dry fog'.
c. 600	Climate change in Peru accelerates the decline of the South American Moche civilization, silting up the canals that irrigate their principal city.
644	The earliest references to the Arab use of windpower dates to Persia at this time. Persian windmills have towers with horizontal sails inside. The wind enters and exits through carefully-positioned inlets.
1040	Water power is employed in hemp mills at Graisivaudan, France.
c. 1125	Tidal mills are built near the mouth of the River Adour, France, to take advantage of the flow of water constantly changing with the tides.
1180	A windmill with vertical sails (not a horizontal post mill) is recorded for the first time in western Europe, at Saint-Sauveur-le-Vicomte, Normandy.
1190	Coal is mined for use in iron forges at Liège, Flanders, apparently the first application for which the mineral was used. Coal mining also begins in France at around this time.
1233	Coal is mined at the English town of Newcastle for the first time and rapidly becomes popular as a fuel, although some complain about the pollution.
1233	The first piped water supply in England takes water from Paddington to Westminster, London.
1273	Complaints in London, England, about pollution produced from the burning of 'seacoal' mined at Newcastle lead to an eventual prohibition on coal burning throughout England.
1594–97	A sharp climatic downturn leads to harvest failures and high grain prices throughout Europe. The shortage causes crises of subsistence, dearth, and famine; unrest, vagabondage, and revolt increase.
1614–42	The Dutch captain Jan May claims Jan Mayen island in the Arctic Ocean as a whaling base; it is deserted after 28 years, when the local whales have been exterminated.
1615	In an effort to conserve the dwindling forests in England, particularly the large (usually) oak trees required for naval ships, timber (large wood) is forbidden as fuel for glass furnaces, which are expanding rapidly as windows are more commonly glazed.
1766	English chemist Henry Cavendish produces nitrogen dioxide by passing electricity through a nitrogen–oxygen mixture; it stimulates experiments on ways to enrich the soil.
1800	Scottish chemist William Cruickshank purifies water by adding chlorine.
1807	British inventor William Cubitt develops 'patent sails'. Used on windmills, they automatically adjust the sails to the speed and direction of the wind.
1818	French scientist Francois de Larderel exploits the energy of geothermal emissions of steam at Larderello, Italy.
1821	The world's first natural gas well is sunk at Fredonia, New York. Lead pipes distribute the gas to consumers for lighting and cooking.
1827	French engineer Benoît Fourneyron builds the world's first water turbine. It generates 6 horsepower.
1840	German chemist Christian Schönbein discovers ozone.
1853	The Forest of Fontainebleau in France becomes the first designated nature reserve.
1856	The world's first oil refinery opens at Ploiesti, Romania.
1859	US engineer Edwin Drake drills the world's first oil well, at Titusville, Pennsylvania. Drilled to a depth of 21 m/69 ft it produces 1,818 l/ 400 gal per day. His success, coinciding with a growing demand for oil products, especially kerosene, leads to further drilling.
1864	Norwegian engineer Sven Foyn invents the gun-launched harpoon with an explosive head. It permits hunting the faster and more

plentiful fin, sei, and blue whales and ushers in the era of modern whaling.

1864 Yosemite National Park, California, is established. It is the first state park and becomes a national park in 1894.

1869 German embryologist Ernst Haeckel coins the word 'ecology'.

1872 Yellowstone National Park is created in Wyoming and Montana. The world's first national park, it is also the USA's largest at 898,000 ha/2.2 million ac.

1885 Banff National Park is established in Canada, the country's first national park.

1892 US naturalist John Muir founds the Sierra Club to preserve scenic resources in the USA.

26 Aug 1895 The first large hydroelectric power station begins operating at Niagara Falls, USA. Fourneyron water turbines are used to turn three 5,000-horsepower Westinghouse electric generators, providing 3,750 kW of alternating current to Buffalo, New York.

1895 The National Trust is founded in Britain to preserve country houses, parks, gardens, and areas of natural beauty.

1896 Swedish chemist Svante August Arrhenius discovers a link between the amount of carbon dioxide in the atmosphere and global temperature.

1898 The 19,500 sq km/7,500 sq mi Kruger National Park is established (in part) in South Africa.

1903 The US president Theodore Roosevelt establishes the first US national wildlife refuge on Pelican Island, off the east coast of Florida.

1905 Roosevelt creates the US Forest Service and establishes millions of acres of additional forest reserves.

1906 The US National Forests Commission is established.

1907 Etosha National Park, one of the world's largest game reserves, covering 22,270 sq km/8,600 sq mi, and Namib Desert National Park (23,400 sq km/9,035 sq mi), containing the only desert in southern Africa, are established in South West Africa (Namibia).

June 1908 Roosevelt appoints a 57-member National Commission for the Conservation of Natural Resources. The Commission will be directed by Gifford Pinchot, who first applied the term 'conservation' to the environment.

1908 Kaziranga National Park, Assam, India is established; it serves as a refuge for the Indian rhinoceros.

1913 French physicist Charles Fabry discovers the ozone layer in the upper atmosphere.

1917 Mount McKinley National Park (now Denali National Park) in Alaska is established; it encompasses Mount McKinley, the highest mountain in North America. Covering 24,419 sq km/9,428 sq mi, it is the largest national park in the USA.

1919 Grand Canyon National Park, Arizona, is established. Covering 4,931 sq km/1,904 sq mi, it preserves 160 km/100 mi of the Grand Canyon.

1921 Ujong-kulon Nature Reserve is established in Java, Indonesia; it contains the last natural forest in Java and is a refuge for the Javanese tiger and Javanese rhinoceros.

1922 Wood Buffalo National Park, Alberta, Canada, is established. The largest national park in the world, covering 44,800 sq km/17,300 sq mi, it contains the only remaining free-ranging buffalo herds.

c. 1925 US chemist Thomas Midgley discovers Freon-12 (dichlorodifluoromethane), a chlorofluorocarbon (CFC) used in refrigeration.

1930 Hundreds of people fall ill and 60 die during a four-day fog in the industrialized Meuse Valley in Belgium. It is the first recorded air pollution disaster.

1930–36 The Boulder Dam (renamed Hoover Dam in 1947) on the Colorado River (on the border of Arizona and Nevada) is constructed to provide hydroelectricity. The dam is 221 m/726 ft high and 379 m/1,244 ft wide, and the lake behind the dam, Lake Mead, is 185 km/115 mi long making it the largest artificial lake in the USA.

1931 The US corporation Du Pont introduces Freon, a chlorofluorocarbon (CFC), as an aerosol propellant and refrigerant; it begins to replace ammonia in refrigerators.

1932 The largest hydroelectric power station in Europe is built at Zaporozhye on the River Dnieper, USSR; it produces 560 megawatts of electricity.

11–13 Nov 1933 A dust storm blows topsoil from South Dakota as far east as New York State; the US Department of Agriculture sets up a Soil Erosion Service to teach farmers tilling methods that minimize erosion.

1933 British biochemist Ernest Kennaway discovers that hydrocarbons produced from incomplete combustion, and found in cigarette smoke, car exhausts, and air pollution, can cause cancer in test animals. These are the first chemical carcinogens to be isolated.

1934 Ecuador makes part of the Galápagos Islands a wildlife sanctuary to protect its unique flora and fauna. It is extended in 1959 and in 1968 becomes the Parque Nacional Galápagos.

1935	A 'Green Belt' scheme is put into operation around London, England, to prevent excessive development.
1940	The UK government creates Serengeti National Park in Tanganyika (modern Tanzania); it covers an area of 14,763 sq km/5,700 sq mi and provides a refuge for the rare black rhinoceros.
1945	The chemical herbicides 2,4-D, 2,4,5,-T and IPC are introduced; they herald a new era in chemical weed control as their high toxicity permits them to be used in dosages as low as one or two pounds per acre.
1947	Everglades National park is established in Florida.
1948	The Fresh Kills landfill site on Staten Island, New York, opens. By the 1990s its volume will exceed that of the Great Wall of China, making it the largest artificial structure on Earth.
1948	Tsavo National Park is created in Kenya; it offers a refuge for the black rhinoceros.
25 Oct 1952	A hydroelectric power station and dam are opened at Donzère-Mondragon in the Rhône Valley of France. The project is part of a larger scheme to produce 25% of France's electricity in the Rhône Valley.
Dec 1952	Smog hits London, England: weather conditions and industrial and domestic pollution combine to produce a haze of toxic pollutants, which limit visibility to a few feet. It lasts for three weeks, and over 4,000 people, mostly elderly, die from respiratory problems caused by poor air quality. The disaster leads to antipollution legislation.
1956	The Clean Air Act is passed by the UK Parliament. It prohibits the burning of untreated coal in London and successfully reduces the emission of sulphur-oxide pollution.
9 Sept 1961	An international conference is held in Tanganyika (now Tanzania) for preserving African wildlife.
1961	Chobe National Park, which contains a fossilized lake bed, and Gemsbok National Park, the largest national park in Africa, covering 24,305 sq km/9,384 sq mi, are established in Bechuanaland (now Botswana).
1961	The World Wildlife Fund (now Worldwide Fund for Nature) is established to promote conservation.
1965	The Gir Lion National Park is established in India. It is the last remaining natural habitat of the Asiatic lion.
1966	The US Department of the Interior issues the first rare and endangered species list; 78 species are included. Congress passes the Rare and Endangered Species Act to protect them.
1967	The International Whaling Commission prohibits hunting blue, right, grey, and humpback whales.

1967	US scientists Syukuvo Manabe and R T Wetherald warn that the increase in carbon dioxide in the atmosphere, produced by human activities, is causing a 'greenhouse effect', which will raise atmospheric temperatures and cause a rise in sea levels.
June 1969	Pesticides spilled into the Rhine by a chemical company in Frankfurt am Main kill millions of fish and contaminate the drinking water of the Netherlands.
15 Jan 1971	The Aswan High Dam in Egypt is officially opened. Its reservoir, Lake Nasser, is 300 km/186 mi long and necessitated the relocation of the Abu Simbel temple complex. The dam allows Egypt to control the annual flooding of the Nile but increases incidence of the disease schistosomiasis.
1971	Greenpeace, the environmental campaign organization, is founded to protest against US nuclear testing at Amchitka Island, Alaska.
1971	The USA bans the importation of all whale products.
June 1972	The United Nations Conference on the Human Environment is held in Stockholm, Sweden; the first international conference on the state of the environment, its aim is to improve the world's environment through monitoring, resource management, and education.
1972	Dennis Meadows's *The Limits to Growth* is published by the Massachusetts Institute of Technology. Based on a Club of Rome report and computer simulation, it predicts environmental catastrophe if the depletion of the Earth's resources, overpopulation, and pollution are not acted upon immediately.
1972	The first section of the Transamazonia Highway is completed in Brazil; designed to assist settlement and exploitation of the underpopulated Amazon River Basin, when completed it will run 15,100 km/3,400 mi from Recife on the Atlantic coast to Cruzeiro do Sul near the Peruvian border.
1972	The UK and several other European countries sign an agreement prohibiting aircraft and ships from dumping toxic and plastic waste into the Atlantic; it is the first major international agreement governing pollution of the sea.
1972	The United Nations Environment Programme (UNEP) is established; its aim is to advise and coordinate environmental activities within the United Nations.
1972	The USA restricts the use of DDT because it is discovered that it thins the eggshells of predatory birds, lowering their reproductive rates.
30 Aug 1973	Kenya bans hunting elephants and trading in ivory.

24 Sept 1973 An accident at the Windscale power station in Cumberland (now Cumbria), northwest England, releases radiation into the surrounding area and raises concern about the safety of nuclear power; the station subsequently changes its name to Sellafield.

1973 An international conference on the environment held in Lebanon warns that the Mediterranean is being turned into a 'dead sea' by the sewage and industrial waste being dumped into it.

1973 Fearing the extinction of kangaroos, Australia bans the sale and export of live animals and hides.

1973 Representatives from 80 nations sign the Convention on International Trade in Endangered Species (CITES) that prohibits trade in 375 endangered species of plants and animals and the products derived from them, such as ivory; the USA does not sign.

1973 The US Fish and Wildlife Bureau issues the first Endangered and Threatened Species List.

1974 Mexican chemist Mario Molina and US chemist F Sherwood Rowland warn that the chlorofluorocarbons (CFCs) used in fridges and as aerosol propellants may be damaging the atmosphere's ozone layer that filters out much of the Sun's ultraviolet radiation.

1974 The Brazilian government begins a $5-billion scheme to replace petrol for cars with an alcohol and gasoline mixture; by 1980, 750,000 cars run on 'gasohol'.

1974 The Parque Nacional da Amazonia is established in Brazil; with an area of 10,000 sq km/4,000 sq mi, it preserves a large area of tropical rainforest.

1975 The Ecology Party is founded in Britain (known since 1985 as the Green Party).

1975 The Worldwatch Institute is founded in the USA to research the interdependence of the world economy and the environment.

17 Dec 1976 More than 180,000 barrels of crude oil are spilled into the Atlantic off Nantucket Island when the Liberian tanker *Argo Merchant* breaks in two after running aground the previous day.

1976 The American Panel on Atmospheric Chemistry warns that the Earth's ozone layer may be being destroyed by chloroflurocarbons (CFCs) from spray cans and refrigeration systems.

20 April 1977 US president Jimmy Carter proposes a radical energy conservation plan to reduce US dependency on imported oil; it involves tax on imported oil, and incentives for purchasing fuel-efficient cars and the discovery of new resources. Congress greatly modifies the programme.

April 1977 A Norwegian oil well in the North Sea blows out of control for eight days spilling about 8.2 million gallons of crude oil into the sea and creating an oil slick 32 km/20 mi long.

4 Aug 1977 Carter signs legislation establishing the Department of Energy, aiming to develop a plan for energy conservation and development after the 'oil shock'. Energy becomes the twelfth cabinet department and the first instituted since 1966.

1977 California enforces strict antipollution legislation compelling car manufacturers to install catalytic converters that reduce exhaust emissions by 90%.

1977 The USA signs the Convention on International Trade in Endangered Species (CITES).

23 Jan 1978 Sweden bans aerosol sprays because of their damaging effect on the environment. It is the first country to do so.

16 March 1978 The US oil tanker *Amoco Cadiz* runs aground off Brittany, France, spilling 1.62 million barrels of oil into the sea and contaminating 177 km/110 mi of coastline.

Aug 1978 Toxic chemicals (PCBs, dioxins, and pesticides) leak into the basements of houses in the Love Canal neighbourhood of Niagara Falls, New York. The site, an abandoned canal, was used as a chemical waste dump by the Hooker Chemicals and Plastics Corporation 1947–53. Residents are evacuated but their long-term exposure results in high rates of chromosomal damage and birth defects. It is the worst environmental disaster involving chemical waste in US history.

5 Nov 1978 A referendum in Austria stops the Zwentersdorf nuclear power station from being switched on.

1978 The USA bans the use of chlorofluorocarbons (CFCs) as spray propellants in order to reduce damage to the ozone layer.

11 July 1979 The International Whaling Commission bans the hunting of sperm whales.

1979 The first International Agreement dealing with transnational air pollution is signed by European members of the United Nations.

1980 A ten-year World Climate Research Programme is launched to study and predict climate changes and human influence on climate change.

1981 The US Committee on the Atmosphere and Biosphere reports evidence linking acid rain to sulphur emissions from power plants.

1981 The US government-commissioned *Global 2000 Report to the President* is published; it predicts global environmental catastrophe if pollution, industrial expansion, and population are not brought under control.

1981 The world's largest solar power generating station, Solar One, is completed in California; it has a capacity of 10 megawatts.

1981 US pilot Stephen Ptacek crosses the English Channel from Paris, France, to Manston, Kent, England (a journey of 368 km /180 mi) in 5.5 hrs in *Solar Challenger*, the first solar-powered aircraft; it uses 16,000 solar cells.

1981 US scientists Adam Heller and Ferdinand Thiel develop a liquid junction cell that converts 11.5% of solar energy into electricity.

30 April 1982 Agreement is reached at the United Nations' Law of the Sea Conference on an international convention governing the use and exploitation of the sea and seabed; the USA and UK do not sign.

1982 The Convention on Conservation of Antarctic Marine Living Organisms comes into effect, establishing a protective oceanic zone around the continent.

1984 The Itaipu power plant on the Paraná River on the border between Brazil and Paraguay starts operating; the largest generating station in the world, it produces 13,200 megawatts of electricity.

1984 The pesticide DDT is banned in Britain.

March 1985 The British Antarctic Survey detects a hole in the ozone layer which opens each year in the spring over Antarctica.

1985 Catalytic converters that require lead-free fuel are made mandatory in Switzerland for private cars. Switzerland is the first country to pass such a law.

1985 The International Whaling Commission bans commercial whaling, in order to prevent the extinction of whales.

26 April 1986 A major accident at the Chernobyl nuclear power station near Kiev, USSR, is announced after abnormally high levels of radiation are reported in Sweden, Denmark, and Finland. Shortly after the accident more than 30 firefighters and plant workers die from radiation exposure and over the next few years an estimated 6,500 to 45,000 people die from cancer. High rates of genetic defects are also reported, and a vast area of land will be uninhabitable for thousands of years.

1 Nov 1986 Water supplies along the Rhine are contaminated and millions of fish are killed when a fire at the Sandoz pharmaceutical company near Basel, Switzerland, results in 1,000 tonnes of toxic chemicals being discharged into the river.

1986 Two wells 4 km/2.5 mi deep and connected at the bottom are used to generate 4 megawatts of geothermal power at the Los Alamos

National Laboratory in New Mexico, USA. Water introduced in one well emerges from the other at a temperature of 190°C/375° F.

15 July–2 Oct 1987 Tax incentives in Brazil encouraging conversion of jungle into ranch land result in an average of over 25,920 sq km/10,000 sq mi of Amazon rainforest being burnt each day during a 79-day period.

1987 At a conference in Montreal, Canada, an international agreement, the Montreal Protocol, is reached to limit the use of ozone-depleting chlorofluorocarbons (CFCs) by 50% by the end of the century; the agreement is later condemned by environmentalists as 'too little, too late'.

1987 The world's most powerful wind-powered generator begins operating in the Orkney Islands, Scotland; it produces 3 megawatts of electricity.

27 March 1989 As the oil spill from the tanker *Exxon Valdez*, aground in Prince William Sound, Alaska, spreads over 100 sq mi/260 sq km, a state of emergency is declared in the area affected.

16 Oct 1989 The Convention on International Trade in Endangered Species (CITES) agrees to a total ban on trading in ivory.

1989 Brazil suspends tax incentives favouring burning in the Amazon jungle in response to worldwide environmentalist opinion; land clearance continues, however.

1990 The British company Imperial Chemical Industries (ICI) develops the first practical biodegradable plastic, Biopal.

15 Nov 1990 A Clean Air Act in the USA raises standards for emissions made by utilities and industrial concerns.

22 Feb–3 Nov 1991 Hundreds of Kuwaiti oil wells are set alight by Iraqi soldiers during the Gulf War; the last fire is extinguished on 3 November.

1991 'Biosphere 2', an experiment that attempts to reproduce the world's biosphere in miniature within a sealed glasshouse, is launched in Arizona, USA. Eight people remain sealed inside for two years.

1991 The Antarctic Treaty is signed by the 39 nations. It imposes a 50-year ban on mineral exploitation in Antarctica.

1991 The European Space Agency's first remote-sensing satellite (*ERS-1*) is launched into polar orbit to monitor the Earth's temperature from space.

1991 The world's first solar-power station connected to a national grid goes on line at Adrano, Sicily. Giant mirrors follow the Sun throughout the day, focusing the rays onto a steam boiler that drives a conventional turbine. The plant generates up to one megawatt.

11 Feb 1992 US president George Bush announces that the USA will phase out CFCs (chlorofluorocarbons)

by 1995; British secretary of state for the environment Michael Heseltine makes a similar announcement for the UK three days later.

3–14 June 1992 The United Nations Conference on Environment and Development is held in Rio de Janeiro, Brazil. It is attended by delegates from 178 countries, most of whom sign binding conventions to combat global warming and to preserve biodiversity (the latter is not signed by the USA).

1993 An ice core drilled in Greenland, providing evidence of climate change over 250,000 years, suggests that sudden fluctuations have been common and that the recent stable climate is unusual.

1993 German shops are obliged to take back the packaging of many of the products they sell for recycling and it becomes illegal to throw away packaging from most large electronic items.

1993 The world's largest array of photovoltaic cells, at Davis, California, is plugged into the local electricity system, producing 479 kilowatts, enough for 125 homes.

14 June 1994 Representatives of 25 European countries and Canada sign a United Nations protocol in Oslo, Norway, to reduce sulphur emissions, a cause of acid rain.

1994 The International Whaling Commission establishes a whale sanctuary in Antarctica.

1994 The United Nations Basel Convention bans the transport of hazardous waste, from the 25 industrialized nations that make up the Organization for Economic Cooperation and Development (OECD), across international boundaries.

April 1995 The European Space Agency's Earth-sensing satellite *ERS-2* is launched successfully. It will work in tandem with *ERS-1*, launched in 1991, to take measurements of global ozone.

10–20 June 1995 The Shell oil company begins towing its disused North Sea oil platform Brent Spa to a dumping site in the Atlantic; on 16 June, Greenpeace activists occupy the platform and on 20 June, following a boycott of Shell petrol stations in Germany and the Netherlands, the company cancels the dumping.

Aug 1995 The world's first commercial wave-powered electricity generator begins operating on the River Clyde, Scotland. Known as 'Osprey', it generates 2 megawatts of electricity.

1995 The Prince Gustav Ice Shelf and the northern Larsen Ice Shelf in Antarctica begin to disintegrate – a result of global warming.

15 Feb 1996 The Liberian-registered *Sea Empress* supertanker runs aground at the entrance to Milford Haven Harbour, south Wales, UK. Salvage attempts fail and 128,000 tonnes of light crude oil are spilled into the sea.

March 1996 The first solar power plant capable of storing heat, 'Solar 2', becomes operational. Located in California's Mojave Desert, it consists of 2,000 motorized mirrors that focus the Sun's rays on to a tower containing molten nitrate salt, which retains its heat for up to 12 hours. The molten salt is used to boil water to drive a 10-megawatt steam turbine.

4 Oct 1996 The World Conservation Union (IUCN) publishes the latest Red List of endangered species. Over 1,000 mammals are listed, far more than on previous lists. The organization believes it has underestimated the risks of habitats from pollution and that the number of endangered species is greater than previously thought.

9–29 June 1997 At the tenth Convention on International Trade in Endangered Species (CITES) convention in Harare, Zimbabwe, the elephant is downlisted to CITES Appendix II (vulnerable) and the ban on ivory exportation in Botswana, Namibia, and Zimbabwe is lifted.

26 June 1997 The second Earth Summit takes place in New York, New York. Delegates report on progress since the 1992 Rio Summit and note that progress on the Rio biodiversity convention has been slower than on the convention on climate. The delegates fail to agree on a deal to address the world's escalating environmental crisis. Dramatic falls in aid to the so-called Third World countries, which the 1992 summit promised to increase, are at the heart of the breakdown.

June–Nov 1997 Plantation owners in Indonesia, burning forests to clear land, cause the worst forest fires in Southeast Asian history. World record levels of atmospheric pollution reach life-threatening levels; up to 20 million people in Indonesia are affected with throat and respiratory inflammations and diarrhoea. By September the fires have consumed between 300,000–600,000 hectares/740,300–1,480,000 acres.

25 July 1997 The US Senate votes unanimously to pass a resolution that urges President Clinton not to agree to an international treaty that limits emissions of greenhouse gases by industrialized countries, warning that such a pact could pose a threat to the US economy.

7 Sep 1997 Australian researcher William de la Mare, using old whaling records which record data on every whale caught since the 1930s, including the ship's latitude, announces the discovery that Antarctic sea-ice could have decreased by up to a quarter between the mid-1950s and the 1970s. The finding has major implications, both for global climate conditions as well as for whaling.

22 Oct 1997 US president Bill Clinton announces a proposal to fight global warming by introducing $5 billion in tax breaks to companies who agree to reduce their greenhouse emissions.

25 Nov 1997 The Federal Energy Regulatory Commission orders the US company Edwards Manufacturing to demolish the 160-year-old Edwards Dam on the Kennebec River, Maine, to give sturgeon and salmon a chance to reach their spawning grounds. It is the first time a working hydroelectric dam has been ordered to be removed.

22 Jan 1998 The Australian government bans Japanese fishing boats from its waters after Japan refuses to limit its fishing of bluefin tuna, a probable endangered species.

Jan 1998 The German Red Cross estimates that 10,000 children a month are dying from malnutrition in North Korea and that two million died in 1997. The famine has been caused by poor agricultural practices that have brought environmental catastrophe.

8 April 1998 The International Union for the Conservation of Nature (IUCN), based in Switzerland, publishes a survey which reports that one in every eight known plant species in the world is in danger of becoming extinct.

17 April 1998 An iceberg 40 km/25 mi long and 4.8 km/3 mi wide breaks off from the Larson B ice shelf in Antarctica. Global warming is thought to be the cause.

Biographies

Carson, Rachel Louise (1907–1964) US biologist, writer, and conservationist. Her book *Silent Spring* (1962), attacking the indiscriminate use of pesticides, especially DDT, inspired the creation of the modern environmental movement. Other publications include *The Sea Around Us* (1951) and *The Edge of the Sea* (1955), an ecological exploration of the seashore. While writing about broad scientific issues of pollution and ecological exploitation, she also raised important issues about the reckless squandering of natural resources by an industrial world.

Elton, Charles Sutherland (1900–1991) British ecologist, a pioneer of the study of animal and plant forms in their natural environments, and of animal behaviour as part of the complex pattern of life. He defined the concept of food chains and originated the concept of the 'pyramid of numbers' as a method of representing the structure of an ecosystem in terms of feeding relationships. He was instrumental in establishing the Nature Conservancy Council in 1949, and was much concerned with the impact of introduced species on natural systems. His books include *Animal Ecology and Evolution* (1930) and *The Pattern of Animal Communities* (1966).

Kelly, Petra (1947–1992) German politician and activist who was a vigorous campaigner against nuclear power and other environmental issues and founded the German Green Party in 1972. Deeply concerned with radioactivity in the environment, she carried out pioneering ecological work in her position with the European Economic Community. She emerged as the German Green Party's most prominent early spokesperson, and the leader of its pragmatic ('realo') wing as the party split into its increasingly bitter divisions. She was Green member of the German parliament 1983–90, and was one of the first Green MPs in the world.

Lovelock, James (1919–) British scientist who began the study of CFCs in the atmosphere in the 1960s (though he did not predict the damage they cause to the ozone layer) and who later elaborated the Gaia hypothesis.

MacArthur, Robert Helmer (1930–1972) Canadian-born US ecologist who did much to change ecology from a descriptive discipline to a quantitative, predictive science. For example, his index of vegetational complexity (foliage height diversity) in 1961 made it possible to compare habitats and predict the diversity of their species in a definite equation. Investigating population biology, he examined how the diversity and relative abundance of species fluctuate over time and how species evolve. In particular, he managed to quantify some of the factors involved in the ecological relationships between species.

Mellanby, Kenneth (1908–) British ecologist and entomologist who in the 1960s drew attention to the environmental effects of pollution, particularly by pesticides. He advocated the use of biological control methods, such as introducing animals that feed on pests.

Muir, John (1838–1914) Scottish-born US conservationist. From 1880 he headed a campaign that led to the establishment of Yosemite National Park, USA, in 1890. He was named adviser to the National Forestry Commission in 1896 and continued to campaign for the preservation of wilderness areas for the rest of his life.

Porritt, Jonathon (1950–) British environmental campaigner, director of Friends of the Earth 1984–90. He has stood in both British and European elections as a Green (formerly Ecology) Party candidate.

Rotblat, Joseph (1908–) Polish physicist. He began working on the atom bomb as part of the Manhattan Project but withdrew in 1944 when he received intelligence that the Germans were not working on a bomb. He was instrumental in the formation of Pugwash, a group of scientists working towards nuclear disarmament. He was awarded the Nobel Peace Prize in 1995 (shared with Pugwash).

Schumacher, Fritz (Ernst Friedrich) (1911–1977) German economist. He believed that the increasing size of institutions, coupled with unchecked economic growth, creates a range of social and environmental problems. He argued his case in books like *Small is Beautiful* (1973), and established the Intermediate Technology Development Group.

Scott, Peter Markham (1909–1989) British naturalist, artist, and explorer, founder of the Wildfowl Trust at Slimbridge, Gloucestershire, England, in 1946, and a founder of the World Wildlife Fund (now World Wide Fund for Nature). The son of Antarctic explorer Robert Falcon Scott, he led ornithological expeditions and published many books on birds, including *Key to the Wild Fowl of the World* (1949) and *Wild Geese and Eskimos* (1951). He was the first president of the World Wildlife Fund 1961–67.

Glossary

abiotic factor nonorganic variable within the ecosystem, affecting the life of organisms. Examples include temperature, light, and soil structure. Abiotic factors can be harmful to the environment, as when sulphur dioxide emissions from power stations produce acid rain.

aerogenerator wind-powered electricity generator. These range from large models used in arrays on wind farms (wind turbines) to battery chargers used on yachts.

aerosol particles of liquid or solid suspended in a gas. Fog is a common natural example. Aerosol cans contain a substance such as scent or cleaner packed under pressure with a device for releasing it as a fine spray. Most aerosols used chlorofluorocarbons (CFCs) as propellants until these were found to cause destruction of the ozone layer in the stratosphere.

afforestation planting of trees in areas that have not previously held forests. (**Reafforestation** is the planting of trees in deforested areas.) Trees may be planted (1) to provide timber and wood pulp; (2) to provide firewood in countries where this is an energy source; (3) to bind soil together and prevent soil erosion; and (4) to act as windbreaks.

agrochemical artificially produced chemical used in modern, intensive agricultural systems. Agrochemicals include nitrate and phosphate fertilizers, pesticides, some animal-feed additives, and pharmaceuticals. Many are responsible for pollution and almost all are avoided by organic farmers.

alkylphenolethoxylate (APEO) chemical used mainly in detergents, but also in herbicides, cleaners, packaging, and paints. Nonylphenol, a breakdown product of APEOs is a significant river pollutant; 60% of APEOs end up in the water. Nonylphenol, and other APEO breakdown products, have a feminizing effect on fish: male fish start to produce yolk protein and the growth of testes is slowed.

animal rights extension of the concept of human rights to animals on the grounds that animals may not be able to reason but can suffer and are easily exploited by humans. The **animal-rights movement** is a general description for a wide range of organizations, both national and local, that take a more radical approach than the traditional **welfare** societies. More radical still is the concept of **animal liberation**, that animals should not be used or exploited in any way at all.

balance of nature in ecology, the idea that there is an inherent equilibrium in most ecosystems, with plants and animals interacting so as to produce a stable and continuing system of life on Earth. The activities of human beings can, and frequently do, disrupt the balance of nature.

beach nourishment the adding of extra sand to the foreshore of a beach to act as a buffer against the sea and reduce erosion.

biodegradable capable of being broken down by living organisms, principally bacteria and fungi. In biodegradable substances, such as food and sewage, the natural processes of decay lead to compaction and liquefaction, and to the release of nutrients that are then recycled by the ecosystem.

biofeedback modification or control of a biological system by its results or effects. For example, a change in the position or ◊trophic level of one species affects all levels above it.

biofouling build-up of barnacles, mussels, seaweed, and other organisms on underwater surfaces, such as ships' hulls. Marine industries worldwide spend at least £1.4 billion controlling biofouling by scraping affected surfaces and painting with antifouling paint.

biomass the total mass of living organisms present in a given area. It may be specified for a particular species (such as earthworm biomass) or for a general category (such as herbivore biomass). Estimates also exist for the entire global plant biomass. Measurements of biomass can be used to study interactions between organisms, the stability of those interactions, and variations in population numbers. Where dry biomass is measured, the material is dried to remove all water before weighing.

bioreactor sealed vessel in which microbial reactions can take place. The simplest bioreactors involve the slow decay of vegetable or animal waste, with the emission of methane that can be used as fuel. Laboratory bioreactors control pH, acidity, and oxygen content and are used in advanced biotechnological operations, such as the production of antibiotics by genetically-engineered bacteria.

BioSphere 2 (BS2) ecological test project, a 'planet in a bottle', in Arizona, USA. Under a sealed glass and metal dome, different habitats are recreated, with representatives of nearly 4,000 species, to test the effects that various environmental factors have on ecosystems. Simulated ecosystems, or 'mesocosms', include savanna, desert, rainforest, marsh and Caribbean reef. The response of such systems to elevated atmospheric concentrations of carbon dioxide gas (CO_2) are among the priorities of Biosphere 2 researchers.

biotic factor organic variable affecting an ecosystem – for example, the changing population of elephants and its effect on the African savanna.

Blueprint for Survival environmental manifesto published in 1972 in the UK by the editors of the *Ecologist* magazine. The statement of support it attracted from a wide range of scientists helped draw attention to the magnitude of environmental problems.

brownfield site site that has previously been developed; for example, a derelict area in the inner city. Before brownfield sites can be redeveloped, site clearance is often necessary, adding to the development cost. The surrounding area may be of poor environmental quality but there is often a surprising level of biodiversity on brownfield sites – for example at one site in London, 300 species of flowering plant were found – and there is often a high proportion of invertebrate species.

Brundtland Report findings of the World Commission on Environment and Development, published in 1987 as *Our Common Future*. It stressed the necessity of environmental protection and popularized the phrase 'sustainable development'. The commission was chaired by the Norwegian prime minister Gro Harlem Brundtland.

bycatch, or **bykill**, in commercial fishing, that part of the catch that is unwanted. Bycatch constitutes approximately 25% of global catch, and consists of a variety of marine life, including fish too small to sell or otherwise without commercial value, seals, dolphins, sharks, turtles, and even seabirds.

carbon sequestration disposal of carbon dioxide waste in solid or liquid form. From 1993 energy conglomerates such as Shell, Exxon, and British Coal have been researching ways to reduce their carbon dioxide emissions by developing efficient technologies to trap the gas and store it securely – for example, by burying it or dumping it in the oceans.

carrying capacity in ecology, the maximum number of animals of a given species that a particular area can support. When the carrying capacity is exceeded, there is insufficient food (or other resources) for the members of the population. The population may then be reduced by emigration, reproductive failure, or death through starvation.

chemical oxygen demand (COD) measure of water and effluent quality, expressed as the amount of oxygen (in parts per million) required to oxidize the reducing substances present.

Chipko Movement Indian grass-roots villagers' movement campaigning against the destruction of their forests. Its broad principles are nonviolent direct action, a commitment to the links between village life and an unplundered environment, and a respect for all living things.

CITES (abbreviation for Convention on International Trade in Endangered Species) international agreement under the auspices of the ◊International Union for the Conservation of Nature (IUCN) with the aim of regulating trade in ◊endangered species of animals and plants. The agreement came into force in 1975 and by 1997 had been signed by 138 states. It prohibits any trade in a category of 8,000 highly endangered species and controls trade in a further 30,000 species. Animals and plants listed in Appendix 1 of CITES are classified endangered; those listed in Appendix 2 are classified vulnerable.

climax community assemblage of plants and animals that is relatively stable in its environment. It is brought about by ecological ◊succession, and represents the point at which succession ceases to occur.

climax vegetation the plants in a ◊climax community.

Club of Rome informal international organization that aims to promote greater understanding of the interdependence of global economic, political, natural, and social systems. Members include industrialists, economists, and research scientists. Membership is limited to 100 people. It was established in 1968.

colonization spread of species into a new habitat, such as a freshly cleared field, a new motorway verge, or a recently flooded valley. The first species to move in are called **pioneers**, and may establish conditions that allow other animals and plants to move in (for example, by improving the condition of the soil or by providing shade). Over time a range of species arrives and the habitat matures; early colonizers will probably be replaced, so that the variety of animal and plant life present changes. This is known as ◊succession.

community assemblage of plants, animals, and other organisms living within a circumscribed area. Communities are usually named by reference to a dominant feature such as characteristic plant species (for example, a beech-wood community), or a prominent physical feature (for example, a freshwater-pond community).

competition interaction between two or more organisms, or groups of organisms (for example, species), that use a common resource which is in short supply. Competition invariably results in a reduction in the numbers of one or both competitors, and in evolution contributes both to the decline of certain species and to the evolution of adaptations.

compost organic material decomposed by bacteria under controlled conditions to make a nutrient-rich natural fertilizer for use in gardening or farming. A well-made compost heap reaches a high temperature during the composting process, killing most weed seeds that might be present.

container habitat small self-contained ecosystem, such as a water pool accumulating in a hole in a tree. Some ecologists believe that much can be learned about larger ecosystems through studying the dynamics of container habitats, which can contain numerous leaf-litter feeders and their predators.

contaminated land land that is considered to pose a health risk to humans because of pollution; usually land that has been the site of industrial activity.

corridor, wildlife route linking areas of similar habitat, or between sanctuaries. For example there is a corridor linking the Masai Mara reserve in Kenya and the Serengeti in Tanzania. On a smaller scale, disused railways provide corridors into urban areas for foxes.

daminozide (trade name **Alar**) chemical formerly used by fruit growers to make apples redder and crisper. In 1989 a report published in the USA found the consumption of daminozide to be linked with cancer, and the US Environment Protection Agency (EPA) called for an end to its use. The makers have now withdrawn it worldwide.

Darwin Initiative UK government conservation initiative announced at the ◊Earth Summit in 1992, pledging an annual budget of £10 million in grants to British researchers. This was reduced to £3 million in 1993.

debt-for-nature swap agreement under which a proportion of a country's debts are written off in exchange for a commitment by the debtor country to undertake projects for environmental protection. Debt-for-nature swaps were set up by environment groups in the 1980s in an attempt to reduce the debt problem of poor countries, while simultaneously promoting conservation.

dioxin any of a family of over 200 organic chemicals, all of which are heterocyclic hydrocarbons. The term is commonly applied, however, to only one member of the family, 2,3,7,8-tetrachlorodibenzo-p-dioxin (2,3,7,8-TCDD), a highly toxic chemical that occurs, for example, as an impurity in the defoliant Agent Orange, used in the Vietnam War (1954–75), and sometimes in the weedkiller 2,4,5-T. It has been associated with a disfiguring skin complaint (chloracne), birth defects, miscarriages, and cancer.

drift net long straight net suspended from the water surface and used by commercial fishermen. Drift nets are controversial as they are indiscriminate in what they catch. Dolphins, sharks, turtles, and other marine animals can drown as a consequence of becoming entangled.

Earth Summit (officially **United Nations Conference on Environment and Development**) international name given to meetings aiming at drawing measures towards environmental protection of the world. The first summit took place in Rio de Janeiro, Brazil, in June 1992. Treaties were made to combat global warming and protect biodiversity (the latter was not signed by the USA). The second Earth Summit was held in New York in June 1997 to review progress on the environment. The meeting agreed to work towards a global forest convention in the year 2000 with the aim of halting the destruction of tropical and old-growth forests.

ecotourism growing trend in tourism to visit sites that are of ecological interest, for example the Galápagos Islands. Ecotourism generates employment and income for local people, providing an incentive for conservation, but if carried out unscrupulously it can lead to damage of environmentally sensitive sites.

effluent liquid discharge of waste from an industrial process, usually into rivers or the sea. Effluent is often toxic but is difficult to control and hard to trace.

electromagnetic pollution electric and magnetic fields set up by high tension power cables, local electric sub-stations, and domestic items such as electric blankets. There have been claims that these electromagnetic fields are linked to increased levels of cancer, especially leukaemia, and to headaches, nausea, dizziness, and depression.

endangered species plant or animal species whose numbers are so few that it is at risk of becoming extinct. Officially designated endangered species are listed by the ◊International Union for the Conservation of Nature.

energy conservation methods of reducing energy use through insulation, increasing energy efficiency, and changes in patterns of use. Profligate energy use by industrialized countries contributes greatly to air pollution and the greenhouse effect when it draws on nonrenewable energy sources.

environmental audit another name for ◊green audit.

eutrophication excessive enrichment of rivers, lakes, and shallow sea areas, primarily by nitrate fertilizers washed from the soil by rain, by phosphates from fertilizers, and from nutrients in municipal sewage, and by sewage itself. These encourage the growth of algae and bacteria which use up the oxygen in the water, thereby making it uninhabitable for fishes and other animal life.

evolutionary toxicology study of the effects of pollution on evolution. A polluted habitat may cause organisms to select for certain traits, as in **industrial melanism** for example, where some insects, such as the peppered moth, are darker in polluted areas, and therefore better camouflaged against predation.

extinction complete disappearance of a species or higher taxon. Extinctions occur when an animal becomes unfit for survival in its natural habitat usually to be replaced by another, better-suited animal. An organism becomes ill-suited for survival because its environment is changed or because its relationship to other organisms is altered. For example, a predator's fitness for survival depends upon the availability of its prey.

field studies study of ecology, geography, geology, history, archaeology, and allied subjects, in the natural environment as opposed to the laboratory.

firewood principal fuel for some 2 billion people, mainly in the Third World. In principle a renewable energy source, firewood is being cut far faster than the trees can regenerate in many areas of Africa and Asia, leading to deforestation.

flue-gas desulphurization process of removing harmful sulphur pollution from gases emerging from a boiler. Sulphur compounds such as sulphur dioxide are commonly produced by burning fossil fuels, especially coal in power stations, and are the main cause of acid rain.

food chain sequence showing the feeding relationships between organisms in a particular ecosystem. Each organism depends on the next lowest member of the chain for its food.

A ◊pyramid of numbers can be used to show the reduction in food energy at each step up the food chain.

forest area where trees have grown naturally for centuries, instead of being logged at maturity (about 150–200 years). A natural, or old-growth, forest has a multistorey canopy and includes young and very old trees (this gives the canopy its range of heights). There are also fallen trees contributing to the very complex ecosystem, which may support more than 150 species of mammals and many thousands of species of insects. Globally forest is estimated to have covered around 68 million sq km/26 million sq mi during prehistoric times. By the late 1990s this is believed to have been reduced by half.

forestation see ◊afforestation.

fur hair of certain animals. Fur is an excellent insulating material and so has been used as clothing. This is, however, vociferously criticized by many groups, as the methods of breeding or trapping animals are often cruel. Mink, chinchilla, and sable are among the most valuable, the wild furs being finer than the farmed.

green accounting inclusion of economic losses caused by environmental degradation in traditional profit and loss accounting systems.

green audit inspection of a company to assess the total environmental impact of its activities or of a particular product or process.

green consumerism marketing term used especially during the 1980s when consumers became increasingly concerned about the environment. Labels such as 'eco-friendly' became a common marketing tool as companies attempted to show that their goods had no negative effect on the environment.

greenhouse effect phenomenon of the Earth's atmosphere by which solar radiation, trapped by the Earth and re-emitted from the surface as infrared radiation, is prevented from escaping by various gases in the air.

green movement collective term for the individuals and organizations involved in efforts to protect the environment. The movement encompasses political parties such as the Green Party and organizations like ◊Friends of the Earth and ◊Greenpeace.

Greenpeace international environmental pressure group, founded in 1971, with a policy of nonviolent direct action backed by scientific research. In 1997 Greenpeace had a membership in 43 'chapters' worldwide.

green tax proposed tax to be levied against companies and individuals causing pollution. For example, a company emitting polluting gases would be obliged to pay a correspondingly significant tax; a company that cleans its emissions, reduces its effluent and uses energy-efficient distribution systems, would be taxed much less.

habitat localized environment in which an organism lives, and which provides for all (or almost all) of its needs. The diversity of habitats found within the Earth's ecosystem is enormous, and they are changing all the time. Many can be considered inorganic or physical; for example, the Arctic ice cap, a cave, or a cliff face. Others are more complex; for instance, a woodland or a forest floor. Some habitats are so precise that they are called **microhabitats**, such as the area under a stone where a particular type of insect lives. Most habitats provide a home for many species.

Hidrovia Project controversial plan to turn the 3,400 km/2,100 mi Paraguay–Paraná river system into a shipping lane. To create a navigable canal up to 50 km/32 mi wide and 4 km/2.5 mi deep the river system will need to be dredged, dammed, and diverted at various stages. In June 1995, the United Nations Environmental Programme called on the governments of the La Plata Basin to halt work on the project until studies on environmental impact had been completed.

hum, environmental disturbing sound of frequency about 40 Hz, heard by individuals sensitive to this range, but inaudible to the rest of the population. It may be caused by industrial noise pollution or have a more exotic origin, such as the jet stream, a fast-flowing high-altitude (about 15,000 m/50,000 ft) mass of air.

indicator species plant or animal whose presence or absence in an area indicates certain environmental conditions, such as soil type, high levels of pollution, or, in rivers, low levels of dissolved oxygen. Many plants show a preference for either alkaline or acid soil conditions, while certain trees require aluminium, and are found only in soils where it is present. Some lichens are sensitive to sulphur dioxide in the air, and absence of these species indicates atmospheric pollution.

integrated pest management (IPM) use of a coordinated array of methods to control pests, including biological control, chemical pesticides, crop rotation, and avoiding monoculture. By cutting back on the level of chemicals used the system can be both economical and beneficial to health and the environment.

International Union for the Conservation of Nature (IUCN) organization established by the United Nations to promote the conservation of wildlife and habitats as part of the national policies of member states.

irrigation artificial water supply for dry agricultural areas by means of dams and channels. Drawbacks are that it tends to concentrate salts at the surface, ultimately causing soil infertility, and that rich river silt is retained at dams, to the impoverishment of the land and fisheries below them.

IUCN abbreviation for ◊International Union for the Conservation of Nature.

ivory hard white substance of which the teeth and tusks of certain mammals are made. Among the most valuable are elephants' tusks, which are of unusual hardness and density. Ivory is used in carving and other decorative work, and is so valuable that poachers continue to illegally destroy the remaining wild elephant herds in Africa to obtain it.

landfill site large holes in the ground used for dumping household and commercial waste. Landfill disposal has been the preferred option in the UK and the USA for many years, with up to 85% of household waste being dumped in this fashion. However, the sites can be dangerous, releasing toxins and other leachates (see ◊leaching) into the soil and the policy is

itself wasteful both in terms of the materials dumped and land usage.

leaching process by which substances are washed through or out of the soil. Fertilizers leached out of the soil drain into rivers, lakes, and ponds and cause water pollution. In tropical areas, leaching of the soil after the destruction of forests removes scarce nutrients and can lead to a dramatic loss of soil fertility. The leaching of soluble minerals in soils can lead to the formation of distinct soil horizons as different minerals are deposited at successively lower levels.

life-cycle analysis assessment of the environmental impact of a product, taking into account all aspects of production (including resources used), packaging, distribution and ultimate end.

methyl bromide pesticide gas used to fumigate soil. It is a major ozone depleter. Industry produces 50,000 tonnes of methyl bromide annually (1995). The European Union proposed a total ban on usage by 2001 at a meeting in July 1998, and the USA intends to ban use by 2001.

Montréal Protocol international agreement, signed in 1987, to stop the production of chemicals that are ozone depleters by the year 2000.

national park land set aside and conserved for public enjoyment. The first was Yellowstone National Park, USA, established in 1872. National parks include not only the most scenic places, but also places distinguished for their historic, prehistoric, or scientific interest, or for their superior recreational assets. They range from areas the size of small countries to pockets of just a few hectares.

nature reserve area set aside to protect a habitat and the wildlife that lives within it, with only restricted admission for the public. A nature reserve often provides a sanctuary for rare species.

niche 'place' occupied by a species in its habitat, including all chemical, physical, and biological components, such as what it eats, the time of day at which the species feeds, temperature, moisture, the parts of the habitat that it uses (for example, trees or open grassland), the way it reproduces, and how it behaves.

nitrate pollution contamination of water by nitrates. Increased use of artificial fertilizers and land cultivation means that higher levels of nitrates are being washed from the soil into rivers, lakes, and aquifers. There they cause an excessive enrichment of the water (◊eutrophication), leading to a rapid growth of algae, which in turn darkens the water and reduces its oxygen content. The water is expensive to purify and many plants and animals die. High levels are now found in drinking water in arable areas. These may be harmful to newborn babies, and it is possible that they contribute to stomach cancer, although the evidence for this is unproven.

noise unwanted sound. Permanent, incurable loss of hearing can be caused by prolonged exposure to high noise levels (above 85 decibels). Over 55 decibels on a daily outdoor basis is regarded as an unacceptable level.

nuclear safety the use of nuclear energy has given rise to concern over safety. Anxiety has been heightened by accidents such as those at Windscale (renamed Sellafield), UK, in 1973; Three Mile Island, USA, in 1979; and Chernobyl, Ukraine, in 1986. There has also been mounting concern about the production and disposal of nuclear waste, the radioactive and toxic by-products of the nuclear energy industry. Burial on land or at sea raises problems of safety, environmental pollution, and security. Nuclear waste has an active half-life of thousands of years and no guarantees exist for the safety of the various methods of disposal.

Nuclear safety is still a controversial subject since governments will not recognize the hazards of radiation and radiation sickness. In 1990 a scientific study revealed an increased risk of leukaemia in children whose fathers had worked at Sellafield between 1950 and 1985. Sellafield is the world's greatest discharger of radioactive waste, followed by Hanford, Washington, USA.

organic farming farming without the use of synthetic fertilizers (such as nitrates and phosphates) or pesticides (herbicides, insecticides, and fungicides) or other agrochemicals (such as hormones, growth stimulants, or fruit regulators). Food produced by genetic engineering cannot be described as organic.

organophosphosphate insecticide insecticidal compounds that cause the irreversible inhibition of the cholinesterase enzymes that break down acetylcholine. As this mechanism of action is very toxic to humans, they should be used with great care. Malathion and permethrin may be used to control lice in humans and have many applications in veterinary medicine and in agriculture.

overfishing fishing at rates that exceed the ◊sustained-yield cropping of fish species, resulting in a net population decline. For example, in the North Atlantic, herring has been fished to the verge of extinction and the cod and haddock populations are severely depleted. In the developing world, use of huge factory ships, often by fisheries from industrialized countries, has depleted stocks for local people who cannot obtain protein in any other way.

oxyfuel fuel enriched with oxygen to decrease carbon monoxide (CO) emissions. Oxygen is added in the form of chemicals such as methyl tertiary butyl ether (MTBE) and ethanol. The use of oxyfuels in winter is compulsory in 35 US cities. There are fears, however, that MTBE can cause health problems, including nausea, headaches, and skin rashes.

ozone layer thin layer of the gas ozone in the upper atmosphere that shields the Earth from harmful ultraviolet rays.

packaging material, usually of metal, paper, plastic, or glass, used to protect products, make them easier to display, and as a form of advertising. Environmentalists have criticized packaging materials as being wasteful of energy and resources. Recycling bins are being placed in residential areas to facilitate the collection of surplus packaging.

pioneer species those species that are the first to colonize and thrive in new areas. Coal tips, recently cleared woodland, and new roadsides are areas where pioneer species will quickly appear. As the habitat matures other species take over, a process known as **succession**.

pitfall trap simple trap for trapping small invertebrates. In its simplest form a beaker or jam jar is buried in the ground so

that the rim of the jar is flush with the soil. Beetles, millipedes, spiders and other arthropods tumble into the jar and are unable to escape.

PM10 (abbreviation for **p**articulate **m**atter less than **10** micrometres across) clusters of small particles, such as carbon particles, in the air that come mostly from vehicle exhausts. There is a link between increase in PM10 levels and a rise in death rate, increased hospital admissions, and asthma incidence. The elderly and those with chronic heart or lung disease are most at risk.

polluter-pays principle the idea that whoever causes pollution is responsible for the cost of repairing any damage. The principle is accepted in British law but has in practice often been ignored; for example, farmers causing the death of fish through slurry pollution have not been fined the full costs of restocking the river.

pooter small device for collecting invertebrates, consisting of a jar to which two tubes are attached. A sharp suck on one of the tubes, while the other is held just above an insect, will propel the animal into the jar. A filter wrapped around the mouth tube, prevents debris or organisms from being swallowed.

population group of animals of one species, living in a certain area and able to interbreed; the members of a given species in a ◊community of living things.

prior informed consent informal policy whereby companies who sell pesticides to developing countries agree to suspend exporting the product if there is an objection from the government of the receiving country and to inform the government of the nature of the pesticide. The situation arises frequently because some pesticides banned in the developed world may be bought by agricultural operations or companies in the Third World, perhaps unaware of any health implications. The policy was adopted by the Food and Agriculture Organization in 1989, and has since been made binding by the European Community on its member states.

pyramid of numbers diagram that shows quantities of plants and animals at different levels of a ◊food chain. This may be measured in terms of numbers (how many animals) or biomass (total mass of living matter), though in terms of showing transfer of food, biomass is a more useful measure. There is always far less biomass at the top of the chain than at the bottom, because only about 10% of the food an animal eats is turned into flesh – the rest is lost through metabolism and excretion. The amount of food flowing through the chain therefore drops with each step up the chain, hence the characteristic 'pyramid' shape.

quadrat in environmental studies, a square structure used to study the distribution of plants in a particular place, for instance a field, rocky shore, or mountainside. The size varies, but is usually 0.5 or 1 m/1.6 or 3.3 ft square, small enough to be carried easily. The quadrat is placed on the ground and the abundance of species estimated. By making such measurements a reliable understanding of species distribution is obtained.

radiation monitoring system network of monitors to detect any rise in background gamma radiation and to warn of a major nuclear accident within minutes of its occurrence. The accident at Chernobyl in Ukraine in 1986 prompted several western European countries to begin installation of such systems locally, and in 1994 work began on a pilot system to provide a **gamma curtain**, a dense net of radiation monitors, throughout eastern and western Europe.

Red Data List report published by the World Conservation Union (IUCN) and regularly updated that lists animal species by their conservation status. Categories of risk include **extinct in the wild**, **critically endangered**, **endangered**, **vulnerable**, and **lower risk** (divided into three subcategories). The list was updated in 1996.

reuse multiple use of a product (often a form of packaging), by returning it to the manufacturer or processor each time. Many such returnable items are sold with a deposit which is reimbursed if the item is returned. Reuse is usually more energy- and resource-efficient than recycling unless there are large transport or cleaning costs.

seabird wreck the washing ashore of significantly larger numbers of seabirds than would be expected for the time of year. In February 1994 around 75,000 birds were washed up along the east coast of Britain. They were mostly fish-eating species, such as guillemots, shags, and razorbills, and appeared to have died of starvation. The cause of seabird wrecks is unknown though overfishing could be a contributory factor.

sere plant ◊succession developing in a particular habitat. A **lithosere** is a succession starting on the surface of bare rock. A **hydrosere** is a succession in shallow freshwater, beginning with planktonic vegetation and the growth of pondweeds and other aquatic plants, and ending with the development of swamp. A **plagiosere** is the sequence of communities that follows the clearing of the existing vegetation.

sewage disposal disposal of human excreta and other waterborne waste products from houses, streets, and factories. Conveyed through sewers to sewage works, sewage has to undergo a series of treatments to be acceptable for discharge into rivers or the sea, according to various local laws and ordinances. Raw sewage, or sewage that has not been treated adequately, is one serious source of water pollution and a cause of ◊eutrophication.

Single European Act 1986 update of the Treaty of Rome (signed in 1957) that provides a legal basis for action by the European Union in matters relating to the environment. The act requires that environmental protection shall be a part of all other Union policies. Also, it allows for agreement by a qualified majority on some legislation, whereas before such decisions had to be unanimous.

site of special scientific interest (SSSI) in the UK, land that has been identified as having animals, plants, or geological features that need to be protected and conserved. From 1991 these sites were designated and administered by English Nature, Scottish Natural Heritage, and the Countryside Council for Wales.

slash and burn simple agricultural method whereby natural vegetation is cut and burned, and the clearing then farmed for a few years until the soil loses its fertility, whereupon farmers move on and leave the area to regrow. Although this is possible with a small, widely dispersed population, it becomes unsustainable with more people and is now a cause of deforestation.

slurry form of manure composed mainly of liquids. Slurry is collected and stored on many farms, especially when large numbers of animals are kept in factory units. When slurry tanks are accidentally or deliberately breached, large amounts can spill into rivers, killing fish and causing ◊eutrophication.

soil depletion decrease in soil quality over time. Causes include loss of nutrients caused by overfarming, erosion by wind, and chemical imbalances caused by acid rain.

soil erosion the wearing away and redistribution of the Earth's soil layer. It is caused by the action of water, wind, and ice, and also by improper methods of agriculture. If unchecked, soil erosion results in the formation of deserts (desertification). It has been estimated that 20% of the world's cultivated topsoil was lost between 1950 and 1990.

SSSI abbreviation for ◊Site of Special Scientific Interest.

standing crop the total number of individuals of a given species alive in a particular area at any moment. It is sometimes measured as the weight (or ◊biomass) of a given species in a sample section.

succession series of changes that occur in the structure and composition of the vegetation in a given area from the time it is first colonized by plants (**primary succession**), or after it has been disturbed by fire, flood, or clearing (**secondary succession**).

sustainable capable of being continued indefinitely. For example, the sustainable yield of a forest is equivalent to the amount that grows back. Environmentalists have made the term a catchword in advocating the sustainable use of resources.

sustained-yield cropping the removal of surplus individuals from a ◊population of organisms so that the population maintains a constant size. This usually requires selective removal of animals of all ages and both sexes to ensure a balanced population structure. Taking too many individuals can result in a population decline, as in overfishing.

TBT abbreviation for ◊tributyl tin, a chemical used in antifouling paints that has become an environmental pollutant.

TRAFFIC the arm of the ◊World Wide Fund for Nature that monitors trade in endangered species.

transboundary pollution pollution generated in one country that affects another country, for example as occurs with acid rain. Natural disasters may also cause pollution; volcanic eruptions, for example, cause ash to be ejected into the atmosphere and deposited on land surfaces.

tributyl tin (TBT) chemical used in antifouling paints on ships' hulls and other submarine structures to deter the growth of barnacles. The tin dissolves in sea water and enters the food chain. It can cause reproductive abnormalities – exposed female whelks develop penises; the use of TBT has therefore been banned in many countries, including the UK.

trophic level the position occupied by a species (or group of species) in a ◊food chain. The main levels are **primary producers** (photosynthetic plants), **primary consumers** (herbivores), **secondary consumers** (carnivores), and **decomposers** (bacteria and fungi).

Tullgren funnel device used to extract mites, springtails, fly larvae, and other small invertebrates from a sample of soil.

urban ecology study of the ecosystems, animal and plant communities, soils, and microclimates found within an urban landscape.

Waldsterben (German 'forest death') tree decline related to air pollution, common throughout the industrialized world. It appears to be caused by a mixture of pollutants; the precise chemical mix varies between locations, but it includes acid rain, ozone, sulphur dioxide, and nitrogen oxides.

Washington Convention alternative name for ◊CITES, the international agreement that regulates trade in endangered species.

weedkiller, or **herbicide**, chemical that kills some or all plants. Selective herbicides are effective with cereal crops because they kill all broad-leaved plants without affecting grasslike leaves. Those that kill all plants include sodium chlorate and paraquat. The widespread use of weedkillers in agriculture has led to an increase in crop yield but also to pollution of soil and water supplies and killing of birds and small animals, as well as creating a health hazard for humans.

wilderness area of uninhabited land that has never been disturbed by humans, usually located some distance from towns and cities. According to estimates by the US group Conservation International, 52% (90 million sq km/35 million sq mi) of the Earth's total land area was still undisturbed in 1994.

wildlife corridor passage between habitats. See ◊corridor, wildlife.

World Wide Fund for Nature (WWF) (formerly the **World Wildlife Fund**) international organization established in 1961 to raise funds for conservation by public appeal. Projects include conservation of particular species, for example, the tiger and giant panda, and special areas, such as the Simen Mountains, Ethiopia.

World Wildlife Fund former and US name of the ◊World Wide Fund for Nature.

WWF abbreviation for ◊World Wide Fund for Nature.

Further Reading

Adams, Douglas, and Carwardine, Mark *Last Chance to See* (1990)

Begon, M; Harper J L; and Townsend, C R *Ecology* (1990)

Blackmore, Roger *Global Environmental Issues* (2nd edition 1996)

Bode, Carl (ed) *The Portable Thoreau* (1980)

Brown, M *The Toxic Cloud* (1987)

Budiansky, Stephen *Nature's Keepers* (1995)

Carson, Rachel *Silent Spring* (1962)

Caufield, Catherine *In the Rainforest* (1984)

Colinvaux, Paul *Why Big Fierce Animals Are Rare* (1980)

Commoner, B *The Closing Circle: Nature, Man, and Technology* (1971)

Crosby, Alfred W *Ecological Imperialism* (1993)

Dickinson, Gordon, and Murphy, Kevin J *Ecosystems* (1998)

Durning, Alan Thein *How Much is Enough?* (1992)

Few, Roger *Atlas of Wild Places* (1997)

Foley, Gerald *The Energy Question* (1992)

Fowler, Cary, and Mooney, Pat *The Threatened Gene: Food, Politics and the Loss of Genetic Diversity* (1990)

Galdikas, Biruté M F *Reflections of Eden* (1995)

George, Susan *How the Other Half Dies – The Real Reasons for World Hunger* (1976)

Goodall, D W (ed) *Ecosystems of the World* (1977)

Groombridge, Brian (ed) *Global Biodiversity: The Status of the Earth's Living Resources* (1992)

Kural, Orhan (ed) *Coal: Resources, Properties, Utilization, Pollution* (1995)

Lear, Linda *Rachel Carson: Witness for Nature* (1997)

Lovelock, James *Gaia: A New Look at Life on Earth* (1979)

Lovins, Amory *Soft Energy Paths: Towards a Durable Peace* (1977)

Luoma, J R *Troubled Skies, Troubled Waters: The Story of Acid Rain* (1984)

MacArthur, Robert, and Wilson, Edward O *The Theory of Island Biogeography* (1967)

McNeely, Jeffrey A; Miller, Kenton R; et al *Conserving the World's Bological Diversity* (1990)

Mann, Charles C, and Plummer, Mark L *Noah's Choice* (1995)

Meadows, Donella L; Meadows, Dennis L; Randers, Jorgen; and Behrens III, William W *The Limits to Growth* (1975)

Myers, Norman *The Sinking Ark* (1979)

Norse, Elliott A (ed) *Global Marine Biological Diversity* (1993)

Noss, Reed F, and Cooperrider, Allen Y *Saving Nature's Legacy* (1994)

Pahl-Wostl, Claudia *The Dynamic nature of Ecosystems: Chaos and Order Entwined* (1995)

Pepper, David *Modern Environmentalism: An Introduction* (1996)

Ponting, Clive *The Green History of the World* (1991)

Priest, Joseph *Energy* (1984)

Pringle, Peter, and Spiegelman, James *The Nuclear Barons* (1981)

Reaka–Kudla, Marjorie (ed) *Biodiversity II* (1997)

Schaller, George D *The Last Panda* (1992)

Schumacher, E F *Small Is Beautiful* (1973)

Shiva, Vandana *Biodiversity: Social and Ecological Consequences* (1992)

Tudge, Colin *Last Animals at the Zoo* (1991)

Whitley, Edward *Gerald Durrell's Army* (1992)

Wilson, Edward O *The Diversity of Life* (1992)

World Commission on Environment and Development, chaired by Gro Harlen Brundtland *Our Common Future* (1987)

BIOLOGY, GENETICS, AND EVOLUTION

Biology is the science of life: the study of millions of different organisms that share that phenomenon – their structure, functioning, interrelationships, and origins. All are able or have the potential to grow, reproduce, move, and respond to such stimuli as light, heat, and sound; and all are sustained by the processes of nutrition, respiration, and excretion.

For all organisms – apart from viruses – the basic unit of life is the cell, each of which also possesses the essential characteristics of life, including the potential to move and reproduce.

origins of life

Life probably originated in the primitive oceans. The original atmosphere, 4 billion years ago, consisted of carbon dioxide, nitrogen, and water. Laboratory experiments have shown that more complex organic molecules, such as ◊amino acids and nucleotides, can be produced from these ingredients by passing electric sparks through a mixture. The climate of the early atmosphere was probably very violent, with lightning a common feature, and these conditions could have resulted in the oceans becoming rich in organic molecules, producing the so-called 'primeval soup'. These molecules may then have organized themselves into clusters capable of reproducing and eventually developing into simple cells. Soon after life developed, photosynthesis would have become the primary source of energy for life. By this process, life would have substantially affected the chemistry of the atmosphere and, in turn, that of its own environment. Once the atmosphere had changed to its present composition, life could only be created by the replication of living organisms (a process called biogenesis).

Biology Timeline
http://www.zoologie.biologie.de/history.html
Chronology of important developments in the biological sciences. It includes items from the mentioning of hand pollination of date palms in 1800 BC to the Nobel prize award for the discovery of site-directed mutagenesis in 1993.

characteristics of living organisms

To remember the major characteristics of living organisms:

Newmarket **r**uns **e**very **g**reat **r**ace **i**n **M**ay.

(**n**utrition, **r**espiration, **e**xcretion, **g**rowth, **r**eproduction, **i**rritability (response to conditions), **m**ovement)

the cell

The cell is the basic structural unit of life. It is the smallest unit capable of independent existence that can reproduce itself exactly. All living organisms – with the exception of viruses – are composed of one or more cells. Single-cell organisms such as bacteria, protozoa, and other microorganisms are termed **unicellular**, while plants and animals that contain many cells are termed **multicellular**. Highly complex organisms such as human beings consist of billions of cells, all of which are adapted to carry out specific functions – for instance, groups of these specialized cells are organized into tissues and organs.

eukaryote and prokaryote cells

Although cells may differ widely in size, appearance, and function, their essential features are similar. Each is composed of a mass of jellylike substance called **cytoplasm**, surrounded by a membrane. The cytoplasm contains (**ribosomes**, which carry out protein synthesis, and ◊DNA, the coded instructions for the behaviour and reproduction of the cell.

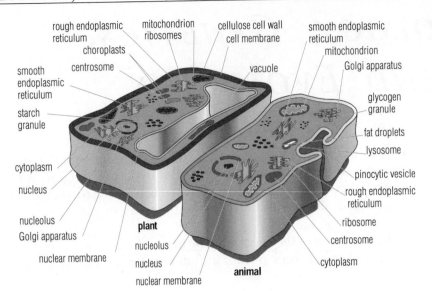

rough endoplasmic reticulum
choroplasts
smooth endoplasmic reticulum
starch granule
cytoplasm
nucleus
nucleolus
Golgi apparatus
nuclear membrane
centrosome
mitochondrion ribosomes
cellulose cell wall
cell membrane
vacuole
plant
nucleolus
nucleus
nuclear membrane
animal
smooth endoplasmic reticulum
mitochondrion
Golgi apparatus
glycogen granule
fat droplets
lysosome
pinocytic vesicle
rough endoplasmic reticulum
ribosome
centrosome
cytoplasm

cell structure Typical plant and animal cell. Plant and animal cells share many structures, such as ribosomes, mitochondria, and chromosomes, but they also have notable differences: plant cells have chloroplasts, a large vacuole, and a cellulose cell wall. Animal cells do not have a rigid cell wall but have an outside cell membrane only.

In **eukaryote** cells (those of protozoa, fungi, plants, and animals) the DNA is organized into chromosomes and is contained within a clearly defined **nucleus**, which is surrounded by a double membrane. (Some cells, however, such as mammalian red blood cells, lose their nuclei as they mature.) The cytoplasm also contains other membrane-bound structures called **organelles**, such as mitochondria and chloroplasts, which carry out specific functions.

In **prokaryote** cells (those of bacteria and cyanobacteria, or blue-green algae) the DNA forms a simple loop and there is no nucleus. The prokaryotic cell also lacks organelles, though it possesses many ribosomes.

eukaryote cell structure

Each eukaryote has a surrounding membrane, which is a thin layer of protein and fat that restricts the flow of substances in and out of the cell and encloses the ◊cytoplasm, a jellylike material containing the ◊nucleus and other structures (organelles) such as mitochondria.

In general, **plant cells** differ from animal cells in that the membrane is surrounded by a cell wall made of cellulose. They also have larger vacuoles (fluid-filled pouches) and contain chloroplasts that convert light energy to chemical energy for the synthesis of glucose. (A fuller description of structures unique to plant cells is given in the Plant Kingdom chapter.)

membrane The membrane is a thin, continuous layer, made up of fat (phospholipid) and protein molecules, that encloses a cell or organelles within a cell. Small molecules, such as water and sugars, can pass through the cell membrane by (osmosis and (diffusion. Large molecules, such as proteins, are transported across the membrane via special channels, a process often involving the input of energy (see (active transport). The Golgi apparatus within the cell is thought to produce certain membranes.

In cell organelles, enzymes may be attached to the membrane at specific positions, often alongside other enzymes involved in the same process, like workers at a conveyor belt. Thus membranes help to make cellular processes more efficient.

cytoplasm This is the part of the cell outside the ◊nucleus. Strictly speaking it includes all the organelles, but often cytoplasm refers to the jellylike matter in which the organelles are embedded (correctly termed the cytosol).

The cytoplasm is permeated with a matrix of protein microfilaments and tubules that, together, form the **cytoskeleton**. This gives the cell a definite shape, keeps the organelles in position, transports vital substances around the cell, and may also be involved in cell movement (cytoplasmic streaming).

In many cells, the cytoplasm is made up of two parts: the **ectoplasm** (or plasmagel), a dense gelatinous outer layer concerned with cell movement, and the **endoplasm** (or plasmasol), a more fluid inner part where most of the organelles are found.

nucleus The nucleus is usually the largest and most prominent structure in the eukaryote cell. Its function is to house and pass on the genetic information to future generations of cells, and to direct and control the activities of the cell according to its own genetic instructions.

The nucleus is bounded by a double membrane, or **nuclear envelope**, with numerous pores that provide

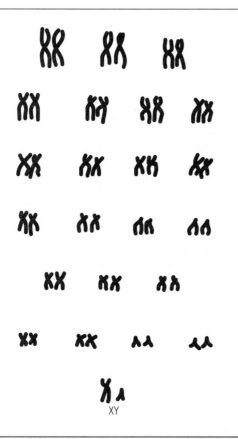

chromosome The 23 pairs of chromosomes of a normal human male.

channels of communication between the nucleus and the rest of the cell. The **nucleoplasm** inside contains the genetic material (DNA.

When the cell is not dividing, the DNA is dispersed in the form of chromatin, but during cell division ((mitosis or (meiosis), it becomes coiled and folded into compact bodies called **chromosomes**, which will carry the genetic code to the next generation. In each species, the number of chromosomes in a cell is constant – for example, in humans it is 46 (23 in ova and sperm).

Also in the nucleoplasm are dense, spherical bodies called **nucleoli**. These contain protein and ribosomal (RNA, and play a role in the manufacture of (ribosomes (structures responsible for synthesizing protein). They disappear during cell division.

Cells Alive
http://www.cellsalive.com/
Lively and attractive collection of microscopic and computer-generated images of living cells and microorganisms. It includes sections on HIV infection, penicillin, and how antibodies are made.

---

ribosome The ribosomes are the protein-making machinery of the cell. They are small, dense bodies composed of protein and a special form of RNA, called ribosomal RNA. They receive messenger RNA (copied from the DNA) and amino acids (the components of protein), and 'translate' the chemically coded instructions on the messenger RNA to link the amino acids in the order required to make a strand of a particular protein (see (protein synthesis).

Ribosomes are located on the endoplasmic reticulum (ER) – creating rough endoplasmic reticulum – or they may be found within other organelles or free in the cytoplasm.

endoplasmic reticulum Endoplasmic reticulum (ER) is a membranous system that forms compartments within the cell. It stores and transports proteins within cells and also carries various enzymes needed for the synthesis of fats. **Rough endoplasmic reticulum** has ribosomes (sites of protein synthesis) attached to its surface.

Under the electron microscope, ER looks like a series of channels and vesicles, but it is in fact a large, sealed, baglike structure crumpled and folded into a convoluted mass. The interior of the 'bag', the ER lumen, stores various proteins needed elsewhere in the cell, then organizes them into transport vesicles formed by a small piece of ER membrane budding from the main membrane.

Golgi apparatus The Golgi apparatus, or Golgi body, is a stack of flattened membranous sacs called **cisternae**. Many proteins and other molecules travel through the Golgi apparatus on their way from the endoplasmic reticulum (ER) to other parts of the cell. Inside the Golgi apparatus, they are modified or fused with other molecules, and then transported away in vesicles (tiny, membrane-bound spheres) that bud off from the tips of the cisternae.

The vesicles may be secretory, migrating to the cell membrane to release their contents to the outside, or they may remain within the cytoplasm, serving as compartments for enzyme reactions (as in, for example, (lysosomes).

lysosome Lysosomes are tiny, membrane-bound sacs, or vesicles, responsible for intracellular digestion. They contain a number of enzymes – known collectively as lysozyme – that can break down proteins and other biological substances. In single-celled organisms they are responsible for digesting foodstuffs, and in white blood cells they destroy ingested bacteria. Lysosomes also play a role in cell death (apoptosis), breaking down the structures of the cell into small fragments that are easily digestible by surrounding cells.

mitochondria Mitochondria (singular 'mitochondrion') are sometimes called the 'powerhouses' of the cell as they are the sites of most of the processes involved in aerobic (respiration, and – in nonphotosynthesizing cells at least – are responsible for producing most of the cell's ATP (the energy-rich molecule that powers cellular activities).

They are rodlike organelles bound in a double membrane, the inner of which is folded into projections called **cristae**. The cristae are covered with proteins that play a role in the electron transport chain (the last stage of respiration), while the viscous matrix inside the mitochondrion contains the enzymes involved in the Krebs cycle (the penultimate stage).

Mitochondria are thought to be derived from free-living bacteria that, at a very early stage in the history of life, invaded larger cells and took up a symbiotic way of life inside. Each still contains its own small loop of DNA, called mitochondrial DNA, and new mitochondria arise by the division of existing ones.

cilia and flagella Cilia (singular 'cilium') and flagella (singular 'flagellum') are small, hairlike organelles found on the surface of certain cells. They are locomotory structures that create motion by beating back and forth. Both are made up of cylinders of protein tubules arranged in a characteristic 'nine plus two' format (nine pairs of tubules around the circumference and two separate central tubules) and covered in cell membrane.

Cilia and flagella differ from each other in that cilia are shorter and occur in large groups, where their action is coordinated so that they beat in a wave. Flagella are longer, occur singly or in pairs, and have a more complex, snakelike action.

Some single-celled organisms move by means of cilia. In mammals they are found in the cells lining the upper respiratory tract, where their wavelike movements waft particles of dust and debris towards the exterior. They also move food in the digestive tracts of some invertebrates.

Flagella are the motile organs of certain protozoa and single-celled algae, and of the sperm cells of multicellular organisms. Water movement inside sponges is also produced by flagella.

cell division

This is the process by which a cell divides to form new cells, either by **mitosis**, which is associated with growth, cell replacement, or repair; or **meiosis**, which

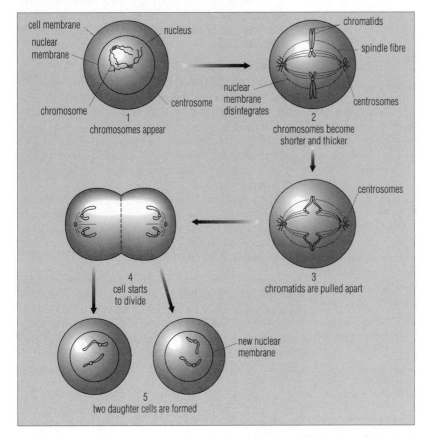

mitosis The stages of mitosis, the process of cell division that takes place when a plant or animal cell divides for growth or repair. The two daughter cells each receive the same number of chromosomes as were in the original cell.

cell membrane
nuclear membrane
nucleus
chromosome
centrosome
1 chromosomes appear

chromatids
spindle fibre
nuclear membrane disintegrates
centrosomes
2 chromosomes become shorter and thicker

centrosomes
3 chromatids are pulled apart

4 cell starts to divide

new nuclear membrane

5 two daughter cells are formed

is associated with sexual reproduction. Both forms involve the duplication of DNA and the splitting of the nucleus.

mitosis Mitosis is the process of cell division by which identical daughter cells are produced. During mitosis the DNA is duplicated and the chromosome number doubled, so new cells contain the same amount of DNA as the original cell.

mitosis

To remember the phases of mitosis:

Prime Minister – a toad!

(Prophase, metaphase, anaphase, telophase)

The genetic material of eukaryotic cells is carried on a number of chromosomes. To control movements of chromosomes during cell division so that both new cells get the correct number, a system of protein tubules, known as the spindle, organizes the chromosomes into position in the middle of the cell before they replicate. The spindle then controls the movement of chromosomes as the cell goes through the stages of division: prophase, metaphase, anaphase, and telophase.

meiosis Meiosis results in the number of chromosomes in the daughter cells being halved. It only occurs in eukaryote cells, and is essential to sexual reproduction because it allows the genes of two parents to be combined without the total number of chromosomes increasing.

Meiosis is the stage in the life cycle where genetic variation arises. This is due to **recombination** mechanisms, which 'shuffle' the genetic material, thus increasing genetic variation in the offspring. The two main mechanisms are: **crossing over**, in which chromosome pairs twist around each other and exchange corresponding segments, and the **random reassortment** of chromosomes that occurs when each gamete (sperm or egg) receives only one of each chromosome pair.

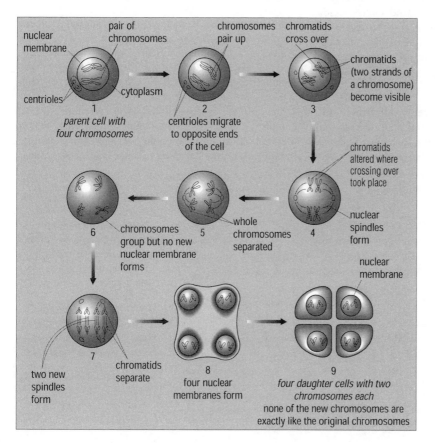

meiosis Meiosis is a type of cell division that produces gametes (sex cells, sperm and egg). This sequence shows an animal cell but only four chromosomes are present in the parent cell (1). There are two stages in the division process. In the first stage (2–6), the chromosomes come together in pairs and exchange genetic material. This is called crossing over. In the second stage (7–9), the cell divides to produce four gamete cells, each with only one copy of each chromosome from the parent cell.

nuclear membrane — pair of chromosomes — chromosomes pair up — chromatids cross over

cytoplasm

centrioles

1
parent cell with four chromosomes

2
centrioles migrate to opposite ends of the cell

3

chromatids (two strands of a chromosome) become visible

chromatids altered where crossing over took place

nuclear spindles form

6
chromosomes group but no new nuclear membrane forms

5
whole chromosomes separated

4

nuclear membrane

7
two new spindles form

chromatids separate

8
four nuclear membranes form

9
four daughter cells with two chromosomes each
none of the new chromosomes are exactly like the original chromosomes

In sexually reproducing diploid animals (having two sets of chromosomes per cell), meiosis occurs during formation of the gametes (sex cells, sperm and egg), so that the gametes are haploid (having only one set of chromosomes). When the gametes unite during fertilization the diploid condition is restored. In plants meiosis occurs just before spore formation. Thus the spores are haploid and in lower plants such as mosses they develop into a haploid plant called a gametophyte which produces the gametes.

the molecules of life

Cells contain many simple, inorganic chemicals, including water, which provide a favourable environment for biochemical reactions and make up 60%–65% of living matter; and inorganic ions, such as those of potassium, sodium, and chlorine, which play an essential role in many cell processes. However, the molecules that form the main functional units of living matter are large, complex organic molecules – they are natural polymers formed by linking together many smaller organic building blocks, or monomers. These **macromolecules** act as structural components – for example, the cell membrane is made up of proteins and lipids, the nucleus of proteins and nucleic acids; or they may provide sources of energy, store genetic information, or speed up biological reactions. They can be classified into four families: proteins, polysaccharides, lipids, and nucleic acids.

proteins

Proteins are complex, biologically important substances composed of long chains of amino acids that are twisted or folded in characteristic shapes. They are essential to all living organisms. As enzymes they reg-

Amino Acids	
name	formula
glycine	$CH_2(NH_2).COOH$
alanine	$CH_3CH.(NH_2).COOH$
phenylalanine	$C_6H_5CH_2CH.(NH_2).COOH$
tyrosine	$C_6H_4OH.CH_2CH.(NH_2).COOH$
valine	$(CH_3)_2CH.CH.(NH_2).COOH$
leucine	$(CH_3)_2CH.CH_2CH.(NH_2).COOH$
iso-leucine	$(CH_3).CH_2CH(CH_3)CH.(NH_2).C\ OOH$
serine	$CH_2OH.CH.(NH_2).COOH$
threonine	$CH_3CHOH.CH.(NH_2).COOH$
cysteine	$CH.CH_2CH.(NH_2).COOH$
methionine	$CH_3.S.(CH_2)_2CH.(NH_2).COOH$
asparagine	$NH_2CO.CH_2CH.(NH_2).COOH$
glutamine	$NH_2CH.(CH_2)_2(CO.NH_2).COOH$
lysine	$NH_2CH_3CH.(NH_2).COOH$
arginine	$NH_2C(NH).NH(CH_2)_3CH.(NH_2).COOH$
aspartic acid	$COOH.CH_2CH.(NH_2).COOH$
glutamic acid	$COOH.(CH_2)_2CH.(NH_2).COOH$
histidine	$C_3H_3N_2.CH_2CH.(NH_2).COOH$
trytophan	$C_4.NH.CH_2CH_2CH.(NH_2).COOH$
proline	$NH.(CH_2)_3CH.COOH$

ulate all aspects of metabolism. Structural proteins such as keratin and collagen make up the skin, claws, bones, tendons, and ligaments; muscle proteins produce movement; haemoglobin transports oxygen; and membrane proteins regulate the movement of substances into and out of cells.

The three-dimensional shapes of proteins are so complex, and so specific to their functioning, that a hierarchy of primary, secondary, tertiary, and even quaternary structures has been created to describe them.

amino acids, where R is one of many possible side chains

peptide – this is one made of just three amino acid units. Proteins consist of very large numbers of amino acid units in long chains, folded up in specific ways

protein A protein molecule is a long chain of amino acids linked by peptide bonds. The properties of a protein are determined by the order, or sequence, of amino acids in its molecule, and by the three-dimensional structure of the molecular chain. The chain folds and twists, often forming a spiral shape.

Molecular Expressions: The Amino Acid Collection
http://micro.magnet.fsu.edu/aminoacids/index.html
Fascinating collection of images showing what all the known amino acids look like when photographed through a microscope. There is also a detailed article about the different amino acids.

amino acids and primary structure Amino acids are water-soluble organic molecules mainly composed of carbon, oxygen, hydrogen, and nitrogen, containing both a basic amino group (NH_2) and an acidic carboxyl (COOH) group. They are the structural units, or monomers, of proteins – linked together by means of **peptide bonds** between the amino group of one amino acid and the carboxyl group of the next to form long, unbranching chains called **polypeptides**.

The creation of polypeptides takes place in the (ribosomes of the cell, where amino acids are joined together in a specific order, according to instructions from the DNA (see (protein synthesis). The primary structure of proteins is the amino acid sequence within their constituent polypeptides. It is this that will determine the shape and properties of the final protein.

All proteins are made up of the same 20 amino acids (although other types of amino acid do occur infrequently in nature). Eight of these, the essential amino acids, cannot be synthesized by humans and must be obtained from the diet. Children need a further two amino acids that are not essential for adults. Other animals also need some preformed amino acids in their diet, but green plants can manufacture all the amino acids they need from simpler molecules, relying on energy from the Sun and minerals (including nitrates) from the soil.

Amino Acids
http://www.chemie.fu-berlin.de/chemistry/bio/
amino-acids_en.html
Small but interesting site giving the names and chemical structures of all the amino acids. The information is available in both English and German.

structure The secondary structure describes the initial repeated folding or organization of the polypeptide chain, which takes place in the rough (endoplasmic reticulum and (Golgi apparatus of the cell. It is basically of two types: the alpha helix or the beta-pleated sheet. In the **alpha helix** a single polypeptide coils up into a regular helix, or spiral, that is cross-linked at intervals by hydrogen bonds. In the **beta-pleated sheet** polypeptides are lined up in parallel and cross-linked to form sheets, with the zigzagging of the peptide bonds giving the impression of pleats.

Fibrous proteins, which need to be strong and insoluble, are predominantly secondary in their structure. For example, structural proteins such as keratin and collagen have considerable alpha-helical content, while silk fibroin, a major constituent of silk, is composed of beta-pleated sheets.

tertiary structure When a protein has adopted its secondary structure, it will fold, coil, or organize itself still further into a complex tertiary structure, such as a helix, sphere, rod, or globule. For example, a length of protein arranged alpha-helically may coil back on itself to create a further helix.

The driving forces behind this are the formation of bonds such as disulphide bridges, hydrogen bonds, and ionic bonds between amino acids, and – most importantly – the tendency for hydrophobic, or water-repelling, amino acids to keep away from the external, aqueous environment. Thus, the protein will fold up with hydrophobic groups in the centre of the protein, and hydrophilic, or water-loving, groups on the outside.

Globular proteins are predominantly tertiary – and sometimes quaternary – in their structure. For example, albumin – a storage protein important in maintaining the blood's osmotic pressure – must be small, compact, and spherical in order to travel in the bloodstream. Hence it has considerable tertiary structure coiling it up into a compact unit. (Enzymes, which have to maintain a highly specific shape in order to function, are always tertiary or quaternary in structure.

quaternary structure Quaternary structure is simply a packing together of two or more tertiary-stage subunits to form a protein. For example, haemoglobin – the oxygen-carrrying protein in red blood cells – is composed of four globular subunits held together with a weak hydrogen bond.

Protein Data Bank WWW Home Page
http://pdb.pdb.bnl.gov/
Expanding scientific site that contains a fully searchable database of molecule images. This is useful for students as well as more in-depth researchers.

polysaccharides
Polysaccharides are long-chain polymers made up of hundreds or thousands of linked simple sugars (monosaccharides) such as glucose. They act as energy-rich food stores in plants (starch) and animals (glycogen), and have structural roles in the plant cell wall (cellulose, pectin) and in the tough outer skeleton of insects and similar creatures (chitin). Polysaccharides and sugars are all types of carbohydrate.

oxygen   CH₂OH   OH   CH₂OH   OH

OH   carbon
hydrogen

OH   CH₂OH   OH   CH₂OH

polysaccharide A molecule of the polysaccharide glycogen (animal starch) is formed from linked glucose ($C_6H_{12}O_6$) molecules. A typical glycogen molecule has 100–1,000 glucose units.

monosaccharide Monosaccharides are the simplest form of carbohydrate: single sugar molecules that form the basic structural units, or monomers, of polysaccharides. They are soluble compounds, some with a sweet taste. Monosaccharides have the general formula $C_nH_{2n}O_n$, where *n* may be any number from three to seven. Monosaccharides that have five carbons in their molecules are called **pentoses** – they include ribose (a constituent of RNA), arabinose, and xylose. **Hexoses** have six carbons in their molecules and include glucose, galactose, and fructose.

Glucose is the most abundant monosaccharide, present in the blood and manufactured by green plants during photosynthesis. The anaerobic respiration reactions inside cells involves the oxidation of glucose to produce ATP, the 'energy molecule' used to drive many of the body's biochemical reactions.

disaccharide Disaccharides are soluble sugars made up of two monosaccharide units linked by a **glycosidic bond**. For example, sucrose, or cane sugar, consists of glucose and fructose; lactose, or milk sugar, consists of glucose and galactose; while maltose, found in germinating seeds, consists of two glucose units.

polysaccharide formation Polysaccharides are made up of many monosaccharides – often of only a single type – joined by glycosidic links. For example,

glycogen, starch, and cellulose are all polymers of glucose, though they are distinctly different in form and function: glycogen and starch being globular, storage polysaccharides, while cellulose is a fibrous, structural material.

The difference is due to the frequency and position of the glycosidic links between monosaccharides, and whether the sugar units are joined together in long chains or in branching structures.

lipids

Lipids are a diverse group of biological molecules that are soluble in alcohol but not in water. They may be divided into **complex lipids**, which are all esters of fatty acids, and **simple lipids**, which do not contain fatty acids. Lipids are the chief constituents of plant and animal waxes, fats, and oils.

complex lipid The **fatty acids** that characterize complex lipids are organic compounds consisting of a hydrocarbon chain, up to 24 carbon atoms long, with a carboxyl group (-COOH) at one end. They are produced in the small intestine when fat is digested. Fatty acids may be described as **saturated** (having only single bonds between their carbon atoms) or **unsaturated** (having one or more double bonds between their carbon atoms).

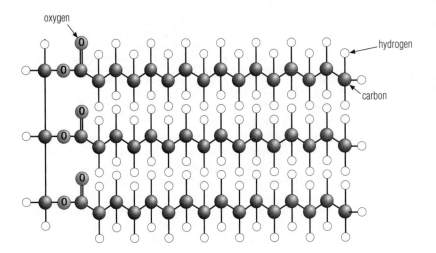

oxygen

hydrogen

carbon

triglyceride The molecular structure of typical fat. The molecule consists of three fatty acid molecules linked to a molecule of glycerol.

The most familiar of the complex lipids are the **triglycerides**, which make up the fats and oils of plants and animals. They are formed as a result of a condensation reaction between a glycerol molecule and three fatty acid molecules.

Phospholipids consist of the same glycerol backbone found in triglycerides, but with one of the fatty acids replaced with a hydrophilic (water-loving) phosphate group. They are major components of cell membranes, forming sandwichlike sheets two molecules thick, with the hydrophilic phosphate 'heads' aligned outwards towards the aqueous cytoplasm and the hydrophobic fatty-acid 'tails' aligning themselves inwards towards the middle of the sandwich.

simple lipid Simple lipids do not have a fatty-acid component; they include sterols and steroids.

Sterols are solid, cyclic, unsaturated alcohols, with a complex structure that includes four carbon rings. They include **cholesterol** – a white, crystalline substance that is an integral part of all cell membranes and is a component of lipoproteins, which transport fats and fatty acids in the blood.

Steroids are derived from sterols, but lack their alcohol (-OH) group. They include the sex hormones (such as testosterone), the corticosteroid hormones produced by the adrenal gland, bile acids, and cholesterol.

nucleic acids

These are complex organic acids made up of a long chain of **nucleotides**, found in the nucleus and sometimes the cytoplasm of the living cell. The two types, known as DNA (deoxyribonucleic acid) and RNA (ribonucleic acid), form the basis of heredity. Each nucleotide is made up of a sugar (deoxyribose or ribose), a phosphate group, and one of four purine or pyrimidine bases. The order of the bases along the nucleic acid strand contains the genetic code.

DNA DNA, or **deoxyribonucleic acid**, is a complex giant molecule that contains, in chemically coded form, the information needed for a cell to make proteins. It is a ladderlike double-stranded nucleic acid that forms the basis of genetic inheritance in all organisms, except for a few viruses that have only ♭RNA. Within the cell, DNA is organized into dense structures called

chromosomes and, in eukaryotes (organisms other than bacteria), it is found only in the cell nucleus.

DNA is made up of two chains of nucleotide subunits, with each nucleotide containing either a purine (**adenine** or **guanine**) or pyrimidine (**cytosine** or

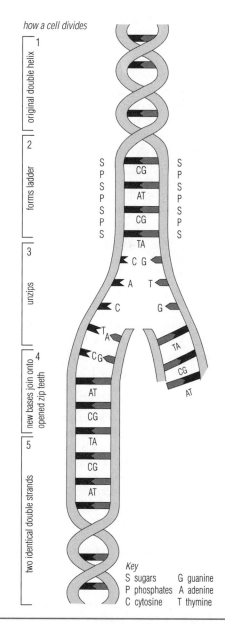

how a cell divides

1 — original double helix
2 — forms ladder
3 — unzips
4 — new bases join onto opened zip teeth
5 — two identical double strands

Key
S sugars   G guanine
P phosphates   A adenine
C cytosine   T thymine

DNA How the DNA molecule divides. The DNA molecule consists of two strands wrapped around each other in a spiral or helix. The main strands consist of alternate sugar (S) and phosphate (P) groups, and attached to each sugar is a nitrogenous base – adenine (A), cytosine (C), guanine (G), or thymine (T). The sequence of bases carries the genetic code which specifies the characteristics of offspring. The strands are held together by weak bonds between the bases, cytosine to guanine, and adenine to thymine. The weak bonds allow the strands to split apart, allowing new bases to attach, forming another double strand.

Crick, Francis Harry Compton
http://kroeber.anthro.mankato.msus.edu/bio/francis_crick.htm
Profile of the life and achievements of the pioneer molecular biologist. It traces his upbringing and education and how he brought his knowledge of X-ray diffraction to his work with James Watson in unravelling the structure of DNA. There is a photo, a listing of Crick's major books and articles, and a bibliography.

thymine) base. The bases link up with each other (adenine linking with thymine, and cytosine with guanine) to form **base pairs** that connect the two strands of the DNA molecule like the rungs of a twisted ladder.

The specific way in which the pairs form means that the base sequence is preserved from generation to generation. Hereditary information is stored as a specific sequence of bases. A set of three bases – known as a **codon** – acts as a blueprint for the manufacture of a particular ◊amino acid, the subunit of a protein molecule.

DNA nucleotides

To remember classes of nucleotides:

In DNA there are four nucleotides (cytosine, thymine, adenine, guanine) divided into two classes (pyrimidines and purines). The easiest way to remember which class they belong to is that the pyrimidines contain the letter y (cytosine and thymine).

The information encoded by the codons is transcribed (see protein synthesis) by messenger RNA and is then translated into amino acids in the ribosomes and cytoplasm. The sequence of codons determines the precise order in which amino acids are linked up during manufacture and, therefore, the kind of protein that is to be produced. Because proteins are the chief structural molecules of living matter and, as enzymes, regulate all aspects of metabolism, it may be seen that the genetic code is effectively responsible for building and controlling the whole organism.

Molecular Expressions: The DNA Collection
http://micro.magnet.fsu.edu/micro/gallery/dna/dna4.html
Spectacular gallery of DNA photographic representations in the laboratory as well as *in vivo*. This site also has links to several other sites offering photographs through a microscope of various substances, including computer chips and various pharmaceutical substances.

RNA

RNA, or **ribonucleic acid**, is responsible for translating the genetic information encoded in the ◊DNA into proteins (see (protein synthesis). It is usually single-stranded, unlike the double-stranded DNA, and consists of a large number of nucleotides strung together, each of which comprises the sugar ribose, a phosphate group, and one of four bases - **uracil** (replacing the thymine of DNA), **cytosine**, **adenine**, or **guanine**.

RNA occurs in three major forms, each with a different function in the synthesis of protein molecules. **Messenger RNA** (mRNA) copies a section of the DNA's genetic code in a process called transcription and then acts as a template for the assembly of amino acids to make proteins. Each codon (a set of three bases) on the mRNA molecule is matched up with the corresponding amino acid, in accordance with the genetic code. This process (translation) takes place in the ribosomes, which are made up of proteins and **ribosomal RNA** (rRNA). **Transfer RNA** (tRNA) is responsible for combining with specific amino acids in the cytoplasm, and then transporting them to the ribosomes to be matched up with the mRNA.

Although RNA is normally associated only with the process of protein synthesis, it makes up the hereditary material itself in some viruses, such as retroviruses.

DNA and RNA

To remember that although DNA and RNA are both nucleic acids, they do different jobs in the cell:

DNA delivers the blueprint, RNA reads it

the chemical processes of life

Living organisms require many hundreds of interrelated chemical processes to enable them to grow and function – the sum of which can be described as an organism's **metabolism**. It involves a constant alternation between building up complex molecules (**anabolism**) and breaking them down (**catabolism**). For example, green plants build up complex organic substances from water, carbon dioxide, and mineral salts (photosynthesis, described in the Plant Kingdom chapter); animals partially break down complex organic substances, ingested as food, and subsequently resynthesize them for use in their own bodies (described in the Human Body chapter); and, within cells, complex molecules are broken down to release their energy in the process of ◊respiration.

Complex processes such as respiration can be described as metabolic pathways, in which the

Discovery of the Structure of DNA

BY JULIAN ROWE

the first announcement

"We wish to suggest a structure for the salt of deoxyribose nucleic acid (DNA). This structure has novel features which are of considerable biological interest." So began a 900-word article that was published in the journal *Nature* in April 1953. Its authors were British molecular biologist Francis Crick (1916–) and US biochemist James Watson (1928–).

The article described the correct structure of DNA, a discovery that many scientists have called the most important since Austrian botanist and monk Gregor Mendel (1822–1884) laid the foundations of the science of genetics.

DNA is the molecule of heredity, and by knowing its structure, scientists can see exactly how forms of life are transmitted from one generation to the next.

the problem of inheritance

The story of DNA really begins with British naturalist Charles Darwin (1809–1882). When, in November 1859, he published *On the Origin of Species by Means of Natural Selection* outlining his theory of evolution, he was unable to explain exactly how inheritance came about. For at that time it was believed that offspring inherited an average of the features of their parents. If this were so, as Darwin's critics pointed out, any remarkable features produced in a living organism by evolutionary processes would, in the natural course of events, soon disappear.

The work of Gregor Mendel, only rediscovered 18 years after Darwin's death, provided a clear demonstration that inheritance was not a "blending" process at all. His description of the mathematical basis to genetics followed years of careful plant-breeding experiments. He concluded that each of the features he studied, such as colour or stem length, was determined by two "factors" of inheritance, one coming from each parent. Each egg or sperm cell contained only one factor of each pair. In this way a particular factor, say for the colour red, would be preserved through subsequent generations.

genes

Today, we call Mendel's factors genes. Through the work of many scientists, it came to be realized that genes are part of the chromosomes located in the nucleus of living cells and that DNA, rather than protein as was first thought, was a hereditary material.

the double helix

In the early 1950s, scientists realized that X-ray crystallography, a method of using X-rays to obtain an exact picture of the atoms in a molecule, could be successfully applied to the large and complex molecules found in living cells.

It had been known since 1946 that genes consist of DNA. At King's College, London, New Zealand–British biophysicist Maurice Wilkins (1916–) had been using X-ray crystallography to examine the structure of DNA, together with his colleague, British X-ray crystallographer Rosalind Franklin (1920–1958), and had made considerable progress.

While in Copenhagen, US scientist James Watson had realized that one of the major unresolved problems of biology was the precise structure of DNA. In 1952, he came as a young postdoctoral student to join the Medical Research Council Unit at the Cavendish Laboratory, Cambridge, where Francis Crick was already working. Convinced that a gene must be some kind of molecule, the two scientists set to work on DNA. Helped by the work of Wilkins, they were able to build an accurate model of DNA. They showed that DNA had a double helical structure, rather like a spiral staircase. Because the molecule of DNA was made from two strands, they envisaged that as a cell divides, the strands unravel, and each could serve as a template as new DNA was formed in the resulting daughter cells. Their model also explained how genetic information might be coded in the sequence of the simpler molecules of which DNA is comprised. Here for the first time was a complete insight into the basis of heredity. James Watson commented that this result was "too pretty not to be true!"

cracking the code

Later, working with South African–British molecular biologist Sidney Brenner (1927–), Crick went on to work out the genetic code, and so ascribe a precise function to each specific region of the molecule of DNA. These triumphant results created a tremendous flurry of scientific activity around the world. The pioneering work of Crick, Wilkins, and Watson was recognized in the award of the Nobel Prize for Physiology or Medicine in 1962.

The unravelling of the structure of DNA lead to a new scientific discipline, molecular biology, and laid the foundation stones for genetic engineering — a powerful new technique that is revolutionizing biology, medicine, and food production through the purposeful adaptation of living organisms.

product of one reaction forms the substrate for another. In each case, the reaction will be accelerated, or catalysed, by the action of a specific enzyme.

enzymes

Enzymes are biological catalysts produced in cells, and capable of speeding up the chemical reactions necessary for life. They are large, complex proteins, and are highly specific – each chemical reaction requiring its own particular enzyme. The enzyme's specificity arises from its **active site**, an area with a shape corresponding to part of the molecule with which it reacts (the substrate). The enzyme and the substrate slot together forming an enzyme–substrate complex that allows the reaction to take place, after which the enzyme falls away unaltered.

The activity and efficiency of enzymes are influenced by various factors, including temperature and pH conditions. Temperatures above 60°C/140°F damage (denature) the intricate structure of enzymes, causing reactions to cease. Each enzyme operates best within a specific pH range, and is denatured by excessive acidity or alkalinity.

Respiration employs about 70 different enzymes to act as catalysts. Digestive enzymes include amylases (which digest starch), lipases (which digest fats), and proteases (which digest protein). Other enzymes play a part in the replication of ◊DNA when a cell divides; the production of hormones; and the control of movement of substances into and out of cells.

transport across cell membranes

The ability to move molecules and ions into and out of cells and cellular structures as required is of vital importance in metabolic reactions. Such transportation may rely on passive processes such as diffusion, or it may involve the expenditure of energy.

diffusion Diffusion is the spontaneous movement of molecules or ions in a fluid from a region in which they are at a high concentration to a region of lower concentration, until a uniform concentration is achieved throughout. In biological systems it plays an essential role in the transport, over short distances, of molecules such as nutrients, respiratory gases, and neurotransmitters. It provides the means by which small molecules pass into and out of individual cells and microorganisms, such as amoebae, that possess no circulatory system.

osmosis Osmosis is the movement of water molecules through a semipermeable membrane from a region in which they are at a high concentration to a region where they are at a lower concentration. Many cell membranes behave as semipermeable membranes, and osmosis is a vital mechanism in the transport of fluids in living organisms – for example, in the transport of water from the roots up the stems of plants.

active transport Active transport requires the expenditure of energy to move molecules or ions across a membrane against a concentration gradient (that is, from where they are at a low concentration to where they are at a higher concentration). The molecule or ion becomes attached to a carrier protein straddling the cell membrane, which then – it is thought – alters its configuration as a result of energy input from ATP

to carry the molecule across the membrane and release it on the other side.

Active transport differs therefore from diffusion and osmosis, which are both passive processes requiring no input of energy.

respiration

This is the process by which food molecules are broken down to release their energy in a form that the organism can use to drive other metabolic processes. The resultant energy is packaged in the energy-carrying molecules ATP (adenosine triphosphate).

aerobic respiration Respiration that involves the presence of oxygen is called **aerobic respiration**. It is the more common form (the other is anaerobic respiration) and takes place largely in the mitochondria of the cell. Glucose is the usual substrate (though other food substances such as fatty acids and pentose sugars are also broken down). The process may be summarized as follows:

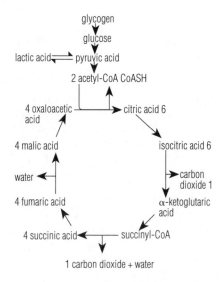

Krebs cycle The purpose of the Krebs (or citric acid) cycle is to complete the biochemical breakdown of food to produce energy-rich molecules, which the organism can use to fuel work. Acetyl coenzyme A (acetyl CoA) – produced by the breakdown of sugars, fatty acids, and some amino acids – reacts with oxaloacetic acid to produce citric acid, which is then converted in a series of enzyme-catalysed steps back to oxaloacetic acid. In the process, molecules of carbon dioxide and water are given off, and the precursors of the energy-rich molecules ATP are formed. (The numbers in the diagram indicate the number of carbon atoms in the principal compounds.)

$$C_6H_{12}O_6 + O_2 \rightarrow 6CO_2 + 6H_2O$$
$$+ \ 686 \ kcal/2,881 \ kJ$$

glucose + oxygen → carbon dioxide
+ water + energy

However, respiration takes place in a number of stages, each of which involves several steps, which proceed by coupling energy-yielding reactions to those that consume energy and are catalysed by the presence of specific enzymes.

The first stage, **glycolysis**, does not require oxygen and is therefore a form of anaerobic respiration. It takes place in the cell cytoplasm and is responsible for converting glucose to acetyl coenzyme A (acetyl CoA), with a net energy gain of two molecules of ATP for each molecule of glucose.

The next stage, the **Krebs cycle** (or citric acid or tricarboxylic acid cycle) takes place in the mitochondrial matrix. The acetyl CoA molecules created by glycolysis diffuse from the cytoplasm to the mitochondria, and are oxidized there to carbon dioxide in a complex series of steps that also result in the release of a further four molecules of ATP.

The last stage, the **electron transport chain**, is the main energy-producing stage and takes place on the mitochondrial walls. Much of the energy released from glucose during glycolysis and the Krebs cycle remains trapped within intermediate products (reduced nucleotides such as NADH, reduced nicotinamide adenine dinucleotide). The electron transport chain brings about the oxidation of the reduced nucleotides, shuttling the resultant electrons along a chain of carriers to bring about the reduction of molecular oxygen to water. It also brings about the phosphorylation of ADP molecules (adenosine diphosphate) to create a further 32 molecules of ATP.

The whole process therefore produces a total of 38 molecules of ATP for each molecule of glucose.

anaerobic respiration Anaerobic respiration takes place in the absence of oxygen. It begins with the process of glycolysis described above, but the acetyl CoA produced is then reduced either to alcohol (as in fermentation by yeast fungi) or to lactic acid (as in muscle cells experiencing oxygen debt). Only the glycolysis stage yields ATP; therefore, the net energy gain per molecule of glucose is only two molecules of ATP (compared with 38 molecules for aerobic respiration).

protein synthesis
This is the process by which cells manufacture the (proteins essential to their functioning, according to instructions encoded in the (DNA. Proteins play a vital role, most importantly as enzymes, which accelerate and control metabolic processes. The structure of proteins, which depends on the exact sequence of their amino-acid building blocks, is closely linked to their correct functioning. It is the genetic information carried in the DNA that controls that sequence.

protein synthesis and the genetic code Each of the nucleotide subunits of DNA is attached to a molecule called a **base**, which may be either adenine, guanine, thymine, or cytosine. A set of three consecutive bases – known as a **codon** – specifies a particular amino acid. A section of DNA that contains the sequence of codons required to encode a particular protein is called a **gene**.

Geneticists identify the codons by the initial letters of their constituent bases – for example, the base sequence of codon CAG is cytosine–adenine–guanine. Because there are four different bases, there must be 4 x 4 x 4 = 64 different codons. Proteins are usually made up of only 20 different amino acids, so many amino acids have more than one codon (for example, GGT, GGC, GGA, and GGG all code for the same amino acid, glycine).

transcription In order for the information represented by the DNA codons to be converted to amino acids, it must be transcribed to another form – **messenger RNA** – that can pass from the cell nucleus to the protein-making machinery in the cytoplasm.

In a section of DNA that represents a particular protein, the two strands of DNA unwind and separate. One of the strands then acts as a template for the production of messenger (RNA (mRNA). An enzyme moves along it, pairing the exposed bases with mRNA nucleotides that have complementary bases (cytosine in the DNA will pair with guanine nucleotides and vice versa, thymine will pair with adenine nucleotides, and adenine with uracil). In this way, a 'mirror image' of the information in the DNA is copied onto the mRNA. The mRNA nucleotides link together to form an mRNA strand, which then detaches itself from the DNA and passes out of the nucleus towards the (ribosomes in the cytoplasm. This is where the information carried by the mRNA will be translated into amino acids and, ultimately, the protein.

translation This is the process by which the information encoded as a sequence of bases in mRNA is transformed into the precise sequence of amino acids that makes up a polypeptide chain (the precursor of a protein molecule). It takes place in the ribosomes, tiny bodies that are either dispersed in the cytoplasm or attached to the (endoplasmic reticulum (a membranous system within the cytoplasm).

During translation, a ribosome binds to an mRNA strand, and then moves along its length. Small, hairpin-shaped strands of RNA, called **transfer RNA** (tRNA), bind to specific amino acids in the cytoplasm and bring these to the ribosome. They possess a special triplet of bases, called an **anticodon**, on their looped end that determines the amino acid that they will collect and also the codon on the mRNA to which they will bring that amino acid.

When the ribosome encounters a codon on the mRNA, the tRNA molecule carrying the correct amino acid is brought into position. The ribosome connects the new amino acid to the preceding amino acid with a peptide bond, and so the polypeptide chain is built up. When completed, the polypeptide is released by the ribosome into the endoplasmic reticulum or cytoplasm, where it may fold itself automatically into a protein or it may experience further modification.

genetics

Genetics is the branch of biology that is concerned with the study of heredity and variation; it attempts to explain how characteristics of living organisms are passed on from one generation to the next.

history of genetics

The science of genetics was initiated by the work of Austrian biologist Gregor Mendel whose experiments with the cross-breeding (hybridization) of peas showed that the inheritance of characteristics and traits takes place by means of discrete 'particles' (which would later be identified as ◊genes). These are present in the cells of all organisms, and are now recognized as being the basic units of heredity. Every organism possesses a **genotype** (a set of variable genes) and a **phenotype** (characteristics produced by certain genes). Modern geneticists investigate the structure, function, and transmission of genes.

Before the publication of Mendel's work in 1865, it had been assumed that the characteristics of both parents were blended during inheritance, but Mendel showed that the genes remain intact, although their combinations change. Since Mendel, the study of genetics has advanced greatly, first through breeding experiments and light-microscope observations (classical genetics), later by means of biochemical and electron microscope studies (molecular genetics).

In 1944 Canadian-born bacteriologist Oswald Avery, together with his colleagues at the Rockefeller Institute, Colin McLeod and Maclyn McCarthy, showed that the genetic material was deoxyribonucleic acid (◊DNA), and not protein as was previously thought. A further breakthrough was made in 1953 when James Watson and Francis Crick published their molecular model for the structure of DNA, the double helix, based on X-ray diffraction photographs. The

following decade saw the cracking of the genetic code. The genetic code is said to be universal since the same code applies to all organisms from bacteria and viruses to higher plants and animals, including humans. Today the deliberate manipulation of genes by biochemical techniques, or ◊genetic engineering, is commonplace.

MendelWeb
**http://www-hpcc.astro.washington.edu/mirrors/
MendelWeb/**
Hefty resource for anyone interested in Gregor Mendel, the origins of classical genetics, and the history and literature of science. View or download Mendel's original paper, with hypertext links to glossaries, biographical information, and exercises, or look up the essays, timeline, bibliography, and statistical tools.

---

genes

The nucleus of each cell of every organism contains a number of chromosomes – long threads made of DNA and protein. The number of chromosomes in each cell is constant for and characteristic of a particular species. Within the DNA of these chromosomes are the genes – units of inherited material that determine the characteristics of each organism, or parts of organisms, and allow these characteristics to be transmitted down from generation to generation. Each gene carries the chemically coded instructions for the production of a particular polypeptide, which in turn forms a protein (see (protein synthesis); it is these proteins that determine the form and function of the organism.

Located on the 23 pairs of chromosomes of each human cell are some 80,000 different genes, which determine features such as hair and skin colour. The presence or absence of specific genes can be a contributing factor in diseases such as haemophilia and cystic fibrosis. Each chromosome of a chromosome pair carries genes for the same characteristics in the same place, or locus. These two kinds of genes defining alternative characteristics are called ◊alleles. If the two alleles match, they are said to be homozygous; if they differ, they are described as heterozygous. Some alleles are dominant, and others are recessive; a

---

Natural History of Genetics
http://raven.umnh.utah.edu/
Through a combination of scientific experts and teachers, this site offers an accessible and well-designed introduction to genetics. It includes several guided projects with experiments and explanations aimed initially at young teenage children. However, this site also includes 'intermediate' and 'expert' sections allowing this page to be used by a wide variety of ages and levels of expertise. In addition to the experiments, the site also includes sections on such topics as 'core genetics', 'teacher workshops', and 'fun stuff'.

---

dominant allele masks the effects of its recessive partner. In other words, the dominant allele is expressed, and the recessive allele is not. A trait that results from a recessive allele is evident only in an individual that has two recessive alleles for that trait.

Human Genome Project

The Human Genome Project is a research scheme, begun in 1988, to map the complete nucleotide (see (nucleic acid) sequence of human DNA. There are approximately 80,000 different genes in the human genome, and one gene may contain more than 2 million nucleotides. The programme aims to collect 10–15,000 genetic specimens from 722 ethnic groups whose genetic make-up is to be preserved for future use and study. The knowledge gained is expected to help prevent or treat many crippling and lethal diseases, but there are potential ethical problems associated with knowledge of an individual's genetic make-up, and fears that it will lead to genetic discrimination.

Only 3% of the genome had been sequenced by mid 1998 but plans were announced to complete a 'rough draft' of 95% of the genome over the next three years. The target date for sequencing the whole genome is 2005, though a private US company announced plans to sequence the entire genome by 2001.

The Human Genome Organization (HUGO) coordinating the project expects to spend $1 billion over the first five years, making this the largest research project ever undertaken in the life sciences. Work is being carried out in more than 20 centres around the world.

Concern that, for example, knowledge of an individual's genes may make that person an unacceptable insurance risk has led to planned legislation on genome privacy in the USA, and 3% of HUGO's funds have been set aside for researching and reporting on the ethical implications of the project.

gene sequencing Each strand of DNA carries a sequence of chemical building blocks, the nucleotides. There are only four different types, but the number of possible combinations is immense. The different combinations of nucleotides produce different proteins in the cell, and thus determine the structure of the body and its individual variations. To establish the nucleotide sequence, DNA strands are broken into fragments, which are duplicated (by being introduced into cells of yeast or the bacterium *Escherichia coli*) and distributed to the research centres.

Genes account for only a small amount of the DNA sequence. Over 90% of DNA appears not to have any function, although it is perfectly replicated each time the cell divides, and handed on to the next generation.

Many higher organisms have large amounts of redundant DNA and it may be that this is an advantage, in that there is a pool of DNA available to form new genes if an old one is lost by mutation.

Human Genome Project Information
http://www.ornl.gov/TechResources/Human_Genome/home.html
US-based site devoted to this mammoth project – with news, progress reports, a molecular genetics primer, and links to other relevant sites.

---

genetic engineering

This is the deliberate manipulation of genetic material by biochemical techniques. It is often achieved by the introduction of new DNA, usually by means of a virus or plasmid. This can be for pure research, gene therapy (curing or alleviating inherited diseases or defects), or to breed functionally specific plants, animals, or bacteria. These organisms with a foreign gene added are said to be **transgenic**. At the beginning of 1995 more than 60 plant species had been genetically engineered, and nearly 3,000 transgenic crops had been field-tested.

practical uses In genetic engineering, the splicing and reconciliation of genes is used to increase knowledge of cell function and reproduction, but it can also achieve practical ends. For example, plants grown for food could be given the ability to fix nitrogen, found in some bacteria, and so reduce the need for expensive fertilizers, or simple bacteria may be modified to produce rare drugs. A foreign gene can be inserted into laboratory cultures of bacteria to generate commercial biological products, such as synthetic insulin, hepatitis-B vaccine, and interferon. Gene splicing was invented in 1973 by the US scientists Stanley Cohen and Herbert Boyer, and patented in the USA in 1984.

new developments Developments in genetic engineering have led to the production of growth hormone, and a number of other bone-marrow stimulating hormones. New strains of animals have also been produced; a new strain of mouse was patented in the USA in 1989 (the application was rejected in the European Patent Office). A vaccine against a sheep parasite (a larval tapeworm) has been developed by genetic

---

Your Genes, Your Choices: Exploring the Issues Raised by Genetic Research
http://www.ornl.gov/hgmis/publicat/ genechoice/index.html
Illustrated electronic book which describes the science of genetic research, as well as the ethical, legal, and social issues that it raises. Detailed and informative in itself, the site also contains an extensive bibliography.

engineering; most existing vaccines protect against bacteria and viruses.

The first genetically engineered food went on sale in 1994; the 'Flavr Savr' tomato, produced by the US biotechnology company Calgene, was available in California and Chicago.

evolution

Evolution is the slow, gradual process by which life has developed from single-celled organisms into the multiplicity of animal and plant life, extinct and existing, that inhabits the Earth. The development of the concept of evolution is usually associated with the English naturalist Charles Darwin, who attributed the main role in evolutionary change to **natural selection** acting on randomly occurring, hereditary variations to produce adaptations that make the organism better suited to its environment. These hereditary variations are now known to be produced by spontaneous changes, or **mutations**, in the genetic material, coupled with the **recombination** of that genetic material during sexual reproduction.

Evolutionary changes have not taken place in a linear manner, but as a branching process of descent from a common ancestor; for example, mammals and the now extinct synapsid reptiles share an ancestor that lived about 225 million years ago, while humans and chimpanzees share an ancestor that lived 5–10 million years ago. Of the 1.5 million identifiable species now existing on Earth, every one is the result of a long line of extinct species. Bacteria are among the earliest known species of life on Earth and are still evolving today.

natural and sexual selection

Evolution depends on the presence, within a population, of **alleles** – or inheritable variations in the (genes – that confer a selective advantage on the individuals possessing them. That is, they are more likely – under the natural pressures of predation, disease, and competition – to succeed than individuals that do not possess these alleles, and so the alleles become more prevalent in the population. The phrase 'survival of the fittest' is misleading since it implies the death of the 'unfit' individuals. From an evolutionary point of view, reproductive success is much more important than survival since if one type regularly leaves more offspring than another, the frequency of the more fertile type in the population is bound to increase. Reproductive success depends on many things including general vigour, the length of the reproductive period, and the ability to mate successfully (sexual selection).

The accumulated effect of natural selection is to produce **adaptations** such as the insulating coat of a polar bear or the spadelike forelimbs of a mole. Natural selection usually takes place over many years, but in fast-breeding organisms it can occur rapidly, for example the spread of antibiotic resistance in some bacteria.

sexual selection Sexual selection is a process similar to natural selection but relating exclusively to success in finding a mate for the purpose of sexual reproduction and producing offspring. Sexual selection occurs when one sex (usually but not always the female) invests more effort in producing young than the other. Members of the other sex compete for access to this limited resource (usually males competing for the chance to mate with females).

Sexual selection often favours features that increase a male's attractiveness to females (such as the pheasant's tail) or enable males to fight with one another (such as a deer's antlers). More subtly, it can produce hormonal effects by which the male makes the female unreceptive to other males, causes the abortion of fetuses already conceived, or removes the sperm of males who have already mated with a female.

speciation

Over the course of time, natural and sexual selection and, perhaps, chance genetic drift may cause a population to diverge so widely from its ancestral population that they are no longer able to interbreed and may therefore be considered a separate species. One cause of speciation is the geographical separation of populations, followed by reproductive isolation. Another cause is assortative mating – selective mating between individuals that are genetically related or have similar characteristics. If sufficiently consistent, assortative mating can theoretically result in the evolution of new species without geographical isolation.

adaptive radiation

Adaptive radiation is the formation of several species, with adaptations to different ways of life, from a single ancestral type. Adaptive radiation is likely to occur whenever members of a species migrate to a new habitat with unoccupied ecological niches. It is thought that the lack of competition in such niches allows sections of the migrant population to develop new adaptations, and eventually to become new species.

The colonization of newly formed volcanic islands has led to the development of many unique species. The 13 species of Darwin's finch on the Galápagos Islands, for example, are probably descended from a single species from the South American mainland. The parent stock evolved into different species that now occupy a range of diverse niches.

evolutionary theory

The idea of evolutionary change can be traced as far back as Lucretius in the 1st century BC, but it did not gain wide acceptance until the 19th century, following the work of Scottish geologist Charles Lyell, French naturalist Jean Baptiste Lamarck (who suggested that characteristics acquired during an organism's lifetime could be inherited by its offspring), and Charles Darwin (in conjunction with Welsh naturalist Alfred Russel Wallace).

The current theory of evolution, **Neo-Darwinism**, combines Darwin's theory of natural selection with the principles of (genetics. Although neither the general concept of evolution nor the importance of natural selection is doubted by biologists, but there still remains dispute over other possible mechanisms involved in evolutionary change. Chance, for example, may play a large part in deciding which genes become characteristic of a population – a phenomenon called **genetic drift**. It is now also clear that evolutionary change does not always occur at a constant rate, but that the process can have long periods of relative stability interspersed with periods of rapid change. This has led to new theories, such as the **punctuated equilibrium model**.

Darwin, Charles
http://www.literature.org/Works/Charles-Darwin
Complete text of Darwin's seminal works *On the Origin of Species* and *Voyage of the Beagle.*

---

evolution and creationism

Some religions deny the theory of evolution, considering it conflicts with their belief that God created all things (creationism). But most scientists accept that there is overwhelming evidence that the diversity of life arose by a process of evolutionary divergence and not by individual acts of divine creation. There are several lines of evidence for this: the fossil record, the existence of similarities – or homologies – between different groups of organisms, embryology, and geographical distribution.

Centre for Scientific Creation
http://www.creationscience.com/
Dedicated to researching the case for creation, rather than evolution, as the origin of living species. Most of the evidence is gathered in a book which is heavily promoted on this page. However, the entire book is available online.

---

fossils, homologies, and embryos Most organisms decompose quite rapidly after death, but sometimes a plant or animal is preserved, usually by being buried soon after death, occasionally by freezing. Burial is generally in peat or mud, although it can also be in volcanic ash or amber (fossilized tree resin) – some remarkably well-preserved fossils of small animals have been found in amber. Even after burial the soft tissues of an organism may decompose so that only the skeleton of an animal or the woody parts of a plant become fossilized.

Evolutionist, The
http://www.lse.ac.uk/depts/cpnss/evolutionist/
Online magazine devoted to evolutionary ideas which includes features, interviews, and comment.

The dating of fossils presents great difficulties. In the first instance they are dated according to the stratum, or layer, in which they are found, and correlated with fossils in the same layer, but radiometric methods have been developed for estimating the absolute ages of fossils. These depend on the fact that natural radioactive isotopes, such as those of carbon and potassium, decay at constant rates so that the amount of isotope remaining in a specimen is proportional to the length of time that has elapsed since its formation or deposition.

Although the fossil record does not actually prove the theory of evolution, a study of a series of fossils can provide a visual record of the evolution of individual species, such as the horse, and their adaptation to changing environments. Certain link fossils provide evidence of a link between species; an example of this is the *Archaeopteryx*, which was a birdlike animal with teeth and this fossil provides corroboration of the reptile ancestry of birds.

Evolution: Theory and History
http://www.ucmp.berkeley.edu/history/evolution.html
Dedicated to the study of the history and theories associated with evolution, this site explores topics on classification, taxonomy, and dinosaur discoveries, and then looks at the key figures in the field and reviews their contributions.

Additional evidence for evolution is found in homologous structures. For instance, a comparison of the limb bones of several different kinds of vertebrate indicates striking similarities in their construction. Such structures are termed homologous and their existence suggests that all these animals have evolved from a common ancestor.

Embryology can also provide important clues as to the ancestry of a group. The vertebrate animals all show very similar embryonic development, and all have embryonic gill slits even though the mature animal has lungs and breathes air. This is taken as evidence that the vertebrates all evolved from an aquatic ancestor breathing through gills like modern fish and the tadpole stage of amphibians. The human embryo has a small tail, the coccyx, which becomes vestigial (functionless) – possible evidence of a common ancestor with a tail.

evolution of humans

The African apes (gorilla and chimpanzee) have been shown by anatomical and molecular comparisons to be the closest living relatives of humans. The oldest known hominids (of the human group), the australopithecines, found in Africa, date from 3.5 to 4.4 million years ago. The first to use tools came 2 million years later, and the first humanoids to use fire and move out of Africa appeared 1.7 million years ago. Neanderthals were not our direct ancestors. Modern humans are all believed to descend from one African female of 200,000 years ago, although there is a rival theory that humans evolved in different parts of the world simultaneously.

Miocene apes Genetic studies indicate that the last common ancestor between chimpanzees and humans lived 5 to 10 million years ago. There are only fragmentary remains of ape and hominid fossils from this period. Dispute continues over the hominid status of *Ramapithecus*, the jaws and teeth of which have been found in India and Kenya in late Miocene deposits, dated between 14 and 10 million years. The lower jaw of a fossil ape found in the Otavi Mountains, Namibia, comes from deposits dated between 10 and 15 million years ago, and is similar to finds from East Africa and Turkey. It is thought to be close to the initial divergence of the great apes and humans.

australopithecines Bones of the earliest known human ancestor, a hominid named *Australopithecus ramidus* were found in Ethiopia in 1994 and dated as 4.4 million years old. *Australopithecus afarensis*, found in Ethiopia and Kenya, date from 3.9 to 4.4 million years ago. These hominids walked upright and they were either direct ancestors or an offshoot of the line that led to modern humans. They may have been the ancestors of *Homo habilis* (considered by some to be a species of *Australopithecus*), who appeared about 2 million years later, had slightly larger bodies and brains, and were probably the first to use stone tools. Also living in Africa at the same time was *Australopithecus africanus*, a possibly carnivorous hominid, and *Australopithecus robustus*, a hominid with robust bones, large teeth, heavy jaws, and thought to be a vegetarian. They are not generally considered to be our ancestors.

Homo erectus Over 1.7 million years ago, *Homo erectus*, believed by some to be descended from *Homo habilis*, appeared in Africa. *Homo erectus* had prominent brow ridges, a flattened cranium, with the widest part of the skull low down, and jaws with a rounded tooth row, but the chin, characteristic of modern humans, is lacking. They also had much larger brains (900-1,200 cu cm), and were probably the first to use fire and the first to move out of Africa. Their remains are found as far afield as China, western Asia, Spain, and southern Britain. Modern humans *Homo sapiens sapiens* and the Neanderthals *Homo sapiens neanderthalensis* are probably descended from *Homo erectus*.

Neanderthals Neanderthals were large brained and heavily built, probably adapted to the cold conditions of the ice ages. They lived in Europe and the Middle East, and disappeared about 40,000 years ago, leaving *Homo sapiens sapiens* as the only remaining species of the hominid group. Possible intermediate forms between Neanderthals and *Homo sapiens sapiens* have been found at Mount Carmel in Israel and at Broken Hill in Zambia, but it seems that *Homo sapiens sapiens* appeared in Europe quite rapidly and either wiped out the Neanderthals or interbred with them.

modern humans There are currently two major views of human evolution: the 'out of Africa' model, according to which *Homo sapiens sapiens* emerged from *Homo erectus*, or a descendant species, in Africa and then spread throughout the world; and the **multiregional model**, according to which selection pressures led to the emergence of similar advanced types of *Homo sapiens sapiens* from *Homo erectus* in different parts of the world at around the same time. Analysis of DNA in recent human populations suggests that *Homo sapiens sapiens* originated about 200,000 years ago in Africa from a single female ancestor, 'Eve'. The oldest known fossils of *Homo sapiens sapiens* also come from Africa, dating from 150,000–100,000 years ago. Separation of human populations would have occurred later, with separation of Asian, European, and Australian populations taking place between 100,000 and 50,000 years ago.

Humans are distinguished from apes by the complexity of their brain and its size relative to body size; by their small jaw, which is situated under the face and is correlated with reduction in the size of the anterior teeth, especially the canines, which no longer project beyond the tooth row; by their bipedalism, which affects the position of the head on the vertebral column, the lumbar and cervical curvature of the vertebral column, and the structure of the pelvis, knee joint, and foot; by their complex language; and by their elaborate culture.

The broad characteristics of human behaviour are a continuation of primate behaviour rather than a departure from it. For example, tool use, once a criterion for human status, has been found regularly in gorillas, orang-utans, and chimpanzees, and sporadically in baboons and macaques. Chimpanzees even make tools. In hominid evolution manual dexterity has increased so that more precise tools can be made. Cooperation in hunting, also once thought to be a unique human characteristic, has been found in chimpanzees, and some gorillas and chimpanzees have been taught to use sign language to communicate.

Fossil Hominids FAQ
http://earth.ics.uci.edu:8080/faqs/fossil-hominids.html
Basic information about hominid species, the most important hominid fossils, and creationist arguments, plus links to related sites.

---

classification and nomenclature

This is the arrangement of organisms into a hierarchy of groups on the basis of their similarities. The basic grouping is a species, several of which may constitute a genus, which in turn are grouped into families, and so on up through orders, classes, phyla (in plants, sometimes called divisions), to kingdoms.

The oldest method of classification, called **phenetic classification**, aims to classify organisms on the basis of as many as possible of their observable characteristics: their morphology, anatomy, physiology, and so on. Greek philosopher Theophrastus adopted this method in the 4th century BC, when he classified plants into trees, shrubs, undershrubs, and herbs. Awareness

---

Journey into Phylogenetic Systematics
http://www.ucmp.berkeley.edu/clad/clad4.html
Online exhibition about evolutionary theory with a specific emphasis on phylogenetic classification: the way that biologists reconstruct the pattern of events that has led to the distribution and diversity of life. The site provides an introduction to the philosophy, methodology, and implication of cladistic analysis, with a separate section on the need for cladistics.

---

of evolutionary theory, however, led to the development of **phylogenetic classification**, which aims to classify groups of organisms on the basis of their common ancestry and their genetic relationship. In practice, most present-day systems of classification compromise between the two approaches.

Cladistics is a controversial phylogenetic method that uses a formal step-by-step procedure for objectively assessing the extent to which organisms share particular characters with a common ancestor, and for assigning them to taxonomic groups called clades.

five-kingdom system

The kingdom is the primary division in biological classification. At one time, only two kingdoms were recognized: animals and plants. Today most biologists prefer a five-kingdom system, even though it still involves grouping together organisms that are probably unrelated. One widely accepted scheme is as follows: Kingdom **Animalia** (all multicellular animals); Kingdom **Plantae** (all plants, excluding seaweeds and other algae); Kingdom **Fungi** (all fungi, including the unicellular yeasts, but not slime moulds); Kingdom **Protista** or Protoctista (multicellular algae, protozoa, diatoms, dinoflagellates, slime moulds, and various

---

The Five Kingdoms of Living Things

Kingdom	Main features of organisms	Number of species
Monera[1]	all are bacteria; single-celled; prokaryotic (lack a membrane-bound nucleus); autotrophic (photosynthesis and chemosynthesis) and heterotrophic; all reproduce asexually, some also reproduce sexually	>10,000
Protista	single-celled or multicellular; eukaryotic (have a membrane-bound nucleus and membrane-bound organelles); autotrophic (photosynthesis in algae and Euglenoids) and heterotrophic; may reproduce asexually or sexually	>65,000
Fungi	single-celled and multicellular; eukaryotic; heterotrophic; form spores at all stages of their life cycle; usually reproduce asexually, many reproduce sexually by conjugation	about 100,000
Plantae	all are multicellular; eukaryotic; most are autotrophic (via photosynthesis); reproduce sexually; in some life cycle includes an alternation of generations (a haploid gametophyte stage and a diploid sporophyte stage)	about 500,000
Animalia	all are multicellular; eukaryotic; all are heterotrophic; reproduce sexually; develop from a blastula; most have tissues organized into organs	>795,000

[1] The Kingdom Monera is sometimes called the Kingdom Prokaryotae.

Evolution: Out of Africa and the Eve Hypothesis

BY CHRIS STRINGER

introduction

Most palaeoanthropologists recognize the existence of two human species during the last million years – *Homo erectus*, now extinct, and *Homo sapiens*, the species which includes recent or 'modern' humans. In general, they believe that *Homo erectus* was the ancestor of *Homo sapiens*.

How did the transition occur?

the multiregional model

There are two opposing views. The multiregional model says that *Homo erectus* gave rise to *Homo sapiens* across its whole range, which, about 700,000 years ago, included Africa, China, Java (Indonesia), and, probably, Europe.

Homo erectus, following an African origin about 1.7 million years ago, dispersed around the Old World, developing the regional variation that lies at the roots of modern 'racial' variation. Particular features in a given region persisted in the local descendant populations of today.

For example, Chinese *Homo erectus* specimens had the same flat faces, with prominent cheekbones, as modern Oriental populations. Javanese *Homo erectus* had robustly built cheekbones and faces that jutted out from the braincase, characteristics found in modern Australian Aborigines. No definite representatives of *Homo erectus* have yet been discovered in Europe. Here, the fossil record does not extend back as far as those of Africa and eastern Asia, although a possible *Homo erectus* jawbone more than a million years old was recently excavated in Georgia.

Nevertheless, the multiregional model claims that European *Homo erectus* did exist, and evolved into a primitive form of *Homo sapiens*. Evolution in turn produced the Neanderthals: the ancestors of modern Europeans. Features of continuity in this European lineage include prominent noses and midfaces.

genetic continuity

The multiregional model was first described in detail by Franz Weidenreich, a German palaeoanthropologist. It was developed further by the American Carleton Coon, who tended to regard the regional lineages as genetically separate. Most recently, the model has become associated with such researchers as Milford Wolpoff (USA) and Alan Thorne (Australia), who have re-emphasized the importance of gene flow between the regional lines. In fact, they regard the continuity in time and space between the various forms of *Homo erectus* and their regional descendants to be so complete that they should be regarded as representing only one species – *Homo sapiens*.

the opposing view

The opposing view is that *Homo sapiens* had a restricted origin in time and space. This is an old idea. Early in the 20th cen-

tury, workers such as Marcellin Boule (France) and Arthur Keith (UK) believed that the lineage of *Homo sapiens* was very ancient, having developed in parallel with that of *Homo erectus* and the Neanderthals. However, much of the fossil evidence used to support their ideas has been re-evaluated, and few workers now accept the idea of a very ancient and separate origin for modern *Homo sapiens*.

the Garden of Eden

Modern proponents of this approach focus on a recent and restricted origin for modern *Homo sapiens*. This was dubbed the 'Garden of Eden' or 'Noah's Ark' model by the US anthropologist William Howells in 1976 because of the idea that all modern human variation had a localized origin from one centre. Howells did not specify the centre of origin, but research since 1976 points to Africa as especially important in modern human origins.

The consequent 'Out of Africa' model claims that *Homo erectus* evolved into modern *Homo sapiens* in Africa about 100,000–150,000 years ago. Part of the African stock of early modern humans spread from the continent into adjoining regions and eventually reached Australia, Europe, and the Americas (probably by 45,000, 40,000, and 15,000 years ago respectively). Regional ('racial') variation only developed during and after the dispersal, so that there is no continuity of regional features between *Homo erectus* and present counterparts in the same regions.

Like the multiregional model, this view accepts that *Homo erectus* evolved into new forms of human in inhabited regions outside Africa, but argues that these non-African lineages became extinct without evolving into modern humans. Some, such as the Neanderthals, were displaced and then replaced by the spread of modern humans into their regions.

... and an African Eve?

In 1987, research on the genetic material called mitochondrial DNA (mtDNA) in living humans led to the reconstruction of a hypothetical female ancestor for all present-day humanity. This 'Eve' was believed to have lived in Africa about 200,000 years ago. Recent re-examination of the 'Eve' research has cast doubt on this hypothesis, but further support for an 'Out of Africa' model has come from genetic studies of nuclear DNA, which also point to a relatively recent African origin for present-day *Homo sapiens*.

Studies of fossil material of the last 50,000 years also seem to indicate that many 'racial' features in the human skeleton have developed only over the last 30,000 years, in line with the 'Out of Africa' model, and at odds with the million-year timespan one would expect from the multiregional model.

other lower organisms with eukaryotic cells); and Kingdom **Monera** or Prokaryotae (the prokaryotes: bacteria and cyanobacteria, or blue-green algae). The first four of these kingdoms make up the eukaryotes.

When only two kingdoms were recognized, any organism with a rigid cell wall was a plant, and so bacteria and fungi were considered plants, despite their many differences. Other organisms, such as the photosynthetic flagellates (euglenoids), were claimed by both kingdoms. The unsatisfactory nature of the two-kingdom system became evident during the 19th century, and the biologist Ernst Haeckel was among the first to try to reform it. High-power microscopes have revealed more about the structure of cells; it has become clear that there is a fundamental difference between cells without a nucleus (prokaryotes) and those with a nucleus (eukaryotes). However, these differences are larger than those between animals and higher plants, and are unsuitable for use as kingdoms. At present there is no agreement on how many king-

doms there are in the natural world. Some schemes have as many as 20.

binomial system of nomenclature

This is the system by which all organisms are identified by a two-part Latinized name. Devised by the biologist Linnaeus, it is also known as the Linnaean system. The first name is capitalized and identifies the genus; the second identifies the species within that genus.

Usually the names are descriptive. Thus, the name of the dog, *Canis familiaris*, means the 'familiar species of the dog genus', *Canis* being Latin for 'dog'. Each species is defined by an officially designated type specimen housed at a particular museum. The rules for naming organisms in this way are specified in a number of International Codes of Taxonomic Nomenclature administered by two International Commissions on Nomenclature, one zoological and one botanical.

Biology, Genetics, and Evolution Chronology

c. 570 BC	Greek philosopher Anaximander argues that life evolved from the sea, and that land animals are descendants of sea animals – the first evolutionary theory.
c. 560 BC	Greek philosopher Xenophanes correctly recognizes the nature of fossils when he suggests that fossil seashells are the result of a great flood that buried them in the mud.
1665	English scientist Robert Hooke publishes *Micrographia*, the first serious scientific work on microscopy, describing the function of the microscope, and coining the name 'cells' to describe cavities he has found in the structure of cork.
1674	Dutch microscopist Anton van Leeuwenhoek develops the single-lens microscope, and begins a series of important discoveries by observing protozoa.
1683	Leeuwenhoek is the first to observe bacteria.
1735	In his *Systema Naturae/System of Nature*, Swedish botanist Carolus Linnaeus introduces a system for classifying plants by genus and species – a taxonomy that will survive the upheavals of evolutionism and remains in use today.
1735	In his *Telliamed*, French scientist Benoit de Maillet puts forward an evolutionary hypothesis.
1745	French naturalist Pierre de Maupertuis attacks the currently favoured theory of reproduction, that the sperm contain a miniature version of the adult. He argues that characteristics of both parents influence the offspring.
1751	French naturalist Georges Buffon is condemned by the Sorbonne (University of Paris, France) for supporting the idea of evolution in his *Natural History*. He is forced to recant, declaring the biblical account of creation correct.
1758	Linnaeus applies the binomial taxonomy he developed for plant classification to animal species.
1780	French chemist Antoine-Laurent Lavoisier demonstrates that respiration is a form of combustion.
1794	English naturalist and physician Erasmus Darwin publishes *Zoonomia, or the Laws of Organic Life*, expressing his ideas on evolution (which he assumes has an environmental cause).
1802	French biologist Jean-Baptiste de Lamarck is the first to use the term 'biology'.
1809	Lamarck theorizes that organs improve with use and degenerate with disuse and that these environmentally adapted traits are inheritable.
1817	French scientist Georges Cuvier breaks away from the view that animals can be arranged in a linear sequence leading to humans and argues instead that they should be classified according to their anatomical organization.
1831	Scottish botanist Robert Brown discovers the nucleus in plant cells.
1831	English naturalist Charles Darwin undertakes a five-year voyage, to South America and the Pacific, as naturalist on the *Beagle*. The voyage convinces him that species have evolved

gradually but he waits over 20 years to publish his findings.

1838 Dutch chemist Gerard Johann Mulder coins the word 'protein'.

1838 German botanist Matthias Jakob Schleiden recognizes that cells are the fundamental units of all plant life. He is thus the first to formulate cell theory.

1838 German chemist Justus von Liebig demonstrates that animal heat is due to respiration.

1839 German physiologist Theodor Schwann argues that all animals and plants are composed of cells. Along with Matthias Schleiden, he thus founds modern cell theory.

1840 Swiss embryologist Rudolf Albert von Kölliker identifies spermatozoa as cells.

1845 German zoologist Carl von Siebold describes the unicellular nature of protozoa and describes the function of the cilia.

1846 German botanist Hugo von Mohl uses the word protoplasm to describe the main living substance in a cell. It leads to the development of cell physiology.

1852 German physician Robert Remak discovers that the growth of tissues involves both the multiplication and division of cells.

1856 German naturalist Johann Fuhrott discovers the first fossil remains of a Neanderthal in Quaternary bed in Feldhofen Cave near Hochdal Cave above the Neander Valley, Germany. They cause immediate debate about whether they are the remains of ancient humans or the deformed bones of a modern human.

1858 English chemists W H Perkin and B F Duppa synthesize glycine, the first amino acid to be manufactured.

1858 English naturalists Charles Darwin and Alfred Russel Wallace contribute a joint paper on the variation of species to the Linnaean Society of London, England, stating their conclusions about natural selection and evolution.

1859 Charles Darwin publishes *On the Origin of Species by Natural Selection*, which expounds his theory of evolution by natural selection, and by implication denies the truth of biblical creation and God's hand in Nature. It sells out immediately and revolutionizes biology.

1860 At the Oxford meeting of the British Association, Bishop Samuel Wilberforce and biologist Thomas Henry Huxley debate creationism versus evolutionism.

1861 German zoologist Max Schultze defines the cell as consisting of protoplasm and a nucleus, a structure he recognizes as fundamental in both plants and animals.

1864 English philosopher Herbert Spencer coins the term 'survival of the fittest' in *Principles of Biology*.

1865 Austrian monk and botanist Gregor Mendel publishes a paper that outlines the fundamental laws of heredity.

1866 German embryologist Ernst Haeckel proposes a third category of living beings intermediate between plants and animals. Called Protista, it consists mostly of microscopic organisms such as protozoans, algae, and fungi.

1868 French geologist Louis Lartet is the first to discover the skeletal remains of anatomically modern humans, in a cave near Cro-Magnon, France. They are 35,000 years old.

1869 Swiss biochemist Johann Miescher discovers a nitrogen and phosphorous material in cell nuclei that he calls nuclein but which is now known as the genetic material DNA.

1876 German cytologist Eduard Adolf Strasburger describes the process of mitosis.

1880 German cytologist Eduard Adolf Strasburger announces that new cell nuclei arise from the division of old nuclei.

1882 German chemist Emil Hermann Fischer shows that proteins are polymers, or large molecules, comprised of amino acids.

1886 German biologist August Friedrich Leopold Weismann states that reproductive cells, or 'germ plasm' cells, remain unchanged from generation to generation, and that they contain some hereditary substance which is now known as chromosomes, DNA, and genes.

1892 Dutch geneticist Hugo Marie de Vries, through a programme of plant breeding, establishes the same laws of heredity discovered by Gregor Mendel in 1865.

1892 German embryologist and anatomist Oscar Hertwig establishes the science of cytology by suggesting that the processes that go on inside the cell are reflections of organismic processes.

1892 Russian microbiologist Dimitry Iosifovich Ivanovsky publishes 'On Two Diseases of Tobacco' in which he announces that mosaic disease in tobacco is caused by microorganisms too small to be seen through a microscope. Now known as viruses, his discovery pioneers the science of virology.

1894 Dutch anatomist Marie Eugène Dubois announces the discovery, in Java, of the remains of the first specimen of *Homo erectus* ('upright man'), which he calls *Pithecanthropus erectus*, and which has a cranial capacity of 900 cc and is 0.5 to 1 million years old.

1899 The amino acid cystine is discovered to be a component of protein.

1900 Dutch geneticist Hugo Marie de Vries, German botanist Carl Erich Correns, and Austrian botanist Erich Tschermak von Seysenegg, simultaneously and independently rediscover the Austrian monk Gregor Mendel's 1865 work on heredity.

1901 English biochemist Frederick Gowland Hopkins isolates the amino acid tryptophan.

1902	German chemists Emil Fischer and Franz Hofmeister discover that proteins are polypeptides consisting of amino acids.
1902–04	US geneticist Walter Sutton and the German zoologist Theodor Boveri found the chromosomal theory of inheritance when they show that cell division is connected with heredity.
1905	Danish botanist Wilhelm Johannsen introduces the terms 'genotype' and 'phenotype' to explain how genetically identical plants differ in external characteristics.
1906	British biochemists Arthur Harden and William Young discover catalysis among enzymes.
1906	English biologist William Bateson introduces the term 'genetics'.
1907	German chemist Emil Fischer describes the synthesis of amino acid chains in proteins.
1907	The Heidelberg jaw is discovered in a sand pit at Mauer, Germany. Belonging to *Homo erectus*, it is the oldest European hominid fossil discovered to date and thought to be 400,000 years old.
1909	Danish botanist Wilhelm Ludvig Johannsen introduces the term 'gene'.
1909	English biologist William Bateson publishes *Mendel's Principles of Genetics*, which introduces Mendelian genetics to the English-speaking world.
1909	German botanist Carl Correns shows that certain hereditary characteristics of plants are determined by factors in the cytoplasm of the female sex cell. It is the first example of non-Mendelian heredity.
1909	Russian-born US chemist Phoebus Levene discovers D-ribose, the five-carbon sugar that forms the basis of RNA.
1910	US geneticist Thomas Hunt Morgan discovers that certain inherited characteristics of the fruit fly *Drosophila melanogaster* are sex linked. He later argues that because all sex-related characteristics are inherited together they are linearly arranged on the X-chromosome.
1914	German biochemist Fritz Albert Lepmann explains the role of adenosine triphosphate (ATP) as the carrier of chemical energy from the oxidation of food to the energy consumption processes in the cells.
1915	US geneticists Thomas Hunt Morgan, Alfred Sturtevant, Calvin Bridges, and Hermann Muller publish *The Mechanism of Mendelian Heredity*, which outlines their work on the fruit fly *Drosophila* demonstrating that genes can be mapped on chromosomes.
1924	Australian-born South African anthropologist Raymond Dart discovers the skull of an early hominid at Tuang, Botswana, which he calls *Australopithecus africanus*. It is now believed to be one of the oldest human ancestors.
1927	Canadian anthropologist Davidson Black discovers the first specimens of 'Beijing man' (*Sinanthropous pekinensis*), a species of *Homo*
	erectus believed to be 300,000 to 400,000 years old, at Choukoutien, China.
1927	US geneticist Hermann Muller uses X-rays to cause mutations in the fruit fly. It permits a greater understanding of the mechanisms of variation.
1930	English geneticist Ronald Fisher publishes *The Genetical Theory of Natural Selection* in which he synthesizes Mendelian genetics and Darwinian evolution.
1932	US anthropologist George Edward Lewis discovers the jaw of a Miocene ape *Ramapithecus* in the Siwalik Hills of India. Living about 8–15 million years ago, it is thought to be the oldest human ancestor.
1935	US biochemist Wendell Meredith Stanley shows that viruses are not submicroscopic organisms but are proteinaceous in nature.
1936	Fossil remains of *Pithecanthropus* (now *Homo erectus*) found in Java indicate that *Homo erectus* lived in the area 500,000–1 million years ago.
1937	German-born British biochemist Hans Krebs describes the citric acid cycle in cells, which converts sugars, fats, and proteins into carbon dioxide, water, and energy – the 'Krebs cycle'.
1944	The role of deoxyribonucleic acid (DNA) in genetic inheritance is first demonstrated by US bacteriologist Oswald Avery, US biologist Colin MacLeod, and US biologist Maclyn McCarthy; it opens the door to the elucidation of the genetic code.
1946	US biologists Max Delbrück and Alfred D Hershey discover recombinant DNA (deoxyribonucleic acid) when they observe that genetic material from different viruses can combine to create new viruses.
1946	US geneticists Joshua Lederberg and Edward Lawrie Tatum pioneer the field of bacterial genetics with their discovery that sexual reproduction occurs in the bacterium *Escherichia coli*.
1948	*Proconsul africanus* is discovered by Kenyan anthropologist Louis Leakey in Kenya. It is a Miocene ape that is a possible ancestor of both apes and monkeys.
1948	US biologist Alfred Mirsky discovers ribonucleic acid (RNA) in chromosomes.
1952	English biophysicist Rosalind Franklin uses X-ray diffraction to study the structure of DNA. She suggests that its sugar-phosphate backbone is on the outside – an important clue that leads to the elucidation of the structure of DNA the following year.
1953	US biochemist Stanley Lloyd Miller shows that amino acids can be formed when simulated lightning is passed through containers of water, methane, ammonia, and hydrogen – conditions under which life may have arisen.

1953	English molecular biologist Francis Crick and US biologist James Watson announce the discovery of the double helix structure of DNA, the basic material of heredity. They also theorize that if the strands are separated then each can form the template for the synthesis of an identical DNA molecule. It is perhaps the most important discovery in biology.
1954	Russian-born US cosmologist George Gamow suggests that the genetic code consists of the order of nucleotide triplets in the DNA molecule.
1955	US geneticists Joshua Lederberg and Norton Zinder discover that some viruses carry part of the chromosome of one bacterium to another; called transduction it becomes an important tool in genetics research.
1956	Romanian-born US biologist George Palade discovers ribosomes, which contain RNA (ribonucleic acid).
1956	Spanish-born US molecular biologist Severo Ochoa discovers polynucleotide phosphorylase, the enzyme responsible for the synthesis of RNA (ribonucleic acid), which allows him to synthesize RNA.
1956	US biochemist Arthur Kornberg, using radioactively tagged nucleotides, discovers that the bacteria *Escherichia coli* uses an enzyme, now known as DNA polymerase, to replicate DNA (deoxyribonucleic acid). It allows him to synthesize DNA in the test tube.
1956	US biologists Mahlon Hoagland and Paul Zamecnik discover transfer RNA (ribonucleic acid) which transfers amino acids, the building blocks of proteins, to the correct site on the messenger RNA.
1961	English molecular biologist Francis Crick and South African chemist Sydney Brenner discover that each base triplet on the DNA strand codes for a specific amino acid in a protein molecule.
1961	French biochemists François Jacob and Jacques Monod discover messenger ribonucleic acid (mRNA), which transfers genetic information to the ribosomes, where proteins are synthesized.
1961	Kenyan anthropologist Louis Leakey and English anthropologist Mary Leakey find the first fossilized remains of *Homo habilis* ('Handy Man') at Olduvai Gorge, Tanganyika (modern Tanzania). Makers of Oldowan stone tools – the oldest stone tools – they lived 1.15 to 1.7 million years ago.
1967	US biochemist Marshall Nirenberg establishes that mammals, amphibians, and bacteria all share a common genetic code.
1967	US scientist Charles Caskey and associates demonstrate that identical forms of messenger RNA produce the same amino acids in a variety of living beings, showing that the genetic code is common to all life forms.
1969	US geneticist Jonathan Beckwith and associates at the Harvard Medical School isolate a single gene for the first time.
1970	US biochemists Howard Temin and David Baltimore separately discover the enzyme reverse transcriptase, which allows some cancer viruses to transfer their RNA to the DNA of their hosts turning them cancerous – a reversal of the common pattern in which genetic information always passes from DNA to RNA.
1970	US geneticist Hamilton Smith discovers type II restriction enzyme that breaks the DNA strand at predictable places, making it an invaluable tool in recombinant DNA technology.
1970	Indian-born US biochemist Har Gobind Khorana assembles an artificial yeast gene from its chemical components.
1972	US palaeontologists Stephen Jay Gould and Nils Eldridge propose the punctuated equilibrium model – the idea that evolution progresses in fits and starts rather than at a uniform rate.
1973	Two fossil skulls, both about 1.85 million years old, are discovered at Koobi Fora, Kenya; they have features typical of *Australopithecus boisei* as well as *Homo habilis*, confounding classification of early hominid species.
1973	US biochemists Stanley Cohen and Herbert Boyer develop the technique of recombinant DNA (deoxyribonucleic acid). Strands of DNA are cut by restriction enzymes from one species and then inserted into the DNA of another; this marks the beginning of genetic engineering.
1974	US anthropologists Donald Johanson and Maurice Taieb discover the 3.2 million-years-old remains of 'Lucy', an adult female hominid classified as *Australopithecus afarensis*, at Hadar in Ethiopia. About 40% of her skeleton is found and it indicates that she was bipedal.
1975	Kenyan field worker Bernard Ngeneo discovers a *Homo erectus* skull at Koobi Fora, Kenya, which is estimated to be 1.7 million years old; discovered in the same stratum as *Australopithecus boisei*, it puts an end to the single species hypothesis, the idea that there has never been more than one hominid species at any point in history.
1975	The gel-transfer hybridization technique for the detection of specific DNA (deoxyribonucleic acid) sequences is developed; it is a key development in genetic engineering.
1976	The first oncogene (cancer-inducing gene) is discovered by US scientists Harold E Varmus and J Michael Bishop.
1976	US biochemist Herbert Boyer and venture capitalist Robert Swanson found Genentech in San Francisco, California, the world's first genetic engineering company.

1976	Har Gobind Khorana and his colleagues announce the construction of the first artificial gene to function naturally when inserted into a bacterial cell. This is a major breakthrough in genetic engineering.
1977	English biochemist Frederick Sanger describes the full sequence of 5,386 bases in the DNA (deoxyribonucleic acid) of virus *phi*X174 in Cambridge, England; the first sequencing of an entire genome.
1977	US scientist Herbert Boyer, of the firm Genentech, fuses a segment of human DNA (deoxyribonucleic acid) into the bacterium *Escherichia coli* which begins to produce the human protein somatostatin; this is the first commercially produced genetically engineered product.
1981	The US Food and Drug Administration grants permission to Eli Lilley and Co to market insulin produced by bacteria, the first genetically engineered product to go on sale.
1981	US geneticists J W Gordon and F H Ruddle of the University of Ohio inject genes from one animal into the fertilized egg of a mouse that develops into mice with the foreign gene in many of the cells; the gene is then passed on to their offspring creating permanently altered (transgenic) animals; it is the first transfer of a gene from one animal species to another.
1982	The Swedish firm Kabivitrum manufactures human growth hormone using genetically engineered bacteria.
1982	US firm Applied Biosystems markets an automated gene sequencer that can sequence 18,000 DNA bases a day, compared with a few hundred a year by hand in the 1970s.
1983	US biologists Andrew Murray and Jack Szostak create the first artificial chromosome; it is grafted onto a yeast cell.
1983	US biochemist Kary Mullis invents the polymerase chain reaction (PCR); a method of multiplying genes or known sections of the DNA molecule a million times without the need for the living cell.
1984	Robert Sinsheimer, the chancellor of the University of California at Santa Cruz, California, proposes that all human genes be mapped; the proposal eventually leads to the development of the Human Genome Project.
1987	The first genetically altered bacteria are released into the environment in the USA; they protect crops against frost.
1987	The US Patent and Trademark Office announces its intention to allow the patenting of animals produced by genetic engineering.
1987	The *New York Times* announces Dr Helen Donis-Keller's mapping of all 23 pairs of human chromosomes, allowing the location of specific genes for the prevention and treatment of genetic disorders.

1988	Fossil remains of a modern *Homo sapiens* are discovered in Israel, dated about 92,000 years ago; they suggest modern humans appeared twice as early as previously thought.
1988	The Human Genome Organization (HUGO) is established in Washington, DC, USA; scientists announce a project to compile a complete 'map' of human genes.
1990	A four-year-old girl in the USA has the gene for adenosine deaminase inserted into her DNA (deoxyribonucleic acid); she is the first human to receive gene therapy.
1991	British geneticists Peter Goodfellow and Robin Lovell-Badge discover the gene on the Y chromosome that determines sex.
1992	The US biotechnology company Agracetus patents transgenic cotton, which has had a foreign gene added to it by genetic engineering.
1992	US biologist Philip Leder receives a patent for the first genetically engineered animal, the oncomouse, which is sensitive to carcinogens.
1994	Bones of the earliest known human ancestor, a hominid named *Australopithecus ramidus*, are found in Ethiopia and dated at 4.4 million years old.
1994	The first genetically engineered food goes on sale in California and Chicago, Illinois. The 'Flavr Savr' tomato is produced by the US biotechnology company Calgene.
1996	US geneticists clone two rhesus monkeys from embryo cells.
1997	Scottish researcher Ian Wilmut of the Roslin Institute in Edinburgh, Scotland, announces that British geneticists have cloned an adult sheep.
1997	Spanish palaeoanthropologist Bermúdez de Castro and his team discover the fossilized remains of six individuals belonging to a new human species in a cave in Spain's Atepuerca Mountains. Named *Homo antecessor*, and about 780,000 years old (the only human fossils found in Europe of that age), they possess a face like *Homo sapiens* and a jaw and brow similar to the Neanderthals, and are believed to be the ancestors of both modern humans and Neanderthals.
1997	US geneticist Huntington F Wilard constructs the first artificial human chromosome.
1997	Teams of researchers from Germany and the USA use mitochondrial DNA (deoxyribonucleic acid) extracted from the original Neanderthal fossils, discovered in the Neander Valley near Düsseldorf, Germany, in 1856, to confirm that Neanderthals and modern humans diverged evolutionarily about 600,000 years ago.
1998	Doctors meeting at the World Medical Association's conference in Hamburg, Germany, call for a worldwide ban on human cloning. US president Clinton calls for legislation banning cloning the following day.

Biographies

Avery, Oswald Theodore (1877–1955) Canadian-born US bacteriologist. His work on transformation in bacteria established in 1944 that DNA is responsible for the transmission of heritable characteristics. Avery proved conclusively that DNA was the transforming principle responsible for the development of polysaccharide capsules in unencapsulated bacteria that had been in contact with dead, encapsulated bacteria. This implicated DNA as the basic genetic material of the cell.

Bateson, William (1861–1926) English geneticist. Bateson was one of the founders of the science of genetics (a term he introduced), and a leading proponent of Austrian biologist Gregor Mendel's work on heredity. In *Material for the Study of Variation* (1894) Bateson put forward his theory of discontinuity to explain the long process of evolution. According to this theory, species do not develop in a predictable sequence of very gradual changes but instead evolve in a series of discontinuous jumps. Mendel's work, which he translated and championed, provided him with supportive evidence. Bateson also carried out breeding experiments, and showed that certain traits are consistently inherited together; this phenomenon (called linkage) is now known to result from genes being situated close together on the same chromosome.

Beadle, George Wells (1903–1989) US biologist. In 1958 he shared a Nobel prize with Edward L ◊Tatum and Joshua ◊Lederberg for his work in biochemical genetics, forming the 'one-gene–one-enzyme' hypothesis (a single gene codes for a single kind of enzyme). This concept found wide applications in biology and virtually created the science of biochemical genetics.

Berg, Paul (1926–) US molecular biologist. In 1972, using gene-splicing techniques developed by others, Berg spliced and combined into a single hybrid the ◊DNA from an animal tumour virus (SV40) and the DNA from a bacterial virus. For his work on recombinant DNA he shared the 1980 Nobel Prize for Chemistry.

In 1956 Berg identified an RNA molecule (later known as a transfer RNA) that is specific to the amino acid methionine. He then perfected a method for making bacteria accept genes from other bacteria. This genetic engineering can be extremely useful for creating strains of bacteria to manufacture specific substances, such as interferon. But there are also dangers: a new, highly virulent pathogenic microorganism might accidentally be created, for example. Berg has therefore advocated restrictions on genetic engineering research.

Brenner, Sidney (1927–) South African scientist, one of the pioneers of genetic engineering. Brenner discovered messenger ◊RNA (a link between ◊DNA and the ◊ribosomes in which proteins are synthesized) in 1960.

Brenner became engaged in one of the most elaborate efforts in anatomy ever attempted: investigating the nervous system of nematode worms and comparing the nervous systems of different mutant forms of the animal. About a hundred genes are involved in constructing the nervous system of a nematode and most of the mutations that occur affect the overall design of a section of the nervous system.

Crick, Francis Harry Compton (1916–) English molecular biologist. From 1949 he researched the molecular structure of ◊DNA, and the means whereby characteristics are transmitted from one generation to another. For this work he was awarded a Nobel prize (with Maurice ◊Wilkins and James ◊Watson) in 1962.

Using Wilkins's and others' discoveries, Crick and Watson postulated that DNA consists of a double helix consisting of two parallel chains of alternate sugar and phosphate groups linked by pairs of organic bases. They built molecular models which also explained how genetic information could be coded – in the sequence of organic bases. Crick and Watson published their work on the proposed structure of DNA in 1953. Their model is now generally accepted as correct.

Darwin, Charles Robert (1809–1882) English naturalist who developed the modern theory of ◊evolution and proposed, with Alfred Russel ◊Wallace, the principle of ◊natural selection.

After research in South America and the Galápagos Islands as naturalist on HMS *Beagle* during 1831–36, Darwin published *On the Origin of Species by Means of Natural Selection or the Preservation of Favoured Races in the Struggle for Life* (1859). This book explained the evolutionary process through the principles of natural selection and aroused bitter controversy because it disagreed with the literal interpretation of the Book of Genesis in the Bible.

On the Origin of Species also refuted earlier evolutionary theories, such as those of French naturalist J B de ◊Lamarck. Darwin himself played little part in the debates, but his *Descent of Man* (1871) added fuel to the theological discussion, in which English scientist T H ◊Huxley and German zoologist Ernst ◊Haeckel took leading parts.

Darwin's theory of natural selection concerned the variation existing between members of a sexually reproducing population. Those members with variations better fitted to the environment would be more likely to survive and breed, subsequently passing on these favourable characteristics to their offspring. He avoided the issue of human evolution, however, remarking at the end of *The Origin of Species* that 'much light will be thrown on the origin of man and his history'. It was not until his publication of *The Descent of Man and Selection in Relation to Sex* (1871) that Darwin argued that people evolved just like other organisms.

Dawkins, (Clinton) Richard (1941–) British zoologist whose book *The Selfish Gene* (1976) popularized the theories of sociobiology (social behaviour in humans and animals in the context of evolution). In *The Blind Watchmaker* (1986) he explained the modern theory of evolution.

In *The Selfish Gene* he argued that genes – not individuals, populations, or species – are the driving force of evolution. He suggested an analogous system of cultural transmission in human societies, and proposed the term 'mimeme', abbreviated to 'meme', as the unit of such a scheme. He considered the idea of God to be a meme with a high survival value.

Dobzhansky, Theodosius (1900–1975), originally Feodosy Grigorevich Dobrzhansky, Ukrainian-born US geneticist who established evolutionary genetics as an independent discipline. He showed that genetic variability between individuals of the same species is very high and that this diversity is vital to the process of evolution.

His book *Genetics and the Origin of Species* (1937) was the first significant synthesis of Darwinian evolutionary theory and Mendelian genetics. Dobzhansky also proved that there is a period when speciation is only partly complete and during which several races coexist.

Fischer, Edmond (1920–) US biochemist who shared the 1992 Nobel Prize for Physiology or Medicine with Edwin Krebs for isolating and describing the action of the enzymes responsible for reversible protein phosphorylation. Reversible phosphorylation is the attachment or detachment of phosphate groups to or from proteins in cells. It is at the heart of a wide range of biological processes ranging from muscle contraction to the regulation of genes.

Fisher, Ronald Aylmer (1890–1962) English statistician and geneticist. He modernized Charles Darwin's theory of evolution, thus securing the key biological concept of genetic change by natural selection. Fisher developed several new statistical techniques and, applying his methods to genetics, published *The Genetical Theory of Natural Selection* (1930).

This classic work established that the discoveries of the geneticist Gregor ◊Mendel could be shown to support Darwin's theory of evolution.

Franklin, Rosalind Elsie (1920–1958) English biophysicist whose research on X-ray diffraction of ◊DNA crystals helped Francis ◊Crick and James D ◊Watson to deduce the chemical structure of DNA.

Gould, Stephen Jay (1941–) US palaeontologist and writer. In 1972 he proposed the theory of punctuated equilibrium, suggesting that the evolution of species did not occur at a steady rate but could suddenly accelerate, with rapid change occurring over a few hundred thousand years. His books include *Ever Since Darwin* (1977), *The Panda's Thumb* (1980), *The Flamingo's Smile* (1985), and *Wonderful Life* (1990).

Hoagland, Mahlon Bush (1921–) US biochemist who was the first to isolate transfer RNA (tRNA), a nucleic acid that plays an essential part in intracellular protein synthesis.

In the late 1950s Hoagland isolated various types of tRNA molecules from cytoplasm and demonstrated that each type of tRNA can combine with only one specific amino acid. Each tRNA molecule has as part of its structure a characteristic triplet of nitrogenous bases that links to a complementary triplet on a messenger RNA (mRNA) molecule. A number of these reactions occur on the ribosome, building up a protein one amino acid at a time.

Holley, Robert William (1922–) US biochemist who established the existence of transfer RNA (tRNA) and its function. For this work he shared the 1968 Nobel Prize for Physiology or Medicine. At Cornell University Holley obtained evidence for the role of tRNAs as acceptors of activated amino acids. In 1958 he succeeded in isolating the alanine-, tyrosine-, and valene-specific tRNAs from baker's yeast, and eventually Holley and his colleagues succeeded in solving the entire nucleotide sequence of this RNA.

Jacob, François (1920–) biochemist who, with Jacques ◊Monod and André Lwoff, pioneered research into molecular genetics and showed how the production of proteins from ◊DNA is controlled. They shared the Nobel Prize for Physiology or Medicine in 1965.

Jacob began his work on the control of gene action in 1958, working with Lwoff and Monod. It was known that the types of proteins produced in an organism are controlled by DNA, and Jacob focused his research on how the amount of protein is controlled. He performed a series of experiments in which he cultured the bacterium *Escherichia coli* in various mediums to discover the effect of the medium on enzyme production. He and his team found that there were three types of gene concerned with the production of each specific protein.

Khorana, Har Gobind (1922–) Indian-born US biochemist who in 1976 led the team that first synthesized a biologically active ◊gene. In 1968 he shared the Nobel Prize for Physiology or Medicine for research on the chemistry of the genetic code and its function in protein synthesis. Khorana's work provides much of the basis for gene therapy and biotechnology.

Khorana systematically synthesized every possible combination of the genetic signals from the four nucleotides known to be involved in determining the genetic code. He showed that a pattern of three nucleotides, called a triplet, specifies a particular amino acid (the building blocks of proteins). He further discovered that some of the triplets provided punctuation marks in the code, marking the beginning and end points of protein synthesis.

Kornberg, Arthur (1918–) US biochemist. In 1956 he discovered the enzyme DNA-polymerase, which enabled molecules of the genetic material DNA to be synthesized for the first time. For this work he shared the 1959 Nobel Prize for Physiology or Medicine. By 1967 he had synthesized a biologically active artificial viral DNA.

Krebs, Hans Adolf (1900–1981) German-born British biochemist. He discovered the citric acid cycle, also known as the **Krebs cycle**, the final pathway by which food molecules are converted into energy in living tissues. For this work he shared the 1953 Nobel Prize for Physiology or Medicine. Knighted in 1958.

Krebs first became interested in the process by which the body degrades amino acids. He discovered that nitrogen atoms are the first to be removed (deamination) and are then excreted as urea in the urine. He then investigated the processes involved in the production of urea from the removed nitrogen atoms, and by 1932 he had worked out the basic steps in the urea cycle.

Lamarck, Jean Baptiste de (1744–1829) French naturalist. His theory of evolution, known as **Lamarckism**, was based on the idea that acquired characteristics (changes acquired in an individual's lifetime) are inherited by the offspring, and that organisms have an intrinsic urge to evolve into better-adapted forms. *Philosophie zoologique/Zoological Philosophy* (1809) tried to show that various parts of the body developed because they were necessary, or disappeared because of disuse when variations in the environment caused a change in habit. If these body changes were inherited over many generations, new species would eventually be produced.

Lederberg, Joshua (1925–) US geneticist who showed that bacteria can reproduce sexually, combining genetic material so that offspring possess characteristics of both parent organisms. In 1958 he shared the Nobel Prize for Physiology or Medicine with George ◊Beadle and Edward ◊Tatum.

Lederberg is a pioneer of genetic engineering, a science that relies on the possibility of artificially shuffling genes from cell to cell. He realized in 1952 that bacteriophages, viruses which invade bacteria, can transfer genes from one bacterium to another, a discovery that led to the deliberate insertion by scientists of foreign genes into bacterial cells.

Lehninger, Albert L(ester) (1917–1986) US biochemist. An authority on cellular energy systems, he made a major contribution to enzymology, the bioenergetics of normal and cancer cells, and the results of calcification. In 1948 Lehninger and E P Kennedy discovered that cellular organelles called mitochondria are the main sites of cell respiration.

McClintock, Barbara (1902–1992) US geneticist who discovered jumping ◊genes (genes that can change their position on a chromosome from generation to generation). This would explain how originally identical cells take on specialized functions as skin, muscle, bone, and nerve, and also how evolution could give rise to the multiplicity of species. She was awarded the Nobel Prize for Physiology or Medicine in 1983.

McClintock's discovery that genes are not stable overturned one of the main tenets of heredity laid down by Gregor ◊Mendel. It had enormous implications and explained, for example, how resistance to antibiotic drugs can be transmitted between entirely different bacterial types.

Mendel, Gregor Johann (1822–1884) Austrian biologist, founder of ◊genetics. His experiments with successive generations of peas gave the basis for his theory of particulate inheritance rather than blending, involving dominant and recessive characters; see ◊Mendelism. His results, published 1865–69, remained unrecognized until the early 20th century.

Mendel formulated two laws now recognized as fundamental laws of heredity: the law of segregation and the law of independent assortment of characters. Mendel concluded that each parent plant contributes a 'factor' to its offspring that determines a particular trait and that the pairs of factors in the offspring do not give rise to a blend of traits.

Monod, Jacques Lucien (1910–1976) French biochemist who shared the 1965 Nobel Prize for Physiology or Medicine with his coworkers André Lwoff and François ◊Jacob for research in genetics and microbiology.

Working on the way in which genes control intracellular metabolism in microorganisms, Monod and his colleagues postulated the existence of a class of genes (which they called operons) that regulate the activities of the genes that actually control the synthesis of enzymes within the cell. They further hypothesized that the operons suppress the activities of the enzyme-synthesizing genes by affecting the synthesis of messenger ◊RNA.

Morgan, Thomas Hunt (1866–1945) US geneticist who helped establish that the ◊genes are located on the chromosomes, discovered sex chromosomes, and invented the techniques of genetic mapping. He was the first to work on the fruit fly *Drosophila*, which has since become a major subject of genetic studies. He was awarded the Nobel Prize for Physiology or Medicine in 1933.

Following the rediscovery of Austrian scientist Gregor ◊Mendel's work, Morgan became interested in the mechanisms involved in heredity, and in 1908 he began his research on the genetics of *Drosophila*. From his findings he postulated that certain characteristics are sex linked, that the X

chromosome carries several discrete hereditary units (genes), and that the genes are linearly arranged on chromosomes. He also demonstrated that sex-linked characters are not invariably inherited together, from which he developed the concept of crossing over and the associated idea that the extent of crossing over is a measure of the spatial separation of genes on chromosomes.

Nirenberg, Marshall Warren (1927–) US biochemist who shared the 1968 Nobel Prize for Physiology or Medicine for his work in deciphering the chemistry of the genetic code. Nirenberg became interested in the way in which the nitrogen bases – adenine (A), cytosine (C), guanine (G), and thymine (T) – specify a particular amino acid. To simplify the task of identifying the RNA triplet (codon) responsible for each amino acid, he used a simple synthetic RNA polymer. He found that certain amino acids could be specified by more than one codon, and that some triplets did not specify an amino acid at all. These 'nonsense' triplets signified the beginning or the end of a sequence. He then worked on finding the orders of the letters in the triplets, and obtained unambiguous results for 60 of the possible codons.

Ochoa, Severo (1905–1993) Spanish-born US biochemist who discovered an enzyme able to assemble units of the nucleic acid RNA in 1955. For his work towards the synthesis of RNA, Ochoa shared the 1959 Nobel Prize for Physiology or Medicine. Ochoa's early work concerned biochemical pathways in the human body, especially those involving carbon dioxide, but his main research was into nucleic acids and how their nucleotide units are linked, either singly (as in RNA) or to form two helically wound strands (as in DNA). In 1955 Ochoa obtained an enzyme from bacteria that was capable of joining together similar nucleotide units to form a nucleic acid, a type of artificial RNA. Nucleic acids containing exactly similar nucleotide units do not occur naturally, but the method of synthesis used by Ochoa was the same as that employed by a living cell.

Sanger, Frederick (1918–) English biochemist. He was the first person to win a Nobel Prize for Chemistry twice: the first in 1958 for determining the structure of the hormone insulin, and the second in 1980 for work on the chemical structure of genes. Sanger's second Nobel prize was shared with two US scientists, Paul Berg and Walter Gilbert, for establishing methods of determining the sequence of nucleotides strung together along strands of RNA and DNA. He also worked out the structures of various enzymes and other proteins.

Schleiden, Matthias Jakob (1804–1881) German botanist who identified the fundamental units of living organisms when, in 1838, he announced that the various parts of plants consist of cells or derivatives of cells. This was extended to animals by Theodor Schwann the following year. The existence of cells had been discovered by British physicist Robert Hooke in 1665, but Schleiden was the first to recognize their importance. He also noted the role of the nucleus in cell division, and the active movement of intracellular material in plant tissues.

Schwann, Theodor (1810–1882) German physiologist who, with Matthias Schleiden, is credited with formulating the cell theory, one of the most fundamental concepts in biology. Schwann also did important work on digestion, fermentation, and the study of tissues.

Sturtevant, Alfred Henry (1891–1970) US geneticist who - working with US biologist Thomas Morgan - was the first, in 1911, to map the position of genes on a chromosome.

Sturtevant developed methods for mapping gene positions on the chromosomes of mutant *Drosophila* (fruit flies). In 1911 he produced the first gene map ever derived, showing the positioning of five genes on a *Drosophila* X chromosome: white-eyed, vermilion-eyed, rudimentary wings, small wings, and yellow body.

Sutton, Walter S(tanborough) (1877–1916) US geneticist and surgeon. He used grasshopper cells to prove that chromosomal behaviour in meiosis is responsible for observed Mendelian phenomena, an achievement still recognized as classic.

Tatum, Edward Lawrie (1909–1975) US microbiologist. For his work on biochemical ◊genetics, he shared the 1958 Nobel Prize for Physiology or Medicine with his coworkers George ◊Beadle and Joshua ◊Lederberg.

Beadle and Tatum used X-rays to cause mutations in bread mould, studying particularly the changes in the enzymes of the various mutant strains. This led them to conclude that for each enzyme there is a corresponding gene. From 1945, with Lederberg, Tatum applied the same technique to bacteria and showed that genetic information can be passed from one bacterium to another. The discovery that a form of sexual reproduction can occur in bacteria led to extensive use of these organisms in genetic research.

Wallace, Alfred Russel (1823–1913) Welsh naturalist who collected animal and plant specimens in South America and Southeast Asia, and independently arrived at a theory of evolution by natural selection similar to that proposed by Charles ◊Darwin.

In 1858 Wallace wrote an essay outlining his ideas on evolution and sent it to Darwin, who had not yet published his. Together they presented a paper to the Linnaean Society that year. Wallace's section, entitled 'On the Tendency of Varieties to Depart Indefinitely from the Original Type', described the survival of the fittest.

Although both thought that the human race had evolved to its present physical form by natural selection, Wallace was of the opinion that humans' higher mental capabilities had arisen from some 'metabiological' agency.

Warburg, Otto Heinrich (1883–1970) German biochemist who in 1923 devised a manometer (pressure gauge) sensitive enough to measure oxygen uptake of respiring tissue. By measuring the rate at which cells absorb oxygen under differing conditions, he was able to show that enzymes called cytochromes enable cells to process oxygen. He was awarded the Nobel Prize for Physiology or Medicine in 1931.

Watson, James Dewey (1928–) US biologist. His research on the molecular structure of ◊DNA and the genetic code, in collaboration with Francis ◊Crick, earned him a shared Nobel prize in 1962. Based on earlier works, they were able to show that DNA formed a double helix of two spiral strands held together by base pairs.

Crick and Watson published their work on the proposed structure of DNA in 1953, and explained how genetic information could be coded.

Crick and Watson envisaged DNA replication occurring by a parting of the two strands of the double helix, each organic base thus exposed linking with a nucleotide (from the free nucleotides within a cell) bearing the complementary base. Thus two complete DNA molecules would eventually be formed by this step-by-step linking of nucleotides, with each of the new DNA molecules comprising one strand from the original DNA and one new strand.

Weismann, August Friedrich Leopold (1834–1914) German biologist, one of the founders of ◊genetics. He postulated that every living organism contains a special hereditary substance, the 'germ plasm', and in 1892 he proposed that changes to the body do not in turn cause an alteration of the genetic material.

This 'central dogma' of biology remains of vital importance to biologists supporting the Darwinian theory of evolution. If the genetic material can be altered only by chance mutation and recombination, then the Lamarckian view that acquired bodily changes can subsequently be inherited becomes obsolete.

Wilkins, Maurice Hugh Frederick (1916–) New Zealand-born British molecular biologist. In 1962 he shared the Nobel Prize for Physiology or Medicine with Francis ◊Crick and James ◊Watson for his work on the molecular structure of nucleic acids, particularly ◊DNA, using X-ray diffraction. Studying the X-ray diffraction pattern of DNA, he discovered that the molecule has a double helical structure and passed on his findings to Crick and Watson.

Wright, Sewall (1889–1988) US geneticist and statistician. During the 1920s he helped modernize Charles ◊Darwin's theory of evolution, using statistics to model the behaviour of populations of ◊genes. Wright's work on genetic drift centred on a phenomenon occurring in small isolated colonies where the chance disappearance of some types of gene leads to evolution without the influence of natural selection.

Glossary

acetyl coenzyme A compound active in processes of metabolism. It is a heat-stable coenzyme with an acetyl group ($-COCH_3$) attached by sulphur linkage. This linkage is a high-energy bond and the acetyl group can easily be donated to other compounds. Acetyl groups donated in this way play an important part in glucose breakdown as well as in fatty acid and steroid synthesis.

acquired character feature of the body that develops during the lifetime of an individual, usually as a result of repeated use or disuse, such as the enlarged muscles of a weightlifter.

active transport in cells, the use of energy to move substances, usually molecules or ions, across a membrane.

adaptation any change in the structure or function of an organism that allows it to survive and reproduce more effectively in its environment. In ◊evolution, adaptation is thought to occur as a result of random variation in the genetic make-up of organisms coupled with ◊natural selection. Species become extinct when they are no longer adapted to their environment – for instance, if the climate suddenly becomes colder.

adaptive radiation in evolution, the formation of several species, with ◊adaptations to different ways of life, from a single ancestral type.

adenosine triphosphate compound present in cells. See ◊ATP.

ADP abbreviation for **adenosine diphosphate**, the chemical product formed in cells when ◊ATP breaks down to release energy.

allele one of two or more alternative forms of a ◊gene at a given position (locus) on a chromosome, caused by a difference in the ◊DNA. Blue and brown eyes in humans are determined by different alleles of the gene for eye colour.

amino acid building block of proteins; a soluble organic molecule, mainly composed of carbon, oxygen, hydrogen, and nitrogen, containing both a basic amino group (NH_2) and an acidic carboxyl (COOH) group.

anabolism process of building up body tissue, promoted by the influence of certain hormones. It is the constructive side of ◊metabolism, as opposed to ◊catabolism.

analogous term describing a structure that has a similar function to a structure in another organism, but not a similar evolutionary path. For example, the wings of bees and of birds have the same purpose – to give powered flight – but have different origins. Compare ◊homologous.

artificial selection the selective breeding of individuals that exhibit the particular characteristics that a plant or animal breeder wishes to develop.

ATP, abbreviation for **adenosine triphosphate**, a nucleotide molecule found in all cells. It can yield large amounts of energy, and is used to drive the thousands of biological processes needed to sustain life, growth, movement, and reproduction. Green plants use light energy to manufacture ATP as part of the process of photosynthesis. In animals, ATP is formed by the breakdown of glucose molecules, usually obtained from the carbohydrate component of a diet, in a series of reactions termed ◊respiration. It is the driving force behind muscle contraction and the synthesis of complex molecules needed by individual cells.

autosome any ◊chromosome in the cell other than a sex chromosome. Autosomes are of the same number and kind in both males and females of a given species.

base pair in biochemistry, the linkage of two base (purine or pyrimidine) molecules in ◊DNA. Bases are found in nucleotides, and form the basis of the genetic code.

biochemistry science concerned with the chemistry of living organisms: the structure and reactions of proteins (such as enzymes), nucleic acids, carbohydrates, and lipids.

biosynthesis synthesis of organic chemicals from simple inorganic ones by living cells – for example, the conversion of carbon dioxide and water to glucose by plants during photosynthesis.

catabolism the destructive part of ◊metabolism where living tissue is changed into energy and waste products. It is the opposite of ◊anabolism.

catalyst substance that alters the speed of, or makes possible, a chemical or biochemical reaction but remains unchanged at the end of the reaction. ◊Enzymes are natural biochemical catalysts. In practice most catalysts are used to speed up reactions.

cell division the process by which a cell divides, either ◊meiosis, associated with sexual reproduction, or ◊mitosis, associated with growth, cell replacement, or repair. Both forms involve the duplication of DNA and the splitting of the nucleus.

cell membrane, or **plasma membrane,** thin layer of protein and fat surrounding cells that controls substances passing between the cytoplasm and the intercellular space. The cell membrane is semipermeable, allowing some substances to pass through and some not.

central dogma in genetics and evolution, the fundamental belief that ◊genes can affect the nature of the physical body, but that changes in the body (◊acquired character, for example, through use or accident) cannot be translated into changes in the genes.

centriole structure found in the ◊cells of animals that plays a role in the processes of ◊meiosis and ◊mitosis (cell division).

centromere part of the ◊chromosome where there are no ◊genes. Under the microscope, it usually appears as a constriction in the strand of the chromosome, and is the point at which the spindle fibres are attached during ◊meiosis and ◊mitosis (cell division).

centrosome cell body that contains the ◊centrioles. During cell division the centrosomes organize the microtubules to form the spindle that divides the chromosomes into daughter cells.

chromosome structure in a cell nucleus that carries the ◊genes. Each chromosome consists of one very long strand of DNA, coiled and folded to produce a compact body. The point on a chromosome where a particular gene occurs is known as its locus. Most higher organisms have two copies of each chromosome, together known as a **homologous pair** (they are ◊diploid) but some have only one (they are ◊haploid).

cilia, singular **cilium,** small hairlike organs on the surface of some cells, particularly the cells lining the upper respiratory tract.

cistron in genetics, the segment of ◊DNA that is required to synthesize a complete polypeptide chain. It is the molecular equivalent of a ◊gene.

citric acid cycle another term for the ◊Krebs cycle.

class in classification, a group of related ◊orders. For example, all mammals belong to the class Mammalia and all birds to the class Aves. Related classes are grouped together in a ◊phylum.

clone an exact replica. In genetics, any one of a group of genetically identical cells or organisms. An identical twin is a clone; so, too, are bacteria living in the same colony.

codominance in genetics, the failure of a pair of alleles, controlling a particular characteristic, to show the classic recessive-dominant relationship. Instead, aspects of both alleles may show in the phenotype.

codon in genetics, a triplet of bases (see ◊base pair) in a molecule of DNA or RNA that directs the placement of a particular amino acid during the process of protein (polypeptide) synthesis. There are 64 codons in the ◊genetic code.

coevolution evolution of those structures and behaviours within a species that can best be understood in relation to another species. For example, insects and flowering plants have evolved together: insects have produced mouthparts suitable for collecting pollen or drinking nectar, and plants have developed chemicals and flowers that will attract insects to them.

concentration gradient change in the concentration of a substance from one area to another.

continuous variation the slight difference of an individual character, such as height, across a sample of the population. Although there are very tall and very short humans, there are also many people with an intermediate height. The same applies to weight. Continuous variation can result from the genetic make-up of a population, or from environmental influences, or from a combination of the two.

convergent evolution the independent evolution of similar structures in species (or other taxonomic groups) that are not closely related, as a result of living in a similar way. Thus, birds and bats have wings, not because they are descended from a common winged ancestor, but because their respective ancestors independently evolved flight.

crossing over exchange of genetic material that occurs during ◊meiosis. While the chromosomes are lying alongside each other in pairs, each partner may twist around the other and exchange corresponding chromosomal segments. It is a form of genetic ◊recombination, which increases variation and thus provides the raw material of evolution.

cytoplasm the part of the cell outside the ◊nucleus. Strictly speaking, this includes all the ◊organelles (mitochondria, chloroplasts, and so on), but often cytoplasm refers to the jellylike matter in which the organelles are embedded.

cytoskeleton matrix of protein filaments and tubules that occurs within the cytoplasm. It gives the cell a definite shape, transports vital substances around the cell, and may also be involved in cell movement.

denaturation irreversible changes occurring in the structure of proteins such as enzymes, usually caused by changes in pH or temperature, by radiation or chemical treatments. An example is the heating of egg albumen resulting in solid egg white.

deoxyribonucleic acid full name of ◊DNA.

diffusion spontaneous and random movement of molecules or particles in a fluid (gas or liquid) from a region in which they are at a high concentration to a region of lower concentration, until a uniform concentration is achieved throughout.

dihybrid inheritance in genetics, a pattern of inheritance observed when two characteristics are studied in succeeding generations. The first experiments of this type, as well as in ◊monohybrid inheritance, were carried out by Austrian biologist Gregor Mendel using pea plants.

diploid having paired ◊chromosomes in each cell. In sexually reproducing species, one set is derived from each parent, the ◊gametes, or sex cells, of each parent being ◊haploid (having only one set of chromosomes) due to ◊meiosis (reduction cell division).

disaccharide sugar made up of two monosaccharides or simple sugars. Sucrose, $C_{12}H_{22}O_{11}$, or table sugar, is a disaccharide.

DNA, abbreviation for **deoxyribonucleic acid**, giant, complex molecule that contains, in chemically coded form, the information needed for a cell to make proteins. DNA is a ladderlike double-stranded ◊nucleic acid, and is organized into ◊chromosomes.

dominance in genetics, the masking of one allele (an alternative form of a gene) by another allele. For example, if a ◊heterozygous person has one allele for blue eyes and one for brown eyes, his or her eye colour will be brown. The allele for blue eyes is described as ◊recessive and the allele for brown eyes as dominant.

endoplasmic reticulum (**ER**) a membranous system of tubes, channels, and flattened sacs that form compartments within eukaryotic cells.

environment in ecology, the sum of conditions affecting a particular organism, including physical surroundings, climate, and influences of other living organisms.

enzyme biological ◊catalyst produced in cells, and capable of speeding up the chemical reactions necessary for life.

eukaryote one of the two major groupings into which all organisms are divided. Included are all organisms, except bacteria and cyanobacteria (blue-green algae), which belong to the ◊prokaryote grouping.

exon in genetics, a sequence of bases in ◊DNA that codes for a protein. Exons make up only 2% of the body's total DNA. The remainder is made up of ◊introns. During RNA processing the introns are cut out of the molecule and the exons spliced together.

family in classification, a group of related genera (see ◊genus). Family names are not printed in italic (unlike genus and species names), and by convention they all have the ending -idae (animals) or -aceae (plants and fungi). Related families are grouped together in an ◊order.

fat in the broadest sense, a mixture of ◊lipids – chiefly triglycerides (lipids containing three ◊fatty acid molecules linked to a molecule of glycerol). More specifically, the term refers to a lipid mixture that is solid at room temperature (20°C); lipid mixtures that are liquid at room temperature are called ◊oils.

fatty acid, or **carboxylic acid,** organic compound consisting of a hydrocarbon chain, up to 24 carbon atoms long, with a carboxyl group (–COOH) at one end.

fitness in genetic theory, a measure of the success with which a genetically determined character can spread in future generations. By convention, the normal character is assigned a fitness of one, and variants (determined by other ◊alleles) are then assigned fitness values relative to this. Those with fitness greater than one will spread more rapidly and will ultimately replace the normal allele; those with fitness less than one will gradually die out.

flagellum small hairlike organ on the surface of certain cells.

gamete cell that functions in sexual reproduction by merging with another gamete to form a zygote. Examples of gametes include sperm and egg cells. In most organisms, the gametes are haploid (they contain half the number of chromosomes of the parent), owing to reduction division or ◊meiosis.

gene unit of inherited material; a strand of ◊DNA that encodes a particular protein. In higher organisms, genes are located on the ◊chromosomes.

gene amplification technique by which selected DNA from a single cell can be duplicated indefinitely until there is a sufficient amount to analyse by conventional genetic techniques.

gene pool total sum of ◊alleles (variants of ◊genes) possessed by all the members of a given population or species alive at a particular time.

genetic engineering deliberate manipulation of genetic material by biochemical techniques. It is often achieved by the introduction of new ◊DNA, usually by means of a virus or ◊plasmid.

genome the full complement of ◊genes carried by a single (haploid) set of ◊chromosomes. The term may be applied to the genetic information carried by an individual or to the range of genes found in a given species. The human genome is made up of about 80,000 genes.

genotype the particular set of ◊alleles (variants of genes) possessed by a given organism. The term is usually used in conjunction with ◊phenotype, which is the product of the genotype and all environmental effects.

genus, plural **genera,** group of ◊species with many characteristics in common. Thus all doglike species (including dogs, wolves, and jackals) belong to the genus *Canis* (Latin 'dog'). Species of the same genus are thought to be descended from a common ancestor species. Related genera are grouped into ◊families.

Golgi apparatus, or **Golgi body,** stack of flattened membranous sacs found in the cells of ◊eukaryotes.

haploid having a single set of ◊chromosomes in each cell. Most higher organisms are ◊diploid – that is, they have two sets – but their gametes (sex cells) are haploid.

heredity the transmission of traits from parent to offspring.

heterozygous term describing an organism that has two different ◊alleles for a given trait. In ◊homozygous organisms, by contrast, both chromosomes carry the same allele.

homologous term describing an organ or structure possessed by members of different taxonomic groups (for example, species, genera, families, orders) that originally derived from the same structure in a common ancestor. The wing of a bat, the arm of a monkey, and the flipper of a seal are homologous because they all derive from the forelimb of an ancestral mammal.

homozygous term describing an organism that has two identical ◊alleles for a given trait. Individuals homozygous for a trait always breed true; that is, they produce offspring that resemble them in appearance when bred with a genetically similar individual. ◊Recessive alleles are only expressed in the homozygous condition. See also ◊heterozygous.

inclusive fitness in genetics, the success with which a given variant (or allele) of a ◊gene is passed on to future generations by a particular individual, after additional copies of the allele in the individual's relatives and their offspring have been taken into account.

intron, or **junk DNA,** in genetics, a sequence of bases in ◊DNA that carries no genetic information. Introns make up 98% of DNA (the rest is made up of ◊exons). Their function is unknown.

karyotype the set of ◊chromosomes characteristic of a given species. It is described as the number, shape, and size of the chromosomes in a single cell of an organism. In humans for example, the karyotype consists of 46 chromosomes, in mice 40, crayfish 200, and in fruit flies 8.

linkage in genetics, the association between two or more genes that tend to be inherited together because they are on the same chromosome. The closer together they are on the chromosome, the less likely they are to be separated by crossing over (one of the processes of ◊recombination) and they are then described as being 'tightly linked'.

lysosome membrane-enclosed structure, or organelle, inside a ◊cell, principally found in animal cells. Lysosomes contain enzymes that can break down proteins and other biological substances. They play a part in digestion, and in the white blood cells known as phagocytes the lysosome enzymes attack ingested bacteria.

meiosis process of cell division in which the number of ◊chromosomes in the cell is halved. It only occurs in ◊eukaryotic cells, and is part of a life cycle that involves sexual reproduction because it allows the genes of two parents to be combined without the total number of chromosomes increasing.

Mendelism in genetics, the theory of inheritance originally outlined by Austrian biologist Gregor Mendel. He suggested that, in sexually reproducing species, all characteristics are inherited through indivisible 'factors' (now identified with ◊genes) contributed by each parent to its offspring.

metabolism the chemical processes of living organisms enabling them to grow and to function. It involves a constant alternation of building up complex molecules (anabolism) and breaking them down (catabolism).

microtubules tiny tubes found in almost all cells with a nucleus. They help to define the shape of a cell by forming scaffolding for cilia and they also form the fibres of mitotic spindle (see ◊mitosis).

mitochondria, singular **mitochondrion**, membrane-enclosed organelles within eukaryotic cells, containing enzymes responsible for energy production during ◊aerobic respiration.

mitosis the process of cell division by which identical daughter cells are produced. During mitosis the DNA is duplicated and the chromosome number doubled, so new cells contain the same amount of DNA as the original cell.

molecular biology study of the molecular basis of life, including the biochemistry of molecules such as DNA, RNA, and proteins, and the molecular structure and function of the various parts of living cells.

monohybrid inheritance pattern of inheritance seen in simple ◊genetics experiments, where the two animals (or two plants) being crossed are genetically identical except for one gene.

monosaccharide, or **simple sugar**, carbohydrate that cannot be hydrolysed (split) into smaller carbohydrate units. Examples are glucose and fructose, both of which have the molecular formula $C_6H_{12}O_6$.

mutation change in the genes produced by a change in the ◊DNA that makes up the hereditary material of all living organisms. Mutations, the raw material of evolution, result from mistakes during replication (copying) of DNA molecules. Only a few improve the organism's performance and are therefore favoured by ◊natural selection. Mutation rates are increased by certain chemicals and by radiation.

natural selection the process whereby gene frequencies in a population change through certain individuals producing more descendants than others because they are better able to survive and reproduce in their environment.

neo-Darwinism the modern theory of ◊evolution, built up since the 1930s by integrating the 19th-century English scientist Charles Darwin's theory of evolution through natural selection with the theory of genetic inheritance founded on the work of the Austrian biologist Gregor Mendel.

nucleic acid complex organic acid made up of a long chain of ◊nucleotides, present in the nucleus and sometimes the cytoplasm of the living cell. The two types, known as ◊DNA (deoxyribonucleic acid) and ◊RNA (ribonucleic acid), form the basis of heredity.

nucleolus structure found in the nucleus of eukaryotic cells. It produces the RNA that makes up the ◊ribosomes, from instructions in the DNA.

nucleotide organic compound consisting of a purine (adenine or guanine) or a pyrimidine (thymine, uracil, or cytosine) base linked to a sugar (deoxyribose or ribose) and a phosphate group. ◊DNA and ◊RNA are made up of long chains of nucleotides.

nucleus the central, membrane-enclosed part of a eukaryotic cell, containing the DNA. The nucleus controls the function of the cell by determining which proteins are produced within it.

oncogene gene carried by a virus that induces a cell to divide abnormally, giving rise to a cancer. Oncogenes arise from mutations in genes (proto-oncogenes) found in all normal cells. They are usually also found in viruses that are capable of transforming normal cells to tumour cells.

oncomouse mouse that has a human ◊oncogene (gene that can cause certain cancers) implanted into its cells by genetic engineering. Such mice are used to test anticancer treatments and were patented within the USA by Harvard University in 1988, thereby protecting its exclusive rights to produce the animal and profit from its research.

operon group of genes that are found next to each other on a chromosome, and are turned on and off as an integrated unit. They usually produce enzymes that control different steps in the same biochemical pathway.

order in classification, a group of related ◊families. Related orders are grouped together in a ◊class.

organelle discrete and specialized structure in a living cell; organelles include mitochondria, chloroplasts, lysosomes, ribosomes, and the nucleus.

osmosis movement of water through a selectively permeable membrane separating solutions of different concentrations.

peptide molecule comprising two or more ◊amino acid molecules (not necessarily different) joined by **peptide bonds**, whereby the acid group of one acid is linked to the amino group of the other (–CO.NH). The number of amino acid molecules in the peptide is indicated by referring to it as a di-, tri-, or polypeptide (two, three, or many amino acids).

phenotype in genetics, visible traits, those actually displayed by an organism. The phenotype is not a direct reflection of the ◊genotype because some alleles are masked by the presence of other, dominant alleles (see ◊dominance). The phenotype is further modified by the effects of the environment (for example, poor nutrition stunts growth).

phospholipid any ◊lipid consisting of a glycerol backbone, a phosphate group, and two long chains. Phospholipids are found everywhere in living systems as the basis for biological membranes.

phylum, plural **phyla**, major grouping in biological classification. Mammals, birds, reptiles, amphibians, fishes, and tunicates belong to the phylum Chordata; the phylum Mollusca consists of snails, slugs, mussels, clams, squid, and octopuses; the phylum Porifera contains sponges; and the phylum Echinodermata includes starfish, sea urchins, and sea cucumbers. Related phyla are grouped together in a ◊kingdom; phyla are subdivided into ◊classes.

pleiotropy process whereby a given gene influences several different observed characteristics of an organism. For example, in the fruit fly *Drosophila* the vestigial gene reduces the size of wings, modifies the halteres, changes the number of egg strings in the ovaries, and changes the direction of certain bristles.

polymerase chain reaction (PCR) technique developed during the 1980s to clone short strands of DNA from the ◊genome of an organism. The aim is to produce enough of the DNA to be able to sequence and identify it.

polymorphism in genetics, the coexistence of several distinctly different types in a population (groups of animals of one species). Examples include the different blood groups in humans, different colour forms in some butterflies, and snail shell size, length, shape, colour, and stripiness.

polyploid in genetics, possessing three or more sets of chromosomes in cases where the normal complement is two sets (◊diploid). Polyploidy arises spontaneously and is common in plants (mainly among flowering plants), but rare in animals. Many crop plants are natural polyploids, including wheat, which has four sets of chromosomes per cell (durum wheat) or six sets (common wheat).

polysaccharide long-chain carbohydrate made up of hundreds or thousands of linked simple sugars (monosaccharides) such as glucose and closely related molecules.

population genetics the branch of genetics that studies the way in which the frequencies of different ◊alleles (alternative forms of a gene) in populations of organisms change, as a result of natural selection and other processes.

prokaryote organism whose cells lack nuclei and other organelles (specialized segregated structures such as mitochondria, and chloroplasts). The prokaryotes comprise only the bacteria and cyanobacteria (blue-green algae); all other organisms are (eukaryotes.

protein complex, biologically important substance composed of amino acids joined by ◊peptide bonds.

punctuated equilibrium model evolutionary theory developed by Niles Eldredge and US palaeontologist Stephen Jay Gould in 1972 to explain discontinuities in the fossil record. It claims that periods of rapid change alternate with periods of relative stability (stasis), and that the appearance of new lineages is a separate process from the gradual evolution of adaptive changes within a species.

recessive gene in genetics, an ◊allele (alternative form of a gene) that will show in the ◊phenotype (observed characteristics of an organism) only if its partner allele on the paired chromosome is similarly recessive. Such an allele will not show if its partner is dominant, that is if the organism is ◊heterozygous for a particular characteristic. Alleles for blue eyes in humans and for shortness in pea plants are recessive.

recombinant DNA in genetic engineering, ◊DNA formed by splicing together genes from different sources into new combinations.

recombination in genetics, any process that recombines, or 'shuffles', the genetic material, thus increasing genetic variation in the offspring.

replication production of copies of the genetic material DNA; it occurs during cell division (◊mitosis and ◊meiosis). Most mutations are caused by mistakes during replication.

restriction enzyme bacterial ◊enzyme that breaks a chain of ◊DNA into two pieces at a specific point; used in ◊genetic engineering. The point along the DNA chain at which the enzyme can work is restricted to places where a specific sequence of base pairs occurs. Different restriction enzymes will break a DNA chain at different points. The overlap

between the fragments is used in determining the sequence of base pairs in the DNA chain.

ribonucleic acid full name of ◊RNA.

ribosome the protein-making machinery of the cell. Ribosomes are located on the endoplasmic reticulum (ER) of eukaryotic cells, and are made of proteins and a special type of ◊RNA, ribosomal RNA.

RNA, abbreviation for **ribonucleic acid**, nucleic acid involved in the process of translating the genetic material ◊DNA into proteins. It is usually single-stranded, unlike the double-stranded DNA, and consists of a large number of nucleotides strung together, each of which comprises the sugar ribose, a phosphate group, and one of four bases (uracil, cytosine, adenine, or guanine).

sequencing in biochemistry, determining the sequence of chemical subunits within a large molecule. Techniques for sequencing amino acids in proteins were established in the 1950s, insulin being the first for which the sequence was completed. The Human Genome Project is attempting to determine the sequence of the 3 billion base pairs within human ◊DNA.

sex chromosome chromosome that differs between the sexes and that serves to determine the sex of the individual. In humans, females have two X chromosomes and males have an X and a Y chromosome.

sex linkage in genetics, the tendency for certain characteristics to occur exclusively, or predominantly, in one sex only. Human examples include red-green colour blindness and haemophilia, both found predominantly in males. In both cases, these characteristics are ◊recessive and are determined by genes on the ◊X chromosome.

species distinguishable group of organisms that resemble each other or consist of a few distinctive types (as in ◊polymorphism), and that can all interbreed to produce fertile offspring. Species are the lowest level in the system of biological classification.

substrate in biochemistry, a compound or mixture of compounds acted on by an enzyme.

systematics science of naming and identifying species, and determining their degree of relatedness.

transcription process by which the information for the synthesis of a protein is transferred from the ◊DNA strand on which it is carried to the messenger ◊RNA strand involved in the actual synthesis.

transduction the transfer of genetic material between cells by an infectious mobile genetic element such as a virus. Transduction is used in ◊genetic engineering to produce new varieties of bacteria.

transgenic organism plant, animal, bacterium, or other living organism which has had a foreign gene added to it by means of ◊genetic engineering.

translation the process by which proteins are synthesized. During translation, the information coded as a sequence of nucleotides in messenger ◊RNA is transformed into a sequence of amino acids in a peptide chain. The process involves the 'translation' of the ◊genetic code. See also ◊transcription.

transposon, or **jumping gene**, segment of DNA able to move within or between chromosomes. Transposons trigger changes in gene expression by shutting off genes or causing insertion ◊mutations.

variation difference between individuals of the same species, found in any sexually reproducing population. Variations may be almost unnoticeable in some cases, obvious in others, and can concern many aspects of the organism. Typically, variations in size, behaviour, biochemistry, or colouring may be found. The cause of the variation is genetic (that is, inherited), environmental, or more usually a combination of the two. The origins of variation can be traced to the recombination of the genetic material during the formation of the gametes, and, more rarely, to mutation.

wild type in genetics, the naturally occurring gene for a particular character that is typical of most individuals of a given species, as distinct from new genes that arise by mutation.

X chromosome the larger of the two sex chromosomes, the smaller being the ◊Y chromosome. These two chromosomes are involved in sex determination. Females have two X chromosomes, males have an X and a Y. Genes carried on the X chromosome produce the phenomenon of ◊sex linkage.

Y chromosome the smaller of the two sex chromosomes. In male mammals it occurs paired with the other type of sex chromosome (X), which carries far more genes. The Y chromosome is the smallest of all the mammalian chromosomes and is considered to be largely inert (that is, without direct effect on the physical body). There are only 20 genes discovered so far on the human Y chromosome, much fewer than on all other human chromosomes. See also ◊sex determination.

Further Reading

Attenborough, David *Life on Earth* (1979)

Baldwin, E *Dynamic Aspects of Biochemistry* (1967)

Bateson, William *Mendel's Principles of Heredity* (1909)

Bernal, J D *The Physical Basis of Life* (1951)

Bloomfield, Molly M *Chemistry and the Living Organism* (1991)

Brent, Peter *Charles Darwin: A Man of Enlarged Curiosity* (1981)

Brooks, James *Origins of Life* (1985)

Carey, George W *Chemistry and Wonders of the Human Body* (1921)

Chaplin, Martin F, and Bucke, Christopher *Enzyme Technology* (1990)

Clark, David, and Russell, Lonnie *Molecular Biology Made Simple* (1997)

Clark, R W *The Survival of Charles Darwin* (1985)

Cronin, Helena *The Ant and the Peacock* (1991)

Darwin, Charles *The Origin of Species* (1859)

Dawkins, Richard *The Blind Watchmaker* (1986)

de Beer. Gavin *Charles Darwin: A Scientific Biography* (1963)

Dennett, Daniel *Darwin's Dangerous Idea* (1995)

Dyson, Freeman John *Origins of Life* (1985)

Dyson, George *Darwin Amongst the Machines* (1998)

Edey, Maitland A, and Johanson, Donald C *Blueprints* (1989)

Fichman, M *Alfred Russel Wallace* (1981)

Fortey, Richard A *Life: An Unauthorized Biography* (1997)

Gersch, Jack *The Matter of Life* (1995)

Goodwin, Brian *How the Leopard Changed Its Spots* (1994)

Gould, Stephen Jay *The Panda's Thumb* (1980)

Johanson, Don, and Edey, Matt *Lucy: The beginnings of humankind* (1981)

Jones, Steve *The Language of the Genes* (1993)

King, Barry (ed) *Cell Biology* (1986)

Leakey, Richard *The Origin of Humankind* (1994)

Lehninger, A L *Principles of Biochemistry* (1993)

Lewin, Roger *Human Evolution: An Illustrated Guide* (1993), *The Origin of Modern Humans* (1993), *Bones of Contention* (1997), *Principles of Human Evolution* (1998)

Lewontin, Richard *The Doctrine of DNA* (1991)

Mason, S *Chemical Evolution* (1993)

Maynard Smith, John *The Theory of Evolution* (1958)

Mayr, Ernst, *The Growth of Biological Thought* (1982)

McKinney, H Lewis *Wallace and Natural Selection* (1972)

Mines, Allan H *Respiratory Physiology* (1993)

Nelkin, Dorothy, and Lindee, M Susan *The DNA Mystique* (1995)

Nisbet, Euan George *Living Earth: A Short History of Life and Its Home* (1991)

Orel, V *Mendel* (1984)

Prange, Henry D *Respiratory Physiology: Understanding Gas Exchange* (1996)

Ridley, Mark *The Problems of Evolution* (1986)

Ritvo, Harriet *The Platypus and the Mermaid and other Figments of the Classifying Imagination* (1997)

Rose, Steven *The Chemistry of Life* (1991)

Rose, Steven; Kamin, Leon; and Lewontin, Richard *Not in our Genes* (1984)

Ruse, Michael *Darwinism Defended* (1982)

Russell, Peter J *Genetics* (1998)

Snedden, Robert *Life* (1994)

Stringer, Chris, and Gamble, Clive *In Search of the Neanderthals* (1993)

Suckling, C J *Enzyme Chemistry: Impact and Applications* (1990)

Tattersall, Ian (ed) and others *Encyclopaedia of Human Evolution and Prehistory* (1988)

Tudge, Colin *The Engineer in the Garden* (1993)

Watson, James D *The Double Helix* (1968)

West, John B *Respiratory Physiology: the Essentials* (1995)

Wilkie, Tom *Perilous Knowledge* (1993)

Williams, George C *Adaptation and Natural Selection* (1966)

THE ANIMAL KINGDOM

Animals (from Latin *anima,* meaning 'breath' or 'life') are members of the kingdom Animalia, one of the major categories of living things, the science of which is zoology. Animals are all heterotrophs (they obtain their energy from organic substances produced by other organisms); they have eukaryotic cells (the genetic material is contained within a distinct nucleus) bounded by a thin cell membrane rather than the thick cell wall of plants. Most animals are capable of moving around for at least part of their life cycle. In the past, it was common to include the single-celled protozoa with the animals, but these are now classified as protists, together with single-celled plants. Thus all animals are multicellular. The oldest land animals known date back 440 million years. Their remains were found in 1990 in a sandstone deposit near Ludlow, Shropshire, England, and included fragments of two centipedes a few centimetres long and a primitive spider measuring about 1 mm.

classification

kingdom

The primary division in biological classification is the kingdom. At one time, only two kingdoms were recognized: animals and plants. Any organism with a rigid cell wall was a plant, and so bacteria and fungi were considered plants, despite their many differences. Today most biologists prefer a five-kingdom system, even though it still involves grouping together organisms that are probably unrelated.

phylum Kingdoms are made up of phyla. The vertebrates (animals with backbones or primitive precursors of these) – mammals, birds, reptiles, amphibians, fishes, and tunicates – belong to the phylum **Chordata**. The invertebrates (animals without backbones) are classified in various phyla including **Echinodermata**, containing starfish, sea urchins, and sea cucumbers; **Mollusca**, consisting of snails, slugs, mussels, clams, squid, and octopuses; **Arthropoda**, containing insects, spiders, millipedes, crabs, and shrimps; **Annelida**, **Nematoda**, and **Platyhelminthes**, containing the segmented worms, unsegmented worms, and flatworms respectively; **Coelenterata**, containing jellyfish, hydra, sea anemones, and coral; and **Porifera**, containing sponges.

There are more than 30 different animal phyla. The most recently identified is Cyclophora, which was described in 1995 and contains only one known species, *Symbion pandora*, a tiny invertebrate that lives in amongst the mouthparts of some lobsters.

class The phyla are made up of classes. For example, all mammals belong to the class Mam-malia, all birds to the class Aves, and all insects to Insecta.

order Each class contains one or more related orders. For example, the horse, rhinoceros, and tapir families are grouped in the order Perissodactyla, the odd-toed ungulates, because they all have either one or three toes on each foot. By convention, the names of orders have the ending '-formes' in birds and fish, and '-a' in mammals, amphibians, reptiles, and other animals.

family An order breaks down into groups of related families. By convention, animal families all have the ending '-idea'. For example, hummingbirds are grouped in the hummingbird family Trochilidae.

genus

Families contain groups of related genera. Each genus groups together species with many characteristics in common. Thus all doglike species (including dogs, wolves, and jackals) belong to the genus *Canis* (Latin 'dog'). Species of the same genus are thought to be descended from a common ancestor species. By convention, genus names are given in italics, with an initial capital letter.

Classification of Living Things

Classification is the grouping of organisms based on similar traits and evolutionary histories. Taxonomy and systematics are the two sciences that attempt to classify living things. In taxonomy, organisms are generally assigned to groups based on their characteristics. In modern systematics, the placement of organisms into groups is based on evolutionary relationships among organisms. Thus, the groupings are based on evolutionary relatedness or family histories called phylogenies.

The groups into which organisms are classified are called taxa (singular, taxon). The taxon that includes the fewest members is the species, which consists of a single organism. Closely related species are placed into a genus (plural, genera). Related genera are placed into families, families into orders, orders into classes, classes into phyla (singular, phylum) or, in the case of plants and fungi, into divisions, and phyla into divisions or kingdoms. The kingdom level, of which five are generally recognized, is the broadest taxonomic group and includes the greatest number of species. The table below provides an example of the classification of an organism representative of the animal kingdom and the plant kingdom.

Taxonomic Species[3] Groups[1]	Common	Kingdom name	Phylum/	Class	Order division[2]	Family	Genus[3]
human	Animalia	Chordata	Mammalia	Primates	Hominoidea	*Homo*	*sapiens*
Douglas fir	Plantae	Tracheophyta	Gymnospermae	Coniferales	Pinaceae	*Pseudotsuga*	*douglasii*

[1] Intermediate taxonomic levels can be created by adding the prefixes 'super-' or 'sub-' to the name of any taxonomic level.

[2] The term division is generally used in place of phylum/phyla for the classification of plants and fungi.

[3] An individual organism is given a two-part name made up of its genus and species names. For example, Douglas fir is correctly known as *Pseudotsuga douglasii*.

species

Species are the lowest level in the system of biological classification. Each species is a distinguishable group of organisms that resemble each other or consist of a few distinctive types (as in polymorphism), and that can all interbreed to produce fertile offspring. A species is designated by the genus name followed by the species name, for example *Canis lupus* (wolf), with both the genus and species name given in italics.

A **native species** is a species that has existed in a country at least from prehistoric times; a **naturalized species** is one known to have been introduced by humans from another country, but which now maintains itself; while an **exotic species** is one that requires human intervention to survive.

Around 1.4 million species of living things have been identified so far, of which 750,000 are insects and 41,000 are vertebrates. It is estimated that one species becomes extinct every day through habitat destruction.

Classification: The Influence of DNA Analysis

DNA analysis is changing the way we classify animals, and throwing new light on how they evolve. A good example is shown by findings concerning whale families. Researchers in Belgium and the USA have completed a detailed analysis of similar genes in 16 species of whales, and the results are at variance with the way these species have been traditionally classified.

Whales with teeth, such as the sperm whale and the dolphins, have previously been thought to be closely related and have been classified as such. However, comparing the DNA that codes for ribosomal RNA in whales shows that the sperm whales are more closely related to the baleen whales, which do not have teeth but which filter food from the sea.

The comparison also showed that the beaked whale, which has a few teeth, is only distantly related to the other whale families. The results were confirmed by comparison of the DNA sequence for myoglobin in different whale species.

the influence of DNA analysis on evolutionary theory

What does this mean? Well, it helps to fill in our understanding of how this important group evolved. One important point, for example, concerns echolocation, the ability of whales to navigate by listening to the echoes of sounds. According to traditional classification, baleen whales never had the ability to echolocate. The new classification implies that baleen whales have evolved from species that were able to echolocate, and have now lost the ability.

mammals

The mammals comprise a large group of warm-blooded vertebrate animals characterized by having mammary glands in the female; these are used for suckling the young. Other features of mammals are hair (very reduced in some species, such as whales); a middle ear formed of three small bones (ossicles); a lower jaw consisting of two bones only; seven vertebrae in the neck; and no nucleus in the red blood cells.

There are over 4,000 species of mammals, adapted to almost every way of life. The smallest shrew weighs only 2 g/0.07 oz, the largest whale up to 140 tonnes. As with every other animal group, a number of mammal species are in danger of extinction from habitat loss and hunting. According to the Red List of endangered species published by the World Conservation Union (IUCN) for 1996, 25% of mammal species are threatened with extinction.

Mammals are divided into three groups: **placental mammals** (the largest group), such as humans, wolves, armadillos, bats, and shrews; **marsupials**, such as kangaroos, koalas, and wombats; and **monotremes**, such as platypuses and echidnas (spiny anteaters).

placental mammals

Most mammals are placental mammals, where the young develop inside the mother's body, in the uterus, receiving nourishment from the blood of the mother via the placenta. There is considerable variation

Major Invertebrate and Vertebrate Groups

Invertebrates

Taxon[1]	Name	Examples
P	Porifera	all sponges
P	Cnidaria	corals, sea anemones, *Hydra*, jellyfishes
P	Ctenophora	sea gooseberries, comb jellies
P	Platyhelminthes	flatworms, flukes, tapeworms
P	Nemertina	nemertine worms, ribbon worms
P	Nematoda	roundworms
P	Mollusca	clams, oysters, snails, slugs, octopuses, squids, cuttlefish
P	Annelida	ringed worms, including lugworms, earthworms, and leeches
P	Arthropoda	(subdivided into classes below)
C	Arachnida	spiders, ticks, scorpions, mites
C	Branchiopoda	water fleas
C	Cirripedia	barnacles
C	Malacostraca	crabs, lobsters, shrimp, woodlice
C	Diplopoda	millipedes
C	Chilopoda	centipedes
C	Insecta	silverfish, dragonflies, mayflies, stoneflies, cockroaches, earwigs, web spinners, termites, booklice, lice, grasshoppers, thrips, lace-wings, scorpion flies, caddis-flies, moths, butterflies, beetles, house flies, fleas, stylopids, ants, bees
P	Echinodermata	sea stars, brittle stars, sea urchins, sand dollars, sea cucumbers
P	Hemichordata	acorn worms, pterobranchs, graptolites

Vertebrates

Taxon[1]	Name	Examples
C	Agnatha	(jawless fishes) lampreys, hagfishes
C	Chondricthyes	(cartilaginous fishes) dogfish, sharks, rays, skates
C	Osteichthyes	(bony fishes) salmon, catfishes, perches, flatfishes including flounder and halibut
C	Amphibia	frogs, toads, newts, salamanders, caecilians
C	Reptilia	tuatara, tortoises, turtles, lizards, snakes
C	Aves	rheas, ostriches, moa, penguins, ducks, pheasants, gulls, swifts, kingfishers, sparrows, woodpeckers, pelicans, flamingoes, herons, falcons, cranes, divers, pigeons, parrots, cuckoos, owls
C	Mammalia	duck-billed platypuses, echidnas, kangaroos, opossums, shrews, bats, dogs, seals, whales, dolphins, rats, rabbits, pigs, camels, deer, horses, tapirs, elephants, hyraxes, anteaters, manatees, pangolins, lemurs, monkeys, humans

[1] P represents phylum; C represents class.

Classification of Mammals

Order	Number of species	Examples
Subclass: Prototheria (egg-laying mammals)		
Monotremata	3	echidna, platypus
Subclass: Theria		
Infraclass: Metatheria (pouched mammals)		
Marsupiala	266	kangaroo, koala, opossum
Infraclass: Eutheria (placental mammals)		
Rodentia	1,700	rat, mouse, squirrel, porcupine
Chiroptera	970	all bats
Insectivora	378	shrew, hedgehog, mole
Carnivora	230	cat, dog, weasel, bear

Infraclass: Eutheria (placental mammals)		
Primates	180	lemur, monkey, ape, human
Artiodactyla	145	pig, deer, cattle, camel, giraffe
Cetacea	79	whale, dolphin
Lagomorpha	58	rabbit, hare, pika
Pinnipedia	33	seal, walrus
Edentata	29	anteater, armadillo, sloth
Perissodactyla	16	horse, rhinoceros, tapir
Hyracoidea	11	hyrax
Pholidota	7	pangolin
Sirenia	4	dugong, manatee
Dermoptera	2	flying lemur
Proboscidea	2	elephant
Tubulidentata	1	aardvark

amongst species as to how long the young remain inside the mother, with larger species tending to have longer gestation periods than smaller species. Placental mammals are found worldwide.

marsupials

Marsupials are mammalian species in which the young are born at an early stage of development and develop further in a pouch on the mother's body where they

Mammals: Average Gestation Times

Mammal	Gestation (days)	Mammal	Gestation (days)
Ass	365	Lion	108
Baboon	187	Llama	330
Bear (black)	210	Mink	40–75
Bear (grizzly)	225	Monkey (rhesus)	164
Bear (polar)	60	Moose	240–250
Beaver	122	Mouse (domestic white)	19
Buffalo (American)	270	Mouse (meadow)	21
Camel (Bactrian)	~410	Muskrat	28–30
Cat (domestic)	58–65	Opossum (American)	12–13
Chimpanzee	23	Oryx	260–5
Chinchilla	110–120	Otter	270–300
Chipmunk	31	Pig (domestic)	112–115
Cow	279–292	Porcupine	112
Deer (white-tailed)	201	Puma	90
Dog (domestic)	58–70	Rabbit (domestic)	30–35
Elephant (Asian)	645	Raccoon	63
Elk (Wapiti)	240–250	Rhinoceros (black)	450
Fox (red)	52	Seal	330
Giraffe	420–450	Sea lion (California)	350
Goat (domestic)	145–155	Sheep (domestic)	144–151
Gorilla	257	Squirrel (grey)	30–40
Guinea pig	68	Tiger	105–113
Hippopotamus	225–250	Whale (sperm)	480–500
Horse	330–342	Wolf	60–68
Kangaroo	42	Zebra (Grant's)	365
Leopard	92–95		

are attached to and fed from a nipple. They occupy many specialized niches in South America and, especially, Australasia.

Marsupials include the kangaroo, wombat, opossum, phalanger, bandicoot, dasyure, and wallaby. The Australian marsupial anteater known as the numbat is unusual in that it has no pouch.

Marsupial Cooperative Research Centre
http://www.newcastle.edu.au/department/bi/
birjt/jrcrc/
Interesting details of latest research into marsupial preservation issues in Australasia. It examines efforts to preserve endangered species, whilst scientifically culling those in excess numbers and restricting introduced predators. For further information there is a link to Australia's Genome Research Network.

---

monotremes

Monotremes are mammals that lay eggs. The young are then nourished with milk. There are only a few types surviving (platypus and two kinds of echidna).

Remarkable Platypus
http://www.anca.gov.au/plants/manageme/
platintr.htm
Comprehensive information on the shy monotreme presented by the Australian Natural Conservation Agency. Findings of latest research into platypuses is presented, together with recommendations to help preserve the species.

echidna The short-nosed echidna is found throughout Australia. Echidnas are egg-laying mammals (monotremes). They are nocturnal and live almost exclusively on ants and termites.

---

birds

Birds make up the class Aves, the biggest group of land vertebrates. They are characterized by warm blood

Classification of Major Bird Groups

Order	Example
Tinamiformes	tinamous
Rheiformes	rhea
Struthioniformes	ostrich
Casuariiformes	cassowary, emu
Apterygiformes	moas, kiwi
Podicipediformes	grebe
Dinornithiformes	moa
Procellariformes	albatross, petrel, shearwater, storm petrel
Sphenisciformes	penguin
Pelecaniformes	pelican, booby, gannet, frigate bird
Anseriformes	duck, goose, swan
Phoenicopteriformes	flamingo
Ciconiiformes	heron, ibis, stork, spoonbill
Falconiformes	falcon, hawk, eagle, buzzard, vulture
Galliformes	grouse, partridge, pheasant, turkey
Gruiformes	crane, rail, bustard, coot
Charadriiformes	wader, gull, auk, oyster-catcher, plover, puffin, tern
Gaviiformes	diver
Columbiformes	dove, pigeon, sandgrouse
Psittaciformes	parrot, macaw, parakeet
Cuculiformes	cuckoo, roadrunner
Strigiformes	owl
Caprimulgiformes	nightjar, oilbird
Apodiformes	swift, hummingbird
Coliiformes	mousebird
Trogoniformes	trogon
Coraciiformes	kingfisher, hoopoe
Piciformes	woodpecker, toucan, puffbird
Passeriformes	finch, crow, warbler, sparrow, weaver, jay, lark, blackbird, swallow, mockingbird, wren, thrush

(usually about 41°C/106°F, higher than that of mammals), feathers, wings, breathing through lungs, and egg-laying by the female. Birds are bipedal (they have two feet); feet are usually adapted for perching and never have more than four toes. Hearing and eyesight are well developed, but the sense of smell is usually poor. No existing species of bird possesses teeth.

Most birds fly, but some groups (such as ostriches) are flightless, and others include flightless members. Many communicate by sounds

bird: names

Rifleman, short-tailed pygmy tyrant, frilled coquette, bobwhite, tawny frogmouth, trembler, wattle-eye, fuscous honeyeater, dickcissel, common grackle, and forktailed drongo are all common names for species of bird.

(nearly half of all known species are songbirds) or by visual displays, in connection with which many species are brightly coloured, usually the males. Birds have highly developed patterns of instinctive behaviour. There are nearly 8,500 species of birds. The study of birds is called **ornithology**.

According to the Red List of endangered species published by the World Conservation Union (IUCN) for 1996, 11% of bird species are threatened with extinction.

wing Birds can fly because of the specialized shape of their wings: a rounded leading edge, flattened underneath and round on top. This aerofoil shape produces lift in the same way that an aircraft wing does. The outline of the wing is related to the speed of flight. Fast birds of prey have a streamlined shape. Larger birds, such as the eagle, have large wings with separated tip feathers which reduce drag and allow slow flight. Insect wings are not aerofoils. They push downwards to produce lift, in the same way that oars are used to push through water.

feather Types of feather. A bird's wing is made up of two types of feather: contour feathers and flight feathers. The primary and secondary feathers are flight feathers. The coverts are contour feathers, used to streamline the bird. Semi-plume and down feathers insulate the bird's body and provide colour.

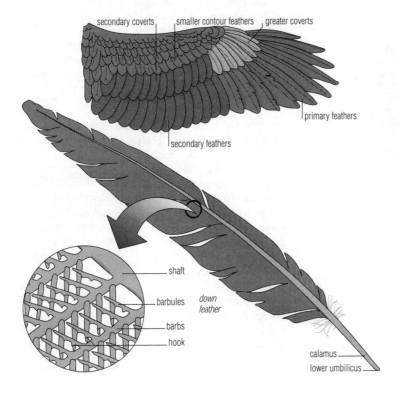

secondary coverts · smaller contour feathers · greater coverts · primary feathers · secondary feathers · shaft · barbules · barbs · hook · down feather · calamus · lower umbilicus

Classification of Dinosaurs

Order	Suborder	Description and examples
Subclass Archosauria	Superorder Dinosauria	
Saurischia	Theropoda	Bipedal, 'beast-footed carnivores: *Tyrannosaurus rex, Dilophosaurus, Velociraptor, Deinonychus*
Sauropodomorpha		Quadripedal, sometimes gigantic herbivores: *Apatosaurus, Brachiosaurus, Diplodocus*
Ornithischia	Ornithopoda	Mostly bipedal, sometimes duck-billed herbivores: *Hadrosaurus, Heterodontosaurus, Iguanodon*
	Thyreophora (Enoplosauria)	armoured herbivores: *Stegosaurus, Ankylosaurus*
	Marginocephalia	horned or 'bone-headed' herbivores: Triceratops, Psittacosaurus

reptiles

Reptiles are vertebrates belonging to the class Reptilia. Unlike amphibians (see below), reptiles have hard-shelled, yolk-filled eggs that are laid on land and from which fully formed young are born. Some snakes and lizards retain their eggs and give birth to live young. Reptiles are cold-blooded, and their skin is usually covered with scales. The metabolism is slow, and in some cases (certain large snakes) intervals between meals may be months. Reptiles date back over 300 million years.

Many extinct forms are known, including the orders Pterosauria, Plesiosauria, Ichthyosauria, and the dinosaur order Saurischia and Ornithschia. The chief living reptile orders are the Testudines or Chelonia (tortoises and turtles), Crocodilia (alligators and crocodiles), and Squamata, divided into three suborders: Sauria or Lacertilia (lizards), Ophidia or Serpentes (snakes), and Amphisbaenia (worm lizards). The order Rhynchocephalia has one surviving genus with two species, the lizardlike tuataras of New Zealand.

crocodile The estuarine or saltwater crocodile, of India, southeast Asia, and Australasia, is one of the largest and most dangerous of its family. It has been known to develop a taste for human flesh. Hunted near to extinction for its leather, it is now protected by restrictions and the trade in skins is controlled.

Classification of Living Reptiles

Order	Suborder	Number of species	Examples
Subclass Anapsida			
Testudines (Chelonia)	Cryptodira	220	turtle, tortoise, terrapin
	Pleurodira	70	sideneck turtle
Subclass Lepidosauria			
	Rhynchocephalia	2	tuatara
Squamata	Sauria (Lacertilia)	4,500	lizards: iguana, gecko, gila monster, skink
	Ophidia (Serpentes)	3,000	snakes: python, viper, cobra, sea snake
	Amphisbaenia	160	worm lizard
Subclass Archosauria			
	Crocodilia	23	crocodile, alligator, gavial, cayman

amphibians

Amphibians belong to the vertebrate class Amphibia (from Greek, meaning 'double life'), which generally spend their larval (tadpole) stage in fresh water, transferring to land at maturity (after metamorphosis) and generally returning to water to breed. Like fish and reptiles, they continue to grow throughout life, and cannot maintain a temperature greatly differing from that of their environment. The class contains 4,553 known species, 4,000 of which are frogs and toads, 390 salamanders, and 163 caecilians (wormlike in appearance).

axolotl The rare axolotl *Ambystoma mexicanum* lives in mountain lakes in Mexico. The capture of specimens as pets and the introduction of predatory fish has led to a decline in numbers. Axolotls spend most of their lives in water in a larval state, breathing by means of three pairs of feathery gills and having undeveloped legs and feet. Occasionally, individuals may develop into the adult, land-living form.

fish

Fish are aquatic vertebrates that use gills to obtain oxygen from fresh or sea water. There are three main groups: the **bony fishes** or Osteichthyes (goldfish, cod, tuna, etc.); the **cartilaginous fishes** or Chondrichthyes (sharks, rays, etc.); and the **jawless fishes** or Agnatha (hagfishes, lampreys, etc.). The majority of fishes are predators, feeding on other fishes and invertebrates.

Fishes of some form are found in virtually every body of water in the world except for the very salty water of the Dead Sea and some of the hot larval springs. Of the 30,000 fish species, approximately 2,500 are freshwater.

The world's largest fish is the whale shark *Rhineodon typus*, more than 20 m/66 ft long; the smallest is the dwarf pygmy goby *Pandaka pygmaea*), 7.5–9.9 mm/0.3–0.4 in long. The study of fishes is called **ichthyology**.

bony fishes

These constitute the majority of living fishes (about 20,000 species). The skeleton is bone, movement is controlled by mobile fins, and the body is usually covered with scales. The gills are covered by a single flap. Many have a swim bladder with which the fish adjusts its buoyancy. Most lay eggs, but some fishes are internally fertilized and retain eggs until hatched inside the body, then giving birth to live young. Most bony fishes are ray-finned fishes, but a few, including lungfishes and coelacanths, are fleshy-finned.

Classification of Living Amphibians

Order	Number of species	Examples
Anura (Salientia)	4,000	frog, toad
Urodela (Caudata)	390	salamander, newt, axolotl, mudpuppy
Apoda (Gymnophiona)	163	caecilian

Classification of Fish

Order	Number of species	Examples
Superclass Agnatha (jawless fishes)		
Class Cyclostomota (scaleless fish with round mouths)		
Petromyzoniformes	30	lamprey
Myxiniformes	15	hagfish
Subphylum: Gnathostomata (jawed vertebrates)		
Superclass Pisces		
Class Chondrichthyes (cartilaginous fishes)		
Subclass Elasmobranchii (sharks and rays)		
Hexanchiformes	6	frilled shark, comb-toothed shark
Heterodontiformes	10	Port Jackson shark
Lamniformes	200	'typical' shark
Rajiformes	300	skate, ray
Subclass Holocephali (rabbitfishes)		
Chimaeriformes	20	chimaera, rabbitfish
Class Osteichthyes (bony fishes)		
Subclass Sarcopterygii (lobe-finned fishes)		
Coelacanthiformes	1	coelacanth
Ceratodiformes	1	Australian lungfish
Lepidosireniformes	4	South American and African lungfish
Subclass Actinopterygii (ray-finned fishes)		
Polypteriformes	11	bichir, reedfish
Acipensiformes	25	paddlefish, sturgeon
Amiiformes	8	bowfin, garpike
Superorder Teleostei		
Elopiformes	12	tarpon, tenpounder
Anguilliformes	300	eel
Notacanthiformes	20	spiny eel
Clupeiformes	350	herring, anchovy
Osteoglossiformes	16	arapaima, African butterfly fish
Mormyriformes	150	elephant-trunk fish, featherback
Salmoniformes	500	salmon, trout, smelt, pike
Gonorhynchiformes	15	milkfish
Cypriniformes	350	carp, barb, characin, loache
Siluriformes	200	catfish
Myctophiformes	300	deep-sea lantern fish, Bombay duck
Percopsiformes	10	pirate perch, cave-dwelling amblyopsid
Batrachoidiformes	10	toadfish
Gobiesociformes	100	clingfish, dragonets
Lophiiformes	150	anglerfish
Gadiformes	450	cod, pollack, pearlfish, eelpout
Atheriniformes	600	flying fish, toothcarp, halfbeak
Lampridiformes	50	opah, ribbonfish
Beryciformes	150	squirrelfish
Zeiformes	60	John Dory, boarfish
Gasterosteiformes	150	stickleback, pipefish, seahorse
Channiformes	5	snakeshead
Synbranchiformes	7	cuchia
Scorpaeniformes	700	gurnard, miller's thumb, stonefish
Dactylopteriformes	6	flying gurnard
Pegasiformes	4	sea-moth
Pleuronectiformes	500	flatfish
Tetraodontiformes	250	puffer fish, triggerfish, sunfish
Perciformes	6,500	perch, cichlid, damsel fish, gobie, wrass, parrotfish, gourami, marlin, mackerel, tuna, swordfish, spiny eel, mullet, barracuda, sea bream, croaker, ice fish, butterfish

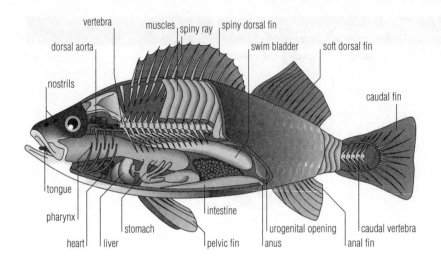

vertebra muscles spiny ray spiny dorsal fin

dorsal aorta swim bladder soft dorsal fin

nostrils caudal fin

tongue

pharynx intestine

heart liver stomach pelvic fin urogenital opening caudal vertebra

anus anal fin

fish anatomy The anatomy of a fish. All fishes move through water using their fins for propulsion. The bony fishes, like the specimen shown here, constitute the largest group of fishes with about 20,000 species.

cartilaginous fishes

These have skeletons made of cartilage, with the mouth generally beneath the head, a large and sensitive nose, and a series of open gill slits along the neck region; they are efficient hunters. Cartilaginous fishes have no swimbladder and, in order to remain buoyant, must keep swimming. They may lay eggs ('mermaid's purses') or bear live young. Some types, such as sharks, retain the shape they had millions of years

shark

A basking shark can sieve more than 1,000 tonnes of water an hour in search of food.

hammerhead shark The hammerhead shark's name derives from the flattened projections at the side of its head. The eyes are on the outer edges of the projections. The advantages of this head design are not known; it may be that the shark's vision is improved by the wide separation of the eyes, or the head may provide extra lift by acting as an aerofoil.

ago. There are fewer than 600 known species of sharks and rays.

Shark
http://www.seaworld.org/animal_bytes/sharkab.html
Illustrated guide to the shark including information about genus, size, life span, habitat, gestation, diet, and a series of fun facts.

---

jawless fishes

Jawless fishes have a body plan like that of some of the earliest vertebrates that existed before true fishes with jaws evolved. There is no true backbone but a notochord. The lamprey attaches itself to the fishes on which it feeds by a suckerlike rasping mouth. Hagfishes are entirely marine, very slimy, and feed on carrion and injured fishes.

Ichthyology Resources
http://muse.bio.cornell.edu/cgi-bin/hl?fish
Everything anyone should need for studying fish, including historical information about development, as well as up-to-date listings of currently endangered species.

---

other major phyla

The mammals, birds, reptiles, amphibians, and fishes all belong to the phylum **Chordata**.

Classification of Molluscs

phylum mollusca

class *Monoplacophora*	primitive marine forms, including Neopilina (2 species)
class *Amphineura*	wormlike marine forms
1 *Aplacophora*	chitons, coat-of-mail
2 *Polyplacophora*	shells (1,150 species)
class *Gastropoda*	snail-like molluscs, with single or no shell (9,000 species)
1 *Prosobranchia*	limpets, winkles, whelks
2 *Opisthobranchia*	seaslugs, land and sea snails
3 *Pulmonata*	freshwater snails, slugs
class *Scaphopoda*	tusk shells, marine burrowers (350 species)
class *Bivalvia*	molluscs with a double (two-valved) shell (15,000 species): mussels, oysters, clams, cockles, scallops, tellins, razor shells, shipworms
class *Cephalopoda*	molluscs with shell generally reduced, arms to capture prey, and beaklike mouth; body bilaterally symmetrical and nervous system well developed (750 species): squids, cuttlefish, octopuses, pearly nautilus, argonaut

Animals belonging to some of the other major phyla are described below.

echinoderms

Echinoderms are marine invertebrates of the phylum **Echinodermata** ('spiny-skinned'), characterized by a five-radial symmetry. Echinoderms have a water-vascular system which transports substances around the body. They include starfishes (or sea stars), brittle-stars, sea lilies, sea urchins, and sea cucumbers. The skeleton is external, made of a series of limy plates. Echinoderms generally move by using tube-feet, small water-filled sacs that can be protruded or pulled back to the body.

molluscs

Molluscs belong to the phylum **Mollusca**. They comprise a group of invertebrate animals, most of which have a body divided into three parts: a head, a central mass containing the main organs, and a foot for movement; the more sophisticated octopuses and related molluscs have arms to capture their prey. Molluscs have

mollusc

The chiton, a species of marine shelled mollusc, uses magnetism to find its way home. The teeth on its tongue-like radula contain magnetite, and enable the chiton to return to exactly the same place on its home rock after night-time feeding.

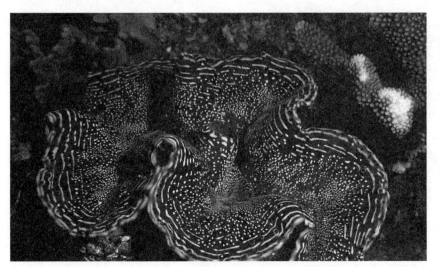

giant clam Giant clam *Tridacna gigus* at Lady Elliott Island, in the Great Barrier Reef, Australia. It obtains its nutrition by a symbiotic association with photosynthetic algae. Giant clams are the largest living bivalves, growing to lengths of around 1 m/ 39 in.

varying diets, the carnivorous species feeding mainly on other molluscs. Some are vegetarian. The majority of molluscs are marine animals, but some live in fresh water, and a few live on dry land. They include clams, mussels, and oysters (bivalves), snails and slugs (gastropods), and cuttlefish, squids, and octopuses (cephalopods). The body is soft, without limbs (except for the cephalopods), and cold-blooded. There is no internal skeleton, but many species have a hard shell covering the body.

arthropods

Arthropods belong to the phylum **Arthropoda**; they are invertebrate animals with jointed legs and a segmented body with a horny or chitinous casing (exoskeleton), which is shed periodically and replaced as the animal grows. The phylum includes **arachnids** such as spiders and mites, as well as **crustaceans, myriapods** such as millipedes and centipedes, and **insects**.

arachnids An arachnid, or arachnoid, is an arthropod of the class Arachnida, which includes spiders, scorpions, ticks, and mites. They differ from insects in possessing only two main body regions, the cephalothorax and the abdomen, and in having eight legs.

crustaceans

A crustacean is one of the class of arthropods that includes crabs, lobsters, shrimps, woodlice, and barnacles. The external skeleton is made of protein and chitin hardened with lime. Each segment bears a pair of appendages that may be modified as sensory feelers (antennae), as mouthparts, or as swimming, walking, or grasping structures.

myriapods

A myriapod is a terrestrial arthropod characterized by a long, segmented body and many legs. Centipedes and millipedes are the most familiar myriapods. The bodies of centipedes are composed of up to almost 200 segments,

termite

Termite nests may house two million termites and reach heights of up to 6 m/ 6.6 yd. A skyscraper built to the same scale would be 8 km/5 mi high.

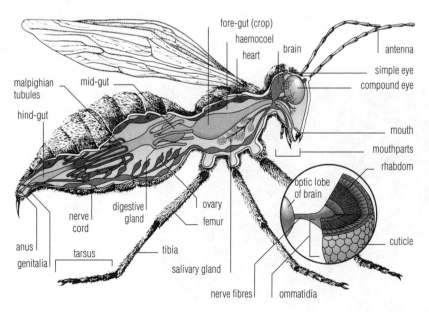

fore-gut (crop)
haemocoel
heart
brain
antenna
simple eye
compound eye
malpighian tubules
mid-gut
hind-gut
mouth
mouthparts
rhabdom
optic lobe of brain
ovary
femur
cuticle
digestive gland
nerve cord
anus
genitalia
tarsus
tibia
salivary gland
nerve fibres
ommatidia

insect, body plan Body plan of an insect. The general features of the insect body include a segmented body divided into head, thorax, and abdomen, jointed legs, feelers or antennae, and usually two pairs of wings. Insects often have compound eyes with a large field of vision.

Classification of Insects

(N/A = not available.)

Order	Number of species	Examples
Apterygota (wingless insects)		
Thysanura	350	three-pronged bristletails, silverfish
Diplura	400	two-pronged bristletails, campodeids, japygids
Protura	50	minute insects living in soil
Collembola[1]	1,500	springtails
Pterygota (winged insects or forms secondary wingless)		
Exopterygota (young resemble adults but have externally-developing wings)		
Ephemeroptera	1,000	mayflies
Odonata	5,000	dragonflies, damselflies
Plecoptera	3,000	stoneflies
Grylloblattodea	12	wingless soil-living insects of North America
Orthoptera	20,000	crickets, grasshoppers, locusts, mantids, roaches
Phasmida	2,000	stick insects, leaf insects
Dermaptera	1,000	earwigs
Embioptera	150	web-spinners
Dictyoptera	5,000	cockroaches, praying mantises
Isoptera	2,000	termites
Zoraptera	16	tiny insects living in decaying plants
Psocoptera	1,600	booklice, barklice, psocids
Mallophaga	2,500	biting lice, mainly parasitic on birds
Anoplura	250	sucking lice, mainly parasitic on mammals
Hemiptera	55,000	true bugs, including shield- and bedbugs, froghoppers, pond skaters, water boatmen
Homoptera	N/A	aphids, cicadas
Thysanoptera	5,000	thrips
Endopterygota (young unlike adults, undergo sudden metamorphosis)		
Neuroptera	4,500	lacewings, alder flies, snake flies, ant lions
Mecoptera	300	scorpion flies
Lepidoptera	165,000	butterflies, moths
Trichoptera	3,000	caddis flies
Diptera	80,000	true flies, including bluebottles, mosquitoes, leatherjackets, midges
Siphonaptera	1,400	fleas
Hymenoptera	100,000	bees, wasps, ants, sawflies, chalcids
Coleoptera	350,000	beetles, including weevils, ladybirds, glow-worms, woodworms, chafers

[1] Some zoologists recognize the Collembola taxon as a class rather than an order.

each of similar form and bearing a single pair of legs. Millipedes have fewer body segments (up to 100) but have two pairs of legs on each.

insects

The insects form a vast group of small invertebrate animals with hard, segmented bodies, three pairs of jointed legs, and, usually, two pairs of wings; they are distributed throughout the world. An insect's body is divided into three segments: head, thorax, and abdomen. On the head is a pair of

housefly

In nine months, a housefly could lay enough eggs to produce a layer of flies that would cover all of Germany to a depth of 14 m/47 ft.

feelers, or antennae. The legs and wings are attached to the thorax, or middle segment of the body. The abdomen, or end segment of the body, is where food is digested and excreted and where the reproductive organs are located.

Insects vary in size from 0.02 cm/0.007 in to 35 cm/13.5 in in length. The world's smallest insect is believed to be a 'fairy fly' wasp in the family Mymaridae, with a wingspan of 0.2 mm/ 0.008 in.

annelids

Annelids are segmented worms belonging to the phylum **Annelida**. Annelids include earthworms, leeches, and marine worms such as lugworms. They have a distinct head and soft body, which is divided into a number of similar segments shut

annelid Annelids are worms with segmented bodies. The ragworm, lugworm, and peacock worm shown here are all marine species. Ragworms commonly live in mucous-lined burrows on muddy shores or under stones, and lugworms occupy U-shaped burrows. The peacock worm, however, builds a smooth, round tube from fine particles of mud.

off from one another internally by membranous partitions, but there are no jointed appendages. Annelids are noted for their ability to regenerate missing parts of their bodies.

nematodes

Nematodes are unsegmented worms belonging to the phylum **Nematoda**. They are pointed at both ends, with a tough, smooth outer skin. They include many free-living species found in soil and water, including the sea, but a large number are parasites, such as the roundworms and pinworms that live in humans, or the eelworms that attack plant roots. They differ from flatworms (the platyhelminths, see below) in that they have two openings to the gut (a mouth and an anus).

Most nematode species are found in deep-sea sediment. Around 13,000

species are known, but a 1995 study by the Natural History Museum, London, based on the analysis of sediment from 17 seabed sites world-wide, estimated that nematodes may make up as much as 75% of all species, with there being an estimated 100 million species. Some are anhydrobiotic, which means they can survive becoming dehydrated, entering a state of suspended animation until they are rehydrated.

platyhelminths

Platyhelminths are flatworms belonging to the phylum **Platyhelminthes**. Some are free-living, but many are parasitic (for example, tapeworms and flukes). The body is simple and bilaterally symmetrical, with one opening to the intestine. Many are hermaphroditic (with both male and female sex organs) and practise self-fertilization.

coelenterates

Coelenterates are freshwater or marine organisms belonging to the phylum **Coelenterata**. They have a body wall composed of two layers of cells. They also possess stinging cells. Examples are jellyfish, hydra, sea anemones, and coral.

protein (as in the bath sponge) or small spikes of silica, or a framework of calcium carbonate.

feeding

As well as being divided up taxonomically, animals are also loosely categorized by what they eat. They can be divided into three feeding types: **herbivores** eat plants and plant products, **carnivores** eat other animals, and **omnivores** eat both. Since few animals can digest cellulose, herbivores have either symbiotic cellulose-digesting bacteria or protozoa in their guts, or grinding mechanisms, such as the large flattened teeth of elephants, to release the plant protoplasm from its

herbivore (sheep)

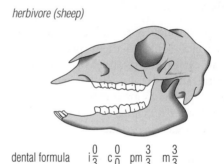

dental formula $i\frac{0}{3}$ $c\frac{0}{0}$ $pm\frac{3}{3}$ $m\frac{3}{3}$

carnivore (dog)

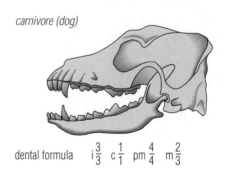

dental formula $i\frac{3}{3}$ $c\frac{1}{1}$ $pm\frac{4}{4}$ $m\frac{2}{3}$

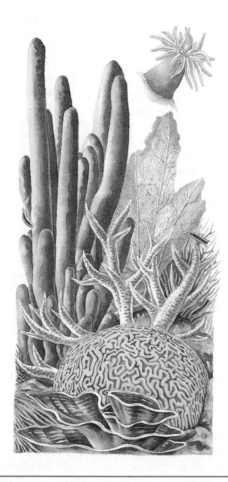

coral Corals are marine animals related to jellyfishes and sea anemones. Most hard or stony corals live as colonies of polyps that secrete a rigid external skeleton of lime. But there are also those that live as solitary individuals. Corals grow in a wide variety of forms as the types here demonstrate. Sea fans, for example, have an internal skeleton linking the polyps, while the polyps of the brain coral are arranged in rows. The stagshorn coral gets its name from the resemblance of the colony to a male deer's antlers. The names of the plate and pillar corals also reflect their appearance.

sponges

Sponges are saclike simple invertebrates belonging to the phylum **Porifera**, usually marine. A sponge has a hollow body, its cavity lined by cells bearing flagella, whose whiplike movements keep water circulating, bringing in a stream of food particles. The body walls are strengthened with

dentition The dentition and dental formulae of a typical herbivore (sheep) and carnivore (dog). The dog has long pointed canines for puncturing and gripping its prey and has modified premolars and molars (carnassials) for shearing flesh. In the sheep, by contrast, there is a wide gap, or diastema, between the incisors, developed for cutting through blades of grass, and the grinding premolars and molars; the canines are absent.

cellulose-walled cells. Carnivores are adapted for hunting and eating flesh, with well-developed sense organs and fast reflexes, and weapons such as sharp fangs, claws, and stings. Omnivores eat whatever they can find, and often scavenge among the remains of carnivores' prey; because of the diversity of their diet, they have more versatile teeth and guts than herbivores or carnivores. Many animals are adapted for a parasitic way of life, living on other animals or plants, and feeding solely by absorbing fluids from their hosts. Some animals absorb food directly into their body cells; others have a digestive system in which food is prepared for absorption by body tissues.

life cycles

All animals go through a sequence of developmental stages, according to their species. Most **vertebrate life cycles** are relatively simple, consisting of **fertilization** of sex cells or gametes, a period of development as an **embryo**, a period of **juvenile growth** after hatching or birth, an **adulthood** including sexual reproduction, and finally **death**.

Invertebrate life cycles are generally more complex and may involve major reconstitution of the individual's appearance (**metamorphosis**) and completely different styles of life. Many insects such as cicadas, dragonflies, and mayflies have a long larval or pupal phase and a short adult phase. Dragonflies live an aquatic life as larvae and an aerial life during the adult phase. In many invertebrates there is a sequence of stages in the life cycle, and in parasites different stages often occur in different host organisms.

reproduction
sexual reproduction Most vertebrate animals reproduce sexually, in a process that requires the union, or fertilization of gametes (eggs and sperm in higher organisms). These are usually produced by two different individuals, a male who produces sperm and a female who produces eggs. Many animals go through some form of courtship before mating. In mammals, birds, reptiles, and terrestrial insects, fertilization

---

monogamy
Less than 3% of mammals are monogamous.

---

occurs internally, within the body of the female; the male introduces sperm into the reproductive tract of the female through a penis or other organ, while most birds transfer sperm by pressing their cloacas (the opening of their reproductive tracts) together. In the majority of amphibians and fishes, and most aquatic invertebrates, fertilization occurs externally, when the female lays her eggs in the water and the male releases sperm onto them. The successful fusion of a male and female gamete produces a zygote, from which a new individual develops. The embryo develops within an egg, where it is nourished by food contained in the yolk, or in mammals in the uterus of the mother. In mammals (except marsupials and monotremes) the embryo is fed through the placenta. After birth or hatching, the parents often continue to care for and feed their young until they are able to survive independently.

other forms of reproduction Other forms of nonsexual reproduction found in the animal kingdom are **parthenogenesis** and **self-fertilization**. Parthenogenesis is the development of an ovum (egg) without any genetic contribution from a male. It is the normal means of reproduction of some fishes. Some sexually reproducing species, such as aphids, show parthenogenesis at some stage in their life cycle to accelerate reproduction to take advantage of good conditions. Self-fertilization occurs in a few hermaphrodites (organisms having both male and female sex organs), such as tapeworms. **Hermaphroditism** is the norm in such species as earthworms and snails, but cross-fertilization is generally the rule among hermaphrodites, with the parents functioning

---

sea horse
In sea horses it is the male that carries the babies. The female deposits her eggs into his brood pouch, where he fertilizes them. They remain there for up to six weeks while he nourishes them till they reach independence.

---

velvet worm
The mating technique of African velvet worms is bizarre. The male places sperm bundles anywhere along the female's body. Beneath the sperm bundle the skin dissolves forming a hole. The sperm can thus enter the female's body cavity, where it makes its way to her ovaries.

Section through a fertilized egg

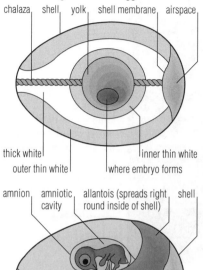

chalaza | shell | yolk | shell membrane | airspace

thick white | inner thin white
outer thin white | where embryo forms

amnion | amniotic cavity | allantois (spreads right round inside of shell) | shell

chorion | yolk sac | chorioallantoic membrane
umbilicus

egg Section through a fertilized bird egg. Inside a bird's egg is a complex structure of liquids and membranes designed to meet the needs of the growing embryo. The yolk, which is rich in fat, is gradually absorbed by the embryo. The white of the egg provides protein and water. The chalaza is a twisted band of protein which holds the yolk in place and acts as a shock absorber. The airspace allows gases to be exchanged through the shell. The allantois contains many blood vessels which carry gases between the embryo and the outside.

as male and female simultaneously, or as one or the other sex at different stages in their development. **Fragmentation** occurs in some invertebrates, such as jellyfish: parts of the organisms break away and subsequently differentiate to form new organisms. Regeneration may sometimes occur before separation, producing chains of offspring budding from the parent organism.

amphibian metamorphosis

Most amphibian species go through an aquatic larval stage as tadpoles, undergoing metamorphosis in the water to emerge as adults that move onto land. In the typical life cycle of a frog, eggs are laid in large masses (spawn) in fresh water. The tadpoles hatch from the eggs in about a fortnight. At first they are fishlike animals with external gills and a long swimming tail, but no limbs. The first change to take place is the disappearance of the external gills and the development of internal gills which are still later supplanted by lungs. The hind legs appear before the front legs, and the last change to occur is the diminution and final disappearance of the tail. The tadpole

aphid During spring and summer female aphids such as this *Macrosiphum cholodkovskyi* produce a continuous succession of offspring by parthenogenesis, but mate and lay eggs before the onset of winter.

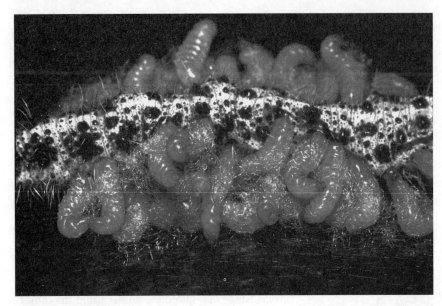

braconid wasp The larvae of many parasitic wasps develop inside the bodies of other insects. These larvae of the braconid wasp *Apanteles glomeratus* are boring their way out through the skin of their host, a caterpillar of the large white butterfly *Pieris brassicae*, in order to pupate. Parasites such as these, which normally kill their host, are sometimes distinguished from 'true' parasites.

stage lasts about three or four months. At the end of this time the adult frog leaves the water.

insect metamorphosis

Many insects hatch out of their eggs as larvae (an immature stage, usually in the form of a caterpillar, grub, or maggot) and have to pass through further major physical changes (metamorphosis) before reaching adulthood. An insect about to go through metamorphosis hides itself or makes a cocoon in which to hide, then rests while the changes take place; at this stage the insect is called a pupa, or a chrysalis if it is a butterfly or moth. When the changes are complete, the adult insect emerges.

animal behaviour

communication

Animals communicate by signalling information, usually with the intention of altering the recipient's behaviour, for example to drive away a rival or to attract a mate. Signals used in communication may be **visual** (such as the colourful plumage in birds, or a submissive or aggressive stance), **auditory** (for example the whines or barks of a dog), **olfactory** (such as the odours released by the scent glands of a deer), **electrical** (as in the pulses emitted by electric fish), or **tactile** (for example, the nuzzling of male and female elephants).

Bees transmit information to each other about

bird of paradise The male blue bird of paradise displays to the female high up in the tree canopy. It swings upside down, exposing its bright blue plumage and tail streamers, at the same time uttering low mechanical-sounding cries.

food sources by 'dances', each movement giving rise to sound impulses which are picked up by tiny hairs on the back of a bee's head, the orientation of the dance also having significance.

courtship

Courtship is the behaviour exhibited by animals as a prelude to mating. The behaviour patterns vary considerably from one species to another, but are often ritualized forms of behaviour not obviously related to courtship or mating (for example, courtship feeding in birds). Courtship rituals include displays of strength or beauty, dances, and so on. Male bowerbirds build elaborate bowers of sticks and grass, decorated with shells, feathers, or flowers, and even painted with the juice of berries, to attract the females. Some frogs, lizards, and even birds (such as the male frigate bird) have inflatable throat sacs, often red in colour, which they puff out to lure a female. Courtship ensures that copulation occurs with a member of the opposite sex of the right species. It also synchronizes the partner's readiness to mate and allows each partner to assess the suitability of the other.

> **courtship**
>
> Birds often use colour in courtship rituals. The kea uses red objects including artificial objects such as buttons; the male satin bower bird uses blue objects; the fairy wren uses a yellow flower.

migration

Certain animals, chiefly birds and fish, undertake long journeys, either as a seasonal movement or as part of a single life cycle, to distant breeding or feeding grounds. The precise methods by which animals navigate and know where to go are still obscure. Birds have much sharper eyesight and better visual memory of ground clues than humans, but in long-distance flights appear to navigate by the Sun and stars, possibly in combination with a 'reading' of the Earth's magnetic field through an inbuilt 'magnetic compass', a tiny mass of tissue between the eye and brain. Similar cells occur in 'homing' honeybees. Leatherback turtles use the contours of underwater mountains and valleys to navigate by. Most striking, however, is the migration of young birds that have never flown a route before and are unaccompanied by adults. It is postulated that they may inherit as part of their genetic code an overall 'sky chart' of their journey that is triggered into use when they become aware of how the local sky pattern, above

> **lobster**
>
> When spiny lobsters migrate, they do so by walking in queues of up to 65 along the sea-bed, each clinging with its claws to the rear of the one in front. Scientific experiments with dead lobsters, weights and pulleys have shown that such an arrangement reduces drag and allows a 25% improvement in speed through the water.

migrating wildebeest During the dry season, herds of tens of thousands of wildebeest embark on migrations of up to 1,600 km/1,000 mi in search of water and grazing. Females and young stay in the centre of the herd, with the males defending the edges.

bonnet macaques Bonnet macaques *Macaca radiata* from southern India usually spend the hottest part of the day grooming each other. This helps not only to keep their skins free of parasites but also to strengthen social relationships.

the place in which they hatch, fits into it. Similar theories have been advanced in the case of fish, such as eels and salmon, with whom vision obviously plays a lesser role, but for whom currents and changes in the composition and temperature of the sea in particular locations may play a part – for example, in enabling salmon to return to the precise river in which they were spawned. Migration also occurs with land animals such as lemmings and antelope.

social behaviour
Social behaviour among animals allows them to live harmoniously in groups by establishing hierarchies of dominance to discourage disabling

fighting. It may be aggressive or submissive (for example cowering and other signals of appeasement), or designed to establish bonds (such as social grooming or preening).

The social behaviour of mammals and birds is generally more complex than that of lower organisms and involves relationships with individually recognized animals. Thus, courtship displays allow individuals to choose appropriate mates and form the bonds necessary for successful reproduction. In the social systems of bees, wasps, ants, and termites, an individual's status and relationships with others are largely determined by its biological form, as a member of a caste of workers, soldiers, or reproductives.

Biodiversity: Number of Species Worldwide

	number identified	% of estimated total number of species
microorganisms	5,800	3–27%
invertebrates	1,021,000	3–27%
plants	322,500	67–100%
fish	19,100	83–100%
birds	9,100	94–100%
reptiles and amphibians	12,000	90–95%
mammals	4,000	90–95%
total	1,393,500	
	number of species	% identified
low estimate of all species	4.4 million	31
high estimate of all species	80 million	2

Selected Animals on the Endangered and Threatened Species List for the World

Source: US Fish and Wildlife Service
(Data as of April 1997.)

Common name	Scientific name	Range	When listed
armadillo, giant	Priodontes maximus (= giganteus)	South America	24 June 1976
bear, brown	Ursus arctos arctos	Europe	24 June 1976
bear, brown	Ursus arctos pruinosus	Asia	24 June 1976
bear, Mexican grizzly	Ursus arctos (= U. a. nelsoni)	North America	2 June 1970
bison, wood	Bison bison athabascae	North America	2 June 1970
bobcat	Felis rufus escuinapae	North America	24 June 1976
camel, Bactrian	Camelus bactrianus (=ferus)	Asia	24 June 1976
cat, leopard	Felis bengalensis bengalensis	Asia	24 June 1976
deer, pampas	Ozotoceros bezoarticus	South America	24 June 1976
dugong	Dugong dugon	Africa	2 December 1970
duiker, Jentink's	Cephalophus jentinki	Africa	25 June 1979
eland, western giant	Taurotragus derbianus derbianus	Africa	25 June 1979
elephant, African	Loxodonta africana	Africa	12 May 1978
elephant, Asian	Elephas maximus	Asia	24 June 1974
gazelle, Clark's (Dibatag)	Ammodorcas clarkei	Africa	2 June 1970
gorilla	Gorilla gorilla	Africa	2 June 1970
kangaroo, Tasmanian forester	Macropus giganteus tasmaniensis	Australia	4 June 1973
lemurs	Lemuridae	Africa	2 June 1970, 14 June 1976, 24 June 1976
leopard	Panthera pardus	Africa, Asia	2 June 1970, 30 March 1972, 28 January 1982
leopard, snow	Panthera uncia	Asia	30 March 1972
manatee, Amazonian	Trichechus inunguis	South America	2 June 1970
manatee, West African	Trichechus senegalensis	Africa	20 July 1979
monkey, black howler	Alouatta pigra	North America	19 October 1976
monkey, red-backed squirrel	Saimiri oerstedii	South America	2 June 1970
monkey, spider	Ateles geoffroyi frontatus	South America	2 June 1970
mouse, Shark Bay	Pseudomys praeconis	Australia	2 December 1970
orangutan	Pongo pygmaeus	Asia	2 June 1970
panda, giant	Ailuropoda melanoleuca	Asia	23 January 1984
rat-kangaroo, Queensland	Bettongia tropica	Australia	2 December 1970
rhinoceros, black	Diceros bicornis	Africa	14 July 1980
rhinoceros, great Indian	Rhinoceros unicornis	Asia	2 December 1970
seal, Saimaa	Phoca hispida saimensis	Europe	28 July 1993
sloth, Brazilian three-toed	Bradypus torquatus	South America	2 June 1970
tapir, Asian	Tapirus indicus	Asia	24 June 1974
tapir, mountain	Tapirus pinchaque	South America	2 June 1970
tiger	Panthera tigris	Asia	2 June 1970, 30 March 1972
tiger, Tasmanian (thylacine)	Thylacinus cynocephalus	Australia	2 June 1970
wallaby, banded hare	Lagostrophus fasciatus	Australia	2 December 1970
whale, blue	Balaenoptera musculus	Oceanic	2 June 1970
whale, bowhead	Balaena mysticetus	Oceanic	2 June 1970
whale, finback	Balaenoptera physalus	Oceanic	2 June 1970
whale, humpback	Megaptera novaeangliae	Oceanic	2 June 1970
whale, right	Balaena glacialis (incl. australis)	Oceanic	2 June 1970
whale, sperm	Physeter macrocephalus (=catodon)	Oceanic	2 June 1970
wombat, hairy-nosed (Barnard's and Queensland hairy-nosed)	Lasiorhinus krefftii (formerly L. Barnardi and L.gillespiei)	Australia	2 December 1970, 4 June 1973
zebra, Grevy's	Equus grevyi	Africa	21 August 1979
zebra, mountain	Equus zebra zebra	Africa	24 June 1976, 10 February 1981
booby, Abbott's	Sula abbotti	Asia	24 June 1976
condor, Andean	Vultur gryphus	South America	2 December 1970
crane, hooded	Grus monacha	Asia	2 December 1970
crane, Japanese	Grus japonenis	Asia	2 June 1970
crane, whooping	Grus americana	North America	11 March 1967, and other dates

Common name	Scientific name	Range	When listed
eagle, harpy	*Harpia harpyja*	South America	24 June 1976
egret, Chinese	*Egretta eulophotes*	Asia	2 June 1970
falcon, American peregrine	*Falco peregrinus anatum*	North America	13 October 1970
hawk, Galapagos	*Buteo galapagoensis*	South America	2 June 1970
ibis, Japanese crested	*Nipponia nippon*	Asia	2 June 1970
ibis, northern bald	*Geronticus eremita*	Africa, Asia, Europe	28 September 1990
macaw, glaucous	*Anodorhynchus glaucus*	South America	24 June 1976
macaw, indigo	*Anodorhynchus leari*	South America	24 June 1976
ostrich, Arabian	*Struthio camelus syriacus*	Asia	2 June 1970
ostrich, West African	*Struthio camelus spatzi*	Africa	2 June 1970
parakeet, gold-shouldered (hooded)	*Psephotus chrysopterygius*	Australia	2 June 1970
parakeet, Norfolk Island	*Cyanoramphus novaezelandiae cookii*	Australia	28 September 1990
parrot, ground	*Pezoporus wallicus*	Australia	4 June 1973
parrot, red-capped	*Pionopsitta pileata*	South America	24 June 1976
stork, oriental white	*Ciconia ciconia boyciana*	Asia	2 June 1970
woodpecker, imperial	*Campephilus imperialis*	North America	2 June 1970
woodpecker, Tristam's	*Drycopus javenis richardsi*	Asia	2 June 1970
alligator, Chinese	*Alligator sinensis*	Asia	24 June 1976
caiman, Apaporis River	*Caiman crocodilus apaporiensis*	South America	24 June 1976
crocodile, African dwarf	*Osteolaemus tetraspis tetraspis*	Africa	24 June 1976
crocodile, African slender-snouted	*Crocodylus cataphractus*	Africa	30 March 1972
crocodile, Morelet's	*Crocodylus moreletii*	South America	2 June 1970
iguana, Barrington land	*Conolophus pallidus*	South America	2 June 1970
lizard, Hierro giant	*Gallotia simonyi simonyi*	Europe	29 February 1984
lizard, Ibiza wall	*Podarcis pityusensis*	Europe	29 February 1984
monitor, Bengal	*Varanus bengalensis*	Asia	24 June 1976
monitor, desert	*Varanus griseus*	Africa	24 June 1976
monitor, yellow	*Varanus flavescens*	Asia	24 June 1976
python, Indian	*Python molurus molurus*	Asia	24 June 1976
tartaruga	*Podocnemis expansa*	South America	2 June 1970
tortoise, Bolson	*Gopherus flavomarginatus*	North America	17 April 1979
tortoise, Galapagos	*Geochelone elephantopus*	South America	2 June 1970
tuatara	*Sphenodon punctatus*	Australia	2 June 1970
turtle, aquatic box	*Terrapene coahuila*	North America	4 June 1973
turtle, green sea	*Chelonia mydas* (incl. *agassizi*)	North America (circumglobal)	28 July 1978
viper, Lar Valley	*Vipera latiffi*	Asia	22 June 1983
frog, Goliath	*Conraua goliath*	Africa	8 December 1994
frog, Israel painted	*Discoglossus nigriventer*	Europe	2 June 1970
frog, Panamanian golden	*Atelopus varius zeteki*	South America	24 June 1976
frog, Stephen Island	*Leiopelma hamiltoni*	Australia	2 June 1970
salamander, Chinese giant	*Andrias davidianus davidianus*	western China	24 June 1976
salamander, Japanese giant	*Andrias davidianus japonicus*	Japan	24 June 1976
toad, African viviparous	*Nectophrynoides* spp.	Africa	24 June 1976
toad, Cameroon	*Bufo superciliaris*	Africa	24 June 1976
toad, Monte Verde	*Bufo periglenes*	South America	24 June 1976

The Animal Kingdom Chronology

c. 1140 BC — Chinese emperor Wen Wang establishes the first zoo. It covers 1,500 acres and is named the *Ling-Yo*/Garden of Intelligence.

c. 1115 BC — King Tiglath-Pileser I of Assyria collects wild animals, including the Egyptian gift of a crocodile.

280 BC — The Egyptian pharaoh Ptolemy II has a large zoological garden at Alexandria, Egypt.

c. 150 — *Physiologus/Naturalist*, a Greek work by an anonymous author, is written in Alexandria. All the medieval 'bestiaries' evolve from this work, which is an encyclopedia of real and imagined natural history.

c. 1115 — King Henry I of England establishes a menagerie at Woodstock, Oxfordshire; it features lions, camels, and leopards.

1220 — Giraffes are brought to Europe and put on public display for the first time, to general amazement.

c. 1297 — The moa, a giant flightless bird of New Zealand, is hunted to extinction by the Maori settlers on North Island, although it is though to have survived a while longer on South Island.

1517 — Taxidermy is developed in Amsterdam, the Netherlands, where naturalists preserve cassowary birds brought from the East Indies.

1551 — Swiss scientist Conrad Gesner publishes the first volume of his *Historia animalium/History of Animals*, a pioneering, highly illustrated, classification of animals.

1599 — The Italian naturalist Ulisse Aldrovandi publishes the first three volumes of his *Natural History*, methodically listing and describing bird species in the first serious zoological study.

1669 — Dutch microscopist Jan Swammerdam writes a *History of Insects*, which lists the reproductive parts of insects and correctly describes metamorphosis.

1669 — Italian anatomist Marcello Malpighi publishes a treatise on the anatomy and development of the silkworm, the first description of the anatomy of an invertebrate.

1681 — The dodo, a form of giant pigeon inhabiting the island of Mauritius in the Indian Ocean, is driven to extinction by the arrival of colonizing Europeans.

1693 — English naturalist John Ray, introduces the first important classification system for animals.

1734 — French scientist René-Antoine Ferchault de Réaumur publishes *History of Insects*, a founding work of entomology.

1740 — Swiss naturalist Abraham Trembley discovers the hydra – a simple water organism combining plant and animal characteristics, and capable of regenerating itself.

1758 — Swedish naturalist Carolus Linnaeus applies the binomial taxonomy he developed for plant classification to animal species.

1760 — Dutch naturalist and engraver Pieter Lyonnet publishes a monograph on the goat-moth caterpillar, containing details and illustrations of dissections. It is one of the best illustrated books on anatomy ever produced and describes over 4,000 muscles.

1760 — French naturalist M Brisson publishes his *Ornithologie/Ornithology*, a classic study of bird life.

1768 — Italian physiologist Lazzaro Spallanzani studies regeneration in animals and shows that the lower animals have a greater capacity to regenerate lost limbs.

1775 — Italian anatomist Johann Fabricius classifies insects according to their mouth structure rather than their wings.

1793 — French naturalist Jean-Baptiste Lamarck argues that fossils are the remains of extremely ancient, extinct species of animals and plants.

1794 — English naturalist and physician Erasmus Darwin publishes *Zoonomia, or the Laws of Organic Life*, expressing his ideas on evolution (which he assumes has an environmental cause).

1797 — The English wood engraver Thomas Bewick publishes *Land Birds*. It is the first part of his work *A History of British Birds*.

1802–1805 — French zoologist Pierre-André Latreille publishes his 14-volume work *Histoire naturele générale et particulière des crustacés et insectes/Comprehensive Natural History of Crustaceans and Insects*.

1812 — French zoologist Georges Cuvier, publishes *Recherches sur les ossements fossiles de quadrupèdes/Research on the Fossil Bones of Quadrupeds*, and establishes comparative vertebrate palaeontology. He theorizes that the extinction of species has been caused by great catastrophes such as sudden land upheavals and floods.

1817 — Cuvier breaks away from the view that animals can be arranged in a linear sequence leading to humans and argues instead that they should be classified according to their anatomical organization.

1826 — English administrator Stamford Raffles founds the Royal Zoological Society in London, England.

1827–1838 — US ornithologist John James Audubon publishes the first volume of his multi-volume work *Birds of America*.

1828 — London Zoo opens in Regent's Park, London, England.

27 Dec 1831 –2 Oct 1836 — The English naturalist Charles Darwin undertakes a five-year voyage, to South

America and the Pacific, as naturalist on the *Beagle*. The voyage convinces him that species have evolved gradually but he waits over 20 years to publish his findings.

1846 English palaeontologist Richard Owen publishes *Lectures on Comparative Anatomy and Physiology of the Vertebrate Animals*, one of the first textbooks on comparative vertebrate anatomy.

1854 English naturalist Philip Henry Gosse builds the first institutional aquarium in England for the protection of marine animals.

1855–1856 Gosse publishes *Manual of Marine Zoology*, the first thorough book on the subject.

24 Nov 1859 Charles Darwin publishes *On the Origin of Species by Natural Selection*, which expounds his theory of evolution by natural selection.

1861 English entomologist H W Bates publishes his paper 'Contributions to the Insect Fauna of the Amazon Valley', in which he describes many of the over 14,000 insects (8,000 of which had previously been unknown) that he has collected.

1868 English naturalist Thomas Henry Huxley makes the first classification of the dinosaurs, creating the order Ornithoscelida and two suborders.

7 Dec 1872– The British ship *Challenger* undertakes the **26 May 1876** world's first major oceanographic survey. Under the command of the Scottish naturalist Wyville Thomson, it discovers hundreds of new marine animals, and finds that at 2,000 fathoms the temperature of the ocean is a constant 2.5°C/36.5°F.

1878 The complete skeletons of several dozen *Iguanodon* are discovered in a coal mine in Belgium. They provide the first evidence that some dinosaurs travelled in herds.

1887 English naturalist H G Seeley classifies the dinosaurs into two groups, those with birdlike pelvises, the *Ornithischia*, and those with reptile-like pelvises, the *Saurischia*.

1915–1916 English naturalist Archibald Thorburn publishes *British Birds*, which describes and catalogues the birds in Britain.

1919 Austrian zoologist Karl von Frisch discovers that bees communicate the location of nectar through wagging body movements and rhythmic dances.

1935 Austrian zoologist Konrad Lorenz founds the discipline of ethology by describing the learning behaviour of young ducklings; visual and auditory stimuli from the parent object

cause them to 'imprint' on the parent.

1938 A coelacanth, an ancient lobe-finned fish assumed to be extinct, is discovered in the Indian Ocean.

1960 English primatologist Jane Goodall discovers that chimpanzees can make tools, something only humans were thought capable of. She watches a chimpanzee fashion a blade of grass into a probe that can be poked into a termite mound to remove termites.

1964 US zoologist William Hamilton recognizes the importance of altruistic behaviour in animals, paving the way for the development of sociobiology.

1967 US biochemist Marshall Nirenberg establishes that mammals, amphibians, and bacteria all share a common genetic code.

1972 The USA restricts the use of DDT because it is discovered that it thins the eggshells of predatory birds, lowering their reproductive rates.

1973 Kenya bans hunting elephants and trading in ivory.

1973 Representatives from 80 nations sign the Convention on International Trade in Endangered Species (CITES) that prohibits trade in 375 endangered species of plants and animals and the products derived from them, such as ivory; the USA does not sign until 1977.

1973 The US Fish and Wildlife Bureau issues the first Endangered and Threatened Species List.

1977 Scientists discover chemosynthetically based animal communities around sulphurous thermal springs deep under the sea near the Galápagos Islands, Ecuador.

1982 Dolphins are discovered to possess magnetized tissues that aid in navigation; they are the first mammals discovered to have such tissues.

1983 The first discovery of a fossil land mammal (a marsupial) in Antarctica is made.

1983 The skull of a creature called *Pakicetus* is discovered in Pakistan; estimated to be 50 million years old, it is intermediate in evolution between whales and land animals.

1984 Allan Wilson and Russell Higuchi of the University of California, Berkeley, USA, clone genes from an extinct animal, the quagga.

1985 An amphibian skeleton dated 340 million years old is discovered in Scottish oil shale. It is the earliest well-preserved amphibian found.

1985 US zoologist Dian Fossey, who tried to protect

endangered gorillas in Rwanda, is murdered. Poachers are suspected.

1989   A lemur, *Allocebus tricholis*, previously thought to be extinct, is discovered in Madagascar.

1989   The United Nations Environment Programme (UNEP) reports that the number of species, and the amount of genetic variation within individual species, is decreasing due to the rapid destruction of natural environments.

1992   The tooth of a 55-million-year-old mammal is discovered at Murgon, Australia, indicating that mammals arrived in Australia at about the same time as marsupials.

1993   US and Pakistani palaeontologists discover in Pakistan a fossil whale which is about the size of an adult male sea lion and has legs. Called *Ambulocetus*, it is 50 million years old and was able to walk on land but spent most of its time in the sea.

1994   A new species of kangaroo is discovered in Papua New Guinea. Known locally as the bondegezou, it weighs 15 kg/7 lb and is 1.2 m/3.9 ft in height.

1995   A fossil chordate *Yunnanozoon lividum* is discovered in Chengjiang, China. It is the first chordate recorded from the early Cambrian period and is 525 million years old.

1996   New Zealand ornithologist Gavin Hunt reveals that crows on the island of New Caledonia in the South Pacific make tools out of leaves and twigs which they use to reach insects in dead wood – something only chimpanzees and humans were thought capable of.

1996   Palaeontologists from the University of Chicago discover, in Morocco, the remains of the largest carnivorous dinosaur known: *Carcharodontosaurus saharicus* ('shark-toothed reptile from the Sahara'). It lived 90 million years ago, weighed 7.8 tonnes/8 tons, ran at a speed of 32 kph/20 mph, and at 13.5 m/44 ft in length was 1 m/3.3 ft longer than *Tyrannosaurus rex*.

1996   The World Conservation Union (IUCN) publishes the latest Red List of endangered species. Over 1,000 mammals are listed, far more than on previous lists. The organization believes it has underestimated the risks of habitats from pollution and that the number of endangered species is greater than previously thought.

1996   British palaeontologist Peter Ward announces the discovery, in South Africa, of the fossil remains of a lystrosaur, a pig-like early mammal which lived during the Permian era about 250 million years ago. The discovery overturns the existing theory that the first mammals were therapsids – small shrew-like creatures which emerged millions of years later during the age of the dinosaurs.

1996   Palaeontologists from the Institute of Vertebrate Palaeontology and Palaeoanthropology in Beijing, China, discover the remains of a 135-million-year-old fossil bird in northeast China. Called the 'Confucius bird' (*Confusciusornis sactus*), it had a modern-looking beak with no teeth, unlike Archaeopteryx.

1997   In a coal mine in southern Thailand, Thai researchers discover fragments of lower and upper jaws and teeth of a medium-sized monkey-like creature that lived about 40 million years ago. Known as *Siamopithecus eocaenus* ('dawn ape from Thailand'), it weighed 6–7 kg/13–15 lb and provides some of the earliest evidence for the evolution of monkeys, apes, and humans.

1997   US zoologists Bill Detrich and Kirk Malloy show that the increased ultraviolet radiation caused by the hole in the ozone layer above Antarctica kills large numbers of fish in the Southern Ocean. Because their transparent eggs and larvae stay near the surface for up to a year, they are exposed to the full force of the ultraviolet rays. It is the first time ozone depletion in the Antarctic has been shown to harm organisms larger than one-celled marine plants.

9–29 June 1997   At the tenth Convention on International Trade in Endangered Species (CITES) convention in Harare, Zimbabwe, the elephant is downlisted to CITES Appendix II (vulnerable) and the ban on ivory exportation in Botswana, Namibia, and Zimbabwe is lifted.

1997   Scientists from the Worldwide Fund for Nature (WWF) announce the discovery of a new species of muntjac deer in Vietnam. A dwarf species weighing only about 16 kg/35 lb, it has antlers the length of a thumbnail and lives at altitudes of 457–914 m/1,500–3,000 ft.

1998   US ornithologists announce the discovery of a new species of bird in Ecuador belonging to the genus *Antpitta*; this is claimed to be the most important species of bird discovered in 50 years.

Biographies

Adamson, Joy Friedericke Victoria (born Gessner) (1910–1985) German-born naturalist whose work with wildlife in Kenya, including the lioness Elsa, is described in the book *Born Free* (1960). She was murdered at her home in Kenya. She worked with her third husband, British game warden **George Adamson (1906–1989)**, who was murdered by bandits.

Andrews, Roy Chapman (1884–1960) US zoologist. He is best known for his discovery of the oldest known mammals and extensive evidence of primitive human life in the central Asiatic plateau. He was the first to find fossilized dinosaur eggs, and the skull and other skeletal parts of the Baluchitherium, the largest known mammal. Andrews proved central Asia to be one of the chief centres of the origin and distribution of reptilian and mammalian life.

Attenborough, David (1926–) English naturalist who has made numerous wildlife films for television. He was the writer and presenter of the television series *Life on Earth* (1979), illustrating evolution; *The Living Planet* (1983), dealing with ecology and the environment; *The Trials of Life* (1990), describing life cycles; *Life in the Freezer* (1993); *The Private Life of Plants* (1995); and *The Life of Birds* (1998).

Audubon, John James (1785–1851) US naturalist and artist. In 1827, after extensive travels and observations of birds, he published the first part of his *Birds of North America*, with a remarkable series of colour plates. Later he produced a similar work on North American quadrupeds.

Bates, H(enry) W(alter) (1825–1892) English naturalist and explorer. He spent 11 years collecting animals and plants in South America and identified 8,000 new species of insects. He made a special study of camouflage in animals, and his observation of insect imitation of species that are unpleasant to predators is known as 'Batesian mimicry'.

Beebe, (Charles) William (1877–1962) US naturalist, explorer, and writer. His interest in deep-sea exploration led to a collaboration with the engineer Otis Barton and the development of a spherical diving vessel, the bathysphere. On 24 August 1934 the two men made a record-breaking dive to 923 m/3,028 ft. Beebe was curator of birds for the New York Zoological Society 1899–1952. He wrote the comprehensive *Monograph of the Pheasants* (1918–22), and his expeditions are described in a series of memoirs.

Buffon, George Louis Leclerc, Comte de (1707–1778) French naturalist and author of the 18th century's most significant work of natural history, the 44-volume *Histoire naturelle* (1749–67). In *The Epochs of Nature*, one of the volumes, he questioned biblical chronology for the first time and raised the Earth's age from the traditional figure of 6,000 year to the seemingly colossal estimate of 75,000 years.

Carter, Herbert James (1858–1940) Australian entomologist, born in England. A schoolteacher, he became interested in entomology, particularly beetles, and was an avid collector and classifier, describing over 1,000 new species. He was joint editor of the first *Australian Encyclopaedia* (1925–27), supervising the science articles. His published work includes 65

papers and the book *Gulliver in the Bush* (1933), recording his field experiences in Australia.

Cousteau, Jacques Yves (1910–1997) French oceanographer who pioneered the invention of the aqualung in 1943 and techniques in underwater filming. In 1951 he began the first of many research voyages in the ship *Calypso*. His film and television documentaries and books established him as a household name.

Darwin, Charles Robert (1809–1882) English naturalist who developed the modern theory of evolution and proposed, with Alfred Russel Wallace, the principle of natural selection. After research in South America and the Galápagos Islands as naturalist on HMS *Beagle* 1831–36, Darwin published *On the Origin of Species by Means of Natural Selection or the Preservation of Favoured Races in the Struggle for Life* (1859). This book explained the evolutionary process through the principles of natural selection and aroused bitter controversy because it disagreed with the literal interpretation of the Book of Genesis in the Bible.

Durrell, Gerald (Malcolm) (1925–1995) English naturalist, writer, and zoo curator. He founded the Jersey Zoological Park in 1958. Critical of the conditions in which most zoos kept animals, the lack of interest in breeding programmes, and the concentration on large species, such as the big cats, rhinos, and elephants, Durrell became determined to build up his own zoo, and to run it so that it could supplement conservation programmes in the wild rather than detract from them. Through his work, Durrell encouraged and inspired a whole generation of naturalists, zoologists, and zoo keepers.

Ehrenberg, Christian Gottfried (1795–1876) German naturalist who developed one of the forerunners of the modern scheme for classification of the animal kingdom. He was the first scientist to study the fossils of microorganisms and can be regarded as the founder of micropalaeontology.

Fabre, Jean Henri Casimir (1823–1915) French entomologist whose studies of wasps, bees, and other insects, particularly their anatomy and behaviour, have become classics. In addition to numerous entomological papers, Fabre wrote the ten-volume *Souvenirs entomologiques* (1879–1907). Based almost entirely on observations made in his small plot of land in Provence, this work is a model of meticulous attention to detail.

Fabricus, Johann Christian (1745–1808) Danish entomologist who developed a classification system for insects. Using the mouth structure as the basis for classification, he named and described over 10,000 insects. He also wrote on evolution, long before Darwin, and was convinced that humans had evolved from the great apes.

Fossey, Dian (1938–1985) US zoologist. Almost completely untrained, Fossey was sent by Louis Leakey into the African wild. From 1975 she studied mountain gorillas in Rwanda and discovered that they committed infanticide and that females were transferred to nearly established groups. Living in close proximity to them, she discovered that they led peaceful family lives. She was murdered by poachers whose snares she had cut.

Frisch, Karl von (1886–1982) Austrian zoologist, founder with Konrad Lorenz of ethology, the study of animal behaviour. He specialized in bees, discovering how they communicate the location of sources of nectar by movements called 'dances'. He was awarded the Nobel Prize for Physiology or Medicine in 1973 together with Lorenz and Nikolaas Tinbergen.

Gesner, Konrad von (1516–1565) Swiss naturalist. He produced an encyclopedia of the animal world, the five-volume *Historia animalium* (1551–58), and is considered the founder of zoology.

Geoffroy Saint-Hilaire, Isidore (1805–1861) French zoologist and anatomist who specialized in the study of apes and developed a system for their classification. He also studied the manner in which animals interact with their environments. In 1859 he wrote a history of the origin of species which put Darwin's proposals in the context of his French predecessors, including Buffon, Lamarck and Etienne Geoffroy Saint-Hilaire.

Goodall, Jane (1934–) English primatologist and conservationist who has studied the chimpanzee community on Lake Tanganyika, Africa, since 1960, and is a world authority on wild chimpanzees. She observed the lifestyles of chimpanzees in their natural habitats, discovering that they are omnivores, not herbivores as originally thought, and that they have highly developed and elaborate forms of social behaviour. In the 1990s most of Goodall's time was devoted to establishing sanctuaries of illegally captured chimpanzees, fundraising, and speaking out against the unnecessary use of animals in research. Her books include *In the Shadow of Man* (1971), *The Chimpanzees of Gombe: Patterns of Behaviour* (1986), and *Through a Window* (1990).

Gosse, Philip Henry (1810–1888) English naturalist who built the first aquarium ever used to house marine animals long-term and wrote many books on marine zoology, including *Manual of Marine Zoology* (1855), *Actinologia Britannica* (1858), a work on sea anemones, *Introduction to Zoology* (1843), and *Evenings at the Microscope* (1859).

Gould, John (1804–1881) English zoologist who with his wife Elizabeth (1804–1841), a natural-history artist, published a successful series of illustrated bird books. They visited Australia 1838–40 and afterwards produced *The Birds of Australia*, issued in 36 parts from 1840. *Mammals of Australia* followed (1845–63), and *Handbook to the Birds of Australia* (1865).

Griffin, Donald Redfield (1915–) US zoologist who discovered that bats use echolocation to navigate and orientate themselves in space, that is they emit ultrasonic sounds that rebound off objects that they are then able to avoid. His later research was mainly in the areas of animal navigation, acoustic orientation and sensory biophysics, and animal consciousness. His writing includes *Listening in the Dark* (1958), *Echoes of Bats and Men* (1959), *Animal Structure and Function* (1962), *Bird Migration* (1964), and *The Question of Animal Awareness* (1976).

Hyman, Libbie Henrietta (1888–1969) US zoologist whose six-volume *The Invertebrates* (1940–68) provided an encyclopedic account of most phyla of invertebrates.

Knight, Charles (1874–1953) US palaeontological artist who was influential in bringing dinosaurs to life in the public imagination. His extensive knowledge of anatomy and his collaborations with palaeontologists, such as Edward Drinker Cope and Henry Fairfield Osborn, mean his paintings accurately reflect scientific thinking of the time.

Lack, David (1910–1973) English ornithologist who used radar to identify groups of migrating birds. He studied and wrote about the robin, the great tit, the swift, and the finches of the Galápagos Islands. He was the director of the Edward Grey Institute of Field Ornithology at Oxford from 1945 and received the Royal Society's Darwin medal in 1972.

Leuckart, Karl Georg Friedrich Rudolf (1822–1898) German zoologist who identified the phylum Coelenterata (now divided into separate phyla, the Cnidaria – jellyfish, sea anemones, and corals – and the Ctenophora – the comb jellies). His research led to the division of the Metazoa (multicellular animals) into Coelenterata, Echinodermata (sea urchins), Annelida (segmented worms), Arthropoda (jointed limbed animals including insects, spiders, crabs, and lobsters), Mollusca (molluscs, including slugs, snails, octopuses, and shellfish), and Vertebrata (animals with backbones, including fish, birds, and mammals).

Lorenz, Konrad Zacharias (1903–1989) Austrian ethologist. He studied the relationship between instinct and behaviour, particularly in birds, and described the phenomenon of imprinting in 1935. His books include *King Solomon's Ring* (1952), on animal behaviour. In 1973 he shared the Nobel Prize for Physiology or Medicine with Nikolaas Tinbergen and Karl von Frisch.

Manton, Sidnie Milana (married name Harding) (1902–1979) English embryologist who specialized in the arthropods (jointed limbed animals including insects, spiders, crabs, and lobsters), concentrating mainly on their embryology and functional morphology in relation to evolution. She summarized her findings in her book *The Arthropoda: Habits, Functional Morphology and Evolution* (1977).

Mayr, Ernst Walter (1904–) German-born US zoologist who was influential in the development of modern evolutionary theories. He led a two-year expedition to New Guinea and the Solomon Islands where he studied the effects of founder populations and speciation on the evolution of the indigenous birds and animals. This research caused him to support neo-Darwinism, a synthesis of the ideas of Darwin and Mendel, being developed at that time. He has written and edited a number of books, including several upon the development of evolutionary thought, which are standard texts on university courses.

Morgan, Ann Haven (1882–1966) US zoologist who promoted the study of ecology and conservation. She particularly studied the zoology of aquatic insects and the comparative physiology of hibernation. Her *Field Book of Ponds and Streams: An Introduction to the Life of Fresh Water* (1930) attracted amateur naturalists as well as providing an authoritative taxonomic guide for professionals.

Morris, Desmond John (1928–) British zoologist, a writer and broadcaster on animal and human behaviour. His book *The Naked Ape* (1967) was a best seller. In his book *The Human Zoo* (1969), Morris compares civilized humans with their captive animal counterparts and shows how confined animals seem to demonstrate the same neurotic behaviour patterns as human beings often do in crowded cities.

Nice, Margaret, born Morse (1883–1974) US ornithologist who made an extensive study of the life history of the sparrow. She also campaigned against the indiscriminate use of pesticides.

Pallas, Peter Simon (1741–1811) German naturalist who classified corals and sponges and whose work in comparative anatomy, including *Zoographia Rosso-Asiatica*, established him as a predecessor of the French comparative anatomist and founder of palaeontology Georges Cuvier.

Pye, John David (1932–) English zoologist who has studied the way bats use echolocation, and also the use of ultrasound in other animals. In 1971 he calculated the resonant frequencies of the drops of water in fog and found that these frequencies coincided with the spectrum of frequencies used by bats for echolocation. In other words, bats cannot navigate in fog. Pye also found that ultrasound seems to be important in the social behaviour of rodents and insects.

Rothschild, Miriam (1908–) English zoologist and entomologist. She studied fleas and was the first to work out the flea jumping mechanism. She also studied fleas' reproductive cycles and linked this, in rabbits, to the hormonal changes within the host. She has written around 350 papers on entomology, zoology, neurophysiology, and chemistry. Still active in her eighties, she worked on plants and studied the telepathic relationship between people and their pets. Her interest in animal consciousness has led her to press for reform in the treatment of animals in agriculture.

Sutton-Pringle, John William (1912–1982) British zoologist who established much of our knowledge of the anatomical mechanisms involved in insect flight.

Swammerdam, Jan (1637–1680) Dutch naturalist who is considered a founder of both comparative anatomy and entomology. Based on their metamorphic development, he classified insects into four main groups, three of which are still used in a modified form in insect classification. He accurately described and illustrated the life cycles and anatomies of many insect species, including bees, mayflies, and dragonflies.

Tinbergen, Niko(laas) (1907–1988) Dutch-born British zoologist. He specialized in the study of instinctive behaviour in animals. One of the founders of ethology, the scientific study of animal behaviour in natural surroundings, he shared a Nobel prize in 1973 with Konrad Lorenz (with whom he worked on several projects) and Karl von Frisch.

Tinbergen investigated other aspects of animal behaviour, such as learning, and aggression. In *The Study of Instinct* (1951), he showed that the aggressive behaviour of the male three-spined stickleback is stimulated by the red coloration on the underside of other males (which develops during the mating season). In *The Herring Gull's World* (1953), he described the social behaviour of gulls, emphasizing the importance of stimulus–response processes in territorial behaviour.

Turner, Charles Henry (1867–1923) US biologist who carried out research into insect behaviour patterns. He was the first to prove that insects can hear and distinguish pitch and that cockroaches learn by trial and error. He published over 50 papers on neurology, animal behaviour, and invertebrate ecology, including his dissertation 'The homing of ants: an experimental study of ant behaviour' (1907).

von Gesner, Konrad Swiss naturalist. See ◊Gesner, Konrad von.

Waite, Edgar Ravenswood (1866–1928) Australian ornithologist and zoologist. He was a member of several expeditions into the subantarctic islands, New Guinea, and central Australia which contributed significantly to scientific knowledge of vertebrates. His published work includes more than 200 scientific papers, *Popular Account of Australian Snakes* (1898), and *The Fishes of South Australia* (1923).

Wilson, Edward Osborne (1929–) US zoologist whose books have stimulated interest in biogeography, the study of the distribution of species, and sociobiology, the evolution of behaviour. He is a world authority on ants. His works include *Sociobiology: The New Synthesis* (1975) and *The Diversity of Life* (1992).

Wynne-Edwards, Vero Copner (1906–) English zoologist who argued that animal behaviour is often altruistic and that animals will behave for the good of the group, even if this entails individual sacrifice. His study *Animal Dispersal in Relation to Social Behaviour* was published in 1962.

The theory that animals are genetically programmed to behave for the good of the species has since fallen into disrepute. From this dispute grew a new interpretation of animal behaviour, seen in the work of biologist Edward Wilson.

Young, J(ohn) Z(achary) (1907–1997) English zoologist who discovered and studied the giant nerve fibres in squids, contributing greatly to knowledge of nerve structure and function. He also did research on the central nervous system of octopuses, demonstrating that memory stores are located in the brain. He published the textbooks *The Life of Vertebrates* (1950) and *The Life of Mammals* (1957).

Zuckerman, Solly, Baron Zuckerman of Burnham Thorpe (1904–1993) South African-born British zoologist, educationalist, and establishment figure. He did extensive research on primates, publishing a number of books that became classics in their field, including *The Social Life of Monkeys and Apes* (1932) and *Functional Affinities of Man, Monkeys and Apes* (1933). He was chief scientific adviser to the British government 1964–71.

Glossary

abdomen in vertebrates, the part of the body containing the digestive organs; in insects and other arthropods, it is the hind part of the body. In mammals, the abdomen is separated from the thorax (containing the heart and lungs, protected by the rib cage) by the diaphragm, a sheet of muscular tissue; in arthropods, commonly by a narrow constriction. In mammals, the female reproductive organs are in the abdomen. In insects and spiders, it is characterized by the absence of limbs.

aestivation a state of inactivity and reduced metabolic activity, similar to hibernation, that occurs during the dry season in species such as lungfish and snails.

afterbirth in mammals, the placenta, umbilical cord, and ruptured membranes, which become detached from the uterus and expelled soon after birth.

air sac in birds, a thin-walled extension of the lungs. There are nine of these and they extend into the abdomen and bones, effectively increasing lung capacity. In mammals, it is another name for the alveoli in the lungs, and in some insects, for widenings of the trachea (airway).

altruism helping another individual of the same species to reproduce more effectively, as a direct result of which the altruist may leave fewer offspring itself. Female honey bees (workers) behave altruistically by rearing sisters in order to help their mother, the queen bee, reproduce, and forgo any possibility of reproducing themselves.

ammonite extinct marine cephalopod mollusc of the order Ammonoidea, related to the modern nautilus. The shell was curled in a plane spiral and made up of numerous gas-filled chambers, the outermost containing the body of the animal. Many species flourished between 200 million and 65 million years ago, ranging in size from that of a small coin to 2 m/6 ft across.

antenna an appendage ('feeler') on the head. Insects, centipedes, and millipedes each have one pair of antennae but there are two pairs in crustaceans, such as shrimps. In insects, the antennae are involved with the senses of smell and touch; they are frequently complex structures with large surface areas that increase the ability to detect scents.

antler 'horn' of a deer, often branched, and made of bone rather than horn. Antlers, unlike true horns, are shed and regrown each year. Reindeer of both sexes grow them, but in all other types of deer, only the males have antlers.

archaeopteryx extinct primitive bird, known from fossilized remains, about 160 million years old, found in limestone deposits in Bavaria, Germany. It is popularly known as 'the first bird', although some earlier bird ancestors are now known. It was about the size of a crow and had feathers and wings, with three clawlike digits at the end of each wing, but in many respects its skeleton is reptilian (teeth and a long, bony tail) and very like some small meat-eating dinosaurs of the time.

axolotl aquatic larval form ('tadpole') of the Mexican salamander *Ambystoma mexicanum*, belonging to the family Ambystomatidae. Axolotls may be up to 30 cm/12 in long. Axolotls are remarkable because they can breed without changing to the adult form, and will metamorphose into adults only in response to the drying-up of their ponds. The adults then migrate to another pond.

bat any mammal of the order Chiroptera, related to the Insectivora (hedgehogs and shrews), but differing from them in being able to fly. Bats are the only true flying mammals. Their forelimbs are developed as wings capable of rapid and sustained flight. There are two main groups of bats: **megabats**, which eat fruit and **microbats**, which mainly eat insects. Although by no means blind, many microbats rely largely on echolocation for navigation and finding prey, sending out pulses of high-pitched sound and listening for the echo.

beak horn-covered projecting jaws of a bird (see **bill**), or other horny jaws such as those of the octopus, platypus, or tortoise.

bee four-winged insect of the superfamily Apoidea in the order Hymenoptera, usually with a sting. There are over 12,000 species, of which fewer than 1 in 20 are social in habit. The **hive bee** or **honeybee** *Apis mellifera* establishes perennial colonies of about 80,000, the majority being infertile females (workers), with a few larger fertile males (drones), and a single very large fertile female (the queen). Worker bees live for no more than a few weeks, while a drone may live a few months, and a queen several years. Queen honeybees lay two kinds of eggs: fertilized, female eggs, which have two sets of chromosomes and develop into workers or queens, and unfertilized, male eggs, which have only one set of chromosomes and develop into drones.

beetle common name of insects in the order Coleoptera with leathery forewings folding down in a protective sheath over the membranous hindwings, which are those used for flight. They pass through a complete metamorphosis. Comprising more than 50% of the animal kingdom, beetles number some 370,000 named species.

bill in birds, the projection of the skull bones covered with a horny sheath. It is not normally sensitive, except in some aquatic birds, rooks, and woodpeckers, where the bill is used to locate food that is not visible. The bills of birds are adapted by shape and size to specific diets, for example, shovellers use their bills to sieve mud in order to extract food; birds of prey have hooked bills adapted to tearing flesh; the bills of the avocet, and the curlew are long and narrow for picking tiny invertebrates out of the mud; and those of woodpeckers are sharp for pecking holes in trees and plucking out insects. The bill is also used by birds for preening, fighting, display, and nest-building.

bioluminescence production of light by living organisms. It is a feature of many deep-sea fishes, crustaceans, and other marine animals. On land, bioluminescence is seen in some nocturnal insects such as glow-worms and fireflies, and in certain bacteria and fungi. Light is usually produced by the oxidation of luciferin, a reaction catalysed by the enzyme luciferase. Animal luminescence is involved in communication, camouflage, or the luring of prey.

birth act of producing live young from within the body of female animals. Both viviparous and ovoviviparous animals give birth to young. In viviparous animals, embryos obtain nourishment from the mother via a placenta or other means.

bivalve marine or freshwater mollusc whose body is enclosed between two shells hinged together by a ligament on the dorsal side of the body.

bone hard connective tissue comprising the skeleton of most vertebrate animals. Bone is composed of a network of collagen fibres impregnated with mineral salts (largely calcium phosphate and calcium carbonate), a combination that gives it great density and strength, comparable in some cases with that of reinforced concrete.

brachiopod, or **lamp shell**, any member of the phylum Brachiopoda, marine invertebrates with two shells, resembling but totally unrelated to bivalves. A single internal organ, the lophophore, handles feeding, aspiration, and excretion. There are about 300 living species.

bug in entomology, an insect belonging to the order Hemiptera. All these have two pairs of wings with forewings partly thickened. They also have piercing mouthparts adapted for sucking the juices of plants or animals, the 'beak' being tucked under the body when not in use.

butterfly insect belonging, like moths, to the order Lepidoptera, in which the wings are covered with tiny scales, often brightly coloured. There are some 15,000 species of butterfly, many of which are under threat throughout the world because of the destruction of habitat.

caecilian tropical amphibian of wormlike appearance. There are about 170 species known in the family Caeciliidae, forming the amphibian order Apoda (also known as Gymnophiona). Caecilians have a grooved skin that gives a 'segmented' appearance; they have no trace of limbs or pelvis. The body is 20–130 cm/8–50 in long, beige to black in colour. The eyes are very small and weak or blind. They eat insects and small worms. Some species bear live young, others lay eggs.

camel large cud-chewing mammal of the even-toed hoofed order Artiodactyla. Unlike typical ruminants, it has a three-chambered stomach. There are two species, the single-humped **Arabian camel** *Camelus dromedarius* and the twin-humped **Bactrian camel** *C. bactrianus* from Asia.

camouflage colours or structures that allow an animal to blend with its surroundings to avoid detection by other animals. Camouflage can take the form of matching the background colour, of countershading (darker on top, lighter below, to counteract natural shadows), or of irregular patters that break up the outline of the animal's body. More elaborate camouflage involves closely resembling a feature of the natural environment, as with the stick insect; this is closely akin to mimicry.

cane toad toad of the genus *Bufo marinus*, family Bufonidae. Also known as the giant or marine toad, the cane toad is the largest in the world. It was introduced to Australia during the 1930s to eradicate the cane beetle, but has now itself become a pest in Australia.

capybara the world's largest rodent *Hydrochoerus hydrochaeris*, up to 1.3 m/4 ft long and 50 kg/110 lb in weight. It is found in South America, and belongs to the guinea-pig family. It inhabits marshes and dense vegetation around water.

carnivore mammal of the order Carnivora. Although its name describes the flesh-eating ancestry of the order, it includes pandas, which are herbivorous, and civet cats, which eat fruit.

cartilage flexible bluish-white connective tissue made up of the protein collagen. In cartilaginous fish it forms the skeleton; in other vertebrates it forms the greater part of the embryonic skeleton, and is replaced by bone in the course of development, except in areas of wear such as bone endings, and the discs between the backbones. It also forms structural tissue in the larynx, nose, and external ear of mammals.

caterpillar larval stage of a butterfly or moth. Wormlike in form, the body is segmented, may be hairy, and often has scent glands. The head has strong biting mandibles, silk glands, and a spinneret.

cephalopod any predatory marine mollusc of the class Cephalopoda, with the mouth and head surrounded by tentacles. Cephalopods are the most intelligent, the fastest-moving, and the largest of all animals without backbones. Examples include squid, octopus, and cuttlefish. Shells are rudimentary or absent in most cephalopods.

chitin complex long-chain compound, or polymer; a nitrogenous derivative of glucose. Chitin is widely found in invertebrates. It forms the exoskeleton of insects and other arthropods. It combines with protein to form a covering that can be hard and tough, as in beetles, or soft and flexible, as in caterpillars and other insect larvae. It is insoluble in water and resistant to acids, alkalis, and many organic solvents. In crustaceans such as crabs, it is impregnated with calcium carbonate for extra strength.

chrysalis pupa of an insect, but especially that of a butterfly or moth. It is essentially a static stage of the creature's life, when the adult insect, benefiting from the large amounts of food laid down by the actively feeding larva, is built up from the disintegrating larval tissues. The chrysalis may be exposed or within a cocoon.

cocoon pupa-case of many insects, especially of moths and silkworms. This outer web or ball is spun from the mouth by caterpillars before they pass into the chrysalis state.

cold-blooded of animals, dependent on the surrounding temperature; see **poikilothermy**.

crustacean one of the class of arthropods that includes crabs, lobsters, shrimps, woodlice, and barnacles. The external skeleton is made of protein and chitin hardened with lime. Each segment bears a pair of appendages that may be modified as sensory feelers (antennae), as mouthparts, or as swimming, walking, or grasping structures.

digestive system the organs and tissues involved in the digestion of food; in animals these consist of the mouth, stomach, intestines, and their associated glands. Birds have two additional digestive organs – the crop and gizzard. In smaller, simpler animals such as jellyfish, the digestive system is simply a cavity (coelenteron or enteric cavity) with a 'mouth' into which food is taken; the digestible portion is dissolved and absorbed in this cavity, and the remains are ejected back through the mouth.

dinosaur (Greek *deinos* 'terrible', *sauros* 'lizard') any of a group (sometimes considered as two separate orders) of extinct reptiles living between 205 million and 65 million years ago. Their closest living relations are crocodiles and birds. Many species of dinosaur evolved during the millions of years they were the dominant large land animals. Most were large (up to 27 m/90 ft), but some were as small as chickens. They disappeared 65 million years ago for reasons not fully understood, although many theories exist.

dodo extinct flightless bird *Raphus cucullatus*, order Columbiformes, formerly found on the island of Mauritius, but exterminated by early settlers around 1681. Although related to the pigeons, it was larger than a turkey, with a bulky body, rudimentary wings, and short curly tail-feathers. The bill was blackish in colour, forming a horny hook at the end.

ecdysis period shedding of the exoskeleton by insects and other arthropods to allow growth. Prior to shedding, a new soft and expandable layer is first laid down underneath the existing one. The old layer then splits, the animal moves free of it, and the new layer expands and hardens.

echolocation method used by certain animals, notably bats, whales, and dolphins, to detect the positions of objects by using sound. The animal emits a stream of high-pitched sounds, generally at ultrasonic frequencies (beyond the range of human hearing), and listens for the returning echoes reflected off objects to determine their exact location.

egg in animals, the ovum, or female gamete (reproductive cell). After fertilization by a sperm cell, it begins to divide to form an embryo. Eggs may be deposited by the female (ovipary) or they may develop within her body (vivipary and ovovivipary). In the oviparous reptiles and birds, the egg is protected by a shell, and well supplied with nutrients in the form of yolk.

embryo early developmental stage of an animal following fertilization of an ovum (egg cell), or activation of an ovum by parthenogenesis.

endoskeleton the internal supporting structure of vertebrates, made up of cartilage or bone. It provides support, and acts as a system of levers to which muscles are attached to provide movement. Certain parts of the skeleton (the skull and ribs) give protection to vital body organs. Sponges are supported by a network of rigid, or semi-rigid, spiky structures called spicules.

ethology comparative study of animal behaviour in its natural setting. Ethology is concerned with the causal mechanisms (both the stimuli that elicit behaviour and the physiological mechanisms controlling it), as well as the development of behaviour, its function, and its evolutionary history.

exoskeleton the hardened external skeleton of insects, spiders, crabs, and other arthropods. It provides attachment for muscles and protection for the internal organs, as well as support. To permit growth it is periodically shed in a process called ecdysis.

extinction the complete disappearance of a species. In the past, extinctions are believed to have occurred because species were unable to adapt quickly enough to a naturally changing environment. Today, most extinctions are due to human activity.

eye the organ of vision. In many animal eyes, as in humans, the light is focused by the combined action of the curved cornea, the internal fluids, and the lens. The insect eye is compound – made up of many separate facets – known as ommatidia, each of which collects light and directs it separately to a receptor to build up an image. Invertebrates have much simpler eyes, with no lenses. Among molluscs, cephalopods have complex eyes similar to those of vertebrates.

feather rigid outgrowth of the outer layer of the skin of birds, made of the protein keratin. Feathers provide insulation and facilitate flight. There are several types, including long quill feathers on the wings and tail, fluffy down feathers for retaining body heat, and contour feathers covering the body. The colouring of feathers is often important in camouflage or in courtship and other displays. Feathers are normally replaced at least once a year.

fin in aquatic animals, flattened extension from the body that aids balance and propulsion through the water.

fur the hair of certain animals. Fur is an excellent insulating material. Fur such as mink is made up of a soft, thick, insulating layer called underfur and a top layer of longer, lustrous guard hairs.

gastropod any member of a very large group of molluscs, having a single shell (in a spiral or modified spiral form) and eyes on stalks, and moving on a flattened, muscular foot. Gastropods have well-developed heads and rough, scraping tongues called radulae. Some are marine, some freshwater, and others land creatures, but they all tend to live in damp places.

gestation in all mammals except the monotremes (platypus and spiny anteaters), the period from the time of implantation of the embryo in the uterus to birth.

gill the main respiratory organ of most fishes and immature amphibians, and of many aquatic invertebrates. In all types, water passes over the gills, and oxygen diffuses across the gill membranes into the circulatory system, while carbon dioxide passes from the system out into the water.

heart muscular organ that rhythmically contracts to force blood around the body of an animal with a circulatory system. Annelid worms and some other invertebrates have simple hearts consisting of thickened sections of main blood vessels that pulse regularly. An earthworm has ten such hearts. Vertebrates have one heart. A fish heart has two chambers – the thin-walled atrium (once called the auricle) that expands to receive blood, and the thick-walled ventricle that pumps it out. Amphibians and most reptiles have two atria and one ventricle; birds and mammals have two atria and two ventricles.

herbivore animal that feeds on green plants (or photosynthetic single-celled organisms) or their products, including seeds, fruit, and nectar. The most numerous type of herbivore is thought to be the zooplankton, tiny invertebrates in the surface waters of the oceans that feed on small photosynthetic algae. Herbivores are more numerous than other animals because their food is the most abundant. They form a vital link in the food chain between plants and carnivores.

hibernation state of dormancy in which certain animals spend the winter. It is associated with a dramatic reduction in all metabolic processes, including body temperature, breathing, and heart rate. It is a fallacy that animals sleep throughout the winter.

homeothermy maintenance of a constant body temperature in endothermic (warm-blooded) animals, by the use of chemical processes to compensate for heat loss or gain when external temperatures change. Such processes include generation of heat by the breakdown of food and the contraction of muscles, and loss of heat by sweating, panting, and other means.

hoof horny covering that protects the sensitive parts of the foot of an animal. The possession of hooves is characteristic of the orders Artiodactyla (even-toed ungulates such as deer and cattle), and Perissodactyla (horses, tapirs, and rhinoceroses).

horn broad term for a hardened processes on the heads of some members of order Artiodactyla: deer, antelopes, cattle, goats, and sheep; and the rhinoceroses in order Perissodactyla. They are used usually for sparring rather than serious fighting, often between members of the same species rather than against predators.

hybrid offspring from a cross between individuals of two different species, or two inbred lines within a species. In most cases, hybrids between species are infertile and unable to reproduce sexually.

hydra any of a group of freshwater polyps, belonging among the coelenterates. The body is a double-layered tube (with six to ten hollow tentacles around the mouth), 1.25 cm/0.5 in long when extended, but capable of contracting to a small knob. Usually fixed to waterweed, hydras feed on minute animals that are caught and paralysed by stinging cells on the tentacles.

imago sexually mature stage of an insect.

insectivore any animal whose diet is made up largely or exclusively of insects. In particular, the name is applied to mammals of the order Insectivora, which includes the shrews, hedgehogs, moles, and tenrecs.

instinct behaviour found in all equivalent members of a given species (for example, all the males, or all the females with young) that is presumed to be genetically determined. Examples include a male robin's tendency to attack other male robins intruding on its territory and the tendency of many female mammals to care for their offspring.

invertebrate animal without a backbone. The invertebrates comprise over 95% of the million or so existing animal species and include sponges, coelenterates, flatworms, nematodes, annelid worms, arthropods, molluscs, echinoderms, and primitive aquatic chordates, such as sea squirts and lancelets.

ivory hard white substance of which the teeth and tusks of certain mammals are made. Among the most valuable are elephants' tusks, which are of unusual hardness and density.

keratin fibrous protein found in the skin of vertebrates and also in hair, nails, claws, hooves, feathers, and the outer coating of horns.

krill any of several Antarctic crustaceans, the most common species being *Euphausia superba*. Similar to a shrimp, it is up to 5 cm/2 in long, with two antennae, five pairs of legs, seven pairs of light organs along the body, and is coloured orange above and green beneath. It is one of the most abundant animals, numbering perhaps 600 trillion (million million) individuals.

lactation secretion of milk in mammals, from the mammary glands. In late pregnancy, the cells lining the lobules inside the mammary glands begin extracting substances from the blood to produce milk. The supply of milk starts shortly after birth and continues practically as long as the young continue to suckle.

larva stage between hatching and adulthood in those species in which the young have a different appearance and way of life from the adults. Examples include tadpoles (frogs) and caterpillars (butterflies and moths). Larvae are typical of the invertebrates, some of which (for example, shrimps) have two or more distinct larval stages. Among vertebrates, it is only the amphibians and some fishes that have a larval stage. The process whereby the larva changes into another stage, such as a pupa (chrysalis) or adult, is known as metamorphosis.

lek a closely spaced set of very small territories each occupied by a single male during the mating season. Leks are found in the mating systems of several ground-dwelling birds (such as grouse) and a few antelopes.

lizard reptile generally distinguishable from snakes, which belong to the same order, by having four legs, moveable eyelids, eardrums, and a fleshy tongue, although some lizards are legless and snakelike in appearance. There are over 3,000 species of lizard worldwide.

maggot soft, plump, limbless larva of flies, a typical example being the larva of the blowfly which is deposited as an egg on flesh.

mammary gland in female mammals, a milk-producing gland derived from epithelial cells underlying the skin, active only after the production of young. In all but monotremes (egg-laying mammals), the mammary glands terminate in teats which aid infant suckling. The number of glands and their position vary between species.

mammoth extinct elephant, remains of which have been found worldwide. Some were 50% taller than modern elephants; others were much samller.

mastodon any of an extinct family of mammals belonging to the elephant order. They differed from elephants and mammoths in the structure of their grinding teeth. There were numerous species, among which the **American mastodon** (*Mastodon americanum*), about 3 m/10 ft high, of the Pleistocene era, is well known.

megapode, or **mound-builder** any of a group of chickenlike birds found in the Malay archipelago and Australia. They pile up large mounds of vegetable matter, earth, and sand 4 m/

13 ft across, in which to deposit their eggs, then cover the eggs and leave them to be incubated by the heat produced by the rotting vegetation. There are 19 species, all large birds, 50—70 cm/20—27 in in length, with very large feet.

metamorphosis period during the life cycle of many invertebrates, most amphibians, and some fish, during which the individual's body changes from one form to another through a major reconstitution of its tissues. For example, adult frogs are produced by metamorphosis from tadpoles, and butterflies are produced from caterpillars following metamorphosis within a pupa.

milk secretion of the mammary glands of female mammals, with which they suckle their young (during lactation). Over 85% is water, the remainder comprising protein, fat, lactose (a sugar), calcium, phosphorus, iron, and vitamins.

mimicry imitation of one species (or group of species) by another. The most common form is **Batesian mimicry** (named after the English naturalist H W Bates), where the mimic resembles a model that is poisonous or unpleasant to eat, and has aposematic, or warning, coloration; the mimic thus benefits from the fact that predators have learned to avoid the model. Hoverflies that resemble bees or wasps are an example. Appearance is usually the basis for mimicry, but calls, songs, scents, and other signals can also be mimicked.

moa any of a group of extinct flightless kiwi-like birds that lived in New Zealand. There were 19 species; they varied from 0.5 to 3.5 m/2 to 12 ft, with strong limbs, a long neck, and no wings. The largest species was *Dinornis maximus*. The last moa was killed in the 1800s.

moth any of a large number of mainly night-flying insects closely related to butterflies. Their wings are covered with microscopic scales. Most moths have a long sucking mouthpart (proboscis) for feeding on the nectar of flowers, but some have no functional mouthparts and rely instead upon stores of fat and other reserves built up during the caterpillar stage. At least 100,000 different species of moth are known.

moulting periodic shedding of the hair or fur of mammals, feathers of birds, or skin of reptiles. In mammals and birds, moulting is usually seasonal and is triggered by changes of day length.

nest place chosen or constructed by a bird or other animal for incubation of eggs, hibernation, and shelter. Nests vary enormously, from saucerlike hollows in the ground, such as the scrapes of hares, to large and elaborate structures, such as the 4-m/13-ft diameter mounds of the megapode birds.

notochord the stiff but flexible rod that lies between the gut and the nerve cord of all embryonic and larval chordates, including the vertebrates. It forms the supporting structure of the adult lancelet, but in vertebrates it is replaced by the vertebral column, or spine.

nymph in entomology, the immature form of insects that do not have a pupal stage; for example, grasshoppers and dragonflies. Nymphs generally resemble the adult (unlike larvae), but do not have fully formed reproductive organs or wings.

omnivore animal that feeds on both plant and animal material. Omnivores have digestive adaptations intermediate between those of herbivores and carnivores, with relatively unspecialized digestive systems and gut microorganisms that can digest a variety of foodstuffs. Omnivores include the chimpanzee, the cockroach, and the ant.

oviparity method of animal reproduction in which eggs are laid by the female and develop outside her body, in contrast to ovoviviparity and viviparity. It is the most common form of reproduction

ovoviviparity method of animal reproduction in which fertilized eggs develop within the female (unlike oviparity), and the embryo gains no nutritional substances from the female (unlike viviparity). It occurs in some invertebrates, fishes, and reptiles.

parasite organism that lives on or in another organism (called the host) and depends on it for nutrition, often at the expense of the host's welfare. Parasites that live inside the host, such as liver flukes and tapeworms, are called **endoparasites**; those that live on the exterior, such as fleas and lice, are called **ectoparasites**.

plankton small, often microscopic, forms of plant and animal life that live in the upper layers of fresh and salt water, and are an important source of food for larger animals. Marine plankton is concentrated in areas where rising currents bring mineral salts to the surface.

poikilothermy the condition in which an animal's body temperature is largely dependent on the temperature of the air or water in which it lives. It is characteristic of all animals except birds and mammals, which maintain their body temperatures by homeothermy (they are 'warm-blooded').

primate any member of the order of mammals that includes monkeys, apes, and humans (together called **anthropoids**), as well as lemurs, bushbabies, lorises, and tarsiers (together called **prosimians**). Generally, they have forward-directed eyes, gripping hands and feet, opposable thumbs, and big toes. They tend to have nails rather than claws, with gripping pads on the ends of the digits, all adaptations to the arboreal, climbing mode of life.

pupa nonfeeding, largely immobile stage of some insect life cycles, in which larval tissues are broken down, and adult tissues and structures are formed. In many insects, the pupa is **exarate**, with the appendages (legs, antennae, wings) visible outside the pupal case; in butterflies and moths, it is called a chrysalis, and is **obtect**, with the appendages developing inside the case.

rodent any mammal of the worldwide order Rodentia, making up nearly half of all mammal species. They are distinguishable by, among other things, a single front pair of incisor teeth in both upper and lower jaw, which continue to grow as they are worn down. They are often subdivided into three suborders: Sciuromorpha, including primitive rodents, with squirrels as modern representatives; Myomorpha, rats and mice and their relatives; and Hystricomorpha, including the Old World and New World porcupines and guinea pigs.

rotifer any of the tiny invertebrates, also called 'wheel animalcules', of the phylum Rotifera. Mainly freshwater, some marine, rotifers have a ring of cilia that carries food to the mouth and also provides propulsion. They are the smallest of multicellular animals – few reach 0.05 cm/0.02 in.

ruminant any even-toed hoofed mammal with a rumen, the 'first stomach' of its complex digestive system. Plant food is stored and fermented before being brought back to the mouth for chewing (chewing the cud) and then is swallowed to the next stomach. Ruminants include cattle, antelopes, goats, deer, and giraffes, all with a four-chambered stomach. Camels are also ruminants, but they have a three-chambered stomach.

shellfish popular name for molluscs and crustaceans, including the whelk and periwinkle, mussel, oyster, lobster, crab, and shrimp.

skeleton the rigid or semirigid framework that supports and gives form to an animal's body, protects its internal organs, and provides anchorage points for its muscles. The skeleton may be composed of bone and cartilage (vertebrates), chitin (arthropods), calcium carbonate (molluscs and other invertebrates), or silica (many protists).

snake reptile of the suborder Serpentes of the order Squamata, which also includes lizards. Snakes are characterized by an elongated limbless body and scaly skin. There are 3,000 species found in the tropical and temperate zones, but none in New Zealand, Ireland, Iceland, and near the poles.

sperm, or **spermatozoon**, the male gamete (reproductive cell) of animals. Each sperm cell has a head capsule containing a nucleus, a middle portion containing mitochondria (which provide energy), and a long tail (flagellum).

spider any arachnid (eight-legged animal) of the order Araneae. There are about 30,000 known species, mostly a few centimetres in size, although a few tropical forms attain great size, for example, some bird-eating spiders attain a body length of 9 cm/3.5 in. Spiders produce silk, and many spin webs to trap their prey. They are found everywhere in the world except Antarctica. Many species are found in woods and dry commons; a few are aquatic. Spiders are predators; they bite their prey, releasing a powerful toxin from poison glands which causes paralysis, together with digestive juices. They then suck out the juices and soft parts.

spiracle in insects, the opening of a trachea (airway), through which oxygen enters the body and carbon dioxide is expelled. In cartilaginous fishes (sharks and rays) the same name is given to a circular opening that marks the remains of the first gill slit.

stomach the first cavity in the digestive system of animals. In mammals it is a bag of muscle situated just below the diaphragm. Food enters it from the oesophagus, is digested by the acid and enzymes secreted by the stomach lining, and then passes into the duodenum. Some plant-eating mammals have multichambered stomachs that harbour bacteria in one of the chambers to assist in the digestion of cellulose. The gizzard is part of the stomach in birds.

tooth in vertebrates, one of a set of hard, bonelike structures in the mouth, used for biting and chewing food, and in defence and aggression. Mammalian teeth have roots surrounded by cementum, which fuses them into their sockets in the jawbones. The neck of the tooth is covered by the gum, while the enamel-covered crown protrudes above the gum line.

trilobite any of a large class (Trilobita) of extinct, marine, invertebrate arthropods of the Palaeozoic era, with a flattened, oval body, 1–65 cm/0.4–26 in long. The hard-shelled body was divided by two deep furrows into three lobes. Some were burrowers, others were swimming and floating forms.

tunicate any marine chordate of the subphylum Tunicata (Urochordata), for example the sea squirt. Tunicates have transparent or translucent tunics made of cellulose. They vary in size from a few millimetres to 30 cm/1 ft in length, and are cylindrical, circular, or irregular in shape. There are more than 1,000 species.

ungulate general name for any hoofed mammal. Included are the odd-toed ungulates (perissodactyls) and the even-toed ungulates (artiodactyls), along with subungulates such as elephants.

vertebrate any animal with a backbone. The 41,000 species of vertebrates include mammals, birds, reptiles, amphibians, and fishes. They include most of the larger animals, but in terms of numbers of species are only a tiny proportion of the world's animals.

viviparity a method of reproduction in which the embryo develops inside the body of the female from which it gains nourishment (in contrast to oviparity and ovoviparity). Viviparity is best developed in placental mammals, but also occurs in some arthropods, fishes, amphibians, and reptiles that have placenta-like structures.

warm-blooded of animals, not dependent on the surrounding temperature; see **homeothermy**.

waterfowl any water bird, but especially any member of the family Anatidae, which consists of ducks, geese, and swans.

whale any marine mammal of the order Cetacea. The only mammals to have adapted to living entirely in water, they have front limbs modified into flippers and no externally visible traces of hind limbs. They have horizontal tail flukes. When they surface to breathe, the hot air they breathe out condenses to form a 'spout' through the blowhole (single or double nostrils) in the top of the head. The largest whales are the baleen whales, with plates of modified mucous membrane called baleen (whalebone) in the mouth; these strain the food, mainly microscopic plankton, from the water. Baleen whales include the finback and right whales, and the blue whale, the largest animal that has ever lived, of length up to 30 m/100 ft.

worm any of various elongated limbless invertebrates belonging to several phyla. Worms include the flatworms, such as flukes and tapeworms; the roundworms or nematodes, such as the eelworm and the hookworm; the marine ribbon worms or nemerteans; and the segmented worms or annelids.

zoo (abbreviation for **zoological gardens**) place where animals are kept in captivity. Originally created purely for visitor entertainment and education, zoos have become major centres for the breeding of endangered species of animals. The Arabian oryx has already been preserved in this way; it was captured in 1962, bred in captivity, and released again in the desert in 1972, where it has flourished.

Further Reading

Berenbaum, May R *Bugs in the System: Insects and Their Impact on Human Affairs* (1995)

Brooke, Michael, and Birkhead, Tim (eds) *The Cambridge Encyclopedia of Ornithology* (1991)

Chapman, R F *The Insects: Structure and Function* (1982)

Cheney, Dorothy L, and Seyfarth, Robert M *How Monkeys See the World* (1990)

Chinery, Michael *A Field Guide to the Insects of Britain and Northern Europe* (1982)

Clarke, P A B, and Linzey, Andrew (eds) *Political Theory and Animal Rights* (1990)

Dawkins, Richard *The Selfish Gene* (1976)

del Hoyo, Josep; Elliott, Andrew; and Sargatal, Jordi (eds) *Handbook of the Birds of the World* (1992, continuing)

Diamond, Anthony *Save the Birds* (1987)

Gadagkar, Raghavendra *Survival Strategies* (1998)

Goodall, Jane *In the Shadow of Man* (1983)

Gould, James, and Gould, Carol Grant (eds) *Life at the Edge* (1989)

Feduccia, Alan *The Age of Birds* (1980)

Halliday, Tim and Alder, K *The Encyclopedia of Reptiles and Amphibians* (1986)

Haraway, Donna *Primate Visions* (1992)

Hölldobler, B, and Wilson, E O *Journey to the Ants* (1994)

Lack, David *The Life of the Robin* (1943)

Linzey, Andrew *Animal Rights* (1976)

McFarland, David (ed) *The Oxford Companion to Animal Behaviour* (1981)

Maeterlink, Maurice *Life of the Bee* (1901)

Mead, Chris *Bird Migration* (1983)

Nelson, Joseph S *Fishes of the World* (1976; 3rd edition 1994)

Norse, Elliott A (ed) *Global Marine Biological Diversity* (1993)

Obee, Bruce, and Ellis, Graeme *Guardians of the Whales – The Quest to Study Whales in the Wild* (1992)

Paxton, John R, and Eschmeyer, William N (eds) *Encyclopedia of Fishes* (1994)

Richards, O W, and Davies, R G *Imm's General Textbook of Entomology*, Volume 2 (10th edition 1977)

Ridley, Matt *The Red Queen* (1993)

Samways, M J *Insect Conservation Biology* (1994)

Springer, Victor G, and Gold, Joy P *Sharks in Question* (1989)

Strattersfield, Alison; Crosby, Michael; et al (eds) *Endemic Bird Areas of the World* (1998)

Tinbergen, Niko *The Study of Instinct* (1951)

Tudge, Colin *Last Animals at the Zoo* (1991)

von Frisch, Karl *The Dance Language and Orientation of Bees* (1967)

Wade, Nicholas (ed) *The Science Times Book of Fish* (1998)

Waller, Geoffrey *Sealife: A Complete Guide to the Marine Environment* (1996)

Wells, Sue, and Hanna, Nick *The Greenpeace Book of Coral Reefs* (1992)

Wheeler, Alwyne *Fishes of the World: An Illustrated Dictionary* (1975)

Wilson, Edward O *Sociobiology* (1975)

THE PLANT KINGDOM

At least a quarter of a million species make up the plant kingdom. From mosses and ferns to the more complex conifers and flowering plants, they are all primarily adapted to life on land, develop from embryos, carry out ∂photosynthesis, have complex cells surrounded by a rigid cellulose wall, and do not move around. A few parasitic plants have lost the ability to photosynthesize but are still considered to be plants.

Plants are **autotrophs,** that is, they are able to nourish themselves by harnessing the energy of sunlight to make carbohydrates from water and carbon dioxide. They and other autotrophs are the primary producers in all food chains since the materials they synthesize and store are the energy sources of all other organisms. Plants also play a vital part in the carbon cycle, removing carbon dioxide from the atmosphere and generating oxygen.

The study of plants is **botany** (from the Greek *botane* for 'herb'). It is divided into a number of specialized studies, such as the identification and classification of plants (taxonomy), their external formation (plant morphology), their internal arrangement (plant anatomy), the microscopic examination of their tissues (plant histology), their functioning and life history (plant physiology), and their distribution over the Earth's surface in relation to their surroundings (plant ecology). Palaeobotany concerns the study of fossil plants, while economic botany deals with the utility of plants. Horticulture, agriculture, and forestry are also branches of botany.

history of botany

Although the study of plants might be said to date back more than 10,000 years, when people first learned to cultivate crops, the earliest known botanical record is that carved on the walls of the temple at Karnak, Egypt, about 1500 BC. The Greeks in the 5th and 4th centuries BC used many plants for medicinal purposes, the first Greek 'Herbal' being drawn up about 350 BC by Diocles of Carystus. Botanical information was collected into the works of Theophrastus of Eresus, a pupil of Aristotle, who founded technical plant nomenclature. Cesalpino in the 16th century sketched out a system of classification based on flowers, fruits, and seeds, while Joachim Jungius (1587–1657) used only flowers as his criterion. The English botanist John Ray arranged plants systematically, based on his findings on fruit, leaf, and flower, and described about 18,600 plants.

The Swedish botanist Carl von Linné, or Linnaeus, who founded systematics in the 18th century, included in his classification all known plants and animals, giving each a two-part descriptive label. His work greatly aided the future study of plants, as botanists found that all plants could be fitted into a systematic classification based on Linnaeus' work. Linnaeus was also the first to recognize the sexual nature of flowers.

Later work revealed the detailed cellular structure of plant tissues and the exact nature of ◊photosynthesis. Julius von Sachs defined the function of ◊chlorophyll and the significance of plant stomata. In the second half of the 20th century much has been learned about cell function, repair, and growth by the hybridization of plant

> **plants: oldest**
>
> The oldest living plant is the Tasmanian king's holly *Loamtia tasmanica*, which has survived for 43,000 years.

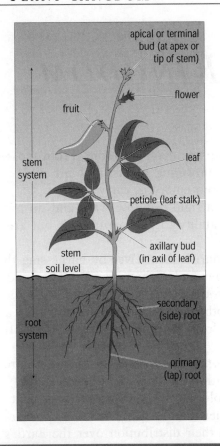

apical or terminal bud (at apex or tip of stem)

flower

fruit

leaf

stem system

petiole (leaf stalk)

axillary bud (in axil of leaf)

stem

soil level

secondary (side) root

root system

primary (tap) root

plant The external anatomy of a typical flowering plant.

cells (the combination of the nucleus of one cell with the cytoplasm of another). With modern tools, such as the electron microscope, the inner structure of plant cells and the function of the intracellular organelles can be studied.

Short Botanical Glossary
http://www.anbg.gov.au/glossary/croft.html
Useful glossary of botanical terms. If you are confused about abaxial, zygote, or anything in between, this glossary will set you straight.

---

classification

The plant kingdom is very varied. Originally it included such organisms as bacteria, fungi, and algae, but these are now more commonly classified in the kingdoms Monera (bacteria and blue-green algae), Fungi, and Protista (protozoa, algae, and slime moulds). The groups that are always classified as plants are the mosses and liverworts; ferns, horsetails, and club mosses; gymnosperms (conifers, cycads, and ginkgos), and angiosperms (flowering plants). How

Botanical Prefixes

Prefix	Meaning	Prefix	Meaning	Prefix	Meaning
acro-	of or towards the top	ex-	without, outwards	ortho-	straight, upright
aero-	concerning the air	gamo-	joined together, fused	penta-	five
amphi-	both	halo-	of or relating to salt	peri-	around, enclosing
andro-	male	gymno-	naked	photo-	concerning light
atro-	dark	hemi-	half	phyllo-	concerning leaves
auto-	self	hetero-	different, other	phyto-	concerning plants
basi-	of or towards the base	hexa-	six	pluri-	several
bi-	two	homo-	similar, same	poly-	many
bryo-	concerning mosses	hydro-	concerning water	proto-	first
cauli-	concerning to stems	hyper-	above	pseudo-	false
centri-	of or towards the centre	hypo-	beneath	pterido-	concerning ferns
		infra-	lower than	quadri-	four
		iso-	identical, equal	re-	backwards
chromo-	colour, coloured	lepto-	thin, slender	rhizo-	concerning roots or rootlike organs
cleisto-	closed	macro-	large		
crypto-	hidden	mega-	large	semi-	half
dendro-, dendri-	concerning trees	meso-	middle	sub-	under
di-	two	micro-	very small	supra-	over
e-	lacking	mono-	single, one	syn-	together, united
ecto-	outside	multi-	many	tetra-	four
endo-	inside	neo-	new	tri-	three
epi-	on, above, outer	ob-	inverted	uni-	single, one
eu-	good, normal	oligo-	few	xero-	dry

Divisions of the Plant Kingdom

Division (Phylum)	Examples
Bryophyta	mosses, liverworts, hornworts
Psilophyta	whisk ferns
Lycopodophyta (Lycophyta)	club mosses
Sphenophyta	horsetails
Filicinophyta (Pterophyta)	ferns
Cycadophyta	cycads
Ginkgophyta	*Ginkgo*
Coniferophyta (Pinophyta)	conifers
Gnetophyta	gnetophytes, such as *Welwitschia*
Angiospermophyta (Magnoliophyta, Anthophyta)	flowering plants
Classes within Angiospermophyta	
Dicotyledoneae (Magnoliopsida)	dicotyledons, such as magnolia, laurel, water lily, buttercup, poppy, pitcher plant, nettle, walnut, cacti, peonies, violet, begonia, willow, primrose, rose, maple, holly, grape, honeysuckle, African violet, daisy
Monocotyledoneae (Liliopsida)	monocotyledons, such as flowering rush, eel grass, lily, iris, banana, orchid, sedge, pineapple, grasses, palms, cat tail

these are related to each other is much debated, and there are at least four systems of classification in common use, which take into account the evolutionary connections between plants, and such characteristics as their life cycles, tissue structure (whether they possess vascular tissue for transporting fluids), and seed structure (whether they produce spores, naked seeds, or covered seeds).

In the system adopted here, plants are classified into ten divisions, or phyla: the non-vascular Bryophyta; the lower vascular Psilophyta, Lycopodophyta, Sphenophyta, and Filicinophyta; the gymnosperm (vascular, producing naked seeds) Cycadophyta, Ginkgophyta, Coniferophyta, and Gnetophyta; and the angiosperm (vascular, producing covered seeds) Angiospermophyta.

bryophyte

Members of the division Bryophyta – **mosses, liverworts**, and **hornworts** – are small, low-growing plants, which chiefly occur in damp habitats and require water for the dispersal of the male gametes (antherozoids). Unlike higher plants, they have no vascular system for conducting fluids. In some liverworts the plant body is a simple and flattened, but in the majority of bryophytes it is differentiated into stemlike, leaflike, and rootlike organs.

The plant exists in two different reproductive forms, sexual and asexual, which appear alternately (described under alternation of generations below). The sexual **gametophyte** generation is dominant, producing a plant body, or **thallus**, which may be flat, green, and lobed like a small leaf, or leafy and mosslike. The asexual **sporophyte** generation is smaller, typically parasitic on the gametophyte, and produces a capsule from which spores are scattered.

liverwort Liverworts belong to a group of plants, called Bryophyta, that also includes the mosses. Neither mosses nor liverworts possess true roots and both require water to enable the male gametes to swim to the female sex organs to fertilize the eggs. Unlike mosses most liverworts have no, or only very frail, leaves.

lower vascular plant The lower vascular plants – ferns and their allies – are mainly terrestrial, though they favour damp habitats. Like the bryophytes, they do not produce seeds, but – like the gymnosperms and angiosperms – they do have special supportive fluid-conducting tissues, which identify them as vascular plants. They tend to possess true stems, roots, and leaves, and show a marked alternation of generations, with the sporophyte (or spore-producing plant) forming the dominant generation in the life cycle.

The lower vascular plants are classified into four divisions: Psilophyta, the whisk ferns; Lycopsida, the club mosses; Sphenopsida, the horsetails; and Filicinophyta, the ferns.

whisk fern The whisk ferns (Psilophyta) are the most primitive of the vascular plants. There are only two living genera, *Psilotum* and *Tmesipteris*, both of which live in the tropics and subtropics. The sporophytes of *Psilotum* are scaly, green, dichotomously branching stems, which lack true roots and leaves.

> ### fern
>
> The fern called bracken is one of the world's six most common plants and also one of the six oldest, being at least 90 million years old.

club moss The club mosses (Lycopsida) have a wide distribution, but were far more numerous in Palaeozoic times, especially the Carboniferous period (362.5–290 million years ago), when members of the group were large trees. The species that exist now are all small in size. Club mosses possess true stems and roots, and have many small leaflike bodies, called microphylls. Special microphylls, which may be modified to form simple cones, bear spore-producing structures, called sporangia.

Common Club Moss
**http://www.botanical.com/botanical/mgmh/m/
mosccl48.html**
Informative resource relating to common club moss. It contains a detailed description of the plant and its habitat. There are also extensive notes on the possible medicinal uses of common club moss, which include acting as a diuretic and treating diarrhoea, eczema, and rheumatism.

---

horsetail Of the horsetails (Sphenopsida), only one genus, *Equisetum*, survives. There are about 35 living species, bearing their sporangia on complex cones at the stem tip. The upright stems are ribbed and jointed, and often have spaced whorls of branches. Today they are of modest size, but hundreds of millions of years ago giant treelike forms existed.

fern The ferns (Filicinophyta), with more than 7,000 species, form the largest and most diverse division.

Most are perennial, spreading by slow-growing rhizomes. Their complex leaves, known as fronds, vary widely in size and shape, and bear sporangia on their undersurface. Some species, such as the tropical tree-ferns, may reach 20 m/66 ft.

Fern Resource Hub
http://www.inetworld.net/~sdfern/
Huge source of fern-related information for the fern hobbyist. Organized by the San Diego Fern Society, this is a clearing house of information for fern societies across the world. There is general information on ferns and how to grow them and information on sources of further information. If you have queries about ferns there are experts to be e-mailed. There is news of upcoming fern events all over the world.

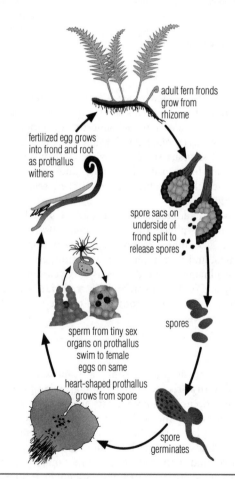

adult fern fronds grow from rhizome

fertilized egg grows into frond and root as prothallus withers

spore sacs on underside of frond split to release spores

spores

sperm from tiny sex organs on prothallus swim to female eggs on same

heart-shaped prothallus grows from spore

spore germinates

---

fern The life cycle of a fern. Ferns have two distinct forms that alternate during their life cycle. For the main – sporophyte – part of its life, a fern consists of a short stem (or rhizome) from which roots and leaves grow. The other part of its life is spent as a small heart-shaped plant called a prothallus.

gymnosperm

The gymnosperms – principally conifers and cycads – produce naked seeds, as opposed to the structurally more advanced angiosperms (see below), where the seeds are contained within an ovary. The gymnosperm seed is usually produced on the scales of the female ♀cone, although some gymnosperms, such as the yew and ginkgo, produce seeds within berrylike structures. Fossil gymnosperms have been found in rocks about 350 million years old.

Gymnosperms are well adapted to dry habitats. Unlike the bryophytes and lower vascular plants, they do not require moisture for the transport of gametes, since the males gametes are transferred to the female ovule by wind ♀pollination. The alternation of generations is hidden, as the female gametophyte generation is contained within the ovule and the male gametophyte within the pollen grain.

The gymnosperm group contains four divisions: Cycadophyta, the cycads; Ginkgophyta, the ginkgo; Coniferophyta, the conifers; and Gnetophyta, the gnetophytes.

cycad

The cycads (Cycadophyta) are superficially similar to either palms, or ferns, depending on species. Their large cones contain fleshy seeds. There are ten genera and about 80 to 100 species, native to tropical and subtropical countries. Cycads were widespread during the Mesozoic era (245–65 million years ago). In 1993 some cycads were discovered to be pollinated by insects, not – as in most other gymnosperms – by wind; their cones produce heat that vaporizes a sweet minty odour to attract insects to a supply of nectarlike liquid.

ginkgo

Ginkgophyta has only one living species, the ginkgo, or maidenhair tree *Ginkgo biloba*. Widespread in the Mesozoic era (245–65 million years ago), it has been cultivated in China and Japan since ancient times, and is planted in many parts of the world. Its leaves are fan-shaped, and it bears fleshy, yellow, foul-smelling berries enclosing edible seeds. It may reach a height of 30 m/100 ft by the time it is 200 years old.

conifer

The conifers (Coniferophyta) comprise the largest gymnosperm division, and include pines,

spruces, firs, yews, junipers, monkey puzzles, and larches. They are cone-bearing trees or shrubs, often pyramid-shaped, with leaves that are either scaled or needle-shaped; most are evergreen. The reproductive organs are small, seasonal male cones and larger female cones, which may be woody or may be modified (as in yews and junipers) to form fleshy berrylike structures. Most conifers grow quickly and can survive in poor soil, on steep slopes, and in short growing seasons. Coniferous forests are widespread in cool temperate regions.

Ancient Bristlecone Pine
http://www.sonic.net/bristlecone/intro.html

All about the bristlecone pines of California, thought to be the world's oldest living inhabitants. This site is a good starting point for information about dendrochronology (the dating of past climatic changes through the study of tree rings).

---

gnetophyte The gnetophytes (Gnetophyta) are, in some respects, intermediate between gymnosperms and angiosperms. The three gnetophyte genera – *Welwitschia*, *Ephedra*, and *Gnetum* – look very different from each other, but are all characterized by producing pollen from stamenlike structures and having vessels within their xylem (water-conducting tissue). Like other gymnosperms, however, they produce naked seeds within cones.

angiosperm

The angiosperms – commonly called **flowering plants** – produce seeds that are enclosed within an ovary. The reproductive organ is the flower, which falls away after fertilization, while the ovary wall ripens into a fruit. Angiosperms, which include the majority of flowers, herbs, grasses, and trees, are the largest, most

---

Angiosperm Anatomy
http://www.botany.uwc.ac.za:80/sci_ed/std8/anatomy/
Good general guide to angiosperms. The differences between monocotyledons and dicotyledons are set out here. The functions of roots, stems, leaves, and flowers are explained by readily understandable text and good accompanying diagrams.

Angiosperms: Differences between Monocotyledons and Dicotyledons

	Monocotyledon	Dicotyledon
embryo	single cotyledon (seed leaf)	two cotyledons
vascular system	scattered vascular bundles; cambium rare	ring of vascular bundles, usually with cambium
leaves	parallel leaf veins; straplike leaves that rarely possess a petiole (stalk)	network of branching veins; broad in shape; usually possessing a petiole
flowers	floral parts in threes or multiples of three	floral parts in fours or fives
roots	taproot or fibrous root system	fibrous root system

dicotyledon The pair of seed leaves typical of this broad group of flowering plants is clearly visible in these tree seedlings, which have germinated in tropical dry forest in Madagascar.

advanced, and most successful group of plants at the present time, occupying a highly diverse range of habitats. There are estimated to be about 230,000 different species. Many have developed highly specialized reproductive structures associated with (pollination by insects, birds, or bats.

There is evidence of fossil angiosperms from the Jurassic period (208–146 million years ago), and genera that seem very similar to modern examples have been found from the Cretaceous period (approximately 144.2–65 million years ago).

Angiosperms are divided into two classes: the dicotyledons and the monocotyledons.

dicotyledon

Dicotyledons are characterized by the presence of two seed leaves, or cotyledons, in the embryo, which is usually surrounded by an ꝑendosperm. They generally have broad leaves with netlike veins. Dicotyledons may be small plants such as daisies and buttercups, shrubs, or trees such as oak and beech.

Grasses
http://www.botanical.com/botanical/mgmh/g/grasse34.html
Informative resource relating to various types of grass, such as couch grass and darnel. It contains detailed descriptions of the plants, including, where relevant, information about the plants' habitat, their constituents, and even their medicinal uses, which, in the case of couch grass, include treating rheumatism and acting as a diuretic.

monocotyledon

Monocotyledons have only one seed leaf, or cotyledon, in the embryo. They usually have narrow leaves with parallel veins and smooth edges, and hollow or soft stems. Their flower parts are arranged in threes. Most are small plants such as grasses, orchids, and lilies, but some are trees such as palms.

the plant cell

The living cells that make up the plant body share a basic structure with those of all other eukaryotic

Ten Largest Angiosperm Families

All the families are dicotyledons, apart from those marked with an asterisk, which are monocotyledons.

Family	Common name	Number of species	Family	Common name	Number of species
Compositae (Asteraceae)	daisy	21,000	Euphorbiaceae	spurge	7,750
Orchidaceae*	orchid	17,500	Labiatae (Lamiaceae)	mint	5,600
Leguminosae (Fabaceae)	bean	16,400	Melastomataceae	melastoma	4,750
Rubiaceae	madder	10,700	Liliaceae*	lily	4,550
Gramineae (Poaceae)*	grass	7,950	Scrophulariaceae	snapdragon	4,500

organisms (organisms other than bacteria and blue-green algae). They possess a clearly defined nucleus, bounded by a membrane, within which the genetic material DNA is formed into distinct chromosomes. They also contain structures called organelles that carry out specialized tasks. For example, the mitochondria are the site of aerobic respiration, and the endoplasmic reticulum takes part in the synthesis of proteins. (A fuller treatment of the eukaryotic cell is given in the *Biology, Genetics, and Evolution* chapter.)

However, there are important differences between plant cells and those of other organisms: notably the possession of a cellulose cell wall (the cell walls of fungi are made of a protein called chitin) and of organelles called chloroplasts, where photosynthesis takes place. The typical cell also contains a single large compartment, called a vacuole.

cell wall The cell wall is the tough outer layer that surrounds the plant cell membrane. It is constructed from a mesh of (cellulose (a complex carbohydrate), and is very strong and relatively inelastic. Most living plant cells are turgid (swollen with water) and develop an internal hydrostatic pressure (wall pressure) that acts against the cellulose wall. The result of this turgor pressure is to give the cell, and therefore the plant, rigidity. Plants that are not woody are particularly reliant on this form of support.

The cellulose in cell walls plays a vital role in global nutrition. No vertebrate is able to produce cellulase, the enzyme necessary for the breakdown of cellulose into sugar. Yet most mammalian herbivores rely on cellulose, using secretions from microorganisms living in the gut to break it down.

chloroplast
Most plant cells that are exposed to light, such as those in leaves, contain organelles called chloroplasts – often in large numbers. Typically, chloroplasts are flattened and disclike, with a double membrane enclosing the **stroma**, a gel-like matrix. Within the stroma are stacks of fluid-containing membranes, or vesicles, on whose surfaces the (**chlorophyll** molecules are bound. It is chlorophyll that gives the chloroplast and the plant its green colour. The light reactions of (photosynthesis (those that require the presence of sunlight and chlorophyll) takes place on the membranes; the dark reactions takes place in the stroma.

It is thought that the chloroplasts were originally free-living cyanobacteria (blue-green algae) which invaded larger, non-photosynthetic cells and developed a symbiotic relationship with them. Like mitochondria, they contain a small amount of DNA and divide by fission.

vacuole
A vacuole is a membrane-bound compartment inside a cell. In living plant cells it is filled with a fluid called **cell sap**, and will take up a large proportion of the cell volume. The vacuole may have a number of functions. In nonwoody plants it may play a role in mechanical support – dissolved substances (solutes) are accumulated in the vacuole at relatively high concentrations, thereby increasing the osmotic flow of water into the cell and making the cell turgid (swollen with water and therefore rigid). The vacuole may act as a reservoir for fluids that the cell will secrete to the outside, or may be filled with excretory products, essential nutrients such as sucrose that the cell needs to store, or with substances that are poisonous to predators.

plant tissues

Plants consist of many different types of cells, which are organized into collections called tissues. Several kinds of tissue can usually be distinguished, each consisting of a particular combination of cells, bound together at the cell walls, which act together to perform specialized functions. **Simple tissues** are made up of only one type of cell, while **complex tissues**, such as those making up the plant's vascular system, contain different cell types. **Meristems** are tissues in which cells are actively dividing to produce new cells and tissues.

simple tissue
Simple tissues are usually named after the cell types of which they are composed.

parenchyma tissue Parenchyma tissue is composed of loosely packed, more or less spherical parenchyma cells, with thin cellulose walls. Although parenchyma often has no specialized function, it is usually present in large amounts, forming a packing or ground tissue. It usually has many intercellular spaces.

collenchyma tissue Collenchyma tissue is composed of relatively elongated cells with thickened cell walls, in particular at the corners where adjacent cells meet. It is a supporting and strengthening tissue found in nonwoody plants, mainly in the stems and leaves.

sclerenchyma tissue The function of sclerenchyma tissue is to strengthen and support. It is composed of thick-walled cells that are heavily lignified (toughened). On maturity the cell inside dies, and only the cell walls remain. There are two types of sclerenchyma tissue: **sclereid cells**, occurring singly or in small clusters, are often found in the hard shells of fruits and in seed coats, bark, and the stem cortex; while the **fibres**, frequently grouped in bundles, are elongated cells, often with pointed ends, associated with the vascular tissue (see below) of the plant.

vascular tissue

A plant's vascular system is a network of complex, fluid-conducting tissues (xylem and phloem) that extend from the roots to the stems and leaves. In young roots, the xylem and phloem are arranged in a **vascular cylinder** at the root's centre; in leaves and young stems they are bound in discrete **vascular bundles**. Typically the phloem is situated nearest to the epidermis and the xylem towards the centre of the cylinder or bundle. In plants exhibiting (secondary growth, the xylem and phloem are separated by a thin layer of vascular cambium, a (meristem that gives rise to new conducting tissues.

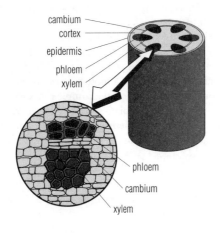

cambium
cortex
epidermis
phloem
xylem

phloem
cambium
xylem

vascular bundle The fluid-carrying tissue of most plants is normally arranged in units called vascular bundles. The vascular tissue is of two types: xylem and phloem. The xylem carries water up through the plant; the phloem distributes food made in the leaves to all parts of the plant.

xylem Xylem is a complex tissue, whose main function is to conduct water and dissolved mineral nutrients from the roots to other parts of the plant. It is composed of a number of different types of cell, and may include long, thin, usually dead cells known as **tracheids**; **fibres** (schlerenchyma); thin-walled **parenchyma** cells; and conducting **vessels**. In most angiosperms (flowering plants) water is moved through these vessels. Most gymnosperms and lower vascular plants lack vessels and depend on tracheids for water

conduction.

Nonwoody plants contain only primary xylem, whereas in trees and shrubs this is replaced for the most part by secondary xylem, formed by secondary growth from the actively dividing vascular cambium. It is this secondary xylem, with its cell walls thickened with lignin, that forms the basis of (wood.

phloem The main function of phloem is to conduct sugars and other food materials from the leaves, where they are produced, to all other parts of the plant. In angiosperms (flowering plants), it is composed of **sieve elements** and their associated **companion cells**, together with some **sclerenchyma** and **parenchyma** cell types. Sieve elements are long, thin-walled cells joined end to end, forming sieve tubes; large pores in the end walls allow the continuous passage of nutrients. Unlike xylem, phloem is a living tissue.

meristem

The meristems are regions of plant tissue containing cells that are actively dividing to produce new cells, which then grow and begin to differentiate to form tissues.

Meristems found in the tip of roots and stems, the **apical meristems**, are responsible for the growth in length of these organs.

The **cambium** is a layer of actively dividing cells (**lateral meristem**), found within stems and roots, that gives rise to secondary growth in perennial plants, causing an increase in girth. There are two main types of cambium: **vascular cambium**, which gives rise to secondary xylem and phloem tissues, and **cork cambium** (or phellogen), which gives rise to secondary cortex and cork tissues (see (bark).

Some plants also have **intercalary meristems**, as in the stems of grasses, for example. These are responsible for their continued growth after cutting or grazing has removed the apical meristems of the shoots.

roots, stems, and leaves

The seed-producing plants – gymnosperms and angiosperms – are usually divided into three parts: root, stem, and leaves. Roots usually grow underground, and serve to anchor the plant in the soil and absorb and conduct water and salts. Stems grow above or below ground. Their cellular structure is designed to carry water and salts from the roots to the leaves in the (xylem, and nutrients from the leaves to the roots in the (phloem. The leaves manufacture the food of the plant by means of photosynthesis, which occurs in the chloroplasts they contain. Flowers (in angiosperms) and cones (in gymnosperms) are modified leaves arranged in groups, enclosing the reproductive organs from which the fruits and seeds result.

root

The root is the part of a plant that is usually underground, and whose primary functions are anchorage and the absorption of water and dissolved mineral salts from the soil. Roots also store foods in the form of starch, and, in some plants, may act as (perennating organs or form symbiotic partnerships with beneficial fungi or nitrogen-fixing microorganisms. Roots usually grow downwards and towards water (that is, they are positively geotropic and hydrotropic; see (tropism).

The root develops from the embryonic root, or radicle, and may form one of two basic types of root system. In the taproot system, a single, robust, main root grows vertically downwards, often to considerable depth. A few smaller lateral roots branch from the main taproot. Taproots are often modified for food storage and are common in biennial plants such as the carrot *Daucus carota*, where they act as perennating organs. In the fibrous system, plants develop several main roots that branch to form a dense network. Such roots are common in grasses.

root structure

The tip of the root plays the greatest role in the absorption of water and mineral salts. This actively growing region is organized into a number of overlapping zones. At the tip is the root apical meristem, a zone of small, continuously dividing cells. A layer of parenchyma cells called a calyptra, or root cap, protects the meristem from abrasion as the root pushes its way through the soil. Above the root apical meristem is the zone of elongation, where the cells no longer divide but begin to grow in size; above this again is the zone of maturation, where the cells begin to mature into tissues such as xylem and phloem. The epidermis (outer surface) of the root at the zone of maturation is characterized by the presence of numerous root hairs, tiny single-celled outgrowths that greatly increase the absorptive area of the roots. These delicate structures survive for a few days only and do not develop into roots.

The tissues of the root include the cortex, a loosely packed layer of parenchyma just beneath the surface cells. The cortex plays a role in conducting water and dissolved salts from the root surface to the vascular cylinder, a solid core of vascular tissue that conducts the water and salts to the stem, and nutrients from the stem to the root body.

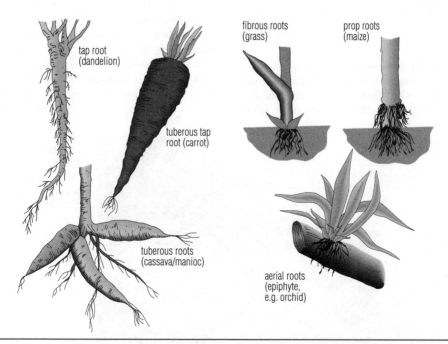

root Types of root. Many flowers (dandelion) and vegetables (carrot) have swollen tap roots with smaller lateral roots. The tuberous roots of the cassava are swollen parts of an underground stem modified to store food. The fibrous roots of the grasses are all of equal size. Prop roots grow out from the stem and then grow down into the ground to support a heavy plant. Aerial roots grow from stems but do not grow into the ground; many absorb moisture from the air.

special roots Special types of root include adventitious roots, which originate from the stem or leaves; contractile roots, which help to position a shoot, corm, or bulb at an appropriate level in the ground; and pneumatophores, erect roots that rise above the soil or water and absorb oxygen from the air.

In **mycorrhizal roots**, found in about 90% of gymnosperms and angiosperms, a symbiotic (mutually beneficial) association is formed with a soil fungus. Such roots take up nutrients more efficiently than do non-mycorrhizal roots, and the fungus benefits by obtaining carbohydrates from the plant or tree.

An ectotrophic mycorrhiza (where the fungus forms a sheath around the root) occurs on many tree species, which usually grow much better, most noticeably in the seeding stage, as a result. Typically the roots become repeatedly branched and coral-like, penetrated by hyphae of a surrounding fungal mycelium. In an endotrophic mycorrhiza, the growth of the fungus is mainly inside the root, as in orchids. Such plants do not usually grow properly, and may not even germinate, unless the appropriate fungus is present.

An Above Grounder's Introduction to Mycorrhiza
http://www.mycorrhiza.com/
General information on mycorrhiza, the 'other half of the root system', including the benefits of mycorrhiza to the plant, how to use mycorrhiza in habitat restoration and revegetation, and current applications of mycorrhiza in agriculture.

Root nodules are growths found on the roots of leguminous plants (members of the pea family), caused by the invasion of *Rhizobium*, a soil bacterium that forms a symbiotic association with the plant.

Rhizobium is a nitrogen-fixing bacterium, that is, it is able to convert gaseous nitrogen in the soil to nitrogenous compounds useful to the plant. The plant thereby obtains valuable nutrients while the bacterium obtains other nutrients and shelter from the plant.

stem

The stem is the main supporting axis of a plant that bears the leaves, buds, and reproductive structures; it may be simple or branched. It contains a continuous vascular system that conducts water and food to and from all parts of the plant. The plant stem usually grows above ground, although some grow underground, including (rhizomes, (corms, and (tubers. The point on a stem from which a leaf or leaves arise is called a **node**, and the space between two successive nodes is the **internode**.

Cactus and Succulent Plant Mall
http://www.cactus-mall.com/index.html
Huge source of information on cacti and how to grow them. There are also reports on conservation of cacti and succulents. There are links to a large number of international cacti associations.

Buds are undeveloped shoots usually enclosed by protective scales; inside is a very short stem and numerous undeveloped leaves, or flower parts, or both. **Terminal buds** are found at the tips of shoots, while **axillary buds** develop in the axils of the leaves, often remaining dormant unless the terminal bud is removed or damaged. **Adventitious buds** may be produced anywhere on the plant, their formation sometimes stimulated by an injury, such as that caused by pruning.

pneumatophore
Pneumatophoric mangrove roots exposed at low tide in Madagascar.

In some plants, the stem is highly modified; for example, it may form a leaflike **cladode** (as in asparagus) or it may be twining (as in many climbing plants), or fleshy and swollen to store water (as in cacti and other succulents).

stem structure Monocotyledons (angiosperms such as grasses that have only one cotyledon in the embryo) and dicotyledons (angiosperms that have two cotyledons in the embryo) differ in the arrangement of tissues in their stems.

In most monocotyledons, the vascular bundles do not form a ring but are scattered throughout the stem. These are surrounded by parenchyma tissue called **ground tissue**. Monocotyledons do not usually show (secondary growth.

Typically, dicotyledons have a central **pith** surrounded by a ring of **vascular bundles** (in which the phloem tissues are outermost) and an outer **cortex**. The outermost cells of the cortex may contain chloroplasts for photosynthesis. The parenchyma cells of the cortex can break down completely to form a hollow core.

secondary growth In woody dicotyledons (trees and shrubs), secondary growth takes place. The xylem and phloem in each vascular bundle becomes separated by a thin layer of vascular cambium, a tissue that divides to form new cells. Cells formed on the outer edge of the cambium become secondary phloem, and cells on the inner edge become secondary xylem (more secondary xylem than phloem is laid down). The laying down of (lignin in the secondary xylem causes the tissue to become woody. A new layer of secondary xylem is laid down each growing season, on the outside of the existing wood, and becomes visible as an **annual ring** (or growth ring) when the tree or shrub is felled, enabling the age of the tree to be estimated.

bark Bark, the protective outer layer on the stems of woody plants, is formed from the tissues external to

leaf margins

entire   serrate   dentate   incised   crenate   sinuate   scalloped   undulate

leaf Leaf shapes and arrangements on the stem are many and varied; in cross section, a leaf is a complex arrangement of cells surrounded by the epidermis. This is pierced by the stomata through which gases enter and leave.

cross section of a leaf

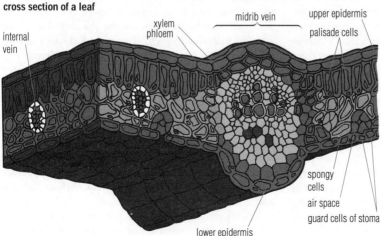

internal vein

xylem
phloem

midrib vein

upper epidermis
palisade cells

spongy cells

air space

guard cells of stoma

lower epidermis

the vascular cambium (the secondary phloem, cortex, and periderm), and its thickness may vary from 2.5 m/0.1 in to 30 cm/12 in or more, as in the giant redwood *Sequoia* where it forms a thick, spongy layer. To allow for expansion of the stem, the bark is continually added to from within, and the outer surface often becomes cracked or is shed as scales.

leaf

A leaf is a lateral outgrowth on the stem of a plant, and in most species the primary organ of (photosynthesis, (transpiration, and gas exchange (the movement of gases between the plant and the atmosphere). The chief leaf types are foliage leaves, cotyledons (seed leaves), scale leaves (on underground stems), and bracts (in the axil of which a flower is produced).

Typically leaves are composed of two parts: the **lamina** or leaf blade, which is flattened and positioned to maximize its exposure to sunlight; and the **petiole** or stalk, which holds the lamina away from the stem. The lamina contains a system of **vascular bundles** or veins, which serve to conduct water and nutrients and also to strengthen the leaf. The leaves of monocotyledons such as grasses often lack petioles and wrap themselves around the stem to form a **sheath**. Leaves that lack petioles are described as **sessile**.

Although the leaves of monocotyledons and dicotyledons serve the same purpose, they differ in basic design. Monocotyledons have long, undivided, straplike leaves, with parallel veins; dicotyledons have a network of veins branching from a single major vein or **midrib**, and the leaf may be greatly divided.

structure of the lamina

In most plants, the lamina is highly adapted for photosynthesis. It contains many chloroplasts (the cell bodies responsible for photosynthesis) and is flattened to provide the greatest surface area for the absorption of sunlight. It also possesses **stomata** – small pores on its epidermis – that allow the exchange of carbon dioxide and oxygen (needed for photosynthesis and respiration) between the internal tissues of the plant and the outside atmosphere. The stomata are also the main route by which water is lost from the plant by transpiration, and they can be closed to conserve water, the movements being controlled by changes in turgidity of its surrounding **guard cells**.

Structurally the lamina is made up of an outer, single-celled layer, called the **epidermis**, which is usually covered with a waxy layer, termed the **cuticle**, which prevents excessive evaporation of water by transpiration. The tissue between the upper and lower epidermis is called **mesophyll**. It consists of parenchyma-like cells containing numerous chloroplasts and, in dicotyledons, is divided into two distinct layers. The **palisade mesophyll** is usually just below the upper epidermis and is

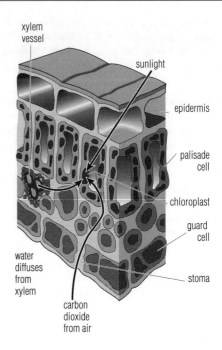

palisade cell Palisade cells are closely packed, columnar cells, lying in the upper surfaces of leaves. They contain many chloroplasts (where photosynthesis takes place) and are well adapted to receive and process the components necessary for photosynthesis – carbon dioxide, water, and sunlight. For instance, their vertical arrangement means that there are few cross-walls to interfere with the passage of sunlight.

composed of regular, tightly packed layers of elongated, chloroplast-rich cells. Lying below them is the **spongy mesophyll**, composed of loosely arranged cells of irregular shape. This layer contains fewer chloroplasts and has many intercellular spaces for the diffusion of carbon dioxide and oxygen, which are linked to the outside by means of stomata in the lower epidermis.

special leaves

The shape, size, and texture of the lamina can vary greatly from species to species. A **simple leaf** is undivided, as in the beech or oak. A **compound leaf** is composed of several leaflets, as in the blackberry, horse-chestnut, or ash tree (the latter being a pinnate leaf). Plants growing in dry situations (xerophytes) frequently have leaves that are reduced in size, or provided with hairs, sunken stomata, and thick cuticles to reduce water loss.

Leaves on the same plant may also be modified to serve different functions. For example, bulbs have very

pinnate leaf The leaves of the mountain ash *Sorbus aucuparia* are pinnate (divided into leaflets along a central midrib).

fleshy leaves that are adapted for storing food and water, while tendrils are often leaves or leaflets adapted for climbing purposes.

Leaves that are shed in the autumn are termed **deciduous**, while evergreen leaves are termed **persistent**.

reproduction in angiosperms (flowering plants)

In angiosperms, the essential steps of reproduction are the formation of male and female gametes, pollination (by which male gametes are transferred from the male reproductive organ to the female reproductive organ), fertilization and seed formation, dispersal of the seed, and germination. The reproductive structure that facilitates gamete production, fertilization, and pollination is the flower.

flower

The site of sexual reproduction in the angiosperm is the flower. This typically consists of four whorls of modified leaves: sepals, petals, stamens, and carpels, borne on the tip of a modified stem or receptacle. The many variations in size, colour, number, and arrangement of parts are closely related to the method of pollination. Flowers adapted for wind pollination typically have reduced or absent petals and sepals and long, feathery stigmas that hang outside the flower to trap airborne pollen. In contrast, the petals of insect-pollinated flowers are usually conspicuous and brightly coloured.

The sepals and petals form the **calyx** and **corolla** respectively and together comprise the **perianth**. In many monocotyledons the sepals and petals are indistinguishable and the segments of the perianth are then known individually as **tepals**.

sepal The sepals are usually green, and surround and protect the flower in bud. In some plants, such as the marsh marigold *Caltha palustris*, where true petals are absent, the sepals are brightly coloured and petal-like, taking over the role of attracting insect pollinators to the flower.

petal The function of the petals is to attract pollinators such as insects or birds. Petals are frequently large

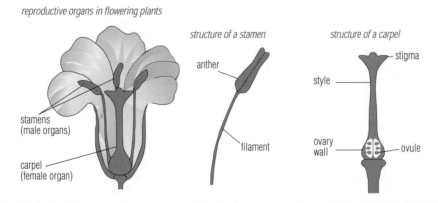

sexual reproduction Reproductive organs in flowering plants. The stamens are the male parts of the plant. Each consists of a stalklike filament topped by an anther. The anther contains four pollen sacs which burst to release tiny grains of pollen, the male sex cells. The carpels are the female reproductive parts. Each carpel has a stigma which catches the pollen grain. The style connects the stigma to the ovary. The ovary contains one or more ovules, which in turn contain the female gametes. Buttercups have many ovaries; the lupin has only one.

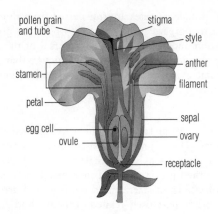

flower Cross-section of a typical flower showing its basic components: sepals, petals, stamens (anthers and filaments), and carpel (ovary and stigma). Flowers vary greatly in the size, shape, colour, and arrangement of these components.

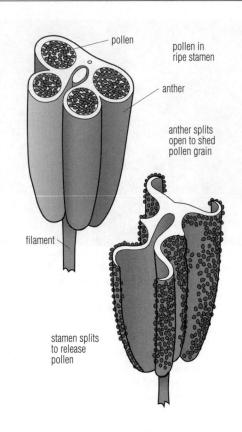

stamen The stamen is the male reproductive organ of a flower. It has a thin stalk called a filament with an anther at the tip. The anther contains pollen sacs, which split to release tiny grains of pollen.

and brightly coloured and may also be scented. Some have a nectary at the base and markings on the petal surface, known as honey guides, to direct pollinators to the source of the nectar. In wind-pollinated plants, however, the petals are usually small and insignificant, and sometimes absent altogether. Some insect-pollinated plants also have inconspicuous petals, with large colourful bracts (leaflike structures) or sepals taking over their role, or strong scents that attract pollinators such as flies.

stamen The stamens, collectively known as the **androecium**, are the male reproductive organs of a flower. A typical stamen consists of a slender stalk, or filament, with an anther, the pollen-bearing organ, at its apex, but in some primitive plants, such as *Magnolia*, the stamen may not be markedly differentiated. The number and position of the stamens are significant in the classification of flowering plants. Generally the more advanced plant families have fewer stamens, but they are often positioned more effectively so that the likelihood of successful pollination is not reduced.

carpel The carpels, collectively known as the **gynoecium**, comprise the female reproductive organs. A flower may have one or more carpels, and they may be separate or fused together. A carpel is usually made up of an **ovary**, a stalk or **style**, and a **stigma** at its top which receives the pollen.

The ovary is the expanded basal portion of the carpel and contains one or more ovules, the structures that develop into seeds after fertilization. Each ovule

consists of an embryo sac containing the female gamete (ovum or egg cell), surrounded by nutritive tissue, the nucellus. Outside this there are one or two coverings that provide protection, developing into the testa, or seed coat, following fertilization. The thick ovary wall provides further protection for the ovules and, following fertilization of the ovum, develops into the fruit wall or pericarp.

flower types Most flowers are **hermaphrodite**, that is they contain both male and female organs. When male and female organs are carried in separate flowers, they are termed **monoecious**; when male and female flowers are on

flower
The world's largest flower smells of rotting flesh. The flower of the parasitic *Rafflesia* has a diameter of up to 1 m/3.3 ft and attracts pollinating insects by mimicking the smell of decomposing corpses. It flowers only once every ten years.

separate plants, the term **dioecious** is used.

In size, flowers range from the tiny blooms of duck-weeds scarcely visible to the naked eye to the gigantic flowers of the Malaysian *Rafflesia*, which can reach over 1 m/3.3 ft across. Flowers may either be borne singly or grouped together in inflorescences. The stalk of the whole inflorescence is termed a **peduncle**, and the stalk of an individual flower is termed a **pedicel**.

pollination
The sacred lotus *Nelumbo nucifera* heats up when it is ready for pollination. For up to four days it maintains steamy temperatures of 30–35°C to attract insects and encourage them to move from one flower to another.

pollination
Pollination is the process by which pollen grains, containing male gametes, are transferred from the male reproductive organ to the female reproductive organ of the plant. In flowering plants, pollen is transferred from the anther to the stigma – the receptive surface at the tip of the carpel. The pollen grain then germinates to form a pollen tube, which grows down towards the ovary and its ovules, where fertilization will take place.

pseudocopulation Close-up of the flower of a bee orchid *Ophrys apifera*. The flower mimics the female bee so exactly that it is even hairy. The orchid will self-pollinate if pseudocopulation does not take place.

types of pollination Self-pollination occurs when pollen is transferred from the anther to a stigma of the same flower, or to another flower on the same plant; **cross-pollination** occurs when pollen is transferred to another plant. This involves external pollen-carrying vectors, such as wind (anemophily), water (hydrophily), insects, birds (ornithophily), bats, and other small mammals.

Animal pollinators carry the pollen on their bodies and are attracted to the flower by scent, or by the sight of the petals. Most flowers are adapted for pollination by one particular agent only. Bat-pollinated flowers tend to smell of garlic, rotting vegetation, or fungus. Those that rely on animals generally produce nectar, a sugary liquid, or surplus pollen, or both, on which the pollinator feeds. Thus the relationship between pollinator and plant is an example of mutualism, in which both benefit. However, in some plants the pollinator receives no benefit – as in **pseudocopulation**, where a flower resembles a female insect so closely that the male

pollination
The flowers of the stapeliads of Africa and southern Asia look very like dung or decaying meat. This is because they rely on carrion-feeding beetles to pollinate them. So convincing are the flowers that they are often covered with maggots that have hatched from eggs mistakenly laid on them.

insect will attempt to mate with it, thereby covering itself with pollen.

fertilization and seed development
In angiosperms (flowering plants), fertilization – the union of two gametes, or sex cells, to form a zygote – takes place in the embryo sac of the ovule (a structure that represents the female gametophyte plant in the alternation of generations; see below). Uniquely, it involves the fertilization not only of the female gamete, or ovum, by a male gamete, but also the fertilization by a second male gamete of *two* polar nuclei within the embryo sac – a process called **double fertilization**.

After (pollination, a pollen tube germinates from the pollen grain and grows down towards the ovule in the carpel's ovary. In the process, a cell within the tube divides by mitosis to form two male gametes. When the tube reaches the ovule and the embryo sac, the male gametes are discharged. One male gamete fertilizes the ovum it encounters there, forming a diploid zygote (having two sets of chromosomes) that will grow into the **embryo**. The second male gamete fertilizes two polar nuclei in the centre of the embryo sac, resulting in the formation of a triploid nucleus (having three sets of chromosomes) that will grow into a

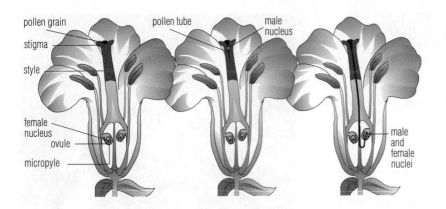

pollen grain   pollen tube   male nucleus
stigma
style
female nucleus
ovule
micropyle
male and female nuclei

fertilization In a flowering plant pollen grains land on the surface of the stigma, and if conditions are acceptable the pollen grain germinates, forming a pollen tube, through which the male gametes pass, entering the ovule in order to reach the female ovum.

specialized tissue, called the **endosperm**, which will provide a food source for the developing embryo.

seed Following this, the ovule matures into a seed, which will comprise the dormant embryo, a food store, and a tough protective seed coat, or testa. The embryo now consists of an embryonic shoot (plumule) and root (radicle), and either one or two seed leaves (cotyledons). The food store is contained either in the endosperm tissue, or in the cotyledons of the embryo.

The number of cotyledons present in a seed is an important character in the classification of angiosperms: monocotyledons (such as grasses, palms, and lilies) have a single cotyledon, whereas dicotyledons (the majority of plant species) have two.

fruit and seed dispersal

In angiosperms (flowering plants), the seed is enclosed within a fruit, which develops from the ripened ovary and serves to protect the seed during its development and to aid in its dispersal. Fruits can be divided into those that are **dry** (such as the capsule, follicle, schizocarp, nut, caryopsis, pod or legume, lomentum, and achene) and those that become **fleshy** (such as the drupe and the berry). The seeds of dry fruits are usually dispersed by the wind or other mechanical means, whereas fleshy fruits are usually dispersed by being eaten by animals.

Seeds of Life
http://www.vol.it/mirror/SeedsOfLife/home.html
Wealth of information about seeds and fruits, with information about the basic structure of a seed, fruit types, how seeds are dispersed, and seeds and humans, plus a mystery seed contest.

The fruit structure consists of the ◊**pericarp**, or fruit wall, which develops from the ovary wall and is usually divided into a number of distinct layers. Sometimes parts other than the ovary are incorporated into the fruit structure, resulting in a false fruit, or **pseudocarp**, such as the apple and strawberry. True fruits include the tomato, orange, melon, and banana.

Fruits may be **dehiscent**, that is, they open to shed their seeds, or **indehiscent**, that is, they remain unopened and are dispersed as a single unit. **Simple fruits** (for example, peaches) are derived from a single

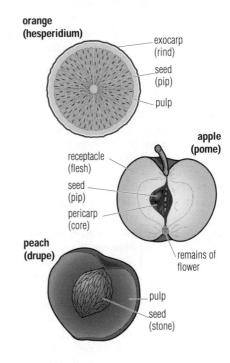

orange (hesperidium)
exocarp (rind)
seed (pip)
pulp

apple (pome)
receptacle (flesh)
seed (pip)
pericarp (core)
remains of flower

peach (drupe)
pulp
seed (stone)

---

fruit A fruit contains the seeds of a plant. Its outer wall is the exocarp, or epicarp; its inner layers are the mesocarp and endocarp. The orange is a hesperidium, a berry having a leathery rind and containing many seeds. The peach is a drupe, a fleshy fruit with a hard seed, or 'stone', at the centre. The apple is a pome, a fruit with a fleshy outer layer and a core containing the seeds.

ovary, whereas **compound fruits** (for example, black-berries) are formed from the ovaries of a number of flowers.

dispersal mechanisms Efficient seed dispersal is essential to avoid overcrowding and enable plants to colonize new areas; the natural function of a fruit is to aid in the dissemination of the seeds which it contains.

A great variety of dispersal mechanisms exist: winged fruits are commonly formed by trees, such as ash and elm, where they are in an ideal position to be carried away by the wind; some wind-dispersed fruits, such as clematis and cotton, have plumes of hairs; others are extremely light, like the poppy, in which the capsule acts like a pepperpot and shakes out the seeds as it is blown about by the wind. Some fruits float on water; the coconut can be dispersed across oceans by means of its buoyant fruit. Geraniums, gorse, and squirting cucumbers have explosive mechanisms, by which seeds are forcibly shot out at dehiscence. Animals often act as dispersal agents either by carrying hooked or sticky fruits (burs) attached to their bodies, or by eating succulent fruits, the seeds passing through the alimentary canal unharmed.

germination

Germination is the initial growth of a seed, spore, or pollen grain. Seeds germinate when they are exposed to favourable external conditions of moisture, light, and temperature, and when any factors causing dormancy have been removed.

The process begins with the uptake of water by the seed. Food reserves, either within the endosperm or from the cotyledons, begin to be broken down to nourish the rapidly growing seedling. The embryonic root, or radicle, is normally the first organ to emerge, its tip protected by a root cap, or calyptra, as it pushes through the soil. The radicle may form the basis of

bur The burs of the great burdock *Arctium lappa*. The tiny hooks will attach to clothing or fur and aid seed dispersal.

the entire root system, or it may be replaced by adventitious roots (positioned on the stem).

The plumule develops into the shoot system of the seedling plant. In most seeds, such as those of the sunflower, the plumule is a small conical structure without any leaf structure, and growth does not occur until

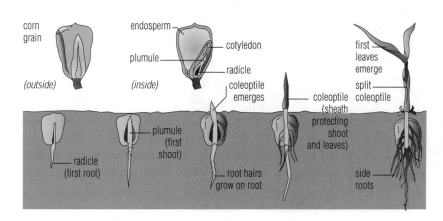

germination The germination of a corn grain. The plumule and radicle emerge from the seed coat and begin to grow into a new plant. The coleoptile protects the emerging bud and the first leaves.

the cotyledons have grown above ground into the first green leaves. This is **epigeal germination**. However, in seeds such as the broad bean, a leaf structure is visible on the plumule in the seed. These seeds develop by the plumule growing up through the soil with the cotyledons remaining below the surface. This is known as **hypogeal germination**.

Germination is considered to have ended with the production of the first true, non-cotyledonous leaves.

alternation of generations

Reproduction in plants takes place in two distinct phases, or generations, occurring alternately: diploid (in which each cell has two sets of chromosomes) and haploid (each cell has one set of chromosomes). The diploid generation produces haploid spores by meiosis, and is called the **sporophyte**. The spores germinate into the haploid generation, which produces gametes (sex cells), and is called the **gametophyte**. The gametes fuse to form a diploid zygote, which develops into a new sporophyte; thus the sporophyte and gametophyte alternate.

In mosses and other bryophytes, the familiar green plant is the gametophyte, while the long-stalked spore capsules growing from it are sporophytes. In lower vascular plants such as ferns, the familiar plant is the sporophyte and the gametophyte, which grows separately from it, is very small and inconspicuous. All higher plants (angiosperms and gymnosperms) are

sporophytes, and the gametophyte is not seen because it completes its life microscopically within the body of the sporophyte.

nutrition

photosynthesis

Photosynthesis is the process by which green plants trap light energy from the Sun. This energy is used to drive a series of chemical reactions that lead to the formation of carbohydrates. The carbohydrates occur in the form of simple sugar, or glucose, which provides the basic food for both plants and animals. For photosynthesis to occur, the plant must possess the green photosynthetic pigment (chlorophyll and must have a supply of carbon dioxide and water. Photosynthesis takes place inside (chloroplasts, which are found mainly in the leaf cells of plants. The by-product of photosynthesis, oxygen, is of great importance to all living organisms, and virtually all atmospheric oxygen has originated by photosynthesis.

chemical process The chemical reactions of photosynthesis occur in two stages. During the **light reactions** sunlight is used to split water (H_2O) into oxygen (O_2), protons (hydrogen ions, H^+), and electrons, and oxygen is given off as a by-product. In the **dark reactions**, for which sunlight is not required, the protons and electrons are used to convert carbon dioxide (CO_2) into carbohydrates ($C_mH_2O)_n$).

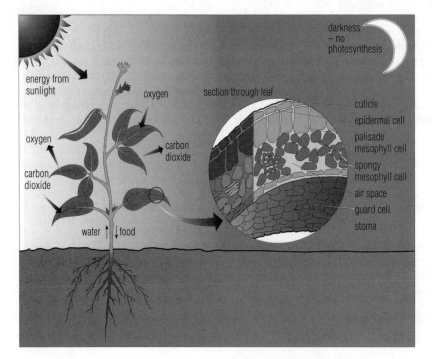

energy from sunlight

oxygen

oxygen

carbon dioxide

carbon dioxide

water ↑ ↓ food

section through leaf

darkness – no photosynthesis

cuticle
epidermal cell
palisade mesophyll cell
spongy mesophyll cell
air space
guard cell
stoma

photosynthesis Process by which green plants manufacture carbohydrates from water and atmospheric carbon dioxide, using the energy of sunlight. Photosynthesis depends on the ability of chlorophyll molecules within plant cells to trap the energy of light to split water molecules, giving off oxygen as a by-product. The hydrogen of the water molecules is then used to reduce carbon dioxide to simple carbohydrates.

photosynthesis and respiration

External respiration in flowering plants is dependent on light intensity. In daylight both photosynthesis and respiration occur. Photosynthesis provides oxygen for respiration, and respiration provides carbon dioxide for photosynthesis. In darkness, only respiration takes place and oxygen is taken in from the outside air. In bright light, the rate of photosynthesis is ten to twenty times the rate of respiration; thus it provides an ample source of oxygen. In decreasing light, the photosynthesis rate drops until a point is reached when the rate of respiration equals the rate of photosynthesis. This is called the **compensation point**. In very dim light, or darkness, the rate of photosynthesis becomes less than the respiratory rate.

Photosynthesis Directory
http://esg-www.mit.edu:8001/esgbio/ps/psdir.html
A wealth of scientific information concerning photosynthesis, its stages and its importance from MIT in Boston, USA. The site discusses issues such as the evolution and discovery of photosynthesis, the chloroplast, and the chlorophyll, and all steps of the light and dark reactions that take place during photosynthesis.

---

mineral salts

Mineral salts are the inorganic salts required by living organisms for growth. For plants, the essential mineral salts (macronutrients) are those of nitrogen, phosphorus, potassium, calcium, magnesium, sulphur, and iron. For example, nitrogen salts in the form of nitrates are required for the manufacture of proteins, nucleic acids, and chlorophyll; phosphates are required for

Venus flytrap The Venus flytrap is native to the swamp lands of Carolina. Charles Darwin described the plant as 'one of the world's most wonderful plants'. The trap is sprung when insects, attracted by the colour and nectar, touch trigger hairs on the faces of the leaves.

proteins; and magnesium salts for chlorophyll. The trace elements (those required only in tiny amounts) are manganese, boron, cobalt, copper, zinc, and chlorine. Plants usually obtain their mineral salts from the soil – either by diffusion into the root cells or by means of an active, energy-requiring process that pumps salts into the cells.

deficiency disease

Shortages of mineral nutrients may lead to deficiency diseases such as discoloration and reduced growth. In general, a deficiency of nitrogen gives rise yellowish lower leaves that are much smaller than normal; a deficiency of phosphorus gives

Classification of Carnivorous Plants

Plants that obtain at least some of their nutrition by capturing and digesting prey are called carnivorous plants. Such plants have adaptations that allow them to attract, catch, and break down or digest prey once it is caught. Estimates of the number of species of carnivorous plants number from 450 to more than 600. Generally, these plants are classified into genera based upon the mechanism they have for trapping and capturing their prey. The major genera of these plants are listed in the table.

Common name	Genus	Scientific name	Trapping mechanism
bladderwort	*Utricularia*	*Utricularia vulgaris*	active trap; shows rapid motion during capture
butterwort	*Pinguicula*	*Pinguicula vulgaris*	semi-active trap; two-stage trap in which prey is initially caught in sticky fluid
calf's head pitcher plant	*Darlingtonia*	*Darlingtonia californica*	passive trap; attracts prey with nectar and then drowns prey in fluid contained within plant
flypaper plant	*Byblis*	*Byblis liniflora*	passive trap; attracts prey with nectar and then drowns prey in fluid contained within plant
sundew	*Drosera*	*Drosera linearis*	semi-active trap; two-stage trap in which prey is initially caught in sticky fluid
Venus flytrap	*Dionaea*	*Dionaea muscipula*	active trap; shows rapid motion during capture

a reddish purple colour to the leaves, whilst one of potassium causes the edges of the leaves to die.

carnivorous plant Carnivorous plants capture and digest live prey (normally insects), to obtain nitrogen compounds that are lacking in their usual marshy habitats. Some are passive traps, for example, the pitcher plants *Nepenthes* and *Sarracenia*. One pitcher-plant species has container-traps holding 1.6 l/3.5 pt of the liquid that 'digests' its food, mostly insects but occasionally even rodents. Others, for example, sundews *Drosera*, butterworts *Pinguicula*, and Venus flytraps *Dionaea muscipula*, have an active trapping mechanism.

Carnivorous plants have adapted to grow in poor soil conditions where the number of microorganisms recycling nitrogen compounds is very much reduced. In these circumstances other plants cannot gain enough nitrates to grow.

Carnivorous Plants FAQ
http://www.indirect.com/www/bazza/cps/faq/faq.html
Well written source of general information about carnivorous plants. Each genus is presented with good text and pictures. There is advice on planting, growing, and feeding carnivorous plants. There is additional information on effort to conserve endangered species.

---

transpiration

Transpiration is the loss of water from a plant by evaporation. Most water is lost from the leaves through pores known as **stomata**, though some loss also occurs from the lenticels (pores in woody stems) and from cracks in the cuticle covering stems and leaves. Transpiration, coupled with the absorption of water and dissolved mineral salts by the roots, causes a continuous upward flow of water from the roots via the xylem – a phenomenon known as the **transpiration stream**. Thus, the water required to maintain turgidity and for photosynthesis, and the mineral salts required for nutrition are transported throughout the plant

The rate of transpiration is affected by atmospheric temperature and humidity, wind, and the availability of water in the soil. Excessive transpiration places the plant in danger of dehydration, and so the plant is able to control the loss of vapour from the leaves by opening and closing guard cells on either side of the stomata.

> **transpiration**
>
> A single maize plant has been estimated to transpire 245 l/54 gal of water in one growing season.

insectivorous plant The insect traps of North American pitcher plants are modified leaves. Insects are lured into the pitchers by sweet secretions and then tumble into the fluid at the base, where they drown and are slowly digested.

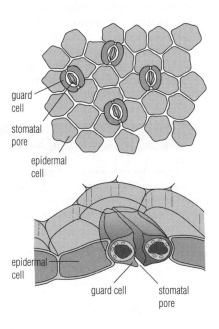

guard cell
stomatal pore
epidermal cell

epidermal cell
guard cell
stomatal pore

stoma The stomata, tiny openings in the epidermis of a plant, are surrounded by pairs of crescent-shaped cells, called guard cells. The guard cells open and close the stoma by changing shape.

transpiration The loss of water from a plant by evaporation is known as transpiration. Most of the water is lost through the surface openings, or stomata, on the leaves. The evaporation produces what is known as the transpiration stream, a tension that draws water up from the roots through the xylem, or water-carrying vessels, in the stem.

transpiration stream

Transpiration from the leaves is the main impetus powering the transpiration stream, generating a strong pulling force on the water column in the plant. Water vapour evaporates from the surfaces of the spongy mesophyll cells within the leaf into the intercellular spaces, and then diffuses through the stomata to the drier air outside the leaf. Water lost from the spongy mesophyll cells is replaced by water carried by the leaf's xylem vessels (veins). The cohesive forces between water molecules mean that the flow of water from the xylem pulls on the rest of the water column in the xylem tube, drawing water up the plant.

Capillary action (the spontaneous movement of liquids through narrow tubes, or capillaries) also plays a role in pulling water through the xylem vessels, though probably only for short distances.

Eventually, the flow of water through the xylem brings about a reduction in the concentration of water in the root tissue. The resultant concentration gradient between the cells of the root cortex and the soil outside the root creates the conditions required from the movement of water by osmosis from the soil into the root. Mineral salts dissolved in the water film surrounding the soil particles pass into the root by diffusion, or are actively absorbed by the root cells. This movement of water and salts into the root creates an upward pressure (root pressure) that plays a small role in driving the transpiration stream.

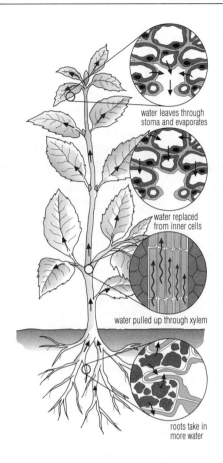

water leaves through stoma and evaporates

water replaced from inner cells

water pulled up through xylem

roots take in more water

Common Drugs Derived from Plants

These plants are poisonous and if swallowed can cause serious illness or unconsciousness. They should only be used if administered by a medically trained professional.

Plant	Drug	Use
Amazonian liana	curare	muscle relaxant
Annual mugwort	artemisinin	antimalarial
Autumn crocus	colchicine	antitumour agent
Coca	cocaine	local anaesthetic
Common thyme	thymol	antifungal
Deadly nightshade (belladonna)	atropine	anticholinergic
Dog button (nux-vomica)	strychnine	central nervous system stimulant
Ergot fungus	ergotamine	analgesic
Foxglove	digitoxin, digitalis	cardiotonic
Indian snakeroot	reserpine	antihypertensive
Meadowsweet	salicylate	analgesic
Mexican yam	diosgenin	birth control pill

Plant	Drug	Use
Opium poppy	codeine, morphine	analgesic (and antitussive)
Pacific yew	taxol	antitumour agent
Recured thornapple	scopolamine	sedative
Rosy periwinkle	vincristine, vinblastine	antileukemia
Velvet bean	L–dopa	antiparkinsonian
White willow	salicylic acid	topical analgesic
Yellow cinchona	quinine	antimalarial, antipyretic

Common Herbs

Common name	Scientific name	Family	Usage
angelica	*Angelica archangelica*	Umbelliferae	cookery, herbal tea
anise	*Pimpinella anisum*	Umbelliferae	cookery, medicine
basil, sweet	*Ocimum basilicum*	Labiatae	cookery
bay, sweet	*Laurus nobilis*	Lauraceae	cookery
bergamot	*Monarda didyma*	Labiatae	perfumery, herbal tea
borage	*Borago officinalis*	Boraginaceae	cookery, herbal tea
burnet, salad	*Sanguisorba minor*	Rosaceae	cookery
camomile	*Chamaemelum nobile*	Compositae	cookery, herbal tea
chervil	*Anthriscus cerefolium*	Umbelliferae	cookery
chive	*Allium schoenoprasum*	Liliaceae	cookery
comfrey	*Symphytum officinale*	Boraginaceae	medicine
coriander (cilantro)	*Coriandrum sativum*	Umbelliferae	cookery
dill	*Anethum graveolens*	Umbelliferae	cookery
fennel	*Foeniculum vulgare*	Umbelliferae	cookery
hyssop	*Hyssopus officinalis*	Labiatae	medicine, cookery
lavender	*Lavandula angustifolia*	Labiatae	perfumery, medicine
lemon grass	*Cymbopogon citradus*	Gramineae	cookery
lemon balm	*Melissa officinalis*	Labiatae	medicine, herbal tea
lovage	*Levisticum officinale*	Umbelliferae	cookery
marjoram	*Origanum majorana*	Labiatae	cookery
oregano	*Origanum vulgare*	Labiatae	cookery
parsley	*Petroselinum crispum*	Umbelliferae	cookery
pennyroyal	*Mentha pulegium*	Labiatae	herbal tea
peppermint	*Mentha piperita*	Labiatae	medicine, cookery
rocket, garden	*Eruca sativa*	Cruciferae	cookery
rosemary	*Rosmarinus officinalis*	Labiatae	cookery
rue	*Ruta graveolens*	Rutaceae	perfumery
sage	*Salvia officinalis*	Labiatae	cookery
sorrel	*Rumex acetosa*	Polygonaceae	cookery
spearmint	*Mentha spicata*	Labiatae	cookery, herbal tea
tansy	*Tanacetum vulgare*	Compositae	cookery
tarragon	*Artemisia dracunculus*	Compositae	cookery
thyme	*Thymus vulgaris*	Labiatae	cookery
thyme, lemon	*Thymus x citriodorus*	Labiatae	cookery, perfumery

The Plant Kingdom Chronology

c. 320 BC The Greek philosopher Theophrastus begins the science of botany with his books *De causis plantarum/The Causes of Plants* and *De historia plantarum/The History of Plants*. In them he classifies 500 plants, develops a scientific terminology for describing biological structures, distinguishes between the internal organs and external tissues of plants, and gives the first clear account of plant sexual reproduction, including how to pollinate the date palm by hand.

170 The Greek physician Claudius Galen develops methods for extracting plant juices, to be used for medicinal purposes.

601 The *Vagbhata*, an Indian medicinal herbal, is compiled around this time.

802 Rose trees from Asia are introduced to Europe and cultivated there for the first time.

1100 The poet and physician Odo of Meung writes *De viribus herbarum/On the Power of Herbs*, a herbal derived from classical sources. Written in verse as an *aide mémoire*, it becomes extremely popular.

1200 The Spanish Muslim agriculturist ibn-al-'Awwam writes his treatise *Kitab al-Filahah*, which describes 585 plants and the many techniques for their cultivation, among other subjects.

1248 The Spanish-born Muslim Al-Baytar, 'chief of botanists' in Cairo, Egypt, writes *Kitab al-jami/Collection of Simple Drugs*, which lists 1,400 different remedies and is the largest and most popular Arab pharmacopoeia.

1317 The physician Matthaeus Sylvaticus writes his *Pandectae*, a dictionary of medicinal herbs.

1406 The Chinese herbalist Chu Hsiao writes *Chiu huang pên ts'ao/Herbal to Relieve Famine*, an illustrated work describing 414 species of plants.

1538 The English naturalist William Turner publishes his *Libellus de re herbaria novus/New Letter on the Properties of Herbs*, the first English herbal to take a scientific approach to botany.

1538 William Turner publishes his *Libellus de re herbaria novus*, the first English herbal to take a scientific approach to botany.

1542 The German botanist Leonhard Fuchs (after whom the fuchsia is named) publishes *De historia stiripium/On the History of Plants*, a pioneering work of plant classification, at Basel.

1543 Europe's first botanical garden is founded at the University of Pisa, in Italy.

1553 The Spanish writer Pedro de Cieza de Leon's *Chronicle of Peru* describes potatoes for the first time.

1583 The Italian botanist and physician Andre Cesalpino attempts to classify plants systematically based on variations in their form, in his *De Plantis*.

1590 The Spanish missionary José de Acosta publishes *Historia natural y moral de las Indias/A Natural and Moral History of the Indies*. Rich in details of the flora and fauna of the New World as well as accounts of Pre-Columbian civilization, it is read throughout Europe. An English translation appears in 1604.

1597 The English naturalist John Gerard publishes his *Herball, or Generall Historie of Plantes*, the first plant catalogue, describing over 1,000 species.

1618 The English botanist John Tradescant visits Russia to collect plant samples, as part of a mission organized by King James I.

1620 John Tradescant visits North Africa, returning with many samples of newly discovered species, including gutta percha and many tropical fruits.

1660 The English naturalist John Ray publishes his *Catalogue of Plants around Cambridge*, giving details of 558 different species.

1662 After the death of John Tradescant the Younger, the English botanist's collection of plants is incorporated into the collection of Elias Ashmole, and ultimately into the Ashmolean Museum (founded in 1683) in Oxford, England.

1663 The English naturalists John Ray and Francis Willughby embark on a three-year tour of Europe to study and collect flora and fauna.

1664 The English writer and diarist John Evelyn writes *Sylva*, a survey of British trees and their management and usefulness.

1672 The English naturalist Nehemiah Grew publishes *The Anatomy of Vegetables Begun*, analysing the structure of bean seeds, and inventing much of the terminology for plant embryology.

1675 The Italian physician and biologist Marcello Malpighi publishes his *Anatome Plantarum/Anatomy of Plants*, a study of plant tissues.

1678 The first chrysanthemums arrive in Europe from Japan.

1682 The English botanist Nehemiah Grew's *Anatomy of Plants* identifies the stamens and pistils as male and female sex organs for the first time.

1686 The English naturalist John Ray publishes the first volume of his *Historia plantarum/Study of Plants*, a catalogue of over 18,000 different plants. It introduces the notion of a species.

1687 The English naturalist Hans Sloane visits Jamaica, collecting 800 new species of plant, and establishing his botanical collection.

1688 John Ray's *Methodus plantarum nova/New Method for Plants* is published, in which he makes a fundamental distinction between monocotyledons and dicotyledons.

1690 Coffee plants are smuggled out of the Arab port of Mocha by Dutch sailors, finding their way to the island of Java and the botanical gardens of Amsterdam, United Netherlands.

1690 The German naturalist Paul Hermann coins the term 'angiosperm' to describe flowering plants.

1694 The German botanist Rudolf Jacob Camerarius identifies the reproductive organs of plants and describes the mechanism of plant fertilization.

1699 The English botanist James Woodward grows flowers in water with a variety of impurities, and discovers they grow better in water from sewers or water mixed with mould.

1704 The English naturalist John Ray completes publication of his three-volume *Historia generalis plantarum/General Study of Plants*, a classification of over 18,000 different plant species.

1710 The Dutch physician Herman Boerhaave publishes his *Index plantarum/Index of Plants*.

1727 The English botanist Stephen Hales's book *Vegetable Staticks* gives the first accurate scientific explanation of the nutrition of plants, and describes numerous experiments in plant physiology.

1728 The Italian mathematician Guido Grandi publishes *Flora geometrica/Geometrical Flowers*, attempting a geometrical definition of the curves of flower petals and leaves.

1733 Stephen Hales describes his investigation of blood flow and sap flow in *Statical Essays*, an expanded version of his *Vegetable Staticks* of 1727

1734 The crocus is introduced into North America from Europe.

1739 The Dutch botanist Jan Frodoronz Gronovius publishes *Flora Virginica/Virginian Flora*, a guide to the plant life of northeastern America.

1741 The Swedish botanist Carolus Linnaeus founds a botanical garden in Uppsala, Sweden.

1753 Carolus Linnaeus publishes his *Species plantarum/The Species of Plants*, an application of his new taxonomic system to thousands of plant species.

1758 Linnaeus applies the binomial taxonomy he developed for plant classification to animal species.

1761 The German botanist Joseph Gottlieb Kölreuter begins to publish the results of his experiments on the artificial fertilization and hybridization of plants, and on the pollination of plants by insects.

1770 The English naturalist John Hill introduces a method of obtaining specimens for microscopic study.

1770 The classification of the gardenia is mistakenly credited to Alexander Garden, for whom its discoverer, Jane Colden, had named it. A New York botanist who catalogued over 300 plants, Colden was denied instruction in Latin by her father because she was female.

1779 The Dutch physician and plant physiologist Jan Ingenhousz discovers two respiratory cycles in plants. He concludes that sunlight is necessary for the production of oxygen by leaves.

1789 The French botanist Antoine-Laurent Jussieu's *Genera Plantarum/The Genera of Plants* refines Carolus Linnaeus's classification of plants.

1793 The German botanist Christian Konrad Sprengel outlines an accurate theory of plant fertilization.

1804 The Swiss plant physiologist Nicolas-Theodre de Saussure demonstrates that plants require nitrogen from the soil, and increase in weight through the absorption of water and carbon dioxide.

1812 Two strains of maize are crossbred in the USA to produce the first hybrid maize.

1813 The Swiss botanist Augustin Pyrame de Candolle publishes *Théorie élémentaire de la botanique/Elementary Theory of Botany* in which he coins the word 'taxonomy', introduces the idea of homology in plants, and argues that the basis of plant classification should be anatomy and not physiology.

1817 The French chemists Joseph Pelletier and Joseph-Bienaimé Caventou isolate chlorophyll.

1823 The Italian botanist Giovanni Battista Amici proves the existence of sexual processes in flowering plants, by observing pollen approaching the plant ovary.

1824 The Swiss botanist Augustin Pyrame de Candolle begins his 17-volume classification of plants *Prodromus systematis naturalis regni vegetabilis/ Treatise on the Classification of the Plant Kingdom*. Completed in 1873 by his son Alphonse, it replaces Lamarck's classification system and serves as the model for future systems.

1831 The Scottish botanist Robert Brown discovers the nucleus in plant cells.

1834 The French chemist Jean-Baptiste Boussingault discovers that plants absorb nitrogen through the soil and carbon from carbon dioxide in the air.

c. 1835 The French chemist Anselme Payen discovers cellulose, the basic component of plant cells.

1836 The US botanist Asa Gray writes *Elements of Botany*, the first textbook on botany.

1837 The French physiologist Henri Dutrochet establishes chlorophyll's essential role in photosynthesis.

1838 The German botanist Matthias Jakob Schleiden publishes the article 'Contributions to Phytogenesis', in which he recognizes that cells are the fundamental units of all plant life. He is thus the first to formulate cell-theory.

1843 The English agronomist John Bennet Lawes and English chemist Joseph Henry Gilbert establish the Rothamsted Experimental Station,

Hertfordshire, England, the world's first agricultural research station. They also discover that nitrogen, potassium, and phosphorus are necessary for plant growth.

1846 The German botanist Hugo von Mohl uses the word protoplasm to describe the main living substance in a cell. It leads to the development of cell physiology.

1846 The Italian botanist Giovanni Battista Amici establishes the circulation of sap in plants.

c. 1850 The French naturalist Antonio Snider-Pellegrini suggests that the similarities between European and North American plant fossils could be explained if the two continents were once in contact.

1851 The German botanist Hugo von Mohl discovers that the secondary walls of plant cells are of a fibrous structure.

1851 The German botanist Wilhelm Friedrich Hofmeister discovers the alternation of generations in ferns and mosses, and establishes the relationships between the conifers (gymnosperms) and flowering plants (angiosperms).

1856 The Austrian monk and botanist Gregor Mendel begins experiments in breeding peas that will lead him to the laws of heredity.

1856 The German botanist Nathaniel Pringsheim is the first to notice fertilization when he observes sperm entering the ovum in plants.

1861 The German zoologist Max Schultze defines the cell as consisting of protoplasm and a nucleus, a structure he recognizes as fundamental in both plants and animals.

1862 The German botanist Julius von Sachs proves that starch is produced by photosynthesis.

1865 Julius von Sachs discusses the transport of water in the roots of plants. He also demonstrates that chlorophyll is contained within chloroplasts, and not diffused throughout the cell.

1868 The German botanists Nathaniel Pringsheim and Julius von Sachs discover the specialized organelles in plant cytoplasm called plastids.

1872 The US plant breeder Luther Burbank develops the Burbank potato. During the next 50 years he develops over 800 new varieties of plants.

1876 The British scientist Henry Wickham collects 70,000 seeds from the rubber plant *Hevea brasiliensis* in the Amazon jungle, and has them planted in Kew Gardens. The saplings are later transported to Ceylon, India, and Malaya where they form the beginning of the rubber industry.

1881 The German botanist Wilhelm Friedrich Philipp Pfeffer publishes *Pflanzenphysiologie: Ein Hanbuch des Stoffwechsels und Kraftwechsels in der Pflanz/The Physiology of Plants: A Treatise Upon the Metabolism and Sources of Energy in Plants*, which becomes the basic handbook on plant physiology.

1888 The Dutch geneticist Hugo Marie de Vries uses the term 'mutation' to describe varieties that arise spontaneously in cultivated primroses.

1888 The German cytologist Eduard Adolf Strasburger determines that germ cell nuclei of flowering plants undergo meiosis.

1892 The Canadian botanist Charles Saunders develops Marquis wheat by crossing Red Fife wheat with an early ripening Indian variety. Made available to farmers in 1900, it dominates spring wheat in the USA and Canada for 50 years.

1892 The Dutch geneticist Hugo Marie de Vries, through a programme of plant breeding, establishes the same laws of heredity discovered by Gregor Mendel in 1865.

1909 The German botanist Carl Correns shows that certain hereditary characteristics of plants are determined by factors in the cytoplasm of the female sex cell. It is the first example of non-Mendelian heredity.

1920 The Russian botanist Nikolay Ivanovich Vavilov states that a plant's place of origin is the region where its greatest diversity is found. He identifies 12 world centres of plant origin.

1921 Dutch elm disease, caused by the fungus *Ceratocystis ulmi* and spread by bark beetles, is first described in the Netherlands. A serious disease of elm trees, it is thought to have come from Asia after World War I.

1962 The Rockefeller and Ford Foundations found the Rice Research Institute in the Philippines and begin cross-breeding more than 10,000 different strains of rice.

1971 C W Ferguson of the University of Arizona, USA, establishes a tree-ring chronology dating back to c. 6000 BC.

1971 The anticancer drug Taxol is isolated from the bark of the Pacific Yew tree.

1973 Representatives from 80 nations sign the Convention on International Trade in Endangered Species (CITES) that prohibits trade in 375 endangered species of plants and animals and the products derived from them; the USA does not sign until 1977.

1974 The Parque Nacional da Amazonia is established in Brazil; with an area of 10,000 sq km/4,000 sq mi, it preserves a large area of tropical rainforest.

1995 A genetically engineered potato is developed that contains the gene for Bt toxin, a natural pesticide produced by a soil bacterium. The potato plant produces Bt within its leaves.

1996 British palaeontologists discover the world's oldest flowering plant, *Bevhalstia pebja*, in southern England. It is a wetland herb about 25 cm/10 in high and it is about 130 million years old.

1997 The Canadian researcher Suzanne W Simard and colleagues announce the discovery that trees use the threadlike growths of fungi called mycorrhizae which infest their roots and connect the trees together underground to exchange food resources. It suggests that forest trees succeed as cooperative communities rather than competing individuals.

Biographies

Alpini, Prospero (1553–1616) Italian botanist and physician, director of the botanical garden in Padua. He studied plants for their therapeutic uses and also out of interest in their structure and function. His publications include *De medicina Aegyptorium/On Egyptian Medicine* (1591) and *De plantis Aegypti liber/Book of Egyptian Plants* (1592), which included the first European descriptions of the coffee bush (coffee arabica) and the banana tree. He also studied the flora of Crete. ◊Linnaeus named the genus *Alpinia* in his honour.

Amici, Giovanni Battista (1786–1863) Italian botanist and microscopist who in the 1820s made a series of observations clarifying the process by which pollen fertilizes the ovule in flowering plants.

Bauhin, Gaspard Casper (1560–1624) Swiss botanist and physician who developed an important early plant classification system. In *Pinax theatri botanica/Illustrated Exposition of Plants* (1623), he attempted to classify all known species of plant, naming 6,000 species. His system was used by both ◊Linnaeus and John ◊Ray, and was to some extent based upon the system which Andrea ◊Cesalpino had outlined earlier in the 16th century.

Blackman, Frederick Frost (1866–1947) English botanist after whom the Blackman reactions of ◊photosynthesis are named. Leading a successful research group, he worked initially upon respiration in plants and showed that the exchange of CO_2 between the leaves and the air occurred via the stomata (the pores in the epidermis of a plant). His later work continued to apply physiochemical concepts to biology.

Brown, Robert (1773–1858) Scottish botanist. He was the first to establish the real basis for the distinction between gymnosperms (conifers) and angiosperms (flowering plants). He also described the organs and mode of reproduction in orchids. In 1831 he discovered that a small body that is fundamental in the creation of plant tissues occurs regularly in plant cells – he called it a 'nucleus', a name that is still used. On an expedition to Australia 1801–05 Brown collected 4,000 plant species and later classified them using the 'natural' system of the French botanist Bernard de Jussieu (1699–1777).

Calvin, Melvin (1911–1997) US chemist who, using radioactive carbon-14 as a tracer, determined the biochemical processes of photosynthesis, in which green plants use chlorophyll to convert carbon dioxide and water into sugar and oxygen. He began work on photosynthesis in 1949, studying how carbon dioxide and water combine to form carbohydrates such as sugar and starch in a single-celled green alga, *Chlorella*. He showed that there is in fact a cycle of reactions (now called the Calvin cycle) involving an enzyme as a catalyst. He was awarded a Nobel prize in 1961.

Candolle, Alphonse Louis Pierre Pyrame de (1806–1893) Swiss botanist who developed a new classification system for plants based on a broad number of their morphological features. He worked extensively on the effects of climatic and other physical variables on the development of distinct species of flora, and published several books including *Introduction à l'étude de la botanique* (1835), *Géographie botanique raisonnée* (1855), and *Historie des sciences et ses savants depuis deux siècles* (1873).

Candolle, Augustin Pyrame de (1778–1841) Swiss botanist who coined the term 'taxonomy' to mean the classification of plants on the basis of their gross anatomy in his book *Théorie élémentaire de la botanique* (1813). He posited that plant relationships can be determined by the symmetry of their sexual organs and introduced the concept of homologous parts, the idea that an organ or structure possessed by different plants may indicate a common ancestor.

Cesalpino, Andrea (1519–1603) Italian botanist who showed that plants could be and should be classified by their anatomy and structure. In *De plantis* (1583) he offered the first remotely modern classification of plants. Before this plants were classed by their location – for example marsh plants, moorland plants, and even foreign plants.

Correns, Carl Franz Joseph Erich (1864–1933) German botanist and geneticist who is credited with rediscovering the Austrian biologist Gregor Mendel's laws of inheritance (Hugo ◊De Vries is similarly credited). His work on the role of pollen in influencing characteristics of fruit and seeds led him to discover ratios like those found by Mendel. He predicted that sex must be inherited in a Mendelian fashion and in 1907 he was able to demonstrate that this was true using experiments on *Bryonia*.

De Vries, Hugo Marie (1848–1935) Dutch botanist who conducted important research on osmosis (the passive diffusion of water from high concentration to low concentration through a semi-permeable membrane) in plant cells and was a pioneer in the study of plant evolution. His work led to the rediscovery of the Austrian biologist Gregor Mendel's laws and the discovery of spontaneously occurring mutations.

Using a plant called *Oenothera lamarckiana*, he began a programme of plant-breeding experiments in 1892. In 1900 he formulated the same laws of heredity that – unknown to De Vries – Mendel had discovered in 1865. De Vries further found that occasionally an entirely new variety of *Oenothera* appeared and that this variety reappeared in subsequent generations; he called these new varieties mutations. He assumed that, in the course of evolution, those mutations that were favourable for the survival of the individual persisted unchanged until other, more favourable mutations occurred. He summarized this work in *Die Mutationstheorie/The Mutation Theory* (1901–03).

Dutrochet, (Rene Joachim) Henri (1776–1847) French physiologist who outlined the process of osmosis (the passive diffusion of water from high concentration to low concentration through a semi-permeable membrane, such as a cell wall) in plants and described various important parts of the plant respiratory mechanism. He was also the first to recognize the role of the pigment chlorophyll in the conversion by plants of carbon dioxide to oxygen (photosynthesis) and to identify stomata (pores) on the surface of leaves, later recognized as important in the exchange of gases between the plant and its surroundings.

Gray, Asa (1810–1888) US botanist and taxonomist. His publications include *Elements of Botany* (1836) and the definitive *Flora of North America* (1838, 1843). He based his revision of the Linnaean system of plant classification on fruit form rather than gross morphology (structure and form). His *Manual of Botany* (1850) remains the standard reference work on flora east of the Rocky Mountains, North America.

Grew, Nehemiah (1641–1712) English botanist and physician who made some of the early microscopical observations of plants. He studied the structure of various plants' anatomy and introduced the term 'parenchyma' to refer to the ground tissue, or unspecialized cells, of a plant. His observations were included in his book *The Anatomy of Plants* (1682).

Hales, Stephen (1677–1761) English scientist who studied the role of water and air in the maintenance of life. He gave accurate accounts of water movement in plants. He demonstrated that plants absorb air, and that some part of that air is involved in their nutrition. His work laid emphasis on measurement and experimentation. His findings were published in *Vegetable Staticks* (1727, enlarged 1733 and retitled *Statical Essays, Containing Haemastaticks, etc.*)

Hedwig, Johannes (1730–1799) Transylvanian-born German botanist whose *Fundamentum historiae naturalis muscorum frondosorum* (1782) led to his establishment as a leading expert on the grouping and early classification of mosses. He was especially interested in the relationship between, and the reproduction of, mosses and liverworts.

Hill, Robert (1899–1991) British biochemist who showed that during photosynthesis, oxygen is produced, and that this derived oxygen comes from water. This process is now known as the Hill reaction. His experiments in 1937 confirmed that the light reactions of photosynthesis (those that require sunlight) occur within the chloroplasts of leaves, as well as elucidating in part the mechanism of the light reactions. To do this, he isolated chloroplasts from leaves and then illuminated them in the presence of an artificial electron-acceptor.

Hofmeister, Wilhelm Friedrich Benedikt (1824–1877) German botanist. He studied plant development and determined how a plant embryo, lying within a seed, is itself formed out of a single fertilized egg (ovule). He also discovered that mosses and ferns display an alternation of generations, in which the plant has two forms, spore-forming and gamete-forming.

Ingenhousz, Jan (1730–1799) Dutch physician and plant physiologist who established in 1779 that in sunlight plants absorb carbon dioxide and give off oxygen. He found that plants, like animals, respire all the time and that respiration occurs in all parts of plants. His *Experiments On Vegetables, Discovering their Great Power of Purifying the Common Air in Sunshine, and of Injuring it in the Shade or at Night* (1779) laid the foundations for the study of photosynthesis.

Jussieu, Antoine Laurent de (1748–1836) French botanist who developed one of the first systems of classification for plants. His study of flowering plants, *Genera plantarum* (1789), became the accepted basis of classification for flowering plants. Building on the foundation laid by ◊Linnaeus, he produced one of the first taxonomies (classifications based on the physical characteristics of plants). Many elements of Jussieu's classification remain in use today.

Linnaeus, Carolus (Latinized form of Carl von Linné) (1707–1778) Swedish naturalist and physician. His botanical work *Systema naturae* (1735) contained his system for classifying plants into groups depending on shared characteristics (such as the number of stamens in flowers), providing a much-needed framework for identification. He also devised the concise and precise system for naming plants and animals, using one Latin (or Latinized) word to represent the genus and a second to distinguish the species. For example, in the Latin name of the daisy, *Bellis perennis*, *Bellis* is the name of the genus to which the plant belongs, and *perennis* distinguishes the species from others of the same genus. The author who first described a particular species is often indicated after the name, for example, *Bellis perennis* Linnaeus, showing that the author was Linnaeus. His system of nomenclature was introduced in *Species plantarum* (1753) and the fifth edition of *Genera plantarum* (1754). In 1758 he applied his binomial system to animal classification.

Linnaeus, Carolus
http://www.ucmp.berkeley.edu/history/linnaeus.html
Profile of the life and legacy of the Swedish 'father of taxonomy'. A biography traces how his childhood interest in plants led to his becoming the greatest botanist of his day.

Mattioli, Pierandrea (1501–1577) Italian physician and botanist. His encyclopedic study of plants *Commentarii a Dioscodie/Commentaries on Dioscorides* (1544) brings together virtually all classical, medieval and Renaissance knowledge of plants and their uses in medicine.

Mitchell, Peter Dennis (1920–1992) English chemist. He received a Nobel prize in 1978 for work on the conservation of energy by plants during respiration and photosynthesis. He showed that the transfer of energy during life processes is not random but directed. It had been believed that the energy absorbed by plants from sunlight was utilized in cells by purely chemical means. The cell was seen as a bag of enzymes in which random and directionless processes took place. Mitchell proved that currents of protons pass through cell walls, which, instead of being simple partitions between cells, are, in fact, full of directional pathways.

Mohl, Hugo von (1805–1872) German botanist who coined the term 'protoplasm' (the living fluid material inside cells), describing it for the first time (although the Czech cell physiologist Johannes Purkinje had already used this term in 1829, he had not applied it to plant cells). Mohl was also the first to describe the cell membrane, nucleus, utricle (bladderlike fruit of certain plants), and the relevance of osmosis with regard to plant cell function. He constructed a prototype of a light microscope, which he used to study and report on the structure of plant cells.

Nageli, Karl Wilhelm von (1817–1891) Swiss botanist and early microscopist. He accurately described cell division, identifying chromosomes as 'transitory cytoblasts'. He was also the first to describe the antheridia (male reproductive organs) and spermatozoids (male gametes) of the fern family. He and Hugo von ◊Mohl were responsible for distinguishing the protoplasm from the cell wall in plants. His own work included the development of the idea of a meristem, a group of formative cells that are always capable of further division.

Pelletier, Pierre-Joseph (1788–1842) French chemist whose extractions of a range of biologically active compounds from plants founded the chemistry of the alkaloids. The most important of his discoveries was quinine, used against malaria. In 1817, together with fellow French chemist Joseph Caventou (1795–1877), he isolated the green pigment in leaves, which they named chlorophyll. In 1818 they turned to plant alkaloids: strychnine in 1818, brucine and veratrine in 1819, and quinine in 1820. Their powerful effects made it possible to specify chemical compounds in pharmacology instead of the imprecise plant extracts and mixtures used previously.

Pfeffer, Wilhelm Friedrich Philipp (1845–1920) German physiological botanist who was the first to measure osmotic pressure, in 1877. He also showed that osmotic pressure varies according to the temperature and concentration of the solute. He also studied respiration, photosynthesis, protein metabolism, and transport in plants. His *Handbuch der Pflanzenphysiologie/Physiology of Plants* (1881) was an important text for many years.

Pringsheim, Nathanael (1823–1894) German botanist who showed that mosses and algae reproduce by sexual union, confirming what Wilhelm (Hofmeister had previously shown to occur in the lower vascular plants such as ferns. Pringsheim also worked on fungi and unsuccessfully on chlorophyll. Like his contemporaries Hugo von ◊Mohl and Hofmeister, he concentrated his study more upon the physiology and dynamics of cell development and life history than upon the traditional classification and collection of specimens.

Ray, John (1627–1705) English naturalist who devised a classification system accounting for some 18,000 plant species. It was the first system to divide flowering plants into monocotyledons and dicotyledons, with additional divisions made on the basis of leaf and flower characters and fruit types. He also established the species as the fundamental unit of classification.

Ray toured Europe 1663–66 with fellow English naturalist Francis Willughby (1635–1672) and they collaborated on a *Catalogus plantarum Angliae/Catalogue of English Plants* (1670). Ray's *Historia generalis plantarum* (1686–1704) contained much information on the morphology, distribution, habitats, and pharmacological uses of individual species as well as general aspects of plant life.

Sachs, Julius von (1832–1897) German botanist and plant physiologist who developed several important experimental techniques and showed that photosynthesis occurs in the chloroplasts (the structure in a plant cell containing the green pigment chlorophyll) and produces oxygen. He was especially gifted in his experimental approach; some of his techniques are still in use today, such as the simple iodine test, which he used to show the existence of starch in a whole leaf.

Saussure, Nicholas Théodore de (1767–1845) Swiss botanist, chemist, and plant physiologist who established the discipline of phytochemistry (the study of the chemistry of plants) and showed that plants gain weight during photosynthesis by converting carbon dioxide to oxygen. Originally, he concluded correctly that this reaction was dependent upon light and incorrectly that carbon and oxygen were the products formed from the carbon dioxide. However, he later realized that more weight was gained than was due to the carbon, and he deduced that water must also be incorporated into the plant's dry weight. He also studied the formation of carbonic acid in plants.

Schimper, Andreas Franz Wilhelm (1856–1901) German botanist and plant geographer who classified the plant life in Africa according to the terrain and climate of the natural habitat, and, in 1880, showed for the first time that starch is an important form of stored energy in plants.

Scott, Dukinfield Henry (1854–1934) English botanist who studied the anatomy of plants and, with William Crawford ◊Williamson, described the evolutionary links between ferns and cycads, research that led to the development of phylogenetic theories of plants. His best known studies were in the field of palaeobotany, including an excellent account of the fruiting bodies of fossil plants in 1904.

Senebier, Jean (1742–1809) Swiss botanist, plant physiologist, and pastor, whose research on photosynthesis showed that 'fixed air' (now known to be carbon dioxide) was converted to 'pure air' (oxygen) in a light-dependent process. He showed that it was the light and not the warmth of sunlight that was necessary for photosynthesis to occur, and that photosynthesis does not occur in boiled water from which the gases have been excluded. His *Action de la lumière sur la végétation* (1779) is an important paper on photosynthesis.

Sprengel, Christian Konrad (1750–1816) German botanist. Writing in 1793, he described the phenomenon of dichogamy, the process whereby stigma and anthers on the same flower ripen at different times and so guarantee cross-fertilization.

Stebbins, George Ledyard (1906–) US botanist and plant geneticist who was the first scientist to apply neo-Darwinism to plants in his *Variation and Evolution in Plants* (1950). With Ernest B Babcock, he developed a technique for doubling the chromosome number of a plant and producing polyploids (plants possessing three or more sets of chromosomes) artificially.

Strasburger, Eduard Adolf (1844–1912) German botanist who discovered that the nucleus of plant cells divides during cell division and clarified the role that chromosomes play in heredity. It had previously been thought that the nucleus disappeared during cell division until Strasburger saw the nucleus of a dividing cell divide. He coined the terms: 'chloroplast' (the structure in a plant cell where photosynthesis occurs), 'cytoplasm' (the part of a cell that is outside the nucleus), 'haploid' (having a single set of chromosomes in each cell), 'diploid' (having two sets of chromosomes in each cell) and 'nucleoplasm' (or karyoplasm, the substance of the nucleus).

Theophrastus (*c.* 372–*c.* 287 BC) Greek philosopher, regarded as the founder of botany. He covered most aspects of botany: descriptions of plants, classification, plant distribution, propagation, germination, and cultivation. He distinguished between two main groups of flowering plants – dicotyledons and monocotyledons in modern terms – and between flowering plants and cone-bearing trees (angiosperms and gymnosperms).

Theophrastus classified plants into trees, shrubs, undershrubs, and herbs. He described and discussed more than 500 species and varieties of plants from lands bordering the

Atlantic and Mediterranean. He noted that some flowers bear petals whereas others do not, and observed the different relative positions of the petals and ovary. In his work on propagation and germination, he described the various ways in which specific plants and trees can grow: from seeds, from roots, from pieces torn off, from a branch or twig, or from a small piece of cleft wood.

Tradescant, John (1570–*c*. 1638) English gardener and botanist who travelled widely in Europe. He was appointed gardener to King Charles I and was succeeded by his son, **John Tradescant the Younger (1608–1662)**, who undertook three plant-collecting trips to Virginia in North America. The Tradescants introduced many new plants to Britain, including the acacia, lilac, and occidental plane. Tradescant senior is generally considered the earliest collector of plants and other natural-history objects. ◊Linnaeus named the genus *Tradescantia* (the spiderworts) after the younger Tradescant.

In 1604 the elder Tradescant became gardener to the Earl of Salisbury, who in 1610 sent him abroad to collect plants. In 1620 he accompanied an official expedition against the North African Barbary pirates and brought back to England gutta-percha and various fruits and seeds. Later, when he became gardener to Charles I, Tradescant set up his own garden and museum in London. In 1624 he published a catalogue of 750 plants grown in his garden.

Van Tiegheim, Phillipe (1839–1914) French botanist and biologist who defined the plant as having three distinct parts – the stem, the root, and the leaf – and studied the origin and differentiation of each type of plant tissue. His best known research included studies of the gross anatomy of the phanerogams (an obsolete term for gymnosperms and angiosperms) and the cryptogams (lower plants, such as mosses and ferns).

Warming, Johannes Eugenius Bülow (1841–1924) Danish botanist whose pioneering studies of the relationships between plants and their natural environments established plant ecology as a new discipline within botany. He investigated the relationships between plants and various environmental conditions, such as light, temperature, and rainfall, and attempted to classify types of plant communities (he defined a plant community as a group of several species that is subject to the same environmental conditions, which he called ecological factors). In *Plantesamfund/Oecology of Plants* (1895) he formulated a programme for future research into the subject.

Williamson, William Crawford (1816–1895) English botanist, surgeon, zoologist, and palaeontologist who was regarded as one of the founders of modern palaeobotany. His research included work on deep-sea deposits, protozoans (single-celled animals), and lower plants such as ferns and mosses. He showed that not all plant fossils containing secondary wood were necessarily seed plants (gymnosperms and angiosperms), but that some were spore-bearing.

Glossary

abscissin, or **abscissic acid**, plant hormone found in all higher plants. It is involved in the process of (abscission and also inhibits stem elongation, germination of seeds, and the sprouting of buds.

abscission the controlled separation of part of a plant from the main plant body – most commonly, the falling of leaves or the dropping of fruit controlled by ◊abscissin.

achene dry, one-seeded ◊fruit that develops from a single ovary and does not split open to disperse the seed. Achenes commonly occur in groups – for example, the fruiting heads of buttercup *Ranunculus* and clematis. The outer surface may be smooth, spiny, ribbed, or tuberculate, depending on the species.

after-ripening process undergone by the seeds of some plants before germination can occur. The length of the after-ripening period in different species may vary from a few weeks to many months.

androecium male part of a flower, comprising a number of ◊stamens.

anemophily type of ◊pollination in which the pollen is carried on the wind. Anemophilous flowers are usually unscented, have either very reduced petals and sepals or lack them altogether, and do not produce nectar.

angiosperm flowering plant in which the seeds are enclosed within an ovary, which ripens into a fruit.

annual plant plant that completes its life cycle within one year, during which time it germinates, grows to maturity, bears flowers, produces seed, and then dies.

anther in a flower, the terminal part of a stamen in which the ◊pollen grains are produced. It is usually borne on a slender stalk or filament, and has two lobes, each containing two chambers, or pollen sacs, within which the pollen is formed.

antheridium organ producing the male gametes, antherozoids, in bryophytes (mosses and liverworts), and lower vascular plants (ferns, club mosses, and horsetails).

archegonium female sex organ found in bryophytes (mosses and liverworts), lower vascular plants (ferns, club mosses, and horsetails), and some gymnosperms.

auxin plant hormone that promotes stem and root growth in plants. Auxins influence many aspects of plant growth and development, including cell enlargement, inhibition of development of axillary buds, ◊tropisms, and the initiation of roots.

axil upper angle between a leaf (or bract) and the stem from which it grows. Organs developing in the axil, such as shoots and buds, are termed axillary, or lateral.

bark protective outer layer on the stems and roots of woody plants, composed mainly of dead cells.

berry fleshy, many-seeded ◊fruit that does not split open to release the seeds. The outer layer of tissue, the exocarp, forms an outer skin that is often brightly coloured to attract birds to eat the fruit and thus disperse the seeds. Examples of berries are the tomato and the grape.

biennial plant plant that completes its life cycle in two years. During the first year it grows vegetatively and the surplus food produced is stored in its ◊perennating organ, usually the root. In the following year these food reserves are used for the production of leaves, flowers, and seeds, after which the plant dies. Many root vegetables are biennials, including the carrot *Daucus carota*.

bract leaflike structure in whose ◊axil a flower or inflorescence develops. Bracts are generally green and smaller than the true leaves. However, in some plants they may be brightly coloured and conspicuous, taking over the role of attracting pollinating insects to the flowers, whose own petals are small; examples include poinsettia *Euphorbia pulcherrima* and bougainvillea.

broad-leaved tree another name for a tree belonging to the ◊angiosperms, such as ash, beech, oak, maple, or birch. The leaves are generally broad and flat, in contrast to the needle-like leaves of most conifers.

bryophyte member of the Bryophyta, a division of the plant kingdom containing the liverworts, mosses, and hornworts.

bud undeveloped shoot usually enclosed by protective scales; inside is a very short stem and numerous undeveloped leaves, or flower parts, or both. Terminal buds are found at the tips of shoots, while axillary buds develop in the ◊axils of the leaves, often remaining dormant unless the terminal bud is removed or damaged.

bulb underground bud with fleshy leaves containing a reserve food supply and with roots growing from its base. Bulbs function in vegetative reproduction and are characteristic of many monocotyledonous plants such as the daffodil, snowdrop, and onion.

bur, or **burr**, type of 'false fruit' or ◊pseudocarp, surrounded by numerous hooks; for instance, that of burdock *Arctium*, where the hooks are formed from bracts surrounding the flowerhead. Burs catch in the feathers or fur of passing animals, and thus may be dispersed over considerable distances.

calyptra in mosses and liverworts, a layer of cells that encloses and protects the young ◊sporophyte (spore-producing generation), forming a sheathlike hood around the capsule. The term is also used to describe the root cap, a layer of ◊parenchyma cells that gives protection to the root tip as it grows through the soil.

calyx collective term for the ◊sepals of a flower, forming the outermost whorl of the ◊perianth.

cambium layer of actively dividing cells (lateral ◊meristem), found within stems and roots, that gives rise to ◊secondary growth in perennial plants, causing an increase in girth.

capsule dry, usually many-seeded fruit formed from an ovary composed of two or more fused ◊carpels, which splits open to release the seeds. The same term is used for the spore-containing structure of mosses and liverworts; this is borne at the top of a long stalk or seta.

carpel female reproductive unit in flowering plants (◊angiosperms).

caryopsis dry, one-seeded ◊fruit in which the wall of the seed becomes fused to the ◊carpel wall during its development. Caryopses are typical of members of the grass family (Gramineae), including the cereals.

cellulose complex carbohydrate composed of long chains of glucose units, joined by chemical bonds called glycosidic links. It is the principal constituent of the plant cell wall.

cell wall the tough outer surface of the plant cell, constructed from a mesh of ◊cellulose.

chemotropism movement by part of a plant in response to a chemical stimulus. The response by the plant is termed 'positive' if the growth is towards the stimulus or 'negative' if the growth is away from the stimulus.

chlorophyll green pigment present in plants; it is responsible for the absorption of light energy during ◊photosynthesis.

chloroplast structure (organelle) within a plant cell containing the green pigment chlorophyll.

cladode flattened stem that is leaflike in appearance and function. It is an adaptation to dry conditions because a stem contains fewer ◊stomata than a leaf, and water loss is thus minimized. Examples of plants with cladodes are butcher's-broom *Ruscus aculeatus* and certain cacti.

coleoptile protective sheath that surrounds the young shoot tip of a grass during its passage through the soil to the surface. Although of relatively simple structure, most coleoptiles are very sensitive to light, ensuring that seedlings grow upwards.

collenchyma plant tissue composed of relatively elongated cells with thickened cell walls, in particular at the corners where adjacent cells meet.

compensation point point at which there is just enough light for a plant to survive. At this point all the food produced by ◊photosynthesis is used up by respiration.

cone, or **strobilus**, reproductive structure of gymnosperms (principally conifers and cycads). It consists of a central axis surrounded by numerous overlapping, scalelike, modified leaves (sporophylls) that bear the reproductive organs. Usually there are separate male and female cones, the former bearing pollen sacs containing pollen grains, and the larger female cones bearing the ovules that contain the ova or egg cells.

contractile root thickened root at the base of a corm, bulb, or other organ that helps position it at an appropriate level in the ground. After they have become anchored in the soil, the upper portion contracts, pulling the plant deeper into the ground.

cork light, waterproof outer layers of the bark covering the branches and roots of almost all trees and shrubs.

corm short, swollen, underground plant stem, surrounded by protective scale leaves, as seen in the genus *Crocus*. It stores food, provides a means of ◊vegetative reproduction, and acts as a ◊perennating organ.

corolla collective name for the petals of a flower.

cotyledon structure in the embryo of a seed plant that may form a 'leaf' after germination and is commonly known as a seed leaf. The number of cotyledons present in an embryo is an important character in the classification of flowering plants (◊angiosperms).

cytokinin plant hormone that stimulates cell division. Cytokinins affect several different aspects of plant growth and development, but only if ◊auxin is also present. They may delay the process of senescence, or ageing, break the dormancy of certain seeds and buds, and induce flowering.

deciduous term describing trees and shrubs that shed their leaves at the end of the growing season or during a dry season to reduce transpiration (the loss of water by evaporation).

dicotyledon major subdivision of the ◊angiosperms, containing the great majority of flowering plants. Dicotyledons are characterized by the presence of two seed leaves, or ◊cotyledons, in the embryo.

dioecious term describing plants with male and female flowers borne on separate individuals of the same species. It is a way of avoiding self-fertilization.

dormancy phase of reduced physiological activity exhibited by certain buds, seeds, and spores. Dormancy can help a plant to survive unfavourable conditions, as in annual plants that pass the cold winter season as dormant seeds, and plants that form dormant buds.

drupe fleshy ◊fruit containing one or more seeds which are surrounded by a hard, protective layer – for example cherry, almond, and plum. The wall of the fruit (◊pericarp) is differentiated into the outer skin (exocarp), the fleshy layer of tissues (mesocarp), and the hard layer surrounding the seed (endocarp).

embryo sac large cell within the ovule of flowering plants that represents the female ◊gametophyte when fully developed.

endosperm nutritive tissue in the seeds of most flowering plants. It surrounds the embryo and is produced by an unusual process that parallels the fertilization of the ovum by a male gamete.

epigeal term describing seed germination in which the ◊cotyledons (seed leaves) are borne above the soil.

epiphyte any plant that grows on another plant or object above the surface of the ground, and has no roots in the soil. An epiphyte does not parasitize the plant it grows on but merely uses it for support.

etiolation form of growth seen in plants receiving insufficient light. It is characterized by long, weak stems, small leaves, and a pale yellowish colour (chlorosis) owing to a lack of chlorophyll. The rapid increase in height enables a plant that is surrounded by others to quickly reach a source of light.

evergreen plant such as pine, spruce, or holly, that bears its leaves all year round. Most conifers are evergreen. Plants that shed their leaves in autumn or during a dry season are described as ◊deciduous.

fern any of a division of plants related to horsetails and clubmosses. Ferns are spore-bearing, not flowering, plants and most are perennial, spreading by slow-growing roots. The leaves, known as fronds, vary widely in size and shape. (Division Filicinophyta.)

flaccidity the loss of rigidity (turgor) in plant cells, caused by loss of water from the central vacuole so that the cytoplasm no longer pushes against the cellulose cell wall. If this condition occurs throughout the plant then wilting is seen.

floret small flower, usually making up part of a larger, composite flower head. There are often two different types present on one flower head: disc florets in the central area, and ray florets around the edge which usually have a single petal known as the ligule. In the common daisy, for example, the disc florets are yellow, while the ligules are white.

flower the reproductive unit of an angiosperm or flowering plant, typically consisting of four whorls of modified leaves: ◊sepals, ◊petals, ◊stamens, and ◊carpels.

flowering plant term generally used for ◊angiosperms, which bear flowers with various parts, including sepals, petals, stamens, and carpels.

follicle dry, usually many-seeded fruit that splits along one side only to release the seeds within. It is derived from a single ◊carpel. Examples include the fruits of the larkspurs *Delphinium* and columbine *Aquilegia*. It differs from a pod, which always splits open (dehisces) along both sides.

forest area where trees have grown naturally for centuries, instead of being logged at maturity (about 150–200 years). A natural, or old-growth, forest has a multistorey canopy and includes young and very old trees (this gives the canopy its range of heights).

frond large leaf or leaflike structure; in ferns it is often pinnately divided. The term is also applied to the leaves of palms and less commonly to the plant bodies of certain seaweeds, liverworts, and lichens.

fruit ripened ovary in flowering plants that develops from one or more seeds or carpels and encloses one or more seeds.

gametophyte the haploid generation in the life cycle of a plant that produces gametes.

germination initial stages of growth in a seed, spore, or pollen grain.

gibberellin plant growth substance (see also ◊auxin) that promotes stem growth and may also affect the breaking of dormancy in certain buds and seeds, and the induction of flowering.

guard cell specialized cell on the undersurface of leaves for controlling gas exchange and water loss through pores called ◊stomata.

gymnosperm any plant whose seeds are exposed, as opposed to the structurally more advanced ◊angiosperms, where they are inside an ovary. The group includes conifers and related plants such as cycads and ginkgos, whose seeds develop in ◊cones.

gynoecium, or **gynaecium**, collective term for the female reproductive organs of a flower, consisting of one or more ◊carpels, either free or fused together.

halophyte plant adapted to live where there is a high concentration of salt in the soil, for example, in salt marshes and mud flats.

herb any plant (usually a flowering plant) tasting sweet, bitter, aromatic, or pungent, used in cooking, medicine, or perfumery; technically, a herb is any plant in which the aerial parts do not remain above ground at the end of the growing season.

herbaceous plant plant with very little or no wood, dying back at the end of every summer. The herbaceous perennials survive winters as underground storage organs such as bulbs and tubers.

heterostyly having ◊styles of different lengths. Certain flowers, such as primroses *Primula vulgaris*, have different-sized ◊anthers and styles to ensure cross-fertilization (through ◊pollination) by visiting insects.

honey guide line or spot on the petals of a flower that indicate to pollinating insects the position of the nectaries (see ◊nectar) within the flower. Sometimes the markings reflect only ultraviolet light, which can be seen by many insects although it is not visible to the human eye.

hornwort nonvascular plant (with no 'veins' to carry water and food), related to the ◊liverworts and ◊mosses. Hornworts are found in warm climates, growing on moist shaded soil. (Division Bryophyta.)

hydrophily type of ◊pollination where the pollen is transported by water. Water-pollinated plants occur in 31 genera in 11 different families. They are found in habitats as diverse as rainforests and seasonal desert pools. Pollen is either dispersed underwater or on the water's surface.

hydrophyte plant adapted to live in water, or in waterlogged soil.

hypogeal term describing seed germination in which the ◊cotyledons remain below ground. It can refer to fruits that develop underground, such as peanuts *Arachis hypogea*.

inflorescence branch, or system of branches, bearing two or more individual flowers.

integument in seed-producing plants, the protective coat surrounding the ovule. In angiosperms (flowering plants) there are two, in gymnosperms only one. A small hole at one end, the micropyle, allows a pollen tube to penetrate through to the egg during fertilization.

lamina in flowering plants (angiosperms), the blade of the leaf on either side of the midrib.

leaf lateral outgrowth on the stem of a plant, and in most species the primary organ of ◊photosynthesis.

lenticel small pore on the stems of woody plants or the trunks of trees. Lenticels are a means of gas exchange between the stem interior and the atmosphere.

lignin naturally occurring substance produced by plants to strengthen their tissues. It is difficult for enzymes to attack lignin, so living organisms cannot digest wood, with the exception of a few specialized fungi and bacteria. Lignin is the essential ingredient of all wood and is, therefore, of great commercial importance.

liverwort nonvascular plant (with no 'veins' to carry water and food), related to ◊hornworts and mosses; it is found growing in damp places. (Division Bryophyta.)

meristem region of plant tissue containing cells that are actively dividing to produce new tissues (or have the potential to do so).

mesophyll the tissue between the upper and lower epidermis of a leaf blade (lamina), consisting of parenchyma-like cells containing numerous ◊chloroplasts.

micropyle in flowering plants, a small hole towards one end of the ovule. At pollination the pollen tube growing down from the ◊stigma eventually passes through this pore.

monocotyledon angiosperm (flowering plant) having an embryo with a single cotyledon, or seed leaf (as opposed to ◊dicotyledons, which have two).

monoecious having separate male and female flowers on the same plant. Monoecism is a way of avoiding self-fertilization.

moss small nonflowering plant of the class Musci, forming with the liverworts and the hornworts the division Bryophyta.

mycorrhiza mutually beneficial (mutualistic) association occurring between plant roots and a soil fungus. Mycorrhizal roots take up nutrients more efficiently than non-mycorrhizal roots, and the fungus benefits by obtaining carbohydrates from the plant or tree.

nastic movement plant movement that is caused by an external stimulus, such as light or temperature, but is directionally independent of its source, unlike ◊tropisms. Nastic movements occur as a result of changes in water pressure within specialized cells or differing rates of growth in parts of the plant.

nectar sugary liquid secreted by some plants from a nectary, a specialized gland usually situated near the base of the flower. Nectar attracts insects, birds, bats, and other animals to the flower for ◊pollination and is the raw material used by bees in the production of honey.

nitrogen fixation process by which nitrogen in the atmosphere is converted into nitrogenous compounds by the action of microorganisms, such as cyanobacteria (blue-green algae) and bacteria, in conjunction with certain legumes (members of the pea family).

nut any dry, single-seeded fruit that does not split open to release the seed, such as the chestnut. A nut is formed from more than one carpel, but only one seed becomes fully formed, the remainder aborting. The wall of the fruit, the pericarp, becomes hard and woody, forming the outer shell.

nutation spiral movement exhibited by the tips of certain stems during growth; it enables a climbing plant to find a suitable support. Nutation sometimes also occurs in tendrils and flower stalks.

ornithophily ◊pollination of flowers by birds.

ovary expanded basal portion of the carpel of flowering plants, containing one or more ovules. It is hollow with a thick wall to protect the ovules. Following fertilization of the ovum, it develops into the fruit wall or pericarp.

ovule structure found in seed plants that develops into a seed after fertilization. It consists of an ◊embryo sac containing the female gamete (ovum or egg cell), surrounded by nutritive tissue, the nucellus.

palisade cell cylindrical cell lying immediately beneath the upper epidermis of a leaf. Palisade cells normally exist as one closely packed row and contain many ◊chloroplasts.

pappus (plural **pappi**) modified ◊calyx comprising a ring of fine, silky hairs, or sometimes scales or small teeth, that persists after fertilization. Pappi are found in members of the daisy family (Compositae) such as the dandelions *Taraxacum*, where they form a parachutelike structure that aids dispersal of the fruit.

parenchyma plant tissue composed of loosely packed, more or less spherical cells, with thin cellulose walls. Although parenchyma often has no specialized function, it is usually present in large amounts, forming a packing or ground tissue.

parthenocarpy formation of fruits without seeds. This phenomenon, of no obvious benefit to the plant, occurs naturally in some plants, such as bananas.

pedicel stalk of an individual flower, which attaches it to the main floral axis, often developing in the axil of a bract.

perennating organ that part of a biennial plant or herbaceous perennial that allows it to survive the winter; usually a root, tuber, rhizome, bulb, or corm.

perennial plant plant that lives for more than two years. Herbaceous perennials have aerial stems and leaves that die each autumn. They survive the winter by means of an underground storage (perennating) organ, such as a bulb or rhizome. Trees and shrubs or woody perennials have stems that persist above ground throughout the year, and may be either ◊deciduous or ◊evergreen.

perianth collective term for the outer whorls of the flower, which protect the reproductive parts during development.

pericarp wall of a ◊fruit. It encloses the seeds and is derived from the ovary wall. In fruits such as the acorn, the pericarp becomes dry and hard, forming a shell around the seed. In fleshy fruits the pericarp is typically made up of three distinct layers. The **epicarp**, or **exocarp**, forms the tough outer skin of the fruit, while the **mesocarp** is often fleshy and forms the middle layers. The innermost layer, or **endocarp**, which surrounds the seeds, may be membranous or thick and hard, as in the drupe (stone) of cherries, plums, and apricots.

petal part of a flower whose function is to attract pollinators such as insects or birds.

petiole stalk attaching the leaf blade, or ◊lamina, to the stem.

phloem tissue found in vascular plants whose main function is to conduct sugars and other food materials from the leaves, where they are produced, to all other parts of the plant.

photosynthesis process by which green plants trap light energy from the Sun. This energy is used to drive a series of chemical reactions which lead to the formation of carbohydrates.

phototropism movement of part of a plant towards or away from a source of light. Leaves are positively phototropic, detecting the source of light and orientating themselves to receive the maximum amount.

pinna the primary division of a ◊pinnate leaf.

pinnate leaf leaf that is divided up into many small leaflets, arranged in rows along either side of a midrib, as in ash trees *Fraxinus*. It is a type of compound leaf. Each leaflet is known as a **pinna**, and where the pinnae are themselves divided, the secondary divisions are known as pinnules.

pistil general term for the female part of a flower, either referring to one single ◊carpel or a group of several fused carpels.

plastid general name for a cell organelle of plants that is enclosed by a double membrane and contains a series of internal membranes and vesicles. Plastids contain DNA and are produced by division of existing plastids. They can be classified into two main groups: the **chromoplasts** (such as chloroplasts), which contain pigments such as carotenes and chlorophyll, and the **leucoplasts**, which are colourless.

plumule part of a seed embryo that develops into the shoot, bearing the first true leaves of the plant.

pneumatophore erect root that rises up above the soil or water and promotes gas exchange.

pod type of fruit that is characteristic of legumes (plants belonging to the Leguminosae family), such as peas and beans. It develops from a single ◊carpel and splits down both sides when ripe to release the seeds.

pollen in angiosperms (flowering plants) and gymnosperms, the grains that contain the male gametes. In angiosperms, pollen is produced within ◊anthers; in most gymnosperms, it is produced in male cones. A pollen grain is typically yellow and, when mature, has a hard outer wall. Pollen of insect-pollinated plants is often sticky and spiny and larger than the smooth, light grains produced by wind-pollinated species.

pollen tube outgrowth from a pollen grain that grows towards the ◊ovule following germination. In ◊angiosperms (flowering plants) the pollen tube reaches the ovule by growing down through the ◊style, carrying the male gametes inside. The gametes are discharged into the ovule and one fertilizes the egg cell.

pollination process by which pollen is transferred from the male reproductive organ to the female reproductive organ of the plant.

pome type of ◊pseudocarp, or 'false fruit', typical of certain plants belonging to the Rosaceae family. The outer skin and fleshy tissues are developed from the ◊receptacle (the enlarged end of the flower stalk) after fertilization, and the five ◊carpels (the true fruit) form the pome's core, which surrounds the seeds. Examples of pomes are apples, pears, and quinces.

prop root, or **stilt root**, modified root that grows from the lower part of a stem or trunk down to the ground, providing a plant with extra support. Prop roots are common on some woody plants, such as mangroves, and also occur on a few herbaceous plants, such as maize.

protandry in a flower, the state where the male reproductive organs reach maturity before those of the female. This is a common method of avoiding self-fertilization. See also ◊protogyny.

prothallus short-lived ◊gametophyte of many ferns and other lower vascular plants (such as horsetails or club mosses). It bears either the male or female sex organs, or both. Typically it is a small, green, flattened structure that is anchored in the soil by several ◊rhizoids (slender, hairlike structures, acting as roots) and needs damp conditions to survive. The reproductive organs are borne on the lower surface close to the soil.

protogyny in a flower, the state where the female reproductive organs reach maturity before those of the male. Like protandry, in which the male organs reach maturity first, this is a method of avoiding self-fertilization, but it is much less common.

pseudocarp fruitlike structure that incorporates tissue that is not derived from the ovary wall. The additional tissues may be derived from floral parts such as the ◊receptacle and ◊calyx. For example, the coloured, fleshy part of a strawberry develops from the receptacle and the true fruits are small ◊achenes – the 'pips' embedded in its outer surface.

radicle part of a plant embryo that develops into the primary root.

receptacle enlarged end of a flower stalk to which the floral parts are attached. Normally the receptacle is rounded, but in some plants it is flattened or cup-shaped.

rhizoid hairlike outgrowth found on the ◊gametophyte generation of ferns, mosses, and liverworts. Rhizoids anchor the plant to the substrate and can absorb water and nutrients. Rhizoids fulfil the same functions as the roots of higher plants but are simpler in construction.

rhizome, or **rootstock**, horizontal underground plant stem. It is a ◊perennating organ in some species, where it is generally thick and fleshy, while in other species it is mainly a means of ◊vegetative reproduction, and is therefore long and slender, with buds all along it that send up new plants. The potato is a rhizome that has two distinct parts, the tuber being the swollen end of a long, cordlike rhizome.

root part of a plant that is usually underground, and whose primary functions are anchorage and the absorption of water and dissolved mineral salts.

root hair tiny hairlike outgrowth on the surface cells of plant roots that greatly increases the area available for the absorption of water and other materials.

rootstock another name for ◊rhizome, an underground plant organ.

runner, or **stolon**, aerial stem that produces new plants.

sap fluids that circulate through ◊vascular plants, especially woody ones. Sap carries water and food to plant tissues. Sap contains alkaloids, protein, and starch; it can be milky (as in rubber trees), resinous (as in pines), or syrupy (as in maples).

schizocarp dry ◊fruit that develops from two or more carpels and splits, when mature, to form separate one-seeded units known as mericarps.

sclerenchyma plant tissue whose function is to strengthen and support, composed of thick-walled cells that are heavily lignified (toughened).

secondary growth, or **secondary thickening**, increase in diameter of the roots and stems of certain plants (notably shrubs and trees) that results from the production of new cells by the ◊cambium. It provides the plant with additional mechanical support and new conducting cells, the secondary ◊xylem and ◊phloem. Secondary growth is generally confined to ◊gymnosperms and, among the ◊angiosperms, to the dicotyledons. With just a few exceptions, the monocotyledons (grasses, lilies) exhibit only primary growth, resulting from cell division at the apical ◊meristems.

seed reproductive structure of higher plants (◊angiosperms and ◊gymnosperms). It develops from a fertilized ovule and consists of an embryo and a food store, surrounded and protected by an outer seed coat, called the testa.

sepal part of a flower, usually green, that surrounds and protects the flower in bud.

sessile term describing a leaf, flower, or fruit that lacks a stalk and sits directly on the stem, as with the sessile acorns of certain oaks.

shoot part of a ◊vascular plant growing above ground, comprising a stem bearing leaves, buds, and flowers.

shrub perennial woody plant that typically produces several separate stems, at or near ground level, rather than the single trunk of most trees. A shrub is usually smaller than a tree, but there is no clear distinction between large shrubs and small trees.

sorus in ferns, a group of sporangia, the reproductive structures that produce ◊spores. They occur on the lower surface of fern fronds.

spadix ◊inflorescence consisting of a long, fleshy axis bearing many small, stalkless flowers. It is partially enclosed by a large bract or ◊spathe. A spadix is characteristic of plants belonging to the family Araceae, including the arum lily *Zantedeschia aethiopica*.

spathe in flowers, the single large bract surrounding the type of inflorescence known as a ◊spadix. It is sometimes brightly coloured and petal-like, as in the brilliant scarlet spathe of the flamingo plant *Anthurium andreanum* from South America; this serves to attract insects.

spikelet one of the units of a grass ◊inflorescence. It comprises a slender axis on which one or more flowers are borne.

sporangium structure in which ◊spores are produced.

spore small reproductive or resting body, usually consisting of just one cell. Unlike a gamete, it does not need to fuse with another cell in order to develop into a new organism. Plant spores are haploid (they have a single set of chromosomes) and are produced by the (sporophyte generation, following meiosis (cell division, in which the number of chromosomes in the cell is halved).

sporophyte diploid (having paired chromosomes in each cell) spore-producing generation in the life cycle of a plant that undergoes alternation of generations.

stamen male reproductive organ of a flower.

stem main supporting axis of a plant that bears the leaves, buds, and reproductive structures.

stigma in a flower, the surface at the tip of a ◊carpel that receives the ◊pollen. It often has short outgrowths, flaps, or hairs to trap pollen and may produce a sticky secretion to which the grains adhere.

stipule outgrowth arising from the base of a leaf or leaf stalk in certain plants. Stipules usually occur in pairs or fused into a single semicircular structure.

stolon type of ◊runner.

stoma (plural **stomata**) pore in the epidermis of a plant. Each stoma is surrounded by a pair of guard cells that are crescent-shaped when the stoma is open but can collapse to an oval shape, thus closing off the opening between them. Stomata allow the exchange of carbon dioxide and oxygen (needed for ◊photosynthesis and respiration) between the internal tissues of the plant and the outside atmosphere. They are also the main route by which water is lost from the plant, and they can be closed to conserve water, the movements being controlled by changes in turgidity of the guard cells.

style in flowers, the part of the ◊carpel bearing the ◊stigma at its tip. In some flowers it is very short or completely lacking, while in others it may be long and slender, positioning the stigma in the most effective place to receive the pollen.

succulent plant thick, fleshy plant that stores water in its tissues; for example, cacti and stonecrops *Sedum*. Succulents live either in areas where water is very scarce, such as deserts, or in places where it is not easily obtainable because of the high concentrations of salts in the soil, as in salt marshes.

suckering reproduction by new shoots (suckers) arising from an existing root system rather than from seed. Plants that produce suckers include elm, dandelion, and members of the rose family.

taproot single, robust, main root that is derived from the embryonic root, or radicle, and grows vertically downwards, often to considerable depth. Taproots are often modified for food storage and are common in biennial plants such as the carrot *Daucus carota*, where they act as ◊perennating organs.

tendril slender, threadlike structure that supports a climbing plant by coiling around suitable supports, such as the stems and branches of other plants. It may be a modified stem, leaf, leaflet, flower, leaf stalk, or stipule (a small appendage on either side of the leaf stalk), and may be simple or branched.

testa outer coat of a seed, formed after fertilization of the ovule. It has a protective function and is usually hard and dry. In some cases the coat is adapted to aid dispersal, for example by being hairy.

thallus any plant body that is not divided into true leaves, stems, and roots. It is often thin and flattened, as in the body of a liverwort, and the gametophyte generation (◊prothallus) of a fern.

tracheid cell found in the water-conducting tissue (◊xylem) of many plants. It is long and thin with pointed ends. The cell walls are thickened by ◊lignin, except for numerous small rounded areas, or pits, through which water and dissolved minerals pass from one cell to another. Once mature, the cell itself dies and only its walls remain.

transpiration loss of water from a plant by evaporation.

tree perennial plant with a woody stem, usually a single stem (trunk), made up of ◊wood and protected by an outer layer of ◊bark.

tropism, or **tropic movement,** directional growth of a plant, or part of a plant, in response to an external stimulus such as gravity or light. If the movement is directed towards the stimulus it is described as positive; if away from it, it is negative. **Geotropism** for example, the response of plants to gravity, causes the root (positively geotropic) to grow downwards, and the stem (negatively geotropic) to grow upwards.

tuber swollen region of an underground stem or root, usually modified for storing food. The potato is a **stem tuber**, as shown by the presence of terminal and lateral buds, the 'eyes' of the potato. **Root tubers**, for example dahlias, developed from adventitious roots (growing from the stem, not from other roots) lack these. Both types of tuber can give rise to new individuals and so provide a means of ◊vegetative reproduction.

turgor rigid condition of a plant caused by the fluid contents of a plant cell exerting a mechanical pressure against the cell wall. Turgor supports plants that do not have woody stems. Plants lacking in turgor visibly wilt. The process of osmosis plays an important part in maintaining the turgidity of plant cells.

vascular bundle strand of primary conducting tissue (a 'vein') in vascular plants, consisting mainly of water-conducting tissues, ◊xylem, and nutrient-conducting tissue, ◊phloem.

vascular plant plant containing vascular bundles. Ferns, horsetails, club mosses, ◊gymnosperms (conifers and cycads), and ◊angiosperms (flowering plants) are all vascular plants.

vegetative reproduction type of asexual reproduction in plants that relies not on spores, but on multicellular structures formed by the parent plant. Some of the main types are ◊stolons and runners, sucker shoots produced from roots (such as in the creeping thistle *Cirsium arvense*), ◊tubers, ◊bulbs, ◊corms, and ◊rhizomes. Vegetative reproduction has long been exploited in horticulture and agriculture, with various methods employed to multiply stocks of plants.

wall pressure in plants, the mechanical pressure exerted by the cell contents against the cell wall. The rigidity (turgor) of a plant often depends on the level of wall pressure found in the cells of the stem. Wall pressure falls if the plant cell loses water.

wilting loss of rigidity (◊turgor) in plants, caused by a decreasing wall pressure within the cells making up the supportive tissues. Wilting is most obvious in plants that have little or no wood.

wood hard tissue beneath the bark of many perennial plants; it is composed of water-conducting cells, or secondary

◊xylem, and gains its hardness and strength from deposits of ◊lignin.

xerophyte plant adapted to live in dry conditions. Common adaptations to reduce the rate of ◊transpiration include a reduction of leaf size, sometimes to spines or scales; a dense covering of hairs over the leaf to trap a layer of moist air (as in edelweiss); water storage cells; sunken ◊stomata; and permanently rolled leaves or leaves that roll up in dry weather (as in marram grass). Many desert cacti are xerophytes.

xylem tissue found in ◊vascular plants, whose main function is to conduct water and dissolved mineral nutrients from the roots to other parts of the plant.

Further Reading

Attenborough, David *The Private Life of Plants* (1995)

Bell, Adrian *Plant Form. An Illustrated Guide to Flowering Plant Morphology* (1991)

Blunt, W *The Complete Naturalist: A Life of Linnaeus* (1971)

Camus, Josephine; Jeremy, Clive; and Thomas, Barry *A World of Ferns* (1991)

Corner, E J H *The Life of Plants* (1981)

Gourlie, N *The Prince of Botanists: Carl Linnaeus* (1953)

Hall, David Oakley, and Rao, K K *Photosynthesis* (1994)

Hecht, Susanna, and Cockburn, Alexander *The Fate of the Forest* (1989)

Heiser, Charles *Seed to Civilization* (1981)

Joyce, Christopher *Earthly Goods* (1994)

Langmead, Clive *A Passion for Plants* (1995)

Lawlor, David W *Photosynthesis: Molecular, Physiological, and Environmental Processes* (1993)

Lewington, Anna *Plants and People* (1990)

Lewis, Walter, and Elvin-Lewis, Memory *Medical Botany. Plants Affecting Man's Health* (1977)

Mabberly, David *The Plant Book* (1990)

Mabey, Richard *The New Age Herbalist* (1988)

Page, C N *The Ferns of Britain and Ireland* (1998)

Phillips, Roger, and Rix, Martin *Vegetables* (1995)

Prance, Ghillean, and Prance, Anne *Bark: The Formation, Characteristics and Uses of Bark Around the World* (1993)

Proctor, Michael, and Yeo, Peter *The Pollination of Plants* (1973)

Raven, Peter; Evert, Ray; and Eichorn, Susan *Biology of Plants* (1992)

Rudall, Paula *Anatomy of Flowering Plants: An Introduction to Structure and Development* (1987)

Slack, Adrian *Carnivorous Plants* (1979)

Stuessy, Tod *Plant Taxonomy* (1990)

Walker, David *Energy, Plants, and Man* (1992)

Weinstock, J *Contemporary Perspectives on Linnaeus* (1985)

Zomlefer, Wendy *Guide to Flowering Plant Families* (1994)

HUMAN BODY

The physical structure of the human being with all its complexities may, for convenience, be considered as a number of systems, such as the circulatory system and the digestive system, all of which are interdependent and interact to maintain life at both a cellular level and a bodily level.

Human Body: Composition

chemical element or substance	body weight (%)	chemical element or substance	body weight (%)
pure elements		magnesium, iron, manganese, copper, iodine, cobalt, zinc	traces
oxygen	65	**water and solid matter**	
carbon	18	water	60–80
hydrogen	10	total solid material	20–40
nitrogen	3	**organic molecules**	
calcium	2	protein	15–20
phosphorus	1.1	lipid	3–20
potassium	0.35	carbohydrate	1–15
sulphur	0.25	small organic particles	0–1
sodium	0.15		
chlorine	0.15		

skeletal system

The skeleton is the rigid framework of bone and cartilage that supports and gives form to the body, protects its internal organs, and provides anchorage points for its muscles. With its flexible (joints acting as fulcrums, the skeleton also forms a system of levers upon which muscles can act to produce movement. The human skeleton is composed of 206 bones, ranging in size from the femur (thigh bone) to the tiny ossicles of the middle ear.

It may be considered in two parts: the axial skeleton and the appendicular skeleton. The **axial skeleton** comprises 80 bones, and consists of the skull, vertebral column (spine), rib cage, and sternum. It supplies the central structure onto which the bones of the appendicular skeleton are joined, and encloses and protects the (central nervous system (brain and spinal cord). The **appendicular skeleton** consists of the limb bones and the pectoral (shoulder) and pelvic (hip) girdles. It comprises 126 bones – 64 in the shoulders and upper limbs, and 62 in the pelvis and lower limbs.

Virtual Body
http://www.medtropolis.com/vbody/
If ever something was worth taking the time to download the 'shockwave' plug-in for, this is it. Authoritative and interactive anatomical animations, complete with voice-overs, guide you round the whole body, with sections on the brain, digestive system, heart, and skeleton.

Human Anatomy Online
http://www.innerbody.com/indexbody.html
Fun, interactive, and educational site on the human body. The site is divided into many informative sections, including hundreds of images, and animations for Java compatible browsers.

special bones
skull The skull is a collection of flat and irregularly shaped bones that enclose the brain and the organs of sight, hearing, and smell, and provide support for the

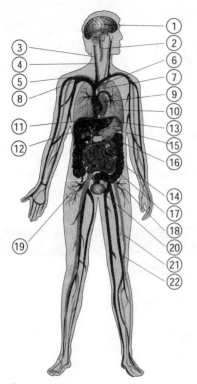

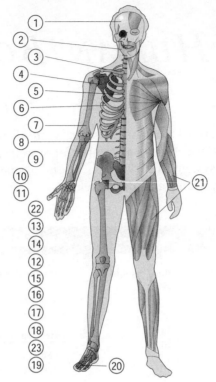

human body The adult human body has approximately 650 muscles, 100 joints, 100,000 km/60,000 mi of blood vessels and 13,000 nerve cells. There are 206 bones in the adult body, nearly half of them in the hands and feet.

Key

1. brain	12. liver
2. spinal cord	13. stomach
3. carotid artery	14. gall bladder
4. jugular vein	15. kidney
5. subclavian artery	16. pancreas
6. superior vena cava	17. small intestine or ileum
7. aorta	18. large intestine or colon
8. subclavian vein	19. appendix
9. heart	20. bladder
10. lungs	21. popliteal artery
11. diaphragm	22. popliteal vein

Key

1. cranium (skull)	13. metacarpals
2. mandible	14. phalanges
3. clavicle	15. femur
4. scapula	16. patella
5. sternum	17. fibula
6. rib cage	18. tibia
7. humerus	19. metatarsals
8. vertebra	20. phalanges
9. ulna	21. superficial (upper)
10. radius	layer of muscles
11. pelvis	22. carpals
12. sacrum	23. tarsals

---

skull

To remember the bones of the skull:

Old people from Texas eat epiders

(**o**ccipital, **p**arietal, **f**rontal, **t**emporal, **e**phnoid, **s**phenoid)

---

jaws. It consists of 22 bones joined by fibrous immobile joints called **sutures**. The floor of the skull is pierced by a large hole (**foramen magnum**) for the spinal cord and a number of smaller apertures through which other nerves and blood vessels pass.

The skull comprises the dome-shaped **cranium** (brain case) and the bones of the face, which include the upper jaw, enclose the sinuses, and form the framework for the nose, eyes, and the roof of the mouth cavity. The lower jaw is hinged to the middle of the skull at its lower edge. The plate at the back of the head is jointed at its lower edge with the upper section of the spine. Inside, the skull has various shallow cavities into which fit different parts of the brain.

Visible Human Project
http://www.nlm.nih.gov/research/visible/visible_gallery.html
Sample images from a long-term US project to collect a complete set of anatomically detailed, three-dimensional representations of the human body.

Bones of the Human Body

Bone	Number
Cranium (Skull)	
Occipital	1
Parietal: 1 pair	2
Sphenoid	1
Ethmoid	1
Inferior nasal conchae	2
Frontal: 1 pair, fused	1
Nasal: 1 pair	2
Lacrimal: 1 pair	2
Temporal: 1 pair	2
Maxilla: 1 pair, fused	1
Zygomatic: 1 pair	2
Vomer	1
Palatine: 1 pair	2
Mandible (jawbone): 1 pair, fused	1
Total	22
Ears	
Malleus (hammer)	2
Incus (anvil)	2
Stapes (stirrups)	2
Total	6
Vertebral Column (Spine)	
Cervical vertebrae	7
Thoracic vertebrae	1
Lumbar vertebrae	5
Sacral vertebrae: 5, fused to form the sacrum	1
Coccygeal vertebrae: between 3 and 5, fused to form the coccyx	1
Total	26
Ribs	
Ribs, 'true': 7 pairs	14
Ribs, 'false': 5 pairs, of which 2 pairs are floating	10
Total	24
Sternum (Breastbone)	
Manubrium	1
Sternebrae	1
Xiphisternum	1
Total	3
Throat	
Hyoid	1
Total	1

Bone		Number
Pectoral Girdle		
Clavicle: 1 pair (collar-bone)		2
Scapula (including coracoid): 1 pair (shoulder blade)		2
Total		4
Upper Extremity (Each Arm)		
Forearm	Humerus	1
	Radius	1
	Ulna	1
Carpus (wrist)	Scaphoid	1
	Lunate	1
	Triquetral	1
	Pisiform	1
	Trapezium	1
	Trapezoid	1
	Capitate	1
	Hamate	1
	Metacarpals	5
Phalanges (fingers)	First digit	2
	Second digit	3
	Third digit	3
	Fourth digit	3
	Fifth digit	3
Total		30
Pelvic Girdle		
Ilium, ischium, and pubis (combined): 1 pair of hip bones, innominate		2
Lower Extremity (Each Leg)		
Leg	Femur (thighbone)	1
	Tibia (shinbone)	1
	Fibula	1
	Patella (kneecap)	1
Tarsus (ankle)	Talus	1
	Calcaneus	1
	Navicular	1
	Cuneiform, medial	1
	Cuneiform, intermediate	1
	Cuneiform, lateral	1
	Cuboid	1
	Metatarsals (foot bones)	5
Phalanges (toes)	First digit	2
	Second digit	3
	Third digit	3
	Fourth digit	3
	Fifth digit	3
Total		30
TOTAL		208

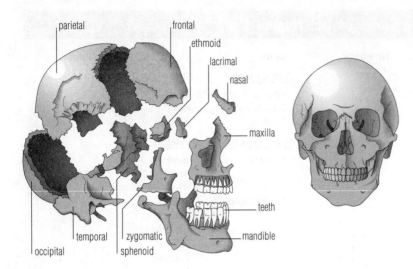

parietal
frontal
ethmoid
lacrimal
nasal
maxilla
teeth
mandible
zygomatic
sphenoid
temporal
occipital

skull The skull is a protective casing for the brain, eyes, and hearing organs. It is also a framework for the teeth and flesh of the face. The cranium has eight bones: occipital, two temporal, two parietal, frontal, sphenoid, and ethmoid. The face has 14 bones, the main ones being two maxillae, two nasal, two zygoma, two lacrimal, and the mandible.

vertebral column The vertebral column, otherwise known as the backbone or spine, is the body's central supporting structure and also serves to protect the spinal cord. It is made up of a series of bones, or **vertebrae**, running from the skull to the coccyx (vestigial tail), with a central canal containing the nerve fibres of the spinal cord. The shape of the vertebrae varies according to position. For example, in the chest region the upper or thoracic vertebrae are shaped to form connections to the ribs. The vertebral column is only slightly flexible to give adequate rigidity to the body.

> ### vertebral column
>
> The vertebral column, or spine, extends every night during sleep. During the day, the cartilage discs between the vertebra are squeezed when the body is in a vertical position, standing or sitting, but at night, with pressure released, the discs swell and the spine lengthens by about 8 mm/0.3 in.

Each vertebra has a body and an arch. The body is a cylinder connected to the adjacent vertebrae by a disk of fibro-cartilage, which absorbs shock. The arch consists of two halves which join together behind, forming a ring. The whole succession of these rings forms the vertebral canal, or channel, containing the main nerve trunk known as the spinal cord. Viewed from the side, the vertebral column has a series of curves. This curvature maintains the strength of the structure and adapts itself to the various movements of the body. The spinal cord occupies the upper two-thirds of the vertebral canal, and 31 pairs of spinal nerves arising from the spinal cord leave the vertebral canal through spaces at the sides between successive vertebrae.

long bone The long bones, such as those of the arms and legs, consist of a shaft, called a **diaphysis**, with an expanded, rounded portion called an **epiphysis** at each end. The whole bone is covered with a tough, fibrous membrane, called the **periosteum**, to which muscles and ligaments are attached.

The diaphysis is a tube of compact bone, resistant to bending, with a core – the **medullary cavity** – filled with a soft matrix called **bone marrow**.

The epiphyses, which are covered in a smooth layer of cartilage, are the regions where the bones meet other bones to form a (joint. They are largely composed of light spongy bone.

> ### bones
>
> To remember the bones of the upper limb:
>
> ### Some criminals have underestimated Royal Canadian Mounted Police.
>
> (**s**capula, **c**lavicle, **h**umerus, **u**lna, **r**adius, **c**arpals, **m**etacarpals, **p**halanges)

skeletal tissues

bone Bone is a living connective tissue, composed of a network of collagen fibres impregnated with mineral salts (largely calcium phosphate and calcium carbonate) – a combination that gives it great density and strength, comparable in some cases with that of reinforced concrete. Enclosed within this solid matrix are bone cells (**osteoblasts**), blood vessels, and nerves.

The osteoblasts, which are housed in small spaces called **lacunae**, are responsible for secreting the matrix.

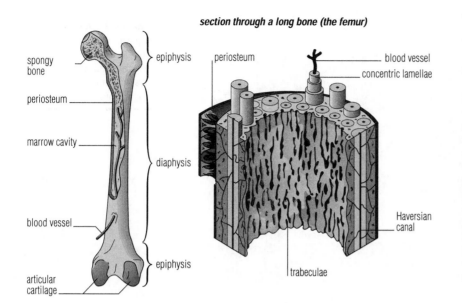

section through a long bone (the femur)

spongy bone

periosteum

marrow cavity

blood vessel

articular cartilage

epiphysis

diaphysis

epiphysis

periosteum

blood vessel

concentric lamellae

Haversian canal

trabeculae

bone Bone is a network of fibrous material impregnated with mineral salts and as strong as reinforced concrete. The upper end of the thighbone or femur is made up of spongy bone, which has a fine lacework structure designed to transmit the weight of the body. The shaft of the femur consists of hard compact bone designed to resist bending. Fine channels carrying blood vessels, nerves, and lymphatics interweave even the densest bone.

Tiny, fluid-filled channels, called canaliculi, penetrate the matrix to supply the cells with oxygen and nutrients and to transfer calcium salts to the exterior.

Compact bone, the dense tissue that forms the outer layer of most bones, is composed of many Haversian systems: tightly packed layers (lamellae) of matrix, formed concentrically around a Haversian canal. This canal acts as a duct for blood and lymph vessels and nerves. **Spongy bone** is a lighter tissue that makes up the rounded ends (epiphyses) of long bones and the inner layer of many other bones. It is composed of a loose, branching network of bone, with many irregularly shaped interconnected spaces that may be filled with bone marrow.

Bone may also be classified according to its origin: whether it developed by replacing ◊cartilage or was formed directly from connective tissue. The latter, which includes the bones of the cranium, are usually platelike in shape and form in the skin of the developing embryo.

cartilage Cartilage is a flexible bluish-white connective tissue made up of the protein collagen. It forms the greater part of the embryonic skeleton, and is replaced by ◊bone in the course of development, except in areas of wear such as bone endings, and the discs between the backbones. It also forms structural tissue in the larynx, nose, and external ear.

joints

A joint is a region where two bones meet. Some joints allow no motion (the sutures of the skull), others allow a very small motion (the sacroiliac joints in the lower back), but most allow a relatively free motion. Of these, some allow a gliding motion (one vertebra of the spine on another), some have a hinge action (elbow and knee), and others allow motion in all directions (hip and shoulder joints) by means of a ball-and-socket arrangement.

The ends of the bones at a moving joint are covered with cartilage for greater elasticity and smoothness, and enclosed in an envelope (capsule) of tough white fibrous tissue lined with a membrane which secretes a lubricating and cushioning synovial fluid. The joint is further strengthened by **ligaments** – strong, flexible bands of connective tissue, that prevent bone dislocation but allow joint flexion.

muscle system

Muscle is a contractile tissue that produces locomotion and power, and maintains the movement of body substances. It is made of long cells that can contract to between one-half and one-third of their relaxed length.

types of muscle The elongated cells or fibres that constitute muscle are classified into three types: striated, cardiac, and smooth.

striated muscle Striated muscle cells are so called because of their striped appearance under the microscope. They comprise the bulk of the body musculature (about 40% of the total body weight). Striated muscle fibres contract as a result of nervous stimulation and they are mostly under voluntary control. The

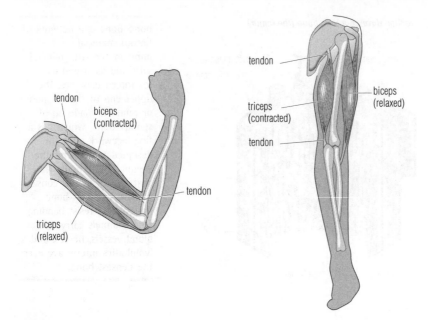

tendon
biceps (contracted)
tendon
triceps (relaxed)

tendon
triceps (contracted)
tendon
biceps (relaxed)

muscles of the arms and legs are examples. Striated muscle is also known as striped, skeletal, or voluntary muscle.

cardiac muscle Cardiac muscle occurs only in the heart. It also has cross-striations, but, unlike striated muscle fibres, its fibres branch so that the mass of muscle tends to function as one unit. It is controlled by the autonomic nervous system, but will continue to contract rhythmically even when its nerve supply is cut.

smooth muscle Smooth muscle lacks the visible cross-striations of striated and cardiac muscle; it is found in the intestinal walls, blood vessels, the iris of the eye, and various ducts. Like cardiac muscle, smooth muscle is normally under the dual control of hormones and the autonomic nervous system, but it also has intrinsic contractility. It is not under voluntary control.

muscle structure

Striated muscles are composed of a large number of parallel cylindrical fibres supported by non-contractile connective tissue. The connective tissue extends at both ends of the muscle to be continuous with the ◊tendons that attach the muscle to ◊bone.

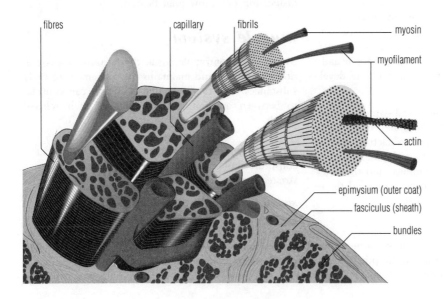

fibres
capillary
fibrils
myosin
myofilament
actin
epimysium (outer coat)
fasciculus (sheath)
bundles

muscle Muscles make up 35–45% of our body weight; there are over 650 skeletal muscles. Muscle cells may be up to 20 cm/0.8 in long. They are arranged in bundles, fibres, fibrils, and myofilaments.

The functional element of the muscle is the **muscle fibre**, which has many fine threads, or **fibrils**, running throughout its length. Although the fibrils themselves can only be observed under an electron microscope, their regular banded structure gives rise to the striped appearance of the muscle fibre. The light bands are called isotropic bands (I-bands), and the dark ones anisotropic bands (A-bands).

A fibril contains two different types of still finer **myofilaments**. The thicker myofilaments consist of the contractile protein **myosin** and each is surrounded by a hexagonal arrangement of thinner contractile protein called **actin**.

muscular contraction

The myosin molecules of the muscle fibrils each have a projecting head that forms a minute cross bridge with an actin myofilament. After nervous stimulation, electrical changes in the membrane surrounding each fibril cause the release of calcium ions, which are normally stored in sacs along the fibril. It is thought that the free calcium ions stimulate the heads of the myosin molecule to swivel, pulling the actin filaments past the myosin filaments. The cross bridges then break temporarily, as the myosin heads swing back to their original angle and attach themselves to the actin filaments again, before repeating the operation. Each cycle of attachment, swivel, and detachment, shortens the muscle by about 1%, so it must be repeated many times in order to produce a significant contraction.

muscles and movement

Usually a striated muscle is linked to bones at both its ends, and spans one or two joints in between. When the muscle fibres contract as a result of nervous stimulation, they tend to bring the bones closer together creating a rotatory movement at the intervening joint(s).

For coordinated movement around a joint, a pair of **antagonistic muscles** is required. The extension of the arm, for example, requires one set of muscles to relax, while another set contracts. The individual components of antagonistic pairs can be classified into **extensors** (muscles that straighten a limb) and **flexors** (muscles that bend a limb).

circulatory system

The circulatory system is the system that transports blood to and from the different parts of the body. It consists principally of a pumping organ – the ◊heart – and a network of blood vessels. In humans and other mammals, there is a double circulation system – the pulmonary (or lung) circulation and the systemic (or body) circulation – whereby blood passes to the lungs and back to the heart before circulating around the remainder of the body. In the **systemic circulation**, blood rich in oxygen is pumped to the tissues of the body, returning to the heart as blood rich in carbon dioxide; in the **pulmonary circulation**, blood rich in carbon dioxide is pumped to the (alveoli of the lungs, where carbon dioxide is exchanged for oxygen, so that oxygenated blood returns to the heart. Valves in the heart, large arteries, and veins ensure that blood flows in one direction only.

The circulatory system performs a number of functions: it supplies the cells of the body with the food and oxygen they need to survive; it carries carbon dioxide and other wastes away from the cells; it helps to regulate the temperature of the body and conveys substances that protect the body from disease. In addition, the system transports hormones, which help to regulate the activities of various parts of the body.

There are three main parts to the human circulatory system. These are the heart, the blood vessels, and the blood itself.

heart

The human heart is more or less conical in shape and is positioned within the chest, behind the breast bone, above the diaphragm, and between the two lungs. It has flattened back and front surfaces and is, in health, the size of an adult's closed fist. It has four chambers: two upper chambers – the left and right **atria** (singular 'atrium') – and two lower chambers – the left and right **ventricles**.

atria and ventricles The atria are thin-walled chambers that act as reservoirs, receiving blood from the veins. The two venae cavae, the major veins bringing back deoxygenated blood from the head, body, and limbs, join the right atrium. This chamber is separated from its respective ventricle by a valve with three flaps, the **tricuspid valve**. The right ventricle is a pyramidal chamber with thicker walls than the atria. The opening of the pulmonary artery, which leaves the right ventricle, has a valve that prevents the ejected blood from flowing back into the ventricle when it relaxes. The left atrium receives blood from the lungs via the four pulmonary veins, and transfers it into the left ventricle. This chamber has the stoutest walls of all, as its contraction should generate sufficient blood pressure to propel the blood into all the arteries of the body. The valve between the left atrium and ventricle has only two flaps and looks somewhat like a bishop's mitre, hence the name **bicuspid** or **mitral valve**.

cardiac cycle The cardiac cycle is the sequence of events during one complete cycle of a heart beat. This

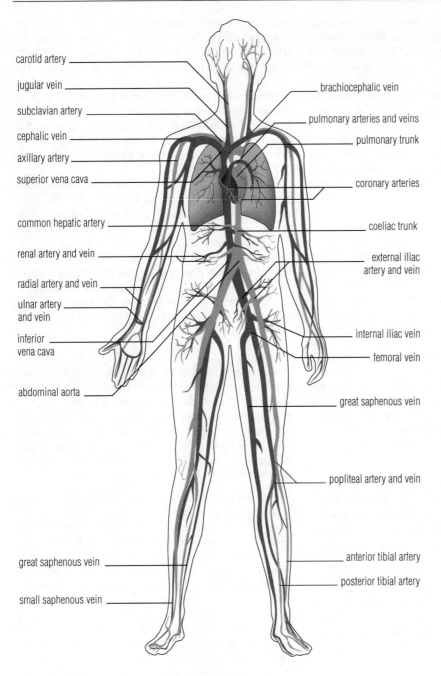

carotid artery

jugular vein

subclavian artery

cephalic vein

axillary artery

superior vena cava

common hepatic artery

renal artery and vein

radial artery and vein

ulnar artery
and vein

inferior
vena cava

abdominal aorta

great saphenous vein

small saphenous vein

brachiocephalic vein

pulmonary arteries and veins

pulmonary trunk

coronary arteries

coeliac trunk

external iliac
artery and vein

internal iliac vein

femoral vein

great saphenous vein

popliteal artery and vein

anterior tibial artery

posterior tibial artery

circulatory system Blood flows through 96,500 km/60,000 mi of arteries and veins, supplying oxygen and nutrients to organs and limbs. Oxygen-poor blood (black) circulates from the heart to the lungs where oxygen is absorbed. Oxygen-rich blood (grey) flows back to the heart and is then pumped round the body through the aorta, the largest artery, to smaller arteries and capillaries. Here oxygen and nutrients are exchanged with carbon dioxide and waste products and the blood returns to the heart via the veins. Waste products are filtered by the liver, spleen, and kidneys, and nutrients are absorbed from the stomach and small intestine.

consists of the simultaneous contraction of the two atria, a short pause, then the simultaneous contraction of the two ventricles, followed by a longer pause while the entire heart relaxes. The contraction phase is called **systole** and the relaxation phase that follows is called **diastole**. The whole cycle is repeated 70–80 times a minute under resting conditions.

When the atria contract, the blood in them enters the two relaxing ventricles, completely filling them. The mitral and tricuspid valves, which were open, now begin to shut and as they do so, they create vibrations in the heart walls and tendons, causing the first heart sound. The ventricles on contraction push open the pulmonary and aortic valves and eject blood into the

Blood: The Discovery of Circulation

BY JULIAN ROWE

background

After completing a preliminary medical course at the University of Cambridge, where would an ambitious young man in the 17th century go to get a really good medical training? To the University of Padua in Italy, where the great Italian anatomist Hieronymous Fabricius (1537–1619) taught. So this is where English physician William Harvey (1578–1657) naturally went.

William Harvey had a consuming interest in the movement of the blood in the body. In 1579, Fabricius had publicly demonstrated the valves, which he termed "sluice gates", in the veins: his principal anatomical work was an accurate and detailed description of them.

Galen's theory

Galen, a Greek physician (c. 130– c. 200), had 1,500 years previously written a monumental treatise covering every aspect of medicine. In this work, he asserted that food turned to blood in the liver, ebbed and flowed in vessels and, on reaching the heart, flowed through pores in the septum (the dividing wall) from the right to left side, and was sent on its way by heart spasms. The blood did not circulate. This doctrine was still accepted and taught well into the 16th century.

one-way flow

Harvey was unconvinced. He had done a simple calculation. He worked out that for each human heart beat, about 60 cm^3 of blood left the heart, which meant that the heart pumped out 259 litres every hour. This is more than three times the weight of the average man.

Harvey examined the heart and blood vessels of 128 mammals and found that the valve which separated the left side of the heart from the right ventricle is a one-way structure, as were the valves in the veins discovered by his tutor Fabricius.

For this reason he decided that the blood in the veins must flow only towards the heart.

Harvey's experiment

Harvey was now in a position to do his famous experiment. He tied a tourniquet round the upper part of his arm. It was just tight enough to prevent the blood from flowing through the veins back into his heart — but not so tight that arterial blood could not enter the arms. Below the tourniquet, the veins swelled up; above it, they remained empty. This showed that the blood could be entering the arm only through the arteries. Further, by carefully stroking the blood out of a short length of vein, Harvey showed that it could fill up only when blood was allowed to enter it from the end that was furthest away from the heart. He had proved that blood in the veins must flow only towards the heart.

a new theory of circulation

Galen's pores in the septum of the heart had never been found. Belgian physician Andreas Versalius (1514–1564) was another alumnus of Padua University. Although brought up in the Galen tradition, he had carried out secret dissections to discover the pores, and had failed. He did, however, show that men and women had the same number of ribs! Harvey clinched his researches into the movement of the blood when he demonstrated that no blood seeps through the septum of the heart. He reasoned that blood must pass from the right side of the heart to the left through the lungs. He had discovered the circulation of the blood, and thus, some 20 years after he left Padua, became the father of modern physiology. In 1628 Harvey published his proof of the circulation of the blood in his classic book *On the Motion of the Heart and Blood in Animals*. A new age in medicine and biology had begun.

---

respective vessels. The closed mitral and tricuspid valves prevent return of blood into the atria during this phase. As the ventricles start to relax, the aortic and pulmonary valves close to prevent backward flow of blood, and their closure causes the second heart sound. By now, the atria have filled once again and are ready to start contracting to begin the next cardiac cycle.

sinoatrial node The heart has a natural pacemaker, called the **sinoatrial node**, which is a group of muscle cells in the heart wall that contracts spontaneously and rhythmically, setting the pace for the contractions of the rest of the heart. The pacemaker's intrinsic rate of contraction is increased or decreased, according to the

needs of the body, by stimulation from the autonomic nervous system.

Heart: An Online Exploration
http://sln.fi.edu/biosci/heart.html
Explore the heart: discover its complexities, development, and structure; follow the blood on its journey through the blood vessels; learn how to maintain a healthy heart; and look back on the history of cardiology.

---

blood vessels There are three major types of blood vessels: **arteries**, which carry blood away from the heart; **veins**, which return blood to the heart; and **capillaries**, which are extremely tiny vessels connecting the arteries and the veins.

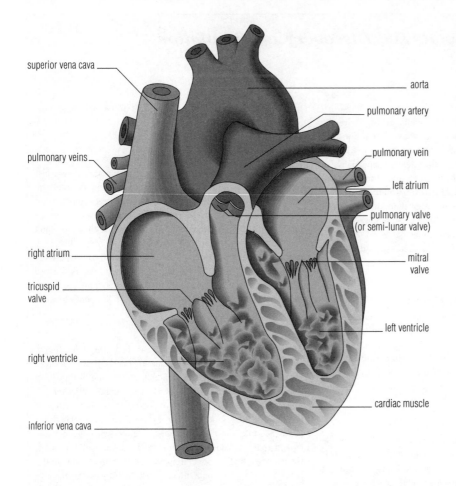

superior vena cava

aorta

pulmonary artery

pulmonary veins

pulmonary vein

left atrium

pulmonary valve
(or semi-lunar valve)

right atrium

mitral
valve

tricuspid
valve

left ventricle

right ventricle

cardiac muscle

inferior vena cava

heart The structure of the human heart. During an average lifetime, the human heart beats more than 2,000 million times and pumps 500 million l/110 million gal of blood. The average pulse rate is 70–72 beats per minute at rest for adult males, and 78–82 beats per minute for adult females.

artery When blood is pumped out of the heart into the arteries, it is forced out at high pressure by contractions of the muscular ventricles. Arteries therefore have a layer of strong, elastic muscle tissue in their walls that can stretch and recoil with the force of the blood; the resulting pulse or pressure wave can be felt at the wrist as the **pulse**. Not all arteries carry oxygenated (oxygen-rich) blood; the pulmonary arteries convey deoxygenated (oxygen-poor) blood from the heart to the lungs.

capillary As the blood travels through the body, the arteries divide into smaller and smaller vessels, finally forming capillaries – the narrowest blood vessels of all, 0.008–0.02 mm in diameter, barely wider than a red blood cell. Capillaries are distributed as beds, complex networks connecting arteries and veins. Their walls are extremely thin, consisting of a single layer of cells, and so nutrients, dissolved gases, and waste products can easily pass through them. This makes the capillaries the main area of exchange between the fluid ((lymph) bathing body tissues and the blood.

vein As the blood continues its passage around the body, capillaries merge again to form veins through which the blood is returned to the heart. By the time blood reaches the veins its pressure has been greatly reduced and its flow is much slower. The walls of veins therefore do not need to be as thick or as elastic as those of arteries. Veins contain valves that prevent the blood from running back when moving against gravity. They always carry deoxygenated blood, with the exception of the pulmonary veins leading from the lungs to the heart, which carry newly oxygenated blood.

blood

In humans, blood makes up 5% of the body weight, occupying a volume of 5.5 l/10 pt in the average adult. It is composed of a colourless, transparent liquid called **plasma**, which is

blood

In 1996 US companies began testing different varieties of artificial blood. All are able to transport oxygen from the lungs round the body and carry carbon dioxide back.

made up mostly of water, but also contains proteins, glucose, amino acids, salts, hormones, and antibodies. Suspended in the plasma are three kinds of microscopic cell: red blood cells, white blood cells, and platelets.

red blood cell The red blood cell, or **erythrocyte**, is the most common type of blood cell, responsible for transporting oxygen around the body. Each is disc-shaped with a depression in the centre and no nucleus. It contains **haemoglobin**, a complex red protein that gives blood its colour and combines with oxygen from the lungs to form oxyhaemoglobin. When transported to the tissues, the red blood cells are able to release the oxygen because the oxyhaemoglobin splits into its original constituents.

Red blood cells are manufactured in the bone marrow – mainly in the ribs, vertebrae, and limbs – at an average rate of 9,000 million per hour. However, they have a relatively short life of only about four months before being destroyed in the liver and spleen.

white blood cell The white blood cell, or **leucocyte**, is a relatively large, nucleated cell found in the blood, (lymph, and elsewhere in the body's tissues. It is colourless, with clear or granulated cytoplasm, and is capable of independent amoeboid movement.

There are a number of different types, each of which plays a part in the body's defences and gives immunity against disease. Some (neutrophils and macrophages) engulf invading microorganisms and infected cells, while lymphocytes produce more specific immune responses.

Human blood contains about 11,000 leucocytes to the cubic millimetre – about one to every 500 red cells.

platelet Platelets are small fragments of cells with no nucleus. They too are produced in the bone marrow and their function is to release substances which enable blood to clot (see (blood clotting). Thus they help to prevent the loss of blood from damaged vessels.

blood group A blood group is one of the types into which blood is classified according to the presence or absence of certain proteins on the surface of its red cells. These proteins act as antigens – that is, they would induce the production of antibodies if they were introduced into the blood of an individual whose red blood cells lack these proteins. Correct typing of blood groups is vital in transfusion, since incompatible types of donor and recipient blood will result in the production of antibodies and the clumping and rupturing of blood cells, with possible death of the recipient.

In the **ABO system**, the two main antigens are designated A and B. These give rise to four blood groups: having A only (A), having B only (B), having both (AB), and having neither (O). People of blood group A, for example, have antibodies against antigen B in their blood plasma, and therefore cannot receive blood transfusions from people possessing this antigen on their red blood cells – that is, from people who are blood group B or AB. People who are blood group AB, on the other hand, have no anti-A or anti-B antibodies in their plasma and so can receive blood of any group.

A further complication is the presence or absence of yet another antigen, called the **rhesus factor** (Rh factor). Most individuals possess the rhesus factor on their red blood cells – that is, they are rhesus positive (Rh+). However, those without this factor are rhesus negative (Rh-) and produce antibodies if they come into contact with it.

blood clotting Blood clotting prevents excessive bleeding after injury. It involves a complex series of events (known as the blood clotting cascade). The result is the formation of a meshwork of protein fibres (fibrin) and trapped blood cells over the cut blood vessels.

When platelets (cell fragments) in the bloodstream come into contact with a damaged blood vessel, they and the vessel wall itself release the enzyme thrombokinase, which brings about the conversion of the inactive enzyme prothrombin into the active thrombin. Thrombin in turn catalyses the conversion of the soluble protein fibrinogen, present in blood plasma, to the insoluble fibrin. This fibrous protein forms a net

Compatibility of Blood Groups

Blood group	Antigen on red blood cell	Antibody in plasma	Blood groups that can be received by this individual	Blood groups that can receive donations from this individual
A	A	anti-B	A, O	A, AB
B	B	anti-A	B, O	B, AB
AB	A and B	none	any	AB
O	neither A nor B	anti-A and anti-B	O	any

over the wound that traps red blood cells and seals the wound; the resulting jellylike clot hardens on exposure to air to form a scab. Calcium, vitamin K, and a variety of enzymes called factors are also necessary for efficient blood clotting.

lymph

Lymph is the fluid that exudes from the capillaries into the tissue spaces between the cells and is similar in composition to blood plasma. It carries some nutrients, and white blood cells to the tissues, and waste matter away from them. It may drain back from the tissues into the blood capillaries and from thence into the veins, or it may drain into a system of lymphatic vessels. These lead to lymph nodes (small, round bodies chiefly situated in the neck, armpit, groin, thorax, and abdomen), which process the lymphocytes produced by the bone marrow, and filter out harmful substances and bacteria. From the lymph nodes, vessels carry the lymph to the thoracic duct and the right lymphatic duct, which drain into the large veins in the neck.

functions of the circulatory system

The circulatory system plays an important role in many of the body's processes including respiration, nutrition, and the removal of wastes and poisons.

In **respiration**, the circulating blood delivers oxygen to the body's cells and removes carbon dioxide from them.

In **nutrition**, it carries digested food substances to the cells. Nutrients from food enter the bloodstream

---

> ### blood
>
> To remember the functions of the blood:
>
> **O**ld **C**harlie **F**oster **h**ates **w**omen **h**aving **d**ull **c**lothes.
>
> (**o**xygen (transport), **c**arbon dioxide (transport), **f**ood, **h**eat, **w**aste, **h**ormones, **d**isease, **c**lotting)

---

by passing through the walls of the small intestine into the capillaries. The blood then carries most of the nutrients to the liver, where some of these are extracted and stored for release back into the blood as and when the body needs them. Other nutrients are converted by the liver into substances which are required in the production of energy, enzymes, and new building materials for the body.

Hormones, which affect or control the activities of various organs and tissues, are produced by the endocrine glands – including the thyroid, pituitary, adrenal, and sex glands – and they too are transported by the blood through the body.

The circulatory system helps to dispose of **waste products** which would prove harmful if allowed to accumulate. In processing food, the liver removes ammonia and other wastes, together with various poisons that enter the body through the digestive system. These are converted into water-soluble substances, which are carried by the blood to the kidneys. The kidneys then filter out these wastes and expel them from the body in urine.

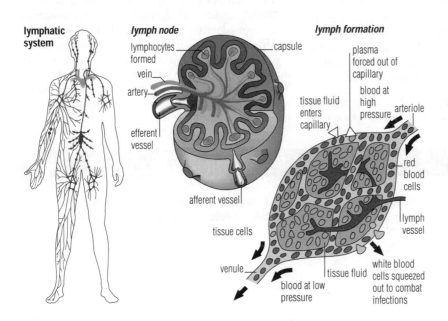

lymphatic system

lymph node
lymphocytes formed
vein
artery
efferent vessel
afferent vessel
capsule

lymph formation
plasma forced out of capillary
tissue fluid enters capillary
blood at high pressure
arteriole
red blood cells
lymph vessel
tissue cells
venule
blood at low pressure
tissue fluid
white blood cells squeezed out to combat infections

lymph Lymph is the fluid that carries nutrients and white blood cells to the tissues. Lymph enters the tissue from the capillaries (right) and is drained from the tissues by lymph vessels. The lymph vessels form a network (left) called the lymphatic system. At various points in the lymphatic system, lymph nodes (centre) filter and clean the lymph.

The circulation also plays a role in **temperature regulation**. Some parts of the body, such as the liver and muscles, produce heat in the course of their activities. This heat is transported by the blood to warm other parts of the body. As the temperature of the body rises, the flow of blood into vessels in the skin increases, and excess heat is conveyed to the surface where it is dispersed. When the temperature of the body drops the flow of blood to the skin is restricted. Thus, the circulatory system acts as a natural thermostat allowing the body to maintain an optimum and stable temperature.

respiratory system

Respiration is a biochemical process by which food substances are broken down to release energy, which the body's cells can use to carry out work. This process requires the presence of oxygen, and humans – like other aerobic organisms – possess special features that enable us both to take in oxygen from the atmosphere and transport it to the respiring cells, and also to eliminate the carbon dioxide that is a waste product of respiration.

lung

The lungs are a pair of large cavities in the chest that are used to exchange oxygen and carbon dioxide between the blood and the atmosphere. The lung tissue, consisting of multitudes of air sacs (alveoli; singular alveolus) and blood vessels, is very light and spongy, and functions to create a huge surface area (of about 80 m²/95 yd²) where inhaled air comes into close contact with the blood so that oxygen can pass into the body and waste carbon dioxide can be passed out. The efficiency of lungs is enhanced by ◊breathing movements, by the thinness and moistness of their surfaces, and by a constant supply of circulating blood.

lung The human lungs contain 300,000 million tiny blood vessels which would stretch for 2,400 km/1,500 mi if laid end to end. A healthy adult at rest breathes 12 times a minute; a baby breathes at twice this rate. Each breath brings 350 millilitres of fresh air into the lungs, and expels 150 millilitres of stale air from the nose and throat.

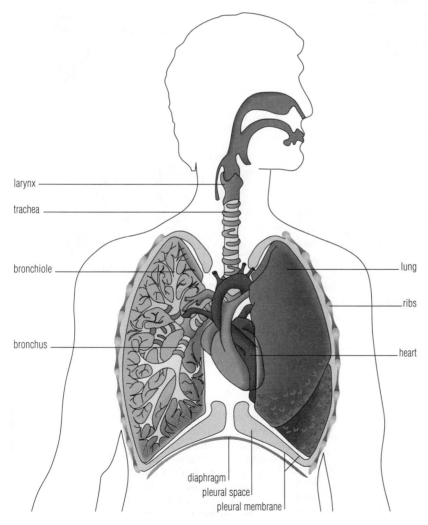

larynx

trachea

bronchiole

bronchus

lung

ribs

heart

diaphragm
pleural space
pleural membrane

The lung may be regarded as a many-chambered elastic bag placed in the air-tight chest cavity and having communication with the exterior only by means of the **trachea** (windpipe) – a flexible tube reinforced by rings of cartilage. Atmospheric pressure acting down the trachea keeps the lungs so far stretched that, together with the heart and great blood vessels, they completely fill the chest cavity. The trachea divides and subdivides into bronchi, bronchioles, and bronchial tubes, which, diverging in all directions, never join up, but terminate separately.

gas exchange at the alveoli

After a certain stage of subdivision, when their diameter is about 1 mm/0.04 in, the bronchial tubes develop into blind, grape-like pouches called **alveoli**, the walls of which consist of a delicate, moist membrane made up of transparent, flattened cells. The diffusion distance between the surrounding mesh of capillaries and the air inside the alveoli is therefore very small.

Oxygen in the alveolar air dissolves into the thin film of moisture that lines the alveolus and diffuses from this region of high concentration across the alveolar and capillary walls into the blood, where it is at a lower concentration. There it combines with haemoglobin in the red blood cells for transport around the body. Carbon dioxide follows a reverse path, diffusing from the blood plasma, where it is at a high concentration, through the capillary and alveolar walls into the alveolar air, where it is at a low concentration and can be exhaled.

breathing

Although lungs are specialized for gas exchange, they are not themselves muscular, consisting of spongy material. In order for fresh supplies of air to be drawn into the alveoli and for stale air to be eliminated, the lungs must be expanded and compressed, respectively, by muscular movement. Air is drawn into the lungs (inhaled) by the contraction of the diaphragm and intercostal muscles (the muscles between the ribs); relaxation of these muscles enables air to be breathed out (exhaled). The rate of breathing is controlled by the brain. High levels of activity lead to a greater demand for oxygen and an increased rate of breathing.

nervous system

The nervous system coordinates the body's activities: it receives and processes information about them and about changes in the external environment, and initiates appropriate responses. It is composed of millions of interconnecting nerve cells, which are organized into a central nervous system, comprising ⊅brain and spinal cord, and a peripheral nervous system, which connects with sensory organs, muscles, and glands.

Neuroscience for Kids
http://weber.u.washington.edu/~chudler/neurok.html
Explore the nervous system – your brain, spinal cord, nerve cells, and senses – by means of this impressive site, designed for primary and secondary school students and teachers.

---

nerve cell

A nerve cell, or **neuron**, is an elongated, branched cell that forms the basic functional unit of the nervous system. It transmits information rapidly – at up to 160 m/525 ft per second – between different parts of the body. Each nerve cell has a cell body, containing the nucleus, from which trail processes called **dendrites**, responsible for receiving incoming signals. The cell's

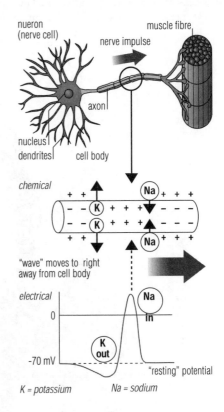

---

nerve cell The anatomy and action of a nerve cell. The nerve cell or neuron consists of a cell body with the nucleus and projections called dendrites which pick up messages. An extension of the cell, the axon, connects one cell to the dendrites of the next. When a nerve cell is stimulated, waves of sodium (Na+) and potassium (K+) ions carry an electrical impulse down the axon.

longest process, called an **axon**, transmits signals away from the cell body.

The unit of information is the **nerve impulse**, a travelling wave of chemical and electrical changes involving the membrane of the nerve cell. The impulse involves the passage of sodium and potassium ions across the nerve-cell membrane. Sequential changes in the permeability of the membrane to positive sodium (Na^+) ions and potassium (K^+) ions produce electrical signals called action potentials. Impulses are received by the cell body and passed, as a pulse of electric charge, along the axon.

The axon terminates at the **synapse**, a specialized area closely linked to the next cell (which may be another nerve cell or a specialized effector cell such as a muscle). On reaching the synapse, the impulse releases a chemical **neurotransmitter**, which diffuses across to the neighbouring cell and there stimulates another impulse or the action of the effector cell.

nerve

A nerve is a bundle of nerve cells enclosed in a sheath of connective tissue (myelin sheath) and transmitting nerve impulses to and from the brain and spinal cord. A single nerve may contain both motor and sensory nerve cells, but they function independently.

The myelin sheath serves to speed up the passage of nerve impulses. It is made up of fats and proteins and is formed from up to a hundred layers, laid down by special cells, the Schwann cells.

central nervous system

The central nervous system (CNS), which consists of the brain and spinal cord, integrates all nervous function.

brain

The brain is a mass of interconnected ⟩nerve cells, forming the anterior part of the central nervous system, the activities of which it coordinates and controls. It is enclosed within three membranes, called the meninges. The thickest is the outermost layer, the dura mater; the middle layer is the arachnoid mater and the innermost layer is the pia mater. Cerebrospinal fluid circulates between the two innermost layers. The brain is further contained and protected by the skull.

The brain may be divided into three parts: forebrain, midbrain, and hindbrain.

The **forebrain** is mainly composed of the **cerebrum** – a pair of outgrowths (cerebral hemispheres) separated by a central fissure. This is the largest and most developed part of the brain, taking up about 70% of its volume. The cerebrum is covered with a deeply infolded layer of grey matter, the **cerebral cortex**, which is responsible for all higher functions, such as speech, emotions, and memory, and for initiating voluntary movement.

Many of the nerve fibres from the two sides of the body cross over as they enter the brain, so that the left cerebral hemisphere is associated with the right side of the body and vice versa. In right-handed people, the left hemisphere seems to play a greater role in controlling verbal and some mathematical skills, whereas the right hemisphere is more involved in spatial perception.

Other structures in the forebrain, lying between the hemispheres, are the thalamus, hypothalamus, and pituitary gland.

The **midbrain** and **hindbrain** together make up the brainstem – the oldest part of the brain in evolutionary terms – with the midbrain acting as a link with the structures of the forebrain. The hindbrain, which connects with the spinal cord, consists of the **medulla oblongata**, a region that contains centres for the control of respiration, heartbeat rate and strength, and blood pressure. Overlying this is the **cerebellum**, which is concerned with coordinating complex muscular processes such as maintaining posture and moving limbs.

spinal cord

The spinal cord is the body's main nerve trunk, consisting of bundles of nerve cells enveloped in three layers of membrane (the meninges) and bathed in cerebrospinal fluid. It is encased and protected by the (vertebral column, lying within the vertebral canal formed by the posterior arches of successive vertebrae. In adults, the spinal cord is about 45 cm/18 in long, extending from the bottom of the skull, where it is continuous with the medulla oblongata, to about waist level.

The spinal cord consists of nerve cell bodies (grey matter) and their myelinated processes or nerve fibres (white matter). In cross-section, the grey matter is arranged in an H-shape around the central canal of the spinal cord, and it is surrounded in turn by the white matter.

Paired spinal nerves arise from the cord at each vertebra. Each is a mixed nerve, consisting of both sensory and motor nerve fibres. The sensory fibres enter the spinal cord at a dorsal root and the motor fibres enter at a ventral root. This arrangement enables the spinal cord to relay impulses coming in and out at the same level, to relay impulses going up and down the cord to other levels, and relay impulses to and from the brain. The first of these involves a **reflex arc**, by which a sensory impulse can create a very rapid, involuntary response to a particular stimulus.

peripheral nervous system

The peripheral nervous system can be classified in many ways. It can be classified anatomically into cranial and **spinal** nerves – those that arise from the brain and spinal cord, respectively. It can be classified into **sensory nerves** – those that carry information from receptor cells or sensory organs to the central nervous system – and **motor nerves** – those that carry information from the central nervous system to the effector organs (muscles and glands). Mixed nerves contain both sensory and motor nerve fibres.

Cranial nerves consist of three sensory nerves, two motor nerves, and seven mixed nerves. The 31 pairs of spinal nerves are all mixed.

The peripheral nervous system can also be classified functionally into the **voluntary nervous system**, which controls voluntary activities, and the **autonomic nervous system**, which controls involuntary functions, such as heart rate, activity of the intestines, and the production of sweat. The autonomic nervous system can be further subdivided into the sympathetic and the parasympathetic systems. The **sympathetic system** responds to stress, when it speeds the heart rate, increases blood pressure, and generally prepares the body for action. The **parasympathetic system** is more important when the body is at rest, since it slows the heart rate, decreases blood pressure, and stimulates the digestive system.

sense organ

Sense organs are used to gain information about our surroundings. They all have specialized receptors (such as light receptors in the eye) and some means of translating their response into a nerve impulse that travels to the brain. The main human sense organs are the eye, which detects light and colour (different wavelengths of light); the ear, which detects sound (vibrations of the air) and gravity; the nose, which detects some of the chemical molecules in the air; and the tongue, which detects some of the chemicals in food, giving a sense of taste. There are also many small sense organs in the skin, including pain, temperature, and pressure sensors, contributing to our sense of touch.

It's All in the Brain
http://www.hhmi.org/senses/a/a110.htm
As part of a much larger site called 'Seeing, Hearing, and Smelling the World', here is a set of pages introducing the way in which we perceive the world through our senses. It is divided into five sections called 'illusions reveal the brain's assumptions', 'sensing change in the environment', 'vision, hearing, and smell: the best-known senses', 'a language the brain can understand', and 'more than the sum of its parts'.

---

ear

ear The ear is primarily the organ of hearing, translating the vibrations that constitute sound into nerve signals that are passed to the brain.

It consists of three parts: outer ear, middle ear, and inner ear. The **outer ear** is a funnel that collects sound, directing it down a tube to the **ear drum** (tympanic membrane), which separates the outer and **middle ears**. Sounds vibrate this membrane, the mechanical movement of which is transferred to a smaller membrane leading to the **inner ear** by three small bones, the ossicles. Vibrations of the inner ear membrane move fluid contained in the snail-shaped cochlea, which vibrates hair cells that stimulate the auditory nerve connected

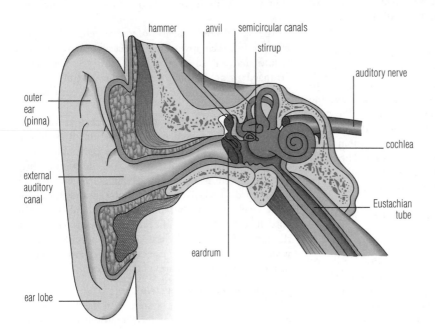

ear The structure of the ear. The three bones (ossicles) of the middle ear – hammer, anvil, and stirrup – vibrate in unison and magnify sounds about 20 times. The spiral-shaped cochlea is the organ of hearing. As sound waves pass down the spiral tube, they vibrate fine hairs lining the tube, which activate the auditory nerve connected to the brain. The semicircular canals are the organs of balance, detecting movements of the head.

to the brain. There are approximately 30,000 sensory hair cells (**stereocilia**). Three fluid-filled canals of the inner ear detect changes of position; this mechanism, with other sensory inputs, is responsible for the sense of balance.

Ear Anatomy
http://weber.u.washington.edu/~otoweb/
ear_anatomy.html
Concise medical information on the anatomy and function of the ear. It includes separate sections on perforated eardrums and their treatment, ear tubes and their use, tinnitus, and a series of different hearing tests.

---

eye The eye is the organ of vision. It is a roughly spherical structure contained in a bony socket called the orbit. Light enters the eye through the cornea, and passes through the circular opening (pupil) in the iris (the coloured part of the eye). The ciliary muscles act on the lens (the rounded transparent structure behind the iris) to change its shape, so that images of objects at different distances can be focused on the **retina**. This is at the back of the eye, and

> **eye**
> A US team of developmental biologists established for the first time in 1997 that the vertebrate eye begins in the embryo as a single structure that later splits in two.

is packed with light-sensitive receptor cells (rods and cones), connected to the brain by the optic nerve.

Breaking the Code of Colour
http://www.hhmi.org/senses/b/b110.htm
As part of a much larger site called 'Seeing, Hearing, and Smelling the World', here is a set of pages examining the way we perceive the world through the sense of sight. It is divided into five sections called 'how do we see colours?', 'red, green, and blue cones', 'colour blindness: more prevalent among males', and 'judging a colour'. This site makes good use of images and animations to help with the explanations, so it is best viewed with an up-to-date browser.

---

endocrine system

Like the nervous system, the endocrine system plays a role in controlling the body's activities. However, its actions are much slower and are mediated via the bloodstream, and its effects – such as growth and development – tend to be more long lasting.

The key players in the endocrine system are the hormones.

hormone

Hormones are chemical secretions, produced mainly by the (endocrine glands, that bring about changes in the functions of distant tissues according to the body's requirements. They maintain homeostasis (a stable

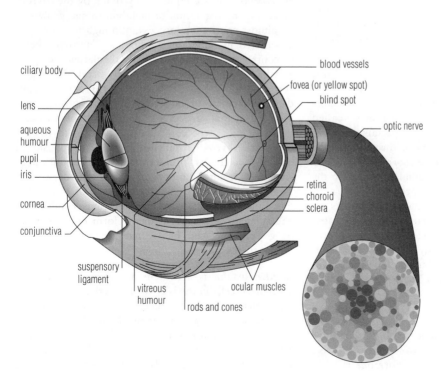

ciliary body
lens
aqueous humour
pupil
iris
cornea
conjunctiva
suspensory ligament
vitreous humour
rods and cones
ocular muscles
blood vessels
fovea (or yellow spot)
blind spot
optic nerve
retina
choroid
sclera

eye The human eye. The retina of the eye contains about 137 million light-sensitive cells in an area of about 650 sq mm/1 sq in. There are 130 million rod cells for black and white vision and 7 million cone cells for colour vision. The optic nerve contains about 1 million nerve fibres. The focusing muscles of the eye adjust about 100,000 times a day. To exercise the leg muscles to the same extent would need an 80 km/50 mi walk.

internal state) and also control tissue development, morphogenesis (the development of the body's form and structure), and reproduction. Many human diseases that are caused by hormone deficiency can be treated with hormone preparations.

Hormones usually act on specific targets. By binding themselves to special receptor sites on cell membranes, they can stimulate or suppress the actions of these cells – either by altering the permeability of the membrane to other chemicals, by activating enzymes, or by activating or inhibiting genes.

hormone interaction The relationships between hormones can be very complex in that some endocrine glands produce hormones that in turn stimulate other glands to produce hormones.

In this way, the anterior pituitary gland controls the thyroid gland, the adrenal cortex, the testis, and the ovary. When stimulated by the relevant pituitary hormone (known as a **trophic hormone**), each one of these glands produces its own hormones which have their own effects. But these hormones do something else that is highly ingenious; they act on the pituitary gland itself. They actually decrease (inhibit) the pituitary's secretion of the original trophic hormone, so that the level of the circulating hormone can never get too high and is kept at an optimum level. This control mechanism is known as **negative feedback**.

Feedback systems are used by the body to keep many of its constituents, such as body temperature, at a constant level; they are analogous to the thermostat system used for keeping the temperature of a house within narrow limits.

neurohormone Hormone secretion is not confined to the endocrine glands. For example, the small intestine secretes a hormone called secretin (the first hormone ever discovered). Other hormones, called **neurohormones**, are secreted by specialized nerve cells, either into the bloodstream or directly into the target tissue. For example, a neurohormone produced by nerve cells in the brain's hypothalamus acts on the pituitary gland, stimulating it to produce ADH (antidiuretic hormone), a hormone that, in turn, acts on the kidney to promote the reabsorption of water. Thus, chemical messages can be relayed to the appropriate part, or parts, of the body in response to stimulants either through nervous impulses via the nerves or through hormones secreted in the blood, or by both together.

endocrine gland
The endocrine glands are ductless glands that have capillaries running through them which provide direct access to the bloodstream. So as a hormone is produced by a gland, it is released directly into the bloodstream and carried to the tissues or organs that it affects. Normal functioning of the endocrine glands enables normal functioning of the body cells and results in general well-being.

The major glands are the pituitary, thyroid, parathyroid, adrenal, pancreas, ovary, and testis. There are also hormone-secreting cells in the kidney, liver, gastrointestinal tract, thymus (in the neck), pineal (in the brain), and placenta.

The anterior pituitary gland and the brain's hypothalamus act as control centres for overall coordination of hormone secretion.

major hormones
adrenaline Adrenaline, or epinephrine, is secreted by the medulla of the adrenal glands. It is synthesized from a closely related substance, noradrenaline, and the two hormones are released into the bloodstream in situations of fear or stress. Adrenaline's action on the liver raises blood-sugar levels by stimulating glucose production and its action on adipose tissue raises blood fatty-acid levels; it also increases the heart rate, increases blood flow to muscles, reduces blood flow to the skin with the production of sweat, widens the smaller breathing tubes (bronchioles) in the lungs, and dilates the pupils of the eyes.

thyroxine The secretion of **thyroxine** by the thyroid gland for the regulation of metabolism and growth is an almost continuous process, although the hormone is secreted in quite small amounts. Thyroxine helps to control metabolism mainly by regulating the speed at which mitochondria in the cells break down glucose in respiration. A deficiency of thyroxine in children will result in abnormal physical growth and retarded mental growth – a condition known as cretinism. This can be cured by injections of thyroxine. In adults, thyroxine deficiency leads to sluggishness and overweight; an overactive thyroid has the opposite effect, resulting in hyperactivity and underweight. Thyroxine contains iodine, and a lack of this in the diet limits production of the hormone, resulting in enlargement of the gland in the form of a swelling, or goitre. This can be treated by the addition of iodine to the diet.

insulin Insulin is a protein hormone, produced by specialized cells in the islets of Langerhans in the pancreas, that regulates the metabolism (rate of activity) of glucose, fats, and proteins. Normally, insulin is secreted in response to rising blood sugar levels (after a meal, for example), stimulating the body's cells to store the excess as glycogen. Failure of this

Hormones and their Functions

Gland	Hormone	Functions
posterior pituitary gland	anti-diuretic hormone (ADH)	water reabsorption from kidney tubules
	oxytocin	contraction of the uterus during birth
anterior pituitary gland	growth hormone (GH)	growth
	prolactin	milk production and secretion
	follice-stimulating hormone (FSH)	in females, maturation of the Graafian follicle; in males, sperm production
	luteininzing hormone (LH)	in females, ovulation, formation of the corpus luteum; in males, testosterone synthesis
	thyroid stimulating hormone (TSH)	stimulates the thyroid to release thyroid hormones
	adrenocorticotrophic hormone (ACTH)	stimulates the adrenal cortext to produce corticosteroid hormones
ovary	oestrogen	female secondary sexual characteristics
ovary and placenta	progesterone	prepares uterus for pregnancy; maintains it during pregnancy
testis	testosterone	male secondary sexual characteristics
adrenal gland (cortex)	corticosteroid hormones	controls salt and water metabolism; regulates use of carbohydrates, proteins, and fats
adrenal gland (medulla)	adrenaline	'fright, flight, or fight': increases heart activity, rate and depth of breathing, blood flow to muscles; inhibits digestion and excretion
thyroid	thyroxine	regulates metabolism and growth
	calcitonin	regulates blood calcium levels by reducing release of calcium from bones
parathyroid	parathormone	regulates blood calcium levels by stimulating release of calcium from bones
pancreas (islets of Langerhans)	insulin	regulates blood glucose levels by stimulating conversion of glucose to glycogen
	glucagon	regulates blood glucose levels by stimulating conversion of glycogen to glucose

regulatory mechanism in diabetes mellitus requires treatment with insulin injections or capsules taken by mouth.

sex hormones The reproductive organs contain endocrine glands: ovaries in females and testes in males. Male sex hormones are called androgens, the most important of which is **testosterone**. The androgens regulate the development of the male sex organs and, at puberty, secondary sexual characteristics such as the growth of facial and pubic hair, breaking of the voice, and muscular development. The female sex hormones, **oestrogens**, regulate the development of female sex organs and, at puberty, breast development and the growth of pubic hair. They also regulate menstruation and pregnancy.

endocrine gland The main human endocrine glands. These glands produce hormones - chemical messengers - which travel in the bloodstream to stimulate certain cells.

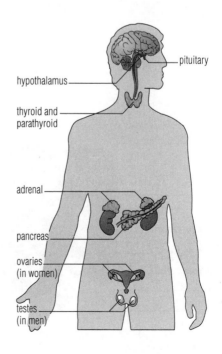

digestive system

The digestive system consists of all the organs and tissues involved in the digestion of food. The process of digestion breaks down the food by physical and chemical means into the different elements that are needed by the body for energy and tissue building and repair. Digestion begins in the mouth and is completed in the stomach and small intestine; most nutrients are absorbed into the small intestine from where they pass through the intestinal wall into the bloodstream; what remains is stored and concentrated into faeces in the large intestine.

The digestive system of humans consists primarily of the **alimentary canal**, a tube that starts at the mouth, continues with the pharynx, oesophagus (or gullet), stomach, large and small intestines, and rectum, and ends at the anus. The food moves through this canal by **peristalsis** whereby waves of involuntary muscular contraction and relaxation produced by the muscles in the wall of the gut cause the food to be ground and mixed with various digestive juices. Most of these juices contain digestive **enzymes**, specialized proteins that speed up reactions involved in the breakdown of food. Other digestive juices empty into the alimentary canal from the salivary glands, gall bladder, and pancreas, which are also part of the digestive system.

digestion and absorption

The fats, proteins, and carbohydrates (starches and sugars) in foods contain very complex molecules that are broken down by the processes of digestion for absorption into the bloodstream: starches and complex sugars are converted to simple sugars; fats are converted to fatty acids and glycerol; and proteins are converted to amino acids and peptides. Foods such as vitamins, minerals, and water do not need to undergo digestion prior to absorption into the bloodstream. The lining of the small intestine, which is the main site of absorption, is covered with small prominences called **villi** which increase the surface area for absorption and allow the digested nutrients to diffuse into small blood-vessels lying immediately under the epithelium.

mouth, pharynx, and oesophagus

The digestive process begins in the mouth with mastication (chewing) and salivation. The purpose of mastication is to crush and grind the food into small

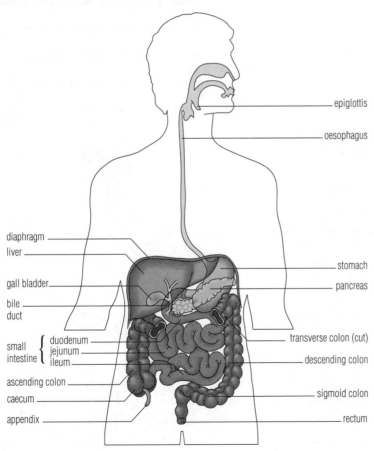

epiglottis

oesophagus

diaphragm
liver

gall bladder

bile
duct

small
intestine { duodenum
jejunum
ileum

ascending colon

caecum

appendix

stomach

pancreas

transverse colon (cut)

descending colon

sigmoid colon

rectum

digestive system The human digestive system. When food is swallowed, it is moved down the oesophagus by the action of muscles (peristalsis) into the stomach. Digestion starts in the stomach as the food is mixed with enzymes and strong acid. After several hours, the food passes to the small intestine. Here more enzymes are added and digestion is completed. After all nutrients have been absorbed, the indigestible parts pass into the large intestine and thence to the rectum. The liver has many functions, such as storing minerals and vitamins and making bile, which is stored in the gall bladder until needed for the digestion of fats. The pancreas supplies enzymes. The appendix appears to have no function in human beings.

Digestive System: Pioneering Experiments on the Digestive System

BY JULIAN ROWE

an army marches on its stomach

On 6 June 1822 at Fort Mackinac, Michigan, USA, an 18-year-old French Canadian was accidentally wounded in the abdomen by the discharge of a musket. He was brought to the army surgeon, US physician William Beaumont (1785–1853), who noted several serious wounds and, in particular, a hole in the abdominal wall and stomach. The surgeon observed that through this hole in the patient 'was pouring out the food he had taken for breakfast'.

The patient, Alexis St Martin, a trapper by profession, was serving with the army as a porter and general servant. Not surprisingly, St Martin was at first unable to keep food in his stomach. As the wound gradually healed, firm dressings were needed to retain the stomach contents. Beaumont tended his patient assiduously and tried during the ensuing months to close the hole in his stomach, without success. After 18 months, a small, protruding fleshy fold had grown to fill the aperture (fistula). This 'valve' could be opened simply by pressing it with a finger.

digestion ... inside and outside

At this point, it occurred to Beaumont that here was an ideal opportunity to study the process of digestion. His patient must have been an extremely tough character to have survived the accident at all. For the next nine years he was the subject of a remarkable series of pioneering experiments, in which Beaumont was able to vary systematically the conditions under which digestion took place and discover the chemical principles involved.

Beaumont attacked the problem of digestion in two ways. He studied how various substances were actually digested in the stomach, and also how they were 'digested' outside the stomach in the digestive juices he extracted from St Martin. He found it was easy enough to drain out the digestive juices from his fortuitously wounded patient 'by placing the subject on his left side, depressing the valve within the aperture, introducing a gum elastic tube and then turning him ... on introducing the tube the fluid soon began to run.'

a typical experiment

Beaumont was basically interested in the rate and temperature of digestion, and also the chemical conditions that favoured different stages of the process of digestion. He describes a typical experiment (he performed hundreds), where (a) digestion in the stomach is contrasted (b) with artificial digestion in glass containers kept at suitable temperatures, like this: (a) 'At 9 o'clock he breakfasted on bread, sausage, and coffee, and kept exercising. 11 o'clock, 30 minutes, stomach two-thirds empty, aspects of weather similar, thermometer 298°F, temperature of stomach 1011/28 and 1003/48. The appearance of contraction and dilation and alternative piston motions were distinctly observed at this examination. 12 o'clock, 20 minutes, stomach empty.' (b) 'February 7. At 8 o'clock, 30 minutes a.m. I put twenty grains of boiled codfish into three drachms of gastric juice and placed them on the bath.' 'At 1 o'clock, 30 minutes, p.m., the fish in the gastric juice on the bath was almost dissolved, four grains only remaining: fluid opaque, white, nearly the colour of milk. 2 o'clock, the fish in the vial all completely dissolved.'

all a matter of chemistry

Beaumont's research showed clearly for the first time just what happens during digestion and that digestion, as a process, can take place independently outside the body. He wrote that gastric juice: 'so far from being inert as water as some authors assert, is the most general solvent in nature of alimentary matter—even the hardest bone cannot withstand its action. It is capable, even out of the stomach, of effecting perfect digestion, with the aid of due and uniform degree of heat (100°Fahrenheit) and gentle agitation ... I am impelled by the weight of evidence ... to conclude that the change effected by it on the aliment, is purely chemical.' Our modern understanding of the physiology of digestion as a process whereby foods are gradually broken down into their basic components follows logically from his work. An explanation of how the digestive juices flowed in the first place came in 1889, when Russian physiologist Ivan Pavlov (1849–1936) showed that their secretion in the stomach was controlled by the nervous system. By preventing the food eaten by a dog from actually entering the stomach, he found that the secretions of gastric juices began the moment the dog started eating, and continued as long as it did so. Since no food had entered the stomach, the secretions must be mediated by the nervous system.

Later, it was found that the further digestion that takes place beyond the stomach was hormonally controlled. But it was Beaumont's careful scientific work, which was published in 1833 with the title Experiments and Observations on the Gastric Juice and Physiology of Digestion, that triggered subsequent research in this field.

fragments that can pass safely into the gut and that will have an increased surface area for the action of digestive enzymes. The teeth play an important role in this process.

tooth A tooth is a hard, bonelike structures in the mouth, used for biting and chewing food. The first teeth – a set of 20 milk teeth – appear from age six months to two and a half years. The permanent dentition

mouth

In human mouths there are over 200 different species of bacteria, fungi, and protozoa. Billions of bacteria still cling to a tooth just brushed. Most species are harmless and some may be even be beneficial.

replaces these from the sixth year onwards, the wisdom teeth (third molars) sometimes not appearing until the age of 25 or 30. Adults have 32 teeth.

Each tooth consists of an enamel coat (hardened calcium deposits), dentine (a thick, bonelike layer), and an inner pulp cavity, housing nerves and blood vessels. The roots are surrounded by cementum, which fuses them into their sockets in the jawbones. The neck of the tooth is covered by the gum, while the enamel-covered crown protrudes above the gum line.

saliva Alkaline saliva is poured into the mouth from three pairs of glands, named parotid, submandibular, and sublingual. The parotid gland secretes a clear saliva, the other two a sticky saliva containing mucin. Saliva

tooth

About one in every 2,000 babies is born with a tooth. Louis XIV of France was born with two teeth, which may explain why he had had eight wet-nurses by the time he moved on to solid foods.

has an important chemical action which, by means of an enzyme called **ptyalin**, converts cooked starch in the food into maltose, a kind of sugar. The saliva also moistens the food so that it can be rolled by the tongue and palate into a soft bolus.

swallowing When the food is sufficiently masticated, the bolus is pushed backwards by the tongue into the pharynx. Here the swallowing reflex causes a series of muscular movements which propel the food into the gullet, or oesophagus, and from there into

the stomach.

A small flap, the **epiglottis**, located behind the root of the tongue, closes off the end of the windpipe during swallowing to prevent food from passing into it and causing choking. The action of the epiglottis is a highly complex reflex process involving two phases. During the first stage a mouthful of chewed food is lifted by the tongue towards the top and back of the mouth. This is accompanied by the cessation of breathing and by the blocking of the nasal areas from the mouth. The second phase involves the epiglottis moving over the larynx while the food passes down into the oesophagus.

stomach

The stomach is a bag of muscle situated just below the diaphragm. It is entered by the cardiac sphincter, which relaxes to admit the food and then closes. The mucous membrane of the stomach is lined with columnar epithelium, in which are embedded little pits called the gastric glands. Gastric juice, containing digestive enzymes and hydrochloric acid, pours from these glands when they are stimulated by the approach of food. The muscular coat of the stomach produces movements that churn the food into a semi-fluid called chyme and tend to urge it towards the intestine, but the pyloric sphincter (opening from the stomach) opens only in response to an acid stimulus. The food received from the mouth is alkaline, owing to the presence of saliva, and so the chyme stays in the stomach until thorough mixing with the hydrochloric acid in the gastric juice has rendered it acid.

Gastric juice contains an enzyme called **pepsin**, which is released as a precursor called pepsinogen; this is a protease (enzyme that acts on protein) that converts the proteins in the food into polypeptides. In babies, there is an enzyme called **rennin** that coagulates milk.

small intestine

The small intestine is 6 m/20 ft long, 4 cm/1.5 in in diameter, and consists of the duodenum, jejunum, and

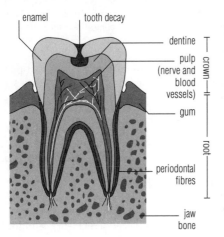

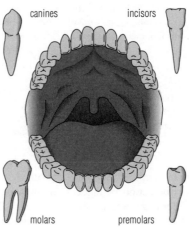

tooth Adults have 32 teeth: two incisors, one canine, two premolars, and three molars on each side of each jaw. Each tooth has three parts: crown, neck, and root. The crown consists of a dense layer of mineral, the enamel, surrounding hard dentine with a soft centre, the pulp.

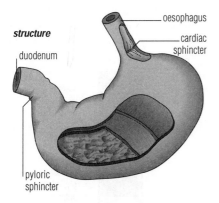

structure

oesophagus

cardiac sphincter

duodenum

pyloric sphincter

detail of stomach wall

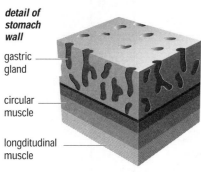

gastric gland

circular muscle

longditudinal muscle

stomach The human stomach can hold about 1.5 l/2.6 pt of liquid. The digestive juices are acidic enough to dissolve metal. To avoid damage, the cells of the stomach lining are replaced quickly – 500,000 cells are replaced every minute, and the whole stomach lining every three days.

---

stomach

The stomach rumbles when empty because of muscle contractions necessary to move the food along. The contractions are controlled by small rhythmic electrical currents generated by cells in the intestine, known as the cells of Cajal.

ileum. It is a muscular tube comprising an inner lining that secretes alkaline digestive juice, a submucous coat containing fine blood vessels and nerves, a muscular coat, and a serous coat covering all, supported by a strong peritoneum, which carries the blood and lymph vessels, and the nerves. The contents are passed along slowly by peristalsis (waves of involuntary muscular action).

duodenum The duodenum is responsible for the digestion of carbohydrates, fats, and proteins, producing smaller molecules that can then be absorbed either by the duodenum itself or the ileum.

Once food has passed into the duodenum from the stomach, it is mixed with intestinal secretions (succus entericus), bile released from the gall bladder, and with a range of enzymes secreted from the pancreas, a digestive gland near the top of the intestine.

Bile consists of bile salts, bile pigments, cholesterol, and lecithin. Bile salts assist in the emulsification of fats – breaking them down into droplets small enough for attack by the enzyme lipase. The bile pigments are the breakdown products of old red blood cells that are passed into the gut to be eliminated with the faeces.

Pancreatic juice contains three enzymes: **trypsin** (released as a precursor, trypsinogen), which attacks proteins more completely than gastric juice, converting them into amino acids; **amylase**, which converts starch into maltose, thus taking over the function of saliva whose activity is stopped in the stomach; and **lipase**, which splits the fats into glycerin and fatty acids.

The succus entericus contains the enzymes **enterokinase**, which is concerned in the production of trypsin, and **aminopeptidases**, which aid trypsin in the breaking up of polypeptides; it also contains enzymes that convert maltose and other sugars into glucose.

jejunum and ileum The jejunum connects the duodenum to the ileum – the part of the intestine that absorbs digested food. The ileum wall is muscular so that waves of contraction (peristalsis) can mix the food and push it forward. Numerous fingerlike projections, or **villi**, point inwards from the wall, increasing the surface area available for absorption. Each villus contains a network of blood vessels, which receives the food molecules diffusing through the wall and transports them to the liver via the hepatic portal vein. It also possess a small lymph vessel, called a **lacteal**.

Glycerin and fatty acids are carried into the lacteal and are again united into small globules of fat (giving the lacteal its characteristic milky appearance). The globules pass into the thoracic duct; from there they pass into the bloodstream and ultimately into the tissues, where they produce energy by oxidation or are stored in the form of adipose tissue.

The amino acids produced by protein digestion are carried in the bloodstream and used to repair and build up the tissues; any excess is converted by the liver into urea, which then passes to the kidneys and excreted via the bladder as urine.

Glucose sugar produced by the digestion of carbohydrates is temporarily stored in the liver as glycogen and converted back to glucose as and when required.

large intestine

The large intestine is 1.5 m/5 ft long, 6 cm/ 2.5 in in diameter, and includes the **caecum**, **colon**, and **rectum**. Materials travel through it slowly, taking from

12 to 18 hours to reach the rectum. During this time, water and mineral salts are absorbed in the colon, and the waste residue is gradually compressed as a compact mass in the rectum, eventually to be egested, or expelled, as faeces at the **anus**.

urinary system

The urinary system is responsible for removing nitrogenous waste products such as urea and excess salts and water from the body. It consists of a pair of kidneys, which produce urine; ureters, which drain the kidneys; and a bladder that stores the urine before its discharge through the urethra.

kidney

The kidneys are a pair of organs situated on the rear wall of the abdomen, which are responsible for osmoregulation (water regulation), excretion of waste products, and maintaining the ionic composition of the blood. Each kidney receives copious quantities of blood via the renal artery, which are processed by more than a million filtering units called **nephrons**, or kidney tubules.

The outer **cortex** of the kidney contains the outermost parts of the nephrons, into which some of the blood's fluid component – plasma – is forced under pressure. The inner **medulla** contains the inner, looping part of the nephron, where about 85% of the plasma – less the waste products and the excess salts and water – is reabsorbed. The remaining fluid in the nephrons drain into the ureters and bladder as urine.

The action of the kidneys is vital, although if one is removed, the other enlarges to take over its function.

nephron Each nephron consists of a knot of blood capillaries called a **glomerulus**, enclosed in a cup-shaped structure called a **Bowman's capsule**, and a long, looping tubule enmeshed with yet more capillaries.

Blood enters the capillaries of the glomerulus at high pressure, forcing about 20% of the blood plasma into the Bowman's capsule. A process of **ultrafiltration** takes place – small molecules in the plasma with a molecular mass of less than 68,000 (such as those of water, salts, urea, and glucose) pass through into the Bowman's capsule, while larger molecules, such as pro-

teins, and blood cells remain in the blood.

The Bowman's capsule leads into the nephron's tubule, which forms a long, U-shaped loop (**loop of**

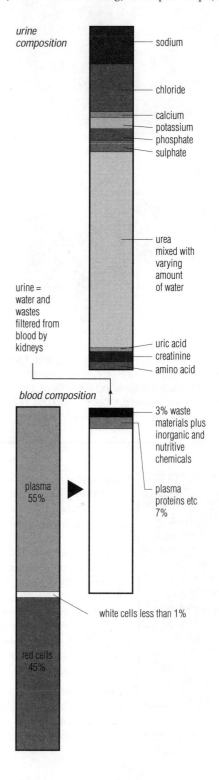

urine composition

— sodium

— chloride

— calcium
— potassium
— phosphate
— sulphate

urine = water and wastes filtered from blood by kidneys

— urea mixed with varying amount of water

— uric acid
— creatinine
— amino acid

blood composition

plasma 55%

red cells 45%

3% waste materials plus inorganic and nutritive chemicals

plasma proteins etc 7%

white cells less than 1%

---

urine Urine consists of excess water and waste products that have been filtered from the blood by the kidneys; it is stored in the bladder until it can be expelled from the body via the urethra. Analysing the composition of an individual's urine can reveal a number of medical conditions, such as poorly functioning kidneys, kidney stones, and diabetes.

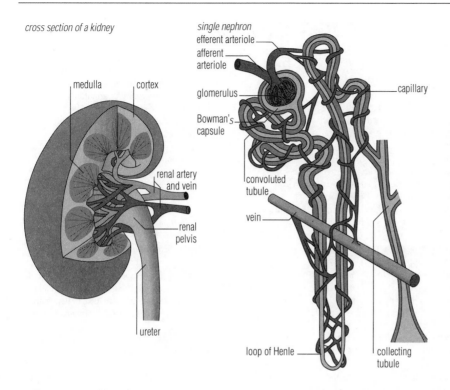

cross section of a kidney

medulla

cortex

renal artery and vein

renal pelvis

ureter

single nephron

efferent arteriole

afferent arteriole

glomerulus

Bowman's capsule

convoluted tubule

vein

loop of Henle

capillary

collecting tubule

nephron The kidney (left) contains more than a million filtering units, or nephrons (right), consisting of the glomerulus, Bowman's capsule, and the loop of Henle. Blood flows through the glomerulus – a tight knot of fine blood vessels from which water and metabolic wastes filter into the tubule. This filtrate flows through the convoluted tubule and loop of Henle where most of the water and useful molecules are reabsorbed into the blood capillaries. The waste materials are passed to the collecting tubule as urine.

Henle) in the kidney's medulla. This is where the selective reabsorption of salts, glucose, and water into the capillaries surrounding the tubule takes place. Additional waste products are also actively secreted into the tubule from the blood.

The antidiuretic hormone (ADH), produced by the pituitary gland, helps to regulate the amount of water reabsorbed from the tubule, depending on whether the overall concentration of the blood is high or low.

The fluid remaining in the tubule – a slightly acidic solution of salts, urea, and other wastes, tinted yellow by a pigment derived from bile – is called urine. It passes into the collecting tubules and ureter to be stored in the bladder, before being discharged to the exterior via the urethra.

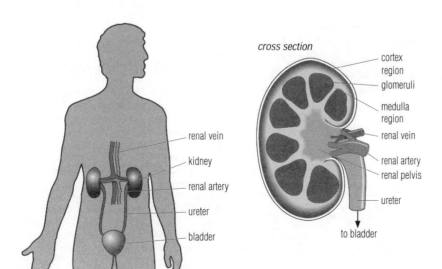

renal vein

kidney

renal artery

ureter

bladder

cross section

cortex region

glomeruli

medulla region

renal vein

renal artery

renal pelvis

ureter

to bladder

kidney Blood enters the kidney through the renal artery. The blood is filtered through the glomeruli to extract the nitrogenous waste products and excess water that make up urine. The urine flows through the ureter to the bladder; the cleaned blood then leaves the kidney via the renal vein.

Human Body Chronology

c. 1600 BC The Edwin Smith papyrus is written. The first medical book, it reveals an accurate understanding of the workings of the heart, stomach, bowels, and larger blood vessels. The papyrus is named after US scientist Edwin Smith, a pioneer in the study of Egyptian science who acquired it in Luxor, Egypt, in 1862.

c. 1550 BC The Ebers papyrus is written. One of the oldest known medical works, it accurately describes the circulatory system, and, for the first time, recognizes the brain's central control function. The papyrus is named after German Egyptologist and novelist George Maurice Ebers, who acquired it in 1873.

c. 550 BC A document from Mohenjo Daro, in the Indus valley, includes information on anatomy, physiology, pathology, and obstetrics.

c. 285 BC Herophilus, an anatomist working at Alexandria, dissects human bodies and compares them with large mammals. He distinguishes the cerebrum and cerebellum, establishes the brain as the seat of thought, writes treatises on the human eye and on general anatomy, and writes a handbook for midwives.

c. 250 BC Greek anatomist Erasistratus of Ceos notes the difference between sensory and motor nerves, and correctly describes the functions of the valves of the heart.

157 Greek physician Galen becomes physician to the gladiators in his native Pergamum, offering him a unique insight into anatomy and the treatment of wounds.

180 Galen, practising at Rome, writes *Methodus medendi/Method of Physicians*, a medical textbook that will become the ultimate authority for medieval medicine.

1110 The text *Anatomia porci/The Anatomy of the Pig* describes the dissection of a pig in the medical school at Salerno, Italy. With human dissection still forbidden by the church, it becomes a valuable reference.

1163 At the council of Tours in France, the Catholic Church issues an edict against the mutilation of dead bodies. Although primarily aimed at the stripping of crusaders' bones for transport back to Europe, it also affects anotomical research.

1214 The Italian physician Marus of the medical school at Salerno, Italy, writes *Anatomia Mauri/Anatomy of a Moor*, one of the earliest Latin texts on anatomy.

1316 The Italian physician Mondino de Liuzzi conducts the first properly recorded dissection of a human corpse at Bologna University, Italy. His book *Anatomia* will become the standard work on anatomy for two centuries.

1527 Swiss physician Paracelsus lectures at Basel University, inviting the public to attend his lectures, and burning the books of Avicenna and Galen – the standard medical works of the day.

1530 Paracelsus writes *Paragranum*, arguing that the body is based on chemical processes, and suggesting specific chemical treatments for different diseases.

1540 Flemish anatomist Andreas Vesalius performs dissections on human cadavers at the University of Bologna. His discoveries contradict the writings of the ancient Greek physician Galen, until now the highest authority.

1542 Vesalius writes *De humani corporis fabrica/On the Fabric of the Human Body*, a highly illustrated, clearly written study of the human body, and effectively the beginning of the science of anatomy.

1552 Italian anatomist Bartolomeo Eustachio, in his *Tabulae anatomicae/Anatomical Writings*, details his discovery of the Eustachian tube between the ear and pharynx, and the Eustachian valve of the heart.

1553 Spanish theologian Michael Servetus, relates his discovery of the pulmonary circulation of the blood.

1561 Italian anatomist Gabriello Falloppio describes the inner ear and female reproductive organs for the first time.

1565 The Royal College of Physicians, London, England, is empowered by Queen Elizabeth I to carry out human dissections.

1568 Italian anatomist Contanzo Varolio publishes his research into the structure of the human brain.

1603 Italian anatomist Hieronymus Fabricius of Acquapendente discovers that the veins contain valves.

1604 Fabricius publishes *De formata foetu/On the Formation of the Fetus*, the first important study of embryology, in which the placenta is identified for the first time.

1614 Italian physician Santorio Santorio (Sanctorius) publishes *De medicina statica/On Medical Statics*, a pioneering study of perspiration and the metabolism.

1619 English physician William Harvey first announces his discovery of the circulation of the blood.

1641 Belgian anatomist Franciscus Sylvius discovers the cerebral or Sylvian fissure that separates the temporal, frontal, and parietal lobes of the brain.

1647 French medical student Jean Pecquet discovers the thoracic duct of the human body, root of the lymphatic system.

1652 Danish physician Thomas Bartholin publishes the first full description of the human lymphatic system.

1667 Italian anatomist Marcello Malpighi identifies the lower layer of the skin known today as the Malpighian layer.

1679   Swiss physician Théophile Bonet collates the results of over 3,000 post mortems in his *Sepulcretum/Cemetery*, founding the study of morbid anatomy.

1684   Dutch microscopist Anton van Leeuwenhoek first describes red blood cells accurately.

1706   Italian anatomist Giovanni Morgagni publishes *Adversaria anatomica/Anatomical Arguments*, establishing him as an important anatomist.

1709   English philosopher and scientist George Berkeley publishes his *New Theory of Vision*. Berkeley maintains that the eye only conveys sensations of colour and that perceptions of form are gathered by touch.

1717   Leeuwenhoek discovers the structure of the nerves.

1745   French naturalist Pierre de Maupertuis attacks the currently favoured theory of reproduction, that the sperm contain a miniature version of the adult. He argues that characteristics of both parents influence the offspring.

1757   Swiss physiologist Albrecht von Haller publishes the first volume of his vast anatomy encyclopedia *Elementa physiologiae corporis humanae/Elements of the Physiology of the Human Body*.

1761   Italian physician Giovanni Morgagni writes *De Sedibus et Causis Morborum per Anatomen Indagatis/On the Seats and Causes of Diseases Investigated by Anatomy*. Based on almost 700 post mortem examinations, it is one of the first books on pathological anatomy.

1766   Albrecht von Haller shows that nerves stimulate muscles to contract, and that all nerves lead to the spinal column and brain. His work lays the foundation of modern neurology.

c. 1780 Italian anatomist Luigi Galvani discovers the electric nature of the nervous impulse.

1787   The German writer and polymath Johann Wolfgang von Goethe discovers the intermaxillary bone.

1802   English physician and physicist Thomas Young postulates that the eye requires only three receptors to see the full spectrum, instead of receptors for each colour as is generally believed.

1811   Scottish anatomist Charles Bell distinguishes between sensory and motor nerves.

1824   French chemists Jean-Baptiste-André Dumas and C Prevost show that sperm is essential to fertilization.

1827   Estonian embryologist Karl von Baer reports the discovery of eggs in mammals and humans. It dispels the idea of the preformation of the embryo.

c. 1833 German physician Gustav Henle describes in detail the structures of the eye and brain.

1834   The English microscopist Joseph Lister discovers the true shape of red blood cells.

1836   German physiologist Theodor Schwann discovers pepsin, the first known animal enzyme to be isolated.

1837   Bohemian physiologist Jan Purkinje discovers large nerve cells in the cerebellum with branching extensions; they are now called Purkinje cells.

1838   German chemist Justus von Liebig demonstrates that animal heat is due to respiration.

1840   Swiss embryologist Rudolf Albert von Kölliker identifies spermatozoa as cells.

1841   German physician Gustav Henle discovers that 'ductless glands' secrete their products directly into the bloodstream.

1842   Polish embryologist Robert Remak discovers that the early embryo consists of three layers: ectoderm, mesoderm, and endoderm.

1846   French physiologist Claude Bernard discovers that pancreatic secretions break down fat molecules into fatty acids and glycerin.

1852   Remak discovers that the growth of tissues, involves both the multiplication and division of cells.

1852   Swiss embryologist and histologist Rudolf Albert von Kölliker publishes *Handbuch der Gewebelehre des Menschen/Handbook of Human Histology*, the first textbook in histology, and the first to discuss tissues in terms of cell theory.

1855   Claude Bernard discovers that ductless glands produce hormones, which he calls 'internal secretions'.

1858   British physician Henry Gray publishes *Anatomy of the Human Body, Descriptive and Surgical* (*Gray's Anatomy*). It remains the standard text in anatomy for over 100 years.

1875   German embryologist Oskar Hertwig discovers that fertilization occurs with the fusion of the nuclei of the sperm and ovum.

1880   The parathyroid gland, which secretes parathormone that regulates calcium levels in the blood, is first described by Swedish physiologist Ivar Sandström.

1889   German physiologists Oskar Minkowski and Joseph von Mering remove the pancreas from a dog, which then develops diabetic symptoms. It leads them to conclude that the pancreas secretes an antidiabetic substance, which is now known as insulin.

1895   Belgian bacteriologist Jules Bordet discovers antibodies.

1896   German biochemist E Baumann discovers iodine in the thyroid gland; it is absent in all other tissues.

1898   Belgian bacteriologist Jules Bordet discovers haemolysis, the rupture of foreign red blood cells in blood serum. It soon leads to the discovery of blood groups.

1900   Austrian immunologist Karl Landsteiner discovers the ABO blood group system.

1901   Japanese-born US biochemist Jokichi Takamine first synthesizes the heart stimulant adrenaline (epinephrine) from the suprarenal gland. It is the first pure hormone to be synthesized from natural sources.

1902   US surgeon Harvey Williams Cushing publishes *The Pituitary Body and its Disorders*, which investigates the pituitary gland and its relationship to various diseases.

1903	Dutch physiologist Willem Einthoven invents the string galvanometer (electrocardiograph), which measures and records the tiny electrical impulses produced by contractions of the heart muscle.
1903	Russian physiologist Ivan Pavlov describes learning by conditioning. He trains dogs to expect food when they hear a bell and eventually they salivate every time the bell rings.
1914	US biochemist Edward Kendall isolates the hormone thyroxine from the thyroid gland. It regulates metabolism by stimulating all cells to consume oxygen.
1921	French physician Jean Athanase Sicard uses a radiopaque iodine substance to X-ray the internal structure of the spinal column. He studies the bronchial tube the next year.
1921	Scottish bacteriologist Alexander Fleming discovers the antibacterial enzyme lysozyme, which is found in tears and saliva.
1922	English biochemist Frederick Gowland Hopkins isolates glutathione and demonstrates its vital role in the cell's utilization of oxygen.
1922	French surgeon Alexis Carrel discovers white blood cells (leucocytes).
1925	US pathologist George Whipple demonstrates that iron is the most important factor involved in the formation of red blood cells.
1929	German biochemist Adolf Butenandt and, simultaneously and independently, US biochemist Edward Doisy isolate an oestrogen hormone, which is involved in the growth and development of females.
1931	Adolf Butenandt isolates the male sex hormone androgen.
1932	German-born British biochemist Hans Krebs discovers the urea cycle, in which ammonia is turned into urea in mammals.
1934	Adolph Butenandt isolates the female sex hormone progesterone.
1935	US biochemist Edward Calvin Kendall isolates the steroid hormone cortisone from the adrenal cortex.
1937	German-born British biochemist Hans Krebs describes the citric acid cycle in cells, which converts sugars, fats, and proteins into carbon dioxide, water, and energy – the 'Krebs cycle'.
1940	Karl Landsteiner and US physician and immunohaematologist Alexander Wiener discover the rhesus (Rh) factor in blood, in the USA.
1940	US physiologist Herbert M Evans uses radioactive iodine to prove that iodine is used by the thyroid gland.
1948	Swiss physiologist Walter Hess describes using fine electrodes to stimulate or destroy specific regions of the brain in cats and dogs; it allows him to discover the role played by various parts of the brain.
1949	US researchers synthesize adrenocorticotropic hormone (ACTH) which the pituitary gland secrets

	to stimulate the adrenal glands.
1951	US biochemist Robert Woodward synthesizes cortisone.
1953	English biochemist Frederick Sanger determines the structure of the insulin molecule. The largest protein molecule to have its chemical structure determined to date, it is essential in the laboratory synthesis.
1957	Interferon, a natural protein that fights viruses, is discovered by Scottish virologist Alick Isaacs and Swiss virologist Jean Lindemann.
1959	Austrian-born British biochemist Max Perutz determines the structure of haemoglobin.
1960	English biochemist John Kendrew, using X-ray diffraction techniques, elucidates the three-dimensional structure of the muscle protein myoglobin.
1971	Li Choh Hao and associates at the University of California Medical Centre announce the synthesis of the human growth hormone somatotrophin.
1971	Polish-born US endocrinologist Andrew Schally isolates the luteinizing hormone-releasing hormone (LH-RH), essential to human ovulation.
1971	Surgeons develop the fibreoptic endoscope, making it possible to view inside the human body by inserting catheters into the arms or legs and manipulating them into organs, such as the heart.
1972	Venezuelan-born US immunologist Baruj Benacerraf and US microbiologist Hugh O'Neill McDevitt show immune response to be genetically determined.
1974	British-born Danish immunologist Niels Jerne proposes a network theory of the immune system.
1975	Swiss scientists publish details of the first chemically directed synthesis of insulin.
1975	US physiologist John Hughes discovers endorphins (morphine-like chemicals) in the brain.
1979	H Goodman and J Baxter of the University of California, Berkeley, together with D V Goeddel of Genentech, announce the biosynthetic production of a human growth hormone.
1988	The Human Genome Organization (HUGO) is established in Washington, DC, USA; scientists announce a project to compile a complete 'map' of human genes.
1992	Sperm cells are discovered by US physician David Garbers of the University of Texas to have odour receptors and may therefore reach eggs by detecting scent.
1996	Two US dentists discover a new muscle running from the jaw to just behind the eye socket. About 3 cm/1 in long, it helps to support and raise the jaw.
1998	Researchers at the University of Texas in conjunction with the British company SmithKline Beecham identify a hormone that triggers hunger in humans.

Biographies

Abel, John Jacob (1857–1938) US biochemist, discoverer of adrenaline. He studied the chemical composition of body tissues, and this led, in 1898, to the discovery of adrenaline, the first hormone to be identified, which Abel called epinephrine. He later became the first to isolate amino acids from blood.

Axelrod, Julius (1912–) US neuropharmacologist who shared the 1970 Nobel Prize for Physiology or Medicine with the biophysicists Bernard Katz and Ulf von Euler (1905–1983) for his work on neurotransmitters (the chemical messengers of the brain). Axelrod wanted to know why the messengers, once transmitted, should stop operating. Through his studies he found a number of specific enzymes that rapidly degraded the neurotransmitters.

Bayliss, William Maddock (1860–1924) English physiologist who discovered the digestive hormone secretin, the first hormone to be found, with Ernest Starling 1902. During World War I, Bayliss introduced the use of saline (salt water) injections to help the injured recover from shock.

By experimenting with the inner lining of the duodenum (first part of the intestines), Bayliss concluded that as hydrochloric acid from the stomach's digestive juices passes into the duodenum during the normal digestive process, the duodenal lining releases a chemical into the bloodstream which, in turn, makes the pancreas secrete its juices. This is the hormone secretin.

Bayliss went on to study the activation of enzymes, particularly the pancreatic enzyme trypsin. Bayliss and Starling also investigated the peristaltic movements of the intestines and their nerve supply, and pressures within the venous and arterial systems.

Cohen, Stanley (1922–) US biochemist who was awarded the Nobel Prize for Physiology or Medicine 1986 jointly with Rita Levi-Montalcini for their work to isolate and characterize growth factors, small proteins that regulate the growth of specific types of cells.

Cohen helped to purify and characterize nerve growth factor, a small protein produced in the male salivary gland that regulates the growth of small nerves and affects the development of the sensory and sympathetic nerve cells. He went on to discover another growth factor, called epidermal growth factor, that affects epithelial cell growth, tooth eruption, and eyelid opening. He then laboured to link epidermal growth factor to the regulation of embryonic growth. Subsequent studies by other scientists have shown that this growth factor also plays a crucial part in the exaggerated growth rate of some cancer cells.

Cori, Carl Ferdinand (1896–1984) and Gerty Theresa (1896–1957) US biochemists born in Austro-Hungary who, together with Argentine physiologist Bernardo Houssay (1887–1971), received a Nobel prize 1947 for their discovery of how glycogen (animal starch) – a derivative of glucose – is broken down and resynthesized in the body, for use as a store and source of energy.

Glycogen is broken down in the muscles into lactic acid, which, when the muscles rest, is reconverted to glycogen. In the 1930s the Coris set out to determine exactly how these changes take place. Gerty Cori found a new substance in muscle tissue, glucose-1-phosphate, now known as Cori ester. Its formation from glycogen involves only a small amount of energy change, so that the balance between the two substances can easily be shifted in either direction. The second step in the reaction chain involves the conversion of glucose-1-phosphate into glucose-6-phosphate. Finally this second phosphate is changed to fructose-1,6-diphosphate, which is eventually converted to lactic acid. The first set of reactions from glycogen to glucose-6-phosphate is now termed glycogenolysis; the second set, from glucose-6-phosphate to lactic acid, is referred to as glycolysis.

Du Bois-Reymond, Emil Heinrich (1818–1896) German physiologist. He showed the existence of electrical currents in nerves, correctly arguing that it would be possible to transmit nerve impulses chemically. His experimental techniques proved the basis for almost all future work in electrophysiology.

Investigating the physiology of muscles and nerves, Du Bois-Reymond demonstrated the presence of electricity in animals, especially researching electric fishes. By 1849 he had evolved a delicate multiplier for measuring nerve currents, enabling him to detect an electric current in ordinary localized muscle tissues, notably contracting muscles. He observantly traced it to individual fibres, finding their interior was negative with regard to the surface.

Erasistratus (c. 304–c. 250 BC) Greek physician and anatomist. Regarded as the founder of physiology, he came close to discovering the true function of several important systems of the body, which were not fully understood until nearly 1,000 years later. For example, the principle of blood circulation, though he had it circulating in the wrong direction. Tracing the network of veins, arteries, and nerves, he postulated that the nerves carry the 'animal' spirit, the arteries the 'vital' spirit, and the veins blood. He did, however, grasp a rudimentary principle of oxygen exchange and condemned bloodletting as a form of treatment.

Erasistratus dissected and examined the human brain, noting the convolutions of the outer surface, and observed that the organ is divided into larger and smaller portions (the cerebrum and cerebellum). He compared the human brain with those of other animals and made the correct hypothesis that the surface area/volume complexity is directly related to the intelligence of the animal.

Fabricius, Geronimo (1537–1619) Latinized name of Girolamo Fabrizio, Italian anatomist and embryologist. He made a detailed study of the veins and discovered the valves that direct the blood flow towards the heart. He also studied the development of chick embryos.

Fabricius also investigated the mechanics of respiration, the action of muscles, the anatomy of the larynx (about which he was the first to give a full description) and the eye (he was the first to correctly describe the location of the lens and the first to demonstrate that the pupil changes size).

Fabricius publicly demonstrated the valves in the veins of the limbs 1579, and in 1603 published the first accurate description, with detailed illustrations, of these valves in *De Venarum Ostiolis/On the Valves of the Veins*. He mistakenly believed, however, that the valves' function was to retard the flow of blood to enable the tissues to absorb nutriment.

Flourens, Pierre Jean Marie (1794–1867) French physiologist who experimented widely on the effects of the removal of various parts of the central nervous system. He determined the function of different parts of the mammalian brain and the role of the semi-circular canals of the inner ear in balance.

Flourens removed the cerebral hemispheres in the brain of a pigeon and observed that this made the bird blind. When one cerebral hemisphere was removed, the bird lost the sight from its opposite eye. Flourens therefore demonstrated that vision depends on the integrity of the cerebral cortex. He next removed only the cerebellum and determined that while the bird could see and hear well, it stood, walked, and flew in an indecisive manner. The bird's equilibrium was almost entirely abolished. Flourens later demonstrated the same results on a dog. Injury to the cerebellum therefore causes loss of coordination. Flourens introduced the idea of nervous coordination to physiology.

Flourens also did important work on the role of the semi-circular canals of the inner ear on balance and demonstrated that the respiratory centre is situated in the brain stem, the medulla oblongata, the area in the brain that is responsible for the involuntary contraction of the respiratory muscles.

Gasser, Herbert Spencer (1888–1963) US physiologist who shared the 1944 Nobel Prize for Physiology or Medicine with Joseph Erlanger for their discoveries regarding the specialized functions of nerve fibres. Gasser was also one of the first to demonstrate the chemical transmission of nerve impulses.

Gasser and Erlanger found that the smaller nerve fibres were responsible for the conduction of pain and that the speed of electrical transmission by a nerve depends upon its diameter. Gasser also performed a great deal of experiments attempting to prove that chemical transmission occurs between nerves. He was one of the first to demonstrate that the injection of acetylcholine into bird muscles or denervated mammalian muscles results in slow contraction. Acetylcholine is now known to be a neurotransmitter, a chemical that carries nerve impulses across synapses between nerves.

Golgi, Camillo (1843–1926) Italian cell biologist who produced the first detailed knowledge of the fine structure of the nervous system. He shared the 1906 Nobel Prize for Physiology or Medicine with Santiago Ramresults in slow contraction. Acetylcholine is now known to be a neurotalts in staining cells proved so effective in showing up the components and fine processes of nerve cells that even the synapses – tiny gaps between the cells – were visible. The **Golgi apparatus**, a series of flattened membranous cavities found in the cytoplasm of cells, was first described by him in 1898.

From his examinations of different parts of the brain, Golgi put forward the theory that there are two types of nerve cells, sensory and motor cells, and that axons are concerned with the transmission of nerve impulses. He discovered tension receptors in the tendons – now called the organs of Golgi.

Graaf, Regnier de (1641–1673) Dutch physician and anatomist who discovered the ovarian follicles, which were later named **Graafian follicles**. He named the ovaries and gave exact descriptions of the testicles. He was also the first to isolate and collect the secretions of the pancreas and gall bladder.

Hall, Marshall (1790–1857) English physician and physiologist who distinguished between voluntary and involuntary reflex muscle contractions, proving that the spinal cord is more than a passive nerve trunk transmitting voluntary signals from the brain and sensory signals to the brain.

Hall is best known for his work on the nervous system of frogs. He showed that if the spinal cord of a frog was severed between the front and back limbs, then the front limbs could still be moved voluntarily but the back limbs were useless. He further showed that the back legs could be stimulated to move artificially, but only once for each stimulus. These were reflex (involuntary) muscle contractions. Pain stimuli applied to the back legs were not felt by the animals. From these experiments Hall deduced that the nervous system is made up of a series of reflex arcs. In the intact spinal cord these reflex arcs are coordinated by the ascending and descending pathways in the cord to form movement patterns.

Hall also demonstrated that stimulus could not be put into the cord through a sensory nerve without it resulting in effects beyond the anatomical segment to which that nerve belongs.

Haller, Albrecht von (1708–1777) Swiss physician and scientist, founder of neurology. He studied the muscles and nerves, and concluded that nerves provide the stimulus that triggers muscle contraction. He also showed that it is the nerves, not muscle or skin, that receive sensation.

Tracing the pathways of nerves, he was able to demonstrate that they always lead to the spinal cord or the brain, suggesting that these regions might be where awareness of sensation and the initiation of answering responses are located. While carrying out his experiments, Haller discovered several processes of the human body, such as the role of bile in digesting fats. He also wrote a report on his study of embryonic development.

Harvey, William (1578–1657) English physician who discovered the circulation of blood. In 1628 he published his book *De motu cordis/On the Motion of the Heart and the Blood in Animals*. He also explored the development of chick and deer embryos. Harvey's discovery marked the beginning of the end of medicine as taught by Greek physician ◊Galen, which had been accepted for 1,400 years.

Examining the heart and blood vessels of mammals, Harvey deduced that the blood in the veins must flow only towards the heart. He also calculated the amount of blood that left the heart at each beat, and realized that the same blood must be circulating continuously around the body. He reasoned that it passes from the right side of the heart to the left through the lungs (pulmonary circulation).

Herophilus of Chalcedon (*c.* 330–*c.* 260 BC) Greek physician, active in Alexandria. His handbooks on anatomy make pioneering use of dissection, which, according to several ancient sources, he carried out on live criminals condemned to death.

Hodgkin, Alan Lloyd (1914–) British physiologist engaged in research with Andrew Huxley on the mechanism of conduction in peripheral nerves 1945–60. He devised techniques for measuring electric currents flowing across a cell membrane. In 1963 they shared the Nobel prize.

Hodgkin and Huxley managed for the first time to record electrical changes across the cell membrane, and Hodgkin then built on these findings working with Bernard ◊Katz, another cell physiologist. They proposed that during the resting phase a nerve membrane allows only potassium ions to diffuse into the cell, but when the cell is excited it allows sodium ions (which are positively charged) to enter and potassium ions to move out. The extrusion of sodium is probably dependent on the metabolic energy supplied either directly or indirectly in the form of ATP (adenosine triphosphate). The amount of sodium flowing in equals that of the potassium flowing out.

Huxley, Andrew Fielding (1917–) English physiologist, awarded the Nobel prize in 1963 with Alan Hodgkin for work on nerve impulses, discovering how ionic mechanisms are used in nerves to transmit impulses.

In 1945 at Cambridge, Hodgkin and Huxley began to measure the electrochemical behaviour of nerve membranes. They experimented on axons of the giant squid – each axon is about 0.7 mm/0.03 in in diameter. They inserted a glass capillary tube filled with sea water into the axon to test the composition of the ions in and surrounding the cell, which also had a microelectrode inserted into it. Stimulating the axon with a pair of outside electrodes, they showed that the inside of the cell was at first negative (the resting potential) and the outside positive, and that during the conduction of the nerve impulse the membrane potential reversed.

Hyrtl, Joseph (1810–1894) Austrian anatomist who, in part, was responsible for the success of the New Vienna School of Medicine in the 19th century. He was the most popular and successful teacher of anatomy in Europe. His textbooks of general anatomy were widely used including his seminal work *The Handbook of Topographical Anatomy* 1845. He is also the author of scholarly works on anatomical terminology and on Hebrew and Arabic elements of anatomy.

Katz, Bernard (1911–) British biophysicist. He shared the 1970 Nobel Prize for Physiology or Medicine for work on the biochemistry of the transmission and control of signals in the nervous system, vital in the search for remedies for nervous and mental disorders.

In the 1940s, Katz joined in the Nobel-prizewinning research of Alan ◊Hodgkin and Andrew ◊Huxley on the electrochemical behaviour of nerve membranes. During the 1950s, Katz found that minute amounts of acetylcholine were randomly released by nerve endings at the neuromuscular junction, giving rise to very small electrical potentials; he also found that the size of the potential was always a multiple of a certain minimum value. These findings led him to suggest that acetylcholine was released in discrete 'packets' (analogous to quanta) of a few thousand molecules each, and that

these packets were released relatively infrequently while a nerve was at rest but very rapidly when an impulse arrived at the neuromuscular junction.

Kuhne, Wilhelme (1837–1900) German physiologist who coined the term 'enzyme' (a substance produced by cells that is capable of speeding up chemical reactions) and was the first to show the reversible effect of light on the retina of the eye.

Kuhne began his research on substances responsible for the breakdown of foodstuffs in the digestive system. After working on the active substances in pancreatic juices, Kuhne coined the term 'enzyme', which he derived from the Greek words **en** which means 'in' and **zyme** which means 'leaven'. He also worked on the photosensitive proteins present in the retina of the frog's eye and was the first to show the reversible effect of light on the activity of a coloured pigment (later called rhodopsin) in the retina.

Langley, John Newport (1852–1925) English physiologist who investigated the structure and function of the autonomic nervous system, the involuntary part of the nervous system, that controls the striated and cardiac muscles and the organs of the gastrointestinal, cardiovascular, excretory, and endocrine systems. He went on to divide up the autonomic nervous system into the sympathetic and parasympathetic branches, with specific functions being apportioned to each.

Langley did a great deal of research on the structure and function of sympathetic nerve fibres and ganglia. Ganglia are clusters of the cell bodies of sensory nerve cells in the peripheral nervous system which lie just outside the spinal cord. Langley blocked nervous impulses by applying various chemicals, such as nicotine, to ganglia. The cell bodies of motor nerve cells in the autonomic nervous system also lie in these ganglia.

Levi-Montalcini, Rita (1909–) Italian neurologist who discovered nerve-growth factor, a substance that controls how many cells make up the adult nervous system. She shared the 1986 Nobel Prize for Physiology or Medicine with her co-worker, US biochemist Stanley ◊Cohen.

Levi-Montalcini first discovered nerve-growth factor in the salivary glands of developing mouse embryos, and later in many tissues. She established that it was chemically a protein, and analysed the mechanism of its action. Her work has contributed to the understanding of some neurological diseases, tissue regeneration, and cancer mechanisms.

Lucas, Keith (1879–1916) English neurophysiologist who investigated the transmission of nerve impulses. He demonstrated that the contraction of muscle fibres follows the 'all or none' law: a certain amount of stimulus is needed in order to induce a nerve impulse and subsequent muscle contraction. Any stimulus below that threshold has no effect regardless of its duration.

Lucas showed that when two successive stimuli are given, the response to the second stimulus cannot be evoked if the first nerve impulse is still in progress. He also demonstrated that following a contraction there is a period of diminished excitability during which the muscle cannot be induced to contract again. This is due to the chemical transmission of impulses over synaptic clefts (the junction between two individual nerve cells).

Malpighi, Marcello (1628–1694) Italian physiologist who made many anatomical discoveries in his pioneering microscope studies of animal and plant tissues. For example, he discovered blood capillaries and indentified the sensory receptors (papillae) of the tongue, which he thought could be nerve endings.

Studying the lungs of a frog, Malpighi found them to consist of thin membranes containing fine blood vessels covering vast numbers of small air sacs. This discovery made it easier to explain how air (oxygen) seeps from the lungs to the blood vessels and is carried around the body.

Mechnikov, Ilya Ilich (1845–1916) Russian-born French zoologist who discovered the function of white blood cells and phagocytes (amoebalike blood cells that engulf foreign bodies). He also described how these 'scavenger cells' can attack the body itself (autoimmune disease). He shared the Nobel Prize for Physiology or Medicine 1908.

While studying the transparent larvae of starfish, Mechnikov observed that certain cells surrounded and engulfed foreign particles that entered the bodies of the larvae. Later he demonstrated that phagocytes exist in higher animals, and form the first line of defence against acute infections.

Meyerhof, Otto (1884–1951) German-born US biochemist who carried out research into the metabolic processes involved in the action of muscles. For this work he shared the 1922 Nobel Prize for Physiology or Medicine.

In 1920 Meyerhof showed that, in anaerobic conditions, the amounts of glycogen metabolized and lactic acid produced in a contracting muscle are proportional to the tension in the muscle. He also demonstrated that 20–25% of the lactic acid is oxidized during the muscle's recovery period and that energy produced by this oxidation is used to convert the remainder of the lactic acid back to glycogen. Meyerhof introduced the term 'glycolysis' to describe the anaerobic degradation of glycogen to lactic acid, and showed the cyclic nature of energy transformations in living cells.

Michaelis, Leonor (1875–1949) German-born, US biochemist who derived a mathematical model to describe the kinetics of how enzymes catalyse (trigger) reactions. The work of Michaelis and German scientist Maude Menton enabled several subsequent generations of biochemists to correctly assess the nature and efficiency of the key enzyme-driven steps in cell metabolism.

In their mathematical calculations they correctly assumed that an enzyme works by rapidly and reversibly binding to a specific molecule (called the substrate) to form an enzyme–substrate complex, triggering a reaction that generates a product molecule. Once the reaction has occurred the enzyme is released unchanged. Michaelis and Menton then correlated the speed of the enzymatic reaction with the concentrations of both the enzyme and the substrate.

Pavlov, Ivan Petrovich (1849–1936) Russian physiologist who studied conditioned reflexes in animals. His work had a great impact on behavioural theory and learning theory. He was awarded Nobel Prize for Physiology or Medicine 1904.

Studying the physiology of the circulatory system and the regulation of blood pressure, Pavlov devised animal experiments such as the dissection of the cardiac nerves of a living dog to show how the nerves that leave the cardiac plexus control heartbeat strength.

Pavlov's work relating to human behaviour and the nervous system also emphasized the importance of conditioning. He deduced that the inhibitive behaviour of a psychotic person is a means of self-protection. The person shuts out the world and, with it, all damaging stimuli. Following this theory, the treatment of psychiatric patients in Russia involved placing a sick person in completely calm and quiet surroundings.

Sakmann, Bert (1942–) German cell physiologist who shared the 1991 Nobel Prize for Physiology or Medicine with Erwin Neher for their studies of the electrical activity of nerve cell membranes. They also determined the role of the neurohormone beta-endorphin.

In 1976 Sakmann worked with Neher to develop a technique called the patch-clamp technique, which greatly enhanced the ability of researchers to measure the electrical activity of nerves and revolutionized the study of ion channels in membranes.

Using the patch-clamp technique Neher and Sakmann also investigated the role of beta-endorphin. Beta-endorphin is a neurohormone which is secreted by the pituitary gland and reaches all body tissues carried in the blood. It is a peptide opiate that has been found to play a clinical role in the perception of pain, behavioural patterns, obesity and diabetes, and psychiatric disorders. They demonstrated that beta-endorphin acts not only to regulate the release of neurotransmitter substances by nerves in the brain, but also, via calcium channels, on the walls of the arteries of the brain.

Samuelsson, Bengt Ingemar (1934–) Swedish biochemist who shared the 1982 Nobel Prize for Physiology or Medicine with Sune Bergström and John Vane for the purification of prostaglandins, chemical messengers produced by the prostate gland.

Ulf Muller originally discovered that human semen and extracts of sheep seminal vesicular glands had peculiar properties. Both substances caused contraction of smooth muscle in vitro (in an artificial environment, such as a test-tube) and sharp decreases in the blood pressure in experimental animals. Muller called the active agents in these substances prostaglandins, because they were primarily made in the prostate gland.

The purification of the prostaglandins was complicated by the very low amounts present in seminal fluid and their extremely short half lives. In 1957, Bergström and Samuelsson managed to obtain crystals from two prostaglandins, alprostadil (PGE1) and PGF1a, which cause the contraction of smooth muscle. They reported the chemical characterization of these two prostaglandins 1962.

Servetus, Michael (Miguel Serveto) (1511–1553) Spanish Christian Anabaptist theologian and physician. He was a pioneer in the study of the circulation of the blood and found that it circulates to the lungs from the right chamber of the heart. He was burned alive by the church reformer Calvin in Geneva, Switzerland, for publishing attacks on the doctrine of the Trinity.

Vesalius, Andreas (1514–1564) Belgian physician who revolutionized anatomy by performing post mortem dissections and making use of illustrations to teach anatomy. Vesalius upset the authority of ◊Galen, and his book – the first real textbook of anatomy – marked the beginning of biology as a science. His dissections of the human body (then illegal) enabled him to discover that ◊Galen's system of medicine was based on fundamental anatomical errors. Vesalius disproved the widely held belief that men had one rib less than women. He also believed, contrary to Aristotle's theory of the heart being the centre of the mind and emotion, that the brain and the nervous system were the centre.

Vesalius's book *De humani corporis fabrica/On the Structure of the Human Body* of 1543 employed talented artists to provide the illustrations and is one of the great books of the 16th century. The quality of anatomical depiction introduced a new standard into all illustrated works, especially into medical books, and highlighted the need to introduce scientific method into the study of anatomy. Together with the main work of astronomer Copernicus, published in the same year, *On the Structure of the Human Body* marked the dawn of modern science.

Glossary

abdomen the part of the body below the ◊thorax, containing the digestive organs and female reproductive organs. It is separated from the thorax by the ◊diaphragm, a sheet of muscular tissue.

accommodation the ability of the ◊eye to focus on near or far objects by changing the shape of the lens.

acetylcholine, ACh, chemical that serves as a ◊neurotransmitter, communicating nerve impulses between the cells of the nervous system. It is largely associated with the transmission of impulses across the ◊synapse (junction) between nerve and muscle cells, causing the muscles to contract.

action potential change in the potential difference (voltage) across the membrane of a nerve cell when an impulse passes along it. A change in potential (from about –60 to +45 millivolts) accompanies the passage of sodium and potassium ions across the membrane.

active transport in cells, the use of energy to move substances, usually molecules or ions, across a membrane.

adenoids masses of lymphoid tissue, similar to ◊tonsils, located in the upper part of the throat, behind the nose. They are part of a child's natural defences against the entry of germs but usually shrink and disappear by the age of ten.

ADH, abbreviation for **antidiuretic hormone**, part of the system maintaining a correct salt/water balance in vertebrates.

adipose tissue ◊connective tissue that serves as an energy reserve, and also pads some organs. It is commonly called fat tissue, and consists of large spherical cells filled with fat. Major layers are in the inner layer of skin and around the kidneys and heart.

adrenal gland, or **suprarenal gland**, triangular gland situated on top of the ◊kidney. The **cortex** (outer part) secretes various steroid hormones and other hormones that control salt and water metabolism and regulate the use of carbohydrates, proteins, and fats. The **medulla** (inner part) secretes the hormones adrenaline and noradrenaline which, during times of stress, prepare the body for 'fight or flight'.

adrenaline, or **epinephrine**, hormone secreted by the medulla of the ◊adrenal glands. Adrenaline is synthesized from a closely related substance, noradrenaline, and the two hormones are released into the bloodstream in situations of fear or stress.

adrenocorticotrophic hormone hormone secreted by the anterior lobe of the ◊pituitary gland.

alimentary canal tube, adapted for (digestion, through which food passes. It is a complex organ, consisting of the mouth cavity, pharynx, oesophagus, stomach, and the small and large intestines.

alveolus (plural **alveoli**) one of the many thousands of tiny air sacs in the ◊lungs in which exchange of oxygen and carbon dioxide takes place between air and the bloodstream.

amylase one of a group of ◊enzymes that break down starches into their component molecules (sugars) for use in the body. It is found in saliva (as ptyalin) and in pancreatic juices.

androgen general name for any male sex hormone, of which ◊testosterone is the most important.

antagonistic muscles pair of muscles allowing coordinated movement of the skeletal joints. The individual components of antagonistic pairs can be classified into extensors (muscles that straighten a limb) and flexors (muscles that bend a limb).

antibody protein molecule produced in the blood by ◊lymphocytes in response to the presence of foreign or invading substances (◊antigens); such substances include the proteins carried on the surface of infecting microorganisms.

antidiuretic hormone pituitary hormone that prevents excessive fluid loss. See ◊ADH.

antigen any substance that causes the production of ◊antibodies by the body's immune system.

anus opening at the end of the alimentary canal that allows undigested food and other waste materials to pass out of the body, in the form of faeces.

appendix a short, blind-ended tube attached to the ◊caecum. It has no known function in humans.

aqueous humour watery fluid found in the chamber between the cornea and lens of the vertebrate eye. Similar to blood plasma in composition, it is constantly renewed.

artery vessel that carries blood from the heart to the rest of the body.

assimilation process by which absorbed food molecules, circulating in the blood, pass into the cells and are used for growth, tissue repair, and other metabolic activities. The actual destiny of each food molecule depends not only on its type, but also on the body requirements at that time.

atrium either of the two upper chambers of the heart.

auditory canal tube leading from the outer ◊ear opening to the eardrum.

autonomic nervous system part of the nervous system that controls involuntary functions, including the heart rate and activity of the intestines.

axon long threadlike extension of a ◊nerve cell that conducts electrochemical impulses away from the cell body towards other nerve cells, or towards an effector organ such as a muscle. Axons terminate in ◊synapses, junctions with other nerve cells, muscles, or glands.

ball-and-socket joint joint allowing considerable movement in three dimensions, for instance the joint between the pelvis and the femur. To facilitate movement, such joints are rimmed with cartilage and lubricated by synovial fluid.

bicuspid valve, or **mitral valve**, in the left side of the ◊heart, a flap of tissue that prevents blood flowing back into the atrium when the ventricle contracts.

bile brownish alkaline fluid produced by the liver. Bile is stored in the gall bladder and is intermittently released into the duodenum (small intestine) to aid digestion.

bladder hollow elastic-walled organ that stores the urine produced in the kidneys. Urine enters the bladder through two ureters, one leading from each kidney, and leaves it through the urethra.

blood transport medium composed of cells and fluid that circulates in the arteries, veins, and capillaries.

blood clotting production of a semisolid mass of protein fibres and blood cells that prevents excessive bleeding after injury.

blood group any of the types into which blood is classified according to the presence or otherwise of certain ◊antigens on the surface of its red cells.

blood pressure pressure, or tension, of the blood against the inner walls of blood vessels, especially the arteries, due to the muscular pumping activity of the heart.

blood vessel tube that conducts blood either away from or towards the heart. The principal types are arteries, veins, and capillaries

bone hard connective tissue, composed of a network of collagen fibres impregnated with mineral salts, that makes up the skeleton.

bone marrow soft tissue in the centre of some large bones that manufactures red and white blood cells.

brain mass of interconnected nerve cells, contained within the skull, that forms the anterior part of the central nervous system, whose activities it coordinates and controls.

brainstem region where the top of the spinal cord merges with the undersurface of the brain, consisting largely of the medulla oblongata and midbrain.

breathing muscular movements whereby air is taken into the lungs and then expelled, a form of gas exchange.

bronchiole small-bore air tube found in the lung responsible for delivering air to the respiratory surface. Bronchioles lead off from the larger bronchus and branch extensively before terminating in the many thousand alveoli that form the bulk of lung tissue.

bronchus one of a pair of large tubes (bronchi) branching off from the windpipe and passing into the lung. Apart from their size, bronchi differ from the bronchioles in possessing cartilaginous rings, which give rigidity and prevent collapse during breathing movements.

caecum in the ◊digestive system, a blind-ending tube branching off from the first part of the large intestine, terminating in the appendix. It has no function in humans.

capillary narrowest type of blood vessel; it forms the main area of exchange between the fluid (◊lymph) bathing body tissues and the blood.

cartilage flexible bluish-white ◊connective tissue that makes up the embryonic skeleton and, in adults, occurs in the discs between the vertebrae, at bone endings, and in the larynx, nose, and external ear.

central nervous system (CNS) the brain and spinal cord, as distinct from other components of the ◊nervous system. The CNS integrates all nervous function.

cerebellum part of the brain that controls muscle tone, movement, balance, and coordination.

cerebrum part of the brain that coordinates all voluntary activity. It is made up of two paired cerebral hemispheres, separated by a central fissure, and is covered with a deeply folded layer of grey matter, the cerebral cortex.

choroid layer found at the rear of the ◊eye beyond the retina. By absorbing light that has already passed through the retina, it stops back-reflection and so prevents blurred vision.

chyme general term for the stomach contents. Chyme resembles a thick creamy fluid and is made up of partly digested food, hydrochloric acid, and a range of enzymes.

ciliary muscle ring of muscle surrounding and controlling the lens inside the eye, used in focusing. Suspensory ligaments, resembling spokes of a wheel, connect the lens to the ciliary muscle and pull the lens into a flatter shape when the muscle relaxes. When the muscle is relaxed the lens has its longest focal length and focuses rays from distant objects. On contraction, the lens returns to its normal spherical state and therefore has a shorter focal length and focuses images of near objects.

clavicle another name for the collar bone.

coagulation another term for ◊blood clotting, the process by which bleeding is stopped in the body.

cochlea part of the inner ◊ear. It is equipped with thousands of hair cells that move in response to sound waves and thus stimulate nerve cells to send messages to the brain. In this way they turn vibrations of the air into electrical signals.

collagen protein that is the main constituent of ◊connective tissue. Collagen is present in skin, cartilage, tendons, and ligaments. Bones are made up of collagen, with the calcium salts providing increased rigidity.

colon the main part of the large intestine, between the caecum and rectum. Water and mineral salts are absorbed from undigested food in the colon, and the residue passes as faeces towards the rectum.

conjunctiva membrane covering the front of the eye. It is continuous with the epidermis of the eyelids, and lies on the surface of the cornea.

connective tissue tissue made up of a noncellular substance, the ◊extracellular matrix, in which some cells are embedded. Skin, bones, tendons, cartilage, and adipose tissue (fat) are the main connective tissues. There are also small amounts of connective tissue in organs such as the brain and liver, where they maintain shape and structure.

cornea transparent front section of the eye. The cornea is curved and behaves as a fixed lens, so that light entering the eye is partly focused before it reaches the lens.

cortex the outer part of a structure such as the brain, kidney, or adrenal gland.

corticosteroid any of several steroid hormones secreted by the cortex of the ◊adrenal glands.

cranium the dome-shaped area of the skull that protects the brain. It consists of eight bony plates fused together by sutures (immovable joints).

dendrite slender filament projecting from the cell body of a ◊nerve cell or neuron. Dendrites receive incoming messages from many other nerve cells and pass them on to the cell body.

diaphragm thin muscular sheet separating the thorax from the abdomen. Arching upwards against the heart and lungs, the diaphragm is important in the mechanics of breathing. It contracts at each inhalation, moving downwards to increase the volume of the chest cavity, and relaxes at exhalation.

diastole the resting period between beats of the heart when blood is flowing into it.

digestion process whereby food is broken down mechanically, and chemically by enzymes, mostly in the stomach and intestines, to make the nutrients available for absorption and cell metabolism.

ductless gland alternative name for an ◊endocrine gland.

duodenum short length of ◊alimentary canal found between the stomach and the ileum.

endocrine gland ductless gland that secretes hormones into the bloodstream to regulate body processes.

endolymph fluid found in the inner ear, filling the central passage of the cochlea as well as the semicircular canals.

endorphin natural substancethat modifies the action of nerve cells. Endorphins are produced by the pituitary gland and hypothalamus. They lower the perception of pain by reducing the transmission of signals between nerve cells.

enzyme biological catalyst produced in cells, and capable of speeding up the chemical reactions necessary for life. Enzymes are large, complex proteins, and are highly specific, each chemical reaction requiring its own particular enzyme.

epiglottis small flap that closes off the end of the trachea (windpipe) during swallowing to prevent food from passing into it and causing choking.

epithelium tissue of closely packed cells that forms a surface or lines a cavity or tube. Epithelium may be protective (as in the skin) or secretory (as in the cells lining the wall of the gut).

erythrocyte another name for ◊red blood cell.

Eustachian tube small air-filled canal connecting the middle ear with the back of the throat. It equalizes the pressure on both sides of the eardrum.

excretion the removal of the waste products of metabolism from the body. It is achieved by specialized excretory organs, principally the kidneys, which excrete nitrogenous wastes such as urea and excess salts and water. Water and metabolic wastes are also excreted in the faeces and through the sweat glands in the skin; carbon dioxide and water are removed via the lungs.

exocrine gland gland that discharges secretions, usually through a tube or a duct, on to a surface. Examples include sweat glands which release sweat on to the skin, and digestive glands which release digestive juices on to the walls of the intestine

extensor muscle that straightens a limb; compare with ◊flexor.

femur another name for the thigh bone.

fibrin insoluble protein involved in blood clotting.

fibula in the leg, the rear lower bone; it is paired with a smaller front bone, the tibia.

flexor any muscle that bends a limb. Flexors usually work in opposition to other muscles, the extensors, an arrangement known as antagonistic.

follicle-stimulating hormone, FSH, a ◊hormone produced by the pituitary gland. It affects the ovaries in women, stimulating the production of an egg cell. In men, FSH stimulates the testes to produce sperm.

gall bladder small muscular sac situated on the underside of the liver and connected to the small intestine by the bile duct. It stores bile from the liver.

ganglion (plural ganglia) solid cluster of nervous tissue containing many cell bodies and ◊synapses, usually enclosed in a tissue sheath.

gas exchange movement of gases between the body and the atmosphere, principally oxygen and carbon dioxide.

gland specialized organ of the body that manufactures and secretes enzymes, hormones, or other chemicals.

glomerulus in the kidney, the cluster of blood capillaries at the threshold of the nephron, or kidney tubule, from which waste products and water pass, ultimately to become urine.

glucagon hormone secreted by the islets of Langerhans in the pancreas, which increases the concentration of glucose in the blood by promoting the breakdown of glycogen in the liver.

grey matter those parts of the brain and spinal cord that are made up of interconnected and tightly packed nerve cell nucleuses. This is in contrast to white matter, which is made of the axons of nerve cells.

growth hormone(GH) or **somatotrophin**, hormone from the anterior ◊pituitary gland that promotes growth of long bones and increases protein synthesis.

haemoglobin protein used for oxygen transport. Oxygen combines readily and reversibly with haemoglobin. It occurs in red blood cells (erythrocytes), giving them their colour.

heart muscular organ that rhythmically contracts to force blood around the body.

hinge joint joint where movement occurs in one plane only. Examples are the elbow and knee.

histology study of tissue by visual examination, usually with a ◊microscope.

hormone chemical secretion of the endocrine glands and other specialized cells that affects the activity of other distant tissues.

humerus the upper bone of the arm.

hypothalamus region of the brain below the cerebrum that regulates rhythmic activity and physiological stability within the body, including water balance and temperature. It regulates the production of the pituitary gland's hormones and controls that part of the ◊nervous system governing the involuntary muscles.

ileum part of the small intestine, between the duodenum and the colon, that absorbs digested food.

incisor one of four sharp teeth found at the front centre of each jaw.Incisors are used for biting.

ingestion process of taking food into the mouth.

insulin hormone, produced by the islets of Langerhans in the pancreas, that stimulates the conversion of glucose to glycogen in the liver, thereby lowering blood glucose levels.

intestine the digestive tract from the stomach outlet to the anus.

iris coloured muscular diaphragm that controls the size of the pupil in the eye. It contains radial muscle that increases the pupil diameter and circular muscle that constricts the pupil diameter. Both types of muscle respond involuntarily to light intensity.

islets of Langerhans groups of cells within the pancreas responsible for the secretion of the hormone insulin. They are sensitive to blood glucose levels, producing more hormone when glucose levels rise.

jaw one of two bony structures that form the framework of the mouth. They consist of the upper jawbone (maxilla), which is fused to the skull, and the lower jawbone (mandible), which is hinged at each side to the bones of the temple by ◊ligaments.

joint point where two bones meet.

kidney one of a pair of organs responsible for excreting waste products such as urea and regulating the water and salt content of blood.

lacteal small lymph vessel responsible for absorbing fat in the small intestine.

larynx cavity at the upper end of the trachea (windpipe) containing the vocal cords. It is stiffened with cartilage and lined with mucous membrane.

leucocyte another name for a white blood cell.

ligament strong, flexible connective tissue, made of the protein ◊collagen, which joins bone to bone at moveable joints and sometimes encloses the joints.

lipase enzyme responsible for breaking down fats into fatty acids and glycerol.

liver large organ, situated in the upper abdomen, that has many regulatory and storage functions. It receives the products of digestion, converts glucose to glycogen (a long-chain carbohydrate used for storage), and then back to glucose when needed. It removes excess amino acids from the blood, converting them to urea, which is excreted by the kidneys. The liver also synthesizes vitamins, produces bile and blood-clotting factors, and removes damaged red cells and toxins such as alcohol from the blood.

lung one of a pair of organs in the chest cavity, used for bringing inhaled air into close contact with the blood so that oxygen can pass into the body and waste carbon dioxide can be passed out.

luteinizing hormone hormone produced by the pituitary gland. In males, it stimulates the testes to produce androgens (male sex hormones). In females, it works together with follicle-stimulating hormone to initiate production of egg cells by the ovary. If fertilization occurs, it plays a part in maintaining the pregnancy by controlling the levels of the hormones oestrogen and progesterone in the body.

lymph fluid found in the lymphatic system of vertebrates. Lymph is drained from the tissues by lymph capillaries, which empty into larger lymph vessels (lymphatics). These lead to lymph nodes (small, round bodies chiefly situated in the neck, armpit, groin, thorax, and abdomen), which process the◊lymphocytes produced by the bone marrow, and filter out harmful substances and bacteria.

lymph node mass of lymphatic tissue that occurs at various points along the major lymphatic vessels. Examples are the tonsils and adenoids. As lymph passes through the nodes, it is filtered, and bacteria and other microorganisms are engulfed by cells known as macrophages.

lymphocyte type of white blood cell with a large nucleus, produced in the bone marrow. B lymphocytes, or B cells, are responsible for producing antibodies; T lymphocytes, or T cells, have several roles in the mechanism of immunity.

macrophage type of white blood cell that specializes in the removal of bacteria and other microorganisms, or of cell debris after injury. Macrophages are found throughout the body, but mainly in the lymph and connective tissues, and especially the lungs, where they ingest dust, fibres, and other inhaled particles.

maltase enzyme that breaks down the disaccharide maltose into glucose.

medulla the central part of a structure such as the kidney, or adrenal gland. In the brain, the medulla oblongata is the posterior region responsible for the coordination of basic activities, such as breathing and temperature control.

molar one of the 12 large teeth, used for crushing food, found at the back of the mouth. The structure of the jaw, and the relation of the muscles, allows a massive force to be applied to molars.

monocyte type of white blood cell. Monocytes are found in the tissues, the lymphatic and circulatory systems where their purpose is to remove foreign particles, such as bacteria and tissue debris, by ingesting them.

motor nerve any nerve that transmits impulses from the central nervous system to muscles or organs.

mucous membrane thin skin lining all body cavities and canals that come into contact with the air (for example, eyelids, breathing and digestive passages, genital tract). It secretes mucus, a moistening, lubricating, and protective fluid.

muscle contractile tissue that produces locomotion and power, and maintains the movement of body substances. It is made up of long fibres that can contract to between one-half and one-third of their relaxed length.

myelin sheath insulating layer that surrounds nerve cells and serves to speed up the passage of nerve impulses.

nephron microscopic unit in the kidney that is responsible for forming urine.

nerve bundle of nerve cells enclosed in a sheath of connective tissue and transmitting nerve impulses to and from the brain and spinal cord.

nerve cell, or **neuron**, elongated cell, the basic functional unit of the ◊nervous system, that transmits information rapidly between different parts of the body.

neuron another name for a ◊nerve cell.

neurotransmitter chemical that diffuses across a synapse, and thus transmits impulses between nerve cells, or between nerve cells and effector organs (such as muscles). Common neurotransmitters are noradrenaline (which also acts as a hormone) and acetylcholine, the latter being most frequent at junctions between nerve and muscle.

noradrenaline, or **norepinephrine**, chemical that acts both as a hormone, produced by the adrenal gland, and as a neurotransmitter, secreted by nerve cells. It acts directly on specific receptors to stimulate the sympathetic nervous system.

nose upper entrance of the respiratory tract; the organ of the sense of smell. The whole nasal cavity is lined with a ◊mucous membrane that warms and moistens the air as it enters and ejects dirt. In the upper parts of the cavity the membrane contains 50 million olfactory receptor cells (cells sensitive to smell).

oesophagus muscular tube by which food travels from mouth to stomach.

oestrogen any of a group of hormones, principally oestradiol, produced by the ◊ovaries. Oestrogens control female sexual development, promote the growth of female secondary sexual characteristics, stimulate egg production, and prepare the lining of the uterus for pregnancy.

optic nerve large nerve passing from the eye to the brain, carrying visual information.

organ part of the body, such as the liver or brain, that has a distinctive function or set of functions. An organ is composed of a group of coordinated ◊tissues.

osmoregulation process whereby the water content is maintained at a constant level. If the water balance is disrupted, the concentration of salts will be too high or too low, and vital functions, such as nerve conduction, will be adversely affected.

ovary in females, the organ that generates the ◊ovum (egg cell) and secretes the hormones responsible for female secondary sexual characteristics, such as smooth, hairless facial skin and enlarged breasts.

oxytocin hormone that stimulates the uterus in late pregnancy to initiate and sustain labour. After birth, it stimulates the uterine muscles to contract, reducing bleeding at the site where the placenta was attached.

pacemaker, or **sinoatrial node** group of muscle cells in the wall of the heart that contracts spontaneously and rhythmically, setting the pace for the contractions of the rest of the heart.

pain sense that gives an awareness of harmful effects on or in the body. It may be triggered by stimuli such as trauma, inflammation, and heat. Pain is transmitted by specialized nerves and also has psychological components controlled by higher centres in the brain.

pancreas accessory gland of the digestive system located close to the duodenum. When stimulated by the hormone secretin, it releases enzymes into the duodenum that digest starches, proteins, and fats. It contains groups of cells called the **islets of Langerhans**, which secrete the hormones insulin and glucagon that regulate the blood sugar level.

parasympathetic nervous system division of the autonomic nervous system responsible for slowing the heart rate, decreasing blood pressure, and stimulating the digestive system.

parathyroid one of a pair of small ◊endocrine glands, located behind the thyroid gland. They secrete parathormone, a hormone that regulates the amount of calcium in the blood.

patella, or **kneecap**, flat bone embedded in the knee tendon, which protects the joint from injury.

pectoral relating to the upper chest; associated with the muscles and bones used in moving the arms.

pelvis lower area of the abdomen featuring the bones and muscles used to move the legs or hindlimbs. The **pelvic girdle** is a set of bones that allows movement of the legs in relation to the rest of the body and provides sites for the attachment of relevant muscles.

pepsin enzyme that breaks down proteins during digestion.

peristalsis wavelike contractions, produced by the contraction of smooth muscle, that pass along tubular organs, such as the intestines.

peritoneum membrane lining the abdominal cavity and digestive organs.

perspiration excretion of water and dissolved substances from the ◊sweat glands of the skin. It has two main functions: body cooling by the evaporation of water from the skin surface, and excretion of waste products such as salts.

phagocyte type of white blood cell, or leucocyte, that can engulf a bacterium or other invading microorganism. Phagocytes are found in blood, lymph, and other body tissues, where they also ingest foreign matter and dead tissue.

phagocytosis the engulfing of foreign bodies or food by white blood cells.

pharynx muscular cavity behind the nose and mouth, extending downwards from the base of the skull. The internal nostrils lead backwards into the pharynx, which continues downwards into the oesophagus and (through the epiglottis) into the windpipe. On each side, a Eustachian tube enters the pharynx from the middle ear cavity.

pituitary gland major endocrine gland, situated in the centre of the brain. The posterior lobe is an extension of the hypothalamus, and is in effect nervous tissue. It stores two hormones synthesized in the hypothalamus: ◊ADH and ◊oxytocin. The anterior lobe secretes six hormones, some of which control the activities of other glands (thyroid, gonads, and adrenal cortex); others are direct-acting hormones affecting milk secretion and controlling growth.

plasma the liquid component of the ◊blood. It is a straw-coloured fluid, largely composed of water (around 90%), in which a number of substances are dissolved. These include a variety of proteins (around 7%) such as fibrinogen (important in blood clotting), inorganic mineral salts such as sodium and calcium, waste products such as urea, traces of hormones, and antibodies to defend against infection.

platelet tiny disc-shaped structure found in the blood, which helps it to clot. Platelets are not true cells, but membrane-bound cell fragments without nuclei that bud off from large cells in the bone marrow.

premolar one of the eight large teeth, used for crushing, found toward the back of the mouth. Unlike molars, they are present in milk dentition as well as in the permanent dentition of adults.

progesterone steroid hormone that regulates the menstrual cycle and pregnancy. Progesterone is secreted by the corpus luteum (the ruptured Graafian follicle of a discharged ovum).

prostaglandin any of a group of complex fatty acids present in the body that act as messenger substances between cells. Effects include stimulating the contraction of smooth muscle (for example, of the uterus during birth), regulating the production of stomach acid, and modifying hormonal activity.

protease general term for a digestive enzyme capable of splitting proteins. Examples include pepsin, found in the stomach, and trypsin, found in the small intestine.

pulmonary pertaining to the ◊lungs.

pylorus the lower opening of the stomach.

radius one of the two bones in the lower arm (the other is the ulna).

receptor discrete area of cell membrane or an area within a cell with which neurotransmitters, hormones, and drugs interact. Such interactions control the activities of the body. For example, adrenaline transmits nervous impulses to receptors in the sympathetic nervous system which initiates the characteristic response to excitement and fear in an individual.

rectum lowest part of the large intestine, which stores faeces prior to elimination (defecation).

red blood cell, or **erythrocyte**, the most common type of blood cell, responsible for transporting oxygen around the body. It contains haemoglobin, which combines with oxygen from the lungs to form oxyhaemoglobin.

reflex very rapid involuntary response to a particular stimulus. A reflex involves only a few nerve cells, unlike the slower but more complex responses produced by the many processing nerve cells of the brain.

renal pertaining to the ◊kidneys.

rennin, or **chymase**, enzyme found in the gastric juice of babies, used in the digestion of milk.

retina light-sensitive area at the back of the eye connected to the brain by the optic nerve. It has several layers and contains over a million rods and cones, sensory cells capable of converting light into nervous messages that pass down the optic nerve to the brain.

rhesus factor group of ◊antigens on the surface of red blood cells that characterize the rhesus blood group system. Most individuals possess the main rhesus factor (Rh+), but those without this factor (Rh−) produce ◊antibodies if they come into contact with it. The name comes from rhesus monkeys, in whose blood rhesus factors were first found.

rib one of 12 pairs of long, usually curved bone that extends laterally from the vertebral column. The ribs protect the lungs and heart, and allow the chest to expand and contract easily.

saliva alkaline secretion from the salivary glands that aids the swallowing and digestion of food in the mouth. It contains the enzyme ptyalin (amylase), which converts starch to sugar.

salivary gland one of three pairs of glands – the parotid, sublingual, and the submandibular – responsible for the manufacture of saliva and its secretion into the mouth.

scapula, or **shoulder blade**, large, flat, triangular bone that lies over the second to seventh ribs on the back, forming part of the pectoral girdle, and assisting in the articulation of the arm with the chest region. Its flattened shape allows a large region for the attachment of muscles.

secretin hormone produced by the small intestine that stimulates the production of digestive secretions by the pancreas and liver.

skin the covering of the body. The outer layer (epidermis) is dead and its cells are constantly being rubbed away and replaced from below; it helps to protect the body from infection and to prevent dehydration. The lower layer (dermis) contains blood vessels, nerves, hair roots, and sweat and sebaceous glands, and is supported by a network of fibrous and elastic cells.

skull collection of flat and irregularly shaped bones that enclose the brain and the organs of sight, hearing, and smell, and provide support for the jaws.

smooth muscle involuntary muscle capable of slow contraction over a period of time. It is present in hollow organs, such as the intestines, stomach, bladder, and blood vessels.

spinal cord major component of the central nervous system, encased in the vertebral column.

spine another word for the vertebral, or spinal column.

spleen organ, situated on the left side of the body, under the stomach, that helps to process ◊lymphocytes. It also regulates the number of red blood cells in circulation by destroying old cells, and stores iron.

sternum or **breastbone**, large flat bone at the front of the chest, joined to the ribs. It gives protection to the heart and lungs.

stomach the first cavity in the digestive system, a bag of muscle situated just below the diaphragm.

suspensory ligament in the eye, a ring of fibre supporting the lens.

synapse junction between two ◊nerve cells, or between a nerve cell and a muscle (a neuromuscular junction), across which a nerve impulse is transmitted. The two cells are separated by a narrow gap, which is bridged by a chemical ◊neurotransmitter, released by the nerve impulse.

synovial fluid viscous colourless fluid that bathes movable joints between the bones. It nourishes and lubricates the ◊cartilage at the end of each bone.

systole contraction of the heart. It alternates with diastole, the resting phase of the heart beat.

taste sense that detects some of the chemical constituents of food. The human tongue can distinguish only four basic tastes (sweet, sour, bitter, and salty) but it is supplemented by the sense of smell.

tears salty fluid exuded by lacrimal glands in the eyes. The fluid contains proteins that are antibacterial, and also absorbs oils and mucus. Apart from cleaning and disinfecting the surface of the eye, the fluid supplies nutrients to the cornea, which does not have a blood supply.

tendon, or **sinew**, cord of very strong, fibrous connective tissue that joins muscle to bone.

testis, plural **testes**, organ that produces ◊sperm in the male. It is one of a pair of oval structures that descend from the body cavity during development, to hang outside the abdomen in a sac called the scrotum. The testes also secrete the male sex hormone testosterone.

testosterone hormone secreted chiefly by the testes, but also by the ovaries and the cortex of the adrenal glands. It promotes the development of secondary sexual characteristics in males.

thorax, or **chest**, the part of the body containing the heart and lungs, and protected by the ribcage.

thrombocyte another name for a ◊platelet.

thymus organ situated in the upper chest cavity. The thymus processes ◊lymphocyte cells to produce T-lymphocytes (T denotes 'thymus-derived'), which are responsible for binding to specific invading organisms and killing them or rendering them harmless.

thyroid endocrine gland, situated in the neck in front of the trachea. It secretes several hormones, principally thyroxine, an iodine-containing hormone that stimulates growth, metabolism, and other functions of the body.

tibia, or **shinbone**, the front bone in the leg between the ankle and the knee.

tissue cellular fabric in the body. Several kinds of tissue can usually be distinguished (for example, nerve and muscle), each consisting of cells of a particular kind bound together by extracellular matrix.

tonsils masses of lymphoid tissue situated at the back of the mouth and throat (palatine tonsils), and on the rear surface of the tongue (lingual tonsils). The tonsils contain many ◊lymphocytes and are part of the body's defence system against infection.

touch sensation produced by specialized nerve endings in the skin. Some respond to light pressure, others to heavy pressure. Temperature detection may also contribute to the overall sensation of touch.

trachea, or **windpipe**, tube that forms part of the airway between the throat and the lungs. It is strong and flexible, and reinforced by rings of ◊cartilage. In the upper chest, the trachea branches into two tubes: the left and right bronchi, which enter the lungs.

tricuspid valve flap of tissue situated on the right side of the ◊heart between the atrium and the ventricle. It prevents blood flowing backwards when the ventricle contracts.

trypsin enzyme responsible for the digestion of protein molecules.

ulna one of the two bones found in the lower arm (the other is the radius).

urea $CO(NH_2)_2$ waste product formed in the liver when nitrogen compounds are broken down. It is filtered from the blood by the kidneys, and stored in the bladder as urine prior to release.

ureter tube connecting the kidney to the bladder. Its wall contains fibres of smooth muscle whose contractions aid the movement of urine out of the kidney.

urethra tube connecting the bladder to the exterior. It carries urine and, in males, semen.

urine amber-coloured fluid filtered out by the kidneys from the blood. It contains excess water, salts, proteins, waste products in the form of urea, a pigment, and some acid.

valve structure for controlling the direction of the blood flow. Valves prevent backflow in the heart, ensuring that its contractions send the blood forward into the arteries. The tendency for low-pressure venous blood to collect at the base of the legs under the influence of gravity is counteracted by a series of small valves within the veins.

vein any vessel that carries blood from the body to the heart.

vena cava either of the two great veins of the trunk, returning deoxygenated blood to the right atrium of the ◊heart. The **superior vena cava**, beginning where the arches of the two innominate veins join high in the chest, receives blood from the head, neck, chest, and arms; the **inferior vena cava**, arising from the junction of the right and left common iliac veins, receives blood from all parts of the body below the diaphragm.

ventricle either of the two lower chambers of the heart that force blood to circulate by contraction of their muscular walls.

vertebra irregularly shaped bone that forms part of the vertebral column.

vertebral column, or **spine**, series of bones, or vertebrae, that support the body and protect the spinal cord.

villus, plural **villi**, small fingerlike projection of the wall of the small intestine. It serves to increase the wall's surface area for the absorption of digested nutrients.

viscera general term for the organs contained in the chest and abdominal cavities.

vitreous humour transparent jellylike substance behind the lens of the eye. It gives rigidity to the spherical form of the eye and allows light to pass through to the retina.

vocal cords the paired folds of tissue within the larynx (voice box). Air constricted between the folds makes them vibrate, producing sounds. Muscles in the larynx change the pitch of the sounds produced, by adjusting the tension of the vocal cords.

windpipe another name for the ◊trachea.

Further Reading

Austin, C R, and Short, R V (eds) *Reproduction in Mammals* 1982

Baker, Robin *Sperm Wars* 1996

Bolt, Robert J *The Digestive System* 1983

Bryan, Jenny *The Pulse of Life: the Circulatory System* 1992

Chaplin, Martin F, and Bucke, Christopher *Enzyme Technology* 1990

Chivers, David John, and Langer, Peter *The Digestive System in Mammals: Food, Form and Nutrition* 1994

Cohen, Jack *Reproduction* 1977

Conn, P Michael, and Melmed, Shlomo *Endocrinology: Basic and Clinical Principles* 1997

Dulbecco, Renato *Encyclopedia of Human Biology* 1991

Friday, Adrian, and Ingram, David S *The Cambridge Encyclopedia of Life Sciences* 1985

Gould, Stephen Jay *The Mismeasure of Man* 1981

Halton, Frances *The Digestive System* 1994

Kimbrell, Andrew *The Human Body Shop* 1993

Miller, Jonathan *The Body in Question* 1978

Miller, Jonathan, and Pelham, David *Human Body* 1994

Mines, Allan H *Respiratory Physiology* 1993

Norman, Anthony W, and Litwack, Gerard *Hormones* 1997

Prange, Henry D *Respiratory Physiology: Understanding Gas Exchange* 1996

Ridley, Matt *The Red Queen* 1993

Ryall, Robert James *The Digestive System* 1984

Smith, James John, and Kampine, John P *Circulatory Physiology: the Essentials* 1990

Suckling, C J *Enzyme Chemistry: Impact and Applications* 1990

Sutcliffe, Margaret *Do You Understand the Circulatory System?* 1991

Vines, Gail *Raging Hormones: Do They Rule Our Lives?* 1993

Vroon, Piet *Smell: The Secret Seducer* 1998

Warnock, Mary *A Question of Life* 1984

West, John B *Respiratory Physiology: the Essentials* 1995

Wolpert, Lewis *The Triumph of the Embryo* 1991

Wong, C H, and Whitesides, G M *Enzymes in Synthetic Organic Chemistry* 1994

HEALTH AND DISEASE

A disease is defined as a condition that disturbs or impairs the normal state of an organism. Diseases can occur in all life forms, and normally affect the functioning of cells, tissues, organs, or systems. Diseases can be classified as infectious or noninfectious. Infectious diseases are caused by microorganisms, such as bacteria and viruses, invading the body; they can be spread across a species, or transmitted between one or more species. All other diseases can be grouped together as noninfectious diseases. These can have many causes, some of which are still unknown today and include such disorders as cardiovascular disease, cancer, and obesity.

Some diseases occur mainly in certain climates or geographical regions of the world. These are **endemic** diseases. For example, African sleeping sickness, which is carried by the tsetse fly, is found mainly in the very hot, humid regions of Africa. Similarly, malaria, a disease spread by mosquitoes, is usually found in or near the marsh or stagnant water which provide breeding grounds for the insect. Other diseases may be seasonal – such as influenza, which tends to occur mainly in winter, or intestinal illnesses that result from food contamination in summer.

Some diseases are more common in certain age groups, such as measles in children, meningitis in young adults, and coronary heart disease in the elderly. Other diseases may tend to occur only in certain racial types and are usually genetic in origin, such as sickle-cell disease which is found mainly among people of black African descent.

More than three centuries ago most people in the Western world died from infection and the average life expectancy is estimated to have been 30–40 years. Since then the pattern of disease aetiology has changed in developed countries, with life expectancy being extended to 70–80 years.

Whilst the main causes of death used to include smallpox, tuberculosis, cholera, typhoid, and dysentery, today's mortality statistics in developed countries are dominated by degenerative diseases such as circulatory disorders, and increasingly, neoplastic diseases.

In developing countries, where the combined problems of famine, war, and poverty boost the mortality figures, dietary deficiencies and infectious diseases are still the main problems being faced.

infectious diseases

introduction

In humans, infections caused by microorganisms (pathogens) are the commonest cause of disease. According to a 1990s World Health Report prepared by the World Health Organization, 17 million deaths (one-third of the total number) occur as a result of infectious diseases. Pathogens are parasites that take over some of the body's cells and tissues, using them for their own growth and reproduction. In the process, the cells and tissues are damaged or destroyed resulting in disease of the host body. These pathogens produce diseases ranging from minor skin infections to

The Leading Causes of Mortality in the World

Source: World Resources Institute

1997

Rank	Cause	Deaths (thousands)	% of overall deaths
1	Infectious and parasitic diseases	17,310	33
2	Diseases of the circulatory system	15,300	29
3	Other and unknown causes	6,250	12
4	Malignant tumours (neoplasms)	6,235	12
5	Perinatal and neonatal causes	3,630	7
6	Chronic lower respiratory diseases	2,890	6
7	Maternal causes	585	1

life-threatening internal disorders.

There are several ways in which pathogens gain access to the body: through the skin, especially through cuts or wounds; through the respiratory system – for instance, cold and influenza viruses are carried through the air in droplets of moisture and are breathed in; in food or water – bacteria causing food poisoning are taken into the alimentary canal in food, poliomyelitis viruses can be transmitted in water; by vectors (organisms that transmit pathogens to host bodies) – such as rabies which is transmitted in the saliva of mammals. (An infected mammal passing on the rabies virus to another is a vector for rabies.)

MedicineNet
http://www.medicinenet.com/
Immense US-based site dealing with all current aspects of medicine in plain language. There is a dictionary of diseases, cures, and medical terms. The site also includes an 'ask the experts' section, lots of current medical news, and last, but not least, some important first aid advice.

Infectious diseases can be classified according to the kind of pathogen that causes them. The most common pathogens are bacteria and viruses, but fungi, protozoans, and worms can also cause infectious diseases. Disease-producing protozoans are found chiefly in tropical areas, and cause such diseases as amoebic dysentery, an intestinal infection. Worm infections also cause many serious tropical diseases, including elephantiasis, river blindness, and schistosomiasis.

The human body has many natural defences against the entry of pathogens. For instance, blood clotting can seal a cut or wound thus preventing the entry of pathogens through the skin. The body also has chemical barriers against infection, such as tears, which not only wash foreign substances from the eyes, but also contain enzymes that destroy many common pathogens. In addition, the mucous membranes release

protective chemicals. The body's own senses of smell and taste can often detect the presence of bacteria in food before harmful quantities are ingested. Any bacteria that reach the stomach may be killed off by the hydrochloric acid in the stomach's digestive juices. Pathogens that manage to penetrate the body's defences will be recognized as foreign and come under attack from antibodies produced by the white cells in the blood.

During the last century the prevalence of old-style infectious diseases has dramatically been reduced in developed countries. This has been the result of:

- improved nutrition and living conditions;
- public health measures, such as the introduction of safe water and drainage facilities;
- immunization and chemotherapy;

Despite these developments, third world countries continue to experience infectious scourges, with acute respiratory infection, gastroenteritis, malaria, and measles being the main cause of early death in children.

And the world burden of infection continues. This is due partly to the emergence of 'new' infectious diseases, such as acquired immune deficiency syndrome (AIDS), increased international travel means that people are more at risk of tropical diseases such as typhoid, malaria, and schistosomiasis. Additionally, certain infectious organisms have become resistant to drugs. These issues are discussed in the following sections.

Acquired Immune Deficiency Syndrome (AIDS)

This is the gravest of the sexually transmitted diseases. It is caused by the human immunodeficiency virus (HIV), now known to be a retrovirus, an organism first identified in 1983. HIV is transmitted in body fluids, mainly blood and genital secretions.

Mortality, Morbidity, and Disability for Selected Infectious Diseases in the World

Source: World Health Organization
(N/A = not available.)

1995

Disease by main mode of transmission	Deaths	New cases (incidence)	All cases (prevalence)	Permanent and long-term affected cases)
Person to Person				
Acute lower respiratory infection (ALR)[1]	4,416	394,750[2]	N/A	N/A
Tuberculosis	3,072	8,888	22,000	N/A
Hepatitis B, viral	1,156	4,149	350,000[3]	N/A
Measles	1,066	N/A	42,000	5,590
AIDS	1,063	1,125	1,538	N/A
Whooping cough (pertussis)	355	N/A	40,000	N/A
Meningococcal meningitis	35	350	N/A	45
Poliomyelitis, acute	9	N/A	82	85
Leprosy	2	561	1,833	3,000
Gonorrhoea	N/A	62,000	N/A	N/A
Syphilis, venereal	N/A	12,000	N/A	N/A
Chancroid	N/A	7,000	N/A	N/A
Trachoma	N/A	20,540	153,832	5,583
Diphtheria[4]	N/A	N/A	35	N/A
Food-, Water-, and Soilborne				
Diarrhoea[5]	3,115	4,002,000[2]	N/A	N/A
Neonatonal tetanus	459	N/A	N/A	N/A
Hookworm diseases	65	N/A	151,000	N/A
Ascariasis	60	N/A	250,000	N/A
Schistosomiasis	20	N/A	200,000	N/A
Cholera[4]	11	384	N/A	N/A
Trichuriasis	10	N/A	45,530	N/A
Trematode infections (foodborne only)	10	N/A	40,000	N/A
Dracunculiasis (guinea-worm infection)	N/A	122	122	N/A
Insect-borne				
Malaria	2,100	approx. 400,000	N/A	N/A
Leishmaniasis	80	1,750	12,000	N/A
Onchocerciasis (river blindness)	47	N/A	18,000	360
Chagas disease (American trypanosomiasis)	45	800	18,000	N/A
Dengue/dengue haemorrhagic fever	24	592	N/A	N/A
Sleeping sickness (African trypanosomiasis)	20	N/A	300	N/A
Japanese encephalitis	11	43	N/A	9
Plague[4]	0.2	2	N/A	N/A
Yellow fever[4]	0.2	0.5	N/A	N/A
Filariasis (lymphatic)	N/A	N/A	120,000	43,000
Animal-borne				
Rabies (dog-mediated)	60	N/A	N/A	N/A
Total	17,312	N/A	N/A	N/A

[1] Figures do not include lower respiratory infections related to measles, pertussis, and HIV infections.
[2] Incidence figure refers to episodes.
[3] Figure refers to chronic HBV carriers.
[4] Officially reported figures only.
[5] Figures relate to acute watery diarrhoea, persistent diarrhoea, and dysentery, but do not include measles- and HIV-related diarrhoea.

Newly Recognized Infectious Diseases

Source: World Health Organization

Year of recognition	Agent	Type	Disease/comments
1973	Rotavirus	virus	major cause of infantile diarrhoea worldwide
1975	Parvovirus B19	virus	aplastic crisis in chronic haemolytic anaemia
1976	Cryptosporidium parvum	parasite	acute and chronic diarrhoea
1977	Ebola virus	virus	Ebola haemorrhagic fever
1977	Legionella pneumophila	bacterium	legionnaires' disease
1977	Hantaan virus	virus	haemorrhagic fever with renal syndrome (HRFS)
1977	Campylobacter jejuni	bacterium	enteric pathogen distributed globally
1980	human T-lymphotropic virus 1 (HTLV-1)	virus	T-cell lymphoma-leukaemia
1981	toxin-producing strains of Staphylococcus aureus	bacterium	toxic shock syndrome
1982	Escherichia coli 0157:H7	bacterium	haemorrhagic colitis; haemolytic uraemic syndrome
1982	HTLV-2	virus	hairy cell leukaemia
1982	Borrelia burgdorferi	bacterium	Lyme disease
1983	human immunodeficiency virus (HIV)	virus	acquired immunodeficiency syndrome (AIDS)
1983	Helicobacter pylori	bacterium	peptic ulcer disease
1985	Enterocytozoon bieneusi	parasite	persistent diarrhoea
1986	Cyclospora cayetanensis	parasite	persistent diarrhoea
1986	BSE agent (uncertain)	non-conventional agent	bovine spongiform encephalopathy in cattle and possibly variant Creutzfeldt-Jakob disease (vCJD) in humans
1988	human herpes virus 6 (HHV-6)	virus	exanthem subitum
1988	hepatitis E virus	virus	enterically transmitted non-A, non-B hepatitis
1989	Ehrlichia chaffeensis	bacterium	human ehrlichiosis
1989	hepatitis C virus	virus	parenterally transmitted non-A, non-B liver hepatitis
1991	Guanarito virus	virus	Venezuelan haemorrhagic fever
1991	Encephalitozoon hellem	parasite	conjunctivitis, disseminated disease
1991	new species of Babesia	parasite	atypical babesiosis
1992	Vibrio cholerae 0139	bacterium	new strain associated with epidemic cholera
1992	Bartonella henselae	bacterium	cat-scratch disease causing flu-like fever; bacillary angiomatosis
1993	Sin Nombre virus	virus	Hantavirus pulmonary syndrome
1993	Encephalitozoon cuniculi	parasite	disseminated disease
1994	Sabia virus	virus	Brazilian haemorrhagic fever
1995	human herpes virus 8	virus	associated with Kaposi's sarcoma in AIDS patients
1995	Hepatitis G	virus	parenterally transmitted non-A, non-B hepatitis
1996	NvCJD	TSE causing agent	New variant Creutzfeldt–Jakob disease
1997	Avian influenza	Type A(9H5N1) virus	Influenza; can cause Reye syndrome

The global count of people infected with HIV continues to increase at an alarming rate. Despite some notable successes with health education campaigns, millions are infected, with the worst infection rates being in the developing world.

Nevertheless, there are signs that the enormous scientific effort that has been mobilized against the disease is now paying dividends. Current drug regimes are able to prolong life and are even allowing scientists to debate the prospect of cure.

When HIV enters the body, the main cell type infected is the CD4+ T lymphocyte, a circulating white cell that plays an important role in controlling immune responses. In addition, the virus can enter a variety of other cell types that have the CD4 molecule on their surface. The net result of infection is a relentless fall in the number of CD4+ T cells and a gradual dismantling of the immune system's ability to fight off

Estimated Number of Deaths from HIV/AIDS in the World

Source: World Health Organization

1997

Region	Estimated deaths	Region	Estimated deaths
East Asia and Pacific	6,200	North America	29,000
Eastern Europe and Central Asia	300	Latin America	81,000
Australia and New Zealand	700	Caribbean	19,000
North Africa and Middle East	13,000	Sub-Saharan Africa	1.8 million
Western Europe	15,000	Total	2.3 million
South and Southeast Asia	250,000		

infectious agents such as bacteria, fungi, and other viruses. This process leaves patients very susceptible to a wide variety of opportunistic infections and the appearance of rare tumours, many of which are virtually never seen in people with a normally functioning immune system. *Pneumocystis carinii* pneumonia, for instance, normally seen only in the malnourished or those whose immune systems have been deliberately suppressed, is common among AIDS patients and, for them, a leading cause of death. One important advance in the last few years has been the use of molecular assays to measure the amount of virus in the blood of infected patients. This procedure is valuable in predicting how rapidly they are likely to progress to an advanced state of the disease.

AIDS and HIV Information
http://www.thebody.com
AIDS/HIV site offering safe sex and AIDS prevention advice, information about treatments and testing, and health/nutritional guidance for those who have contracted the disease.

early treatment The initial successes in the drug treatment of HIV infection were with agents that could treat or prevent the infectious complications of the disease. These drugs remain very valuable but do not have any activity against HIV itself. The first drug with proven activity against HIV was zidovudine (AZT). Zidovudine resembles one of the building blocks of DNA, and when HIV undergoes replication,

HIV/AIDS: Regional Statistics and Features

Source: UNAIDS/WHO

1997

	Sub-Saharan Africa	South and South-East Asia	Latin America	Established Market Economies[1]	Caribbean	Eastern Europe–Central Asia	East Asia–Pacific	North Africa–Middle East[2]
Epidemic started	Late 1970s–early 1980s	Late 1980s	Late 1970s–early 1980s	Late 1970s–early 1980s	Late 1970s–early 1980s	Early 1990s	Late 1980s	Late 1980s
Adults and children living with HIV/AIDS	20.8 million	6.0 million	1.3 million	1.4 million	310,000	150,000	440,000	210,000
Prevalence (%)	7.4	0.6	0.5	0.3	1.9	0.07	0.05	0.1
Women (%)	50	25	19	20	33	25	11	20
Main mode(s) of transmission for those living with HIV/AIDS[2]	Hetero	Hetero-IDU	MSM-IDU-Hetero	MSM-IDU-Hetero	Hetero	IDU-MSM	IDU-Hetero-MSM	IDU-Hetero

[1] North America, Western Europe, Australia, New Zealand, Japan.

[2] Hetero: heterosexual transmission; IDU: transmission through injecting drug use; MSM: men who have sex with men.

Current Total of Reported AIDS Cases in the World

Source: UNAIDS/WHO

Category	Group	Number (millions)
New infections in 1997	adults	5.21
	children	0.59
	Total	5.80
People living with HIV/AIDS as of end 1997	adults	29.50
	children	1.10
	Total	30.60
HIV/AIDS associated deaths during 1997	adults	1.84
	children	0.46
	Total	2.30
Cumulative HIV/AIDS deaths	adults	9.00
	children	2.70
	Total	11.70
Cumulative number of AIDS orphans[1]	children	8.20

[1] Orphans are defined as HIV-negative children having lost their mother or both parents due to AIDS.

zidovudine can bind to an essential HIV enzyme and prevent the virus from completing its life cycle. Zidovudine can improve the symptoms of HIV infection, is valuable in asymptomatic patients with low CD4+ T lymphocyte counts, and is effective at reducing the rate of transmission of HIV from pregnant women to their babies. However, when zidovudine is used alone, the virus is usually able to escape from the effects of zidovudine by mutating its DNA sequence. There is now an increasing appreciation of the need to use zidovudine in combination with some new antiviral drugs. This combination treatment of opportunistic infections extended the average length of survival with AIDS (in Western countries) from about 11 months in 1985 to 23 months in 1994.

new developments At the moment, probably the most exciting class of drugs that inhibit HIV replication is the protease inhibitors. When HIV replicates inside a cell it has to make a copy of its DNA, and then this genetic message is decoded into a protein. Some HIV proteins need to be broken down into smaller pieces in order to function, and this is done by a protease molecule. Normal function of the HIV protease appears to be vital for efficient replication of the virus, and over the last few years researchers have spent a great deal of effort in developing drugs that can block its function.

At least four protease inhibitors have been tried in clinical practice: saquinavir, ritonavir, nelfinavir, and indinavir. All have slightly different properties and different side effects. In clinical trials, these drugs have demonstrated a spectacular ability to reduce the amount of virus in the body. Sensitive molecular assays such as the polymerase chain reaction (PCR) have shown that the amount of HIV in blood can be reduced by over a thousandfold and sometimes may reach undetectable levels. Although effective on their own, most of the current drive in HIV therapeutics is to use these agents in combination with other anti-HIV agents. Typically this would include zidovudine, a protease inhibitor, and another agent such as didanosine.

In a recent trial this combination led to the virus being undetectable in 60% of patients after 24 weeks of treatment. A very encouraging observation with protease inhibitors is that they can be used at a very advanced state of the disease. It seems that the drugs should be used at quite large doses, to avoid the development of a resistant virus, and unfortunately they do have several side effects. Although most of these effects are not serious, many patients are unable to tolerate a particular drug combination; in these cases a change to another combination is indicated.

Antiviral Agents Bulletin
http://www.bioinfo.com/antiviral.html
News articles, abstracts, and patent details about antiviral treatments – including interferon and HIV-infection and AIDS-related therapeutics.

---

the role of combination therapy The exact role of combination therapy in the overall management of HIV infection is a subject of considerable debate at present. In an attempt to achieve consensus, a panel of the International AIDS Society-USA met in 1996. After results from many clinical trials, the group suggested that combination therapy was now the treatment of choice. It remains unclear, however, whether or not protease inhibitors should be used with all patients or just those at particular risk of rapid progression to full-blown AIDS based on measurement of the amount of HIV in their blood. Patients with symptoms should be offered treatment, but for those who are asymptomatic the situation is less clear and the decision will be based on the CD4+ count and the viral load in the blood. There are relatively little data to recommend how to treat patients who have been infected in the last month or two and are suffering from the typical symptoms of acute infection: fever, swollen lymph nodes, and headaches. As this is a time of intense viral replication, there is a theoretical advantage in using the strongest available treatment to limit the initial multiplication of the virus. This may also reduce the chance of the virus making mutations that would allow it to resist drug treatment.

The last few years have seen valuable advances in the treatment of HIV infection. Several powerful drugs are now available and are being tested in trials around the world. The human immunodeficiency virus has an astounding ability to mutate itself in order to evade drugs, and there are likely to be setbacks ahead. It is too soon to say whether or not some patients may be offered the prospect of complete cure. Nevertheless, many AIDS researchers are hoping that they can now maintain the upper hand in the battle against this formidable virus.

HIV Info Web
http://www.infoweb.org/
Up-to-date information about HIV and AIDS treatments and other important resources for patients and their families, such as housing services and legal assistance.

---

costs The cumulative direct and indirect costs of HIV and AIDS in the 1980s have been conservatively estimated at $240 billion. The global cost – direct and indirect – of HIV and AIDS by the year 2000 could be as high as $500 billion a year – equivalent to more than 2% of global GDP.

tropical diseases

The concept of 'tropical disease' first arose when Victorian Europeans explored the tropics and encountered diseases, in areas such as the 'white man's grave' in West Africa, with which they were unfamiliar. However, it now usually encompasses all the diseases of the tropics, including those, such as tuberculosis, which are also common in temperate regions.

It is difficult for most of us to grasp the impact of tropical diseases on the human race because the numbers are so large. Of the 52 million deaths in 1996, for example, 40 million occurred in the developing world and 17 million of these (nearly three times the number killed by cancers worldwide) were caused by infectious or parasitic disease. Such disease accounts for 45% of all deaths in developing countries, compared with just over 1% in the rest of the world.

Disease in the tropics may be caused by poor nutrition, environmental factors and/or by living organisms such as fungi (ringworm, athlete's foot), viruses (polio, hepatitis A and B, yellow fever, some forms of meningitis), bacteria (travellers' diarrhoea, typhoid, cholera, tetanus, tuberculosis, leprosy, legionnaires' disease, other forms of meningitis), the single-celled organisms known as protozoa (malaria, sleeping sickness, amoebic dysentery, Chagas's disease, leishmaniasis) and the worms known as nematodes (elephantiasis, river blindness, Guinea worm, hookworm, *Strongyloides*), trematodes (bilharzia) and cestodes (tapeworms, hydatid disease).

Yellow Fever
http://www.outbreak.org/cgi-unreg/dynaserve.exe/YellowFever/index.html
Outbreak page on yellow fever. The Web site contains a section on frequently asked questions about the disease, and a section on current and recent outbreaks. The outbreaks section is updated regularly, often giving numbers of casualties and descriptions of operations by the World Health Organization to curb the spread of any outbreak.

Each of the living organisms has to get below the surface of the skin to survive. Some of the fungi live in the keratin just below the surface but the other pathogens need to get deeper, often into the bloodstream or gut. Several of them (the bacteria causing typhoid, cholera, other causes of diarrhoea, the amoebae causing dysentery, the *Ascaris* roundworm, Guinea worm, hepatitis A virus and tapeworms) rely on us to swallow them in food or water or to ingest them accidentally as we touch our mouths with unclean hands. Other pathogens (some hepatitis B and human immunodeficiency viruses) rely on openings in the skin, caused by accident or injection, to gain access to the bloodstream, although HIV, hepatitis B and gonorrhoea are mainly spread by sexual contact. As 30% of the populations of some towns in the tropics now carry HIV, sex (particularly unprotected sex) with a new partner is a life-threatening gamble.

The bacteria causing tuberculosis are drawn into our lungs from the air and those causing leprosy are mostly passed on in sneezes. A few parasites have stages in their life cycles which simply burrow into our skin, either from the soil (hookworm, *Strongyloides*) or water (bilharzia). Some of the most successful parasites of man (those causing malaria, sleeping sickness, river blindness, elephantiasis, yellow fever and leishmaniasis) spend part of their life cycle in insects (called the vectors of the pathogens) which distribute the parasites and later transmit them to humans as the insects feed.

The parasites causing malaria and elephantiasis use mosquitoes in this way, the nematodes of river blindness use blackflies (which breed in rivers, hence the name) and the trypanosomes of sleeping sickness and Chagas's disease use tsetse flies and bugs, respectively. Clean insects acquire the parasites when they feed on an infected individual. Most of the parasites do not simply use the vectors for distribution but also multiply and develop within them, the vectors acting as intermediate hosts of the parasites and humans as the definitive hosts.

malaria Malaria remains one of the most important parasitic diseases of humans. More than two-fifths of the world's population live in the 100 countries where transmission of malaria occurs and there are about 400 million clinical cases of malaria each year (1.5–2.7 million of them fatal). Human malaria may be caused by any of four species: *Plasmodium falciparum*, *P. vivax* and the rarer *P. ovale* and *P. malariae*. Although only one of these species (*P. falciparum*) causes fatal infections in humans, it is this species which is the most common and widespread, being present in 92 of the 100 countries were malaria transmission occurs. In fact, *P. falciparum* may cause more illness and deaths in humans than any other pathogen on Earth.

The malarial parasite injected by a feeding mosquito passes to the liver and multiplies there for a week or two. It then emerges, invades the red blood cells and multiplies again until the blood cells rupture and the released parasites invade more cells. It is the rupture of the infected red cells, which occurs more or less at the same time (every 2–3 days), that causes the first symptom of malaria: intermittent fever. After a while, some of the parasites breaking out of the blood cells develop into the sexual forms which are capable of infecting mosquitoes. Liver and kidney failure, severe anaemia, convulsions, coma and/or death may develop in falciparum malaria. As malarial parasites in the liver are not affected by the majority of anti-malarial drugs

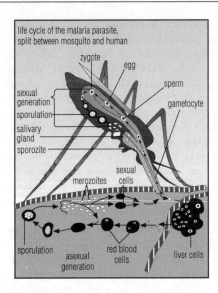

Life cycle of malaria parasite The life cycle of the malaria parasite is split between mosquito and human hosts. The parasites are injected into the human bloodstream by an infected Anopheles mosquito and carried to the liver. Here they attack red blood cells, and multiply asexually. The infected blood cells burst, producing spores, or merozoites, which reinfect the bloodstream. After several generations, the parasite develops into a sexual form. If the human host is bitten at this stage, the sexual form of the parasite is sucked into the mosquito's stomach. Here fertilization takes place, the zygotes formed reproduce asexually and migrate to the salivary glands ready to be injected into another human host, completing the cycle.

in the bloodstream, any traveller who is infected in the last few days of their trip returns home with viable parasites. All travellers at risk of malaria must therefore continue to take anti-malarials for 4 weeks after leaving a malarious area, so that parasites emerging from the liver to infect the blood cells are killed. Two species causing human malaria (*P. vivax* and *P. ovale*) have special stages, called hypnozoites, which can lie dormant in the liver for months and sometimes years, and cause attacks of malaria when they eventually become active.

The relationship between an individual's natural resistance to disease (their level of immunity) and the severity of their suffering is an important one. Those who live in countries where tropical diseases are common are often exposed to the diseases many times and those that survive may have high levels of immunity to them. One individual infected with *P. falciparum* may stay healthy whereas another, infected in exactly

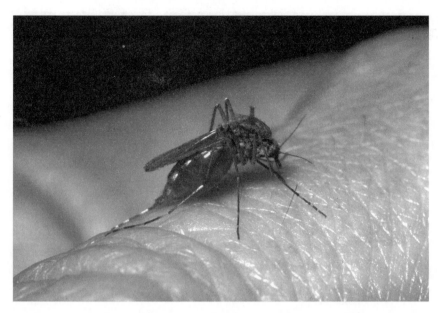

Mosquito Though most of the more than 2,400 species of mosquito are tropical, they are found worldwide. The largest mosquito in the British Isles is the banded-legged *Culiseta annulata*, seen here.

the same way but without any immunity to the infection, may die within 24 hours. This is why a traveller to such countries may suffer when all around appear healthy.

world travel The recent increase in the numbers of visitors to the tropics has meant that more of those from the developed world are putting themselves at risk of acquiring a tropical disease and of taking it home. Although there is usually little risk of the returned traveller passing on their disease, there is a major risk that it will not be diagnosed or treated correctly, simply because local physicians are unfamiliar with its signs and symptoms or fail to consider it. The symptoms and signs of malaria are so nonspecific that local physicians often take them to indicate influenza or gastritis unless the sick traveller mentions that they have visited a malarious area in the last year or more. Of those returning to the UK or USA and eventually dying of falciparum malaria, for example, almost 50% are misdiagnosed as cases of influenza or of another viral infection when first examined by their physicians. Fatal infectious or parasitic disease is fortunately rare in travellers to the tropics and accounts for only a small proportion of deaths in this group (most of the deaths are caused by accidents, especially road traffic accidents, or by cardiovascular disease in the older traveller).

The commonest afflictions the traveller to a developing country is likely to suffer are, in descending frequency, sunburn, diarrhoea and vomiting (usually caused by *Escherichia coli* and of a few days'

duration), malaria, acute infection of the respiratory tract (with fever) and hepatitis A. The chances of acquiring some of the perhaps better known diseases, such as leprosy, are virtually nil.

Anyone from a temperate country planning to visit a tropical country can take many steps to reduce their chances of developing a health problem as a result (prevention is always better than cure, particularly when the disease involved may be rapidly fatal and difficult to treat). It is important to be aware of the risks and how to avoid or minimize them. Travellers can educate themselves by contacting local travel health clinics, reading recent guides or exploring the World Wide Web; one of the best Web sites on travel health is run by the 'Centers for Disease Control' in the USA (**http://www.cdc.gov**). Just in case a severe problem arises, travel insurance should be enough to cover the expenses of repatriation.

Travellers must take appropriate advice on what drugs and other measures they should take to protect themselves against malaria. As the malarial parasites continue to develop resistance to the drugs used against them, it is important to seek up-to-date information on the best drug or drugs to use in any particular area. Use of a bednet (preferably one impregnated with insecticide), keeping the skin covered and using an insect repellent (preferably one containing DEET) and insecticides (in mosquito coils or from electronic vaporizers) not only reduces the risk of being bitten by mosquitoes carrying malarial parasites but also limits the nuisance and loss of sleep caused by other biting insects.

Immunizations for Travellers

Source: Travel Health Online

Disease	Immunization	Timing	Reaction	Protection	Duration of protection	Other precautions	Notes
cholera	2 injections not less than week apart	1 week to 1 month before departure	soreness where injected, fever, headache, fatigue	50–60%	6 months	avoid food or water that may be dirty	low risk in reasonable tourist accommodation; infants under 6 months should not be vaccinated
hepatitis A	(a) injection of immunoglobulin or (b) a vaccine consisting of 2 injections 1 month apart, then a 3rd injection 6–12 months later	(a) just before travel (b) 2 months before	(a) immunoglobulin – some may experience soreness and swelling at the injection site and, in some cases, hives (b) vaccine – soreness where injected, sometimes headache and fatigue soreness where injected	(a) prevents illness (b) lessens its severity	(a) 3 months (b) 10 years	as typhoid	
hepatitis B	2 injections of vaccine 1 month apart, then booster 4 months later	last injection 1 month before travel		80–95%	perhaps 5 years		given to those at high risk, such as health workers, and is now also part of the recommended childhood vaccination series some antimalarial drugs are not recommended for pregnant mothers or children under 1 year
malaria	none; take preventative tablets from 1 week before to 4 weeks after leaving malaria area	order tablets 2 weeks before travel	side effects are rare	90%	only while tablets are taken	use anti-mosquito sprays, mosquito nets; keep arms and legs covered after sunset	
polio	(a) oral vaccine – 3–4 doses (b) injection – 3–4 shots, the best way to be protected is to get 4 doses of polio vaccine, immunized adults: 1 booster dose	for travellers who are not up-to-date with their vaccination it may be necessary to allow as much as 7 months for the full recommended vaccination schedule, depending upon other vaccines that may be necessary for the trip	very rare cases develop polio	95+%	10 years		(a) should not be given to pregnant mothers (b) (injections) recommended only for people 18 years and older who have not yet been vaccinated, and for people who cannot take the oral vaccine because of health reasons; should not be given to a person who has had an allergy problem with the antibiotics neomycin or streptomycin the overall risk to travellers is low
rabies	3-dose series of injections, usually given on days 0, 7, and 21 or 28	5 weeks before travel	swelling and itching where injected, headache, abdominal pain, muscle aches, nausea	opinion divided as to whether vaccine prevents rabies or promotes a faster response to treatment	3 months	avoid bites, scratches, or licks from any animal; wash any bite or scratch with antiseptic or soap as quickly as possible and get immediate medical	

Disease	Immunization	Timing	Reaction	Protection	Duration of protection	Other precautions	Notes
tetanus	normally given in childhood with booster every 10 years; unimmunized adults: 2 injections 1 month apart then 3rd injection 6 months later	not critical	headache, lethargy in rare cases	>90%	about 10 years	wash any wounds with antiseptic	Recommended for those visiting areas in developing countries with poor food and water sanitation
typhoid	3 types of vaccine: (a) 1–2 injections 4–6 weeks apart (b) single injection (c) 4 oral doses, every-other-day series on days 1, 2, 4, and 6	5–7 weeks before departure	soreness where injected, nausea, headache (worst in those over 35 and on repeat immunizations) may last 36 hours	50–70%	(a) 1–3 years (b) 3 years (c) 1 year	avoid food, milk, or water that may be contaminated by sewage or by flies	may only be available from special centres
yellow fever	1 injection	at least 10 days before departure, but not more than 10 years before (arrival)	possible slight headache and low fever 5–10 days later, muscle ache	almost 100%	10 years	against mosquitoes, as for malaria	

details correct at end August 1998

It is possible to be vaccinated against several tropical diseases (hepatitis A and B, yellow fever, polio, typhoid) before travel. Every traveller should be immunized against tetanus but the most common cholera vaccine is no longer generally recommended as it is not very effective (a new, more effective one is now on the market). As vaccination may take several weeks it is important to seek health advice early. Once in the tropics, personal hygiene is important and care should be taken with any food or drink. Any food should be cooked adequately and preferably eaten immediately after cooking (shellfish, ice cream and any raw foods, including salads and unpasteurized milk, are particularly hazardous). Any water to be drunk, if from a local source, is best boiled or sterilized chemically. Ice cubes (which may be made with contaminated water) are best avoided. Skin contact with fresh water should be avoided as much as possible as the water may contain the schistosome stages (cercariae) which can burrow into skin and lying directly on sand or walking barefoot should also be avoided. Most bouts of diarrhoea only last a few days even if untreated, but plenty of safe fluids should be drunk (and the use of oral rehydration salts which can be bought and taken with the traveller should also be considered) to prevent dehydration.

antibiotic resistance

When, in 1969, the US Surgeon-General announced that we could soon 'close the book' on infectious diseases, he was speaking prematurely. For two years previously, first reports had surfaced of penicillin resistance developing in *Pneumococcus*, a bacterium which causes a number of potentially fatal diseases, including pneumonia and meningitis. Within little more than a decade, epidemics of pneumococcal disease were breaking out in many countries.

In 1995, when workers from the Centers for Disease Control in Atlanta looked at samples taken from patients with severe pneumonia, they found that a quarter had been hit by pneumococcal (*Staphylococcus pneumonia*) strains resistant to penicillin; some 15% were also resistant to the antibiotic erythromycin. Today, growing resistance of this and other organisms to antibiotics is a global public health problem.

rise of the superbugs Bacteria vary greatly in their sensitivity to antibiotics, and there is no such thing as a 'magic bullet' with blanket activity against all pathogens. The trouble is that, after more than half a century of antibiotic use, bacteria are mutating faster than new drugs can be found. The growth of super-resistance is being hastened by the indiscriminate use of antibiotics, including over-prescribing.

In the developed world, the danger is greatest in hospitals, where people are already very sick. Currently thousands of Britons are dying each year from hospital-acquired infections caused by resistant bacteria. Most at risk are people whose immune systems are in some way impaired, including the very young or the frail elderly, AIDS patients, organ transplant recipients, and patients receiving chemotherapy for cancer.

Most notorious of the so-called 'superbugs' stalking hospital floors is methicillin-resistant *Staphylococcus aureus* (MRSA), a pathogen with an awesome talent for acquiring resistance traits from its microscopic neighbours. So far outbreaks of MRSA, which can cause temporary closure of operating rooms and intensive care units, have been met with vancomycin, a 'last resort' antibiotic normally reserved for life-threatening infections. But in spring of 1997, the Japanese reported the most convincing evidence yet of the appearance of vancomycin-resistant strains.

Antibiotics: How Do Antibiotics Work?
http://ericir.syr.edu/Projects/Newton/12/Lessons/antibiot.html
Introduction to the use and importance of antibiotics in easy to understand language. The site also includes a glossary of scientific and difficult terms, a further reading list, and an activities sheet.

The Japanese report brings one step closer the spectre of the unstoppable 'superbug' overrunning hospitals. In fact, microbiologists have been predicting just this scenario, not least because strains of the intestinal bacterium *Enterococcus* have for some time been defying all existing antibiotics, including vancomycin and the closely related teicoplanin. In the laboratory, it has been shown that genes for resistance can pass from *Enterococcus* to the more deadly *S. aureus* by plasmid transfer.

It was fear of an epidemic of untreatable infections that prompted the recent European ban on avoparcin, a drug administered to farm animals to promote growth. The rationale for antibiotic use in this context is that it improves feeding efficiency. However, avoparcin is close in chemical structure to vancomycin, and many scientists argue that its use as a growth promoter in livestock creates a potential reservoir of vancomycin-resistant bacteria that would be transmissible to human beings.

return of old-time diseases
The phenomenon of super-resistance means also that many one-time killer diseases, including tuberculosis (TB), typhoid, cholera, and diphtheria, are returning in force. TB, always a major problem in the Third World, is the biggest threat, since now it is making a comeback in countries where previously it had been brought under control. Moreover, some strains of the bacterium have become resistant. Parts of the country worst affected are deprived inner-city areas where resistance is widespread.

A big factor in the spread of multi-drug resistant tuberculosis (MDRTB) has been failure of control programs in the industrialized countries. This has meant that many patients starting out on medication lapse before the six-month course is completed. If a patient carrying a resistant strain takes only one drug instead of the prescribed combination, or fails to complete the course, the effect is to promote resistant strains.

fresh strategies
Bacteria developing the ability to foil an antibiotic can become permanently resistant, according to researchers at Emory University in Atlanta. They demonstrated this in another rising superbug, *Escherichia coli*, which causes gastrointestinal infections. This finding implies that, contrary to what many doctors previously believed, reducing antibiotic use will not eliminate resistant strains.

While the quest for new drugs to fight infection is now paramount, many researchers are developing fresh strategies. These include: tinkering with the structure of antibiotics to add in helper molecules; seeking to disable genes for resistance; and developing laser-activated chemical compounds to blitz the superbugs. Some teams, too, are reviving the old idea of turning bacteriophages (bacteria-eating viruses) loose on resistant bacteria.

All these strategies and more may be needed to overcome the rising toll exacted by antibiotic resistance. Certainly, with infectious diseases claiming more than 17 million lives a year worldwide, we are still no nearer to 'closing the book'.

mechanisms of resistance
Bacteria may be naturally resistant to some antibiotics, or resistance may be acquired, mostly by the phenomenon of plasmid transfer. A plasmid is a free-floating fragment of DNA adrift in the cell cytoplasm. It carries some genetic material, including data governing the cell's resistance to antibiotics. Plasmids can be transmitted from one bacterium to another.

Occasionally resistance may be due to spontaneous mutation, which is the result of an error in replication of the cell's nuclear material during reproduction. Further reproduction causes the development of a resistant strain.

Bacteria demonstrate resistance in two ways. One way is by producing enzymes that disable drugs. A drug-defying enzyme is not always expressed by the organism targeted by therapy. The normally innocuous *Staphylococcus epidermidis* produces an enzyme that disables penicillin before it can act against harmful staphylococcal species; or some bacteria can contrive metabolic changes that foil the action of drugs. Sulphonamides – antibacterials introduced before the discovery of antibiotics – are often defeated by these metabolic readjustments on the part of bacteria.

noninfectious diseases

This is a broad term that groups together all those disorders not caused by pathogens. It includes degenerative diseases, metabolic disorders, diet-related diseases, neurological and mental disorders, and birth defects. The following sections examine some of the most common noninfectious diseases prevalent today.

cardiovascular diseases

Cardiovascular diseases are those affecting the heart and blood vessels. This includes atherosclerosis, high blood pressure, heart attacks, and strokes.

Cardiovascular disease is one of the most common causes of death worldwide, but the distribution of the various types of cardiac diseases varies.

- In the UK about 30% of all deaths among men and 20% of those among women are due to ischaemic heart disease.
- Hypertension is prevalent in the Caribbean and West Africa, and is the major cause of premature death from stroke.
- In countries where malnutrition occurs, heart failure develops due to dietary deficiency diseases such as protein deficiency (kwashiorkor and marasmus) and beriberi (vitamin B_1 deficiency).
- Rheumatic heart disease is common in countries with overcrowded and inadequate housing combined with poor hygiene, but is becoming much less common in the West.

atherosclerosis Atherosclerosis consists of accumulation of a variable combination of lipids, complex carbohydrates, blood and blood products, fibrous tissue, and calcium deposits in the lining of the arteries; this leads to thickening and hardening of the artery walls.

The main clinical forms of atherosclerotic disease arise from narrowing of the coronary arteries, the cerebral arteries, and the femoral artery and its branches.

hypertension Hypertension is abnormally high blood pressure due to a variety of causes, leading to excessive contraction of the smooth muscle cells of the walls of the arteries. It increases the risk of kidney disease, stroke, and heart attack.

Hypertension is one of the major public health problems of the developed world, affecting 15–20% of

---

American Heart Association
http://www.amhrt.org/
Home page of the American Heart Association offers a risk assessment test, information about healthy living, including the effects of diet, and access to resources for both patients and carers.

---

adults in industrialized countries (1996). It may be of unknown cause (essential hypertension), or it may occur in association with some other condition, such as kidney disease (secondary or symptomatic hypertension). It is controlled with a low-salt diet and drugs.

Preventing Recurrent Coronary Heart Disease
http://www.bmj.com/cgi/content/full/316/7142/1400
Part of a collection of articles maintained by the British Medical Journal, this page features the full text of a 9 May 1998 editorial concerning coronary heart disease, in which the authors explain that lowering the heart rate may reduce mortality. In addition to the editorial, you will find a list of related articles from the BMJ's collection and the extensive PubMed collection. You can also search the collections under which this article appears, search other articles written by the authors of this article, and submit or read responses to this article.

---

coronary heart disease or ischaemic heart disease

This is a disorder caused by reduced perfusion of the coronary arteries due to atherosclerosis. It is the commonest cause of death in the Western world, leading to more than a million deaths each year in the USA and about 160,000 in the UK.

Early symptoms of ischaemic heart disease include angina or palpitations, but sometimes a heart attack is the first indication that a person is affected.

Risk of developing ischaemic heart disease is particularly high in the UK. Scotland has the highest mortality from cardiovascular disease of any country in the world, followed closely by England and Wales. In the USA, deaths from this cause declined steeply from 650 per 100,000 in 1968 to just over 300 in 1985, reflecting widespread public concern about the risks of atherosclerosis, but in England and Wales the rates have remained almost static.

heart attack or myocardial infarction This involves a sudden onset of gripping central chest pain, often accompanied by sweating and vomiting, caused by death of a portion of the heart muscle following obstruction of a coronary artery by thrombosis (formation of a blood clot). Half of all heart attacks result in death within the first two hours, but in the remainder survival has improved following the widespread use of thrombolytic (clot-buster) drugs.

After a heart attack, most people remain in hospital for

> **heart disease**
>
> Deposits of calcium in diseased heart valves are sometimes actual bone tissue, possibly caused when heart cells follow a new developmental pathway as a result of inflammation.

Death Rates for Cardiovascular Disease (CVD) in the World

Source: American Heart Association
(Figures indicate death rate per 100,000 population.)

1995

Country Men[1]	CVD deaths	Total deaths	Women[1]		
			Russia	581	1,123
Russia	1,310	2,957	Romania[3]	491	899
Hungary	899	2,246	Bulgaria[2]	466	827
Bulgaria[2]	938	1,757	Hungary	408	980
Romania[3]	850	1,765	Poland	334	754
Poland	761	1,769	Czech Republic[4]	345	754
Czech Republic[4]	765	1,621	Colombia[2]	286	748
Scotland[2]	567	1,268	China (urban)[2]	296	716
Argentina[4]	595	1,335	Scotland[2]	281	774
Northern Ireland[2]	522	1,117	Argentina[4]	260	667
Finland	500	1,158	Northern Ireland[2]	253	670
Ireland, Republic of[4]	553	1,199	China (rural)[2]	275	868
England and Wales[2]	445	1,019	Ireland, Republic of[4]	240	669
New Zealand[4]	440	1,022	England and Wales[2]	204	613
Austria[2]	433	1,080	New Zealand[4]	215	630
Colombia[2]	397	1,166	USA	201	649
Norway[2]	393	957	Israel	191	565
USA	413	1,122	Portugal	177	539
Germany	411	1,105	Denmark[4]	197	757
China (urban)[2]	409	1,079	Austria[2]	188	536
Denmark[4]	419	1,196	Finland	171	494
Portugal	356	1,159	Mexico[4]	194	770
Sweden	354	828	Puerto Rico[3]	189	616
Israel	354	886	Germany	179	551
Puerto Rico[3]	384	1,323	Greece	173	435
China (rural)[2]	393	1,357	Norway[2]	147	506
Netherlands	354	958	Australia[2]	140	486
Australia[2]	321	885	Sweden	133	462
Greece	352	882	Netherlands	148	518
Canada[2]	312	912	Italy[4]	131	457
Italy[4]	304	953	Canada[2]	135	514
Switzerland[2]	273	852	Spain[2]	111	403
Spain[2]	268	952	Japan[2]	99	362
Mexico[4]	274	1,191	Switzerland[2]	105	423
France[2]	225	998	France[2]	84	412
Japan[2]	201	774			

[1] Data are for persons aged 35–74.
[2] Data as of 1994.
[3] Data as of 1992.
[4] Data as of 1993

seven to ten days, and may make a gradual return to normal activity over the following months. How soon a patient is able to return to work depends on the physical and mental demands of their job. Despite widespread fears to the contrary, it is safe to return to normal sexual activity within about a month of the attack.

angina, or angina pectoris This is severe pain in the chest due to impaired blood supply to the heart muscle because a coronary artery is narrowed. Faintness and difficulty in breathing accompany the pain. Treatment is by drugs or bypass surgery.

stroke A stroke is an interruption of the blood supply to part of the brain due to a sudden bleed in the brain (cerebral haemorrhage) or embolism or thrombosis. Strokes vary in severity from producing almost no symptoms to proving rapidly fatal. In between are those (often recurring) that leave a wide range of impaired function, depending on the size and location of the event. Strokes involving the right side of the brain, for example, produce weakness of the left side of the body. Some affect speech. Around 80% of strokes are ischaemic strokes, caused by a blood clot blocking an artery transporting blood to the brain. Transient ischaemic attacks, or 'mini-strokes', with effects lasting only briefly (less than 24 hours), require investigation to try to forestall the possibility of a subsequent full-blown stroke.

The disease of the arteries that predisposes to stroke is atherosclerosis. High blood pressure (hypertension) is also a precipitating factor – a worldwide study in 1995 estimated that high blood pressure before middle age gives a tenfold increase in the chance of having a stroke later in life.

Stroke Mini Fact Sheet
http://www.ninds.nih.gov/healinfo/disorder/stroke/
strkmini.HTM
Mini fact sheet and handy reference tool on strokes. It offers a brief description of the diagnostic symptoms, a list of the warning signs of the onset of the attack and suggested resources for further information on strokes, ways to prevent them, and also how to deal with them.

Strokes can sometimes be prevented by surgery (as in the case of some aneurysms), or by use of anticoagulant drugs or vitamin E or daily aspirin to minimize the risk of stroke due to blood clots. According to the results of a US trial announced in December 1995, the clot-buster drug tPA, if administered within three hours of a stroke, can cut the number of stroke victims experiencing lasting disability by 50%.

In the US around 500,000 people a year suffer a stroke, and it is the third commonest cause of death (after heart attack and cancer), and the main cause of disability.

heart failure This condition arises when the heart's pumping mechanism is inadequate. It results in back pressure of blood, causing congestion of the liver and lungs, failure of the peripheral blood supply and oedema. It may be a consequence of hypertension or

various types of heart disease. Treatment is with a diuretic and heart drugs, especially digoxin.

Drug Treatment in Heart Failure
http://www.bmj.com/cgi/content/full/316/7131/567
Part of a collection of articles maintained by the *British Medical Journal*, this page features the full text of a 14 February 1998 editorial concerning the treatment of heart failure, in which the author advocates the need for implementing 'the new evidence on preventing coronary heart disease effectively in general practice'.

---

rheumatic heart disease Rheumatic fever results from a Lancefield group A streptococcal infection causing tonsillitis and scarlet fever. The most important complication of rheumatic fever is damage to the heart and its valves, producing rheumatic heart disease many years later, which may lead to disability and death.

The various forms of cardiovascular disease are by no means always fatal. Many individuals survive, but with varying levels of disability. Cardiovascular diseases therefore account for a large proportion of the resources (both time and money) of a health service.

cancer

Cancer is a group of diseases characterized by abnormal proliferation of cells. Cancer (malignant) cells are usually degenerate, capable only of reproducing themselves (tumour formation). Malignant cells tend to spread from their site of origin by travelling through the bloodstream or lymphatic system. Cancer kills about 6 million people a year worldwide.

In some families there is a genetic tendency towards a particular type of cancer. In 1993 researchers isolated the first gene that predisposes individuals to cancer. About 1 in 200 people in the West carry the gene. If the gene mutates, those with the altered gene have a 70% chance of developing colon cancer, and female carriers have a 50% chance of developing cancer of the uterus. This accounts for an estimated 10% of all colon cancer.

In 1994 a gene that triggers breast cancer was identified. *BRCA1* was found to be responsible for almost half the cases of inherited breast cancer, and most cases of ovarian cancer. In 1995 a link between *BRCA1* and

---

Breast Cancer Awareness
http://avon.avon.com/showpage.asp?thepage=crusade
Promotes awareness of this disease through a library of frequently asked questions about breast cancer and mammograms, as well as a glossary of common terms and access to support groups. There is also information here about their fundraising activities and recipients of awards for people and groups seen to be contributing the most to the fight against this disease. It is sponsored by the make-up company Avon.

non-inherited breast cancer was discovered. Women with the gene have an 85% chance of developing breast or ovarian cancer during their lifetime. A second breast cancer gene *BRCA2* was identified later in 1995. In Britain 28,000 women develop breast cancer annually. Of these about 5% have an inherited form of the disease.

The commonest cancer in young men is testicular cancer, the incidence of which has been rising by 3% a year since 1974 (1998).

types of cancer There are over 100 different types of cancer. The following section details some of these.

leukaemia

Leukaemia is one of a group of cancers of the blood cells, with widespread involvement of the bone marrow and other blood-forming tissue. The central feature of leukaemia is runaway production of white blood cells that are immature or in some way abnormal. These rogue cells, which lack the defensive capacity of healthy white cells, overwhelm the normal ones, leaving the victim vulnerable to infection. Treatment is with radiotherapy and cytotoxic drugs to suppress replication of abnormal cells, or by bone-marrow transplantation.

Abnormal functioning of the bone marrow also suppresses production of red blood cells and blood platelets, resulting in anaemia and a failure of the blood to clot.

Leukaemias are classified into acute or chronic, depending on their known rates of progression. They are also grouped according to the type of white cell involved.

lung cancer

The main risk factor for lung cancer is smoking, with almost nine out of ten cases attributed to it. Other risk factors include workspace exposure to carcinogenic substances such as asbestos, and radiation. Warning symptoms include a persistent and otherwise unexplained cough, breathing difficulties, and pain in the chest or shoulder. Treatment is with chemotherapy, radiotherapy, and surgery.

colon cancer

Incidence rates for carcinoma of the large intestine tend to be low in most African countries (0–5 per 100,000), higher in Asian countries (5–15), and much higher in Europe, North America and Australia (15–35). The disease is associated with affluence and coronary heart disease. There is also a positive correlation with the following dietary factors: high fat intake, high meat intake, and low fibre intake.

melanoma

This is a highly malignant tumour of the melanin-forming cells (melanocytes) of the skin. It develops from an existing mole in up to two thirds of cases, but can also arise in the eye or mucous membranes.

Malignant melanoma is the most dangerous of the skin cancers; it is associated with brief but excessive exposure to sunlight. It is easily treated if caught early but deadly once it has spread. There is a genetic factor in some cases.

Once rare, this disease is increasing at the rate of 7% in most countries with a predominantly fair-skinned population, owing to the increasing popularity of holidays in the sun. Most at risk are those with fair hair and light skin, and those who have had a severe sunburn in childhood. Cases of melanoma are increasing by 4% a year worldwide. It strikes about 3,000 people a year in the UK, killing 1,250.

Melanoma Skin Cancer Information
http://www.cancer.org/cidSpecificCancers/
melanomaskin/index.html
Comprehensive information on melanoma from the American Cancer Society. Written in easily understandable language, this guide explains the normal function of the skin and non-malignant cancers before turning to melanoma. There is information on risk factors, causes, diagnosis, treatment, latest research news, and prognosis for those with these highly malignant tumours.

---

factors that contribute to an individual's risk of developing cancer Current research shows that attention to diet and quitting smoking reduce an individual's risk of developing cancer.

smoking

Cigarette smoking is responsible for 90% of lung cancers among men and 79% among women – about 87% overall. Smoking accounts for about 30% of all cancer deaths. Those who smoke two or more packs of cigarettes a day have lung cancer mortality rates 12–25 times greater than those of nonsmokers.

nutrition and diet

Nutrition plays an important role in preventing cancer. Research indicates that people may reduce their cancer risk by observing these nutrition guidelines:

- Maintain a desirable weight. For people who are obese, weight reduction is a good way to lower cancer risk. Weight maintenance can be accomplished by reducing intake of calories and by maintaining a physically active lifestyle.
- Eat a varied diet. A varied diet eaten in moderation offers the best hope for lowering the risk of cancer.
- Include a variety of vegetables and fruits in the daily diet. Studies have shown that daily consumption of vegetables and fresh fruits is associated with a decreased risk of lung, prostate, bladder, oesophagus, colorectal, and stomach cancers.

World Mortality from Cancer and Preventive Measures

Source: Worldwatch Institute

1997

Cancers	Deaths	Dietary and lifestyle preventive measures
Trachea, bronchus, and lung	1,050,000	eliminate smoking
Stomach	765,000	increase fruit and vegetable consumption; reduce salt intake
Colon and rectum	525,000	reduce fat and protein consumption; increase vegetable consumption
Liver	505,000	reduce alcohol consumption; vaccinate against hepatitis B
Breast (female)	385,000	reduce fat and animal protein consumption; avoid obesity
Oesophagus	355,000	eliminate smoking; reduce alcohol consumption
Mouth and pharynx	260,000	Eliminate smoking; reduce alcohol consumption
Bladder	140,000	eliminate smoking; reduce alcohol consumption

- Eat more high-fibre foods, such as whole grain cereals, breads, pasta, vegetables, and fruits. High-fibre diets are a healthy substitute for fatty foods and may reduce the risk of colon cancer.
- Cut down on total fat intake. A diet high in fat may be a factor in the development of certain cancers, particularly breast, colon, and prostate cancers. The American Cancer Society recommends reducing total fat intake to 30% or less of total calorie intake.
- Limit consumption of alcohol, if you drink at all. Heavy drinking, especially when accompanied by cigarette smoking or smokeless tobacco use, increases risk of cancers of the mouth, larynx, throat, oesophagus, and liver.

- Limit consumption of salt-cured, smoked, and nitrite-cured foods. In areas of the world where salt-cured and smoked foods are eaten frequently, there is higher incidence of cancer of the oesophagus and stomach. Modern methods of food processing and preserving appear to avoid the cancer-causing by-products associated with older methods of food treatment.

sunlight

Almost all of the over 800,000 cases of basal and squamous cell skin cancer diagnosed each year in the USA are sun-related (ultraviolet radiation). Epidemiological evidence shows that sun exposure is a major factor in the development of melanoma and

Recommendations for Early Detection of Cancer in People with No Symptoms

Source: American Cancer Society

Test	Sex	Age		Frequency
Sigmoidoscopy, preferably flexible	both	>50		every 3–5 years
Faecal occult blood test	both	>50		every year
Digital rectal examination	both	>40		every year
Prostate examination[1]	male	>50		every year
Breast self-examination	female	>20		every month
Breast clinical examination	female	20–40		every 3 years
		>40	every year	
Mammography[2]	female	40–49		every 1–2 years
		>50	every year	
Pap test	female	all women who are, or who have been, sexually active, or have reached age 18, should have an annual Pap test and pelvic examination. After a woman has had 3 or more consecutive satisfactory normal annual examinations, the Pap test may be performed less frequently at the discretion of her physician		

[1] Annual digital rectal examination and prostate-specific antigen should be performed on men 50 years and older. If either result is abnormal, further evaluation should be considered.

[2] Screening mammography should begin by age 40.

that incidence increases for those living near the equator.

smokeless tobacco

Use of chewing tobacco or snuff increases the risk of cancer of the mouth, larynx, throat, and oesophagus, and is a highly addictive habit.

oestrogen

Oestrogen treatment to control menopausal symptoms can increase risk of endometrial cancer. However, using progesterone with the oestrogen helps to minimize this risk. Consultation with a physician will help each woman to assess personal risks and benefits. Continued research is needed in the area of oestrogen use and breast cancer.

occupational hazards

Exposure to several different industrial agents, such as nickel, chromate, asbestos, vinyl chloride, increases risk of various cancers. Risk of lung cancer from asbestos is greatly increased when combined with cigarette smoking.

ionizing radiation

Excessive exposure to ionizing radiation can increase cancer risk. Most medical and dental X-rays are adjusted to deliver the lowest dose possible without sacrificing image quality. Excessive radon exposure (emitted by rocks in the Earth's crust) in homes may increase risk of lung cancer, especially in cigarette smokers. If levels are found to be too high, remedial actions should be taken.

OncoLink – The University of Pennsylvania Cancer Centre Resource
http://cancer.med.upenn.edu/
As the title of this page says, a broad resource of cancer-related information for sufferers and their families. It includes sections on 'causes, screening, and prevention', 'clinical trials', and 'conferences and meetings'.

---

treatment Cancer is one of the leading causes of death in the industrialized world, yet it is by no means incurable, particularly in the case of certain tumours, including Hodgkin's disease, acute leukaemia, and testicular cancer. Cures are sometimes achieved with specialized treatments, such as surgery, chemotherapy with cytotoxic drugs, and irradiation, or a combination of all three. Monoclonal antibodies have been used therapeutically against some cancers, with limited success. There is also hope of combining a monoclonal antibody with a drug that will kill the cancer cell to produce a highly specific magic bullet drug. In 1990 it was discovered that the presence in some patients of a particular protein, p-glycoprotein, actively protects the cancer cells from drugs intended to destroy

them. If this action can be blocked, the cancer should become far easier to treat.

Cancer: The Facts
http://www.icnet.uk/research/factsheet/index.html
Information on cancer from the Imperial Cancer Research Fund. There is a simple explanation of what cancer is, followed by links to different types of cancer – bowel, leukaemia, lung, pancreatic, skin, and multiple myeloma cancers, as well as cancers specific to either men or women, or prevalent within families. Cancer statistics are also given.

---

diabetes

Diabetes mellitus is a disorder of the islets of Langerhans in the pancreas which prevents the body producing the hormone insulin, so that sugars cannot be used properly. Treatment is by strict dietary control and oral or injected insulin, depending on the type of diabetes.

There are two forms of diabetes: Type 1, or insulin-dependent diabetes, which usually begins in childhood (early onset) and is an autoimmune condition; and Type 2, or noninsulin-dependent diabetes, which occurs in later life (late onset).

In diabetes, sugar accumulates first in the blood and then in the urine. The patient experiences thirst, weight loss, and copious voiding, along with degenerative changes in the capillary system. Without treatment, the patient may go blind, ulcerate, lapse into diabetic coma, and die. Early-onset diabetes tends to be more severe than that developing in later years, and its inci-

---

Symptoms of Diabetes

Source: American Diabetes Association

Type of diabetes	Symptoms
Type I (IDDM)[1]	frequent urination
	excessive thirst
	extreme hunger
	dramatic weight loss
	weakness and fatigue
	nausea and vomiting
	irritability
	high levels of sugar in the blood
	high levels of sugar in the urine
Type II (NIDDM)	any of the IDDM symptoms
	recurring, hard-to-heal skin or
	bladder infections
	drowsiness
	blurred vision
	tingling or numbness in extremities
	itching

[1] Symptoms usually occur suddenly.

dence has almost doubled since the 1970s; it now affects 1 in 500 people. Before the discovery of insulin by Frederick Banting and Charles Best, severe diabetics did not survive.

Careful management of diabetes, including control of high blood pressure, can delay some of the serious complications associated with the condition, which include blindness, disease of the peripheral blood vessels and kidney failure. A continuous infusion of insulin can be provided via a catheter implanted under the skin, which is linked to an electric pump. This more accurately mimics the body's natural secretion of insulin than injections or oral doses, and can provide better control of diabetes. It can, however, be very dangerous if the pump malfunctions.

In 1997, there were about 135 million sufferers worldwide.

risk factors In the majority of patients diabetes is a primary disorder, but it may arise secondary to other diseases that impair the function of the pancreas. An individual's risk of developing diabetes is affected by:
- genetic factors
- dietary factors such as obesity, dietary restriction, sugar intake and fibre intake
- infections
- stress
- other diseases that destroy the pancreas or impair insulin secretion

Insulin

To remember the role of insulin:

Remember that **in**sulin gets **in**to cells. Without insulin, a person can have excess sugar in the blood yet die of lack of sugar.

treatment for diabetes Treatment emphasizes control of blood glucose through blood glucose monitoring, regular physical activity, meal planning, and attention to relevant medical and psychosocial factors. In many patients, oral medications and/or insulin injections are also required for appropriate blood sugar (glucose) control. Treatment of diabetes is an ongoing process that is planned and regularly reassessed by the health-care team, the person with diabetes, and his or her family. Patient and family education are important parts of the process.

Type I diabetes
The goal of the Type I diabetes treatment plan is to keep the blood-sugar level as close to normal as possible (good blood-sugar control). A treatment plan will probably include:

Insulin to lower blood sugar. A health-care practitioner will prescribe for each patient the dosage, frequency, and type of insulin. Insulin must be injected into the fat below the skin to be effective. It cannot be taken as a pill because the stomach juices would destroy the insulin before it could work. Scientists are looking for new ways to give insulin.

Specified foods to increase blood sugar. Most people with Type I diabetes have a personal meal plan made up by a registered dietician, detailing the amount of food and when it should be eaten. Most diets include three meals and at least two snacks a day.

Exercise to help the body use and lower blood sugar. Generally, a health-care practitioner prescribes exercise as part of the daily routine.

Blood and urine testing to determine blood-sugar levels. These tests are simple. The blood test requires only a drop of blood from a finger-prick, which patients can learn to do at home, as well as how to use the results. Urine tests, which may be needed to test for substances in the body called ketones, can also be done by the diabetic at home. Ketones in the urine may indicate that the diabetes is not being adequately controlled.

Type II diabetes
The goal of the Type II diabetes treatment plan is to keep the blood-sugar level as close to normal as possible (good blood-sugar control). A treatment plan will probably include:

Healthy diet and regular exercise Type II diabetes can often be controlled through diet and exercise alone. But some people also need medicine – either diabetes pills or insulin shots. Many people find their diabetes gets better when they follow their treatment plan.

Losing weight to help some overweight people to bring their blood sugars into the normal range. People who have a tendency to get Type II diabetes can try to avoid the condition by losing weight or not becoming overweight. (The health-care practitioner may allow some overweight people to stop their medication so long as they lose weight and follow a good meal plan.)

Recent Advances: Diabetes
http://www.bmj.com/cgi/content/full/316/7139/1221
Part of a collection of articles maintained by the *British Medical Journal*, this page features the full full text of an 18 April 1998 clinical review regarding the alarming rise in the number of patients with diabetes around the world. The text includes diagrams, and there is a list of references.

Blood and urine testing: the health-care practitioner may also want the patient to test blood-sugar levels regularly, to check that the diabetes is under control.

complications with diabetes

short-term complications

Hypoglycaemia, or low blood sugar, sometimes called an insulin reaction, occurs when the blood sugar drops too low. This problem can be corrected by eating some sugar, for example glucose tablets or chocolate.

Hyperglycaemia, or high blood sugar, occurs when the blood sugar is too high. It can be a sign that diabetes is not well controlled.

Ketoacidosis, or ketone poisoning, occurs when the metabolism of fat is not being controlled by the treatment. It is very serious and may lead to coma.

long-term complications

Heart disease (cardiovascular disease) is more common among people with diabetes.

Stroke the risk of stroke is higher among people with diabetes.

High blood pressure is more common amongst diabetics.

Blindness diabetes is the leading cause of new cases of blindness among adults. Diabetes affects the retina in the eye and causes many new cases of blindness each year.

Kidney disease (treatment by dialysis or transplantation) diabetes is the leading cause of end-stage renal disease.

The efficient management of patients with diabetes presents an opportunity for good medical and dietetic practice, and unlike other chronic diseases a great difference can be made to the patient's life. The aetiology of diabetes and restriction of its long-term complications continues to be a challenge to medical research today.

diet and diseases

Nutrition has had an important role in promoting health and in preventing and treating diseases for thousands of years. It is well documented that Hippocrates frequently gave advice to his patients on the types of food they should be eating. However, the concerns of modern day reflect not only the importance of receiving sufficient amounts of food and nutrients, but also problems that have arisen through the lifestyle factors of affluent societies. Such diseases as obesity, cancer, and cardiovascular diseases can be directly linked to diet.

Dietary requirements may vary over the lifespan of an animal, according to whether it is growing, reproducing, highly active, or approaching death. An adequate diet for humans is one that supplies the body's daily nutritional needs and provides sufficient energy to meet individual levels of activity. The average daily requirement for men is 2,500 calories, but this will vary with age, occupation, and weight; in general, women need fewer calories than men. The energy requirements of active children increase steadily with age, reaching a peak in the late teens. At present, about 450 million people in the world – mainly living in famine or poverty stricken areas, especially in Third World countries – subsist on fewer than 1,500 calories per day. The average daily intake in developed countries is 3,300 calories.

A well-balanced diet is essential to ensure the body's peak performance. There are many factors that determine an individual's energy requirements – such as age, sex, occupation, and general lifestyle – and so it is important that diets provide the right amount of energy to match individual needs. In addition to energy-producing carbohydrates, however, the body has many other nutritional requirements, and so the emphasis of a healthy, well-balanced diet is not just a matter of calorie-counting, but ensuring that all the components (nutrients) that together comprise a well-balanced diet are present in adequate and correct amounts.

nutrients

Essential nutrients are those substances that are necessary for growth, normal functioning, and maintenance of life; they must be supplied by food because they cannot be made by the body. The main nutrients known to be essential for humans are:

- amino acids (proteins)
- carbohydrates
- essential fatty acids (fats)
- vitamins
- minerals and trace elements
- water

proteins These are made up of smaller units called amino acids. There are about twenty different kinds of amino acids. The primary function of dietary protein is to provide the amino acids required for growth and maintenance of body tissues and to regulate metabolism. Proteins are also needed by the white cells in the blood to produce antibodies; these assist the body's immune system to ward off attacks by bacteria and viruses. All fruits and vegetables contain some protein; good sources include peas, beans, lentils, grains, nuts, seeds, and potatoes. Animal proteins include milk, cheese, meat, eggs, and fish.

carbohydrates Carbohydrates are composed of carbon, hydrogen, and oxygen. The major groups are starches, sugars, and cellulose and related material (or 'roughage'). The prime function of the carbohydrates is to provide energy for the body; they also serve as efficient sources of glucose, which the body requires for brain functioning, utilization of foods, and maintenance of body temperature. Roughage includes the stiff structural materials of vegetables, fruits, and cereal products.

fats Fats are composed of glycerol and fatty acids and, like carbohydrates, contain carbon, hydrogen, and oxygen atoms. There are two types of fatty acids: saturated and unsaturated. Unsaturated fatty acids are found in fish oils and vegetable oils – coconut oil and palm oil are the only saturated vegetable oils. Most saturated fats are animal fats, and are solid, such as lard and butter. Margarine is saturated by the process of hydrogenation which forces hydrogen gas through vegetable oil. Fats provide the most concentrated form of energy: in other words, when they are burned in the body, they supply more than twice the number of calories per gram available from carbohydrates. They are also high in cholesterol. Hence the need to control fat intake where obesity and cholesterol levels present a health problem. Heart disease has been linked to the consumption of hydrogenated fats. However, fats are a necessary part of any well-balanced diet. They provide insulation, build cells, and facilitate metabolism. Unsaturated fatty acids are essential for healthy skin, circulation, bone, brain, and nerves.

Most fatty acids can be synthesized by the body, with the exception of three, the essential fatty acids (EFAs): linoleic acid, linolenic acid, and arachidonic acid. These have to be supplied from food; vegetable oils, particularly if they unrefined and cold pressed, are the best sources of EFAs. Sunflower and safflower oils are among the richest sources, containing up to 90%. EFAs are vital for the maintenance of good health. Among their many uses, they help to prevent atheroschlerosis (coronary heart disease) and the formation of blood clots in arteries, and they regulate such diverse reactions as stomach secretions, hormone release, and pancreatic function.

Your Body and Nutrition
http://www.ilcnet.com/~nutrition/body.htm
Information and pictures on the way the human body works. There are pages on different cell functions and the importance of free radicals. Each page is accompanied by a picture and a brief explanation. It is a commercially sponsored site, but there is a lot of good and free biological information here.

---

vitamins These are organic substances which are used by the body in very small amounts but are vital for normal body chemistry. They are all obtained from food, but vitamin D is also produced by the action of daylight on the skin, and vitamin K is produced by microorganisms in the bowel. An insufficient intake of one or more vitamins can result in a wide range of deficiency diseases. For instance, vitamin A, which can be found in eggs, milk, dairy products, fish liver oil, and animal liver, helps to maintain the cells lining the respiratory system and the mucous membranes of the eyes, ear, nose throat, and bladder; it helps to fight colds and a deficiency of the vitamin can lead to respiratory infections.

Some sixteen different B vitamins have been isolated, and as they usually occur together, they are known as vitamin B complex. They can be found in vegetables and animal foods, such as organ meats (particularly liver), wholemeal bread, yeast extract, and brown rice, and are vital for converting carbohydrates to glucose and food into energy. When B vitamins are lacking in the body, carbohydrates are not fully utilized and this can result in stress, nervousness, constipation, fatigue, and indigestion. Most of the complex are concerned with various processes in the liver, eyes, skin, and hair, and have a wide range of effects from alleviating stress to preventing atherosclerosis.

Vitamin C (ascorbic acid) is a very unstable, water-soluble vitamin, which is easily lost in food preparation – not only in cooking, but also in peeling, stoning, and soaking fruits and vegetables. Among its best natural sources are citrus fruits, peppers, broccoli, tomatoes, cabbage, green leafy vegetables, and potatoes. Vitamin C has many functions in the body, among the best known of which are the prevention of scurvy, a skin condition, and fighting the symptoms of the common cold. In addition, it helps to form collagen (a sub-skin 'cement'), increases immune responses to infectious diseases, and has been found to lower the risk of cancers of the mouth, oesophagus, lung, stomach, colon, cervix, and breast. It has also been shown by several studies to lower cholesterol levels. Human beings are one of very few mammals that cannot synthesize vitamin C, and a regular daily intake from a food source is necessary.

Vitamin D is a fat-soluble vitamin which is supplied from both food, especially milk and dairy foods, and exposure to the sun. It is stored mainly in the liver, but also in smaller quantities in the skin, brain, and bones. Vitamin D promotes absorption of calcium and phosphorus which are both vital for strong teeth and bones, and for the prevention of rickets in children. It also helps to maintain a healthy nervous system, normal heartbeat, and efficient blood clotting. Since vitamin D is scarce in vegetables, people who do not drink milk may need to supplement their diet with cod liver oil and fish such as sardines, herring, salmon, and

Vitamins

Vitamin	Name	Main dietary sources	Established benefit	Deficiency symptoms
A	retinol	dairy products, egg yolk, liver; also formed in body from ß-carotene, a pigment present in some leafy vegetables	aids growth; prevents night blindness and xerophthalmia (a common cause of blindness among children in developing countries); helps keep the skin and mucous membranes resistant to infection	night blindness; rough skin; impaired bone growth
B_1	thiamin	germ and bran of seeds and grains, yeast	essential for carbohydrate metabolism and health of nervous system	beriberi; Korsakov's syndrome
B_2	riboflavin	eggs, liver, milk, poultry, broccoli, mushrooms	involved in energy metabolism; protects skin, mouth, eyes, eyelids, mucous membranes	inflammation of tongue and lips; sores in corners of the mouth
B_6	pyridoxine/ pantothenic acid/biotin	meat, poultry, fish, fruits, nuts, whole grains, leafy vegetables, yeast extract	important in the regulation of the central nervous system and in protein metabolism; helps prevent anaemia, skin lesions, nerve damage	dermatitis; neurological problems; kidney stones
B_{12}	cyanocobalamin	liver, meat, fish, eggs, dairy products, soybeans	involved in synthesis of nucleic acids, maintenance of myelin sheath around nerve fibres; efficient use of folic acid	anaemia; neurological disturbance
	folic acid	green leafy vegetables, liver, peanuts; cooking and processing can cause serious losses in food	involved in synthesis of nucleic acids; helps protect against cervical dysplasia (precancerous changes in the cells of the uterine cervix)	megaloblastic anaemia
	nicotinic acid (or niacin)	meat, yeast extract, some cereals; also formed in the body from the amino acid tryptophan	maintains the health of the skin, tongue, and digestive system	pellagra
C	ascorbic acid	citrus fruits, green vegetables, tomatoes, potatoes; losses occur during storage and cooking	prevents scurvy, loss of teeth; fights haemorrhage; important in synthesis of collagen (constituent of connective tissue); aids in resistance to some types of virus and bacterial infections	scurvy
D	calciferol, cholecalciferol	liver, fish oil, dairy products, eggs; also produced when skin is exposed to sunlight	promotes growth and mineralization of bone	rickets in children; osteomalacia in adults
E	tocopherol	vegetable oils, eggs, butter, some cereals, nuts	prevents damage to cell membranes	anaemia
K	phytomenadione, menaquinone	green vegetables, cereals, fruits, meat, dairy products	essential for blood clotting	haemorrhagic problems

tuna. Overdoses of the vitamin can lead to toxicity symptoms such as diarrhoea, nausea, excessive urination, and kidney damage.

minerals and trace elements These are inorganic substances vital to normal development; calcium and iron are particularly important as they are required in relatively large amounts. Minerals required by the body in trace amounts include chromium, copper, fluoride, iodine, iron, magnesium, manganese, molybdenum, phosphorus, potassium, selenium, sodium, and zinc. The relative proportions of these nutrients in the diet can influence health. If a particular nutrient is present in insufficient quantities then deficiency symptoms can result. Likewise, if too much of a certain nutrient is present in the diet then toxic symptoms may occur or this may lead to other disease-associated manifestations. For instance, calcium, which is found in milk, cheese, and bread, is necessary for healthy bones and teeth. A deficiency of this mineral can result in brittle bones and teeth. Similarly, phosphorus, which is contained in milk, is also needed in the formation of bones and teeth. Iron, which is present in liver and egg yolk, is used by the body in the manufacture of haemo-

Nutritive Value of Foods

Source: UK Ministry of Agriculture, Fisheries and Food
The energy value of each food is given in kilojoules (kJ) and kilocalories (kcal), and both have been calculated from the protein, fat, and carbohydrate content per 100 g of edible portion.

Food	Energy kJ	kcal	Protein (g)	Fat (g)	Saturated fat (g)	Carbohydrate (g)
Cereal and Cereal Products						
Bread, brown	927	218	8.5	2.0	0.4	44.3
Bread, white	1,002	235	8.4	1.9	0.4	49.3
Flour, plain, white	1,450	341	9.4	1.3	0.2	77.7
Flour, wholemeal	1,318	310	12.7	2.2	0.3	63.9
Oats, porridge, raw	1,587	375	11.2	9.2	1.6	66.0
Rice, brown, boiled	597	141	2.6	1.1	0.3	32.1
Rice, white, boiled	587	138	2.6	1.3	0.3	30.9
Spaghetti, white, boiled	442	104	3.6	0.7	0.1	22.2
Dairy Products						
Butter	3,031	737	0.5	81.7	54.0	0.0
Cheddar cheese	1,708	412	25.5	34.4	21.7	0.1
Cottage cheese	413	98	13.8	3.9	2.4	2.1
Cream, fresh, heavy	1,849	449	1.7	48.0	30.0	2.7
Cream, fresh	817	198	2.6	19.1	11.9	4.1
Eggs, boiled	612	147	12.5	10.8	3.1	0.0
Low-fat spread	1,605	390	5.8	40.5	11.2	0.5
Margarine, polyunsaturated	3,039	739	0.2	81.6	16.2	1.0
Milk, semi-skimmed	195	46	3.3	1.6	1.0	5.0
Milk, skimmed	140	33	3.3	0.1	0.1	5.0
Milk, whole	275	66	3.2	3.9	2.4	4.8
Yoghurt, whole milk, plain	333	79	5.7	3.0	1.7	7.8
Fish						
White fish, steamed, flesh only	417	98	22.8	0.8	0.2	0.0
Shrimps, boiled	451	107	22.6	1.8	0.4	0.0
Fruit						
Apples	199	47	0.4	0.1	0.0	11.8
Apricots	674	158	4.0	0.6	0.0	36.5
Avocados	784	190	1.9	19.5	4.1	1.9
Bananas	403	95	1.2	0.3	0.1	23.2
Cherries	203	48	0.9	0.1	0.0	11.5
Grapefruit	126	30	0.8	0.1	0.0	6.8
Grapes	257	60	0.4	0.1	0.0	15.4
Mangoes	245	57	0.7	0.2	0.1	14.1
Melon	119	28	0.6	0.1	0.0	6.6
Oranges	158	37	1.1	0.1	0.0	8.5
Peaches	142	33	1.0	0.1		
Pears	169	40	0.3	0.1	0.0	10.0
Plums	155	36	0.6	0.1	0.0	8.8
Raspberries	109	25	1.4	0.3	0.1	4.6
Strawberries	113	27	0.8	0.1	0.0	6.0
Meat						
Beef, lean only, raw	517	123	20.3	4.6	1.9	0.0
Chicken, meat and skin, raw	954	230	17.6	17.7	5.9	0.0
Lamb, lean only, raw	679	162	20.8	8.8	4.2	0.0
Pork, lean only, raw	615	147	20.7	7.1	2.5	0.0

Food	Energy		Protein (g)	Fat (g)	Saturated fat (g)	Carbohydrate (g)
	kJ	kcal				
Vegetables						
Aubergine	64	15	0.9	0.4	0.1	2.2
Beetroot	195	46	2.3	0.0	0.0	9.5
Cabbage	109	26	1.7	0.4	0.1	4.1
Celery	32	7	0.5	0.2	0.0	0.9
Courgettes	74	18	1.8	0.4	0.1	1.8
Cucumber	40	10	0.7	0.1	0.0	1.5
Lettuce	59	14	0.8	0.5	0.1	1.7
Mushrooms	55	13	1.8	0.5	0.1	0.4
Onions	150	36	1.2	0.2	0.0	7.9
Parsnips	278	66	1.6	1.2	0.2	12.9
Peas	291	69	6.0	0.9	0.2	9.7
Peppers	65	15	0.8	0.3	0.1	2.6
Potatoes, new, flesh only	298	70	1.7	0.3	0.1	16.1
Potatoes, old, flesh only	318	75	2.1	0.2	0.0	17.2
Spinach	90	21	301	0.8	0.1	0.5
Sweetcorn kernels	519	122	2.9	1.2	0.2	26.6
Sweet potatoes	358	84	1.1	0.3	0.1	20.5
Tofu, soya bean, steamed	304	73	8.1	4.2	0.5	0.7
Watercress	94	22	3.0	1.0	0.3	0.4

globin; a deficiency of this mineral can result in various forms of anaemia.

water Water is involved in nearly every body process. Animals and humans will succumb to water deprivation sooner than to starvation.

the Western diet

During the twentieth century with the advent of food processing techniques, people have been able to 'refine', 'process', and 'add' to food. Whilst this has increased the quantity and range of food that an

Dietary Components of Traditional, Mediterranean, and Western Diets

Source: Worldwatch Institute
(N/A = not available.)

Dietary components	Diet			WHO recommendations
	Traditional	Mediterranean	Western	
Carbohydrates				
Complex (starch) (% of dietary energy)	60–75	N/A	28	45–55
Simple (sugar) (% of dietary energy)	5	N/A	22	10
Total (% of dietary energy)	65–80	>50	50	55–65
Fats and Oils				
Saturated fats (% of dietary energy)	approx. 10–15	<8	18–30	10–15
Total (% of dietary energy)	<20	30–37	38–43	20–30
Protein				
Protein (% of dietary energy)	10	8–12	12	8–12
Cholesterol				
Cholesterol (mg per day)	<100	approx. 300–500	approx. 500	<300
Salt				
Salt (g per day)	5–15	N/A	15	5
Fibre				
Fibre (g per day)	60–120	7–11	7	>30

Recommended Weight Tables

Source: Metropolitan Life Insurance Company

Height		Small frame weight		Medium frame weight		Large frame weight	
m	ft/in	kg	lbs	kg	lbs	kg	lbs
Men Aged 25 and Over							
1.55	5'1"	48–51	105–13	50–55	111–22	54–61	119–34
1.57	5'2"	49–53	108–16	52–57	114–26	55–62	122–37
1.60	5'3"	50–54	111–19	53–59	117–29	57–64	125–41
1.63	5'4"	52–55	114–22	54–60	120–32	58–66	128–45
1.65	5'5"	53–57	117–26	56–62	123–36	59–68	131–49
1.68	5'6"	55–59	121–30	58–64	127–40	61–70	135–54
1.70	5'7"	57–61	125–34	59–66	131–45	64–72	140–59
1.73	5'8"	59–68	129–49	61–68	135–49	65–74	144–63
1.75	5'9"	60–65	133–43	63–69	139–53	67–76	148–67
1.78	5'10"	62–67	137–47	65–72	143–58	69–78	152–72
1.80	5'11"	64–69	141–51	67–74	147–63	71–80	157–77
1.83	6'0"	66–70	145–55	69–76	151–68	73–83	161–82
1.85	6'1"	68–73	149–60	70–79	155–73	76–85	168–87
1.88	6'2"	69–74	153–64	73–81	160–78	78–87	171–92
1.91	6'3"	71–76	157–68	75–83	165–83	79–89	175–97
Women Aged 25 and Over							
1.45	4'9"	41–44	90–97	43–48	94–106	46–54	102–18
1.47	4'10"	42–45	92–100	44–49	97–109	48–55	106–21
1.50	4'11"	43–47	95–103	45–51	100–12	49–56	108–24
1.52	5'0"	45–48	98–106	47–53	103–16	50–58	111–27
1.55	5'1"	46–49	101–09	48–54	106–18	52–59	114–30
1.57	5'2"	47–51	104–12	49–51	109–12	53–61	117–34
1.60	5'3"	49–52	107–15	51–57	112–26	55–63	121–38
1.63	5'4"	50–54	110–19	53–59	116–31	57–64	125–42
1.65	5'5"	52–56	114–23	54–62	120–36	59–66	129–46
1.68	5'6"	54–58	118–27	56–63	124–39	60–68	133–50
1.70	5'7"	55–59	122–31	58–65	128–43	62–70	137–54
1.73	5'8"	57–62	126–36	60–67	132–47	64–72	141–59
1.75	5'9"	59–64	130–40	62–69	136–51	66–74	145–64
1.78	5'10"	61–65	134–44	64–70	140–55	68–77	149–69

individual can sample, in some cases it has also had a deleterious effect on the nutritional quality of food, with direct consequences for health and well-being.

The so-called 'Western diet' of affluent societies leaves much to be desired. Unlike traditional diets and diets of those people living in Mediterranean countries it either lacks certain beneficial nutrients or contains too much of others. Compared to a traditional diet, a Western diet contains:

- too much refined carbohydrates and sugar;
- too much saturated fats, mostly animal fats;
- too much cholesterol;
- too much salt;
- insufficient fibre;
- too many processed foods;

- too much tea, coffee, and alcoholic beverages;
- pesticide residues and other potentially harmful food additives;
- adopting such a diet can lead to diseases such as obesity.

obesity Obesity is the most prevalent nutritional disorder in prosperous communities and is the result of an energy imbalance. An individual is considered to be obese when they have a Body Mass Index (BMI) in excess of 30. The BMI is calculated by taking the individual's weight in kilogrammes and dividing this by the height in metres. Obesity occurs when energy intake exceeds actual energy expended, so the unused energy is stored as body fat. A BMI of over 30 is a hazard to health, and is now common enough in the

Western world to constitute one of the most important medical and public health problems, putting the individual at high risk of degenerative diseases or even death, and these risks increase as the BMI increases.

Obesity has been attributed to numerous causes including excessive intake, reactive eating due to stress or anxiety, low resting metabolic rate influencing the caloric value of energy expenditure, environmental effects, family behaviour problems, and genetic factors. Other influences include human biology, endocrine alterations, society, culture, and physical activity. It is a complex issue.

genetic influence

An individual's genetic composition (their genotype) has a specific influence on their sex, height, eye colour, and skeleton, and may also influence their body and muscular mass, as well as body fat and how this is regionally distributed. The variation in body fat distribution may also result from a complex interaction of physical, environmental, and social features, as well as from the genotype. Each gene may have a varying or specific influence with many genes each exerting a small effect, or perhaps a single rarer gene that plays a larger role over time, with inherited differences in the likelihood of developing obesity.

Each genotype may have its own particular influence, with a single gene defect influencing obesity, or it may exert several independent (polygenic) influences, including regulation of energy expenditure (or metabolism). For example, there may be a form of thrifty genotype that has developed over time, and which influences the more efficient use of energy intake. The body fat itself may consist mainly of enlarged fat cells, or there may be a larger number of smaller fat cells, each of which may have a specific genotype influence. Perhaps there is no particular gene that influences the development and/or maintenance of obesity, but probably several genes and aspects of human physiology that determine many aspects of weight, appetite, satiety, metabolism, and a predisposition to physical activity. Families may have shared genes, and they often share the same environment and particular eating patterns – also influential factors that need to be examined separately. With regard to body fat variation, the additive genetic effect on the amount of subcutaneous fat is considered to be quite low, but has been reported as highest for fat mass and regional distribution of the fat, which suggests that the genotype may be more influential on the visceral (deep) fat than the subcutaneous store. From the limited data available it appears that the genotype could account for up to 40% of the individual differences in the resting metabolic rate, the thermic effect of food, and the energy cost of light exercise, which would explain why some people become obese.

studies

Studies of obesity in families show there is possibly a familial influence; for example, in a series of studies of the families of obese children, it was found that a child with one obese parent was at greater risk of developing obesity, but surprisingly also that those with two obese parents were at a lesser risk. This shows how complex this issue is and demonstrates the need for much more investigation.

Studies of adopted children and both their biological and adoptive parents suggest that inheritance plays an important role in the risk of developing obesity. Research has shown that there was no relationship between the BMI of the child and the adoptive parents, but that the BMI of the biological parents actually increased with that of the child whose weight was increasing.

Further evidence for the effect of the genotype and its influence has come from work examining the body weight of twins, where nearly two-thirds of the variability in BMI was attributed to genetic factors. The body weight of twins of the same sex (monozygotic) has been found to have identical genetic features and is more similar than the body weight of twins of different sexes (dizygotic). The genetic features of dizygotic twins are similar to those of other siblings.

Obesity appears to be more prevalent in the families of obese patients seeking surgical treatment for their condition than in a group of nonobese patients treated with other abdominal surgery. In approximately 10% of the families of the patients seeking help, more than one other member sought treatment. The incidence of obesity in the families of these patients was compared with that in a group of nonobese patients, and it was found that there were significantly more obese family members in the obese group. These included mothers and daughters, fathers and daughters, sisters and brothers, all with a BMI of over 35. This would suggest a genetic influence, and a need for further investigations into these factors.

There are also certain inherited syndromes where obesity is one of the characteristics. These include, in particular, Prader–Willi Syndrome, where children develop a ravenous appetite and a variety of food-seeking behaviours, with parents fighting a losing battle to control the child's intake. The Bardet–Biedl Syndrome is transmitted via an autosomal recessive trait, with obesity occurring in up to 80% of the cases.

There appears to be some single and polygenic influence on the transmission of obesity, although the genetic influence is more important in determining body fat distribution, and studies show that it may run in families, but one needs to examine separately the influences of the environment and the genes.

anorexia In the wealthier developed countries of the world, many instances of primary deficiency diseases are self-inflicted and the direct result of weight-loss and slimming regimes, whereby essential nutrients are omitted from the basic diet.

Anorexia is a lack of desire to eat, or refusal to eat, and can result in the pathological condition of **anorexia nervosa**, most often found in adolescent girls and young women. Anorexia nervosa is characterized by severe self-imposed restriction of food intake. The consequent weight loss may lead, in women, to absence of menstruation. Anorexic patients sometimes commit suicide. Anorexia nervosa is often associated with increased physical activity and symptoms of mental disorders. Psychotherapy is an important part of the treatment.

The mental Health Foundation estimated in 1997 that as many as 1 in 20 people (of which the vast majority are women) display symptoms of anorexia, although most of them are never formally diagnosed. About 1 sufferer in 100 needs long-term treatment, and of these one-fifth die, half from starvation and half from suicide.

The causes of anorexia nervosa are not known. Teenage pressures and family rivalries and hostilities may be contributive factors. The anorectic may be trying to attain an 'ideal' figure, or may be resisting becoming a grown-up woman like her mother. She often thinks of herself as without an identity of her own. The condition also occurs with older women and, very rarely, with men.

deficiency diseases Deficiency diseases are widespread in developing countries. In many areas the combined effects of drought and civil war mean that harvests do not provide sufficient food to feed the people and many die from starvation.

The World Health Organization estimated in 1995 that one-third of the world's children are undernourished. According to UNICEF, in 1996 around 86 million children under the age of five (50% of all under-fives) in South Asia were malnourished, compared to 32 million (25%) in sub-Saharan Africa.

In a report released at the World Food Summit in Rome in 1996, the World Bank warned of an impending international food crisis. In 1996, more than 800 million people were unable to get enough food to meet their basic needs. Eighty-two countries, half of them in Africa, did not grow enough food for their own people; nor could they afford to import it. The World Bank calculated that food production would have to double over the next 30 years as the world population increases. Contrary to earlier predictions, food stocks and particularly grain stocks have fallen during the 1990s.

Nutritional deficiency diseases can be divided into two categories: primary deficiency diseases which are a direct result of an inadequate supply of an essential nutrient; and secondary deficiency diseases which result from the body's failure to make adequate use of an essential nutrient. This can be due either to an inability to absorb the nutrient, or an inability to metabolize the nutrient once it has been absorbed.

In the poorer countries of the Third World and developing nations primary deficiency diseases are, for the most part, endemic to regions where poverty and famine are common, or where essential nutrients are lacking in the staple diet.

protein-energy malnutrition
Protein-energy malnutrition describes a range of clinical disorders. At one end of the spectrum lies **kwashiorkor**, which is caused by a quantitative and qualitative lack of protein in the diet, but where energy intake may be adequate. At the other end is marasmus, which is the result of a continual restriction of both energy and protein, as well as other nutrients. Both conditions arise as a result of poverty and ignorance concerning infant feeding practices, where breast feeding may have ended too early and unsuitable infant foods introduced.

As a result of protein-energy malnutrition there is a failure in the functioning of certain body organs and impairment of the immune system, a drastic alteration in body composition, and disorders of metabolism. If the disease is so severe as to warrant treatment in hospital, the prognosis of the patient is uncertain.

iodine deficiency
Low iodine intakes are one cause of endemic goitre, and cretinism also occurs in areas in which severe iodine deficiency has been prevalent for several generations. Endemic cretinism is known to occur in parts of Nepal, the Andes, the Democratic Republic of Congo (formerly Zaire), and New Guinea, and in some communities may affect 1–5% of the population. Prevention of endemic goitre and cretinism can be established by effective iodine prophylaxis. This can be achieved by iodinization of table salt or iodized oil injections.

iron deficiency anaemia
Iron deficiency anaemia, in which the red blood cells contain abnormally low amounts of haemoglobin, occurs in all countries of the world. In the Middle East, Africa, and Asia as much as 20% of the population is affected. It is still common in the developed world, particularly in women of childbearing age. Iron deficiency may be due to an iron-poor diet, poor absorption of the mineral due to disease, or an unreplaced loss of iron from menstruation or other.

vitamin deficiencies

Vitamins can be found in a wide variety of dietary sources, and in developed countries deficiencies tend to be rare. However, this is not the case in developing countries, where insufficient food supplies combined with infectious diseases mean that vitamin deficiencies are a common occurrence. The following table lists the deficiency symptoms of fat- and water-soluble vitamins.

mental disorders

Psychiatric disorders are a common, if somewhat stigmatized, problem. The main areas dealt with include:
- schizophrenia and other psychoses;
- mood disorders (bipolar disorders, depressive disorders, manic depression);
- anxiety disorders (panic disorders, phobias, obsessive-compulsive disorder, post-traumatic stress disorder);
- personality disorders.

schizophrenia

Schizophrenia is a mental disorder, a psychosis of unknown origin, which can lead to profound changes in personality, behaviour, and perception, including delusions and hallucinations. It is more common in males and the early-onset form is more severe than when the illness develops in later life. Modern treatment approaches include drugs, family therapy, stress reduction, and rehabilitation.

Schizophrenia implies a severe divorce from reality in the patient's thinking. Although the causes are poorly understood, it is now recognized as an organic disease, associated with structural anomalies in the brain. There is some evidence that early trauma, either in the womb or during delivery, may play a part in causation. There is also a genetic contribution.

There is an enormous intercountry variation in the symptoms of schizophrenia and in the incidence of the main forms of the disease, according to a 1997 report by US investigators. Paranoid schizophrenia, characterized by feelings of persecution, is 50% more common in developed countries, whereas catatonic schizophrenia, characterized by total immobility, is six times more frequent in developing countries. Hebephrenic schizophrenia, characterized by disorganized behaviour and speech and emotional bluntness, is four times more prevalent in developed countries overall but is rare in the USA.

Canadian researchers in 1995 identified a protein in the brain, PSA-NCAM, that plays a part in filtering sensory information. The protein is significantly reduced in the brains of schizophrenics, supporting the idea that schizophrenia occurs when the brain is overwhelmed by sensory information. In 1997, US researchers linked schizophrenia to a mutation in the gene that codes for an acetylcholine receptor. The receptor, α7-nictotinic receptor, is also stimulated by nicotine.

The prevalence of schizophrenia in Europe is about 2–5 cases per 1,000 of the population.

Schizophrenia
http://www.pslgroup.com/SCHIZOPHR.HTM
Facts about schizophrenia – causes and symptoms, the different types of the disease, how it affects sufferers' family members, available treatments, and new developments. It includes a list of available support resources.

---

paranoia

This is a mental disorder marked by delusions of grandeur or persecution. In popular usage, paranoia means baseless or exaggerated fear and suspicion.

In chronic paranoia, patients exhibit a rigid system of false beliefs and opinions, believing themselves, for example, to be followed by the secret police, to be loved by someone at a distance, or to be of great importance or in special relation to God. There are no hallucinations and patients are in other respects normal. In paranoid states, the delusions of persecution or grandeur are present but not systematized. In paranoid schizophrenia, the patient suffers from many unsystematized and incoherent delusions, is extremely suspicious, and experiences hallucinations and the feeling that external reality has altered.

depression

Depression is an emotional state characterized by sadness, unhappy thoughts, apathy, and dejection. It is the most common reason in the UK for people consulting a general practitioner. After childbirth, postnatal depression is common. However, clinical depression, which is prolonged or unduly severe, often requires treatment, such as antidepressant medication, cognitive therapy, or, in very rare cases, electroconvulsive therapy (ECT), in which an electrical current is passed through the brain.

Periods of depression may alternate with periods of high optimism, over-enthusiasm, and confidence. This

---

Depression Central
http://www.psycom.net/depression.central.html
Jumping-off point for information on all types of depressive disorders and on their most effective treatment. This site is divided into many useful sections, including 'bipolar disorder', 'depression in the elderly', and 'electroconvulsive therapy'.

---

is the manic phase in a disorder known as manic depression or bipolar disorder.

manic depression or bipolar disorder

This mental disorder is characterized by recurring periods of either depression or mania (inappropriate elation, agitation, and rapid thought and speech) in which a person switches repeatedly from one extreme to the other. Each mood can last for weeks or months. Typically, the depressive state lasts longer than the manic phase.

Sufferers may be genetically predisposed to the condition. Some cases have been improved by taking prescribed doses of lithium. Some manic-depressive patients have only manic attacks, others only depressive, and in others the alternating, or circular, form exists. The episodes may be of varying severity, from mild to psychotic (when the patient loses touch with reality and may experience hallucinations), and sometimes continue for years without interruption.

Long Term Pharmacotherapy of Depression
http://www.bmj.com/cgi/content/full/316/7139/1180
Part of a collection of articles maintained by the *British Medical Journal*, this page features the full text of a 18 April 1998 editorial addressing the 'importance of long term psychological and pharmacological treatment'.

---

anxiety

Anxiety is a subjective experience of fear and apprehension accompanied by an overwhelming feeling of impending doom. It is commonly associated with symptoms such as breathlessness, sweating, palpitations, nausea, and chest pain.

obsessive-compulsive disorder

This is an anxiety disorder that manifests itself in the need to check constantly that certain acts have been performed 'correctly'. Sufferers may, for example, feel compelled to repeatedly wash themselves or return home again and again to check that doors have been locked and appliances switched off. They may also hoard certain objects and insist in these being arranged in a precise way or be troubled by intrusive and unpleasant thoughts. In extreme cases normal life is disrupted through the hours devoted to compulsive actions. Treatment involves cognitive therapy and drug therapy with serotonin-blocking drugs such as Prozac.

phobia

A phobia is an excessive irrational fear of an object or situation – for example, agoraphobia (fear of open spaces and crowded places), acrophobia (fear of heights), and claustrophobia (fear of enclosed places). Behaviour therapy is one form of treatment.

personality disorders

This involves a maladjusted, often antisocial pattern of behaviour that is usually well in evidence by the teenage years. The person concerned lacks insight and tends to be a liability (if not a danger) to himself and others. The condition remains curiously intractable, but some people mature out of their maladaptive behaviour and become better adjusted in middle life.

Psychopathy is a personality disorder characterized by chronic antisocial behaviour (violating the rights of others, often violently) and an absence of feelings of guilt about the behaviour. Because the term 'psychopathy' has been misused to refer to any severe mental disorder, many psychologists now prefer the term 'antisocial personality disorder', though this also includes cases in which absence or a lesser degree of guilt is not a characteristic feature.

maternal and fetal health

complications of pregnancy and childbirth

Every day, at least 1,600 women die from the complications of pregnancy and childbirth. A high proportion of these deaths – almost 90% – occur in Asia and sub-Saharan Africa; approximately 10% in other developing regions; and less than 1% developed countries.

Maternal mortality is about 18 times higher in the developing world than in developed countries. These maternal mortality ratios reflect a woman's risk of dying each time she becomes pregnant; because women in developing countries bear many children and obstetric care is poor, their lifetime risk of maternal death is much higher – approximately 40 times higher than in the developed world.

In addition to maternal mortality, half of all perinatal deaths are due primarily to inadequate maternal care during pregnancy and infant delivery. Each year, 8 million neonatal deaths and stillbirths occur, largely the result of those factors that also cause the death and disability of their mothers – poor maternal health, poor hygiene and inappropriate management of infant delivery, as well as lack of neonatal care.

abortion Each year, about 20 million unsafe abortions occur around the world, resulting in some 80,000 maternal deaths, and hundreds of thousands of disabilities. Most of these deaths take place in developing countries. Unsafe abortion accounts for at least 13% of global maternal mortality, and in some countries is the most common cause of maternal mortality and morbidity.

Phobias

Fear	Name of phobia	Fear	Name of phobia
Animals	zoophobia	Fever	febriphobia
Bacteria	bacteriophobia, bacillophobia	Fire	pyrophobia
Beards	pogonophobia	Fish	ichthyophobia
Bees	apiphobia, melissophobia	Flying, the air	aerophobia
Being alone	monophobia, autophobia, eremophobia	Fog	homichlophobia
		Food	sitophobia
Being buried alive	taphophobia	Foreign languages	xenoglossophobia
Being seen by others	scopophobia	Freedom	eleutherophobia
Being touched	haphephobia, aphephobia	Fun	cherophobia
Birds	ornithophobia	Germs	spermophobia, bacillophobia
Blood	h(a)ematophobia, hemophobia	Ghosts	phasmophobia
Blushing	ereuthrophobia, e(y)rythrophobia	Glass	hyalophobia
Books	bibliophobia	God	theophobia
Cancer	cancerophobia, carcinophobia	Going to bed	clinophobia
Cats	ailurophobia, gatophobia	Graves	taphophobia
Chickens	alektorophobia	Hair	chaetophobia, trichophobia, hypertrichophobia
Childbirth	tocophobia, parturiphobia		
Children	paediphobia	Heart conditions	cardiophobia
Cold	cheimatophobia, frigophobia	Heat	thermophobia
Colour	chromatophobia, chromophobia, psychrophobia	Heaven	ouranophobia
		Heights	acrophobia, altophobia
Comets	cometophobia	Hell	hadephobia, stygiophobia
Computers	computerphobia, cyberphobia	Home	domatophobia, oikophobia
Contamination	misophobia, coprophobia	Homosexuality	homophobia
Criticism	enissophobia	Horses	hippophobia
Crossing bridges	gephyrophobia	Human beings	anthrophobia
Crossing streets	dromophobia	Ice, frost	cryophobia
Crowds	demophobia, ochlophobia	Ideas	ideophobia
Darkness	achulophobia, nyctophobia, scotophobia	Illness	nosemaphobia, nosophobia
		Imperfection	atelophobia
Dawn	eosophobia	Infection	mysophobia
Daylight	phengophobia	Infinity	apeirophobia
Death, corpses	necrophobia, thanatophobia	Injustice	dikephobia
Defecation	rhypophobia	Inoculations, injections	trypanophobia
Deformity	dysmorphophobia	Insanity	lyssophobia, maniaphobia
Demons	demonophobia	Insects	entomophobia
Dirt	mysophobia	Itching	acarophobia, scabiophobia
Disease	nosophobia, pathophobia	Jealousy	zelophobia
Disorder	ataxiophobia	Knowledge	epistemophobia
Dogs	cynophobia	Lakes	limnophobia
Draughts	anemophobia	Large objects	macrophobia
Dreams	oneirophobia	Leaves	phyllophobia
Drinking	dipsophobia	Left side	levophobia
Drugs	pharmacophobia	Leprosy	leprophobia
Duration	chronophobia	Lice	pediculophobia
Dust	amathophobia, koniphobia	Lightning	astraphobia
Eating	phagophobia	Machinery	mechanophobia
Enclosed spaces	claustrophobia	Many things	polyphobia
Everything	pan(t)ophobia	Marriage	gamophobia
Facial hair	trichopathophobia	Meat	carnophobia
Faeces	coprophobia	Men	androphobia
Failure	kakorrphiaphobia	Metals	metallophobia
Fatigue	kopophobia, ponophobia	Meteors	meteorophobia
Fears	phobophobia	Mice	musophobia

Fear	Name of phobia	Fear	Name of phobia
Mind	psychophobia	Small objects	microphobia
Mirrors	eisoptrophobia, catotrophobia	Smell	olfactophobia
Money	chrometophobia	Smothering, choking	pnigerophobia
Monsters, monstrosities	teratophobia	Snakes	ophidiophobia, ophiophobia
Motion	kinesophobia, kinetophobia	Snow	chionophobia
Music	musicophobia	Soiling	rypophobia
Names	onomatophobia	Solitude	eremitophobia, eremophobia
Narrowness	anginaphobia	Sound	akousticophobia
Needles	belonophobia	Sourness	acerophobia
Night, darkness	achluophobia	Speaking aloud	phonophobia
Noise	phonophobia	Speed	tachophobia
Novelty	cainophobia, cenotophobia, neophobia	Spiders	arachn(e)ophobia
		Standing	stasiphobia
Nudity	gymnotophobia	Standing erect	stasibasiphobia
Number 13	triskaidekaphobia, terdekaphobia	Stars	siderophobia
Odours	osmophobia	Stealing	kleptophobia
Open spaces	agoraphobia	Stillness	eremophobia
Pain	algophobia, odynophobia	Stings	cnidophobia
Parasites	parasitophobia	Strangers	xenophobia
Physical love	erotophobia	Strong light	photophobia
Pins	enetophobia	Stuttering	laliophobia, lalophobia
Places	topophobia	Suffocation	anginophobia
Pleasure	hedonophobia	Sun	heliophobia
Pointed instruments	aichmophobia	Symbols	symbolophobia
Poison	toxiphobia, toxophobia, iophobia	Taste	geumaphobia
Poverty	peniaphobia	Teeth	odontophobia
Precipices	cremnophobia	Thinking	phronemophobia
Pregnancy	maieusiophobia	Thrown objects	ballistophobia
Punishment	poinephobia	Thunder	astraphobia, brontophobia, keraunophobia
Rain	ombrophobia		
Reptiles	batrachophobia	Touch	aphephobia, haptophobia, haphephobia
Responsibility	hypegiaphobia		
Ridicule	katagalophobia	Travel	hodophobia
Rivers	potamophobia	Travelling by train	siderodromophobia
Robbery	harpaxophobia	Trees	dendrophobia
Ruin	atephobia	Trembling	tremophobia
Rust	iophobia	Vehicles	amaxophobia, ochophobia
Sacred things	hierophobia	Venereal disease	cypridophobia
Satan	satanophobia	Void	kenophobia
School	scholionophobia	Vomiting	emetophobia
Sea	thalassophobia	Walking	basiphobia
Semen	spermatophobia	Wasps	spheksophobia
Sex	genophobia	Water	hydrophobia, aquaphobia
Sexual intercourse	coitophobia	Weakness	asthenophobia
Shadows	sciophobia	Wind	ancraophobia
Sharp objects	belonephobia	Women	gynophobia
Shock	hormephobia	Words	logophobia
Sin	hamartiophobia	Work	ergophobia, ergasiophobia
Sinning	peccatophobia	Worms	helminthophobia
Skin	dermatophobia	Wounds, injury	traumatophobia
Sleep	hypnophobia	Writing	graphophobia

Maternal Deaths, Maternal Mortality Ratio, and Lifetime Risk of Maternal Death, by Region.

Source: WHO and Unicef

1996

Region	Annual number of maternal deaths	Maternal mortality ratio[1]	Lifetime risk of maternal death (one in:)
World	585,000	430	60
Developed countries	4,000	27	1,800
Developing countries	582,000	480	48
Africa	235,000	870	16
Eastern Africa	97,000	1,060	12
Middle Africa	31,000	950	14
Northern Africa	16,000	340	55
Southern Africa	3,600	260	75
Western Africa	87,000	1,020	12
Asia[2]	323,000	390	65
Eastern Asia	24,000	95	410
South-central Asia	227,000	560	35
South-eastern Asia	56,000	440	55
Western Asia	16,000	320	55
Europe	3,200	36	1,400
Eastern Europe	2,500	62	730
Latin America and the Caribbean	23,000	190	130
Central America	4,700	140	170
South America	15,000	200	140
North America	500	11	3,700
Oceania[2]	1,400	680	26

[1] Maternal Mortality Ratio (MMR) = total number of maternal deaths for every 100, 000 live births. This ratio measures the risk of death a woman faces each time she becomes pregnant.
[2] Australia, New Zealand, and Japan have been excluded from the regional; totals, but are included in the total for developed countries.

Unsafe Abortion: Regional Estimates of Mortality and Risk of Death

Source: World Health Organization

Region	No. of maternal deaths due to unsafe abortion	Risk of dying after unsafe abortion (one in:)	Maternal deaths due to unsafe abortion (%)
Africa	33,000	150	13
Asia[1]	37,500	250	12
Latin America and the Caribbean	4,600	900	21
Europe	500	1900	17

[1] Japan, Australia, and New Zealand excluded.

Unsafe abortion is now recognized by many governments as a major public health issue.

complementary therapy

The term 'complementary therapy' is used to encompass any practice or system of beliefs about treatment that is not included in what is generally understood as Western scientific medicine. Science and much of Western medical practice is incompatible with many of the practices of complementary therapy, as such practices do not often have a scientific basis, though this should not be seen as an obstacle to its understanding and acceptance. Practitioners of complementary therapy are seeing their role increasingly as offering something that Western scientific medicine does not, and something that adds to it without denying that scientific medicine has an essential role.

Medical intervention is restricted to qualified

Diseases and Disorders

Source: World Health Organization

Group	Subgroup	Examples
Diseases of the blood and blood-forming organs and disorders involving the immune mechanism	nutritional anaemias	iron deficiency anaemia, vitamin B12 deficiency anaemia
	haemolytic anaemias	thalassaemia, sickle-cell disorder
	aplastic and other anaemias	acquired pure red cell aplasia, acute posthaemorrhagic anaemia
	coagulation defects, purpura, and other haemorrhagic conditions	hereditary factor VIII deficiency, hereditary factor IX deficiency
	other diseases of blood and blood-forming organs	agranulocytosis, diseases of spleen, methaemoglobinaemia
Endocrine and metabolic diseases	disorders involving the immune mechanism	immunodeficiencies, sarcoidosis
	disorders of thyroid gland	iodine-deficiency syndrome, hypothyroism, thyroditis
	Disorders of glucose regulation and pancreatic internal secretion	non-diabetic hypoglycaemic coma
	disorders of other endocrine glands	hyper- or hypofunction of pituitary gland, Cushing's syndrome, diseases of thymus, polyglandular dysfunction
	metabolic disorders	lactose intolerance, cystic fibrosis, amyloidosis
Diseases of the nervous system	inflammatory diseases of the central nervous system	bacterial meningitis, encephalitis, myelitis, encephalomyelitis
	systemic atrophies primarily affecting the central nervous system	Huntington's disease, hereditary ataxia
	extrapyramidal and movement disorders	Parkinson's disease, dystonia
	other degenerative metabolic disorders of the nervous system	Alzheimer's disease
	demyelinating diseases of the central nervous system	multiple sclerosis
	episodic and paroxysmal disorders	epilepsy, migraine, other headache syndromes, sleep disorders
	nerve, nerve root, and plexus disorders	of trigeminal nerve, facial, cranial nerve, mononeuropathies
	polyneuropathies and other disorders of the perypheral nervous system	hereditary, idiopathic, inflammatory
	diseases of myoneural junction and muscle	myasthenia gravis
	cerebral palsy and other paralytic syndromes	infantile cerebral palsy, hemiplegia, paraplegia, tetraplegia
	other disorders of the nervous system	hydrocephalus, toxic encephalopathy
Diseases of the eye and adnexa	disorders of eyelid, lacrimal system, and orbit	hordeolum and chalazion
	disorders of conjunctiva	conjunctivitis
	disorders of sclera, cornea, iris, and ciliary body	keratitis, iridocyclitis
	disorders of lens	cataract
	disorders of choroid and retina	retinal detachments and breaks
	glaucoma	glaucoma
	disorders of vitreous body and globe	various
	disorders of optic nerve and visual pathways	optic neuritis
	disorders of ocular muscles, binocular movement, accommodation, and refraction	strabismus
	visual disturbances and blindness	low vision
	other disorders of eye and adnexa	nystagmus
Diseases of the ear and mastoid process	diseases of external ear	otitis externa
	diseases of middle ear and mastoid	mastoiditis, ostitis media, disorders of tympanic membrane
	diseases of inner ear	otosclerosis
	other disorders of ear	otalgia and effusion of ear, hearing loss

Group	Subgroup	Examples
Diseases of the respiratory system	acute upper respiratory infections	acute nasopharyngitis (common cold), acute sinisitis, acute laryngitis, acute obstructive laryngitis (croup)
	influenza and pneumonia	influenza, pneumonia
	other acute lower respiratory infections	acute bronchitis
	other diseases of upper respiratory tract	chronic rhinitis, chronic sinusitis, nasal polyp, chronic laryngitis
	chronic lower respiratory diseases	bronchitis (non-acute and non-chronic), emphysema, asthma
	lung diseases due to external agents	coalworker's pneumoconiosis; pneumoconiosis due to asbestos, other mineral fibres, dust containing silica; conditions due to inhalation of chemicals, gases, fumes, and vapours
	other respiratory diseases primarily affecting the interstitium	pulmonary oedema
	suppurative and necrotic conditions of lower respiratory tract	abscess of lung and mediastinum, pyothorax
	other diseases of pleura	pleural effusion, pleural plaque
	other diseases of the respiratory system	respiratory failure, not elsewhere classified
Diseases of the digestive system	diseases of oral cavity, salivary glands, and jaws	dental caries, gingivitis and periodontal diseases, stomatitis
	diseases of oesophagus, stomach, and duodenum	oesophagitis, gastric ulcer, duodenal ulcer, peptic ulcer, dyspepsia
	diseases of appendix	acute appendicitis
	hernia	inguinal, femoral, umbilical, ventral, diaphragmatic hernia
	noninfective enteritis and colitis	Crohn's disease
	other diseases of intestines	irritable bowel syndrome
	diseases of peritoneum	peritonitis
	diseases of liver	alcoholic liver disease, toxic liver disease, fibrosis and cirrhosis of liver
	disorders of gallbladder, biliary tract, and pancreas	cholelithiasis, cholecystitis, pancreatitis
	other diseases of the digestive system	intestinal malabsorption
Diseases of skin and subcutaneous tissue	infections of skin and subcutaneous tissue	impetigo, cutaneous abscess, furuncle, carbuncle, cellulitis
	bullous disorders	pemphigus
	dermatitis and eczema	dermatitis, eczema, pruritus
	papulosquamous disorders	psoriasis, pityriasis rosea, lichen planus
	urticaria and erythema	urticaria, erythema
	radiation-related disorders of the skin and subcutaneous tissue	sunburn, radiothermatitis
	disorders of skin appendages	nail disorders, hair disorders, acne, rosacea, sweat disorders
Diseases of the musculoskeletal system and connective tissue	other disorders of skin and subcutaneous tissue	vitiligo, seborrhoeic keratosis, corns and callosities, lupus erythematosus
	arthropathies	infectious arthropathies, inflammmatory polyarthropathies, arthrosis
	systemic connective tissue disorders	polyarteritis nodosa, dermatopolymyositis
	dorsopathies	deforming dorsopathies, spondylopathies
	soft tissue disorders	of muscles, of synovium and tendon
	osteopathies and chondropathies	disorders of bone density and structure (including osteoporosis); osteomyelitis, osteonecrosis
	other disorders of the musculoskeletal system and connective tissue	other acquired deformities, postprocedural disorders, not elsewhere classified

Group	Subgroup	Examples
Diseases of the genito-urinary system	glomerular diseases	nephritis syndrome
	renal tubulo-interstitial diseases	nephritis
	renal failure	acute, chronic
	urolithiasis	calculus of kidney, ureter, lower urinary tract
	other disorders of kidney and ureter	unspecified contracted kidney, small kidney of unknown cause
	other diseases of the urinary system	cystitis, urethritis
	diseases of male genital organs	disorders of prostate, male infertility, disorders of penis
	disorders of breast	benign mammary dysplasia, hypertrophy of breast
	inflammatory diseases of female pelvic organs	uterus, cervix uteri, vagina, vulva
	noninflammatory disorders of female genital tract	endometriosis, conditions related to menstruation and menstrual cycle, female infertility
	other disorders of the genito-urinary system	postprocedural disorders, not elsewhere classified
Congenital malformations, deformations, and chromosomal abnormalities	congenital malformations of the nervous system	anencephaly, microcephaly, spina bifida
	congenital malformations of eye, ear, face, and neck	anophthalmos, micro- and macrophthalmos
	congenital malformations of the circulatory system	of cardiac chambers, connections, septa; of great arteries; of great veins
	congenital malformations of the respiratory system	of nose, larynx, trachea, bronchus, lung
	cleft lip and cleft palate	either or both
	other congenital malformations of the digestive system	of tongue, mouth, oesophagus, intestines
	congenital malformations of genital organs	of ovaries, fallopian tubes, uterus, cervix, female gemnitalia; undescended testicle, hypospadias, malformations of male genitalia; indeterminate sex and pseudohermaphroditism
	congenital malformations of the urinary system	renal agenesis, cystic kidney disease
	congenital malformations and deformations of the musculoskeletal system	polydactyly, syndactyly, osteochondrodysplasias
	other congenital malformations	congenital ichthyosis, epidermolysis bullosa
	chromosomal abnormalities, not elsewhere classified	Down's syndrome, Edward's syndrome, Patau's syndrome, Turner's syndrome
Symptoms, signs, and abnormal clinical and laboratory findings, not elsewhere classified	symptoms and signs involving the circulatory and respiratory systems	gangrene, not elsewhere classified; cardial murmurs; cough; pain in throat and chest
	symptoms and signs involving the digestive system and abdomen	abdominal and pelvic pain, nausea and vomiting, heartburn, dysphagia, faecal incontinence
	symptoms and signs involving the skin and subcutaneous tissue	rash, localized swelling; disturbances of skin sensation
	symptoms and signs involving the nervous and musculoskeletal systems	abnormal involuntary movements, lack of coordination
	symptoms and signs involving the urinary system	unspecified urinary incontinence, retention of urine
	symptoms and signs involving cognition, perception, emotional state, and behaviour	somnolence, stupor, and coma; dizziness and giddiness; disturbances of smell and taste; symptoms and signs involving appearance
	symptoms and signs involving speech and voice	dyslexia, speech disturbances (not elsewhere classified)
	general symptoms and signs	fever, headache, malaise and fatigue, senility, syncope and collapse, enlarged lymph nodes, lack of expected physiological development, other general symptoms and signs, unknown and unspecified causes of morbidity

Group	Subgroup	Examples
	abnormal findings on examination of blood, without diagnosis	abnormalities of plasma viscosity, red blood cells, white blood cells (not elsewhere specified), elevated blood glucose levels, findings of drugs or other substances not normally found in blood
	abnormal findings on examination of urine, without diagnosis	isolated proteinuria, glycosuria
	abnormal findings on examination of other body fluids, substances and tissues, without diagnosis	findings in cerebrospinal fluid and tissues
	abnormal findings on diagnostic imaging and in function studies, without diagnosis	imaging of central nervous system, lung, breast
	ill-defined and unknown causes of mortality	sudden infant death syndrome
	injuries to the head	includes, where applicable, superficial injury, open wound, fracture, dislocation, sprain, strain, crushing injury, injury to nerves in the area, injury to blood vessels in the area, traumatic amputation of (part of) organ or body part
Injury, poisoning, and certain other consequences of external causes	injuries to the neck	
	injuries to the thorax	
	injuries to the abdomen, lower back, lumbar spine, and pelvis	
	injuries to the shoulder and upper arm	
	injuries to the elbow and forearm	
	injuries to the wrist and hand	
	injuries to the hip and thigh	
	injuries to the knee and lower leg	
	injuries to the ankle and foot	
	injuries involving multiple body regions	
	injuries to unspecified part(s) of trunk, limb, or body region	
	effects of foreign body entering through natural orifice	foreign body on external eye, in ear, in respiratory tract
	burns and corrosions	of external body surface; confined to eye and internal organs; of multiple and unspecified body regions
	frostbite	superficial; with tissue necrosis
	poisoning by drugs, medicaments, and biological substances	poisoning (various forms)
	toxic effects of substances chiefly unmedical as to source	alcohol, organic solvents, corrosive substances, soaps and detergents, metals, pesticides, venomous animals
	other and unspecified effects of external causes	hypothermia, asphyxiation, maltreatment syndromes, unspecified effects of radiation
	certain early complications of trauma	complications not elsewhere classified
	complications of surgical and medical care, not elsewhere classified	following procedures, implants, prosthetic devices, failure and rejection of transplanted organs and tissues, complications peculiar to reattachment and amputation
	sequelae of injuries, of poisoning, and of other consequences of external causes	sequela of conditions as above

medical practitioners in many Western countries and a significant proportion of them use complementary therapy. Not all complementary practitioners are regulated and this has caused concern as complementary therapy grows in popularity. However, organizations concerned with the training and registration of complementary practitioners have developed at a rapid rate recently to meet the growing demands of the public for a pluralist, holistic, effective, and safe approach to health care.

homoeopathy

The roots of homoeopathy lie in the work of Samuel Hahnemann (1755–1843), an 18th-century German physician. The brutalities of medical practice, which included purging, bleeding, and the use of poisons, were instrumental in his decision to explore other avenues in which he could use his skills. He was prompted to investigate the use of cinchona bark in the treatment of fever while he was working on a *Materia Medica*. Cinchona bark produced many of the symptoms associated with fever without inducing pyrexia. This led to him to speculate that a substance that was effective against a disease would produce symptoms resembling those of that disease if it was given to a healthy person.

Homoeopathy recognizes a need for balance and harmony. Disturbances in the harmony of the body due to illness can precede the appearance of symptoms as the body reacts to the illness. Symptoms of the same illness differ between individuals, and the establishment of a detailed profile of the patient and a complete symptom picture are an essential part of the diagnosis. Homoeopathic remedies are intended to stimulate the resources of the body to restore its natural harmony, following Hahnemann's principle that the appropriate treatment is one that produces the same symptoms in a healthy person. Many of these principles pose a problem to the scientist, but the greatest obstacle to scientific acceptance is the observation that homoeopathic remedies become more potent when subject to serial dilution and mechanical shock. Theories have been developed in an attempt to explain why more dilute preparations are more powerful, but none of these have yet been proven.

herbal medicine

A wide range of medical practices that use unrefined and refined plant materials for treatment is encompassed by the term 'herbal medicine'. They do not necessarily have a common belief system and they range from traditional herbal medicine to many ethnic medical systems, such as Chinese and Ayurvedic medicine.

Plants were used medicinally in China, Egypt, Greece, Rome, and other ancient civilizations up to 5,000 years ago. The central position of herbalism in health care was challenged by changes in medical thinking in Renaissance Europe, and by the consequent emphasis on the management of symptoms with specific remedies by the emerging medical establishment.

Samuel Thomson, an American physician, documented much of this early herbalist knowledge in the early 19th century, and he is now credited as the founder of Western herbal medicine. He developed a theory resembling that of the four humours, in which life and health were represented by heat; illness and death by cold; motion by air; and energy or life force by fire. He believed that the fever associated with infection was a healthy sign and, unlike his contemporaries, he wanted to facilitate it rather than to suppress it. He considered that coughing, vomiting, and diarrhoea were healthy signs of the body removing toxins.

The underlying belief in herbalism is that health depends on maintaining the natural state of the body. The natural state of the body, or 'vital pulse', is represented by the rhythmic variation in tissue and cell activity, which protects, regulates, and renews the body. A herbal remedy is a preparation of the entire plant rather than an active ingredient extracted from it. The constituent components may have many different activities that restore the balance of the body, which is lost with the onset of illness. They are not specific for a particular symptom and they are thought to act by stimulating the natural defences of the body and enhancing the elimination of toxins. Attention to diet is often used as an adjunct to treatment by herbalists.

aromatherapy

The therapeutic use of volatile oils can be traced back to the ancient civilizations of India, China, Egypt, Babylon, and Greece, and their use in medicine survived until challenged by the advent of scientific chemistry in the 17th century.

Aromatherapists believe that health depends on a balance of mental, emotional, and physical processes which is disturbed in illness. A holistic approach is taken to diagnosis and treatment. The volatile oils used in aromatherapy are complex mixtures of chemicals that occur naturally in plants. Specific disorders are treated with particular oils. They are believed to be absorbed through the skin or by inhalation before exerting their subtle effects on physical symptoms and emotional well being. The pleasant smell and application of the oils by massage can enhance the beneficial effects of aromatherapy through reduction of tension and pain, increased relaxation, and improved circulation. Aromatherapy is widely used as an adjunct to conventional care in patients who are terminally ill.

acupuncture

Traditional Chinese medicine is based on the fundamental principle that a life force known as *qi* flows around the body in 12 channels known as meridians. The flow of *qi* is important for good health. The balance between *yin* and *yang*, qualities possessed by all things including the internal functions and processes of the body and the meridians, is also vital to health. *Yin* is cold, dark, passive, and negative; *yin* organs include the heart, spleen, kidney, and liver. *Yang* is warm, light, active, and positive; *yang* organs include the small intestine, stomach, bladder, and gall bladder. Illness occurs when either *yin* or *yang* is dominant, and treatment helps to restore their balance.

Acupuncture
http://acupuncture.com/
Thorough introduction to the alternative world of acupuncture with descriptions of the main notions and powers of herbology, yoga, Qi Gong, and Chinese nutrition. Consumers are given access to lists of practitioners and an extensive section on acupuncture research. Practitioners can browse through journal listings and the latest industry news and announcements.

Acupuncture is one form of treatment used in traditional Chinese medicine. It involves the insertion of fine needles into the skin at particular points on the body to correct the imbalances between *yin* and *yang*. There are about 2,000 acupoints that lie on the meridians through which *qi* flows. The imbalance between *yin* and *yang* may be associated with many factors, such as stress, emotion, diet, or injury, and these are considered in conjunction with the medical history of the patient, and examination of the 12 pulses and the condition of the tongue before a diagnosis is made. The acupoints to be stimulated are then selected and needles are inserted to a depth of about $1/4$ of an inch (6 mm) and rotated. The direction of insertion and rotation of the needles regulates the flow of *qi* and the collection or dispersion of energy.

Acupuncture can be used to treat a wide range of acute and chronic illnesses, such as pain, anxiety, asthma, migraine, menstrual disorders, and gastrointestinal complaints. It can also be used as an aid to dieting and giving up smoking.

osteopathy

Osteopaths perceive the function of muscles and the skeletal system to be central to a range of health problems. Osteopathy was conceived as a system of diagnosis and treatment by Andrew Taylor Still in the USA in the 19th century. He believed that a lack of balance in the mechanical functioning of the body, such as muscle groups being too tense or joints moving incorrectly, may cause illness. He developed a range of manipulative techniques to correct the imbalances and claimed therapeutic success. These include massage, passive movement, and stretching of the limbs. Despite opposition from the medical profession, osteopathy has slowly achieved success because it can produce a dramatic improvement in disorders that are difficult to treat by conventional medicine, such as chronic back pain and sciatica.

These are just a few of the more popular forms of complementary therapy that are available today. There are many others: chiropractic, reflexology, and relaxation techniques such as hypnotherapy, to name but a few.

Osteopathic Home Page
http://www.am-osteo-assn.org/general.htm
Well organized site, from the American Association of Osteopathy, dedicated to increasing public understanding of osteopathy. The theory behind the osteopaths' belief in the body's innate healing power and some of the healing techniques are fully explained. There is also an account of the origins and development of osteopathy and a selection of case histories attest to its success.

Health and Disease Chronology

c. 2800 BC   Chinese emperor Shen Nong describes the therapeutic powers of numerous medicinal plants.

c. 2500 BC   The practice of acupuncture is developed in China.

c. 1600 BC   The Edwin Smith papyrus is written. The first medical book, it contains clinical descriptions of the examination, diagnosis, and treatment of injuries, and reveals an accurate understanding of the workings of the heart, stomach, bowels, and larger blood vessels. The papyrus is named after US scientist Edwin Smith, a pioneer in the study of Egyptian science who acquired it in Luxor, Egypt, in 1862.

c. 1122 BC   Smallpox is first described, in China. Pharaoh Ramses V, who dies in 1157 BC, is considered the first known victim of the disease.

c. 650 BC   Tuberculosis, leprosy, and gonorrhoea are first described with accuracy.

437 BC   The world's first hospital is established in Ceylon.

c. 400 BC   Greek physician Hippocrates of Cos begins the corpus of the Hippocratic Collection of about 70 medical treatises which cover topics such as epidemics and epilepsy. There is no evidence that he wrote any of them himself, but by recognizing that disease has natural causes he begins the science of medicine.

c. 200 BC   The Greek physician Aretaeus of Cappadocia describes manic-depressive psychosis.

c. 90 BC   The Roman scholar Marcus Terentius Varro writes that disease is caused by the entry of imperceptible particles into the body – the first enunciation of germ theory.

180   The Greek physician Claudius Galen, practising at Rome, writes *Methodus medendi/Method of Physicians*, a medical textbook that will become the ultimate authority for medieval medicine.

900   The Muslim physician al-Razi describes diseases such as plague, consumption, smallpox, and rabies.

1443   Quarantine laws are introduced in England for the first time.

1489   The first major typhus epidemic in Europe occurs when Spanish soldiers returning from Cyprus introduce the disease.

1493   Syphilis appears in Europe for the first time, brought back from South America by sailors returning with the explorer Christopher Columbus. The disease is first reported in Barcelona, Spain.

1598   The physician known as G W publishes *The Cure of the Diseased in Remote Regions*, the earliest treatise on tropical medicine.

1601   Captain James Lancaster of the East India Company provides his crew with lemon juice and citrus fruits, avoiding an outbreak of scurvy, the deficiency disease that devastates the crews of other ships on his trade mission.

1625   German chemist Johann Glauber prepares hydrochloric acid by the action of sulphuric acid on common salt. He discovers the laxative properties of the by-product, sodium sulphate, and markets it as a patent medicine, Glauber's salt, or *sal mirabile*.

1635   French physician Duterte describes yellow fever for the first time, during an outbreak on the islands of St Kitts and Guadeloupe.

1642   Seville physician Pedro Barba recounts his use of 'Peruvian bark', a source of quinine, for treating the Countess of Chinchou's malaria.

1650   English physician Francis Glisson writes his *Treatise on Rickets*, a detailed clinical description of rickets.

1659   Typhoid fever is given a detailed description for the first time by British physician Thomas Willis.

1665   The last major outbreak of the Black Death (a form of bubonic plague) affects London, England, in an epidemic known as the 'Great Plague', which reaches a peak in September, and around 70,000 people die.

1717   Lady Mary Wortley Montagu writes *Inoculation Against Smallpox*, reporting the method of immunization known in the East for centuries and introducing the practice of inoculation for smallpox into England; the inoculation of the Princess of Wales makes it fashionable.

1736   North American physician William Douglass publishes the first clinical description of scarlet fever, in a paper on the outbreak that hit the city the previous year.

1761   English naturalist John Hill writes that excessive use of snuff may lead to cancer; this is the first association of tobacco and cancer.

1763   English clergyman Edmund Stone describes the effective treatment of fever using willow bark, from which the active ingredient of aspirin is later derived.

1768   English physician William Heberden gives a full and correct clinical description of angina pectoris as pains of the chest precipitated by effort and originating from the heart.

1775   The first evidence that environmental and occupational factors can cause cancer is provided by English surgeon Percivall Pott, who suggests that chimney sweeps' exposure to soot causes cancer of the scrotum and nasal cavity.

1783   English physician Thomas Cawley correctly diagnoses diabetes mellitus by demonstrating the presence of sugar in a patient's urine.

1796   English physician Edward Jenner performs the first vaccination against smallpox.

1800   Scottish chemist William Cruickshank purifies water by adding chlorine.

1806	German chemist Friedrich Wilhelm Saturner isolates morphine – the first painkiller – from opium.
1816	French physician René Laënnec invents the stethoscope to detect cardiovascular diseases.
1817	English surgeon James Parkinson describes the central nervous system disorder now known as Parkinson's disease.
1818	French chemist Jean-Baptiste-André Dumas uses iodine to treat goitres.
1819	British surgeon James Blundell makes the first human-to-human blood transfusion. The patient survives for 56 hours.
1819	French physician Jean-Louis-Marie Poiseuille is the first to use a mercury manometer to measure blood pressure.
1820	French chemists Joseph Pelletier and Joseph-Bienaimé Caventou isolate the antimalarial drug quinine, as well as the other alkaloids brucine, cinchonine, colchicine, strychnine, and veratrine.
1832	English physician Thomas Hodgkin first describes the lymphatic cancer, Hodgkin's disease.
1832	French chemist Pierre-Jean Robiquet isolates the analgesic, codeine, from opium.
1835	English physician James Paget discovers the parasitic worm *Trichina spiralis* which causes trichinosis.
1838	Italian chemist R Piria synthesizes salicylic acid, the basic ingredient in aspirin, from willow bark.
1840	German physician Gustav Jacob Henle suggests that infectious diseases are caused by living microscopic organisms.
1840	Swiss chemist Charles Choss demonstrates that calcium is needed for proper bone development.
1842	US surgeon Crawford Williamson Long is the first to use ether as an anaesthetic in an operation. He does not publish his findings until 1849.
1851–1854	Over 250,000 people (approximately 2% of the population) die from tuberculosis in Britain.
1852	German physician Karl von Vierordt develops a method of counting red blood cells that becomes important in diagnosing anaemia.
1853	Nearly 8,000 die in a yellow fever epidemic in New Orleans, Louisiana.
Nov 1854	English nurse Florence Nightingale arrives in Scutari, Ottoman Empire, and introduces sanitary measures in an effort to reduce deaths from cholera, dysentery, and typhus during the Crimean War.
1856	German scientist Theodore Bilharz discovers the parasitic worm that causes schistosomiasis (bilharzia).
1860	French physician Etienne Lancereaux suggests that diabetes is due to a disorder of the pancreas.
c. 1860	French chemist and microbiologist Louis Pasteur develops the process of pasteurization: sterilizing milk and other beverages by heating to a high temperature for a few minutes to kill microorganisms.
1861	Pasteur develops the germ theory of disease.
1863	French parasitologist Casimir-Joseph Davaine shows that anthrax is due to the presence of rodlike microorganisms in the blood. It is the first disease of animals and humans to be shown to be caused by a specific microorganism.
1863	French physiologist Etienne-Jules Marey invents the sphygmomanometer to record blood pressure graphically. It is still used today.
12 Aug 1865	English surgeon Joseph Lister first uses phenol (carbolic acid) as an antiseptic during surgery, to kill germs.
1869	Norwegian physician Gerhard Henrik Armauer Hansen discovers the leprosy bacillus *Mycobacterium leprae*.
1871	German surgeon F Steiner is the first to use electrical cardiac stimulation to successfully restart a patient's heart.
1873	English physician William Budd publishes *Typhoid Fever, Its Nature, Mode of Spreading, and Prevention*, in which he establishes the infectious nature of the disease.
1877	British physician Patrick Manson shows how insects can be the carriers of infectious diseases, and demonstrates that the embryo of the *Filaria* worm, which causes elephantiasis, is transmitted by mosquito.
1879	Pasteur discovers that chickens infected with weakened cholera bacteria are immune to the normal form of the disease. It leads to the development of vaccines.
1880	Pasteur discovers the bacterial genus *Streptococcus* that causes diseases such as scarlet fever and rheumatic fever.
1880	French pathologist Alphonse Laveran discovers the malaria parasite, in Algeria.
1880	German bacteriologist Karl Joseph Eberth discovers the *Salmonella typhii* bacteria responsible for typhoid.
5 May 1881	Pasteur vaccinates sheep against anthrax. It is the first infectious disease to be treated effectively with an antibacterial vaccine, and his success lays the foundations of immunology.
1881	Austrian surgeon Theodor Christian Albert Billroth initiates modern abdominal surgery by removing the cancerous lower part of a patient's stomach.
1881	US Army physician George Miller Sternberg discovers the bacillus responsible for pneumonia.
24 March 1882	German physician Robert Koch announces the discovery of *Mycobacterium tuberculosis*, the

bacillus responsible for tuberculosis. This is the first time a microorganism has been definitively associated with a human disease.

1883   Koch discovers preventive inoculation against anthrax.

1883   German bacteriologists Edwin Klebs and Friedrich August Johannes Löffler discover *Corynebacterium diptheriae*, the diphtheria bacillus.

1883   Koch discovers the cholera bacillus.

25 Nov 1884   German-born British surgeon Rickman John Godlee performs the first operation to remove a brain tumour.

1884   German physician Arthur Nikolaier discovers the tetanus bacillus, *Clostridium tetani*.

6 July 1885   Pasteur develops a vaccine against rabies and uses it to save the life of a young boy, Joseph Meister, who has been bitten by a rabid dog.

1889   German physiologists Oskar Minkowski and Joseph von Mering remove the pancreas from a dog, which then develops diabetic symptoms. It leads them to conclude that the pancreas secretes an antidiabetic substance, which is now known as insulin.

1891   German medical scientist Paul Ehrlich treats malaria with methylene blue. Its use marks the beginning of chemotherapy.

10 July 1893   US surgeon Daniel Hale Williams performs the first open-heart surgery, on a patient who has been wounded by a knife. The patient lives for another 20 years.

1894   During an epidemic of bubonic plague in Hong Kong, Japanese bacteriologist Shibasaburo Kitasato and French bacteriologist Alexandre Yersin, simultaneously and independently, discover the bacillus *Yersinia pestis* which is responsible.

1895   Belgian bacteriologist Jules Bordet discovers antibodies.

1895   The protozoa *Trypanosoma gambiense* is discovered to be the cause of African sleeping sickness.

1897   British bacteriologist Ronald Ross discovers the malaria parasite in the gastrointestinal tract of the *Anopheles* mosquito, and realizes that the insect is responsible for the transmission of the disease.

1897   English bacteriologist Almroth Edward Wright introduces a vaccine against typhoid. It is successfully tested on 3,000 troops sent to India.

1898   Martinus Willem Beijerinck identifies the first virus; it is the cause of tobacco mosaic disease in plants.

1900   US army pathologist Walter Reed establishes that yellow fever is caused by the bite of an *Aëdes aegypti* mosquito infected with the yellow fever parasite. His discovery leads to the creation of a vaccine and makes possible the completion of the Panama Canal.

1901   Dutch physician Gerrit Grijns demonstrates that

beriberi is caused by a nutritional deficiency of vitamin B_1.

c. 1901   British soldiers are immunized against typhoid fever during the Boer War.

1903   German surgeon Georg Clems Perthes discovers that X-rays inhibit the growth of cancerous tumours and suggests they be used as a treatment.

1903   In New York, New York, a cook called Mary Mallon – 'Typhoid Mary' – is discovered to be a carrier of typhoid, unwittingly carrying and spreading the disease through handling food.

c. 1905   English biochemist Frederick Gowland Hopkins shows that the amino acid tryptophan and other essential amino acids cannot be manufactured from other nutrients but must be supplied in the diet.

1906   Belgian bacteriologists Jules Bordet and Octave Gengou discover the bacterium responsible for whooping cough, *Bordetella pertussis*.

1906   English biochemist Frederick Gowland Hopkins suggests that necessary 'accessory factors' (vitamins) are contained in foods in addition to carbohydrates, fats, minerals, and water.

1910   US physician James Herrick is the first to describe the genetic disease sickle-cell anaemia.

1912   US entomologist Leland Ossian Howard, publishes *The House Fly, Disease Carrier*, in which he identifies the common housefly as a major carrier of disease.

1912   US physician James Herrick is the first to describe a coronary thrombosis – a blood clot in the coronary artery that causes damage to the heart (heart attack).

1913   German bacteriologist Emil von Behring introduces a toxin–antitoxin vaccine against diphtheria.

1914   Polish–American biochemist Casimir Funk isolates vitamin B, a vital discovery in the treatment of beriberi.

1918   English pharmacologist Edward Mellanby discovers that a vitamin (vitamin D) in cod-liver oil cures rickets.

1921   Canadian physiologists Frederick Banting, Charles Best, and John James MacLeod isolate insulin. A diabetic patient in Toronto, Canada, receives the first insulin injection.

1923   French bacteriologists Albert Calmette and Camille Guérin develop the tuberculosis vaccine, known as Bacillus Calmette-Guérin (BCG), and use it to vaccinate newborns at a hospital in Paris, France.

1923   US physicians George and Gladys Dick isolate the microorganism responsible for scarlet fever (*Streptococcus pyogenes*) and develop an antitoxin.

1926   US biochemist Elmer McCollum isolates vitamin D and uses it to successfully treat rickets.

1926   US physicians George Richards Minot and William Parry Murphy use raw-liver extract for

	treating the previously fatal disease pernicious anaemia.
1928	Greek-born US physician George Papanicolaou develops the Pap smear to test for uterine cancers.
1928	Scottish bacteriologist Alexander Fleming discovers penicillin when he notices that the mould *Penicillium notatum*, which has invaded a culture of staphylococci, inhibits the bacteria's growth.
1932	English physician Cecily Williams describes the protein deficiency disease kwashiorkor.
1933	Austrian-born German organic chemist R Kuhn, Hungarian-born US biochemist A von Szent-Györgyi, and J Wagner-Jauregg discover vitamin B_2 (riboflavin) in Hungary.
1933	Swiss biochemist Paul Karrer establishes the structure of vitamin A (retinol). and US biologist George Wald demonstrates its important in preventing night blindness.
1937	French microbiologist Max Theiler develops a vaccine against yellow fever; it is the first antiviral vaccine.
1937	US biochemist Conrad Arnold Elvehjem finds that vitamin B_3 (niacin) prevents pellagra, a vitamin deficiency disease.
1938–1940	Five independent researchers isolate and synthesize vitamin B_6 in Germany.
1939	US microbiologist René J Dubos is the first to search systematically for, and discover, natural antibiotics. He looks for soil bacteria that kill other bacteria and discovers the antibiotics gramicidin and tyrocidine.
1942	French physician André Loubatière leads the development of oral drugs for diabetics with his discovery that sulfa drugs lower blood sugar levels.
1943	US biologist Selman A Waksman discovers the antibiotic streptomycin, which is used as a treatment for tuberculosis; he coins the term 'antibiotic'.
1944	Swiss pharmacologist Daniel Bovet discovers pyrilamine, the first antihistamine.
Oct 1945	The Food and Agriculture Organization (FAO) is established, the first of the UN's specialized agencies. Its objective is to eliminate hunger and improve nutrition worldwide.
1945	US physician Alton Ochsner points to the relationship between cigarette smoking and lung cancer.
c. 1945	The first effective vaccine against influenza is developed.
1946	Penicillin is synthesized by US chemist Vincent du Vigneaud.
1947	The poliomyelitis virus is isolated by US physician Jonas E Salk.
5 July 1948	The British National Health Service comes into effect.
1948	British biochemist Dorothy Hodgkin analyses the complex structure of vitamin B_{12} and makes the first X-ray photographs of it.

1948	The World Health Organization (WHO) is established with its headquarters in Geneva, Switzerland; its aim is to improve world health conditions.
1948	US chemist Karl Folkers and British chemist Alexander Todd isolate vitamin B_{12}.
1949	British biochemist Dorothy Hodgkin works out the chemical structure of penicillin.
1949	Lithium is first used to treat mental patients.
1949	The antibiotic chloramphenicol (Chloromycetin) is introduced; it is the first effective treatment for typhoid fever.
1949	The antibiotic tetracycline is discovered.
6 May 1953	US physician John Gibbon performs the first successful open-heart operation. He uses a heart-lung machine to oxygenate the blood during the operation.
1957	Interferon, a natural protein that fights viruses, is discovered by Scottish virologist Alick Isaacs and Swiss virologist Jean Lindemann.
1959	French researcher Jérôme Lejeune discovers that Down's syndrome is due to an extra chromosome 21. It is the first chromosomal disorder discovered.
1963	US surgeon James Hardy performs the first lung transplant.
1963	US surgeon Thomas Starlz performs the first liver transplant.
1966	US virologists Harry Meyer and Paul Parman develop a live virus vaccine for rubella (German measles), which reduces the incidence of the disease.
3 Dec 1967	South African surgeon Christiaan Barnard performs the first heart transplant operation. The patient survives for 18 days.
1967	Greek-born US neurologist George Cotzias begins to use L-dopa, a precursor of dopamine, to successfully treat patients with Parkinson's disease.
1967	US surgeon Rene Favaloro develops the coronary bypass operation.
15 Sept 1969	The world's first heart and lung transplant is performed at the Stanford Medical Center in California.
1973	US biochemists Stanley Cohen and Herbert Boyer develop the technique of recombinant DNA (deoxyribonucleic acid). Strands of DNA are cut by restriction enzymes from one species and then inserted into the DNA of another; this marks the beginning of genetic engineering.
1976	The first oncogene (cancer-inducing gene) is discovered by US scientists Harold E Varmus and J Michael Bishop.
1977	Two homosexual men in New York, New York, are diagnosed as having the rare cancer Kaposi's sarcoma. They are later thought to be the first victims of AIDS.
May 1980	The World Health Organization announces the eradication of smallpox.

1981	The US Food and Drug Administration grants permission to Eli Lilley and Co to market insulin produced by bacteria, the first genetically engineered product to go on sale.
Oct1983	UK and US studies indicate that long-term use of oral contraceptives may increase the risk of breast and cervical cancer.
1983	US medical researcher Robert Gallo at the US National Cancer Institute, Maryland, and French medical researcher Luc Montagnier at the Pasteur Institute in Paris, France, isolate the virus thought to cause AIDS; it becomes known as the HIV virus (human immunodeficiency virus).
1984	UK researchers develop the first vaccine against leprosy.
1985	An epidemic of bovine spongiform encephalopathy (BSE) is reported in beef cattle in Britain; it is later traced to cattle feed containing sheep carcasses infected with scrapie; in following years there are fears that beef consumption could lead to Creutzfeld-Jakob disease in humans.
1985	Researchers locate gene markers on chromosomes for cystic fibrosis and polycystic kidney disease.
1985	Screening blood donations for the AIDS virus begins in Britain.
1985	US researcher Steven Rosenberg discovers interleukin 2, a crucial protein in the immune system involved in the activation of lymphocytes; researchers soon begin experimenting on its anticancer properties.
17 Dec 1986	British surgeons John Wallwork and Roy Calne perform the first triple transplant – heart, lung, and liver.
20 March 1987	The AIDS treatment drug AZT is given approval by the US Federal Drug Administration. Treatment costs $10,000 per year per patient and does not cure the disease although it relieves some symptoms and does extend victims' lives.
1987	A three-year-old girl in the USA receives a new liver, pancreas, small intestine, and parts of the stomach and colon; this is the first successful five-organ transplant.
1987	German-born British geneticist Walter Bodmer and associates announce the discovery of a marker for a gene that causes cancer of the colon.
1988	A US survey indicates that the risk of a heart attack can be halved by taking aspirin daily.
1989	Researchers in Toronto, Canada, identify a gene responsible for cystic fibrosis.
1990	A four-year-old girl in the USA has the gene for adenosine deaminase inserted into her DNA (deoxyribonucleic acid); she is the first human to receive gene therapy.
1991	Australian and British studies show that passive smoking is a significant cause of lung cancer. Children whose parents smoke suffer an increased risk of asthma and respiratory infections.
1992	A vaccine for hepatitis A becomes available.
1993	AIDS becomes the leading cause of death among men aged between 25 and 44 in the USA.
1993	The gene responsible for Huntington's chorea is identified. It makes it easier to test individuals for the disease and increases the chances of developing a cure.
Sep 1994	A gene that triggers breast cancer is identified and is found to be responsible for almost half the cases of inherited breast cancer and most cases of ovarian cancer.
1995	Trials begin in the USA to treat breast cancer by gene therapy. The women are injected with a virus genetically engineered to destroy their tumours.
Jan 1997	The World Health Organization (WHO) estimates that 22.6 million men, women, and children have to date been infected by HIV, the virus responsible for causing AIDS. Approximately 42% of adult sufferers are female, with the proportion of women infected steadily increasing.
8 May 1997	US AIDS researcher David D Ho and colleagues show how aggressive treatment of HIV-1 infection with a cocktail of three antiviral drugs can drive the virus to below the limits of conventional clinical detection within eight weeks.
17 May 1997 –20 May 1997	At a meeting of the American Society of Clinical Oncology, researchers announce the development of vaccines which cause the immune system to shrink certain cancers, such as those attacking the skin, breast, prostate, and ovaries. Unlike the normal preventative vaccines, the new vaccines fight tumours that already exist. They use components of the cancer to provoke white blood cells to attack the invader.
27 June 1997	US scientists at the National Human Genome Research Institute in Bethesda, Maryland, announce the discovery of a gene that causes Parkinson's disease.
2 Oct 1997	UK scientists Moira Bruce and, independently, John Collinge and their colleagues show that the new variant form of the brain-wasting Creutzfeldt-Jakob disease (CJD) is the same disease as bovine spongiform encephalopathy (BSE or 'mad cow disease') in cows.
22 April 1998	Scientists at the Public Health Laboratory Service in London, England, report the discovery of a bacterium Pseudonas aeruginosa that is resistant to all known antibiotics. It causes a wide range of infections in people with impaired immune systems.

Biographies

Anderson, Elizabeth Garrett (1836–1917) English physician, the first English woman to qualify in medicine. Unable to attend medical school, Anderson studied privately and was licensed by the Society of Apothecaries in London in 1865. She was physician to the Marylebone Dispensary for Women and Children (later renamed the Elizabeth Garrett Anderson Hospital), a London hospital now staffed by women and serving female patients.

Banting, Frederick Grant (1891–1941) Canadian physician. He discovered a technique for isolating the hormone insulin 1921 when he and his colleague Charles Best tied off the ducts of the pancreas to determine the function of the cells known as the islets of Langerhans. This made possible the treatment of diabetes. Banting and John J R Macleod (1876–1935), his mentor, shared the 1923 Nobel Prize for Physiology or Medicine, and Banting divided his prize with Best. He was knighted in 1934

Discovery of Insulin
http://web.idirect.com/~discover/
Details the discovery of the protein hormone insulin. Included here are brief descriptions of the lives of the discovery members as well as an article on the discoverer Dr F G Banting's death. An overview of insulin and diabetes is also included here, describing the hormone and the disease, the research experiments, and current developments.

Barnard, Christiaan Neethling (1922–) South African surgeon who performed the first human heart transplant in 1967. The 54-year-old patient lived for 18 days.

Barnard also discovered that intestinal artresia – a congenital deformity in the form of a hole in the small intestine – is the result of an insufficient supply of blood to the fetus during pregnancy. It was a fatal defect before he developed the corrective surgery.

Behring, Emil von (1854–1917) German physician who discovered that the body produces antitoxins, substances able to counteract poisons released by bacteria. Using this knowledge, he developed new treatments for such diseases as diphtheria. He won the first Nobel Prize for Physiology or Medicine, 1901.

Behring discovered the diphtheria antitoxin and developed serum therapy together with Japanese bacteriologist Shibasaburo Kitasato, and they went on to apply the technique to tetanus. Behring also introduced early vaccination techniques against diphtheria and tuberculosis.

Best, Charles H(erbert) (1899–1978) Canadian physiologist. He was one of the team of Canadian scientists including Frederick Banting whose research resulted 1922 in the discovery of insulin as a treatment for diabetes. Best also discovered the vitamin choline and the enzyme histaminase, and introduced the use of the anticoagulant heparin.

Breuer, Josef (1842–1925) Viennese physician, one of the pioneers of psychoanalysis. He applied it successfully to cases of hysteria, and collaborated with Freud on *Studien über Hysterie/Studies in Hysteria* 1895.

Carrel, Alexis (1873–1944) US surgeon born in France, whose experiments paved the way for organ transplantation. Working at the Rockefeller Institute, New York City, he devised a way of joining blood vessels end to end (anastomosing). This was a key move in the development of transplant surgery, as was his work on keeping organs viable outside the body, for which he was awarded the Nobel Prize for Physiology or Medicine 1912.

Chain, Ernst Boris (1906–1979) German-born British biochemist. After the discovery of penicillin by Alexander Fleming, Chain worked to isolate and purify it. For this work, he shared the 1945 Nobel Prize for Medicine with Fleming and Howard Florey. Chain also discovered penicillinase, an enzyme that destroys penicillin. Knighted 1969.

Charcot, Jean-Martin (1825–1893) French neurologist who studied hysteria, sclerosis, locomotor ataxia, and senile diseases. Among his pupils was the founder of psychiatry, Sigmund Freud. One of the most influential neurologists of his day, Charcot exhibited hysterical women at weekly public lectures, which became fashionable events. He was also fascinated by the relations between hysteria and hypnotic phenomena.

Cushing, Harvey Williams (1869–1939) US neurologist who pioneered neurosurgery. He developed a range of techniques for the surgical treatment of brain tumours, and also studied the link between the pituitary gland and conditions such as dwarfism. He first described the chronic wasting disease now known as **Cushing's syndrome.**

Doll, (William) Richard Shaboe (1912–) British physician who, working with Bradford Hill (1897–), provided the first statistical proof of the link between smoking and lung cancer in 1950. In a later study of the smoking habits of doctors, they were able to show that stopping smoking immediately reduces the risk of cancer. Knighted 1971.

Domagk, Gerhard Johannes Paul (1895–1964) German pathologist, discoverer of antibacterial sulphonamide drugs. He found in 1932 that a coal-tar dye called Prontosil red contains chemicals with powerful antibacterial properties. Sulphanilamide became the first of the sulphonamide drugs, used – before antibiotics were discovered – to treat a wide range of conditions, including pneumonia and septic wounds. Nobel prize 1939.

Ehrlich, Paul (1854–1915) German bacteriologist and immunologist who produced the first cure for syphilis. He developed the arsenic compounds, in particular Salvarsan, that were used in the treatment of syphilis before the discovery of antibiotics. He shared the 1908 Nobel Prize for Physiology or Medicine with Ilya Mechnikov, awarded for his work on immunity.

Ehrlich also founded chemotherapy – the use of a chemical substance to destroy disease organisms in the body. He was also one of the earliest workers on immunology, and through his studies on blood samples the discipline of haematology was recognized.

Fleming, Alexander (1881–1955) Scottish bacteriologist who discovered the first antibiotic drug, penicillin, in 1928. In

1922 he had discovered lysozyme, an antibacterial enzyme present in saliva, nasal secretions, and tears. While studying this, he found an unusual mould growing on a culture dish, which he isolated and grew into a pure culture; this led to his discovery of penicillin. It came into use in 1941. In 1945 he won the Nobel Prize for Physiology or Medicine with Howard W Florey and Ernst B Chain, whose research had brought widespread realization of the value of penicillin.

Florey, Howard Walter, Baron Florey (1898–1968) Australian pathologist whose research into lysozyme, an antibacterial enzyme discovered by Alexander Fleming, led him to study penicillin (another of Fleming's discoveries), which he and Ernst Chain isolated and prepared for widespread use. With Fleming, they were awarded the Nobel Prize for Physiology or Medicine 1945.

Freud, Sigmund (1856–1939) Austrian physician who pioneered the study of the unconscious mind. He developed the methods of free association and interpretation of dreams that are basic techniques of psychoanalysis. The influence of unconscious forces on people's thoughts and actions was Freud's discovery, as was his controversial theory of the repression of infantile sexuality as the root of neuroses in the adult. His influence has permeated the world to such an extent that it may be discerned today in almost every branch of thought.

Galen (c. 129–c. 200) Greek physician and anatomist whose ideas dominated Western medicine for almost 1,500 years. Central to his thinking were the theories of humours and the threefold circulation of the blood. He remained the highest medical authority until Andreas Vesalius and William Harvey exposed the fundamental errors of his system.

Galen postulated a circulation system in which the liver produced the natural spirit, the heart the vital spirit, and the brain the animal spirit. He also wrote about philosophy and believed that Nature expressed a divine purpose, a belief that became increasingly popular with the rise of Christianity (Galen himself was not a Christian). This helped to account for the enormous influence of his ideas.

On the Natural Faculties
http://classics.mit.edu/Galen/natfac.html
Text of this work by the Greek physician Galen.

Gallo, Robert Charles (1937–) US scientist credited with identifying the virus responsible for AIDS. Gallo discovered the virus, now known as human immunodeficiency virus (HIV), in 1984; the French scientist Luc Montagnier (1932–) of the Pasteur Institute, Paris, discovered the virus, independently, in 1983. The sample in which Gallo discovered the virus was supplied by Montagnier, and it has been alleged that this may have been contaminated by specimens of the virus isolated by Montagnier a few months earlier.

Guérin, Camille (1872–1961) French bacteriologist who, with Albert Calmette, developed the BCG vaccine for tuberculosis 1921.

Hippocrates (c. 460–c. 377 BC) Greek physician, often called the founder of medicine. Important Hippocratic ideas include cleanliness (for patients and physicians), moderation in eating and drinking, letting nature take its course, and living where the air is good. He believed that health was the result of the 'humours' of the body being in balance; imbalance caused disease. These ideas were later adopted by Galen.

He is known to have discovered aspirin in willow bark. The *Corpus Hippocraticum/Hippocratic Collection*, a group of some 70 works, is attributed to him but was probably not written by him, although the works outline his approach to medicine. They include *Aphorisms* and the **Hippocratic Oath**, which embodies the essence of medical ethics.

Works by Hippocrates
http://classics.mit.edu/Browse/
browse-Hippocrates.html
Translations of the works of Hippocrates by Francis Adams in downloadable form. Seventeen works by the father of medicine are to be found here, ranging from *On Ancient Medicine* (this was written in 400BC!) to *On The Surgery*. The site also includes the philosophers best known work *The Oath*. The text files are very large (with some exceptions, notably *The Oath*), but the server is fast so don't be too put off downloading.

Hunter, John (1728–1793) Scottish surgeon, pathologist, and comparative anatomist who insisted on rigorous scientific method. He was the first to understand the nature of digestion.

Jenner, Edward (1749–1823) English physician who pioneered vaccination. In Jenner's day, smallpox was a major killer. His discovery in 1796 that inoculation with cowpox gives immunity to smallpox was a great medical breakthrough.

Jenner observed that people who worked with cattle and contracted cowpox from them never subsequently caught smallpox. In 1798 he published his findings that a child inoculated with cowpox, then two months later with smallpox, did not get smallpox.

Kitasato, Shibasaburo (1852–1931) Japanese bacteriologist who discovered the plague bacillus while investigating an outbreak of plague in Hong Kong. He was the first to grow the tetanus bacillus in pure culture. He and German bacteriologist Emil von Behring discovered that increasing nonlethal doses of tetanus toxin give immunity to the disease.

Koch, (Heinrich Hermann) Robert (1843–1910) German bacteriologist. Koch and his assistants devised the techniques for culturing bacteria outside the body, and formulated the rules for showing whether or not a bacterium is the cause of a disease. Nobel Prize for Physiology or Medicine 1905.

His techniques enabled him to identify the bacteria responsible for tuberculosis (1882), cholera (1883), and other diseases. He investigated anthrax bacteria in the 1870s and showed that they form spores which spread the infection.

Lister, Joseph (1827–1912) 1st Baron Lister, English surgeon. He was the founder of antiseptic surgery, influenced by Louis Pasteur's work on bacteria. He introduced dressings soaked in carbolic acid and strict rules of hygiene to combat wound sepsis in hospitals. Baronet 1883, Baron 1897.

Nicolle, Charles Jules Henri (1866–1936) French bacteriologist whose discovery in 1909 that typhus is transmitted by

Louis Pasteur

More germs are transmitted when shaking hands than when kissing. Louis Pasteur, the pioneer of hygienic methods, refused to shake hands with acquaintances for fear of infection.

the body louse made the armies of World War I introduce delousing as a compulsory part of the military routine. Nobel Prize for Physiology or Medicine 1928.

Pasteur, Louis (1822–1895) French chemist and microbiologist who discovered that fermentation is caused by microorganisms and developed the germ theory of disease. He also created a vaccine for rabies, which led to the foundation of the Pasteur Institute in Paris in 1888.

Ross, Ronald (1857–1932) British physician and bacteriologist, born in India. From 1881 to 1899 he served in the Indian Medical Service, and during 1895–98 identified mosquitoes of the genus *Anopheles* as being responsible for the spread of malaria. Nobel prize 1902. KCB 1911.

Salk, Jonas Edward (1914–1995) US physician and microbiologist. In 1954 he developed the original vaccine that led to virtual eradication of paralytic polio in industrialized countries. He was director of the Salk Institute for Biological Studies, University of California, San Diego, 1963–75.

Sydenham, Thomas (1624–1689) English physician, the first person to describe measles and to recommend the use of quinine for relieving symptoms of malaria. His original reputation as the 'English Hippocrates' rested upon his belief that careful observation is more useful than speculation.

Glossary

abortion ending of a pregnancy before the fetus is developed sufficiently to survive outside the uterus. Loss of a fetus at a later gestational age is termed premature stillbirth. Abortion may be accidental (◊miscarriage) or deliberate (termination of pregnancy).

acquired immune deficiency syndrome full name for the disease ◊AIDS.

acupuncture in alternative medicine, a system of inserting long, thin metal needles into the body at predetermined points to relieve pain, as an anaesthetic in surgery, and to assist healing. The method, developed in ancient China and increasingly popular in the West, is thought to work by stimulating the brain's own painkillers, the endorphins.

AIDS acronym for **acquired immune deficiency syndrome,** the gravest of the sexually transmitted diseases, or ◊STDs. It is caused by the human immunodeficiency virus (◊HIV), now known to be a ◊retrovirus. HIV is transmitted in body fluids, mainly blood and genital secretions.

allopathy in ◊homoeopathy, a term used for orthodox medicine, using therapies designed to counteract the manifestations of the disease. In strict usage, allopathy is the opposite of homoeopathy.

alternative medicine any form of medical treatment that does not use synthetic drugs or surgery in response to the symptoms of a disease, but aims to treat the patient as a whole. The emphasis is on maintaining health (with diet and exercise) and on dealing with the underlying causes rather than just the symptoms of illness.

Alzheimer's disease common manifestation of ◊dementia, thought to afflict one in 20 people over 65. Attacking the brain's 'grey matter', it is a disease of mental processes rather than physical function, characterized by memory loss and progressive intellectual impairment. It was first described by Alois Alzheimer 1906. It affects up to 4 million people in the USA and around 600,000 in Britain.

amniocentesis sampling the amniotic fluid surrounding a fetus in the womb for diagnostic purposes. It is used to detect Down's syndrome and other genetic abnormalities. The procedure carries a 1 in 200 risk of miscarriage.

anaemia condition caused by a shortage of haemoglobin, the oxygen-carrying component of red blood cells. The main symptoms are fatigue, pallor, breathlessness, palpitations, and poor resistance to infection. Treatment depends on the cause.

analgesic agent for relieving ◊pain. Opiates alter the perception or appreciation of pain and are effective in controlling 'deep' visceral (internal) pain. Non-opiates, such as ◊aspirin, ◊paracetamol, and NSAIDs (nonsteroidal anti-inflammatory drugs), relieve musculoskeletal pain and reduce inflammation in soft tissues.

aneurysm weakening in the wall of an artery, causing it to balloon outwards with the risk of rupture and serious, often fatal, blood loss. If detected in time, some accessible aneurysms can be repaired by bypass surgery, but such major surgery carries a high risk for patients in poor health.

angina or **angina pectoris** severe pain in the chest due to impaired blood supply to the heart muscle because a coronary artery is narrowed. Faintness and difficulty in breathing accompany the pain. Treatment is by drugs or bypass surgery.

anorexia lack of desire to eat, or refusal to eat, especially the pathological condition of **anorexia nervosa,** most often found in adolescent girls and young women. Compulsive eating, or ◊bulimia, distortions of body image, and depression often accompany anorexia.

anthrax disease of livestock, occasionally transmitted to humans, usually via infected hides and fleeces. It may develop as black skin pustules or severe pneumonia. Treatment is with antibiotics. Vaccination is effective.

antibiotic drug that kills or inhibits the growth of bacteria and fungi. It is derived from living organisms such as fungi or bacteria, which distinguishes it from synthetic antimicrobials.

antibody protein molecule produced in the blood by ◊lymphocytes in response to the presence of foreign or invading substances (◊antigens); such substances include the proteins carried on the surface of infecting microorganisms. Antibody production is only one aspect of ◊immunity in vertebrates.

antigen any substance that causes the production of ◊antibodies by the body's immune system. Common antigens include the proteins carried on the surface of bacteria, viruses, and pollen grains. The proteins of incompatible blood groups or tissues also act as antigens, which has to be taken into account in medical procedures such as blood transfusions and organ transplants.

antiseptic any substance that kills or inhibits the growth of microorganisms. The use of antiseptics was pioneered by Joseph ◊Lister.

arteriosclerosis hardening of the arteries, with thickening and loss of elasticity. It is associated with smoking, ageing, and a diet high in saturated fats. The term is used loosely as a synonym for ◊atherosclerosis.

artificial respiration emergency procedure to restart breathing once it has stopped; in cases of electric shock or apparent drowning, for example, the first choice is the expired-air method, the **kiss of life** by mouth-to-mouth breathing until natural breathing is restored.

ascorbic acid $C_6H_8O_6$ or **vitamin C** a relatively simple organic acid found in citrus fruits and vegetables. Lack of ascorbic acid results in scurvy.

asepsis practice of ensuring that bacteria are excluded from open sites during surgery, wound dressing, blood sampling, and other medical procedures. Aseptic technique is a first line of defence against infection.

aspirin acetylsalicylic acid, a popular pain-relieving drug (◊analgesic) developed in the late 19th century as a household remedy for aches and pains. It relieves pain and reduces inflammation and fever. It is derived from the white willow tree *Salix alba*, and is the world's most widely used drug.

asthma chronic condition characterized by difficulty in breathing due to spasm of the bronchi (air passages) in the lungs. Attacks may be provoked by allergy, infection, and stress. The incidence of asthma may be increasing as a result of air pollution and occupational hazard. Treatment is with bronchodilators to relax the bronchial muscles and thereby ease the breathing, and in severe cases by inhaled ◊steroids that reduce inflammation of the bronchi.

atherosclerosis thickening and hardening of the walls of the arteries, associated with atheroma.

autism, infantile rare disorder, generally present from birth, characterized by a withdrawn state and a failure to develop normally in language or social behaviour. Special education may bring about some improvement.

autoimmunity condition where the body's immune responses are mobilized not against 'foreign' matter, such as invading germs, but against the body itself. Diseases considered to be of autoimmune origin include myasthenia gravis, rheumatoid arthritis, and ◊lupus erythematosus.

AZT drug used in the treatment of AIDS; see ◊zidovudine.

bacille Calmette–Guérin tuberculosis vaccine ◊BCG.

basal metabolic rate, BMR, minimum amount of energy needed by the body to maintain life. It is measured when the subject is awake but resting, and includes the energy required to keep the heart beating, sustain breathing, repair tissues, and keep the brain and nerves functioning. Measuring the subject's consumption of oxygen gives an accurate value for BMR, because oxygen is needed to release energy from food.

BCG abbreviation for **bacille Calmette–Guérin**, bacillus injected as a vaccine to confer active immunity to ◊tuberculosis (TB).

behaviour therapy in psychology, the application of behavioural principles, derived from learning theories, to the treatment of clinical conditions such as ◊phobias, ◊obsessions, and sexual and interpersonal problems.

Benzedrine trade name for ◊amphetamine, a stimulant drug.

beta-blocker any of a class of drugs that block impulses that stimulate certain nerve endings (beta receptors) serving the heart muscle. This reduces the heart rate and the force of contraction, which in turn reduces the amount of oxygen (and therefore the blood supply) required by the heart. Beta-blockers may be useful in the treatment of angina, arrhythmia (abnormal heart rhythms), and raised blood pressure, and following heart attacks. They must be withdrawn from use gradually.

biopsy removal of a living tissue sample from the body for diagnostic examination.

blood poisoning presence in the bloodstream of quantities of bacteria or bacterial toxins sufficient to cause serious illness.

blood pressure pressure, or tension, of the blood against the inner walls of blood vessels, especially the arteries, due to the muscular pumping activity of the heart. Abnormally high blood pressure (◊hypertension) may be associated with various conditions or arise with no obvious cause; abnormally low blood pressure (hypotension) occurs in shock and after excessive fluid or blood loss from any cause.

blood test laboratory evaluation of a blood sample. There are numerous blood tests, from simple typing to establish the ◊blood group to sophisticated biochemical assays of substances, such as hormones, present in the blood only in minute quantities.

bovine spongiform encephalopathy, (BSE) or **mad cow disease**, disease of cattle, related to ◊scrapie in sheep, which attacks the nervous system, causing aggression, lack of coordination, and collapse. First identified in 1986, it is almost entirely confined to the UK. By 1996 it had claimed 158,000 British cattle.

bronchitis inflammation of the bronchi (air passages) of the lungs, usually caused initially by a viral infection, such as a cold or flu. It is aggravated by environmental pollutants, especially smoking, and results in a persistent cough, irritated mucus-secreting glands, and large amounts of sputum.

bulimia eating disorder in which large amounts of food are consumed in a short time ('binge'), usually followed by depression and self-criticism. The term is often used for **bulimia nervosa**, an emotional disorder in which eating is followed by deliberate vomiting and purging. This may be a chronic stage in ◊anorexia nervosa.

Caesarean section surgical operation to deliver a baby by way of an incision in the mother's abdominal and uterine walls. It may be recommended for almost any obstetric complication implying a threat to mother or baby.

cancer group of diseases characterized by abnormal proliferation of cells. Cancer (malignant) cells are usually degenerate, capable only of reproducing themselves (tumour formation). Malignant cells tend to spread from their site of origin by travelling through the bloodstream or lymphatic system (see metastasis). Cancer kills about 6 million people a year worldwide.

carbohydrate chemical compound composed of carbon, hydrogen, and oxygen, with the basic formula $C_m(H_2O)_n$, and related compounds with the same basic structure but modified functional groups. As sugar and starch, carbohydrates are an important part of a balanced human diet, providing energy for life processes including growth and movement. Excess carbohydrate intake can be converted into fat and stored in the body.

carcinogen any agent that increases the chance of a cell becoming cancerous (see ◊cancer), including various chemical compounds, some viruses, X-rays, and other forms of ionizing radiation. The term is often used more narrowly to mean chemical carcinogens only.

carcinoma malignant ◊tumour arising from the skin, the glandular tissues, or the mucous membranes that line the gut and lungs..

carrier anyone who harbours an infectious organism without ill effects but can pass the infection to others. The term is also applied to those who carry a recessive gene for a disease or defect without manifesting the condition.

casein main protein of milk used as a protein supplement in the treatment of malnutrition.

chemotherapy any medical treatment with chemicals. It usually refers to treatment of cancer with cytotoxic and other drugs.

chickenpox or **varicella** common, usually mild disease, caused by a virus of the ◊herpes group and transmitted by airborne droplets. Chickenpox chiefly attacks children under the age of ten. The incubation period is two to three weeks. One attack normally gives immunity for life.

chiropractic in alternative medicine, technique of manipulation of the spine and other parts of the body, based on the principle that physical disorders are attributable to aberrations in the functioning of the nervous system, which manipulation can correct.

chlamydia viruslike bacteria which live parasitically in animal cells, and cause disease in humans and birds. Chlamydiae are thought to be descendants of bacteria that have lost certain metabolic processes. In humans, a strain of chlamydia causes ◊trachoma, a disease found mainly in the tropics (a leading cause of blindness); venereally transmitted chlamydiae cause genital and urinary infections.

cholera disease caused by infection with various strains of the bacillus *Vibrio cholerae*, transmitted in contaminated water and characterized by violent diarrhoea and vomiting. It is prevalent in many tropical areas.

cholesterol white, crystalline sterol found throughout the body, especially in fats, blood, nerve tissue, and bile; it is also provided in the diet by foods such as eggs, meat, and butter. A high level of cholesterol in the blood is thought to contribute to atherosclerosis (hardening of the arteries).

chronic term used to describe a condition that is of slow onset and then runs a prolonged course, such as rheumatoid arthritis or chronic bronchitis. In contrast, an **acute** condition develops quickly and may be of relatively short duration.

clinical psychology branch of psychology dealing with the understanding and treatment of health problems, particularly mental disorders. The main problems dealt with include anxiety, phobias, depression, obsessions, sexual and marital problems, drug and alcohol dependence, childhood behavioural problems, psychoses (such as schizophrenia), mental disability, and brain disease (such as dementia) and damage. Other areas of work include forensic psychology (concerned with criminal behaviour) and health psychology.

cold, common minor disease of the upper respiratory tract, caused by a variety of viruses. Symptoms are headache, chill, nasal discharge, sore throat, and occasionally cough. Research indicates that the virulence of a cold depends on psychological factors and either a reduction or an increase of social or work activity, as a result of stress, in the previous six months.

colitis inflammation of the colon (large intestine) with diarrhoea (often bloody). It is usually due to infection or some types of bacterial dysentery.

complementary medicine systems of care based on methods of treatment or theories of disease that differ from those taught in most western medical schools. See ◊ alternative medicine.

congenital disease disease present at birth. It is not necessarily genetic in origin; for example, congenital herpes may be acquired by the baby as it passes through the mother's birth canal.

corticosteroid any of several steroid hormones secreted by the cortex of the adrenal glands; also synthetic forms with similar properties. Corticosteroids have anti-inflammatory and immunosuppressive effects and may be used to treat a number of conditions, including rheumatoid arthritis, severe allergies, asthma, some skin diseases, and some cancers. Side effects can be serious, and therapy must be withdrawn very gradually.

cot death or **sudden infant death syndrome** (SIDS) death of an apparently healthy baby, almost always during sleep. It is most common in the winter months, and strikes more boys than girls. The cause is not known but risk factors that have been identified include prematurity, respiratory infection, overheating, and sleeping position.

Creutzfeldt–Jakob disease (CJD) rare brain disease that causes progressive physical and mental deterioration, leading to death usually within a year of onset. It claims one person in every million and is universally fatal. It has been linked with ◊bovine spongiform encephalopathy (BSE), and there have also been occurrences in people treated with pituitary hormones derived from cows for growth or fertility problems. Research published by British pathologists in 1997 proved that the new variant of CJD (vCJD) is caused by the same agent that causes BSE, indicating that the disease has jumped species, from cattle to humans.

croup inflammation of the larynx in small children, with harsh, difficult breathing and hoarse coughing. Croup is most often associated with viral infection of the respiratory tract.

cyanocobalamin chemical name for vitamin B_{12}, which is normally produced by microorganisms in the gut. The richest sources are liver, fish, and eggs. It is essential to the replacement of cells, the maintenance of the myelin sheath, which insulates nerve fibres, and the efficient use of folic acid, another vitamin in the B complex. Deficiency can result in pernicious anaemia (defective production of red blood cells), and possible degeneration of the nervous system.

dementia mental deterioration as a result of physical changes in the brain. It may be due to degenerative change, circulatory disease, infection, injury, or chronic poisoning. **Senile dementia**, a progressive loss of mental faculties such as memory and orientation, is typically a disease process of old age, and can be accompanied by ◊depression.

depression emotional state characterized by sadness, unhappy thoughts, apathy, and dejection. Sadness is a normal response to major losses such as bereavement or unemployment. After childbirth, postnatal depression is common. However, clinical depression, which is prolonged or unduly severe, often requires treatment, such as antidepressant medication, cognitive therapy, or, in very rare cases, electroconvulsive therapy (ECT), in which an electrical current is passed through the brain.

developmental psychology study of development of cognition and behaviour from birth to adulthood.

diabetes disease *diabetes mellitus* in which a disorder of the islets of Langerhans in the ◊pancreas prevents the body producing the hormone ◊insulin, so that sugars cannot be used properly. Treatment is by strict dietary control and oral or injected insulin, depending on the type of diabetes.

diarrhoea frequent or excessive action of the bowels so that the faeces are liquid or semiliquid. It is caused by intestinal irritants (including some drugs and poisons), infection with harmful organisms (as in dysentery, salmonella, or cholera), or allergies.

diphtheria acute infectious disease in which a membrane forms in the throat (threatening death by asphyxia), along with the production of a powerful toxin that damages the heart and nerves. The organism responsible is a bacterium (*Corynebacterium diphtheriae*). It is treated with antitoxin and antibiotics.

drug any of a range of substances, natural or synthetic, administered to humans and animals as therapeutic agents: to diagnose, prevent, or treat disease, or to assist recovery from injury. Traditionally many drugs were obtained from plants or animals; some minerals also had medicinal value. Today, increasing numbers of drugs are synthesized in the laboratory.

drug misuse illegal use of drugs for nontherapeutic purposes. Drugs used illegally include: narcotics, such as heroin, morphine, and the synthetic opioids; barbiturates; amphetamines and related substances; benzodiazepine tranquillizers; cocaine, LSD, and cannabis.

dysentery infection of the large intestine causing abdominal cramps and painful ◊diarrhoea with blood. There are two kinds of dysentery: **amoebic** (caused by a protozoan), common in the tropics, which may lead to liver damage; and **bacterial**, the kind most often seen in the temperate zones.

evening primrose oil plant oil rich in gamma-linoleic acid (GLA). The body converts GLA into substances which resemble hormones, and evening primrose oil is beneficial in relieving the symptoms of premenstrual tension. It is also used in treating eczema and chronic fatigue syndrome.

food poisoning any acute illness characterized by vomiting and diarrhoea and caused by eating food contaminated with harmful bacteria (for example, listeriosis), poisonous food (for example, certain mushrooms, puffer fish), or poisoned food (such as lead or arsenic introduced accidentally during processing). A frequent cause of food poisoning is ◊Salmonella bacteria. Salmonella comes in many forms, and strains are found in cattle, pigs, poultry, and eggs.

gangrene death and decay of body tissue (often of a limb) due to bacterial action; the affected part gradually turns black and causes blood poisoning.

gastroenteritis inflammation of the stomach and intestines, giving rise to abdominal pain, vomiting, and diarrhoea. It may be caused by food or other poisoning, allergy, or infection. Dehydration may be severe and it is a particular risk in infants.

gene therapy medical technique for curing or alleviating inherited diseases or defects; certain infections, and several kinds of cancer in which affected cells from a sufferer would be removed from the body, the ◊DNA repaired in the laboratory (genetic engineering), and the functioning cells reintroduced. In 1990 a genetically engineered gene was used for the first time to treat a patient.

German measles or **rubella** mild, communicable virus disease, usually caught by children. It is marked by a sore throat, pinkish rash, and slight fever, and has an incubation period of two to three weeks. If a woman contracts it in the first three months of pregnancy, it may cause serious damage to the unborn child.

ginseng plant with a thick forked aromatic root used in alternative medicine as a tonic.

glandular fever or **infectious mononucleosis** viral disease characterized at onset by fever and painfully swollen lymph nodes; there may also be digestive upset, sore throat, and skin rashes. Lassitude persists for months and even years, and recovery can be slow. It is caused by the Epstein–Barr virus.

glue-sniffing or **solvent misuse** inhalation of the fumes from organic solvents of the type found in paints, lighter fuel, and glue, for their hallucinatory effects. As well as being addictive, solvents are dangerous for their effects on the user's liver, heart, and lungs.

goitre enlargement of the thyroid gland seen as a swelling on the neck. It is most pronounced in simple goitre, which is caused by iodine deficiency. More common is toxic goitre or hyperthyroidism, caused by overactivity of the thyroid gland.

gonorrhoea common sexually transmitted disease arising from infection with the bacterium *Neisseria gonorrhoeae*, which causes inflammation of the genito-urinary tract. After an incubation period of two to ten days, infected men experience pain while urinating and a discharge from the penis; infected women often have no external symptoms.

gynaecology medical speciality concerned with disorders of the female reproductive system.

haematology medical speciality concerned with disorders of the blood.

health, world the health of people worldwide is monitored by the World Health Organization (WHO). Outside the industrialized world in particular, poverty and degraded environmental conditions mean that easily preventable diseases are widespread: WHO estimated in 1990 that 1 billion people, or 20% of the world's population, were diseased, in poor health, or malnourished. In North Africa and the Middle East, 25% of the population were ill.

heart attack or **myocardial infarction** sudden onset of gripping central chest pain, often accompanied by sweating and vomiting, caused by death of a portion of the heart muscle following obstruction of a coronary artery by thrombosis (formation of a blood clot). Half of all heart attacks result in death within the first two hours, but in the remainder survival has improved following the widespread use of thrombolytic (clot-buster) drugs.

hepatitis any inflammatory disease of the liver, usually caused by a virus. Other causes include alcohol, drugs, gallstones, lupus erythematous, and amoebic dysentery. Symptoms include weakness, nausea, and jaundice.

herbalism prescription and use of plants and their derivatives for medication.

herpes any of several infectious diseases caused by viruses of the herpes group. **Herpes simplex I** is the causative agent of a common inflammation, the cold sore. **Herpes simplex II** is responsible for genital herpes, a highly contagious, sexually transmitted disease characterized by painful blisters in the genital area. It can be transmitted in the birth canal from mother to newborn. **Herpes zoster** causes shingles; another herpes virus causes chickenpox.

HIV abbreviation for **human immunodeficiency virus**, the infectious agent that is believed to cause ◊AIDS.

holistic medicine umbrella term for an approach that virtually all alternative therapies profess, which considers the overall health and lifestyle profile of a patient, and treats specific ailments not primarily as conditions to be alleviated but rather as symptoms of more fundamental disease.

homoeopathy or **homoeopathy** system of alternative medicine based on the principle that symptoms of disease are part of the body's self-healing processes, and on the practice of administering extremely diluted doses of natural substances found to produce in a healthy person the symptoms manifest in the illness being treated.

hypertension abnormally high ◊blood pressure due to a variety of causes, leading to excessive contraction of the smooth muscle cells of the walls of the arteries. It increases the risk of kidney disease, stroke, and heart attack.

hysterectomy surgical removal of all or part of the uterus (womb). The operation is performed to treat fibroids (benign tumours growing in the uterus) or cancer; also to relieve heavy menstrual bleeding. A woman who has had a hysterectomy will no longer menstruate and cannot bear children.

immunity protection that organisms have against foreign microorganisms, such as bacteria and viruses, and against cancerous cells (see ◊cancer). The cells that provide this protection are called white blood cells, or leucocytes, and make up the immune system. They include neutrophils and macrophages, which can engulf invading organisms and other unwanted material, and natural killer cells that destroy cells infected by viruses and cancerous cells. Some of the most important immune cells are the B cells and T cells. Immune cells coordinate their activities by means of chemical messengers or lymphokines, including the antiviral messenger interferon. The lymph nodes play a major role in organizing the immune response.

immunization conferring immunity to infectious disease by artificial methods. The most widely used technique is ◊vaccination.

infection invasion of the body by disease-causing organisms (pathogens, or germs) that become established, multiply, and produce symptoms. Bacteria and viruses cause most diseases, but diseases are also caused by other microorganisms, protozoans, and other parasites.

influenza any of various viral infections primarily affecting the air passages, accompanied by systemic effects such as fever, chills, headache, joint and muscle pains, and lassitude. Treatment is with bed rest and analgesic drugs such as aspirin or paracetamol.

inoculation injection into the body of dead or weakened disease-carrying organisms or their toxins (◊vaccine) to produce immunity by inducing a mild form of a disease.

insulin protein hormone, produced by specialized cells in the islets of Langerhans in the pancreas, that regulates the metabolism (rate of activity) of glucose, fats, and proteins. It is used in treating ◊diabetes.

kwashiorkor severe protein deficiency in children under five years, resulting in retarded growth, lethargy, oedema, diarrhoea, and a swollen abdomen. It is common in Third World countries with a high incidence of malnutrition.

lactation secretion of milk in mammals, from the mammary glands. In late pregnancy, the cells lining the lobules inside the mammary glands begin extracting substances from the blood to produce milk. The supply of milk starts shortly after birth with the production of colostrum, a clear fluid consisting largely of water, protein, antibodies, and vitamins. The production of milk continues practically as long as the baby continues to suckle.

Lassa fever acute disease caused by an arenavirus, first detected 1969, and spread by a species of rat found only in West Africa. It is classified as a haemorrhagic fever and characterized by high fever, headache, muscle pain, and internal bleeding. There is no known cure, the survival rate being less than 50%.

leprosy or **Hansen's disease** chronic, progressive disease caused by a bacterium *Mycobacterium leprae* closely related to that of tuberculosis. The infection attacks the skin and nerves. It is controlled with drugs. In 1998 there were an estimated 1.5 million cases of leprosy, with 60% of these being in India.

leukaemia any one of a group of cancers of the blood cells, with widespread involvement of the bone marrow and other blood-forming tissue. The central feature of leukaemia is runaway production of white blood cells that are immature or in some way abnormal. These rogue cells, which lack the defensive capacity of healthy white cells, overwhelm the normal ones, leaving the victim vulnerable to infection. Treatment is with radiotherapy and cytotoxic drugs to suppress replication of abnormal cells, or by bone-marrow transplantation.

lithium soft, ductile, silver-white, metallic element used in medicine to treat manic depression.

louse parasitic insect that lives on mammals. It has a flat, segmented body without wings, and a tube attached to the head, used for sucking blood from its host.

lung cancer cancer of the lung. The main risk factor is smoking, with almost nine out of ten cases attributed to it. Other risk factors include workspace exposure to carcinogenic substances such as asbestos, and radiation. Warning symptoms include a persistent and otherwise unexplained cough, breathing difficulties, and pain in the chest or shoulder. Treatment is with chemotherapy, radiotherapy, and surgery.

lupus any of various diseases characterized by lesions of the skin. One form (lupus vulgaris) is caused by the tubercle bacillus (see tuberculosis). The organism produces ulcers that spread and eat away the underlying tissues. Treatment is primarily with standard antituberculous drugs, but ultraviolet light may also be used.

malaria infectious parasitic disease of the tropics transmitted by mosquitoes, marked by periodic fever and an enlarged spleen. Malaria affects about 267 million people in 103 countries, and in 1995 around 2.1 million people died of the disease.

malnutrition condition resulting from a defective diet where certain important food nutrients (such as protein, vitamins, or carbohydrates) are absent. It can lead to deficiency diseases.

manic depression or **bipolar disorder** mental disorder characterized by recurring periods of either ◊depression or mania (inappropriate elation, agitation, and rapid thought and speech) or both.

ME abbreviation for **myalgic encephalomyelitis**, a popular name for chronic fatigue syndrome.

measles acute virus disease (rubeola), spread by airborne infection. Symptoms are fever, severe catarrh, small spots inside the mouth, and a raised, blotchy red rash appearing for about a week after two weeks' incubation. Prevention is by vaccination.

melanoma highly malignant tumour of the melanin-forming cells (melanocytes) of the skin. It develops from an existing mole in up to two thirds of cases, but can also arise in the eye or mucous membranes. Malignant melanoma is the most dangerous of the skin cancers; it is associated with brief but excessive exposure to sunlight. It is easily treated if caught early but deadly once it has spread. There is a genetic factor in some cases.

meningitis inflammation of the meninges (membranes) surrounding the brain, caused by bacterial or viral infection. Bacterial meningitis, though treatable by antibiotics, is the more serious threat. Diagnosis is by lumbar puncture.

mental illness disordered functioning of the mind. It is broadly divided into two categories: ◊neurosis, in which the patient remains in touch with reality; and ◊psychosis, in which perception, thought, and belief are disordered.

miscarriage spontaneous expulsion of a fetus from the womb before it is capable of independent survival. Often, miscarriages are due to an abnormality in the developing fetus.

morphine narcotic alkaloid $C_{17}H_{19}NO_3$ derived from opium and prescribed only to alleviate severe pain. Its use produces serious side effects, including nausea, constipation, tolerance, and addiction, but it is highly valued for the relief of the terminally ill.

mumps or **infectious parotitis** virus infection marked by fever, pain, and swelling of one or both parotid salivary glands (situated in front of the ears). It is usually shortlived in children, although meningitis is a possible complication. In adults the symptoms are more serious and it may cause sterility in men.

narcotic pain-relieving and sleep-inducing drug. The term is usually applied to heroin, morphine, and other opium derivatives, but may also be used for other drugs which depress brain activity, including anaesthetic agents and hypnotics.

neoplasm any lump or tumour, which may be benign or malignant (cancerous).

neurosis in psychology, a general term referring to emotional disorders, such as anxiety, depression, and phobias. The main disturbance tends to be one of mood; contact with reality is relatively unaffected, in contrast to ◊psychosis.

obesity condition of being overweight (generally, 20% or more above the desirable weight for one's sex, build, and height). Obesity increases susceptibility to disease, strains the vital organs, and reduces life expectancy; it is usually remedied by controlled weight loss, healthy diet, and exercise.

obstetrics medical speciality concerned with the management of pregnancy, childbirth, and the immediate postnatal period.

oncology medical speciality concerned with the diagnosis and treatment of ◊neoplasms, especially cancer.

osteopathy system of alternative medical practice that relies on physical manipulation to treat mechanical stress.

paediatrics or **pediatrics** medical speciality concerned with the care of children.

paranoia mental disorder marked by delusions of grandeur or persecution. In popular usage, paranoia means baseless or exaggerated fear and suspicion.

parasite organism that lives on or in another organism (called the host) and depends on it for nutrition, often at the expense of the host's welfare. Parasites that live inside the host, such as liver flukes and tapeworms, are called **endoparasites**; those that live on the exterior, such as fleas and lice, are called **ectoparasites**.

pathogen any microorganism that causes disease. Most pathogens are ◊parasites, and the diseases they cause are incidental to their search for food or shelter inside the host. Nonparasitic organisms, such as soil bacteria or those living in the human gut and feeding on waste foodstuffs, can also become pathogenic to a person whose immune system or liver is damaged.

pathology medical speciality concerned with the study of disease processes and how these provoke structural and functional changes in the body.

penicillin any of a group of ◊antibiotic (bacteria killing) compounds obtained from filtrates of moulds of the genus *Penicillium* (especially *P. notatum*) or produced synthetically. Penicillin was the first antibiotic to be discovered (by Alexander Fleming); it kills a broad spectrum of bacteria, many of which cause disease in humans.

phobia excessive irrational fear of an object or situation – for example, agoraphobia (fear of open spaces and crowded places), acrophobia (fear of heights), and claustrophobia (fear of enclosed places).

phytomenadione one form of vitamin K, a fat-soluble chemical found in green vegetables. It is involved in the production of prothrombin, which is essential in blood clotting.

placenta organ that attaches the developing embryo or fetus to the uterus in placental mammals (mammals other than marsupials, platypuses, and echidnas). Composed of maternal and embryonic tissue, it links the blood supply of the embryo to the blood supply of the mother, allowing the exchange of oxygen, nutrients, and waste products. The two blood systems are not in direct contact, but are separated by thin membranes, with materials diffusing across from one system to the other. The placenta also produces hormones that maintain and regulate pregnancy. It is shed as part of the afterbirth.

pneumonia inflammation of the lungs, generally due to bacterial or viral infection but also to particulate matter or gases. It is characterized by a build-up of fluid in the alveoli, the clustered air sacs (at the ends of the air passages) where oxygen exchange takes place.

polio, poliomyelitis viral infection of the central nervous system affecting nerves that activate muscles. Two kinds of vaccine are available, one injected and one given by mouth. In 1997 the World Health Organization reported that causes of polio had dropped by nearly 90% since 1988. Most cases remain in Africa and southeast Asia (in 1998 India accounted for over 30% of the world's polio cases).

prophylaxis any measure taken to prevent disease, including exercise and ◊vaccination. Prophylactic (preventive) medicine is an aspect of public-health provision that is receiving increasing attention.

psychiatry branch of medicine dealing with the diagnosis and treatment of mental disorder, normally divided into the areas of **neurotic conditions**, including anxiety, depression, and hysteria, and **psychotic disorders**, such as schizophrenia. Psychiatric treatment consists of drugs, analysis, or electroconvulsive therapy (ECT).

psychoanalysis theory and treatment method for neuroses, developed by Sigmund Freud in the 1890s. Psychoanalysis asserts that the impact of early childhood sexuality and experiences, stored in the ◊unconscious, can lead to the development of adult emotional problems. The main treatment method involves the free association of ideas, and their interpretation by patient and analyst, in order to discover these long-buried events and to grasp their significance to the patient, linking aspects of the patient's historical past with the present relationship to the analyst.

psychosis or **psychotic disorder** general term for a serious mental disorder where the individual commonly loses contact with reality and may experience hallucinations or delusions (fixed false beliefs). For example, in a paranoid psychosis, an individual may believe that others are plotting against him or her. A major type of psychosis is schizophrenia.

psychosomatic of a physical symptom or disease thought to arise from emotional or mental factors.

psychotherapy any treatment for psychological problems that involves talking rather than surgery or drugs. Examples include cognitive therapy and psychoanalysis.

pus yellowish fluid that forms in the body as a result of bacterial infection; it includes white blood cells (leucocytes), living and dead bacteria, dead tissue, and serum. An enclosed collection of pus is called an abscess.

quinine antimalarial drug extracted from the bark of the cinchona tree. It is a bitter alkaloid, with the formula $C_{20}H_{24}N_2O_2$.

rabies or **hydrophobia** viral disease of the central nervous system that can afflict all warm-blooded creatures. It is caused by a lyssavirus. It is almost invariably fatal once symptoms have developed. Its transmission to humans is generally by a bite from an infected animal. Rabies continues to kill hundreds of thousands of people every year; almost all these deaths occur in Asia, Africa, and South America.

radiotherapy treatment of disease by radiation from X-ray machines or radioactive sources. Radiation, which reduces the activity of dividing cells, is of special value for its effect on malignant tissues, certain nonmalignant tumours, and some diseases of the skin.

retinol or **vitamin A** fat-soluble chemical derived from ß-carotene and found in milk, butter, cheese, egg yolk, and liver.

Lack of retinol in the diet leads to the eye disease **xerophthalmia**.

retrovirus any of a family of viruses (Retroviridae) containing the genetic material RNA rather than the more usual DNA.

rheumatic fever or **acute rheumatism** acute or chronic illness characterized by fever and painful swelling of joints. Some victims also experience involuntary movements of the limbs and head, a form of chorea. It is now rare in the developed world.

riboflavin or **vitamin B$_2$** vitamin of the B complex important in cell respiration. It is obtained from eggs, liver, and milk. A deficiency in the diet causes stunted growth.

rickets defective growth of bone in children due to an insufficiency of calcium deposits. The bones, which do not harden adequately, are bent out of shape. It is usually caused by a lack of vitamin D and insufficient exposure to sunlight. Renal rickets, also a condition of malformed bone, is associated with kidney disease.

ringworm any of various contagious skin infections due to related kinds of fungus, usually resulting in circular, itchy, discoloured patches covered with scales or blisters. The scalp and feet (athlete's foot) are generally involved. Treatment is with antifungal preparations.

rubella technical term for German measles.

salmonella any of a very varied group of bacteria, genus *Salmonella* that colonize the intestines of humans and some animals. Some strains cause typhoid and paratyphoid fevers, while others cause salmonella food poisoning, which is characterized by stomach pains, vomiting, diarrhoea, and headache. It can be fatal in elderly people, but others usually recover in a few days without antibiotics. Most cases are caused by contaminated animal products, especially poultry meat.

sarcoma malignant tumour arising from the fat, muscles, bones, cartilage, or blood and lymph vessels and connective tissues. Sarcomas are much less common than carcinomas.

scabies contagious infection of the skin caused by the parasitic itch mite *Sarcoptes scabiei*, which burrows under the skin to deposit eggs. Treatment is by antiparasitic creams and lotions.

scarlet fever or **scarlatina** acute infectious disease, especially of children, caused by the bacteria in the *Streptococcus pyogenes* group. It is marked by fever, vomiting, sore throat, and a bright red rash spreading from the upper to the lower part of the body. The rash is followed by the skin peeling in flakes. It is treated with antibiotics.

schizophrenia mental disorder, a psychosis of unknown origin, which can lead to profound changes in personality, behaviour, and perception, including delusions and hallucinations. It is more common in males and the early-onset form is more severe than when the illness develops in later life. Modern treatment approaches include drugs, family therapy, stress reduction, and rehabilitation.

scurvy disease caused by deficiency of vitamin C (ascorbic acid), which is contained in fresh vegetables and fruit. The signs are weakness and aching joints and muscles, progressing to bleeding of the gums and other spontaneous haemorrhage, and drying-up of the skin and hair. It is reversed by giving the vitamin.

sepsis general term for infectious change in the body caused by bacteria or their toxins.

septicaemia general term for any form of ◊blood poisoning.

serum clear fluid that separates out from clotted blood. It is blood plasma with the anticoagulant proteins removed, and contains ◊antibodies and other proteins, as well as the fats and sugars of the blood. It can be produced synthetically, and is used to protect against disease.

sexually transmitted disease (STD) any disease transmitted by sexual contact. STDs include venereal disease, ◊AIDS, scabies, and viral ◊hepatitis. The WHO estimate that there are 356,000 new cases of STDs daily worldwide (1995).

shingles common name for ◊herpes zoster, a disease characterized by infection of sensory nerves, with pain and eruption of blisters along the course of the affected nerves.

sleeping sickness infectious disease of tropical Africa, a form of ◊trypanosomiasis. Early symptoms include fever, headache, and chills, followed by anaemia and joint pains. Later, the disease attacks the central nervous system, causing drowsiness, lethargy, and, if left untreated, death. Sleeping sickness is caused by either of two trypanosomes, *Trypanosoma gambiense* or *T. rhodesiense*. Control is by eradication of the tsetse fly, which transmits the disease to humans.

smallpox acute, highly contagious viral disease, marked by aches, fever, vomiting, and skin eruptions leaving pitted scars. Widespread vaccination programmes have eradicated this often fatal disease.

STD abbreviation for ◊sexually transmitted disease.

stroke or **cerebrovascular accident** or **apoplexy** interruption of the blood supply to part of the brain due to a sudden bleed in the brain (cerebral haemorrhage) or embolism or thrombosis. Strokes vary in severity from producing almost no symptoms to proving rapidly fatal. In between are those (often recurring) that leave a wide range of impaired function, depending on the size and location of the event.

sulphonamide any of a group of compounds containing the chemical group sulphonamide (SO_2NH_2) or its derivatives, which were, and still are in some cases, used to treat bacterial diseases.

superbug popular name given to an infectious bacterium that has developed resistance to most or all known antibiotics.

surgery branch of medicine concerned with the treatment of disease, abnormality, or injury by operation. Traditionally it has been performed by means of cutting instruments, but today a number of technologies are used to treat or remove lesions, including ultrasonic waves and laser surgery.

syphilis sexually transmitted disease caused by the spiral-shaped bacterium (spirochete) *Treponema pallidum*. Untreated, it runs its course in three stages over many years, often starting with a painless hard sore, or chancre, developing within a month on the area of infection (usually the genitals). The second stage, months later, is a rash with arthritis, hepatitis, and/or meningitis. The third stage, years later, leads eventually to paralysis, blindness, insanity, and death. The Wassermann test is a diagnostic blood test for syphilis.

tapeworm any of various parasitic flatworms of the class Cestoda. They lack digestive and sense organs, can reach 15 m/50 ft in length, and attach themselves to the host's intestines by means of hooks and suckers. The larvae of tapeworms usually reach humans in imperfectly cooked meat or fish, causing anaemia and intestinal disorders.

TB abbreviation for the infectious disease ◊tuberculosis.

tetanus or **lockjaw** acute disease caused by the toxin of the bacillus *Clostridium tetani*, which usually enters the body through a wound. The bacterium is chiefly found in richly manured soil. Untreated, in seven to ten days tetanus produces muscular spasm and rigidity of the jaw spreading to other parts of the body, convulsions, and death. There is a vaccine, and the disease may be treatable with tetanus antitoxin and antibiotics.

tetracycline one of a group of antibiotic compounds having in common the four-ring structure of chlortetracycline, the first member of the group to be isolated. They are prepared synthetically or obtained from certain bacteria of the genus *Streptomyces*. They are broad-spectrum antibiotics, effective against a wide range of disease-causing bacteria.

thiamine or **vitamin B$_1$** a water-soluble vitamin of the B complex. It is found in seeds and grain. Its absence from the diet causes the disease beriberi.

thrombosis condition in which a blood clot forms in a vein or artery, causing loss of circulation to the area served by the vessel. If it breaks away, it often travels to the lungs, causing pulmonary embolism.

thymus organ in vertebrates, situated in the upper chest cavity in humans. The thymus processes lymphocyte cells to produce T-lymphocytes (T denotes 'thymus-derived'), which are responsible for binding to specific invading organisms and killing them or rendering them harmless.

toxaemia another term for ◊blood poisoning; **toxaemia of pregnancy** is another term for pre-eclampsia.

trypanosomiasis any of several debilitating long-term diseases caused by a trypanosome (protozoan of the genus *Trypanosoma*). They include sleeping sickness in Africa, transmitted by the bites of tsetse flies, and Chagas's disease in Central and South America, spread by assassin bugs.

tuberculosis (TB) formerly known as **consumption** or **phthisis**, infectious disease caused by the bacillus *Mycobacterium tuberculosis*. It takes several forms, of which pulmonary tuberculosis is by far the most common. A vaccine, BCG, was developed around 1920 and the first antituberculosis drug, streptomycin, in 1944. The bacterium is mostly kept in check by the body's immune system; about 5% of those infected develop the disease. Treatment of patients with a combination of anti-TB medicines for 6–8 months produces a cure rate of 80%. There are 7 million new cases of TB annually worldwide (1998) and 3 million deaths.

tumour overproduction of cells in a specific area of the body, often leading to a swelling or lump. Tumours are classified as **benign** or **malignant**. Benign tumours grow more slowly, do not invade surrounding tissues, do not spread to other parts of the body, and do not usually recur after removal. However, benign tumours can be dangerous in areas such as the brain. The most familiar types of benign tumour are warts on the skin. In some cases, there is no sharp dividing line between benign and malignant tumours.

typhoid fever acute infectious disease of the digestive tract, caused by the bacterium *Salmonella typhi*, and usually contracted through a contaminated water supply. It is characterized by bowel haemorrhage and damage to the spleen. Treatment is with antibiotics.

typhus any one of a group of infectious diseases caused by bacteria transmitted by lice, fleas, mites, and ticks. Symptoms include fever, headache, and rash. The most serious form is epidemic typhus, which also affects the brain, heart, lungs, and kidneys and is associated with insanitary overcrowded conditions. Treatment is by antibiotics.

vaccine any preparation of modified pathogens (viruses or bacteria) that is introduced into the body, usually either orally or by a hypodermic syringe, to induce the specific antibody reaction that produces ◊immunity against a particular disease.

valvular heart disease damage to the heart valves, leading to either narrowing of the valve orifice when it is open (stenosis) or leaking through the valve when it is closed (regurgitation).

virus infectious particle consisting of a core of nucleic acid (DNA or RNA) enclosed in a protein shell. Viruses are acellular and able to function and reproduce only if they can invade a living cell to use the cell's system to replicate themselves. In the process they may disrupt or alter the host cell's own DNA. The healthy human body reacts by producing an antiviral protein, ◊interferon, which prevents the infection spreading to adjacent cells.

vitamin any of various chemically unrelated organic compounds that are necessary in small quantities for the normal functioning of the human body. Vitamins must be supplied by the diet because the body cannot make them. They are normally present in adequate amounts in a balanced diet. Deficiency of a vitamin may lead to a metabolic disorder ('deficiency disease'), which can be remedied by sufficient intake of the vitamin. They are generally classified as **water-soluble** (B and C) or **fat-soluble** (A, D, E, and K).

WHO acronym for **World Health Organization**.

whooping cough or **pertussis** acute infectious disease, seen mainly in children, caused by colonization of the air passages by the bacterium *Bordetella pertussis*. There may be catarrh, mild fever, and loss of appetite, but the main symptom is violent coughing, associated with the sharp intake of breath that is the characteristic 'whoop', and often followed by vomiting and severe nose bleeds. The cough may persist for weeks.

yellow fever or **yellow jack** acute tropical viral disease, prevalent in the Caribbean area, Brazil, and on the west coast of Africa. The yellow fever virus is an arbovirus transmitted by mosquitoes. Its symptoms include a high fever, headache, joint and muscle pains, vomiting, and yellowish skin (jaundice, possibly leading to liver failure); the heart and kidneys may also be affected. The mortality rate is 25%, with 91% of all cases occurring in Africa.

zidovudine, formerly **AZT**, antiviral drug used in the treatment of ◊AIDS. It is not a cure for AIDS but is effective in prolonging life; it does not, however, delay the onset of AIDS in people carrying the virus.

Further Reading

Ackerknecht, Erwin H *A Short History of Medicine* (1968)

Altman, Lawrence K *Who Goes First: The Story of Self-Experimentation in Medicine* (1988)

American Medical Association *International Classification of Diseases: Clinical Modification* (1998)

Asher, Richard *Talking Sense* (1972)

Blaue, David, Brunner, Eric, and Wilkinson, Richard G *Health and Social Organization: Towards a Health Policy in the Twenty-First Century* (1996)

Bliss, Michael *The Discovery of Insulin* (1982)

Briant, Keith *Passionate Paradox* (1962)

Bynum, W F and Porter, Roy *Companion Encyclopedia of the History of Medicine* (1993)

Chalmers, Irena *Foods that Harm, Foods that Heal* (1996)

Clare, Anthony W *Psychiatry in Dissent* (1980)

Conrad, Lawrence; Neve, Michael; Nuttonand, Vivian; and Porter, Roy *The Western Medical Tradition: 800 BC to AD 1800* (1995)

Dally, P and Gomez, J *Anorexia Nervosa* (1979)

Daniel, Thomas and Robbins, Frederick (eds) *Polio* (1998)

Desowitz, Robert *Tropical Diseases: from 50,000 BC to 2500 AD* (1998)

Dubos, René *Louis Pasteur: Free Lance of Science* (1986)

Duncan, R and Weston-Smith, M *The Encyclopaedia of Medical Ignorance* (1984)

Dworkins, Ronald *Life's Dominion: An Argument about Abortion and Euthanasia* (1993)

Eastwood, Martin Anthony *Principles of Human Nutrition* (1997)

Ellenberger, Henri F *The Discovery of the Unconscious* (1970)

Fabrega, Horacio *Evolution of Sickness and Healing* (1997)

Fisher, Richard *Joseph Lister, 1827–1912* (1977)

Fosbery, Richard *Human Health and Disease* (1997)

Ganellin, C R and Roberts, S M *Medicinal Chemistry: The Role of Organic Chemistry in Drug Research* (1993)

Gay, Peter *Freud: A Life for Our Time* (1988)

Gellner, Ernest *The Psychoanalytic Movement* (1985)

Gillon, Raanan *Philosophical Medical Ethics* (1986)

Goldman, Martin *Lister Ward* (1987)

Goodwin, Frederick K and Jamison, Kay Redfield *Manic-Depressive Illness* (1990)

Gordon, Richard A *Anorexia and Bulimia* (1990)

Grunbaum, Adolf *The Foundations of Psychoanalysis* (1984)

Guyton, A C *Human Physiology and Mechanism of Disease* (1992)

Harris, Seale *Banting's Miracle* (1946)

Holtzmann Kevles, Bettyann *Naked to the Bone* (1996)

Jaspers, Karl *General Psychopathology* (1946)

Kellock, Brian *The Fibre Man, the Life Story of Dr Denis Burkitt* (1985)

Kinoy, Barbara (ed) *Eating Disorders: New Directions in Treatment and Recovery* (1994)

Knight, David *Robert Koch: Father of Bacteriology* (1961)

Le Fanu, James (ed) *Preventionitis: The Exaggerated Claims of Health Promotion* (1994)

Leone, Daniel A *The Spread of AIDS* (1997)

Levine, Israel *The Discoverer of Insulin* (1959)

MacFarlane, Gwyn *Alexander Fleming* (1985)

Maxcy, Kenneth Fuller, Rosenau, Milton Joseph, Last, John M, Wallace, Robert B *Public Health and Preventative Medicine* (1992)

McKenna, P J *Schizophrenia and Related Syndromes* (1994)

Nicolle, J *Louis Pasteur: The Story of His Major Discoveries* (1961)

Phillips, Jonathan *The Biology of Disease* (1995)

Pinker, Steven *How the Mind Works* (1998)

Reid, Robert *Microbes and Men* (1975)

Rowland, John *The Penicillin Man* (1957)

Rowland, John *The Insulin Man* (1966)

Rycroft, Charles *Anxiety and Neurosis* (1968)

Ryle, J A *The Natural History of Disease* (1988)

Sournia, Jean-Charles *The Illustrated History of Medicine* (1992)

Sulloway, Frank J *Freud: Biologist of the Mind* (1979)

COMPUTER SCIENCE

A computer is a programmable electronic device that processes data and performs calculations and other symbol manipulation tasks. There are three types: the **digital computer**, which manipulates information coded as binary numbers (the digits 0 and 1, representing two different signals, on and off, with an individual digit being known as a binary digit or bit); the **analogue computer**, which works with continuously varying quantities; and the **hybrid computer**, which has characteristics of both analogue and digital computers.

There are four types of digital computer, corresponding roughly to their size and intended use. **Microcomputers** are the smallest and most common, used in small businesses, at home, and in schools. They are usually single-user machines. **Minicomputers** are found in medium-sized businesses and university departments. They may support from around ten to two hundred users at once. **Mainframes,** which can often service several hundred users simultaneously, are found in large organizations, such as national companies and government departments. **Supercomputers** are mostly used for highly complex scientific tasks, such as analysing the results of nuclear physics experiments and weather forecasting.

The first mechanical computer was conceived by Charles Babbage in 1835. He designed a general-purpose mechanical computing device for performing different calculations according to a program input on punched cards. In the 1880s Herman Hollerith devised the first device for high-volume data processing, a mechanical tabulating machine. In 1943 Thomas Flowers built Colossus, the first electronic computer. John Von Neumann's computer, EDVAC, built in 1949, was the first to use binary arithmetic and to store its operating instructions internally. His design still forms the basis of today's computers.

parts of a computer

central processing unit (CPU)

At the heart of a computer is the central processing unit, which executes individual program instructions and controls the operation of other parts. It is sometimes called the central processor or, when contained on a single integrated circuit, a microprocessor.

The CPU has three main components: the **arithmetic and logic unit** (ALU), where all calculations and logical operations are carried out; a **control unit**, which decodes, synchronizes, and executes program instruc-

tions; and the **immediate access memory**, which stores the data and programs on which the computer is currently working. All these components contain registers, which are memory locations reserved for specific purposes.

The central processing unit communicates with the component parts of the computer and/or its peripherals via a number of electrical pathways, known as buses. Physically, a bus is a set of parallel tracks that can carry digital signals; it may take the form of copper tracks laid down on the computer's printed circuit boards (PCBs), or of an external cable or connection.

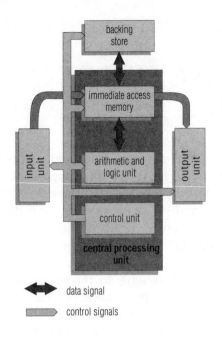

data signal

control signals

central processing unit The relationship between the three main areas of a computer's central processing unit. The arithmetic and logic unit (ALU) does the arithmetic, using the registers to store intermediate results, supervised by the control unit. Input and output circuits connect the ALU to external memory, input, and output devices.

discs

Discs are the most common medium for storing large volumes of computer data. A **magnetic disc** is rotated at high speed in a disc-drive unit in the computer as a read/write (playback or record) head passes over its surfaces to read or record the magnetic variations that encode the data on the disc's surface. The more recent **optical discs**, which have a much larger capacity, are read by a laser-scanning device.

hard disc The hard disc is a rigid metal disc coated with a magnetic material. A hard disc may be permanently fixed into the disc drive (a **fixed hard disc**) or in the form of a disc pack that can be removed and exchanged with a different pack (a **removable hard**

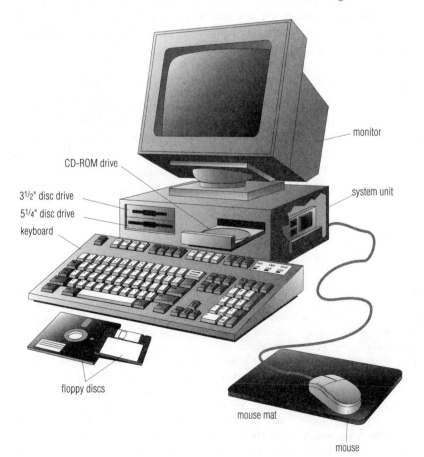

monitor

CD-ROM drive

3½" disc drive

5¼" disc drive

keyboard

system unit

floppy discs

mouse mat

mouse

microcomputer The component parts of the microcomputer: the system unit contains the hub of the system, including the central processing unit (CPU), information on all of the computer's peripheral devices, and often a fixed disc drive. The monitor (or visual display unit) displays text and graphics, the keyboard and mouse are used to input data, and the floppy disc and CD-ROM drives read data stored on discs.

read-write heads
locate data by cylinder,
sector and surface location

drive spindle

hard discs

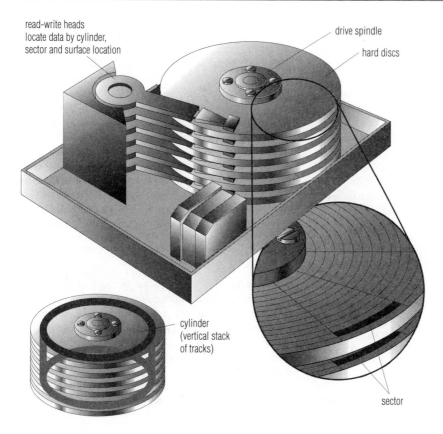

cylinder
(vertical stack
of tracks)

sector

disc A hard disc. Data is stored in sectors within cylinders and is read by a head which passes over the spinning surface of each disc.

read-write head
moves to locate
specific track

floppy disc

access cover moves
to expose
disc surface

plastic casing

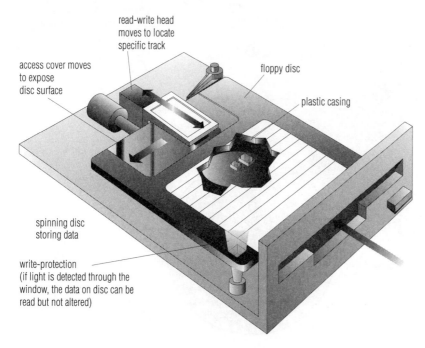

spinning disc
storing data

write-protection
(if light is detected through the
window, the data on disc can be
read but not altered)

disc drive A floppy disc drive. As the disc is inserted into the drive, its surface is exposed to the read/write head, which moves over the spinning disc surface to locate a specific track.

disc). A fixed disc cannot be removed: once it is full, data must be deleted in order to free space or a complete new disc drive must be added to the computer system in order to increase storage capacity. Removable discs can be taken out of the drive unit and kept for later use. By swapping such discs around, a single hard disc drive can be made to provide a potentially infinite storage capacity. However, access speeds and capacities tend to be lower than those associated with large fixed hard discs.

Hard discs vary from large units with capacities of more than 3,000 megabytes, intended for use with mainframe computers, to small units with capacities as low as 20 megabytes, intended for use with microcomputers.

floppy disc The floppy disc is a light, flexible disc enclosed in a plastic jacket. Floppy discs are the most common form of backing store for microcomputers. They are much smaller in size and capacity than hard discs, either 13.13 cm/5.25 in or 8.8 cm/3.5 in in diameter, and they normally hold 0.5–2 megabytes of data. Floppy discs are inexpensive, and light enough to send through the post, but have slower access speeds and are more fragile than hard discs.

optical disc Plastic-coated metal discs, such as a **CD-ROM** (compact-disc read-only memory) and **WORM**

(write once, read many times), are optical discs. Data are recorded as binary digital information etched on the disc surface in the form of microscopic pits; this is read by a laser-scanning device. Optical discs have an enormous capacity – about 550 megabytes on a compact disc, and thousands of megabytes on a full-size optical disc. CD-ROMs are used in distributing large amounts of text, graphics, audio, and video, such as encyclopedias, catalogues, technical manuals, and games.

Standard CD-ROMs cannot have information written onto them by computer, but must be manufactured from a master, although recordable CDs, called CD-R discs, have been developed for use as computer discs. A compact disc, CD-RW, that can be overwritten repeatedly by a computer has also been developed.

memory

The memory of a computer is the part of the system used to store data and programs either permanently or temporarily. There are two main types: **immediate access memory** and **backing storage** (see central processing unit illustration). Memory capacity is measured in bytes (sufficient memory to store a single character of data) or, more conveniently, in kilobytes (units of 1,024 bytes) or megabytes (units of 1,024 kilobytes).

Immediate access memory, or **internal memory**,

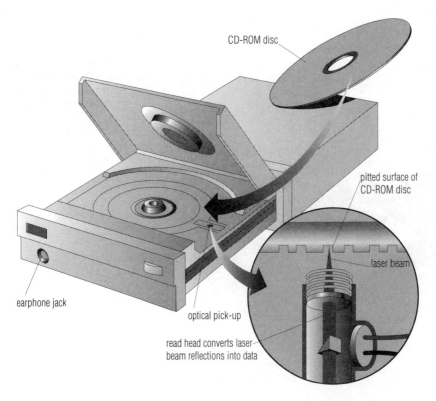

CD-ROM drive Data is obtained by the CD-ROM drive by converting the reflections from a disc's surface into digital form.

CD-ROM disc

pitted surface of CD-ROM disc

laser beam

earphone jack

optical pick-up

read head converts laser beam reflections into data

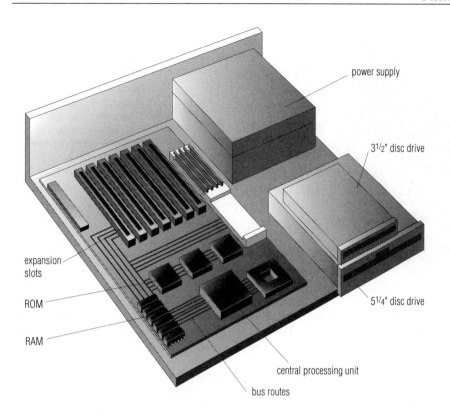

describes the memory locations that can be addressed directly and individually by the central processing unit. It is either read-only (stored in ROM, PROM, and EPROM chips) or read/write (stored in RAM chips). Read-only memory stores information that must be constantly available and is unlikely to be changed. It is nonvolatile – that is, it is not lost when the computer is switched off. Read/write memory is volatile – it stores programs and data only while the computer is switched on.

Backing storage, or **external memory**, is nonvolatile memory, located outside the central processing unit, used to store programs and data that are not in current use. Backing storage is provided by such devices as magnetic discs (floppy and hard discs), magnetic tape (tape streamers and cassettes), optical discs (such as CD-ROM), and bubble memory. By rapidly switching blocks of information between the backing storage and the immediate-access memory, the limited size of the immediate-access memory may be increased artificially. When this technique is used to give the appearance of a larger internal memory than physically exists, the additional capacity is referred to as virtual memory.

motherboard

The motherboard is a large **printed circuit board** (PCB,

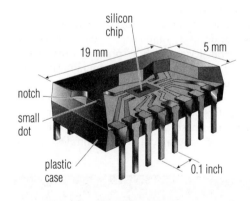

integrated circuit An integrated circuit (IC), or silicon chip. The IC is a piece of silicon, about the size of a child's fingernail, on which the components of an electrical circuit are etched. The IC is packed in a plastic container with metal legs that connect it to the printed circuit board.

an internal electrical circuit laid out on insulating board) that contains the main components of a microcomputer. The power, memory capacity, and capability of the microcomputer may be enhanced by adding expansion

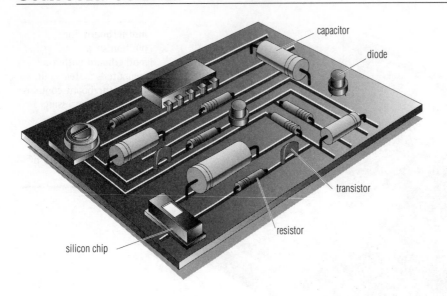

capacitor

diode

transistor

resistor

silicon chip

printed circuit board A typical microcomputer printed circuit board (PCB). The PCB contains sockets for the integrated circuits, or chips, and the connecting tracks.

boards to the motherboard. A computer may contain several printed circuit boards for various purposes; expansion boards and adaptors are examples of PCBs. Each PCB, in turn, may accommodate one or more **integrated circuits** (ICs), or **silicon chips**. These are miniaturized electronic circuits produced on a single crystal, or chip, of a semiconducting material, usually silicon.

operating system (OS)

The operating system is a program that controls the basic operation of a computer. A typical OS controls the peripheral devices such as printers, organizes the filing system, provides a means of communicating with the operator, and runs other programs.

Many operating systems were written to run on specific computers, but some are available from third-party software houses and will run on machines from a variety of manufacturers. Examples include

Microware's OS/9, Microsoft's MS-DOS, and UNIX. UNIX is the standard on workstations, minicomputers, and supercomputers; it is also used on desktop PCs and mainframes.

A **DOS** (disc operating system) is a computer operating system specifically designed for use with disc storage; also used as an alternative name for Microsoft's operating system, MS-DOS.

peripheral devices

A peripheral device is any item connected to a computer's central processing unit (CPU). Typical peripherals include the keyboard, mouse, visual display unit (VDU), and printer. Users who enjoy playing games might add a joystick or a trackball; others might connect a modem, scanner, or Integrated Services Digital Network (ISDN) terminal to their machines.

Major Operating Systems				
Operating system	**Developer**	**Interface**	**Use**	**Features**
DOS	Microsoft and IBM 1981	command line	IBM-compatible PCs	most widely used system
OS/2	IBM and Microsoft late 1980s	GUI	IBM-compatible computers	multitasking
Mac Operating System	Apple 1984	pioneered GUI	Macintosh	allows plug and play peripherals
UNIX	AT&T late 1960s	complex commands	from mainframes to PCs	multiuser, multitasking
Windows NT	Microsoft	GUI	predominantly PCs	networking and multitasking
Windows 95/98	Microsoft 1995/98	GUI	dominates the PC market	multitasking and multimedia

trackball
(alternative to mouse)

printer

computer projector

microphone
and speaker
(if computer has
soundcard)

scanner

modem

mouse

peripheral device Some
of the types of peripheral
device that may be
connected to a computer
include printers, scanners,
and modems.

input device
An input device is used for entering information into
a computer. Input devices include keyboards, joysticks,
mice, light pens, touch-sensitive screens, scanners,
graphics tablets, speech-recognition devices, and vision
systems.

keyboard
The keyboard is the most frequently
used input device, resembling a typewriter keyboard.
Keyboards are used to enter instructions and data
via keys. There are many variations on the layout
and labelling of keys. Extra numeric keys may
be added, as may special-purpose function keys,

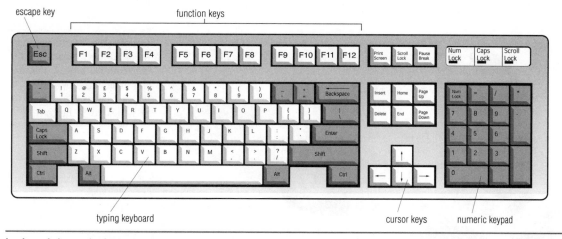

escape key   function keys

typing keyboard   cursor keys   numeric keypad

keyboard A standard 102-key keyboard. As well as providing a QWERTY typing keyboard, the function keys (labelled
F1–F12) may be assigned tasks specific to a particular system.

whose effects can be defined by programs in the computer.

graphics tablet The graphics tablet is an input device, also known as a **bit pad**, in which a stylus or cursor is moved by hand over a flat surface. The computer can keep track of the position of the stylus, enabling the operator to input drawings or diagrams into the computer. A graphics tablet is often used with a form overlaid for users to mark boxes in positions that relate to specific registers in the computer, although recent developments in handwriting recognition may increase its future versatility.

joystick The joystick is a hand-held lever that signals to a computer the direction and extent of displacement required by the user. It is similar to the joystick used to control the flight of an aircraft. Joysticks are sometimes used to control the movement of a cursor (marker) across a display screen, but are much more frequently used to provide fast and direct input for moving the characters and symbols that feature in computer games. Unlike a mouse, which can move a pointer in any direction, simple games joysticks are often capable only of moving an object in one of eight different directions.

light pen Light pens, resembling ordinary pens, are used to indicate locations on a computer screen. With certain computer-aided design (CAD) programs, a light pen can be used to instruct the computer to change the shape, size, position, and colours of sections of a screen image.

mouse Perhaps the most frequently used input device is the mouse, used to control a pointer on a computer screen. It is a feature of graphical user interface (GUI) systems. The mouse is about the size of a pack of play-

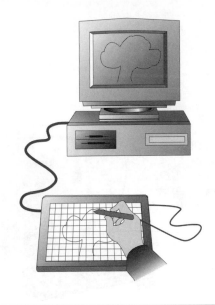

graphics tablet A graphics tablet enables images drawn freehand to be translated directly to the computer screen.

ing cards. It is connected to the computer by a wire, and incorporates one or more buttons that can be pressed. Moving the mouse across a flat surface causes a corresponding movement of the pointer. In this way, the operator can manipulate objects on the screen and make menu selections.

Mice work either mechanically (with electrical contacts to sense the movement in two planes of a ball on a level surface), or optically (photocells detecting movement by recording light reflected from a grid on which the mouse is moved).

joystick

'fire' buttons

joy pad

'fire' buttons

'direction' buttons

joystick The directional and other controls on a conventional joystick may be translated to a joy pad, which enables all controls to be activated by buttons.

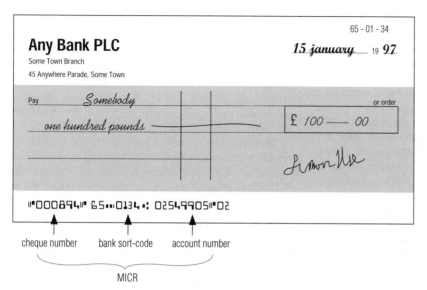

Any Bank PLC
Some Town Branch
45 Anywhere Parade, Some Town

65 - 01 - 34

15 january 19 **97**

Pay   *Somebody*   or order

one hundred pounds   £ *100 — 00*

⑈000894⑈ 65⑈0134⑈: 025499051⑈02

cheque number   bank sort-code   account number

MICR

magnetic-ink character recognition An example of one of the uses of magnetic ink in automatic character recognition. Because of the difficulties in forging magnetic-ink characters, and the speed with which they can be read by computer systems, MICR is used extensively in banking.

scanner The scanner is a peripheral device that produces a digital image of a document for input and storage in a computer, using technology similar to that of a photocopier. Small scanners can be passed over the document surface by hand; larger versions have a flat bed, like that of a photocopier, on which the input document is placed and scanned. Scanners are widely used to input graphics for use in desktop publishing.

Various forms of scanners are used to read and capture large volumes of data very rapidly. They include **document readers** for magnetic-ink character recognition (MICR), optical character recognition (OCR), and optical mark recognition (OMR).

modem A modem (**modulator/demodulator**) is a device for transmitting computer data over telephone

external modem

external modem for a notebook computer

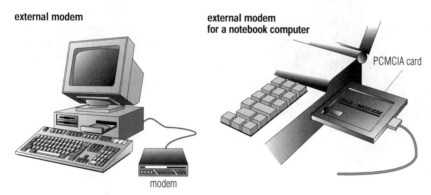

PCMCIA card

modem

modem Modems are available in various forms: microcomputers may use an external device connected through a communications port, or an internal device, which takes the form of an expansion board inside the computer. Notebook computers use an external modem connected via a special interface card.

internal modem

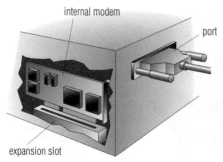

internal modem

port

expansion slot

lines. Such a device is necessary because the digital signals produced by computers cannot, at present, be transmitted directly over the telephone network, which uses analogue signals. The modem converts the digital signals to analogue, and back again.

Modems are used for linking remote terminals to central computers and enable computers to communicate with each other anywhere in the world. In 1997 the fastest standard modems transmitted data at a nominal rate of about 33,600 bps (bits per second), often abbreviated to 33.6 K. 56 K modems launched in 1997 achieve higher speeds by using a digital connection to the user's computer, while using a conventional analogue connection in the other direction. In theory the downstream link can transfer data at 56 Kbps but in practice, speeds are usually 45–50 K or less, depending on the quality of the phone line and other factors.

output device

An output device is used for displaying, in a form intelligible to the user, the results of processing carried out by a computer. The most common output devices are the VDU (visual display unit, or screen) and the printer. Other output devices include graph plotters, speech synthesizers, and COM (computer output on microfilm/microfiche).

port The socket that enables a computer processor to communicate with an external device is called the port. It may be an **input port** (such as a joystick port), or an **output port** (such as a printer port), or both (an **i/o port**).

Microcomputers may provide ports for cartridges, televisions and/or monitors, printers, and modems, and sometimes for hard discs and musical instruments (MIDI, the musical instrument digital interface). Ports may be serial or parallel.

printer The printer is an output device for producing printed copies of text or graphics. Types include the **daisywheel printer**, which produces good-quality text but no graphics; the **dot matrix printer**, which produces text and graphics by printing a pattern of small dots; the **ink jet printer**, which creates text and graphics by spraying a fine jet of quick-drying ink onto the paper; and the **laser printer**, which uses electrostatic technology very similar to that used by a photocopier to produce high-quality text and graphics.

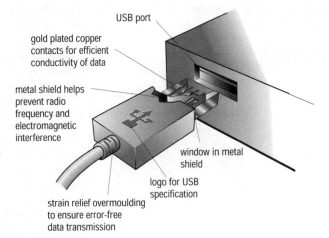

USB port

gold plated copper contacts for efficient conductivity of data

metal shield helps prevent radio frequency and electromagnetic interference

window in metal shield

logo for USB specification

strain relief overmoulding to ensure error-free data transmission

port The two types of communications port in a microcomputer. The parallel port enables up to eight bits of data to travel through it at any one time; the serial port enables only one.

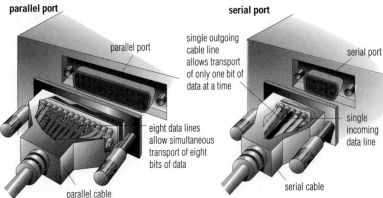

parallel port

parallel port

eight data lines allow simultaneous transport of eight bits of data

parallel cable

serial port

single outgoing cable line allows transport of only one bit of data at a time

serial port

single incoming data line

serial cable

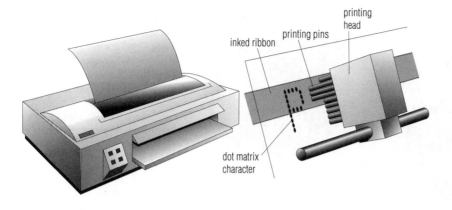

dot matrix printer Characters and graphics printed by a dot matrix printer are produced by a block of pins which strike the ribbon and make up a pattern using many small dots.

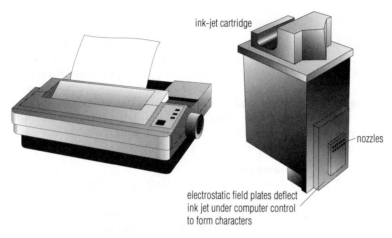

ink-jet printer High-quality print images are produced by ink-jet printers by squirting ink through a series of nozzles.

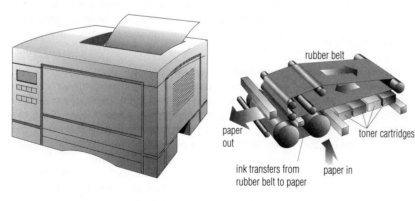

laser printer A laser printer works by transferring tiny ink particles contained in a toner cartridge to paper via a rubber belt. The image is produced by laser on a light-sensitive drum within the printer.

Printers may be classified as **impact printers** (such as daisywheel and dot matrix printers), which form characters by striking an inked ribbon against the paper, and **non-impact printers** (such as ink jet and laser printers), which use a variety of techniques to produce characters without physical impact on the paper.

A further classification is based on the basic unit of printing, and categorizes printers as character printers, line printers, or page printers, according to whether they print one character, one line, or a complete page at a time.

visual display unit (VDU) A VDU is a computer terminal consisting of a keyboard for input data and a screen, or monitor, for displaying output. The oldest and most popular type of VDU screen is the cathode ray tube (CRT), which uses essentially the same tech-

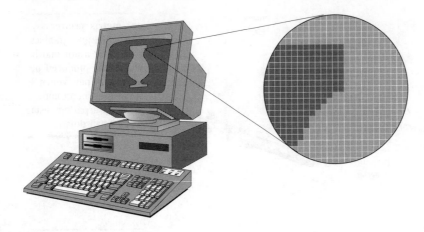

pixel Computer screen images are made of a number of pixels ('dots'). The greater the number of pixels the greater the resolution of the image; most computer screens are set at 640 x 480 pixels, although higher resolutions are available.

nology as a television screen. Other types use plasma display technology and liquid crystal displays.

Screen resolution (the quality of the display) is dependent on the number of **pixels** (picture elements, or single dots on a computer screen) used; the smaller the pixel, the higher the resolution. Within the limits of the resolution supported by the screen itself, the display quality can be altered by the user to suit particular applications. In the same way, the number of **screen colours** supported by a VDU can be selected by the user.

types of computer

microcomputer or personal computer
A microcomputer is a small desktop or portable computer, typically designed to be used by one person at a time, although individual computers can be linked in a network so that users can share data and programs. Its central processing unit is a **microprocessor**, contained on a single integrated circuit, or chip. The first microprocessor appeared in 1971, designed for a pocket calculator manufacturer; it led to a dramatic fall in the size and cost of computers and heralded the introduction of the microcomputer.

Microcomputers are the smallest of the four classes of computer. Since the appearance in 1975 of the first commercially available microcomputer, the Altair 8800, micros have become ubiquitous in commerce, industry, and education.

laptops, notebooks, or portables
Laptops are portable microcomputers that are small enough to be used on the operator's lap. The laptop consists of a single unit, incorporating a keyboard, floppy disc and hard disc drives, and a screen. The screen often forms a lid that folds back in use. It uses a liquid crystal or gas plasma display, rather than the bulkier and heavier

cathode ray tubes found in most display terminals. A typical laptop computer measures about 210 x 297 mm/8.3 x 11.7 in (A4), is 5 cm/2 in in depth, and weighs less than 3 kg/6 lb 9 oz.

In the 1990s the notebook format became the standard for portable PCs and Apple PowerBooks. Some manufacturers also offer smaller portable PCs called subnotebooks or palmtops.

minicomputer
A minicomputer is a multiuser computer with a size and processing power between those of a mainframe and a microcomputer. Nowadays almost all minicomputers are based on microprocessors. Minicomputers are often used in medium-sized businesses and in university departments handling database or other commercial programs and running scientific or graphical applications.

mainframe computer
A mainframe computer is a large computer used for commercial data processing and other large-scale operations. Because of the general increase in computing power, the differences between the mainframe, supercomputer, minicomputer, and microcomputer (personal computer) are becoming less marked.

Mainframe manufacturers include IBM, Amdahl, Fujitsu, and Hitachi. Typical mainframes have from 128 MB to 4 GB of memory and hundreds of gigabytes of disc storage.

supercomputer
The supercomputer is the fastest, most powerful type of computer, capable of performing its basic operations in picoseconds (thousand-billionths of a second), rather than nanoseconds (billionths of a second), like most other computers.

To achieve these extraordinary speeds, supercomputers use several processors working together and

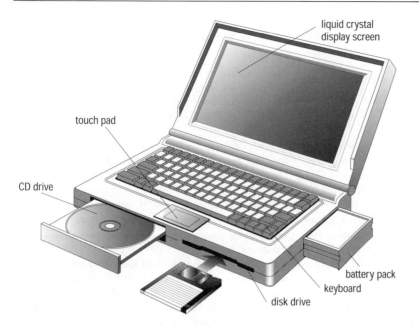

notebook computer The component parts of a notebook computer. Although as powerful as a microcomputer, the battery pack enables the notebook to be used while travelling.

techniques such as cooling processors down to nearly absolute zero temperature, so that their components conduct electricity many times faster than normal. Supercomputers are used in weather forecasting, fluid dynamics, and aerodynamics. Manufacturers include Cray Research, Fujitsu, and NEC.

Of the world's 500 most powerful supercomputers 232 are in the USA, 109 in Japan, and 140 in Europe, with 23 in the UK.

computer programming

Computer programmers write instructions in a **programming language** (see below) that is translated into **machine code** (a set of instructions that a computer's central processing unit can understand and obey directly) by means of a **compiler** or **interpreter program** to enable the user to operate the computer and perform specific tasks.

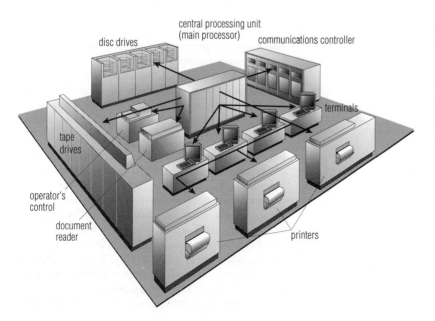

computer A mainframe computer. Functionally, it has the same component parts as a microcomputer, but on a much larger scale. The central processing unit is at the hub, and controls all the attached devices.

computer coding systems

All modern digital computers are directly operated by **machine codes** based on binary numbers (using combinations of the digits 1 and 0), which represent instructions and data. The values of the binary digits, or 'bits', are stored or transmitted as, for example, open/closed switches or high/low voltages in circuits.

ASCII (American standard code for information interchange)

Almost all mini- and microcomputers use ASCII, a coding system in which numbers are assigned to letters, digits, and punctuation symbols. Although computers work in code based on the binary number system, ASCII numbers are usually quoted as decimal or hexadecimal numbers. For example, the decimal number 45 (binary 0101101) represents a hyphen, and 65 (binary 1000001) a capital A. The first 32 codes are used for control functions, such as carriage return and backspace.

Strictly speaking, ASCII is a 7-bit binary code, allowing 128 different characters to be represented, but an eighth bit is often used to provide parity (the state of being even or odd, see below) or to allow for extra characters. The system is widely used for the storage of text and for the transmission of data between computers.

parity

Parity is the term referring to the number of 1s in the binary codes used to represent data. A binary representation has **even parity** if it contains an even number of 1s and **odd parity** if it contains an odd number of 1s.

For example, the binary code 1000001, commonly used to represent the character 'A', has even parity because it contains two 1s, and the binary code 1000011, commonly used to represent the character 'C', has odd parity because it contains three 1s. A **parity bit** (0 or 1) is sometimes added to each binary representation to give all the same parity so that a validation check can be carried out each time data are transferred from one part of the computer to another. So, for example, the codes 1000001 and 1000011 could have parity bits added and become 01000001 and 11000011, both with even parity. If any bit in these codes should be altered in the course of processing the parity would change and the error would be quickly detected.

ASCII Codes

character	binary code
A	1000001
B	1000010
C	1000011
D	1000100
E	1000101
F	1000110
G	1000111
H	1001000
I	1001001
J	1001010
K	1001011
L	1001100
M	1001101
N	1001110
O	1001111
P	1010000
Q	1010001
R	1010010
S	1010011
T	1010100
U	1010101
V	1010110
W	1010111
X	1011000
Y	1011001
Z	1011010

Parity

character	binary code	parity	base-ten representation
A	1000001	even	65
B	1000010	even	66
C	1000011	odd	67
D	1000100	even	68

computer programming languages

Programming languages are designed to be easy for people to write and read, but must be capable of being mechanically translated (by a compiler or an interpreter) into the machine code that the computer can execute. Programming languages may be classified as high-level languages or low-level languages.

assembly language

This is a low-level computer-programming language closely related to a computer's internal codes. It consists chiefly of a set of short sequences of letters (mnemonics), which are translated, by a program called an **assembler**, into machine code for the computer's central processing unit (CPU) to follow directly. In assembly language, for example, 'JMP' means 'jump' and 'LDA' means 'load accumulator'. Assembly code is used by programmers who need to write very fast or efficient programs.

Because they are much easier to use, high-level lan-

Programming Languages

Language	Main uses	Description
Ada	defence applications	high-level
assembly languages	jobs needing detailed control of the hardware, fast execution, and small program size	fast and efficient but require considerable effort and skill
ALGOL (algorithmic language)	mathematical work	high-level with an algebraic style; no longer in current use, but has influenced languages such as Ada an PASCAL
BASIC (beginners' all-purpose symbolic instruction code)	mainly in education, business, and the home, and among non-professional programmers, such as engineers	easy to learn; early versions lacked the features of other languages
C	systems and general programming	fast and efficient; widely used as a general-purpose language; especially popular among professional programmers
C++	systems and general programming; commercial software development	developed from C, adding the advantages of object-oriented programming
COBOL (common business-oriented language)	business programming	strongly oriented towards commercial work; easy to learn but very verbose; widely used on mainframes
FORTH	control applications	reverse Polish notation language
FORTRAN (formula translation)	scientific and computational work	based on mathematical formulae; popular among engineers, scientists, and mathematicians
Java	developed for consumer electronics; used for many interactive Web sites	multipurpose, cross-platform, object-oriented language with similar features to C and C++ but simpler
LISP (list processing)	artificial intelligence	symbolic language with a reputation for being hard to learn; popular in the academic and research communities
LOGO	teaching of mathematical concepts	high-level; popular with schools and home computer users
Modula-2	systems and real-time programming; general programming highly-structured	intended to replace PASCAL for 'real-world' applications
OBERON	general programming	small, compact language incorporating many of the features of PASCAL and Modula-2
PASCAL (program appliqué à la sélection et la compilation automatique de la littérature)	general-purpose language	highly-structured; widely used for teaching programming in universities
Perl (pathological eclectic rubbish lister)	systems programming and Web development	easy manipulation of text, files, and processes, especially in UNIX environment
PROLOG (programming in logic)	artificial intelligence	symbolic-logic programming system, originally intended for theorem solving but now used more generally in artificial intelligence

guages are normally used in preference to assembly languages. An assembly language may still be used in some cases, however, particularly when no suitable high-level language exists or where a very efficient machine-code program is required.

compiler A compiler is a computer program that translates programs written in a high-level language into machine code (the form in which they can be run by the computer). The compiler translates each high-level instruction into several machine-code instructions – in a process called **compilation** – and produces a complete independent program that can be run by the computer as often as required, without the original source program being present.

Different compilers are needed for different high-level languages and for different computers. In contrast to using an interpreter, using a compiler adds

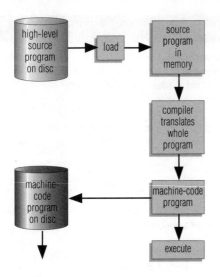

compiler The process of compilation; a program written in a high-level language is translated into a program that can be run without the original source being present.

slightly to the time needed to develop a new program because the machine-code program must be recompiled after each change or correction. Once compiled, however, the machine-code program will run much faster than an interpreted program.

fourth-generation language This is a type of programming language designed for the rapid programming of applications but often lacking the ability to control the individual parts of the computer. Such a language typically provides easy ways of designing screens and reports, and of using databases.

high-level language This is a programming language designed to suit the requirements of the programmer; it is independent of the internal machine code of any particular computer. High-level languages are used to solve problems and are often described as **problem-oriented languages** – for example, BASIC was designed to be easily learnt by first-time programmers; COBOL is used to write programs solving business problems; and FORTRAN is used for programs solving scientific and mathematical problems. In contrast, low-level languages, such as assembly languages, closely reflect the machine codes of specific computers, and are therefore described as **machine-oriented languages**.

Unlike low-level languages, high-level languages are relatively easy to learn because the instructions bear a close resemblance to everyday language, and because the programmer does not require a detailed knowledge of the internal workings of the computer. Each instruc-

tion in a high-level language is equivalent to several machine-code instructions. High-level programs are therefore more compact than equivalent low-level programs. However, each high-level instruction must be translated into machine code – by either a compiler or an interpreter program – before it can be executed by a computer. High-level languages are designed to be **portable** – programs written in a high-level language can be run on any computer that has a compiler or interpreter for that particular language.

interpreter This is a computer program that translates and executes a program written in a high-level language. Unlike a compiler, which produces a complete machine-code translation of the high-level program in one operation, an interpreter translates the source program, instruction by instruction, each time that program is run.

Because each instruction must be translated each time the source program is run, interpreted programs run far more slowly than do compiled programs. However, unlike compiled programs, they can be executed immediately without waiting for an intermediate compilation stage.

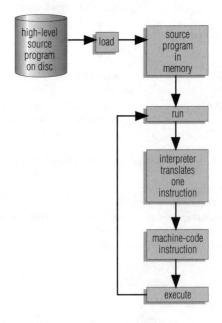

interpreter The sequence of events when running an interpreter on a high-level language program. Instructions are translated one at a time, making the process a slow one; however, interpreted programs do not need to be compiled and may be executed immediately.

low-level language This is a programming language designed for a particular computer and reflecting its internal machine code; low-level languages are therefore often described as **machine-oriented languages**. They cannot easily be converted to run on a computer with a different central processing unit, and they are relatively difficult to learn because a detailed knowledge of the internal working of the computer is required. Since they must be translated into machine code by an assembler program, low-level languages are also called **assembly languages.**

A mnemonic-based low-level language replaces binary machine-code instructions, which are very hard to remember, write down, or correct, with short codes chosen to remind the programmer of the instructions they represent. For example, the binary-code instruction that means 'store the contents of the accumulator' may be represented with the mnemonic STA.

computer programs

A computer program is a set of instructions that controls the operation of a computer. There are two main kinds: **applications programs**, which carry out tasks for the benefit of the user, for example, word processing; and **systems programs**, which control the internal workings of the computer. A **utility program** is a systems program that carries out specific tasks for the user, for example converting the format of a data file so that it can be accessed by a different applications program. Programs can be written in any of a number of programming languages but are always translated into machine code before they can be executed by the computer.

communications programs or comms programs

These are general-purpose programs for accessing older online systems and bulletin board systems which use a command-line interface; also known as terminal emulators.

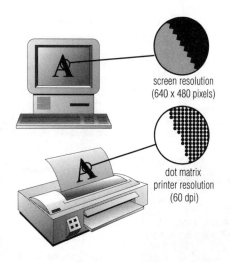

resolution An example of typical resolutions of screens and printers. The resolution of a screen image when printed can only be as high as the resolution supported by the printer itself.

screen resolution
(640 x 480 pixels)

dot matrix
printer resolution
(60 dpi)

Most operating systems include a trimmed-down comms program, but full-featured programs include facilities to store phone numbers and settings for frequently called services, address books, and the ability to write scripts to automate logging on. Popular comms programs include ProComm, Smartcom, Qmodem, and Odyssey.

computer graphics programs These are application programs that enable a computer to display and manipulate information in pictorial form. Input may be achieved by scanning an image, by drawing with a mouse or stylus on a graphics tablet, or by drawing directly on the screen with a light pen.

The output may be as simple as a pie chart, or as

bit map The difference in close-up between a bit-mapped and vector font. As separate sets of bit maps are required for each different type size, scaleable vector graphics (outline) is the preferred medium for fonts.

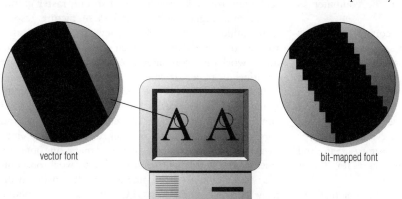

vector font

bit-mapped font

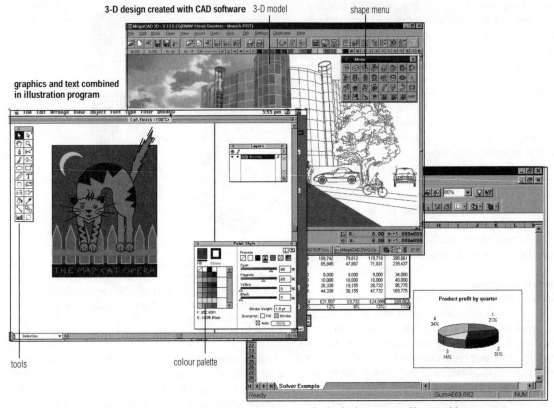

3-D design created with CAD software   3-D model

shape menu

graphics and text combined
in illustration program

tools

colour palette

simple pie chart generared by spreadsheet program

---

computer graphics Some examples of the kinds of graphic design that can be achieved using computers. Text and graphics may be combined within an illustration package, and sophisticated three-dimensional drawings can be created using a computer-aided design (CAD) system.

---

complex as an animated sequence in a science-fiction film, or a seemingly three-dimensional engineering blueprint. The drawing is stored in the computer as raster graphics or vector graphics.

Vector graphics are stored in the computer memory by using geometric formulae. They can be transformed (enlarged, rotated, stretched, and so on) without loss of picture resolution. It is also possible to select and transform any of the components of a vector-graphics display because each is separately defined in the computer memory. In these respects vector graphics are superior to raster graphics. They are typically used for drawing applications, allowing the user to create and modify technical diagrams such as designs for houses or cars.

Raster graphics are stored in the computer memory by using a map to record data (such as colour and

intensity) for every pixel (picture element, or individual dot) that makes up the image. When transformed (enlarged, rotated, stretched, and so on), raster graphics become ragged and suffer loss of picture resolution, unlike vector graphics. They are typically used for painting applications, which allow the user to create artwork on a computer screen much as if they were painting on paper or canvas.

Computer graphics are increasingly used in computer-aided design (CAD), and to generate models and simulations in engineering, meteorology, medicine and surgery, and other fields of science. Recent developments in software mean that designers on opposite sides of the world will soon be able to work on complex three-dimensional computer models using ordinary personal computers (PCs) linked by telephone lines rather then powerful graphics workstations.

Major Graphics and Design Programs

Software	Manufacturer	Description
Adobe Illustrator	Adobe	professional Draw program with PostScript graphics
Adobe Photoshop	Adobe	professional Paint program offers image creation and manipulation
AutoCAD	AutoDesk	industry standard and leading CAD software
ClarisDraw	Claris	paint and draw with smart tools; strong on presentations
Corel Draw	Corel	paint and draw with OCR, animation and presentation
DesignCAD	PMS (Instruments)	range of programs for CAD and modelling
FreeHand	Macromedia	established Draw program with built in effects and colour support
Painter	Fractal Design	suite of creative artist's tools; provides tutorials and stunning results
Paint Shop Pro	JASC	popular, easy-to-use image-editing program; with shareware option
Simply 3D	Visual Software	paint program with full camera animation

database programs These are programs used to create databases; they enable a user to define the database structure by selecting the number of fields, naming those fields, and allocating the type and amount of data that is valid for each field. To sort records within a database, one or more **sort fields** may be selected, so that when the data is sorted, it is ordered according to the contents of these fields. A **key field** is used to give a unique identifier to a particular record. Data programs also determine how data can be viewed on screen or extracted into files.

A database-management system (DBMS) program ensures that the integrity of the data is maintained by controlling the degree of access of the applications programs using the data.

Major Database Programs

Software	Manufacturer	Description
Access	Microsoft	features wizards and macros; included in Microsoft Office Professional
Approach	Lotus	easy-to-use; requires no programming; with multiple database formats
dBASE	Borland	powerful and flexible relational database
Filemaker Pro	Claris	versatile and easy-to-use application with relational features
FoxPro	Microsoft	relational database with programming language

database table containing customer details

data format options

database containing customer details

database An example of the type of information that may be stored on a database. The information may be stored in various formats, enabling it to be sorted and output to other software programs.

Major Desktop Publishing Software Programs

Software	Manufacturer	Description
FrameMaker	Frame Technologies	strong on technical reports and book production
PageMaker	Adobe	powerful professional tool; strong layout and colour capabilities
PagePlus	Serif	good value; includes vector drawing program and bitmap editor
Publisher	Microsoft	low-level for beginners; provides wizards and clip art gallery
QuarkXpress	Quark	industry standard; numerous enhancement modules available

---

Ultimate Electronic Publishing Resource
http://desktoppublishing.com/
Includes reviews of a number of DTP packages, as well as a
large library of clip art and a number of links to sites on
related topics, including Java, fonts, and graphics programs.

---

desktop publishing (DTP) programs These are
application programs that enable small-scale typesetting and page make-up to be performed on a microcomputer. DTP packages use a graphical interface to
import text and graphics from other packages; run text
as columns, over pages, and around artwork and other
insertions; enable a wide range of fonts; and allow
accurate positioning of all elements required to make
a page.

DTP systems are capable of producing camera-ready
pages (pages ready for photographing and printing),
made up of text and graphics, with text set in different typefaces and sizes. The page can be previewed on
the screen before final printing on a laser printer.

spreadsheets These are application programs that
mimic a sheet of ruled paper, divided into columns
down the page, and rows across. The user enters values into cells within the sheet, then instructs the program to perform some operation on them, such as
totalling a column or finding the average of a series
of numbers. Highly complex numerical analyses may
be built up from these simple steps.

Columns and rows in a spreadsheet are labelled;
although different programs use different methods,
columns are often labelled with alpha characters, and
rows with numbers. This way, each cell has its own
reference, unique within that spreadsheet. For example, A5 would be the cell reference for the fifth row
in the first column. Cells can also be grouped using
references; the range H9:H30 groups together all the
cells in column H between (and including) rows 9 and
30. Single references or cell ranges may be used when
inputting formulae into cells.

When a cell containing a formula is copied and
pasted within a spreadsheet, the formula is said to be
relative, meaning the cell references it takes its values

from are relative to its new position. An **absolute** reference does not change.

The pages of a spreadsheet can be formatted to
make them easier to read; the height of rows, the width
of columns, and the typeface of the text may all be
changed. Number formats may also be changed to display, for example, fractions as decimals or numbers as
integers.

Spreadsheets are widely used in business for forecasting and financial control. The first spreadsheet program, Software Arts' VisiCalc, appeared in 1979. The
best known include Lotus 1–2–3 and Microsoft Excel.

word processors These are programs that allow the
input, amendment, manipulation, storage, and retrieval
of text; also computer systems that run such software.
Since word-processing programs became available to
microcomputers, the method has largely replaced the
typewriter for producing letters or other text. Typical
facilities include insert, delete, cut and paste, reformat,
search and replace, copy, print, mail merge, and
spelling check.

The leading word-processing programs include
Microsoft Word, the market leader, Lotus WordPro,
and Corel WordPerfect.

errors
A fault or mistake, either in the software or on the
part of the user, can cause a program to stop running
(crash) or produce unexpected results. Program errors,
or bugs, are largely eliminated in the course of the programmer's initial testing procedure, but some will
remain in most programs. All computer operating systems are designed to produce an **error message** (on the
display screen, or in an error file or printout) whenever an error is detected, reporting that an error has
taken place and, wherever possible, diagnosing its
cause.

Errors can be categorized into several types: **syntax
errors** are caused by the incorrect use of the programming language, and include spelling and keying
mistakes. These errors are detected when the compiler
or interpreter fails to translate the program into
machine code (instructions that a computer can understand directly); **logical errors** are faults in the program

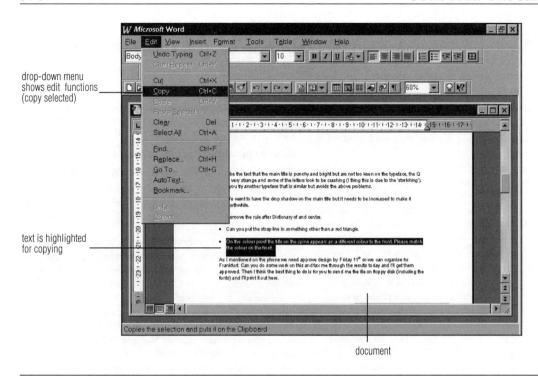

drop-down menu shows edit functions (copy selected)

text is highlighted for copying

document

word processing A word processing software package enables text to manipulated in a variety of ways, such as copying and pasting and changing its size and typeface.

design – for example, in the order of instructions. They may cause a program to respond incorrectly to the user's requests or to crash completely; **execution errors**, or **run-time errors**, are caused by combinations of data that the programmer did not anticipate. A typical execution error is caused by attempting to divide a number by zero. This is impossible, and so the program stops running at this point. Execution errors occur only when a program is running, and cannot be detected by a compiler or interpreter.

Computers are designed to deal with a set range of numbers to a given range of accuracy. Many errors are caused by these limitations: **overflow error** occurs when a number is too large for the computer to deal with; an **underflow error** occurs when a number is too small; **rounding** and **truncation errors** are caused by the need to round off decimal numbers, or to cut them off (truncate them) after the maximum number of decimal places allowed by the computer's level of accuracy.

flow chart

A flow chart is a diagram often used in computing to show the possible paths that data can take through a system or program.

A **system flow chart**, or **data flow chart**, is used to

describe the flow of data through a complete data-processing system. Different graphic symbols represent the clerical operations involved and the different input, storage, and output equipment required. Although the flow chart may indicate the specific programs used, no details are given of how the programs process the data.

A **program flow chart** is used to describe the flow of data through a particular computer program, showing the exact sequence of operations performed by that program in order to process the data. Different graphic symbols are used to represent data input and output, decisions, branches, and subroutines.

viruses

A computer virus is a piece of software that can replicate and transfer itself from one computer to another, without the user being aware of it. Some viruses are relatively harmless, but others can damage or destroy data.

Viruses are written by anonymous programmers, often maliciously, and are spread on floppy discs, CD-ROMs, and via networks. Most viruses hide in the boot sectors of floppy or hard discs or infect program files, but more recently there has been a huge growth in macro viruses that infect Microsoft Word or Excel.

Millennium Bug: Preparing Computers for the Year 2000

BY SCOTT KIRSNER

digital disarray

It sounds like a bad riddle: how will two missing digits create a $600 billion industry when the calendar flips from 1999 to 2000?

Unfortunately, it's not a riddle; it's reality. As the 20th century draws to a close, an expensive computer problem dubbed 'the millennium bug' has emerged. In brief, most computers handle dates using a two-digit shorthand: 98 instead of the four-digit 1998. When presented with a date like 00, they become hopelessly confused. Since they're missing the two digits that indicate what millennium and century the date is in (19 or 20), computers tend to assume that 00 is actually 1900. So they'll either begin making errors of calculation or they'll stop working altogether. Reprogramming them to be capable of comprehending dates in the new millennium is expected to cost as much as $600 billion dollars worldwide.

The origins of the problem are simple. First, the programmers who wrote software in the 1960s for the first generation of commercial mainframe computers were shortsighted. They didn't imagine that the programs they were creating – or the machines they were creating them for – would still be in service in the far-off year of 2000. So they conserved the computers' memory by using a two-digit shorthand for the year. Every byte of memory was precious in those days, and lopping off 19 from dates was an obvious way to save a few bytes here and there.

Banks were among the first institutions to notice the downside to that approach. When they began writing long-term mortgages and approving loans that lasted past 1999, they were forced to confront the millennium bug. But the problem received little widespread attention until the mid-1990s. Technology consultants began writing articles and giving speeches about 'the year 2000 problem'. Business executives began to take notice, and even consumers couldn't ignore the problem when credit card issuers and driver's licence organizations began to renew cards and licences for shorter periods of time, because their systems couldn't handle an expiration date past 1999.

How might businesses and consumers be affected by the millennium bug? Computers, both new and old, in every industry are vulnerable. They might shut down as a result of being asked to process the date 00. Some say that is the best-case scenario, because at least businesses will know something is wrong. Worse would be if computers continued to operate, making numerous date-related errors that would be difficult to identify and fix.

The areas of greatest concern are defence, health care, transportation, telecommunications, financial services, and national and local governments. Technology experts warn of the hazards of air travel if the Federal Aviation Administration's computers can't manage data properly, the danger of hospital stays if the computers that monitor patients go awry, and the possibility of social unrest if the federal government can't provide services in 2000.

date expansion or windowing?

Fixing the millennium bug is a labour-intensive endeavour. An organization can opt to replace its systems entirely with new ones that can function in the 21st century, or it may pursue one of two basic repair strategies – 'date expansion' or 'windowing'.

Date expansion involves changing the two-digit dates to four. That entails converting all of the data an organization has stored from one format to the other, and reprogramming systems to handle four-digit dates like 2001.

Windowing is considered a simpler, less expensive solution, but it's only a temporary patch. Rather than converting all of a company's data, the windowing approach merely adds logic to a program to help it determine whether a two-digit date belongs in the 20th century or the 21st. Programmers might create a 'window' of time – from 00 to 30, for example – and then instruct the computer to assume that those dates should all be preceded by 20, whereas dates between 31 and 99 should be preceded by 19. But when 2031 rolls around, that hypothetical company would have a new problem on its hands. Windowing assumes that an organization will either replace its older systems before the window of time closes, or reprogram them yet again.

Eradicating the millennium bug is a multistage process. An organization must first assess which of its systems will be unable to handle dates in the 21st century. Then, it must convert those systems, either through expansion or windowing. Finally, it has to test the systems to ensure that they will work after the clock ticks past midnight on 31 December 1999.

ripple effects

But even if companies successfully repair their own systems, they're still vulnerable to what has been dubbed 'the ripple effect'. One of their suppliers or customers, or a government regulator, could send them unconverted data and contaminate their systems. Or even worse, a key supplier might be unable to provide services or raw materials as a result of the bug, hamstringing its customers. For those reasons, organizations must make sure that everyone else with whom they do business is solving their own year 2000 problems. Certain sectors of the economy, like the financial services arena, are even coordinating massive, interorganizational tests to make sure that stock exchanges, banks, regulators, and clearing houses will be able to work together in the new millennium.

And waiting in the wings are the lawyers. If software or hardware fails, they'll be scrutinizing contracts to see who is liable.

If a conversion project turns out to have been defective, they may bring litigation against the service provider that was contracted to perform the fix. And if a company's stock takes a dive as a result of year 2000-related failures, lawyers may file negligence lawsuits against the Board of Directors. Once litigation and damages are figured into the cost of the millennium bug, some analysts believe the total worldwide cost could skyrocket to as much as $3.6 trillion.

The sudden emergence of the year 2000 problem has created an entire mini-economy. Programmers and technology managers are finding that they can demand and receive higher salaries, computer consultants have more work than they can handle, and software companies have begun to market tools aimed at making assessment, conversion, and testing more efficient. There are dozens of Web sites and books devoted to the problem. The American Stock Exchange has even created an options index that enables investors to speculate on the fortunes of 18 companies selling software or services intended to solve the year 2000 problem.

Few participants in this mini-economy are willing to speculate about the extent to which the world will be affected by the millennium bug. Will 1 January 2000 arrive without a hitch, or will, as some technology experts predict, the front pages of every major newspaper be filled with stories about date-related computer crises? All that's certain is that programmers and their technology managers won't be among the celebrants on New Year's Eve 1999; they'll be huddled over their mainframes, fingers crossed.

Antivirus software can be used to detect and destroy well-known viruses, but new viruses continually appear and these may bypass existing antivirus programs.

Computer viruses may be programmed to operate on a particular date, such as Dark Avenger's Michelangelo Virus, which was triggered on 6 March 1992 (the anniversary of the birth of Italian artist Michelangelo) and erased hard discs. An estimated 5,000–10,000 PCs were affected.

the computer—human interface

user interface

The term 'user interface' describes the procedures and methods through which the user operates a computer program. These might include menus, input forms, error messages, and keyboard procedures. A graphical user interface (GUI or WIMP) is one that makes use of icons (small pictures) and allows the user to make menu selections with a mouse.

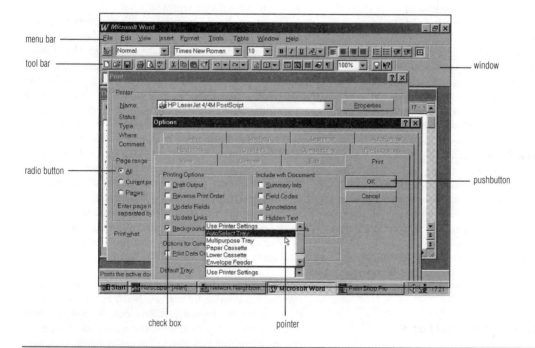

graphical user interface A typical graphical user interface (GUI), where the user moves around the system by clicking on representative buttons or icons using the mouse.

icon · wallpaper

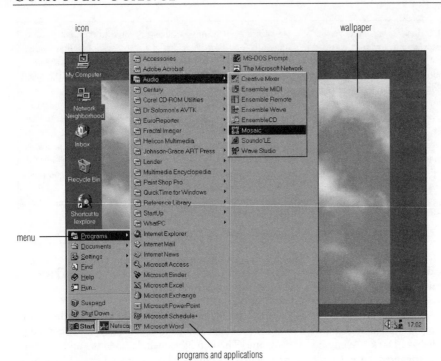

desktop A typical graphical desktop, showing the menu system, icons, programs, and applications available to the user.

menu

programs and applications

A **command line interface** is a character-based interface in which a prompt is displayed on the screen at which the user types a command, followed by carriage return, at which point the command, if valid, is executed. An example of a command line interface is the **DOS prompt**.

A **menu-driven interface** presents various options to the user in the form of a list, from which commands may be selected. Types of menu include the **menu bar**, which displays the top level options available to the user as a single line across the top of the screen; selecting one of these options displays a **pull-down menu**. Programs such as Microsoft Word use menus in this way.

In a **graphical user interface** programs and files appear as icons, user options are selected from pull-down menus, and data are displayed in windows (rectangular areas), which the operator can manipulate in various ways. The operator uses a pointing device, typically a mouse, to make selections and initiate actions. The graphical user interface is now available on many types of computer – most notably as Windows, an operating system for IBM PC-compatible microcomputers developed by the software company Microsoft.

computer applications

artificial intelligence (AI)

Artificial intelligence is concerned with creating computer programs that can perform actions comparable with those of an intelligent human. Current AI research covers such areas as planning (for robot behaviour), language understanding, pattern recognition, and knowledge representation.

The possibility of artificial intelligence was first proposed by the English mathematician Alan Turing in 1950. Early AI programs, developed in the 1960s, attempted simulations of human intelligence or were aimed at general problem-solving techniques. By the mid-1990s, scientists were concluding that AI was more difficult to create than they had imagined. It is now thought that intelligent behaviour depends as much on the knowledge a system possesses as on its reasoning power. Present emphasis is on knowledge-based systems, such as expert systems, while research projects focus on neural networks, which attempt to mimic the structure of the human brain.

On the Internet, small bits of software that automate common routines or attempt to predict human likes or behaviour based on past experience are called intelligent agents or bots.

fuzzy logic Fuzzy logic is a form of knowledge representation suitable for notions (such as 'hot' or 'loud') that cannot be defined precisely but depend on their context. Fuzzy logic enables computerized devices to reason more like humans, receiving complex messages from their control panels and sensors and responding effectively to these messages.

Fuzzy logic has been largely ignored in Europe and the USA, but was taken up by Japanese manufacturers in the mid-1980s and has since been applied to hundreds of electronic goods and industrial machines. For example, a vacuum cleaner launched in 1992 by Matsushita uses fuzzy logic to adjust its sucking power in response to messages from its sensors about the type of dirt on the floor, its distribution, and its depth.

neural network This is an artificial network of processors that attempts to mimic the structure of nerve cells (neurons) in the human brain. Neural networks may be electronic, optical, or simulated by computer software.

A basic network has three layers of processors: an input layer, an output layer, and a 'hidden' layer in between. Each processor is connected to every other in the network by a system of 'synapses'; every processor in the top layer connects to every one in the hidden layer, and each of these connects to every processor in the output layer. This means that each nerve cell in the middle and bottom layers receives input from several different sources; only when the amount of input exceeds a critical level does the cell fire an output signal.

The chief characteristic of neural networks is their ability to sum up large amounts of imprecise data and decide whether they match a pattern or not. Networks of this type may be used in developing robot vision, matching fingerprints, and analysing fluctuations in stock-market prices. However, it is thought unlikely by scientists that such networks will ever be able to accurately imitate the human brain, which is very much more complicated; it contains around ten billion nerve cells, whereas current artificial networks contain only a few hundred processors.

robotics This term is applied to any computer-controlled machine that can be programmed to move or carry out work. Robots are often used in industry to transport materials or to perform repetitive tasks. For instance, robotic arms, fixed to a floor or workbench, may be used to paint machine parts or assemble electronic circuits. Other robots are designed to work in situations that would be dangerous to humans - for example, in defusing bombs or in space and deep-sea exploration. Some robots are equipped with sensors, such as touch sensors and video cameras, and can be programmed to make simple decisions based on the sensory data received.

computer-aided design and manufacturing

CAD (computer-aided design) The use of computers in creating and editing design drawings. CAD also allows such things as automatic testing of designs and multiple or animated three-dimensional views of designs. CAD systems are widely used in architecture, electronics, and engineering, for example in the motor-vehicle industry, where cars designed with the assistance of computers are now commonplace. With a CAD system, picture components are accurately positioned using grid lines. Pictures can be resized, rotated, or mirrored without loss of quality or proportion.

CAM (computer-aided manufacturing) The use of computers to control production processes; in particular, the control of machine tools and robots in factories. In some factories, the whole design and production system has been automated by linking CAD (computer-aided design) to CAM.

Linking flexible CAD/CAM manufacturing to computer-based sales and distribution methods makes it possible to produce semicustomized goods cheaply and in large numbers.

computer games or video games

Computers can be used for leisure and entertainment as well as more serious purposes, and computer games represent an important and fast-growing area of use. There are a wide variety of computer-controlled games in which the computer (sometimes) opposes the human player. Computer games typically employ fast, animated graphics on a VDU (visual display unit) and synthesized sound.

Doomgate – Where it all Begins
http://doomgate.cs.buffalo.edu/doomgate/
Clearing house of information for the computer game *Doom*. Frequently updated, it contains advice, frequently asked questions, technical details about graphics and specifications, add-on utilities, and links to the large number of *Doom* newsgroups.

Commercial computer games became possible with the advent of the microprocessor in the mid-1970s and rapidly became popular as amusement arcade games, using dedicated chips. Available games range from chess to fighter-plane simulations.

Some of the most popular computer games in the early 1990s were id Software's *Wolfenstein 3D* and *Doom*, which were designed to be played across networks including the Internet. A whole subculture built up around those particular games, as users took advantage of id's help to create their own additions to the game.

Kasparov v. Deep Blue – The Rematch
http://www.chess.ibm.com/
Official site of the team that produced the first computer able to beat a world chess champion. This is a complete account of the tussle between Deep Blue and Gary Kasparov. There are some thought provoking articles on the consequences of Deep Blue's victory. There is also some video footage of the games.

---

computer simulation

In this type of application, a real-life situation is represented in a computer program. For example, the program might simulate the flow of customers arriving at a bank. The user can alter variables, such as the number of cashiers on duty, and see the effect.

More complex simulations can model the behaviour of chemical reactions or even nuclear explosions. The behaviour of solids and liquids at high temperatures can be simulated using quantum simulation. Computers also control the actions of machines – for example, a flight simulator models the behaviour of real aircraft and allows training to take place in safety. Computer simulations are very useful when it is too dangerous, time consuming, or simply impossible to carry out a real experiment or test.

virtual reality Virtual reality is an advanced form of computer simulation, in which a participant has the illusion of being part of an artificial environment. The participant views the environment through two tiny television screens (one for each eye) built into a visor. Sensors detect movements of the participant's head or body, causing the apparent viewing position to change. Gloves (datagloves) fitted with sensors may be worn, which allow the participant seemingly to pick up and move objects in the environment.

What Is Virtual Reality?
http://www.cms.dmu.ac.uk/~cph/VR/whatisvr.html
Text-based introduction to VR and an information resource list. The site covers all major aspects of the subject, and also provides a great many literature and Internet references for further reading.

The technology is still under development but is expected to have widespread applications, for example in military and surgical training, architecture, and home entertainment.

databases

Databases are structured collections of data, which may be manipulated to select and sort desired items of information. For example, an accounting system might be built around a database containing details of customers and suppliers. A telephone directory stored as a database might allow all those people whose names start with the letter B to be selected by one program, and all those living in Chicago by another. Databases are normally used by large organizations as an effective and fast way of handling large amounts of information in various ways via mainframes or minicomputers.

There are three main types (or 'models') of database: hierarchical, network, and relational, of which relational is the most widely used. In a **relational database** data are viewed as a collection of linked tables. A **free-text database** is one that holds the unstructured text of articles or books in a form that permits rapid searching. A collection of databases is known as a **databank**.

communications and the Internet

e-mail

E-mails are messages sent electronically from computer to computer via network connections such as Ethernet or the Internet, or via telephone lines to a host system. Messages once sent are stored on the network or by the host system until the recipient picks them up. As well as text, messages may contain enclosed text files, artwork, or multimedia clips.

Subscribers to an electronic mail system type messages in ordinary letter form on a word processor, or microcomputer, and 'drop' the letters into a central computer's memory bank by means of a computer/telephone connector (a modem). The recipient 'collects' the letter by calling up the central computer and feeding a unique password into the system. Due to the high speed of delivery electronic mail is cheaper than an equivalent telephone call or fax.

Integrated Services Digital Network (ISDN)

The ISDN is an internationally developed telecommunications system for sending signals in digital format. It involves converting the 'local loop' – the link between the user's telephone (or private automatic branch exchange) and the digital telephone exchange – from an analogue system into a digital system, thereby greatly increasing the amount of information that can be carried. The first large-scale use of ISDN began in Japan in 1988.

ISDN has advantages in higher voice quality, better quality faxes, and the possibility of data transfer between computers faster than current modems. With ISDN's **Basic Rate Access** a multiplexer divides one voice telephone line into three channels: two B bands and a D band. Each B band offers 64 kilobits per second and can carry one voice conversation or 50 simultaneous data calls at 1,200 bits per second. The D band is a data-signalling channel operating at 16 kilobits per second. With **Primary Rate Access** ISDN provides 30 B channels.

the Internet

The Internet is a global computer network connecting governments, companies, universities, and many other networks and users. Electronic mail, conferencing, and

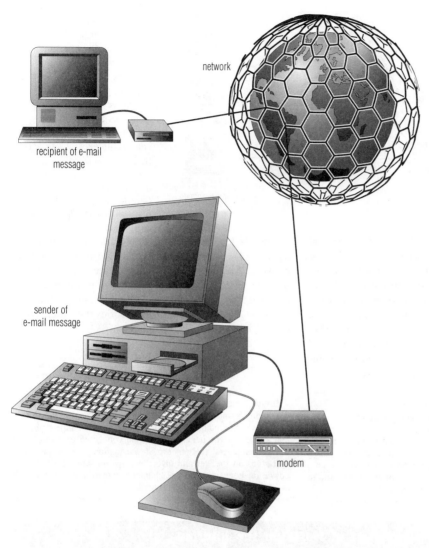

network

recipient of e-mail
message

sender of
e-mail message

modem

e-mail system The basic
structure of an electronic mail
system. A message is sent via
a telephone line and stored in
a central computer. The
message remains there until
the recipient calls up the
central computer and collects
the message.

chat services are all supported across the network, as is the ability to access remote computers and send and retrieve files. In 1997 around 55 million adults had access to the Internet in the USA alone.

The technical underpinnings of the Internet were developed as a project funded by the Advanced Research Project Agency (ARPA) to research how to build a network that would withstand bomb damage. The Internet itself began in 1984 with funding from the US National Science Foundation as a means to allow

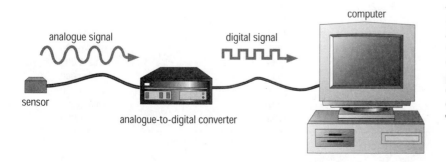

analogue signal

digital signal

computer

sensor

analogue-to-digital converter

analogue-to-digital converter
An analogue-to-digital
converter, or ADC, converts a
continuous analogue signal
produced by a sensor to a
digital ('off and on') signal for
computer processing.

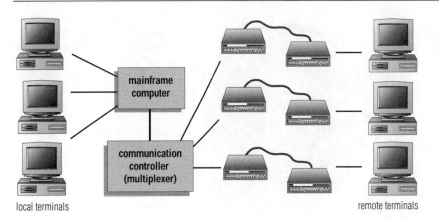

local terminals   remote terminals

remote terminal Remote computer terminals communicate with the central mainframe via modems and telephone lines. The controller allocates computer time to the terminals according to predetermined priority rules. The multiplexor allows more than one terminal to use the same communications link at the same time (multiplexing).

US universities to share the resources of five regional supercomputing centres. The number of users grew quickly, and in the early 1990s access became cheap enough for domestic users to have their own links on home personal computers. As the amount of information available via the Internet grew, indexing and search services such as Gopher, Archie, Veronica, and WAIS were created by Internet users to help both themselves and others. The newer World Wide Web allows seamless browsing across the Internet via hypertext.

The US vice president Al Gore announced plans, in April 1998, for Internet2 that will run on a second network Abilene, operated by private contractors, to provide a high-speed data communications backbone to serve the main US research universities, enabling them to bypass the congestion on the Internet. It should be operational by 1999.

Internet Starter Kit
http://206.246.131.227/resources/geninternet/
iskm/iskw2/index.html
If you are struggling online then there are worse places to start than this – the full text of a recent book that covers everything from advice on getting connected to recommendations about the best shareware. It also includes some information on how to start creating Web pages yourself.

Parents' Guide to the Internet
http://www.ed.gov/pubs/parents/internet/
Useful electronic booklet aiming to bridge the gap between children's and parents' knowledge of the Internet. The site introduces the main features of the 'information superhighway' and argues for the benefits of getting connected to the Internet at home. Several sections provide navigation to assist parents' first steps on the Net and give them tips on safe travelling, and advice on how to encourage children's activities at home and at school.

Categories of Hosts on the Internet

Number of Hosts on the Internet by Category

Source: Key Note (As of July 1996.)

Category	Hosts
Commercial	3,323,647
Educational	2,114,851
Networks	1,232,902
Government	361,065
Organizations	327,148
International organizations	1,930
Other	5,519,156
Total	12,880,699
% of which UK	4.5

Number of Hosts on the Internet by Year

Source: Key Note

Year	Number of hosts
1981	213
1985	1,961
1990	313,000
1991	535,000
1992	992,000
1993	1,776,000
1994	3,212,000
1995	6,642,000
1996	12,881,000
1997	19,540,000
1998	36,739,000

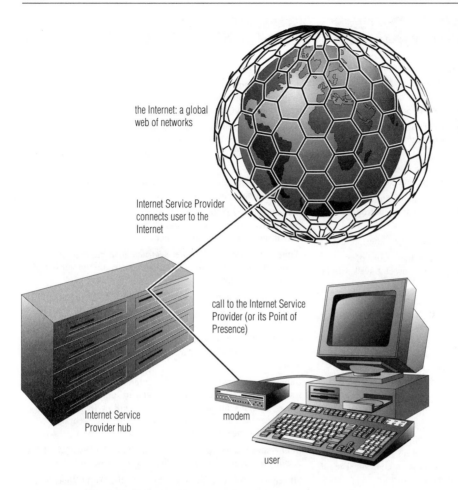

the Internet: a global web of networks

Internet Service Provider connects user to the Internet

Internet Service Provider hub

call to the Internet Service Provider (or its Point of Presence)

modem

user

Internet The Internet is accessed by users via a modem to the service provider's hub, which handles all connection requests. Once connected, the user can access a whole range of information from many different sources, including the World Wide Web.

Internet Service Provider (ISP) Dial-up access to the Internet is sold by an Internet Service Provider. Several types of company provide Internet access, including online information services such as CompuServe and America Online (AOL), electronic conferencing systems such as the WELL and Compulink Information eXchange, and local bulletin board systems (BBSs). Most recently founded ISPs, such as Demon Internet and PIPEX, offer only direct access to the Internet without the burden of running services of their own just for their members.

Such companies typically work out cheaper for their users, as they charge a low flat rate for unlimited usage. By contrast, commercial online services typically charge by the hour or minute.

World Wide Web (WWW)
hypertext system for publishing information on the Internet. World Wide Web documents ('**Web pages**') are text files coded using HTML (hypertext mark-up language) to include text and graphics, and are stored on a Web server connected to the Internet. Web pages may also contain dynamic objects and Java applets for enhanced animation, video, sound, and interactivity. The Web server can be any computer, from the simplest Apple Macintosh to the largest mainframe, if Web server software is available.

Every Web page has a URL (Uniform Resource Locator) – a unique address (usually starting with http://www) which tells a browser program (such as Netscape Navigator or Microsoft Internet Explorer) where to find it. An important feature of the World Wide Web is that most documents contain links enabling readers to follow whatever aspects of a subject interest them most. These links may connect to different computers all over the world. Interlinked or

---

Worldwide Web Workbook
http://sln.fi.edu/primer/setup.html
Guide for novice Web surfers (limited to users of PCs and MS Windows). Topics include hypertext, graphics, hypergraphics, imagemaps, and thumbnails. Once the basics have been covered, users are offered a short tour with the help of Spot, the mascot webdog.

Some Internet Terms

acceptable use set of rules enforced by a service provider or backbone network restricting the use to which their facilities may be put

access provider another term for Internet Service Provider

ack radio-derived term for 'acknowledge', used on the Internet as a brief way of indicating agreement with or receipt of a message or instruction

alt hierarchy 'alternative' set of newsgroups on USENET, set up so that anyone can start a newsgroup on any topic

anonymous remailer service that allows Internet users to post to USENET and send e-mail without revealing their true identity or e-mail address

Archie software tool for locating information on the Internet

bang path list of routing that appears in the header of a message sent across the Internet, showing how it travelled from the sender to its destination

Big Seven hierarchies original seven hierarchies of newsgroups on USENET. They are: **comp.** – computing; **misc.** – miscellaneous; **news.** – newsgroups; **rec.** – recreation; **sci.** – science; **soc.** – social issues; and **talk.** – debate

blocking software any of various software programs that work on the World Wide Web to block access to categories of information considered offensive or dangerous

blue–ribbon campaign campaign for free speech launched to protest against moves towards censorship on the Internet

'bot (short for robot) automated piece of software that performs specific tasks on the Internet. 'Bots are commonly found on multi-user dungeons (MUDs) and other multi-user role-playing game sites, where they maintain a constant level of activity even when few human users are logged on

bozo filter facility to eliminate messages from irritating users

browser any program that allows the user to search for and view data; Web browsers allow access to the World Wide Web

bulletin board centre for the electronic storage of messages; bulletin board systems are usually dedicated to specific interest groups, and may carry public and private messages, notices, and programs

cancelbot automated software program that cancels messages on USENET; Cancelbot is activated by the CancelMoose, an anonymous individual who monitors newsgroups for complaints about spamming

crawler automated indexing software that scours the Web for new or updated sites

crossposting practice of sending a message to more than one newsgroup on USENET

cybersex online sexual fantasy spun by two or more participants via live, online chat

cyberspace the imaginary, interactive 'worlds' created by networked computers; often used interchangeably with 'virtual world'

cypherpunk passionate believer in the importance of free access to strong encryption on the Internet, in the interests of guarding privacy and free speech

digital city area in cyberspace, either text-based or graphical, that uses the model of a city to make it easy for visitors and residents to find specific types of information

e–zine (contraction of **electronic magazine**) periodical sent by e-mail. E-zines can be produced very cheaply as there are no production costs for design and layout, and minimal costs for distribution

FAQ (abbreviation for **frequently asked questions**) file of answers to commonly asked questions on any topic

firewall security system built to block access to a particular computer or network while still allowing some types of data to flow in and out onto the Internet

flame angry public or private electronic mail message used to express disapproval of breaches of netiquette or the voicing of an unpopular opinion

follow–up post publicly posted reply to a USENET message; unlike a personal e-mail reply, follow-up post can be read by anyone

FurryMUCK popular MUD site where the players take on the imaginary shapes and characters of furry, anthropomorphic animals

Gopher menu-based server on the Internet that indexes resources and retrieves them according to user choice via any one of several built-in methods such as FTP or Telnet. Gopher servers can also be accessed via the World Wide Web and searched via special servers called Veronicas

Gopherspace name for the knowledge base composed of all the documents indexed on all the Gophers in the world

hit request sent to a file server. Sites on the World Wide Web often measure their popularity in numbers of hits

home page opening page on a particular site on the World Wide Web

hop intermediate stage of the journey taken by a message travelling from one site to another on the Internet

HTTP (abbreviation for **Hypertext Transfer Protocol**) protocol used for communications between client (the Web browser) and server on the World Wide Web

hypermedia system that uses links to lead users to related graphics, audio, animation, or video files in the same way that hypertext systems link related pieces of text

in–line graphics images included in Web pages that are displayed automatically by Web browsers without any action required by the user

Internet Relay Chat (IRC) service that allows users connected to the Internet to chat with each other over many channels

Internet Service Provider (ISP) any company that sells dial-up access to the Internet

Jughead (acronym for **Jonzy's Universal Gopher Hierarchy Excavation and Display**) search engine enabling users of the Internet server Gopher to find keywords in Gopherspace directories

killfile file specifying material that you do not wish to see when accessing a newsgroup. By entering names, subjects or phrases into a killfile, users can filter out tedious threads, offensive subject headings, spamming, or contributions from other subscribers

link image or item of text in a World Wide Web document that acts as a route to another Web page or file on the Internet

lurk read a USENET newsgroup without making a contribution

MBONE (contraction of **multicast backbone**) layer of the Internet designed to deliver packets of multimedia data, enabling video and audio communication

MIME (acronym for **Multipurpose Internet Mail Extensions**) standard for transferring multimedia e-mail messages and World Wide Web hypertext documents over the Internet

moderator person or group of people that screens submissions to certain newsgroups and mailing lists before passing them on for wider circulation

MUD (acronym for **multi-user dungeon**) interactive multi-player game, played via the Internet or modem connection to one of the participating computers. MUD players typically have to solve puzzles, avoid traps, fight other participants, and carry out various tasks to achieve their goals

MUSE (abbreviation for **multi-user shared environment**) type of MUD

MUSH (acronym for **multi-user shared hallucination**) a MUD (multi-user dungeon) that can be altered by the players

netiquette behaviour guidelines evolved by users of the Internet including: no messages typed in upper case (considered to be the equivalent of shouting); new users, or new members of a newsgroup, should read the frequently asked questions (FAQ) file before asking a question; and no advertising via USENET newsgroups

net police USENET readers who monitor and 'punish' postings which they find offensive or believe to be in breach of netiquette. Many newsgroups are policed by these self-appointed guardians

newbie insulting term for a new user of a USENET newsgroup

newsgroup discussion group on the Internet's USENET. Newsgroups are organized in seven broad categories: **comp.** – computers and programming; **news.** – newsgroups themselves; **rec.** – sports and hobbies; **sci.** – scientific research and ideas; **talk.** – discussion groups; **soc.** – social issues and **misc.** – everything else. In addition, there are alternative hierarchies such as the wide-ranging and anarchic **alt.** (alternative). Within these categories there is a hierarchy of subdivisions

newsreader program that gives access to USENET newsgroups, interpreting the standard commands understood by news servers in a simple, user-friendly interface

news server computer that stores USENET messages for access by users. Most Internet Service Providers (ISPs) offer a news server as part of the service

off-line browser program that downloads and copies Web pages onto a computer so that they can be viewed without being connected to the Internet

off-line reader program that downloads information from newsgroups, FTP servers, or other Internet resources, storing it locally on a hard disc so that it can be read without running up a large phone bill

Pretty Good Privacy (PGP) strong encryption program that runs on personal computers and is distributed on the Internet free of charge

proxy server server on the World Wide Web that 'stands in' for another server, storing and forwarding files on behalf of a computer which might be slower or too busy to deal with the request itself

pseudonym name adopted by someone on the Internet, especially to participate in USENET or discussions using IRC (Internet Relay Chat)

signature (or **.sig**) personal information appended to a message by the sender of an e-mail message or USENET posting in order to add a human touch

spamming advertising on the Internet by broadcasting to many or all newsgroups regardless of relevance

spider program that combs the Internet for new documents such as Web pages and FTP files. Spiders start their work by retrieving a document such as a Web page and then following all the links and references contained in it

surfing exploring the Internet. The term is rather misleading: the glitches, delays, and complexities of the system mean the experience is more like wading through mud

sysop (contraction of **system operator**) the operator of a bulletin board system (BBS)

trolling mischievously posting a deliberately erroneous or obtuse message to a newsgroup in order to tempt others to reply – usually in a way that makes them appear gullible, intemperate, or foolish

URL (abbreviation for **Uniform Resource Locator**) series of letters and/or numbers specifying the location of a document on the World Wide Web. Every URL consists of a domain name, a description of the document's location within the host computer, and the name of the document itself, separated by full stops and backslashes

USENET (acronym for **users' network**) the world's largest bulletin board system, which brings together people with common interests to exchange views and information. It consists of e-mail messages and articles organized into newsgroups

vertical spam on USENET, spam which consists of many, often repetitive, messages per day posted to the same newsgroup or small set of newsgroups. The effect is to drown out other, more useful, conversation in the newsgroup

wAreZ slang for pirated games or other applications that can be downloaded using FTP

Web authoring tool software for creating Web pages. The basic Web authoring tool is HTML, the source code that determines how a Web page is constructed and how it looks

Web browser client software that allows you to access the World Wide Web

Webmaster system administrator for a server on the World Wide Web

Web page hypertext document on the World Wide Web

webzine magazine published on the World Wide Web, instead of on paper

WWW address (URL)

icons link to required audio
and video plug-ins

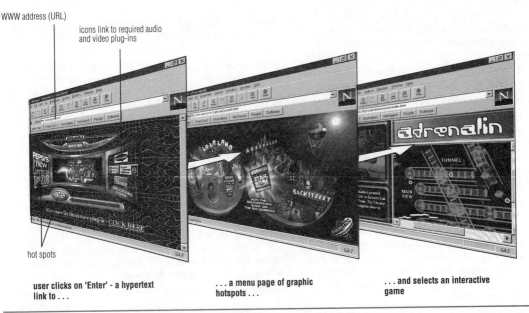

hot spots

user clicks on 'Enter' - a hypertext link to . . .   **. . . a menu page of graphic hotspots . . .**   **. . . and selects an interactive game**

web page An example of how pages on the World Wide Web may be linked to take the user to additional pages of information.

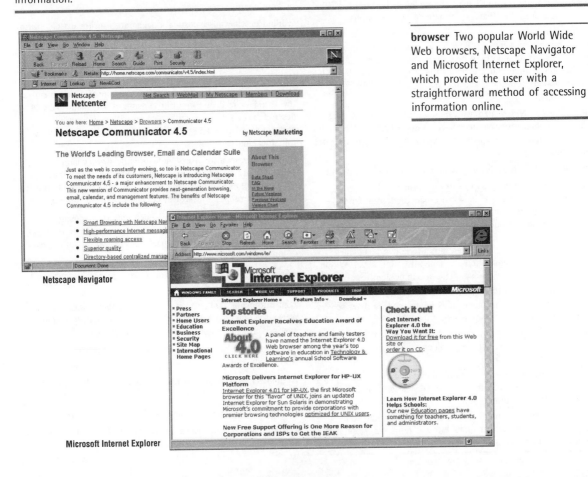

Netscape Navigator

Microsoft Internet Explorer

browser Two popular World Wide Web browsers, Netscape Navigator and Microsoft Internet Explorer, which provide the user with a straightforward method of accessing information online.

nested Web pages belonging to a single organization are known as a **Web site**.

Web browser Web browsers allow access to the World Wide Web. Netscape Navigator and Microsoft's Internet Explorer were the leading Web browsers in 1996–97. They act as a graphical interface to information available on the Internet – they read HTML documents and display them as graphical documents which may include images, video, sound, and hypertext links to other documents.

Browsers using graphical user interfaces became widely available from 1993 with the release of Mosaic, written by Marc Andreessen and Eric Bina. For some specialist applications such as viewing the virtual reality sites beginning to appear on the Web, a special virtual reality modelling language (VRML) browser is needed.

networking

Networking is the term for connecting computers so that they can share data and peripheral devices, such as printers. The main types of network are classified by the pattern of the connections – star or ring network, for example – or by the degree of geographical spread allowed; for example, **local area networks** (LANs) for communication within a room or building, and **wide area networks** (WANs) for more remote systems. The **Internet** is the computer network that connects major English-speaking institutions throughout the world, with more than 12 million users. **Janet** (Joint Academic Network), a variant of the Internet, is used in Britain. **SuperJanet**, launched in 1992, is an extension of this that can carry 1,000 million bits of information per second.

One of the most common networking systems is **Ethernet**, developed in 1973 (released in 1980) at Xerox's Palo Alto Research Center, California, by R M Metcalfe and D R Boggs.

computers and the future

fifth-generation computer

The fifth-generation computer is an anticipated new type of computer based on emerging microelectronic technologies with high computing speeds and parallel processing (see below). The development of very large-scale integration (VLSI) technology, which can put many more circuits on to an integrated circuit (chip) than is currently possible, and developments in computer hardware and software design may produce computers far more powerful than those in current use.

It has been predicted that such a computer will be able to communicate in natural spoken language with its user; store vast knowledge databases; search rapidly through these databases, making intelligent inferences and drawing logical conclusions; and process images and 'see' objects in the way that humans do.

In 1981 Japan's Ministry of International Trade and Industry launched a ten-year project to build the first fifth-generation computer, the 'parallel inference machine', consisting of over a thousand microprocessors operating in parallel with each other. By 1992, however, the project was behind schedule and had only produced 256 processor modules. It has since been suggested that research into other technologies, such as neural networks, may present more promising approaches to artificial intelligence.

parallel processing

Parallel processing is an emerging computer technology that allows more than one computation at the same time. Although in the 1980s this technology enabled only a small number of computer processor units to work in parallel, in theory thousands or millions of processors could be used at the same time.

Parallel processing, which involves breaking down computations into small parts and performing thousands of them simultaneously, rather than in a linear sequence, offers the prospect of a vast improvement in working speed for certain repetitive applications.

smart card

A smart card is a plastic card with an embedded microprocessor and memory. It can store, for example, personal data, identification, and bank account details, to enable it to be used as a credit or debit card. The card can be loaded with credits, which are then spent electronically, and reloaded as needed. Possible other uses range from hotel door 'keys' to passports.

The smart card was invented by French journalist Juan Moreno in 1974. It is expected that by the year 2000 it will be possible to make cards with as much computing power as the leading personal computers of 1990.

Computer Museum
http://www.net.org/
Well designed interactive museum, examining the history, development, and future of computer technology. As well as plenty of illustrations and detailed explanations, it is possible to change your route through the museum by indicating whether you are a kid, student, adult, or educator.

Computer Science Chronology

100	Greek mathematician and inventor Hero of Alexandria devises a method of representing numbers and performing simple calculating tasks using a train of gears – a primitive computer.
1623	German inventor William Schickhard of Tübingen builds an early adding machine.
1642	French mathematician Blaise Pascal invents the first calculating machine.
1673	German mathematician Gottfried von Leibniz presents a calculating machine to the Royal Society. It is the most advanced yet, capable of multiplication, division, and extracting roots.
1679	Leibniz introduces binary arithmetic, in which only two symbols are used to represent all numbers. It will eventually pave the way for computers.
1805	French inventor Joseph-Marie Jacquard develops a loom that uses punched cards to control the weaving of cloth.
c. 1835	English inventor Charles Babbage devises his analytical engine: 'a mechanical device designed to combine basic arithmetic operations'. Never completed because the production techniques were not yet available, it embodies most of the basic elements of modern computers: program control, memory, arithmetic processing, and automatic printout.
1843	English mathematician Ada Byron, Countess Lovelace, writes a programme for Charles Babbage's analytical engine – the first computer program.
1847	The English mathematician George Boole publishes *The Mathematical Analysis of Logic*, in which he shows that the rules of logic can be treated mathematically. Boole's work lays the foundation of computer logic.
1876	Scottish physicist William Thomson (Lord Kelvin), develops the first analogue computer. Called the 'Harmonic Analyser', he uses it to solve differential equations to predict tides.
1907	US physicist Lee De Forest invents the 'audion tube', a triode vacuum tube with a third electrode, shaped like a grid, between the cathode and anode that controls the flow of electrons and permits the amplification of sound. It is an essential element in the development of computers.
1915	US physicist Manson Benedicks discovers that a germanium crystal can convert alternating current to direct current. It leads to the development of the microchip.
1930	US electrical engineer Vannevar Bush builds the differential analyser. The first analogue computer, it is used to solve differential equations. It is the forerunner of modern computers.

1936	British mathematician Alan Turing supplies the theoretical basis for digital computers by describing a machine, now known as the Turing machine, capable of universal rather than special-purpose problem solving.
1937	US mathematician Georges Stibitz builds the first binary circuit that can add two binary numbers based on Boolean algebra. Consisting of batteries, lights, and wires, it is instrumental in the development of subsequent electromechanical computers.
1937–39	US mathematician and physicist John V Atanasoff invents an electromechanical digital computer for solving systems of linear equations. It uses punched cards and is the first electronic calculator using electronic vacuum tubes.
1938	German inventor Konrad Zuse constructs the first binary calculator using a binary code (Boolean algebra); it is the first working computer.
April 1939	US physicists Georges Stibitz and Samuel B Williams of Bell Laboratories build a computer consisting of over 400 relays connected to a teletype machine for input and output of data, thus introducing the idea of operating a computer via a terminal. Called a Complex Number Calculator, it is demonstrated on 8 January 1940.
1943	Colossus, the first electronic computer and code-breaker, is developed at Bletchley Park, England, to break German codes. Designed by Thomas Flowers, M H A Newman, and English mathematician Alan Turing, it has 1,500 vacuum tubes and is the first all-electronic calculating device.
1944	US mathematician Howard Aitken builds the Harvard University Mark I, or Automatic Sequence Controlled Calculator. The first programme-controlled computer, it is 15 m/ 50 ft long and 2.4 m/8 ft high, and its operations are controlled by a sequence of instruction codes on punched paper that operate electromechanical switches. Simple multiplication takes 4 seconds and division 11 seconds.
1946	British scientist Maurice Wilkes writes the first assembly language – a mnemonic code using alphabetic symbols that translates instructions into computer machine language.
1946	ENIAC (acronym for Electronic Numerical Integrator, Analyser, and Calculator), the first general-purpose, fully electronic digital computer, is completed at the University of Pennsylvania for use in military research. It uses 18,000 vacuum tubes instead of mechanical relays, and can make 4,500

calculations a second. It is 24 m/80 ft long and is built by electrical engineers John Presper Eckert and John Mauchly, with input from John V Atanasoff.

1947 US–Hungarian mathematician John Von Neumann introduces the idea of a stored program computer, in which both instruction codes and data are stored.

1948 A magnetic drum for storage of computer data is introduced; data are recorded on magnetic tape on a rapidly spinning drum.

1948 Manchester University in Manchester, England, demonstrates a computer with a simple memory, which permits some software development. The stored-program electronic computer, Mark I, designed by Tom Kilurn, is the first to use Von Neumann architecture and stores data in a type of cathode ray tube (Williams tube).

Aug 1949 BINAC (acronym for Binary Automatic Computer) is built by US scientists John W Mauchly and John Presper Eckert. It is the first electronic stored-program computer to store data on magnetic tape.

1949 EDSAC (acronym for Electronic Delay Storage Automatic Calculator) is constructed at Cambridge University, England; one of the first stored-program computers, it uses 3,000 vacuum tubes and is nearly six times faster than other computers; data are stored in mercury delay lines.

1949 US engineer John W Mauchly develops the Short Code, the first high-level programming language, which allows computers to recognize two-digit mathematical codes.

1950 Dr Yoshiro Nakamata of the Imperial University, Tokyo, Japan, develops the floppy disc and licenses it to International Business Machines (IBM).

1950 EDVAC (Electronic, Discrete, Variable, Automatic Computer) is constructed at Princeton University in Princeton, New Jersey. Its instructions, or programs, are stored within the computer in numerical form.

June 1951 US engineers John Mauchly and John Eckert build UNIVAC 1 (Universal Automated Computer), the first commercially available electronic digital computer, in Philadelphia, Pennsylvania. Built for the US Bureau of the Census by the Remington Rand corporation, it uses vacuum tubes, is the first to handle both numeric and alphabetic information easily, has a memory of 1.5 kilobytes, and is the first to store data on magnetic tape.

1951 US computer scientist Grace Hopper develops the first compiler. It translates programmers' codes into the binary machine codes used by computers.

1953 IBM introduces its first computer, the IBM 701, which competes with Remington Rand's UNIVAC. It has a memory of 4 kilobytes.

Feb 1955 IBM introduces the IBM 705 computer, the first commercially successful business computer to use magnetic core memory.

1955 US firm IBM develops SABRE (Semi-Automated Business Related Environment) for American Airlines passenger reservations. It consists of more than 1,000 teletypewriters connected to a central database – the first computer network.

1956 IBM introduce RAMAC (Random Access Method of Accounting and Control), the first hard disc storage of data. Indexes are used to locate the data on the disc.

1956 Univac initiates the second generation of computers when it introduces the first commercially successful computer using transistors instead of vacuum tubes.

1956 US computer programmer Jack Backus at IBM invents FORTRAN (formula translation), the first computer-programming language. It is used primarily by scientists and mathematicians.

12 Sep 1958 US electrical engineer Jack Kilby demonstrates the first integrated circuit. It consists of transistors, resistors, and capacitors contained within a silicon substrate. It leads to the third generation of computers.

1958 The US telecommunications company Bell Laboratories invents the first modem, which allows telephone lines to transmit binary data.

1959 US computer programmer Grace Hopper invents COBOL (Common Business Oriented Language), a computer language for business use.

1959 US computer programmer John McCarthy develops the List Processor (LISP) computer language that is used in artificial intelligence applications.

1959 US engineer Jean Hoerni of Fairchild Semiconductor Corporation designs the planar or 'flat' transistor and US engineer Robert Noyce discovers a way to join the circuits by printing, eliminating hundreds of hours in their production. Their work leads to the creation of the first microchip, which stimulates the computer industry with its sharply reduced size and cost and leads to the third generation of computers.

Nov 1960 US computer scientist Kenneth Olsen, at Digital Equipment Corporation, introduces the PDP-1 computer. It has a memory of 26 megabytes and is the first to use a monitor and keyboard. It is the forerunner of the minicomputer.

1962 IBM builds the 7030 computer for the Los Alamos Laboratories, New Mexico. It contains 169,100 transistors and is 30 times faster than IBM's 704 mainframe computer.

1962 Magnetic discs begin to replace magnetic tape as the main means of storing computer data.

1964 IBM introduces the system 2250 computer, the first CAD (computer-aided design) computer.

1965 US computer scientists John Kemeny and Thomas Kurtz develop BASIC (Beginners All-purpose Symbolic Instruction Code), a simplified computer programming language used in schools, businesses, and microcomputers.

1965	US Digital Equipment Corporation (DEC) introduces the PDP-8 (Programmed Data Processor) computer. The first minicomputer, it has 4 kilobytes of memory, is easy to use, and costs $18,000. It stimulates the growth of computers in business and education.
1966	The seven-level American Standard Code for Information Interchange (ASCII) receives widespread acceptance as a means of transmitting the high volumes of data generated by business machines.
1967	US scientist Gene Amdahl proposes the use of parallel processors in computers to produce faster processing speeds.
1968	US computer scientist Douglas Engelbart demonstrates the first computer mouse.
1968	US firms Control Data, NCR, and Burroughs introduce the first commercial computers that use integrated circuits. The first of the 'third generation' of computers, they are faster and have a greater capacity than previous ones.
1968	US scientist Edward Feigenbaum and US geneticist Joshua Lederberg develop DENDRAL, an expert system (which duplicates human decision-making processes) for identifying chemical substances in compounds based on the results of mass-spectrographic results. Its success spurs the development of other expert systems, especially in medicine.
1969	The US Department of Defense establishes a computer network that is the basis of the Internet.
1970	The US firm IBM develops the floppy disc for storing computer data.
1970	US computer programmers Kenneth Thomson and Dennis Ritchie develop the UNIX computer operating system. It becomes the standard operating system for computer systems with multiple tasking and multiple users.
1971	Dot matrix printers are first introduced.
1971	Swiss programmer Niklaus Wirth develops the computer language PASCAL. It is designed as a teaching tool for computer programming and allows errors to be discovered quickly.
c. 1971	A technique known as large-scale integration (LSI) is developed in the USA which makes it possible to pack thousands of transistors, diodes, and resistors on a silicon chip less than 5 mm/0.2 in square; it makes possible the development of microprocessors and microcomputers.
c. 1971	The programming language C is developed by Dennis Ritchie and Kenneth Thompson at Bell Laboratories; it is the preferred language of professional programmers and is widely used for writing software packages.
1972	SMALLTALK, one of the first object-oriented computer languages, is developed by US scientist Alan Kay. It is especially adapted to graphics and uses icons.
1972	The computer language PROLOG (Programming-in-Logic) is developed by French computer scientist Alain Colmeraurer; it has applications in artificial intelligence.
1972	US company Telnet Communications Corporation establishes a worldwide computer network.
1972	US computer scientist Nolan Bushnell invents 'Pong', the first computer game.
April 1974	Intel introduces the 8-bit 8080 microprocessor; it has 5,000 transistors.
1974	The first word processors are introduced by the Xerox corporation.
1975	The first 'personal computer', the Altair 8800, is marketed in the USA; it has no keyboard or screen but uses toggle switches to input data and flashing lights for output.
1975	The US firm IBM introduces the laser printer.
1976	Daisywheel printers are introduced; they can print between 30 and 55 characters per second.
1976	IBM computers are built with chips with 16 kilobits (16,384 bits) of memory.
1976	IBM develops the ink-jet printer.
1977	Apple Computers launches the Apple II personal computer; owners must use their own television screens and store data on audiocassette tapes. It is the first mass-produced personal computer in assembled form.
1977	The US firms Commodore Business Machines and Tandy Corporation introduce computers with built-in monitors although data and programmes are still stored on tape cassettes; other computers use separate television screens.
1978	Apple Computers introduces personal computers with disc drives.
1978	Intel introduce the 16-bit 8086 microprocessor starting the x86 line of microprocessors. It has 29,000 transistors and runs at 10 MHz.
1978	The US company DEC introduces the VAX (virtual address extension) computer; able to run very large programmes, it becomes an industry standard for scientific and technical applications.
1978	US computer programmer John Barnaby develops the word-processing program 'Wordstar'; it becomes the most popular word processor in the early 1980s.
1979	Motorola introduces the 8 MHz 68000 microprocessor; the first with a 32-bit register, it becomes the basis of the Macintosh computer.
1979	The Dutch company Philips and the Japanese company Sony work collaboratively to develop the compact disc (CD); tiny pits on the plastic are read by laser to reproduce sound or other information. CDs are first marketed in 1982.
1979	The first spreadsheet program for personal computers, VisiCalc, leads to the expansion in business use of PCs.

1980	The database software package dBase II is developed by US computer scientist Wayne Ratliffe; later versions of it become the principal filing system for personal computers.	**1985**	A chip that operates on fuzzy logic is developed at AT&T Bell Labs by Masaki Togai and Hiroyuki Watanabe.
12 Aug 1981	IBM launches its personal computer, using the Microsoft disc-operating system (MS-DOS).	**1985**	US computer chip manufacturer Intel launches the 32-bit 20MHz 80386 microprocessor; it has 275,000 transistors.
1981	The Japanese introduce computer chips with 64 kilobytes of memory.	**1985**	US firm Cray Research introduce the Cray 2, a supercomputer with 4 processors and a 2 billion-byte memory that can perform 1 billion floating point operations per second.
1981	US firm 3M develops the erasable optical disc, enabling discs to be reused.		
1982	US company Intel introduces the 16-bit 80286 microprocessor; it has 130,000 transistors and runs at speeds up to 12 MHz.	**1985**	US firm Microsoft develops Windows for the IBM PC.
1982	US firms Columbia Data Products and Compaq produce the first 'clones' of an IBM personal computer; they use the same operating system as the IBM personal computer.	**1986**	The first laptop computer is introduced in the USA.
Feb 1983	IBM introduces the PC-XT personal computer, the first to have a built-in hard disc drive. The hard disc can store 10 megabytes of information even when the machine is turned off. It is supplied with DOS 2 which allows an unlimited number of files and subdirectories to be created.	**2 Nov 1988**	Serious damage is done to more than 6,000 computer systems worldwide when the 'Internet worm' computer virus, developed by Cornell University graduate student Robert Morris, is implanted in the Internet computer network.
		1988	At Fujitsu laboratories in Japan, T Kotani and coworkers develop a microprocessor that incorporates a Josephson junction; it works hundreds of times faster than conventional computer chips.
29 Mar1983	The Tandy Corporation markets the first laptop computer in the USA. The TRS-80 Model 100 weighs less than 2 kg/4 lb and runs on 4 small batteries; prices range from $799 to $999.	**1988**	US computer scientists John Gustafson, Gary Montry, and Robert Benner develop a method of parallel processing that speeds up the processing of complex problems by a factor of 1,000; 100 times was thought to be the limit of this method.
1983	British computer company Inmos develops a computer capable of parallel processing: several operations such as memory, logic, and control are processed simultaneously, considerably speeding up overall processing speed.		
		1988	US firm Motorola introduces a RISC (Reduced Instruction Set Computing) microprocessor; it processes fewer instructions and can operate much faster – processing up to 17 million instructions per second – than other processors.
1983	Japan launches the 'fifth generation' computer project, aimed at producing a machine capable of a billion computations per second.		
1983	US computer manufacturer Apple introduces the 'Lisa', the first computer to use a mouse and pull-down menus.	**1988**	US researcher Dana Anderson invents a holographic computer, capable of generating three-dimensional images.
20 Dec 1984	The development of a 1 MB random access memory (RAM) chip is announced by Bell Laboratories; it is capable of storing four times as much data as any currently available.	**1989**	US computer microchip manufacturer Intel launches the 25MHz 80486 microprocessor, it has 1.2 million transistors and includes a maths coprocessor and cache memory.
1984	Computers are used to generate the 25 minutes of space-battle scenes in the film *The Last Starfighter*; it is the first film to make extensive use of computers.	**1989**	US computing innovator Jaron Lanier makes the experience of virtual reality possible with his design of a headset and special gloves, which will allow a user to experience and manipulate a computer-generated world.
1984	Japanese firm NEC produces computer chips with 256 kilobits of memory; similar ones are manufactured in the USA the following year.		
		c. 1989	Developments in desktop publishing make high-quality print production more generally accessible.
1984	The Dutch company Philips and Japanese firm Sony introduce the CD-ROM, a laser-read, read-only disc.	**29 Jan 1990**	US scientist Alan Huang and his colleagues at Bell Laboratories demonstrate the first all-optical processor; calculations are performed optically using lasers, lenses, and fast light switches.
1984	US computer manufacturer Apple launches the Macintosh personal computer in the USA; it is the first successful graphic-based microcomputer using icons and a mouse.		
		1991	British firm Virtuality launches its first commercial virtual reality products: games machines in arcades where players wear head-mounted displays.
c. 1984	Computer 'viruses' such as 'Friday 13th', 'Trojan Horse', 'Holland Girl', and 'Christmas Tree' begin to appear.	**1991**	British Telecom begins offering Integrated Services Digital Network (ISDN) to businesses in

	the UK. Introduced in Japan in 1988, it provides the fast transfer of computerized information. New services include computer conferencing, teleshopping, home banking, and services where both voice and computer communications take place simultaneously.	**1993**	Personal computers based on the first 64-bit processor, the Intel Pentium chip, go on sale in the USA.
1991	Japanese electronics companies Sega and Nintendo compete for the lucrative console games market. Sega's 'Sonic the Hedgehog' is matched against Nintendo's 'Super Mario Brothers'.	**1994**	The World Wide Web, a computer network that allows users to utilize graphical interfaces through Web 'browsers', makes the Internet much more accessible to general users and permits a freedom of information distribution not previously possible.
1991	Several US companies introduce local area networks (LANs), which use nondirectional microwaves to transmit data as fast as fibre optic cables.	**Nov 1995**	The Pentium Pro is launched by Intel. It is a 64-bit microprocessor containing 5.5 million transistors, compared with Pentium's 3.1 million, and can execute 166 million instructions per second.
1991	The British Science Museum constructs Charles Babbage's second difference engine, demonstrating that it would have worked had the materials then been available. It evaluates polynomials up to the seventh power, with 30-figure accuracy.	**1995**	US firm Sun Microsystems develops the computer-programming language 'Java', which is used to construct World Wide Web sites.
1991	US computer manufacturer Apple introduces 'System 7', an intuitive, easy-to-use interface, with icons, windows, and a mouse.	**10 Feb 1996**	IBM's Deep Blue computer beats Russian grand master Gary Kasparov at a chess match, the first computer to defeat a grand master. However, this is the first of a six-match series that Kasparov goes on to win 4–2.
1991	US firm Cray Research introduces the Cray Y-MP 90 computer, which is capable of 16 billion calculations a second.		
Nov 1992	The US national on-line information service Delphi becomes the first national US service to open a gateway to the Internet.	**3 June 1997**	US computer scientists announce the construction of logic gates from DNA (deoxyribonucleic acid) which simulate the functions of an **OR** gate and an **AND** gate. Rather than responding to an electronic signal, the DNA gates respond to nucleotide sequences.
1992	The Japanese firm Fujitsu announces the launch of the first computer capable of performing 300 billion calculations a second.		
1993	Fujitsu Corporation announces the development of a 256-megabit memory chip.	**1997**	An attempt in the USA to bring in legislation to control the Internet, intended to prevent access to sexual material, is rejected as unconstitutional.
1993	Mosaic, the first graphical browser that allows pictures from the Internet to be seen, is developed at the National Center for Supercomputing Applications at the University of Illinois, USA.	**14 April 1998**	US vice president Al Gore announces plans for Internet2, a high-speed data communications network which will serve the main US research universities, and bypass the congestion on the Internet. It should be operational by 1999.

Biographies

Aiken, Howard Hathaway (1900–1973) US mathematician and computer pioneer. In 1939, in conjunction with engineers from IBM, he started work on the design of an automatic calculator using standard business-machine components. In 1944 the team completed one of the first computers, the Automatic Sequence Controlled Calculator (known as the Harvard Mark I), a programmable computer controlled by punched paper tape and using punched cards.

The Harvard Mark I was principally a mechanical device, although it had a few electronic features; it was 15 m/49 ft long and 2.5 m/8 ft high, and weighed more than 30 tonnes. Addition took 0.3 seconds, multiplication 4 seconds. It was able to manipulate numbers of up to 23 decimal places and to store 72 of them. The Mark II, completed in 1947, was a fully electronic machine, requiring only 0.2 seconds for

addition and 0.7 seconds for multiplication. It could store 100 ten-digit figures and their signs.

Andreessen, Marc (1972–) US systems developer and coauthor of the first widely available graphical browser for the World Wide Web, Mosaic. He wrote Mosaic with fellow researcher Eric Bina while working at the National Center for Supercomputing Applications (NCSA), based at the University of Illinois. In 1994 both moved to the start-up company Netscape Communications Corporation to work on the next generation of browser software. This included Netscape Navigator, which was made freely available on the Internet and contributed to the explosive growth of the World Wide Web in the mid-1990s.

Babbage, Charles (1792–1871) English mathematician who devised a precursor of the computer. He designed an analytical engine, a general-purpose mechanical computing device for performing different calculations according to a program input on punched cards (an idea borrowed from the Jacquard loom). This device was never built, but it embodied many of the principles on which digital computers are based.

In 1822 he began early work on a mechanical calculator, or difference engine, which could compute squares to six places of decimals, and was commissioned by the British Admiralty to work on an expanded version of the engine. But this project was abandoned in favour of the analytical engine, which he worked on for the rest of his life. The difference engine could perform only one function once it was set up. The analytical engine was intended to perform many functions; it was to store numbers and be capable of working to a program.

Babbage, Charles
http://ei.cs.vt.edu/~history/Babbage.html
Extended biography of the visionary mathematician, industrialist, and misanthrope. This very full and entertaining account of his life is supported by a number of pictures of Babbage, his plan of the difference engine, and the actual difference engine built by the British Science Museum in 1991. There is a full bibliography.

Barlow, John Perry (1948–) US writer and cofounder in 1991 of the Electronic Frontier Foundation, a non-profit-making organization concerned with protecting civil liberties, in particular freedom of speech, on the Internet. His writings about cyberspace issues, such as 'Crime and Puzzlement' (1991) and 'A Declaration of the Independence of Cyberspace' (1996), have circulated widely and influentially on the Net.

Electronic Frontier Foundation
http://www.eff.org/
US-based non-profit organization that aims to protect free speech on the Internet. This site includes a lot of technical legal jargon and the full text of Supreme Court decisions relating to their campaigns. However, there is also a lot of news-style pieces on issues such as encryption, privacy, and free speech which are more accessible to the casual browser.

Berners-Lee, Tim(othy) (1955–) English inventor of the World Wide Web in 1990. He developed the Web whilst working as a consultant at CERN. He currently serves as director of the W3 Consortium, a neutral body that manages the Web. In 1996 the British Computing Society (BCS) gave him a Distinguished Fellow award.

His parents, both mathematicians, worked on the England's first commercial computer, the Ferranti Mark 1, in the 1950s.

Boole, George (1815–1864) English mathematician. His work *The Mathematical Analysis of Logic* (1847) established the basis of modern mathematical logic, and his **Boolean algebra** can be used in designing computers. His system is essentially two-valued. By subdividing objects into separate classes, each

with a given property, his algebra makes it possible to treat different classes according to the presence or absence of the same property. Hence it involves just two numbers, 0 and 1 – the binary system used in the computer.

Boole, George
http://www-history.mcs.st-and.ac.uk/~history/
Mathematicians/Boole.html
Extensive biography of the mathematician. The site contains a clear description of his working relationship with his contemporaries, and also includes the title page of his famous book *Investigation of the Laws of Thought*. Several literature references for further reading on the mathematician are also listed, and the Web site also features a portrait of Boole.

Byron, (Augusta) Ada, Countess of Lovelace (1815–1852) English mathematician, a pioneer in writing programs for Charles Babbage's analytical engine. In 1983 a new, high-level computer language, Ada, was named after her.

Cerf, Vinton (1943–) US inventor of part of the TCP/IP protocols on which the Internet is based. Known throughout the industry as the 'Father of the Internet', Cerf is president of the Internet Society and was a principal developer of the ARPANET.

Cerf is senior vice president of data architecture for MCI Telecommunications Corporation's Data Services Division in Reston, Virginia. He was awarded the US National Medal of Technology in 1997.

Clark, Jim (James) US founder of Silicon Graphics Inc. in 1982 and the Netscape Communications Corporation in 1994. As an associate professor at Stanford University, California, he and a team of graduate students developed the initial technology upon which Silicon Graphics's first products were built. He resigned as chair of Silicon Graphics early in 1994 to start up Netscape, of which he is chair.

Cray, Seymour
http://www.si.edu/resource/tours/comphist/cray.htm
Lengthy interview with Seymour Cray, carried out on behalf of the National Museum of American History, at the Smithsonian Institution.

Cray, Seymour Roger (1925–1996) US computer scientist and pioneer in the field of supercomputing. He designed one of the earliest computers to contain transistors in 1960. In 1972 he formed Cray Research to build the first supercomputer, the Cray-1, released in 1976. Its success led to the production of further supercomputers, including the Cray-2 in 1985, the Cray Y-MP, a multiprocessor design, in 1988, and the Cray-3 in 1989.

Eckert, John Presper
http://www-groups.dcs.st-
and.ac.uk/~history/Mathematicians/Eckert_John.html
Part of an archive containing the biographies of the world's greatest mathematicians, this site is devoted to the life and contributions of John Eckert.

Eckert, John Presper Jr (1919–1995) US electronics engineer and mathematician who collaborated with John Mauchly on the development of the early ENIAC (1946) and UNIVAC 1 (1951) computers.

The ENIAC (Electronic Numerical Integrator, Analyser, and Calculator) weighed many tonnes and lacked a memory, but could store a limited amount of information and perform mathematical functions. It was used for calculating ballistic firing tables and for meteorological and research problems. It was superseded by BINAC (Binary Automatic Computer), also designed in part by Eckert, and in the early 1950s Eckert's group began to produce computers for the commercial market with the construction of the UNIVAC 1 (Universal Automated Computer). Its chief advance was the capacity to store programs.

The Eckert–Mauchly Computer Corporation, formed in 1947, was incorporated in Remington Rand in 1950 and subsequently came under the control of the Sperry Rand Corporation.

Gates, Bill (William) Henry, III (1955–) US businessman and computer programmer. He cofounded the Microsoft Corporation in 1975 and was responsible for supplying MS-DOS, the operating system and the Basic language that IBM used in the IBM PC. In 1997 Gates controlled a $39.8 billion shareholding in Microsoft, making him the world's richest individual.

When the IBM deal was struck in 1980, Microsoft did not actually have an operating system, but Gates bought one from another company, renamed it MS-DOS, and modified it to suit IBM's new computer. Microsoft also retained the right to sell MS-DOS to other computer manufacturers, and because the IBM PC was not only successful but easily copied by other manufacturers, MS-DOS found its way onto the vast majority of PCs. The revenue from MS-DOS helped Microsoft to expand into other areas of software, guided by Gates.

In 1994 he invested $10 million into a biotechnology company, Darwin Molecular, with Microsoft cofounder Paul Allen. In 1997 Gates and his wife, Melinda French Gates, formed the non-profitmaking Gates Library Foundation whose aim is to provide computers and software to public libraries. This initiative expanded Microsoft's program Libraries Online supporting libraries in the USA and Canada.

Herzog, Bertram (1929–) German-born computer scientist, one of the pioneers in the use of computer graphics in engineering design. He has alternated academic posts with working in industry. In 1963 he joined the Ford Motor Company as engineering methods manager, where he extensively applied computers to tasks involved in planning and design. In 1965 he became professor of industrial engineering at the University of Michigan. Two years later he became professor of electrical engineering and computer science at the University of Colorado.

Hollerith, Herman (1860–1929) US inventor of a mechanical tabulating machine, the first device for high-volume data processing. Hollerith's tabulator was widely publicized after being successfully used in the 1890 census. The firm he established, the Tabulating Machine Company, was later one of the founding companies of IBM.

While working on the 1880 US census, he saw the need for an automated recording process for data, and had the idea of punching holes in cards or rolls of paper. By 1889

he had developed machines for recording, counting, and collating census data. The system was used in 1891 for censuses in several countries, and was soon adapted to the needs of government departments and businesses that handled large quantities of data.

Hollerith, Herman
http://www-groups.dcs.st-and.ac.uk/~history/Mathematicians/Hollerith.html
Part of an archive containing the biographies of the world's greatest mathematicians, this site is devoted to the life and contributions of inventor Herman Hollerith.

Hopper, Grace (1906–1992) US computer pioneer who created the first compiler and helped invent the computer language COBOL. She also coined the term 'debug'.

In 1945 she was ordered to Harvard University to assist Howard Aiken in building a computer. One day a breakdown of the machine was found to be due to a moth that had flown into the computer. Aiken came into the laboratory as Hopper was dealing with the insect. 'Why aren't you making numbers, Hopper?' he asked. Hopper replied: 'I am debugging the machine!'

Hopper's main contribution was to create the first computer language, together with the compiler needed to translate the instructions into a form that the computer could work with. In 1959 she was invited to join a Pentagon team attempting to create and standardize a single computer language for commercial use. This led to the development of COBOL, still one of the most widely used languages.

Jacquard, Joseph Marie (1752–1834) French textile manufacturer. He invented a punched-card system for programming designs on a carpetmaking loom. In 1801 he constructed looms that used a series of punched cards to control the pattern of longitudinal warp threads depressed before each sideways passage of the shuttle. On later machines the punched cards were joined to form an endless loop that represented the 'program' for the repeating pattern of a carpet. Jacquard-style punched cards were used in the early computers of the 1940s–60s.

Jobs, Steven Paul (1955–) US computer entrepreneur. He cofounded Apple Computer Inc with Steve Wozniak in 1976, and founded NeXT Technology Inc in 1985. In 1986 he bought Pixar Animation Studios, the computer animation studio spin-off from George Lucas's LucasFilm.

Jobs has been involved with the creation of three different types of computer: the Apple II personal computer in 1977, the Apple Macintosh in 1984 – marketed as 'the computer for the rest of us' – and the NeXT workstation in 1988.

The NeXT was technically the most sophisticated and powerful design, but it was a commercial disaster, and in 1993 NeXT abandoned hardware manufacturing to concentrate on its highly-regarded UNIX-based object-oriented operating system, NextStep. Apple Computer bought NeXT at the end of 1996 to obtain NextStep, and Jobs returned to Apple in an advisory capacity. However, he soon took over as acting chief executive officer of the struggling firm in 1997.

Kahle, Brewster (1960–) US computing entrepreneur who is best known for inventing the software tool WAIS system for publishing material on the Internet. Early in his career, Kahle founded Thinking Machines Corporation, a company that designed supercomputers. He sold his second company, WAIS Inc, to America Online in 1995. In 1996 he continued to coordinate WAIS, and set up an Internet archive which aims to keep a copy of every item on the Net.

Kapor, Mitchell (1951–) US entrepreneur and software designer who founded Lotus Development Corporation, a leading business software company, in 1982. In 1991 he cofounded the Electronic Frontier Foundation, a non-profit-making organization concerned with protecting civil liberties, in particular freedom of speech on the Internet. Kapor is also a professor of media arts and sciences at the Massachussetts Institute of Technology.

Kay, Alan US computing expert and a key figure in the development of graphical user interfaces (later popularized by the Apple Macintosh) and object-oriented languages (Smalltalk) while working at Xerox's Palo Alto Research Centre (Parc) throughout the 1970s. Kay also came up with the inspirational idea of the DynaBook, a sort of computer-based personal digital assistant. Kay spent 1984–96 as an Apple Fellow, working mainly on future-oriented projects with children. In 1996 he joined Walt Disney Imagineering as a Disney Fellow.

Mauchly, John William (1907–1980) US physicist and engineer who, in 1946, constructed the first general-purpose computer, the ENIAC (Electronic Numerical Integrator, Analyser, and Calculator), in collaboration with John Eckert. Their company was bought by Remington Rand (later Sperry Rand) in 1950, and they built the UNIVAC 1 computer (Universal Automated Computer) in 1951 for the US census. Mauchly was a consultant to Remington Rand 1950–59 and again from 1973, after setting up his own consulting company in 1959.

The work on ENIAC was carried out by Mauchly and Eckert during World War II, and was commissioned to automate the calculation of artillery firing tables for the US Army. In 1949 the two partners designed a small-scale binary computer, BINAC, which was faster and cheaper to use. Punched cards were replaced with magnetic tape, and the computer stored programs internally.

Mitnick, Kevin (1963–) US computer criminal, known as 'the world's most wanted hacker' during the three years he spent on the run before being caught in 1994. He was a compulsive hacker who specialized in penetrating communications systems including MCI, Pacific Bell, the Manhattan telephone system, and a Pentagon defence computer.

Moore, Gordon (1928–) US cofounder, with the late Robert Noyce, of the microchip manufacturer Intel in 1968. In 1965, when writing an article for the 35th anniversary edition of *Electronics* magazine, Moore formulated what has since been named **Moore's Law**: the number of components that could be squeezed onto a silicon chip would double every year. Moore updated this prediction in 1975 from doubling every year to doubling every two years. These observations proved remarkably accurate – the processing technology of 1996, for example, was some 8 million times more powerful than that of 1966 – partly because chip manufacturers tried to keep up with Moore's Law so as to avoid falling behind their rivals.

Negroponte, Nicholas US founder and the director of the MIT Media Lab and columnist for *Wired* magazine. In 1996 he published *Being Digital*, in which he made an analogy between bits of data ('the DNA of information') and atoms of matter. Negroponte also predicted that mass media such as newspapers and television will give way to consumer-led electronic media in which people will take only the information they need.

Nelson, Ted (Theodore) (1937–) US computer scientist who coined the term hypertext in 1965 to propose a type of literature that used links embedded in text to connect readers to sources of further information. He went on to develop a global electronic publishing project called Xanadu, and was appointed professor of environmental information at Keio University, Japan, in 1996.

Noyce, Robert Norton (1927–1990) US scientist and inventor, with Jack Kilby, of the integrated circuit (chip), which revolutionized the computer and electronics industries in the 1970s and 1980s. In 1968 he and six colleagues founded the Intel Corporation, which became one of the USA's leading semiconductor manufacturers.

Noyce was awarded a patent for the integrated circuit in 1959. In 1961 he founded his first company, Fairchild Camera and Instruments Corporation, around which Silicon Valley was to grow. The company was the first in the world to understand and exploit the commercial potential of the integrated circuit. It quickly became the basis for such products as the personal computer, the pocket calculator, and the programmable microwave oven. At the time of his death, he was president of Sematech Incorporated, a government–industry research consortium created to help US firms regain a lead in semiconductor technology that they had lost to Japanese manufacturers.

Sinclair, Clive Marles (1940–) British electronics engineer. He produced the first widely available pocket calculator, pocket and wristwatch televisions, a series of home computers, and the innovative but commercially disastrous C5 personal transport (a low cyclelike three-wheeled vehicle powered by a washing-machine motor).

Turing, Alan Mathison (1912–1954) English mathematician and logician. In 1936 he described a 'universal computing machine' that could theoretically be programmed to solve any problem capable of solution by a specially designed machine. This concept, now called the **Turing machine**, foreshadowed the digital computer.

Turing is believed to have been the first to suggest (in 1950) the possibility of machine learning and artificial intelligence. His test for distinguishing between real (human) and simulated (computer) thought is known as the **Turing test**: with a person in one room and the machine in another, an interrogator in a third room asks questions of both to try to identify them. When the interrogator cannot distinguish between them by questioning, the machine will have reached a state of humanlike intelligence.

During World War II Turing worked on the Ultra project in the team that cracked the German Enigma cipher code. After the war he worked briefly on the project to design the general computer known as the Automatic Computing Engine, or ACE, and was involved in the pioneering computer developed at Manchester University from 1948.

Alan Turing Home Page
http://www.turing.org.uk/turing/
Authoritative illustrated biography of the computer pioneer, plus links to related sites. This site contains information on his origins and his code-breaking work during World War II, as well as several works written by Turing himself.

Von Neumann, John (originally Johann) (1903–1957)
Hungarian-born US scientist and mathematician, a pioneer of computer design. He invented his 'rings of operators' (called Von Neumann algebras) in the late 1930s, and also contributed to set theory, game theory, quantum mechanics, cybernetics (with his theory of self-reproducing automata, called **Von Neumann machines**), and the development of the atomic and hydrogen bombs.

He designed and supervised the construction of the first computer able to use a flexible stored program (named MANIAC-1) at the Institute for Advanced Study at Princeton 1940–52. This work laid the foundations for the design of all subsequent programmable computers.

Von Neumann, John
http://ei.cs.vt.edu/~history/VonNeumann.html
Biographical feature on this pioneer of computer design, plus quotations, and a bibliography. It is a text-based site, but contains plenty of information about Von Neumann.

Wang, An (1920–1990) Chinese-born US engineer, founder of Wang Laboratories in 1951, one of the world's largest computer companies in the 1970s. In 1948 he invented the computer memory core, the most common device used for storing computer data before the invention of the integrated circuit (chip).

Wang emigrated to the USA in 1945. He developed his own company with the $500,000 he received from IBM from the sale of his patent. His company took off in 1964 with the introduction of a desktop calculator. Later, Wang switched with great success to the newly emerging market for word-processing systems based on cheap silicon chips, turning Wang Laboratories into a multibillion-dollar company. However, with the advent of the personal computer, the company fell behind and had to seek protection from its creditors. It staged a comeback, doubling in size during 1994–97 to achieve annual revenues of $1.3 billion.

Wiener, Norbert (1894–1964) US mathematician, credited with the establishment of the science of cybernetics in his book *Cybernetics* (1948). In mathematics, he laid the foundation of the study of stochastic processes (those dependent on random events), particularly Brownian motion. He devoted much of his efforts to methodology, developing mathematical approaches that could usefully be applied to continuously changing processes.

During World War II, Wiener worked on the control of anti-aircraft guns (which required him to consider factors such as the machinery itself, the gunner, and the unpredictable evasive action on the part of the target's pilot), on filtering 'noise' from useful information for radar, and on coding and decoding. His investigations stimulated his interest in information transfer and processes such as information feedback.

Wilkes, Maurice Vincent (1913–) English mathematician who led the team at Cambridge University that built the EDSAC (Electronic Delay Storage Automatic Calculator) in 1949, one of the earliest British electronic computers. He chose the serial mode, in which the information in the computer is processed in sequence (and not several parts at once, as in the parallel type). This design incorporated mercury delay lines (developed at the Massachusetts Institute of Technology, USA) as the elements of the memory.

In May 1949 the EDSAC ran its first program and became the first delay-line computer in the world. From early 1950 it offered a regular computing facility to the members of Cambridge University, the first general-purpose computer service. Much time was spent by the research group on programming and on the compilation of a library of programs. The EDSAC was in operation until 1958.

EDSAC II came into service in 1957. This was a parallel-processing machine and the delay line was abandoned in favour of magnetic storage methods.

Glossary

AI abbreviation for artificial intelligence.

algorithm the logical sequence of operations to be performed by a program. A flow chart is a visual representation of an algorithm.

analogue (of a quantity or device) changing continuously; by contrast a digital quantity or device varies in a series of distinct steps. For example, an analogue clock measures time by means of a continuous movement of hands around a dial, whereas a digital clock measures time with a numerical display that changes in a series of discrete steps.

applet mini-software application. Examples of applets include Microsoft WordPad, the simple word processor in Windows 95 or the single-purpose applications that in 1996 were beginning to appear on the World Wide Web, written in Java. These include small animations such as a moving ticker tape of stock prices.

application a program or job designed for the benefit of the end user. Examples of **general purpose** application programs include word processors, desktop publishing programs, databases, spreadsheet packages, and graphics programs. **Application-specific** programs include payroll and stock-control systems. Applications may also be **custom designed** to solve a specific problem not catered for in other types of application.

The term is used to distinguish such programs from those that control the computer (systems programs) or assist the programmer, such as a compiler.

baud a unit of electrical signalling speed equal to one pulse per second, measuring the rate at which signals are sent between electronic devices such as telegraphs and computers; 300 baud is about 300 words a minute.

binary number system a system of numbers to base two, using combinations of the digits 1 and 0. Codes based on binary numbers are used to represent instructions and data in all modern digital computers, the values of the binary digits (contracted to 'bits') being stored or transmitted as, for example, open/closed switches, magnetized/unmagnetized discs and tapes, and high/low voltages in circuits.

bit (contraction of **binary digit**) a single binary digit, either 0 or 1. A bit is the smallest unit of data stored in a computer; all other data must be coded into a pattern of individual bits. A ◊byte represents sufficient computer memory to store a single character of data, and usually contains eight bits. For example, in the ASCII code system used by most microcomputers the capital letter A would be stored in a single byte of memory as the bit pattern 01000001.

boot, or **bootstrap**, the process of starting up a computer. Most computers have a small, built-in boot program that starts automatically when the computer is switched on – its only task is to load a slightly larger program, usually from a hard disc, which in turn loads the main operating system.

browser any program that allows the user to search for and view data. Browsers are usually limited to a particular type of data, so, for example, a graphics browser will display graphics files stored in many different file formats. Browsers usually do not permit the user to edit data, but are sometimes able to convert data from one file format to another.

bubble memory a memory device based on the creation of small 'bubbles' on a magnetic surface. Bubble memories typically store up to 4 megabits (4 million bits) of information. They are not sensitive to shock and vibration, unlike other memory devices such as disc drives, yet, like magnetic discs, they are nonvolatile and do not lose their information when the computer is switched off.

bug an error in a program. It can be an error in the logical structure of a program or a syntax error, such as a spelling mistake. Some bugs cause a program to fail immediately; others remain dormant, causing problems only when a particular combination of events occurs. The process of finding and removing errors from a program is called **debugging**.

byte sufficient computer memory to store a single character of data. The character is stored in the byte of memory as a pattern of ◊bits (binary digits), using a code such as ASCII. A byte usually contains eight bits – for example, the capital letter F can be stored as the bit pattern 01000110.

cache memory a reserved area of the immediate access memory used to increase the running speed of a computer program.

CAD acronym for computer-aided design.

CAL (acronym for computer-assisted learning) the use of computers in education and training: the computer displays instructional material to a student and asks questions about the information given; the student's answers determine the sequence of the lessons.

CAM acronym for computer-aided manufacturing.

CD-ROM (abbreviation for compact-disc read-only memory) a computer storage device developed from the technology of the audio compact disc.

chat real-time exchange of messages between users of a particular system. Chat allows people who are geographically far apart to type messages to each other which are sent and received instantly. The biggest chat system is Internet Relay Chat (IRC), which is used for the exchange of information and software as well as for social interaction.

chip, or **silicon chip**, another name for an ◊integrated circuit, a complete electronic circuit on a slice of silicon (or other semiconductor) crystal only a few millimetres square.

client–server architecture a system in which the mechanics of looking after data are separated from the programs that use the data. For example, the 'server' might be a central database, typically located on a large computer that is reserved for this purpose. The 'client' would be an ordinary program that requests data from the server as needed.

clip art small graphics used to liven up documents and presentations. Many software packages such as word processors and presentation graphics packages come with a selection of clip art.

clone copy of hardware or software that may not be identical to the original design but provides the same functions. All personal computers (PCs) are to some extent clones of the original IBM PC and PC AT launched by IBM in 1981 and 1984, respectively – including IBM's current machines. Cloning a disc drive or workstation, however, means making an exact copy of all the files or software so that the new drive or machine functions identically to the original one.

command language a set of commands and the rules governing their use, by which users control a program. For example, an operating system may have commands such as SAVE and DELETE, or a payroll program may have commands for adding and amending staff records.

computer art art produced with the help of a computer. Since the 1950s the aesthetic use of computers has been increasingly evident in most artistic disciplines, including film animation, architecture, and music. Computer graphics has been the most developed area, with the 'paint-box' computer liberating artists from the confines of the canvas. It is now also possible to programme computers in advance to generate graphics, music, and sculpture, according to 'instructions' which may include a preprogrammed element of unpredictability.

computer-assisted learning use of computers in education and training; see ◊CAL.

computer crime broad term applying to any type of crime committed via a computer, including unauthorized access to files. Most computer crime is committed by disgruntled former employees or subcontractors. Examples include the releasing of viruses, hacking, and computer fraud. Many countries, including the USA and the UK, have specialized law enforcement units to supply the technical knowledge needed to investigate computer crime.

computer generation any of the five broad groups into which computers may be classified: **first generation** the earliest computers, developed in the 1940s and 1950s, made from valves and wire circuits; **second generation** from the early 1960s, based on transistors and printed circuits; **third generation** from the late 1960s, using integrated circuits and often sold as families of computers, such as the IBM 360 series; **fourth generation** using microprocessors, large-scale integration (LSI), and sophisticated programming languages, still in use in the 1990s; and **fifth generation** based on parallel processing and very large-scale integration, currently under development.

control character any character produced by depressing the control key (Ctrl) on a keyboard at the same time as another (usually alphabetical) key. The control characters form the first 32 ASCII characters and most have specific meanings according to the operating system used. They are also used in combination to provide formatting control in many word processors, although the user may not enter them explicitly.

corruption of data introduction or presence of errors in data. Most computers use a range of verification and validation routines to prevent corrupt data from entering the computer system or detect corrupt data that are already present.

CPU abbreviation for central processing unit.

data (singular **datum**) facts, figures, and symbols, especially as stored in computers. The term is often used to mean raw, unprocessed facts, as distinct from information, to which a meaning or interpretation has been applied.

data compression techniques for reducing the amount of storage needed for a given amount of data. They include word tokenization (in which frequently used words are stored as shorter codes), variable bit lengths (in which common characters are represented by fewer bits than less common ones), and run-length encoding (in which a repeated value is stored once along with a count).

digit any of the numbers from 0 to 9 in the decimal system. Different bases have different ranges of digits. For example, the ◊hexadecimal system has digits 0 to 9 and A to F, whereas the binary system has two digits (or ◊bits), 0 and 1.

digital recording technique whereby the pressure of sound waves is sampled more than 30,000 times a second and the values converted by computer into precise numerical values. These are recorded and, during playback, are reconverted to sound waves.

directory a list of file names, together with information that enables a computer to retrieve those files from backing storage. The computer operating system will usually store and update a directory on the backing storage to which it refers. So, for example, on each disc used by a computer a directory file will be created listing the disc's contents.

The term is also used to refer to the area on a disc where files are stored; the main area, the **root** directory, is at the topmost level, and may contain several separate **subdirectories**.

document reader an input device that reads marks or characters, usually on preprepared forms and documents. Such devices are used to capture data by (optical mark recognition (OMR), optical character recognition (OCR), and mark sensing.

DOS (acronym for disc operating system) a computer operating system specifically designed for use with disc storage; also used as an alternative name for a particular operating system, MS-DOS.

DTP abbreviation for desktop publishing.

Dynamic HTML the fourth version of hypertext mark-up language (HTML), the language used to create Web pages. It is called Dynamic HTML because it enables dynamic effects to be incorporated in pages without the delays involved in downloading Java applets and without referring back to the server.

EBCDIC (abbreviation for extended binary-coded decimal interchange code) a code used for storing and communicating alphabetic and numeric characters. It is an 8-bit code, capable of holding 256 different characters, although only 85 of these are defined in the standard version. It is still used in many mainframe computers, but almost all mini-and microcomputers now use ASCII code.

EEPROM (abbreviation for electrically erasable programmable read-only memory) computer memory that can record data and retain it indefinitely. The data can be erased with an electrical charge and new data recorded. Some EEPROM must be removed from the computer and erased and reprogrammed using a special device. Other EEPROM, called **flash memory**, can be erased and reprogrammed without removal from the computer.

e-mail abbreviation for electronic **mail**.

EPROM (abbreviation for erasable programmable read-only memory) computer memory device in the form of an ◊integrated circuit (chip) that can record data and retain it indefinitely. The data can be erased by exposure to ultraviolet light, and new data recorded. Other kinds of computer memory chips are ◊ROM (read-only memory), ◊PROM (programmable read-only memory), and ◊RAM (random-access memory).

expansion board, or **expansion card**, printed circuit board that can be inserted into a computer in order to enhance its capabilities (for example, to increase its memory) or to add facilities (such as graphics).

expert system computer program for giving advice (such as diagnosing an illness or interpreting the law) that incorporates knowledge derived from human expertise. It is a kind of ◊knowledge-based system containing rules that can be applied to find the solution to a problem. It is a form of artificial intelligence.

fax, common name for **facsimile transmission** or **telefax**, the transmission of images over a telecommunications link, usually the telephone network. When placed on a fax machine, the original image is scanned by a transmitting device and converted into coded signals, which travel via the telephone lines to the receiving fax machine, where an image is created that is a copy of the original. Photographs as well as printed text and drawings can be sent. The standard transmission takes place at 4,800 or 9,600 bits of information per second.

feedback general principle whereby the results produced in an ongoing reaction become factors in modifying or changing the reaction; it is the principle used in self-regulating control systems, from a simple thermostat and steam-engine governor to automatic computer-controlled machine tools. A fully computerized control system, in which there is no operator intervention, is called a **closed-loop feedback** system. A system that also responds to control signals from an operator is called an **open-loop feedback** system.

font, or **fount**, complete set of printed or display characters of the same typeface, size, and style (bold, italic, underlined, and so on). Fonts used in computer setting are of two main types: bit-mapped and outline. **Bit-mapped fonts** are stored in the computer memory as the exact arrangement of pixels or printed dots required to produce the characters in a particular size on a screen or printer. **Outline fonts** are stored in the computer memory as a set of instructions for drawing the circles, straight lines, and curves that make up the outline of each character. In the UK, font sizes are measured in points, a point being approximately 0.3 mm.

gigabyte in computing, a measure of memory capacity, equal to 1,024 megabytes. It is also used, less precisely, to mean 1,000 billion ◊bytes.

hacking unauthorized access to a computer, either for fun or for malicious or fraudulent purposes. Hackers generally use microcomputers and telephone lines to obtain access. In computing, the term is used in a wider sense to mean using software for enjoyment or self-education, not necessarily involving unauthorized access. The most destructive form of hacking is the introduction of a computer virus.

hardware the mechanical, electrical, and electronic components of a computer system, as opposed to the various programs, which constitute ◊software.

hertz SI unit (symbol Hz) of frequency (the number of repetitions of a regular occurrence in one second). Radio waves are often measured in megahertz (MHz), millions of hertz, and the clock rate of a computer is usually measured in megahertz. The unit is named after Heinrich Hertz.

hexadecimal number system, or **hex**, a number system to the base 16, used in computing. In hex the decimal numbers 0–15 are represented by the characters 0, 1, 2, 3, 4, 5, 6, 7, 8, 9, A, B, C, D, E, F.

Hexadecimal numbers are easy to convert to the computer's internal binary code and are more compact than binary numbers.

holography method of producing three-dimensional (3-D) images, called holograms, by means of laser light. Holography uses a photographic technique (involving the splitting of a laser beam into two beams) to produce a picture, or hologram, that contains 3-D information about the object photographed. Some holograms show meaningless patterns in ordinary light and produce a 3-D image only when laser light is projected through them, but reflection holograms produce images when ordinary light is reflected from them (as found on credit cards).

HTML (Hypertext Markup Language) the standard for structuring and describing a document on the World Wide Web. The HTML standard provides labels for constituent parts of a document (for example headings and paragraphs) and permits the inclusion of images, sounds, and 'hyperlinks' to other documents. A browser program is then used to convert this information into a graphical document on screen. The specifications for HTML version 4, called Dynamic HTML, were adopted at the end of 1997.

Lightning HTML Editor
http://www.lightningsp.com/HTML_Editor/
HTML editor that teaches you about HTML as you use it. This is a well organized guide to the intricacies of designing Web pages. There is a good list of frequently asked questions and tips and tricks.

---

hypertext system for viewing information (both text and pictures) on a computer screen in such a way that related items of information can easily be reached. For example, the program might display a map of a country; if the user clicks (with a mouse) on a particular city, the program will display information about that city.

icon a small picture on the computer screen, or VDU, representing an object or function that the user may manipulate or otherwise use. It is a feature of graphical user interface (GUI) systems. Icons make computers easier to use by allowing the user to point to and click with a mouse on pictures, rather than type commands.

information technology (IT) collective term for the various technologies involved in processing and transmitting information. They include computing, telecommunications, and microelectronics.

integrated circuit (IC), popularly called **silicon chip**, a miniaturized electronic circuit produced on a single crystal, or chip, of a semiconducting material – usually silicon. It may contain many millions of components and yet measure only 5 mm/0.2 in square and 1 mm/0.04 in thick. The IC is encapsulated within a plastic or ceramic case, and linked via gold wires to metal pins with which it is connected to a ◊printed circuit board and the other components that make up such electronic devices as computers and calculators.

interactive video (IV) computer-mediated system that enables the user to interact with and control information (including text, recorded speech, or moving images) stored on video disc. IV is most commonly used for training purposes, using analogue video discs, but has wider applications with digital video systems such as CD-I (Compact Disc Interactive, from Philips and Sony) which are based on the CD-ROM format derived from audio compact discs.

interface the point of contact between two programs or pieces of equipment. The term is most often used for the physical connection between the computer and a peripheral device, which is used to compensate for differences in such operating characteristics as speed, data coding, voltage, and power consumption. For example, a **printer interface** is the cabling and circuitry used to transfer data from a computer to a printer, and to compensate for differences in speed and coding.

ISDN (abbreviation for Integrated Services Digital Network) a telecommunications system.

kilobyte (K or KB) a unit of memory equal to 1,024 ◊bytes. It is sometimes used, less precisely, to mean 1,000 bytes.

knowledge-based system (KBS) computer program that uses an encoding of human knowledge to help solve problems. It was discovered during research into artificial intelligence that adding heuristics (rules of thumb) enabled programs to tackle problems that were otherwise difficult to solve by the usual techniques of computer science.

LCD abbreviation for ◊liquid-crystal display.

LED abbreviation for ◊light-emitting diode.

light-emitting diode (LED) an electronic component that converts electrical energy into light or infrared radiation in the range of 550 nm (green light) to 1300 nm (infrared). They are used for displaying symbols in electronic instruments and devices. An LED is a diode made of semiconductor material, such as gallium arsenide phosphide, that glows when electricity is passed through it. The first digital watches and calculators had LED displays, but many later models use ◊liquid-crystal displays.

liquid-crystal display (LCD) display of numbers (for example, in a calculator) or pictures (such as on a pocket television screen) produced by molecules of a substance in a semiliquid state with some crystalline properties, so that clusters of molecules align in parallel formations. The display is a blank until the application of an electric field, which 'twists' the molecules so that they reflect or transmit light falling on them. There two main types of LCD are **passive matrix** and **active matrix**.

LSI (abbreviation for **large-scale integration**) the technology that enables whole electrical circuits to be etched into a piece of semiconducting material just a few millimetres square.

machine code a set of instructions that a computer's central processing unit (CPU) can understand and obey directly, without any translation. Each type of CPU has its own machine code.

magnetic-ink character recognition (MICR) a technique that enables special characters printed in magnetic ink to be read and input rapidly to a computer. MICR is used extensively in banking because magnetic-ink characters are difficult to forge and are therefore ideal for marking and identifying cheques.

magnetic tape narrow plastic ribbon coated with an easily magnetizable material on which data can be recorded. It is used in sound recording, audiovisual systems (videotape), and computing. For mass storage on commercial mainframe computers, large reel-to-reel tapes are still used, but cartridges are becoming popular. Various types of cartridge are now standard on minis and PCs, while audio cassettes are sometimes used with home computers.

megabyte (MB) a unit of memory equal to 1,024 ◊kilobytes. It is sometimes used, less precisely, to mean 1 million bytes.

menu a list of options, displayed on screen, from which the user may make a choice – for example, the choice of services offered to the customer by a bank cash dispenser: withdrawal, deposit, balance, or statement. Menus are used extensively in graphical user interface (GUI) systems, where the menu options are often selected using a mouse.

Action 2000
http://www.open.gov.uk/bug2000/index2.html
Practical UK government advice on how to avoid computer chaos at midnight on 31 December1999. The 'busting the bug' campaign warns of the consequences of inaction and advises how to check your software and take preventive measures. Further sources of information are provided.

Millennium Bug crisis facing some computer systems in the year 2000 that will arise because computers may be unable to operate normally when faced with the unfamiliar date format. Information about the year has typically been stored in a two-digit instead of a four-digit field in order to save memory space, which may mean that after the year 1999 ends the year will appear as '00'. Systems may consider this to mean 1900, or they may not recognize it at all and will crash, resulting in data corruption.

monitor, or **screen**, output device on which a computer displays information for the benefit of the operator user – usually in the form of a graphical user interface such as Windows. The commonest type is the cathode-ray tube (CRT), which is similar to a television screen. Portable computers often use liquid crystal display (LCD) screens. These are harder to read than CRTs, but require less power, making them suitable for battery operation.

multimedia computerized method of presenting information by combining audio and video components using text, sound, and graphics (still, animated, and video sequences). For example, a multimedia database of musical instruments may allow a user not only to search and retrieve text about a particular instrument but also to see pictures of it and hear it play a piece of music. Multimedia applications emphasize interactivity between the computer and the user.

multitasking, or **multiprogramming**, a system in which one processor appears to run several different programs (or different parts of the same program) at the same time. All the programs are held in memory together and each is allowed to run for a certain period.

online system originally a system that allows the computer to work interactively with its users, responding to each instruction as it is given and prompting users for information when necessary. Since almost all the computers used now work this way, 'online system' is now used to refer to large database, electronic mail, and conferencing systems accessed via a dial-up modem. These often have tens or hundreds of users from different places – sometimes from different countries – 'on line' at the same time.

optical character recognition (OCR) a technique for inputting text to a computer by means of a document reader. First, a scanner produces a digital image of the text; then character-recognition software makes use of stored knowledge about the shapes of individual characters to convert the digital image to a set of internal codes that can be stored and processed by computer.

optical fibre very fine, optically pure glass fibre through which light can be reflected to transmit images or data from one end to the other. Although expensive to produce and install, optical fibres can carry more data than traditional cables, and are less susceptible to interference. Standard optical fibre transmitters can send up to 10 billion bits of information per second by switching a laser beam on and off.

optical mark recognition (OMR) a technique that enables marks made in predetermined positions on computer-input forms to be detected optically and input to a computer. An **optical mark reader** shines a light beam onto the input document and is able to detect the marks because less light is reflected back from them than from the paler, unmarked paper.

personal computer (PC) another name for a microcomputer. The term is also used, more specifically, to mean the IBM Personal Computer and computers compatible with it. The first IBM PC was introduced in 1981; it had 64 kilobytes of random access memory (RAM) and one floppy-disc drive. It was followed in 1983 by the XT (with a hard-disc drive) and in 1984 by the AT (based on a more powerful microprocessor). Many manufacturers have copied the basic design, which is now regarded as a standard for business microcomputers. Computers designed to function like an IBM PC are **IBM-compatible computers**.

pixel (derived from **picture element**) a single dot on a computer screen. All screen images are made up of a collection of pixels, with each pixel being either off (dark) or on (illuminated, possibly in colour). The number of pixels available determines the screen's resolution. Typical resolutions of microcomputer screens vary from 320 x 200 pixels to 640 x 480 pixels, but screens with 1,024 x 768 pixels are now common for high-quality graphic (pictorial) displays.

printed circuit board (PCB) electrical circuit created by laying (printing) 'tracks' of a conductor such as copper on one or both sides of an insulating board. The PCB was invented in 1936 by Austrian scientist Paul Eisler, and was first used on a large scale in 1948.

processor another name for the central processing unit or microprocessor of a computer.

PROM (abbreviation for **programmable read-only memory**) a memory device in the form of an integrated circuit (chip) that can be programmed after manufacture to hold information permanently. PROM chips are empty of information when manufactured, unlike ROM (read-only memory) chips, which have information built into them. Other memory devices are ◊EPROM (erasable programmable read-only memory) and ◊RAM (random-access memory).

protocol an agreed set of **standards** for the transfer of data between different devices. They cover transmission speed, format of data, and the signals required to synchronize the transfer.

RAM (acronym for **random-access memory**) a memory device in the form of a collection of integrated circuits (chips), frequently used in microcomputers. Unlike ◊ROM (read-only memory) chips, RAM chips can be both read from and written to by the computer, but their contents are lost when the power is switched off.

real-time system a program that responds to events in the world as they happen. For example, an automatic-pilot program in an aircraft must respond instantly in order to cor-rect deviations from its course. Process control, robotics, games, and many military applications are examples of real-time systems.

repetitive strain injury (RSI) inflammation of tendon sheaths, mainly in the hands and wrists, which may be disabling. It is found predominantly in factory workers involved in constant repetitive movements, and in those who work with computer keyboards. The symptoms include aching muscles, weak wrists, tingling fingers and in severe cases, pain and paralysis. Some victims have successfully sued their employers for damages.

Repetitive Strain Injury
http://engr-www.unl.edu/ee/eeshop/rsi.html
Unofficial, but very informative and extensively-linked, site produced by a sufferer of RSI. It includes a section on the symptoms, as well as diagrams and photos of how to type without straining your back or hands. The site is largely focused upon prevention rather than cure.

RISC (abbreviation for **reduced instruction-set computer**) a microprocessor (processor on a single chip) that carries out fewer instructions than other (CISC) microprocessors in common use in the 1990s. Because of the low number and the regularity of machine code instructions, the processor carries out those instructions very quickly.

ROM (abbreviation for **read-only memory**) a memory device in the form of a collection of integrated circuits (chips), frequently used in microcomputers. ROM chips are loaded with data and programs during manufacture and, unlike ◊RAM (random-access memory) chips, can subsequently only be read, not written to, by computer. However, the contents of the chips are not lost when the power is switched off, as happens in RAM.

RSI (abbreviation for ◊**repetitive strain injury**) a condition affecting workers, such as typists, who repeatedly perform certain movements with their hands and wrists.

search engine a remotely accessible program to help users find information on the Internet. Commercial search engines such as AltaVista and Lycos comprise databases of documents, URLs, USENET articles and more, which can be searched by keying in a key word or phrase. The databases are compiled by a mixture of automated agents (spiders) and webmasters registering their sites.

SGML (Standard Generalized Markup Language) International Standards Organization standard describing how the structure (features such as headers, columns, margins, and tables) of a text can be identified so that it can be used, probably via filters, in applications such as desktop publishing and electronic publishing. HTML and VRML are both types of SGML.

shareware software distributed free via the Internet or on discs given away with magazines. Users have the opportunity to test its functionality and ability to meet their requirements before paying a small registration fee directly to the author.

silicon chip ◊**integrated circuit** with microscopically small electrical components on a piece of silicon crystal only a few millimetres square.

software a collection of programs and procedures for making a computer perform a specific task, as opposed to ◊hardware, the physical components of a computer system. Software is created by programmers and is either distributed on a suitable medium, such as the floppy disc, or built into the computer in the form of firmware. Examples of software include operating systems, compilers, and applications programs such as payrolls or word processors. No computer can function without some form of software.

source language the language in which a program is written, as opposed to machine code, which is the form in which the program's instructions are carried out by the computer. Source languages are classified as either high-level languages or low-level languages, according to whether each notation in the source language stands for many or only one instruction in machine code.

speech recognition, or **voice input**, any technique by which a computer can understand ordinary speech. Spoken words are divided into 'frames', each lasting about one-thirtieth of a second, which are converted to a wave form. These are then compared with a series of stored frames to determine the most likely word. Research into speech recognition started in 1938, but the technology did not become sufficiently developed for commercial applications until the late 1980s.

speech synthesis, or **voice output**, computer-based technology for generating speech. A speech synthesizer is controlled by a computer, which supplies strings of codes representing basic speech sounds (phonemes); together these make up words. Speech-synthesis applications include children's toys, car and aircraft warning systems, and talking books for the blind.

standards any agreed system or protocol that helps different pieces of software or different computers to work together. If computers are to communicate over a network, standards must be coordinated: the World Wide Web, for example, works because everybody who uses it agrees to follow the same conventions, such as using HTML to build Web documents. Other standards, like SMTP – the procedure for sending e-mail – exist to make cross-platform communication (for example between a UNIX machine and a Macintosh) possible. Bodies involved with this process include: the Internet Architecture Board, the W3 Consortium, and the International Standards Organization.

systems analysis the investigation of a business activity or clerical procedure, with a view to deciding if and how it can be computerized. The analyst discusses the existing procedures with the people involved, observes the flow of data through the business, and draws up an outline specification of the required computer system. The next step is ◊systems design.

systems design the detailed design of an applications package. The designer breaks the system down into component programs, and designs the required input forms, screen layouts, and printouts. Systems design forms a link between systems analysis and programming.

teletext broadcast system of displaying information on a television screen. The information – typically about news items, entertainment, sport, and finance – is constantly updated. Teletext is a form of ◊videotext, pioneered in Britain by the British Broadcasting Corporation (BBC) with Ceefax and by Independent Television with Teletext.

terminal a device consisting of a keyboard and display screen (VDU) to enable the operator to communicate with the computer. The terminal may be physically attached to the computer or linked to it by a telephone line (remote terminal). A 'dumb' terminal has no processor of its own, whereas an 'intelligent' terminal has its own processor and takes some of the processing load away from the main computer.

tree-and-branch filing system a filing system where all files are stored within directories, like folders in a filing cabinet. These directories may in turn be stored within further directories. The root directory contains all the other directories and may be thought of as equivalent to the filing cabinet. Another way of picturing the system is as a tree with branches from which grow smaller branches, ending in leaves (individual files).

UNIX multiuser operating system designed for minicomputers but becoming increasingly popular on microcomputers, workstations, mainframes, and supercomputers.

VDU abbreviation for visual display unit.

video adapter an expansion board that allows display of graphics and colour. Commonly used video adapters for IBM PC-based systems are Hercules, CGA, EGA, VGA, XGA, and SVGA.

videotext system in which information (text and simple pictures) is displayed on a television (video) screen. There are two basic systems, known as ◊teletext and viewdata. In the teletext system information is broadcast with the ordinary television signals, whereas in the viewdata system information is relayed to the screen from a central data bank via the telephone network. Both systems require the use of a television receiver (or a connected VTR) with special decoder.

voice modem a modem which handles voice as well as data communications, so that it can be used to add the capabilities of a voice mail system to a personal computer. Primarily aimed at small and home-based businesses, voice modems typically also include fax facilities.

VRAM (video random-access memory) a form of ◊RAM that allows simultaneous access by two different devices, so that graphics can be handled at the same time as data are updated. VRAM improves graphic display performance.

VRML (Virtual Reality Modelling Language) method of displaying three-dimensional images on a Web page. VRML, which functions as a counterpart to HTML, is a platform-independent language that creates a virtual reality scene which users can 'walk' through and follow links much like a conventional Web page. In some contexts, VRML can replace conventional computer interfaces with their icons, menus, files, and folders.

Windows graphical user interface (GUI) from Microsoft that has become the standard for IBM PCs and clones. **Windows 95**, updated to **Windows 98**, is designed for homes and offices and retains maximum compatibility with programs written for the MS-DOS operating system. **Windows NT** is a 32-bit multiuser and multitasking operating system designed for business use, especially on workstations and server computers, where it is seen as a rival to UNIX. **Windows CE** is a small operating system that supports a subset of the Windows applications programming interface. It is designed for hand-held personal computers (HPCs) and consumer electronics products.

World Wide Web a hypertext system for publishing information on the ◊Internet. The original program was created in 1990 for internal use at CERN, the Geneva-based physics research centre, by Tim Berners-Lee and Robert Cailliau. The system was released on the Internet in 1991, but only caught on in 1993, following the release of Mosaic, an easy-to-use PC-compatible browser. From the 600-odd Web servers in existence in December 1993, the number grew to around 2,000,000 by the end of 1997. According to an estimate by the search engine Alta Vista, there were 100–150 million Web pages in 1997.

History of the Web
http://dbhs.wvusd.k12.ca.us/Chem-History/
Hist-of-Web.html
Transcript of Birthplace of the Web by Eric Berger, Office of Public Affairs at FermiLab. The text covers the origins of the web as a means of communication between scientists at CERN and at FermiLab, and describes how one person's idea in 1991 has brought about a social and cultural revolution in just a few years.

Further Reading

Angelides, Marios C, and Dustdar, Schahram *Multimedia Information Systems* (1997)

Beynon-Davies, Paul *Database Systems* (1996)

Bines, W J (ed) *Microcomputer Applications Handbook* (1989)

Campbell-Kelly, Martin, and Aspray, William *Computer: A History of the Information Machine* (1996)

Clements, A *Principles of Computer Hardware* (1992, second edition)

Collin, Simon *E-Mail: a Practical Guide* (1995)

Dorf, Richard Carl *Modern Control Systems* (1995, seventh edition)

Edwards, John, and Finlay, Paul N *Decision Making with Computers: the Spreadsheet and Beyond* (1997)

Farkas, Bart, and Breen, Christopher *The Macintosh Bible Guide to Games* (1996)

Foley, James D *Introduction to Computer Graphics* (1994)

Galitz, Wilbert O *The Essential Guide to User Interface Design: An Introduction to GUI Design Principles and Techniques* (1997)

Hennessy, John L, and Patterson, David A *Computer Organization and Design: the Hardware/Software Interface* (1998, second edition)

Herz, J C *Surfing on the Internet* (1995)

Hillman, David *Multimedia Technology and Applications* (1998)

Horstmann, Cay *Computing Concepts with Java Essentials* (1998)

Kamin, Jonathan *Understanding Hard Disc Management on the PC* (1989)

Kaplan, Randy M *Intelligent Multimedia Systems: A Handbook for Creating Applications* (1997)

Knuth, Donald E *The Art of Computer Programming* (1998)

Korolenko, Michael *Writing for Multimedia: A Guide and Sourcebook for the Digital Writer* (1997)

Lathrop, Olin *The Way Computer Graphics Works* (1997)

Mandel, Theo *The Elements of User Interface Design* (1997)

Maybury, Mark T *Intelligent Multimedia Information Retrieval* (1997)

Mulholland, Dawn *Desktop Publishing: A Complete Course* (1996)

Nutt, Gary J *Operating Systems: A Modern Perspective* (1997)

Oxborrow, Elizabeth A *Databases and Database Systems: Concepts and Issues* (1989, second edition)

Parker, Roger C *Desktop Publishing and Design for Dummies* (1995)

Peck, Dave D *Multimedia: A Hands-On Introduction* (1997)

Pimentel, Ken, and Teixeira, Kevin *Virtual Reality: Through the New Looking Glass* (1993)

Rathbone, Andy *Multimedia and CD ROMs for Dummies* (1995, second edition)

Reid, T R *Microchip: The Story of a Revolution and the Men who Made It* (1985)

Ritchie, Colin *Operating Systems* (1997, third edition)

Ross, Sheldon M *Simulation* (1997, second edition)

Slade, Robert M *Robert Slade's Guide to Computer Viruses: How to Avoid Them, How to Get Rid of Them, How to Get Help* (1994)

Wood, John M *Desktop Magic: Electronic Publishing, Document Management, and Workgroups* (1995)

Zorkoczy, Peter *Information Technology: An Introduction* (1995, fourth edition)

TECHNOLOGY

Technology is the use of tools, power, and materials, generally for the purposes of production. Almost every human process for getting food and shelter depends on complex technological systems developed over a two and half million-year period when *Homo habilis*, the ancestors of modern humans, first began to use tools.

muscle to microelectronics

In human prehistory, the only power available was muscle power, augmented by primitive tools such as the wedge or lever. The domestication of animals about 8500 BC and invention of the wheel about 2000 BC paved the way for the water mill (1st century BC) and later the windmill (12th century AD). Not until 1712 did an alternative source of power appear in the form of the first working steam engine, constructed by the English inventor Thomas Newcomen. The English chemist and physicist Michael Faraday's demonstration of the dynamo in 1831 revealed the potential of the electrical motor, and in 1876 the German scientist Nikolaus Otto introduced the four-stroke cycle used in the modern internal combustion engine. The 1940s saw the explosion of the first atomic bomb and the subsequent development of the nuclear power industry. Recent concern over the use of nonrenewable power sources and the pollution caused by the burning of fossil fuels has caused technologists to turn increasingly to exploring renewable sources of energy, in particular solar energy, wind energy, and wave power.

The earliest materials used by humans were wood, bone, horn, shell, and stone. Metals were rare and/or difficult to obtain, although forms of bronze and iron were in use from 6000 BC and 1000 BC respectively. The introduction of the blast furnace in the 15th century enabled cast iron to be extracted, but this process remained expensive until the English ironmaker Abraham Darby substituted coke for charcoal in 1709, thus ensuring a plentiful supply of cheap iron at the start of the Industrial Revolution. Rubber, glass, leather, paper, bricks, and porcelain underwent similar processes of trial and error before becoming readily available. From the mid-1800s entirely new materials, synthetics, appeared. First dyes, then plastic and the more versatile celluloid, and later drugs were synthesized, a process continuing into the 1980s with the

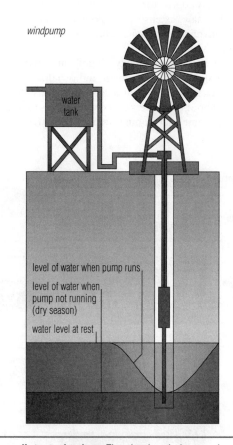

windpump

water tank

level of water when pump runs

level of water when pump not running (dry season)

water level at rest

intermediate technology The simple wind pump is an example of intermediate technology if it utilizes local materials and traditional design. In this way, there is no need for complex maintenance and repair, nor expensive spare parts.

growth of genetic engineering, which enabled the production of synthetic insulin and growth hormones.

The utilization of power sources and materials for production frequently lagged behind their initial discovery. The lathe, known in antiquity in the form of a pole powered by a foot treadle, was not fully developed until the 18th century when it was used to produce objects of great precision, ranging from astronomical instruments to mass-produced screws. The realization that gears, cranks, cams, and wheels could operate in harmony to perform complex motion made mechanization possible. With the perfection of the programmable electronic computer in the 1960s, the way lay open for fully automatic plants. The 1960s–90s saw extensive developments in the electronic and microelectronic industries and in the field of communications.

The **advanced technology** (highly automated and specialized) on which modern industrialized society depends is frequently contrasted with the **low technology** (labour-intensive and unspecialized) that characterizes some developing countries. **Intermediate technology** is an attempt to adapt scientifically advanced inventions and technologies to less developed areas by using local materials and methods of manufacture.

technology of energy conversion

Energy is the capacity to do work. It can exist in many different forms and can be converted from one form to another. So-called energy resources are stores of convertible energy. Harnessing resources generally implies converting their energy into electrical form, because electrical energy is easy to convert to other forms and to transmit from place to place, though not to store.

Nonrenewable resources include the fossil fuels (coal, oil, and gas) and nuclear-fission fuels, for example, uranium 235. **Renewable resources**, such as wind, tidal, and geothermal power, have so far been less

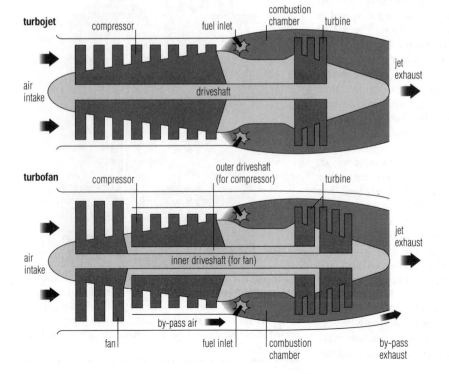

jet propulsion Two forms of jet engine. In the turbojet, air passing into the air intake is compressed by the compressor and fed into the combustion chamber where fuel burns. The hot gases formed are expelled at high speed from the rear of the engine, driving the engine forwards and turning a turbine which drives the compressor. In the turbofan, some air flows around the combustion chamber and mixes with the exhaust gases. This arrangement is more efficient and quieter than the turbojet.

exploited, although hydroelectric projects are well established, and wind turbines and tidal systems are being developed. The ultimate nonrenewable, but almost inexhaustible energy source, would be nuclear fusion (the way in which energy is generated in the Sun), but controlled fusion is a long way off.

conversion of nonrenewable energy sources

fossil fuels The energy stored in such fossil fuels as coal, oil, and natural gas is converted into useful work or movement in devices known as engines. The fuel is burnt to produce heat energy – hence the name 'heat engine' – which is then converted into movement. Heat engines can be classified according to the fuel they use such as petrol or diesel; or according to whether the fuel is burnt inside, as in an internal combustion engine; or externally, as in a steam engine; or according to whether they produce a reciprocating or rotary motion. The ◊diesel engine and ◊petrol engine are both internal-combustion engines. Gas ◊turbines and ◊jet and rocket engines are also considered to be internal-combustion engines because they burn their fuel inside their combustion chambers.

Steam Engine
http://www.easystreet.com/pnwc/museum/
Steam_Locomotive.html
Basic guide to the principles behind the steam engine. It includes a labelled diagram and explanatory text of the engine in its most well known form – the steam locomotive

---

steam power In conventional power stations boilers convert water into steam to feed steam turbines, which drive the electricity generators. Every boiler has a furnace in which fuel (coal, oil, or gas) is burned to produce hot gases, and a system of tubes in which heat is transferred from the gases to the water. Boilers are also used in steamships, which are propelled by steam turbines, and in steam locomotives.

chemical energy Chemical energy can be converted into electrical energy in an **electrical cell**, more com-

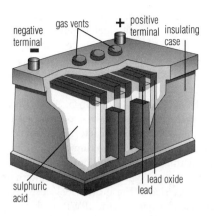

accumulator The lead–acid car battery is a typical example of an accumulator. The battery has a set of grids immersed in a sulphuric acid electrolyte. One set of grids is made of lead (Pb) and acts as the anode and the other set made of lead oxide (PbO_2) acts as the cathode.

---

monly known as a battery. The reactions that take place in a simple cell depend on the fact that some metals are more reactive than others. If an electrolyte and a wire join two different metals, the more reactive metal loses electrons to form ions. The ions pass into solution in the electrolyte, while the electrons flow down the wire to the less reactive metal. At the less reactive metal the electrons are taken up by the positive ions in the electrolyte, which completes the circuit.

A **fuel cell** converts chemical energy directly to electrical energy. It works on the same principle as a battery but is continually fed with fuel, usually hydrogen. Fuel cells are silent and reliable (no moving parts) but expensive to produce.

nuclear energy Nuclear reactors are used to produce nuclear energy in a controlled manner. There are various types of reactor in use, all using nuclear fission.

---

diesel engine In a diesel engine, fuel is injected on the power stroke into hot compressed air at the top of the cylinder, where it ignites spontaneously. The four stages are exactly the same as those of the four-stroke or Otto cycle.

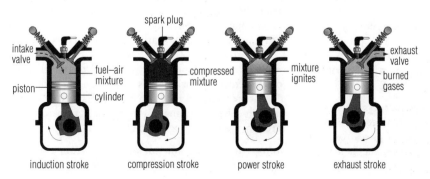

induction stroke   compression stroke   power stroke   exhaust stroke

In a **gas-cooled reactor**, a circulating gas under pressure (such as carbon dioxide) removes heat from the core of the reactor, which usually contains natural uranium. Slowing neutrons in the core by using a moderator such as carbon increases the efficiency of the fission process. The reaction is controlled with neutron-absorbing rods made of boron. An **advanced gas-cooled reactor** (AGR) generally has enriched uranium as its fuel. A **water-cooled reactor**, such as the steam-generating heavy water (deuterium oxide) reactor, has water circulating through the hot core. The water is converted to steam, which drives turbo-alternators for generating electricity. The most widely used reactor is the **pressurized-water reactor** (PWR), which contains a sealed system of pressurized water that is heated to form steam in heat exchangers in an external circuit. The **fast reactor** or **fast breeder reactor** has no moderator and uses fast neutrons to bring about fission. It uses a mixture of plutonium and uranium oxide as fuel. When operating, uranium is converted to plutonium, which can be extracted and used later as fuel. It is also called the fast breeder because it produces more plutonium than it consumes. Heat is removed from the reactor by a coolant of liquid sodium. Nuclear power generates around 17% of the world's electricity.

Advanced Reactors
http://www.uic.com.au/nip16.htm
Overview of the features of the next generation of nuclear reactors currently being developed around the world. This 'nuclear issues briefing paper', from Australia, focuses on the reactors currently being developed in USA, Japan, France, Germany, and Canada.

---

conversion of renewable energy sources

hydroelectric power In a typical hydroelectric scheme, whereby electricity is generated by moving water, water stored in a reservoir, often created by damming a river, is piped into water ◊turbines, coupled to electricity generators. In pumped storage plants, water flowing through the turbines is recycled. A tidal power station exploits the rise and fall of the tides. About one-fifth of the world's electricity comes from hydroelectric power. Hydroelectric plants have prodigious generating capacities. The Grand Coulee plant in Washington State, USA, has a power output of around 10,000 megawatts.

wave power Various schemes to harness the energy of water waves have been advanced since 1973 when an energy shortage threatened and oil prices rose dramatically. In 1974 the British engineer Stephen Salter developed the 'duck' – a floating boom, the segments of which nod up and down with the waves. The nodding motion can be used to drive pumps and spin generators. Another device uses an oscillating water column to harness wave power. A major breakthrough will be required if wave power is ever to contribute significantly to the world's energy needs, although several ideas have reached prototype stage.

A UK government adviser on wave power concluded in a 1998 report that wave power devices have been improved to such a degree as to have become economically viable. The duck, for example, can generate electricity at 2.6 pence per kilowatt hour, compared with 2.5 pence for a gas-fired power station and 4.5 pence for a nuclear-powered one.

wind power The wind has long been used as a source of energy: sailing ships and windmills are ancient inventions. After the energy crisis of the 1970s wind turbines began to be used to produce electricity on a large scale. Wind turbines are windmills of advanced aerodynamic design connected to an electricity generator and used in wind-power installations. They can be either large propeller-type rotors mounted on a tall tower, or flexible metal strips fixed to a vertical axle at top and bottom. The world's largest wind turbine is on Hawaii, in the Pacific Ocean. It has two blades 50 m/160 ft long on top of a tower 20 storeys high. Other machines use novel rotors, such as the 'egg-beater' design developed at Sandia Laboratories in New Mexico, USA. Worldwide, wind turbines on land produce only the energy equivalent of a single nuclear power station.

Wind Turbines
http://www.nrel.gov/wind/turbines.html
Reports on latest wind turbine research. There are technical details of a variety of experimental turbines, information on the work of companies active in research and development, and speculation about the configurations of the kinds of turbines likely to become significant contributors to energy resources in the next century.

---

geothermal energy Energy extracted from natural steam, hot water, or hot dry rocks in the Earth's crust for heating and electricity generation is an important source of energy in volcanically active areas such as Iceland and New Zealand. Water is pumped down through an injection well where it passes through joints in the hot rocks. It rises to the surface through a recovery well and may be converted to steam or run through a heat exchanger. Dry steam may be directed through turbines to produce electricity.

solar energy The amount of energy from the Sun's radiation falling on just 1 sq km/0.3861 sq mi is about

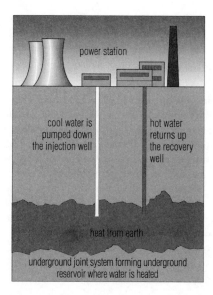

geothermal energy Geothermal energy is derived from the natural heat present below the surface of the Earth. Cool water is pumped down where it is heated up in large underground reservoirs before being pumped back to the surface.

4,000 megawatts: enough to heat and light a small town. **Solar heaters** have industrial or domestic uses. They usually consist of a black (heat-absorbing) panel containing pipes through which air or water, heated by the Sun, is circulated, either by thermal convection or by a pump. Solar energy may also be harnessed indirectly using **solar cells** (photovoltaic cells) made of panels of semiconductor material (usually silicon), which generate electricity when illuminated by sunlight. Although it is difficult to generate a high output from solar energy compared to sources such as nuclear or fossil fuels, it is a major nonpolluting and renewable energy source.

engineering and technological science

Engineering is the application of science to the design, construction, and maintenance of works, machinery, roads, railways, bridges, harbour installations, engines, ships, aircraft and airports, spacecraft and space stations, and the generation, transmission, and use of electrical power. The main divisions of engineering are aerospace, chemical, civil, computer, electrical, electronic, gas, marine, materials, mechanical, mining, production, radio, and structural.

civil engineering Civil engineering is concerned with the construction of roads, bridges, airports, aqueducts, waterworks, tunnels, canals, irrigation works, and harbours. The term is thought to have been used for the first time by the British engineer John Smeaton in about 1750 to distinguish civilian from military engineering projects.

electronic engineering Electronic engineering deals with the construction of electronic devices. The first electronic device was the thermionic valve, or vacuum tube, in which electrons moved in a vacuum, and led to such inventions as radio, television, ⟩radar, and the digital computer. Replacement of valves with the comparatively tiny and reliable transistor from 1948 revolutionized electronic development. Modern electronic devices are based on minute integrated circuits (silicon chips), wafer-thin crystal slices holding tens of thousands of electronic components. By using solid-state devices such as integrated circuits, extremely complex electronic circuits can be constructed, leading to the development of digital watches, pocket calculators, powerful microcomputers, and word processors.

hydraulic engineering Hydraulic engineering is concerned with utilizing the properties of water and other liquids, in particular the way they flow and transmit pressure. It applies the principles of hydrostatics and hydrodynamics. The oldest type of hydraulic machine is the **hydraulic press**, invented by Joseph Bramah in England in 1795. The hydraulic principle of pressurized liquid increasing a force is commonly used on

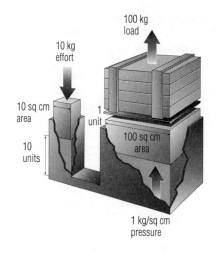

hydraulics The hydraulic jack transmits the pressure on a small piston to a larger one. The larger piston develops a larger total force but it moves a smaller distance than the small piston.

vehicle braking systems, the forging press, and the hydraulic systems of aircraft and excavators.

aeronautics Aeronautics is concerned with travel through the Earth's atmosphere, and includes aerodynamics, aircraft structures, jet and rocket propulsion, and aerial navigation. For all flight speeds streamlining is necessary to reduce the effects of air resistance. In **subsonic aeronautics** (below the speed of sound), aerodynamic forces increase at the rate of the square of the speed. **Transonic aeronautics** covers the speed range from just below to just above the speed of sound and is crucial to aircraft design. **Supersonic aeronautics** concerns speeds above that of sound and in one sense may be considered a much older study than aeronautics itself, since the study of the flight of bullets, known as ballistics, was undertaken soon after the introduction of firearms. **Hypersonics** is the study of airflows and forces at speeds above five times that of sound (Mach 5); for example, for guided missiles, space rockets, and advanced concepts such as HOTOL (horizontal takeoff and landing). Aeronautics is distinguished from astronautics, which is the science of travel through space.

American Institute of Aeronautics and Astronautics Home Page
http://www.aiaa.org/
Access to the *AIAA Bulletin*. This site also includes details of the institute's research departments, recent conferences, technical activities, and project updates. If the extensive front page doesn't have what you need, the site is also fully-searchable.

---

bionics Bionics (from 'biological electronics') deals with the design and development of electronic or mechanical artificial systems that imitate those of living things. The bionic arm, for example, is an artificial limb (prosthesis) that uses electronics to amplify minute electrical signals generated in body muscles to work electric motors, which operate the joints of the fingers and wrist.

Chemisty and Industry Magazine
http://biotech.mond.org/
Latest developments in the fast-changing world of biotechnology. This online edition of Chemistry and Industry has well written and readily understandable articles on biotechnology, chemistry, and the pharmaceutical industry.

---

biotechnology Biotechnology is the industrial use of living organisms to manufacture food, drugs, or other products. The brewing and baking industries have long relied on the yeast microorganism for fermentation purposes, while the dairy industry employs a range of

bacteria and fungi to convert milk into cheeses and yoghurts. Enzymes, whether extracted from cells or produced artificially, are central to most biotechnological applications.

metallurgy Metallurgy is the science and technology of producing metals, which includes extraction, alloying, and hardening. **Extractive** or **process metallurgy** is concerned with the extraction of metals from their ores and refining and adapting them for use. Metals can be extracted from their ores in three main ways: **dry processes**, such as smelting, volatilization, or amalgamation (treatment with mercury); **wet processes**, involving chemical reactions; and **electrolytic processes**, which work on the principle of electrolysis (using electricity conducted by a solution or melt to effect chemical changes). **Physical metallurgy** is concerned with the properties of metals and their applications.

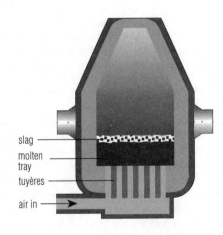

Bessemer process In a Bessemer converter, a blast of high-pressure air oxidizes impurities in molten iron and converts it to steel.

---

nanotechnology Nanotechnology is an experimental technology which uses individual atoms or molecules as the components of minute machines, measured by the nanometre, or millionth of a millimetre. Nanotechnology research in the 1990s focused on testing molecular structures and refining ways to manipulate atoms using a scanning tunnelling microscope. The ultimate aim is to create very small computers and molecular machines that can perform vital engineering or medical tasks. The scanning electron microscope can be used to see and position single atoms and molecules, and to drill holes a nanometre across in a variety of materials.

optoelectronics Optoelectronics is the branch of electronics concerned with the development of devices (based on the semiconductor gallium arsenide) that respond not only to the electrons of electronic data transmission, but also to photons. In 1989, scientists at IBM in the USA built a gallium arsenide microprocessor ('chip') containing 8,000 transistors and four photodetectors. The densest optoelectronic chip yet produced, this can detect and process data at a speed of 1 billion bits per second.

radiography Radiography is concerned with the use of radiation (particularly X-rays) to produce images on photographic film or fluorescent screens. X-rays penetrate matter according to its nature, density, and thickness. In doing so they can cast shadows on photographic film, producing a radiograph. Radiography is widely used in medicine for examining bones and tissues and in industry for examining solid materials; for example, to check welded seams in pipelines.

technology of machines

A machine is a device that allows a small force (the effort) to overcome a larger one (the load). There are three basic machines: the **inclined plane** (ramp), the **lever**, and the **wheel and axle**. All other machines are combinations of these three basic types. Simple machines or machine components derived from the inclined plane include the wedge, the gear, the pulley, and the screw. The principal features of a machine are its mechanical advantage, which is the ratio of load to effort, its velocity ratio, and its efficiency, which is the work done by the load divided by the work done by the effort; the latter is expressed as a percentage. In a perfect machine, with no friction, the efficiency would be 100%. All practical machines have efficiencies of less than 100%, otherwise perpetual motion would be possible.

simple machines

inclined plane An inclined plane is a slope that allows a load to be raised gradually using a smaller effort than would be needed if it were lifted vertically upwards. It is a force multiplier, possessing a mechanical advantage greater than one. Bolts and screws are based on the principle of the inclined plane.

lever A lever is a simple machine consisting of a rigid rod pivoted at a fixed point called the fulcrum, used for shifting or raising a heavy load or applying force. Levers are classified into orders according to where the effort is applied, and the load-moving force developed, in relation to the position of the fulcrum. A **first-order lever** has the load and the effort on opposite sides of the fulcrum – for example, a see-saw or pair of scissors. A **second-order lever** has the load and the effort on the same side of the fulcrum, with the load nearer the fulcrum – for example, nutcrackers or a wheelbarrow. A **third-order lever** has the effort nearer the fulcrum than the load, with both on the same side of it – for example, a pair of tweezers or tongs. The mechanical advantage of a lever is the ratio of load to effort, equal to the perpendicular distance of the effort's line of action from the fulcrum divided by the distance to the load's line of action. Thus tweezers, for instance, have a mechanical advantage of less than one.

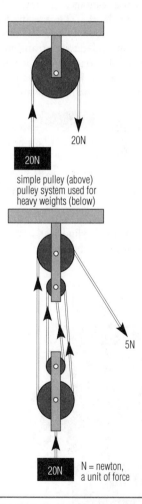

pulley The mechanical advantage of a pulley increases with the number of rope strands. If a pulley system has four ropes supporting the load, the mechanical advantage is four, and a 5 Newton force will lift a 20 Newton load.

wheel and axle A wheel and axle consists of a rope wound round an axle connected to a larger wheel with another rope attached to its rim. Pulling on the wheel rope (applying an effort) lifts a load attached to the axle rope. The velocity ratio of the machine (distance moved by load divided by distance moved by effort) is equal to the ratio of the wheel radius to the axle radius. A wheel is defined as a circular disc that supports an object and allows its easy movement.

pulley A pulley consists of a fixed, grooved wheel, sometimes in a block, around which a rope or chain can be run. A simple pulley serves only to change the direction of the applied effort (as in a simple hoist for raising loads). The use of more than one pulley results in a mechanical advantage, so that a given effort can raise a heavier load. The mechanical advantage depends on the arrangement of the pulleys. For instance, a block and tackle arrangement with three ropes supporting the load will lift it with one-third of the effort needed to lift it directly (if friction is ignored), giving a mechanical advantage of three.

engineering

To remember that to tighten a bolt or nut you turn it clockwise (right), and to take it off you turn it anti-clockwise (left):

Righty tighty, lefty loosie

screw A screw is a cylindrical or tapering piece of metal or plastic with a helical (spiral) groove cut into it. Each turn of a screw moves it forward or backwards by a distance equal to the pitch (the spacing between neighbouring threads). Its mechanical advantage equals $2rP$ where P is the pitch and r is the radius of the thread. Thus the mechanical advantage of a tapering wood screw, for example, increases as it is rotated into the wood. The thread is comparable to an inclined plane (wedge) wrapped around a cylinder or cone.

wedge A wedge is a block of triangular cross-section that can be used as a simple machine. An axe is a wedge that splits wood by redirecting the energy of the downward blow sideways, where it exerts the force needed to split the wood.

major machine components
bearing Bearings are used in a machine to allow free movement between two parts, typically the rotation of a shaft in a housing. **Ball bearings** consist of two rings, one fixed to a housing, one to the rotating shaft. Between them is a set, or race, of steel balls. They are widely used to support shafts, as in the spindle in the hub of a bicycle wheel. The **sleeve**, or **journal bearing**, is the simplest bearing. It is a hollow cylinder, split into two halves. It is used for the big-end and main bearings on a car crankshaft. In some machinery the balls of ball bearings are replaced by cylindrical rollers or thinner **needle bearings**.

cam A cam converts circular motion to linear motion or vice versa. The **edge cam** in a car engine is in the form of a rounded projection on a shaft, the camshaft.

clutch The clutch consists of two main plates: a drive plate connected to the engine crankshaft and a driven plate connected to the wheels. When the clutch is disengaged, the drive plate does not press against the driven plate. When the clutch is engaged, the two plates are pressed into contact and the rotation of the crankshaft is transmitted to the wheels.

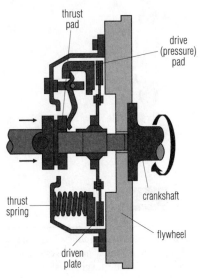

disengaged (pedal pressed down)

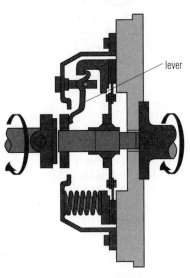

engaged (pedal up)

When the camshaft turns, the cams press against linkages (plungers or followers) that open the valves in the cylinders.

clutch A clutch is any device for disconnecting rotating shafts, used especially in a car's transmission system. In a car with a manual gearbox, the driver depresses the clutch when changing gear, thus disconnecting the engine from the gearbox.

gear A gear is a toothed wheel that transmits the turning movement of one shaft to another shaft. Gear wheels may be used in pairs, or in threes if both shafts are to turn in the same direction. The gear ratio – the ratio of the number of teeth on the two wheels – determines the torque ratio, the turning force on the output shaft compared with the turning force on the input shaft. The ratio of the angular velocities of the shafts is the inverse of the gear ratio.

shaft A shaft is a bar, usually cylindrical, used to support rotating pieces of machinery or to transmit power or motion by rotation.

spring A spring is a device, usually a metal coil, that returns to its original shape after being stretched or compressed. Springs are used in some machines (such as clocks) to store energy, which can be released at a controlled rate. In other machines (such as engines) they are used to close valves.

technology of industry

Industry is the extraction and conversion of raw materials, the manufacture of goods, and the provision of services. Industry can be either low technology, unspecialized, and labour-intensive, as in countries with a large unskilled labour force, or highly automated, mechanized, and specialized, using advanced technology, as in the industrialized countries. Major recent trends in industrial activity have been the growth of electronic, robotic, and microelectronic technologies.

mining Mining entails the extraction of minerals from under the land or sea for industrial or domestic uses. Exhaustion of traditionally accessible resources has led to development of new mining techniques; for example, extraction of oil from offshore deposits and from land shale reserves. Technology is also under development for the exploitation of minerals from entirely new sources such as mud deposits and mineral nodules from the sea bed.

refining Refining is any process that purifies or converts something into a more useful form. Metals usually need refining after they have been extracted from their ores by such processes as smelting. Petroleum, or crude oil, needs refining before it can be used; the process involves fractional distillation, the separation of the substance into separate components or 'fractions'. Electrolytic metal-refining methods use the principle of electrolysis (conducting electricity in a solution or melt to effect chemical changes) to obtain pure metals. When refining petroleum, or crude oil, further refinery processes after fractionation convert the heavier fractions into more useful lighter products. The most important of these processes is cracking; others include polymerization, hydrogenation, and reforming.

Oil Industry: Panoramic Photographs, 1851–1991
http://lcweb2.loc.gov/cgi-bin/query/r?ammem/
pan:@field(SUBJ+@band(+Oil+Petroleum+))
Part of the Panoramic Photo Collection of the US Library of Congress, this page features approximately 50 panoramic photographs of the US oil industry taken between 1851 and 1991. Most images are of Texas and California.

---

automation The widespread use of self-regulating machines in industry is known as automation. It involves the addition of control devices, using electronic sensing and computing techniques, which often follow the pattern of human nervous and brain functions, to already mechanized physical processes of production and distribution; for example, steel processing, mining, chemical production, and road, rail, and air control. Automation builds on the process of mechanization to improve manufacturing efficiency.

robotics Robotics is the application of any computer-controlled machine that can be programmed to move or carry out work. Robots are often used in industry to transport materials or to perform repetitive tasks. For instance, robotic arms, fixed to a floor or workbench, may be used to paint machine parts or assemble electronic circuits. Other robots are designed to work in situations that would be dangerous to humans – for example, in defusing bombs or in space and deep-sea exploration. Some robots are equipped with sensors, such as touch sensors and video cameras, and can be programmed to make simple decisions based on the sensory data received. As robots do not suffer from fatigue or become distracted, researchers in robotics aim to produce robots that can carry out sophisticated tasks more efficiently than humans.

refrigeration Refrigeration is the use of technology to transfer heat from cold to warm, against the normal temperature gradient, so that a refrigeration unit can remain substantially colder than its surroundings. Refrigeration equipment is used for the chilling and deep-freezing of food in food technology, and in air conditioners and industrial processes. Refrigeration is commonly achieved by a vapour-compression cycle, in which a suitable chemical (the refrigerant) travels through a long circuit of tubing, during which it changes from a vapour to a liquid and back again. A compression chamber makes it condense, and thus give out heat. In another part of the circuit, called the evaporator coils, the pressure is much lower, so the refrigerant evaporates, absorbing heat as it does so. The evaporation process takes place near the central part of the refrigerator, which therefore becomes colder, while the compression process takes place near a ventilation grille, transferring the heat to the air outside. The most commonly used refrigerants in modern systems were formerly chlorofluorocarbons, but these are now being replaced by coolants that do not damage the ozone layer.

food technology The US Food and Drug Administration (FDA) has guidelines for acceptable levels of insect contamination in food. For example it is acceptable to have up to 3 fruitfly maggots per 200 g/7 oz of tomato juice; 100 insect fragments per 25 g/0.8 oz of curry powder; and up to 13 insect heads per 100 g/3.5 oz of fig paste.

technology of construction industries

bridge A bridge is a structure that provides a continuous path or road over water, valleys, ravines, or above other roads. The basic designs and composites of these are based on the way they bear the weight of the structure and its load. **Beam** or **girder bridges** are supported at each end by the ground with the weight thrusting downwards. **Cantilever bridges** are a complex form of girder in which only one end is supported. **Arch** bridges thrust outwards and downwards at their ends. **Suspension bridges** use cables under tension to pull inwards against anchorages on either side of the span, so that the roadway hangs from the main cables by the network of vertical cables. The **cable-stayed bridge** relies on diagonal cables connected directly between the bridge deck and supporting towers at each end. Some bridges are too low to allow traffic to pass beneath easily, so they are designed with movable parts, like **swing and draw bridges**.

The Longest Bridges by Span in the World

Bridge	Location	Date opened	Length m	ft
Suspension spans				
Akashi–Kaikyo	Honshu–Awaji Islands, Japan	1998	1,990	6,527
Store Baelt	Zealand–Funen, Denmark	1997	1,600	5,248
Humber Bridge	Kingston-upon-Hull, UK	1973–81	1,410	4,626
Verrazono Narrows	Brooklyn–Staten Island, New York Harbor (NY), USA	1959–64	1,298	4,260
Golden Gate	San Francisco (CA), USA	1937	1,280	4,200
Mackinac Straits	Michigan (MI), USA	1957	1,158	3,800
Bosporus	Golden Horn, Istanbul, Turkey	1973	1,074	3,524
George Washington	Hudson River, New York (NY), USA	1927–31	1,067	3,500
Ponte 25 Abril (Salazar)	Tagus River, Lisbon, Portugal	1966	1,013	3,323
Firth of Forth (road)	South Queensferry, UK	1958–64	1,006	3,300
Severn Bridge	Beachley, UK	1961–66	988	3,240
Cable-stayed spans				
Pont de Normandie	Seine Estuary, France	1995	2,200	7,216
Skarnsundet	near Trondheim, Norway	1991	530	1,740
Cantilever spans				
Howrah (railroad)	Hooghly River, Calcutta, India	1936–43	988	3,240
Pont de Québec (railroad)	St Lawrence, Canada	1918	549	1,800
Ravenswood	Ravenswood (WV) USA	1981	525	1,723
Firth of Forth (rail)	South Queensferry, UK	1882–90	521	1,710
Commodore Barry	Chester (PA), USA	1974	494	1,622
Greater New Orleans	Mississippi River (LA), USA	1958	480	1,575
Steel arch spans				
New River Gorge	Fayetteville (WV), USA	1977	518	1,700
Bayonne (Killvan Kull)	New Jersey–Staten Island (NY), USA	1932	504	1,652
Sydney Harbour	Sydney, Australia	1923–32	503	1,500

The Tallest Buildings in the World

(N/A = not available.)

Building/structure	City	Height m	ft	Storeys
Inhabited buildings				
Miglin-Beitler Tower[1]	Chicago (IL), USA	609	1,999	N/A
Chongqing Tower	Chongqing, China	460[2]	1,509[2]	114
Petronas Tower[3]	Kuala Lumpur, Malaysia	452[2]	1,483[2]	113
Sears Tower	Chicago (IL), USA	443[2]	1,454[2]	110
World Trade Center[3]	New York (NY), USA	417[2]	1,368[2]	110
Empire State Building	New York (NY), USA	381[2]	1,250[2]	102
Bank of China	Hong Kong, China	368	1,209	72
Amoco Building	Chicago (IL), USA	346	1,136	80
John Hancock Centre	Chicago (IL), USA	344	1,127	100
Chrysler Building	New York (NY), USA	319	1,046	77
Nations Bank Tower	Atlanta (GA), USA	312	1,023	55
First Interstate World Center	Los Angeles (CA), USA	310	1,017	73
Stratosphere Tower	Las Vegas (NV), USA	308	1,012	114
Texas Commerce Tower	Houston (TX), USA	305	1,002	75
Allied Bank Plaza	Houston (TX), USA	302	992	71
Two Prudential Plaza	Chicago (IL), USA	298	978	64
311 South Waker Drive	Chicago (IL), USA	295	969	65
First Canadian Place	Toronto, Ontario, Canada	290	952	72
American International	New York (NY), USA	290	952	66
Bay/Adelaide Centre	Toronto, Ontario, Canada	288	945	53
One Liberty Place	Philadelphia (PA), USA	288	945	62
Columbia Seafirst Center	Seattle (WA), USA	287	943	76
40 Wall Tower	New York (NY), USA	283	927	70
Nations Bank Plaza	Dallas (TX), USA	281	921	72
Citicorp Center	New York (NY), USA	279	915	59
Scotia Plaza	Toronto, Ontario, Canada	275	902	68
One Peach Tree Center	Atlanta (GA), USA	275	902	60
Transco Tower	Houston (TX), USA	274	901	64
Society Center	Cleveland (OH), USA	271	888	57
Two Union Square	Seattle (WA), USA	270	886	56
AT&T Corporate Center	Chicago (IL), USA	270	885	60
Mellon Bank Center	Philadelphia (PA), USA	268	880	56
Nations Bank Corporate Center	Charlotte (NC), USA	267	875	60
900 North Michigan	Chicago (IL), USA	265	871	66
Canada Trust Tower	Toronto, Ontario, Canada	263	863	51
Water Tower Place	Chicago (IL), USA	262	859	74
First Interstate Bank	Los Angeles (CA), USA	261	858	62
Transamerica Pyramid	San Francisco (CA), USA	260	853	61
G E Building, Rockefeller Center	New York (NY), USA	259	850	70
One First National Plaza	Chicago (IL), USA	259	851	60
Two Liberty Place	Philadelphia (PA), USA	258	845	52
USX Towers	Pittsburgh (PA), USA	256	841	64
One Atlantic Center	Atlanta (GA), USA	251	825	50
Cityspire	New York (NY), USA	248	814	72
One Chase Manhattan	New York (NY), USA	248	813	60
Metlife Building	New York (NY), USA	246	808	59
John Hancock Tower	Boston (MA), USA	244	800	60
Tallest structures				
Warszawa Radio Maszt[4]	Konstantynadio Maszt	646	2,120	–
KTHI-TV Mast	Fargo (ND), USA	629	2,063	–
CN Tower	Toronto, Ontario, Canada	555	1,822	–

[1] Planned; this will become the tallest inhabited building when completed.
[2] Excluding TV antennas.
[3] Tallest tower in building listed.
[4] Collapsed during renovation, August 1991.

dam A dam is a structure built to hold back water in order to prevent flooding, to provide water for irrigation and storage, and to provide hydroelectric power. The biggest dams are of the earth- and rock-fill type, also called **embankment dams**. Such dams are generally built on broad valley sites. Earth dams have a watertight core wall, formerly made of puddle clay but nowadays constructed of concrete. Their construction is very economical even for very large structures. Rock-fill dams are a variant of the earth dam in which dumped rock takes the place of compacted earth fill.

Deep, narrow gorges dictate a concrete dam, where the strength of reinforced concrete can withstand the water pressures involved. Many concrete dams are triangular in cross section, with their vertical face pointing upstream. Their sheer weight holds them in position, and so they are called **gravity dams**. They are no longer favoured for very large dams, however, because they are expensive and time-consuming to build. Other concrete dams are built in the shape of an arch, which transfers the horizontal force into the sides of the river valley: the **arch dam** derives its strength from the arch shape, just as an arch bridge does, and has been widely used in the 20th century.

They require less construction material than other dams and are the strongest type. **Buttress dams** are used when economy of construction is important or foundation conditions preclude any other type. The upstream portion of a buttress dam may comprise series of cantilevers, slabs, arches, or domes supported from the back by a line of buttresses. They are usually made from reinforced and pre-stressed concrete. In 1997 there were approximately 40,000 large dams (more than 15 m/49 ft in height) and 800,000 small ones worldwide.

If it were not for the world's dams holding back river water, sea levels would be 3 cm/1.2 in higher than they are.

canal Canals are artificial waterways constructed for drainage, irrigation, or navigation. **Irrigation canals** carry water for irrigation from rivers, reservoirs, or wells, and are designed to maintain an even flow of water over the whole length. **Navigation and ship canals** are constructed at one level between locks, and frequently link with rivers or sea inlets to form a waterway system. Where speed is not a prime factor, the cost-effectiveness of transporting goods by canal has encouraged a revival; Belgium, France, Germany, and the states of the former USSR are among countries that have extended and streamlined their canals.

tunnel A tunnel is a passageway through a mountain, under a body of water, or underground. The difficulties of tunnelling naturally increase with the size, length, and depth of tunnel, but with the mechanical appliances now available no serious limitations are

The Highest Dams in the World

Source: Institute of Civil Engineers, London

Dam	Location	Height above lowest formation		Dam	Location	Height above lowest formation	
		m	ft			m	ft
Rogun[1]	Tajikistan	335	1,099	Ertan[1]	China	240	787
Nurek	Tajikistan	300	984	Mauvoisin	Switzerland	237	778
Grand Dixence	Switzerland	285	935	Chivor	Colombia	237	778
Inguri	Georgia	272	892	El Cajon	Honduras	234	768
Boruca[1]	Costa Rica	267	875	Chirkey	Russia	233	765
Chicoasen	Mexico	261	856	Oroville	USA	230	754
Tehri[1]	India	261	856	Bekhme[1]	Iraq	230	754
Kambaratinsk[1]	Kyrgyzstan	255	836	Bhakra	India	226	741
Kishau[1]	India	253	830	Hoover	USA	225	738
Sayano-Shushensk[1]	Russia	245	804	Contra	Switzerland	235	772
Guavio	Colombia	243	797	Mratinje	Yugoslavia	235	772
Mica	Canada	242	794				

[1] Under construction.

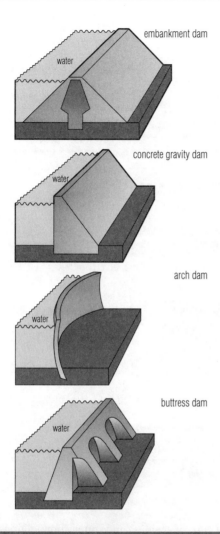

embankment dam

water

concrete gravity dam

water

arch dam

water

buttress dam

water

dam There are two basic types of dam: the gravity dam and the arch dam. The gravity dam relies upon the weight of its material to resist the forces imposed upon it; the arch dam uses an arch shape to take the forces in a horizontal direction into the sides of the river valley. The largest dams are usually embankment dams. Buttress dams are used to hold back very wide rivers or lakes.

imposed. In recent years there have been notable developments in linings (for example, concrete segments and steel liner plates), and in the use of rotary diggers and cutters and explosives. Small-section tunnels are usually driven from one end to the other at their full dimensions. Large-section tunnels are often driven in two stages; a pilot heading is excavated in advance which is afterwards enlarged to the full section of the main tunnel. The normal procedure in tunnelling in rock is as follows. Power drills are used to bore successive rounds of holes in the face. Each round is fired and the broken rock removed by hand shovels or mechanical loaders. In hard rock, blasting is necessary to break down the material. The section is trimmed to its proper size by further blasting or by pneumatic picks, and timber or steel supports are erected. Sometimes side and top lagging boards are required. In loose ground the top laggings are driven in advance of the last supporting set (fore-poling) before the debris is removed. In sand or gravel the problem is one of support rather than excavation, and fore-poling is necessary. The poling pieces are driven along the sides and top of the tunnel to protect the workers from sudden falls or 'runs' of ground. In soft ground, tunnelling

Canals and Waterways

Name	Country	Opened	Length (km/mi)
Amsterdam	Netherlands	1876	26.6/16.5
Baltic–Volga	Russian Federation, Belarus, Ukraine	1964	2,430/1,510
Baltic–White Sea	Russian Federation	1933	235/146
Corinth	Greece	1893	6.4/4
Elbe and Trave	Germany	1900	66/41
Erie	USA	1825	580/360
Göta	Sweden	1832	185/115
Grand Canal	China	485 BC–AD 1972	1,050/650
Kiel	Germany	1895	98/61
Manchester	England	1894	57/35.5
Panama	Panama (US zone)	1914	81/50.5
Princess Juliana	Netherlands	1935	32/20
St Lawrence	Canada	1959	3,770/2,342
Sault Ste Marie	USA	1855	2.6/1.6
Sault Ste Marie	Canada	1895	1.8/1.1
Welland	Canada	1929	45/28
Suez	Egypt	1869	166/103

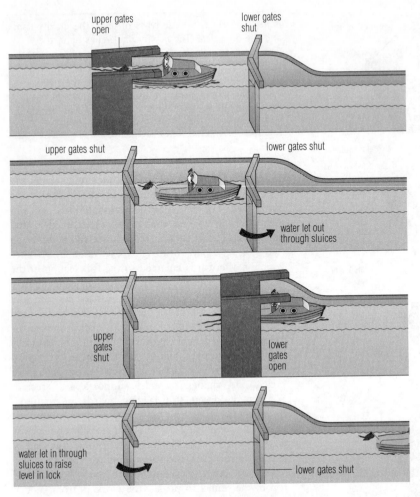

upper gates open

lower gates shut

upper gates shut

lower gates shut

water let out through sluices

upper gates shut

lower gates open

water let in through sluices to raise level in lock

lower gates shut

lock Travelling downstream, a boat enters the lock with the lower gates closed. The upper gates are then shut and the water level lowered by draining through sluices. When the water level in the lock reaches the downstream level, the lower gates are opened.

shields provide overhead protection, and the support given to the face helps to reduce air losses when air pressure is used to keep water out of the workings. The shield is more easily pushed ahead when the ground immediately adjacent to the shield is replaced with soft, puddled clay. Alignment is laser-controlled to keep the tunnels within a tolerance on position of 30 mm/1.18 in vertically and 300 mm/11.8 in laterally.

technology of transportation

land

car Originally the automated version of the horse-drawn carriage, a car is a small, driver-guided, passenger-carrying motor vehicle designed to transport people. Most are four-wheeled and have water-cooled, piston-type internal-combustion engines fuelled by petrol or diesel. Variations have existed for decades

that use ingenious and often nonpolluting power plants; steam was an attractive form of power to the English pioneers. A typical present-day medium-sized saloon car has a semi-monocoque construction in which the body panels, suitably reinforced, support the road loads through independent front and rear sprung suspension, with seats located within the wheelbase for comfort.

A car is usually powered by a petrol engine using a carburettor to mix petrol and air for feeding to the engine cylinders (typically four or six), and the engine is usually water cooled. In the 1980s high-performance diesel engines were being developed for use in private cars, and it is anticipated that this trend will continue for reasons of economy. From the engine, power is transmitted through a clutch to a four- or five-speed gearbox and from there, in a front-engine rear-drive car, through a drive (propeller) shaft to a differential gear, which drives the rear wheels. In a front-engine,

The Longest Vehicular Tunnels in the World

Tunnel	Location	Year opened	Length	
			km	mi
Saint Gotthard	Switzerland	1980	16.3	10.1
Arlberg	Austria	1978	14.0	8.7
Fréjus	France–Italy	1980	12.9	8.0
Mont Blanc	France–Italy	1965	11.7	7.3
Gran Sasso	Italy	1976	10.0	6.2
Seelisberg	Switzerland	1979	9.3	5.8
Mount Ena	Japan	1976	8.5	5.3
Rokko 11	Japan	1974	6.9	4.3
San Bernardino	Switzerland	1967	6.6	4.1
Tauren	Austria	1974	6.4	4.0

front-wheel drive car, clutch, gearbox, and final drive are incorporated with the engine unit.

Since cars are responsible for almost a quarter of the world's carbon dioxide emissions, technologists have experimented with alternative power sources developing steam cars, solar-powered cars, and hybrid cars using both electricity (in town centres) and petrol (on the open road). To achieve weight reduction in the body, aluminium and plastics are being used and fuel consumption is being improved through aerodynamic body designs to reduce air resistance. Microprocessors are also being developed to measure temperature, engine speed, pressure, and oxygen/carbon dioxide content of exhaust gases, and to readjust the engine accordingly.

locomotive A locomotive is an engine for hauling railway trains. In a steam locomotive, fuel (usually coal, sometimes wood) is burned in a furnace. The hot gases and flames produced are drawn through tubes running

locomotive The drive of an electric locomotive is provided by powerful electric motors (traction motors) in the bogies beneath the body of the locomotive. The motors are controlled by equipment inside the locomotive. Both AC and DC power supplies are used, although most modern systems use a 2500 V supply.

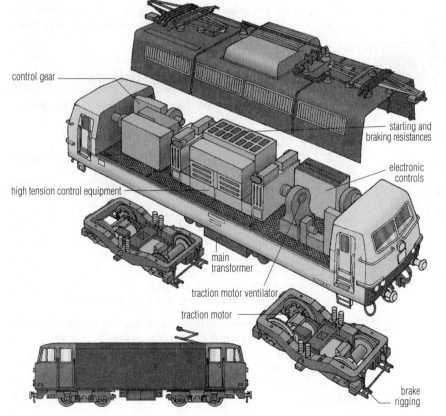

control gear

high tension control equipment

starting and braking resistances

electronic controls

main transformer

traction motor ventilator

traction motor

brake rigging

through a huge water-filled boiler and heat up the water to steam. The steam is then fed to the cylinders, where it forces the pistons back and forth. Movement of the pistons is conveyed to the wheels by cranks and connecting rods. Today most locomotives are diesel or electric.

Diesel locomotives have a powerful diesel engine, burning oil, and **electric locomotives** draw their power from either an overhead cable or a third rail alongside the ordinary track. The engine may drive a generator to produce electricity to power electric motors that turn the wheels, or the engine drives the wheels mechanically or through a hydraulic link. A number of **gas-turbine locomotives** are in use, in which a turbine spun by hot gases provides the power to drive the wheels.

maglev Maglev (an acronym for **magnetic levitation**) is a high-speed surface transport using the repellent force of superconductive magnets to propel and support, for example, a train above a track. The train is levitated by electromagnets and forward thrust is provided by linear motors aboard the cars, propelling the

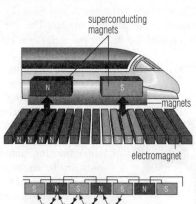

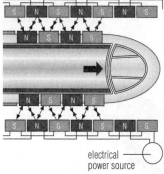

maglev The repulsion of superconducting magnets and electromagnets in the track keeps a maglev train suspended above the track. By varying the strength and polarity of the track electromagnets, the train can be driven forward.

train along a reaction plate Technical trials on a maglev train track began in Japan in the 1970s, and a speed of 500 kph/310 mph has been reached, with a cruising altitude of 10 cm/4 in.

sea

ship A ship is a large seagoing vessel. The invention of the stern rudder during the 12th century, together with the developments made in sailing during the Crusades, enabled the use of sails to almost completely supersede that of oars. One of the finest of the tea clippers, the *Sir Launcelot*, was built in 1865 and marked the highest development of the sailing ship. Early **steamers** depended partly on sails for auxiliary power. The opening of the Suez Canal in 1869, together with the simultaneous introduction of the compound engine, raised steamships to superiority over sailing ships. This was followed by the introduction of the internal combustion engine. The **tanker** was developed after World War II to carry oil supplies to the areas of consumption. The prolonged closure of the Suez Canal after 1967 and the great increase in oil consumption led to the development of the very large tanker, or 'supertanker'. More recently hovercraft and hydrofoil boats have been developed for specialized purposes, particularly as short-distance ferries. Sailing ships in automated form for cargo purposes, and maglev ships, were in development in the early 1990s.

---

port and starboard

To associate port/starboard, left/right and red/green navigation lights:

'There's no **red port** wine **left**'

(knowing this, you can work out that green, starboard and right go together)

---

hovercraft A hovercraft is a vehicle that rides on a cushion of high-pressure air, free from all contact with the surface beneath, invented by the British engineer Christopher Cockerell in 1959. Hovercraft need a smooth terrain when operating overland and are best adapted to use on waterways. They are useful in places where harbours have not been established.

hydrofoil Hydrofoils are wings that develop lift in the water in much the same way that an aeroplane wing develops lift in the air. A hydrofoil boat is a seagoing vessel in which the hull rises out of the water , the boat skimming along the surface due to the lift created by its hydrofoils. The first hydrofoil was fitted to

compass points

To remember the points of the compass, in the correct order:

Nobody **e**ver **s**wallows **w**hales.

(Place the first letter of each word in a clockwise circle starting at the 12 o'clock (north) position)

---

a boat in 1906. The first commercial hydrofoil went into operation in 1956.

submarine A submarine is an underwater warship. The conventional submarine of World War I was driven by diesel engine on the surface and by battery-powered electric motors underwater. The diesel engine also drove a generator that produced electricity to charge the batteries. In 1954 the USA launched the first **nuclear-powered submarine**, the *Nautilus*. Operating depth is usually up to 300 m/1,000 ft, and nuclear-powered speeds of 30 knots (55 kph/34 mph) are reached. As in all nuclear submarines, propulsion is by steam turbine driving a propeller. The steam is raised using the heat given off by the nuclear reactor.

U-475: Soviet Foxtrot Class Submarine
http://www.wtj.com/artdocs/u-475.htm
This page provides a room-by-room tour of a Cold War-era Soviet submarine, the U-475. The text is complimented by numerous colour photographs of the submarine's interior. Since being sold to a private owner in 1994, the submarine has been moored near London and is open to the public.

---

sailing and depth of water

If sailing in unfamiliar waters, remember:

Brown brown, run aground
White white, you might
Green green, nice and clean
Blue blue, run right through

This allows you to estimate the depth of the water from shallowest to deepest, based on its colour

---

air

aeroplane Aeroplanes are powered heavier-than-air craft supported in flight by fixed wings. They are propelled by the thrust of a jet engine or airscrew (propeller). They must be designed aerodynamically, since efficient streamlining prevents the formation of shock waves over the body surface and wings, which would cause instability and power loss. Streamlining the plane also reduces the drag resulting in higher speeds and reduced fuel consumption for a given power. The wing of an aeroplane has the cross-sectional shape of an airfoil, being broad and curved at the front, flat underneath, curved on top, and tapered to a sharp point at the rear. It is so shaped that air passing above it is speeded up, reducing pressure below atmospheric pressure. This follows from Bernoulli's principle and results in a force acting vertically upwards, called lift, which counters the plane's weight.

The shape of a plane is dictated principally by the

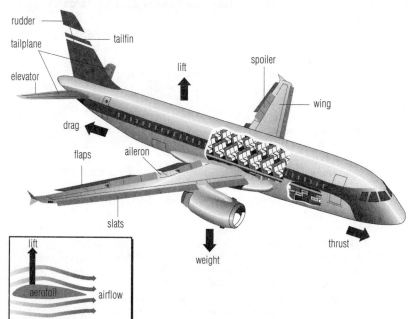

---

aeroplane In flight, the forces on an aeroplane are lift, weight, drag, and thrust. The lift is generated by the air flow over the wings, which have the shape of an aerofoil. The engine provides the thrust. The drag results from the resistance of the air to the aeroplane's passage through it. Various moveable flaps on the wings and tail allow the aeroplane to be controlled. The rudder is moved to turn the aeroplane. The elevators allow the craft to climb or dive. The ailerons are used to bank the aeroplane while turning. The flaps, slats, and spoilers are used to reduce lift and speed during landing.

Aviation Firsts

(– = not applicable.)

Date	Name	Achievement
1804	George Cayley (UK)	first modern-configuration aeroplane prototype in history ('Cayley's glider')
1891	Otto Lilienthal (Germany)	first successful crewed glider flight
1896	Samuel Pierpont Langley (USA)	first steam-driven pilotless aeroplane (12,082 m/24,000 ft over the Potomac River)
1903	Orville and Wilbur Wright (USA)	first successful flight of a powered aircraft, Kitty Hawk (NC) (59 seconds; 259 m/850 ft)
1906	Alberto Santos-Dumos (Brazil)	first successful European flight (Bagatelle Field, Paris, France)
1908	Frank P Lahm (USA)	first aeroplane passenger (with Wilbur Smith piloting)
	Samuel Franklin Cody (UK)	first powered flight in the UK
1909	Louis Blériot (France)	first successful flight across the English Channel
1910	Raymonde de la Roche (France)	first licensed female pilot
	Henri Faber (France)	first take-off from water
1911	Harriet Quimby (USA)	first US female pilot
1912	Harriet Quimby (USA)	first crossing of the English Channel by female pilot
	–	first automatic pilot in service
1913	Ivan Sikorsky (Russia)	first multi-engined aircraft
1914	–	first scheduled passenger airline service (St Petersburg–Tampa, Florida)
1914–18	–	first military use of aircraft; developments in speed and power of aircrafts triggered by World War I
1918	–	first airmail service established in the USA (Washington–New York City)
1919	Walter Hinton (USA)	first transatlantic flight (Trepassy Bay, Newfoundland–Lisbon, Portugal via Horta, Azores, and Ponta Delgada)
	John W Alcock, Arthur Whitten Brown (UK)	first nonstop transatlantic flight (St John's, Newfoundland–Clifden, Ireland; 3,058 km/1,900 mi; 16 hr 12 min)
	Ross Smith, Keith Smith (Australia)	first flight from the UK to Australia
	–	first scheduled London, UK–Paris, France passenger service
	–	first airline food served
1923	John A Macready, Oakley Kelly (USA)	first nonstop transcontinental flight (New York–San Diego, USA; 4,023 km/2,500 mi; 26 hr 50 min)
1924	Lowell Smith, Erik Nelson (USA)	first round-the-world flight
1926	Richard E Byrd, Floyd Bennett (USA)	first flight over the North Pole
1927	Charles Augustus Lindbergh (USA)	first solo nonstop transatlantic flight
1928	Charles K Smith, C T P Ulm	first transpacific flight (San Francisco, USA–Brisbane, Australia)
1929	James H Doolittle (USA)	first take-off and landing relying solely on instruments ('blind')
	Richard E Byrd, Nernt Balchen, Harold I June, A C McKinley (USA)	first flight over the South Pole
1930	Frank Whittle (UK)	jet engine patented
	Amy Johnson (UK)	first female pilot to fly from the UK to Australia
	Ellen Church (USA)	first flight attendant
1931	Hugh Herndon, Clyde Pangborn (USA)	first nonstop transpacific flight (Sabishiro Beach, Japan–near Wenatchee (WA), USA; 41 hr 13 min)
1932	Amelia Earhart (USA)	first transatlantic solo by a female pilot (Harbor Grace, Newfoundland–Ireland; around 15 hr)
1933	Wiley Post (USA)	first round-the-world solo flight
1937	Amelia Earhart (USA)	first attempt at a round-the-world flight by a female pilot (Earhart disappeared in the Pacific, between New Guinea and Howland Island)
	–	first fully pressurized aircraft (*Lockheed XC-35*)
1939	Erich Warsitz (Germany)	first turbojet flight, Germany (*Heinkel He 178*)
1939–45	–	developments related to World War II (*Hawker Hurricane* and *Supermarine Spitfire* fighters; *Avro Lancaster* and *Boeing Flying Fortress* bombers)

Date	Name	Achievement
1941	–	first UK jet aircraft (Gloster e.28/39)
1942	Robert Stanley (USA)	first US jet plane flight
1944	–	first rocket-engine fighter plane operational, Germany (Messerschmitt Me 163B Komet)
1947	Charles E ('Chuck') Yeager (USA)	first piloted supersonic flight (Bell X-1 rocket-powered aircraft)
1949	James Gallagher (USA)	first round-the-world nonstop flight
	–	first jet airliner in service (de Havilland Comet)
1950	David C Schilling (USA)	first nonstop transatlantic jet flight (UK to Limestone, Maine)
1952	–	first jetliner service (London, UK–Johannesburg, South Africa)
1953	–	first vertical take-off aircraft tested (Rolls-Royce Flying Bedstead)
	Jacqueline Cochran (USA)	first female pilot to break the sound barrier
1957	Archie J Old Jr (USA)	first nonstop round-the-world jet plane flight (45 hr 19 min)
1958	–	first domestic jet passenger service (New York–Miami, USA); first transatlantic jet passenger service (New York, USA–London, UK, and Paris, France)
1960	–	first supersonic bomber (Convair B-58)
1968	–	first supersonic plane (Tupolev Tu-144; supersonic speed achieved: Mach 2, c. 1,924 kph/1,200 mph)
1970	–	Mach 2 exceeded (Tupolev Tu-144; 2,140 kph/1,335 mph)
	–	Boeing 747 jumbo jet in service
1976	–	first commercial supersonic flights (Concorde)
1977	Paul MacCready (USA)	first succcessful human-powered aircraft (Gossamer Condor)
1979	Bryan Allen (USA)	first crossing of the English Channel by a human-powered aircraft (Gossamer Albatross)
1980	Janice Brown (USA)	first successful long-distance solar-powered flight (Solar Challenger; 10 km/6 mi in 22 min)
1986	Dick Rutan, Jeana Yeager (USA)	first nonstop around-the-world flight without refuelling (Voyager; 216 hr 3 min 44 sec)
1988	Kaneilos Kanellopoulos (Greece)	first flight of a human-powered aircraft across the Aegean Sea
1992	–	first radio-controlled ornithopter (aircraft propelled and manoeuvred by flapping wings; model demonstrated in the USA)
1993	Barbara Harmer (UK)	first female co-pilot of a commercial supersonic plane
	–	automatic on-board collision avoidance system (TCAS-2) mandatory in US airspace

speed at which it will operate. A low-speed plane operating at well below the speed of sound (about 965 kph/600 mph) need not be particularly well streamlined, and it can have its wings broad and projecting at right angles from the fuselage. An aircraft operating close to the speed of sound must be well streamlined and have swept-back wings. Supersonic planes (faster than sound) need to be severely streamlined, and require a needle nose, extremely swept-back wings, and what is often termed a 'Coke-bottle' (narrow-waisted) fuselage, in order to pass through the sound barrier without suffering undue disturbance. To give great flexibility of operation at low as well as high speeds, some supersonic planes are designed with variable geometry, or swing wings. For low-speed flight the wings are outstretched; for high-speed flight they are swung close to the fuselage to form an efficient **delta wing** configuration.

Wings by themselves are unstable in flight, and a plane requires a tail to provide stability. The tail comprises a horizontal tailplane and vertical tailfin, called the horizontal and vertical stabilizer respectively. The tail plane has hinged flaps at the rear called elevators to control pitch (attitude). Raising the elevators depresses the tail and inclines the wings upwards (increases the angle of attack). This speeds the airflow above the wings until lift exceeds weight and the plane climbs. However, the steeper attitude increases drag, so more power is needed to maintain speed and the engine throttle must be opened up. Moving the elevators in the opposite direction produces the reverse effect. The angle of attack is reduced, and the plane descends. Speed builds up rapidly if the engine is not throttled back. Turning (changing direction) is effected by moving the rudder hinged to the rear of the tailfin, and by banking (rolling) the plane. It is banked by moving the ailerons, interconnected flaps at the rear of the wings which move in opposite directions, one up, the other down.

helicopter A helicopter is a powered aircraft that achieves both lift and propulsion by means of a rotary wing, or rotor, on top of the fuselage. It can take off and land vertically, move in any direction, or remain stationary in the air. It can be powered by a piston or jet engine. The rotor of a helicopter has two or more blades of aerofoil cross-section like an aeroplane's wings. Lift and propulsion are achieved by angling the blades as they rotate. A single-rotor helicopter must also have a small tail rotor to counter the torque, or tendency of the body to spin in the opposite direction to the main rotor. Twin-rotor helicopters, like the Boeing Chinook, have their rotors turning in opposite directions to prevent the body from spinning.

airship, or dirigible Any aircraft that is lighter than air and power-driven is known as an airship or dirigible. It consists of an ellipsoidal balloon that forms the streamlined envelope or hull and has below it the propulsion system (propellers), steering mechanism, and space for crew, passengers, and/or cargo. The balloon section is filled with lighter-than-air gas, either the nonflammable helium or, before helium was industrially available in large enough quantities, the easily ignited and flammable hydrogen. The envelope's form is maintained by internal pressure in the nonrigid (blimp) and semirigid (zeppelin) types. The zeppelin type maintains its form using an internal metal framework.

technology of telecommunications

Telecommunications is simply communications over a distance, generally by electronic means. The Scottish scientist Alexander Graham Bell pioneered long-distance voice communication in 1876 when he invented the telephone. Today it is possible to communicate internationally by telephone cable or by satellite or

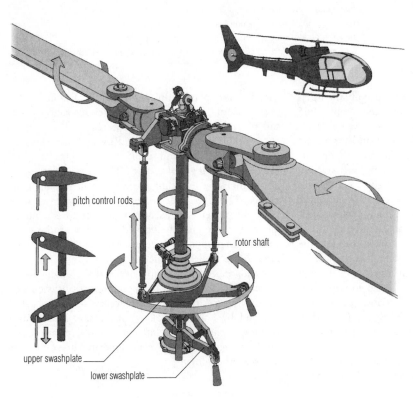

pitch control rods

rotor shaft

upper swashplate

lower swashplate

helicopter The helicopter is controlled by varying the rotor pitch (the angle of the rotor blade as it moves through the air). For backwards flight, the blades in front of the machine have greater pitch than those behind the craft. This means that the front blades produce more lift and a backwards thrust. For forwards flight, the situation is reversed. In level flight, the blades have unchanging pitch.

microwave link, with over 100,000 simultaneous conversations and several television channels being carried by the latest satellites.

radio Radio is the transmission and reception of radio waves. In radio transmission a microphone converts sound waves (pressure variations in the air) into electromagnetic waves that are then picked up by a receiving aerial and fed to a loudspeaker, which converts them back into sound waves. To carry the transmitted electrical signal, an oscillator produces a carrier wave of high frequency. Different stations are allocated different transmitting carrier frequencies. A modulator superimposes the audiofrequency signal on the carrier. There are two main ways of doing this: **amplitude modulation** (AM), used for long- and medium-wave broadcasts, in which the strength of the carrier is made to fluctuate in time with the audio signal; and **frequency modulation** (FM), as used for VHF broadcasts, in which the frequency of the carrier is made to fluctuate. The transmitting aerial emits the modulated electromagnetic waves, which travel outwards from it.

In radio reception a receiving aerial picks up minute voltages in response to the waves sent out by a transmitter. A tuned circuit selects a particular frequency, usually by means of a variable capacitor connected across a coil of wire. A demodulator disentangles the audio signal from the carrier, which is now discarded, having served its purpose. An amplifier boosts the audio signal for feeding to the loudspeaker. In a superheterodyne receiver, the incoming signal is mixed with an internally-generated signal of fixed frequency so that the amplifier circuits can operate near their optimum frequency.

telegraphy Telegraphy is the transmission of messages along wires by means of electrical signals. The first modern form of telecommunication, it now uses printers for the transmission and receipt of messages. Telex is an international telegraphy network.

fax Fax is the common name for **facsimile transmission** or **telefax**, the transmission of images over a telecommunications link, usually the telephone net-

work. When placed on a fax machine, the original image is scanned by a transmitting device and converted into coded signals, which travel via the telephone lines to the receiving fax machine, where an image is created that is a copy of the original. Photographs as well as printed text and drawings can be sent.

telephone Developed by Scottish inventor Alexander Graham Bell in 1876, the telephone is an instrument for communicating by voice along wires. The transmitter (mouthpiece) consists of a carbon microphone, with a diaphragm that vibrates when a person speaks into it. The diaphragm vibrations compress grains of carbon to a greater or lesser extent, altering their resistance to an electric current passing through them. This sets up variable electrical signals, which travel along the telephone lines to the receiver of the person being called. There they cause the magnetism of an electromagnet to vary, making a diaphragm above the electromagnet vibrate and give out sound waves, which mirror those that entered the mouthpiece originally.

The standard instrument has a handset, which houses the transmitter (mouthpiece), and receiver (earpiece), resting on a base, which has a dial or pushbutton mechanism for dialling a telephone number. Some telephones combine a push-button mechanism and mouthpiece and earpiece in one unit. A cordless telephone is of this kind, connected to a base unit not by wires but by radio. It can be used at distances up to about 100 m/330 ft from the base unit.

television (TV) Television is the reproduction of visual images at a distance using radio waves. For transmission, a television camera converts the pattern

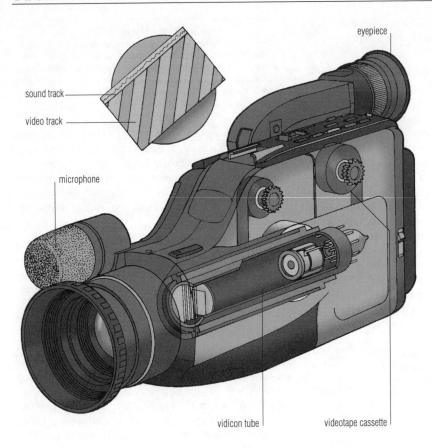

eyepiece

sound track

video track

microphone

vidicon tube

videotape cassette

video camera The heart of the video camera is the vidicon tube which converts light entering the front lens to an electrical signal. An image is formed on a light-sensitive surface at the front of the tube. The image is then scanned by an electron beam, to give an output signal corresponding to the image brightness. The signal is recorded as a magnetic track traversing the tape diagonally. The sound track, which records the sounds picked up by the microphone, runs along the edge of the tape.

of light it takes in, into a pattern of electrical charges. This is scanned line by line by a beam of electrons from an electron gun, resulting in variable electrical signals that represent the picture. These signals are combined with a radio carrier wave and broadcast as electromagnetic waves. The TV aerial picks up the wave and feeds it to the receiver (TV set). This separates out the vision signals, which pass to a cathode-ray tube where a beam of electrons is made to scan across the screen line by line, mirroring the action of the electron gun in the TV camera. The result is a recreation of the pattern of light that entered the camera. In Europe, 25 pictures (30 in North America) are built up each second with interlaced scanning with a total of 625 lines (525 lines in North America and Japan).

In colour television, signals indicate the amounts of red, green, and blue light to be generated at the receiver. To transmit each of these three signals in the same way as the single brightness signal in black and white television, three times the normal band width would be needed which would reduce the number of possible stations and programmes to one-third of that

possible with monochrome television. The three signals are therefore coded into one complex signal, which is transmitted as a more or less normal black and white signal and produces a satisfactory – or compatible – picture on black and white receivers. A fraction of each primary red, green, and blue signal is added together to produce the normal brightness, or luminance, signal. The minimum of extra colouring information is then sent by a special subcarrier signal, which is superimposed on the brightness signal. This extra colouring information corresponds to the hue and saturation of the transmitted colour, but without any of the fine detail of the picture. The impression of

Big Dream, Small Screen
http://www.pbs.org/wgbh/pages/amex/
technology/bigdream/index.html
Companion to the US Public Broadcasting Service (PBS) television programme The American Experience, this page tells the story of Philo Farnsworth, David Sarnoff, and the invention of the television. You will also find an interesting section on TV milestones that features numerous photographs, a list of quotes about television, a biography of David Sarnoff, and a list of sources for further reading.

sharpness is conveyed only by the brightness signal, the colouring being added as a broad colour wash. The colour receiver has to amplify the complex signal and decode it back to the basic red, green, and blue signals; these primary signals are then applied to a colour cathode-ray tube.

High-definition television (HDTV) offers a significantly greater number of scanning lines, and therefore a clearer picture, than the 525/625 lines of established television systems. In 1989 the Japanese broadcasting station NHK and a consortium of manufacturers launched the Hi-Vision HDTV system, with 1,125 lines and a wide-screen format. **Digital television** is a system of transmitting television programmes in digital codes. Until the late 1980s it was considered impossible to convert a TV signal into digital code because of the sheer amount of information needed to represent a visual image. However, data compression techniques have been developed to reduce the number of bits that need to be transmitted each second. As a result, digital technology is being developed that will offer sharper pictures on wider screens, and HDTV with image quality comparable to a cinema.

communications satellites The chief method of relaying long-distance calls on land is microwave radio transmission. This has the drawback that the transmissions follow a straight line from tower to tower, so that over the sea the system becomes impracticable. A system of communications satellites in an orbit 35,900 km/22,300 mi above the Equator, where they circle the Earth in exactly 24 hours and thus appear fixed in the sky, is

now in operation internationally, by Intelsat. The satellites are called geostationary satellites (syncoms). Telegraphy, telephony, and television transmissions are carried simultaneously by high-frequency radio waves. They are beamed to the satellites from large dish antennae or Earth stations, which connect with international networks.

key technological inventions

photographic and print technology

camera A camera is an apparatus used in photography, consisting of a lens system set in a light-proof box

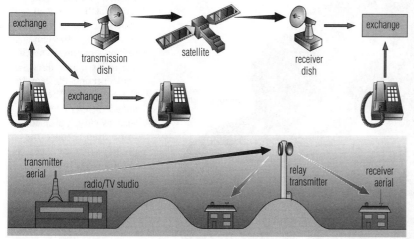

exchange

transmission dish

satellite

receiver dish

exchange

exchange

exchange

transmitter aerial

radio/TV studio

relay transmitter

receiver aerial

telecommunications The international telecommunications system relies on microwave and satellite links for long-distance international calls. Cable links are increasingly made of optical fibres. The capacity of these links is enormous. The TDRS-C (tracking data and relay satellite communications) satellite, the world's largest and most complex satellite, can transmit in a single second the contents of a 20-volume encyclopedia, with each volume containing 1,200 pages of 2,000 words. A bundle of optical fibres, no thicker than a finger, can carry 10,000 phone calls – more than a copper wire as thick as an arm.

Technology Timeline
http://www.pbs.org/wgbh/pages/amex/technology/techtimeline/
index.html

Companion to the US Public Broadcasting Service (PBS) television programme The American Experience, this page follows the course of technological progress and invention in the USA from the time of Benjamin Franklin's experiments in the 1750s to the creation of the Hubble Telescope in the 1990s. The information is divided into decades along a timeline. Along the way, you can learn about such inventions as the cotton gin, steamboat, revolver, false teeth, passenger elevator, burglar alarm, barbed wire, skyscraper, dishwasher, zipper, frozen food, electric guitar, atomic bomb, optic fibre, video game, bar code, space shuttle, personal computer, and artificial heart.

inside of which a sensitized film or plate can be placed. The lens collects rays of light reflected from the subject and brings them together as a sharp image on the film. The opening or hole at the front of the camera, through which light enters, is called an aperture. The aperture size controls the amount of light that can enter. A shutter controls the amount of time light has to affect the film. There are small-, medium-, and large-format cameras; the format refers to the size of recorded image and the dimensions of the image obtained.

Cameras: The Technology of Photographic Imaging
http://www.mhs.ox.ac.uk/cameras/index.htm

General presentation of the camera collection at the Museum of the History of Science, Oxford, UK. Of the many highlights perhaps the most distinguished are some very early photographs, the photographic works of Sarah A Acland, and the cameras of T E Lawrence.

A simple camera has a fixed shutter speed and aperture, chosen so that on a sunny day the correct amount of light is admitted. More complex cameras allow the shutter speed and aperture to be adjusted; most have a built-in exposure meter to help choose the correct combination of shutter speed and aperture for the ambient conditions and subject matter. The most versatile camera is the single lens reflex (SLR) which allows the lens to be removed and special lenses attached. A pin-hole camera has a small (pin-sized) hole instead of a lens. It must be left on a firm support during exposures, which are up to ten seconds with slow film, two seconds with fast film, and five minutes for paper negatives in daylight. The pin-hole camera gives sharp images from close-up to infinity.

Daguerreian Society
http://abell.austinc.edu/dag/home.html

Information about the process involved in this early form of photography, 19th- and 20th-century texts about it, an extensive bibliography of related literature, plenty of daguerreotypes to look at, and information about this society itself can be found here.

film, photographic Photographic film is a strip of transparent material (usually cellulose acetate) coated with a light-sensitive emulsion, used in cameras to take pictures. The emulsion contains a mixture of light-sensitive silver halide salts (for example, bromide or iodide) in gelatine. When the emulsion is exposed to

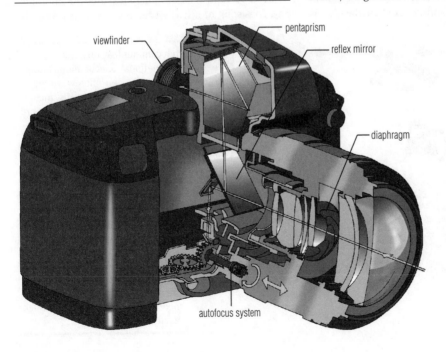

camera The single-lens reflex (SLR) camera in which an image can be seen through the lens before a picture is taken. The reflex mirror directs light entering the lens to the viewfinder. The SLR allows different lenses, such as close-up or zoom, to be used because the photographer can see exactly what is being focused on.

light, the silver salts are invisibly altered, giving a latent image, which is then made visible by the process of developing. Films differ in their sensitivities to light, this being indicated by their speeds.

Colour film consists of several layers of emulsion, each of which records a different colour in the light falling on it. In colour film the front emulsion records blue light, then comes a yellow filter, followed by layers that record green and red light respectively. In the developing process the various images in the layers are dyed yellow, magenta (red), and cyan (blue), respectively. When they are viewed, either as a transparency or as a colour print, the colours merge to produce the true colour of the original scene photographed.

four-colour process The four-colour process is a colour printing process using four printing plates, based on the principle that any colour is made up of differing proportions of the primary colours blue, red, and green. The first stage in preparing a colour picture for printing is to produce separate films, one each for the blue, red, and green respectively in the picture (colour separations). From these separations three printing plates are made, with a fourth plate for black (for shading or outlines and type). Ink colours complementary to those represented on the plates are used

for printing – yellow for the blue plate, cyan for the red, and magenta for the green.

Timeline of Photography
http://www.eastman.org/menu.html
Created by the George Eastman International Museum of Photography & Film, this page provides a detailed timeline of the history of photography and film, which begins as far back as the 5th century with Aristotle's description of optical principles. You can scroll the timeline or jump to specific periods of interest. Underlined text will take you to more involved descriptions. The bottom of the page contains a list of sources used to create the timeline.

---

halftone process The halftone process is a technique used in printing to reproduce the full range of tones in a photograph or other illustration. The intensity of the printed colour is varied from full strength to the lightest shades, even if one colour of ink is used. The picture to be reproduced is photographed through a screen ruled with a rectangular mesh of fine lines, which breaks up the tones of the original into areas of dots that vary in frequency according to the intensity of the tone. In the darker areas the dots run together; in the lighter areas they have more space between them. Colour pictures are broken down into a pattern of dots in the same way, the

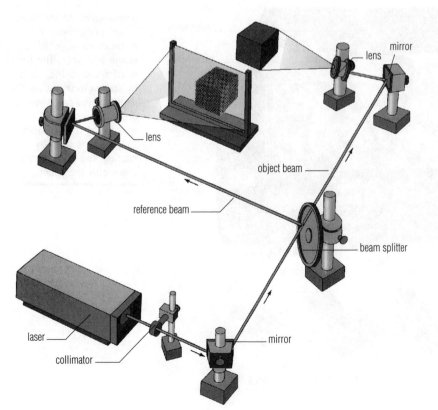

holography Recording a transmission hologram. Light from a laser is divided into two beams. One beam goes directly to the photographic plate. The other beam reflects off the object before hitting the photographic plate. The two beams combine to produce a pattern on the plate which contains information about the 3-D shape of the object. If the exposed and developed plate is illuminated by laser light, the pattern can be seen as a 3-D picture of the object.

original being photographed through a number of colour filters. The process is known as **colour separation**. Plates made from the separations are then printed in sequence, yellow, magenta (blue-red), cyan (blue-green), and black, which combine to give the full colour range.

holography Holography is a method of producing three-dimensional (3-D) images, called holograms, by means of laser light. Holography uses a photographic technique (involving the splitting of a laser beam into two beams) to produce a picture, or hologram, that contains 3-D information about the object photographed. Some holograms show meaningless patterns in ordinary light and produce a 3-D image only when laser light is projected through them, but reflection holograms produce images when ordinary light is reflected from them (as found on credit cards).

Holographic techniques have applications in storing dental records, detecting stresses and strains in construction and in retail goods, detecting forged paintings and documents, and producing three-dimensional body scans. The technique of detecting strains is of widespread application. It involves making two different holograms of an object on one plate, the object being stressed between exposures. If the object has dis-torted during stressing, the hologram will be greatly changed, and the distortion readily apparent.

printing Printing is the reproduction of multiple copies of text or illustrative material on paper, as in books or newspapers, or on an increasing variety of materials; for example, on plastic containers. The first printing used woodblocks, followed by carved wood type or moulded metal type and hand-operated presses. Modern printing is effected by electronically controlled machinery. Current printing processes include electronic phototypesetting with ◊offset printing, and ◊gravure print. Offset printing, prints from an inked flat surface, while the gravure method (used for high-circulation magazines), uses recessed plates.

The introduction of electronic phototypesetting machines in the 1960s allowed the entire process of setting and correction to be done in the same way that a typist operates, thus eliminating the hot-metal composing room and leaving only the making of plates and the running of the presses to be done traditionally. By the 1970s some final steps were taken to plate-less printing, using various processes, such as a computer-controlled laser beam, or continuous jets of ink acoustically broken up into tiny equal-sized drops, which are electrostatically charged under computer

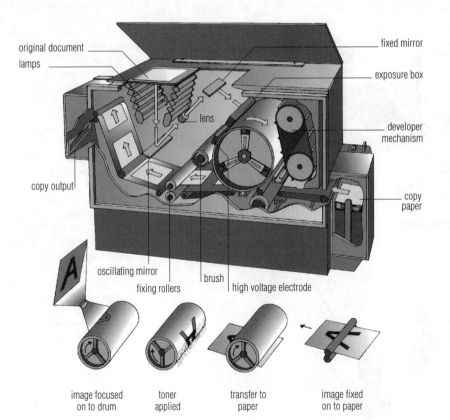

original document
lamps
fixed mirror
exposure box
lens
developer mechanism
copy output
copy paper
oscillating mirror
fixing rollers
brush
high voltage electrode

image focused on to drum

toner applied

transfer to paper

image fixed on to paper

photocopier At the heart of the photocopier is a metal drum on which an image is formed. The toner is attracted to the image by static electricity. As the paper moves past the drum, the image is transferred to the paper. The image is fixed by heating and pressing the toner into the paper.

control. Pictures can be fed into computer typesetting systems by optical scanners.

lithography Lithography is a printmaking technique invented in 1798 by Aloys Senefelder, based on the mutual repulsion of grease and water. A drawing is made with greasy crayon on an absorbent stone, which is then wetted. The wet stone repels ink (which is greasy) applied to the surface and the crayon absorbs it, so that the drawing can be printed. Lithographic printing is used in book production, posters, and prints, and this basic principle has developed into complex processes.

silk-screen printing, or serigraphy Silk-screening is a method of printing based on stencilling. It can be used to print on most surfaces, including paper, plastic, cloth, and wood. An impermeable stencil (either paper or photosensitized gelatine plate) is attached to a finely meshed silk screen that has been stretched on a wooden frame, so that the ink passes through to the area beneath only where an image is required. The design can also be painted directly on the screen with varnish. A series of screens can be used to add successive layers of colour to the design.

photocopier A photocopier is a machine that uses some form of photographic process to reproduce copies of documents or illustrations. Most modern photocopiers, as pioneered by the Xerox Corporation, use electrostatic photocopying, or xerography ('dry writing'). The Japanese company Ricoh produced a prototype for a 'recycle copier' in 1994. It reheats the toner on the paper to 100°C and adds a 'peel off' solution to reduce the adhesion of the toner. The paper is then passed through a roller removing all toner. By this method paper can be used up to 10 times.

Technology Chronology

c. 2600000 BC	The first tools, simple pebble choppers, are used in Africa near Lake Rudolph in present-day Kenya. They remain essentially unchanged for the next 1.5 million years.
c. 8000 BC– **c. 2700** BC	The Mesolithic or Middle Stone Age begins in western Europe. It is characterized by the use of microliths (very small stone tools mounted on a shaft), chipped stone tools, and bone, antler, and wooden tools. Important inventions include the barbed fish-hook, harpoon, woven basket, clay cooking pot, and the comb.
c. 5500 BC	Copper is smelted in Persia, the first metal to be smelted.
c. 4400 BC	The weaving loom is invented in Egypt. It consists of a frame that holds two sets of alternating parallel threads in place (the warp). By raising one set of threads it is possible to run a cross-thread (the weft) between them using a shuttle.
c. 3500 BC	The gnomon – the first clock – is invented, probably in Egypt. It consists of a vertical stick or pillar inserted in the ground, the length of its shadow giving an idea of the time.
c. 3500 BC	The Sumerians invent the wheel. Consisting of two or three wooden segments held together by transverse struts that rotate on a wooden pole, its invention transforms transportation, warfare, and industry.
c. 3200 BC	The Bronze Age begins when metallurgists in the Middle East discover that the addition of about 10% tin to copper in the molten state produces bronze. It lasts about 2,000 years.
c. 3000 BC	Papyrus, derived from reed, is invented in Egypt.
c. 3000 BC	The abacus, which uses rods and beads for making calculations, is developed in the Middle East and adopted throughout the Mediterranean. A form of the abacus is also used in China at this time.
c. 1700 BC	Windmills are used in Babylon to pump water for irrigation.
c. 1200 BC	The Iron Age begins as iron displaces bronze as the most important metal in Egypt and elsewhere.
c. 1100 BC	The spinning wheel is invented in China, derived from the machines used to draw out silk from the silkworm. It subsequently spreads to India and reaches Europe about the 13th century AD.
c. 350 BC	Work begins on Shan-yang Canal in China; it later forms the Southern Grand Canal.
c. 270 BC	The Greek physicist and inventor Ctesibius of Alexandria lays the foundations for the development of modern pumps with his invention of a small pipe organ, the hydraulis, which is supplied with air by a piston pump.
c. 250 BC	The Greek mathematician and inventor Archimedes invents the Archimedes screw for removing water from the hold of a large ship. A similar device is already in use in Egypt for irrigation.

c. 200 BC	The Romans invent concrete.
850	The earliest Chinese reference to gunpowder, a mixture of saltpetre, sulphur and charcoal, dates to this time. At first, it appears to have been used only for fireworks.
1600	Around this time, the compound microscope, which uses two lenses to magnify objects, is invented – probably by Hans Lippershey or Hans Jansen and his son Zacharias, both spectacle makers from Middelburg in the Netherlands.
1643	The Italian scientist Evangelista Torricelli invents the mercury barometer that records air pressure by the changes in the level of mercury within a glass tube sealed at the top.
1674	The Dutch scientist and instrumentmaker Christiaan Huygens makes a watch using a balance wheel controlled by the oscillations of a spring to keep time.
1679	The French Huguenot refugee Denis Papin makes a 'steam digester' – an early form of pressure cooker – for softening bones, and demonstrates it to the Royal Society in England. It uses a safety valve which will find widespread use in the coming industrial revolution.
1712	The first recorded practical steam-engine to use a piston and cylinder, constructed by English inventor Thomas Newcomen, is installed at Dudley Castle, near Wolverhampton, England, where it is used for pumping out underground mineworkings.
1728	The English clockmaker John Harrison completes construction of a clock for measuring longitude at sea, the H-1. After presenting it to English astronomer Edmund Halley and watchmaker George Graham, he makes some adjustments.
1758	The English engineer John Wilkinson installs the first blast furnace at his works in Bilston, Staffordshire, England.
1761	The English engineer James Brindley completes the Duke of Bridgewater's Canal between Manchester and the Worsley collieries in northern England; it is the first British canal of major economic importance.
1764	The English inventor James Hargreaves invents the spinning jenny, which allows one individual to spin several threads simultaneously.
1765	The Scottish inventor James Watt invents a condenser to condense exhaust steam from Newcomen steam engines; it reduces the loss of latent heat, making them more efficient.
1769	The English inventor Richard Arkwright patents a spinning machine (or 'water frame' because it operates by water) that produces cotton yarn suitable for warp; it is one of the key inventions of Britain's Industrial Revolution.
1769	The French engineer Nicolas-Joseph Cugnot's steam road carriage carries four people at speeds of 3.6 kph/2.25 mph.
c. 1770	The English inventor Richard Arkwright operates a number of factories using his water-powered spinning machines and in 1773 begins to manufacture calico, establishing textile manufacture as the leading industry in northern England.
1773	T F Pritchard's bridge, the first made completely of iron, is completed at Coalbrookdale, Shropshire, England. It is built for the industrialist Abraham Darby.
1775	The Italian physicist Alessandro Volta invents the electrophorus, a device for generating and storing static electricity; it later leads to modern electrical condensers.
1776	The American engineer David Bushnell builds a hand-powered wooden-hulled submarine named the *Turtle*. It is used in an unsuccessful attempt to attach an explosive device to a British ship.
1779	The English inventor Samuel Crompton devises the spinning mule, a cross between a spinning jenny and a water-frame spinning machine; it makes possible the large-scale manufacture of thread.
1779	The French inventors Jean-Pierre-François Blanchard and M Masurier construct a velocipede, a type of early bicycle, in Paris, France.
1781	The Scottish engineer James Watt discovers how to convert the up-and-down motion of his steam engine into rotary motion which can then turn a shaft.
1782	James Watt patents the double-acting steam engine, which provides power on both the upstroke and the downstroke of the piston.
1782	The distributing mechanism of English agriculturist Jethro Tull's seed-planting machine is improved with the addition of gears.
1784	Henry Cort discovers the 'puddling process' of converting pig iron into wrought iron by stirring to burn off impurities. It revolutionizes the manufacture of iron, production of which quadruples over the next 20 years.
1784	The Swiss inventor Aimé Argand invents an oil-burner consisting of a cylindrical wick, two concentric metal tubes to provide air, and a glass chimney to increase the draught. It gives a light ten times brighter than previous lamps and the principle is later used in gas-burners.
1784	The American engineer Oliver Evans invents an automated process for grinding grain and

	sifting flour; it marks the beginning of automation in America.
1785	The English inventor Edmund Cartwright develops a steam-powered loom.
1788	The Scottish millwright Andrew Meikle patents a threshing machine for separating the grain from the straw.
1789	The English inventor Edmund Cartwright patents a wool-combing machine.
1794	The English engineer Robert Street patents the first practical internal-combustion engine.
1795	The English inventor Joseph Bramah invents a hydraulic press capable of exerting a force of several thousand tonnes.
1795	The French chef Nicolas Appert begins experiments on preserving food in hermetically sealed containers.
1797	The English engineer Henry Maudslay invents the carriage lathe, which permits the operator to use the lathe without holding the metal-cutting tool.
1798	The Frenchman Nicholas-Louis Robert invents a paper-making machine.
1800	The English engineer Henry Maudslay improves his screw-cutting lathe so that it can cut screws of varying pitches.
1800	The Italian physicist Alessandro Volta invents the voltaic pile made of discs of silver and zinc – the first battery.
1800	The US engineer James Finley builds the first suspension bridge in the USA.
24 Dec 1801	The English engineer Richard Trevithick builds a steam-powered carriage that he successfully drives up a hill in Camborne, Cornwall, England.
1803–1822	The Scottish engineer Thomas Telford constructs the 100 km/60 mi Caledonian Canal, linking Scotland's east and west coasts.
21 Feb 1804	The English engineer Richard Trevithick builds the first steam railway locomotive, and on a wager runs it on a 16 km/10 mi track at the Pen-y-darren Ironworks in South Wales carrying 10 tons of iron and 70 passengers.
1 Dec 1804	The English aviation pioneer George Cayley develops an instrument to measure wind resistance. About this time he also begins to construct models of gliders with fixed wings, fuselage, elevators and a rudder – the basic configuration of the modern aeroplane.
1805	The French inventor Joseph-Marie Jacquard develops a loom that uses punched cards to control the weaving of cloth.
1807	The English inventors Henry and Sealy Fourdrinier receive a patent for an improved version of Nicolas-Louis Robert's papermaking machine. Their new 'Fourdrinier machine' allows production of paper in continuous sheets.
1807	The German promoter Frederick Albert Winsor's National Light and Heat Company lights one

	side of Pall Mall, London, England, with gas lamps – the first street-lighting in the world.
1808	The German artist Ferdinand Piloty produces the first colour lithographs.
c. 1812	The Baltimore clipper ship is introduced by US shipbuilders; its revolutionary design – it has a great expanse of sail and hull that offers little water resistance – makes it one of the fastest ships afloat.
1813	The British engineer Bryan Donkin develops a rotary printing press. This is an improvement over the flatbed press, as it allows faster printing, and spurs the development of newspapers.
25 July 1814	The English engineer George Stephenson constructs the first effective steam locomotive. Called the 'Bulcher' it hauls up to 30 tonnes of coal at 6.4 kph/4 mph, out of mines at Killingorth Collier, Newcastle upon Tyne, England.
1815	The English chemist Humphry Davy invents the miner's safety lamp. It does not ignite marsh gas in the mines which causes explosions.
1815	The Scottish inventor John Loudon McAdam begins building roads around Bristol, southwest England. Comprised of two grades of large crushed stone for good drainage and to support the load, and covered by a surface of compacted smaller stones to form a pavement to withstand wear and tear and to shed water to the drainage ditches, they are the most advanced roads built to date.
1816	The French photography pioneer and inventor Joseph-Nicéphore Niépce invents the 'celeripede'. A two-wheeled ancestor of the bicycle, it is propelled by pushing the feet against the ground, but cannot be steered.
1816	The Scottish clergyman Robert Stirling patents the Stirling hot-air engine. It is powered by the expansion and displacement of air inside an externally heated cylinder. It is forgotten until the Dutch company Philips becomes interested in it in the 1970s.
1821	The English physicist Michael Faraday builds an apparatus that transforms electrical energy into mechanical energy – the principle of the electric motor.
1821	The world's first natural gas well is sunk at Fredonia, New York. Lead pipes distribute the gas to consumers for lighting and cooking.
1822	The US inventor William Church patents the world's first typesetting machine in Britain.
1824	The English mason Joseph Aspdin patents Portland cement. More water-resistant than other cements, it quickly becomes widely used.
27 Sep 1825	The Stockton to Darlington railway line in England opens. Built by George Stephenson, it is the world's first public railway to carry steam trains. Stephenson's locomotive *Active*

carries 450 passengers at 24 kph/15 mph over the 43 km/27 mi track.

25 Oct 1825 The canal boat *Seneca Chief* opens the Erie Canal. Linking the Great Lakes with the Hudson River, it opens the Midwest to settlement.

1825 The US engineer John Stevens constructs the first steam locomotive to run on rails in the USA. It runs on a short circular track at his home in Hoboken, New Jersey.

1825–1843 The English engineer Isambard Kingdom Brunel's Thames Tunnel, the first under a river, is constructed. Its success on opening to pedestrian traffic prompts British solicitor Charles Pearson to propose the construction of a subway system for London, England.

1826 The world's first railway tunnel is built on the Liverpool–Manchester railway in England.

1827 The English chemist John Walker invents the friction match ('Lucifer'); made with antimony sulphide and potassium chloride it is ignited by drawing it through sandpaper.

10 Oct 1829– George and Robert Stephenson's *Rocket* wins
14 Oct 1829 the Liverpool and Manchester Railway competition. Using a multiple fire-tube boiler, rather than the single flue boiler other contestants use, its design sets the pattern for future railway locomotives.

1830 The French inventor Barthélemy Thimonnier patents the first sewing machine.

1832 The French engineer Benoît Fourneyron develops a water turbine capable of 50 horsepower.

1832 The French inventor Hippolyte Pixii builds the first magneto or magneto-electric generator; it is the first machine to convert mechanical energy into electrical energy.

1832 The US engineer William Avery builds the first practical steam turbine; it is used to power sawmills.

1834 The German chemist Justus von Liebig develops melamine – the basis for synthetic resins such as 'Formica' and 'Melmac'.

1834 The US blacksmith Thomas Davenport constructs the first battery-powered electric motor. He uses it to operate a small car on a short section of track – the first streetcar.

1834 The US inventor Jacob Perkins develops, in Britain, a compression machine that, by alternate compression and expansion of gases freezes water – the beginning of gas refrigeration.

1835 The English scientist William Henry Fox Talbot publishes a paper describing the paper negative. He exposes paper infused with silver chloride to light, which then separates into fine silver and dark tones from which he can take positive prints.

1835 The US manufacturer Samuel Colt patents a six-shot revolver with a rotating cartridge cylinder. Each time the trigger is pulled a new bullet moves in front of the barrel. Its effective range is 23 m/75 ft.

1837 The French artist Louis-Jacques-Mandé Daguerre produces a detailed photograph of his studio on a silvered copper plate.

1837 The French engineer Benoît Fourneyron builds a water turbine which rotates at 2,300 revolutions per minute and generates 60 horsepower. Weighing only 18 kg/40 lb, and with a wheel only 0.3 m/1.0 ft in diameter it is far more productive than the waterwheel and is used to power factories, especially the textile industry, in Europe and the USA.

1837 The German scientist Moritz Hermann von Jacobi, in Russia, develops electroplating; first with silver and then with nickel and chrome.

1839 The British engineer James Nasmyth designs the steam hammer; an important tool for forging heavy machinery in the Industrial Revolution. He patents it on 24 November 1842.

1839 The US inventor Charles Goodyear vulcanizes rubber by adding sulphur and then heating it.

1844 The German engineer Gottlob Keller develops an effective process for making paper from wood pulp, which reduces the cost of newspaper production and helps the growth of mass media.

1845 The English inventor Thomas Wright obtains the first patent for an arc lamp.

1845 The Scottish inventor Robert Thomson patents the pneumatic tyre. Although used for 1,931 km/1,200 mi on a horsedrawn brougham carriage, pneumatic tyres are not used again until the end of the century.

10 Sep 1846 The US inventor Elias Howe patents a practical sewing machine; it revolutionizes garment manufacture in both the factory and home.

1849 The French inventor Eugène Bourdon invents the Bourdon tube, the most commonly used industrial pressure gauge for measuring the pressures of liquids and gases.

1 May 1851– The Great Exhibition is held in Hyde Park,
18 Sep 1851 London, England. Devised by Prince Albert, it is the first exhibition to display the latest technical innovations in industry, from both Britain and Europe.

12 Aug 1851 The US inventor Isaac Merritt Singer patents the first practical domestic sewing machine for general use, in Boston, Massachusetts. His design, which enables continuous and curved stitching, and allows any part of the material to be worked on, sets the pattern for all subsequent sewing machines.

1852 The US inventor and machine-shop owner Elisha Otis installs, in a factory in Albany, New York, a freight lift equipped with an automatic safety device that prevents it from falling if the lifting chain or rope breaks.

1854 The French scientist Henri-Etienne St Claire Deville develops a new process for making aluminium. Through the action of metallic sodium on aluminium chloride he produces marble-sized lumps of the metal.

1855 The US manufacturer Samuel Colt opens an armoury at Hartford, Connecticut. Using 1,400 machine tools, he develops the use of interchangeable parts to a high degree and revolutionizes the manufacture of small arms.

1856 The English inventor Henry Bessemer obtains a patent for the Bessemer converter which converts cast iron into steel by injecting air into molten iron to remove carbon and increase the temperature of the molten mass. It allows iron to be poured and thus shaped and brings down prices.

1857 The Scottish-born Australian inventor James Harrison develops a vapour-compression machine using ether as the refrigerant. It is the first to be used in the brewing industry and for freezing meat for shipment.

31 Jan 1858 The English engineer Isambard Kingdom Brunel's steamship *Great Eastern* is launched. With a displacement of 19,222 tonnes/18,918 tons, and 211 m/692 ft long, it is the largest ship in the world. It has two sets of engines that drive two screw propellers and two paddlewheels, and is the first steamship with a double iron hull. Its design serves as the prototype for modern ocean liners.

16 Aug 1858 Queen Victoria of Britain and US president James Buchanan are the first to exchange messages on the first successful Atlantic telegraph cable laid between Valentia, Ireland and Newfoundland, Canada. The cable lasts for only 27 days.

1859 The Belgian inventor (Jean-Joseph-) Etienne Lenoir builds the first internal combustion engine in Paris, France. Operating on coal gas it has only a 4% efficiency.

1861 The Belgian chemist Ernest Solvay patents a method for the economic production of sodium carbonate (washing soda) from sodium chloride, ammonia, and carbon dioxide. Used to make paper, glass, and bleach, and to treat water and refine petroleum, it is a key development in the Industrial Revolution.

1861 The French inventor Pierre Michaux and his son Ernst construct the first successful bicycle with pedals. The pedals are attached to the front wheel, and because it has steel tyres and no springs it is called the 'bone-shaker'.

1861 The German engineer Nikolaus August Otto constructs an internal combustion engine that runs on gasoline.

1862 The Belgian inventor (Jean-Joseph-) Etienne Lenoir constructs the first car with an internal combustion engine and makes a 10 km/6 mi trip.

1864 The French engineers Pierre and Emile Martin, and British engineer William Siemens, simultaneously develop the open-hearth process for making steel using a regenerative gas-fired furnace. By using hot waste gases to heat the furnace, high quality steel is produced in bulk, and scrap steel can be melted and reused.

1867 The French gardener Joseph Monier patents reinforced concrete by adding steel rods, bars, or mesh to the concrete. It dramatically increases the tensile strength of the concrete, making it capable of sustaining heavy stresses.

1867 The US printer Christopher Latham Sholes constructs the first practical typewriter.

17 Nov 1869 The French engineer Ferdinand de Lesseps completes the 168 km/105 mi long Suez Canal in Egypt that links the Mediterranean and the Red Sea, and which reduces the route from Europe to Asia by 8,000 km/5,000 mi.

1869 The first chain-driven bicycle is built by the firm of A Guilmet and Meyer.

1869 The US scientist John Wesley Hyatt, in an effort to find a substitute for the ivory in billiard balls invents (independently of Alexander Parkes) celluloid. The first artificial plastic, it can be produced cheaply in a variety of colours and is resistant to water, oil, and weak acids.

1874 The English inventor H J Lawson develops the 'safety bicycle'. Because it has two equal-sized wheels, rubber tyres, and is powered by an endless chain between the pedals and the rear wheel, it has greater stability and is easier to brake than other bicycles.

1875 The Austrian engineer Siegfried Marcus builds one of the first cars powered by petrol. It is the oldest existing automobile.

1876 The German engineer Karl von Linde develops the first really efficient refrigerator, replacing the potentially explosive methyl ether with ammonia. It opens the way for refrigerated railway cars and ships.

1876 The German engineer Nikolaus Otto patents the four-stroke internal combustion engine, the prototype of modern engines. Its development marks the beginning of the end of the age of steam. More than 30,000 are built in the following decade.

18 Dec 1878 The English physicist Joseph Swan demonstrates his electric 'glow lamp', in Newcastle, England. It is the first practical carbon-filament incandescent light bulb.

1878 The German-born British inventor Charles Siemens invents the electric arc furnace, the first to use electricity to make steel.

21 Oct 1879 The US inventor Thomas Edison demonstrates his carbon-filament incandescent lamp light. He lights his Menlo Park power station with

30 lamps that burn for two days; later filaments burn for several hundred hours. Each light can be turned on or off separately in the first demonstration of parallel circuit.

1879 The US manufacturer George Eastman invents a machine that applies photographic emulsion to a gelatin plate, which allows him to mass-produce photographic dry-plates.

1883 The English physicist and chemist Joseph Wilson Swan patents a method of creating nitrocellulose (cellulose nitrate) fibre by squeezing it though small holes. It becomes a basic process in the artificial textile industry.

1883 The US architect William Lebaron Jenney completes construction of the ten=storey Home Insurance Building in Chicago, Illinois. The world's first true skyscraper, it consists of a steel=girder framework on which the outer covering of masonry hangs. It sparks a boom in the construction of skyscrapers in Chicago.

1884 The German-born US inventor Ottmar Mergenthaler patents the first Linotype typesetting machine. Characters are cast as metal type in complete lines rather than as individual letters as in a monotype machine.

26 Jan 1885 The German mechanical engineer Karl Friedrich Benz patents a three-wheeled vehicle powered by a two-cycle, single-cylinder internal combustion engine, pioneering the development of the motorcar. His car achieves a speed of 14.4 kph/9 mph.

1885 The British inventor John Starley builds the 'Rover' safety bicycle. The forerunner of modern bicycles, its wheels are of equal size.

1885 The German mechanical engineers Gottlieb Daimler and Wilhelm Maybach develop a successful lightweight high-speed internal combustion petrol engine, and fit it to a bicycle to create the prototype of the present-day motorcycle.

1885 The US inventor Tolbert Lanston invents the Monotype typesetting machine. Type is cast in individual letters, using a 120-key keyboard.

1886 The English inventor Thomas Crapper invents the modern flush toilet.

1886 The US chemist Charles Martin Hall and French chemist Paul–Louis–Toussaint Héroult, working independently, each develop a method for the production of aluminium by the electrolysis of aluminium oxide. The process reduces the price of the metal dramatically and brings it into widespread use.

31 Oct 1888 The Scottish veterinary surgeon John Boyd Dunlop patents the pneumatic bicycle tyre.

1888 The Serbian-born US inventor Nikola Tesla invents the first alternating current (AC) electric motor, which serves as the model for most modern electric motors.

1890 The US inventor and statistician Herman Hollerith uses punched cards to automate counting the US census. The holes, which represent numerical data, are sorted and tabulated by an electric machine, the forerunner of modern computers. In 1896 Hollerith forms the Tabulating Machine Company, which later changes its name to International Business Machines (IBM).

1892 The French chemist Ferdinand-Frédéric Henri Moissan invents the first electric-arc furnace. He uses it to vaporize and fuse different elements and create new materials.

1892 The German engineer Rudolf Diesel patents the diesel engine, a new type of internal combustion engine which runs on a fuel cheaper than petrol and which, because ignition of the fuel is achieved by compression rather than electric spark, is simpler in construction.

1893 At the World's Columbian Exhibition, at Chicago, Illinois, the US inventor Whitcomb L Judson exhibits a 'clasp locker or unlocker for shoes', now known as a zipper.

1893 The German mechanical engineer Karl Friedrich Benz constructs his first four-wheeled car.

1893 The US manufacturer Edward Drummond Libbey produces fabric woven from a mixture of glass fibre and silk, creating the first fibreglass.

1901 The German engineer Carl von Linde separates liquid oxygen from liquid air. It leads to the widespread use of oxygen in industry.

1901 The mass production of cars in Detroit, Michigan, begins when US car manufacturer Ransom Eli Olds produces the three-horsepower Oldsmobile buggy. The first car with a curved dash, it is also the first to be made using assembly line techniques and the first commercially successful car in the USA.

17 Dec 1903 The US aviator Orville Wright makes the first successful flight in an aeroplane with a petrol engine at Kitty Hawk, North Carolina, covering 37 m/120ft in a flight lasting just 12 seconds. During the day, Orville and his brother Wilbur make a number of flights, the longest covering 260 m/852 ft and lasting 59 seconds.

1903 The 16-storey Ingalls Building in Cincinnati, Ohio, is completed. It is the first skyscraper built of reinforced concrete.

1906 George Albert Smith of the Charles Urban Trading Co. develops Kinemacolour, the first commercially successful colour process for film: it uses two colour filters and two reels of film.

12 Aug 1908 The US car manufacturer Henry Ford of the Ford Motor Company introduces the Model T.

Inexpensive (sold for $850), easy to maintain, and mass-produced after 1913, it revolutionizes transportation.

1908 The Belgian-born US chemist Leo H Baekeland invents the plastic Bakelite: its insulating and malleable properties, combined with the fact that it does not bend when heated, ensures it has many uses.

1912 The Edison film studio produces the first film with sound. It is a 15 minute musical based on nursery rhymes in which the sound is roughly synchronized on a phonograph with the image.

1913 The world's first diesel-electric locomotives begin running in Sweden.

15 Aug 1914 The Panama Canal opens to traffic. One of the world's greatest engineering feats, it is 81.6 km/50.7 mi long and saves 12,800 km/8,000 mi on the trip around South America. It cost $366,650,000 and around 6,000 workers died during its construction.

1916 A passive form of sonar is developed in the USA and Britain to detect submarines. It consists of a microphone towed behind ships.

1918 The Anti-Submarine Detection Investigation Device (ASDIC) is developed by British and US naval scientists. An active sonar system, it uses the echo of a pulsed sound to detect submarines.

1920 The US inventor James Smathers pioneers the electric typewriter as an office machine.

16 Mar 1926 The US inventor Robert Hutchings Goddard achieves the first flight of a liquid-propelled rocket, at Auburn, Michigan. It reaches an altitude of 12 m/41 ft.

1926 The US research company Bell Laboratories issues the first synchronous sound motion picture system.

1928 Polyvinyl chloride (PVC) is developed, simultaneously, by the US companies Carbide and Carbon Corporation and Du Pont and the German firm I G Farben.

1930 The English inventor Frank Whittle patents a turbo-jet engine. It is later used on the first jet aeroplane.

1932 Pioneering radar equipment at the US Naval Research Laboratory is able to detect aircraft 80 km/50 mi away from its transmitter, but unable to locate them.

April 1936 US scientists produce a workable radar with a range of 4 km/2.5 mi. By the end of the year this is extended to 11 km/7 mi.

2 Nov 1936 The British Broadcasting Corporation (BBC) starts the world's first public high-definition television service from its transmitter at Alexandra Palace, London, England, using Logie Baird's mechanical system and EMI's electronic system.

12 April 1937 The English engineer Frank Whittle tests the first prototype jet engine. A similar engine is developed in Germany at the same time.

27 May 1937 The Golden Gate Bridge in San Francisco, California, opens; it is the longest suspension bridge to date at 1,965 m/6,450 ft.

1937 Nylon, developed by W H Carothers, is patented by the US chemicals company Du Pont and is commercially available the following year in the form of toothbrush bristles; nylon stockings become widely available in the USA in May 1940.

22 Oct 1938 In the USA, Chester Carlson produces the first example of xerography ('dry writing'), which is to develop into modern photocopying.

27 Aug 1939 The Heinkel He 178 makes a test flight in Germany, achieving a speed of 500 kph/360 mph; it is the first jet aeroplane to fly.

3 Oct 1942 The V2 rocket, the ancestor of modern space rockets, is first launched, in Germany; weighing 40 tons it is 12 m/40 ft long, burns an alcohol-liquid oxygen mixture, can reach a distance of 200 km/125 mi, a height of 97 km/60 mi, and travels at 5,300kph/3,300 mph.

1943 The French oceanographer Jacques Cousteau invents the aqualung (or self-contained underwater breathing apparatus, 'scuba'), the first fully automatic compressed-air breathing apparatus. It allows him to dive to a depth of 64 m/210 ft.

8 Oct 1945 In Waltham, Massachusetts, USA, Percy LeBaron Spencer patents the first microwave, which is used in restaurants and institutions.

1948 The first atomic clock is installed at the National Bureau of Standards, Washington, DC; it is based on the oscillation of the ammonia molecule and operates using the natural vibrations of atoms. It is extremely accurate, with an error margin of 2 seconds in every 2 million years.

1948 The US physicists John Bardeen, William Bradley Shockley, and Walter Brattain develop the transistor in research at Bell Telephone Laboratories in the USA.

1949 The Swiss engineer Georges de Mestral invents velcro after getting the idea from the burs that stick to his socks. The name is formed from the first letters of velvet and crochet.

1949 The Intertype Fotosetter Photographic Line Composing Machine is introduced in the USA; it is the first typesetting machine that does not use metal type.

1950 The first Xerox photocopying machine is produced by the Haloid Company (later to become the Xerox Corporation) in Rochester, New York.

1952 A method of converting iron to steel, known as the basic—oxygen process, is developed in Linz and Donawitz, Austria. By blowing a supersonic jet of oxygen over the surface of the molten material the nitrogen content is reduced, producing a better quality steel. It replaces the Bessemer process.

1954	The silicon transistor is developed by Texas Instruments engineer Gordon Teal. Cheaper and more resistant to higher temperatures than germanium, the development of silicon transistors stimulates the growth of solid state components in computers, aeroplanes, and missiles.
12 Sep 1955	The English engineer Christopher Cockerell patents the first hovercraft.
1956	The German engineer Felix Wankel invents the Wankel engine, an internal combustion engine that uses a triangular-shaped rotor instead of pistons.
1959	The US engineer Jean Hoerni of Fairchild Semiconductor Corporation designs the planar or 'flat' transistor and US engineer Robert Noyce discovers a way to join the circuits by printing, eliminating hundreds of hours in their production. Their work leads to the creation of the first microchip, which stimulates the computer industry with its sharply reduced size and cost and leads to the third generation of computers.
1960	The halogen lamp is introduced. The halogen gas in the lamp regenerates the filament, permitting it to burn at higher temperatures, and thus burn brighter.
1961	The US inventor George Devol and US businessman Joseph Engelberger develop the first true robot, a programmable manipulator called 'Programmed Article Transfer'. Installed at General Motors by their company Unimation, it is used to unload parts from a die-casting operation.
1963	The US chemist Leslie Phillips and colleagues at the Royal Aircraft Establishment, Farnborough, develop carbon fibre. It is used for strength in bridges and turbine blades.
1964	Xerox develops the first office fax in the USA. It can only operate on dedicated phone lines.
1965	The Japanese electronic company Sony launches the Sony CV-2000, the first home video recorder, using Sony's Betamax format. The first colour video recorder is available the following year.

1970	Canon Business Machines markets the first pocket calculator, in Japan.
1970	The US scientist Charles Burns invents the light-emitting diode (LED).
1972	The English engineer Godfrey Hounsfield performs the first successful CAT (computerized axial tomography) scan, which provides cross-sectional X-rays of the human body.
1972	The Japanese researcher Hideki Shirakawa attempts to make the polymer polyacetylene but accidentally adds a thousand times too much catalyst and discovers electrically conductive plastics; they have a metallic appearance.
1973	The US chemist and medical information scientist Paul Lauterbur obtains the first NMR (nuclear magnetic resonance) image, in Britain. Radio waves are beamed through a patient's body while subjected to a powerful magnetic field; an image is generated because different atoms absorb radio waves at different frequencies under the influence of a magnetic field.
1975	Liquid crystals are first used for display purposes in electronic devices such as watches and calculators.
21 Jan 1976	Two Concorde aircraft make their first commercial flights, from London, England, to Bahrain and from Paris, France, to Rio de Janeiro, Brazil.
1976	The home video cassette recorder is introduced into the US market, with two incompatible models. The Japanese electronics company Sony markets the Betamax system, released in 1975, and fellow Japanese electronics company Japanese Matsushita Corporation (JVC) markets the Video Home System (or VHS), which eventually dominates the trade.
1979	The Dutch company Philips and the Japanese company Sony work collaboratively to develop the compact disc (CD); tiny pits on the plastic are read by laser to reproduce sound or other information. CDs are first marketed in 1982.

22 Sep 1981	French railways introduce the TGV (*train à grande vitesse*, 'high-speed train'); electrically powered and capable of cruising at 290 kph/180 mph, it is Europe's first super high speed passenger train. Later in the year achieves a record speed of 380 kph/236 mph.
Oct 1982	The Japanese company Sony launches the first compact disc (CD) players in Japan, working with Philips, the Dutch manufacturer of the compact disc.
April 1986	The first large plant to irradiate food is set up in New Jersey, USA, to process fruits and vegetables arriving from tropical countries; other plants soon open in the Netherlands, Japan, Canada, and other countries.
1986	The first digital audio tape (DAT) recorders are demonstrated in Japan.
1986	The M25, the world's longest ring road at 195 km/121 mi, is completed around London, England.
12 Feb 1987	The Chinese physicist Paul Ching-Wu Chu and associates at the University of Huston, Texas, USA, make a material that is superconducting at the temperature of liquid nitrogen – 77K or –196°C/–321°F.
1987	The Japanese firm Nippon Zeon discovers memory plastics – plastics that change their shape at one temperature and then return to their original shape at another. Applications are envisioned in the car industry.
1988	Fujitsu laboratories in Japan develop monocrystalline superconductors – crystal superconductors only one layer of crystal thick. They retain heat better than several layers and thus allow the construction of large integrated circuits with little energy loss and unequalled speed.
1988	Researchers at IBM's Almaden Research Center in San José, California, using a scanning tunnelling microscope, produce the first image of the ring structure of benzene, the simplest aromatic hydrocarbon.
1988	The Seikan Tunnel under Tsugaru Strait between Honshu and Hokkaido islands, Japan, is completed; 54 km/86.9 mi long, it is the longest undersea railway tunnel in the world.
1 Dec 1990	British and French tunnelling engineers, working from opposite sides of the English Channel to build the Channel Tunnel, break through the last few yards of ground separating their excavations.
1991	Several US companies introduce local area networks (LANs), which use nondirectional microwaves to transmit data as fast as fibre optic cables.
1992	The electronics companies Matsushita and Philips launch the Digital Compact Cassette.
1993	French and Russian chemists create a superhard material by crystallizing buckminsterfullerenes at very high pressure. The material is able to scratch diamond.
June 1994	The Dutch electronics company Philips launches films on digital video disc, using technology developed jointly by Philips and Japanese electronics company Sony.
2 July 1996	The US aerospace company Lockheed Martin unveils plans for the X-33, a $1-billion wedge-shaped rocket ship. Called the *Venture Star*, it will be built and operated by Lockheed Martin and will replace the US space shuttle fleet by the year 2012.
July 1996	The US engineers Theodore O Poehler and Peter C Searson announce the invention of the first all-plastic battery. It uses polymers instead of conventional electrode materials and has implications for military and space applications as well as its use in consumer devices such as hearing aids and wristwatches.
5 Dec 1996	General Motors launches the Saturn EV1 in California and Arizona: the first mass-market electric car.
1997	A credit card-sized version of the plastic battery is introduced by its US inventors in Baltimore, Maryland. It produces 2.5 volts of electricity.
5 April 1998	The world's largest suspension bridge, linking Kobe and Awaji Island in Japan, opens to traffic. It costs £2.2. billion and is 3.9 km/2.4 mi long.

Biographies

Arkwright, Richard (1732–1792) English inventor and manufacturing pioneer who, in 1768, developed a machine for spinning cotton, called a 'water frame', which was the first machine capable of producing sufficiently strong cotton thread to be used as warp. In 1773 Arkwright produced the yarn 'water twist', the first cloth made entirely from cotton; previously, the warp had been of linen and only the weft was cotton.

Baird, John Logie (1888–1946) Scottish electrical engineer who pioneered television. In 1925 he gave the first public demonstration of television, transmitting an image of a recognizable human face. Baird used a mechanical scanner which temporarily changed an image into a sequence of electronic signals that could then be reconstructed on a screen as a pattern of half-tones. His first pictures were formed of only 30 lines repeated approximately 10 times a second. The following year, he gave the world's first demonstration of true television before an audience of about 50 scientists at the Royal Institution, London. By 1928 Baird had succeeded in demonstrating colour television.

Bell, Alexander Graham (1847–1922) Scottish-born US scientist and inventor. He was the first person ever to transmit speech from one point to another by electrical means. This invention – the telephone – was made in 1876. Later, Bell experimented with aeronautics. Bell also invented a type of phonograph, the tricycle undercarriage, and the photophone, which used selenium crystals to apply the telephone principle to transmitting words in a beam of light. He thus achieved the first wireless transmission of speech.

Benz, Karl Friedrich (1844–1929) German automobile engineer. He produced the world's first petrol-driven motor vehicle. He produced a two-stroke engine of his own design in 1878, and in 1885, the first vehicle successfully propelled by an internal-combustion engine. It achieved a speed of up to 5 kph/3 mph. The production model Tri-car appeared 1886–87 and had a 1 kW/1.5 hp single-cylinder engine. Benz made his first four-wheeled prototype in 1891 and by 1895, he was building a range of four-wheeled vehicles that were light, strong, inexpensive, and simple to operate. These vehicles ran at speeds of about 24 kph/15 mph.

Bessemer, Henry (1813–1898) British engineer and inventor who developed a method of converting molten pig iron into steel (the **Bessemer process**) in 1856. By modifying the standard process, he found a way to produce steel without an intermediate wrought-iron stage, reducing its cost dramatically. However, to obtain high-quality steel, phosphorus-free ore was required. In 1860 Bessemer erected his own steelworks in Sheffield, importing phosphorus-free iron ore from Sweden.

Biró, Lazlo Hungarian-born Argentine who invented a ballpoint pen in 1944. His name became generic for ballpoint pens in the UK.

Bridgewater, Francis Egerton, 3rd Duke of Bridgewater (1736–1803) Pioneer of British inland navigation. With James Brindley as his engineer, he constructed 1762–72 the Bridgewater Canal from Worsley to Manchester and on to the Mersey, a distance of 67.5 km/42 mi. Initially built to carry coal, the canal crosses the Irwell Valley on an aqueduct.

Brunel, Isambard Kingdom (1806–1859) British engineer and inventor. The son of Marc Brunel, he made major contributions in shipbuilding and bridge construction, and assisted his father in the Thames tunnel project. He built the Clifton Suspension Bridge over the River Avon at Bristol and the Saltash Bridge over the River Tamar near Plymouth. His shipbuilding designs include the *Great Western* (1837), the first steamship to cross the Atlantic regularly, and the *Great Britain* (1843), the first large iron ship to have a screw propeller. Brunel's last ship, the *Great Eastern* (1858), was to remain the largest ship in service until the end of the 19th century. With over ten times the tonnage of his first ship, it was the first ship to be built with a double iron hull, was driven by both paddles and a screw propeller, and laid the first transatlantic telegraph cable.

Brunel, Marc Isambard (1769–1849) French-born British engineer and inventor, father of Isambard Kingdom Brunel. In 1799 he moved to England to mass-produce marine blocks, which were needed by the navy. Brunel demonstrated that with specially designed machine tools 10 men could do the work of 100, more quickly, more cheaply, and yield a better product. He constructed the tunnel under the River Thames in London from Wapping to Rotherhithe 1825–43.

Cartwright, Edmund (1743–1823) British inventor. He patented the water-driven power loom in 1785, built a weaving mill in 1787, and patented a wool-combing machine in 1789 which did the work of 20 hand-combers.

Cockerell, Christopher Sydney (1910–) English engineer who invented the hovercraft in the 1950s.

Crompton, Samuel (1753–1827) British inventor. He invented the 'spinning mule' in 1779. Called the mule because it combined the ideas of Richard ◊Arkwright's water frame and James ◊Hargreaves's spinning jenny, it spun a fine, continuous yarn and revolutionized the production of high-quality cotton textiles.

Cugnot, Nicolas-Joseph (1725–1804) French engineer who produced the first high-pressure steam engine and, in 1769, the first self-propelled road vehicle. The three-wheeled, high-pressure carriage was capable of carrying 1,800 litres/400 gallons of water and four passengers at a speed of 5 kph/3 mph. Although it proved the viability of steam-powered traction, the problems of water supply and pressure maintenance severely handicapped the vehicle.

Daguerre, Louis Jacques Mandé (1787–1851) French pioneer of photography. Together with his fellow French pioneer Joseph-Nicéphore Niépce (1765–1833), he is credited with the invention of photography (though others were reaching the same point simultaneously). In 1838 he invented the daguerreotype, a single image process superseded ten years later by Fox ◊Talbot's negative/positive process.

Daimler, Gottlieb Wilhelm (1834–1900) German engineer who pioneered the car and the internal-combustion engine together with Wilhelm Maybach. In 1885 he produced a motor bicycle and in 1889 his first four-wheeled motor vehicle. He combined the vaporization of fuel with the high-speed four-stroke petrol engine.

De Havilland, Geoffrey (1882–1965) British aircraft designer who designed and whose company produced the Moth biplane, the Mosquito fighter-bomber of World War II, and in 1949 the Comet, the world's first jet-driven airliner to enter commercial service.

Diesel, Rudolf Christian Karl (1858–1913) German engineer who patented the diesel engine. He began his career as a refrigerator engineer and, like many engineers of the period, sought to develop a better power source than the conventional steam engine. Able to operate with greater efficiency and economy, the diesel engine soon found a ready market.

Dunlop, John Boyd (1840–1921) Scottish inventor who founded the rubber company that bears his name. In 1888, to help his child win a tricycle race, he bound an inflated rubber hose to the wheels. The same year he developed commercially practical pneumatic tyres, first patented by Robert William Thomson (1822–1873) in 1845 for bicycles and cars. Dunlop's first simple design consisted of a rubber inner tube, covered by a jacket of linen tape with an outer tread also of rubber. The inner tube was inflated using a football pump and the tyre was attached by flaps in the jacket which were rubber-cemented to the wheel. Later, he incorporated a wire through the edge of the tyre which secured it to the rim of the wheel.

Eastman, George (1854–1932) US entrepreneur and inventor who founded the Eastman Kodak photographic company in 1892. He patented flexible film in 1884, invented the Kodak box camera in 1888, and followed this up in 1889 with the first commercially available transparent nitrocellulose (celluloid) roll films. He introduced daylight-loading film in 1892. By 1900 his company was selling a pocket camera for as little as one dollar ushering in the era of press-button photography.

Edison, Thomas Alva (1847–1931) US scientist and inventor, whose work in the fields of communications and electrical power greatly influenced the world in which we live. His first invention was an automatic repeater for telegraphic messages. He then invented a tape machine called a 'ticker', which communicated stock exchange prices across the country. Edison improved Bell and Gray's telephone with his invention of the carbon transmitter, which so increased the volume of the telephone signal that it was used as a microphone in the Bell telephone. In 1877 he launched the era of recorded sound by inventing the phonograph, a device in which the vibrations of the human voice were engraved by a needle on a revolving cylinder coated with tin foil. In 1879, using carbonized sewing cotton mounted on an electrode in a vacuum (one millionth of an atmosphere), he invented the electric light bulb. He also constructed a system of electric power distribution for consumers, the telephone transmitter, the megaphone, and the kinetoscopic camera, an early cine camera. With more than 1,000 patents, Edison produced his most important inventions in Menlo Park, New Jersey, between 1876 and 1887.

Ford, Henry (1863–1947) US automobile manufacturer. He built his first car in 1896 and founded the Ford Motor Company in 1903. His Model T (1908–27) was the first car to be constructed solely by assembly-line methods and to be mass-marketed; 15 million of these cars were sold. In 1928 he launched the Model A, a stepped-up version of the Model T.

Giffard, Henri (1825–1882) French inventor of the first passenger-carrying powered and steerable airship, called a dirigible, built in 1852. The hydrogen-filled airship was 43 m/144 ft long, had a 2,200-W/3-hp steam engine that drove a three-bladed propeller, and was steered using a sail-like rudder. It flew at an average speed of 5 kph/3 mph.

Goddard, Robert Hutchings (1882–1945) US rocket pioneer. He developed the principle of combining liquid fuels in a rocket motor, the technique used subsequently in every practical space vehicle. His first liquid-fuelled rocket was launched at Auburn, Massachusetts, in 1926. In 1929, instruments, and a camera to record them, were carried aloft for the first time, and by 1935 his rockets had gyroscopic control. Two years later a Goddard rocket gained the world altitude record with an ascent of 3 km/1.9 mi. He was the first to prove by actual test that a rocket will work in a vacuum and he was the first to fire a rocket faster than the speed of sound.

Goodyear, Charles (1800–1860) US inventor who developed rubber coating in 1837 and vulcanized rubber in 1839, a method of curing raw rubber to make it strong and elastic.

Gutenberg, Johannes (Gensfleisch) (c. 1398–1468) German printer, the inventor of printing from movable metal type based on the Chinese wood-block-type method. Gutenberg began work on the process in the 1440s and in 1450 set up a printing business in Mainz. By 1456 he had produced the first printed Bible (known as the Gutenberg Bible). He punched and engraved a steel character (letter shape) into a piece of copper to form a mould that he filled with molten metal.

Hargreaves, James (c. 1720–1778) English inventor who co-invented a carding machine for combing wool in 1760. About 1764 he invented the 'spinning jenny' (patented in 1770), which enabled a number of threads to be spun simultaneously by one person. The spinning jenny multiplied eightfold the output of the spinner and could be worked easily by children.

Hooke, Robert (1635–1703) English scientist and inventor, originator of **Hooke's law**. His inventions included a telegraph system, the spirit level, marine barometer, and sea gauge. He coined the term 'cell' in biology. He studied elasticity, furthered the sciences of mechanics and microscopy, invented the hairspring regulator in timepieces, perfected the air pump, and helped improve such scientific instruments as microscopes, telescopes, and barometers. His work on gravitation and in optics contributed to the achievements of his contemporary the English physicist and mathematician Isaac Newton.

Jacquard, Joseph Marie (1752–1834) French textile manufacturer. He invented a punched-card system for programming designs on a carpetmaking loom. In 1801 he constructed looms that used a series of punched cards to control the pattern of longitudinal warp threads depressed before each sideways passage of the shuttle. On later machines the punched cards were joined to form an endless loop that represented the 'program' for the repeating pattern of a carpet. Jacquard-style punched cards were used in the early computers of the 1940s–1960s.

Lesseps, Ferdinand Marie, Vicomte de Lesseps (1805–1894) French engineer. He designed and built the Suez Canal 1859–69, shortening the route between Britain and India by 9,700 km/6,000 mi. He began work on the Panama Canal in 1881, but withdrew after failing to construct it without locks.

Lilienthal, Otto (1848–1896) German aviation pioneer who inspired the US aviators Orville and Wilbur Wright (see ◊Wright brothers). Lilienthal demonstrated the superiority of cambered wings over flat wings – the principle of the aerofoil and from 1891 he made and successfully flew many gliders, including two biplanes. In his planes the pilot was suspended by the arms, as in a modern hang-glider. He achieved glides of more than 300 m/1,000 ft.

Lumière, Auguste Marie Louis Nicolas (1862–1954) and Louis Jean (1864–1948) French brothers who pioneered cinematography. In February 1895 they patented their cinematograph, a combined camera and projector operating at 16 frames per second, screening short films for the first time on 22 March, and in December opening the world's first cinema in Paris.

McAdam, John Loudon (1756–1836) Scottish engineer, inventor of the **macadam** road surface. McAdam introduced a method of road building that raised the road above the surrounding terrain, compounding a surface of small stones bound with gravel on a firm base of large stones. A camber, making the road slightly convex in section, ensured that rainwater rapidly drained off the road and did not penetrate the foundation. It originally consisted of broken granite bound together with slag or gravel, raised for drainage. Today, it is bound with tar or asphalt.

Marconi, Guglielmo (1874–1937) Italian electrical engineer and pioneer in the invention and development of radio. In 1895 he achieved radio communication over more than a mile. In 1898 he successfully transmitted signals across the English Channel, and in 1901 established communication with St John's, Newfoundland, from Poldhu in Cornwall, and in 1918 with Australia. Marconi's later inventions included the magnetic detector (1902), horizontal direction telegraphy (1905), and the continuous wave system (1912).

Montgolfier, Joseph Michel (1740–1810) and Jacques Etienne (1745–1799) French brothers whose hot-air balloon was used for the first successful human flight on 21 November 1783. The Montgolfier experiments greatly stimulated scientific interest in aviation.

Morse, Samuel Finley Breese (1791–1872) US inventor. In 1835 he produced the first adequate electric telegraph. The signal current was sent in an intermittent coded pattern and would cause an electromagnet to attract intermittently to the same pattern on a piece of soft iron to which a pencil or pen would be attached and which in turn would make marks on a moving strip of paper. In 1843 he was granted $30,000 by Congress for an experimental line between Washington, DC, and Baltimore. With his assistant Alexander Bain (1810–1877) he invented the Morse code.

Muybridge, Eadweard (1830–1904) Adopted name of Edward James Muggeridge, English-born US photographer. He made a series of animal locomotion photographs in the USA in the 1870s and proved that, when a horse trots, there are times when all its feet are off the ground. He also explored motion in birds and humans.

Otis, Elisha Graves (1811–1861) US engineer who developed a lift that incorporated a safety device, making it acceptable for passenger use in the first skyscrapers. The device, invented in 1852, consisted of vertical ratchets on the sides of the lift shaft into which spring-loaded catches would engage and lock the lift in position in the event of cable failure. In 1857 the first public passenger lift was installed in New York. Otis also invented and patented railway trucks and brakes, a steam plough, and a baking oven.

Remington, Philo (1816–1889) US inventor. He designed the breech-loading rifle that bears his name. He began manufacturing typewriters in 1873, using the patent of the US printer and newspaper editor Christopher Sholes (1819–1890), and made improvements that resulted five years later in the first machine with a shift key, thus providing lower-case letters as well as capital letters.

Sikorsky, Igor Ivan (1889–1972) Ukrainian-born US engineer. He built the first successful helicopter in 1939 (commercially produced from 1943). His first biplane flew in 1910, and in 1929 he began to construct multi-engined flying boats.

Singer, Isaac Merrit (1811–1875) US inventor of domestic and industrial sewing machines. Within a few years of opening his first factory in 1851, he became the world's largest sewing-machine manufacturer and by the late 1860s more than 100,000 Singer sewing machines were in use in the USA alone. Singer used the best of the US inventor Elias Howe's sewing machine design and altered some of the other features. The basic mechanism was the same: as the handle turned, the needle paused at a certain point in its stroke so that the shuttle could pass through the loop formed in the cotton. When the needle continued, the threads were tightened, forming a secure stitch.

Stephenson, George (1781–1848) English engineer who built the first successful steam locomotive. He introduced a system by which exhaust steam was redirected into the chimney through a blast pipe, bringing in air with it and increasing the draught through the fire. This development made the locomotive truly practical. He also invented a safety lamp independently of the English chemist Humphrey Davy in 1815. In 1821 he was appointed engineer of the Stockton and Darlington Railway, the world's first public railway, setting the gauge at 1.4 m/4 ft 8 in, which became the standard gauge for railways in most of the world. In 1829 he won a prize with his locomotive *Rocket*. He also advocated the use of malleable iron rails instead of cast iron.

Stephenson, Robert (1803–1859) English civil engineer. He constructed railway bridges such as the high-level bridge at Newcastle-upon-Tyne, England, and the Menai and Conway tubular bridges in Wales. He was the son of George Stephenson. The successful *Rocket* steam locomotive was built under his direction in 1829, as were subsequent improvements to it.

Swan, Joseph Wilson (1828–1914) English inventor. Swan invented an incandescent-filament electric lamp using a filament of cotton thread partly dissolved by sulphuric acid. He patented the process in 1880 and began manufacturing lamps. He also developed bromide paper for use in developing photographs, and a miner's electric safety lamp, which was the ancestor of the modern miner's lamp, and in the course of this invention he devised a new lead cell (battery) which would not spill acid. He also attempted to make an early type of fuel cell.

Talbot, William Henry Fox (1800–1877) English pioneer of photography. In 1841 he invented the paper-based calotype process – the first negative/positive method. Writing paper was coated successively with solutions of silver nitrate and potassium iodide, forming silver iodide. The iodized paper was made more sensitive by brushing with solutions of gallic acid and silver nitrate, and then it was exposed (either moist or dry). The latent image was developed with an application of gallo-silver nitrate solution, and when the image became visible, the paper was warmed for one to two minutes. It was fixed with a solution of potassium bromide (later replaced by sodium hyposulphite). Calotypes did not have the sharp definition of daguerreotypes and were generally considered inferior. Talbot made photograms several years before Louis ◊Daguerre's invention was announced. In 1851 he made instantaneous photographs by electric light and in 1852 photo engravings. *The Pencil of Nature* (1844–46) by Talbot was the first book illustrated with photographs to be published.

Telford, Thomas (1757–1834) Scottish civil engineer. He opened up northern Scotland by building roads and waterways. He constructed many aqueducts and canals, including the Caledonian Canal (1802–23), and erected the Menai road suspension bridge between Wales and Anglesey (1819–26), a type of structure scarcely tried previously in the UK. In Scotland he constructed over 1,600 km/1,000 mi of road and 1,200 bridges, churches, and harbours.

Tesla, Nikola (1856–1943) Serbian-born US physicist and electrical engineer who invented fluorescent lighting, the Tesla induction motor (1882–87), and the Tesla coil, and developed the alternating current (AC) electrical supply system.

Trevithick, Richard (1771–1833) English engineer, constructor of a steam road locomotive, the *Puffing Devil*, in 1801. By 1804 he had produced the first railway locomotive to run on rails and the first to carry passengers. Able to haul 10 tonnes and 70 people for 15 km/9.5 mi it was set up on rails used by horse-drawn trains at a mine in Wales.

von Braun, Wernher Magnus Maximilian (1912–1977) German rocket engineer responsible for Germany's rocket development programme in World War II (V1 and V2). He later worked for the space agency NASA in the USA. He also invented the Saturn rocket (*Saturn V*) that sent the Apollo spacecraft to the Moon in 1969.

Watt, James (1736–1819) Scottish engineer who developed the steam engine in the 1760s, making the English inventor

Thomas Newcomen's earlier engine vastly more efficient by cooling the used steam in a condenser separate from the main cylinder. He eventually made a double-acting machine that supplied power with both directions of the piston and developed rotary motion. During the period 1775–90, Watt invented an automatic centrifugal governor, which cut off the steam when the engine began to work too quickly and turned it on again when it had slowed sufficiently. He also devised a steam engine indicator that showed steam pressure and the degree of vacuum within the cylinder. Watt devised a rational method to rate the capability of his engines by considering the rate at which horses worked. After many experiments, he concluded that a 'horsepower' was 33,000 lb (15,000 kg) raised through 1 ft (0.3 m) each minute. The modern unit of power, the watt, is named after him.

Westinghouse, George (1846–1914) US inventor and founder of the Westinghouse Corporation in 1886. Westinghouse helped to standardize railway components, including the development of a completely new signalling system. He also developed a system of gas mains, and patented a powerful air brake for trains in 1869, which allowed trains to run more safely with greater loads at higher speeds. In the 1880s he turned his attention to the generation of electricity. Unlike Thomas ◊Edison, Westinghouse introduced alternating current (AC) into his power stations. In 1895 the Westinghouse Electric Company harnessed Niagara Falls to generate electricity for the lights and trams of the nearby town of Buffalo.

Whitney, Eli (1765–1825) US inventor who in 1794 patented the cotton gin, a device for separating cotton fibre from its seeds. He also used machine tools to make firearms with fully interchangeable parts creating a standardization system that was the precursor of the assembly line.

Whittle, Frank (1907–1996) British engineer. He patented the basic design for the turbojet engine in 1930. In May 1941 the Gloster E 28/39 aircraft first flew with the Whittle jet engine.

Wright brothers, Orville (1871–1948) and Wilbur (1867–1912) US inventors, brothers, who pioneered piloted, powered flight. Inspired by Otto ◊Lilienthal's gliding, they perfected their piloted glider in 1902. In 1903 they built a powered machine, a 12-hp 341-kg/750-lb plane, and became the first to make a successful powered flight, near Kitty Hawk, North Carolina. Orville flew 36.6 m/120 ft in 12 seconds; Wilbur, 260 m/852 ft in 59 seconds.

Zworykin, Vladimir Kosma (1889–1982) Russian-born US electronics engineer who invented a television camera tube, the iconoscope, which uses an electron beam to scan the charge pattern on a signal plate, which corresponds to the pattern of light and dark of an image focused on it by a lens. He also developed the electron microscope, an early form of electric eye, an electronic image tube sensitive to infrared light, which was the basis for World War II inventions for seeing in the dark, and in 1929 he demonstrated an improved electronic television system.

Glossary

accumulator a storage ◊battery – that is, a group of rechargeable secondary cells. A familiar example is the lead–acid car battery.

aerial, or **antenna**, in radio and television broadcasting, a conducting device that radiates or receives electromagnetic waves. The design of an aerial depends principally on the wavelength of the signal.

alloy metal blended with some other metallic or nonmetallic substance to give it special qualities, such as resistance to corrosion, greater hardness, or tensile strength. Useful alloys include bronze, brass, cupronickel, duralumin, German silver, gunmetal, pewter, solder, steel, and stainless steel.

alternator electricity generator that produces an alternating current.

altimeter instrument used in aircraft that measures altitude, or height above sea level. The common type is a form of aneroid barometer, which works by sensing the differences in air pressure at different altitudes. This must continually be recalibrated because of the change in air pressure with changing weather conditions. The ◊radar altimeter measures the height of the aircraft above the ground, measuring the time it takes for radio pulses emitted by the aircraft to be reflected.

annealing controlled cooling of a material to increase ductility and strength. The process involves first heating a material (usually glass or metal) for a given time at a given temperature, followed by slow cooling.

anodizing process that increases the resistance to corrosion of a metal, such as aluminium, by building up a protective oxide layer on the surface. The natural corrosion resistance of aluminium is provided by a thin film of aluminium oxide; anodizing increases the thickness of this film and thus the corrosion protection.

aqualung, or **scuba**, underwater breathing apparatus worn by divers, developed in the early 1940s by the French diver Jacques Cousteau. Compressed-air cylinders strapped to the diver's back are regulated by a valve system and by a mouth tube to provide air to the diver at the same pressure as that of the surrounding water (which increases with the depth).

aqueduct any artificial channel or conduit for water, originally applied to water supply tunnels, but later used to refer to elevated structures of stone, wood, or iron carrying navigable canals across valleys. One of the first great aqueducts was built in 691 BC, carrying water for 80 km/50 mi to Ninevah, capital of the ancient Assyrian Empire.

Archimedes screw one of the earliest kinds of pump, associated with the Greek mathematician Archimedes. It consists of an enormous spiral screw revolving inside a close-fitting cylinder. It is used, for example, to raise water for irrigation and for land drainage. Of robust and simple construction, it has the advantage of being able to shift water containing mud, sand, gravel, and even larger debris.

arc lamp, or **arc light**, electric light that uses the illumination of an electric arc maintained between two electrodes. The lamp consists of two carbon electrodes, between which a very high voltage is maintained. Electric current arcs (jumps) between the two electrolytes, creating a brilliant light. Its main use in recent years has been in cinema projectors.

armature in a motor or generator, the wire-wound coil that carries the current and rotates in a magnetic field. (In alternating-current machines, the armature is sometimes stationary.) The pole piece of a permanent magnet or electromagnet and the moving, iron, part of a solenoid, especially if the latter acts as a switch, may also be referred to as armatures.

atomic clock timekeeping device regulated by various periodic processes occurring in atoms and molecules, such as atomic vibration or the frequency of absorbed or emitted radiation. The first atomic clock, the **ammonia clock**, was invented at the US National Bureau of Standards in 1948, and was regulated by measuring the speed at which the nitrogen atom in an ammonia molecule vibrated back and forth. The rate of molecular vibration is not affected by temperature, pressure, or other external influences, and can be used to regulate an electronic clock. Atomic clocks are so accurate that minute adjustments must be made periodically to the length of the year to keep the calendar exactly synchronized with the Earth's rotation, which has a tendency to slow down.

autogiro, or **autogyro**, heavier-than-air craft that supports itself in the air with a rotary wing, or rotor. The autogiro's rotor provides only lift and not propulsion; it has been superseded by the helicopter, in which the rotor provides both. The autogiro is propelled by an orthodox propeller.

Bakelite the first synthetic plastic, created by the Belgian-born US chemist Leo Baekeland in 1909. Bakelite is hard, tough, and heatproof, and is used as an electrical insulator. It is made by the reaction of phenol with formaldehyde, producing a powdery resin that sets solid when heated. Objects are made by subjecting the resin to compression moulding (simultaneous heat and pressure in a mould).

ball valve valve that works by the action of external pressure raising a ball and thereby opening a hole. An example is the valve used in lavatory cisterns to cut off the water supply when it reaches the correct level.

basic–oxygen process the most widely used method of steelmaking, involving the blasting of oxygen at supersonic speed into molten pig iron.

battery any energy-storage device allowing release of electricity on demand. It is made up of one or more electrical cells. Primary-cell batteries are disposable; secondary-cell batteries, or ◊accumulators, are rechargeable. Primary-cell batteries are an extremely uneconomical form of energy, since they produce only 2% of the power used in their manufacture.

Bessemer process the first cheap method of making ◊steel, invented by the engineer and inventor Henry Bessemer in England in 1856. In the Bessemer process compressed air is blown into the bottom of a converter, a furnace shaped like a cement mixer, containing molten pig iron. The excess carbon in the iron burns off, other impurities form a slag, and the furnace is emptied by tilting. The process has since been superseded by the ◊basic–oxygen process.

binoculars optical instrument for viewing an object in magnification with both eyes; for example, field glasses and opera glasses. Binoculars consist of two telescopes containing lenses and prisms, which produce a stereoscopic effect as well as magnifying the image.

blast furnace smelting furnace used to extract metals from their ores, chiefly pig iron from iron ore. The temperature is raised by the injection of an air blast.

Bourdon gauge instrument for measuring pressure, patented by the French watchmaker Eugène Bourdon in 1849. The gauge contains a C-shaped tube, closed at one end. When the pressure inside the tube increases, the tube uncurls slightly causing a small movement at its closed end. A system of levers and gears magnifies this movement and turns a pointer, which indicates the pressure on a circular scale. Bourdon gauges are often fitted to cylinders of compressed gas used in industry and hospitals.

cable television distribution of broadcast signals through cable relay systems. Systems using coaxial and fibreoptic cable, are increasingly used for distribution and development of home-based interactive services, typically telephones.

caesium clock extremely accurate type of ◊atomic clock. Because of its internal structure, a caesium atom produces or absorbs radiation of a very precise frequency (9,192,631,770 Hz) that varies by less than one part in 10 billion. This frequency has been used to define the second, and is the basis of atomic clocks used in international timekeeping.

cantilever beam or structure that is fixed at one end only, though it may be supported at some point along its length; for example, a diving board. The cantilever principle, widely used in construction engineering, eliminates the need for a second main support at the free end of the beam, allowing for more elegant structures and reducing the amount of materials required. Many large-span bridges have been built on the cantilever principle.

carbon fibre fine, black, silky filament of pure carbon produced by heat treatment from a special grade of Courtelle acrylic fibre and used for reinforcing plastics. The resulting composite is very stiff and, weight for weight, has four times the strength of high-tensile steel. It is used in the aerospace industry, cars, and electrical and sports equipment.

carburation mixing of a gas, such as air, with a volatile hydrocarbon fuel, such as petrol, kerosene, or fuel oil, in order to form an explosive mixture. The process, which ensures that the maximum amount of heat energy is released during combustion, is used in internal-combustion engines. In most petrol engines the liquid fuel is atomized and mixed with air by means of a device called a **carburettor.**

cast iron cheap but invaluable constructional material, most commonly used for car engine blocks. Cast iron is partly refined pig (crude) iron, which is very fluid when molten and highly suitable for shaping by casting; it contains too many impurities (for example, carbon) to be readily shaped in any other way. Solid cast iron is heavy and can absorb great shock but is very brittle.

catalytic converter device fitted to the exhaust system of a motor vehicle in order to reduce toxic emissions from the engine. It converts harmful exhaust products to relatively harmless ones by passing the exhaust gases over a mixture of catalysts coated on a metal or ceramic honeycomb (a structure that increases the surface area and therefore the amount of active catalyst with which the exhaust gases will come into contact).

cellular phone, or **cellphone,** mobile radio telephone, one of a network connected to the telephone system by a computer-controlled communication system. Service areas are divided into small 'cells', about 5 km/3 mi across, each with a separate low-power transmitter.

cement any bonding agent used to unite particles in a single mass or to cause one surface to adhere to another. **Portland cement** is a powder which when mixed with water and sand or gravel turns into mortar or concrete. Cement sets by losing water. The term 'cement' covers a variety of materials, such as fluxes and pastes, and also bituminous products obtained from tar.

charge-coupled device (CCD) device for forming images electronically, using a layer of silicon that releases electrons when struck by incoming light. The electrons are stored in pixels and read off into a computer at the end of the exposure. CCDs have now almost entirely replaced photographic film for applications such as astrophotography where extreme sensitivity to light is paramount.

cine camera camera that takes a rapid sequence of still photographs called frames. When the frames are projected one after the other onto a screen, they appear to show movement, because our eyes hold onto the image of one picture until the next one appears.

coaxial cable electric cable that consists of a solid or stranded central conductor insulated from and surrounded by a solid or braided conducting tube or sheath. It can transmit the high-frequency signals used in television, telephones, and other telecommunications transmissions.

compact disc (CD) disc for storing digital information, about 12 cm/4.5 in across, mainly used for music, when it can have over an hour's playing time. A laser beam etches the compact disc with microscopic pits that carry a digital code representing the sounds; the pitted surface is then coated with aluminium. During playback, a laser beam reads the code and produces signals that are changed into near-exact replicas of the original sounds.

concrete building material composed of cement, stone, sand, and water. It has been used since Egyptian and Roman times. Since the late 19th century it has been combined with steel to increase its tension capacity. Reinforced concrete and prestressed concrete are strengthened by combining concrete with another material, such as steel rods or glass fibres. The addition of carbon fibres to concrete increases its conductivity. The electrical resistance of the concrete changes with increased stress or fracture, so this 'smart concrete' can be used as an early indicator of structural damage.

delta wing aircraft wing shaped like the Greek letter delta. Its design enables an aircraft to pass through the sound barrier with little effect. The supersonic airliner Concorde and the US space shuttle have delta wings.

diesel engine an internal-combustion engine that burns a light-weight fuel oil and operates by compressing air until it becomes sufficiently hot to ignite the fuel. It is a piston-in-cylinder engine, like the ◊petrol engine, but only air (rather than an air-and-fuel mixture) is taken into the cylinder on the first piston stroke (down). The piston moves up and compresses the air until it is at a very high temperature. The fuel oil is then injected into the hot air, where it burns, driving the piston down on its power stroke. For this reason the engine is called a compression-ignition engine.

digital audio tape (DAT) digitally recorded audio tape produced in cassettes that can carry up to two hours of sound on each side and are about half the size of standard cassettes. DAT players/recorders were developed in 1987. The first DAT for computer data was introduced in 1988.

digital camera camera that uses a ◊ charge-coupled device to take pictures which are stored as digital data rather than on film. The output from digital cameras can be downloaded onto a computer for retouching or storage, and can be readily distributed as computer files.

digital compact cassette (DCC) digitally recorded audio cassette that is roughly the same size as a standard cassette. It cannot be played on a normal tape recorder.

electron microscope instrument that produces a magnified image by using a beam of electrons instead of light rays, as in an optical ◊microscope. An electron lens – an arrangement of electromagnetic coils – control and focus the beam. Electrons are not visible to the eye, so instead of an eyepiece there is a fluorescent screen or a photographic plate on which the electrons form an image. The wavelength of the electron beam is much shorter than that of light, so much greater magnification and resolution (ability to distinguish detail) can be achieved. A **transmission electron microscope** passes the electron beam through a very thin slice of a specimen. A **scanned-probe microscope**, which has a probe with a tip so fine that it may consist only of a single atom, runs across the surface of the specimen. A **scanning electron microscope** looks at the exterior of a specimen. In the **scanning tunnelling microscope**, an electric current flows through the probe to construct an image of the specimen. In the **atomic force microscope**, the force felt by the probe is measured and used to form the image. A **scanning transmission electron microscope** (STEM) can produce a magnification of 90 million times.

fibreglass glass that has been formed into fine fibres, either as long continuous filaments or as a fluffy, short-fibred glass wool. Fibreglass is heat- and fire-resistant and a good electrical insulator. It has applications in the field of fibre optics and as a strengthener for plastics in GRP (glass-reinforced plastics).

four-stroke cycle the engine-operating cycle of most petrol and ◊diesel engines. The 'stroke' is an upward or downward movement of a piston in a cylinder. In a petrol engine the cycle begins with the induction of a fuel mixture as the piston goes down on its first stroke. On the second stroke (up) the piston compresses the mixture in the top of the cylinder. An electric spark then ignites the mixture, and the gases produced force the piston down on its third, power, stroke. On the fourth stroke (up) the piston expels the burned gases from the cylinder into the exhaust. The four-stroke cycle is also called the **Otto cycle**. The diesel engine cycle works in a slightly different way to that of the petrol engine on the first two strokes.

gravure one of the three main printing methods, in which printing is done from a plate etched with a pattern of recessed cells in which the ink is held. The greater the depth of a cell, the greater the strength of the printed ink.

heat treatment in industry, the subjection of metals and alloys to controlled heating and cooling after fabrication to relieve internal stresses and improve their physical properties. Methods include ◊annealing, quenching, and tempering.

hydraulic press two liquid-connected pistons in cylinders, one of narrow bore, one of large bore. A force applied to the narrow piston applies a certain pressure (force per unit area) to the liquid, which is transmitted to the larger piston. Because the area of this piston is larger, the force exerted on it is larger. Thus the original force has been magnified, although the smaller piston must move a great distance to move the larger piston only a little.

hydrogen maser clock the most accurate type of ◊atomic clock based on the radiation from hydrogen atoms. The hydrogen maser clock at the US Naval Research Laboratory, Washington, DC, is estimated to lose one second in 1,700,000 years. Cooled hydrogen maser clocks could theoretically be accurate to within one second in 300 million years.

hydroplane on a submarine, a movable horizontal fin angled downwards or upwards when the vessel is descending or ascending. It is also a highly manoeuvrable motorboat with its bottom rising in steps to the stern, or a hydrofoil boat that skims over the surface of the water when driven at high speed.

inductor device included in an electrical circuit because of its inductance.

interactive video (IV) computer-mediated system that enables the user to interact with and control information (including text, recorded speech, or moving images) stored on video disc. IV is most commonly used for training purposes, using analogue video discs, but has wider applications with digital video systems such as CD-I (Compact Disc Interactive, from Philips and Sony) which are based on the CD-ROM format derived from audio compact discs.

jet engine a kind of gas ◊turbine. Air, after passing through a forward-facing intake, is compressed by a compressor, or fan, and fed into a combustion chamber. Fuel (usually kerosene) is sprayed in and ignited. The hot gas produced expands rapidly rearwards, spinning a turbine that drives the compressor before being finally ejected from a rearward-facing tail pipe, or nozzle, at very high speed. Reaction to the jet of gases streaming backwards produces a propulsive thrust forwards, which acts on the aircraft through its engine-mountings, not from any pushing of the hot gas stream against the static air.

jetfoil advanced type of hydrofoil boat built by Boeing, propelled by water jets. It features horizontal, fully submerged hydrofoils fore and aft and has a sophisticated computerized control system to maintain its stability in all waters.

laser (acronym for **l**ight **a**mplification by **s**timulated **e**mission of **r**adiation) device for producing a narrow beam of light, capable of travelling over vast distances without dispersion, and of being focused to give enormous power densities. The uses of lasers include communications (a laser beam can carry much more information than can radio waves), cutting, drilling, welding, satellite tracking, medical and biological research, and surgery. Any substance in which the majority of atoms or molecules can be put into an excited energy state can be used as laser material. Many solid, liquid, and gaseous substances have been used, including synthetic ruby crystal, carbon dioxide gas, a helium–neon gas mixture, and complex organic dyes.

lie detector instrument that records graphically certain body activities, such as thoracic and abdominal respiration, blood pressure, pulse rate, and galvanic skin response (changes in electrical resistance of the skin). Marked changes in these activities when a person answers a question may indicate that the person is lying.

light-emitting diode (LED) electronic component that converts electrical energy into light or infrared radiation in the range of 550 nm (green light) to 1300 nm (infrared). It is used for displaying symbols in electronic instruments and devices. An LED is a diode made of semiconductor material, such as gallium arsenide phosphide, that glows when electricity is passed through it. The first digital watches and calculators had LED displays, but many later models use ◊liquid-crystal displays.

linear motor type of electric motor, an induction motor in which the fixed stator and moving armature are straight and parallel to each other (rather than being circular and one inside the other as in an ordinary induction motor). Linear motors are used, for example, to power sliding doors. There is a magnetic force between the stator and armature; this force has been used to support a vehicle, as in the experimental maglev linear motor train.

liquid-crystal display (LCD) display of numbers (for example, in a calculator) or pictures (such as on a pocket television screen) produced by molecules of a substance in a semiliquid state with some crystalline properties, so that clusters of molecules align in parallel formations. The display is a blank until the application of an electric field, which 'twists' the molecules so that they reflect or transmit light falling on them.

loom any machine for weaving yarn or thread into cloth. A loom is a frame on which a set of lengthwise threads (warp) is strung. A second set of threads (weft), carried in a shuttle, is inserted at right angles over and under the warp. In most looms the warp threads are separated by a device called a treddle to create a gap, or shed, through which the shuttle can be passed in a straight line. A kind of comb called a reed presses each new line of weave tight against the previous ones.

loudspeaker electromechanical device that converts electrical signals into sound waves, which are radiated into the air. The most common type of loudspeaker is the **moving-coil speaker**. Electrical signals from a radio, for example, are fed to a coil of fine wire wound around the top of a cone. A magnet surrounds the coil. When signals pass through it, the coil becomes an electromagnet, which by moving causes the cone to vibrate, setting up sound waves.

magnetic tape narrow plastic ribbon coated with an easily magnetizable material on which data can be recorded. It is used in sound recording, audiovisual systems (videotape), and computing.

metal detector electronic device for detecting metal, usually below ground, developed from the wartime mine detector. In the head of the metal detector is a coil, which is part of an electronic circuit. The presence of metal causes the frequency of the signal in the circuit to change, setting up an audible note in the headphones worn by the user.

meter any instrument used for measurement. The term is often compounded with a prefix to denote a specific type of meter: for example, ammeter, voltmeter, flowmeter, or pedometer.

microphone primary component in a sound-reproducing system, whereby the mechanical energy of sound waves is converted into electrical signals by means of a transducer. One of the simplest is the telephone receiver mouthpiece, invented by Alexander Graham Bell in 1876; other types of microphone are used with broadcasting and sound-film apparatus. Telephones have a **carbon microphone**, which reproduces only a narrow range of frequencies. For live music a **moving-coil microphone** is often used. In it, a diaphragm that vibrates with sound waves moves a coil through a magnetic field, thus generating an electric current. The **ribbon microphone** combines the diaphragm and coil. The **condenser microphone** is most commonly used in recording and works by a capacitor.

microscope instrument for forming magnified images with high resolution for detail. Optical and electron microscopes are the ones chiefly in use; other types include acoustic, scanning tunnelling, and atomic force microscopes.

missile rocket-propelled weapon, which may be nuclear-armed. Modern missiles are often classified as **surface-to-surface missiles (SSM)**, **air-to-air missiles (AAM)**, **surface-to-air missiles (SAM)**, or **air-to-surface missiles (ASM)**. A **cruise missile** is in effect a pilotless, computer-guided aircraft; it can be sea-launched from submarines or surface ships, or launched from the air or the ground.

monorail railway that runs on a single rail; the cars can be balanced on it or suspended from it. It was invented in 1882 to carry light loads, and when run by electricity was called a **telpher**.

motor anything that produces or imparts motion; a machine that provides mechanical power – for example, an electric motor.

motorcycle, or **motorbike**, two-wheeled vehicle propelled by a ◊petrol engine.

nylon synthetic long-chain polymer similar in chemical structure to protein. Nylon was the first all-synthesized fibre, made from petroleum, natural gas, air, and water by the Du Pont firm in 1938. It is used in the manufacture of moulded articles, textiles, and medical sutures. Nylon fibres are stronger and more elastic than silk and are relatively insensitive to moisture and mildew.

offset printing the most common method of printing, which uses smooth (often rubber) printing plates. It works on the principle of lithography: that grease and water repel one another.

optical fibre very fine, optically pure glass fibre through which light can be reflected to transmit images or data from one end to the other. Although expensive to produce and install, optical fibres can carry more data than traditional cables, and are less susceptible to interference. Standard optical fibre transmitters can send up to 10 billion bits of information per second by switching a laser beam on and off. Optical fibres are increasingly being used to replace metal communications cables, the messages being encoded as digital pulses of light rather than as fluctuating electric current. Current research is investigating how optical fibres could replace wiring inside computers. Bundles of optical fibres are also used in endoscopes to inspect otherwise inaccessible parts of machines or of the living body.

Otto cycle alternative name for the ◊four-stroke cycle, introduced by the German engineer Nikolaus Otto (1832–1891) in 1876. It improved on existing piston engines by compressing the fuel mixture in the cylinder before it was ignited.

oxyacetylene torch gas torch that burns ethyne (acetylene) in pure oxygen, producing a high-temperature flame (3,000°C/5,400°F). It is widely used in welding to fuse metals. In the cutting torch, a jet of oxygen burns through metal already melted by the flame.

pacemaker a medical device implanted under the skin of a patient whose heart beats inefficiently. It delivers minute electric shocks to stimulate the heart muscles at regular intervals and restores normal heartbeat.

petrol engine a reciprocating piston engine in which a number of pistons move up and down in cylinders. A mixture of petrol and air is introduced to the space above the pistons and ignited. The gases produced force the pistons down, generating power. The engine-operating cycle is repeated every four strokes (upward or downward movement) of the piston, this being known as the ◊four-stroke cycle. The motion of the pistons rotate a crankshaft, at the end of which is a heavy flywheel. From the flywheel the power is transferred to the car's driving wheels via the transmission system of clutch, gearbox, and final drive.

piston barrel-shaped device used in reciprocating engines (steam, petrol, diesel oil) to harness power. Pistons are driven up and down in cylinders by expanding steam or hot gases. They pass on their motion via a connecting rod and crank to a crankshaft, which turns the driving wheels. In a pump or compressor, the role of the piston is reversed, being used to move gases and liquids.

pitch in mechanics, the distance between the adjacent threads of a screw or bolt. When a screw is turned through one full turn it moves a distance equal to the pitch of its thread.

Pitot tube instrument that measures fluid (gas and liquid) flow. It is used to measure the speed of aircraft, and works by sensing pressure differences in different directions in the airstream.

pneumatic drill drill operated by compressed air, used in mining and tunnelling, for drilling shot holes (for explosives), and in road repairs for breaking up pavements. It contains an air-operated piston that delivers hammer blows to the drill bit many times a second.

Polaroid camera instant-picture camera, invented by Edwin Land in the USA in 1947. The original camera produced black-and-white prints in about one minute. Modern cameras can produce black-and-white prints in a few seconds, and colour prints in less than a minute. A Polaroid camera ejects a piece of film on paper immediately after the picture has been taken. The film consists of layers of emulsion and colour dyes together with a pod of chemical developer. When the film is ejected the pod bursts and processing occurs in the light, producing a paper-backed print.

potentiometer an electrical resistor that can be divided so as to compare, measure, or control voltages. In radio circuits, any rotary variable resistance (such as volume control) is referred to as a potentiometer.

propeller screwlike device used to propel some ships and aeroplanes. A propeller has a number of curved blades that describe a helical path as they rotate with the hub, and accelerate fluid (liquid or gas) backwards during rotation. Reaction to this backward movement of fluid sets up a propulsive thrust forwards.

prosthesis artificial device used to substitute for a body part which is defective or missing. Prostheses include artificial limbs, hearing aids, false teeth and eyes, heart ◊pacemakers and plastic heart valves and blood vessels.

pump any device for moving liquids and gases, or compressing gases. Some pumps, such as the traditional **lift pump** used to raise water from wells, work by a reciprocating (up-and-down) action. Movement of a piston in a cylinder with a one-way valve creates a partial vacuum in the cylinder, thereby sucking water into it. **Gear pumps**, used to pump oil in a car's lubrication system, have two meshing gears that rotate inside a housing, and the teeth move the oil. **Rotary pumps** contain a rotor with vanes projecting from it inside a casing, sweeping the liquid round as they move.

radar (acronym for **ra**dio **d**irection **a**nd **r**anging) device for locating objects in space, direction finding, and navigation by means of transmitted and reflected high-frequency radio waves. The direction of an object is ascertained by transmitting a beam of short-wavelength (1–100 cm/0.5–40 in), short-pulse radio waves, and picking up the reflected beam. Distance is determined by timing the journey of the radio waves (travelling at the speed of light) to the object and back again. Radar is essential to navigation in darkness, cloud, and fog, and is widely used in warfare to detect enemy aircraft and missiles. Radar is also used to detect objects underground, for example service pipes, and in archaeology.

rayon any of various shiny textile fibres and fabrics made from cellulose. It is produced by pressing whatever cellulose solution is used through very small holes and solidifying the resulting filaments. A common type is ◊viscose, which consists of regenerated filaments of pure cellulose.

reinforced concrete material formed by casting concrete in timber or metal formwork around a cage of steel reinforcement. The steel gives added strength by taking up the tension stresses, while the concrete takes up the compression stresses.

reflex camera camera that uses a mirror and prisms to reflect light passing through the lens into the viewfinder, showing the photographer the exact scene that is being shot. When the shutter button is released the mirror springs out of the way, allowing light to reach the film. The most common type is the single-lens reflex (SLR) camera. The twin-lens reflex (TLR) camera has two lenses: one has a mirror for viewing, the other is used for exposing the film.

relay in electrical engineering, an electromagnetic switch. A small current passing through a coil of wire wound around an iron core attracts an ◊armature whose movement closes a pair of sprung contacts to complete a secondary circuit, which may carry a large current or activate other devices. The solid-state equivalent is a thyristor switching device.

rocket projectile driven by the reaction of gases produced by a fast-burning fuel. Unlike jet engines, which are also reaction engines, modern rockets carry their own oxygen supply to burn their fuel and do not require any surrounding atmosphere.

rolling common method of shaping metal. Rolling is carried out by giant mangles, consisting of several sets, or stands, of heavy rollers positioned one above the other. Red-hot metal slabs are rolled into sheet and also (using shaped rollers) girders and rails. Metal sheets are often cold-rolled finally to impart a harder surface.

scuba (acronym for **s**elf-**c**ontained **u**nderwater **b**reathing **a**pparatus) another name for ◊aqualung.

servomechanism automatic control system used in aircraft, motor cars, and other complex machines. A specific input, such as moving a lever or joystick, causes a specific output, such as feeding current to an electric motor that moves, for example, the rudder of the aircraft. At the same time, the position of the rudder is detected and fed back to the central control, so that small adjustments can continually be made to maintain the desired course.

sextant navigational instrument for determining latitude by measuring the angle between some heavenly body and the horizon. It can be used only in clear weather.

solder any of various alloys used when melted for joining metals such as copper, its common alloys (brass and bronze), and tin-plated steel, as used for making food cans. Soft solders (usually alloys of tin and lead, sometimes with added antimony) melt at low temperatures (about 200°C/392°F), and are widely used in the electrical industry for joining copper wires. Hard (or brazing) solders, such as silver solder (an alloy of copper, silver, and zinc), melt at much higher temperatures and form a much stronger joint. Printed circuit boards for computers are assembled by soldering.

solid-state circuit electronic circuit where all the components (resistors, capacitors, transistors, and diodes) and interconnections are made at the same time, and by the same processes, in or on one piece of single-crystal silicon. The small size of this construction accounts for its use in electronics for space vehicles and aircraft.

sonar (acronym for **so**und **na**vigation and **r**anging) method of locating underwater objects by the reflection of ultrasonic waves. The time taken for an acoustic beam to travel to the object and back to the source enables the distance to be found since the velocity of sound in water is known. Sonar devices, or **echo sounders**, were developed in 1920, and are the commonest means of underwater navigation.

soundtrack band at one side of a cine film on which the accompanying sound is recorded. Usually it takes the form of an optical track (a pattern of light and shade). The pattern is produced on the film when signals from the recording microphone are made to vary the intensity of a light beam. During playback, a light is shone through the track on to a photocell, which converts the pattern of light falling on it into appropriate electrical signals. These signals are then fed to loudspeakers to recreate the original sounds.

spark plug plug that produces an electric spark in the cylinder of a petrol engine to ignite the fuel mixture. It consists essentially of two electrodes insulated from one another. High-voltage (18,000 V) electricity is fed to a central electrode via the distributor. At the base of the electrode, inside the cylinder, the electricity jumps to another electrode earthed to the engine body, creating a spark.

stainless steel widely used ◊alloy of iron, chromium, and nickel that resists rusting. Its chromium content also gives it a high tensile strength. It is used for cutlery and kitchen fittings, and in surgical instruments.

steel alloy or mixture of iron and up to 1.7% carbon, sometimes with other elements, such as manganese, phosphorus, sulphur, and silicon. Steel has innumerable uses, including ship and car manufacture, skyscraper frames, and machinery of all kinds.

submersible vessel designed to operate under water, especially a small submarine used by engineers and research scientists as a ferry craft to support diving operations. The most advanced submersibles are the so-called lock-out type, which have two compartments: one for the pilot, the other to carry divers. The diving compartment is pressurized and provides access to the sea.

synthetic any material made from chemicals. Since the 1900s, more and more of the materials used in everyday life are synthetics, including plastics (polythene, polystyrene), synthetic fibres (nylon, acrylics, polyesters), synthetic resins, and synthetic rubber. Most naturally occurring organic substances are now made synthetically, especially pharmaceuticals.

tachograph combined speedometer and clock that records a vehicle's speed (on a small card disc, magnetic disc, or tape) and the length of time the vehicle is moving or stationary. It is used to monitor a lorry driver's working hours.

tape recording, magnetic method of recording electric signals on a layer of iron oxide, or other magnetic material, coating a thin plastic tape. The electrical signals from the microphone are fed to the electromagnetic recording head, which magnetizes the tape in accordance with the frequency and amplitude of the original signal. The impulses may be audio (for sound recording), video (for television), or data (for computer). For playback, the tape is passed over the same, or another, head to convert magnetic into electrical signals, which are then amplified for reproduction. Tapes are easily demagnetized (erased) for reuse, and come in cassette, cartridge, or reel form.

teleprinter, or **teletypewriter**, transmitting and receiving device used in telecommunications to handle coded messages. Teleprinters are automatic typewriters keyed telegraphically to convert typed words into electrical signals (using a five-unit Baudot code) at the transmitting end, and signals into typed words at the receiving end.

telex (acronym for **tele**printer **ex**change) international telecommunications network that handles telegraph messages in the form of coded signals. It uses ◊teleprinters for transmitting and receiving, and makes use of land lines (cables) and radio and satellite links to make connections between subscribers.

Tesla coil air core transformer with the primary and secondary windings tuned in resonance to produce high-frequency, high-voltage electricity.

theodolite instrument for the measurement of horizontal and vertical angles, used in surveying. It consists of a small telescope mounted so as to move on two graduated circles, one horizontal and the other vertical, while its axes pass through the centre of the circles.

thermocouple electric temperature measuring device consisting of a circuit having two wires made of different metals welded together at their ends. A current flows in the circuit when the two junctions are maintained at different temperatures (Seebeck effect). The electromotive force generated – measured by a millivoltmeter – is proportional to the temperature difference.

thermometer instrument for measuring temperature. There are many types, designed to measure different temperature ranges to varying degrees of accuracy. Each makes use of a different physical effect of temperature. Expansion of a liquid is employed in common **liquid-in-glass thermometers**, such as those containing mercury or alcohol. The more accurate **gas thermometer** uses the effect of temperature on the pressure of a gas held at constant volume. A **resistance thermometer** takes advantage of the change in resistance of a conductor (such as a platinum wire) with variation in temperature.

thermostat temperature-controlling device that makes use of feedback. It employs a temperature sensor (often a bimetallic strip) to operate a switch or valve to control electricity or fuel supply. Thermostats are used in central heating, ovens, and car engines.

transformer device in which, by electromagnetic induction, an alternating current (AC) of one voltage is transformed to another voltage, without change of frequency. Transformers are widely used in electrical apparatus of all kinds, and in particular in power transmission where high voltages and low currents are utilized.

transistor solid-state electronic component, made of semiconductor material, with three or more electrodes, that can regulate a current passing through it. A transistor can act as an amplifier, oscillator, photocell, or switch, and (unlike earlier thermionic valves) usually operates on a very small amount of power. Transistors commonly consist of a tiny sandwich of germanium or silicon, alternate layers having different electrical properties because they are impregnated with minute amounts of different impurities.

turbine an engine in which steam, water, gas, or air is made to spin a rotating shaft by pushing on angled blades, like a fan. Turbines are among the most powerful machines. **Steam turbines** are used to drive generators in power stations and ships' propellers; **water turbines** spin the generators in hydroelectric power plants; and **gas turbines** power most aircraft and drive machines in industry. A steam turbine consists of a shaft, or rotor, which rotates inside a fixed casing (stator). The rotor carries 'wheels' consisting of blades, or vanes. The stator has vanes set between the vanes of the rotor, which direct the steam through the rotor vanes at the optimum angle. When steam expands through the turbine, it spins the rotor by reaction and is called a **reaction turbine**. The **impulse turbine** works by directing a jet of steam at blades on a rotor. Impulse water turbines work on the same principle as the water wheel and consist of sets of buckets arranged around the edge of a wheel; reaction turbines look much like propellers and are fully immersed in the water. In a **gas turbine** a compressed mixture of air and gas, or vaporized fuel, is ignited, and the hot gases produced expand through the turbine blades, spinning the rotor. In the industrial gas turbine, the rotor shaft drives machines. In the jet engine, the turbine drives the compressor, which supplies the compressed air to the engine, but most of the power developed comes from the jet exhaust in the form of propulsive thrust.

turbocharger turbine-driven device fitted to engines to force more air into the cylinders, producing extra power. The turbocharger consists of a 'blower', or compressor, driven by a turbine, which in most units is driven by the exhaust gases leaving the engine.

turbofan jet engine of the type used by most airliners, so called because of its huge front fan. The fan sends air not only into the engine for combustion but also around the engine for additional thrust. This results in a faster and more fuel-efficient propulsive jet.

turboprop jet engine that derives its thrust partly from a jet of exhaust gases, but mainly from a propeller powered by a turbine in the jet exhaust. Turboprops are more economical than turbojets but can be used only at relatively low speeds.

two-stroke cycle operating cycle for internal combustion piston engines. The engine cycle is completed after just two strokes (up or down) of the piston, which distinguishes it from the more common ◊four-stroke cycle. In a typical two-stroke engine, fuel mixture is drawn into the crankcase as the piston moves up on its first stroke to compress the mixture above it. Then the compressed mixture is ignited, and hot gases are produced, which drive the piston down on its second stroke. As it moves down, it uncovers an opening (port) that allows the fresh fuel mixture in the crankcase to flow into the combustion space above the piston. At the same time, the exhaust gases leave through another port. Power mowers, most marine diesel engines, and lightweight motorcycles use two-stroke petrol engines, which are cheaper and simpler than four-strokes.

typesetting means by which text, or copy, is prepared for printing, now usually carried out by computer. Text is keyed on a typesetting machine in a similar way to typing. Laser or light impulses are projected on to light-sensitive film that, when developed, can be used to make plates for printing.

typewriter keyboard machine that produces characters on paper. Recent typewriters work electronically, are equipped with a memory, and can be given an interface that enables them to be connected to a computer. The word processor has largely replaced the typewriter for typing letters and text.

vacuum flask, or **Dewar flask** or **Thermos flask,** container for keeping things either hot or cold. It has two silvered glass walls with a vacuum between them, in a metal or plastic outer case. This design reduces the three forms of heat transfer: radiation (prevented by the silvering), conduction, and convection (both prevented by the vacuum). A vacuum flask is therefore equally efficient at keeping cold liquids cold or hot liquids hot.

video camera, or **camcorder,** portable television camera that records moving pictures electronically on magnetic tape. It produces an electrical output signal corresponding to rapid line-by-line scanning of the field of view. The output is recorded on video cassette and is played back on a television screen via a video cassette recorder.

video cassette recorder (VCR) device for recording on and playing back video cassettes; see ◊videotape recorder.

video disc disc with pictures and sounds recorded on it, played back by laser. The video disc is a type of ◊compact disc. The video disc is chiefly used to provide commercial films for private viewing. Most systems use a 30 cm/12 in rotating vinyl disc coated with a reflective material. Laser scanning recovers picture and sound signals from the surface where they are recorded as a spiral of microscopic pits.

videotape recorder (VTR) device for recording pictures and sound on cassettes or spools of magnetic tape. Video recording works in the same way as audio ◊tape recording: the picture information is stored as a line of varying magnetism, or track, on a plastic tape covered with magnetic material. The main difficulty – the huge amount of information needed to reproduce a picture – is overcome by arranging the video track diagonally across the tape. During recording, the tape is wrapped around a drum in a spiral fashion. The recording head rotates inside the drum. The combination of the forward motion of the tape and the rotation of the head produces a diagonal track. The audio signal accompanying the video signal is recorded as a separate track along the edge of the tape.

videotext system in which information (text and simple pictures) is displayed on a television (video) screen. There are two basic systems, known as teletext and viewdata. In the teletext system information is broadcast with the ordinary television signals, whereas in the viewdata system information is relayed to the screen from a central data bank via the telephone network. Both systems require the use of a television receiver (or a connected videotape recorder) with special decoder.

viscose yellowish, syrupy solution made by treating cellulose with sodium hydroxide and carbon disulphide. The solution is then regenerated as continuous filament for the making of ◊rayon and as cellophane.

Wankel engine rotary petrol engine developed by the German engineer Felix Wankel (1902–1988) in the 1950s. It operates according to the same stages as the ◊four-stroke petrol engine cycle, but these stages take place in different sectors of a figure-eight chamber in the space between the chamber walls and a triangular rotor. Power is produced once on every turn of the rotor. The Wankel engine is simpler in construction than the four-stroke piston petrol engine, and produces rotary power directly (instead of via a crankshaft). Problems with rotor seals have prevented its widespread use.

welding joining pieces of metal (or nonmetal) at faces rendered plastic or liquid by heat or pressure (or both). The principal processes today are gas and arc welding, in which the heat from a gas flame or an electric arc melts the faces to be joined. Additional 'filler metal' is usually added to the joint. Forge (or hammer) welding, employed by blacksmiths since early times, was the only method available until the late 19th century. Resistance welding is another electric method in which the weld is formed by a combination of pressure and resistance heating from an electric current. Recent developments include electric-slag, electron-beam, high-energy laser, and the still experimental radio-wave energy-beam welding processes.

wind tunnel test tunnel in which air is blown over, for example, a stationary model aircraft, motor vehicle, or locomotive to simulate the effects of movement. Lift, drag, and airflow patterns are observed by the use of special cameras and sensitive instruments. Wind-tunnel testing assesses aerodynamic design, prior to full-scale construction.

Further Reading

Adler, Michael H *The Writing Machine* (1973)

Alexander, William, and Street, Arthur *Metals in the Service of Man* (1972)

American Society for Testing and Materials *Chemical and Spectrometric Test Methods for Steel* (1992)

Ashby, M F, and Jones, D R H *Engineering Materials: An Introduction to their Properties and Application* (1980)

Ball, P *Designing the Molecular World: Chemistry at the Frontier* (1994)

Batty, Peter *The House of Krupp* (1966)

Beeching, Wilfred A *Century of the Typewriter* (1990)

Bruce, R V *Alexander Graham Bell and the Conquest of Solitude* (1973)

Bud, R *The Uses of Life: A History of Biotechnology* (1993)

Buchanan, R A *The Power of the Machine* (1992)

Burton, Anthony *The Canal Builders* (1981)

Cahn, Robert W *Artifice and Artifacts: 100 Essays in Materials Science* (1992)

Cahn, Robert W, and Haasen, Peter (eds) *Physical Metallurgy* (1974; 3rd revised edition 1993)

Carr, Marilyn *The AT Reader: Theory and Practice in Appropriate Technology* (1985)

Connot, R *Thomas A Edison* (1987)

Crouch, T D *The Bishop's Boys: A Life of Wilbur and Orville Wright* (1989)

Davies, Hunter *A Biographical Study of the Father of the Railways: George Stephenson* (1977)

Dennis, W H *A Hundred Years of Metallurgy* (1963)

Dickinson, H W and Vowles, H P *James Watt and the Industrial Revolution* (1943)

Dudley, Eric *The Critical Villager* (1993)

Dunn, P *Appropriate Technology: Technology with a Human Face* (1978)

Eber, Dorothy *Genius at Work: Images of Alexander Graham Bell* (1982)

Edison, Thomas (ed D Runes) *Diary and Sundry Observations* (1949)

Faraday, Michael *Experimental Researches in Electricity* (1839–55)

Forester, T (ed) *The Materials Revolution* (1988)

Fussell, G E *Jethro Tull* (1973)

Gamser, Matthew; Appleton, Helen; and Carter, Nicola (eds) *Tinker, Tiller, Technical Change: Technologies from the People* (1990)

Hadfield, Charles *The Canal Age* (1968)

Harrison, Paul *The Third World Tomorrow* (1980)

Herndon, B *Ford* (1969)

Ing, Janet *Johann Gutenberg and His Bible* (1988)

Josephson, M *Edison: A Biography* (1959)

Kaplinsky, Raphael *The Economies of Small: Appropriate Technology in a Changing World* (1990)

Kaufman, M *The First Century of Plastics* (1963)

Klass, Gert von *Krupps: The Story of an Industrial Empire* (trs 1954)

Lacey, R *Ford: The Man and the Machine* (1986)

Lindbergh, Charles *The Spirit of St Louis* (1953–75)

McNeil, Ian (ed) *An Encyclopaedia of the History of Technology* (1990)

Manchester, William *The Arms of Krupp* (1959)

Millard, A *Edison and the Business of Invention* (1990)

Newcomb, Horace (ed) *Encyclopedia of Television* (1997)

Nye, D *Henry Ford: Ignorant Idealist* (1979)

Pearce, R M *Thomas Telford* (1973)

Penfold, A (ed) *Thomas Telford: Engineer* (1980)

Psaras, Peter A, and Langford, A Dale *Advancing Materials Research* (1987)

Rae, J B *Henry Ford* (1969)

Richards, George Tilghman *The History and Development of Typewriters* (1964)

Robinson, E H, and Musson, James (eds) *James Watt and the Steam Revolution: A Documentary History* (1969)

Rolt, L T C *Thomas Telford* (1958), *George Stephenson* (1960), *The Railway Revolution: George and Robert Stephenson* (1962)

Scholderer, Victor *Johann Gutenberg: The Inventor of Printing* (1963)

Schumacher, E F *Small is Beautiful: A Study of Economics as if People Mattered* (1973)

Seymour, Raymond B (ed) *Pioneers in Polymer Science* (1989)

Taylor, Michael J *History of Helicopters* (1984)

Wachhorst, W *Thomas Alva Edison: An American Myth* (1981)

DISCOVERIES, INVENTIONS, AND PRIZES

Scientific Discoveries

Discovery	Date	Discoverer	Nationality
Absolute zero, concept	1851	William Thomson, 1st Baron Kelvin	Irish
Adrenalin, isolation	1901	Jokichi Takamine	Japanese
Alizarin, synthesized	1869	William Perkin	English
Allotropy (in carbon)	1841	Jöns Jakob Berzelius	Swedish
Alpha rays	1899	Ernest Rutherford	New Zealand-born British
Alternation of generations (ferns and mosses)	1851	Wilhelm Hofmeister	German
Aluminium, extraction by electrolysis of aluminium oxide	1886	Charles Hall, Paul Héroult	US, French
Aluminium, improved isolation of	1827	Friedrich Wöhler	German
Anaesthetic, first use (ether)	1842	Crawford Long	US
Anthrax vaccine	1881	Louis Pasteur	French
Antibacterial agent, first specific (Salvarsan for treatment of syphilis)	1910	Paul Ehrlich	German
Antiseptic surgery (using phenol)	1865	Joseph Lister	English
Argon	1892	William Ramsay	Scottish
Asteroid, first (Ceres)	1801	Giuseppe Piazzi	Italian
Atomic theory	1803	John Dalton	English
Australopithecus	1925	Raymond Dart	Australian-born South African
Avogadro's hypothesis	1811	Amedeo Avogadro	Italian
Bacteria, first observation	1683	Anton van Leeuwenhoek	Dutch
Bacteriophages	1916	Felix D'Herelle	Canadian
Bee dance	1919	Karl von Frisch	Austrian
Benzene, isolation	1825	Michael Faraday	English
Benzene, ring structure	1865	Friedrich Kekulé	German
Beta rays	1899	Ernest Rutherford	New Zealand-born British
Big-Bang theory	1948	Ralph Alpher, George Gamow	US
Binary arithmetic	1679	Gottfried Leibniz	German
Binary stars	1802	William Herschel	German-born English
Binomial theorem	1665	Isaac Newton	English
Blood, circulation	1619	William Harvey	English
Blood groups, ABO system	1900	Karl Landsteiner	Austrian-born US
Bode's law	1772	Johann Bode, Johann Titius	German
Bohr atomic model	1913	Niels Bohr	Danish
Boolean algebra	1854	George Boole	English
Boyle's law	1662	Robert Boyle	Irish
Brewster's law	1812	David Brewster	Scottish
Brownian motion	1827	Robert Brown	Scottish
Cadmium	1817	Friedrich Strohmeyer	German
Caesium	1861	Robert Bunsen	German
Carbon dioxide	1755	Joseph Black	Scottish
Charles' law	1787	Jacques Charles	French
Chlorine	1774	Karl Scheele	Swedish

Discovery	Date	Discoverer	Nationality
Complex numbers, theory	1746	Jean d'Alembert	French
Conditioning	1902	Ivan Pavlov	Russian
Continental drift	1912	Alfred Wegener	German
Coriolis effect	1834	Gustave-Gaspard Coriolis	French
Cosmic radiation	1911	Victor Hess	Austrian
Decimal fractions	1576	François Viète	French
Dinosaur fossil, first recognized	1822	Mary Ann Mantell	English
Diphtheria bacillus, isolation	1883	Edwin Krebs	US
DNA	1869	Johann Frederick Miescher	Swiss
DNA, components of	1909	Phoebus Levene	Russian-born US
DNA, double-helix structure	1953	Francis Crick, James Watson	English, US
Doppler effect	1842	Christian Doppler	Austrian
Earth's magnetic pole	1546	Gerardus Mercator	Flemish
Earth's molten core, proof	1906	Richard Oldham	Welsh
Earth's rotation, demonstration	1851	Léon Foucault	French
Eclipse, prediction	585 BC	Thales of Miletus	Greek
Electrolysis, laws	1833	Michael Faraday	English
Electromagnetic induction	1831	Michael Faraday	English
Electromagnetism	1819	Hans Christian Oersted	Danish
Electron	1897	J J Thomson	English
Electroweak unification theory	1967	Sheldon Lee Glashow, Abdus Salam, Steven Weinberg	US, Pakistani, US
Endorphins	1975	John Hughes	US
Enzyme, first animal (pepsin)	1836	Theodor Schwann	German
Enzyme, first (diastase from barley)	1833	Anselme Payen	French
Enzymes, 'lock and key' hypothesis	1899	Emil Fischer	German
Ether, first anaesthetic use	1842	Crawford Long	US
Eustachian tube	1552	Bartolomeo Eustachio	Italian
Evolution by natural selection	1859	Charles Darwin	English
Exclusion principle	1925	Wolfgang Pauli	Austrian-born Swiss
Fallopian tubes	1561	Gabriello Fallopius	Italian
Fluorine, preparation	1886	Henri Moissan	French
Fullerines	1985	Harold Kroto, David Walton	English
Gay-Lussac's law	1808	Joseph-Louis Gay-Lussac	French
Geometry, Euclidean	300 BC	Euclid	Greek
Germanium	1886	Clemens Winkler	German
Germ theory	1861	Louis Pasteur	French
Global temperature and link with atmospheric carbon dioxide	1896	Svante Arrhenius	Swedish
Gravity, laws	1687	Isaac Newton	English
Groups, theory	1829	Evariste Galois	French
Gutenberg discontinuity	1914	Beno Gutenberg	German-born US
Helium, production	1896	William Ramsay	Scottish
Homo erectus	1894	Marie Dubois	Dutch
Homo habilis	1961	Louis Leakey, Mary Leakey	Kenyan, English
Hormones	1902	William Bayliss, Ernest Starling	English
Hubble's law	1929	Edwin Hubble	US
Hydraulics, principles	1642	Blaise Pascal	French
Hydrogen	1766	Henry Cavendish	English
Iapetus	1671	Giovanni Cassini	Italian-born French
Infrared solar rays	1801	William Herschel	German-born English
Insulin, isolation	1921	Frederick Banting, Charles Best	Canadian

Discovery	Date	Discoverer	Nationality
Insulin, structure	1969	Dorothy Hodgkin	English
Interference of light	1801	Thomas Young	English
Irrational numbers	450 BC	Hipparcos	Greek
Jupiter's satellites	1610	Galileo	Italian
Kinetic theory of gases	1850	Rudolf Clausius	German
Krypton	1898	William Ramsay, Morris Travers	Scottish, English
Lanthanum	1839	Carl Mosander	Swedish
Lenses, how they work	1039	Ibn al-Haytham Alhazen	Arabic
Light, finite velocity	1675	Ole Römer	Danish
Light, polarization	1678	Christiaan Huygens	Dutch
Linnaean classification system	1735	Linnaeus	Swedish
'Lucy', hominid	1974	Donald Johanson	US
Magnetic dip	1576	Robert Norman	English
Malarial parasite in *Anopheles* mosquito	1897	Ronald Ross	British
Mars, moons	1877	Asaph Hall	US
Mendelian laws of inheritance	1866	Gregor Mendel	Austrian
Messenger RNA	1960	Sydney Brenner, François Jacob	South African, French
Microorganisms as cause of fermentation	1856	Louis Pasteur	French
Monoclonal antibodies	1975	César Milstein, George Köhler	Argentinean-born British, German
Motion, laws	1687	Isaac Newton	English
Neon	1898	William Ramsay, Morris Travers	Scottish, English
Neptune	1846	Johann Galle	German
Neptunium	1940	Edwin McMillan, Philip Abelson	US
Nerve impulses, electric nature	1771	Luigi Galvani	Italian
Neutron	1932	James Chadwick	English
Nitrogen	1772	Daniel Rutherford	Scottish
Normal distribution curve	1733	Abraham De Moivre	French
Nuclear atom, concept	1911	Ernest Rutherford	New Zealand-born British
Nuclear fission	1938	Otto Hahn, Fritz Strassman	German
Nucleus, plant cell	1831	Robert Brown	Scottish
Ohm's law	1827	Georg Ohm	German
Organic substance, first synthesis (urea)	1828	Friedrich Wöhler	German
Oxygen	1774	Joseph Priestley	English
Oxygen, liquefaction	1894	James Dewar	Scottish
Ozone layer	1913	Charles Fabry	French
Palladium	1803	William Hyde Wollaston	English
Pallas (asteroid)	1802	Heinrich Olbers	German
Pendulum, principle	1581	Galileo	Italian
Penicillin	1928	Alexander Fleming	Scottish
Penicillin, widespread preparation	1940	Ernst Chain, Howard Florey	German, Australian
Pepsin	1836	Theodor Schwann	German
Periodic law for elements	1869	Dmitri Mendeleyev	Russian
Period–luminosity law	1912	Henrietta Swan	US
Phosphorus	1669	Hennig Brand	German
Piezoelectric effect	1880	Pierre Curie	French
Pi meson (particle)	1947	Cecil Powell, Giuseppe Occhialini	English, Italian
Pistils, function	1676	Nehemiah Grew	English
Planetary nebulae	1790	William Herschel	German-born English
Planets, orbiting Sun	1543	Copernicus	Polish

Discovery	Date	Discoverer	Nationality
Pluto	1930	Clyde Tombaugh	US
Polarization of light by reflection	1808	Etienne Malus	French
Polio vaccine	1952	Jonas Salk	US
Polonium	1898	Marie and Pierre Curie	French
Positron	1932	Carl Anderson	US
Potassium	1806	Humphry Davy	English
Probability theory	1654	Blaise Pascal, Pierre de Fermat	French
Probability theory, expansion	1812	Pierre Laplace	French
Proton	1914	Ernest Rutherford	New Zealand-born British
Protoplasm	1846	Hugo von Mohl	German
Pulsar	1967	Jocelyn Bell Burnell	English
Pythagoras' theorem	550 BC	Pythagoras	Greek
Quantum chromodynamics	1972	Murray Gell-Mann	US
Quantum electrodynamics	1948	Richard Feynman, Seymour Schwinger, Shin'chiro Tomonaga	US, US, Japanese
Quark, first suggested existence	1963	Murray Gell-Mann, George Zweig	US
Quasar	1963	Maarten Schmidt	Dutch-born US
Rabies vaccine	1885	Louis Pasteur	French
Radioactivity	1896	Henri Becquerel	French
Radio emissions, from Milky Way	1931	Karl Jansky	US
Radio waves, production	1887	Heinrich Hertz	German
Radium	1898	Marie and Pierre Curie	French
Radon	1900	Friedrich Dorn	German
Refraction, laws	1621	Willibrord Snell	Dutch
Relativity, general theory	1915	Albert Einstein	German-born US
Relativity, special theory	1905	Albert Einstein	German-born US
Rhesus factor	1940	Karl Landsteiner, Alexander Wiener	Austrian, US
Rubidium	1861	Robert Bunsen	German
Sap flow in plants	1733	Stephen Hales	English
Saturn, 18th moon	1990	Mark Showalter	US
Saturn's satellites	1656	Christiaan Huygens	Dutch
Smallpox inoculation	1796	Edward Jenner	English
Sodium	1806	Humphry Davy	English
Stamens, function	1676	Nehemiah Grew	English
Stars, luminosity sequence	1905	Ejnar Hertzsprung	Danish
Stereochemistry, foundation	1848	Louis Pasteur	French
Stratosphere	1902	Léon Teisserenc	French
Sunspots	1611	Galileo, Christoph Scheiner	Italian, German
Superconductivity	1911	Heike Kamerlingh-Onnes	Dutch
Superconductivity, theory	1957	John Bardeen, Leon Cooper, John Schrieffer	US
Thermodynamics, second law	1834	Benoit-Pierre Clapeyron	French
Thermodynamics, third law	1906	Hermann Nernst	German
Thermoelectricity	1821	Thomas Seebeck	German
Thorium-X	1902	Ernest Rutherford, Frederick Soddy	New Zealand-born British, English
Titius–Bode law	1772	Johan Bode, Johann Titius	German
Tranquillizer, first (reserpine)	1956	Robert Woodward	US
Transformer	1831	Michael Faraday	English
Troposphere	1902	Léon Teisserenc	French
Tuberculosis bacillus, isolation	1883	Robert Koch	German

Discovery	Date	Discoverer	Nationality
Tuberculosis vaccine	1923	Albert Calmette, Camille Guérin	French
Uranus	1781	William Herschel	German-born English
Urea cycle	1932	Hans Krebs	German
Urease, isolation	1926	James Sumner	US
Urea, synthesis	1828	Friedrich Wöhler	German
Valves, in veins	1603	Geronimo Fabricius	Italian
Van Allen radiation belts	1958	James Van Allen	US
Virus, first identified (tobacco mosaic disease)	1898	Martinus Beijerinck	Dutch
Vitamin A, isolation	1913	Elmer McCollum	US
Vitamin A, structure	1931	Paul Karrer	Russian-born Swiss
Vitamin B, composition	1955	Dorothy Hodgkin	English
Vitamin B, isolation	1925	Joseph Goldberger	Austrian-born US
Vitamin C	1928	Charles Glen King, Albert Szent-Györgi	US, Hungarian-born US
Vitamin C, isolation	1932	Charles Glen King	US
Vitamin C, synthesis	1933	Tadeus Reichstein	Polish-born Swiss
Wave mechanics	1926	Erwin Schrödinger	Austrian
Xenon	1898	William Ramsay, Morris Travers	Scottish, English
X-ray crystallography	1912	Max von Laue	German
X-rays	1895	Wilhelm Röntgen	German

Inventions

Invention	Date	Inventor	Nationality
Achromatic lens	1733	Chester Moor Hall	English
Adding machine	1642	Blaise Pascal	French
Aeroplane, powered	1903	Orville and Wilbur Wright	US
Air conditioning	1902	Willis Carrier	US
Air pump	1654	Otto Guericke	German
Airship, first successful	1852	Henri Giffard	French
Airship, rigid	1900	Ferdinand von Zeppelin	German
Amniocentesis test	1952	Douglas Bevis	English
Aqualung	1943	Jacques Cousteau	French
Arc welder	1919	Elihu Thomson	US
Armillary ring	125	Zhang Heng	Chinese
Aspirin	1899	Felix Hoffman	German
Assembly line	1908	Henry Ford	US
Autogiro	1923	Juan de la Cierva	Spanish
Automatic pilot	1912	Elmer Sperry	US
Babbitt metal	1839	Isaac Babbitt	US
Bakelite, first synthetic plastic	1909	Leo Baekeland	US
Ballpoint pen	1938	Lazlo Biró	
Barbed wire	1874	Joseph Glidden	US
Bar code system	1970	Monarch Marking, Plessey Telecommunications	US, English
Barometer	1642	Evangelista Torricelli	Italian

Invention	Date	Inventor	Nationality
Bathysphere	1934	Charles Beebe	US
Bessemer process	1856	Henry Bessemer	British
Bicycle	1839	Kirkpatrick Macmillan	Scottish
Bifocal spectacles	1784	Benjamin Franklin	US
Binary calculator	1938	Konrad Zuse	German
Bottling machine	1895	Michael Owens	US
Braille	1837	Louis Braille	French
Bunsen burner	1850	Robert Bunsen	German
Calculator, pocket	1971	Texas Instruments	US
Camera film (roll)	1888	George Eastman	US
Camera obscura	1560	Battista Porta	Italian
Carbon fibre	1963	Leslie Phillips	English
Carbon-zinc battery	1841	Robert Bunsen	German
Carburettor	1893	Wilhelm Maybach	German
Car, four-wheeled	1887	Gottlieb Daimler	German
Car, petrol-driven	1885	Karl Benz	German
Carpet sweeper	1876	Melville Bissell	US
Cash register	1879	James Ritty	US
Cassette tape	1963	Philips	Dutch
Catapult	c. 400 BC	Dionysius of Syracuse	Greek
Cathode ray oscilloscope	1897	Karl Braun	German
CD-ROM	1984	Sony, Fujitsu, Philips	Japanese, Japanese, Dutch
Cellophane	1908	Jacques Brandenberger	Swiss
Celluloid	1869	John Wesley Hyatt	US
Cement, Portland	1824	Joseph Aspidin	English
Centigrade scale	1742	Anders Celsius	Swedish
Chemical symbols	1811	Jöns Jakob Berzelius	Swedish
Chronometer, accurate	1762	John Harrison	English
Cinematograph	1895	Auguste and Louis Lumière	French
Clock, pendulum	1656	Christiaan Huygens	Dutch
Colt revolver	1835	Samuel Colt	US
Compact disc	1972	RCA	US
Compact disc player	1984	Sony, Philips	Japanese, Dutch
Compass, simple	1088	Shen Kua	Chinese
Computer, bubble memory	1967	A H Bobeck, Bell Telephone Laboratories team	US
Computer, first commercially available (UNIVAC 1)	1951	John Mauchly, John Eckert	US
Computerized axial tomography (CAT) scanning	1972	Godfrey Hounsfield	English
Contraceptive pill	1954	Gregory Pincus	US
Cotton gin	1793	Eli Whitney	US
Cream separator	1878	Carl de Laval	Swedish
Crookes tube	1878	William Crookes	English
Cyclotron	1931	Ernest O Lawrence	US
DDT	1940	Paul Müller	Swiss
Diesel engine	1892	Rudolf Diesel	German
Difference engine	1822	Charles Babbage	English
Diode valve	1904	Ambrose Fleming	English

Invention	Date	Inventor	Nationality
Dynamite	1866	Alfred Nobel	Swedish
Dynamo	1831	Michael Faraday	English
Electric cell	1800	Alessandro Volta	Italian
Electric fan	1882	Schuyler Wheeler	US
Electric generator, first commercial	1867	Zénobe Théophile Gramme	French
Electric light bulb	1879	Thomas Edison	US
Electric motor	1821	Michael Faraday	English
Electric motor, alternating current	1888	Nikola Tesla	Croatian-born US
Electrocardiography	1903	Willem Einthoven	Dutch
Electroencephalography	1929	Hans Berger	German
Electromagnet	1824	William Sturgeon	English
Electron microscope	1933	Ernst Ruska	German
Electrophoresis	1930	Arne Tiselius	Swedish
Fahrenheit scale	1714	Gabriel Fahrenheit	Polish-born Dutch
Felt-tip pen	1955	Esterbrook	English
Floppy disc	1970	IBM	US
Flying shuttle	1733	John Kay	English
FORTRAN	1956	John Backus, IBM	US
Fractal images	1962	Benoit Mandelbrot	Polish-born French
Frozen food	1929	Clarence Birdseye	US
Fuel cell	1839	William Grove	Welsh
Galvanometer	1820	Johann Schwiegger	German
Gas mantle	1885	Carl Welsbach	Austrian
Geiger counter	1908	Hans Geiger, Ernest Rutherford	German, New Zealand-born British
Genetic fingerprinting	1985	Alec Jeffreys	British
Glider	1877	Otto Lilienthal	German
Gramophone	1877	Thomas Edison	US
Gramophone (flat discs)	1887	Emile Berliner	German
Gyrocompass	1911	Elmer Sperry	US
Gyroscope	1852	Jean Foucault	French
Heart, artificial	1982	Robert Jarvik	US
Heart-lung machine	1953	John Gibbon	US
Helicopter	1939	Igor Sikorsky	US
Holography	1947	Dennis Gabor	Hungarian-born British
Hovercraft	1955	Christopher Cockerell	English
Hydrogen bomb	1952	US government scientists	US
Hydrometer	1675	Robert Boyle	Irish
Iconoscope	1923	Vladimir Zworykin	Russian-born US
Integrated circuit	1958	Jack Kilby, Texas Instruments	US
Internal-combustion engine, four-stroke	1877	Nikolaus Otto	German
Internal-combustion engine, gas-fuelled	1860	Etienne Lenoir	Belgian
In vitro fertilization	1969	Robert Edwards	Welsh
Jet engine	1930	Frank Whittle	English
Jumbo jet	1969	Joe Sutherland, Boeing team	US

Invention	Date	Inventor	Nationality
Laser, prototype	1960	Theodore Maiman	US
Lightning rod	1752	Benjamin Franklin	US
Linoleum	1860	Frederick Walton	English
Liquid crystal display (LCD)	1971	Hoffmann-LaRoche Laboratories	Swiss
Lock (canal)	980	Ciao Wei-yo	Chinese
Lock, Yale	1851	Linus Yale	US
Logarithms	1614	John Napier	Scottish
Loom, power	1785	Edmund Cartwright	English
Machine gun	1862	Richard Gatling	US
Magnifying glass	1250	Roger Bacon	English
Map	c. 510 BC	Hecataeus	Greek
Map, star	c. 350 BC	Eudoxus	Greek
Maser	1953	Charles Townes, Arthur Schawlow	US
Mass-spectrograph	1918	Francis Aston	English
Microscope	1590	Zacharias Janssen	Dutch
Miners' safety lamp	1813	Humphry Davy	English
Mohs' scale for mineral hardness	1822	Frederick Mohs	German
Morse code	1838	Samuel Morse	US
Motorcycle	1885	Gottlieb Daimler	German
Neutron bomb	1977	US military	US
Nylon	1934	Wallace Carothers	US
Paper chromatography	1944	Archer Martin, Richard Synge	English
Paper, first	105	Ts'ai Lun	Chinese
Particle accelerator	1932	John Cockcroft, Ernest Walton	English, Irish
Pasteurization (wine)	1864	Louis Pasteur	French
Pen, fountain	1884	Lewis Waterman	US
Photoelectric cell	1904	Johann Elster	German
Photograph, first colour	1881	Frederic Ives	US
Photograph, first (on a metal plate)	1827	Joseph Niepce	French
Piano	1704	Bartelommeo Cristofori	Italian
Planar transistor	1959	Robert Noyce	US
Plastic, first (Parkesine)	1862	Alexander Parkes	English
Plough, cast iron	1785	Robert Ransome	English
Punched-card system for carpet-making loom	1805	Joseph-Marie Jacquard	French
Radar, first practical equipment	1935	Robert Watson-Watt	Scottish
Radio	1901	Guglielmo Marconi	Italian
Radio interferometer	1955	Martin Ryle	English
Radio, transistor	1952	Sony	Japanese
Razor, disposable safety	1895	King Gillette	US
Recombinant DNA, technique	1973	Stanley Cohen, Herbert Boyer	US
Refrigerator, domestic	1918	Nathaniel Wales, E J Copeland	US
Richter scale	1935	Charles Richter	US
Road locomotive, steam	1801	Richard Trevithick	English
Road vehicle, first self-propelled (steam)	1769	Nicolas-Joseph Cugnot	French

Invention	Date	Inventor	Nationality
Rocket, powered by petrol and liquid oxygen	1926	Robert Goddard	US
Rubber, synthetic	1909	Karl Hoffman	German
Scanning tunnelling microscope	1980	Heinrich Rohrer, Gerd Binning	Swiss, German
Seed drill	1701	Jethro Tull	English
Seismograph	1880	John Milne	English
Shrapnel shell	1784	Henry Shrapnel	English
Silicon transistor	1954	Gordon Teal	US
Silk, method of producing artificial	1887	Hilaire, Comte de Chardonnet	French
Spinning frame	1769	Richard Arkwright	English
Spinning jenny	1764	James Hargreaves	English
Spinning mule	1779	Samuel Crompton	English
Stainless steel	1913	Harry Brearley	English
Steam engine	50 BC	Hero of Alexandria	Greek
Steam engine, first successful	1712	Thomas Newcomen	English
Steam engine, improved	1765	James Watt	Scottish
Steam locomotive, first effective	1814	George Stephenson	English
Steam turbine, first practical	1884	Charles Parsons	English
Steel, open-hearth production	1864	William Siemens, Pierre Emile Martin	German, French
Submarine	1620	Cornelius Drebbel	Dutch
Superheterodyne radio receiver	1918	Edwin Armstrong	US
Tank	1914	Ernest Swinton	English
Telephone	1876	Alexander Graham Bell	Scottish-born US
Telescope, binocular	1608	Johann Lippershey	Dutch
Telescope, reflecting	1668	Isaac Newton	English
Television	1926	John Logie Baird	Scottish
Terylene (synthetic fibre)	1941	John Whinfield, J T Dickson	English
Thermometer	1607	Galileo	Italian
Thermometer, alcohol	1730	Réné Antoine Ferchault de Réaumur	French
Thermometer, mercury	1714	Gabriel Fahrenheit	Polish-born Dutch
TNT	1863	J Willbrand	German
Toaster, pop-up	1926	Charles Strite	US
Toilet, flushing	1778	Joseph Bramah	English
Transistor	1948	John Bardeen, Walter Brattain, William Shockley	US
Triode valve	1906	Lee De Forest	US
Tunnel diode	1957	Leo Esaki, Sony	Japanese
Tupperware	1944	Earl Tupper	US
Type, movable earthenware	1045	Pi Shêng	Chinese
Type, movable metal	1440	Johannes Gutenberg	German
Ultrasound, first use in obstetrics	1958	Ian Donald	Scottish
Velcro	1948	Georges de Mestral	Swiss
Video, home	1975	Matsushita, JVC, Sony	Japanese
Viscose	1892	Charles Cross	English
Vulcanization of rubber	1839	Charles Goodyear	US
Wind tunnel	1932	Ford Motor Company	US
Wireless telegraphy	1895	Guglielmo Marconi	Italian
Word processor	1965	IBM	US
Zinc-carbon battery	1868	George Leclanché	French
Zip	1891	Whitcombe Judson	US

Nobel Prize: Introduction

The Nobel Prizes were first awarded in 1901 under the will of Alfred B Nobel (1833–1896), a Swedish chemist, who invented dynamite. The interest on the Nobel endowment fund is divided annually among the persons who have made the greatest contributions in the fields of physics, chemistry, medicine, literature, and world peace. The first four are awarded by academic committees based in Sweden, while the peace prize is awarded by a committee of the Norwegian parliament. A sixth prize, for economics, financed by the Swedish National Bank, was first awarded in 1969. The prizes have a large cash award and are given to organizations – such as the United Nations peacekeeping forces, which received the Nobel Peace Prize in 1988 – as well as to individuals.

Nobel Prize for Chemistry

Year	Winner(s)[1]	Awarded for
1901	Jacobus van't Hoff (Netherlands)	laws of chemical dynamics and osmotic pressure
1902	Emil Fischer (Germany)	sugar and purine syntheses
1903	Svante Arrhenius (Sweden)	theory of electrolytic dissociation
1904	William Ramsay (UK)	discovery of inert gases in air and their locations in the periodic table
1905	Adolf von Baeyer (Germany)	work in organic dyes and hydroaromatic compounds
1906	Henri Moissan (France)	isolation of fluorine and adoption of electric furnace
1907	Eduard Buchner (Germany)	biochemical research and discovery of cell-free fermentation
1908	Ernest Rutherford (UK)	work in atomic disintegration, and the chemistry of radioactive substances
1909	Wilhelm Ostwald (Germany)	work in catalysis, and principles of equilibria and rates of reaction
1910	Otto Wallach (Germany)	work in alicyclic compounds
1911	Marie Curie (France)	discovery of radium and polonium, and the isolation and study of radium
1912	Victor Grignard (France)	discovery of Grignard reagent
	Paul Sabatier (France)	finding method of catalytic hydrogenation of organic compounds
1913	Alfred Werner (Switzerland)	work in bonding of atoms within molecules
1914	Theodore Richards (USA)	accurate determination of the atomic masses of many elements
1915	Richard Willstäter (Germany)	research into plant pigments, especially chlorophyll
1916	no award	
1917	no award	
1918	Fritz Haber (Germany)	synthesis of ammonia from its elements
1919	no award	
1920	Walther Nernst (Germany)	work in thermochemistry
1921	Frederick Soddy (UK)	work in radioactive substances, especially isotopes
1922	Francis Aston (UK)	work in mass spectrometry of isotopes of radioactive elements, and enunciation of the whole-number rule
1923	Fritz Pregl (Austria)	method of microanalysis of organic substances
1924	no award	
1925	Richard Zsigmondy (Austria)	elucidation of heterogeneity of colloids
1926	Theodor Svedberg (Sweden)	investigation of dispersed systems
1927	Heinrich Wieland (Germany)	research on constitution of bile acids and related substances
1928	Adolf Windaus (Germany)	research on constitution of sterols and related vitamins
1929	Arthur Harden (UK) and Hans von Euler-Chelpin (Sweden)	work on fermentation of sugar, and fermentative enzymes
1930	Hans Fischer (Germany)	analysis of haem (the iron-bearing group in haemoglobin) and chlorophyll, and the synthesis of haemin (a compound of haem)
1931	Carl Bosch (Germany) and Friedrich Bergius (Germany)	invention and development of chemical high-pressure methods
1932	Irving Langmuir (USA)	discoveries and investigations in surface chemistry
1933	no award	
1934	Harold Urey (USA)	discovery of deuterium (heavy hydrogen)
1935	Irène and Frédéric Joliot-Curie (France)	synthesis of new radioactive elements

Year	Winner(s)[1]	Awarded for
1936	Peter Debye (Netherlands)	work in molecular structures by investigation of dipole moments and the diffraction of X-rays and electrons in gases
1937	Norman Haworth (UK)	work in carbohydrates and ascorbic acid (vitamin C)
	Paul Karrer (Switzerland)	work in carotenoids, flavins, retinol (vitamin A) and riboflavin (vitamin B_2)
1938	Richard Kuhn (Germany) (declined)	carotenoids and vitamins research
1939	Adolf Butenandt (Germany) (declined)	work in sex hormones
	Leopold Ruzicka (Switzerland)	polymethylenes and higher terpenes
1940	no award	
1941	no award	
1942	no award	
1943	Georg von Hevesy (Hungary)	use of isotopes as tracers in chemical processes
1944	Otto Hahn (Germany)	discovery of nuclear fission
1945	Artturi Virtanen (Finland)	work in agriculture and nutrition, especially fodder preservation
1946	James Sumner (USA)	discovery of crystallization of enzymes
	John Northrop (USA) and Wendell Stanley (USA)	preparation of pure enzymes and virus proteins
1947	Robert Robinson (UK)	investigation of biologically important plant products, especially alkaloids
1948	Arne Tiselius (Sweden)	researches in electrophoresis and adsorption analysis, and discoveries concerning serum proteins
1949	William Giauque (USA)	work in chemical thermodynamics, especially at very low temperatures
1950	Otto Diels (West Germany) and Kurt Alder (West Germany)	discovery and development of diene synthesis
1951	Edwin McMillan (USA) and Glenn Seaborg (USA)	discovery and work in chemistry of transuranic elements
1952	Archer Martin (UK) and Richard Synge (UK)	development of partition chromatography
1953	Hermann Staudinger (West Germany)	discoveries in macromolecular chemistry
1954	Linus Pauling (USA)	study of nature of chemical bonds, especially in complex substances
1955	Vincent du Vigneaud (USA)	investigations into biochemically important sulphur compounds, and the first synthesis of a polypeptide hormone
1956	Cyril Hinshelwood (UK) and Nikolay Semenov (USSR)	work in mechanism of chemical reactions
1957	Alexander Todd (UK)	work in nucleotides and nucleotide coenzymes
1958	Frederick Sanger (UK)	determination of the structure of proteins, especially insulin
1959	Jaroslav Heyrovsky (Czechoslovakia)	discovery and development of polarographic methods of chemical analysis
1960	Willard Libby (USA)	development of radiocarbon dating in archaeology, geology, and geography
1961	Melvin Calvin (USA)	study of assimilation of carbon dioxide by plants
1962	Max Perutz (UK) and John Kendrew (UK)	determination of structures of globular proteins
1963	Karl Ziegler (West Germany) and Giulio Natta (Italy)	chemistry and technology of high polymers
1964	Dorothy Crowfoot Hodgkin (UK)	crystallographic determination of the structures of biochemical compounds, notably penicillin and cyanocobalamin (vitamin B_{12})
1965	Robert Woodward (USA)	organic synthesis
1966	Robert Mulliken (USA)	molecular orbital theory of chemical bonds and structures
1967	Manfred Eigen (West Germany), Ronald Norrish (UK), and George Porter (UK)	investigation of rapid chemical reactions by means of very short pulses of energy
1968	Lars Onsager (USA)	discovery of reciprocal relations, fundamental for the thermodynamics of irreversible processes
1969	Derek Barton (UK) and Odd Hassel (Norway)	concept and applications of conformation
1970	Luis Federico Leloir (Argentina)	discovery of sugar nucleotides and their role in carbohydrate biosynthesis

Year	Winner(s)[1]	Awarded for
1971	Gerhard Herzberg (Canada)	research on electronic structure and geometry of molecules, particularly free radicals
1972	Christian Anfinsen (USA), Stanford Moore (USA), and William Stein (USA)	work in amino-acid structure and biological activity of the enzyme ribonuclease
1973	Ernst Fischer (West Germany) and Geoffrey Wilkinson (UK)	work in chemistry of organometallic sandwich compounds
1974	Paul Flory (USA)	studies of physical chemistry of macromolecules
1975	John Cornforth (UK)	work in stereochemistry of enzyme-catalysed reactions
	Vladimir Prelog (Switzerland)	work in stereochemistry of organic molecules and their reactions
1976	William Lipscomb (USA)	study of structure and chemical bonding of boranes (compounds of boron and hydrogen)
1977	Ilya Prigogine (Belgium)	work in thermodynamics of irreversible and dissipative processes
1978	Peter Mitchell (UK)	formulation of a theory of biological energy transfer and chemiosmotic theory
1979	Herbert Brown (USA) and Georg Wittig (West Germany)	use of boron and phosphorus compounds, respectively, in organic syntheses
1980	Paul Berg (USA)	biochemistry of nucleic acids, especially recombinant DNA
	Walter Gilbert (USA) and Frederick Sanger (UK)	base sequences in nucleic acids
1981	Kenichi Fukui (Japan) and Roald Hoffmann (USA)	theories concerning chemical reactions
1982	Aaron Klug (UK)	determination of crystallographic electron microscopy: structure of biologically important nucleic-acid–protein complexes
1983	Henry Taube (USA)	study of electron-transfer reactions in inorganic chemical reactions
1984	Bruce Merrifield (USA)	development of chemical syntheses on a solid matrix
1985	Herbert Hauptman (USA) and Jerome Karle (USA)	development of methods of determining crystal structures
1986	Dudley Herschbach (USA), Yuan Lee (USA), and John Polanyi (Canada)	development of dynamics of chemical elementary processes
1987	Donald Cram (USA), Jean-Marie Lehn (France), and Charles Pedersen (USA)	development of molecules with highly selective structure-specific interactions
1988	Johann Deisenhofer (West Germany), Robert Huber (West Germany), and Hartmut Michel (West Germany)	discovery of three-dimensional structure of the reaction centre of photosynthesis
1989	Sidney Altman (USA) and Thomas Cech (USA)	discovery of catalytic function of RNA
1990	Elias James Corey (USA)	new methods of synthesizing chemical compounds
1991	Richard Ernst (Switzerland)	improvements in the technology of nuclear magnetic resonance (NMR) imaging
1992	Rudolph Marcus (USA)	theoretical discoveries relating to reduction and oxidation reactions
1993	Kary Mullis (USA)	invention of the polymerase chain reaction technique for amplifying DNA
	Michael Smith (Canada)	invention of techniques for splicing foreign genetic segments into an organism's DNA in order to modify the proteins produced
1994	George Olah (USA)	development of technique for examining hydrocarbon molecules
1995	F Sherwood Rowland (USA), Mario Molina (USA), and Paul Crutzen (Netherlands)	explaining the chemical process of the ozone layer
1996	Robert Curl, Jr (USA), Harold Kroto (UK), and Richard Smalley (USA)	discovery of fullerenes
1997	John Walker (UK), Paul Boyer (USA), and Jens Skou (Denmark)	study of the enzymes involved in the production of adenosine triphospate (ATP), which acts as a store of energy in bodies called mitochondria inside cells
1998	Walter Kohn (USA) and John Pople (USA)	contribution to quantum chemistry

[1] Nationality given is the citizenship of recipient at the time award was made.

Nobel Prize for Physics

Year	Winner(s)[1]	Awarded for
1901	Wilhelm Röntgen (Germany)	discovery of X-rays
1902	Hendrik Lorentz (Netherlands) and Pieter Zeeman (Netherlands)	influence of magnetism on radiation phenomena
1903	Henri Becquerel (France)	discovery of spontaneous radioactivity
	Pierre Curie (France) and Marie Curie (France)	research on radiation phenomena
1904	John Strutt (Lord Rayleigh, UK)	densities of gases and discovery of argon
1905	Philipp von Lenard (Germany)	work on cathode rays
1906	Joseph J Thomson (UK)	theoretical and experimental work on the conduction of electricity by gases
1907	Albert Michelson (USA)	measurement of the speed of light through the design and application of precise optical instruments such as the interferometer
1908	Gabriel Lippmann (France)	photographic reproduction of colours by interference
1909	Guglielmo Marconi (Italy) and Karl Ferdinand Braun (Germany)	development of wireless telegraphy
1910	Johannes van der Waals (Netherlands)	equation describing the physical behaviour of gases and liquids
1911	Wilhelm Wien (Germany)	laws governing radiation of heat
1912	Nils Dalén (Sweden)	invention of light-controlled valves, which allow lighthouses and buoys to operate automatically
1913	Heike Kamerlingh Onnes (Netherlands)	studies of properties of matter at low temperatures
1914	Max von Laue (Germany)	discovery of diffraction of X-rays by crystals
1915	William Bragg (UK) and Lawrence Bragg (UK)	X-ray analysis of crystal structures
1916	no award	
1917	Charles Barkla (UK)	discovery of characteristic X-ray emission of the elements
1918	Max Planck (Germany)	formulation of quantum theory
1919	Johannes Stark (Germany)	discovery of Doppler effect in rays of positive ions, and splitting of spectral lines in electric fields
1920	Charles Guillaume (Switzerland)	discovery of anomalies in nickel-steel alloys
1921	Albert Einstein (Switzerland)	theoretical physics, especially law of photoelectric effect
1922	Niels Bohr (Denmark)	discovery of the structure of atoms and radiation emanating from them
1923	Robert Millikan (USA)	discovery of the electric charge of an electron, and study of the photoelectric effect
1924	Karl Siegbahn (Sweden)	X-ray spectroscopy
1925	James Franck (Germany) and Gustav Hertz (Germany)	discovery of laws governing the impact of an electron upon an atom
1926	Jean Perrin (France)	confirmation of the discontinuous structure of matter
1927	Arthur Compton (USA)	transfer of energy from electromagnetic radiation to a particle
	Charles Wilson (UK)	invention of the Wilson cloud chamber, by which the movement of electrically charged particles may be tracked
1928	Owen Richardson (UK)	work on thermionic phenomena and associated law
1929	Louis Victor de Broglie (France)	discovery of the wavelike nature of electrons
1930	Chandrasekhara Raman (India)	discovery of the scattering of single-wavelength light when it is passed through a transparent substance
1931	no award	
1932	Werner Heisenberg (Germany)	creation of quantum mechanics
1933	Erwin Schrödinger (Austria) and Paul Dirac (UK)	development of quantum mechanics
1934	no award	
1935	James Chadwick (UK)	discovery of the neutron
1936	Victor Hess (Austria)	discovery of cosmic radiation
	Carl Anderson (USA)	discovery of the positron
1937	Clinton Davisson (USA) and George Thomson (UK)	diffraction of electrons by crystals

Year	Winner(s)[1]	Awarded for
1938	Enrico Fermi (Italy)	use of neutron irradiation to produce new elements, and discovery of nuclear reactions induced by slow neutrons
1939	Ernest Lawrence (USA)	invention and development of the cyclotron, and production of artificial radioactive elements
1940	no award	
1941	no award	
1942	no award	
1943	Otto Stern (USA)	molecular-ray method of investigating elementary particles, and discovery of magnetic moment of proton
1944	Isidor Isaac Rabi (USA)	resonance method of recording the magnetic properties of atomic nuclei
1945	Wolfgang Pauli (Austria)	discovery of the exclusion principle
1946	Percy Bridgman (USA)	development of high-pressure physics
1947	Edward Appleton (UK)	physics of the upper atmosphere
1948	Patrick Blackett (UK)	application of the Wilson cloud chamber to nuclear physics and cosmic radiation
1949	Hideki Yukawa (Japan)	theoretical work predicting existence of mesons
1950	Cecil Powell (UK)	use of photographic emulsion to study nuclear processes, and discovery of pions (pi mesons)
1951	John Cockcroft (UK) and Ernest Walton (Ireland)	transmutation of atomic nuclei by means of accelerated subatomic particles
1952	Felix Bloch (USA) and Edward Purcell (USA)	precise nuclear magnetic measurements
1953	Frits Zernike (Netherlands)	invention of phase-contrast microscope
1954	Max Born (UK)	statistical interpretation of wave function in quantum mechanics
	Walther Bothe (West Germany)	coincidence method of detecting the emission of electrons
1955	Willis Lamb (USA)	structure of hydrogen spectrum
	Polykarp Kusch (USA)	determination of magnetic moment of the electron
1956	William Shockley (USA), John Bardeen (USA), and Walter Houser Brattain (USA)	study of semiconductors, and discovery of the transistor effect
1957	Tsung-Dao Lee (China) and Chen Ning Yang (China)	investigations of weak interactions between elementary particles
1958	Pavel Cherenkov (USSR), Ilya Frank (USSR), and Igor Tamm (USSR)	discovery and interpretation of Cherenkov radiation
1959	Emilio Segrè (USA) and Owen Chamberlain (USA)	discovery of the antiproton
1960	Donald Glaser (USA)	invention of the bubble chamber
1961	Robert Hofstadter (USA)	scattering of electrons in atomic nuclei, and structure of protons and neutrons
	Rudolf Mössbauer (West Germany)	resonance absorption of gamma radiation
1962	Lev Landau (USSR)	theories of condensed matter, especially liquid helium
1963	Eugene Wigner (USA)	discovery and application of symmetry principles in atomic physics
	Maria Goeppert-Mayer (USA) and Hans Jensen (Germany)	discovery of the shell-like structure of atomic nuclei
1964	Charles Townes (USA), Nikolai Basov (USSR), and Aleksandr Prokhorov (USSR)	work on quantum electronics leading to construction of oscillators and amplifiers based on maser–laser principle
1965	Shin'ichiro Tomonaga (Japan), Julian Schwinger (USA), and Richard Feynman (USA)	basic principles of quantum electrodynamics
1966	Alfred Kastler (France)	development of optical pumping, whereby atoms are raised to higher energy levels by illumination
1967	Hans Bethe (USA)	theory of nuclear reactions, and discoveries concerning production of energy in stars
1968	Luis Alvarez (USA)	elementary-particle physics, and discovery of resonance states, using hydrogen bubble chamber and data analysis
1969	Murray Gell-Mann (USA)	classification of elementary particles, and study of their interactions
1970	Hannes Alfvén (Sweden)	work in magnetohydrodynamics and its applications in plasma physics
	Louis Néel (France)	work in antiferromagnetism and ferromagnetism in solid-state physics

Year	Winner(s)[1]	Awarded for
1971	Dennis Gabor (UK)	invention and development of holography
1972	John Bardeen (USA), Leon Cooper (USA), and John Robert Schrieffer (USA)	theory of superconductivity
1973	Leo Esaki (Japan) and Ivar Giaever (USA)	tunnelling phenomena in semiconductors and superconductors
	Brian Josephson (UK)	theoretical predictions of the properties of a supercurrent through a tunnel barrier
1974	Martin Ryle (UK) and Antony Hewish (UK)	development of radioastronomy, particularly the aperture-synthesis technique, and the discovery of pulsars
1975	Aage Bohr (Denmark), Ben Mottelson (Denmark), and James Rainwater (USA)	discovery of connection between collective motion and particle motion in atomic nuclei, and development of theory of nuclear structure
1976	Burton Richter (USA) and Samuel Ting (USA)	discovery of the psi meson
1977	Philip Anderson (USA), Nevill Mott (UK), and John Van Vleck (USA)	contributions to understanding electronic structure of magnetic and disordered systems
1978	Pyotr Kapitsa (USSR)	invention and application of low-temperature physics
	Arno Penzias (USA) and Robert Wilson (USA)	discovery of cosmic background radiation
1979	Sheldon Glashow (USA), Abdus Salam (Pakistan), and Steven Weinberg (USA)	unified theory of weak and electromagnetic fundamental forces, and prediction of the existence of the weak neutral current
1980	James W Cronin (USA) and Val Fitch (USA)	violations of fundamental symmetry principles in the decay of neutral kaon mesons
1981	Nicolaas Bloembergen (USA) and Arthur Schawlow (USA)	development of laser spectroscopy
	Kai Siegbahn (Sweden)	high-resolution electron spectroscopy
1982	Kenneth Wilson (USA)	theory for critical phenomena in connection with phase transitions
1983	Subrahmanyan Chandrasekhar (USA)	theoretical studies of physical processes in connection with structure and evolution of stars
	William Fowler (USA)	nuclear reactions involved in the formation of chemical elements in the universe
1984	Carlo Rubbia (Italy) and Simon van der Meer (Netherlands)	contributions to the discovery of the W and Z particles (weakons)
1985	Klaus von Klitzing (West Germany)	discovery of the quantized Hall effect
1986	Erns Ruska (West Germany)	electron optics, and design of the first electron microscope
	Gerd Binnig (West Germany) and Heinrich Rohrer (Switzerland)	design of scanning tunnelling microscope
1987	Georg Bednorz (West Germany) and Alex Müller (Switzerland)	superconductivity in ceramic materials
1988	Leon M Lederman (USA), Melvin Schwartz (USA), and Jack Steinberger (USA)	neutrino-beam method, and demonstration of the doublet structure of leptons through discovery of muon neutrino
1989	Norman Ramsey (USA)	measurement techniques leading to discovery of caesium atomic clock
	Hans Dehmelt (USA) and Wolfgang Paul (Germany)	ion-trap method for isolating single atoms
1990	Jerome Friedman (USA), Henry Kendall (USA), and Richard Taylor (Canada)	experiments demonstrating that protons and neutrons are made up of quarks
1991	Pierre-Gilles de Gennes (France)	work on disordered systems including polymers and liquid crystals; development of mathematical methods for studying the behaviour of molecules in a liquid on the verge of solidifying
1992	Georges Charpak (France)	invention and development of detectors used in high-energy physics
1993	Joseph Taylor (USA) and Russell Hulse (USA)	discovery of first binary pulsar (confirming the existence of gravitational waves)
1994	Clifford Shull (USA) and Bertram Brockhouse (Canada)	development of technique known as 'neutron scattering' which led to advances in semiconductor technology
1995	Frederick Reines (USA)	discovery of the neutrino
	Martin Perl (USA)	discovery of the tau lepton
1996	David Lee (USA), Douglas Osheroff (USA), and Robert Richardson (USA)	discovery of superfluidity in helium-3

| 1997 | Claude Cohen-Tannoudji (France), William Phillips (USA), and Steven Chu (USA) | discovery of a way to slow down individual atoms using lasers for study in a near-vacuum |
| 1998 | Robert Laughlin (USA), Horst Störmer (USA), and Daniel Tsui (USA) | discovery of quasi particles |

[1] Nationality given is the citizenship of recipient at the time award was made.

Nobel Prize for Physiology or Medicine

Year	Winner(s)[1]	Awarded for
1901	Emil von Behring (Germany)	discovery that the body produces antitoxins, and development of serum therapy for diseases such as diphtheria
1902	Ronald Ross (UK)	work on the role of the *Anopheles* mosquito in transmitting malaria
1903	Niels Finsen (Denmark)	discovery of the use of ultraviolet light to treat skin diseases
1904	Ivan Pavlov (Russia)	discovery of the physiology of digestion
1905	Robert Koch (Germany)	investigations and discoveries in relation to tuberculosis
1906	Camillo Golgi (Italy) and Santiago	Ramtion to tuberculosidiscovery of the fine structure of the nervous system
1907	Charles Laveran (France)	discovery that certain protozoa can cause disease
1908	Ilya Mechnikov (Russia) and Paul Ehrlich (Germany)	work on immunity
1909	Emil Kocher (Switzerland)	work on the physiology, pathology, and surgery of the thyroid gland
1910	Albrecht Kossel (Germany)	study of cell proteins and nucleic acids
1911	Allvar Gullstrand (Sweden)	work on the refraction of light through the different components of the eye
1912	Alexis Carrel (France)	work on the techniques for connecting severed blood vessels and transplanting organs
1913	Charles Richet (France)	work on allergic responses
1914	Robert Bárány (Austria-Hungary)	work on the physiology and pathology of the equilibrium organs of the inner ear
1915	no award	
1916	no award	
1917	no award	
1918	no award	
1919	Jules Bordet (Belgium)	work on immunity
1920	August Krogh (Denmark)	discovery of the mechanism regulating the dilation and constriction of blood capillaries
1921	no award	
1922	Archibald Hill (UK)	work in the production of heat in contracting muscle
	Otto Meyerhof (Germany)	work in the relationship between oxygen consumption and metabolism of lactic acid in muscle
1923	Frederick Banting (Canada) and John Macleod (UK)	discovery and isolation of the hormone insulin
1924	Willem Einthoven (Netherlands)	invention of the electrocardiograph
1925	no award	
1926	Johannes Fibiger (Denmark)	discovery of a parasite *Spiroptera carcinoma* that causes cancer
1927	Julius Wagner-Jauregg (Austria)	use of induced malarial fever to treat paralysis caused by mental deterioration
1928	Charles Nicolle (France)	work on the role of the body louse in transmitting typhus
1929	Christiaan Eijkman (Netherlands)	discovery of a cure for beriberi, a vitamin-deficiency disease
	Frederick Hopkins (UK)	discovery of trace substances, now known as vitamins, that stimulate growth
1930	Karl Landsteiner (USA)	discovery of human blood groups
1931	Otto Warburg (Germany)	discovery of respiratory enzymes that enable cells to process oxygen
1932	Charles Sherrington (UK) and Edgar Adrian (UK)	discovery of function of neurons (nerve cells)

Year	Winner(s)[1]	Awarded for
1933	Thomas Morgan (USA)	work on the role of chromosomes in heredity
1934	George Whipple (USA), George Minot (USA), and William Murphy (USA)	work on treatment of pernicious anaemia by increasing the amount of liver in the diet
1935	Hans Spemann (Germany)	organizer effect in embryonic development
1936	Henry Dale (UK) and Otto Loewi (Germany)	chemical transmission of nerve impulses
1937	Albert Szent-Györgyi (Hungary)	investigation of biological oxidation processes and of the action of ascorbic acid (vitamin C)
1938	Corneille Heymans (Belgium)	mechanisms regulating respiration
1939	Gerhard Domagk (Germany)	discovery of the first antibacterial sulphonamide drug
1940	no award	
1941	no award	
1942	no award	
1943	Henrik Dam (Denmark)	discovery of vitamin K
	Edward Doisy (USA)	chemical nature of vitamin K
1944	Joseph Erlanger (USA) and Herbert Gasser (USA)	transmission of impulses by nerve fibres
1945	Alexander Fleming (UK)	discovery of the bactericidal effect of penicillin
	Ernst Chain (UK) and Howard Florey (Australia)	isolation of penicillin and its development as an antibiotic drug
1946	Hermann Muller (USA)	discovery that X-ray irradiation can cause mutation
1947	Carl Cori (USA) and Gerty Cori (USA)	production and breakdown of glycogen (animal starch)
	Bernardo Houssay (Argentina)	function of the pituitary gland in sugar metabolism
1948	Paul Müller (Switzerland)	discovery of the first synthetic contact insecticide DDT
1949	Walter Hess (Switzerland)	mapping areas of the midbrain that control the activities of certain body organs
	Antonio Egas Moniz (Portugal)	therapeutic value of prefrontal leucotomy in certain psychoses
1950	Edward Kendall (USA), Tadeus Reichstein (Switzerland), and Philip Hench (USA)	structure and biological effects of hormones of the adrenal cortex
1951	Max Theiler (South Africa)	discovery of a vaccine against yellow fever
1952	Selman Waksman (USA)	discovery of streptomycin, the first antibiotic effective against tuberculosis
1953	Hans Krebs (UK)	discovery of the Krebs cycle
	Fritz Lipmann (USA)	discovery of coenzyme A, a nonprotein compound that acts in conjunction with enzymes to catalyse metabolic reactions leading up to the Krebs cycle
1954	John Enders (USA), Thomas Weller (USA), and Frederick Robbins (USA)	cultivation of the polio virus in the laboratory
1955	Hugo Theorell (Sweden)	work on the nature and action of oxidation enzymes
1956	André Cournand (USA), Werner Forssmann (West Germany), and Dickinson Richards (USA)	work on the technique for passing a catheter into the heart for diagnostic purposes
1957	Daniel Bovet (Italy)	discovery of synthetic drugs used as muscle relaxants in anaesthesia
1958	George Beadle (USA) and Edward Tatum (USA)	discovery that genes regulate precise chemical effects
	Joshua Lederberg (USA)	work on genetic recombination and the organization of bacterial genetic material
1959	Severo Ochoa (USA) and Arthur Kornberg (USA)	discovery of enzymes that catalyse the formation of RNA (ribonucleic acid) and DNA (deoxyribonucleic acid)
1960	Macfarlane Burnet (Australia) and Peter Medawar (UK)	acquired immunological tolerance of transplanted tissues
1961	Georg von Békésy (USA)	investigations into the mechanism of hearing within the cochlea of the inner ear
1962	Francis Crick (UK), James Watson (USA), and Maurice Wilkins (UK)	discovery of the double-helical structure of DNA and of the significance of this structure in the replication and transfer of genetic information

Year	Winner(s)[1]	Awarded for
1963	John Eccles (Australia), Alan Hodgkin (UK), and Andrew Huxley (UK)	ionic mechanisms involved in the communication or inhibition of impulses across neuron (nerve cell) membranes
1964	Konrad Bloch (USA) and Feodor Lynen (West Germany)	work on the cholesterol and fatty-acid metabolism
1965	François Jacob (France), André Lwoff (France), and Jacques Monod (France)	genetic control of enzyme and virus synthesis
1966	Peyton Rous (USA)	discovery of tumour-inducing viruses
	Charles Huggins (USA)	hormonal treatment of prostatic cancer
1967	Ragnar Granit (Sweden), Haldan Hartline (USA), and George Wald (USA)	physiology and chemistry of vision
1968	Robert Holley (USA), Har Gobind Khorana (USA), and Marshall Nirenberg (USA)	interpretation of genetic code and its function in protein synthesis
1969	Max Delbrück (USA), Alfred Hershey (USA), and Salvador Luria (USA)	replication mechanism and genetic structure of viruses
1970	Bernard Katz (UK), Ulf von Euler (Sweden), and Julius Axelrod (USA)	work on the storage, release, and inactivation of neurotransmitters
1971	Earl Sutherland (USA)	discovery of cyclic AMP, a chemical messenger that plays a role in the action of many hormones
1972	Gerald Edelman (USA) and Rodney Porter (UK)	work on the chemical structure of antibodies
1973	Karl von Frisch (Austria), Konrad Lorenz (Austria), and Nikolaas Tinbergen (UK)	work in animal behaviour patterns
1974	Albert Claude (USA), Christian de Duve (Belgium), and George Palade (USA)	work in structural and functional organization of the cell
1975	David Baltimore (USA), Renato Dulbecco (USA), and Howard Temin (USA)	work on interactions between tumour-inducing viruses and the genetic material of the cell
1976	Baruch Blumberg (USA) and Carleton Gajdusek (USA)	new mechanisms for the origin and transmission of infectious diseases
1977	Roger Guillemin (USA) and Andrew Schally (USA)	discovery of hormones produced by the hypothalamus region of the brain
	Rosalyn Yalow (USA)	radioimmunoassay techniques by which minute quantities of hormone may be detected
1978	Werner Arber (Switzerland), Daniel Nathans (USA), and Hamilton Smith (USA)	discovery of restriction enzymes and their application to molecular genetics
1979	Allan Cormack (USA) and Godfrey Hounsfield (UK)	development of the computed axial tomography (CAT) scan
1980	Baruj Benacerraf (USA), Jean Dausset (France), and George Snell (USA)	work on genetically determined structures on the cell surface that regulate immunological reactions
1981	Roger Sperry (USA)	functional specialization of the brain's cerebral hemispheres
	David Hubel (USA) and Torsten Wiesel (Sweden)	work on visual perception
1982	Sune Bergström (Sweden), Bengt Samuelsson (Sweden), and John Vane (UK)	discovery of prostaglandins and related biologically active substances
1983	Barbara McClintock (USA)	discovery of mobile genetic elements
1984	Niels Jerne (Denmark-UK), Georges Köhler (West Germany), and César Milstein (Argentina)	work on immunity and discovery of a technique for producing highly specific, monoclonal antibodies
1985	Michael Brown (USA) and Joseph L Goldstein (USA)	work on the regulation of cholesterol metabolism
1986	Stanley Cohen (USA) and Rita Levi-Montalcini (USA-Italy)	discovery of factors that promote the growth of nerve and epidermal cells
1987	Susumu Tonegawa (Japan)	work on the process by which genes alter to produce a range of different antibodies
1988	James Black (UK), Gertrude Elion (USA), and George Hitchings (USA)	work on the principles governing the design of new drug treatment

Year	Winner(s)[1]	Awarded for
1989	Michael Bishop (USA) and Harold Varmus (USA)	discovery of oncogenes, genes carried by viruses that can trigger cancerous growth in normal cells
1990	Joseph Murray (USA) and Donnall Thomas (USA)	pioneering work in organ and cell transplants
1991	Erwin Neher (Germany) and Bert Sakmann (Germany)	discovery of how gatelike structures (ion channels) regulate the flow of ions into and out of cells
1992	Edmond Fisher (USA) and Edwin Krebs (USA)	isolating and describing the action of the enzyme responsible for reversible protein phosphorylation, a major biological control mechanism
1993	Phillip Sharp (USA) and Richard Roberts (UK)	discovery of split genes (genes interrupted by nonsense segments of DNA)
1994	Alfred Gilman (USA) and Martin Rodbell (USA)	discovery of a family of proteins (G-proteins) that translate messages – in the form of hormones or other chemical signals – into action inside cells
1995	Edward Lewis (USA), Eric Wieschaus (USA), and Christiane Nüsslein-Volhard (Germany)	discovery of genes which control the early stages of the body's development
1996	Peter Doherty (Australia) and Rolf Zinkernagel (Switzerland)	discovery of how the immune system recognizes virus-infected cells
1997	Stanley Prusiner (USA)	discoveries, including the 'prion' theory, that could lead to new treatments of dementia-related diseases, including Alzheimer's and Parkinson's diseases
1998	Robert Furchgott (USA), Ferid Murad (USA), and Louis Ignarro (USA)	discovery that nitric oxide (NO) acts as a messenger between cells

[1] Nationality given is the citizenship of recipient at the time award was made.

Ig Nobel Prizes

These awards are a spoof on science awards and the Nobel prizes and are announced at Harvard University annually, honouring people whose work 'cannot or should not be reproduced'.

Biology T Yagyu and colleagues from the University Hospital of Zurich, Switzerland, from Kansai Medical University in Osaka, Japan, and from Neuroscience Technology Research in Prague, Czech Republic, for their report on measuring people's brainwaves while chewing different flavours of chewing gum. 'Chewing gum flavor affects measures of global complexity of multichannel EEG' was published in 1997.

Entomology Mark Hostetler of the University of Florida, for his scholarly book, *That Gunk on Your Car*, which identifies the insect splats that appear on automobile windows.

Astronomy Richard Hoagland of New Jersey, for identifying artificial features on the moon and on Mars, including a human face on Mars and ten-mile high buildings on the far side of the moon.

Communications Sanford Wallace, president of Cyber Promotions of Philadelphia, for delivering electronic junk mail around the world.

Physics John Bockris of Texas A&M University, for achievements in cold fusion, in the transmutation of base elements into gold, and in the electrochemical incineration of domestic rubbish.

Medicine Carl J Charnetski and Francis X Brennan Jr of Wilkes University, and James F Harrison of Muzak Ltd in Seattle, Washington, for their discovery that listening to elevator Muzak stimulates immunoglobulin A (IgA) production, and thus may help prevent the common cold.

Meteorology Bernard Vonnegut of the State University of Albany, for his report 'Chicken Plucking as Measure of Tornado Wind Speed', published in 1975.

Fields Medal

This international prize for achievement in the field of mathematics is awarded every four years by the International Mathematical Union.

Year	Winner(s)
1936	Lars Ahlfors (Finland); Jesse Douglas (USA)
1950	Atle Selberg (USA); Laurent Schwartz (France)
1954	Kunihiko Kodaira (USA); Jean-Pierre Serre (France)
1958	Klaus Roth (UK); René Thom (France)
1962	Lars Hörmander (Sweden); John Milnor (USA)
1966	Michael Atiyah (UK); Paul J Cohen (USA); Alexander Grothendieck (France); Stephen Smale (USA)
1970	Alan Baker (UK); Heisuke Hironaka (USA); Sergei Novikov (USSR); John G Thompson (USA)
1974	Enrico Bombieri (Italy); David Mumford (USA)
1978	Pierre Deligne (Belgium); Charles Fefferman (USA); G A Margulis (USSR); Daniel Quillen (USA)
1982	Alain Connes (France); William Thurston (USA); S T Yau (USA)
1986	Simon Donaldson (UK); Gerd Faltings (West Germany); Michael Freedman (USA)
1990	Vladimir Drinfeld (USSR); Vaughan F R Jones (USA); Shigefumi Mori (Japan); Edward Witten (USA)
1994	L J Bourgain (USA/France); P-L Lions (France); J-C Yoccoz (France); E I Zelmanov (USA)
1998	Richard E Borcherds (UK); W Timothy Gowers (UK); Maxim Kontsevich (Russia); Curtis T McMullen (USA)

INDEX

A–Z glossaries of useful terms are also included towards the end of each chapter.